DIFFERENTIAL EQUATIONS & LINEAR ALGEBRA

C. Henry Edwards • David E. Penney

CUSTOM EDITION FOR VILLANOVA UNIVERSITY

Taken from:
Differential Equations & Linear Algebra, Third Edition
by C. Henry Edwards and David E. Penney

Custom Publishing

New York Boston San Francisco
London Toronto Sydney Tokyo Singapore Madrid
Mexico City Munich Paris Cape Town Hong Kong Montreal

Cover art courtesy of Photodisc/Getty Images.

Taken from:

Differential Equations & Linear Algebra, Third Edition
by C. Henry Edwards and David E. Penney

Published by Prentice Hall
Upper Saddle River, New Jersey 07458

This special edition published in cooperation with Pearson Custom Publishing.

Printed in the United States of America

10 9 8 7 6 5 4 3 2 1

2009360110

LB

Pearson
Custom Publishing
is a division of

PEARSON

www.pearsonhighered.com

ISBN 10: 0-558-23099-7
ISBN 13: 978-0-558-23099-9

CONTENTS

CHAPTER 4 Vector Spaces 227

CHAPTER 5 Higher-Order Linear Differential Equations 286

CHAPTER 6 Eigenvalues and Eigenvectors 366

CHAPTER 7 Linear Systems of Differential Equations 396

APPLICATION MODULES

The modules listed here follow the indicated sections in the text. Most provide computing projects that illustrate the corresponding text sections. *Maple*, *Mathematica*, and MATLAB versions of these investigations are included in the Applications Manual that accompanies this textbook.

PREFACE

The evolution of the present text is based on experience teaching introductory differential equations and linear algebra with an emphasis on conceptual ideas and the use of applications and projects to involve students in active problem-solving experiences. Technical computing environments like *Maple*, *Mathematica*, and MATLAB® are widely available and are now used extensively by practicing engineers and scientists. This change in professional practice motivates a shift from the traditional concentration on manual symbolic methods to coverage also of qualitative and computer-based methods that employ numerical computation and graphical visualization to develop greater conceptual understanding. A bonus of this more comprehensive approach is accessibility to a wider range of more realistic applications of differential equations.

Both the conceptual and the computational aspects of such a course depend heavily on the perspective and techniques of linear algebra. Consequently, the study of differential equations and linear algebra in tandem reinforces the learning of both subjects. In this book we therefore have combined core topics in elementary differential equations with those concepts and methods of elementary linear algebra that are needed for a contemporary introduction to differential equations.

Principal Features of This Revision

While successful features of the preceding editions have been retained, the exposition has been enhanced in every chapter and in many individual sections of the text. Both new graphics and new text have been inserted where needed for improved student understanding of key concepts. However, the solid class-tested chapter and section structure of the book is unchanged, so class notes and syllabi will not require revision for use of this new edition. The following examples of this revision illustrate the way the local structure of the text has been augmented and polished for this edition.

Chapter 1: New Figures 1.3.9 and 1.3.10 showing direction fields that illustrate failure of existence and uniqueness of solutions (page 24); new Problems 34 and 35 showing that small changes in initial conditions can make big differences in results, but big changes in initial conditions may sometimes make only small differences in results (page 30); new Remarks 1 and 2 clarifying the concept of implicit solutions (page 35); new example of a concrete physical situation described by an initial value problem having non-unique solutions (page 43); new Remark clarifying the meaning of homogeneity for first-order equations (page 62).

Chapter 2: Additional details inserted in the derivation of the rocket propulsion equation (page 110), and new Problem 5 inserted to investigate the liftoff pause of a rocket on the launch pad sometimes observed before blastoff (page 112).

Chapter 3: Clearer outline of the method of elimination (page 149); new Remark on different parametrizations of the same solution of a linear system (pageb 151); new interpretation of problem as the 2-dimensional case of the Cayley-Hamilton theorem (page 187).

Chapter 4: New Remark explaining uniqueness of linear combinations of linearly independent vectors (page 249); clarification of row matrices, row vectors, and row rank (page 261); converse of the Pythagorean theorem included (page 273).

Chapter 5: New explanation of signs and directions of internal forces in mass-spring systems (page 287); new introduction of differential operators and clarification of the algebra of polynomial operators (page 316); new introduction and illustration of polar exponential forms of complex numbers (page 321); fuller explanation of method of undetermined coefficients in Examples 1 and 3 (pages 339–340).

Chapter 6: New problems on characteristic polynomials of low-dimensional matrices (page 375).

Chapter 7: New subsection explaining the graphic on the cover of this edition (page 429); new Problems 20–23 for student exploration of three-railway-cars systems with different initial velocity conditions (page 443); new discussion with new Figures 7.6.11 and 7.6.12 clarifying the difference between rotating and non-rotating coordinate systems in moon-earth orbit problems (page 474).

Chapter 8: New Remark illustrating the relation between matrix exponential methods and the generalized eigenvalue methods discussed previously (page 489); new exposition inserted at end of section to explain the connection between matrix variation of parameters here and (scalar) variation of parameters for second-order equations discussed previously in Chapter 5 (page 501).

Chapter 9: New remarks on phase plane portraits, autonomous systems, and critical points (pages 519–520); new introduction of linearized systems (page 532).

Chapter 10: New discussion clarifying functions of exponential order and existence of Laplace transforms (page 583); new Remark discussing the mechanics of partial-fraction decomposition (page 590); new much-expanded discussion of the proof of the Laplace-transform existence theorem and its extension to include the jump discontinuities that play an important role in many practical applications (pages 596–597); added details in the solutions of Examples 1, 3, 4, and 6 in Section 10.3 (pages 601–607).

Chapter 11: New Problem 35 on determination of radii of convergence of power series solutions of differential equations (page 653).

Computing Features

The following features highlight the flavor of computing technology that distinguishes much of our exposition.

- Almost 600 *computer-generated figures* show students vivid pictures of direction fields, solution curves, and phase plane portraits that bring symbolic solutions of differential equations to life.
- About 3 dozen *application modules* follow key sections throughout the text. Most of these applications outline "technology neutral" investigations illustrating the use of technical computing systems and seek to actively engage students in the application of new technology.

- A fresh *numerical emphasis* that is afforded by the early introduction of numerical solution techniques in Chapter 2 (on mathematical models and numerical methods). Here and in Section 7.6, where numerical techniques for systems are treated, a concrete and tangible flavor is achieved by the inclusion of numerical algorithms presented in parallel fashion for systems ranging from graphing calculators to MATLAB.

Modeling Features

Mathematical modeling is a goal and constant motivation for the study of differential equations. To sample the range of applications in this text, take a look at the following questions:

- What explains the commonly observed time lag between indoor and outdoor daily temperature oscillations? (Section 1.5)
- What makes the difference between doomsday and extinction in alligator populations? (Section 2.1)
- How do a unicycle and a two-axle car react differently to road bumps? (Sections 5.6 and 7.4)
- Why might an earthquake demolish one building and leave standing the one next door? (Section 7.4)
- How can you predict the time of next perihelion passage of a newly observed comet? (Section 7.6)
- What determines whether two species will live harmoniously together, or whether competition will result in the extinction of one of them and the survival of the other? (Section 9.3)

Organization and Content

We have reshaped the usual approach and sequence of topics to accommodate new technology and new perspectives. For instance:

- After a précis of first-order equations in Chapter 1 (though with the coverage of certain traditional symbolic methods streamlined a bit), Chapter 2 offers an early introduction to mathematical modeling, stability and qualitative properties of differential equations, and numerical methods—a combination of topics that frequently are dispersed later in an introductory course.
- Chapters 3 (Linear Systems and Matrices), 4 (Vector Spaces), and 6 (Eigenvalues and Eigenvectors) provide concrete and self-contained coverage of the elementary linear algebra concepts and techniques that are needed for the solution of linear differential equations and systems. Chapter 4 concludes with sections 4.5 (row and column spaces) and 4.6 (orthogonal vectors in $\mathbf{R}^n$). Chapter 6 concludes with applications of diagonalizable matrices and a proof of the Cayley-Hamilton theorem for such matrices.
- Chapter 5 exploits the linear algebra of Chapters 3 and 4 to present efficiently the theory and solution of single linear differential equations. Chapter 7 is based on the eigenvalue approach to linear systems, and includes (in Section 7.5) the Jordan normal form for matrices and its application to the general Cayley-Hamilton theorem. This chapter includes an unusual number of applications (ranging from railway cars to earthquakes) of the various cases of the

eigenvalue method, and concludes in Section 7.6 with numerical methods for systems.

- Chapter 8 is devoted to matrix exponentials with applications to linear systems of differential equations. The spectral decomposition method of Section 8.3 offers students an especially concrete approach to the computation of matrix exponentials. Our treatment of this material owes much to advice and course notes provided by Professor Dar-Veig Ho of the Georgia Institute of Technology.
- Chapter 9 exploits linear methods for the investigation of nonlinear systems and phenomena, and ranges from phase plane analysis to applications involving ecological and mechanical systems.
- Chapter 10 treats Laplace transform methods for the solution of constant-coefficient linear differential equations with a goal of handling the piecewise continuous and periodic forcing functions that are common in physical applications. Chapter 11 treats power series methods with a goal of discussing Bessel's equation with sufficient detail for the most common elementary applications.

Applications and Solutions Manuals

The answer section has been expanded considerably to increase its value as a learning aid. It now includes the answers to most odd-numbered problems plus a good many even-numbered ones. The almost 600-page **Instructor's Solutions Manual** (0-13-605426-9) accompanying this book provides worked-out solutions for most of the problems in the book, and the over 300-page **Student Solutions Manual** (0-13-605427-7) contains solutions for most of the odd-numbered problems.

The approximately 3 dozen application modules in the text contain additional problem and project material designed largely to engage students in the exploration and application of computational technology. These investigations are expanded considerably in the 250-page **Applications Manual** (0-321-61525-5) that accompanies the text and supplements it with additional and sometimes more challenging investigations. Each section in this manual has parallel subsections **Using *Maple***, **Using *Mathematica***, and **Using MATLAB** that detail the applicable methods and techniques of each system, and will afford student users an opportunity to compare the merits and styles of different computational systems.

Acknowledgments

In preparing this revision we profited greatly from the advice and assistance of the following very capable and perceptive reviewers:

Miklos Bona
University of Florida
Mark Bridger
Northeastern University
Raymond A. Claspadle
University of Memphis
Sigal Gottlieb
University of Massachusetts, Dartmouth
Grant Gustafson
University of Utah
Semion Gutman
University of Oklahoma
Carl Lutzer
Rochester Institute of Technology
Ifran Ul-Haq
University of Wisconsin-Platteville

It is a pleasure to (once again) credit Dennis Kletzing and his extraordinary TEXpertise for the attractive presentation of both the text and the art in this book.

Thanks are due to Paul Lorczak for meticulously checking every example, problem, and answer in the text. Finally, but far from least, I am especially happy to acknowledge a new contributor to this effort, David Calvis, who assisted in every aspect of this revision and contributed tangibly to the improvement of every chapter in the book.

C. H. E.
h.edwards@mindspring.com

DIFFERENTIAL EQUATIONS & LINEAR ALGEBRA

1 First-Order Differential Equations

1.1 Differential Equations and Mathematical Models

The laws of the universe are written in the language of mathematics. Algebra is sufficient to solve many static problems, but the most interesting natural phenomena involve change and are described by equations that relate changing quantities.

Because the derivative $dx/dt = f'(t)$ of the function f is the rate at which the quantity $x = f(t)$ is changing with respect to the independent variable t, it is natural that equations involving derivatives are frequently used to describe the changing universe. An equation relating an unknown function and one or more of its derivatives is called a **differential equation**.

Example 1 The differential equation

$$\frac{dx}{dt} = x^2 + t^2$$

involves both the unknown function $x(t)$ and its first derivative $x'(t) = dx/dt$. The differential equation

$$\frac{d^2y}{dx^2} + 3\frac{dy}{dx} + 7y = 0$$

involves the unknown function y of the independent variable x and the first two derivatives y' and y'' of y. ■

The study of differential equations has three principal goals:

1. To discover the differential equation that describes a specified physical situation.
2. To find—either exactly or approximately—the appropriate solution of that equation.
3. To interpret the solution that is found.

In algebra, we typically seek the unknown *numbers* that satisfy an equation such as $x^3+7x^2-11x+41=0$. By contrast, in solving a differential equation, we are challenged to find the unknown *functions* $y = y(x)$ for which an identity such as $y'(x) = 2xy(x)$—that is, the differential equation

$$\frac{dy}{dx} = 2xy$$

—holds on some interval of real numbers. Ordinarily, we will want to find *all* solutions of the differential equation, if possible.

Example 2 If C is a constant and

$$y(x) = Ce^{x^2}, \tag{1}$$

then

$$\frac{dy}{dx} = C\left(2xe^{x^2}\right) = (2x)\left(Ce^{x^2}\right) = 2xy.$$

Thus every function $y(x)$ of the form in Eq. (1) *satisfies*—and thus is a solution of—the differential equation

$$\frac{dy}{dx} = 2xy \tag{2}$$

for all x. In particular, Eq. (1) defines an *infinite* family of different solutions of this differential equation, one for each choice of the arbitrary constant C. By the method of separation of variables (Section 1.4) it can be shown that every solution of the differential equation in (2) is of the form in Eq. (1). ■

Differential Equations and Mathematical Models

The following three examples illustrate the process of translating scientific laws and principles into differential equations. In each of these examples the independent variable is time t, but we will see numerous examples in which some quantity other than time is the independent variable.

Example 3 Newton's law of cooling may be stated in this way: The *time rate of change* (the rate of change with respect to time t) of the temperature $T(t)$ of a body is proportional to the difference between T and the temperature A of the surrounding medium (Fig. 1.1.1). That is,

$$\frac{dT}{dt} = -k(T - A), \tag{3}$$

where k is a positive constant. Observe that if $T > A$, then $dT/dt < 0$, so the temperature is a decreasing function of t and the body is cooling. But if $T < A$, then $dT/dt > 0$, so that T is increasing.

Thus the physical law is translated into a differential equation. If we are given the values of k and A, we should be able to find an explicit formula for $T(t)$, and then—with the aid of this formula—we can predict the future temperature of the body. ■

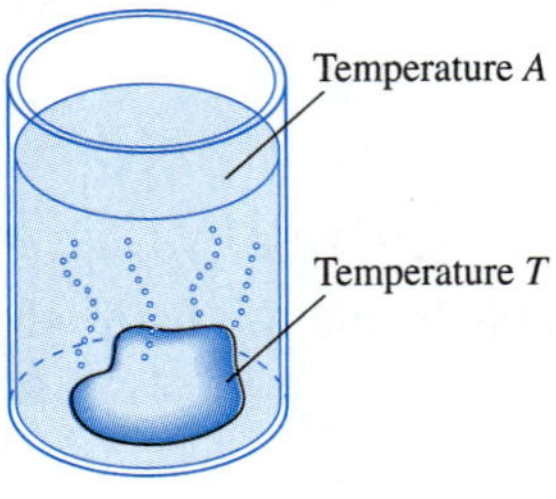

FIGURE 1.1.1. Newton's law of cooling, Eq. (3), describes the cooling of a hot rock in water.

Example 4 Torricelli's law implies that the *time rate of change* of the volume V of water in a draining tank (Fig. 1.1.2) is proportional to the square root of the depth y of water in the tank:

$$\frac{dV}{dt} = -k\sqrt{y}, \tag{4}$$

where k is a constant. If the tank is a cylinder with vertical sides and cross-sectional area A, then $V = Ay$, so $dV/dt = A \cdot (dy/dt)$. In this case Eq. (4) takes the form

$$\frac{dy}{dt} = -h\sqrt{y}, \tag{5}$$

where $h = k/A$ is a constant. ■

Example 5 The *time rate of change* of a population $P(t)$ with constant birth and death rates is, in many simple cases, proportional to the size of the population. That is,

$$\frac{dP}{dt} = kP, \tag{6}$$

where k is the constant of proportionality. ■

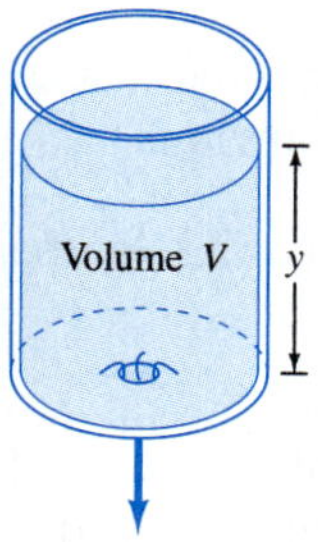

FIGURE 1.1.2. Torricelli's law of draining, Eq. (4), describes the draining of a water tank.

Let us discuss Example 5 further. Note first that each function of the form

$$P(t) = Ce^{kt} \tag{7}$$

is a solution of the differential equation

$$\frac{dP}{dt} = kP$$

in (6). We verify this assertion as follows:

$$P'(t) = Cke^{kt} = k\left(Ce^{kt}\right) = kP(t)$$

for all real numbers t. Because substitution of each function of the form given in (7) into Eq. (6) produces an identity, all such functions are solutions of Eq. (6).

Thus, even if the value of the constant k is known, the differential equation $dP/dt = kP$ has *infinitely many* different solutions of the form $P(t) = Ce^{kt}$, one for each choice of the "arbitrary" constant C. This is typical of differential equations. It is also fortunate, because it may allow us to use additional information to select from among all these solutions a particular one that fits the situation under study.

Example 6 Suppose that $P(t) = Ce^{kt}$ is the population of a colony of bacteria at time t, that the population at time $t = 0$ (hours, h) was 1000, and that the population doubled after 1 h. This additional information about $P(t)$ yields the following equations:

$$1000 = P(0) = Ce^0 = C,$$
$$2000 = P(1) = Ce^k.$$

It follows that $C = 1000$ and that $e^k = 2$, so $k = \ln 2 \approx 0.693147$. With this value of k the differential equation in (6) is

$$\frac{dP}{dt} = (\ln 2)P \approx (0.693147)P.$$

Substitution of $k = \ln 2$ and $C = 1000$ in Eq. (7) yields the particular solution

$$P(t) = 1000e^{(\ln 2)t} = 1000(e^{\ln 2})^t = 1000 \cdot 2^t \qquad (\text{because } e^{\ln 2} = 2)$$

that satisfies the given conditions. We can use this particular solution to predict future populations of the bacteria colony. For instance, the predicted number of bacteria in the population after one and a half hours (when $t = 1.5$) is

$$P(1.5) = 1000 \cdot 2^{3/2} \approx 2828.$$ ■

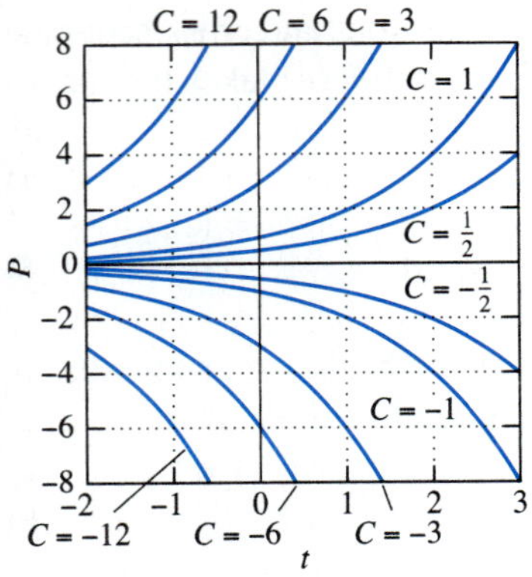

FIGURE 1.1.3. Graphs of $P(t) = Ce^{kt}$ with $k = \ln 2$.

The condition $P(0) = 1000$ in Example 6 is called an **initial condition** because we frequently write differential equations for which $t = 0$ is the "starting time." Figure 1.1.3 shows several different graphs of the form $P(t) = Ce^{kt}$ with $k = \ln 2$. The graphs of all the infinitely many solutions of $dP/dt = kP$ in fact fill the entire two-dimensional plane, and no two intersect. Moreover, the selection of any one point P_0 on the P-axis amounts to a determination of $P(0)$. Because exactly one solution passes through each such point, we see in this case that an initial condition $P(0) = P_0$ determines a unique solution agreeing with the given data.

Mathematical Models

Our brief discussion of population growth in Examples 5 and 6 illustrates the crucial process of *mathematical modeling* (Fig. 1.1.4), which involves the following:

1. The formulation of a real-world problem in mathematical terms; that is, the construction of a mathematical model.
2. The analysis or solution of the resulting mathematical problem.
3. The interpretation of the mathematical results in the context of the original real-world situation—for example, answering the question originally posed.

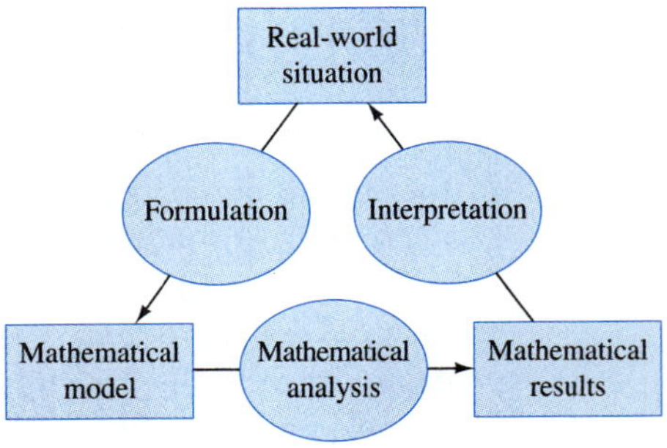

FIGURE 1.1.4. The process of mathematical modeling.

In the population example, the real-world problem is that of determining the population at some future time. A **mathematical model** consists of a list of variables (P and t) that describe the given situation, together with one or more equations relating these variables ($dP/dt = kP$, $P(0) = P_0$) that are known or are assumed to hold. The mathematical analysis consists of solving these equations (here, for P as a function of t). Finally, we apply these mathematical results to attempt to answer the original real-world question.

As an example of this process, think of first formulating the mathematical model consisting of the equations $dP/dt = kP$, $P(0) = 1000$, describing the bacteria population of Example 6. Then our mathematical analysis there consisted of solving for the solution function $P(t) = 1000e^{(\ln 2)t} = 1000 \cdot 2^t$ as our mathematical result. For an interpretation in terms of our real-world situation—the actual bacteria population—we substituted $t = 1.5$ to obtain the predicted population of $P(1.5) \approx 2828$ bacteria after 1.5 hours. If, for instance, the bacteria population is growing under ideal conditions of unlimited space and food supply, our prediction may be quite accurate, in which case we conclude that the mathematical model is adequate for studying this particular population.

On the other hand, it may turn out that no solution of the selected differential equation accurately fits the actual population we're studying. For instance, for *no* choice of the constants C and k does the solution $P(t) = Ce^{kt}$ in Eq. (7) accurately describe the actual growth of the human population of the world over the past few centuries. We must conclude that the differential equation $dP/dt = kP$ is inadequate for modeling the world population—which in recent decades has "leveled off" as compared with the steeply climbing graphs in the upper half ($P > 0$) of Fig. 1.1.3. With sufficient insight, we might formulate a new mathematical model including a perhaps more complicated differential equation, one that takes into account such factors as a limited food supply and the effect of increased population on birth and death rates. With the formulation of this new mathematical model, we may attempt to traverse once again the diagram of Fig. 1.1.4 in a counterclockwise manner. If we can solve the new differential equation, we get new solution functions to compare with the real-world population. Indeed, a successful population analysis may require refining the mathematical model still further as it is repeatedly measured against real-world experience.

But in Example 6 we simply ignored any complicating factors that might affect our bacteria population. This made the mathematical analysis quite simple, perhaps unrealistically so. A satisfactory mathematical model is subject to two contradictory requirements: It must be sufficiently detailed to represent the real-world situation with relative accuracy, yet it must be sufficiently simple to make the mathematical analysis practical. If the model is so detailed that it fully represents the physical situation, then the mathematical analysis may be too difficult to carry out. If the model is too simple, the results may be so inaccurate as to be useless. Thus there is an inevitable tradeoff between what is physically realistic and what is mathematically possible. The construction of a model that adequately bridges this gap between realism and feasibility is therefore the most crucial and delicate step in the process. Ways must be found to simplify the model mathematically without sacrificing essential features of the real-world situation.

Mathematical models are discussed throughout this book. The remainder of this introductory section is devoted to simple examples and to standard terminology used in discussing differential equations and their solutions.

Examples and Terminology

Example 7 If C is a constant and $y(x) = 1/(C - x)$, then

$$\frac{dy}{dx} = \frac{1}{(C - x)^2} = y^2$$

if $x \neq C$. Thus

$$y(x) = \frac{1}{C - x} \tag{8}$$

defines a solution of the differential equation

$$\frac{dy}{dx} = y^2 \tag{9}$$

on any interval of real numbers not containing the point $x = C$. Actually, Eq. (8) defines a *one-parameter family* of solutions of $dy/dx = y^2$, one for each value of the arbitrary constant or "parameter" C. With $C = 1$ we get the particular solution

$$y(x) = \frac{1}{1 - x}$$

that satisfies the initial condition $y(0) = 1$. As indicated in Fig. 1.1.5, this solution is continuous on the interval $(-\infty, 1)$ but has a vertical asymptote at $x = 1$. ■

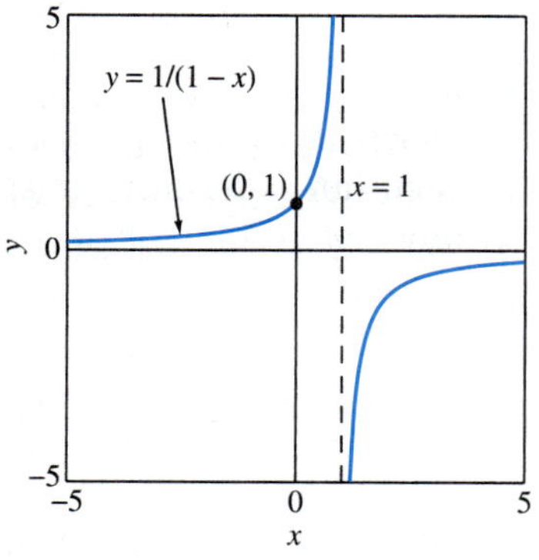

FIGURE 1.1.5. The solution of $y' = y^2$ defined by $y(x) = 1/(1 - x)$.

Example 8 Verify that the function $y(x) = 2x^{1/2} - x^{1/2}\ln x$ satisfies the differential equation

$$4x^2y'' + y = 0 \tag{10}$$

for all $x > 0$.

Solution First we compute the derivatives

$$y'(x) = -\tfrac{1}{2}x^{-1/2}\ln x \quad \text{and} \quad y''(x) = \tfrac{1}{4}x^{-3/2}\ln x - \tfrac{1}{2}x^{-3/2}.$$

Then substitution into Eq. (10) yields

$$4x^2y'' + y = 4x^2\left(\tfrac{1}{4}x^{-3/2}\ln x - \tfrac{1}{2}x^{-3/2}\right) + 2x^{1/2} - x^{1/2}\ln x = 0$$

if x is positive, so the differential equation is satisfied for all $x > 0$. ■

The fact that we can write a differential equation is not enough to guarantee that it has a solution. For example, it is clear that the differential equation

$$(y')^2 + y^2 = -1 \tag{11}$$

has *no* (real-valued) solution, because the sum of nonnegative numbers cannot be negative. For a variation on this theme, note that the equation

$$(y')^2 + y^2 = 0 \tag{12}$$

obviously has only the (real-valued) solution $y(x) \equiv 0$. In our previous examples any differential equation having at least one solution indeed had infinitely many.

The **order** of a differential equation is the order of the highest derivative that appears in it. The differential equation of Example 8 is of second order, those in Examples 2 through 7 are first-order equations, and

$$y^{(4)} + x^2y^{(3)} + x^5y = \sin x$$

is a fourth-order equation. The most general form of an ***n*th-order** differential equation with independent variable x and unknown function or dependent variable $y = y(x)$ is

$$F\left(x, y, y', y'', \dots, y^{(n)}\right) = 0, \tag{13}$$

where F is a specific real-valued function of $n + 2$ variables.

Our use of the word *solution* has been until now somewhat informal. To be precise, we say that the continuous function $u = u(x)$ is a **solution** of the differential equation in (13) **on the interval** I provided that the derivatives $u', u'', \dots, u^{(n)}$ exist on I and

$$F\left(x, u, u', u'', \dots, u^{(n)}\right) = 0$$

for all x in I. For the sake of brevity, we may say that $u = u(x)$ **satisfies** the differential equation in (13) on I.

Remark: Recall from elementary calculus that a differentiable function on an open interval is necessarily continuous there. This is why only a continuous function can qualify as a (differentiable) solution of a differential equation on an interval. ■

Example 7 *Continued*

Figure 1.1.5 shows the two "connected" branches of the graph $y = 1/(1-x)$. The left-hand branch is the graph of a (continuous) solution of the differential equation $y' = y^2$ that is defined on the interval $(-\infty, 1)$. The right-hand branch is the graph of a *different* solution of the differential equation that is defined (and continuous) on the different interval $(1, \infty)$. So the single formula $y(x) = 1/(1-x)$ actually defines two different solutions (with different domains of definition) of the same differential equation $y' = y^2$. ■

Example 9

If A and B are constants and

$$y(x) = A\cos 3x + B\sin 3x, \tag{14}$$

then two successive differentiations yield

$$y'(x) = -3A\sin 3x + 3B\cos 3x,$$
$$y''(x) = -9A\cos 3x - 9B\sin 3x = -9y(x)$$

for all x. Consequently, Eq. (14) defines what it is natural to call a *two-parameter family* of solutions of the second-order differential equation

$$y'' + 9y = 0 \tag{15}$$

on the whole real number line. Figure 1.1.6 shows the graphs of several such solutions. ■

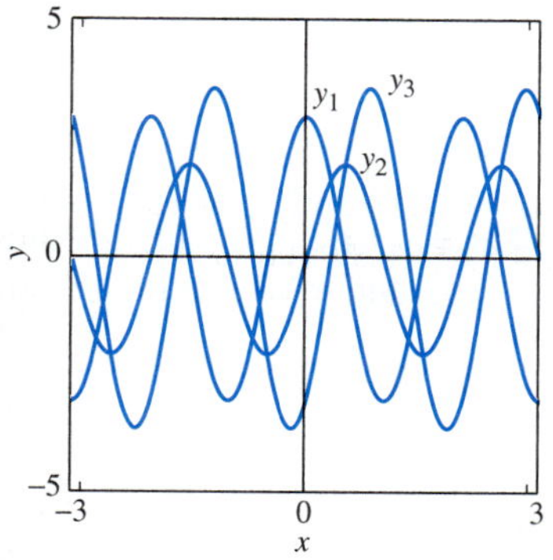

FIGURE 1.1.6. The three solutions $y_1(x) = 3\cos 3x$, $y_2(x) = 2\sin 3x$, and $y_3(x) = -3\cos 3x + 2\sin 3x$ of the differential equation $y'' + 9y = 0$.

Although the differential equations in (11) and (12) are exceptions to the general rule, we will see that an nth-order differential equation ordinarily has an n-parameter family of solutions—one involving n different arbitrary constants or parameters.

In both Eqs. (11) and (12), the appearance of y' as an implicitly defined function causes complications. For this reason, we will ordinarily assume that any differential equation under study can be solved explicitly for the highest derivative that appears; that is, that the equation can be written in the so-called *normal form*

$$y^{(n)} = G\left(x, y, y', y'', \dots, y^{(n-1)}\right), \tag{16}$$

where G is a real-valued function of $n+1$ variables. In addition, we will always seek only real-valued solutions unless we warn the reader otherwise.

All the differential equations we have mentioned so far are **ordinary** differential equations, meaning that the unknown function (dependent variable) depends on only a *single* independent variable. If the dependent variable is a function of two or more independent variables, then partial derivatives are likely to be involved; if they are, the equation is called a **partial** differential equation. For example, the temperature $u = u(x, t)$ of a long thin uniform rod at the point x at time t satisfies (under appropriate simple conditions) the partial differential equation

$$\frac{\partial u}{\partial t} = k\frac{\partial^2 u}{\partial x^2},$$

where k is a constant (called the *thermal diffusivity* of the rod). In this book we will be concerned only with *ordinary* differential equations and will refer to them simply as differential equations.

In this chapter we concentrate on *first-order* differential equations of the form

$$\frac{dy}{dx} = f(x, y). \tag{17}$$

We also will sample the wide range of applications of such equations. A typical mathematical model of an applied situation will be an **initial value problem**, consisting of a differential equation of the form in (17) together with an **initial condition** $y(x_0) = y_0$. Note that we call $y(x_0) = y_0$ an initial condition whether or not $x_0 = 0$. To **solve** the initial value problem

$$\frac{dy}{dx} = f(x, y), \quad y(x_0) = y_0 \tag{18}$$

means to find a differentiable function $y = y(x)$ that satisfies both conditions in Eq. (18) on some interval containing x_0.

Example 10 Given the solution $y(x) = 1/(C - x)$ of the differential equation $dy/dx = y^2$ discussed in Example 7, solve the initial value problem

$$\frac{dy}{dx} = y^2, \quad y(1) = 2.$$

Solution We need only find a value of C so that the solution $y(x) = 1/(C - x)$ satisfies the initial condition $y(1) = 2$. Substitution of the values $x = 1$ and $y = 2$ in the given solution yields

$$2 = y(1) = \frac{1}{C - 1},$$

so $2C - 2 = 1$, and hence $C = \frac{3}{2}$. With this value of C we obtain the desired solution

$$y(x) = \frac{1}{\frac{3}{2} - x} = \frac{2}{3 - 2x}.$$

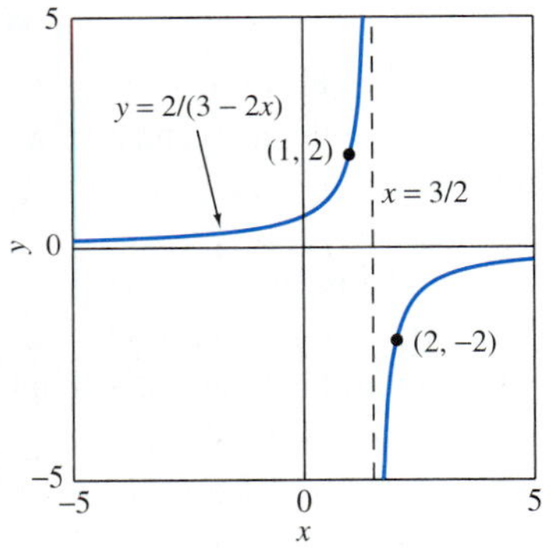

FIGURE 1.1.7. The solutions of $y' = y^2$ defined by $y(x) = 2/(3 - 2x)$.

Figure 1.1.7 shows the two branches of the graph $y = 2/(3 - 2x)$. The left-hand branch is the graph on $(-\infty, \frac{3}{2})$ of the solution of the given initial value problem $y' = y^2$, $y(1) = 2$. The right-hand branch passes through the point $(2, -2)$ and is therefore the graph on $(\frac{3}{2}, \infty)$ of the solution of the different initial value problem $y' = y^2$, $y(2) = -2$. ■

The central question of greatest immediate interest to us is this: If we are given a differential equation known to have a solution satisfying a given initial condition, how do we actually *find* or *compute* that solution? And, once found, what can we do with it? We will see that a relatively few simple techniques—separation of variables (Section 1.4), solution of linear equations (Section 1.5), elementary substitution methods (Section 1.6)—are enough to enable us to solve a variety of first-order equations having impressive applications.

1.1 Problems

In Problems 1 through 12, verify by substitution that each given function is a solution of the given differential equation. Throughout these problems, primes denote derivatives with respect to x.

1. $y' = 3x^2$; $y = x^3 + 7$
2. $y' + 2y = 0$; $y = 3e^{-2x}$
3. $y'' + 4y = 0$; $y_1 = \cos 2x$, $y_2 = \sin 2x$
4. $y'' = 9y$; $y_1 = e^{3x}$, $y_2 = e^{-3x}$
5. $y' = y + 2e^{-x}$; $y = e^x - e^{-x}$
6. $y'' + 4y' + 4y = 0$; $y_1 = e^{-2x}$, $y_2 = xe^{-2x}$
7. $y'' - 2y' + 2y = 0$; $y_1 = e^x \cos x$, $y_2 = e^x \sin x$
8. $y'' + y = 3\cos 2x$, $y_1 = \cos x - \cos 2x$, $y_2 = \sin x - \cos 2x$
9. $y' + 2xy^2 = 0$; $y = \dfrac{1}{1 + x^2}$
10. $x^2y'' + xy' - y = \ln x$; $y_1 = x - \ln x$, $y_2 = \dfrac{1}{x} - \ln x$
11. $x^2y'' + 5xy' + 4y = 0$; $y_1 = \dfrac{1}{x^2}$, $y_2 = \dfrac{\ln x}{x^2}$

12. $x^2y'' - xy' + 2y = 0$; $y_1 = x\cos(\ln x)$, $y_2 = x\sin(\ln x)$

In Problems 13 through 16, substitute $y = e^{rx}$ into the given differential equation to determine all values of the constant r for which $y = e^{rx}$ is a solution of the equation.

13. $3y' = 2y$ **14.** $4y'' = y$
15. $y'' + y' - 2y = 0$ **16.** $3y'' + 3y' - 4y = 0$

In Problems 17 through 26, first verify that $y(x)$ satisfies the given differential equation. Then determine a value of the constant C so that $y(x)$ satisfies the given initial condition. Use a computer or graphing calculator (if desired) to sketch several typical solutions of the given differential equation, and highlight the one that satisfies the given initial condition.

17. $y' + y = 0$; $y(x) = Ce^{-x}$, $y(0) = 2$
18. $y' = 2y$; $y(x) = Ce^{2x}$, $y(0) = 3$
19. $y' = y + 1$; $y(x) = Ce^{x} - 1$, $y(0) = 5$
20. $y' = x - y$; $y(x) = Ce^{-x} + x - 1$, $y(0) = 10$
21. $y' + 3x^2y = 0$; $y(x) = Ce^{-x^3}$, $y(0) = 7$
22. $e^y y' = 1$; $y(x) = \ln(x + C)$, $y(0) = 0$
23. $x\dfrac{dy}{dx} + 3y = 2x^5$; $y(x) = \frac{1}{4}x^5 + Cx^{-3}$, $y(2) = 1$
24. $xy' - 3y = x^3$; $y(x) = x^3(C + \ln x)$, $y(1) = 17$
25. $y' = 3x^2(y^2 + 1)$; $y(x) = \tan(x^3 + C)$, $y(0) = 1$
26. $y' + y\tan x = \cos x$; $y(x) = (x + C)\cos x$, $y(\pi) = 0$

In Problems 27 through 31, a function $y = g(x)$ is described by some geometric property of its graph. Write a differential equation of the form $dy/dx = f(x, y)$ having the function g as its solution (or as one of its solutions).

27. The slope of the graph of g at the point (x, y) is the sum of x and y.
28. The line tangent to the graph of g at the point (x, y) intersects the x-axis at the point $(x/2, 0)$.
29. Every straight line normal to the graph of g passes through the point $(0, 1)$. Can you *guess* what the graph of such a function g might look like?
30. The graph of g is normal to every curve of the form $y = x^2 + k$ (k is a constant) where they meet.
31. The line tangent to the graph of g at (x, y) passes through the point $(-y, x)$.

In Problems 32 through 36, write—in the manner of Eqs. (3) through (6) of this section—a differential equation that is a mathematical model of the situation described.

32. The time rate of change of a population P is proportional to the square root of P.
33. The time rate of change of the velocity v of a coasting motorboat is proportional to the square of v.
34. The acceleration dv/dt of a Lamborghini is proportional to the difference between 250 km/h and the velocity of the car.
35. In a city having a fixed population of P persons, the time rate of change of the number N of those persons who have heard a certain rumor is proportional to the number of those who have not yet heard the rumor.
36. In a city with a fixed population of P persons, the time rate of change of the number N of those persons infected with a certain contagious disease is proportional to the product of the number who have the disease and the number who do not.

In Problems 37 through 42, determine by inspection at least one solution of the given differential equation. That is, use your knowledge of derivatives to make an intelligent guess. Then test your hypothesis.

37. $y'' = 0$ **38.** $y' = y$
39. $xy' + y = 3x^2$ **40.** $(y')^2 + y^2 = 1$
41. $y' + y = e^x$ **42.** $y'' + y = 0$

43. **(a)** If k is a constant, show that a general (one-parameter) solution of the differential equation

$$\frac{dx}{dt} = kx^2$$

is given by $x(t) = 1/(C - kt)$, where C is an arbitrary constant.

(b) Determine by inspection a solution of the initial value problem $x' = kx^2$, $x(0) = 0$.

44. **(a)** Continuing Problem 43, assume that k is positive, and then sketch graphs of solutions of $x' = kx^2$ with several typical positive values of $x(0)$.

(b) How would these solutions differ if the constant k were negative?

45. Suppose a population P of rodents satisfies the differential equation $dP/dt = kP^2$. Initially, there are $P(0) = 2$ rodents, and their number is increasing at the rate of $dP/dt = 1$ rodent per month when there are $P = 10$ rodents. How long will it take for this population to grow to a hundred rodents? To a thousand? What's happening here?

46. Suppose the velocity v of a motorboat coasting in water satisfies the differential equation $dv/dt = kv^2$. The initial speed of the motorboat is $v(0) = 10$ meters per second (m/s), and v is decreasing at the rate of 1 m/s^2 when $v = 5$ m/s. How long does it take for the velocity of the boat to decrease to 1 m/s? To $\frac{1}{10}$ m/s? When does the boat come to a stop?

47. In Example 7 we saw that $y(x) = 1/(C - x)$ defines a one-parameter family of solutions of the differential equation $dy/dx = y^2$. **(a)** Determine a value of C so that $y(10) = 10$. **(b)** Is there a value of C such that $y(0) = 0$? Can you nevertheless find by inspection a solution of $dy/dx = y^2$ such that $y(0) = 0$? **(c)** Figure 1.1.8 shows typical graphs of solutions of the form $y(x) = 1/(C - x)$. Does it appear that these solution curves fill the entire xy-plane? Can you conclude that, given any point (a, b) in the plane, the differential equation $dy/dx = y^2$ has exactly one solution $y(x)$ satisfying the condition $y(a) = b$?

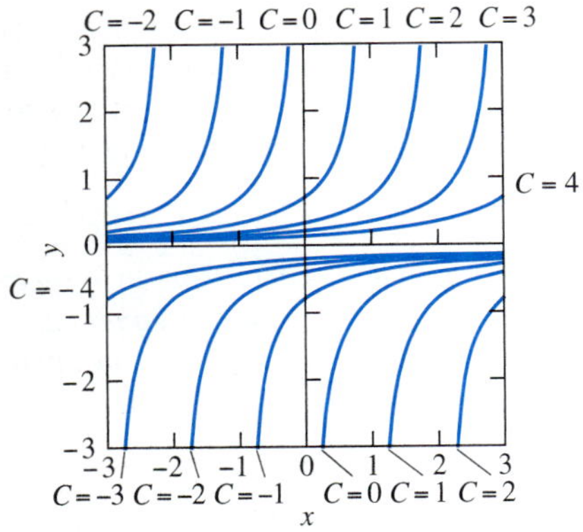

FIGURE 1.1.8. Graphs of solutions of the equation $dy/dx = y^2$.

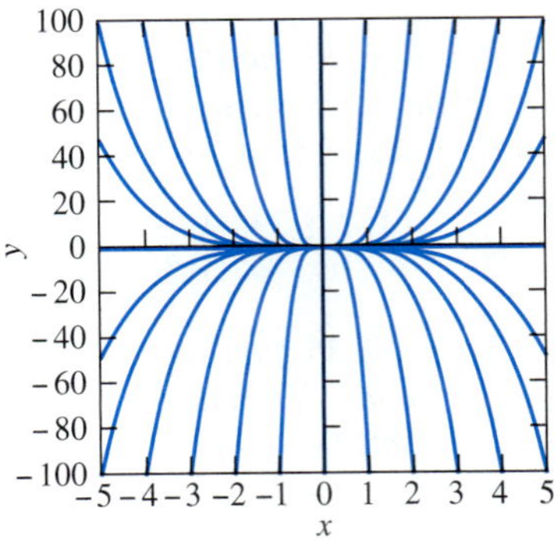

FIGURE 1.1.9. The graph $y = Cx^4$ for various values of C.

48. **(a)** Show that $y(x) = Cx^4$ defines a one-parameter family of differentiable solutions of the differential equation $xy' = 4y$ (Fig. 1.1.9). **(b)** Show that

$$y(x) = \begin{cases} -x^4 & \text{if } x < 0, \\ x^4 & \text{if } x \geqq 0 \end{cases}$$

defines a differentiable solution of $xy' = 4y$ for all x, but is not of the form $y(x) = Cx^4$. **(c)** Given any two real numbers a and b, explain why—in contrast to the situation in part (c) of Problem 47—there exist infinitely many differentiable solutions of $xy' = 4y$ that all satisfy the condition $y(a) = b$.

1.2 Integrals as General and Particular Solutions

The first-order equation $dy/dx = f(x, y)$ takes an especially simple form if the right-hand-side function f does not actually involve the dependent variable y, so

$$\frac{dy}{dx} = f(x). \tag{1}$$

In this special case we need only integrate both sides of Eq. (1) to obtain

$$y(x) = \int f(x)\,dx + C. \tag{2}$$

This is a **general solution** of Eq. (1), meaning that it involves an arbitrary constant C, and for every choice of C it is a solution of the differential equation in (1). If $G(x)$ is a particular antiderivative of f—that is, if $G'(x) \equiv f(x)$—then

$$y(x) = G(x) + C. \tag{3}$$

The graphs of any two such solutions $y_1(x) = G(x) + C_1$ and $y_2(x) = G(x) + C_2$ on the same interval I are "parallel" in the sense illustrated by Figs. 1.2.1 and 1.2.2. There we see that the constant C is geometrically the vertical distance between the two curves $y(x) = G(x)$ and $y(x) = G(x) + C$.

To satisfy an initial condition $y(x_0) = y_0$, we need only substitute $x = x_0$ and $y = y_0$ into Eq. (3) to obtain $y_0 = G(x_0) + C$, so that $C = y_0 - G(x_0)$. With this choice of C, we obtain the **particular solution** of Eq. (1) satisfying the initial value problem

$$\frac{dy}{dx} = f(x), \quad y(x_0) = y_0.$$

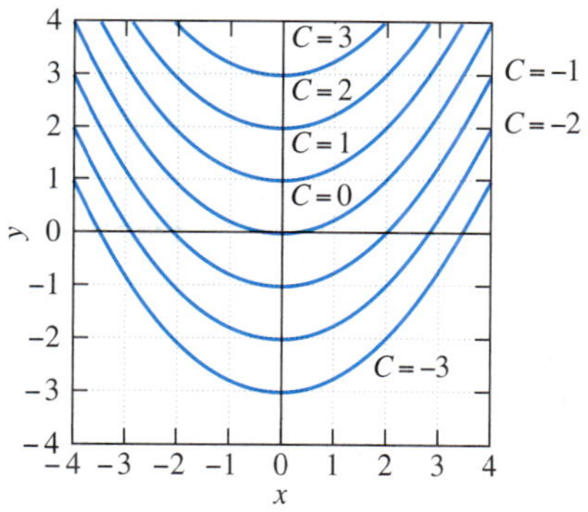

FIGURE 1.2.1. Graphs of $y = \frac{1}{4}x^2 + C$ for various values of C.

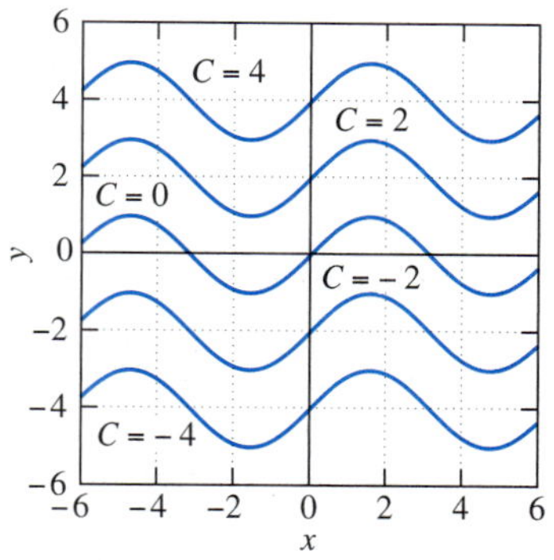

FIGURE 1.2.2. Graphs of $y = \sin x + C$ for various values of C.

We will see that this is the typical pattern for solutions of first-order differential equations. Ordinarily, we will first find a *general solution* involving an arbitrary constant C. We can then attempt to obtain, by appropriate choice of C, a *particular solution* satisfying a given initial condition $y(x_0) = y_0$.

Remark: As the term is used in the previous paragraph, a *general solution* of a first-order differential equation is simply a one-parameter family of solutions. A natural question is whether a given general solution contains *every* particular solution of the differential equation. When this is known to be true, we call it **the** general solution of the differential equation. For example, because any two antiderivatives of the same function $f(x)$ can differ only by a constant, it follows that every solution of Eq. (1) is of the form in (2). Thus Eq. (2) serves to define **the** general solution of (1). ■

Example 1 Solve the initial value problem

$$\frac{dy}{dx} = 2x + 3, \quad y(1) = 2.$$

Solution Integration of both sides of the differential equation as in Eq. (2) immediately yields the general solution

$$y(x) = \int (2x + 3)\,dx = x^2 + 3x + C.$$

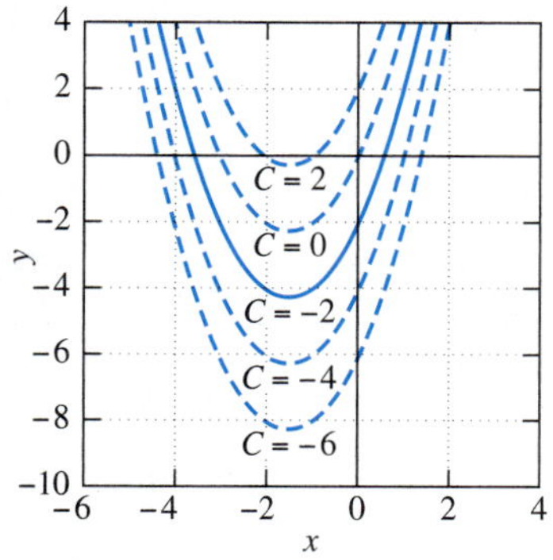

FIGURE 1.2.3. Solution curves for the differential equation in Example 1.

Figure 1.2.3 shows the graph $y = x^2 + 3x + C$ for various values of C. The particular solution we seek corresponds to the curve that passes through the point $(1, 2)$, thereby satisfying the initial condition

$$y(1) = (1)^2 + 3 \cdot (1) + C = 2.$$

It follows that $C = -2$, so the desired particular solution is

$$y(x) = x^2 + 3x - 2.$$ ■

Second-order equations. The observation that the special first-order equation $dy/dx = f(x)$ is readily solvable (provided that an antiderivative of f can be found) extends to second-order differential equations of the special form

$$\frac{d^2y}{dx^2} = g(x), \tag{4}$$

in which the function g on the right-hand side involves neither the dependent variable y nor its derivative dy/dx. We simply integrate once to obtain

$$\frac{dy}{dx} = \int y''(x)\,dx = \int g(x)\,dx = G(x) + C_1,$$

where G is an antiderivative of g and C_1 is an arbitrary constant. Then another integration yields

$$y(x) = \int y'(x)\,dx = \int [G(x) + C_1]\,dx = \int G(x)\,dx + C_1x + C_2,$$

where C_2 is a second arbitrary constant. In effect, the second-order differential equation in (4) is one that can be solved by solving successively the *first-order* equations

$$\frac{dv}{dx} = g(x) \quad \text{and} \quad \frac{dy}{dx} = v(x).$$

Velocity and Acceleration

Direct integration is sufficient to allow us to solve a number of important problems concerning the motion of a particle (or *mass point*) in terms of the forces acting on it. The motion of a particle along a straight line (the x-axis) is described by its **position function**

$$x = f(t) \tag{5}$$

giving its x-coordinate at time t. The **velocity** of the particle is defined to be

$$v(t) = f'(t); \quad \text{that is,} \quad v = \frac{dx}{dt}. \tag{6}$$

Its **acceleration** $a(t)$ is $a(t) = v'(t) = x''(t)$; in Leibniz notation,

$$a = \frac{dv}{dt} = \frac{d^2x}{dt^2}. \tag{7}$$

Equation (6) is sometimes applied either in the indefinite integral form $x(t) = \int v(t)\,dt$ or in the definite integral form

$$x(t) = x(t_0) + \int_{t_0}^{t} v(s)\,ds,$$

which you should recognize as a statement of the fundamental theorem of calculus (precisely because $dx/dt = v$).

Newton's *second law of motion* says that if a force $F(t)$ acts on the particle and is directed along its line of motion, then

$$ma(t) = F(t); \quad \text{that is,} \quad F = ma, \tag{8}$$

where m is the mass of the particle. If the force F is known, then the equation $x''(t) = F(t)/m$ can be integrated twice to find the position function $x(t)$ in terms of two constants of integration. These two arbitrary constants are frequently determined by the **initial position** $x_0 = x(0)$ and the **initial velocity** $v_0 = v(0)$ of the particle.

Constant acceleration. For instance, suppose that the force F, and therefore the acceleration $a = F/m$, are *constant*. Then we begin with the equation

$$\frac{dv}{dt} = a \quad (a \text{ is a constant}) \tag{9}$$

and integrate both sides to obtain

$$v(t) = \int a\,dt = at + C_1.$$

We know that $v = v_0$ when $t = 0$, and substitution of this information into the preceding equation yields the fact that $C_1 = v_0$. So

$$v(t) = \frac{dx}{dt} = at + v_0. \tag{10}$$

A second integration gives

$$x(t) = \int v(t)\,dt = \int (at + v_0)\,dt = \tfrac{1}{2}at^2 + v_0 t + C_2,$$

and the substitution $t = 0$, $x = x_0$ gives $C_2 = x_0$. Therefore,

$$x(t) = \tfrac{1}{2}at^2 + v_0 t + x_0. \tag{11}$$

Thus, with Eq. (10) we can find the velocity, and with Eq. (11) the position, of the particle at any time t in terms of its *constant* acceleration a, its initial velocity v_0, and its initial position x_0.

Example 2 A lunar lander is falling freely toward the surface of the moon at a speed of 450 meters per second (m/s). Its retrorockets, when fired, provide a constant deceleration of 2.5 meters per second per second (m/s^2) (the gravitational acceleration produced by the moon is assumed to be included in the given deceleration). At what height above the lunar surface should the retrorockets be activated to ensure a "soft touchdown" ($v = 0$ at impact)?

Solution We denote by $x(t)$ the height of the lunar lander above the surface, as indicated in Fig. 1.2.4. We let $t = 0$ denote the time at which the retrorockets should be fired. Then $v_0 = -450$ (m/s, negative because the height $x(t)$ is decreasing), and $a = +2.5$, because an upward thrust increases the velocity v (although it decreases the *speed* $|v|$). Then Eqs. (10) and (11) become

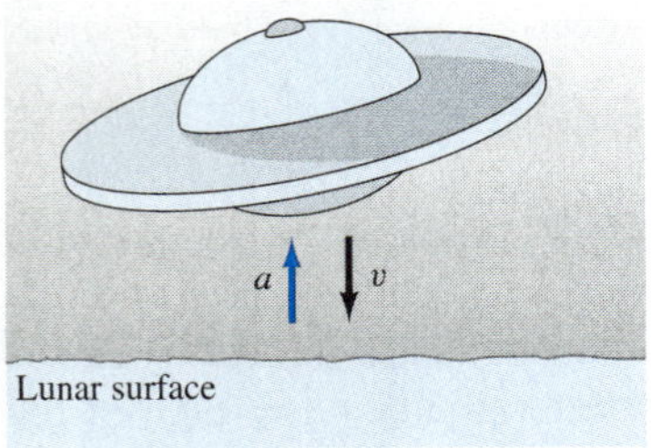

FIGURE 1.2.4. The lunar lander of Example 2.

$$v(t) = 2.5t - 450 \tag{12}$$

and

$$x(t) = 1.25t^2 - 450t + x_0, \tag{13}$$

where x_0 is the height of the lander above the lunar surface at the time $t = 0$ when the retrorockets should be activated.

From Eq. (12) we see that $v = 0$ (soft touchdown) occurs when $t = 450/2.5 = 180$ s (that is, 3 minutes); then substitution of $t = 180$, $x = 0$ into Eq. (13) yields

$$x_0 = 0 - (1.25)(180)^2 + 450(180) = 40{,}500$$

meters—that is, $x_0 = 40.5$ km $\approx 25\frac{1}{6}$ miles. Thus the retrorockets should be activated when the lunar lander is 40.5 kilometers above the surface of the moon, and it will touch down softly on the lunar surface after 3 minutes of decelerating descent. ■

Physical Units

Numerical work requires units for the measurement of physical quantities such as distance and time. We sometimes use ad hoc units—such as distance in miles or kilometers and time in hours—in special situations (such as in a problem involving an auto trip). However, the foot-pound-second (fps) and meter-kilogram-second (mks) unit systems are used more generally in scientific and engineering problems. In fact, fps units are commonly used only in the United States (and a few other countries), while mks units constitute the standard international system of scientific units.

	fps units	**mks units**
Force	pound (lb)	newton (N)
Mass	slug	kilogram (kg)
Distance	foot (ft)	meter (m)
Time	second (s)	second (s)
g	32 ft/s^2	9.8 m/s^2

The last line of this table gives values for the gravitational acceleration g at the surface of the earth. Although these approximate values will suffice for most examples and problems, more precise values are 9.7805 m/s^2 and 32.088 ft/s^2 (at sea level at the equator).

Both systems are compatible with Newton's second law $F = ma$. Thus 1 N is (by definition) the force required to impart an acceleration of 1 m/s^2 to a mass of 1 kg. Similarly, 1 slug is (by definition) the mass that experiences an acceleration of 1 ft/s^2 under a force of 1 lb. (We will use mks units in all problems requiring mass units and thus will rarely need slugs to measure mass.)

Inches and centimeters (as well as miles and kilometers) also are commonly used in describing distances. For conversions between fps and mks units it helps to remember that

$$1 \text{ in.} = 2.54 \text{ cm (exactly)} \quad \text{and} \quad 1 \text{ lb} \approx 4.448 \text{ N}.$$

For instance,

$$1 \text{ ft} = 12 \text{ in.} \times 2.54 \frac{\text{cm}}{\text{in.}} = 30.48 \text{ cm},$$

and it follows that

$$1 \text{ mi} = 5280 \text{ ft} \times 30.48 \frac{\text{cm}}{\text{ft}} = 160934.4 \text{ cm} \approx 1.609 \text{ km}.$$

Thus a posted U.S. speed limit of 50 mi/h means that—in international terms—the legal speed limit is about $50 \times 1.609 \approx 80.45$ km/h.

Vertical Motion with Gravitational Acceleration

The **weight** W of a body is the force exerted on the body by gravity. Substitution of $a = g$ and $F = W$ in Newton's second law $F = ma$ gives

$$W = mg \tag{14}$$

for the weight W of the mass m at the surface of the earth (where $g \approx 32 \text{ ft/s}^2 \approx 9.8$ m/s^2). For instance, a mass of $m = 20$ kg has a weight of $W = (20 \text{ kg})(9.8 \text{ m/s}^2) = 196$ N. Similarly, a mass m weighing 100 pounds has mks weight

$$W = (100 \text{ lb})(4.448 \text{ N/lb}) = 444.8 \text{ N},$$

so its mass is

$$m = \frac{W}{g} = \frac{444.8 \text{ N}}{9.8 \text{ m/s}^2} \approx 45.4 \text{ kg}.$$

To discuss vertical motion it is natural to choose the y-axis as the coordinate system for position, frequently with $y = 0$ corresponding to "ground level." If we choose the *upward* direction as the positive direction, then the effect of gravity on a vertically moving body is to decrease its height and also to decrease its velocity $v = dy/dt$. Consequently, if we ignore air resistance, then the acceleration $a = dv/dt$ of the body is given by

$$\frac{dv}{dt} = -g. \tag{15}$$

This acceleration equation provides a starting point in many problems involving vertical motion. Successive integrations (as in Eqs. (10) and (11)) yield the velocity and height formulas

$$v(t) = -gt + v_0 \tag{16}$$

and

$$y(t) = -\tfrac{1}{2}gt^2 + v_0 t + y_0. \tag{17}$$

Here, y_0 denotes the initial ($t = 0$) height of the body and v_0 its initial velocity.

Example 3 **(a)** Suppose that a ball is thrown straight upward from the ground ($y_0 = 0$) with initial velocity $v_0 = 96$ (ft/s, so we use $g = 32$ ft/s^2 in fps units). Then it reaches its maximum height when its velocity (Eq. (16)) is zero,

$$v(t) = -32t + 96 = 0,$$

and thus when $t = 3$ s. Hence the maximum height that the ball attains is

$$y(3) = -\tfrac{1}{2} \cdot 32 \cdot 3^2 + 96 \cdot 3 + 0 = 144 \text{ (ft)}$$

(with the aid of Eq. (17)).
(b) If an arrow is shot straight upward from the ground with initial velocity $v_0 = 49$ (m/s, so we use $g = 9.8$ m/s^2 in mks units), then it returns to the ground when

$$y(t) = -\tfrac{1}{2} \cdot (9.8)t^2 + 49t = (4.9)t(-t + 10) = 0,$$

and thus after 10 s in the air. ■

A Swimmer's Problem

Figure 1.2.5 shows a northward-flowing river of width $w = 2a$. The lines $x = \pm a$ represent the banks of the river and the y-axis its center. Suppose that the velocity v_R at which the water flows increases as one approaches the center of the river, and indeed is given in terms of distance x from the center by

$$v_R = v_0\left(1 - \frac{x^2}{a^2}\right). \tag{18}$$

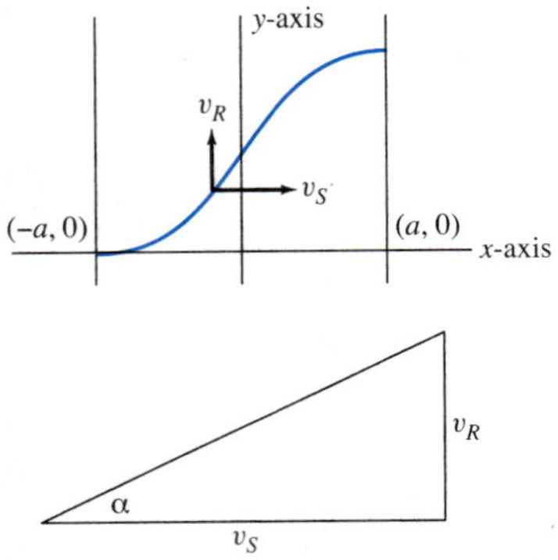

FIGURE 1.2.5. A swimmer's problem (Example 4).

You can use Eq. (18) to verify that the water does flow the fastest at the center, where $v_R = v_0$, and that $v_R = 0$ at each riverbank.

Suppose that a swimmer starts at the point $(-a, 0)$ on the west bank and swims due east (relative to the water) with constant speed v_S. As indicated in Fig. 1.2.5, his velocity vector (relative to the riverbed) has horizontal component v_S and vertical component v_R. Hence the swimmer's direction angle α is given by

$$\tan\alpha = \frac{v_R}{v_S}.$$

Because $\tan\alpha = dy/dx$, substitution using (18) gives the differential equation

$$\frac{dy}{dx} = \frac{v_0}{v_S}\left(1 - \frac{x^2}{a^2}\right) \tag{19}$$

for the swimmer's trajectory $y = y(x)$ as he crosses the river.

Example 4 Suppose that the river is 1 mile wide and that its midstream velocity is $v_0 = 9$ mi/h. If the swimmer's velocity is $v_S = 3$ mi/h, then Eq. (19) takes the form

$$\frac{dy}{dx} = 3(1 - 4x^2).$$

Integration yields

$$y(x) = \int (3 - 12x^2)\,dx = 3x - 4x^3 + C$$

for the swimmer's trajectory. The initial condition $y\left(-\frac{1}{2}\right) = 0$ yields $C = 1$, so

$$y(x) = 3x - 4x^3 + 1.$$

Then

$$y\left(\tfrac{1}{2}\right) = 3\left(\tfrac{1}{2}\right) - 4\left(\tfrac{1}{2}\right)^3 + 1 = 2,$$

so the swimmer drifts 2 miles downstream while he swims 1 mile across the river. ■

1.2 Problems

In Problems 1 through 10, find a function $y = f(x)$ satisfying the given differential equation and the prescribed initial condition.

1. $\dfrac{dy}{dx} = 2x + 1;\ y(0) = 3$

2. $\dfrac{dy}{dx} = (x-2)^2;\ y(2) = 1$

3. $\dfrac{dy}{dx} = \sqrt{x};\ y(4) = 0$

4. $\dfrac{dy}{dx} = \dfrac{1}{x^2};\ y(1) = 5$

5. $\dfrac{dy}{dx} = \dfrac{1}{\sqrt{x+2}};\ y(2) = -1$

6. $\dfrac{dy}{dx} = x\sqrt{x^2+9};\ y(-4) = 0$

7. $\dfrac{dy}{dx} = \dfrac{10}{x^2+1};\ y(0) = 0$

8. $\dfrac{dy}{dx} = \cos 2x;\ y(0) = 1$

9. $\dfrac{dy}{dx} = \dfrac{1}{\sqrt{1-x^2}};\ y(0) = 0$

10. $\dfrac{dy}{dx} = xe^{-x};\ y(0) = 1$

In Problems 11 through 18, find the position function $x(t)$ of a moving particle with the given acceleration $a(t)$, initial position $x_0 = x(0)$, and initial velocity $v_0 = v(0)$.

11. $a(t) = 50,\ v_0 = 10,\ x_0 = 20$

12. $a(t) = -20,\ v_0 = -15,\ x_0 = 5$

13. $a(t) = 3t,\ v_0 = 5,\ x_0 = 0$

14. $a(t) = 2t + 1,\ v_0 = -7,\ x_0 = 4$

15. $a(t) = 4(t+3)^2,\ v_0 = -1,\ x_0 = 1$

16. $a(t) = \dfrac{1}{\sqrt{t+4}},\ v_0 = -1,\ x_0 = 1$

17. $a(t) = \dfrac{1}{(t+1)^3},\ v_0 = 0,\ x_0 = 0$

18. $a(t) = 50 \sin 5t,\ v_0 = -10,\ x_0 = 8$

In Problems 19 through 22, a particle starts at the origin and travels along the x-axis with the velocity function $v(t)$ whose graph is shown in Figs. 1.2.6 through 1.2.9. Sketch the graph of the resulting position function $x(t)$ for $0 \leqq t \leqq 10$.

19.

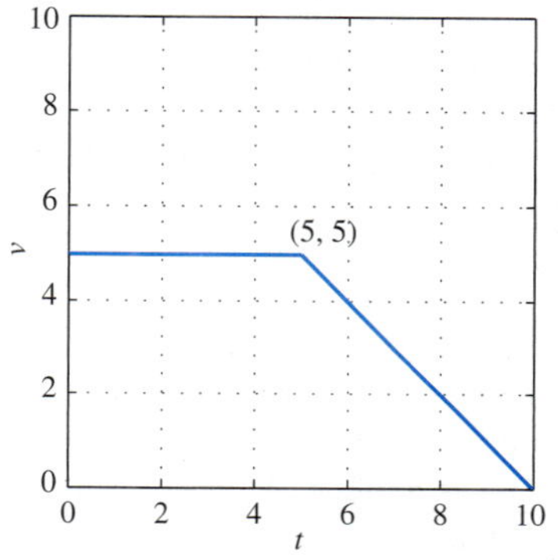

FIGURE 1.2.6. Graph of the velocity function $v(t)$ of Problem 19.

20.

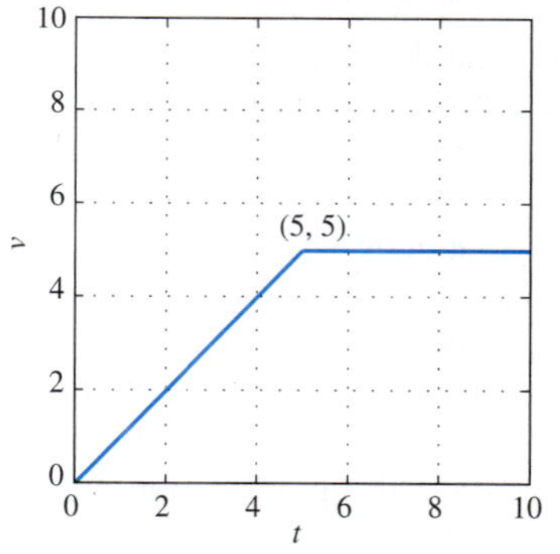

FIGURE 1.2.7. Graph of the velocity function $v(t)$ of Problem 20.

21.

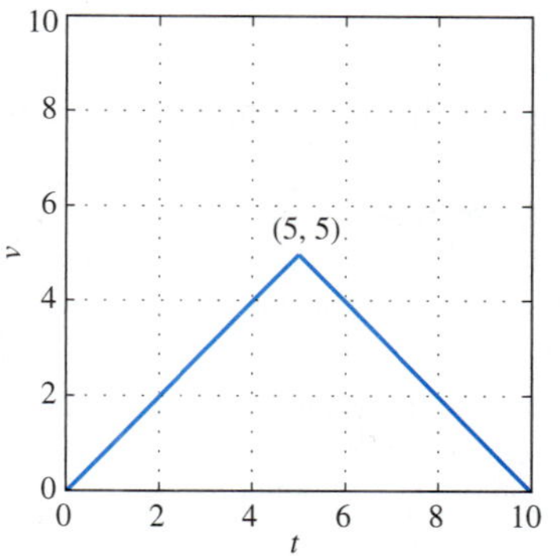

FIGURE 1.2.8. Graph of the velocity function $v(t)$ of Problem 21.

22.

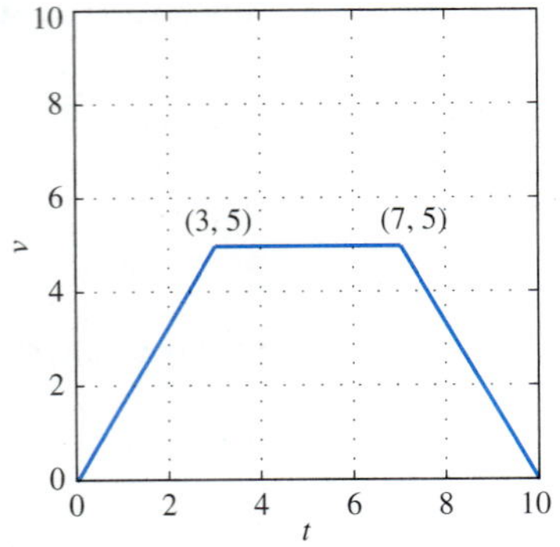

FIGURE 1.2.9. Graph of the velocity function $v(t)$ of Problem 22.

23. What is the maximum height attained by the arrow of part (b) of Example 3?

24. A ball is dropped from the top of a building 400 ft high. How long does it take to reach the ground? With what speed does the ball strike the ground?

25. The brakes of a car are applied when it is moving at 100 km/h and provide a constant deceleration of 10 meters per second per second (m/s^2). How far does the car travel before coming to a stop?

26. A projectile is fired straight upward with an initial velocity of 100 m/s from the top of a building 20 m high and falls to the ground at the base of the building. Find (a) its maximum height above the ground; (b) when it passes the top of the building; (c) its total time in the air.

27. A ball is thrown straight downward from the top of a tall building. The initial speed of the ball is 10 m/s. It strikes the ground with a speed of 60 m/s. How tall is the building?

28. A baseball is thrown straight downward with an initial speed of 40 ft/s from the top of the Washington Monument (555 ft high). How long does it take to reach the ground, and with what speed does the baseball strike the ground?

29. A diesel car gradually speeds up so that for the first 10 s its acceleration is given by

$$\frac{dv}{dt} = (0.12)t^2 + (0.6)t \quad (\text{ft/s}^2).$$

If the car starts from rest ($x_0 = 0$, $v_0 = 0$), find the distance it has traveled at the end of the first 10 s and its velocity at that time.

30. A car traveling at 60 mi/h (88 ft/s) skids 176 ft after its brakes are suddenly applied. Under the assumption that the braking system provides constant deceleration, what is that deceleration? For how long does the skid continue?

31. The skid marks made by an automobile indicated that its brakes were fully applied for a distance of 75 m before it came to a stop. The car in question is known to have a constant deceleration of 20 m/s^2 under these conditions. How fast—in km/h—was the car traveling when the brakes were first applied?

32. Suppose that a car skids 15 m if it is moving at 50 km/h when the brakes are applied. Assuming that the car has the same constant deceleration, how far will it skid if it is moving at 100 km/h when the brakes are applied?

33. On the planet Gzyx, a ball dropped from a height of 20 ft hits the ground in 2 s. If a ball is dropped from the top of a 200-ft-tall building on Gzyx, how long will it take to hit the ground? With what speed will it hit?

34. A person can throw a ball straight upward from the surface of the earth to a maximum height of 144 ft. How high could this person throw the ball on the planet Gzyx of Problem 33?

35. A stone is dropped from rest at an initial height h above the surface of the earth. Show that the speed with which it strikes the ground is $v = \sqrt{2gh}$.

36. Suppose a woman has enough "spring" in her legs to jump (on earth) from the ground to a height of 2.25 feet. If she jumps straight upward with the same initial velocity on the moon—where the surface gravitational acceleration is (approximately) 5.3 ft/s^2—how high above the surface will she rise?

37. At noon a car starts from rest at point A and proceeds at constant acceleration along a straight road toward point B. If the car reaches B at 12:50 P.M. with a velocity of 60 mi/h, what is the distance from A to B?

38. At noon a car starts from rest at point A and proceeds with constant acceleration along a straight road toward point C, 35 miles away. If the constantly accelerated car arrives at C with a velocity of 60 mi/h, at what time does it arrive at C?

39. If $a = 0.5$ mi and $v_0 = 9$ mi/h as in Example 4, what must the swimmer's speed v_S be in order that he drifts only 1 mile downstream as he crosses the river?

40. Suppose that $a = 0.5$ mi, $v_0 = 9$ mi/h, and $v_S = 3$ mi/h as in Example 4, but that the velocity of the river is given by the fourth-degree function

$$v_R = v_0\left(1 - \frac{x^4}{a^4}\right)$$

rather than the quadratic function in Eq. (18). Now find how far downstream the swimmer drifts as he crosses the river.

41. A bomb is dropped from a helicopter hovering at an altitude of 800 feet above the ground. From the ground directly beneath the helicopter, a projectile is fired straight upward toward the bomb, exactly 2 seconds after the bomb is released. With what initial velocity should the projectile be fired, in order to hit the bomb at an altitude of exactly 400 feet?

42. A spacecraft is in free fall toward the surface of the moon at a speed of 1000 mph (mi/h). Its retrorockets, when fired, provide a constant deceleration of 20,000 mi/h^2. At what height above the lunar surface should the astronauts fire the retrorockets to insure a soft touchdown? (As in Example 2, ignore the moon's gravitational field.)

43. Arthur Clarke's *The Wind from the Sun* (1963) describes Diana, a spacecraft propelled by the solar wind. Its aluminized sail provides it with a constant acceleration of $0.001g = 0.0098$ m/s^2. Suppose this spacecraft starts from rest at time $t = 0$ and simultaneously fires a projectile (straight ahead in the same direction) that travels at one-tenth of the speed $c = 3 \times 10^8$ m/s of light. How long will it take the spacecraft to catch up with the projectile, and how far will it have traveled by then?

44. A driver involved in an accident claims he was going only 25 mph. When police tested his car, they found that when its brakes were applied at 25 mph, the car skidded only 45 feet before coming to a stop. But the driver's skid marks at the accident scene measured 210 feet. Assuming the same (constant) deceleration, determine the speed he was actually traveling just prior to the accident.

1.3 Slope Fields and Solution Curves

Consider a differential equation of the form

$$\frac{dy}{dx} = f(x, y), \tag{1}$$

where the right-hand function $f(x, y)$ involves both the independent variable x and the dependent variable y. We might think of integrating both sides in (1) with respect to x, and hence write $y(x) = \int f(x, y(x))\,dx + C$. However, this approach does not lead to a solution of the differential equation, because the indicated integral involves the *unknown* function $y(x)$ itself, and therefore cannot be evaluated explicitly. Actually, there exists *no* straightforward procedure by which a general differential equation can be solved explicitly. Indeed, the solutions of such a simple-looking differential equation as $y' = x^2 + y^2$ cannot be expressed in terms of the ordinary elementary functions studied in calculus textbooks. Nevertheless, the graphical and numerical methods of this and later sections can be used to construct *approximate* solutions of differential equations that suffice for many practical purposes.

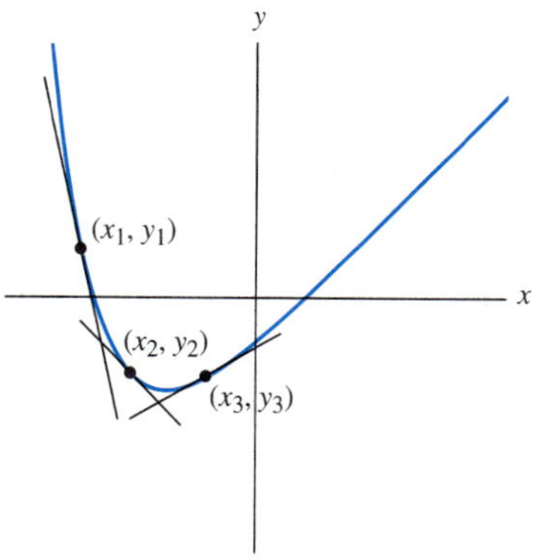

FIGURE 1.3.1. A solution curve for the differential equation $y' = x - y$ together with tangent lines having

- slope $m_1 = x_1 - y_1$ at the point (x_1, y_1);
- slope $m_2 = x_2 - y_2$ at the point (x_2, y_2); and
- slope $m_3 = x_3 - y_3$ at the point (x_3, y_3).

Slope Fields and Graphical Solutions

There is a simple geometric way to think about solutions of a given differential equation $y' = f(x, y)$. At each point (x, y) of the xy-plane, the value of $f(x, y)$ determines a slope $m = f(x, y)$. A solution of the differential equation is simply a differentiable function whose graph $y = y(x)$ has this "correct slope" at each point $(x, y(x))$ through which it passes—that is, $y'(x) = f(x, y(x))$. Thus a **solution curve** of the differential equation $y' = f(x, y)$—the graph of a solution of the equation—is simply a curve in the xy-plane whose tangent line at each point (x, y) has slope $m = f(x, y)$. For instance, Fig. 1.3.1 shows a solution curve of the differential equation $y' = x - y$ together with its tangent lines at three typical points.

This geometric viewpoint suggests a *graphical method* for constructing *approximate* solutions of the differential equation $y' = f(x, y)$. Through each of a representative collection of points (x, y) in the plane we draw a short line segment having the proper slope $m = f(x, y)$. All these line segments constitute a **slope field** (or a **direction field**) for the equation $y' = f(x, y)$.

Example 1 Figures 1.3.2 (a)–(d) show slope fields and solution curves for the differential equation

$$\frac{dy}{dx} = ky \tag{2}$$

with the values $k = 2, 0.5, -1$, and -3 of the parameter k in Eq. (2). Note that each slope field yields important qualitative information about the set of all solutions

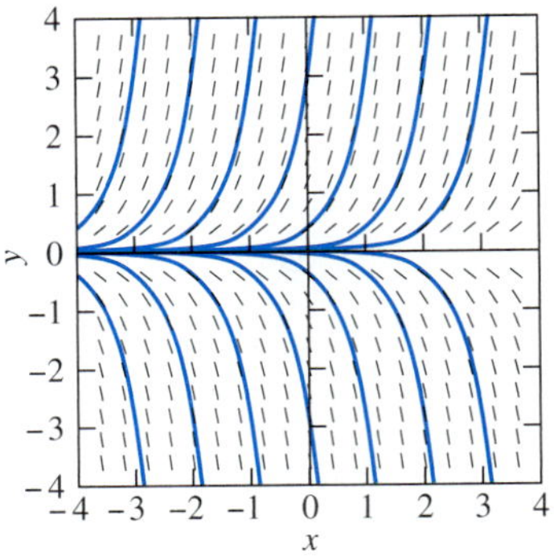

FIGURE 1.3.2(a) Slope field and solution curves for $y' = 2y$.

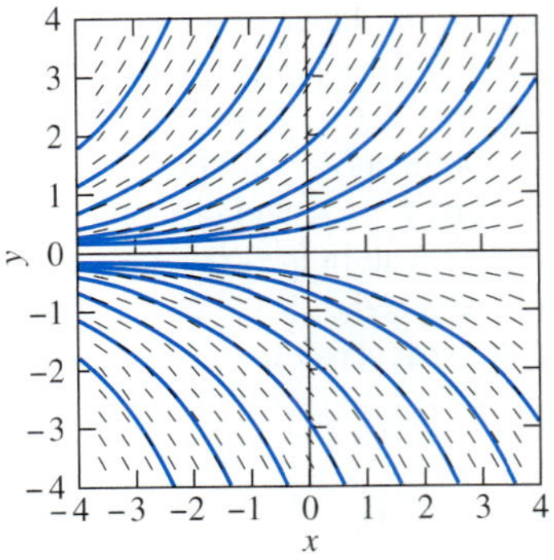

FIGURE 1.3.2(b) Slope field and solution curves for $y' = (0.5)y$.

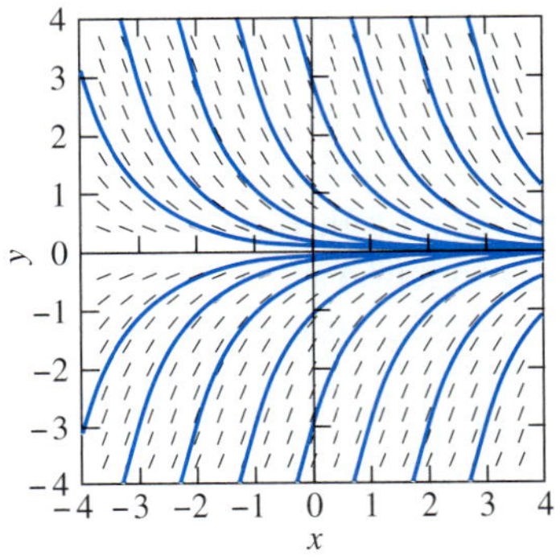

FIGURE 1.3.2(c) Slope field and solution curves for $y' = -y$.

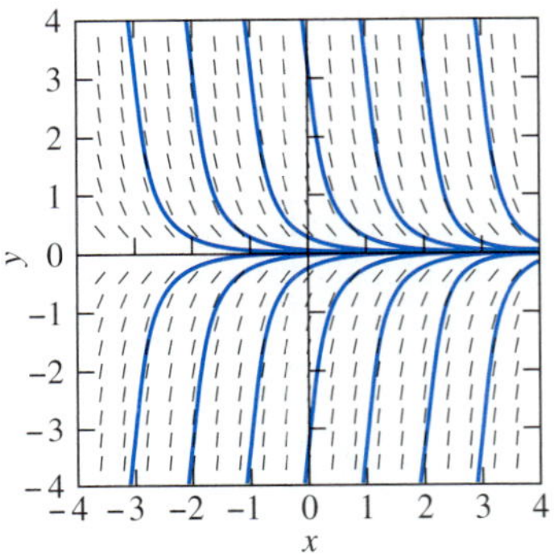

FIGURE 1.3.2(d) Slope field and solution curves for $y' = -3y$.

of the differential equation. For instance, Figs. 1.3.2(a) and (b) suggest that each solution $y(x)$ approaches $\pm\infty$ as $x \to +\infty$ if $k > 0$, whereas Figs. 1.3.2(c) and (d) suggest that $y(x) \to 0$ as $x \to +\infty$ if $k < 0$. Moreover, although the sign of k determines the *direction* of increase or decrease of $y(x)$, its absolute value $|k|$ appears to determine the *rate of change* of $y(x)$. All this is apparent from slope fields like those in Fig. 1.3.2, even without knowing that the general solution of Eq. (2) is given explicitly by $y(x) = Ce^{kx}$. ■

A slope field suggests visually the general shapes of solution curves of the differential equation. Through each point a solution curve should proceed in such a direction that its tangent line is nearly parallel to the nearby line segments of the slope field. Starting at any initial point (a, b), we can attempt to sketch freehand an approximate solution curve that threads its way through the slope field, following the visible line segments as closely as possible.

Example 2 Construct a slope field for the differential equation $y' = x - y$ and use it to sketch an approximate solution curve that passes through the point $(-4, 4)$.

Solution Figure 1.3.3 shows a table of slopes for the given equation. The numerical slope $m = x - y$ appears at the intersection of the horizontal x-row and the vertical y-column of the table. If you inspect the pattern of upper-left to lower-right diagonals in this table, you can see that it was easily and quickly constructed. (Of

$x \backslash y$	−4	−3	−2	−1	0	1	2	3	4
−4	0	−1	−2	−3	−4	−5	−6	−7	−8
−3	1	0	−1	−2	−3	−4	−5	−6	−7
−2	2	1	0	−1	−2	−3	−4	−5	−6
−1	3	2	1	0	−1	−2	−3	−4	−5
0	4	3	2	1	0	−1	−2	−3	−4
1	5	4	3	2	1	0	−1	−2	−3
2	6	5	4	3	2	1	0	−1	−2
3	7	6	5	4	3	2	1	0	−1
4	8	7	6	5	4	3	2	1	0

FIGURE 1.3.3. Values of the slope $y' = x - y$ for $-4 \leqq x,\ y \leqq 4$.

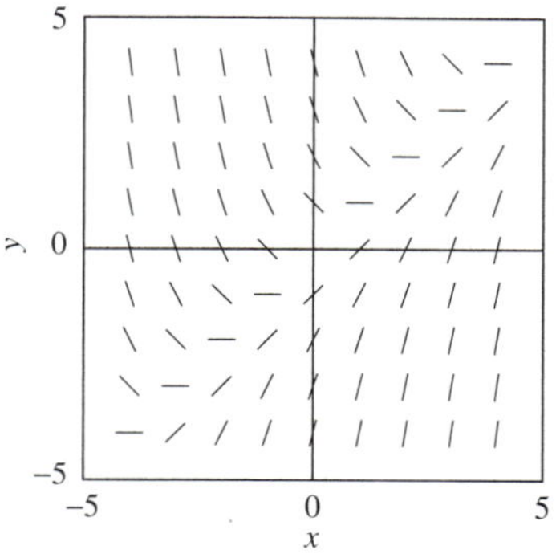

FIGURE 1.3.4. Slope field for $y' = x - y$ corresponding to the table of slopes in Fig. 1.3.3.

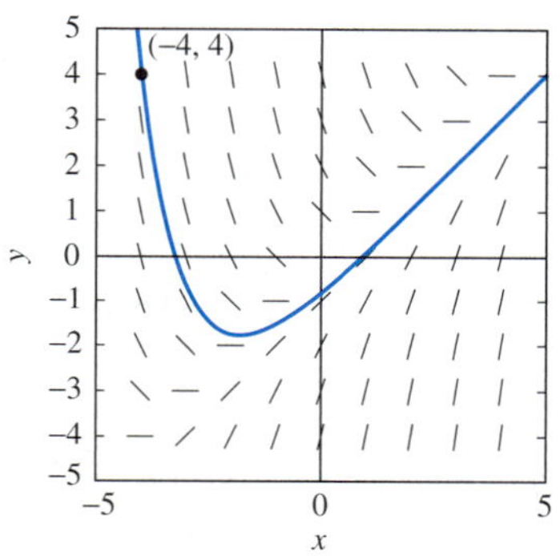

FIGURE 1.3.5. The solution curve through (−4, 4).

course, a more complicated function $f(x, y)$ on the right-hand side of the differential equation would necessitate more complicated calculations.) Figure 1.3.4 shows the corresponding slope field, and Fig. 1.3.5 shows an approximate solution curve sketched through the point (−4, 4) so as to follow this slope field as closely as possible. At each point it appears to proceed in the direction indicated by the nearby line segments of the slope field. ■

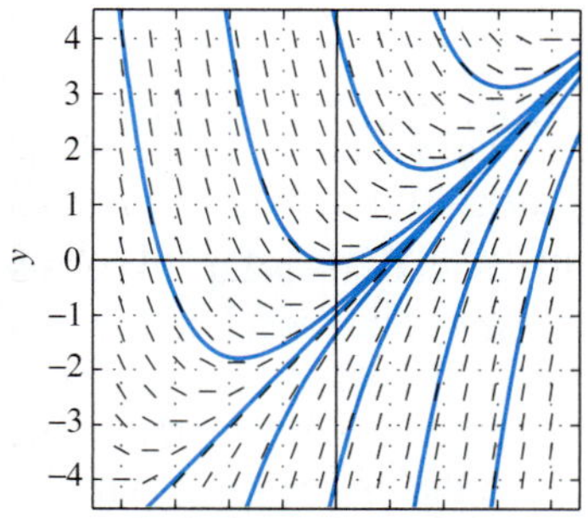

FIGURE 1.3.6. Slope field and typical solution curves for $y' = x - y$.

Although a spreadsheet program (for instance) readily constructs a table of slopes as in Fig. 1.3.3, it can be quite tedious to plot by hand a sufficient number of slope segments as in Fig. 1.3.4. However, most computer algebra systems include commands for quick and ready construction of slope fields with as many line segments as desired; such commands are illustrated in the application material for this section. The more line segments are constructed, the more accurately solution curves can be visualized and sketched. Figure 1.3.6 shows a "finer" slope field for the differential equation $y' = x - y$ of Example 2, together with typical solution curves treading through this slope field.

If you look closely at Fig. 1.3.6, you may spot a solution curve that appears to be a straight line! Indeed, you can verify that the linear function $y = x - 1$ is a solution of the equation $y' = x - y$, and it appears likely that the other solution curves approach this straight line as an asymptote as $x \to +\infty$. This inference illustrates the fact that a slope field can suggest tangible information about solutions that is not at all evident from the differential equation itself. Can you, by tracing the appropriate solution curve in this figure, infer that $y(3) \approx 2$ for the solution $y(x)$ of the initial value problem $y' = x - y$, $y(-4) = 4$?

Applications of Slope Fields

The next two examples illustrate the use of slope fields to glean useful information in physical situations that are modeled by differential equations. Example 3 is based on the fact that a baseball moving through the air at a moderate speed v (less than about 300 ft/s) encounters air resistance that is approximately proportional to v. If the baseball is thrown straight downward from the top of a tall building or from a hovering helicopter, then it experiences both the downward acceleration of gravity and an upward acceleration of air resistance. If the y-axis is directed *downward*, then the ball's velocity $v = dy/dt$ and its gravitational acceleration $g = 32$ ft/s^2 are both positive, while its acceleration due to air resistance is negative. Hence its total acceleration is of the form

$$\frac{dv}{dt} = g - kv. \tag{3}$$

A typical value of the air resistance proportionality constant might be $k = 0.16$.

Example 3 Suppose you throw a baseball straight downward from a helicopter hovering at an altitude of 3000 feet. You wonder whether someone standing on the ground below could conceivably catch it. In order to estimate the speed with which the ball will land, you can use your laptop's computer algebra system to construct a slope field for the differential equation

$$\frac{dv}{dt} = 32 - 0.16v. \tag{4}$$

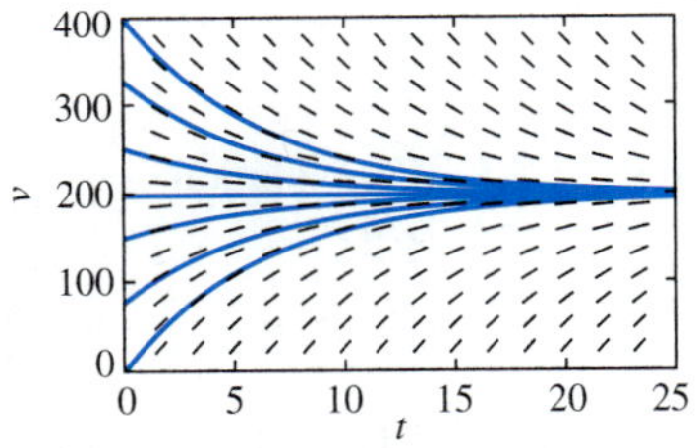

FIGURE 1.3.7. Slope field and typical solution curves for $v' = 32 - 0.16v$.

The result is shown in Fig. 1.3.7, together with a number of solution curves corresponding to different values of the initial velocity $v(0)$ with which you might throw the baseball downward. Note that all these solution curves appear to approach the horizontal line $v = 200$ as an asymptote. This implies that—however you throw it—the baseball should approach the *limiting velocity* $v = 200$ ft/s instead of accelerating indefinitely (as it would in the absence of any air resistance). The handy fact that 60 mi/h = 88 ft/s yields

$$v = 200\frac{\text{ft}}{\text{s}} \times \frac{60 \text{ mi/h}}{88 \text{ ft/s}} \approx 136.36 \frac{\text{mi}}{\text{h}}.$$

Perhaps a catcher accustomed to 100 mi/h fastballs would have some chance of fielding this speeding ball. ■

Comment: If the ball's initial velocity is $v(0) = 200$, then Eq. (4) gives $v'(0) = 32 - (0.16)(200) = 0$, so the ball experiences *no* initial acceleration. Its velocity therefore remains unchanged, and hence $v(t) \equiv 200$ is a constant "equilibrium solution" of the differential equation. If the initial velocity is greater than 200, then the initial acceleration given by Eq. (4) is negative, so the ball slows down as it falls. But if the initial velocity is less than 200, then the initial acceleration given by (4) is positive, so the ball speeds up as it falls. It therefore seems quite reasonable that, because of air resistance, the baseball will approach a limiting velocity of 200 ft/s—whatever initial velocity it starts with. You might like to verify that—in the absence of air resistance—this ball would hit the ground at over 300 mi/h. ■

In Section 2.1 we will discuss in detail the logistic differential equation

$$\frac{dP}{dt} = kP(M - P) \tag{5}$$

that often is used to model a population $P(t)$ that inhabits an environment with *carrying capacity* M. This means that M is the maximum population that this environment can sustain on a long-term basis (in terms of the maximum available food, for instance).

Example 4 If we take $k = 0.0004$ and $M = 150$, then the logistic equation in (5) takes the form

$$\frac{dP}{dt} = 0.0004P(150 - P) = 0.06P - 0.0004P^2. \tag{6}$$

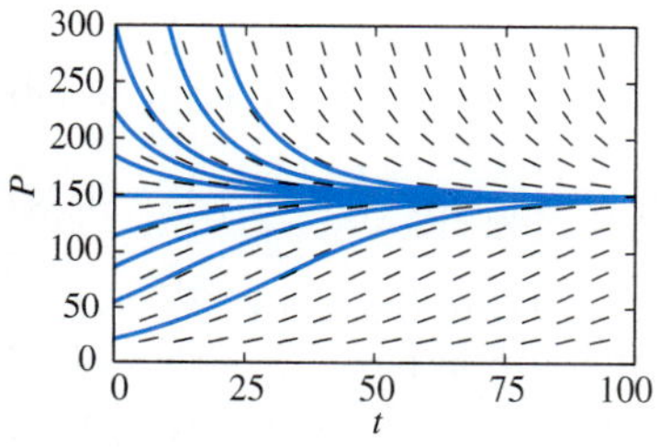

FIGURE 1.3.8. Slope field and typical solution curves for $P' = 0.06P - 0.0004P^2$.

The positive term $0.06P$ on the right in (6) corresponds to natural growth at a 6% annual rate (with time t measured in years). The negative term $-0.0004P^2$ represents the inhibition of growth due to limited resources in the environment.

Figure 1.3.8 shows a slope field for Eq. (6), together with a number of solution curves corresponding to possible different values of the initial population $P(0)$. Note that all these solution curves appear to approach the horizontal line $P = 150$ as an asymptote. This implies that—whatever the initial population—the population $P(t)$ approaches the *limiting population* $P = 150$ as $t \to \infty$. ■

Comment: If the initial population is $P(0) = 150$, then Eq. (6) gives

$$P'(0) = 0.0004(150)(150 - 150) = 0,$$

so the population experiences *no* initial (instantaneous) change. It therefore remains unchanged, and hence $P(t) \equiv 150$ is a constant "equilibrium solution" of the differential equation. If the initial population is greater than 150, then the initial rate of change given by (6) is negative, so the population immediately begins to decrease. But if the initial population is less than 150, then the initial rate of change given by (6) is positive, so the population immediately begins to increase. It therefore seems quite reasonable to conclude that the population will approach a limiting value of 150—whatever the (positive) initial population. ■

Existence and Uniqueness of Solutions

Before one spends much time attempting to solve a given differential equation, it is wise to know that solutions actually *exist*. We may also want to know whether there is only one solution of the equation satisfying a given initial condition—that is, whether its solutions are *unique*.

Example 5 **(a)** [Failure of existence] The initial value problem

$$y' = \frac{1}{x}, \quad y(0) = 0 \tag{7}$$

has *no* solution, because no solution $y(x) = \int(1/x)\,dx = \ln|x| + C$ of the differential equation is defined at $x = 0$. We see this graphically in Fig. 1.3.9, which shows a direction field and some typical solution curves for the equation $y' = 1/x$. It is apparent that the indicated direction field "forces" all solution curves near the y-axis to plunge downward so that none can pass through the point $(0, 0)$.

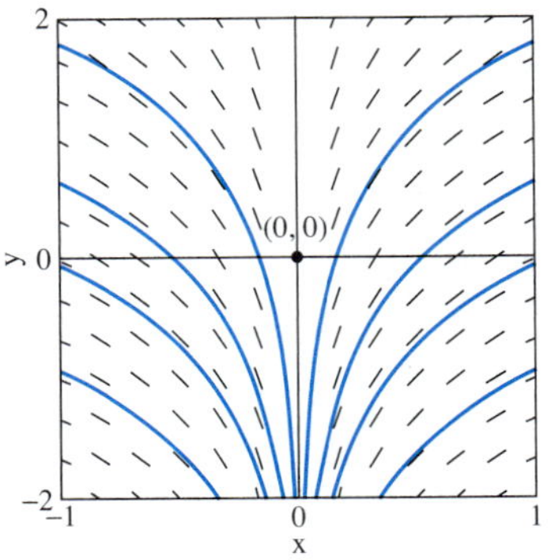

FIGURE 1.3.9. Direction field and typical solution curves for the equation $y' = 1/x$.

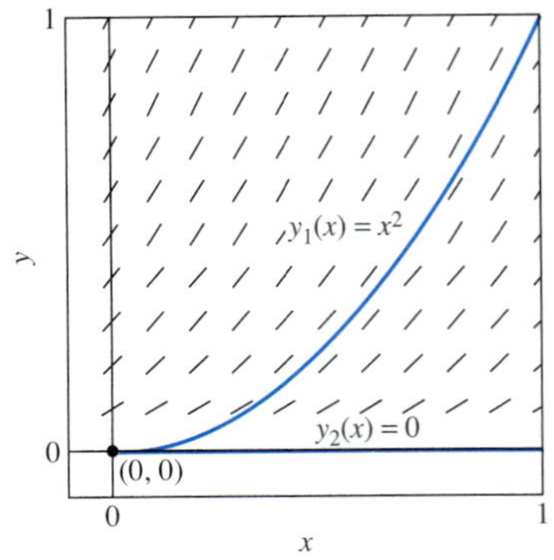

FIGURE 1.3.10. Direction field and two different solution curves for the initial value problem $y' = 2\sqrt{y}$, $y(0) = 0$.

(b) [Failure of uniqueness] On the other hand, you can readily verify that the initial value problem

$$y' = 2\sqrt{y}, \quad y(0) = 0 \tag{8}$$

has the *two* different solutions $y_1(x) = x^2$ and $y_2(x) \equiv 0$ (see Problem 27). Figure 1.3.10 shows a direction field and these two different solution curves for the initial value problem in (8). We see that the curve $y_1(x) = x^2$ threads its way through the indicated direction field, whereas the differential equation $y' = 2\sqrt{y}$ specifies slope $y' = 0$ along the x-axis $y_2(x) = 0$. ■

Example 5 illustrates the fact that, before we can speak of "the" solution of an initial value problem, we need to know that it has *one and only one* solution. Questions of existence and uniqueness of solutions also bear on the process of mathematical modeling. Suppose that we are studying a physical system whose behavior is completely determined by certain initial conditions, but that our proposed mathematical model involves a differential equation *not* having a unique solution satisfying those conditions. This raises an immediate question as to whether the mathematical model adequately represents the physical system.

The theorem stated below implies that the initial value problem $y' = f(x, y)$, $y(a) = b$ has one and only one solution defined near the point $x = a$ on the x-axis, provided that both the function f and its partial derivative $\partial f/\partial y$ are continuous near the point (a, b) in the xy-plane. Methods of proving existence and uniqueness theorems are discussed in Appendix A.

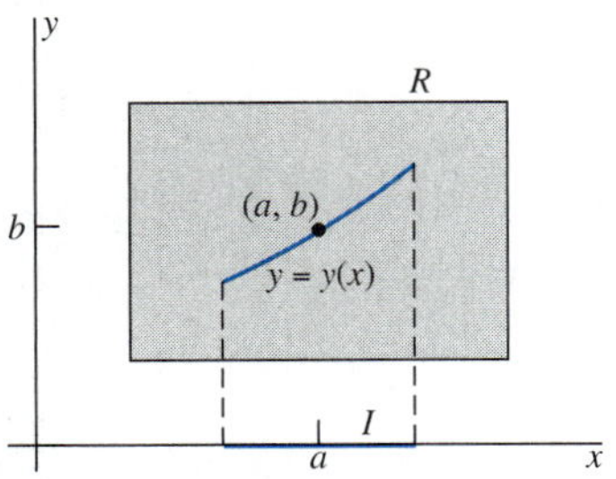

FIGURE 1.3.11. The rectangle R and x-interval I of Theorem 1, and the solution curve $y = y(x)$ through the point (a, b).

THEOREM 1 Existence and Uniqueness of Solutions

Suppose that both the function $f(x, y)$ and its partial derivative $D_y f(x, y)$ are continuous on some rectangle R in the xy-plane that contains the point (a, b) in its interior. Then, for some open interval I containing the point a, the initial value problem

$$\frac{dy}{dx} = f(x, y), \quad y(a) = b \tag{9}$$

has one and only one solution that is defined on the interval I. (As illustrated in Fig. 1.3.11, the solution interval I may not be as "wide" as the original rectangle R of continuity; see Remark 3 on the next page.)

Remark 1: In the case of the differential equation $dy/dx = -y$ of Example 1 and Fig. 1.3.2(c), both the function $f(x, y) = -y$ and the partial derivative $\partial f/\partial y = -1$ are continuous everywhere, so Theorem 1 implies the existence of a unique solution for any initial data (a, b). Although the theorem ensures existence only on some open interval containing $x = a$, each solution $y(x) = Ce^{-x}$ actually is defined for all x.

Remark 2: In the case of the differential equation $dy/dx = 2\sqrt{y}$ of Example 5(b) and Eq. (8), the function $f(x, y) = 2\sqrt{y}$ is continuous wherever $y > 0$, but the partial derivative $\partial f/\partial y = 1/\sqrt{y}$ is discontinuous when $y = 0$, and hence at the point $(0, 0)$. This is why it is possible for there to exist two different solutions $y_1(x) = x^2$ and $y_2(x) \equiv 0$, each of which satisfies the initial condition $y(0) = 0$.

Remark 3: In Example 7 of Section 1.1 we examined the especially simple differential equation $dy/dx = y^2$. Here we have $f(x, y) = y^2$ and $\partial f/\partial y = 2y$. Both of these functions are continuous everywhere in the xy-plane, and in particular on the rectangle $-2 < x < 2$, $0 < y < 2$. Because the point $(0, 1)$ lies in the interior of this rectangle, Theorem 1 guarantees a unique solution—necessarily a continuous function—of the initial value problem

$$\frac{dy}{dx} = y^2, \quad y(0) = 1 \tag{10}$$

on some open x-interval containing $a = 0$. Indeed this is the solution

$$y(x) = \frac{1}{1 - x}$$

that we discussed in Example 7. But $y(x) = 1/(1 - x)$ is discontinuous at $x = 1$, so our unique continuous solution does not exist on the entire interval $-2 < x < 2$. Thus the solution interval I of Theorem 1 may not be as wide as the rectangle R where f and $\partial f/\partial y$ are continuous. Geometrically, the reason is that the solution curve provided by the theorem may leave the rectangle—wherein solutions of the differential equation are guaranteed to exist—before it reaches the one or both ends of the interval (see Fig. 1.3.12). ■

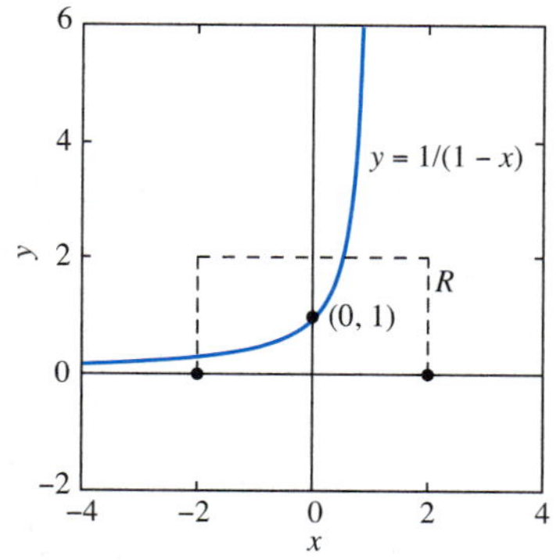

FIGURE 1.3.12. The solution curve through the initial point (0, 1) leaves the rectangle R before it reaches the right side of R.

The following example shows that, if the function $f(x, y)$ and/or its partial derivative $\partial f/\partial y$ fail to satisfy the continuity hypothesis of Theorem 1, then the initial value problem in (9) may have *either* no solution *or* many—even infinitely many—solutions.

Example 6 Consider the first-order differential equation

$$x\frac{dy}{dx} = 2y. \tag{11}$$

Applying Theorem 1 with $f(x, y) = 2y/x$ and $\partial f/\partial y = 2/x$, we conclude that Eq. (11) must have a unique solution near any point in the xy-plane where $x \neq 0$. Indeed, we see immediately by substitution in (11) that

$$y(x) = Cx^2 \tag{12}$$

satisfies Eq. (11) for any value of the constant C and for all values of the variable x. In particular, the initial value problem

$$x\frac{dy}{dx} = 2y, \quad y(0) = 0 \tag{13}$$

has infinitely many different solutions, whose solution curves are the parabolas $y = Cx^2$ illustrated in Fig. 1.3.13. (In case $C = 0$, the "parabola" is actually the x-axis $y = 0$.)

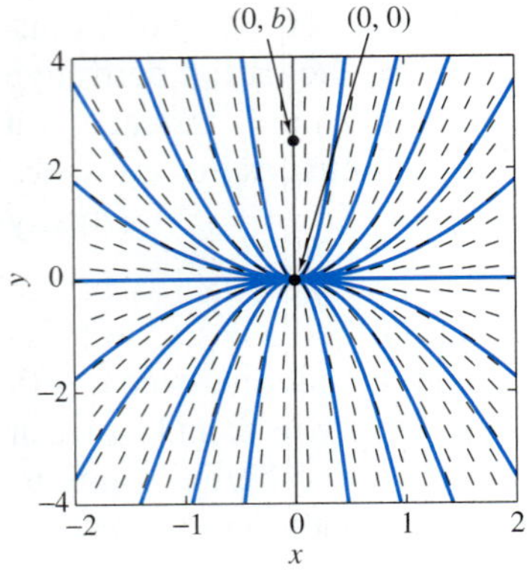

FIGURE 1.3.13. There are infinitely many solution curves through the point (0, 0), but no solution curves through the point $(0, b)$ if $b \neq 0$.

Observe that all these parabolas pass through the origin (0, 0), but none of them passes through any other point on the y-axis. It follows that the initial value problem in (13) has infinitely many solutions, but the initial value problem

$$x\frac{dy}{dx} = 2y, \quad y(0) = b \tag{14}$$

has no solution if $b \neq 0$.

Finally, note that through any point off the y-axis there passes only one of the parabolas $y = Cx^2$. Hence, if $a \neq 0$, then the initial value problem

$$x\frac{dy}{dx} = 2y, \quad y(a) = b \tag{15}$$

has a unique solution on any interval that contains the point $x = a$ but not the origin $x = 0$. In summary, the initial value problem in (15) has

- a unique solution near (a, b) if $a \neq 0$;
- no solution if $a = 0$ but $b \neq 0$;
- infinitely many solutions if $a = b = 0$. ■

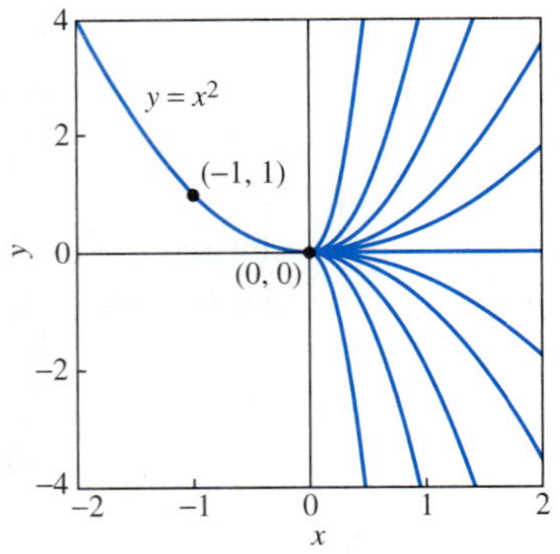

FIGURE 1.3.14. There are infinitely many solution curves through the point $(1, -1)$.

Still more can be said about the initial value problem in (15). Consider a typical initial point off the y-axis—for instance, the point $(-1, 1)$ indicated in Fig. 1.3.14. Then for any value of the constant C the function defined by

$$y(x) = \begin{cases} x^2 & \text{if } x \leq 0, \\ Cx^2 & \text{if } x > 0 \end{cases} \tag{16}$$

is continuous and satisfies the initial value problem

$$x\frac{dy}{dx} = 2y, \quad y(-1) = 1. \tag{17}$$

For a particular value of C, the solution curve defined by (16) consists of the left half of the parabola $y = x^2$ and the right half of the parabola $y = Cx^2$. Thus the unique solution curve near $(-1, 1)$ branches at the origin into the infinitely many solution curves illustrated in Fig. 1.3.14.

We therefore see that Theorem 1 (if its hypotheses are satisfied) guarantees uniqueness of the solution near the initial point (a, b), but a solution curve through (a, b) may eventually branch elsewhere so that uniqueness is lost. Thus a solution may exist on a larger interval than one on which the solution is unique. For instance, the solution $y(x) = x^2$ of the initial value problem in (17) exists on the whole x-axis, but this solution is unique only on the negative x-axis $-\infty < x < 0$.

1.3 Problems

In Problems 1 through 10, we have provided the slope field of the indicated differential equation, together with one or more solution curves. Sketch likely solution curves through the additional points marked in each slope field.

1. $\dfrac{dy}{dx} = -y - \sin x$

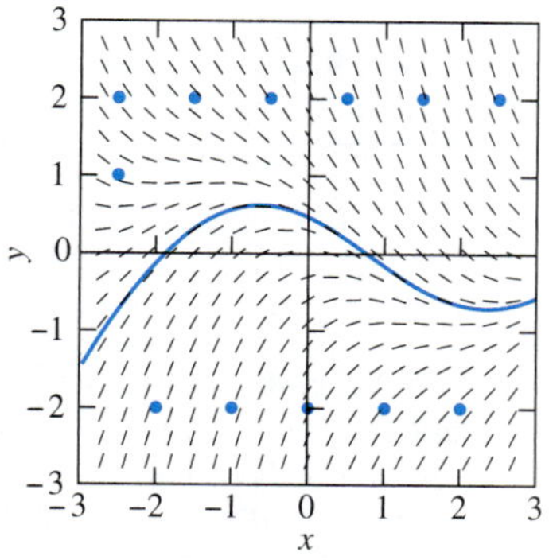

FIGURE 1.3.15

2. $\dfrac{dy}{dx} = x + y$

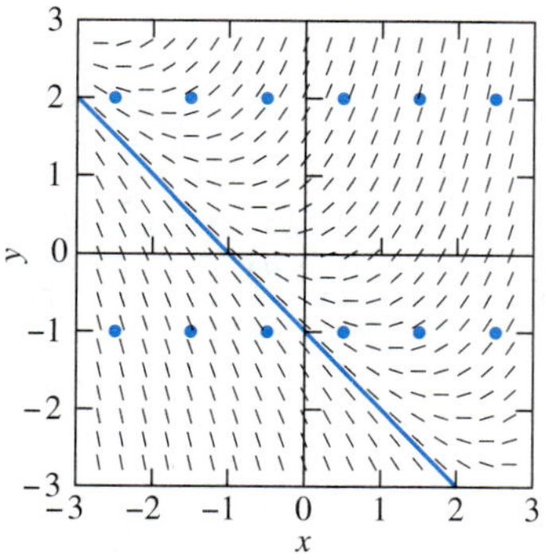

FIGURE 1.3.16

3. $\dfrac{dy}{dx} = y - \sin x$

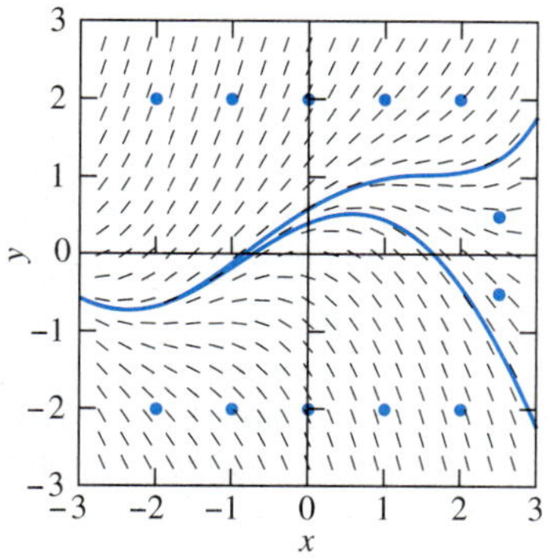

FIGURE 1.3.17

4. $\dfrac{dy}{dx} = x - y$

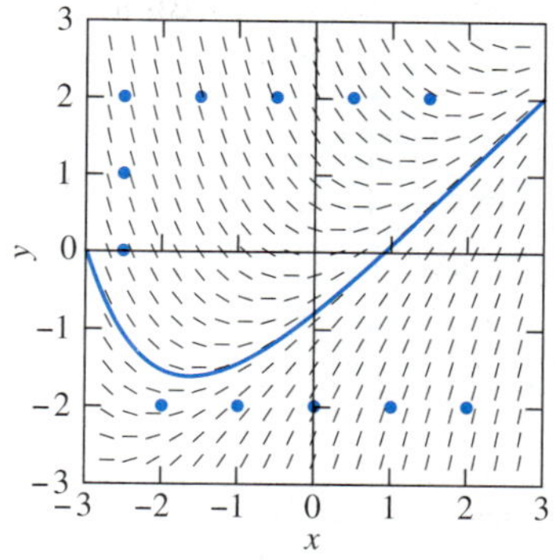

FIGURE 1.3.18

5. $\dfrac{dy}{dx} = y - x + 1$

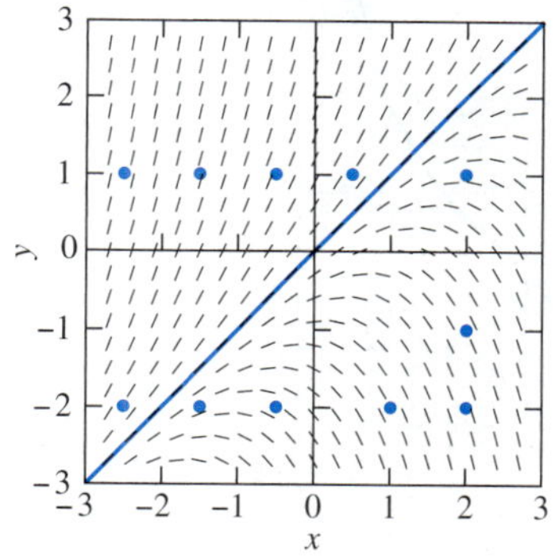

FIGURE 1.3.19

6. $\dfrac{dy}{dx} = x - y + 1$

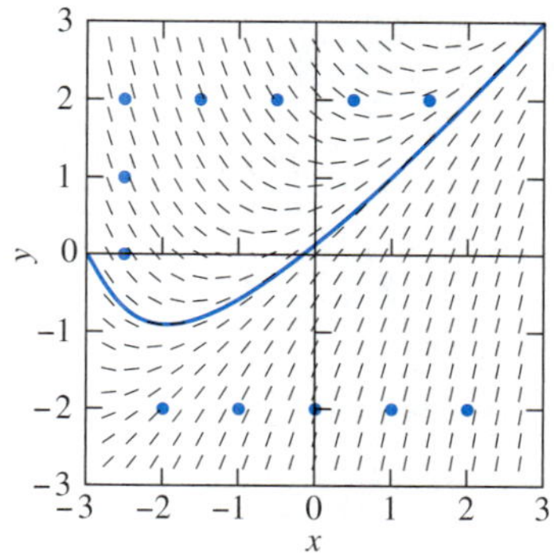

FIGURE 1.3.20

7. $\dfrac{dy}{dx} = \sin x + \sin y$

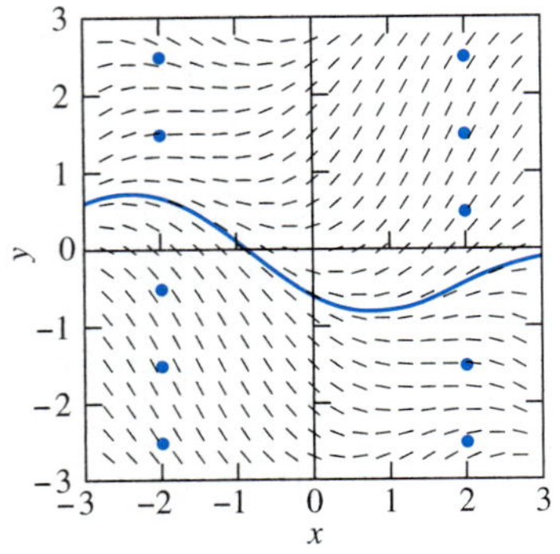

FIGURE 1.3.21

8. $\dfrac{dy}{dx} = x^2 - y$

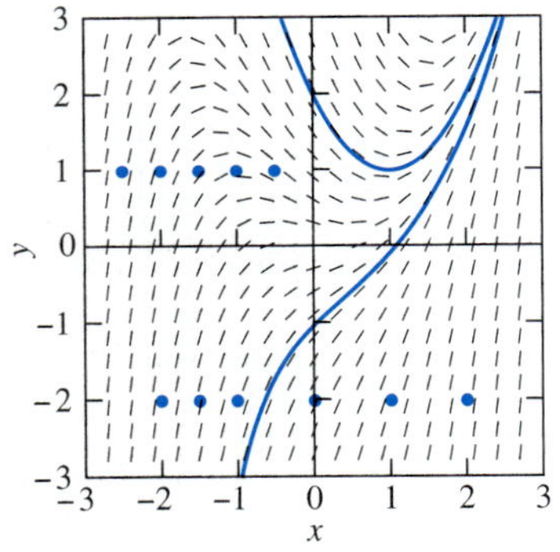

FIGURE 1.3.22

9. $\dfrac{dy}{dx} = x^2 - y - 2$

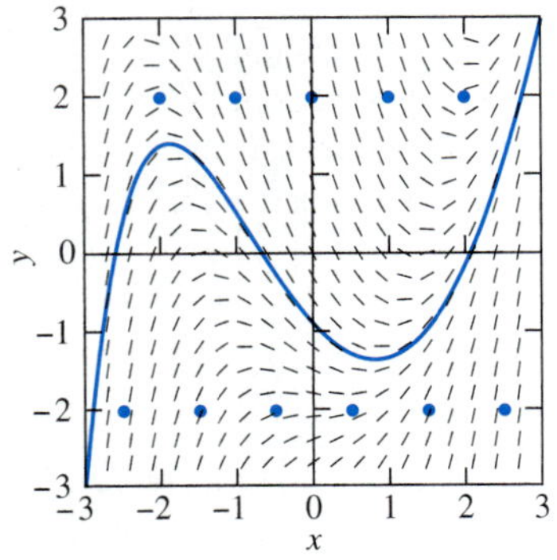

FIGURE 1.3.23

10. $\dfrac{dy}{dx} = -x^2 + \sin y$

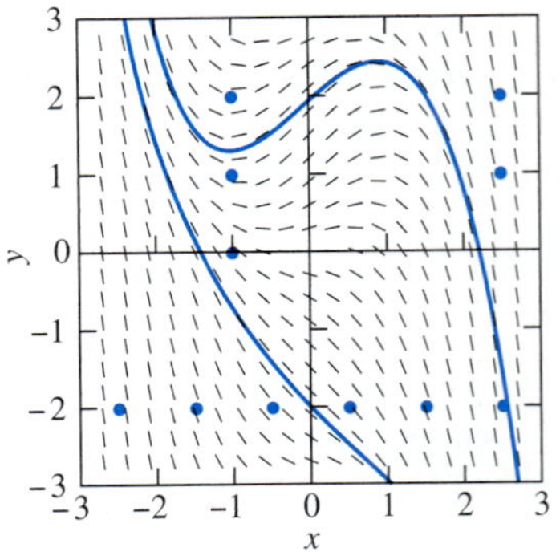

FIGURE 1.3.24

A more detailed version of Theorem 1 says that, if the function $f(x, y)$ is continuous near the point (a, b), then at least one solution of the differential equation $y' = f(x, y)$ exists on some open interval I containing the point $x = a$ and, moreover, that if in addition the partial derivative $\partial f/\partial y$ is continuous near (a, b), then this solution is unique on some (perhaps smaller) interval J. In Problems 11 through 20, determine whether existence of at least one solution of the given initial value problem is thereby guaranteed and, if so, whether uniqueness of that solution is guaranteed.

11. $\dfrac{dy}{dx} = 2x^2y^2;\quad y(1) = -1$

12. $\dfrac{dy}{dx} = x \ln y;\quad y(1) = 1$

13. $\dfrac{dy}{dx} = \sqrt[3]{y};\quad y(0) = 1$

14. $\dfrac{dy}{dx} = \sqrt[3]{y};\quad y(0) = 0$

15. $\dfrac{dy}{dx} = \sqrt{x - y};\quad y(2) = 2$

16. $\dfrac{dy}{dx} = \sqrt{x - y};\quad y(2) = 1$

17. $y\dfrac{dy}{dx} = x - 1;\quad y(0) = 1$

18. $y\dfrac{dy}{dx} = x - 1;\quad y(1) = 0$

19. $\dfrac{dy}{dx} = \ln(1 + y^2);\quad y(0) = 0$

20. $\dfrac{dy}{dx} = x^2 - y^2;\quad y(0) = 1$

In Problems 21 and 22, first use the method of Example 2 to construct a slope field for the given differential equation. Then sketch the solution curve corresponding to the given initial condition. Finally, use this solution curve to estimate the desired value of the solution $y(x)$.

21. $y' = x + y,\quad y(0) = 0;\quad y(-4) = ?$

22. $y' = y - x,\quad y(4) = 0;\quad y(-4) = ?$

Problems 23 and 24 are like Problems 21 and 22, but now use a computer algebra system to plot and print out a slope field for the given differential equation. If you wish (and know how), you can check your manually sketched solution curve by plotting it with the computer.

23. $y' = x^2 + y^2 - 1, \quad y(0) = 0; \quad y(2) = ?$

24. $y' = x + \frac{1}{2}y^2, \quad y(-2) = 0; \quad y(2) = ?$

25. You bail out of the helicopter of Example 3 and pull the ripcord of your parachute. Now $k = 1.6$ in Eq. (3), so your downward velocity satisfies the initial value problem

$$\frac{dv}{dt} = 32 - 1.6v, \quad v(0) = 0.$$

In order to investigate your chances of survival, construct a slope field for this differential equation and sketch the appropriate solution curve. What will your limiting velocity be? Will a strategically located haystack do any good? How long will it take you to reach 95% of your limiting velocity?

26. Suppose the deer population $P(t)$ in a small forest satisfies the logistic equation

$$\frac{dP}{dt} = 0.0225P - 0.0003P^2.$$

Construct a slope field and appropriate solution curve to answer the following questions: If there are 25 deer at time $t = 0$ and t is measured in months, how long will it take the number of deer to double? What will be the limiting deer population?

The next seven problems illustrate the fact that, if the hypotheses of Theorem 1 are not satisfied, then the initial value problem $y' = f(x, y)$, $y(a) = b$ may have either no solutions, finitely many solutions, or infinitely many solutions.

27. **(a)** Verify that if c is a constant, then the function defined piecewise by

$$y(x) = \begin{cases} 0 & \text{for } x \leqq c, \\ (x-c)^2 & \text{for } x > c \end{cases}$$

satisfies the differential equation $y' = 2\sqrt{y}$ for all x (including the point $x = c$). Construct a figure illustrating the fact that the initial value problem $y' = 2\sqrt{y}$, $y(0) = 0$ has infinitely many different solutions. **(b)** For what values of b does the initial value problem $y' = 2\sqrt{y}$, $y(0) = b$ have (i) no solution, (ii) a unique solution that is defined for all x?

28. Verify that if k is a constant, then the function $y(x) \equiv kx$ satisfies the differential equation $xy' = y$ for all x. Construct a slope field and several of these straight line solution curves. Then determine (in terms of a and b) how many different solutions the initial value problem $xy' = y$, $y(a) = b$ has—one, none, or infinitely many.

29. Verify that if c is a constant, then the function defined piecewise by

$$y(x) = \begin{cases} 0 & \text{for } x \leqq c, \\ (x-c)^3 & \text{for } x > c \end{cases}$$

satisfies the differential equation $y' = 3y^{2/3}$ for all x. Can you also use the "left half" of the cubic $y = (x-c)^3$ in piecing together a solution curve of the differential equation? (See Fig. 1.3.25.) Sketch a variety of such solution curves. Is there a point (a, b) of the xy-plane such that the initial value problem $y' = 3y^{2/3}$, $y(a) = b$ has either no solution or a unique solution that is defined for all x? Reconcile your answer with Theorem 1.

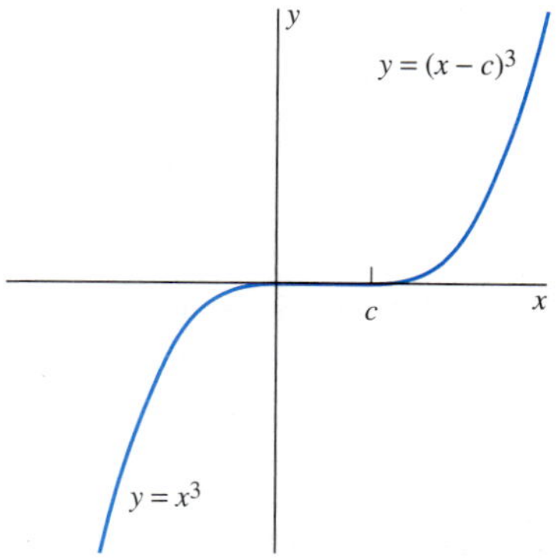

FIGURE 1.3.25. A suggestion for Problem 29.

30. Verify that if c is a constant, then the function defined piecewise by

$$y(x) = \begin{cases} +1 & \text{if } x \leqq c, \\ \cos(x-c) & \text{if } c < x < c + \pi, \\ -1 & \text{if } x \geqq c + \pi \end{cases}$$

satisfies the differential equation $y' = -\sqrt{1-y^2}$ for all x. (Perhaps a preliminary sketch with $c = 0$ will be helpful.) Sketch a variety of such solution curves. Then determine (in terms of a and b) how many different solutions the initial value problem $y' = -\sqrt{1-y^2}$, $y(a) = b$ has.

31. Carry out an investigation similar to that in Problem 30, except with the differential equation $y' = +\sqrt{1-y^2}$. Does it suffice simply to replace $\cos(x-c)$ with $\sin(x-c)$ in piecing together a solution that is defined for all x?

32. Verify that if $c > 0$, then the function defined piecewise by

$$y(x) = \begin{cases} 0 & \text{if } x^2 \leqq c, \\ (x^2-c)^2 & \text{if } x^2 > c \end{cases}$$

satisfies the differential equation $y' = 4x\sqrt{y}$ for all x. Sketch a variety of such solution curves for different values of c. Then determine (in terms of a and b) how many different solutions the initial value problem $y' = 4x\sqrt{y}$, $y(a) = b$ has.

33. If $c \neq 0$, verify that the function defined by $y(x) = x/(cx-1)$ (with graph illustrated in Fig. 1.3.26) satisfies the differential equation $x^2y' + y^2 = 0$ if $x \neq 1/c$. Sketch a variety of such solution curves for different values of c. Also, note the constant-valued function $y(x) \equiv 0$ that does not result from any choice of the constant c. Finally, determine (in terms of a and b) how many different solutions the initial value problem $x^2y' + y^2 = 0$, $y(a) = b$ has.

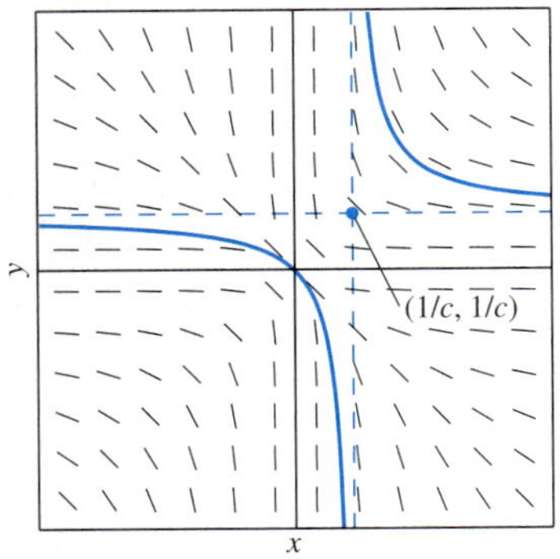

FIGURE 1.3.26. Slope field for $x^2y' + y^2 = 0$ and graph of a solution $y(x) = x/(cx-1)$.

34. **(a)** Use the direction field of Problem 5 to estimate the values at $x = 1$ of the two solutions of the differential equation $y' = y - x + 1$ with initial values $y(-1) = -1.2$ and $y(-1) = -0.8$.

(b) Use a computer algebra system to estimate the values at $x = 3$ of the two solutions of this differential equation with initial values $y(-3) = -3.01$ and $y(-3) = -2.99$.

The lesson of this problem is that small changes in initial conditions can make big differences in results.

35. **(a)** Use the direction field of Problem 6 to estimate the values at $x = 2$ of the two solutions of the differential equation $y' = x - y + 1$ with initial values $y(-3) = -0.2$ and $y(-3) = +0.2$.

(b) Use a computer algebra system to estimate the values at $x = 2$ of the two solutions of this differential equation with initial values $y(-3) = -0.5$ and $y(-3) = +0.5$.

The lesson of this problem is that big changes in initial conditions may make only small differences in results.

1.3 Application Computer-Generated Slope Fields and Solution Curves

Widely available computer algebra systems and technical computing environments include facilities to automate the construction of slope fields and solution curves, as do some graphing calculators (see Fig. 1.3.27).

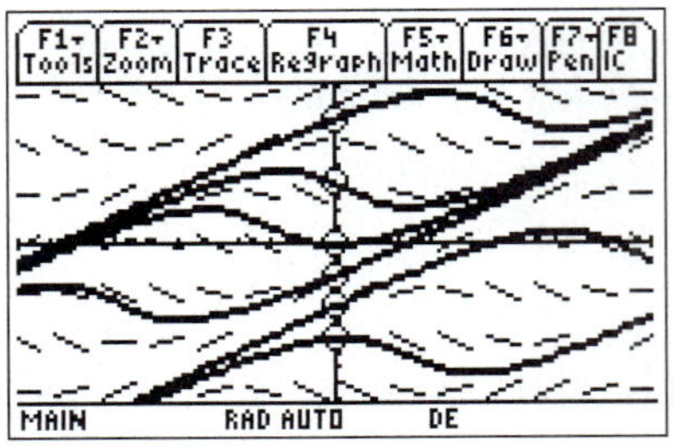

FIGURE 1.3.27. Slope field and solution curves for the differential equation

$$\frac{dy}{dx} = \sin(x-y)$$

with initial points $(0, b)$, $b = -3, -1, -2, 0, 2, 4$ and window $-5 \leq x, y \leq 5$ on a TI-89 graphing calculator.

The applications manual accompanying this textbook includes discussion of *Maple*™, *Mathematica*™, and MATLAB™ resources for the investigation of differential equations. For instance, the *Maple* command

```
with(DEtools):
DEplot(diff(y(x),x)=sin(x-y(x)), y(x), x=-5..5, y=-5..5);
```

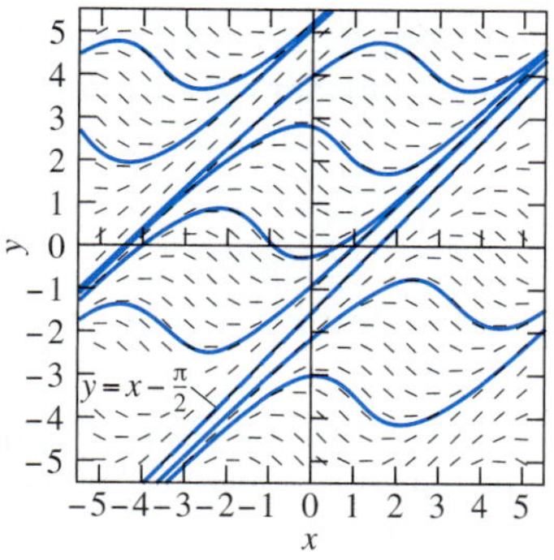

FIGURE 1.3.28. Computer-generated slope field and solution curves for the differential equation $y' = \sin(x - y)$.

and the *Mathematica* command

```
<< Graphics\PlotField.m
PlotVectorField[{1, Sin[x-y]}, {x, -5, 5}, {y, -5, 5}]
```

produce slope fields similar to the one shown in Fig. 1.3.28. Figure 1.3.28 itself was generated with the MATLAB program **dfield** [John Polking and David Arnold, *Ordinary Differential Equations Using* MATLAB, 3rd edition, Upper Saddle River, NJ: Prentice Hall, 2004] that is freely available for educational use (**math.rice.edu/~dfield**). When a differential equation is entered in the **dfield** setup menu (Fig. 1.3.29), you can (with mouse button clicks) plot both a slope field and the solution curve (or curves) through any desired point (or points). Another freely available and user-friendly MATLAB-based ODE package with impressive graphical capabilities is **Iode** (**www.math.uiuc.edu/iode**).

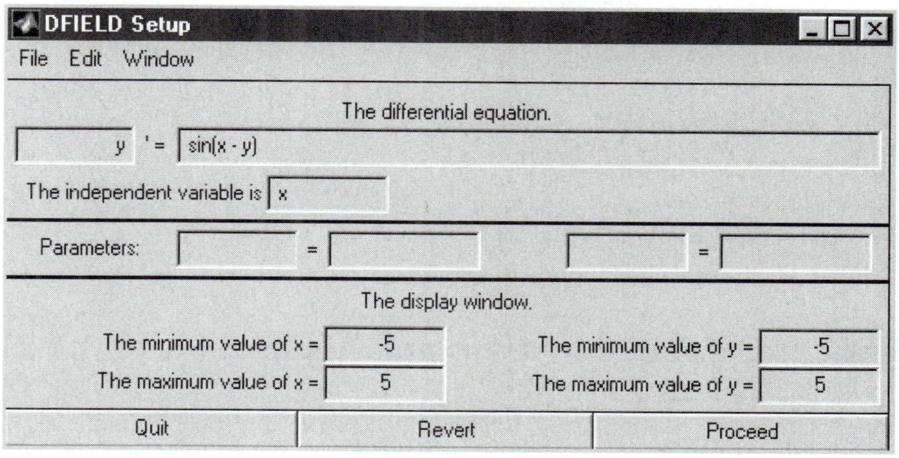

FIGURE 1.3.29. MATLAB **dfield** setup to construct slope field and solution curves for $y' = \sin(x - y)$.

Use a graphing calculator or computer system in the following investigations. You might warm up by generating the slope fields and some solution curves for Problems 1 through 10 in this section.

INVESTIGATION A: Plot a slope field and typical solution curves for the differential equation $dy/dx = \sin(x - y)$, but with a larger window than that of Fig. 1.3.28. With $-10 \leqq x \leqq 10$, $-10 \leqq y \leqq 10$, for instance, a number of apparent straight line solution curves should be visible.

(a) Substitute $y = ax + b$ in the differential equation to determine what the coefficients a and b must be in order to get a solution.

(b) A computer algebra system gives the general solution

$$y(x) = x - 2\tan^{-1}\left(\frac{x - 2 - C}{x - C}\right).$$

Plot this solution with selected values of the constant C to compare the resulting solution curves with those indicated in Fig. 1.3.28. Can you see that *no* value of C yields the linear solution $y = x - \pi/2$ corresponding to the initial condition $y(\pi/2) = 0$? Are there any values of C for which the corresponding solution curves lie close to this straight line solution curve?

INVESTIGATION B: For your own personal investigation, let n be the *smallest* digit in your student ID number that is greater than 1, and consider the differential equation

$$\frac{dy}{dx} = \frac{1}{n}\cos(x - ny).$$

(a) First investigate (as in part (a) of Investigation A) the possibility of straight line solutions.

(b) Then generate a slope field for this differential equation, with the viewing window chosen so that you can picture some of these straight lines, plus a sufficient number of nonlinear solution curves that you can formulate a conjecture about what happens to $y(x)$ as $x \to +\infty$. State your inference as plainly as you can. Given the initial value $y(0) = y_0$, try to predict (perhaps in terms of y_0) how $y(x)$ behaves as $x \to +\infty$.

(c) A computer algebra system gives the general solution

$$y(x) = \frac{1}{n}\left[x + 2\tan^{-1}\left(\frac{1}{x - C}\right)\right].$$

Can you make a connection between this symbolic solution and your graphically generated solution curves (straight lines or otherwise)?

1.4 Separable Equations and Applications

The first-order differential equation

$$\frac{dy}{dx} = H(x, y) \tag{1}$$

is called **separable** provided that $H(x, y)$ can be written as the product of a function of x and a function of y:

$$\frac{dy}{dx} = g(x)h(y) = \frac{g(x)}{f(y)},$$

where $h(y) = 1/f(y)$. In this case the variables x and y can be *separated*—isolated on opposite sides of an equation—by writing informally the equation

$$f(y)\,dy = g(x)\,dx,$$

which we understand to be concise notation for the differential equation

$$f(y)\frac{dy}{dx} = g(x). \tag{2}$$

It is easy to solve this special type of differential equation simply by integrating both sides with respect to x:

$$\int f(y(x))\frac{dy}{dx}\,dx = \int g(x)\,dx + C;$$

equivalently,

$$\int f(y)\,dy = \int g(x)\,dx + C. \tag{3}$$

All that is required is that the antiderivatives

$$F(y) = \int f(y)\,dy \quad \text{and} \quad G(x) = \int g(x)\,dx$$

can be found. To see that Eqs. (2) and (3) are equivalent, note the following consequence of the chain rule:

$$D_x[F(y(x))] = F'(y(x))y'(x) = f(y)\frac{dy}{dx} = g(x) = D_x[G(x)],$$

which in turn is equivalent to

$$F(y(x)) = G(x) + C, \tag{4}$$

because two functions have the same derivative on an interval if and only if they differ by a constant on that interval.

Example 1 Solve the initial value problem

$$\frac{dy}{dx} = -6xy, \quad y(0) = 7.$$

Solution Informally, we divide both sides of the differential equation by y and multiply each side by dx to get

$$\frac{dy}{y} = -6x\,dx.$$

Hence

$$\int \frac{dy}{y} = \int (-6x)\,dx;$$

$$\ln|y| = -3x^2 + C.$$

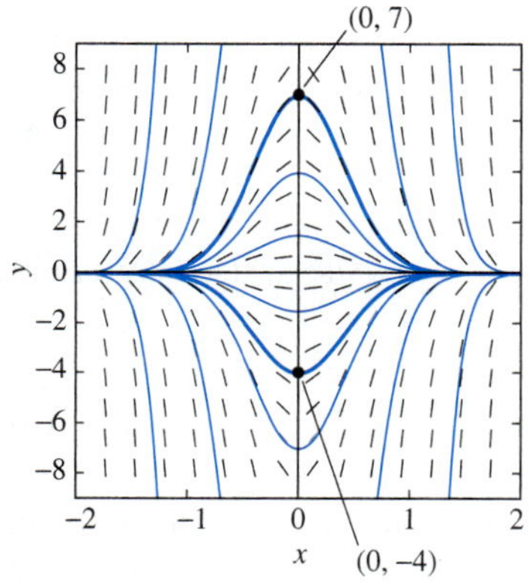

FIGURE 1.4.1. Slope field and solution curves for $y' = -6xy$ in Example 1.

We see from the initial condition $y(0) = 7$ that $y(x)$ is positive near $x = 0$, so we may delete the absolute value symbols:

$$\ln y = -3x^2 + C,$$

and hence

$$y(x) = e^{-3x^2+C} = e^{-3x^2}e^C = Ae^{-3x^2},$$

where $A = e^C$. The condition $y(0) = 7$ yields $A = 7$, so the desired solution is

$$y(x) = 7e^{-3x^2}.$$

This is the upper emphasized solution curve shown in Fig. 1.4.1. ■

Remark: Suppose, instead, that the initial condition in Example 1 had been $y(0) = -4$. Then it would follow that $y(x)$ is *negative* near $x = 0$. We should therefore replace $|y|$ with $-y$ in the integrated equation $\ln|y| = -3x^2 + C$ to obtain

$$\ln(-y) = -3x^2 + C.$$

The initial condition then yields $C = \ln 4$, so $\ln(-y) = -3x^2 + \ln 4$, and hence

$$y(x) = -4e^{-3x^2}.$$

This is the lower emphasized solution curve in Fig. 1.4.1. ■

Example 2 Solve the differential equation

$$\frac{dy}{dx} = \frac{4 - 2x}{3y^2 - 5}. \tag{5}$$

Solution When we separate the variables and integrate both sides, we get

$$\int (3y^2 - 5)\,dy = \int (4 - 2x)\,dx;$$

$$y^3 - 5y = 4x - x^2 + C. \tag{6}$$

This equation is not readily solved for y as an explicit function of x. ■

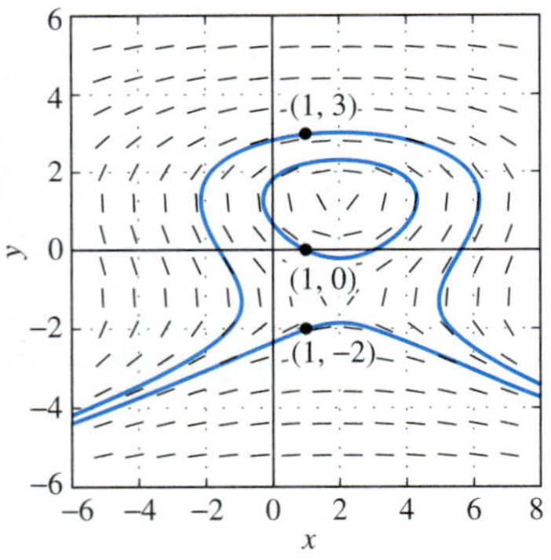

FIGURE 1.4.2. Slope field and solution curves for $y' = (4 - 2x)/(3y^2 - 5)$ in Example 2.

As Example 2 illustrates, it may or may not be possible or practical to solve Eq. (4) explicitly for y in terms of x. If not, then we call (4) an *implicit solution* of the differential equation in (2). Thus Eq. (6) gives an implicit solution of the differential equation in (5). Although it is not convenient to solve Eq. (6) explicitly in terms of x, we see that each solution curve $y = y(x)$ lies on a contour (or level) curve where the function

$$H(x, y) = x^2 - 4x + y^3 - 5y$$

is constant. Figure 1.4.2 shows several of these contour curves.

Example 3 To solve the initial value problem

$$\frac{dy}{dx} = \frac{4 - 2x}{3y^2 - 5}, \quad y(1) = 3, \tag{7}$$

we substitute $x = 1$ and $y = 3$ in Eq. (6) and get $C = 9$. Thus the desired particular solution $y(x)$ is defined implicitly by the equation

$$y^3 - 5y = 4x - x^2 + 9. \tag{8}$$

The corresponding solution curve $y = y(x)$ lies on the upper contour curve in Fig. 1.4.2—the one passing through $(1, 3)$. Because the graph of a differentiable solution cannot have a vertical tangent line anywhere, it appears from the figure that this particular solution is defined on the interval $(-1, 5)$ but not on the interval $(-3, 7)$. ■

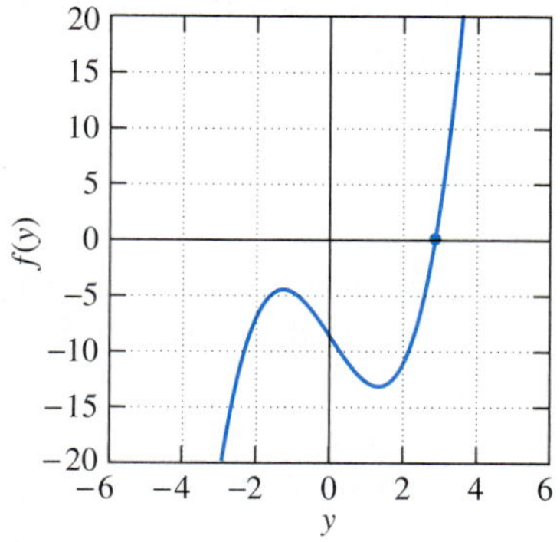

FIGURE 1.4.3. Graph of $f(y) = y^3 - 5y - 9$.

Remark 1: When a specific value of x is substituted in Eq. (8), we can attempt to solve numerically for y. For instance, $x = 4$ yields the equation

$$f(y) = y^3 - 5y - 9 = 0.$$

Figure 1.4.3 shows the graph of f. With a graphing calculator we can solve for the single real root $y \approx 2.8552$. This yields the value $y(4) \approx 2.8552$ of the particular solution in Example 3.

Remark 2: If the initial condition in (7) is replaced with the condition $y(1) = 0$, then the resulting particular solution of the differential equation in (5) lies on the lower "half" of the oval contour curve in Fig. 1.4.2. It appears that this particular solution through $(1, 0)$ is defined on the interval $(0, 4)$ but not on the interval $(-1, 5)$. On the other hand, with the initial condition $y(1) = -2$ we get the lower contour curve in Fig. 1.4.2. This particular solution is defined for all x. Thus the initial condition can determine whether a particular solution is defined on the whole real line or only on some bounded interval. With a computer algebra system one can readily calculate a table of values of the y-solutions of Eq. (8) for x-values at desired increments from $x = -1$ to $x = 5$ (for instance). Such a table of values serves effectively as a "numerical solution" of the initial value problem in (7). ■

Implicit, General, and Singular Solutions

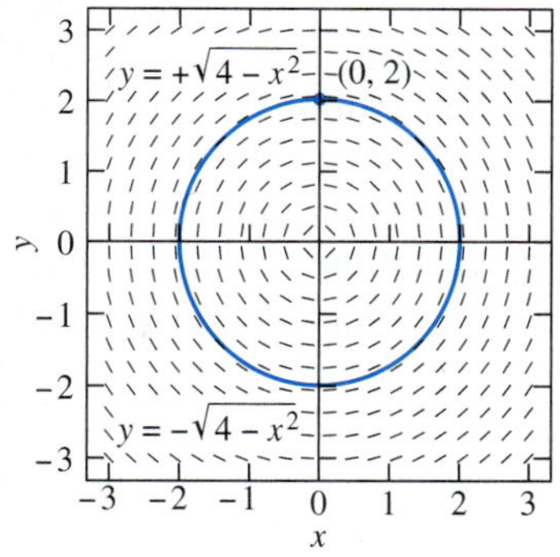

FIGURE 1.4.4. Slope field and solution curves for $y' = -x/y$.

The equation $K(x, y) = 0$ is commonly called an **implicit solution** of a differential equation if it is satisfied (on some interval) by some solution $y = y(x)$ of the differential equation. But note that a particular solution $y = y(x)$ of $K(x, y) = 0$ may or may not satisfy a given initial condition. For example, differentiation of $x^2 + y^2 = 4$ yields

$$x + y\frac{dy}{dx} = 0,$$

so $x^2 + y^2 = 4$ is an implicit solution of the differential equation $x + yy' = 0$. But only the first of the two explicit solutions

$$y(x) = +\sqrt{4 - x^2} \quad \text{and} \quad y(x) = -\sqrt{4 - x^2}$$

satisfies the initial condition $y(0) = 2$ (Fig. 1.4.4).

Remark 1: You should not assume that every possible algebraic solution $y = y(x)$ of an implicit solution satisfies the same differential equation. For instance, if we multiply the implicit solution $x^2 + y^2 - 4 = 0$ by the factor $(y - 2x)$, then we get the new implicit solution

$$(y - 2x)(x^2 + y^2 - 4) = 0$$

that yields (or "contains") not only the previously noted explicit solutions $y = +\sqrt{4 - x^2}$ and $y = -\sqrt{4 - x^2}$ of the differential equation $x + yy' = 0$, but also the additional function $y = 2x$ that does *not* satisfy this differential equation.

Remark 2: Similarly, solutions of a given differential equation can be either gained or lost when it is multiplied or divided by an algebraic factor. For instance, consider the differential equation

$$(y - 2x)y\frac{dy}{dx} = -x(y - 2x) \tag{9}$$

having the obvious solution $y = 2x$. But if we divide both sides by the common factor $(y - 2x)$, then we get the previously discussed differential equation

$$y\frac{dy}{dx} = -x, \quad \text{or} \quad x + y\frac{dy}{dx} = 0, \tag{10}$$

of which $y = 2x$ is *not* a solution. Thus we "lose" the solution $y = 2x$ of Eq. (9) upon its division by the factor $(y - 2x)$; alternatively, we "gain" this new solution when we multiply Eq. (10) by $(y - 2x)$. Such elementary algebraic operations to simplify a given differential equation before attempting to solve it are common in practice, but the possibility of loss or gain of such "extraneous solutions" should be kept in mind. ■

A solution of a differential equation that contains an "arbitrary constant" (like the constant C in the solution of Examples 1 and 2) is commonly called a **general solution** of the differential equation; any particular choice of a specific value for C yields a single particular solution of the equation.

The argument preceding Example 1 actually suffices to show that *every* particular solution of the differential equation $f(y)y' = g(x)$ in (2) satisfies the equation $F(y(x)) = G(x) + C$ in (4). Consequently, it is appropriate to call (4) not merely ***a*** general solution of (2), but ***the*** general solution of (2).

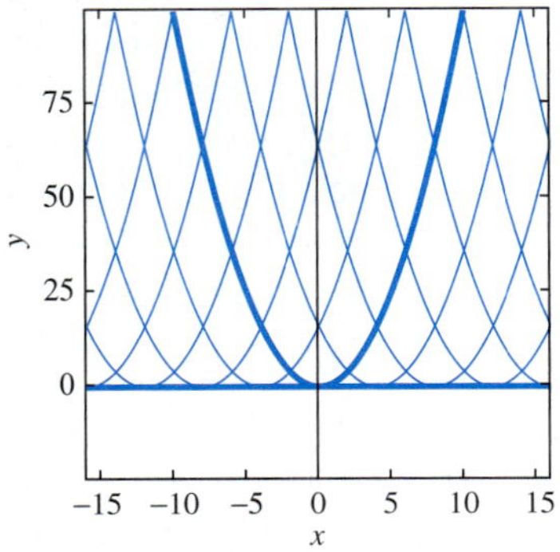

FIGURE 1.4.5. The general solution curves $y = (x - C)^2$ and the singular solution curve $y = 0$ of the differential equation $(y')^2 = 4y$.

In Section 1.5 we shall see that every particular solution of a *linear* first-order differential equation is contained in its general solution. By contrast, it is common for a nonlinear first-order differential equation to have both a general solution involving an arbitrary constant C and one or several particular solutions that cannot be obtained by selecting a value for C. These exceptional solutions are frequently called **singular solutions**. In Problem 30 we ask you to show that the general solution of the differential equation $(y')^2 = 4y$ yields the family of parabolas $y = (x - C)^2$ illustrated in Fig. 1.4.5, and to observe that the constant-valued function $y(x) \equiv 0$ is a singular solution that cannot be obtained from the general solution by any choice of the arbitrary constant C.

Example 4 Find all solutions of the differential equation

$$\frac{dy}{dx} = 6x(y - 1)^{2/3}.$$

Solution Separation of variables gives

$$\int \frac{1}{3(y-1)^{2/3}}\,dy = \int 2x\,dx;$$

$$(y-1)^{1/3} = x^2 + C;$$

$$y(x) = 1 + (x^2 + C)^3.$$

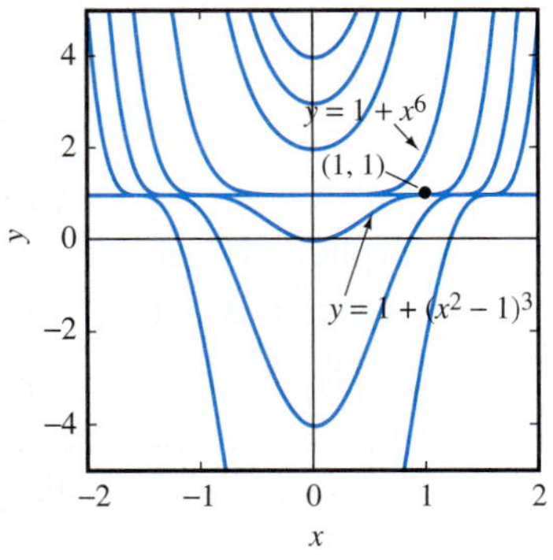

FIGURE 1.4.6. General and singular solution curves for $y' = 6x(y - 1)^{2/3}$.

Positive values of the arbitrary constant C give the solution curves in Fig. 1.4.6 that lie above the line $y = 1$, whereas negative values yield those that dip below it. The value $C = 0$ gives the solution $y(x) = 1 + x^6$, but *no* value of C gives the singular solution $y(x) \equiv 1$ that was lost when the variables were separated. Note that the two different solutions $y(x) \equiv 1$ and $y(x) = 1 + (x^2 - 1)^3$ both satisfy the initial condition $y(1) = 1$. Indeed, the whole singular solution curve $y = 1$ consists of points where the solution is not unique and where the function $f(x, y) = 6x(y - 1)^{2/3}$ is not differentiable. ■

Natural Growth and Decay

The differential equation

$$\frac{dx}{dt} = kx \quad (k \text{ a constant}) \tag{11}$$

serves as a mathematical model for a remarkably wide range of natural phenomena—any involving a quantity whose time rate of change is proportional to its current size. Here are some examples.

POPULATION GROWTH: Suppose that $P(t)$ is the number of individuals in a population (of humans, or insects, or bacteria) having *constant* birth and death rates β and δ (in births or deaths per individual per unit of time). Then, during a short time interval Δt, approximately $\beta P(t)\,\Delta t$ births and $\delta P(t)\,\Delta t$ deaths occur, so the change in $P(t)$ is given approximately by

$$\Delta P \approx (\beta - \delta) P(t)\,\Delta t,$$

and therefore

$$\frac{dP}{dt} = \lim_{\Delta t \to 0} \frac{\Delta P}{\Delta t} = kP, \tag{12}$$

where $k = \beta - \delta$.

COMPOUND INTEREST: Let $A(t)$ be the number of dollars in a savings account at time t (in years), and suppose that the interest is *compounded continuously* at an annual interest rate r. (Note that 10% annual interest means that $r = 0.10$.) Continuous compounding means that during a short time interval Δt, the amount of interest added to the account is approximately $\Delta A = rA(t)\,\Delta t$, so that

$$\frac{dA}{dt} = \lim_{\Delta t \to 0} \frac{\Delta A}{\Delta t} = rA. \tag{13}$$

RADIOACTIVE DECAY: Consider a sample of material that contains $N(t)$ atoms of a certain radioactive isotope at time t. It has been observed that a constant fraction of those radioactive atoms will spontaneously decay (into atoms of another element or into another isotope of the same element) during each unit of time. Consequently, the sample behaves exactly like a population with a constant death rate and no births. To write a model for $N(t)$, we use Eq. (12) with N in place of P, with $k > 0$ in place of δ, and with $\beta = 0$. We thus get the differential equation

$$\frac{dN}{dt} = -kN. \tag{14}$$

The value of k depends on the particular radioactive isotope.

The key to the method of *radiocarbon dating* is that a constant proportion of the carbon atoms in any living creature is made up of the radioactive isotope ^{14}C of carbon. This proportion remains constant because the fraction of ^{14}C in the atmosphere remains almost constant, and living matter is continuously taking up carbon from the air or is consuming other living matter containing the same constant ratio of ^{14}C atoms to ordinary ^{12}C atoms. This same ratio permeates all life, because organic processes seem to make no distinction between the two isotopes.

The ratio of ^{14}C to normal carbon remains constant in the atmosphere because, although ^{14}C is radioactive and slowly decays, the amount is continuously replenished through the conversion of ^{14}N (ordinary nitrogen) to ^{14}C by cosmic rays bombarding the upper atmosphere. Over the long history of the planet, this decay and replenishment process has come into nearly steady state.

Of course, when a living organism dies, it ceases its metabolism of carbon and the process of radioactive decay begins to deplete its ^{14}C content. There is no replenishment of this ^{14}C, and consequently the ratio of ^{14}C to normal carbon begins to drop. By measuring this ratio, the amount of time elapsed since the death of the organism can be estimated. For such purposes it is necessary to measure the **decay constant** k. For ^{14}C, it is known that $k \approx 0.0001216$ if t is measured in years.

(Matters are not as simple as we have made them appear. In applying the technique of radiocarbon dating, extreme care must be taken to avoid contaminating the sample with organic matter or even with ordinary fresh air. In addition, the cosmic ray levels apparently have not been constant, so the ratio of ^{14}C in the atmosphere has varied over the past centuries. By using independent methods of dating samples, researchers in this area have compiled tables of correction factors to enhance the accuracy of this process.)

DRUG ELIMINATION: In many cases the amount $A(t)$ of a certain drug in the bloodstream, measured by the excess over the natural level of the drug, will decline at a rate proportional to the current excess amount. That is,

$$\frac{dA}{dt} = -\lambda A, \tag{15}$$

where $\lambda > 0$. The parameter λ is called the **elimination constant** of the drug.

The Natural Growth Equation

The prototype differential equation $dx/dt = kx$ with $x(t) > 0$ and k a constant (either negative or positive) is readily solved by separating the variables and integrating:

$$\int \frac{1}{x}\,dx = \int k\,dt;$$

$$\ln x = kt + C.$$

Then we solve for x:

$$e^{\ln x} = e^{kt+C}; \quad x = x(t) = e^C e^{kt} = Ae^{kt}.$$

Because C is a constant, so is $A = e^C$. It is also clear that $A = x(0) = x_0$, so the particular solution of Eq. (11) with the initial condition $x(0) = x_0$ is simply

$$x(t) = x_0 e^{kt}. \tag{16}$$

Because of the presence of the natural exponential function in its solution, the differential equation

$$\frac{dx}{dt} = kx \tag{17}$$

is often called the **exponential** or **natural growth equation.** Figure 1.4.7 shows a typical graph of $x(t)$ in the case $k > 0$; the case $k < 0$ is illustrated in Fig. 1.4.8.

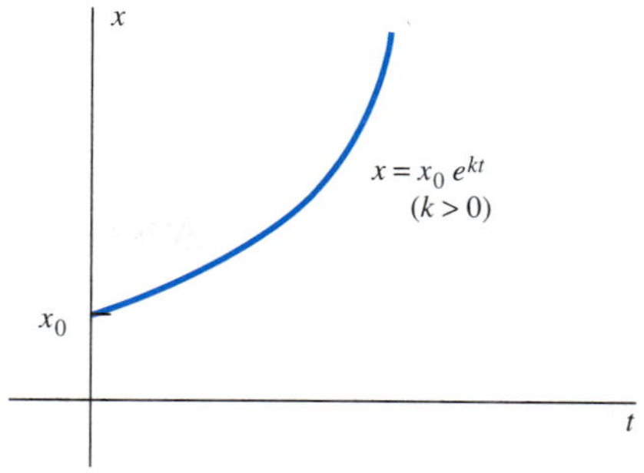

FIGURE 1.4.7. Natural growth.

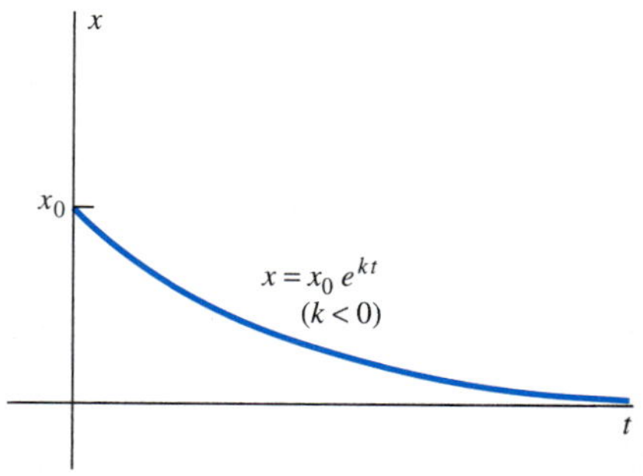

FIGURE 1.4.8. Natural decay.

Example 5 According to data listed at **www.census.gov**, the world's total population reached 6 billion persons in mid-1999, and was then increasing at the rate of about 212 thousand persons each day. Assuming that natural population growth at this rate continues, we want to answer these questions:

(a) What is the annual growth rate k?

(b) What will be the world population at the middle of the 21st century?

(c) How long will it take the world population to increase tenfold—thereby reaching the 60 billion that some demographers believe to be the maximum for which the planet can provide adequate food supplies?

Solution **(a)** We measure the world population $P(t)$ in billions and measure time in years. We take $t = 0$ to correspond to (mid) 1999, so $P_0 = 6$. The fact that P is increasing by 212,000, or 0.000212 billion, persons per day at time $t = 0$ means that

$$P'(0) = (0.000212)(365.25) \approx 0.07743$$

billion per year. From the natural growth equation $P' = kP$ with $t = 0$ we now obtain

$$k = \frac{P'(0)}{P(0)} \approx \frac{0.07743}{6} \approx 0.0129.$$

Thus the world population was growing at the rate of about 1.29% annually in 1999. This value of k gives the world population function

$$P(t) = 6e^{0.0129t}.$$

(b) With $t = 51$ we obtain the prediction

$$P(51) = 6e^{(0.0129)(51)} \approx 11.58 \text{ (billion)}$$

for the world population in mid-2050 (so the population will almost have doubled in the just over a half-century since 1999).

(c) The world population should reach 60 billion when

$$60 = 6e^{0.0129t}; \quad \text{that is, when } t = \frac{\ln 10}{0.0129} \approx 178;$$

and thus in the year 2177. ■

Note: Actually, the rate of growth of the world population is expected to slow somewhat during the next half-century, and the best current prediction for the 2050 population is "only" 9.1 billion. A simple mathematical model cannot be expected to mirror precisely the complexity of the real world.

The decay constant of a radioactive isotope is often specified in terms of another empirical constant, the *half-life* of the isotope, because this parameter is more convenient. The **half-life** τ of a radioactive isotope is the time required for *half* of it to decay. To find the relationship between k and τ, we set $t = \tau$ and $N = \frac{1}{2}N_0$ in the equation $N(t) = N_0 e^{-kt}$, so that $\frac{1}{2}N_0 = N_0 e^{-k\tau}$. When we solve for τ, we find that

$$\tau = \frac{\ln 2}{k}. \tag{18}$$

For example, the half-life of ^{14}C is $\tau \approx (\ln 2)/(0.0001216)$, approximately 5700 years.

Example 6 A specimen of charcoal found at Stonehenge turns out to contain 63% as much ^{14}C as a sample of present-day charcoal of equal mass. What is the age of the sample?

Solution We take $t = 0$ as the time of the death of the tree from which the Stonehenge charcoal was made and N_0 as the number of ^{14}C atoms that the Stonehenge sample contained then. We are given that $N = (0.63)N_0$ now, so we solve the equation $(0.63)N_0 = N_0 e^{-kt}$ with the value $k = 0.0001216$. Thus we find that

$$t = -\frac{\ln(0.63)}{0.0001216} \approx 3800 \text{ (years)}.$$

Thus the sample is about 3800 years old. If it has any connection with the builders of Stonehenge, our computations suggest that this observatory, monument, or temple—whichever it may be—dates from 1800 B.C. or earlier. ■

Cooling and Heating

According to Newton's law of cooling (Eq. (3) of Section 1.1), the time rate of change of the temperature $T(t)$ of a body immersed in a medium of constant temperature A is proportional to the difference $A - T$. That is,

$$\frac{dT}{dt} = k(A - T), \tag{19}$$

where k is a positive constant. This is an instance of the linear first-order differential equation with constant coefficients:

$$\frac{dx}{dt} = ax + b. \tag{20}$$

It includes the exponential equation as a special case ($b = 0$) and is also easy to solve by separation of variables.

Example 7 A 4-lb roast, initially at 50°F, is placed in a 375°F oven at 5:00 P.M. After 75 minutes it is found that the temperature $T(t)$ of the roast is 125°F. When will the roast be 150°F (medium rare)?

Solution We take time t in minutes, with $t = 0$ corresponding to 5:00 P.M. We also assume (somewhat unrealistically) that at any instant the temperature $T(t)$ of the roast is uniform throughout. We have $T(t) < A = 375$, $T(0) = 50$, and $T(75) = 125$. Hence

$$\frac{dT}{dt} = k(375 - T);$$

$$\int \frac{1}{375 - T}\,dT = \int k\,dt;$$

$$-\ln(375 - T) = kt + C;$$

$$375 - T = Be^{-kt}.$$

Now $T(0) = 50$ implies that $B = 325$, so $T(t) = 375 - 325e^{-kt}$. We also know that $T = 125$ when $t = 75$. Substitution of these values in the preceding equation yields

$$k = -\tfrac{1}{75}\ln\left(\tfrac{250}{325}\right) \approx 0.0035.$$

Hence we finally solve the equation

$$150 = 375 - 325e^{(-0.0035)t}$$

for $t = -[\ln(225/325)]/(0.0035) \approx 105$ (min), the total cooking time required. Because the roast was placed in the oven at 5:00 P.M., it should be removed at about 6:45 P.M. ■

Torricelli's Law

Suppose that a water tank has a hole with area a at its bottom, from which water is leaking. Denote by $y(t)$ the depth of water in the tank at time t, and by $V(t)$ the volume of water in the tank then. It is plausible—and true, under ideal conditions—that the velocity of water exiting through the hole is

$$v = \sqrt{2gy}, \tag{21}$$

which is the velocity a drop of water would acquire in falling freely from the surface of the water to the hole (see Problem 35 of Section 1.2). One can derive this formula beginning with the assumption that the sum of the kinetic and potential energy of the system remains constant. Under real conditions, taking into account the constriction of a water jet from an orifice, $v = c\sqrt{2gy}$, where c is an empirical constant between 0 and 1 (usually about 0.6 for a small continuous stream of water). For simplicity we take $c = 1$ in the following discussion.

As a consequence of Eq. (21), we have

$$\frac{dV}{dt} = -av = -a\sqrt{2gy}; \tag{22a}$$

equivalently,

$$\frac{dV}{dt} = -k\sqrt{y} \quad \text{where} \quad k = a\sqrt{2g}. \tag{22b}$$

This is a statement of Torricelli's law for a draining tank. Let $A(y)$ denote the horizontal cross-sectional area of the tank at height y. Then, applied to a thin horizontal slice of water at height $\overline{y}$ with area $A(\overline{y})$ and thickness $d\overline{y}$, the integral calculus method of cross sections gives

$$V(y) = \int_0^y A(\overline{y})\, d\overline{y}.$$

The fundamental theorem of calculus therefore implies that $dV/dy = A(y)$ and hence that

$$\frac{dV}{dt} = \frac{dV}{dy} \cdot \frac{dy}{dt} = A(y)\frac{dy}{dt}. \tag{23}$$

From Eqs. (22) and (23) we finally obtain

$$A(y)\frac{dy}{dt} = -a\sqrt{2gy} = -k\sqrt{y}, \tag{24}$$

an alternative form of Torricelli's law. ■

Example 8 A hemispherical bowl has top radius 4 ft and at time $t = 0$ is full of water. At that moment a circular hole with diameter 1 in. is opened in the bottom of the tank. How long will it take for all the water to drain from the tank?

Solution From the right triangle in Fig. 1.4.9, we see that

$$A(y) = \pi r^2 = \pi\left[16 - (4 - y)^2\right] = \pi(8y - y^2).$$

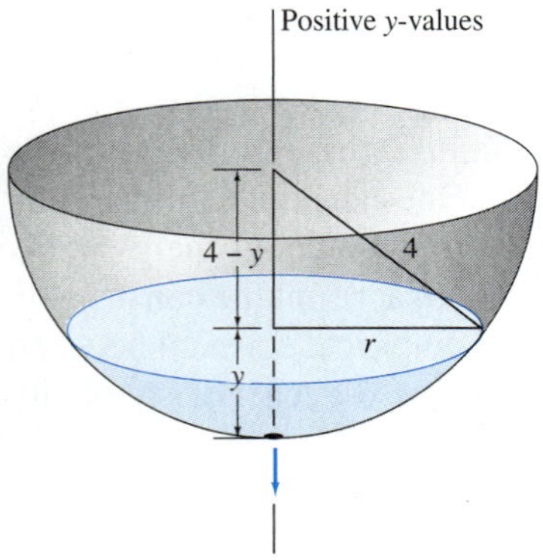

FIGURE 1.4.9. Draining a hemispherical tank.

With $g = 32$ ft/s^2, Eq. (24) becomes

$$\pi(8y - y^2)\frac{dy}{dt} = -\pi\left(\tfrac{1}{24}\right)^2\sqrt{2 \cdot 32y};$$

$$\int (8y^{1/2} - y^{3/2})\, dy = -\int \tfrac{1}{72}\, dt;$$

$$\tfrac{16}{3}y^{3/2} - \tfrac{2}{5}y^{5/2} = -\tfrac{1}{72}t + C.$$

Now $y(0) = 4$, so

$$C = \tfrac{16}{3} \cdot 4^{3/2} - \tfrac{2}{5} \cdot 4^{5/2} = \tfrac{448}{15}.$$

The tank is empty when $y = 0$, thus when

$$t = 72 \cdot \tfrac{448}{15} \approx 2150 \text{ (s)};$$

that is, about 35 min 50 s. So it takes slightly less than 36 min for the tank to drain. ■

Example 9 In the case of an upright cylindrical tank with constant cross-sectional area A, Torricelli's law in Eq. (24) takes the form

$$\frac{dy}{dt} = -c\sqrt{y}$$

with $c = k/A$. With initial condition $y(0) = 0$ we routinely separate variables and integrate to get

$$y = \tfrac{1}{4}c^2t^2.$$

However, this formula if taken at face value implies that $y > 0$ if $t > 0$. But if the tank is empty at time $t = 0$ as prescribed by the initial condition $y(0) = 0$, then certainly the tank remains empty thereafter, so $y(t) \equiv 0$ for $t > 0$.

To see what is going on here, note that the right-hand side function $f(t, y) = -c\sqrt{y}$ in our differential equation $y' = -c\sqrt{y}$ does *not* satisfy the condition that $\partial f/\partial y$ be continuous at $(0, 0)$, so the existence-uniqueness theorem of Section 1.3 does *not* guarantee uniqueness of a solution near $t = 0$. Indeed, we note the two different but physically meaningful solutions

$$y_1(t) \equiv 0 \quad \text{for all } t, \qquad \text{and} \qquad y_2(t) = \begin{cases} \frac{1}{4}c^2t^2 & \text{for } t < 0, \\ 0 & \text{for } t \geqq 0 \end{cases}$$

of the initial value problem $y' = -c\sqrt{y}$, $y(0) = 0$. The constant solution $y_1(t) \equiv 0$ corresponds to a tank that always has been and always will be empty, while $y_2(t)$ corresponds to a tank draining while $t < 0$ that empties precisely at time $t = 0$ and remains empty thereafter.

Thus this example provides a concrete physical situation described by an initial value problem with non-unique solutions. ■

1.4 Problems

Find general solutions (implicit if necessary, explicit if convenient) of the differential equations in Problems 1 through 18. Primes denote derivatives with respect to x.

1. $\dfrac{dy}{dx} + 2xy = 0$

2. $\dfrac{dy}{dx} + 2xy^2 = 0$

3. $\dfrac{dy}{dx} = y \sin x$

4. $(1 + x)\dfrac{dy}{dx} = 4y$

5. $2\sqrt{x}\,\dfrac{dy}{dx} = \sqrt{1 - y^2}$

6. $\dfrac{dy}{dx} = 3\sqrt{xy}$

7. $\dfrac{dy}{dx} = (64xy)^{1/3}$

8. $\dfrac{dy}{dx} = 2x \sec y$

9. $(1 - x^2)\dfrac{dy}{dx} = 2y$

10. $(1 + x)^2\dfrac{dy}{dx} = (1 + y)^2$

11. $y' = xy^3$

12. $yy' = x(y^2 + 1)$

13. $y^3\dfrac{dy}{dx} = (y^4 + 1)\cos x$

14. $\dfrac{dy}{dx} = \dfrac{1 + \sqrt{x}}{1 + \sqrt{y}}$

15. $\dfrac{dy}{dx} = \dfrac{(x - 1)y^5}{x^2(2y^3 - y)}$

16. $(x^2 + 1)(\tan y)y' = x$

17. $y' = 1 + x + y + xy$ (*Suggestion*: Factor the right-hand side.)

18. $x^2y' = 1 - x^2 + y^2 - x^2y^2$

Find explicit particular solutions of the initial value problems in Problems 19 through 28.

19. $\dfrac{dy}{dx} = ye^x, \quad y(0) = 2e$

20. $\dfrac{dy}{dx} = 3x^2(y^2 + 1), \quad y(0) = 1$

21. $2y\dfrac{dy}{dx} = \dfrac{x}{\sqrt{x^2 - 16}}, \quad y(5) = 2$

22. $\dfrac{dy}{dx} = 4x^3y - y, \quad y(1) = -3$

23. $\dfrac{dy}{dx} + 1 = 2y, \quad y(1) = 1$

24. $(\tan x)\dfrac{dy}{dx} = y, \quad y\left(\tfrac{1}{2}\pi\right) = \tfrac{1}{2}\pi$

25. $x\dfrac{dy}{dx} - y = 2x^2y, \quad y(1) = 1$

26. $\dfrac{dy}{dx} = 2xy^2 + 3x^2y^2, \quad y(1) = -1$

27. $\dfrac{dy}{dx} = 6e^{2x-y}, \quad y(0) = 0$

28. $2\sqrt{x}\dfrac{dy}{dx} = \cos^2 y, \quad y(4) = \pi/4$

29. **(a)** Find a general solution of the differential equation $dy/dx = y^2$. **(b)** Find a singular solution that is not included in the general solution. **(c)** Inspect a sketch of typical solution curves to determine the points (a, b) for which the initial value problem $y' = y^2$, $y(a) = b$ has a unique solution.

30. Solve the differential equation $(dy/dx)^2 = 4y$ to verify the general solution curves and singular solution curve that are illustrated in Fig. 1.4.5. Then determine the points (a, b) in the plane for which the initial value problem $(y')^2 = 4y$, $y(a) = b$ has **(a)** no solution, **(b)** infinitely many solutions that are defined for all x, **(c)** on some neighborhood of the point $x = a$, only finitely many solutions.

31. Discuss the difference between the differential equations $(dy/dx)^2 = 4y$ and $dy/dx = 2\sqrt{y}$. Do they have the same solution curves? Why or why not? Determine the points (a, b) in the plane for which the initial value problem $y' = 2\sqrt{y}$, $y(a) = b$ has **(a)** no solution, **(b)** a unique solution, **(c)** infinitely many solutions.

32. Find a general solution and any singular solutions of the differential equation $dy/dx = y\sqrt{y^2 - 1}$. Determine the points (a, b) in the plane for which the initial value problem $y' = y\sqrt{y^2 - 1}$, $y(a) = b$ has **(a)** no solution, **(b)** a unique solution, **(c)** infinitely many solutions.

33. (Population growth) A certain city had a population of 25,000 in 1960 and a population of 30,000 in 1970. Assume that its population will continue to grow exponentially at a constant rate. What population can its city planners expect in the year 2000?

34. (Population growth) In a certain culture of bacteria, the number of bacteria increased sixfold in 10 h. How long did it take for the population to double?

35. (Radiocarbon dating) Carbon extracted from an ancient skull contained only one-sixth as much ^{14}C as carbon extracted from present-day bone. How old is the skull?

36. (Radiocarbon dating) Carbon taken from a purported relic of the time of Christ contained 4.6×10^{10} atoms of ^{14}C per gram. Carbon extracted from a present-day specimen of the same substance contained 5.0×10^{10} atoms of ^{14}C per gram. Compute the approximate age of the relic. What is your opinion as to its authenticity?

37. (Continuously compounded interest) Upon the birth of their first child, a couple deposited \$5000 in an account that pays 8% interest compounded continuously. The interest payments are allowed to accumulate. How much will the account contain on the child's eighteenth birthday?

38. (Continuously compounded interest) Suppose that you discover in your attic an overdue library book on which your grandfather owed a fine of 30 cents 100 years ago. If an overdue fine grows exponentially at a 5% annual rate compounded continuously, how much would you have to pay if you returned the book today?

39. (Drug elimination) Suppose that sodium pentobarbital is used to anesthetize a dog. The dog is anesthetized when its bloodstream contains at least 45 milligrams (mg) of sodium pentobarbitol per kilogram of the dog's body weight. Suppose also that sodium pentobarbitol is eliminated exponentially from the dog's bloodstream, with a half-life of 5 h. What single dose should be administered in order to anesthetize a 50-kg dog for 1 h?

40. The half-life of radioactive cobalt is 5.27 years. Suppose that a nuclear accident has left the level of cobalt radiation in a certain region at 100 times the level acceptable for human habitation. How long will it be until the region is again habitable? (Ignore the probable presence of other radioactive isotopes.)

41. Suppose that a mineral body formed in an ancient cataclysm—perhaps the formation of the earth itself—originally contained the uranium isotope ^{238}U (which has a half-life of 4.51×10^9 years) but no lead, the end product of the radioactive decay of ^{238}U. If today the ratio of ^{238}U atoms to lead atoms in the mineral body is 0.9, when did the cataclysm occur?

42. A certain moon rock was found to contain equal numbers of potassium and argon atoms. Assume that all the argon is the result of radioactive decay of potassium (its half-life is about 1.28×10^9 years) and that one of every nine potassium atom disintegrations yields an argon atom. What is the age of the rock, measured from the time it contained only potassium?

43. A pitcher of buttermilk initially at 25°C is to be cooled by setting it on the front porch, where the temperature is 0°C. Suppose that the temperature of the buttermilk has dropped to 15°C after 20 min. When will it be at 5°C?

44. When sugar is dissolved in water, the amount A that remains undissolved after t minutes satisfies the differential equation $dA/dt = -kA$ $(k > 0)$. If 25% of the sugar dissolves after 1 min, how long does it take for half of the sugar to dissolve?

45. The intensity I of light at a depth of x meters below the surface of a lake satisfies the differential equation $dI/dx = (-1.4)I$. **(a)** At what depth is the intensity half the intensity I_0 at the surface (where $x = 0$)? **(b)** What is the intensity at a depth of 10 m (as a fraction of I_0)? **(c)** At what depth will the intensity be 1% of that at the surface?

46. The barometric pressure p (in inches of mercury) at an altitude x miles above sea level satisfies the initial value problem $dp/dx = (-0.2)p$, $p(0) = 29.92$. **(a)** Calculate the barometric pressure at 10,000 ft and again at 30,000 ft. **(b)** Without prior conditioning, few people can survive when the pressure drops to less than 15 in. of mercury. How high is that?

47. A certain piece of dubious information about phenylethylamine in the drinking water began to spread one day in a city with a population of 100,000. Within a week, 10,000 people had heard this rumor. Assume that the rate of increase of the number who have heard the rumor is proportional to the number who have not yet heard it. How long will it be until half the population of the city has heard the rumor?

48. According to one cosmological theory, there were equal amounts of the two uranium isotopes ^{235}U and ^{238}U at the creation of the universe in the "big bang." At present there are 137.7 atoms of ^{238}U for each atom of ^{235}U. Using the half-lives 4.51×10^9 years for ^{238}U and 7.10×10^8 years for ^{235}U, calculate the age of the universe.

49. A cake is removed from an oven at 210°F and left to cool at room temperature, which is 70°F. After 30 min the temperature of the cake is 140°F. When will it be 100°F?

50. The amount $A(t)$ of atmospheric pollutants in a certain mountain valley grows naturally and is tripling every 7.5 years.

(a) If the initial amount is 10 pu (pollutant units), write a formula for $A(t)$ giving the amount (in pu) present after t years.

(b) What will be the amount (in pu) of pollutants present in the valley atmosphere after 5 years?

(c) If it will be dangerous to stay in the valley when the amount of pollutants reaches 100 pu, how long will this take?

51. An accident at a nuclear power plant has left the surrounding area polluted with radioactive material that decays naturally. The initial amount of radioactive material present is 15 su (safe units), and 5 months later it is still 10 su.

(a) Write a formula giving the amount $A(t)$ of radioactive material (in su) remaining after t months.

(b) What amount of radioactive material will remain after 8 months?

(c) How long—total number of months or fraction thereof—will it be until $A = 1$ su, so it is safe for people to return to the area?

52. There are now about 3300 different human "language families" in the whole world. Assume that all these are derived from a single original language, and that a language family develops into 1.5 language families every 6 thousand years. About how long ago was the single original human language spoken?

53. Thousands of years ago ancestors of the Native Americans crossed the Bering Strait from Asia and entered the western hemisphere. Since then, they have fanned out across North and South America. The single language that the original Native Americans spoke has since split into many Indian "language families." Assume (as in Problem 52) that the number of these language families has been multiplied by 1.5 every 6000 years. There are now 150 Native American language families in the western hemisphere. About when did the ancestors of today's Native Americans arrive?

54. A tank is shaped like a vertical cylinder; it initially contains water to a depth of 9 ft, and a bottom plug is removed at time $t = 0$ (hours). After 1 h the depth of the water has dropped to 4 ft. How long does it take for all the water to drain from the tank?

55. Suppose that the tank of Problem 54 has a radius of 3 ft and that its bottom hole is circular with radius 1 in. How long will it take the water (initially 9 ft deep) to drain completely?

56. At time $t = 0$ the bottom plug (at the vertex) of a full conical water tank 16 ft high is removed. After 1 h the water in the tank is 9 ft deep. When will the tank be empty?

57. Suppose that a cylindrical tank initially containing V_0 gallons of water drains (through a bottom hole) in T minutes. Use Torricelli's law to show that the volume of water in the tank after $t \leqq T$ minutes is $V = V_0\,[1 - (t/T)]^2$.

58. A water tank has the shape obtained by revolving the curve $y = x^{4/3}$ around the y-axis. A plug at the bottom is removed at 12 noon, when the depth of water in the tank is 12 ft. At 1 P.M. the depth of the water is 6 ft. When will the tank be empty?

59. A water tank has the shape obtained by revolving the parabola $x^2 = by$ around the y-axis. The water depth is 4 ft at 12 noon, when a circular plug in the bottom of the tank is removed. At 1 P.M. the depth of the water is 1 ft. **(a)** Find the depth $y(t)$ of water remaining after t hours. **(b)** When will the tank be empty? **(c)** If the initial radius of the top surface of the water is 2 ft, what is the radius of the circular hole in the bottom?

60. A cylindrical tank with length 5 ft and radius 3 ft is situated with its axis horizontal. If a circular bottom hole with a radius of 1 in. is opened and the tank is initially half full of xylene, how long will it take for the liquid to drain completely?

61. A spherical tank of radius 4 ft is full of gasoline when a circular bottom hole with radius 1 in. is opened. How long will be required for all the gasoline to drain from the tank?

62. Suppose that an initially full hemispherical water tank of radius 1 m has its flat side as its bottom. It has a bottom hole of radius 1 cm. If this bottom hole is opened at 1 P.M., when will the tank be empty?

63. Consider the initially full hemispherical water tank of Example 8, except that the radius r of its circular bottom hole is now unknown. At 1 P.M. the bottom hole is opened and at 1:30 P.M. the depth of water in the tank is 2 ft. **(a)** Use Torricelli's law in the form $dV/dt = -(0.6)\pi r^2\sqrt{2gy}$ (taking constriction into account) to determine when the tank will be empty. **(b)** What is the radius of the bottom hole?

64. (The *clepsydra*, or water clock) A 12-h water clock is to be designed with the dimensions shown in Fig. 1.4.10, shaped like the surface obtained by revolving the curve $y = f(x)$ around the y-axis. What should be this curve, and what should be the radius of the circular bottom hole, in order that the water level will fall at the *constant* rate of 4 inches per hour (in./h)?

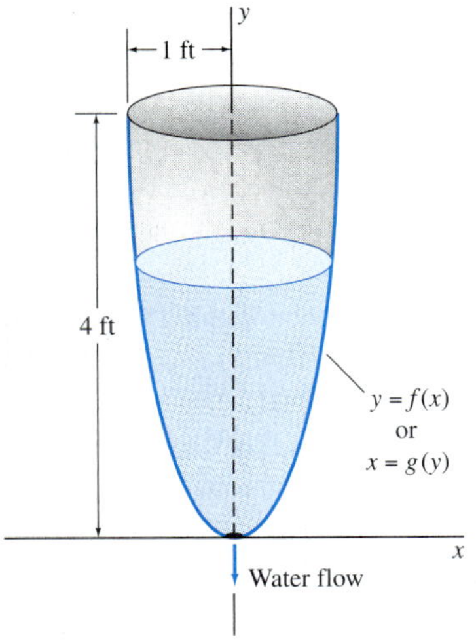

FIGURE 1.4.10. The clepsydra.

65. Just before midday the body of an apparent homicide victim is found in a room that is kept at a constant temperature of 70°F. At 12 noon the temperature of the body is 80°F and at 1 P.M. it is 75°F. Assume that the temperature of the body at the time of death was 98.6°F and that it has cooled in accord with Newton's law. What was the time of death?

66. Early one morning it began to snow at a constant rate. At 7 A.M. a snowplow set off to clear a road. By 8 A.M. it had traveled 2 miles, but it took two more hours (until 10 A.M.) for the snowplow to go an additional 2 miles. (a) Let $t = 0$ when it began to snow and let x denote the distance traveled by the snowplow at time t. Assuming that the snowplow clears snow from the road at a constant rate (in cubic feet per hour, say), show that

$$k\frac{dx}{dt} = \frac{1}{t},$$

where k is a constant. (b) What time did it start snowing? (*Answer*: 6 A.M.)

67. A snowplow sets off at 7 A.M. as in Problem 66. Suppose now that by 8 A.M. it had traveled 4 miles and that by 9 A.M. it had moved an additional 3 miles. What time did it start snowing? This is a more difficult snowplow problem because now a transcendental equation must be solved numerically to find the value of k. (*Answer*: 4:27 A.M.)

68. Figure 1.4.11 shows a bead sliding down a frictionless wire from point P to point Q. The *brachistochrone problem* asks what shape the wire should be in order to minimize the bead's time of descent from P to Q. In June of 1696, John Bernoulli proposed this problem as a public challenge, with a 6-month deadline (later extended to Easter 1697 at George Leibniz's request). Isaac Newton, then retired from academic life and serving as Warden of the Mint in London, received Bernoulli's challenge on January 29, 1697. The very next day he communicated his own solution—the curve of minimal descent time is an arc of an inverted cycloid—to the Royal Society of London. For a modern derivation of this result, suppose the bead starts from rest at the origin P and let $y = y(x)$ be the equation of the desired curve in a coordinate system with the y-axis pointing downward. Then a mechanical analogue of Snell's law in optics implies that

$$\frac{\sin\alpha}{v} = \text{constant}, \qquad \text{(i)}$$

where α denotes the angle of deflection (from the vertical) of the tangent line to the curve—so $\cot\alpha = y'(x)$ (why?)—and $v = \sqrt{2gy}$ is the bead's velocity when it has descended a distance y vertically (from KE $= \frac{1}{2}mv^2 = mgy = -$PE).

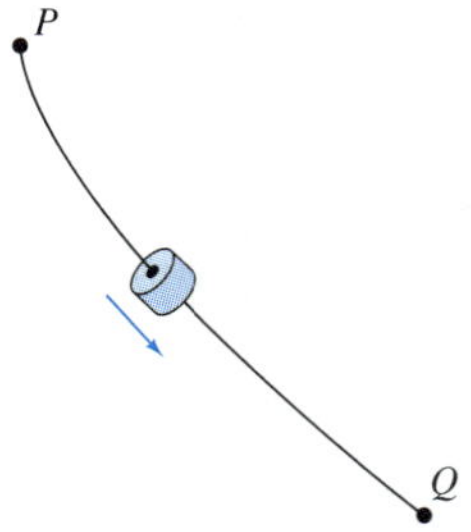

FIGURE 1.4.11. A bead sliding down a wire—the brachistochrone problem.

(a) First derive from Eq. (i) the differential equation

$$\frac{dy}{dx} = \sqrt{\frac{2a - y}{y}}, \qquad \text{(ii)}$$

where a is an appropriate positive constant.

(b) Substitute $y = 2a\sin^2 t$, $dy = 4a \sin t \cos t \, dt$ in (ii) to derive the solution

$$x = a(2t - \sin 2t), \quad y = a(1 - \cos 2t) \qquad \text{(iii)}$$

for which $t = y = 0$ when $x = 0$. Finally, the substitution of $\theta = 2a$ in (iii) yields the standard parametric equations $x = a(\theta - \sin\theta)$, $y = a(1 - \cos\theta)$ of the cycloid that is generated by a point on the rim of a circular wheel of radius a as it rolls along the x-axis. [See Example 5 in Section 9.4 of Edwards and Penney, *Calculus: Early Transcendentals*, 7th edition (Upper Saddle River, NJ: Prentice Hall, 2008).]

69. Suppose a uniform flexible cable is suspended between two points $(\pm L, H)$ at equal heights located symmetrically on either side of the x-axis (Fig. 1.4.12). Principles of physics can be used to show that the shape $y = y(x)$ of the hanging cable satisfies the differential equation

$$a\frac{d^2y}{dx^2} = \sqrt{1 + \left(\frac{dy}{dx}\right)^2},$$

where the constant $a = T/\rho$ is the ratio of the cable's tension T at its lowest point $x = 0$ (where $y'(0) = 0$) and its (constant) linear density ρ. If we substitute $v = dy/dx$,

$dv/dx = d^2y/dx^2$ in this second-order differential equation, we get the first-order equation

$$a\frac{dv}{dx} = \sqrt{1+v^2}.$$

Solve this differential equation for $y'(x) = v(x) = \sinh(x/a)$. Then integrate to get the shape function

$$y(x) = a\cosh\left(\frac{x}{a}\right) + C$$

of the hanging cable. This curve is called a *catenary*, from the Latin word for *chain*.

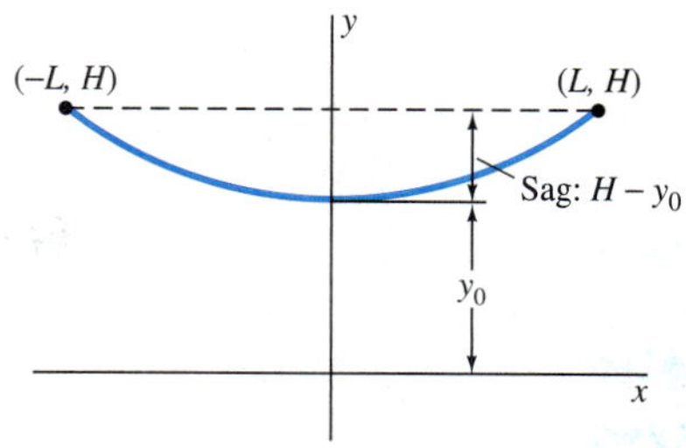

FIGURE 1.4.12. The catenary.

1.4 Application The Logistic Equation

As in Eq. (3) of this section, the solution of a separable differential equation reduces to the evaluation of two indefinite integrals. It is tempting to use a symbolic algebra system for this purpose. We illustrate this approach using the *logistic differential equation*

$$\frac{dx}{dt} = ax - bx^2 \tag{1}$$

that models a population $x(t)$ with births (per unit time) proportional to x and deaths proportional to x^2. Here we concentrate on the solution of Eq. (1) and defer discussion of population applications to Section 2.1.

If $a = 0.01$ and $b = 0.0001$, for instance, Eq. (1) is

$$\frac{dx}{dt} = (0.01)x - (0.0001)x^2 = \frac{x}{10000}(100 - x). \tag{2}$$

Separation of variables leads to

$$\int \frac{1}{x(100-x)}\,dx = \int \frac{1}{10000}\,dt = \frac{t}{10000} + C. \tag{3}$$

We can evaluate the integral on the left by using the *Maple* command

```
int(1/(x*(100 - x)), x);
```

the *Mathematica* command

```
Integrate[ 1/(x*(100 - x)), x ]
```

or the MATLAB command

```
syms x;   int(1/(x*(100 - x)))
```

Any computer algebra system gives a result of the form

$$\frac{1}{100}\ln x - \frac{1}{100}\ln(x - 100) = \frac{t}{10000} + C \tag{4}$$

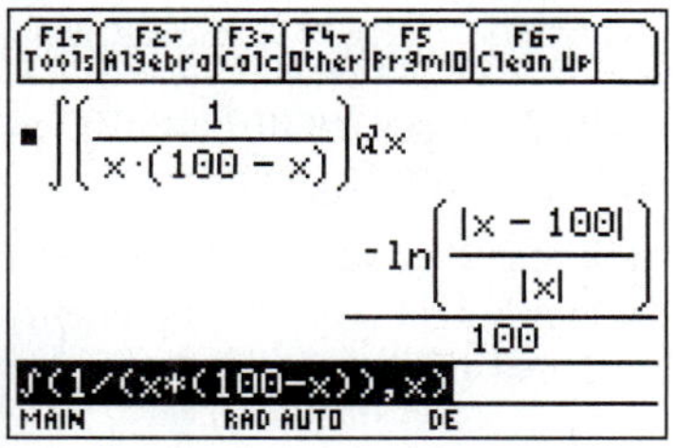

FIGURE 1.4.13. TI-89 screen showing the integral in Eq. (3).

equivalent to the graphing calculator result shown in Fig. 1.4.13.

You can now apply the initial condition $x(0) = x_0$, combine logarithms, and finally exponentiate to solve Eq. (4) for the particular solution

$$x(t) = \frac{100x_0e^{t/100}}{100 - x_0 + x_0e^{t/100}} \tag{5}$$

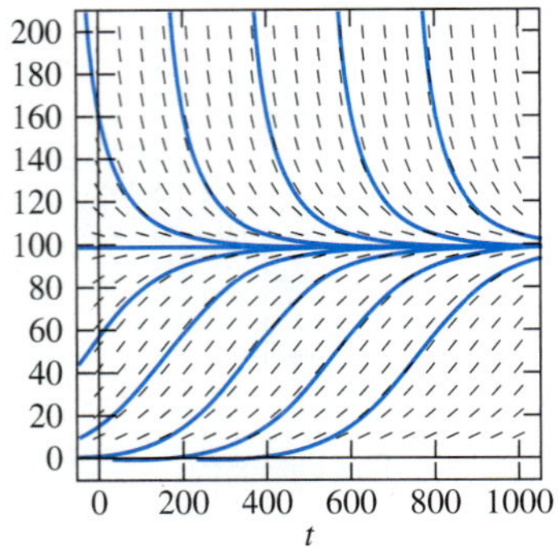

FIGURE 1.4.14. Slope field and solution curves for $x' = (0.01)x - (0.0001)x^2$.

of Eq. (2). The slope field and solution curves shown in Fig. 1.4.14 suggest that, whatever is the initial value x_0, the solution $x(t)$ approaches 100 as $t \to +\infty$. Can you use Eq. (5) to verify this conjecture?

INVESTIGATION: For your own personal logistic equation, take $a = m/n$ and $b = 1/n$ in Eq. (1), with m and n being the *largest* two distinct digits (in either order) in your student ID number.

(a) First generate a slope field for your differential equation and include a sufficient number of solution curves that you can see what happens to the population as $t \to +\infty$. State your inference plainly.

(b) Next use a computer algebra system to solve the differential equation symbolically; then use the symbolic solution to find the limit of $x(t)$ as $t \to +\infty$. Was your graphically based inference correct?

(c) Finally, state and solve a numerical problem using the symbolic solution. For instance, how long does it take x to grow from a selected initial value x_0 to a given target value x_1?

1.5 Linear First-Order Equations

In Section 1.4 we saw how to solve a separable differential equation by integrating *after* multiplying both sides by an appropriate factor. For instance, to solve the equation

$$\frac{dy}{dx} = 2xy \quad (y > 0), \tag{1}$$

we multiply both sides by the factor $1/y$ to get

$$\frac{1}{y} \cdot \frac{dy}{dx} = 2x; \quad \text{that is,} \quad D_x(\ln y) = D_x\left(x^2\right). \tag{2}$$

Because each side of the equation in (2) is recognizable as a *derivative* (with respect to the independent variable x), all that remains are two simple integrations, which yield $\ln y = x^2 + C$. For this reason, the function $\rho(y) = 1/y$ is called an *integrating factor* for the original equation in (1). An **integrating factor** for a differential equation is a function $\rho(x, y)$ such that the multiplication of each side of the differential equation by $\rho(x, y)$ yields an equation in which each side is recognizable as a derivative.

With the aid of the appropriate integrating factor, there is a standard technique for solving the **linear first-order equation**

$$\frac{dy}{dx} + P(x)y = Q(x) \tag{3}$$

on an interval on which the coefficient functions $P(x)$ and $Q(x)$ are continuous. We multiply each side in Eq. (3) by the integrating factor

$$\rho(x) = e^{\int P(x)\,dx}. \tag{4}$$

The result is

$$e^{\int P(x)\,dx}\frac{dy}{dx} + P(x)e^{\int P(x)\,dx}y = Q(x)e^{\int P(x)\,dx}. \tag{5}$$

Because

$$D_x\left[\int P(x)\,dx\right] = P(x),$$

the left-hand side is the derivative of the *product* $y(x)\cdot e^{\int P(x)\,dx}$, so Eq. (5) is equivalent to

$$D_x\left[y(x)\cdot e^{\int P(x)\,dx}\right] = Q(x)e^{\int P(x)\,dx}.$$

Integration of both sides of this equation gives

$$y(x)e^{\int P(x)\,dx} = \int \left(Q(x)e^{\int P(x)\,dx}\right)dx + C.$$

Finally, solving for y, we obtain the general solution of the linear first-order equation in (3):

$$y(x) = e^{-\int P(x)\,dx}\left[\int \left(Q(x)e^{\int P(x)\,dx}\right)dx + C\right]. \tag{6}$$

This formula should **not** be memorized. In a specific problem it generally is simpler to use the *method* by which we developed the formula. That is, in order to solve an equation that can be written in the form in Eq. (3) with the coefficient functions $P(x)$ and $Q(x)$ displayed explicitly, you should attempt to carry out the following steps.

METHOD: SOLUTION OF FIRST-ORDER EQUATIONS

1. Begin by calculating the integrating factor $\rho(x) = e^{\int P(x)\,dx}$.
2. Then multiply both sides of the differential equation by $\rho(x)$.
3. Next, recognize the left-hand side of the resulting equation as the derivative of a product:

$$D_x\,[\rho(x)y(x)] = \rho(x)Q(x).$$

4. Finally, integrate this equation,

$$\rho(x)y(x) = \int \rho(x)Q(x)\,dx + C,$$

then solve for y to obtain the general solution of the original differential equation.

Remark 1: Given an initial condition $y(x_0) = y_0$, you can (as usual) substitute $x = x_0$ and $y = y_0$ into the general solution and solve for the value of C yielding the particular solution that satisfies this initial condition.

Remark 2: You need not supply explicitly a constant of integration when you find the integrating factor $\rho(x)$. For if we replace

$$\int P(x)\,dx \quad \text{with} \quad \int P(x)\,dx + K$$

in Eq. (4), the result is

$$\rho(x) = e^{K+\int P(x)\,dx} = e^K e^{\int P(x)\,dx}.$$

But the constant factor e^K does not affect materially the result of multiplying both sides of the differential equation in (3) by $\rho(x)$, so we might as well take $K = 0$. You may therefore choose for $\int P(x)\,dx$ *any* convenient antiderivative of $P(x)$, without bothering to add a constant of integration. ■

Example 1 Solve the initial value problem

$$\frac{dy}{dx} - y = \tfrac{11}{8}e^{-x/3}, \quad y(0) = -1.$$

Solution Here we have $P(x) \equiv -1$ and $Q(x) = \frac{11}{8}e^{-x/3}$, so the integrating factor is

$$\rho(x) = e^{\int(-1)\,dx} = e^{-x}.$$

Multiplication of both sides of the given equation by e^{-x} yields

$$e^{-x}\frac{dy}{dx} - e^{-x}y = \tfrac{11}{8}e^{-4x/3}, \tag{7}$$

which we recognize as

$$\frac{d}{dx}\left(e^{-x}y\right) = \tfrac{11}{8}e^{-4x/3}.$$

Hence integration with respect to x gives

$$e^{-x}y = \int \tfrac{11}{8}e^{-4x/3}\,dx = -\tfrac{33}{32}e^{-4x/3} + C,$$

and multiplication by e^x gives the general solution

$$y(x) = Ce^x - \tfrac{33}{32}e^{-x/3}. \tag{8}$$

Substitution of $x = 0$ and $y = -1$ now gives $C = \frac{1}{32}$, so the desired particular solution is

$$y(x) = \tfrac{1}{32}e^x - \tfrac{33}{32}e^{-x/3} = \tfrac{1}{32}\left(e^x - 33e^{-x/3}\right). \qquad ■$$

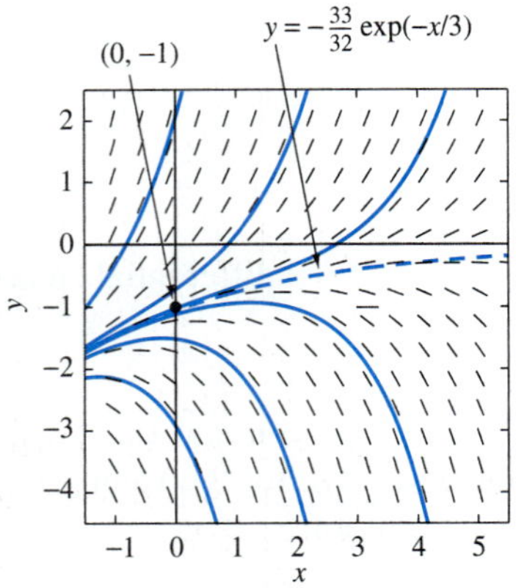

FIGURE 1.5.1. Slope field and solution curves for $y' = y + \frac{11}{8}e^{-x/3}$.

Remark: Figure 1.5.1 shows a slope field and typical solution curves for Eq. (7), including the one passing through the point $(0, -1)$. Note that some solutions grow rapidly in the positive direction as x increases, while others grow rapidly in the negative direction. The behavior of a given solution curve is determined by its initial condition $y(0) = y_0$. The two types of behavior are separated by the particular solution $y(x) = -\frac{33}{32}e^{-x/3}$ for which $C = 0$ in Eq. (8), so $y_0 = -\frac{33}{32}$ for the solution curve that is dashed in Fig. 1.5.1. If $y_0 > -\frac{33}{32}$, then $C > 0$ in Eq. (8), so the term e^x eventually dominates the behavior of $y(x)$, and hence $y(x) \to +\infty$ as $x \to +\infty$. But if $y_0 < -\frac{33}{32}$, then $C < 0$, so both terms in $y(x)$ are negative and therefore $y(x) \to -\infty$ as $x \to +\infty$. Thus the initial condition $y_0 = -\frac{33}{32}$ is *critical* in the sense that solutions that start above $-\frac{33}{32}$ on the y-axis grow in the positive direction, while solutions that start lower than $-\frac{33}{32}$ grow in the negative direction as $x \to +\infty$. The interpretation of a mathematical model often hinges on finding such a critical condition that separates one kind of behavior of a solution from a different kind of behavior. ■

Example 2 Find a general solution of

$$(x^2+1)\frac{dy}{dx}+3xy=6x. \tag{9}$$

Solution After division of both sides of the equation by x^2+1, we recognize the result

$$\frac{dy}{dx}+\frac{3x}{x^2+1}y=\frac{6x}{x^2+1}$$

as a first-order linear equation with $P(x)=3x/(x^2+1)$ and $Q(x)=6x/(x^2+1)$. Multiplication by

$$\rho(x)=\exp\left(\int\frac{3x}{x^2+1}\,dx\right)=\exp\left(\tfrac{3}{2}\ln(x^2+1)\right)=(x^2+1)^{3/2}$$

yields

$$(x^2+1)^{3/2}\frac{dy}{dx}+3x(x^2+1)^{1/2}y=6x(x^2+1)^{1/2},$$

and thus

$$D_x\left[(x^2+1)^{3/2}y\right]=6x(x^2+1)^{1/2}.$$

Integration then yields

$$(x^2+1)^{3/2}y=\int 6x(x^2+1)^{1/2}\,dx=2(x^2+1)^{3/2}+C.$$

Multiplication of both sides by $(x^2+1)^{-3/2}$ gives the general solution

$$y(x)=2+C(x^2+1)^{-3/2}. \tag{10}$$

■

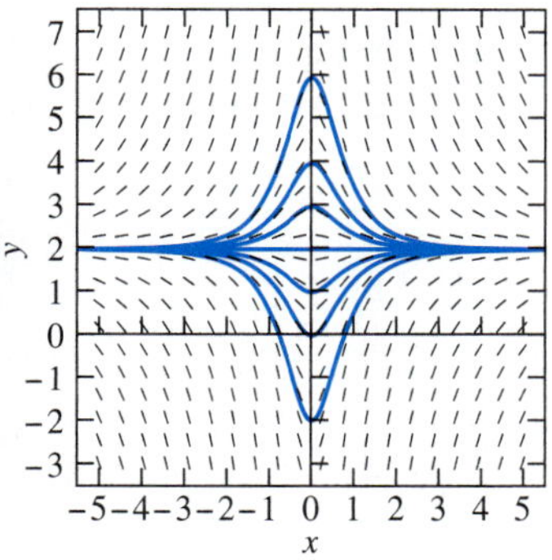

FIGURE 1.5.2. Slope field and solution curves for the differential equation in Eq. (9).

Remark: Figure 1.5.2 shows a slope field and typical solution curves for Eq. (9). Note that, as $x\to+\infty$, all other solution curves approach the constant solution curve $y(x)\equiv 2$ that corresponds to $C=0$ in Eq. (10). This constant solution can be described as an *equilibrium solution* of the differential equation, because $y(0)=2$ implies that $y(x)=2$ for all x (and thus the value of the solution remains forever where it starts). More generally, the word "equilibrium" connotes "unchanging," so by an equilibrium solution of a differential equation is meant a constant solution $y(x)\equiv c$, for which it follows that $y'(x)\equiv 0$. Note that substitution of $y'=0$ in the differential equation (9) yields $3xy=6x$, so it follows that $y=2$ if $x\neq 0$. Hence we see that $y(x)\equiv 2$ is the only equilibrium solution of this differential equation, as seems visually obvious in Fig. 1.5.2. ■

A Closer Look at the Method

The preceding derivation of the solution in Eq. (6) of the linear first-order equation $y'+Py=Q$ bears closer examination. Suppose that the coefficient functions $P(x)$ and $Q(x)$ are continuous on the (possibly unbounded) open interval I. Then the antiderivatives

$$\int P(x)\,dx \quad\text{and}\quad \int\left(Q(x)e^{\int P(x)\,dx}\right)dx$$

exist on I. Our derivation of Eq. (6) shows that *if* $y = y(x)$ is a solution of Eq. (3) on I, *then* $y(x)$ is given by the formula in Eq. (6) for some choice of the constant C. Conversely, you may verify by direct substitution (Problem 31) that the function $y(x)$ given in Eq. (6) satisfies Eq. (3). Finally, given a point x_0 of I and any number y_0, there is—as previously noted—a unique value of C such that $y(x_0) = y_0$. Consequently, we have proved the following existence-uniqueness theorem.

THEOREM 1 The Linear First-Order Equation

If the functions $P(x)$ and $Q(x)$ are continuous on the open interval I containing the point x_0, then the initial value problem

$$\frac{dy}{dx} + P(x)y = Q(x), \quad y(x_0) = y_0 \tag{11}$$

has a unique solution $y(x)$ on I, given by the formula in Eq. (6) with an appropriate value of C.

Remark 1: Theorem 1 gives a solution on the *entire* interval I for a *linear* differential equation, in contrast with Theorem 1 of Section 1.3, which guarantees only a solution on a possibly smaller interval.

Remark 2: Theorem 1 tells us that every solution of Eq. (3) is included in the general solution given in Eq. (6). Thus a *linear* first-order differential equation has *no* singular solutions.

Remark 3: The appropriate value of the constant C in Eq. (6)—as needed to solve the initial value problem in Eq. (11)—can be selected "automatically" by writing

$$\begin{aligned} \rho(x) &= \exp\left(\int_{x_0}^{x} P(t)\,dt\right), \\ y(x) &= \frac{1}{\rho(x)}\left[y_0 + \int_{x_0}^{x} \rho(t)Q(t)\,dt\right]. \end{aligned} \tag{12}$$

The indicated limits x_0 and x effect a choice of indefinite integrals in Eq. (6) that guarantees in advance that $\rho(x_0) = 1$ and that $y(x_0) = y_0$ (as you can verify directly by substituting $x = x_0$ in Eqs. (12)). ■

Example 3 Solve the initial value problem

$$x^2\frac{dy}{dx} + xy = \sin x, \quad y(1) = y_0. \tag{13}$$

Solution Division by x^2 gives the linear first-order equation

$$\frac{dy}{dx} + \frac{1}{x}y = \frac{\sin x}{x^2}$$

with $P(x) = 1/x$ and $Q(x) = (\sin x)/x^2$. With $x_0 = 1$ the integrating factor in (12) is

$$\rho(x) = \exp\left(\int_1^x \frac{1}{t}\,dt\right) = \exp(\ln x) = x,$$

so the desired particular solution is given by

$$y(x) = \frac{1}{x}\left(y_0 + \int_1^x \frac{\sin t}{t}\,dt\right). \tag{14}$$

In accord with Theorem 1, this solution is defined on the whole positive x-axis. ■

Comment: In general, an integral such as the one in Eq. (14) would (for given x) need to be approximated numerically—using Simpson's rule, for instance—to find the value $y(x)$ of the solution at x. In this case, however, we have the sine integral function

$$\text{Si}(x) = \int_0^x \frac{\sin t}{t}\,dt,$$

which appears with sufficient frequency in applications that its values have been tabulated. A good set of tables of special functions is Abramowitz and Stegun, *Handbook of Mathematical Functions* (New York: Dover, 1965). Then the particular solution in Eq. (14) reduces to

$$y(x) = \frac{1}{x}\left(y_0 + \int_0^x \frac{\sin t}{t}\,dt - \int_0^1 \frac{\sin t}{t}\,dt\right) = \frac{1}{x}\left[y_0 + \text{Si}(x) - \text{Si}(1)\right]. \tag{15}$$

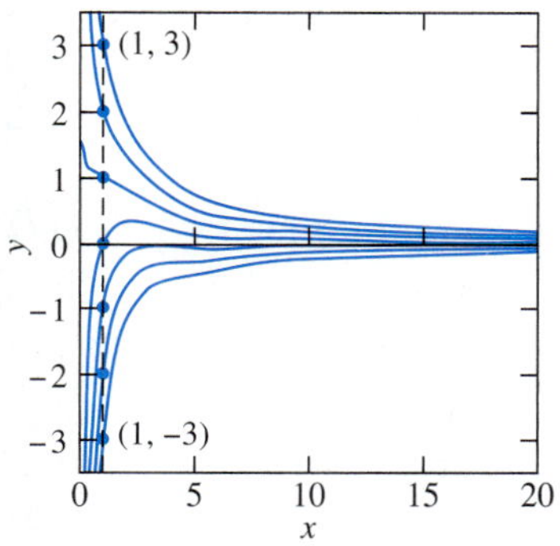

FIGURE 1.5.3. Typical solution curves defined by Eq. (15).

The sine integral function is available in most scientific computing systems and can be used to plot typical solution curves defined by Eq. (15). Figure 1.5.3 shows a selection of solution curves with initial values $y(1) = y_0$ ranging from $y_0 = -3$ to $y_0 = 3$. It appears that on each solution curve, $y(x) \to 0$ as $x \to +\infty$, and this is in fact true because the sine integral function is bounded. ■

In the sequel we will see that it is the exception—rather than the rule—when a solution of a differential equation can be expressed in terms of elementary functions. We will study various devices for obtaining good approximations to the values of the nonelementary functions we encounter. In Chapter 2 we will discuss numerical integration of differential equations in some detail.

Mixture Problems

As a first application of linear first-order equations, we consider a tank containing a solution—a mixture of solute and solvent—such as salt dissolved in water. There is both inflow and outflow, and we want to compute the *amount* $x(t)$ of solute in the tank at time t, given the amount $x(0) = x_0$ at time $t = 0$. Suppose that solution with a concentration of c_i grams of solute per liter of solution flows into the tank at the constant rate of r_i liters per second, and that the solution in the tank—kept thoroughly mixed by stirring—flows out at the constant rate of r_o liters per second.

To set up a differential equation for $x(t)$, we estimate the change Δx in x during the brief time interval $[t, t + \Delta t]$. The amount of solute that flows into the tank during Δt seconds is $r_i c_i\,\Delta t$ grams. To check this, note how the cancellation of dimensions checks our computations:

$$\left(r_i \frac{\text{liters}}{\text{second}}\right)\left(c_i \frac{\text{grams}}{\text{liter}}\right)(\Delta t \text{ seconds})$$

yields a quantity measured in grams.

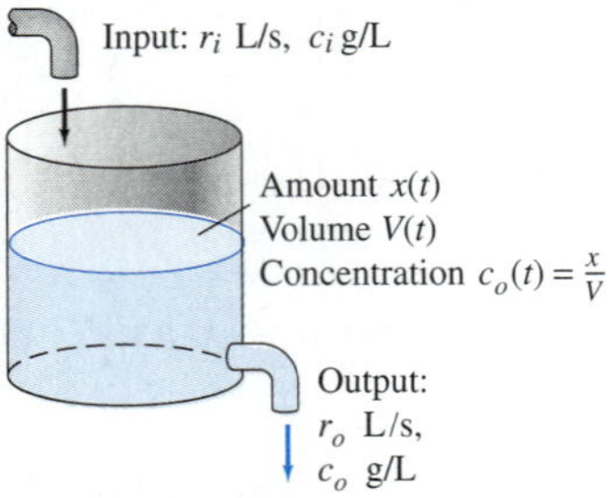

FIGURE 1.5.4. The single-tank mixture problem.

The amount of solute that flows out of the tank during the same time interval depends on the concentration $c_o(t)$ of solute in the solution at time t. But as noted in Fig. 1.5.4, $c_o(t) = x(t)/V(t)$, where $V(t)$ denotes the volume (not constant unless $r_i = r_o$) of solution in the tank at time t. Then

$$\Delta x = \{\text{grams input}\} - \{\text{grams output}\} \approx r_i c_i \,\Delta t - r_o c_o \,\Delta t.$$

We now divide by Δt:

$$\frac{\Delta x}{\Delta t} \approx r_i c_i - r_o c_o.$$

Finally, we take the limit as $\Delta t \to 0$; if all the functions involved are continuous and $x(t)$ is differentiable, then the error in this approximation also approaches zero, and we obtain the differential equation

$$\frac{dx}{dt} = r_i c_i - r_o c_o, \tag{16}$$

in which r_i, c_i, and r_o are constants, but c_o denotes the variable concentration

$$c_o(t) = \frac{x(t)}{V(t)} \tag{17}$$

of solute in the tank at time t. Thus the amount $x(t)$ of solute in the tank satisfies the differential equation

$$\frac{dx}{dt} = r_i c_i - \frac{r_o}{V} x. \tag{18}$$

If $V_0 = V(0)$, then $V(t) = V_0 + (r_i - r_o)t$, so Eq. (18) is a linear first-order differential equation for the amount $x(t)$ of solute in the tank at time t.

***Important*:** Equation (18) need not be committed to memory. It is the *process* we used to obtain that equation—examination of the behavior of the system over a short time interval $[t, t + \Delta t]$—that you should strive to understand, because it is a very useful tool for obtaining all sorts of differential equations.

Remark: It was convenient for us to use g/L mass/volume units in deriving Eq. (18). But any other consistent system of units can be used to measure amounts of solute and volumes of solution. In the following example we measure both in cubic kilometers. ■

Example 4 Assume that Lake Erie has a volume of 480 km^3 and that its rate of inflow (from Lake Huron) and outflow (to Lake Ontario) are both 350 km^3 per year. Suppose that at the time $t = 0$ (years), the pollutant concentration of Lake Erie—caused by past industrial pollution that has now been ordered to cease—is five times that of Lake Huron. If the outflow henceforth is perfectly mixed lake water, how long will it take to reduce the pollution concentration in Lake Erie to twice that of Lake Huron?

Solution Here we have

$$V = 480 \text{ (km}^3\text{)},$$
$$r_i = r_o = r = 350 \text{ (km}^3\text{/yr)},$$
$$c_i = c \text{ (the pollutant concentration of Lake Huron), and}$$
$$x_0 = x(0) = 5cV,$$

and the question is this: When is $x(t) = 2cV$? With this notation, Eq. (18) is the separable equation

$$\frac{dx}{dt} = rc - \frac{r}{V}x, \tag{19}$$

which we rewrite in the linear first-order form

$$\frac{dx}{dt} + px = q \tag{20}$$

with constant coefficients $p = r/V$, $q = rc$, and integrating factor $\rho = e^{pt}$. You can either solve this equation directly or apply the formula in (12). The latter gives

$$\begin{aligned} x(t) &= e^{-pt}\left[x_0 + \int_0^t qe^{pt}\,dt\right] = e^{-pt}\left[x_0 + \frac{q}{p}\left(e^{pt} - 1\right)\right] \\ &= e^{-rt/V}\left[5cV + \frac{rc}{r/V}\left(e^{rt/V} - 1\right)\right]; \\ x(t) &= cV + 4cVe^{-rt/V}. \end{aligned} \tag{21}$$

To find when $x(t) = 2cV$, we therefore need only solve the equation

$$cV + 4cVe^{-rt/V} = 2cV \quad \text{for} \quad t = \frac{V}{r}\ln 4 = \frac{480}{350}\ln 4 \approx 1.901 \text{ (years).}$$ ■

Example 5 A 120-gallon (gal) tank initially contains 90 lb of salt dissolved in 90 gal of water. Brine containing 2 lb/gal of salt flows into the tank at the rate of 4 gal/min, and the well-stirred mixture flows out of the tank at the rate of 3 gal/min. How much salt does the tank contain when it is full?

Solution The interesting feature of this example is that, due to the differing rates of inflow and outflow, the volume of brine in the tank increases steadily with $V(t) = 90 + t$ gallons. The change Δx in the amount x of salt in the tank from time t to time $t + \Delta t$ (minutes) is given by

$$\Delta x \approx (4)(2)\,\Delta t - 3\left(\frac{x}{90 + t}\right)\Delta t,$$

so our differential equation is

$$\frac{dx}{dt} + \frac{3}{90 + t}x = 8.$$

An integrating factor is

$$\rho(x) = \exp\left(\int \frac{3}{90 + t}\,dt\right) = e^{3\ln(90+t)} = (90 + t)^3,$$

which gives

$$\begin{aligned} D_t\left[(90 + t)^3 x\right] &= 8(90 + t)^3; \\ (90 + t)^3 x &= 2(90 + t)^4 + C. \end{aligned}$$

Substitution of $x(0) = 90$ gives $C = -(90)^4$, so the amount of salt in the tank at time t is

$$x(t) = 2(90+t) - \frac{90^4}{(90+t)^3}.$$

The tank is full after 30 min, and when $t = 30$, we have

$$x(30) = 2(90+30) - \frac{90^4}{120^3} \approx 202 \text{ (lb)}$$

of salt in the tank. ■

1.5 Problems

Find general solutions of the differential equations in Problems 1 through 25. If an initial condition is given, find the corresponding particular solution. Throughout, primes denote derivatives with respect to x.

1. $y' + y = 2$, $y(0) = 0$

2. $y' - 2y = 3e^{2x}$, $y(0) = 0$

3. $y' + 3y = 2xe^{-3x}$

4. $y' - 2xy = e^{x^2}$

5. $xy' + 2y = 3x$, $y(1) = 5$

6. $xy' + 5y = 7x^2$, $y(2) = 5$

7. $2xy' + y = 10\sqrt{x}$

8. $3xy' + y = 12x$

9. $xy' - y = x$, $y(1) = 7$

10. $2xy' - 3y = 9x^3$

11. $xy' + y = 3xy$, $y(1) = 0$

12. $xy' + 3y = 2x^5$, $y(2) = 1$

13. $y' + y = e^x$, $y(0) = 1$

14. $xy' - 3y = x^3$, $y(1) = 10$

15. $y' + 2xy = x$, $y(0) = -2$

16. $y' = (1 - y)\cos x$, $y(\pi) = 2$

17. $(1 + x)y' + y = \cos x$, $y(0) = 1$

18. $xy' = 2y + x^3 \cos x$

19. $y' + y \cot x = \cos x$

20. $y' = 1 + x + y + xy$, $y(0) = 0$

21. $xy' = 3y + x^4 \cos x$, $y(2\pi) = 0$

22. $y' = 2xy + 3x^2 \exp(x^2)$, $y(0) = 5$

23. $xy' + (2x - 3)y = 4x^4$

24. $(x^2 + 4)y' + 3xy = x$, $y(0) = 1$

25. $(x^2 + 1)\dfrac{dy}{dx} + 3x^3 y = 6x \exp\left(-\frac{3}{2}x^2\right)$, $y(0) = 1$

Solve the differential equations in Problems 26 through 28 by regarding y as the independent variable rather than x.

26. $(1 - 4xy^2)\dfrac{dy}{dx} = y^3$

27. $(x + ye^y)\dfrac{dy}{dx} = 1$

28. $(1 + 2xy)\dfrac{dy}{dx} = 1 + y^2$

29. Express the general solution of $dy/dx = 1 + 2xy$ in terms of the **error function**

$$\operatorname{erf}(x) = \frac{2}{\sqrt{\pi}} \int_0^x e^{-t^2}\,dt.$$

30. Express the solution of the initial value problem

$$2x\frac{dy}{dx} = y + 2x\cos x, \quad y(1) = 0$$

as an integral as in Example 3 of this section.

Problems 31 and 32 illustrate—for the special case of first-order linear equations—techniques that will be important when we study higher-order linear equations in Chapter 3.

31. **(a)** Show that

$$y_c(x) = Ce^{-\int P(x)\,dx}$$

is a general solution of $dy/dx + P(x)y = 0$. **(b)** Show that

$$y_p(x) = e^{-\int P(x)\,dx}\left[\int \left(Q(x)e^{\int P(x)\,dx}\right)dx\right]$$

is a particular solution of $dy/dx + P(x)y = Q(x)$. **(c)** Suppose that $y_c(x)$ is any general solution of $dy/dx + P(x)y = 0$ and that $y_p(x)$ is any particular solution of $dy/dx + P(x)y = Q(x)$. Show that $y(x) = y_c(x) + y_p(x)$ is a general solution of $dy/dx + P(x)y = Q(x)$.

32. **(a)** Find constants A and B such that $y_p(x) = A\sin x + B\cos x$ is a solution of $dy/dx + y = 2\sin x$. **(b)** Use the result of part (a) and the method of Problem 31 to find the general solution of $dy/dx + y = 2\sin x$. **(c)** Solve the initial value problem $dy/dx + y = 2\sin x$, $y(0) = 1$.

33. A tank contains 1000 liters (L) of a solution consisting of 100 kg of salt dissolved in water. Pure water is pumped into the tank at the rate of 5 L/s, and the mixture—kept uniform by stirring— is pumped out at the same rate. How long will it be until only 10 kg of salt remains in the tank?

34. Consider a reservoir with a volume of 8 billion cubic feet (ft^3) and an initial pollutant concentration of 0.25%. There is a daily inflow of 500 million ft^3 of water with a pollutant concentration of 0.05% and an equal daily outflow of the well-mixed water in the reservoir. How long will it take to reduce the pollutant concentration in the reservoir to 0.10%?

35. Rework Example 4 for the case of Lake Ontario, which empties into the St. Lawrence River and receives inflow from Lake Erie (via the Niagara River). The only differences are that this lake has a volume of 1640 km^3 and an inflow-outflow rate of 410 km^3/year.

36. A tank initially contains 60 gal of pure water. Brine containing 1 lb of salt per gallon enters the tank at 2 gal/min, and the (perfectly mixed) solution leaves the tank at 3 gal/min; thus the tank is empty after exactly 1 h. **(a)** Find the amount of salt in the tank after t minutes. **(b)** What is the maximum amount of salt ever in the tank?

37. A 400-gal tank initially contains 100 gal of brine containing 50 lb of salt. Brine containing 1 lb of salt per gallon enters the tank at the rate of 5 gal/s, and the well-mixed brine in the tank flows out at the rate of 3 gal/s. How much salt will the tank contain when it is full of brine?

38. Consider the *cascade* of two tanks shown in Fig. 1.5.5, with $V_1 = 100$ (gal) and $V_2 = 200$ (gal) the volumes of brine in the two tanks. Each tank also initially contains 50 lb of salt. The three flow rates indicated in the figure are each 5 gal/min, with pure water flowing into tank 1. **(a)** Find the amount $x(t)$ of salt in tank 1 at time t. **(b)** Suppose that $y(t)$ is the amount of salt in tank 2 at time t. Show first that

$$\frac{dy}{dt} = \frac{5x}{100} - \frac{5y}{200},$$

and then solve for $y(t)$, using the function $x(t)$ found in part (a). **(c)** Finally, find the maximum amount of salt ever in tank 2.

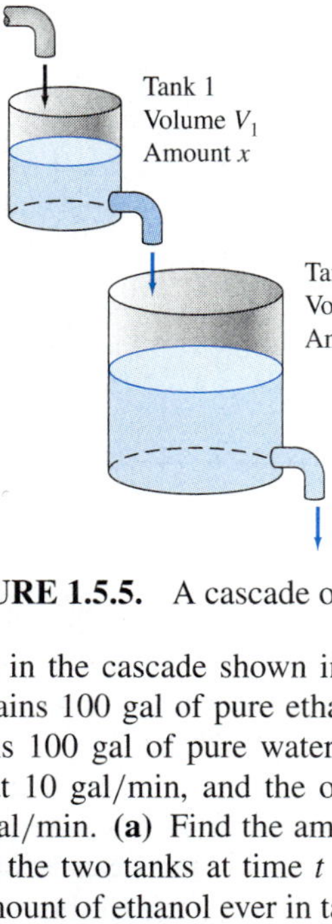

FIGURE 1.5.5. A cascade of two tanks.

39. Suppose that in the cascade shown in Fig. 1.5.5, tank 1 initially contains 100 gal of pure ethanol and tank 2 initially contains 100 gal of pure water. Pure water flows into tank 1 at 10 gal/min, and the other two flow rates are also 10 gal/min. **(a)** Find the amounts $x(t)$ and $y(t)$ of ethanol in the two tanks at time $t \geqq 0$. **(b)** Find the maximum amount of ethanol ever in tank 2.

40. A multiple cascade is shown in Fig. 1.5.6. At time $t = 0$, tank 0 contains 1 gal of ethanol and 1 gal of water; all the remaining tanks contain 2 gal of pure water each. Pure water is pumped into tank 0 at 1 gal/min, and the varying mixture in each tank is pumped into the one below it at the same rate. Assume, as usual, that the mixtures are kept perfectly uniform by stirring. Let $x_n(t)$ denote the amount of ethanol in tank n at time t.

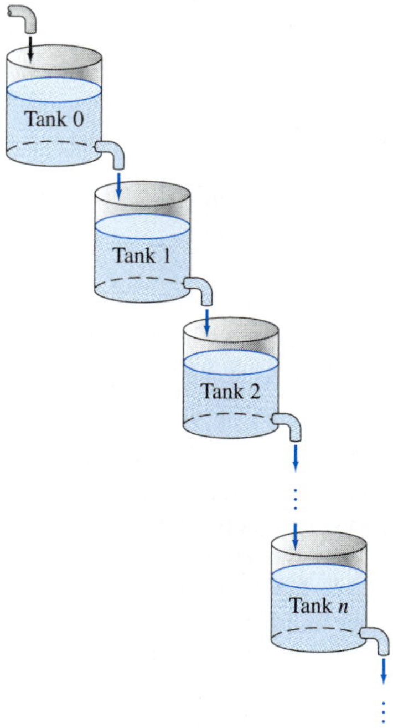

FIGURE 1.5.6. A multiple cascade.

(a) Show that $x_0(t) = e^{-t/2}$. **(b)** Show by induction on n that

$$x_n(t) = \frac{t^n e^{-t/2}}{n!\,2^n} \quad \text{for } n > 0.$$

(c) Show that the maximum value of $x_n(t)$ for $n > 0$ is $M_n = x_n(2n) = n^n e^{-n}/n!$. **(d)** Conclude from **Stirling's approximation** $n! \approx n^n e^{-n}\sqrt{2\pi n}$ that $M_n \approx (2\pi n)^{-1/2}$.

41. A 30-year-old woman accepts an engineering position with a starting salary of \$30,000 per year. Her salary $S(t)$ increases exponentially, with $S(t) = 30e^{t/20}$ thousand dollars after t years. Meanwhile, 12% of her salary is deposited continuously in a retirement account, which accumulates interest at a continuous annual rate of 6%. **(a)** Estimate ΔA in terms of Δt to derive the differential equation satisfied by the amount $A(t)$ in her retirement account after t years. **(b)** Compute $A(40)$, the amount available for her retirement at age 70.

42. Suppose that a falling hailstone with density $\delta = 1$ starts from rest with negligible radius $r = 0$. Thereafter its radius is $r = kt$ (k is a constant) as it grows by accretion during its fall. Use Newton's second law—according to which the net force F acting on a possibly variable mass m equals the time rate of change dp/dt of its momentum $p = mv$—to set up and solve the initial value problem

$$\frac{d}{dt}(mv) = mg, \quad v(0) = 0,$$

where m is the variable mass of the hailstone, $v = dy/dt$ is its velocity, and the positive y-axis points downward. Then show that $dv/dt = g/4$. Thus the hailstone falls as though it were under *one-fourth* the influence of gravity.

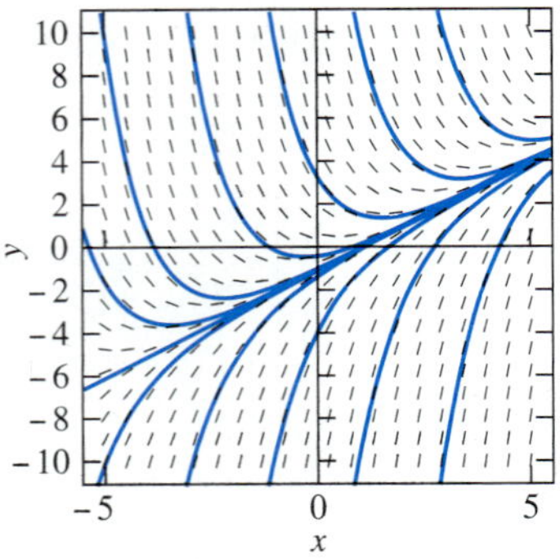

FIGURE 1.5.7. Slope field and solution curves for $y' = x - y$.

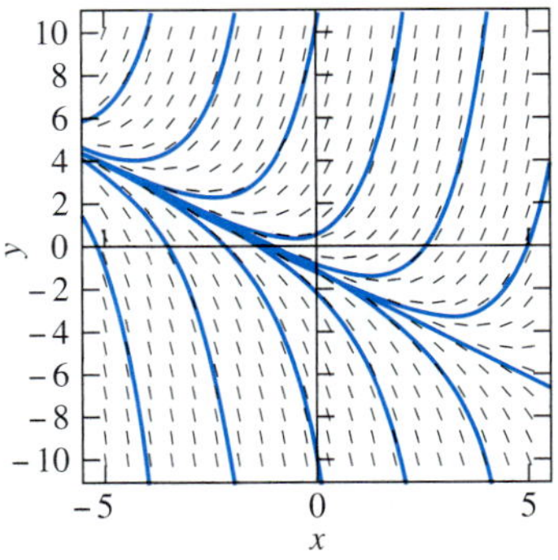

FIGURE 1.5.8. Slope field and solution curves for $y' = x + y$.

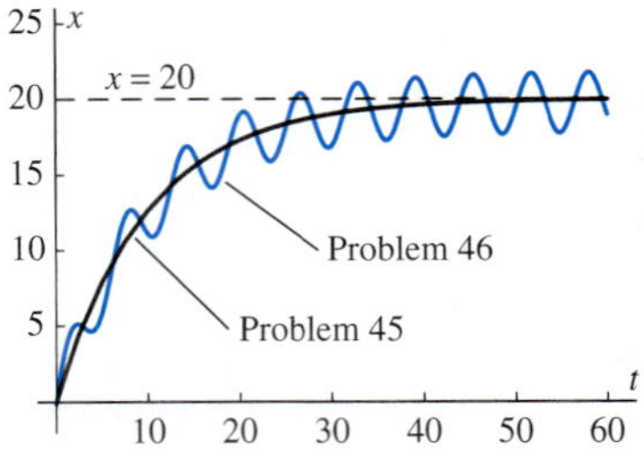

FIGURE 1.5.9. Graphs of solutions in Problems 45 and 46.

43. Figure 1.5.7 shows a slope field and typical solution curves for the equation $y' = x - y$. **(a)** Show that every solution curve approaches the straight line $y = x - 1$ as $x \to +\infty$. **(b)** For each of the five values $y_1 = 3.998$, 3.999, 4.000, 4.001, and 4.002, determine the initial value y_0 (accurate to four decimal places) such that $y(5) = y_1$ for the solution satisfying the initial condition $y(-5) = y_0$.

44. Figure 1.5.8 shows a slope field and typical solution curves for the equation $y' = x + y$. **(a)** Show that every solution curve approaches the straight line $y = -x - 1$ as $x \to -\infty$. **(b)** For each of the five values $y_1 = -10$, -5, 0, 5, and 10, determine the initial value y_0 (accurate to five decimal places) such that $y(5) = y_1$ for the solution satisfying the initial condition $y(-5) = y_0$.

Problems 45 and 46 deal with a shallow reservoir that has a one-square-kilometer water surface and an average water depth of 2 meters. Initially it is filled with fresh water, but at time $t = 0$ water contaminated with a liquid pollutant begins flowing into the reservoir at the rate of 200 thousand cubic meters per month. The well-mixed water in the reservoir flows out at the same rate. Your first task is to find the amount $x(t)$ of pollutant (in millions of liters) in the reservoir after t months.

45. The incoming water has a pollutant concentration of $c(t) = 10$ liters per cubic meter (L/m^3). Verify that the graph of $x(t)$ resembles the steadily rising curve in Fig. 1.5.9, which approaches asymptotically the graph of the equilibrium solution $x(t) \equiv 20$ that corresponds to the reservoir's long-term pollutant content. How long does it take the pollutant concentration in the reservoir to reach 10 L/m^3?

46. The incoming water has pollutant concentration $c(t) = 10(1 + \cos t)$ L/m^3 that varies between 0 and 20, with an average concentration of 10 L/m^3 and a period of oscillation of slightly over $6\frac{1}{4}$ months. Does it seem predictable that the lake's pollutant content should ultimately oscillate periodically about an average level of 20 million liters? Verify that the graph of $x(t)$ does, indeed, resemble the oscillatory curve shown in Fig. 1.5.9. How long does it take the pollutant concentration in the reservoir to reach 10 L/m^3?

1.5 Application Indoor Temperature Oscillations

For an interesting applied problem that involves the solution of a linear differential equation, consider indoor temperature oscillations that are driven by outdoor temperature oscillations of the form

$$A(t) = a_0 + a_1 \cos \omega t + b_1 \sin \omega t. \tag{1}$$

If $\omega = \pi/12$, then these oscillations have a period of 24 hours (so that the cycle of outdoor temperatures repeats itself daily) and Eq. (1) provides a realistic model for the temperature outside a house on a day when no change in the overall day-to-day weather pattern is occurring. For instance, for a typical July day in Athens, Georgia with a minimum temperature of 70°F when $t = 4$ (4 A.M.) and a maximum of 90°F when $t = 16$ (4 P.M.), we would take

$$A(t) = 80 - 10 \cos \omega(t - 4) = 80 - 5 \cos \omega t - 5\sqrt{3} \sin \omega t. \tag{2}$$

We derived Eq. (2) by using the identity $\cos(\alpha - \beta) = \cos\alpha\cos\beta + \sin\alpha\sin\beta$ to get $a_0 = 80$, $a_1 = -5$, and $b_1 = -5\sqrt{3}$ in Eq. (1).

If we write Newton's law of cooling (Eq. (3) of Section 1.1) for the corresponding indoor temperature $u(t)$ at time t, but with the outside temperature $A(t)$ given by Eq. (1) instead of a constant ambient temperature A, we get the linear first-order differential equation

$$\frac{du}{dt} = -k(u - A(t));$$

that is,

$$\frac{du}{dt} + ku = k(a_0 + a_1\cos\omega t + b_1\sin\omega t) \tag{3}$$

with coefficient functions $P(t) \equiv k$ and $Q(t) = kA(t)$. Typical values of the proportionality constant k range from 0.2 to 0.5 (although k might be greater than 0.5 for a poorly insulated building with open windows, or less than 0.2 for a well-insulated building with tightly sealed windows).

SCENARIO: Suppose that our air conditioner fails at time $t_0 = 0$ one midnight, and we cannot afford to have it repaired until payday at the end of the month. We therefore want to investigate the resulting indoor temperatures that we must endure for the next several days.

Begin your investigation by solving Eq. (3) with the initial condition $u(0) = u_0$ (the indoor temperature at the time of the failure of the air conditioner). You may want to use the integral formulas in 49 and 50 of the endpapers, or possibly a computer algebra system. You should get the solution

$$u(t) = a_0 + c_0 e^{-kt} + c_1\cos\omega t + d_1\sin\omega t, \tag{4}$$

where

$$c_0 = u_0 - a_0 - \frac{k^2 a_1 - k\omega b_1}{k^2 + \omega^2},$$

$$c_1 = \frac{k^2 a_1 - k\omega b_1}{k^2 + \omega^2}, \qquad d_1 = \frac{k\omega a_1 + k^2 b_1}{k^2 + \omega^2}$$

with $\omega = \pi/12$.

With $a_0 = 80$, $a_1 = -5$, $b_1 = -5\sqrt{3}$ (as in Eq. (2)), $\omega = \pi/12$, and $k = 0.2$ (for instance), this solution reduces (approximately) to

$$u(t) = 80 + e^{-t/5}(u_0 - 82.3351) + (2.3351)\cos\frac{\pi t}{12} - (5.6036)\sin\frac{\pi t}{12}. \tag{5}$$

Observe first that the "damped" exponential term in Eq. (5) approaches zero as $t \to +\infty$, leaving the long-term "steady periodic" solution

$$u_{sp}(t) = 80 + (2.3351)\cos\frac{\pi t}{12} - (5.6036)\sin\frac{\pi t}{12}. \tag{6}$$

Consequently, the long-term indoor temperatures oscillate every 24 hours around the same average temperature 80°F as the average outdoor temperature.

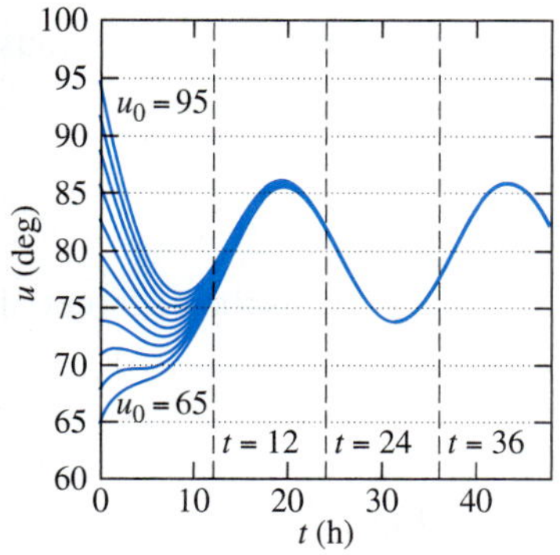

FIGURE 1.5.10. Solution curves given by Eq. (5) with $u_0 = 65, 68, 71, \ldots, 92, 95$.

Figure 1.5.10 shows a number of solution curves corresponding to possible initial temperatures u_0 ranging from 65°F to 95°F. Observe that—whatever the initial temperature—the indoor temperature "settles down" within about 18 hours to a periodic daily oscillation. But the amplitude of temperature variation is less indoors than outdoors. Indeed, using the trigonometric identity mentioned earlier, Eq. (6) can be rewritten (verify this!) as

$$\begin{aligned} u(t) &= 80 - (6.0707)\cos\left(\frac{\pi t}{12} - 1.9656\right) \\ &= 80 - (6.0707)\cos\frac{\pi}{12}(t - 7.5082). \end{aligned} \tag{7}$$

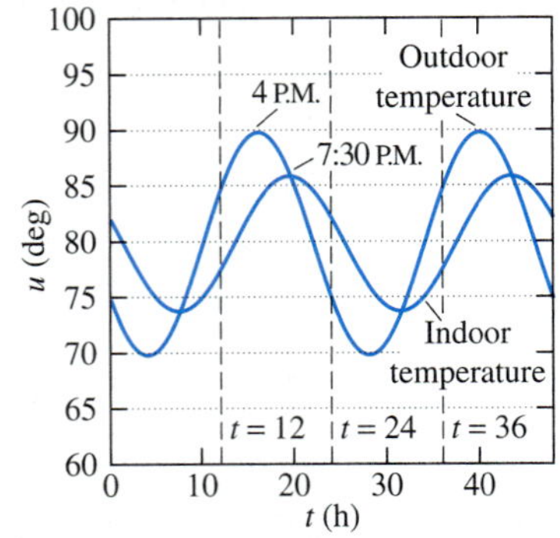

FIGURE 1.5.11. Comparison of indoor and outdoor temperature oscillations.

Do you see that this implies that the indoor temperature varies between a minimum of about 74°F and a maximum of about 86°F?

Finally, comparison of Eqs. (2) and (7) indicates that the indoor temperature lags behind the outdoor temperature by about $7.5082 - 4 \approx 3.5$ hours, as illustrated in Fig. 1.5.11. Thus the temperature inside the house continues to rise until about 7:30 P.M. each evening, so the hottest part of the day inside is early evening rather than late afternoon (as outside).

For a personal problem to investigate, carry out a similar analysis using average July daily maximum/minimum figures for your own locale and a value of k appropriate to your own home. You might also consider a winter day instead of a summer day. (What is the winter-summer difference for the indoor temperature problem?) You may wish to explore the use of available technology both to solve the differential equation and to graph its solution for the indoor temperature in comparison with the outdoor temperature.

1.6 Substitution Methods and Exact Equations

The first-order differential equations we have solved in the previous sections have all been either separable or linear. But many applications involve differential equations that are neither separable nor linear. In this section we illustrate (mainly with examples) substitution methods that sometimes can be used to transform a given differential equation into one that we already know how to solve.

For instance, the differential equation

$$\frac{dy}{dx} = f(x, y), \tag{1}$$

with dependent variable y and independent variable x, may contain a conspicuous combination

$$v = \alpha(x, y) \tag{2}$$

of x and y that suggests itself as a new independent variable v. Thus the differential equation

$$\frac{dy}{dx} = (x + y + 3)^2$$

practically demands the substitution $v = x + y + 3$ of the form in Eq. (2).

If the substitution relation in Eq. (2) can be solved for

$$y = \beta(x, v), \tag{3}$$

then application of the chain rule—regarding v as an (unknown) function of x—yields

$$\frac{dy}{dx} = \frac{\partial\beta}{\partial x}\frac{dx}{dx} + \frac{\partial\beta}{\partial v}\frac{dv}{dx} = \beta_x + \beta_v\frac{dv}{dx}, \tag{4}$$

where the partial derivatives $\partial\beta/\partial x = \beta_x(x, v)$ and $\partial\beta/\partial v = \beta_v(x, v)$ are *known* functions of x and v. If we substitute the right-hand side in (4) for dy/dx in Eq. (1) and then solve for dv/dx, the result is a *new* differential equation of the form

$$\frac{dv}{dx} = g(x, v) \tag{5}$$

with *new* dependent variable v. If this new equation is either separable or linear, then we can apply the methods of preceding sections to solve it.

If $v = v(x)$ is a solution of Eq. (5), then $y = \beta(x, v(x))$ will be a solution of the original Eq. (1). The trick is to select a substitution such that the transformed Eq. (5) is one we can solve. Even when possible, this is not always easy; it may require a fair amount of ingenuity or trial and error.

Example 1 Solve the differential equation

$$\frac{dy}{dx} = (x + y + 3)^2.$$

Solution As indicated earlier, let's try the substitution

$$v = x + y + 3; \qquad \text{that is,} \qquad y = v - x - 3.$$

Then

$$\frac{dy}{dx} = \frac{dv}{dx} - 1,$$

so the transformed equation is

$$\frac{dv}{dx} = 1 + v^2.$$

This is a separable equation, and we have no difficulty in obtaining its solution

$$x = \int \frac{dv}{1 + v^2} = \tan^{-1} v + C.$$

So $v = \tan(x - C)$. Because $v = x + y + 3$, the general solution of the original equation $dy/dx = (x + y + 3)^2$ is $x + y + 3 = \tan(x - C)$; that is,

$$y(x) = \tan(x - C) - x - 3.$$ ■

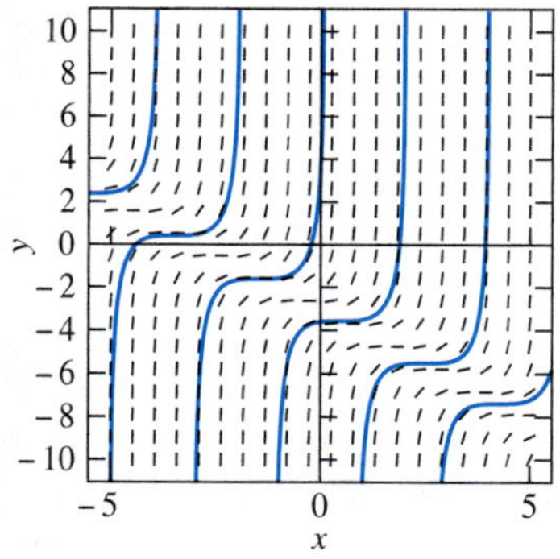

FIGURE 1.6.1. Slope field and solution curves for $y' = (x + y + 3)^2$.

Remark: Figure 1.6.1 shows a slope field and typical solution curves for the differential equation of Example 1. We see that, although the function $f(x, y) = (x+y+3)^2$ is continuously differentiable for all x and y, each solution is continuous only on a bounded interval. In particular, because the tangent function is continuous on the open interval $(-\pi/2, \pi/2)$, the particular solution with arbitrary constant value C is continuous on the interval where $-\pi/2 < x - C < \pi/2$; that is, $C - \pi/2 < x < C + \pi/2$. This situation is fairly typical of nonlinear differential equations, in contrast with linear differential equations, whose solutions are continuous wherever the coefficient functions in the equation are continuous. ■

Example 1 illustrates the fact that any differential equation of the form

$$\frac{dy}{dx} = F(ax + by + c) \tag{6}$$

can be transformed into a separable equation by use of the substitution $v = ax + by + c$ (see Problem 55). The paragraphs that follow deal with other classes of first-order equations for which there are standard substitutions that are known to succeed.

Homogeneous Equations

A **homogeneous** first-order differential equation is one that can be written in the form

$$\frac{dy}{dx} = F\left(\frac{y}{x}\right). \tag{7}$$

If we make the substitutions

$$v = \frac{y}{x}, \quad y = vx, \quad \frac{dy}{dx} = v + x\frac{dv}{dx}, \tag{8}$$

then Eq. (7) is transformed into the *separable* equation

$$x\frac{dv}{dx} = F(v) - v.$$

Thus every homogeneous first-order differential equation can be reduced to an integration problem by means of the substitutions in (8).

Remark: A dictionary definition of "homogeneous" is "of a similar kind or nature." Consider a differential equation of the form

$$Ax^m y^n \frac{dy}{dx} = Bx^p y^q + Cx^r y^s \tag{$*$}$$

whose polynomial coefficient functions are "homogeneous" in the sense that each of their terms has the same total degree, $m + n = p + q = r + s = K$. If we divide each side of ($*$) by x^K, then the result—because $x^m y^n / x^{m+n} = (y/x)^n$, and so forth—is the equation

$$A\left(\frac{y}{x}\right)^n \frac{dy}{dx} = B\left(\frac{y}{x}\right)^q + C\left(\frac{y}{x}\right)^s,$$

which evidently can be written (by another division) in the form of Eq. (7). More generally, a differential equation of the form $P(x, y)y' = Q(x, y)$ with polynomial coefficients P and Q is homogeneous if the terms in these polynomials all have the same total degree K. The differential equation in the following example is of this form with $K = 2$. ■

Example 2 Solve the differential equation

$$2xy\frac{dy}{dx} = 4x^2 + 3y^2.$$

Solution This equation is neither separable nor linear, but we recognize it as a homogeneous equation by writing it in the form

$$\frac{dy}{dx} = \frac{4x^2 + 3y^2}{2xy} = 2\left(\frac{x}{y}\right) + \frac{3}{2}\left(\frac{y}{x}\right).$$

The substitutions in (8) then take the form

$$y = vx, \quad \frac{dy}{dx} = v + x\frac{dv}{dx}, \quad v = \frac{y}{x}, \quad \text{and} \quad \frac{1}{v} = \frac{x}{y}.$$

These yield

$$v + x\frac{dv}{dx} = \frac{2}{v} + \frac{3}{2}v,$$

and hence

$$x\frac{dv}{dx} = \frac{2}{v} + \frac{v}{2} = \frac{v^2+4}{2v};$$

$$\int \frac{2v}{v^2+4}\,dv = \int \frac{1}{x}\,dx;$$

$$\ln(v^2+4) = \ln|x| + \ln C.$$

We apply the exponential function to both sides of the last equation to obtain

$$v^2 + 4 = C|x|;$$

$$\frac{y^2}{x^2} + 4 = C|x|;$$

$$y^2 + 4x^2 = kx^3.$$

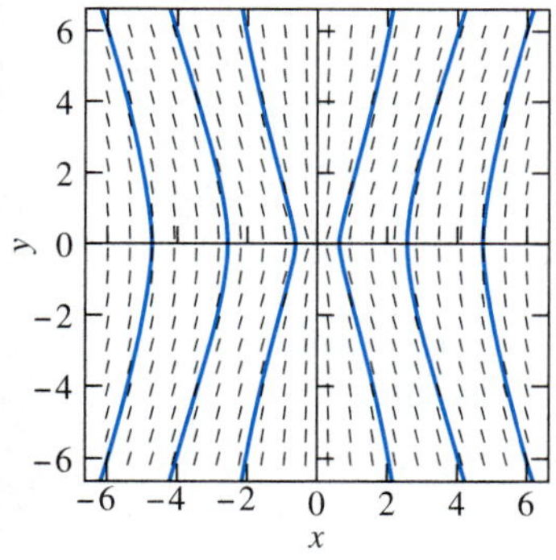

FIGURE 1.6.2. Slope field and solution curves for $2xyy' = 4x^2 + 3y^2$.

Note that the left-hand side of this equation is necessarily nonnegative. It follows that $k > 0$ in the case of solutions that are defined for $x > 0$, while $k < 0$ for solutions where $x < 0$. Indeed, the family of solution curves illustrated in Fig. 1.6.2 exhibits symmetry about both coordinate axes. Actually, there are positive-valued and negative-valued solutions of the forms $y(x) = \pm\sqrt{kx^3 - 4x^2}$ that are defined for $x > 4/k$ if the constant k is positive, and for $x < 4/k$ if k is negative. ■

Example 3 Solve the initial value problem

$$x\frac{dy}{dx} = y + \sqrt{x^2 - y^2}, \quad y(x_0) = 0,$$

where $x_0 > 0$.

Solution We divide both sides by x and find that

$$\frac{dy}{dx} = \frac{y}{x} + \sqrt{1 - \left(\frac{y}{x}\right)^2},$$

so we make the substitutions in (8); we get

$$v + x\frac{dv}{dx} = v + \sqrt{1 - v^2};$$

$$\int \frac{1}{\sqrt{1 - v^2}}\,dv = \int \frac{1}{x}\,dx;$$

$$\sin^{-1} v = \ln x + C.$$

We need not write $\ln|x|$ because $x > 0$ near $x = x_0 > 0$. Now note that $v(x_0) = y(x_0)/x_0 = 0$, so $C = \sin^{-1} 0 - \ln x_0 = -\ln x_0$. Hence

$$v = \frac{y}{x} = \sin(\ln x - \ln x_0) = \sin\left(\ln \frac{x}{x_0}\right),$$

and therefore

$$y(x) = x \sin\left(\ln \frac{x}{x_0}\right)$$

is the desired particular solution. Figure 1.6.3 shows some typical solution curves. Because of the radical in the differential equation, these solution curves are confined to the indicated triangular region $x \geqq |y|$. You can check that the boundary lines $y = x$ and $y = -x$ (for $x > 0$) are singular solution curves that consist of points of tangency with the solution curves found earlier. ■

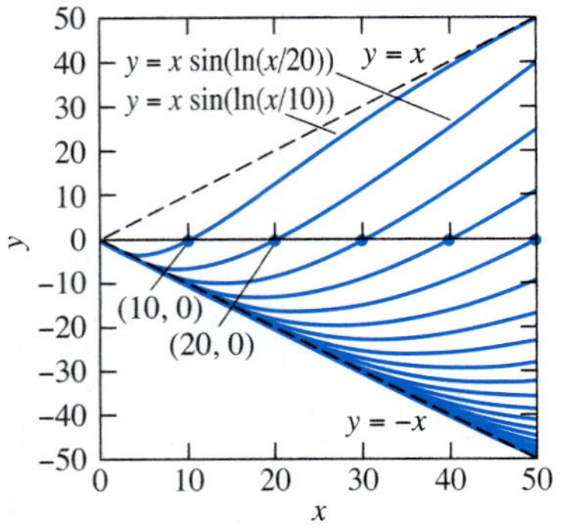

FIGURE 1.6.3. Solution curves for $xy' = y + \sqrt{x^2 - y^2}$.

Bernoulli Equations

A first-order differential equation of the form

$$\frac{dy}{dx} + P(x)y = Q(x)y^n \tag{9}$$

is called a **Bernoulli equation.** If either $n = 0$ or $n = 1$, then Eq. (9) is linear. Otherwise, as we ask you to show in Problem 56, the substitution

$$v = y^{1-n} \tag{10}$$

transforms Eq. (9) into the linear equation

$$\frac{dv}{dx} + (1 - n)P(x)v = (1 - n)Q(x).$$

Rather than memorizing the form of this transformed equation, it is more efficient to make the substitution in Eq. (10) explicitly, as in the following examples.

Example 4 If we rewrite the homogeneous equation $2xyy' = 4x^2 + 3y^2$ of Example 2 in the form

$$\frac{dy}{dx} - \frac{3}{2x}y = \frac{2x}{y},$$

we see that it is also a Bernoulli equation with $P(x) = -3/(2x)$, $Q(x) = 2x$, $n = -1$, and $1 - n = 2$. Hence we substitute

$$v = y^2, \quad y = v^{1/2}, \quad \text{and} \quad \frac{dy}{dx} = \frac{dy}{dv}\frac{dv}{dx} = \frac{1}{2}v^{-1/2}\frac{dv}{dx}.$$

This gives

$$\frac{1}{2}v^{-1/2}\frac{dv}{dx} - \frac{3}{2x}v^{1/2} = 2xv^{-1/2}.$$

Then multiplication by $2v^{1/2}$ produces the linear equation

$$\frac{dv}{dx} - \frac{3}{x}v = 4x$$

with integrating factor $\rho = e^{\int(-3/x)\,dx} = x^{-3}$. So we obtain

$$\begin{aligned} D_x(x^{-3}v) &= \frac{4}{x^2}; \\ x^{-3}v &= -\frac{4}{x} + C; \\ x^{-3}y^2 &= -\frac{4}{x} + C; \\ y^2 &= -4x^2 + Cx^3. \end{aligned}$$

■

Example 5 The equation

$$x\frac{dy}{dx} + 6y = 3xy^{4/3}$$

is neither separable nor linear nor homogeneous, but it is a Bernoulli equation with $n = \frac{4}{3}$, $1 - n = -\frac{1}{3}$. The substitutions

$$v = y^{-1/3}, \quad y = v^{-3}, \quad \text{and} \quad \frac{dy}{dx} = \frac{dy}{dv}\frac{dv}{dx} = -3v^{-4}\frac{dv}{dx}$$

transform it into

$$-3xv^{-4}\frac{dv}{dx} + 6v^{-3} = 3xv^{-4}.$$

Division by $-3xv^{-4}$ yields the linear equation

$$\frac{dv}{dx} - \frac{2}{x}v = -1$$

with integrating factor $\rho = e^{\int(-2/x)\,dx} = x^{-2}$. This gives

$$D_x(x^{-2}v) = -\frac{1}{x^2}; \quad x^{-2}v = \frac{1}{x} + C; \quad v = x + Cx^2;$$

and finally,

$$y(x) = \frac{1}{(x + Cx^2)^3}.$$

■

Example 6 The equation

$$2xe^{2y}\frac{dy}{dx} = 3x^4 + e^{2y} \tag{11}$$

is neither separable, nor linear, nor homogeneous, nor is it a Bernoulli equation. But we observe that y appears only in the combinations e^{2y} and $D_x(e^{2y}) = 2e^{2y}y'$. This prompts the substitution

$$v = e^{2y}, \quad \frac{dv}{dx} = 2e^{2y}\frac{dy}{dx}$$

that transforms Eq. (11) into the linear equation $xv'(x) = 3x^4 + v(x)$; that is,

$$\frac{dv}{dx} - \frac{1}{x}v = 3x^3.$$

After multiplying by the integrating factor $\rho = 1/x$, we find that

$$\frac{1}{x}v = \int 3x^2\,dx = x^3 + C, \quad \text{so} \quad e^{2y} = v = x^4 + Cx,$$

and hence

$$y(x) = \tfrac{1}{2}\ln\left|x^4 + Cx\right|. \qquad ■$$

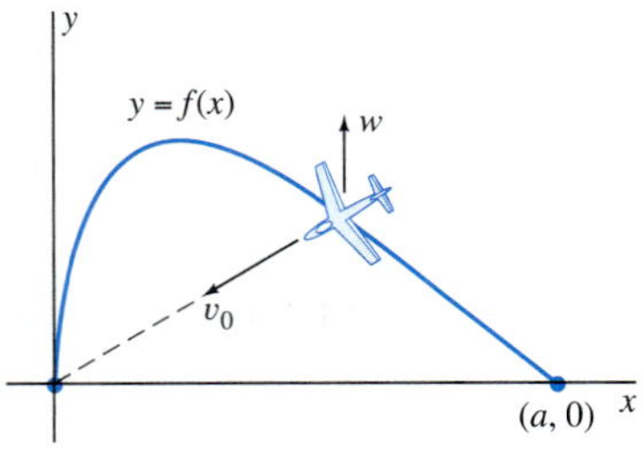

FIGURE 1.6.4. The airplane headed for the origin.

Flight Trajectories

Suppose that an airplane departs from the point $(a, 0)$ located due east of its intended destination—an airport located at the origin $(0, 0)$. The plane travels with constant speed v_0 relative to the wind, which is blowing due north with constant speed w. As indicated in Fig. 1.6.4, we assume that the plane's pilot maintains its heading directly toward the origin.

Figure 1.6.5 helps us derive the plane's velocity components relative to the ground. They are

$$\frac{dx}{dt} = -v_0\cos\theta = -\frac{v_0 x}{\sqrt{x^2+y^2}},$$

$$\frac{dy}{dt} = -v_0\sin\theta + w = -\frac{v_0 y}{\sqrt{x^2+y^2}} + w.$$

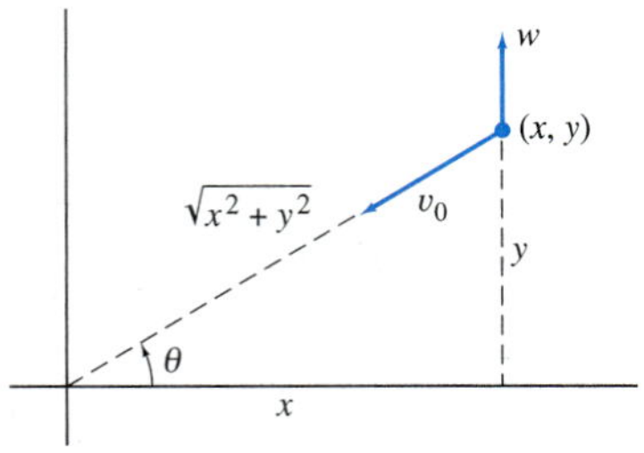

FIGURE 1.6.5. The components of the velocity vector of the airplane.

Hence the trajectory $y = f(x)$ of the plane satisfies the differential equation

$$\frac{dy}{dx} = \frac{dy/dt}{dx/dt} = \frac{1}{v_0 x}\left(v_0 y - w\sqrt{x^2+y^2}\right). \tag{12}$$

If we set

$$k = \frac{w}{v_0}, \tag{13}$$

the ratio of the windspeed to the plane's airspeed, then Eq. (12) takes the homogeneous form

$$\frac{dy}{dx} = \frac{y}{x} - k\left[1 + \left(\frac{y}{x}\right)^2\right]^{1/2}. \tag{14}$$

The substitution $y = xv$, $y' = v + xv'$ then leads routinely to

$$\int \frac{dv}{\sqrt{1+v^2}} = -\int \frac{k}{x}\,dx. \tag{15}$$

By trigonometric substitution, or by consulting a table for the integral on the left, we find that

$$\ln\left(v + \sqrt{1+v^2}\right) = -k \ln x + C, \tag{16}$$

and the initial condition $v(a) = y(a)/a = 0$ yields

$$C = k \ln a. \tag{17}$$

As we ask you to show in Problem 68, the result of substituting (17) in Eq. (16) and then solving for v is

$$v = \frac{1}{2}\left[\left(\frac{x}{a}\right)^{-k} - \left(\frac{x}{a}\right)^{k}\right]. \tag{18}$$

Because $y = xv$, we finally obtain

$$y(x) = \frac{a}{2}\left[\left(\frac{x}{a}\right)^{1-k} - \left(\frac{x}{a}\right)^{1+k}\right] \tag{19}$$

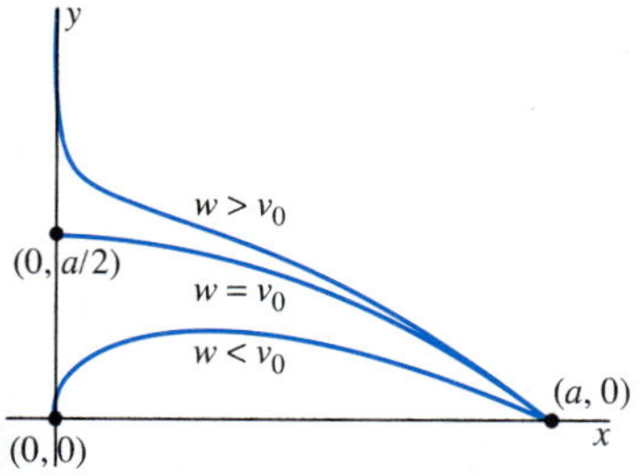

FIGURE 1.6.6. The three cases $w < v_0$ (plane velocity exceeds wind velocity), $w = v_0$ (equal velocities), and $w > v_0$ (wind is greater).

for the equation of the plane's trajectory.

Note that only in the case $k < 1$ (that is, $w < v_0$) does the curve in Eq. (19) pass through the origin, so that the plane reaches its destination. If $w = v_0$ (so that $k = 1$), then Eq. (19) takes the form $y(x) = \frac{1}{2}a(1 - x^2/a^2)$, so the plane's trajectory approaches the point $(0, a/2)$ rather than $(0, 0)$. The situation is even worse if $w > v_0$ (so $k > 1$)—in this case it follows from Eq. (19) that $y \to +\infty$ as $x \to 0$. The three cases are illustrated in Fig. 1.6.6.

Example 7 If $a = 200$ mi, $v_0 = 500$ mi/h, and $w = 100$ mi/h, then $k = w/v_0 = \frac{1}{5}$, so the plane will succeed in reaching the airport at $(0, 0)$. With these values, Eq. (19) yields

$$y(x) = 100\left[\left(\frac{x}{200}\right)^{4/5} - \left(\frac{x}{200}\right)^{6/5}\right]. \tag{20}$$

Now suppose that we want to find the maximum amount by which the plane is blown off course during its trip. That is, what is the maximum value of $y(x)$ for $0 \leqq x \leqq 200$?

Solution Differentiation of the function in Eq. (20) yields

$$\frac{dy}{dx} = \frac{1}{2}\left[\frac{4}{5}\left(\frac{x}{200}\right)^{-1/5} - \frac{6}{5}\left(\frac{x}{200}\right)^{1/5}\right],$$

and we readily solve the equation $y'(x) = 0$ to obtain $(x/200)^{2/5} = \frac{2}{3}$. Hence

$$y_{\max} = 100\left[\left(\frac{2}{3}\right)^2 - \left(\frac{2}{3}\right)^3\right] = \frac{400}{27} \approx 14.81.$$

Thus the plane is blown almost 15 mi north at one point during its westward journey. (The graph of the function in Eq. (20) is the one used to construct Fig. 1.6.4. The vertical scale there is exaggerated by a factor of 4.) ■

Exact Differential Equations

We have seen that a general solution $y(x)$ of a first-order differential equation is often defined implicitly by an equation of the form

$$F(x, y(x)) = C, \tag{21}$$

where C is a constant. On the other hand, given the identity in (21), we can recover the original differential equation by differentiating each side with respect to x. Provided that Eq. (21) implicitly defines y as a differentiable function of x, this gives the original differential equation in the form

$$\frac{\partial F}{\partial x} + \frac{\partial F}{\partial y}\frac{dy}{dx} = 0;$$

that is,

$$M(x, y) + N(x, y)\frac{dy}{dx} = 0, \tag{22}$$

where $M(x, y) = F_x(x, y)$ and $N(x, y) = F_y(x, y)$.

It is sometimes convenient to rewrite Eq. (22) in the more symmetric form

$$M(x, y)\,dx + N(x, y)\,dy = 0, \tag{23}$$

called its **differential** form. The general first-order differential equation $y' = f(x, y)$ can be written in this form with $M = f(x, y)$ and $N \equiv -1$. The preceding discussion shows that, *if* there exists a function $F(x, y)$ such that

$$\frac{\partial F}{\partial x} = M \quad \text{and} \quad \frac{\partial F}{\partial y} = N,$$

then the equation

$$F(x, y) = C$$

implicitly defines a general solution of Eq. (23). In this case, Eq. (23) is called an **exact differential equation**—the differential

$$dF = F_x\,dx + F_y\,dy$$

of $F(x, y)$ is exactly $M\,dx + N\,dy$.

Natural questions are these: How can we determine whether the differential equation in (23) is exact? And if it is exact, how can we find the function F such that $F_x = M$ and $F_y = N$? To answer the first question, let us recall that if the mixed second-order partial derivatives F_{xy} and F_{yx} are continuous on an open set in the xy-plane, then they are equal: $F_{xy} = F_{yx}$. If Eq. (23) is exact and M and N have continuous partial derivatives, it then follows that

$$\frac{\partial M}{\partial y} = F_{xy} = F_{yx} = \frac{\partial N}{\partial x}.$$

Thus the equation

$$\frac{\partial M}{\partial y} = \frac{\partial N}{\partial x} \tag{24}$$

is a *necessary condition* that the differential equation $M\,dx + N\,dy = 0$ be exact. That is, if $M_y \neq N_x$, then the differential equation in question is not exact, so we need not attempt to find a function $F(x, y)$ such that $F_x = M$ and $F_y = N$—there is no such function.

Example 8 The differential equation

$$y^3\,dx + 3xy^2\,dy = 0 \tag{25}$$

is exact because we can immediately see that the function $F(x, y) = xy^3$ has the property that $F_x = y^3$ and $F_y = 3xy^2$. Thus a general solution of Eq. (25) is

$$xy^3 = C;$$

if you prefer, $y(x) = kx^{-1/3}$. ■

But suppose that we divide each term of the differential equation in Example 8 by y^2 to obtain

$$y\,dx + 3x\,dy = 0. \tag{26}$$

This equation is not exact because, with $M = y$ and $N = 3x$, we have

$$\frac{\partial M}{\partial y} = 1 \neq 3 = \frac{\partial N}{\partial x}.$$

Hence the necessary condition in Eq. (24) is not satisfied.

We are confronted with a curious situation here. The differential equations in (25) and (26) are essentially equivalent, and they have exactly the same solutions, yet one is exact and the other is not. In brief, whether a given differential equation is exact or not is related to the precise form $M\,dx + N\,dy = 0$ in which it is written.

Theorem 1 tells us that (subject to differentiability conditions usually satisfied in practice) the necessary condition in (24) is also a *sufficient* condition for exactness. In other words, if $M_y = N_x$, then the differential equation $M\,dx + N\,dy = 0$ is exact.

THEOREM 1 Criterion for Exactness

Suppose that the functions $M(x, y)$ and $N(x, y)$ are continuous and have continuous first-order partial derivatives in the open rectangle $R: a < x < b$, $c < y < d$. Then the differential equation

$$M(x, y)\,dx + N(x, y)\,dy = 0 \tag{23}$$

is exact in R if and only if

$$\frac{\partial M}{\partial y} = \frac{\partial N}{\partial x} \tag{24}$$

at each point of R. That is, there exists a function $F(x, y)$ defined on R with $\partial F/\partial x = M$ and $\partial F/\partial y = N$ if and only if Eq. (24) holds on R.

***Proof*:** We have seen already that it is necessary for Eq. (24) to hold if Eq. (23) is to be exact. To prove the converse, we must show that if Eq. (24) holds, then we can construct a function $F(x, y)$ such that $\partial F/\partial x = M$ and $\partial F/\partial y = N$. Note first that, for *any* function $g(y)$, the function

$$F(x, y) = \int M(x, y)\,dx + g(y) \tag{27}$$

satisfies the condition $\partial F/\partial x = M$. (In Eq. (27), the notation $\int M(x, y)\,dx$ denotes an antiderivative of $M(x, y)$ with respect to x.) We plan to choose $g(y)$ so that

$$N = \frac{\partial F}{\partial y} = \left(\frac{\partial}{\partial y}\int M(x, y)\,dx\right) + g'(y)$$

as well; that is, so that

$$g'(y) = N - \frac{\partial}{\partial y}\int M(x, y)\,dx. \tag{28}$$

To see that there *is* such a function of y, it suffices to show that the right-hand side in Eq. (28) is a function of y alone. We can then find $g(y)$ by integrating with respect to y. Because the right-hand side in Eq. (28) is defined on a rectangle, and hence on an interval as a function of x, it suffices to show that its derivative with respect to x is identically zero. But

$$\begin{aligned}
\frac{\partial}{\partial x}\left(N - \frac{\partial}{\partial y}\int M(x, y)\,dx\right) &= \frac{\partial N}{\partial x} - \frac{\partial}{\partial x}\frac{\partial}{\partial y}\int M(x, y)\,dx \\
&= \frac{\partial N}{\partial x} - \frac{\partial}{\partial y}\frac{\partial}{\partial x}\int M(x, y)\,dx \\
&= \frac{\partial N}{\partial x} - \frac{\partial M}{\partial y} = 0
\end{aligned}$$

by hypothesis. So we can, indeed, find the desired function $g(y)$ by integrating Eq. (28). We substitute this result in Eq. (27) to obtain

$$F(x, y) = \int M(x, y)\,dx + \int\left(N(x, y) - \frac{\partial}{\partial y}\int M(x, y)\,dx\right)dy \tag{29}$$

as the desired function with $F_x = M$ and $F_y = N$. ■

Instead of memorizing Eq. (29), it is usually better to solve an exact equation $M\,dx + N\,dy = 0$ by carrying out the process indicated by Eqs. (27) and (28). First we integrate $M(x, y)$ with respect to x and write

$$F(x, y) = \int M(x, y)\,dx + g(y),$$

thinking of the function $g(y)$ as an "arbitrary constant of integration" as far as the variable x is concerned. Then we determine $g(y)$ by imposing the condition that $\partial F/\partial y = N(x, y)$. This yields a general solution in the implicit form $F(x, y) = C$.

Example 9 Solve the differential equation

$$(6xy - y^3)\,dx + (4y + 3x^2 - 3xy^2)\,dy = 0. \tag{30}$$

Solution Let $M(x, y) = 6xy - y^3$ and $N(x, y) = 4y + 3x^2 - 3xy^2$. The given equation is exact because

$$\frac{\partial M}{\partial y} = 6x - 3y^2 = \frac{\partial N}{\partial x}.$$

Integrating $\partial F/\partial x = M(x, y)$ with respect to x, we get

$$F(x, y) = \int (6xy - y^3)\,dx = 3x^2y - xy^3 + g(y).$$

Then we differentiate with respect to y and set $\partial F/\partial y = N(x, y)$. This yields

$$\frac{\partial F}{\partial y} = 3x^2 - 3xy^2 + g'(y) = 4y + 3x^2 - 3xy^2,$$

and it follows that $g'(y) = 4y$. Hence $g(y) = 2y^2 + C_1$, and thus

$$F(x, y) = 3x^2y - xy^3 + 2y^2 + C_1.$$

Therefore, a general solution of the differential equation is defined implicitly by the equation

$$3x^2y - xy^3 + 2y^2 = C \tag{31}$$

(we have absorbed the constant C_1 into the constant C). ■

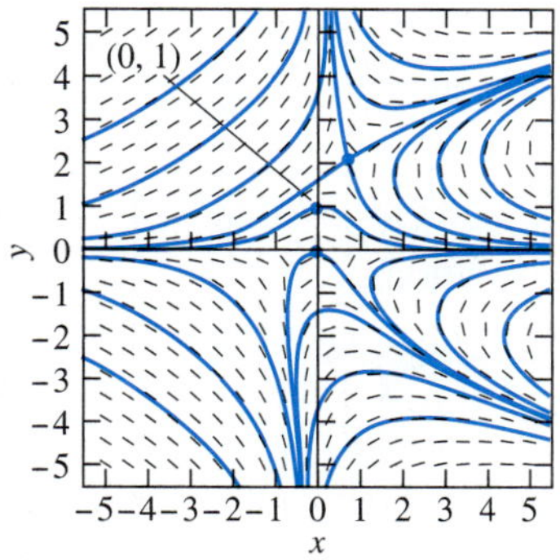

FIGURE 1.6.7. Slope field and solution curves for the exact equation in Example 9.

Remark: Figure 1.6.7 shows a rather complicated structure of solution curves for the differential equation of Example 9. The solution satisfying a given initial condition $y(x_0) = y_0$ is defined implicitly by Eq. (31), with C determined by substituting $x = x_0$ and $y = y_0$ in the equation. For instance, the particular solution satisfying $y(0) = 1$ is defined implicitly by the equation $3x^2y - xy^3 + 2y^2 = 2$. The other two special points in the figure—at (0, 0) and near (0.75, 2.12)—are ones where both coefficient functions in Eq. (30) vanish, so the theorem of Section 1.3 does not guarantee a unique solution. ■

Reducible Second-Order Equations

A *second-order differential equation* involves the second derivative of the unknown function $y(x)$, and thus has the general form

$$F(x, y, y', y'') = 0. \tag{32}$$

If *either* the dependent variable y *or* the independent variable x is missing from a second-order equation, then it is easily reduced by a simple substitution to a first-order equation that may be solvable by the methods of this chapter.

Dependent variable y missing. If y is missing, then Eq. (32) takes the form

$$F(x, y', y'') = 0. \tag{33}$$

Then the substitution

$$p = y' = \frac{dy}{dx}, \quad y'' = \frac{dp}{dx} \tag{34}$$

results in the *first-order* differential equation

$$F(x, p, p') = 0.$$

If we can solve this equation for a general solution $p(x, C_1)$ involving an arbitrary constant C_1, then we need only write

$$y(x) = \int y'(x)\,dx = \int p(x, C_1)\,dx + C_2$$

to get a solution of Eq. (33) that involves two arbitrary constants C_1 and C_2 (as is to be expected in the case of a second-order differential equation).

Example 10 Solve the equation $xy'' + 2y' = 6x$ in which the dependent variable y is missing.

Solution The substitution defined in (34) gives the first-order equation

$$x\frac{dp}{dx} + 2p = 6x; \quad \text{that is,} \quad \frac{dp}{dx} + \frac{2}{x}p = 6.$$

Observing that the equation on the right here is linear, we multiply by its integrating factor $\rho = \exp\left(\int (2/x)\,dx\right) = e^{2\ln x} = x^2$ and get

$$D_x(x^2 p) = 6x^2,$$
$$x^2 p = 2x^3 + C_1,$$
$$p = \frac{dy}{dx} = 2x + \frac{C_1}{x^2}.$$

A final integration with respect to x yields the general solution

$$y(x) = x^2 - \frac{C_1}{x} + C_2$$

of the second-order equation $xy'' + 2y' = 6x$. Solution curves with $C_1 = 0$ but $C_2 \neq 0$ are simply vertical translates of the parabola $y = x^2$ (for which $C_1 = C_2 = 0$). Figure 1.6.8 shows this parabola and some typical solution curves with $C_2 = 0$ but $C_1 \neq 0$. Solution curves with C_1 and C_2 both nonzero are vertical translates of those (other than the parabola) shown in Fig. 1.6.8. ■

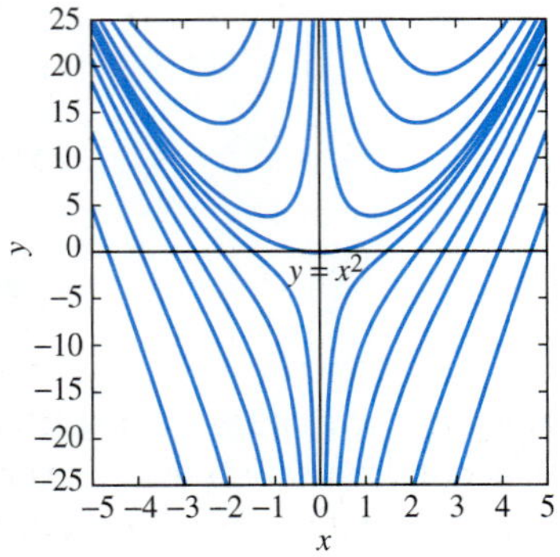

FIGURE 1.6.8. Solution curves of the form $y = x^2 - \frac{C_1}{x}$ for $C_1 = 0, \pm 3, \pm 10, \pm 20, \pm 35, \pm 60, \pm 100$.

Independent variable x missing. If x is missing, then Eq. (32) takes the form

$$F(y, y', y'') = 0. \tag{35}$$

Then the substitution

$$p = y' = \frac{dy}{dx}, \quad y'' = \frac{dp}{dx} = \frac{dp}{dy}\frac{dy}{dx} = p\frac{dp}{dy} \tag{36}$$

results in the *first-order* differential equation

$$F\left(y, p, p\frac{dp}{dy}\right) = 0$$

for p as a function of y. If we can solve this equation for a general solution $p(y, C_1)$ involving an arbitrary constant C_1, then (assuming that $y' \neq 0$) we need only write

$$x(y) = \int \frac{dx}{dy}\,dy = \int \frac{1}{dy/dx}\,dy = \int \frac{1}{p}\,dy = \int \frac{dy}{p(y, C_1)} + C_2.$$

If the final integral $P = \int (1/p)\,dy$ can be evaluated, the result is an implicit solution $x(y) = P(y, C_1) + C_2$ of our second-order differential equation.

Example 11 Solve the equation $yy'' = (y')^2$ in which the independent variable x is missing.

Solution We assume temporarily that y and y' are both nonnegative, and then point out at the end that this restriction is unnecessary. The substitution defined in (36) gives the first-order equation

$$yp\frac{dp}{dy} = p^2.$$

Then separation of variables gives

$$\int \frac{dp}{p} = \int \frac{dy}{y},$$

$$\ln p = \ln y + C \quad (\text{because } y > 0 \text{ and } p = y' > 0),$$

$$p = C_1 y$$

where $C_1 = e^C$. Hence

$$\frac{dx}{dy} = \frac{1}{p} = \frac{1}{C_1 y},$$

$$C_1 x = \int \frac{dy}{y} = \ln y + C_2.$$

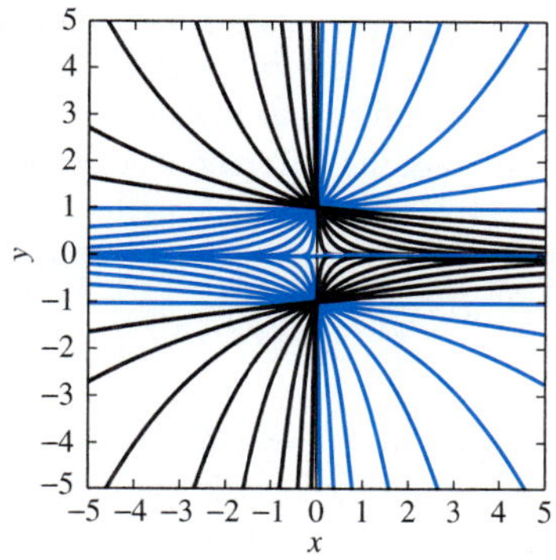

FIGURE 1.6.9. The solution curves $y = Ae^{Bx}$ with $B = 0$ and $A = 0, \pm 1$ are the horizontal lines $y = 0, \pm 1$. The exponential curves with $B > 0$ and $A = \pm 1$ are in color, those with $B < 0$ and $A = \pm 1$ are black.

The resulting general solution of the second-order equation $yy'' = (y')^2$ is

$$y(x) = \exp(C_1 x - C_2) = Ae^{Bx},$$

where $A = e^{-C_2}$ and $B = C_1$. Despite our temporary assumptions, which imply that the constants A and B are both positive, we readily verify that $y(x) = Ae^{Bx}$ satisfies $yy'' = (y')^2$ for *all* real values of A and B. With $B = 0$ and different values of A, we get all horizontal lines in the plane as solution curves. The upper half of Fig. 1.6.9 shows the solution curves obtained with $A = 1$ (for instance) and different positive values of B. With $A = -1$ these solution curves are reflected in the x-axis, and with negative values of B they are reflected in the y-axis. In particular, we see that we get solutions of $yy'' = (y')^2$, allowing both positive and negative possibilities for both y and y'. ■

1.6 Problems

Find general solutions of the differential equations in Problems 1 through 30. Primes denote derivatives with respect to x throughout.

1. $(x+y)y' = x - y$
2. $2xyy' = x^2 + 2y^2$
3. $xy' = y + 2\sqrt{xy}$
4. $(x-y)y' = x + y$
5. $x(x+y)y' = y(x-y)$
6. $(x+2y)y' = y$
7. $xy^2y' = x^3 + y^3$
8. $x^2y' = xy + x^2e^{y/x}$
9. $x^2y' = xy + y^2$
10. $xyy' = x^2 + 3y^2$
11. $(x^2 - y^2)y' = 2xy$
12. $xyy' = y^2 + x\sqrt{4x^2 + y^2}$
13. $xy' = y + \sqrt{x^2 + y^2}$
14. $yy' + x = \sqrt{x^2 + y^2}$
15. $x(x+y)y' + y(3x+y) = 0$
16. $y' = \sqrt{x+y+1}$
17. $y' = (4x+y)^2$
18. $(x+y)y' = 1$
19. $x^2y' + 2xy = 5y^3$
20. $y^2y' + 2xy^3 = 6x$
21. $y' = y + y^3$
22. $x^2y' + 2xy = 5y^4$
23. $xy' + 6y = 3xy^{4/3}$
24. $2xy' + y^3e^{-2x} = 2xy$
25. $y^2(xy' + y)(1 + x^4)^{1/2} = x$
26. $3y^2y' + y^3 = e^{-x}$
27. $3xy^2y' = 3x^4 + y^3$
28. $xe^yy' = 2(e^y + x^3e^{2x})$
29. $(2x \sin y \cos y)y' = 4x^2 + \sin^2 y$
30. $(x + e^y)y' = xe^{-y} - 1$

In Problems 31 through 42, verify that the given differential equation is exact; then solve it.

31. $(2x + 3y)\,dx + (3x + 2y)\,dy = 0$
32. $(4x - y)\,dx + (6y - x)\,dy = 0$
33. $(3x^2 + 2y^2)\,dx + (4xy + 6y^2)\,dy = 0$
34. $(2xy^2 + 3x^2)\,dx + (2x^2y + 4y^3)\,dy = 0$
35. $\left(x^3 + \dfrac{y}{x}\right)dx + (y^2 + \ln x)\,dy = 0$
36. $(1 + ye^{xy})\,dx + (2y + xe^{xy})\,dy = 0$
37. $(\cos x + \ln y)\,dx + \left(\dfrac{x}{y} + e^y\right)dy = 0$
38. $(x + \tan^{-1} y)\,dx + \dfrac{x+y}{1+y^2}\,dy = 0$
39. $(3x^2y^3 + y^4)\,dx + (3x^3y^2 + y^4 + 4xy^3)\,dy = 0$
40. $(e^x \sin y + \tan y)\,dx + (e^x \cos y + x \sec^2 y)\,dy = 0$
41. $\left(\dfrac{2x}{y} - \dfrac{3y^2}{x^4}\right)dx + \left(\dfrac{2y}{x^3} - \dfrac{x^2}{y^2} + \dfrac{1}{\sqrt{y}}\right)dy = 0$
42. $\dfrac{2x^{5/2} - 3y^{5/3}}{2x^{5/2}y^{2/3}}\,dx + \dfrac{3y^{5/3} - 2x^{5/2}}{3x^{3/2}y^{5/3}}\,dy = 0$

Find a general solution of each reducible second-order differential equation in Problems 43 through 54. Assume x, y, and/or y′ positive where helpful (as in Example 11).

43. $xy'' = y'$
44. $yy'' + (y')^2 = 0$
45. $y'' + 4y = 0$
46. $xy'' + y' = 4x$
47. $y'' = (y')^2$
48. $x^2y'' + 3xy' = 2$
49. $yy'' + (y')^2 = yy'$
50. $y'' = (x + y')^2$
51. $y'' = 2y(y')^3$
52. $y^3y'' = 1$
53. $y'' = 2yy'$
54. $yy'' = 3(y')^2$

55. Show that the substitution $v = ax + by + c$ transforms the differential equation $dy/dx = F(ax + by + c)$ into a separable equation.

56. Suppose that $n \neq 0$ and $n \neq 1$. Show that the substitution $v = y^{1-n}$ transforms the Bernoulli equation $dy/dx + P(x)y = Q(x)y^n$ into the linear equation

$$\frac{dv}{dx} + (1-n)P(x)v(x) = (1-n)Q(x).$$

57. Show that the substitution $v = \ln y$ transforms the differential equation $dy/dx + P(x)y = Q(x)(y \ln y)$ into the linear equation $dv/dx + P(x) = Q(x)v(x)$.

58. Use the idea in Problem 57 to solve the equation

$$x\frac{dy}{dx} - 4x^2y + 2y \ln y = 0.$$

59. Solve the differential equation

$$\frac{dy}{dx} = \frac{x - y - 1}{x + y + 3}$$

by finding h and k so that the substitutions $x = u + h$, $y = v + k$ transform it into the homogeneous equation

$$\frac{dv}{du} = \frac{u - v}{u + v}.$$

60. Use the method in Problem 59 to solve the differential equation

$$\frac{dy}{dx} = \frac{2y - x + 7}{4x - 3y - 18}.$$

61. Make an appropriate substitution to find a solution of the equation $dy/dx = \sin(x - y)$. Does this general solution contain the linear solution $y(x) = x - \pi/2$ that is readily verified by substitution in the differential equation?

62. Show that the solution curves of the differential equation

$$\frac{dy}{dx} = -\frac{y(2x^3 - y^3)}{x(2y^3 - x^3)}$$

are of the form $x^3 + y^3 = 3Cxy$.

63. The equation $dy/dx = A(x)y^2 + B(x)y + C(x)$ is called a **Riccati equation.** Suppose that one particular solution $y_1(x)$ of this equation is known. Show that the substitution

$$y = y_1 + \frac{1}{v}$$

transforms the Riccati equation into the *linear* equation

$$\frac{dv}{dx} + (B + 2Ay_1)v = -A.$$

Use the method of Problem 63 to solve the equations in Problems 64 and 65, given that $y_1(x) = x$ is a solution of each.

64. $\dfrac{dy}{dx} + y^2 = 1 + x^2$

65. $\dfrac{dy}{dx} + 2xy = 1 + x^2 + y^2$

66. An equation of the form

$$y = xy' + g(y') \tag{37}$$

is called a **Clairaut equation**. Show that the one-parameter family of straight lines described by

$$y(x) = Cx + g(C) \tag{38}$$

is a general solution of Eq. (37).

67. Consider the Clairaut equation

$$y = xy' - \tfrac{1}{4}(y')^2$$

for which $g(y') = -\frac{1}{4}(y')^2$ in Eq. (37). Show that the line

$$y = Cx - \tfrac{1}{4}C^2$$

is tangent to the parabola $y = x^2$ at the point $\left(\frac{1}{2}C, \frac{1}{4}C^2\right)$. Explain why this implies that $y = x^2$ is a singular solution of the given Clairaut equation. This singular solution and the one-parameter family of straight line solutions are illustrated in Fig. 1.6.10.

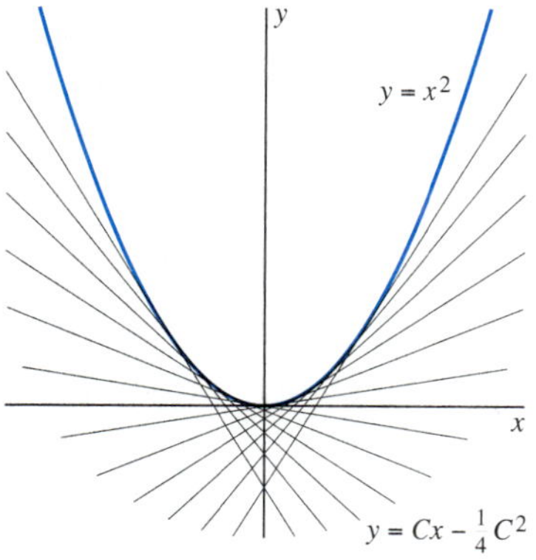

FIGURE 1.6.10. Solutions of the Clairaut equation of Problem 67. The "typical" straight line with equation $y = Cx - \frac{1}{4}C^2$ is tangent to the parabola at the point $(\frac{1}{2}C, \frac{1}{4}C^2)$.

68. Derive Eq. (18) in this section from Eqs. (16) and (17).

69. In the situation of Example 7, suppose that $a = 100$ mi, $v_0 = 400$ mi/h, and $w = 40$ mi/h. Now how far northward does the wind blow the airplane?

70. As in the text discussion, suppose that an airplane maintains a heading toward an airport at the origin. If $v_0 = 500$ mi/h and $w = 50$ mi/h (with the wind blowing due north), and the plane begins at the point (200, 150), show that its trajectory is described by

$$y + \sqrt{x^2 + y^2} = 2(200x^9)^{1/10}.$$

71. A river 100 ft wide is flowing north at w feet per second. A dog starts at (100, 0) and swims at $v_0 = 4$ ft/s, always heading toward a tree at (0, 0) on the west bank directly across from the dog's starting point. **(a)** If $w = 2$ ft/s, show that the dog reaches the tree. **(b)** If $w = 4$ ft/s, show that the dog reaches instead the point on the west bank 50 ft north of the tree. **(c)** If $w = 6$ ft/s, show that the dog never reaches the west bank.

72. In the calculus of plane curves, one learns that the *curvature* κ of the curve $y = y(x)$ at the point (x, y) is given by

$$\kappa = \frac{|y''(x)|}{[1 + y'(x)^2]^{3/2}},$$

and that the curvature of a circle of radius r is $\kappa = 1/r$. [See Example 3 in Section 11.6 of Edwards and Penney, *Calculus: Early Transcendentals*, 7th edition (Upper Saddle River, NJ: Prentice Hall, 2008).] Conversely, substitute $\rho = y'$ to derive a general solution of the second-order differential equation

$$ry'' = [1 + (y')^2]^{3/2}$$

(with r constant) in the form

$$(x - a)^2 + (y - b)^2 = r^2.$$

Thus a circle of radius r (or a part thereof) is the *only* plane curve with constant curvature $1/r$.

1.6 Application Computer Algebra Solutions

Computer algebra systems typically include commands for the "automatic" solution of differential equations. But two different such systems often give different results whose equivalence is not clear, and a single system may give the solution in an overly complicated form. Consequently, computer algebra solutions of differential equations often require considerable "processing" or simplification by a human user in order to yield concrete and applicable information. Here we illustrate these issues using the interesting differential equation

$$\frac{dy}{dx} = \sin(x - y) \tag{1}$$

that appeared in the Section 1.3 Application. The *Maple* command

```
dsolve( D(y)(x) = sin(x - y(x)), y(x));
```

yields the simple and attractive result

$$y(x) = x - 2\tan^{-1}\left(\frac{x - 2 - C1}{x - C1}\right) \tag{2}$$

that was cited there. But the supposedly equivalent *Mathematica* command

```
DSolve[ y'[x] == Sin[x - y[x]], y[x], x]
```

yields a considerably more complicated result from which—with a fair amount of effort in simplification—one can extract the quite different looking solution

$$y(x) = 2\cos^{-1}\left(\frac{2\cos\frac{x}{2} + (x - c)\left(\cos\frac{x}{2} + \sin\frac{x}{2}\right)}{\sqrt{2 + 2(x - c + 1)^2}}\right). \tag{3}$$

This apparent disparity is not unusual; different symbolic algebra systems, or even different versions of the same system, often yield different forms of a solution of the same differential equation. As an alternative to attempted reconciliation of such seemingly disparate results as in Eqs. (2) and (3), a common tactic is simplification of the differential equation before submitting it to a computer algebra system.

EXERCISE 1: Show that the plausible substitution $v = x - y$ in Eq. (1) yields the separable equation

$$\frac{dv}{dx} = 1 - \sin v. \tag{4}$$

Now the *Maple* command `int(1/(1-sin(v)),v)` yields

$$\int \frac{dv}{1 - \sin v} = \frac{2}{1 - \tan\frac{v}{2}} \tag{5}$$

(omitting the constant of integration, as symbolic computer algebra systems often do).

EXERCISE 2: Use simple algebra to deduce from Eq. (5) the integral formula

$$\int \frac{dv}{1 - \sin v} = \frac{1 + \tan\frac{v}{2}}{1 - \tan\frac{v}{2}} + C. \tag{6}$$

EXERCISE 3: Deduce from (6) that Eq. (4) has the general solution

$$v(x) = 2\tan^{-1}\left(\frac{x-1+C}{x+1+C}\right),$$

and hence that Eq. (1) has the general solution

$$y(x) = x - 2\tan^{-1}\left(\frac{x-1+C}{x+1+C}\right). \tag{7}$$

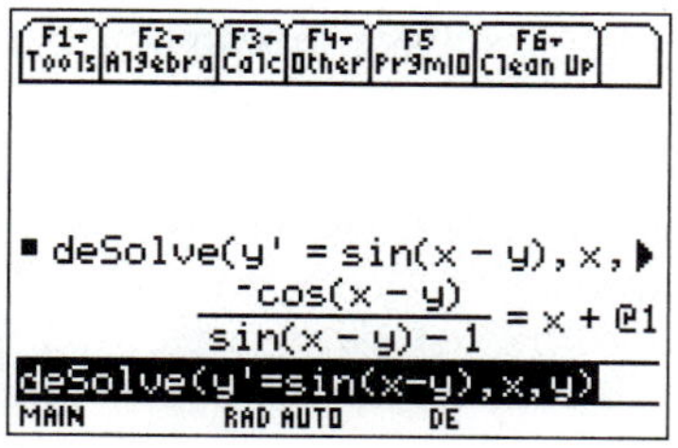

FIGURE 1.6.11. Implicit solution of $y' = \sin(x - y)$ generated by a TI-89 graphing calculator.

EXERCISE 4: Finally, reconcile the forms in Eq. (2) and Eq. (7). What is the relation between the constants C and $C1$?

EXERCISE 5: Show that the integral in Eq. (5) yields immediately the graphing calculator implicit solution shown in Fig. 1.6.11.

INVESTIGATION: For your own personal differential equation, let p and q be two distinct nonzero digits in your student ID number, and consider the differential equation

$$\frac{dy}{dx} = \frac{1}{p}\cos(x - qy). \tag{8}$$

(a) Find a symbolic general solution using a computer algebra system and/or some combination of the techniques listed in this project.

(b) Determine the symbolic particular solution corresponding to several typical initial conditions of the form $y(x_0) = y_0$.

(c) Determine the possible values of a and b such that the straight line $y = ax + b$ is a solution curve of Eq. (8).

(d) Plot a direction field and some typical solution curves. Can you make a connection between the symbolic solution and your (linear and nonlinear) solution curves?

Chapter 1 Summary

In this chapter we have discussed applications of and solution methods for several important types of first-order differential equations, including those that are separable (Section 1.4), linear (Section 1.5), or exact (Section 1.6). In Section 1.6 we also discussed substitution techniques that can sometimes be used to transform a given first-order differential equation into one that is either separable, linear, or exact.

Lest it appear that these methods constitute a "grab bag" of special and unrelated techniques, it is important to note that they are all versions of a single idea. Given a differential equation

$$f(x, y, y') = 0, \tag{1}$$

we attempt to write it in the form

$$\frac{d}{dx}\left[G(x, y)\right] = 0. \tag{2}$$

It is precisely to obtain the form in Eq. (2) that we multiply the terms in Eq. (1) by an appropriate integrating factor (even if all we are doing is separating the variables).

But once we have found a function $G(x, y)$ such that Eqs. (1) and (2) are equivalent, a general solution is defined implicitly by means of the equation

$$G(x, y) = C \tag{3}$$

that one obtains by integrating Eq. (2).

Given a specific first-order differential equation to be solved, we can attack it by means of the following steps:

- Is it *separable*? If so, separate the variables and integrate (Section 1.4).
- Is it *linear*? That is, can it be written in the form

$$\frac{dy}{dx} + P(x)y = Q(x)?$$

  If so, multiply by the integrating factor $\rho = \exp\left(\int P\,dx\right)$ of Section 1.5.
- Is it *exact*? That is, when the equation is written in the form $M\,dx + N\,dy = 0$, is $\partial M/\partial y = \partial N/\partial x$ (Section 1.6)?
- If the equation as it stands is not separable, linear, or exact, is there a plausible substitution that will make it so? For instance, is it homogeneous (Section 1.6)?

Many first-order differential equations succumb to the line of attack outlined here. Nevertheless, many more do not. Because of the wide availability of computers, numerical techniques are commonly used to *approximate* the solutions of differential equations that cannot be solved readily or explicitly by the methods of this chapter. Indeed, most of the solution curves shown in figures in this chapter were plotted using numerical approximations rather than exact solutions. Several numerical methods for the appropriate solution of differential equations will be discussed in Chapter 2.

Chapter 1 Review Problems

Find general solutions of the differential equations in Problems 1 through 30. Primes denote derivatives with respect to x.

1. $x^3 + 3y - xy' = 0$
2. $xy^2 + 3y^2 - x^2y' = 0$
3. $xy + y^2 - x^2y' = 0$
4. $2xy^3 + e^x + (3x^2y^2 + \sin y)y' = 0$
5. $3y + x^4y' = 2xy$
6. $2xy^2 + x^2y' = y^2$
7. $2x^2y + x^3y' = 1$
8. $2xy + x^2y' = y^2$
9. $xy' + 2y = 6x^2\sqrt{y}$
10. $y' = 1 + x^2 + y^2 + x^2y^2$
11. $x^2y' = xy + 3y^2$
12. $6xy^3 + 2y^4 + (9x^2y^2 + 8xy^3)y' = 0$
13. $4xy^2 + y' = 5x^4y^2$
14. $x^3y' = x^2y - y^3$
15. $y' + 3y = 3x^2e^{-3x}$
16. $y' = x^2 - 2xy + y^2$
17. $e^x + ye^{xy} + (e^y + xe^{yx})y' = 0$
18. $2x^2y - x^3y' = y^3$
19. $3x^5y^2 + x^3y' = 2y^2$
20. $xy' + 3y = 3x^{-3/2}$
21. $(x^2 - 1)y' + (x - 1)y = 1$
22. $xy' = 6y + 12x^4y^{2/3}$
23. $e^y + y\cos x + (xe^y + \sin x)y' = 0$
24. $9x^2y^2 + x^{3/2}y' = y^2$
25. $2y + (x + 1)y' = 3x + 3$
26. $9x^{1/2}y^{4/3} - 12x^{1/5}y^{3/2} + (8x^{3/2}y^{1/3} - 15x^{6/5}y^{1/2})y' = 0$
27. $3y + x^3y^4 + 3xy' = 0$
28. $y + xy' = 2e^{2x}$
29. $(2x + 1)y' + y = (2x + 1)^{3/2}$
30. $y' = \sqrt{x + y}$

Each of the differential equations in Problems 31 through 36 is of two different types considered in this chapter—separable, linear, homogeneous, Bernoulli, exact, etc. Hence, derive general solutions for each of these equations in two different ways; then reconcile your results.

31. $\dfrac{dy}{dx} = 3(y + 7)x^2$
32. $\dfrac{dy}{dx} = xy^3 - xy$
33. $\dfrac{dy}{dx} = -\dfrac{3x^2 + 2y^2}{4xy}$
34. $\dfrac{dy}{dx} = \dfrac{x + 3y}{y - 3x}$
35. $\dfrac{dy}{dx} = \dfrac{2xy + 2x}{x^2 + 1}$
36. $\dfrac{dy}{dx} = \dfrac{\sqrt{y} - y}{\tan x}$

2 Mathematical Models and Numerical Methods

2.1 Population Models

In Section 1.4 we introduced the exponential differential equation $dP/dt = kP$, with solution $P(t) = P_0e^{kt}$, as a mathematical model for natural population growth that occurs as a result of constant birth and death rates. Here we present a more general population model that accommodates birth and death rates that are not necessarily constant. As before, however, our population function $P(t)$ will be a *continuous* approximation to the actual population, which of course changes only by integral increments—that is, by one birth or death at a time.

Suppose that the population changes only by the occurrence of births and deaths—there is no immigration or emigration from outside the country or environment under consideration. It is customary to track the growth or decline of a population in terms of its *birth rate* and *death rate* functions defined as follows:

- $\beta(t)$ is the number of births per unit of population per unit of time at time t;
- $\delta(t)$ is the number of deaths per unit of population per unit of time at time t.

Then the numbers of births and deaths that occur during the time interval $[t, t + \Delta t]$ is given (approximately) by

$$\text{births:} \quad \beta(t) \cdot P(t) \cdot \Delta t, \qquad \text{deaths:} \quad \delta(t) \cdot P(t) \cdot \Delta t.$$

Hence the change ΔP in the population during the time interval $[t, t + \Delta t]$ of length Δt is

$$\Delta P = \{\text{births}\} - \{\text{deaths}\} \approx \beta(t) \cdot P(t) \cdot \Delta t - \delta(t) \cdot P(t) \cdot \Delta t,$$

so

$$\frac{\Delta P}{\Delta t} \approx [\beta(t) - \delta(t)]\, P(t).$$

The error in this approximation should approach zero as $\Delta t \to 0$, so—taking the limit—we get the differential equation

$$\frac{dP}{dt} = (\beta - \delta)P, \tag{1}$$

in which we write $\beta = \beta(t)$, $\delta = \delta(t)$, and $P = P(t)$ for brevity. Equation (1) is the **general population equation.** If β and δ are constant, Eq. (1) reduces to the natural growth equation with $k = \beta - \delta$. But it also includes the possibility that β and δ are variable functions of t. The birth and death rates need not be known in advance; they may well depend on the unknown function $P(t)$.

Example 1 Suppose that an alligator population numbers 100 initially, and that its death rate is $\delta = 0$ (so none of the alligators is dying). If the birth rate is $\beta = (0.0005)P$—and thus increases as the population does—then Eq. (1) gives the initial value problem

$$\frac{dP}{dt} = (0.0005)P^2, \quad P(0) = 100$$

(with t in years). Then upon separating the variables we get

$$\int \frac{1}{P^2}\,dP = \int (0.0005)\,dt;$$

$$-\frac{1}{P} = (0.0005)t + C.$$

Substitution of $t = 0$, $P = 100$ gives $C = -1/100$, and then we readily solve for

$$P(t) = \frac{2000}{20 - t}.$$

For instance, $P(10) = 2000/10 = 200$, so after 10 years the alligator population has doubled. But we see that $P \to +\infty$ as $t \to 20$, so a real "population explosion" occurs in 20 years. Indeed, the direction field and solution curves shown in Fig. 2.1.1 indicate that a population explosion always occurs, whatever the size of the (positive) initial population $P(0) = P_0$. In particular, it appears that the population always becomes unbounded in a *finite* period of time. ■

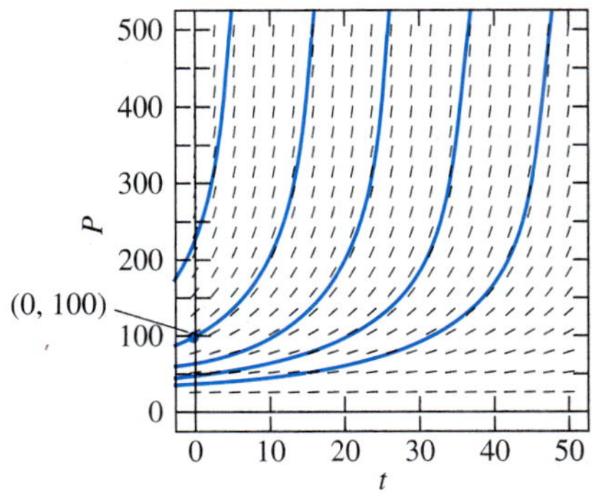

FIGURE 2.1.1. Slope field and solution curves for the equation $dP/dt = (0.0005)P^2$ in Example 1.

Bounded Populations and the Logistic Equation

In situations as diverse as the human population of a nation and a fruit fly population in a closed container, it is often observed that the birth rate decreases as the population itself increases. The reasons may range from increased scientific or cultural sophistication to a limited food supply. Suppose, for example, that the birth rate β is a *linear* decreasing function of the population size P, so that $\beta = \beta_0 - \beta_1 P$, where β_0 and β_1 are positive constants. If the death rate $\delta = \delta_0$ remains constant, then Eq. (1) takes the form

$$\frac{dP}{dt} = (\beta_0 - \beta_1 P - \delta_0)P;$$

that is,

$$\frac{dP}{dt} = aP - bP^2, \tag{2}$$

where $a = \beta_0 - \delta_0$ and $b = \beta_1$.

If the coefficients a and b are both positive, then Eq. (2) is called the **logistic equation.** For the purpose of relating the behavior of the population $P(t)$ to the values of the parameters in the equation, it is useful to rewrite the logistic equation in the form

$$\frac{dP}{dt} = kP(M - P), \tag{3}$$

where $k = b$ and $M = a/b$ are constants.

Example 2 In Example 4 of Section 1.3 we explored graphically a population that is modeled by the logistic equation

$$\frac{dP}{dt} = 0.0004P(150 - P) = 0.06P - 0.0004P^2. \tag{4}$$

To solve this differential equation symbolically, we separate the variables and integrate. We get

$$\int \frac{dP}{P(150 - P)} = \int 0.0004\,dt,$$

$$\frac{1}{150}\int \left(\frac{1}{P} + \frac{1}{150 - P}\right) dP = \int 0.0004\,dt \qquad \text{[partial fractions]},$$

$$\ln|P| - \ln|150 - P| = 0.06t + C,$$

$$\frac{P}{150 - P} = \pm e^C e^{0.06t} = Be^{0.06t} \qquad [\text{where } B = \pm e^C].$$

If we substitute $t = 0$ and $P = P_0 \neq 150$ into this last equation, we find that $B = P_0/(150 - P_0)$. Hence

$$\frac{P}{150 - P} = \frac{P_0 e^{0.06t}}{150 - P_0}.$$

Finally, this equation is easy to solve for the population

$$P(t) = \frac{150P_0}{P_0 + (150 - P_0)e^{-0.06t}} \tag{5}$$

at time t in terms of the initial population $P_0 = P(0)$. Figure 2.1.2 shows a number of solution curves corresponding to different values of the initial population ranging from $P_0 = 20$ to $P_0 = 300$. Note that all these solution curves appear to approach the horizontal line $P = 150$ as an asymptote. Indeed, you should be able to see directly from Eq. (5) that $\lim_{t\to\infty} P(t) = 150$, whatever the initial value $P_0 > 0$. ■

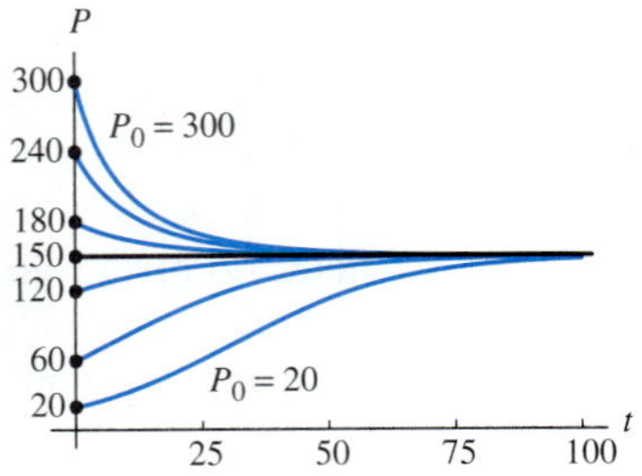

FIGURE 2.1.2. Typical solution curves for the logistic equation $P' = 0.06P - 0.0004P^2$.

Limiting Populations and Carrying Capacity

The finite limiting population noted in Example 2 is characteristic of logistic populations. In Problem 32 we ask you to use the method of solution of Example 2 to show that the solution of the logistic initial value problem

$$\frac{dP}{dt} = kP(M - P), \quad P(0) = P_0 \tag{6}$$

is

$$P(t) = \frac{MP_0}{P_0 + (M - P_0)e^{-kMt}}. \tag{7}$$

Actual animal populations are positive valued. If $P_0 = M$, then (7) reduces to the unchanging (constant-valued) "equilibrium population" $P(t) \equiv M$. Otherwise, the behavior of a logistic population depends on whether $0 < P_0 < M$ or $P_0 > M$. If $0 < P_0 < M$, then we see from (6) and (7) that $P' > 0$ and

$$P(t) = \frac{MP_0}{P_0 + (M - P_0)e^{-kMt}} = \frac{MP_0}{P_0 + \{\text{pos. number}\}} < \frac{MP_0}{P_0} = M.$$

However, if $P_0 > M$, then we see from (6) and (7) that $P' < 0$ and

$$P(t) = \frac{MP_0}{P_0 + (M - P_0)e^{-kMt}} = \frac{MP_0}{P_0 + \{\text{neg. number}\}} > \frac{MP_0}{P_0} = M.$$

In either case, the "positive number" or "negative number" in the denominator has absolute value less than P_0 and—because of the exponential factor—approaches 0 as $t \to +\infty$. It follows that

$$\lim_{t\to+\infty} P(t) = \frac{MP_0}{P_0 + 0} = M. \tag{8}$$

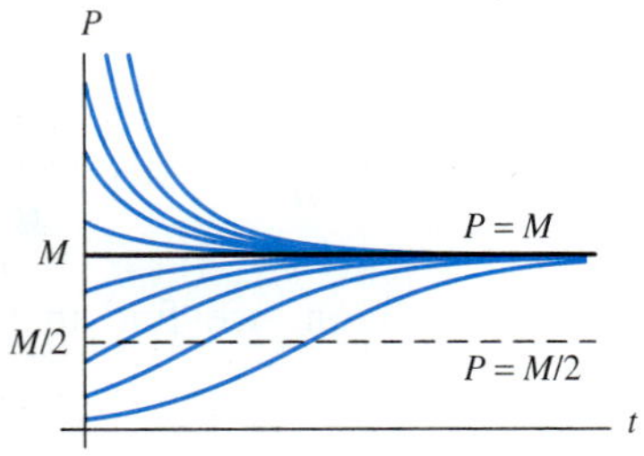

FIGURE 2.1.3. Typical solution curves for the logistic equation $P' = kP(M - P)$. Each solution curve that starts below the line $P = M/2$ has an inflection point on this line. (See Problem 34.)

Thus a population that satisfies the logistic equation does *not* grow without bound like a naturally growing population modeled by the exponential equation $P' = kP$. Instead, it approaches the finite **limiting population** M as $t \to +\infty$. As illustrated by the typical logistic solution curves in Fig. 2.1.3, the population $P(t)$ steadily increases and approaches M from below if $0 < P_0 < M$, but steadily decreases and approaches M from above if $P_0 > M$. Sometimes M is called the **carrying capacity** of the environment, considering it to be the maximum population that the environment can support on a long-term basis.

Example 3 Suppose that in 1885 the population of a certain country was 50 million and was growing at the rate of 750,000 people per year at that time. Suppose also that in 1940 its population was 100 million and was then growing at the rate of 1 million per year. Assume that this population satisfies the logistic equation. Determine both the limiting population M and the predicted population for the year 2000.

Solution We substitute the two given pairs of data in Eq. (3) and find that

$$0.75 = 50k(M - 50), \qquad 1.00 = 100k(M - 100).$$

We solve simultaneously for $M = 200$ and $k = 0.0001$. Thus the limiting population of the country in question is 200 million. With these values of M and k, and with $t = 0$ corresponding to the year 1940 (in which $P_0 = 100$), we find that—according to Eq. (7)—the population in the year 2000 will be

$$P(60) = \frac{100 \cdot 200}{100 + (200 - 100)e^{-(0.0001)(200)(60)}},$$

about 153.7 million people. ■

Historical Note

The logistic equation was introduced (around 1840) by the Belgian mathematician and demographer P. F. Verhulst as a possible model for human population growth. In the next two examples we compare natural growth and logistic model fits to the 19th-century U.S. population census data, then compare projections for the 20th century.

Example 4 The U.S. population in 1800 was 5.308 million and in 1900 was 76.212 million. If we take $P_0 = 5.308$ (with $t = 0$ in 1800) in the natural growth model $P(t) = P_0 e^{rt}$ and substitute $t = 100$, $P = 76.212$, we find that

$$76.212 = 5.308e^{100r}, \quad \text{so} \quad r = \frac{1}{100} \ln \frac{76.212}{5.308} \approx 0.026643.$$

Thus our natural growth model for the U.S. population during the 19th century is

$$P(t) = (5.308)e^{(0.026643)t} \tag{9}$$

(with t in years and P in millions). Because $e^{0.026643} \approx 1.02700$, the average population growth between 1800 and 1900 was about 2.7% per year. ■

Example 5 The U.S. population in 1850 was 23.192 million. If we take $P_0 = 5.308$ and substitute the data pairs $t = 50$, $P = 23.192$ (for 1850) and $t = 100$, $P = 76.212$ (for 1900) in the logistic model formula in Eq. (7), we get the two equations

$$\begin{aligned} \frac{(5.308)M}{5.308 + (M - 5.308)e^{-50kM}} &= 23.192, \\ \frac{(5.308)M}{5.308 + (M - 5.308)e^{-100kM}} &= 76.212 \end{aligned} \tag{10}$$

in the two unknowns k and M. Nonlinear systems like this ordinarily are solved numerically using an appropriate computer system. But with the right algebraic trick (Problem 36 in this section) the equations in (10) can be solved manually for $k = 0.000167716$, $M = 188.121$. Substitution of these values in Eq. (7) yields the logistic model

$$P(t) = \frac{998.546}{5.308 + (182.813)e^{-(0.031551)t}}. \tag{11}$$

The table in Fig. 2.1.4 compares the actual 1800–1990 U.S. census population figures with those predicted by the exponential growth model in (9) and the logistic model in (11). Both agree well with the 19th-century figures. But the exponential model diverges appreciably from the census data in the early decades of the 20th century, whereas the logistic model remains accurate until 1940. By the end of the 20th century the exponential model vastly overestimates the actual U.S. population—predicting over a billion in the year 2000—whereas the logistic model somewhat underestimates it.

Year	Actual U.S. Pop.	Exponential Model	Exponential Error	Logistic Model	Logistic Error
1800	5.308	5.308	0.000	5.308	0.000
1810	7.240	6.929	0.311	7.202	0.038
1820	9.638	9.044	0.594	9.735	−0.097
1830	12.861	11.805	1.056	13.095	−0.234
1840	17.064	15.409	1.655	17.501	−0.437
1850	23.192	20.113	3.079	23.192	0.000
1860	31.443	26.253	5.190	30.405	1.038
1870	38.558	34.268	4.290	39.326	−0.768
1880	50.189	44.730	5.459	50.034	0.155
1890	62.980	58.387	4.593	62.435	0.545
1900	76.212	76.212	0.000	76.213	−0.001
1910	92.228	99.479	−7.251	90.834	1.394
1920	106.022	129.849	−23.827	105.612	0.410
1930	123.203	169.492	−46.289	119.834	3.369
1940	132.165	221.237	−89.072	132.886	−0.721
1950	151.326	288.780	−137.454	144.354	6.972
1960	179.323	376.943	−197.620	154.052	25.271
1970	203.302	492.023	−288.721	161.990	41.312
1980	226.542	642.236	−415.694	168.316	58.226
1990	248.710	838.308	−589.598	173.252	75.458
2000	281.422	1094.240	−812.818	177.038	104.384

FIGURE 2.1.4. Comparison of exponential growth and logistic models with U.S. census populations (in millions).

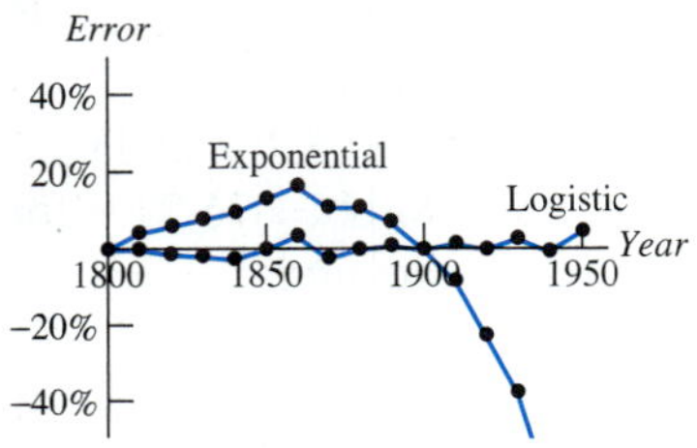

FIGURE 2.1.5. Percentage errors in the exponential and logistic population models for 1800–1950.

The two models are compared in Fig. 2.1.5, where plots of their respective errors—as a *percentage* of the actual population—are shown for the 1800–1950 period. We see that the logistic model tracks the actual population reasonably well throughout this 150-year period. However, the exponential error is considerably larger during the 19th century and literally goes off the chart during the first half of the 20th century.

In order to measure the extent to which a given model fits actual data, it is customary to define the **average error** (in the model) as *the square root of the average of the squares of the individual errors* (the latter appearing in the fourth and sixth columns of the table in Fig. 2.1.4). Using only the 1800–1900 data, this definition gives 3.162 for the average error in the exponential model, while the average error in the logistic model is only 0.452. Consequently, even in 1900 we might well have anticipated that the logistic model would predict the U.S. population growth during the 20th century more accurately than the exponential model. ■

The moral of Examples 4 and 5 is simply that one should not expect too much of models that are based on severely limited information (such as just a pair of data points). Much of the science of *statistics* is devoted to the analysis of large "data sets" to formulate useful (and perhaps reliable) mathematical models.

More Applications of the Logistic Equation

We next describe some situations that illustrate the varied circumstances in which the logistic equation is a satisfactory mathematical model.

1. *Limited environment situation.* A certain environment can support a population of at most M individuals. It is then reasonable to expect the growth rate $\beta - \delta$ (the combined birth and death rates) to be proportional to $M - P$, because we may think of $M - P$ as the potential for further expansion. Then $\beta - \delta = k(M - P)$, so that

$$\frac{dP}{dt} = (\beta - \delta)P = kP(M - P).$$

   The classic example of a limited environment situation is a fruit fly population in a closed container.

2. *Competition situation.* If the birth rate β is constant but the death rate δ is proportional to P, so that $\delta = \alpha P$, then

$$\frac{dP}{dt} = (\beta - \alpha P)P = kP(M - P).$$

   This might be a reasonable working hypothesis in a study of a cannibalistic population, in which all deaths result from chance encounters between individuals. Of course, competition between individuals is not usually so deadly, nor its effects so immediate and decisive.

3. *Joint proportion situation.* Let $P(t)$ denote the number of individuals in a constant-size susceptible population M who are infected with a certain contagious and incurable disease. The disease is spread by chance encounters. Then $P'(t)$ should be proportional to the product of the number P of individuals having the disease and the number $M - P$ of those not having it, and therefore $dP/dt = kP(M - P)$. Again we discover that the mathematical model is the logistic equation. The mathematical description of the spread of a rumor in a population of M individuals is identical.

Example 6 Suppose that at time $t = 0$, 10 thousand people in a city with population $M = 100$ thousand people have heard a certain rumor. After 1 week the number $P(t)$ of those who have heard it has increased to $P(1) = 20$ thousand. Assuming that $P(t)$ satisfies a logistic equation, when will 80% of the city's population have heard the rumor?

Solution Substituting $P_0 = 10$ and $M = 100$ (thousand) in Eq. (7), we get

$$P(t) = \frac{1000}{10 + 90e^{-100kt}}. \tag{12}$$

Then substitution of $t = 1$, $P = 20$ gives the equation

$$20 = \frac{1000}{10 + 90e^{-100k}}$$

that is readily solved for

$$e^{-100k} = \tfrac{4}{9}, \quad \text{so} \quad k = \tfrac{1}{100} \ln \tfrac{9}{4} \approx 0.008109.$$

With $P(t) = 80$, Eq. (12) takes the form

$$80 = \frac{1000}{10 + 90e^{-100kt}},$$

which we solve for $e^{-100kt} = \frac{1}{36}$. It follows that 80% of the population has heard the rumor when

$$t = \frac{\ln 36}{100k} = \frac{\ln 36}{\ln \frac{9}{4}} \approx 4.42,$$

thus after about 4 weeks and 3 days. ■

Doomsday versus Extinction

Consider a population $P(t)$ of unsophisticated animals in which females rely solely on chance encounters to meet males for reproductive purposes. It is reasonable to expect such encounters to occur at a rate that is proportional to the product of the number $P/2$ of males and the number $P/2$ of females, hence at a rate proportional to P^2. We therefore assume that births occur at the rate kP^2 (per unit time, with k constant). The birth rate (births/time/population) is then given by $\beta = kP$. If the death rate δ is constant, then the general population equation in (1) yields the differential equation

$$\frac{dP}{dt} = kP^2 - \delta P = kP(P - M) \tag{13}$$

(where $M = \delta/k > 0$) as a mathematical model of the population.

Note that the right-hand side in Eq. (13) is the *negative* of the right-hand side in the logistic equation in (3). We will see that the constant M is now a **threshold population**, with the way the population behaves in the future depending critically on whether the initial population P_0 is less than or greater than M.

Example 7 Consider an animal population $P(t)$ that is modeled by the equation

$$\frac{dP}{dt} = 0.0004P(P - 150) = 0.0004P^2 - 0.06P. \tag{14}$$

We want to find $P(t)$ if (a) $P(0) = 200$; (b) $P(0) = 100$.

Solution To solve the equation in (14), we separate the variables and integrate. We get

$$\int \frac{dP}{P(P-150)} = \int 0.0004\,dt,$$

$$-\frac{1}{150}\int\left(\frac{1}{P} - \frac{1}{P-150}\right)dP = \int 0.0004\,dt \qquad \text{[partial fractions]},$$

$$\ln|P| - \ln|P-150| = -0.06t + C,$$

$$\frac{P}{P-150} = \pm e^C e^{-0.06t} = Be^{-0.06t} \qquad [\text{where } B = \pm e^C]. \qquad (15)$$

(a) Substitution of $t = 0$ and $P = 200$ into (15) gives $B = 4$. With this value of B we solve Eq. (15) for

$$P(t) = \frac{600e^{-0.06t}}{4e^{-0.06t} - 1}. \qquad (16)$$

Note that, as t increases and approaches $T = \ln(4)/0.06 \approx 23.105$, the positive denominator on the right in (16) decreases and approaches 0. Consequently $P(t) \to +\infty$ as $t \to T^-$. This is a *doomsday* situation—a real population *explosion*.

(b) Substitution of $t = 0$ and $P = 100$ into (15) gives $B = -2$. With this value of B we solve Eq. (15) for

$$P(t) = \frac{300e^{-0.06t}}{2e^{-0.06t} + 1} = \frac{300}{2 + e^{0.06t}}. \qquad (17)$$

Note that, as t increases without bound, the positive denominator on the right in (16) approaches $+\infty$. Consequently, $P(t) \to 0$ as $t \to +\infty$. This is an (eventual) *extinction* situation. ■

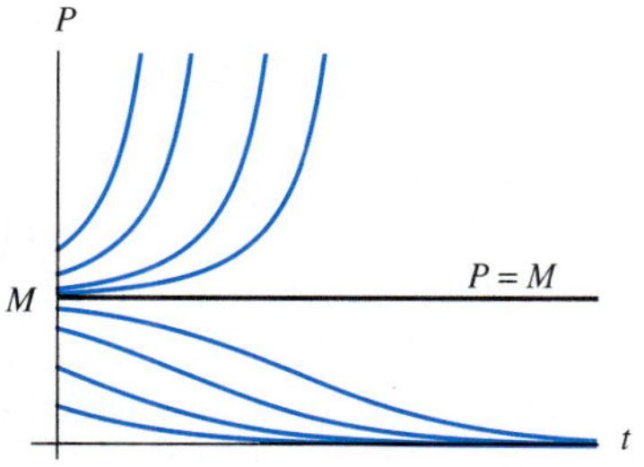

FIGURE 2.1.6. Typical solution curves for the explosion/extinction equation $P' = kP(P - M)$.

Thus the population in Example 7 either explodes or is an endangered species threatened with extinction, depending on whether or not its initial size exceeds the threshold population $M = 150$. An approximation to this phenomenon is sometimes observed with animal populations, such as the alligator population in certain areas of the southern United States.

Figure 2.1.6 shows typical solution curves that illustrate the two possibilities for a population $P(t)$ satisfying Eq. (13). If $P_0 = M$ (exactly!), then the population remains constant. However, this equilibrium situation is very unstable. If P_0 exceeds M (even slightly), then $P(t)$ rapidly increases without bound, whereas if the initial (positive) population is less than M (however slightly), then it decreases (more gradually) toward zero as $t \to +\infty$. See Problem 33.

2.1 Problems

Separate variables and use partial fractions to solve the initial value problems in Problems 1 through 8. Use either the exact solution or a computer-generated slope field to sketch the graphs of several solutions of the given differential equation, and highlight the indicated particular solution.

1. $\dfrac{dx}{dt} = x - x^2,\ x(0) = 2$

2. $\dfrac{dx}{dt} = 10x - x^2,\ x(0) = 1$

3. $\dfrac{dx}{dt} = 1 - x^2,\ x(0) = 3$

4. $\dfrac{dx}{dt} = 9 - 4x^2,\ x(0) = 0$

5. $\dfrac{dx}{dt} = 3x(5 - x)$, $x(0) = 8$

6. $\dfrac{dx}{dt} = 3x(x - 5)$, $x(0) = 2$

7. $\dfrac{dx}{dt} = 4x(7 - x)$, $x(0) = 11$

8. $\dfrac{dx}{dt} = 7x(x - 13)$, $x(0) = 17$

9. The time rate of change of a rabbit population P is proportional to the square root of P. At time $t = 0$ (months) the population numbers 100 rabbits and is increasing at the rate of 20 rabbits per month. How many rabbits will there be one year later?

10. Suppose that the fish population $P(t)$ in a lake is attacked by a disease at time $t = 0$, with the result that the fish cease to reproduce (so that the birth rate is $\beta = 0$) and the death rate δ (deaths per week per fish) is thereafter proportional to $1/\sqrt{P}$. If there were initially 900 fish in the lake and 441 were left after 6 weeks, how long did it take all the fish in the lake to die?

11. Suppose that when a certain lake is stocked with fish, the birth and death rates β and δ are both inversely proportional to $\sqrt{P}$. **(a)** Show that

$$P(t) = \left(\tfrac{1}{2}kt + \sqrt{P_0}\right)^2,$$

where k is a constant. **(b)** If $P_0 = 100$ and after 6 months there are 169 fish in the lake, how many will there be after 1 year?

12. The time rate of change of an alligator population P in a swamp is proportional to the square of P. The swamp contained a dozen alligators in 1988, two dozen in 1998. When will there be four dozen alligators in the swamp? What happens thereafter?

13. Consider a prolific breed of rabbits whose birth and death rates, β and δ, are each proportional to the rabbit population $P = P(t)$, with $\beta > \delta$. **(a)** Show that

$$P(t) = \frac{P_0}{1 - kP_0t}, \qquad k \text{ constant.}$$

Note that $P(t) \to +\infty$ as $t \to 1/(kP_0)$. This is doomsday. **(b)** Suppose that $P_0 = 6$ and that there are nine rabbits after ten months. When does doomsday occur?

14. Repeat part (a) of Problem 13 in the case $\beta < \delta$. What now happens to the rabbit population in the long run?

15. Consider a population $P(t)$ satisfying the logistic equation $dP/dt = aP - bP^2$, where $B = aP$ is the time rate at which births occur and $D = bP^2$ is the rate at which deaths occur. If the initial population is $P(0) = P_0$, and B_0 births per month and D_0 deaths per month are occurring at time $t = 0$, show that the limiting population is $M = B_0P_0/D_0$.

16. Consider a rabbit population $P(t)$ satisfying the logistic equation as in Problem 15. If the initial population is 120 rabbits and there are 8 births per month and 6 deaths per month occurring at time $t = 0$, how many months does it take for $P(t)$ to reach 95% of the limiting population M?

17. Consider a rabbit population $P(t)$ satisfying the logistic equation as in Problem 15. If the initial population is 240 rabbits and there are 9 births per month and 12 deaths per month occurring at time $t = 0$, how many months does it take for $P(t)$ to reach 105% of the limiting population M?

18. Consider a population $P(t)$ satisfying the extinction-explosion equation $dP/dt = aP^2 - bP$, where $B = aP^2$ is the time rate at which births occur and $D = bP$ is the rate at which deaths occur. If the initial population is $P(0) = P_0$ and B_0 births per month and D_0 deaths per month are occurring at time $t = 0$, show that the threshold population is $M = D_0P_0/B_0$.

19. Consider an alligator population $P(t)$ satisfying the extinction-explosion equation as in Problem 18. If the initial population is 100 alligators and there are 10 births per month and 9 deaths per months occurring at time $t = 0$, how many months does it take for $P(t)$ to reach 10 times the threshold population M?

20. Consider an alligator population $P(t)$ satisfying the extinction/explosion equation as in Problem 18. If the initial population is 110 alligators and there are 11 births per month and 12 deaths per month occurring at time $t = 0$, how many months does it take for $P(t)$ to reach 10% of the threshold population M?

21. Suppose that the population $P(t)$ of a country satisfies the differential equation $dP/dt = kP(200 - P)$ with k constant. Its population in 1940 was 100 million and was then growing at the rate of 1 million per year. Predict this country's population for the year 2000.

22. Suppose that at time $t = 0$, half of a "logistic" population of 100,000 persons have heard a certain rumor, and that the number of those who have heard it is then increasing at the rate of 1000 persons per day. How long will it take for this rumor to spread to 80% of the population? (*Suggestion*: Find the value of k by substituting $P(0)$ and $P'(0)$ in the logistic equation, Eq. (3).)

23. As the salt KNO_3 dissolves in methanol, the number $x(t)$ of grams of the salt in a solution after t seconds satisfies the differential equation $dx/dt = 0.8x - 0.004x^2$.

(a) What is the maximum amount of the salt that will ever dissolve in the methanol?

(b) If $x = 50$ when $t = 0$, how long will it take for an additional 50 g of salt to dissolve?

24. Suppose that a community contains 15,000 people who are susceptible to Michaud's syndrome, a contagious disease. At time $t = 0$ the number $N(t)$ of people who have developed Michaud's syndrome is 5000 and is increasing at the rate of 500 per day. Assume that $N'(t)$ is proportional to the product of the numbers of those who have caught the disease and of those who have not. How long will it take for another 5000 people to develop Michaud's syndrome?

25. The data in the table in Fig. 2.1.7 are given for a certain population $P(t)$ that satisfies the logistic equation in (3). **(a)** What is the limiting population M? (*Suggestion*: Use the approximation

$$P'(t) \approx \frac{P(t+h) - P(t-h)}{2h}$$

with $h = 1$ to estimate the values of $P'(t)$ when $P = 25.00$ and when $P = 47.54$. Then substitute these values in the logistic equation and solve for k and M.) **(b)** Use the values of k and M found in part (a) to determine when $P = 75$. (*Suggestion*: Take $t = 0$ to correspond to the year 1925.)

Year	P (millions)
1924	24.63
1925	25.00
1926	25.38
⋮	⋮
1974	47.04
1975	47.54
1976	48.04

FIGURE 2.1.7. Population data for Problem 25.

26. A population $P(t)$ of small rodents has birth rate $\beta = (0.001)P$ (births per month per rodent) and *constant* death rate δ. If $P(0) = 100$ and $P'(0) = 8$, how long (in months) will it take this population to double to 200 rodents? (*Suggestion*: First find the value of δ.)

27. Consider an animal population $P(t)$ with constant death rate $\delta = 0.01$ (deaths per animal per month) and with birth rate β proportional to P. Suppose that $P(0) = 200$ and $P'(0) = 2$. **(a)** When is $P = 1000$? **(b)** When does doomsday occur?

28. Suppose that the number $x(t)$ (with t in months) of alligators in a swamp satisfies the differential equation $dx/dt = 0.0001x^2 - 0.01x$.

(a) If initially there are 25 alligators in the swamp, solve this differential equation to determine what happens to the alligator population in the long run.

(b) Repeat part (a), except with 150 alligators initially.

29. During the period from 1790 to 1930, the U.S. population $P(t)$ (t in years) grew from 3.9 million to 123.2 million. Throughout this period, $P(t)$ remained close to the solution of the initial value problem

$$\frac{dP}{dt} = 0.03135P - 0.0001489P^2, \quad P(0) = 3.9.$$

(a) What 1930 population does this logistic equation predict?

(b) What limiting population does it predict?

(c) Has this logistic equation continued since 1930 to accurately model the U.S. population?

[This problem is based on a computation by Verhulst, who in 1845 used the 1790–1840 U.S. population data to predict accurately the U.S. population through the year 1930 (long after his own death, of course).]

30. A tumor may be regarded as a population of multiplying cells. It is found empirically that the "birth rate" of the cells in a tumor decreases exponentially with time, so that $\beta(t) = \beta_0 e^{-\alpha t}$ (where α and β_0 are positive constants), and hence

$$\frac{dP}{dt} = \beta_0 e^{-\alpha t} P, \quad P(0) = P_0.$$

Solve this initial value problem for

$$P(t) = P_0 \exp\left(\frac{\beta_0}{\alpha}(1 - e^{-\alpha t})\right).$$

Observe that $P(t)$ approaches the finite limiting population $P_0 \exp(\beta_0/\alpha)$ as $t \to +\infty$.

31. For the tumor of Problem 30, suppose that at time $t = 0$ there are $P_0 = 10^6$ cells and that $P(t)$ is then increasing at the rate of 3×10^5 cells per month. After 6 months the tumor has doubled (in size and in number of cells). Solve numerically for α, and then find the limiting population of the tumor.

32. Derive the solution

$$P(t) = \frac{MP_0}{P_0 + (M - P_0)e^{-kMt}}$$

of the logistic initial value problem $P' = kP(M - P)$, $P(0) = P_0$. Make it clear how your derivation depends on whether $0 < P_0 < M$ or $P_0 > M$.

33. **(a)** Derive the solution

$$P(t) = \frac{MP_0}{P_0 + (M - P_0)e^{kMt}}$$

of the extinction-explosion initial value problem $P' = kP(P - M)$, $P(0) = P_0$.

(b) How does the behavior of $P(t)$ as t increases depend on whether $0 < P_0 < M$ or $P_0 > M$?

34. If $P(t)$ satisfies the logistic equation in (3), use the chain rule to show that

$$P''(t) = 2k^2 P(P - \tfrac{1}{2}M)(P - M).$$

Conclude that $P'' > 0$ if $0 < P < \frac{1}{2}M$; $P'' = 0$ if $P = \frac{1}{2}M$; $P'' < 0$ if $\frac{1}{2}M < P < M$; and $P'' > 0$ if $P > M$. In particular, it follows that any solution curve that crosses the line $P = \frac{1}{2}M$ has an inflection point where it crosses that line, and therefore resembles one of the lower S-shaped curves in Fig. 2.1.3.

35. Consider two population functions $P_1(t)$ and $P_2(t)$, both of which satisfy the logistic equation with the same limiting population M but with different values k_1 and k_2 of the constant k in Eq. (3). Assume that $k_1 < k_2$. Which population approaches M the most rapidly? You can reason *geometrically* by examining slope fields (especially if appropriate software is available), *symbolically* by analyzing the solution given in Eq. (7), or *numerically* by substituting successive values of t.

36. To solve the two equations in (10) for the values of k and M, begin by solving the first equation for the quantity $x = e^{-50kM}$ and the second equation for $x^2 = e^{-100kM}$. Upon equating the two resulting expressions for x^2 in terms of M, you get an equation that is readily solved for M. With M now known, either of the original equations is readily solved for k. This technique can be used to "fit" the logistic equation to any three population values P_0, P_1, and P_2 corresponding to *equally spaced* times $t_0 = 0$, t_1, and $t_2 = 2t_1$.

37. Use the method of Problem 36 to fit the logistic equation to the actual U.S. population data (Fig. 2.1.4) for the years 1850, 1900, and 1950. Solve the resulting logistic equation and compare the predicted and actual populations for the years 1990 and 2000.

38. Fit the logistic equation to the actual U.S. population data (Fig. 2.1.4) for the years 1900, 1930, and 1960. Solve the resulting logistic equation, then compare the predicted and actual populations for the years 1980, 1990, and 2000.

39. Birth and death rates of animal populations typically are not constant; instead, they vary periodically with the passage of seasons. Find $P(t)$ if the population P satisfies the differential equation

$$\frac{dP}{dt} = (k + b\cos 2\pi t)P,$$

where t is in years and k and b are positive constants. Thus the growth-rate function $r(t) = k + b\cos 2\pi t$ varies periodically about its mean value k. Construct a graph that contrasts the growth of this population with one that has the same initial value P_0 but satisfies the natural growth equation $P' = kP$ (same constant k). How would the two populations compare after the passage of many years?

2.1 Application Logistic Modeling of Population Data

These investigations deal with the problem of fitting a logistic model to given population data. Thus we want to determine the numerical constants a and b so that the solution $P(t)$ of the initial value problem

$$\frac{dP}{dt} = aP + bP^2, \qquad P(0) = P_0 \tag{1}$$

approximates the given values $P_0, P_1, \ldots, P_n$ of the population at the times $t_0 = 0$, $t_1, \ldots, t_n$. If we rewrite Eq. (1) (the logistic equation with $kM = a$ and $k = -b$) in the form

$$\frac{1}{P}\frac{dP}{dt} = a + bP, \tag{2}$$

then we see that the points

$$\left(P(t_i), \frac{P'(t_i)}{P(t_i)}\right), \qquad i = 0, 1, 2, \ldots, n,$$

should all lie on the straight line with y-intercept a and slope b (as determined by the function of P on the right-hand side in Eq. (2)).

This observation provides a way to find a and b. If we can determine the approximate values of the derivatives $P_1', P_2', \ldots$ corresponding to the given population data, then we can proceed with the following agenda:

- First plot the points $(P_1, P_1'/P_1)$, $(P_2, P_2'/P_2)$, $\ldots$ on a sheet of graph paper with horizontal P-axis.
- Then use a ruler to draw a straight line that appears to approximate these points well.

- Finally, measure this straight line's y-intercept a and slope b.

But where are we to find the needed values of the derivative $P'(t)$ of the (as yet) unknown function P? It is easiest to use the approximation

$$P_i' = \frac{P_{i+1} - P_{i-1}}{t_{i+1} - t_{i-1}} \tag{3}$$

suggested by Fig. 2.1.8. For instance, if we take $i = 0$ corresponding to the year 1790, then the U.S. population data in Fig. 2.1.9 give

$$P_1' = \frac{P_2 - P_0}{t_2 - t_0} = \frac{7.240 - 3.929}{20} \approx 0.166$$

for the slope at (t_1, P_1) corresponding to the year 1800.

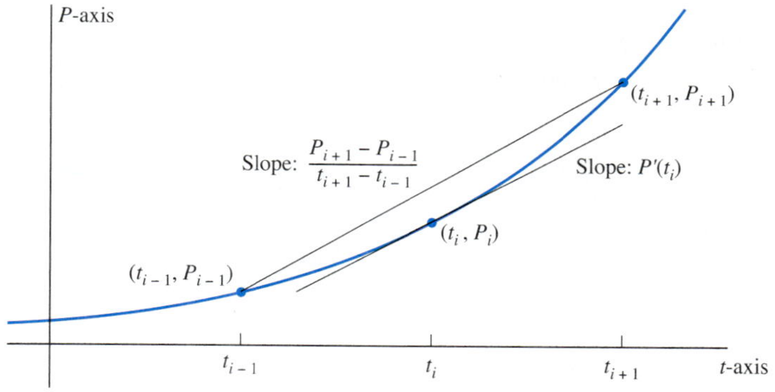

FIGURE 2.1.8. The symmetric difference approximation $\frac{P_{i+1} - P_{i-1}}{t_{i+1} - t_{i-1}}$ to the derivative $P'(t_i)$.

INVESTIGATION A: Use Eq. (3) to verify the slope figures shown in the final column of the table in Fig. 2.1.9, then plot the points $(P_1, P_1'/P_1), \ldots, (P_{11}, P_{11}'/P_{11})$ indicated by the dots in Fig. 2.1.10. If an appropriate graphing calculator, spreadsheet, or computer program is available, use it to find the straight line $y = a + bP$ as in (2) that best fits these points. If not, draw your own straight line approximating these points, and then measure its intercept a and slope b as accurately as you can. Next, solve the logistic equation in (1) with these numerical parameters, taking $t = 0$ corresponding to the year 1800. Finally, compare the predicted 20th-century U.S. population figures with the actual data listed in Fig. 2.1.4.

INVESTIGATION B: Repeat Investigation A, but take $t = 0$ in 1900 and use only 20th-century population data. Do you get a better approximation for the U.S. population during the final decades of the 20th century?

INVESTIGATION C: Model similarly the world population data shown in Fig. 2.1.11. The Population Division of the United Nations predicts a world population of 8.177 billion in the year 2025. What do you predict?

Year	i	t_i	Population P_i	Slope P'_i
1790	0	−10	3.929	
1800	1	0	5.308	0.166
1810	2	10	7.240	0.217
1820	3	20	9.638	0.281
1830	4	30	12.861	0.371
1840	5	40	17.064	0.517
1850	6	50	23.192	0.719
1860	7	60	31.443	0.768
1870	8	70	38.558	0.937
1880	9	80	50.189	1.221
1890	10	90	62.980	1.301
1900	11	100	76.212	1.462
1910	12	110	92.228	

FIGURE 2.1.9. U.S. population data (in millions) and approximate slopes.

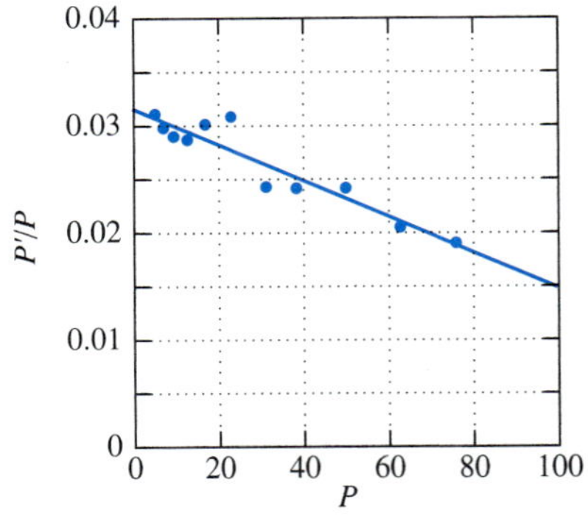

FIGURE 2.1.10. Points and approximating straight line for U.S. population data from 1800 to 1900.

Year	World Population (billions)
1960	3.049
1965	3.358
1970	3.721
1975	4.103
1980	4.473
1985	4.882
1990	5.249
1995	5.679
2000	6.127

FIGURE 2.1.11. World population data.

2.2 Equilibrium Solutions and Stability

In previous sections we have often used explicit solutions of differential equations to answer specific numerical questions. But even when a given differential equation is difficult or impossible to solve explicitly, it often is possible to extract *qualitative* information about general properties of its solutions. For example, we may be able to establish that every solution $x(t)$ grows without bound as $t \to +\infty$, or approaches a finite limit, or is a periodic function of t. In this section we introduce—mainly by consideration of simple differential equations that *can* be solved explicitly—some of the more important qualitative questions that can sometimes be answered for equations that are difficult or impossible to solve.

Example 1 Let $x(t)$ denote the temperature of a body with initial temperature $x(0) = x_0$. At time $t = 0$ this body is immersed in a medium with constant temperature A. Assuming Newton's law of cooling,

$$\frac{dx}{dt} = -k(x - A) \qquad (k > 0 \quad \text{constant}), \tag{1}$$

we readily solve (by separation of variables) for the explicit solution

$$x(t) = A + (x_0 - A)e^{-kt}.$$

It follows immediately that

$$\lim_{t\to\infty} x(t) = A, \tag{2}$$

so the temperature of the body approaches that of the surrounding medium (as is evident to one's intuition). Note that the constant function $x(t) \equiv A$ is a solution of Eq. (1); it corresponds to the temperature of the body when it is in thermal equilibrium with the surrounding medium. In Fig. 2.2.1 the limit in (2) means that every other solution curve approaches the equilibrium solution curve $x = A$ asymptotically as $t \to +\infty$. ■

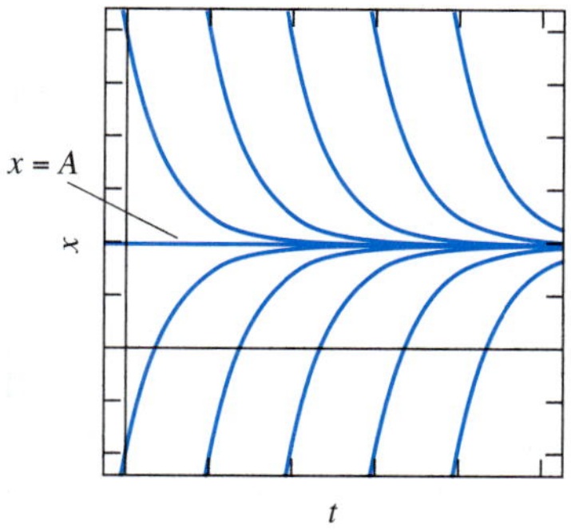

FIGURE 2.2.1. Typical solution curves for the equation of Newton's law of cooling, $dx/dt = -k(x - A)$.

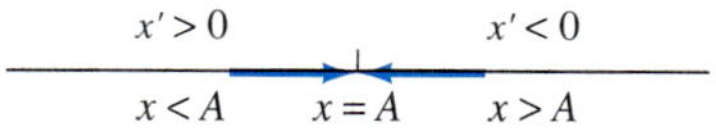

FIGURE 2.2.2. Phase diagram for the equation $dx/dt = f(x) = k(A - x)$.

Remark: The behavior of solutions of Eq. (1) is summarized briefly by the **phase diagram** in Fig. 2.2.2—which indicates the direction (or "phase") of change in x as a function of x itself. The right-hand side $f(x) = -k(x - A) = k(A - x)$ is positive if $x < A$, negative if $x > A$. This observation corresponds to the fact that solutions starting above the line $x = A$ and those starting below it both approach the limiting solution $x(t) \equiv A$ as t increases (as indicated by the arrows). ■

In Section 2.1 we introduced the general population equation

$$\frac{dx}{dt} = (\beta - \delta)x, \tag{3}$$

where β and δ are the birth and death rates, respectively, in births or deaths per individual per unit of time. The question of whether a population $x(t)$ is bounded or unbounded as $t \to +\infty$ is of evident interest. In many situations—like the logistic and explosion/extinction populations of Section 2.1—the birth and death rates are known functions of x. Then Eq. (3) takes the form

$$\frac{dx}{dt} = f(x). \tag{4}$$

This is an **autonomous** first-order differential equation—one in which the independent variable t does not appear explicitly (the terminology here stemming from the Greek word *autonomos* for "independent," e.g., of the time t). As in Example 1, the solutions of the equation $f(x) = 0$ play an important role and are called **critical points** of the autonomous differential equation $dx/dt = f(x)$.

If $x = c$ is a critical point of Eq. (4), then the differential equation has the constant solution $x(t) \equiv c$. A constant solution of a differential equation is sometimes called an **equilibrium solution** (one may think of a population that remains constant because it is in "equilibrium" with its environment). Thus the critical point $x = c$, a number, corresponds to the equilibrium solution $x(t) \equiv c$, a constant-valued function.

Example 2 illustrates the fact that the qualitative behavior (as t increases) of the solutions of an autonomous first-order equation can be described in terms of its critical points.

Example 2 Consider the logistic differential equation

$$\frac{dx}{dt} = kx(M - x) \tag{5}$$

(with $k > 0$ and $M > 0$). It has two critical points—the solutions $x = 0$ and $x = M$ of the equation

$$f(x) = kx(M - x) = 0.$$

In Section 2.1 we discussed the logistic-equation solution

$$x(t) = \frac{Mx_0}{x_0 + (M - x_0)e^{-kMt}} \tag{6}$$

satisfying the initial condition $x(0) = x_0$. Note that the initial values $x_0 = 0$ and $x_0 = M$ yield the equilibrium solutions $x(t) \equiv 0$ and $x(t) \equiv M$ of Eq. (5).

We observed in Section 2.1 that if $x_0 > 0$, then $x(t) \to M$ as $t \to +\infty$. But if $x_0 < 0$, then the denominator in Eq. (6) initially is positive, but vanishes when

$$t = t_1 = \frac{1}{kM} \ln \frac{M - x_0}{-x_0} > 0.$$

Because the numerator in (6) is negative in this case, it follows that

$$\lim_{t \to t_1^-} x(t) = -\infty \quad \text{if} \quad x_0 < 0.$$

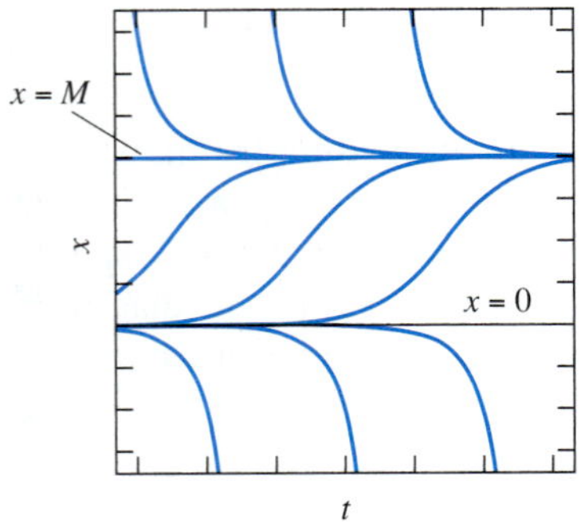

FIGURE 2.2.3. Typical solution curves for the logistic equation $dx/dt = kx(M - x)$.

It follows that the solution curves of the logistic equation in (5) look as illustrated in Fig. 2.2.3. Here we see graphically that every solution either approaches the equilibrium solution $x(t) \equiv M$ as t increases, or (in a visually obvious sense) diverges away from the other equilibrium solution $x(t) \equiv 0$. ■

Stability of Critical Points

Figure 2.2.3 illustrates the concept of *stability*. A critical point $x = c$ of an autonomous first-order equation is said to be *stable* provided that, if the initial value x_0 is sufficiently close to c, then $x(t)$ remains close to c for all $t > 0$. More precisely, the critical point c is **stable** if, for each $\epsilon > 0$, there exists $\delta > 0$ such that

$$|x_0 - c| < \delta \quad \text{implies that} \quad |x(t) - c| < \epsilon \tag{7}$$

for all $t > 0$. The critical point $x = c$ is **unstable** if it is not stable.

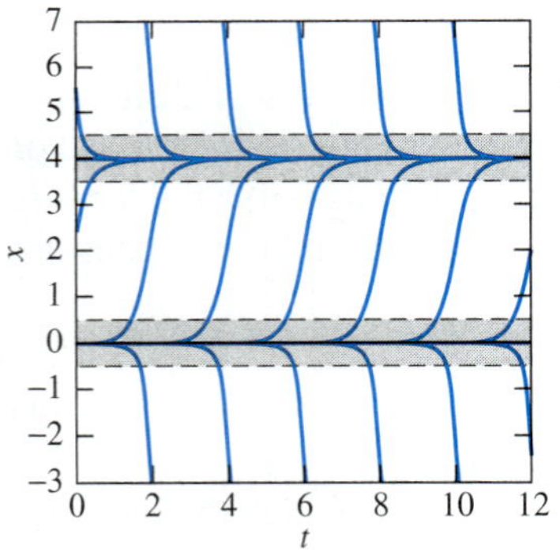

FIGURE 2.2.4. Solution curves, funnel, and spout for $dx/dt = 4x - x^2$.

Figure 2.2.4 shows a "wider view" of the solution curves of a logistic equation with $k = 1$ and $M = 4$. Note that the strip $3.5 < x < 4.5$ enclosing the stable equilibrium curve $x = 4$ acts like a *funnel*—solution curves (moving from left to right) enter this strip and thereafter remain within it. By contrast, the strip $-0.5 < x < 0.5$ enclosing the unstable solution curve $x = 0$ acts like a *spout*—solution curves leave this strip and thereafter remain outside it. Thus the critical point $x = M$ is stable, whereas the critical point $x = 0$ is unstable.

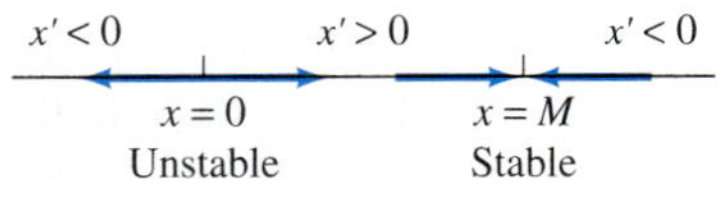

FIGURE 2.2.5. Phase diagram for the logistic equation $dx/dt = f(x) = kx(M - x)$.

Remark 1: We can summarize the behavior of solutions of the logistic equation in (5)—in terms of their initial values—by means of the phase diagram shown in Fig. 2.2.5. It indicates that $x(t) \to M$ as $t \to +\infty$ if either $x_0 > M$ or $0 < x_0 < M$, whereas $x(t) \to -\infty$ as t increases if $x_0 < 0$. The fact that M is a stable critical point would be important, for instance, if we wished to conduct an experiment with a population of M bacteria. It is impossible to count precisely M bacteria for M large, but any initially positive population will approach M as t increases.

Remark 2: Related to the stability of the limiting solution $M = a/b$ of the logistic equation

$$\frac{dx}{dt} = ax - bx^2 \tag{8}$$

is the "predictability" of M for an actual population. The coefficients a and b are unlikely to be known precisely for an actual population. But if they are replaced with close approximations $a^\star$ and $b^\star$—derived perhaps from empirical measurements—then the approximate limiting population $M^\star = a^\star/b^\star$ will be close to the actual limiting population $M = a/b$. We may therefore say that the value M of the limiting population predicted by a logistic equation not only is a stable critical point of the differential equation; this value also is "stable" with respect to small perturbations of the constant coefficients in the equation. (Note that one of these two statements involves changes in the initial value x_0; the other involves changes in the coefficients a and b.) ■

Example 3 Consider now the explosion/extinction equation

$$\frac{dx}{dt} = kx(x - M) \tag{9}$$

of Eq. (10) in Section 2.1. Like the logistic equation, it has the two critical points $x = 0$ and $x = M$ corresponding to the equilibrium solutions $x(t) \equiv 0$ and $x(t) \equiv M$. According to Problem 33 in Section 2.1, its solution with $x(0) = x_0$ is given by

$$x(t) = \frac{Mx_0}{x_0 + (M - x_0)e^{kMt}} \tag{10}$$

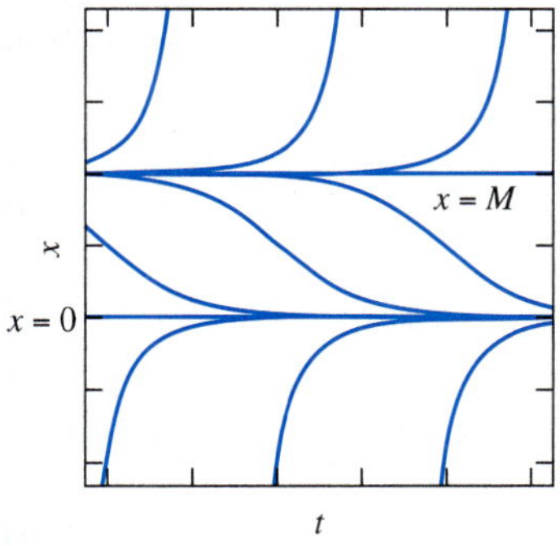

FIGURE 2.2.6. Typical solution curves for the explosion/extinction equation $dx/dt = kx(x - M)$.

(with only a single difference in sign from the logistic solution in (6)). If $x_0 < M$, then (because the coefficient of the exponential in the denominator is positive) it follows immediately from Eq. (10) that $x(t) \to 0$ as $t \to +\infty$. But if $x_0 > M$, then the denominator in (10) initially is positive, but vanishes when

$$t = t_1 = \frac{1}{kM} \ln \frac{x_0}{x_0 - M} > 0.$$

Because the numerator in (10) is positive in this case, it follows that

$$\lim_{t \to t_1^-} x(t) = +\infty \quad \text{if} \quad x_0 > M.$$

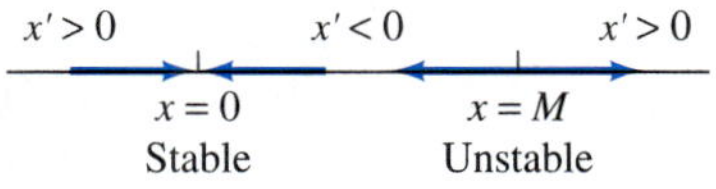

FIGURE 2.2.7. Phase diagram for the explosion/extinction equation $dx/dt = f(x) = kx(x - M)$.

Therefore, the solution curves of the explosion/extinction equation in (9) look as illustrated in Fig. 2.2.6. A narrow band along the equilibrium curve $x = 0$ (as in Fig. 2.2.4) would serve as a funnel, while a band along the solution curve $x = M$ would serve as a spout for solutions. The behavior of the solutions of Eq. (9) is summarized by the phase diagram in Fig. 2.2.7, where we see that the critical point $x = 0$ is stable and the critical point $x = M$ is unstable. ■

Harvesting a Logistic Population

The autonomous differential equation

$$\frac{dx}{dt} = ax - bx^2 - h \tag{11}$$

(with a, b, and h all positive) may be considered to describe a logistic population *with harvesting*. For instance, we might think of the population of fish in a lake from which h fish per year are removed by fishing.

Example 4 Let us rewrite Eq. (11) in the form

$$\frac{dx}{dt} = kx(M - x) - h, \tag{12}$$

which exhibits the limiting population M in the case $h = 0$ of no harvesting. Assuming hereafter that $h > 0$, we can solve the quadratic equation $-kx^2+kMx-h = 0$ for the two critical points

$$H, N = \frac{kM \pm \sqrt{(kM)^2 - 4hk}}{2k} = \frac{1}{2}\left(M \pm \sqrt{M^2 - 4h/k}\right), \tag{13}$$

assuming that the harvesting rate h is sufficiently small that $4h < kM^2$, so both roots H and N are real with $0 < H < N < M$. Then we can rewrite Eq. (12) in the form

$$\frac{dx}{dt} = k(N - x)(x - H). \tag{14}$$

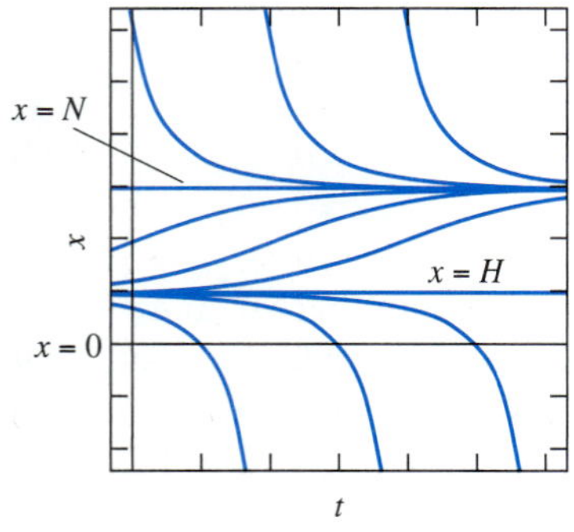

FIGURE 2.2.8. Typical solution curves for the logistic harvesting equation $dx/dt = k(N - x)(x - H)$.

For instance, the number of critical points of the equation may change abruptly as the value of a parameter is changed. In Problem 24 we ask you to solve this equation for the solution

$$x(t) = \frac{N(x_0 - H) - H(x_0 - N)e^{-k(N-H)t}}{(x_0 - H) - (x_0 - N)e^{-k(N-H)t}} \tag{15}$$

in terms of the initial value $x(0) = x_0$.

Note that the exponent $-k(N - H)t$ is negative for $t > 0$. If $x_0 > N$, then each of the coefficients within parentheses in Eq. (15) is positive; it follows that

$$\text{If} \quad x_0 > N \quad \text{then} \quad x(t) \to N \quad \text{as} \quad t \to +\infty. \tag{16}$$

In Problem 25 we ask you to deduce also from Eq. (15) that

$$\text{If} \quad H < x_0 < N \quad \text{then} \quad x(t) \to N \quad \text{as} \quad t \to +\infty, \quad \text{whereas} \tag{17}$$

$$\text{if} \quad x_0 < H \quad \text{then} \quad x(t) \to -\infty \quad \text{as} \quad t \to t_1 \tag{18}$$

for a positive value t_1 that depends on x_0. It follows that the solution curves of Eq. (12)—still assuming that $4h < kM^2$—look as illustrated in Fig. 2.2.8. (Can you visualize a funnel along the line $x = N$ and a spout along the line $x = H$?) Thus the constant solution $x(t) \equiv N$ is an equilibrium *limiting solution,* whereas $x(t) \equiv H$ is a *threshold solution* that separates different behaviors—the population approaches N if $x_0 > H$, while it becomes extinct because of harvesting if $x_0 < H$. Finally, the stable critical point $x = N$ and the unstable critical point $x = H$ are illustrated in the phase diagram in Fig. 2.2.9. ■

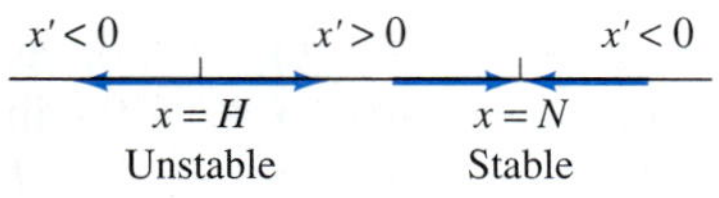

FIGURE 2.2.9. Phase diagram for the logistic harvesting equation $dx/dt = f(x) = k(N-x)(x-H)$.

Example 5 For a concrete application of our stability conclusions in Example 4, suppose that $k = 1$ and $M = 4$ for a logistic population $x(t)$ of fish in a lake, measured in hundreds after t years. Without any fishing at all, the lake would eventually contain nearly 400 fish, whatever the initial population. Now suppose that $h = 3$, so that 300 fish are "harvested" annually (at a constant rate throughout the year). Equation (12) is then $dx/dt = x(4 - x) - 3$, and the quadratic equation

$$-x^2 + 4x - 3 = (3 - x)(x - 1) = 0$$

has solutions $H = 1$ and $N = 3$. Thus the threshold population is 100 fish and the (new) limiting population is 300 fish. In short, if the lake is stocked initially with more than 100 fish, then as t increases, the fish population will approach a limiting value of 300 fish. But if the lake is stocked initially with fewer than 100 fish, then the lake will be "fished out" and the fish will disappear entirely within a finite period of time. ■

Bifurcation and Dependence on Parameters

A biological or physical system that is modeled by a differential equation may depend crucially on the numerical values of certain coefficients or parameters that appear in the equation. For instance, the number of critical points of the equation may change abruptly as the value of a parameter is changed.

Example 6 The differential equation

$$\frac{dx}{dt} = x(4 - x) - h \tag{19}$$

(with x in hundreds) models the harvesting of a logistic population with $k = 1$ and limiting population $M = 4$ (hundred). In Example 5 we considered the case of harvesting level $h = 3$, and found that the new limiting population is $N = 3$ hundred and the threshold population is $H = 1$ hundred. Typical solution curves, including the equilibrium solutions $x(t) \equiv 3$ and $x(t) \equiv 1$, then look like those pictured in Fig. 2.2.8.

Now let's investigate the dependence of this picture upon the harvesting level h. According to Eq. (13) with $k = 1$ and $M = 4$, the limiting and threshold populations N and H are given by

$$H, N = \frac{1}{2}\left(4 \pm \sqrt{16 - 4h}\right) = 2 \pm \sqrt{4 - h}. \tag{20}$$

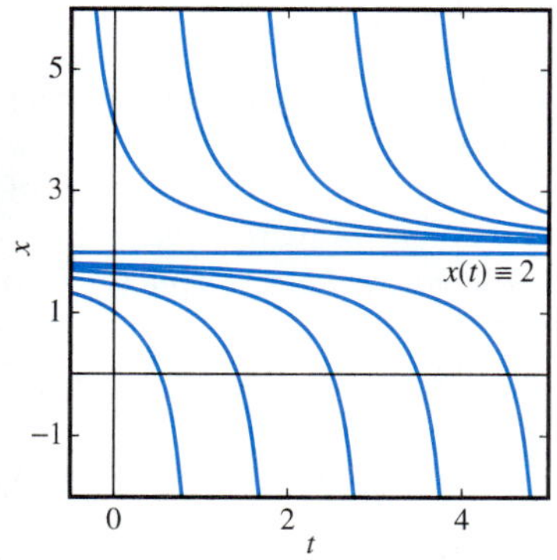

FIGURE 2.2.10. Solution curves of the equation $x' = x(4 - x) - h$ with critical harvesting $h = 4$.

If $h < 4$—we can consider negative values of h to describe stocking rather than harvesting the fish—then there are distinct equilibrium solutions $x(t) \equiv N$ and $x(t) \equiv H$ with $N > H$ as in Fig. 2.2.8.

But if $h = 4$, then Eq. (20) gives $N = H = 2$, so the differential equation has only the single equilibrium solution $x(t) \equiv 2$. In this case the solution curves of the equation look like those illustrated in Fig. 2.2.10. If the initial number x_0 (in hundreds) of fish exceeds 2, then the population approaches a limiting population of 2 (hundred fish). However, any initial population $x_0 < 2$ (hundred) results in extinction with the fish dying out as a consequence of the harvesting of 4 hundred fish annually. The critical point $x = 2$ might therefore be described as "semistable"—it looks stable on the side $x > 2$ where solution curves approach the equilibrium solution $x(t) \equiv 2$ as t increases, but unstable on the side $x < 2$ where solution curves instead diverge away from the equilibrium solution.

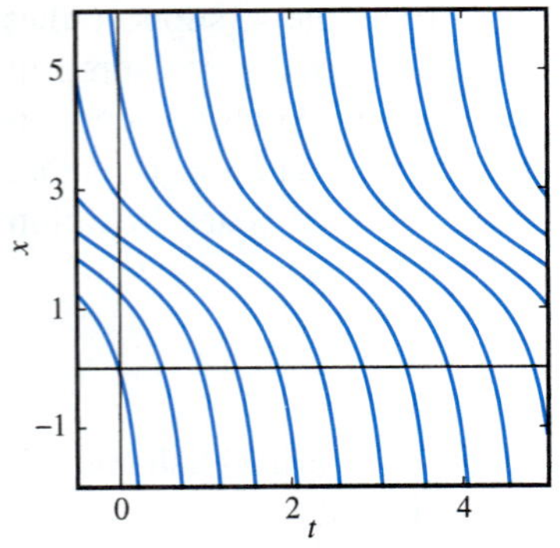

FIGURE 2.2.11. Solution curves of the equation $x' = x(4 - x) - h$ with excessive harvesting $h = 5$.

If, finally, $h > 4$, then the quadratic equation corresponding to (20) has no real solutions and the differential equation in (19) has no equilibrium solutions. The solution curves then look like those illustrated in Fig. 2.2.11, and (whatever the initial number of fish) the population dies out as a result of the excessive harvesting. ■

If we imagine turning a dial to gradually increase the value of the parameter h in Eq. (19), then the picture of the solution curves changes from one like Fig. 2.2.8 with $h < 4$, to Fig. 2.2.10 with $h = 4$, to one like Fig. 2.2.11 with $h > 4$. Thus the differential equation has

- two critical points if $h < 4$;
- one critical point if $h = 4$;
- no critical point if $h > 4$.

The value $h = 4$—for which the qualitative nature of the solutions changes as h increases—is called a **bifurcation point** for the differential equation containing the parameter h. A common way to visualize the corresponding "bifurcation" in the solutions is to plot the **bifurcation diagram** consisting of all points (h, c), where c is a critical point of the equation $x' = x(4 - x) + h$. For instance, if we rewrite Eq. (20) as

$$c = 2 \pm \sqrt{4 - h},$$
$$(c - 2)^2 = 4 - h,$$

where either $c = N$ or $c = H$, then we get the equation of the parabola that is shown in Fig. 2.2.12. This parabola is then the bifurcation diagram for our differential equation that models a logistic fish population with harvesting at the level specified by the parameter h.

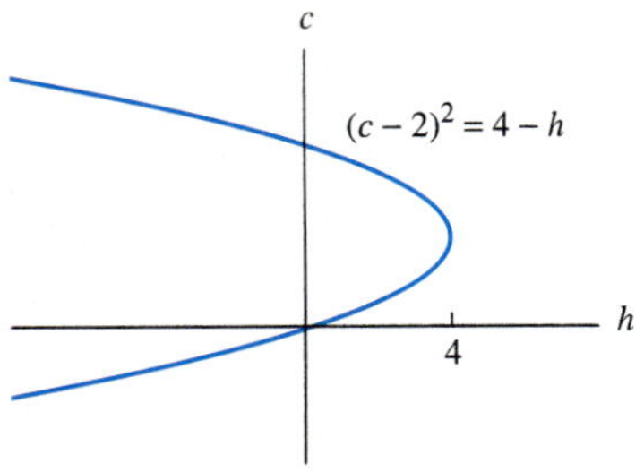

FIGURE 2.2.12. The parabola $(c - 2)^2 = 4 - h$ is the bifurcation diagram for the differential equation $x' = x(4 - x) - h$.

2.2 Problems

In Problems 1 through 12, first solve the equation $f(x) = 0$ to find the critical points of the given autonomous differential equation $dx/dt = f(x)$. Then analyze the sign of $f(x)$ to determine whether each critical point is stable or unstable, and construct the corresponding phase diagram for the differential equation. Next, solve the differential equation explicitly for $x(t)$ in terms of t. Finally, use either the exact solution or a computer-generated slope field to sketch typical solution curves for the given differential equation, and verify visually the stability of each critical point.

1. $\dfrac{dx}{dt} = x - 4$

2. $\dfrac{dx}{dt} = 3 - x$

3. $\dfrac{dx}{dt} = x^2 - 4x$

4. $\dfrac{dx}{dt} = 3x - x^2$

5. $\dfrac{dx}{dt} = x^2 - 4$

6. $\dfrac{dx}{dt} = 9 - x^2$

7. $\dfrac{dx}{dt} = (x - 2)^2$

8. $\dfrac{dx}{dt} = -(3 - x)^2$

9. $\dfrac{dx}{dt} = x^2 - 5x + 4$

10. $\dfrac{dx}{dt} = 7x - x^2 - 10$

11. $\dfrac{dx}{dt} = (x - 1)^3$

12. $\dfrac{dx}{dt} = (2 - x)^3$

In Problems 13 through 18, use a computer system or graphing calculator to plot a slope field and/or enough solution curves to indicate the stability or instability of each critical point of the given differential equation. (Some of these critical points may be semistable in the sense mentioned in Example 6.)

13. $\dfrac{dx}{dt} = (x + 2)(x - 2)^2$

14. $\dfrac{dx}{dt} = x(x^2 - 4)$

15. $\dfrac{dx}{dt} = (x^2 - 4)^2$

16. $\dfrac{dx}{dt} = (x^2 - 4)^3$

17. $\dfrac{dx}{dt} = x^2(x^2 - 4)$

18. $\dfrac{dx}{dt} = x^3(x^2 - 4)$

19. The differential equation $dx/dt = \frac{1}{10}x(10 - x) - h$ models a logistic population with harvesting at rate h. Determine (as in Example 6) the dependence of the number of critical points on the parameter h, and then construct a bifurcation diagram like Fig. 2.2.12.

20. The differential equation $dx/dt = \frac{1}{100}x(x - 5) + s$ models a population with stocking at rate s. Determine the dependence of the number of critical points c on the parameter s, and then construct the corresponding bifurcation diagram in the sc-plane.

21. Consider the differential equation $dx/dt = kx - x^3$. **(a)** If $k \leqq 0$, show that the only critical value $c = 0$ of x is stable. **(b)** If $k > 0$, show that the critical point $c = 0$ is now unstable, but that the critical points $c = \pm\sqrt{k}$ are stable. Thus the qualitative nature of the solutions changes at $k = 0$ as the parameter k increases, and so $k = 0$ is a bifurcation point for the differential equation with parameter k. The plot of all points of the form (k, c), where c is a critical point of the equation $x' = kx - x^3$, is the "pitchfork diagram" shown in Fig. 2.2.13.

22. Consider the differential equation $dx/dt = x + kx^3$ containing the parameter k. Analyze (as in Problem 21) the dependence of the number and nature of the critical points on the value of k, and construct the corresponding bifurcation diagram.

23. Suppose that the logistic equation $dx/dt = kx(M - x)$ models a population $x(t)$ of fish in a lake after t months during which no fishing occurs. Now suppose that, because of fishing, fish are removed from the lake at the rate of hx fish per month (with h a positive constant). Thus fish are "harvested" at a rate proportional to the existing fish population, rather than at the constant rate of Example 4. **(a)** If $0 < h < kM$, show that the population is still logistic. What is the new limiting population? **(b)** If $h \geqq kM$, show that $x(t) \to 0$ are $t \to +\infty$, so the lake is eventually fished out.

24. Separate variables in the logistic harvesting equation $dx/dt = k(N - x)(x - H)$ and then use partial fractions to derive the solution given in Eq. (15).

25. Use the alternative forms

$$x(t) = \frac{N(x_0 - H) + H(N - x_0)e^{-k(N-H)t}}{(x_0 - H) + (N - x_0)e^{-k(N-H)t}}$$

$$= \frac{H(N - x_0)e^{-k(N-H)t} - N(H - x_0)}{(N - x_0)e^{-k(N-H)t} - (H - x_0)}$$

of the solution in (15) to establish the conclusions stated in (17) and (18).

Example 4 dealt with the case $4h > kM^2$ in the equation $dx/dt = kx(M - x) - h$ that describes constant-rate harvesting of a logistic population. Problems 26 and 27 deal with the other cases.

26. If $4h = kM^2$, show that typical solution curves look as illustrated in Fig. 2.2.14. Thus if $x_0 \geqq M/2$, then $x(t) \to M/2$ as $t \to +\infty$. But if $x_0 < M/2$, then $x(t) = 0$ after a finite period of time, so the lake is fished out. The critical point $x = M/2$ might be called *semistable,* because it looks stable from one side, unstable from the other.

27. If $4h > kM^2$, show that $x(t) = 0$ after a finite period of time, so the lake is fished out (whatever the initial population). [*Suggestion*: Complete the square to rewrite the differential equation in the form $dx/dt = -k[(x - a)^2 + b^2]$. Then solve explicitly by separation of variables.] The results of this and the previous problem (together with Example 4) show that $h = \frac{1}{4}kM^2$ is a critical harvesting rate for a logistic population. At any lesser harvesting rate the population approaches a limiting population N that is less than M (why?), whereas at any greater harvesting rate the population reaches extinction.

28. This problem deals with the differential equation $dx/dt = kx(x - M) - h$ that models the harvesting of an unsophisticated population (such as alligators). Show that this equation can be rewritten in the form $dx/dt = k(x - H)(x - K)$, where

$$H = \tfrac{1}{2}\left(M + \sqrt{M^2 + 4h/k}\right) > 0,$$

$$K = \tfrac{1}{2}\left(M - \sqrt{M^2 + 4h/k}\right) < 0.$$

Show that typical solution curves look as illustrated in Fig. 2.2.15.

29. Consider the two differential equations

$$\frac{dx}{dt} = (x - a)(x - b)(x - c) \tag{21}$$

and

$$\frac{dx}{dt} = (a - x)(b - x)(c - x), \tag{22}$$

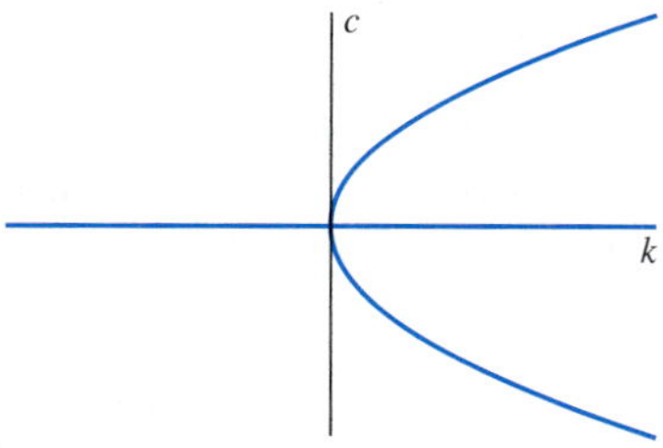

FIGURE 2.2.13. Bifurcation diagram for $dx/dt = kx - x^3$.

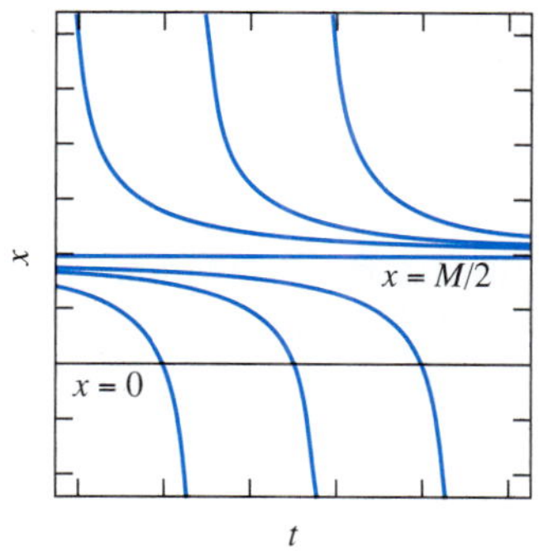

FIGURE 2.2.14. Solution curves for harvesting a logistic population with $4h = kM^2$.

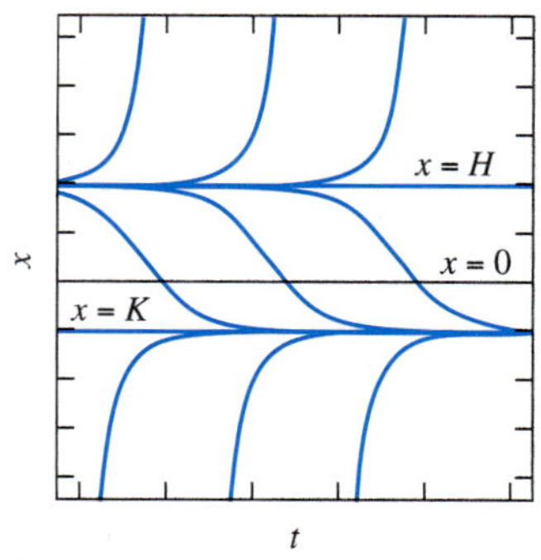

FIGURE 2.2.15. Solution curves for harvesting a population of alligators.

each having the critical points a, b, and c; suppose that $a < b < c$. For one of these equations, only the critical point b is stable; for the other equation, b is the only unstable critical point. Construct phase diagrams for the two equations to determine which is which. Without attempting to solve either equation explicitly, make rough sketches of typical solution curves for each. You should see two funnels and a spout in one case, two spouts and a funnel in the other.

2.3 Acceleration–Velocity Models

In Section 1.2 we discussed vertical motion of a mass m near the surface of the earth under the influence of constant gravitational acceleration. If we neglect any effects of air resistance, then Newton's second law ($F = ma$) implies that the velocity v of the mass m satisfies the equation

$$m\frac{dv}{dt} = F_G, \tag{1}$$

where $F_G = -mg$ is the (downward-directed) force of gravity, where the gravitational acceleration is $g \approx 9.8\ \text{m/s}^2$ (in mks units; $g \approx 32\ \text{ft/s}^2$ in fps units).

Example 1 Suppose that a crossbow bolt is shot straight upward from the ground ($y_0 = 0$) with initial velocity $v_0 = 49$ (m/s). Then Eq. (1) with $g = 9.8$ gives

$$\frac{dv}{dt} = -9.8, \quad \text{so} \quad v(t) = -(9.8)t + v_0 = -(9.8)t + 49.$$

Hence the bolt's height function $y(t)$ is given by

$$y(t) = \int [-(9.8)t + 49]\,dt = -(4.9)t^2 + 49t + y_0 = -(4.9)t^2 + 49t.$$

The bolt reaches its maximum height when $v = -(9.8)t + 49 = 0$, hence when $t = 5$ (s). Thus its maximum height is

$$y_{\text{max}} = y(5) = -(4.9)(5^2) + (49)(5) = 122.5\ \text{(m)}.$$

The bolt returns to the ground when $y = -(4.9)t(t - 10) = 0$, and thus after 10 seconds aloft. ■

Now we want to take account of air resistance in a problem like Example 1. The force F_R exerted by air resistance on the moving mass m must be added in Eq. (1), so now

$$m\frac{dv}{dt} = F_G + F_R. \tag{2}$$

Newton showed in his *Principia Mathematica* that certain simple physical assumptions imply that F_R is proportional to the *square* of the velocity: $F_R = kv^2$. But empirical investigations indicate that the actual dependence of air resistance on velocity can be quite complicated. For many purposes it suffices to assume that

$$F_R = kv^p,$$

where $1 \leqq p \leqq 2$ and the value of k depends on the size and shape of the body, as well as the density and viscosity of the air. Generally speaking, $p = 1$ for relatively low speeds and $p = 2$ for high speeds, whereas $1 < p < 2$ for intermediate speeds. But how slow "low speed" and how fast "high speed" are depend on the same factors that determine the value of the coefficient k.

Thus air resistance is a complicated physical phenomenon. But the simplifying assumption that F_R is exactly of the form given here, with either $p = 1$ or $p = 2$, yields a tractable mathematical model that exhibits the most important qualitative features of motion with resistance.

Resistance Proportional to Velocity

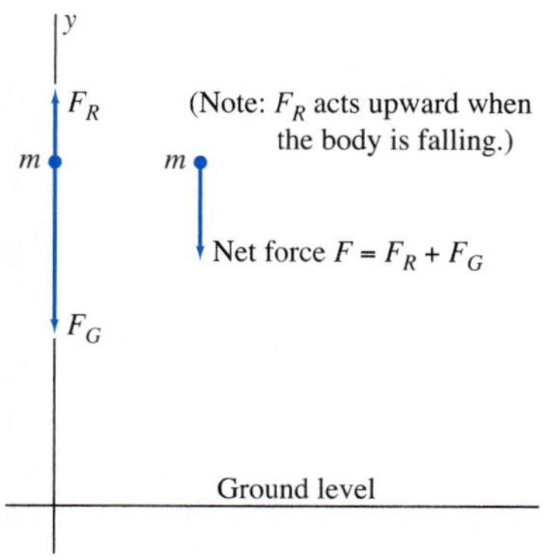

FIGURE 2.3.1. Vertical motion with air resistance.

Let us first consider the vertical motion of a body with mass m near the surface of the earth, subject to two forces: a downward gravitational force F_G and a force F_R of air resistance that is proportional to velocity (so that $p = 1$) and of course directed opposite the direction of motion of the body. If we set up a coordinate system with the positive y-direction upward and with $y = 0$ at ground level, then $F_G = -mg$ and

$$F_R = -kv, \tag{3}$$

where k is a positive constant and $v = dy/dt$ is the velocity of the body. Note that the minus sign in Eq. (3) makes F_R positive (an upward force) if the body is falling (v is negative) and makes F_R negative (a downward force) if the body is rising (v is positive). As indicated in Fig. 2.3.1, the net force acting on the body is then

$$F = F_R + F_G = -kv - mg,$$

and Newton's law of motion $F = m(dv/dt)$ yields the equation

$$m\frac{dv}{dt} = -kv - mg.$$

Thus

$$\frac{dv}{dt} = -\rho v - g, \tag{4}$$

where $\rho = k/m > 0$. You should verify for yourself that if the positive y-axis were directed downward, then Eq. (4) would take the form $dv/dt = -\rho v + g$.

Equation (4) is a separable first-order differential equation, and its solution is

$$v(t) = \left(v_0 + \frac{g}{\rho}\right)e^{-\rho t} - \frac{g}{\rho}. \tag{5}$$

Here, $v_0 = v(0)$ is the initial velocity of the body. Note that

$$v_\tau = \lim_{t\to\infty} v(t) = -\frac{g}{\rho}. \tag{6}$$

Thus the speed of a body falling with air resistance does *not* increase indefinitely; instead, it approaches a *finite* limiting speed, or **terminal speed,**

$$|v_\tau| = \frac{g}{\rho} = \frac{mg}{k}. \tag{7}$$

This fact is what makes a parachute a practical invention; it even helps explain the occasional survival of people who fall without parachutes from high-flying airplanes.

We now rewrite Eq. (5) in the form

$$\frac{dy}{dt} = (v_0 - v_\tau)e^{-\rho t} + v_\tau. \tag{8}$$

Integration gives

$$y(t) = -\frac{1}{\rho}(v_0 - v_\tau)e^{-\rho t} + v_\tau t + C.$$

We substitute 0 for t and let $y_0 = y(0)$ denote the initial height of the body. Thus we find that $C = y_0 + (v_0 - v_\tau)/\rho$, and so

$$y(t) = y_0 + v_\tau t + \frac{1}{\rho}(v_0 - v_\tau)(1 - e^{-\rho t}). \tag{9}$$

Equations (8) and (9) give the velocity v and height y of a body moving vertically under the influence of gravity and air resistance. The formulas depend on the initial height y_0 of the body, its initial velocity v_0, and the *drag coefficient* ρ, the constant such that the acceleration due to air resistance is $a_R = -\rho v$. The two equations also involve the terminal velocity v_τ defined in Eq. (6).

For a person descending with the aid of a parachute, a typical value of ρ is 1.5, which corresponds to a terminal speed of $|v_\tau| \approx 21.3$ ft/s, or about 14.5 mi/h. With an unbuttoned overcoat flapping in the wind in place of a parachute, an unlucky skydiver might increase ρ to perhaps as much as 0.5, which gives a terminal speed of $|v_\tau| \approx 65$ ft/s, about 44 mi/h. See Problems 10 and 11 for some parachute-jump computations.

Example 2 We again consider a bolt shot straight upward with initial velocity $v_0 = 49$ m/s from a crossbow at ground level. But now we take air resistance into account, with $\rho = 0.04$ in Eq. (4). We ask how the resulting maximum height and time aloft compare with the values found in Example 1.

Solution We substitute $y_0 = 0$, $v_0 = 49$, and $v_\tau = -g/\rho = -245$ in Eqs. (5) and (9), and obtain

$$v(t) = 294e^{-t/25} - 245,$$
$$y(t) = 7350 - 245t - 7350e^{-t/25}.$$

To find the time required for the bolt to reach its maximum height (when $v = 0$), we solve the equation

$$v(t) = 294e^{-t/25} - 245 = 0$$

for $t_m = 25\ln(294/245) \approx 4.558$ (s). Its maximum height is then $y_{\max} = v(t_m) \approx 108.280$ meters (as opposed to 122.5 meters without air resistance). To find when the bolt strikes the ground, we must solve the equation

$$y(t) = 7350 - 245t - 7350e^{-t/25} = 0.$$

Using Newton's method, we can begin with the initial guess $t_0 = 10$ and carry out the iteration $t_{n+1} = t_n - y(t_n)/y'(t_n)$ to generate successive approximations to the root. Or we can simply use the **`Solve`** command on a calculator or computer. We find that the bolt is in the air for $t_f \approx 9.411$ seconds (as opposed to 10 seconds without air resistance). It hits the ground with a reduced speed of $|v(t_f)| \approx 43.227$ m/s (as opposed to its initial velocity of 49 m/s).

Thus the effect of air resistance is to decrease the bolt's maximum height, the total time spent aloft, and its final impact speed. Note also that the bolt now spends more time in descent ($t_f - t_m \approx 4.853$ s) than in ascent ($t_m \approx 4.558$ s). ■

Resistance Proportional to Square of Velocity

Now we assume that the force of air resistance is proportional to the *square* of the velocity:

$$F_R = \pm kv^2, \tag{10}$$

with $k > 0$. The choice of signs here depends on the direction of motion, which the force of resistance always opposes. Taking the positive y-direction as upward, $F_R < 0$ for upward motion (when $v > 0$) while $F_R > 0$ for downward motion (when $v < 0$). Thus the sign of F_R is always opposite that of v, so we can rewrite Eq. (10) as

$$F_R = -kv|v|. \tag{10$'$}$$

Then Newton's second law gives

$$m\frac{dv}{dt} = F_G + F_R = -mg - kv|v|;$$

that is,

$$\frac{dv}{dt} = -g - \rho v|v|, \tag{11}$$

where $\rho = k/m > 0$. We must discuss the cases of upward and downward motion separately.

UPWARD MOTION: Suppose that a projectile is launched straight upward from the initial position y_0 with initial velocity $v_0 > 0$. Then Eq. (11) with $v > 0$ gives the differential equation

$$\frac{dv}{dt} = -g - \rho v^2 = -g\left(1 + \frac{\rho}{g}v^2\right). \tag{12}$$

In Problem 13 we ask you to make the substitution $u = v\sqrt{\rho/g}$ and apply the familiar integral

$$\int \frac{1}{1+u^2}\,du = \tan^{-1} u + C$$

to derive the projectile's velocity function

$$v(t) = \sqrt{\frac{g}{\rho}}\tan\left(C_1 - t\sqrt{\rho g}\right) \quad \text{with} \quad C_1 = \tan^{-1}\left(v_0\sqrt{\frac{\rho}{g}}\right). \tag{13}$$

Because $\int \tan u\,du = -\ln|\cos u| + C$, a second integration (see Problem 14) yields the position function

$$y(t) = y_0 + \frac{1}{\rho}\ln\left|\frac{\cos\left(C_1 - t\sqrt{\rho g}\right)}{\cos C_1}\right|. \tag{14}$$

DOWNWARD MOTION: Suppose that a projectile is launched (or dropped) straight downward from the initial position y_0 with initial velocity $v_0 \leqq 0$. Then Eq. (11) with $v < 0$ gives the differential equation

$$\frac{dv}{dt} = -g + \rho v^2 = -g\left(1 - \frac{\rho}{g}v^2\right). \tag{15}$$

In Problem 15 we ask you to make the substitution $u = v\sqrt{\rho/g}$ and apply the integral

$$\int \frac{1}{1-u^2}\,du = \tanh^{-1} u + C$$

to derive the projectile's velocity function

$$v(t) = \sqrt{\frac{g}{\rho}}\tanh\left(C_2 - t\sqrt{\rho g}\right) \quad \text{with} \quad C_2 = \tanh^{-1}\left(v_0\sqrt{\frac{\rho}{g}}\right). \tag{16}$$

Because $\int \tanh u\,du = \ln|\cosh u| + C$, another integration (Problem 16) yields the position function

$$y(t) = y_0 - \frac{1}{\rho}\ln\left|\frac{\cosh\left(C_2 - t\sqrt{\rho g}\right)}{\cosh C_2}\right|. \tag{17}$$

(Note the analogy between Eqs. (16) and (17) and Eqs. (13) and (14) for upward motion.)

If $v_0 = 0$, then $C_2 = 0$, so $v(t) = -\sqrt{g/\rho}\tanh\left(t\sqrt{\rho g}\right)$. Because

$$\lim_{x\to\infty}\tanh x = \lim_{x\to\infty}\frac{\sinh x}{\cosh x} = \lim_{x\to\infty}\frac{\frac{1}{2}(e^x - e^{-x})}{\frac{1}{2}(e^x + e^{-x})} = 1,$$

it follows that in the case of downward motion the body approaches the terminal speed

$$|v_\tau| = \sqrt{\frac{g}{\rho}} \tag{18}$$

(as compared with $|v_\tau| = g/\rho$ in the case of downward motion with linear resistance described by Eq. (4)).

Example 3 We consider once more a bolt shot straight upward with initial velocity $v_0 = 49$ m/s from a crossbow at ground level, as in Example 2. But now we assume air resistance proportional to the square of the velocity, with $\rho = 0.0011$ in Eqs. (12) and (15). In Problems 17 and 18 we ask you to verify the entries in the last line of the following table.

Air Resistance	Maximum Height (ft)	Time Aloft (s)	Ascent Time (s)	Descent Time (s)	Impact Speed (ft/s)
0.0	122.5	10	5	5	49
$(0.04)v$	108.28	9.41	4.56	4.85	43.23
$(0.0011)v^2$	108.47	9.41	4.61	4.80	43.49

Comparison of the last two lines of data here indicates little difference—for the motion of our crossbow bolt—between linear air resistance and air resistance proportional to the square of the velocity. And in Fig. 2.3.2, where the corresponding

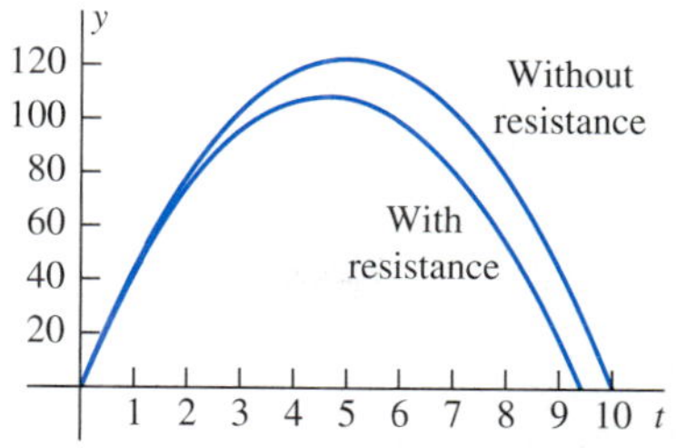

FIGURE 2.3.2. The height functions in Example 1 (without air resistance), Example 2 (with linear air resistance), and Example 3 (with air resistance proportional to the square of the velocity) are all plotted. The graphs of the latter two are visually indistinguishable.

height functions are graphed, the difference is hardly visible. However, the difference between linear and nonlinear resistance can be significant in more complex situations—such as, for instance, the atmospheric reentry and descent of a space vehicle. ■

Variable Gravitational Acceleration

Unless a projectile in vertical motion remains in the immediate vicinity of the earth's surface, the gravitational acceleration acting on it is not constant. According to Newton's law of gravitation, the gravitational force of attraction between two point masses M and m located at a distance r apart is given by

$$F = \frac{GMm}{r^2}, \tag{19}$$

where G is a certain empirical constant ($G \approx 6.6726 \times 10^{-11}$ N·(m/kg)^2 in mks units). The formula is also valid if either or both of the two masses are homogeneous spheres; in this case, the distance r is measured between the centers of the spheres.

The following example is similar to Example 2 in Section 1.2, but now we take account of lunar gravity.

Example 4 A lunar lander is free-falling toward the moon, and at an altitude of 53 kilometers above the lunar surface its downward velocity is measured at 1477 km/h. Its retrorockets, when fired in free space, provide a deceleration of $T = 4$ m/s^2. At what height above the lunar surface should the retrorockets be activated to ensure a "soft touchdown" ($v = 0$ at impact)?

Solution Let $r(t)$ denote the lander's distance from the center of the moon at time t (Fig. 2.3.3). When we combine the (constant) thrust acceleration T and the (negative) lunar acceleration $F/m = GM/r^2$ of Eq. (19), we get the (acceleration) differential equation

$$\frac{d^2r}{dt^2} = T - \frac{GM}{r^2}, \tag{20}$$

where $M = 7.35 \times 10^{22}$ (kg) is the mass of the moon, which has a radius of $R = 1.74 \times 10^6$ meters (or 1740 km, a little over a quarter of the earth's radius). Noting that this second-order differential equation does not involve the independent variable t, we substitute

$$v = \frac{dr}{dt}, \quad \frac{d^2r}{dt^2} = \frac{dv}{dt} = \frac{dv}{dr} \cdot \frac{dr}{dt} = v\frac{dv}{dr}$$

(as in Eq. (36) of Section 1.6) and obtain the first-order equation

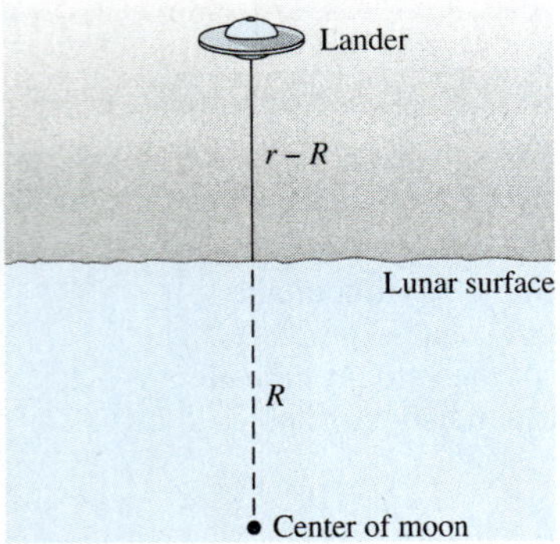

FIGURE 2.3.3. The lunar lander descending to the surface of the moon.

$$v\frac{dv}{dr} = T - \frac{GM}{r^2}$$

with the new independent variable r. Integration with respect to r now yields the equation

$$\frac{1}{2}v^2 = Tr + \frac{GM}{r} + C \tag{21}$$

that we can apply both before ignition ($T = 0$) and after ignition ($T = 4$).

Before ignition: Substitution of $T = 0$ in (21) gives the equation

$$\frac{1}{2}v^2 = \frac{GM}{r} + C_1, \tag{21a}$$

where the constant is given by $C_1 = v_0^2/2 - GM/r_0$ with

$$v_0 = -1477\,\frac{\text{km}}{\text{h}} \times 1000\,\frac{\text{m}}{\text{km}} \times \frac{1\text{ h}}{3600\text{ s}} = -\frac{14770}{36}\,\frac{\text{m}}{\text{s}}$$

and $r_0 = (1.74 \times 10^6) + 53{,}000 = 1.793 \times 10^6$ m (from the initial velocity–position measurement).

After ignition: Substitution of $T = 4$ and $v = 0$, $r = R$ (at touchdown) into (21) gives

$$\frac{1}{2}v^2 = 4r + \frac{GM}{r} + C_2 \tag{21b}$$

where the constant $C_2 = -4R - GM/R$ is obtained by substituting the values $v = 0$, $r = R$ at touchdown.

At the instant of ignition the lunar lander's position and velocity satisfy both (21a) and (21b). Therefore, we can find its desired height h above the lunar surface at ignition by equating the right-hand sides in (21a) and (21b). This gives $r = \frac{1}{4}(C_1 - C_2) = 1.78187 \times 10^6$ and finally $h = r - R = 41{,}870$ meters (that is, 41.87 kilometers—just over 26 miles). Moreover, substitution of this value of r in (21a) gives the velocity $v = -450$ m/s at the instant of ignition. ■

Escape Velocity

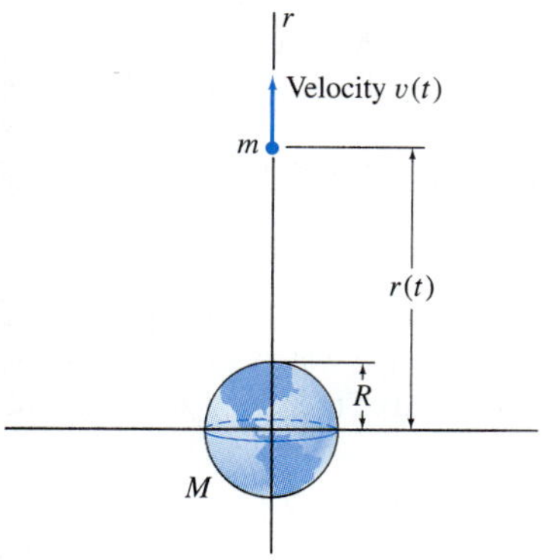

FIGURE 2.3.4. A mass m at a great distance from the earth.

In his novel *From the Earth to the Moon* (1865), Jules Verne raised the question of the initial velocity necessary for a projectile fired from the surface of the earth to reach the moon. Similarly, we can ask what initial velocity v_0 is necessary for the projectile to escape from the earth altogether. This will be so if its velocity $v = dr/dt$ remains *positive* for all $t > 0$, so it continues forever to move away from the earth. With $r(t)$ denoting the projectile's distance from the earth's center at time t (Fig. 2.3.4), we have the equation

$$\frac{dv}{dt} = \frac{d^2r}{dt^2} = -\frac{GM}{r^2}, \tag{22}$$

similar to Eq. (20), but with $T = 0$ (no thrust) and with $M = 5.975 \times 10^{24}$ (kg) denoting the mass of the earth, which has an equatorial radius of $R = 6.378 \times 10^6$ (m). Substitution of the chain rule expression $dv/dt = v(dv/dr)$ as in Example 4 gives

$$v\frac{dv}{dr} = -\frac{GM}{r^2}.$$

Then integration of both sides with respect to r yields

$$\frac{1}{2}v^2 = \frac{GM}{r} + C.$$

Now $v = v_0$ and $r = R$ when $t = 0$, so $C = \frac{1}{2}v_0^2 - GM/R$, and hence solution for v^2 gives

$$v^2 = v_0^2 + 2GM\left(\frac{1}{r} - \frac{1}{R}\right). \tag{23}$$

This implicit solution of Eq. (22) determines the projectile's velocity v as a function of its distance r from the earth's center. In particular,

$$v^2 > v_0^2 - \frac{2GM}{R},$$

so v will remain positive provided that $v_0^2 \geqq 2GM/R$. Therefore, the **escape velocity** from the earth is given by

$$v_0 = \sqrt{\frac{2GM}{R}}. \tag{24}$$

In Problem 27 we ask you to show that, if the projectile's initial velocity exceeds $\sqrt{2GM/R}$, then $r(t) \to \infty$ as $t \to \infty$, so it does, indeed, "escape" from the earth. With the given values of G and the earth's mass M and radius R, this gives $v_0 \approx 11{,}180$ (m/s) (about 36,680 ft/s, about 6.95 mi/s, about 25,000 mi/h).

Remark: Equation (24) gives the escape velocity for any other (spherical) planetary body when we use *its* mass and radius. For instance, when we use the mass M and radius R for the moon given in Example 4, we find that escape velocity from the lunar surface is $v_0 \approx 2375$ m/s. This is just over one-fifth of the escape velocity from the earth's surface, a fact that greatly facilitates the return trip ("From the Moon to the Earth"). ■

2.3 Problems

1. The acceleration of a Maserati is proportional to the difference between 250 km/h and the velocity of this sports car. If this machine can accelerate from rest to 100 km/h in 10 s, how long will it take for the car to accelerate from rest to 200 km/h?
2. Suppose that a body moves through a resisting medium with resistance proportional to its velocity v, so that $dv/dt = -kv$. **(a)** Show that its velocity and position at time t are given by

$$v(t) = v_0 e^{-kt}$$

and

$$x(t) = x_0 + \left(\frac{v_0}{k}\right)(1 - e^{-kt}).$$

(b) Conclude that the body travels only a finite distance, and find that distance.
3. Suppose that a motorboat is moving at 40 ft/s when its motor suddenly quits, and that 10 s later the boat has slowed to 20 ft/s. Assume, as in Problem 2, that the resistance it encounters while coasting is proportional to its velocity. How far will the boat coast in all?
4. Consider a body that moves horizontally through a medium whose resistance is proportional to the *square* of the velocity v, so that $dv/dt = -kv^2$. Show that

$$v(t) = \frac{v_0}{1 + v_0 kt}$$

and that

$$x(t) = x_0 + \frac{1}{k}\ln(1 + v_0 kt).$$

Note that, in contrast with the result of Problem 2, $x(t) \to +\infty$ as $t \to +\infty$. Which offers less resistance when the body is moving fairly slowly—the medium in this problem or the one in Problem 2? Does your answer seem consistent with the observed behaviors of $x(t)$ as $t \to \infty$?
5. Assuming resistance proportional to the square of the velocity (as in Problem 4), how far does the motorboat of Problem 3 coast in the first minute after its motor quits?
6. Assume that a body moving with velocity v encounters resistance of the form $dv/dt = -kv^{3/2}$. Show that

$$v(t) = \frac{4v_0}{\left(kt\sqrt{v_0} + 2\right)^2}$$

and that

$$x(t) = x_0 + \frac{2}{k}\sqrt{v_0}\left(1 - \frac{2}{kt\sqrt{v_0} + 2}\right).$$

Conclude that under a $\frac{3}{2}$-power resistance a body coasts only a finite distance before coming to a stop.
7. Suppose that a car starts from rest, its engine providing an acceleration of 10 ft/s^2, while air resistance provides 0.1 ft/s^2 of deceleration for each foot per second of the car's velocity. **(a)** Find the car's maximum possible (limiting) velocity. **(b)** Find how long it takes the car to attain 90% of its limiting velocity, and how far it travels while doing so.
8. Rework both parts of Problem 7, with the sole difference that the deceleration due to air resistance now is $(0.001)v^2$ ft/s^2 when the car's velocity is v feet per second.
9. A motorboat weighs 32,000 lb and its motor provides a thrust of 5000 lb. Assume that the water resistance is 100 pounds for each foot per second of the speed v of the boat. Then

$$1000\frac{dv}{dt} = 5000 - 100v.$$

If the boat starts from rest, what is the maximum velocity that it can attain?
10. A woman bails out of an airplane at an altitude of 10,000 ft, falls freely for 20 s, then opens her parachute. How long will it take her to reach the ground? Assume linear air resistance ρv ft/s^2, taking $\rho = 0.15$ without the parachute and $\rho = 1.5$ with the parachute. (*Suggestion*: First determine her height above the ground and velocity when the parachute opens.)
11. According to a newspaper account, a paratrooper survived a training jump from 1200 ft when his parachute failed to open but provided some resistance by flapping unopened in the wind. Allegedly he hit the ground at 100 mi/h after falling for 8 s. Test the accuracy of this account. (*Suggestion*: Find ρ in Eq. (4) by assuming a terminal velocity of 100 mi/h. Then calculate the time required to fall 1200 ft.)
12. It is proposed to dispose of nuclear wastes—in drums with weight $W = 640$ lb and volume 8 ft^3—by dropping them into the ocean ($v_0 = 0$). The force equation for a drum falling through water is

$$m\frac{dv}{dt} = -W + B + F_R,$$

where the buoyant force B is equal to the weight (at 62.5 lb/ft^3) of the volume of water displaced by the drum (Archimedes' principle) and F_R is the force of water resistance, found empirically to be 1 lb for each foot per second of the velocity of a drum. If the drums are likely to burst upon an impact of more than 75 ft/s, what is the maximum depth to which they can be dropped in the ocean without likelihood of bursting?
13. Separate variables in Eq. (12) and substitute $u = v\sqrt{\rho/g}$ to obtain the upward-motion velocity function given in Eq. (13) with initial condition $v(0) = v_0$.
14. Integrate the velocity function in Eq. (13) to obtain the upward-motion position function given in Eq. (14) with initial condition $y(0) = y_0$.
15. Separate variables in Eq. (15) and substitute $u = v\sqrt{\rho/g}$ to obtain the downward-motion velocity function given in Eq. (16) with initial condition $v(0) = v_0$.
16. Integrate the velocity function in Eq. (16) to obtain the downward-motion position function given in Eq. (17) with initial condition $y(0) = y_0$.

17. Consider the crossbow bolt of Example 3, shot straight upward from the ground ($y = 0$) at time $t = 0$ with initial velocity $v_0 = 49$ m/s. Take $g = 9.8$ m/s^2 and $\rho = 0.0011$ in Eq. (12). Then use Eqs. (13) and (14) to show that the bolt reaches its maximum height of about 108.47 m in about 4.61 s.

18. Continuing Problem 17, suppose that the bolt is now dropped ($v_0 = 0$) from a height of $y_0 = 108.47$ m. Then use Eqs. (16) and (17) to show that it hits the ground about 4.80 s later with an impact speed of about 43.49 m/s.

19. A motorboat starts from rest (initial velocity $v(0) = v_0 = 0$). Its motor provides a constant acceleration of 4 ft/s^2, but water resistance causes a deceleration of $v^2/400$ ft/s^2. Find v when $t = 10$ s, and also find the *limiting velocity* as $t \to +\infty$ (that is, the maximum possible speed of the boat).

20. An arrow is shot straight upward from the ground with an initial velocity of 160 ft/s. It experiences both the deceleration of gravity and deceleration $v^2/800$ due to air resistance. How high in the air does it go?

21. If a ball is projected upward from the ground with initial velocity v_0 and resistance proportional to v^2, deduce from Eq. (14) that the maximum height it attains is

$$y_{\max} = \frac{1}{2\rho}\ln\left(1 + \frac{\rho v_0^2}{g}\right).$$

22. Suppose that $\rho = 0.075$ (in fps units, with $g = 32$ ft/s^2) in Eq. (15) for a paratrooper falling with parachute open. If he jumps from an altitude of 10,000 ft and opens his parachute immediately, what will be his terminal speed? How long will it take him to reach the ground?

23. Suppose that the paratrooper of Problem 22 falls freely for 30 s with $\rho = 0.00075$ before opening his parachute. How long will it now take him to reach the ground?

24. The mass of the sun is 329,320 times that of the earth and its radius is 109 times the radius of the earth. **(a)** To what radius (in meters) would the earth have to be compressed in order for it to become a *black hole*—the escape velocity from its surface equal to the velocity $c = 3 \times 10^8$ m/s of light? **(b)** Repeat part (a) with the sun in place of the earth.

25. **(a)** Show that if a projectile is launched straight upward from the surface of the earth with initial velocity v_0 less than escape velocity $\sqrt{2GM/R}$, then the maximum distance from the center of the earth attained by the projectile is

$$r_{\max} = \frac{2GMR}{2GM - Rv_0^2},$$

where M and R are the mass and radius of the earth, respectively. **(b)** With what initial velocity v_0 must such a projectile be launched to yield a maximum altitude of 100 kilometers above the surface of the earth? **(c)** Find the maximum distance from the center of the earth, expressed in terms of earth radii, attained by a projectile launched from the surface of the earth with 90% of escape velocity.

26. Suppose that you are stranded—your rocket engine has failed—on an asteroid of diameter 3 miles, with density equal to that of the earth with radius 3960 miles. If you have enough spring in your legs to jump 4 feet straight up on earth while wearing your space suit, can you blast off from this asteroid using leg power alone?

27. **(a)** Suppose a projectile is launched vertically from the surface $r = R$ of the earth with initial velocity $v_0 = \sqrt{2GM/R}$ so $v_0^2 = k^2/R$, where $k^2 = 2GM$. Then solve the differential equation $dr/dt = k/\sqrt{r}$ (from Eq. (23) in this section) explicitly to deduce that $r(t) \to \infty$ as $t \to \infty$.

(b) If the projectile is launched vertically with initial velocity $v_0 > \sqrt{2GM/R}$, deduce that

$$\frac{dr}{dt} = \sqrt{\frac{k^2}{r} + \alpha} > \frac{k}{\sqrt{r}}.$$

Why does it again follow that $r(t) \to \infty$ as $t \to \infty$?

28. **(a)** Suppose that a body is dropped ($v_0 = 0$) from a distance $r_0 > R$ from the earth's center, so its acceleration is $dv/dt = -GM/r^2$. Ignoring air resistance, show that it reaches the height $r < r_0$ at time

$$t = \sqrt{\frac{r_0}{2GM}}\left(\sqrt{rr_0 - r^2} + r_0\cos^{-1}\sqrt{\frac{r}{r_0}}\right).$$

(*Suggestion*: Substitute $r = r_0\cos^2\theta$ to evaluate $\int\sqrt{r/(r_0 - r)}\,dr$.) **(b)** If a body is dropped from a height of 1000 km above the earth's surface and air resistance is neglected, how long does it take to fall and with what speed will it strike the earth's surface?

29. Suppose that a projectile is fired straight upward from the surface of the earth with initial velocity $v_0 < \sqrt{2GM/R}$. Then its height $y(t)$ above the surface satisfies the initial value problem

$$\frac{d^2y}{dt^2} = -\frac{GM}{(y+R)^2}; \quad y(0) = 0, \quad y'(0) = v_0.$$

Substitute $dv/dt = v(dv/dy)$ and then integrate to obtain

$$v^2 = v_0^2 - \frac{2GMy}{R(R+y)}$$

for the velocity v of the projectile at height y. What maximum altitude does it reach if its initial velocity is 1 km/s?

30. In Jules Verne's original problem, the projectile launched from the surface of the earth is attracted by both the earth and the moon, so its distance $r(t)$ from the center of the earth satisfies the initial value problem

$$\frac{d^2r}{dt^2} = -\frac{GM_e}{r^2} + \frac{GM_m}{(S-r)^2}; \quad r(0) = R, \quad r'(0) = v_0,$$

where M_e and M_m denote the masses of the earth and the moon, respectively; R is the radius of the earth; and $S = 384{,}400$ km is the distance between the centers of the earth and the moon. To reach the moon, the projectile must only just pass the point between the moon and earth where its net acceleration vanishes. Thereafter it is "under the control" of the moon, and falls from there to the lunar surface. Find the *minimal* launch velocity v_0 that suffices for the projectile to make it "From the Earth to the Moon."

2.3 Application Rocket Propulsion

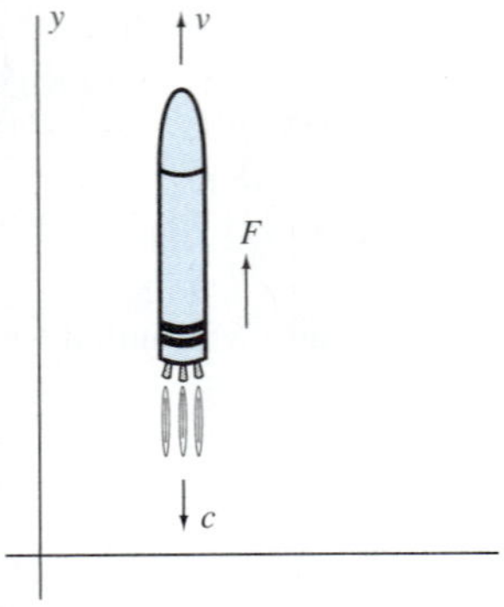

FIGURE 2.3.5. An ascending rocket.

Suppose that the rocket of Fig. 2.3.5 blasts off straight upward from the surface of the earth at time $t = 0$. We want to calculate its height y and velocity $v = dy/dt$ at time t. The rocket is propelled by exhaust gases that exit (rearward) with constant speed c (relative to the rocket). Because of the combustion of its fuel, the mass $m = m(t)$ of the rocket is variable.

To derive the equation of motion of the rocket, we use Newton's second law in the form

$$\frac{dP}{dt} = F, \tag{1}$$

where P is momentum (the product of mass and velocity) and F denotes net external force (gravity, air resistance, etc.). If the mass m of the rocket is constant so $m'(t) \equiv 0$—when its rockets are turned off or burned out, for instance—then Eq. (1) gives

$$F = \frac{d(mv)}{dt} = m\frac{dv}{dt} + \frac{dm}{dt}v = m\frac{dv}{dt},$$

which (with $dv/dt = a$) is the more familiar form $F = ma$ of Newton's second law.

But here m is not constant. Suppose m changes to $m + \Delta m$ and v to $v + \Delta v$ during the short time interval from t to $t + \Delta t$. Then the change in the momentum *of the rocket itself* is

$$\Delta P \approx (m + \Delta m)(v + \Delta v) - mv = m\,\Delta v + v\,\Delta m + \Delta m\,\Delta v.$$

But the system also includes the exhaust gases expelled during this time interval, with mass $-\Delta m$ and approximate velocity $v - c$. Hence the total change in momentum during the time interval Δt is

$$\begin{aligned}\Delta P &\approx (m\,\Delta v + v\,\Delta m + \Delta m\,\Delta v) + (-\Delta m)(v - c)\\ &= m\,\Delta v + c\,\Delta m + \Delta m\,\Delta v.\end{aligned}$$

Now we divide by Δt and take the limit as $\Delta t \to 0$, so $\Delta m \to 0$, assuming continuity of $m(t)$. The substitution of the resulting expression for dP/dt in (1) yields the **rocket propulsion equation**

$$m\frac{dv}{dt} + c\frac{dm}{dt} = F. \tag{2}$$

If $F = F_G + F_R$, where $F_G = -mg$ is a constant force of gravity and $F_R = -kv$ is a force of air resistance proportional to velocity, then Eq. (2) finally gives

$$m\frac{dv}{dt} + c\frac{dm}{dt} = -mg - kv. \tag{3}$$

Constant Thrust

Now suppose that the rocket fuel is consumed at the constant "burn rate" β during the time interval $[0, t_1]$, during which time the mass of the rocket decreases from m_0 to m_1. Thus

$$\begin{aligned} m(0) &= m_0, & m(t_1) &= m_1,\\ m(t) &= m_0 - \beta t, & \frac{dm}{dt} &= -\beta \quad \text{for } t \leq t_1, \end{aligned} \tag{4}$$

with burnout occurring at time $t = t_1$.

PROBLEM 1 Substitute the expressions in (4) into Eq. (3) to obtain the differential equation

$$(m_0 - \beta t)\frac{dv}{dt} + kv = \beta c - (m_0 - \beta t)g. \tag{5}$$

Solve this linear equation for

$$v(t) = v_0 M^{k/\beta} - \frac{g\beta t}{\beta - k} + \left(\frac{\beta c}{k} + \frac{g m_0}{\beta - k}\right)(1 - M^{k/\beta}), \tag{6}$$

where $v_0 = v(0)$ and

$$M = \frac{m(t)}{m_0} = \frac{m_0 - \beta t}{m_0}$$

denotes the rocket's **fractional mass** at time t.

No Resistance

PROBLEM 2 For the case of no air resistance, set $k = 0$ in Eq. (5) and integrate to obtain

$$v(t) = v_0 - gt + c \ln \frac{m_0}{m_0 - \beta t}. \tag{7}$$

Because $m_0 - \beta t_1 = m_1$, it follows that the velocity of the rocket at burnout ($t = t_1$) is

$$v_1 = v(t_1) = v_0 - g t_1 + c \ln \frac{m_0}{m_1}. \tag{8}$$

PROBLEM 3 Start with Eq. (7) and integrate to obtain

$$y(t) = (v_0 + c)t - \frac{1}{2}gt^2 - \frac{c}{\beta}(m_0 - \beta t)\ln \frac{m_0}{m_0 - \beta t}. \tag{9}$$

It follows that the rocket's altitude at burnout is

$$y_1 = y(t_1) = (v_0 + c)t_1 - \frac{1}{2}g t_1^2 - \frac{c m_1}{\beta} \ln \frac{m_0}{m_1}. \tag{10}$$

PROBLEM 4 The V-2 rocket that was used to attack London in World War II had an initial mass of 12,850 kg, of which 68.5% was fuel. This fuel burned uniformly for 70 seconds with an exhaust velocity of 2 km/s. Assume it encounters air resistance of 1.45 N per m/s of velocity. Then find the velocity and altitude of the V-2 at burnout under the assumption that it is launched vertically upward from rest on the ground.

PROBLEM 5 Actually, our basic differential equation in (3) applies without qualification only when the rocket is already in motion. However, when a rocket is sitting on its launch pad stand and its engines are turned on initially, it is observed that a certain time interval passes before the rocket actually "blasts off" and begins to ascend. The reason is that if $v = 0$ in (3), then the resulting initial acceleration

$$\frac{dv}{dt} = \frac{c}{m}\frac{dm}{dt} - g$$

of the rocket may be *negative*. But the rocket does not descend into the ground; it just "sits there" while (because m is decreasing) this calculated acceleration increases until it reaches 0 and (thereafter) positive values so the rocket can begin to ascend. With the notation introduced to describe the constant-thrust case, show that the rocket initially just "sits there" if the exhaust velocity c is less than $m_0 g/\beta$, and that the time t_B which then elapses before actual blastoff is given by

$$t_B = \frac{m_0 g - \beta c}{\beta g}.$$

Free Space

Suppose finally that the rocket is accelerating in free space, where there is neither gravity nor resistance, so $g = k = 0$. With $g = 0$ in Eq. (8) we see that, as the mass of the rocket decreases from m_0 to m_1, its increase in velocity is

$$\Delta v = v_1 - v_0 = c \ln \frac{m_0}{m_1}. \tag{11}$$

Note that Δv depends only on the exhaust gas speed c and the initial-to-final mass ratio m_0/m_1, but does not depend on the burn rate β. For example, if the rocket blasts off from rest ($v_0 = 0$) and $c = 5$ km/s and $m_0/m_1 = 20$, then its velocity at burnout is $v_1 = 5 \ln 20 \approx 15$ km/s. Thus if a rocket initially consists predominantly of fuel, then it can attain velocities significantly greater than the (relative) velocity of its exhaust gases.

2.4 Numerical Approximation: Euler's Method

It is the exception rather than the rule when a differential equation of the general form

$$\frac{dy}{dx} = f(x, y)$$

can be solved exactly and explicitly by elementary methods like those discussed in Chapter 1. For example, consider the simple equation

$$\frac{dy}{dx} = e^{-x^2}. \tag{1}$$

A solution of Eq. (1) is simply an antiderivative of e^{-x^2}. But it is known that every antiderivative of $f(x) = e^{-x^2}$ is a **nonelementary** function—one that cannot be expressed as a finite combination of the familiar functions of elementary calculus. Hence no particular solution of Eq. (1) is finitely expressible in terms of elementary functions. Any attempt to use the symbolic techniques of Chapter 1 to find a simple explicit formula for a solution of (1) is therefore doomed to failure.

As a possible alternative, an old-fashioned computer plotter—one that uses an ink pen to draw curves mechanically—can be programmed to draw a solution curve that starts at the initial point (x_0, y_0) and attempts to thread its way through the slope field of a given differential equation $y' = f(x, y)$. The procedure the plotter carries out can be described as follows.

- The plotter pen starts at the initial point (x_0, y_0) and moves a tiny distance along the slope segment though (x_0, y_0). This takes it to the point (x_1, y_1).
- At (x_1, y_1) the pen changes direction, and now moves a tiny distance along the slope segment through this new starting point (x_1, y_1). This takes it to the next starting point (x_2, y_2).
- At (x_2, y_2) the pen changes direction again, and now moves a tiny distance along the slope segment through (x_2, y_2). This takes it to the next starting point (x_3, y_3).

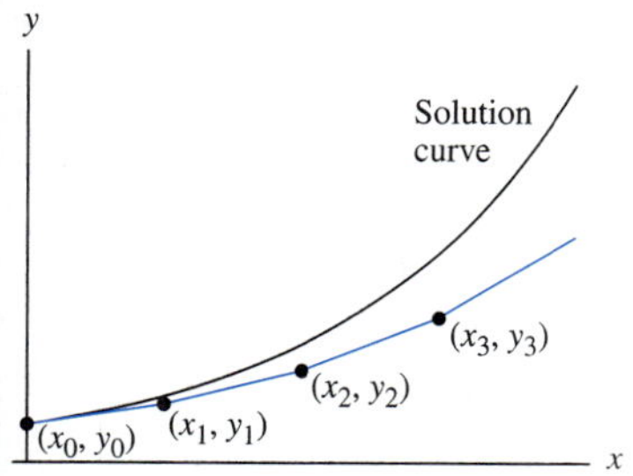

FIGURE 2.4.1. The first few steps in approximating a solution curve.

Figure 2.4.1 illustrates the result of continuing in this fashion—by a sequence of discrete straight-line steps from one starting point to the next. In this figure we see a polygonal curve consisting of line segments that connect the successive points (x_0, y_0), (x_1, y_1), (x_2, y_2), (x_3, y_3), ... However, suppose that each "tiny distance" the pen travels along a slope segment—before the midcourse correction that sends it along a fresh new slope segment—is so very small that the naked eye cannot distinguish the individual line segments constituting the polygonal curve. Then the resulting polygonal curve looks like a smooth, continuously turning solution curve of the differential equation. Indeed, this is (in essence) how most of the solution curves shown in the figures of Chapter 1 were computer generated.

Leonhard Euler—the great 18th-century mathematician for whom so many mathematical concepts, formulas, methods, and results are named—did not have a computer plotter, and his idea was to do all this numerically rather than graphically. In order to approximate the solution of the initial value problem

$$\frac{dy}{dx} = f(x, y), \quad y(x_0) = y_0, \tag{2}$$

we first choose a fixed (horizontal) **step size** h to use in making each step from one point to the next. Suppose we've started at the initial point (x_0, y_0) and after n steps have reached the point (x_n, y_n). Then the step from (x_n, y_n) to the next point (x_{n+1}, y_{n+1}) is illustrated in Fig. 2.4.2. The slope of the direction segment through (x_n, y_n) is $m = f(x_n, y_n)$. Hence a horizontal change of h from x_n to x_{n+1} corresponds to a vertical change of $m \cdot h = h \cdot f(x_n, y_n)$ from y_n to y_{n+1}. Therefore the coordinates of the new point (x_{n+1}, y_{n+1}) are given in terms of the old coordinates by

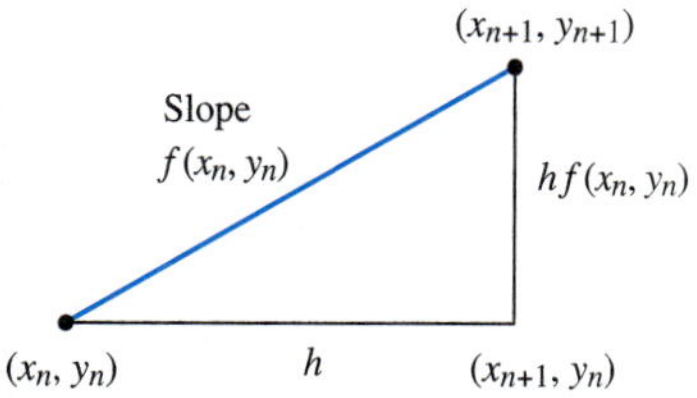

FIGURE 2.4.2. The step from (x_n, y_n) to (x_{n+1}, y_{n+1}).

$$x_{n+1} = x_n + h, \quad y_{n+1} = y_n + h \cdot f(x_n, y_n).$$

Given the initial value problem in (2), **Euler's method** with step size h consists of starting with the initial point (x_0, y_0) and applying the formulas

$$\begin{aligned}
x_1 &= x_0 + h & y_1 &= y_0 + h \cdot f(x_0, y_0) \\
x_2 &= x_1 + h & y_2 &= y_1 + h \cdot f(x_1, y_1) \\
x_3 &= x_2 + h & y_3 &= y_2 + h \cdot f(x_2, y_2) \\
&\vdots & &\vdots \\
&\vdots & &\vdots
\end{aligned}$$

to calculate successive points (x_1, y_1), (x_2, y_2), (x_3, y_3), ... on an approximate solution curve.

However, we ordinarily do not sketch the corresponding polygonal approximation. Instead, the numerical result of applying Euler's method is the sequence of *approximations*

$$y_1, y_2, y_3, \dots, y_n, \dots$$

to the *true values*

$$y(x_1), y(x_2), y(x_3), \dots, y(x_n), \dots$$

at the points $x_1, x_2, x_3, \dots, x_n, \dots$ of the *exact* (though unknown) solution $y(x)$ of the initial value problem. These results typically are presented in the form of a table of approximate values of the desired solution.

ALGORITHM The Euler Method

Given the initial value problem

$$\frac{dy}{dx} = f(x, y), \quad y(x_0) = y_0, \tag{2}$$

Euler's method with step size h consists of applying the iterative formula

$$y_{n+1} = y_n + h \cdot f(x_n, y_n) \quad (n \geq 0) \tag{3}$$

to calculate successive approximations $y_1, y_2, y_3, \dots$ to the (true) values $y(x_1)$, $y(x_2)$, $y(x_3)$, ... of the (exact) solution $y = y(x)$ at the points $x_1, x_2, x_3, \dots$, respectively.

The iterative formula in (3) tells us how to make the typical step from y_n to y_{n+1} and is the heart of Euler's method. Although the most important applications of Euler's method are to nonlinear equations, we first illustrate the method with a simple initial value problem whose exact solution is available, just for the purpose of comparison of approximate and actual solutions.

Example 1 Apply Euler's method to approximate the solution of the initial value problem

$$\frac{dy}{dx} = x + \frac{1}{5}y, \quad y(0) = -3, \tag{4}$$

(a) first with step size $h = 1$ on the interval $[0, 5]$,
(b) then with step size $h = 0.2$ on the interval $[0, 1]$.

Solution **(a)** With $x_0 = 0$, $y_0 = -3$, $f(x, y) = x + \frac{1}{5}y$, and $h = 1$, the iterative formula in (3) yields the approximate values

$$\begin{aligned}
y_1 &= y_0 + h \cdot [x_0 + \tfrac{1}{5}y_0] = (-3) + (1)[0 + \tfrac{1}{5}(-3)] = -3.6,\\
y_2 &= y_1 + h \cdot [x_1 + \tfrac{1}{5}y_1] = (-3.6) + (1)[1 + \tfrac{1}{5}(-3.6)] = -3.32,\\
y_3 &= y_2 + h \cdot [x_2 + \tfrac{1}{5}y_2] = (-3.32) + (1)[2 + \tfrac{1}{5}(-3.32)] = -1.984,\\
y_4 &= y_3 + h \cdot [x_3 + \tfrac{1}{5}y_3] = (-1.984) + (1)[3 + \tfrac{1}{5}(-1.984)] = 0.6192, \quad \text{and}\\
y_5 &= y_4 + h \cdot [x_4 + \tfrac{1}{5}y_4] = (0.6912) + (1)[4 + \tfrac{1}{5}(0.6912)] \approx 4.7430
\end{aligned}$$

at the points $x_1 = 1$, $x_2 = 2$, $x_3 = 3$, $x_4 = 4$, and $x_5 = 5$. Note how the result of each calculation feeds into the next one. The resulting table of approximate values is

x	0	1	2	3	4	5
Approx. y	−3	−3.6	−3.32	−1.984	0.6912	4.7430

Figure 2.4.3 shows the graph of this approximation, together with the graphs of the Euler approximations obtained with step sizes $h = 0.2$ and 0.05, as well as the graph of the exact solution

$$y(x) = 22e^{x/5} - 5x - 25$$

that is readily found using the linear-equation technique of Section 1.5. We see that decreasing the step size increases the accuracy, but with any single approximation, the accuracy decreases with distance from the initial point.

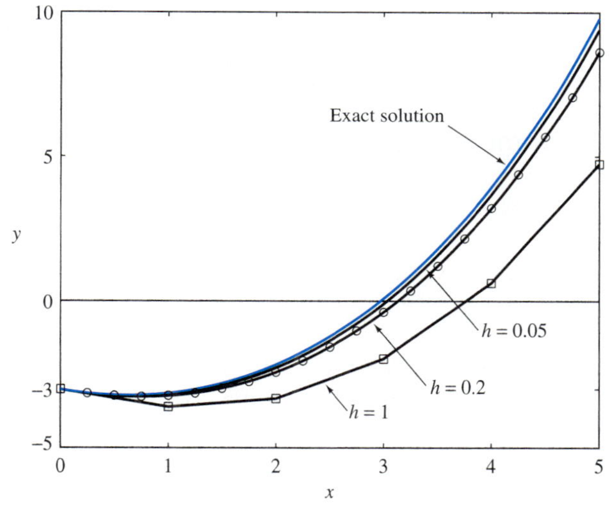

FIGURE 2.4.3. Graphs of Euler approximations with step sizes $h = 1$, $h = 0.2$, and $h = 0.05$.

(b) Starting afresh with $x_0 = 0$, $y_0 = -3$, $f(x, y) = x + \frac{1}{5}y$, and $h = 0.2$, we get the approximate values

$$\begin{aligned}
y_1 &= y_0 + h \cdot [x_0 + \tfrac{1}{5}y_0] = (-3) + (0.2)[0 + \tfrac{1}{5}(-3)] = -3.12,\\
y_2 &= y_1 + h \cdot [x_1 + \tfrac{1}{5}y_1] = (-3.12) + (0.2)[0.2 + \tfrac{1}{5}(-3.12)] \approx -3.205,\\
y_3 &= y_2 + h \cdot [x_2 + \tfrac{1}{5}y_2] \approx (-3.205) + (0.2)[0.4 + \tfrac{1}{5}(-3.205)] \approx -3.253,\\
y_4 &= y_3 + h \cdot [x_3 + \tfrac{1}{5}y_3] \approx (-3.253) + (0.2)[0.6 + \tfrac{1}{5}(-3.253)] \approx -3.263,\\
y_5 &= y_4 + h \cdot [x_4 + \tfrac{1}{5}y_4] \approx (-3.263) + (0.2)[0.8 + \tfrac{1}{5}(-3.263)] \approx -3.234
\end{aligned}$$

at the points $x_1 = 0.2$, $x_2 = 0.4$, $x_3 = 0.6$, $x_4 = 0.8$, and $x_5 = 1$. The resulting table of approximate values is

x	0	0.2	0.4	0.6	0.8	1
Approx. y	−3	−3.12	−3.205	−3.253	−3.263	−3.234

■

High accuracy with Euler's method usually requires a very small step size and hence a larger number of steps than can reasonably be carried out by hand. The application material for this section contains calculator and computer programs for automating Euler's method. One of these programs was used to calculate the table entries shown in Fig. 2.4.4. We see that 500 Euler steps (with step size $h = 0.002$) from $x = 0$ to $x = 1$ yield values that are accurate to within 0.001.

x	Approx y with $h = 0.2$	Approx y with $h = 0.02$	Approx y with $h = 0.002$	Actual value of y
0	−3.000	−3.000	−3.000	−3.000
0.2	−3.120	−3.104	−3.102	−3.102
0.4	−3.205	−3.172	−3.168	−3.168
0.6	−3.253	−3.201	−3.196	−3.195
0.8	−3.263	−3.191	−3.184	−3.183
1	−3.234	−3.140	−3.130	−3.129

FIGURE 2.4.4. Euler approximations with step sizes $h = 0.2$, $h = 0.02$, and $h = 0.002$.

Example 2 Suppose the baseball of Example 3 in Section 1.3 is simply dropped (instead of being thrown downward) from the helicopter. Then its velocity $v(t)$ after t seconds satisfies the initial value problem

$$\frac{dv}{dt} = 32 - 0.16v, \qquad v(0) = 0. \tag{5}$$

We use Euler's method with $h = 1$ to track the ball's increasing velocity at 1-second intervals for the first 10 seconds of fall. With $t_0 = 0$, $v_0 = 0$, $F(t, v) = 32 - 0.16v$, and $h = 1$, the iterative formula in (3) yields the approximate values

$$\begin{aligned}
v_1 &= v_0 + h \cdot [32 - 0.16v_0] = (0) + (1)[32 - 0.16(0)] = 32, \\
v_2 &= v_1 + h \cdot [32 - 0.16v_1] = (32) + (1)[32 - 0.16(32)] = 58.88, \\
v_3 &= v_2 + h \cdot [32 - 0.16v_2] = (58.88) + (1)[32 - 0.16(58.88)] \approx 81.46, \\
v_4 &= v_3 + h \cdot [32 - 0.16v_3] = (81.46) + (1)[32 - 0.16(81.46)] \approx 100.43, \quad \text{and} \\
v_5 &= v_4 + h \cdot [32 - 0.16v_4] = (100.43) + (1)[32 - 0.16(100.43)] \approx 116.36.
\end{aligned}$$

Continuing in this fashion, we complete the $h = 1$ column of v-values shown in the table of Fig. 2.4.5—where we have rounded off velocity entries to the nearest foot per second. The values corresponding to $h = 0.1$ were calculated using a computer, and we see that they are accurate to within about 1 ft/s. Note also that after 10 seconds the falling ball has attained about 80% of its limiting velocity of 200 ft/s. ■

Local and Cumulative Errors

There are several sources of error in Euler's method that may make the approximation y_n to $y(x_n)$ unreliable for large values of n, those for which x_n is not sufficiently close to x_0. The error in the linear approximation formula

$$y(x_{n+1}) \approx y_n + h \cdot f(x_n, y_n) = y_{n+1} \tag{6}$$

t	Approx v with $h = 1$	Approx v with $h = 0.1$	Actual value of v
1	32	30	30
2	59	55	55
3	81	77	76
4	100	95	95
5	116	111	110
6	130	124	123
7	141	135	135
8	150	145	144
9	158	153	153
10	165	160	160

FIGURE 2.4.5. Euler approximations in Example 2 with step sizes $h = 1$ and $h = 0.1$.

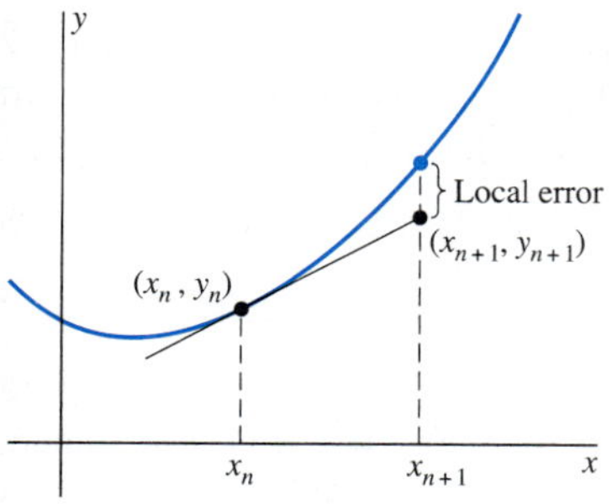

FIGURE 2.4.6. The local error in Euler's method.

is the amount by which the tangent line at (x_n, y_n) departs from the solution curve through (x_n, y_n), as illustrated in Fig. 2.4.6. This error, introduced at each step in the process, is called the **local error** in Euler's method.

The local error indicated in Fig. 2.4.6 *would be* the total error in y_{n+1} *if* the starting point y_n in (6) were an exact value, rather than merely an approximation to the actual value $y(x_n)$. But y_n itself suffers from the accumulated effects of all the local errors introduced at the previous steps. Thus the tangent line in Fig. 2.4.6 is tangent to the "wrong" solution curve—the one through (x_n, y_n) rather than the actual solution curve through the initial point (x_0, y_0). Figure 2.4.7 illustrates this **cumulative error** in Euler's method; it is the amount by which the polygonal stepwise path from (x_0, y_0) departs from the actual solution curve through (x_0, y_0).

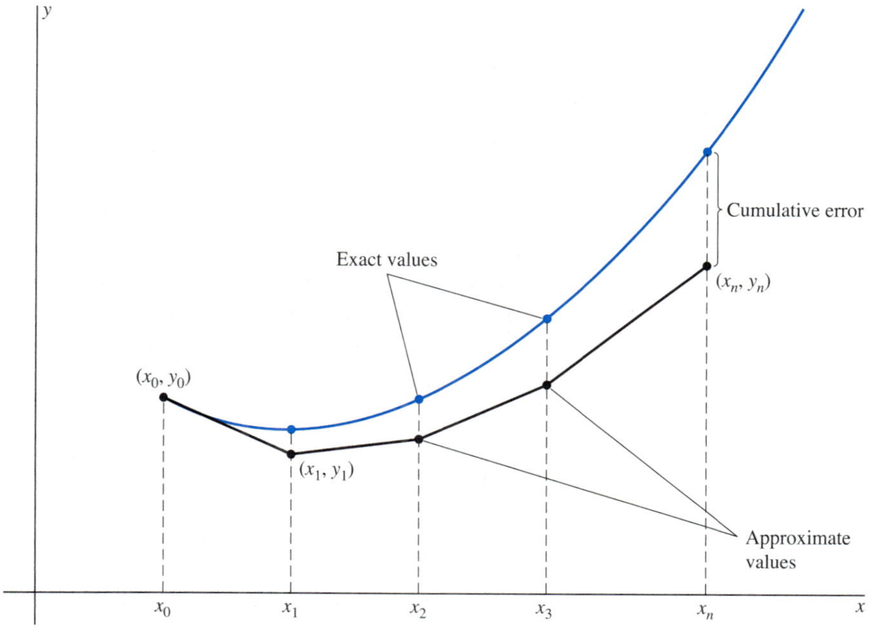

FIGURE 2.4.7. The cumulative error in Euler's method.

x	y with $h = 0.1$	y with $h = 0.02$	y with $h = 0.005$	y with $h = 0.001$	Actual y
0.1	1.1000	1.1082	1.1098	1.1102	1.1103
0.2	1.2200	1.2380	1.2416	1.2426	1.2428
0.3	1.3620	1.3917	1.3977	1.3993	1.3997
0.4	1.5282	1.5719	1.5807	1.5831	1.5836
0.5	1.7210	1.7812	1.7933	1.7966	1.7974
0.6	1.9431	2.0227	2.0388	2.0431	2.0442
0.7	2.1974	2.2998	2.3205	2.3261	2.3275
0.8	2.4872	2.6161	2.6422	2.6493	2.6511
0.9	2.8159	2.9757	3.0082	3.0170	3.0192
1.0	3.1875	3.3832	3.4230	3.4338	3.4366

FIGURE 2.4.8. Approximating the solution of $dy/dx = x + y$, $y(0) = 1$ with successively smaller step sizes.

The usual way of attempting to reduce the cumulative error in Euler's method is to decrease the step size h. The table in Fig. 2.4.8 shows the results obtained in approximating the exact solution $y(x) = 2e^x - x - 1$ of the initial value problem

$$\frac{dy}{dx} = x + y, \quad y(0) = 1,$$

using the successively smaller step sizes $h = 0.1$, $h = 0.02$, $h = 0.005$, and $h = 0.001$. We show computed values only at intervals of $\Delta x = 0.1$. For instance, with $h = 0.001$, the computation required 1000 Euler steps, but the value y_n is shown only when n is a multiple of 100, so that x_n is an integral multiple of 0.1.

By scanning the columns in Fig. 2.4.8, we observe that, for each fixed step size h, the error $y_{\text{actual}} - y_{\text{approx}}$ increases as x gets farther from the starting point $x_0 = 0$. But by scanning the rows of the table we see that for each fixed x, the error decreases as the step size h is reduced. The percentage errors at the final point $x = 1$ range from 7.25% with $h = 0.1$ down to only 0.08% with $h = 0.001$. Thus the smaller the step size, the more slowly does the error grow with increasing distance from the starting point.

The column of data for $h = 0.1$ in Fig. 2.4.8 requires only 10 steps, so Euler's method can be carried out with a hand-held calculator. But 50 steps are required to reach $x = 1$ with $h = 0.02$, 200 steps with $h = 0.005$, and 1000 steps with $h = 0.001$. A computer is almost always used to implement Euler's method when more than 10 or 20 steps are required. Once an appropriate computer program has been written, one step size is—in principle—just as convenient as another; after all, the computer hardly cares how many steps it is asked to carry out.

Why, then, do we not simply choose an exceedingly small step size (such as $h = 10^{-12}$), with the expectation that very great accuracy will result? There are two reasons for not doing so. The first is obvious: the time required for the computation. For example, the data in Fig. 2.4.8 were obtained using a hand-held calculator that carried out nine Euler steps per second. Thus it required slightly over one second to approximate $y(1)$ with $h = 0.1$ and about 1 min 50 s with $h = 0.001$. But with $h = 10^{-12}$ it would require over 3000 years!

The second reason is more subtle. In addition to the local and cumulative errors discussed previously, the computer itself will contribute **roundoff error** at each stage because only finitely many significant digits can be used in each calculation.

An Euler's method computation with $h = 0.0001$ will introduce roundoff errors 1000 times as often as one with $h = 0.1$. Hence with certain differential equations, $h = 0.1$ might actually produce more accurate results than those obtained with $h = 0.0001$, because the cumulative effect of roundoff error in the latter case might exceed combined cumulative and roundoff error in the case $h = 0.1$.

The "best" choice of h is difficult to determine in practice as well as in theory. It depends on the nature of the function $f(x, y)$ in the initial value problem in (2), on the exact code in which the program is written, and on the specific computer used. With a step size that is too large, the approximations inherent in Euler's method may not be sufficiently accurate, whereas if h is too small, then roundoff errors may accumulate to an unacceptable degree or the program may require too much time to be practical. The subject of *error propagation* in numerical algorithms is treated in numerical analysis courses and textbooks.

The computations in Fig. 2.4.8 illustrate the common strategy of applying a numerical algorithm, such as Euler's method, several times in succession, beginning with a selected number n of subintervals for the first application, then doubling n for each succeeding application of the method. Visual comparison of successive results often can provide an "intuitive feel" for their accuracy. In the next two examples we present graphically the results of successive applications of Euler's method.

Example 3 The exact solution of the logistic initial value problem

$$\frac{dy}{dx} = \tfrac{1}{3}y(8 - y), \quad y(0) = 1$$

is $y(x) = 8/(1 + 7e^{-8x/3})$. Figure 2.4.9 shows both the exact solution curve and approximate solution curves obtained by applying Euler's method on the interval $0 \leqq x \leqq 5$ with $n = 5$, $n = 10$, and $n = 20$ subintervals. Each of these "curves" actually consists of line segments joining successive points (x_n, y_n) and (x_{n+1}, y_{n+1}). The Euler approximation with 5 subintervals is poor, and the approximation with 10 subintervals also overshoots the limiting value $y = 8$ of the solution before leveling off, but with 20 subintervals we obtain fairly good qualitative agreement with the actual behavior of the solution. ■

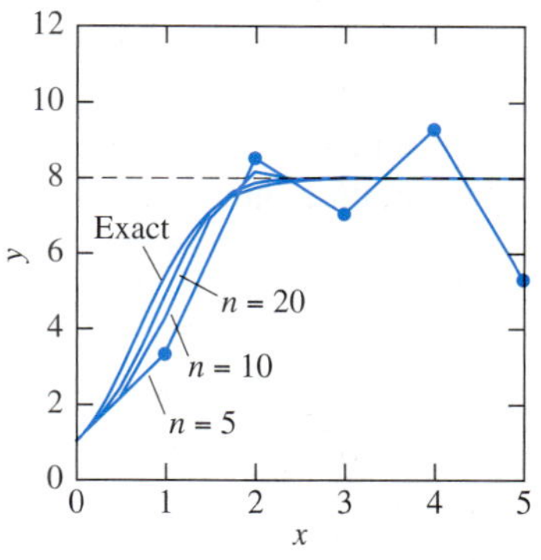

FIGURE 2.4.9. Approximating a logistic solution using Euler's method with $n = 5$, $n = 10$, and $n = 20$ subintervals.

Example 4 The exact solution of the initial value problem

$$\frac{dy}{dx} = y \cos x, \quad y(0) = 1$$

is the periodic function $y(x) = e^{\sin x}$. Figure 2.4.10 shows both the exact solution curve and approximate solution curves obtained by applying Euler's method on the interval $0 \leqq x \leqq 6\pi$ with $n = 50$, $n = 100$, $n = 200$, and $n = 400$ subintervals. Even with this many subintervals, Euler's method evidently has considerable difficulty keeping up with the oscillations in the actual solution. Consequently, the more accurate methods discussed in succeeding sections are needed for serious numerical investigations. ■

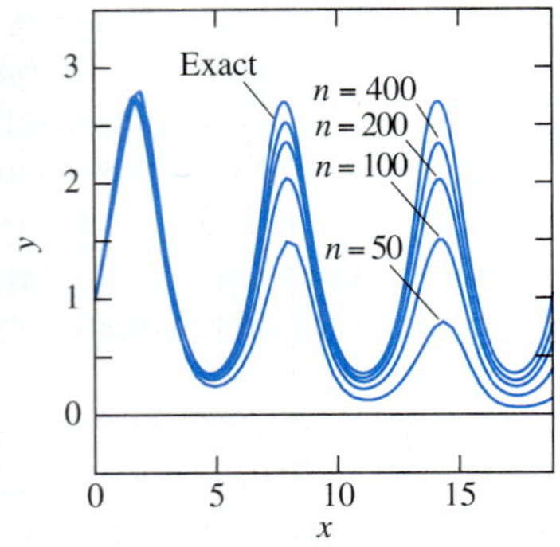

FIGURE 2.4.10. Approximating the exact solution $y = e^{\sin x}$ using Euler's method with 50, 100, 200, and 400 subintervals.

A Word of Caution

The data shown in Fig. 2.4.8 indicate that Euler's method works well in approximating the solution of $dy/dx = x + y$, $y(0) = 1$ on the interval [0, 1]. That is, for each fixed x it appears that the approximate values approach the actual value of $y(x)$ as the step size h is decreased. For instance, the approximate values in the rows corresponding to $x = 0.3$ and $x = 0.5$ suggest that $y(0.3) \approx 1.40$ and $y(0.5) \approx 1.80$, in accord with the actual values shown in the final column of the table.

Example 5, in contrast, shows that some initial value problems are not so well behaved.

Example 5 Use Euler's method to approximate the solution of the initial value problem

$$\frac{dy}{dx} = x^2 + y^2, \quad y(0) = 1 \tag{7}$$

on the interval [0, 1].

Solution Here $f(x, y) = x^2 + y^2$, so the iterative formula of Euler's method is

$$y_{n+1} = y_n + h \cdot (x_n^2 + y_n^2). \tag{8}$$

With step size $h = 0.1$ we obtain

$$\begin{aligned} y_1 &= 1 + (0.1) \cdot [(0)^2 + (1)^2] = 1.1, \\ y_2 &= 1.1 + (0.1) \cdot [(0.1)^2 + (1.1)^2] = 1.222, \\ y_3 &= 1.222 + (0.1) \cdot [(0.2)^2 + (1.222)^2] \approx 1.3753, \end{aligned}$$

and so forth. Rounded to four decimal places, the first ten values obtained in this manner are

$$\begin{aligned} y_1 &= 1.1000 & y_6 &= 2.1995 \\ y_2 &= 1.2220 & y_7 &= 2.7193 \\ y_3 &= 1.3753 & y_8 &= 3.5078 \\ y_4 &= 1.5735 & y_9 &= 4.8023 \\ y_5 &= 1.8371 & y_{10} &= 7.1895. \end{aligned}$$

But instead of naively accepting these results as accurate approximations, we decided to use a computer to repeat the computations with smaller values of h. The

x	y with $h = 0.1$	y with $h = 0.02$	y with $h = 0.005$
0.1	1.1000	1.1088	1.1108
0.2	1.2220	1.2458	1.2512
0.3	1.3753	1.4243	1.4357
0.4	1.5735	1.6658	1.6882
0.5	1.8371	2.0074	2.0512
0.6	2.1995	2.5201	2.6104
0.7	2.7193	3.3612	3.5706
0.8	3.5078	4.9601	5.5763
0.9	4.8023	9.0000	12.2061
1.0	7.1895	30.9167	1502.2090

FIGURE 2.4.11. Attempting to approximate the solution of $dy/dx = x^2 + y^2$, $y(0) = 1$.

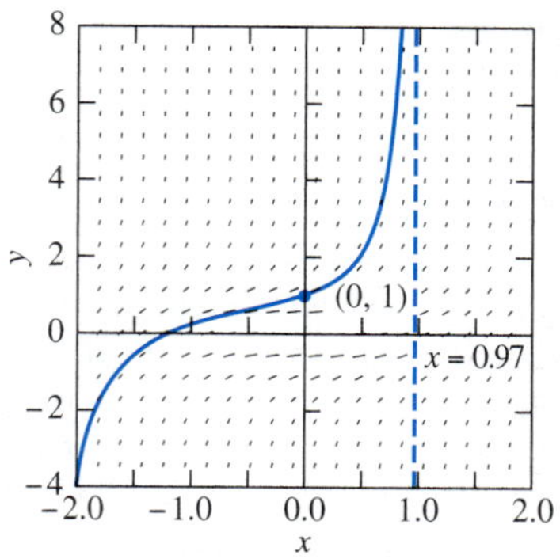

FIGURE 2.4.12. Solution of $dy/dx = x^2 + y^2$, $y(0) = 1$.

table in Fig. 2.4.11 shows the results obtained with step sizes $h = 0.1$, $h = 0.02$, and $h = 0.005$. Observe that now the "stability" of the procedure in Example 1 is missing. Indeed, it seems obvious that something is going wrong near $x = 1$.

Figure 2.4.12 provides a graphical clue to the difficulty. It shows a slope field for $dy/dx = x^2 + y^2$, together with a solution curve through (0, 1) plotted using one of the more accurate approximation methods of the following two sections. It appears that this solution curve may have a vertical asymptote near $x = 0.97$. Indeed, an exact solution using Bessel functions (see Problem 34 in Section 11.4) can be used to show that $y(x) \to +\infty$ as $x \to 0.969811$ (approximately). Although Euler's method gives values (albeit spurious ones) at $x = 1$, the actual solution does not exist on the entire interval [0, 1]. Moreover, Euler's method is unable to "keep up" with the rapid changes in $y(x)$ that occur as x approaches the infinite discontinuity near 0.969811. ■

The moral of Example 5 is that there are pitfalls in the numerical solution of certain initial value problems. Certainly it's pointless to attempt to approximate a solution on an interval where it doesn't even exist (or where it is not unique, in which case there's no general way to predict which way the numerical approximations will branch at a point of nonuniqueness). One should never accept as accurate the results of applying Euler's method with a single fixed step size h. A second "run" with smaller step size ($h/2$, say, or $h/5$, or $h/10$) may give seemingly consistent results, thereby suggesting their accuracy, or it may—as in Example 5—reveal the presence of some hidden difficulty in the problem. Many problems simply require the more accurate and powerful methods that are discussed in the final two sections of this chapter.

2.4 Problems

In Problems 1 through 10, an initial value problem and its exact solution $y(x)$ are given. Apply Euler's method twice to approximate to this solution on the interval $[0, \frac{1}{2}]$, first with step size $h = 0.25$, then with step size $h = 0.1$. Compare the three-decimal-place values of the two approximations at $x = \frac{1}{2}$ with the value $y(\frac{1}{2})$ of the actual solution.

1. $y' = -y$, $y(0) = 2$; $y(x) = 2e^{-x}$
2. $y' = 2y$, $y(0) = \frac{1}{2}$; $y(x) = \frac{1}{2}e^{2x}$
3. $y' = y + 1$, $y(0) = 1$; $y(x) = 2e^{x} - 1$
4. $y' = x - y$, $y(0) = 1$; $y(x) = 2e^{-x} + x - 1$
5. $y' = y - x - 1$, $y(0) = 1$; $y(x) = 2 + x - e^{x}$
6. $y' = -2xy$, $y(0) = 2$; $y(x) = 2e^{-x^2}$

7. $y' = -3x^2y$, $y(0) = 3$; $y(x) = 3e^{-x^3}$
8. $y' = e^{-y}$, $y(0) = 0$; $y(x) = \ln(x + 1)$
9. $y' = \frac{1}{4}(1 + y^2)$, $y(0) = 1$; $y(x) = \tan \frac{1}{4}(x + \pi)$
10. $y' = 2xy^2$, $y(0) = 1$; $y(x) = \dfrac{1}{1 - x^2}$

Note: *The application following this problem set lists illustrative calculator/computer programs that can be used in the remaining problems.*

A programmable calculator or a computer will be useful for Problems 11 through 16. In each problem find the exact solution of the given initial value problem. Then apply Euler's method twice to approximate (to four decimal places) this solution on the given interval, first with step size $h = 0.01$, then with step size $h = 0.005$. Make a table showing the approximate values and the actual value, together with the percentage error in the more accurate approximation, for x an integral multiple of 0.2. Throughout, primes denote derivatives with respect to x.

11. $y' = y - 2$, $y(0) = 1$; $0 \leqq x \leqq 1$
12. $y' = \frac{1}{2}(y - 1)^2$, $y(0) = 2$; $0 \leqq x \leqq 1$
13. $yy' = 2x^3$, $y(1) = 3$; $1 \leqq x \leqq 2$
14. $xy' = y^2$, $y(1) = 1$; $1 \leqq x \leqq 2$
15. $xy' = 3x - 2y$, $y(2) = 3$; $2 \leqq x \leqq 3$
16. $y^2y' = 2x^5$, $y(2) = 3$; $2 \leqq x \leqq 3$

A computer with a printer is required for Problems 17 through 24. In these initial value problems, use Euler's method with step sizes $h = 0.1$, 0.02, 0.004, and 0.0008 to approximate to four decimal places the values of the solution at ten equally spaced points of the given interval. Print the results in tabular form with appropriate headings to make it easy to gauge the effect of varying the step size h. Throughout, primes denote derivatives with respect to x.

17. $y' = x^2 + y^2$, $y(0) = 0$; $0 \leqq x \leqq 1$
18. $y' = x^2 - y^2$, $y(0) = 1$; $0 \leqq x \leqq 2$
19. $y' = x + \sqrt{y}$, $y(0) = 1$; $0 \leqq x \leqq 2$
20. $y' = x + \sqrt[3]{y}$, $y(0) = -1$; $0 \leqq x \leqq 2$
21. $y' = \ln y$, $y(1) = 2$; $1 \leqq x \leqq 2$
22. $y' = x^{2/3} + y^{2/3}$, $y(0) = 1$; $0 \leqq x \leqq 2$
23. $y' = \sin x + \cos y$, $y(0) = 0$; $0 \leqq x \leqq 1$
24. $y' = \dfrac{x}{1 + y^2}$, $y(-1) = 1$; $-1 \leqq x \leqq 1$

25. You bail out of the helicopter of Example 2 and immediately pull the ripcord of your parachute. Now $k = 1.6$ in Eq. (5), so your downward velocity satisfies the initial value problem

$$\frac{dv}{dt} = 32 - 1.6v, \quad v(0) = 0$$

(with t in seconds and v in ft/sec). Use Euler's method with a programmable calculator or computer to approximate the solution for $0 \leqq t \leqq 2$, first with step size $h = 0.01$ and then with $h = 0.005$, rounding off approximate v-values to one decimal place. What percentage of the limiting velocity 20 ft/sec has been attained after 1 second? After 2 seconds?

26. Suppose the deer population $P(t)$ in a small forest initially numbers 25 and satisfies the logistic equation

$$\frac{dP}{dt} = 0.0225P - 0.0003P^2$$

(with t in months). Use Euler's method with a programmable calculator or computer to approximate the solution for 10 years, first with step size $h = 1$ and then with $h = 0.5$, rounding off approximate P-values to integral numbers of deer. What percentage of the limiting population of 75 deer has been attained after 5 years? After 10 years?

Use Euler's method with a computer system to find the desired solution values in Problems 27 and 28. Start with step size $h = 0.1$, and then use successively smaller step sizes until successive approximate solution values at $x = 2$ agree rounded off to two decimal places.

27. $y' = x^2 + y^2 - 1$, $y(0) = 0$; $y(2) = ?$
28. $y' = x + \frac{1}{2}y^2$, $y(-2) = 0$; $y(2) = ?$

29. Consider the initial value problem

$$7x\frac{dy}{dx} + y = 0, \quad y(-1) = 1.$$

(a) Solve this problem for the exact solution

$$y(x) = -\frac{1}{x^{1/7}},$$

which has an infinite discontinuity at $x = 0$. **(b)** Apply Euler's method with step size $h = 0.15$ to approximate this solution on the interval $-1 \leqq x \leqq 0.5$. Note that, from these data alone, you might not suspect any difficulty near $x = 0$. The reason is that the numerical approximation "jumps across the discontinuity" to another solution of $7xy' + y = 0$ for $x > 0$. **(c)** Finally, apply Euler's method with step sizes $h = 0.03$ and $h = 0.006$, but still printing results only at the original points $x = -1.00$, -0.85, -0.70, ..., 1.20, 1.35. and 1.50. Would you now suspect a discontinuity in the exact solution?

30. Apply Euler's method with successively smaller step sizes on the interval [0, 2] to verify empirically that the solution of the initial value problem

$$\frac{dy}{dx} = x^2 + y^2, \quad y(0) = 0$$

has a vertical asymptote near $x = 2.003147$. (Contrast this with Example 2, in which $y(0) = 1$.)

31. The general solution of the equation

$$\frac{dy}{dx} = (1 + y^2)\cos x$$

is $y(x) = \tan(C + \sin x)$. With the initial condition $y(0) = 0$, the solution $y(x) = \tan(\sin x)$ is well behaved. But with $y(0) = 1$ the solution $y(x) = \tan\left(\frac{1}{4}\pi + \sin x\right)$ has a vertical asymptote at $x = \sin^{-1}(\pi/4) \approx 0.90334$. Use Euler's method to verify this fact empirically.

2.4 Application Implementing Euler's Method

Construction of a calculator or computer program to implement a numerical algorithm can sharpen one's understanding of the algorithm. Figure 2.4.13 lists TI-85 and BASIC programs implementing Euler's method to approximate the solution of the initial value problem

$$\frac{dy}{dx} = x + y, \quad y(0) = 1$$

considered in this section. The comments provided in the final column should make these programs intelligible even if you have little familiarity with the BASIC and TI calculator programming languages. Indeed, the BASIC language is no longer widely used for programming computers but is still useful (as in Fig. 2.4.13 and subsequent ones in this text) for brief description of mathematical algorithms in a transparent form intermediate between English and higher programming languages. (Appropriately, the name BASIC is an acronym describing the **B**eginner's **A**ll-purpose **S**ymbolic **I**nstruction **C**ode introduced in 1963, intially for instructional use at Dartmouth College.)

TI-85	BASIC	Comment
`PROGRAM:EULER`	`Program EULER`	`Program title`
`:10→N`	`N = 10`	`Number of steps`
`:0→X`	`X = 0`	`Initial x`
`:1→Y`	`Y = 1`	`Initial y`
`:1→X1`	`X1 = 1`	`Final x`
`:(X1-X)/N→H`	`H = (X1-X)/N`	`Step size`
`:For(I,1,N)`	`FOR I=1 TO N`	`Begin loop`
`:X+Y→F`	`F = X + Y`	`Function value`
`:Y+H*F→Y`	`Y = Y + H*F`	`Euler iteration`
`:X+H→X`	`X = X + H`	`New x`
`:Disp X,Y`	`PRINT X,Y`	`Display results`
`:End`	`NEXT I`	`End loop`

FIGURE 2.4.13. TI-85 and BASIC Euler's method programs.

To increase the number of steps (and thereby decrease the step size), you need only change the value of **`N`** specified in the first line of the program. To apply Euler's method to a different equation $dy/dx = f(x, y)$, you need change only the single line that calculates the function value **`F`**.

Any other procedural programming language (such as FORTRAN or Pascal) would follow the pattern illustrated by the parallel lines of TI-85 and BASIC code in Fig. 2.4.13. Some of the modern functional programming languages mirror standard mathematical notation even more closely. Figure 2.4.14 shows a MATLAB implementation of Euler's method. The **`euler`** function takes as input the initial value **`x`**, the initial value **`y`**, the final value **`x1`** of x, and the desired number **`n`** of subintervals. For instance, the MATLAB command

```
[X, Y] = euler(0, 1, 1, 10)
```

then generates the x_n- and y_n-data shown in the first two columns of the table of Fig. 2.4.8.

```
function yp = f(x,y)
yp = x + y;                        % yp = y'

function [X,Y] = euler(x,y,x1,n)
h = (x1 - x)/n;                    % step size
X = x;                             % initial x
Y = y;                             % initial y
for i = 1:n                        % begin loop
     y = y + h*f(x,y);             % Euler iteration
     x = x + h;                    % new x
     X = [X;x];                    % update x-column
     Y = [Y;y];                    % update y-column
     end                           % end loop
```

FIGURE 2.4.14. MATLAB implementation of Euler's method.

You should begin this project by implementing Euler's method with your own calculator or computer system. Test your program by first applying it to the initial value problem in Example 1, then to some of the problems for this section.

Famous Numbers Investigation

The following problems describe the numbers $e \approx 2.71828$, $\ln 2 \approx 0.69315$, and $\pi \approx 3.14159$ as specific values of solutions of certain initial value problems. In each case, apply Euler's method with $n = 50, 100, 200, \ldots$ subintervals (doubling n each time). How many subintervals are needed to obtain—twice in succession—the correct value of the target number rounded to three decimal places?

1. The number $e = y(1)$, where $y(x)$ is the solution of the initial value problem $dy/dx = y$, $y(0) = 1$.
2. The number $\ln 2 = y(2)$, where $y(x)$ is the solution of the initial value problem $dy/dx = 1/x$, $y(1) = 0$.
3. The number $\pi = y(1)$, where $y(x)$ is the solution of the initial value problem $dy/dx = 4/(1 + x^2)$, $y(0) = 0$.

Also explain in each problem what the point is—why the indicated famous number is the expected numerical result.

2.5 A Closer Look at the Euler Method

The Euler method as presented in Section 2.4 is not often used in practice, mainly because more accurate methods are available. But Euler's method has the advantage of simplicity, and a careful study of this method yields insights into the workings of more accurate methods, because many of the latter are extensions or refinements of the Euler method.

To compare two different methods of numerical approximation, we need some way to measure the accuracy of each. Theorem 1 tells what degree of accuracy we can expect when we use Euler's method.

THEOREM 1 The Error in the Euler Method

Suppose that the initial value problem

$$\frac{dy}{dx} = f(x, y), \quad y(x_0) = y_0 \tag{1}$$

has a unique solution $y(x)$ on the closed interval $[a, b]$ with $a = x_0$, and assume that $y(x)$ has a continuous second derivative on $[a, b]$. (This would follow from the assumption that f, f_x, and f_y are all continuous for $a \leqq x \leqq b$ and $c \leqq y \leqq d$, where $c \leqq y(x) \leqq d$ for all x in $[a, b]$.) Then there exists a constant C such that the following is true: If the approximations $y_1, y_2, y_3, \ldots, y_k$ to the actual values $y(x_1), y(x_2), y(x_3), \ldots, y(x_k)$ at points of $[a, b]$ are computed using Euler's method with step size $h > 0$, then

$$|y_n - y(x_n)| \leqq Ch \tag{2}$$

for each $n = 1, 2, 3, \ldots, k$.

Remark: The **error**

$$y_{\text{actual}} - y_{\text{approx}} = y(x_n) - y_n$$

in (2) denotes the (cumulative) error in Euler's method after n steps in the approximation, *exclusive* of roundoff error (as though we were using a perfect machine that made no roundoff errors). The theorem can be summarized by saying that *the error in Euler's method is of order h*; that is, the error is bounded by a (predetermined) constant C multiplied by the step size h. It follows, for instance, that (on a given closed interval) halving the step size cuts the maximum error in half; similarly, with step size $h/10$ we get 10 times the accuracy (that is, 1/10 the maximum error) as with step size h. Consequently, we can—in principle—get any degree of accuracy we want by choosing h sufficiently small. ■

We will omit the proof of this theorem, but one can be found in Chapter 7 of G. Birkhoff and G.-C. Rota, *Ordinary Differential Equations,* 4th ed. (New York: John Wiley, 1989). The constant C deserves some comment. Because C tends to increase as the maximum value of $|y''(x)|$ on $[a, b]$ increases, it follows that C must depend in a fairly complicated way on y, and actual computation of a value of C such that the inequality in (2) holds is usually impractical. In practice, the following type of procedure is commonly employed.

1. Apply Euler's method to the initial value problem in (1) with a reasonable value of h.
2. Repeat with $h/2$, $h/4$, and so forth, at each stage halving the step size for the next application of Euler's method.
3. Continue until the results obtained at one stage agree—to an appropriate number of significant digits—with those obtained at the previous stage. Then the approximate values obtained at this stage are considered likely to be accurate to the indicated number of significant digits.

Example 1 Carry out this procedure with the initial value problem

$$\frac{dy}{dx} = -\frac{2xy}{1+x^2}, \quad y(0) = 1 \tag{3}$$

to approximate accurately the value $y(1)$ of the solution at $x = 1$.

Solution Using an Euler method program, perhaps one of those listed in Figs. 2.4.13 and 2.4.14, we begin with a step size $h = 0.04$ requiring $n = 25$ steps to reach $x = 1$. The table in Fig. 2.5.1 shows the approximate values of $y(1)$ obtained with successively smaller values of h. The data suggest that the true value of $y(1)$ is exactly 0.5. Indeed, the exact solution of the initial value problem in (3) is $y(x) = 1/(1+x^2)$, so the true value of $y(1)$ is exactly $\frac{1}{2}$. ■

h	Approximate $y(1)$	Actual $y(1)$	$\lvert\text{Error}\rvert/h$
0.04	0.50451	0.50000	0.11
0.02	0.50220	0.50000	0.11
0.01	0.50109	0.50000	0.11
0.005	0.50054	0.50000	0.11
0.0025	0.50027	0.50000	0.11
0.00125	0.50013	0.50000	0.10
0.000625	0.50007	0.50000	0.11
0.0003125	0.50003	0.50000	0.10

FIGURE 2.5.1. Table of values in Example 1.

The final column of the table in Fig. 2.5.1 displays the ratio of the magnitude of the error to h; that is, $|y_{\text{actual}} - y_{\text{approx}}|/h$. Observe how the data in this column substantiate Theorem 1—in this computation, the error bound in (2) appears to hold with a value of C slightly larger than 0.1.

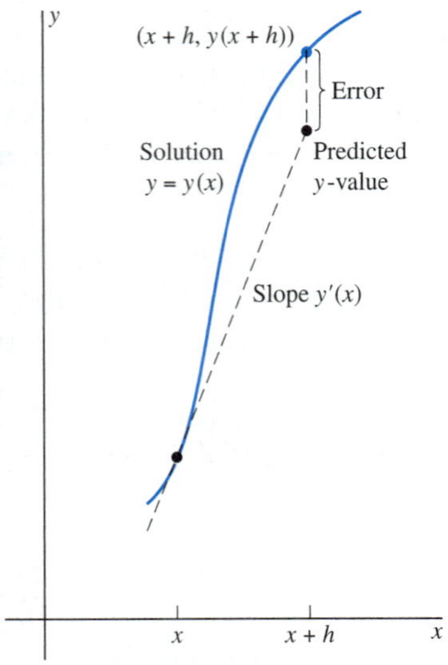

FIGURE 2.5.2. True and predicted values in Euler's method.

An Improvement in Euler's Method

As Fig. 2.5.2 shows, Euler's method is rather unsymmetrical. It uses the predicted slope $k = f(x_n, y_n)$ of the graph of the solution at the left-hand endpoint of the interval $[x_n, x_n + h]$ as if it were the actual slope of the solution over that entire interval. We now turn our attention to a way in which increased accuracy can easily be obtained; it is known as the *improved Euler method.*

Given the initial value problem

$$\frac{dy}{dx} = f(x, y), \quad y(x_0) = y_0, \tag{4}$$

suppose that after carrying out n steps with step size h we have computed the approximation y_n to the actual value $y(x_n)$ of the solution at $x_n = x_0 + nh$. We can use the Euler method to obtain a first estimate—which we now call u_{n+1} rather than y_{n+1}—of the value of the solution at $x_{n+1} = x_n + h$. Thus

$$u_{n+1} = y_n + h \cdot f(x_n, y_n) = y_n + h \cdot k_1.$$

Now that $u_{n+1} \approx y(x_{n+1})$ has been computed, we can take

$$k_2 = f(x_{n+1}, u_{n+1})$$

as a second estimate of the slope of the solution curve $y = y(x)$ at $x = x_{n+1}$.

Of course, the approximate slope $k_1 = f(x_n, y_n)$ at $x = x_n$ has already been calculated. Why not *average* these two slopes to obtain a more accurate estimate of the average slope of the solution curve over the entire subinterval $[x_n, x_{n+1}]$? This idea is the essence of the *improved* Euler method. Figure 2.5.3 shows the geometry behind this method.

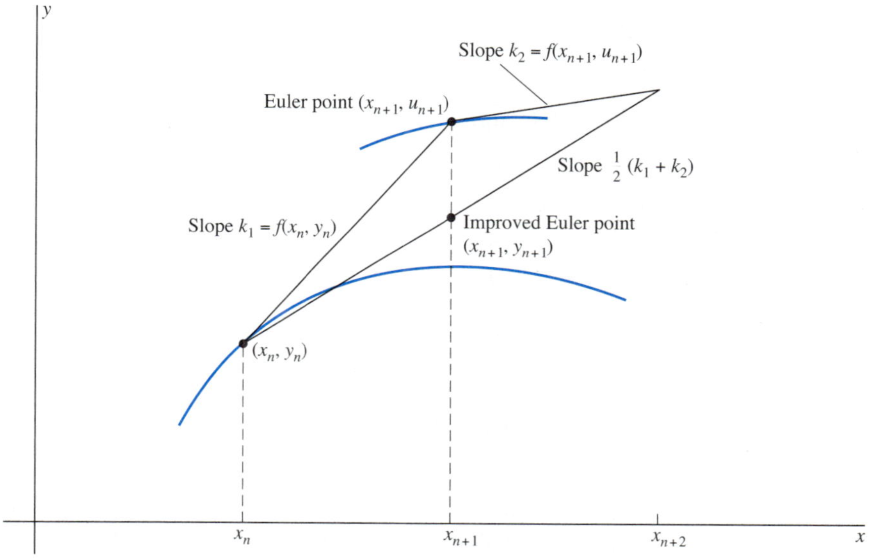

FIGURE 2.5.3. The improved Euler method: Average the slopes of the tangent lines at (x_n, y_n) and (x_{n+1}, u_{n+1}).

ALGORITHM The Improved Euler Method

Given the initial value problem

$$\frac{dy}{dx} = f(x, y), \quad y(x_0) = y_0,$$

the **improved Euler method with step size** h consists in applying the iterative formulas

$$\begin{aligned} k_1 &= f(x_n, y_n), \\ u_{n+1} &= y_n + h \cdot k_1, \\ k_2 &= f(x_{n+1}, u_{n+1}), \\ y_{n+1} &= y_n + h \cdot \tfrac{1}{2}(k_1 + k_2) \end{aligned} \tag{5}$$

to compute successive approximations $y_1, y_2, y_3, \ldots$ to the (true) values $y(x_1)$, $y(x_2)$, $y(x_3)$, $\ldots$ of the (exact) solution $y = y(x)$ at the points $x_1, x_2, x_3, \ldots$, respectively.

Remark: The final formula in (5) takes the "Euler form"

$$y_{n+1} = y_n + h \cdot k$$

if we write

$$k = \frac{k_1 + k_2}{2}$$

for the approximate *average slope* on the interval $[x_n, x_{n+1}]$. ■

The improved Euler method is one of a class of numerical techniques known as **predictor-corrector** methods. First a predictor u_{n+1} of the next y-value is computed; then it is used to correct itself. Thus the **improved Euler method** with step size h consists of using the **predictor**

$$u_{n+1} = y_n + h \cdot f(x_n, y_n) \tag{6}$$

and the **corrector**

$$y_{n+1} = y_n + h \cdot \tfrac{1}{2}\left[f(x_n, y_n) + f(x_{n+1}, u_{n+1})\right] \tag{7}$$

iteratively to compute successive approximations $y_1, y_2, y_2, \ldots$ to the values $y(x_1)$, $y(x_2), y(x_3), \ldots$ of the actual solution of the initial value problem in (4).

Remark: Each improved Euler step requires two evaluations of the function $f(x, y)$, as compared with the single function evaluation required for an ordinary Euler step. We naturally wonder whether this doubled computational labor is worth the trouble.

Answer: Under the assumption that the exact solution $y = y(x)$ of the initial value problem in (4) has a continuous third derivative, it can be proved—see Chapter 7 of Birkhoff and Rota—that the error in the improved Euler method is of order h^2. This means that on a given bounded interval $[a, b]$, each approximate value y_n satisfies the inequality

$$|y(x_n) - y_n| \leqq Ch^2, \tag{8}$$

where the constant C does not depend on h. Because h^2 is much smaller than h if h itself is small, this means that the improved Euler method is more accurate than Euler's method itself. This advantage is offset by the fact that about twice as many computations are required. But the factor h^2 in (8) means that halving the step size results in 1/4 the maximum error, and with step size $h/10$ we get 100 times the accuracy (that is, 1/100 the maximum error) as with step size h. ■

Example 2 Figure 2.4.8 shows results of applying Euler's method to the initial value problem

$$\frac{dy}{dx} = x + y, \quad y(0) = 1 \tag{9}$$

with exact solution $y(x) = 2e^x - x - 1$. With $f(x, y) = x + y$ in Eqs. (6) and (7), the predictor-corrector formulas for the improved Euler method are

$$u_{n+1} = y_n + h \cdot (x_n + y_n),$$
$$y_{n+1} = y_n + h \cdot \tfrac{1}{2}\left[(x_n + y_n) + (x_{n+1} + u_{n+1})\right].$$

With step size $h = 0.1$ we calculate

$$
\begin{aligned}
u_1 &= 1 + (0.1) \cdot (0 + 1) = 1.1, \\
y_1 &= 1 + (0.05) \cdot [(0 + 1) + (0.1 + 1.1)] = 1.11, \\
u_2 &= 1.11 + (0.1) \cdot (0.1 + 1.11) = 1.231, \\
y_2 &= 1.11 + (0.05) \cdot [(0.1 + 1.11) + (0.2 + 1.231)] = 1.24205,
\end{aligned}
$$

and so forth. The table in Fig. 2.5.4 compares the results obtained using the improved Euler method with those obtained previously using the "unimproved" Euler method. When the same step size $h = 0.1$ is used, the error in the Euler approximation to $y(1)$ is 7.25%, but the error in the improved Euler approximation is only 0.24%.

x	Euler Method, $h = 0.1$ Values of y	Euler Method, $h = 0.005$ Values of y	Improved Euler, $h = 0.1$ Values of y	Actual y
0.1	1.1000	1.1098	1.1100	1.1103
0.2	1.2200	1.2416	1.2421	1.2428
0.3	1.3620	1.3977	1.3985	1.3997
0.4	1.5282	1.5807	1.5818	1.5836
0.5	1.7210	1.7933	1.7949	1.7974
0.6	1.9431	2.0388	2.0409	2.0442
0.7	2.1974	2.3205	2.3231	2.3275
0.8	2.4872	2.6422	2.6456	2.6511
0.9	2.8159	3.0082	3.0124	3.0192
1.0	3.1875	3.4230	3.4282	3.4366

FIGURE 2.5.4. Euler and improved Euler approximations to the solution of $dy/dx = x + y$, $y(0) = 1$.

Indeed, the improved Euler method with $h = 0.1$ is more accurate (in this example) than the original Euler method with $h = 0.005$. The latter requires 200 evaluations of the function $f(x, y)$, but the former requires only 20 such evaluations, so in this case the improved Euler method yields greater accuracy with only about one-tenth the work.

Figure 2.5.5 shows the results obtained when the improved Euler method is applied to the initial value problem in (9) using step size $h = 0.005$. Accuracy of five significant figures is apparent in the table. This suggests that, in contrast with the original Euler method, the improved Euler method is sufficiently accurate for certain practical applications—such as plotting solution curves. ■

x	Improved Euler, Approximate y	Actual y
0.0	1.00000	1.00000
0.1	1.11034	1.11034
0.2	1.24280	1.24281
0.3	1.39971	1.39972
0.4	1.58364	1.58365
0.5	1.79744	1.79744
0.6	2.04423	2.04424
0.7	2.32749	2.32751
0.8	2.65107	2.65108
0.9	3.01919	3.01921
1.0	3.43654	3.43656

FIGURE 2.5.5. Improved Euler approximation to the solution of Eq. (9) with step size $h = 0.005$.

An improved Euler program (similar to the ones listed in the project material for this section) was used to compute approximations to the exact value $y(1) = 0.5$ of the solution $y(x) = 1/(1 + x^2)$ of the initial value problem

$$\frac{dy}{dx} = -\frac{2xy}{1 + x^2}, \quad y(0) = 1 \tag{3}$$

of Example 1. The results obtained by successively halving the step size appear in the table in Fig. 2.5.6. Note that the final column of this table impressively corroborates the form of the error bound in (8), and that each halving of the step size

h	Improved Euler Approximation to $y(1)$	Error	$\lvert\text{Error}\rvert/h^2$
0.04	0.500195903	−0.000195903	0.12
0.02	0.500049494	−0.000049494	0.12
0.01	0.500012437	−0.000012437	0.12
0.005	0.500003117	−0.000003117	0.12
0.0025	0.500000780	−0.000000780	0.12
0.00125	0.500000195	−0.000000195	0.12
0.000625	0.500000049	−0.000000049	0.12
0.0003125	0.500000012	−0.000000012	0.12

FIGURE 2.5.6. Improved Euler approximation to $y(1)$ for $dy/dx = -2xy/(1+x^2)$, $y(0) = 1$.

reduces the error by a factor of almost exactly 4, as should happen if the error is proportional to h^2.

In the following two examples we exhibit graphical results obtained by employing this strategy of successively halving the step size, and thus doubling the number of subintervals of a fixed interval on which we are approximating a solution.

Example 3 In Example 3 of Section 2.4 we applied Euler's method to the logistic initial value problem

$$\frac{dy}{dx} = \tfrac{1}{3}y(8-y), \quad y(0) = 1.$$

Figure 2.4.9 shows an obvious difference between the exact solution $y(x) = 8/(1+7e^{-8x/3})$ and the Euler approximation on $0 \leqq x \leqq 5$ using $n = 20$ subintervals. Figure 2.5.7 shows approximate solution curves plotted using the improved Euler's method.

The approximation with five subintervals is still bad—perhaps worse! It appears to level off considerably short of the actual limiting population $M = 8$. You should carry out at least the first two improved Euler steps manually to see for yourself how it happens that, after increasing appropriately during the first step, the approximate solution *decreases* in the second step rather than continuing to increase (as it should). In the project for this section we ask you to show empirically that the improved Euler approximate solution with step size $h = 1$ levels off at $y \approx 4.3542$.

In contrast, the approximate solution curve with $n = 20$ subintervals tracks the exact solution curve rather closely, and with $n = 40$ subintervals the exact and approximate solution curves are indistinguishable in Fig. 2.5.7. The table in Fig. 2.5.8 indicates that the improved Euler approximation with $n = 200$ subintervals is accurate rounded to three decimal places (that is, four significant digits) on the interval $0 \leqq x \leqq 5$. Because discrepancies in the fourth significant digit are not visually apparent at the resolution of an ordinary computer screen, the improved Euler method (using several hundred subintervals) is considered adequate for many graphical purposes. ■

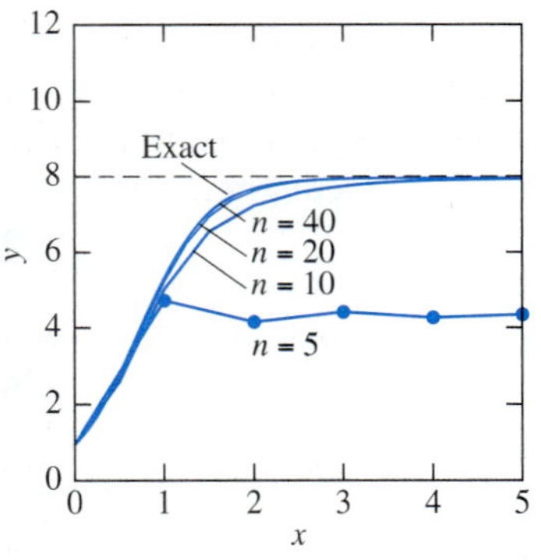

FIGURE 2.5.7. Approximating a logistic solution using the improved Euler method with $n = 5$, $n = 10$, $n = 20$, and $n = 40$ subintervals.

x	Actual $y(x)$	Improved Euler with $n = 200$
0	1.0000	1.0000
1	5.3822	5.3809
2	7.7385	7.7379
3	7.9813	7.9812
4	7.9987	7.9987
5	7.9999	7.9999

FIGURE 2.5.8. Using the improved Euler method to approximate the actual solution of the initial value problem in Example 3.

Example 4 In Example 4 of Section 2.4 we applied Euler's method to the initial value problem

$$\frac{dy}{dx} = y \cos x, \quad y(0) = 1.$$

Figure 2.4.10 shows obvious visual differences between the periodic exact solution $y(x) = e^{\sin x}$ and the Euler approximations on $0 \leqq x \leqq 6\pi$ with as many as $n = 400$ subintervals.

Figure 2.5.9 shows the exact solution curve and approximate solution curves plotted using the improved Euler method with $n = 50$, $n = 100$, and $n = 200$ subintervals. The approximation obtained with $n = 200$ is indistinguishable from the exact solution curve, and the approximation with $n = 100$ is only barely distinguishable from it. ■

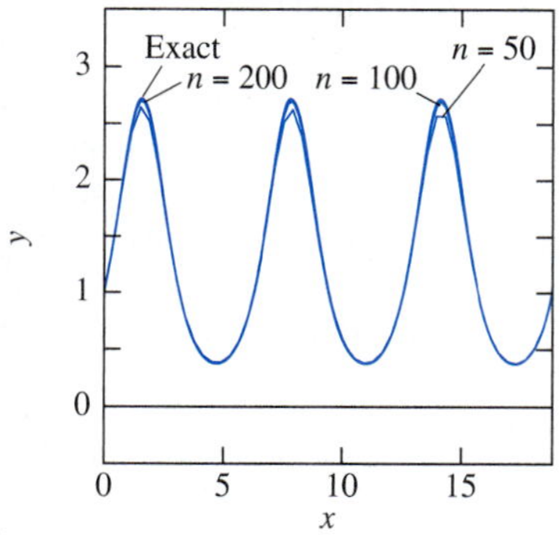

FIGURE 2.5.9. Approximating the exact solution $y = e^{\sin x}$ using the improved Euler method with $n = 50$, 100, and 200 subintervals.

Although Figs. 2.5.7 and 2.5.9 indicate that the improved Euler method can provide accuracy that suffices for many graphical purposes, it does not provide the higher-precision numerical accuracy that sometimes is needed for more careful investigations. For instance, consider again the initial value problem

$$\frac{dy}{dx} = -\frac{2xy}{1+x^2}, \quad y(0) = 1$$

of Example 1. The final column of the table in Fig. 2.5.6 suggests that, if the improved Euler method is used on the interval $0 \leqq x \leqq 1$ with n subintervals and step size $h = 1/n$, then the resulting error E in the final approximation $y_n \approx y(1)$ is given by

$$E = |y(1) - y_n| \approx (0.12)h^2 = \frac{0.12}{n^2}.$$

If so, then 12-place accuracy (for instance) in the value $y(1)$ would require that $(0.12)n^{-2} < 5 \times 10^{-13}$, which means that $n \geqq 489{,}898$. Thus, roughly half a million steps of length $h \approx 0.000002$ would be required. Aside from the possible impracticality of this many steps (using available computational resources), the roundoff error resulting from so many successive steps might well overwhelm the cumulative error predicted by theory (which assumes exact computations in each separate step). Consequently, still more accurate methods than the improved Euler method are needed for such high-precision computations. Such a method is presented in Section 2.6.

2.5 Problems

A hand-held calculator will suffice for Problems 1 through 10, where an initial value problem and its exact solution are given. Apply the improved Euler method to approximate this solution on the interval $[0, 0.5]$ *with step size* $h = 0.1$*. Construct a table showing four-decimal-place values of the approximate solution and actual solution at the points* $x = 0.1, 0.2, 0.3, 0.4, 0.5$.

1. $y' = -y$, $y(0) = 2$; $y(x) = 2e^{-x}$
2. $y' = 2y$, $y(0) = \frac{1}{2}$; $y(x) = \frac{1}{2}e^{2x}$
3. $y' = y + 1$, $y(0) = 1$; $y(x) = 2e^{x} - 1$
4. $y' = x - y$, $y(0) = 1$; $y(x) = 2e^{-x} + x - 1$
5. $y' = y - x - 1$, $y(0) = 1$; $y(x) = 2 + x - e^{x}$
6. $y' = -2xy$, $y(0) = 2$; $y(x) = 2e^{-x^2}$
7. $y' = -3x^2y$, $y(0) = 3$; $y(x) = 3e^{-x^3}$
8. $y' = e^{-y}$, $y(0) = 0$; $y(x) = \ln(x + 1)$
9. $y' = \frac{1}{4}(1 + y^2)$, $y(0) = 1$; $y(x) = \tan\frac{1}{4}(x + \pi)$
10. $y' = 2xy^2$, $y(0) = 1$; $y(x) = \dfrac{1}{1 - x^2}$

Note: *The application following this problem set lists illustrative calculator/computer programs that can be used in Problems 11 through 24.*

A programmable calculator or a computer will be useful for Problems 11 through 16. In each problem find the exact solution of the given initial value problem. Then apply the improved Euler method twice to approximate (to five decimal places) this solution on the given interval, first with step size $h = 0.01$*, then with step size* $h = 0.005$*. Make a table showing the approximate values and the actual value, together with the percentage error in the more accurate approximations, for* x *an integral multiple of* 0.2*. Throughout, primes denote derivatives with respect to* x.

11. $y' = y - 2$, $y(0) = 1$; $0 \leqq x \leqq 1$
12. $y' = \frac{1}{2}(y - 1)^2$, $y(0) = 2$; $0 \leqq x \leqq 1$
13. $yy' = 2x^3$, $y(1) = 3$; $1 \leqq x \leqq 2$
14. $xy' = y^2$, $y(1) = 1$; $1 \leqq x \leqq 2$
15. $xy' = 3x - 2y$, $y(2) = 3$; $2 \leqq x \leqq 3$
16. $y^2y' = 2x^5$, $y(2) = 3$; $2 \leqq x \leqq 3$

A computer with a printer is required for Problems 17 through 24. In these initial value problems, use the improved Euler method with step sizes $h = 0.1$, 0.02, 0.004, *and* 0.0008 *to approximate to five decimal places the values of the solution at ten equally spaced points of the given interval. Print the results in tabular form with appropriate headings to make it easy to gauge the effect of varying the step size* h*. Throughout, primes denote derivatives with respect to* x.

17. $y' = x^2 + y^2$, $y(0) = 0$; $0 \leqq x \leqq 1$
18. $y' = x^2 - y^2$, $y(0) = 1$; $0 \leqq x \leqq 2$
19. $y' = x + \sqrt{y}$, $y(0) = 1$; $0 \leqq x \leqq 2$
20. $y' = x + \sqrt[3]{y}$, $y(0) = -1$; $0 \leqq x \leqq 2$
21. $y' = \ln y$, $y(1) = 2$; $1 \leqq x \leqq 2$
22. $y' = x^{2/3} + y^{2/3}$, $y(0) = 1$; $0 \leqq x \leqq 2$
23. $y' = \sin x + \cos y$, $y(0) = 0$; $0 \leqq x \leqq 1$
24. $y' = \dfrac{x}{1 + y^2}$, $y(-1) = 1$; $-1 \leqq x \leqq 1$

25. As in Problem 25 of Section 2.4, you bail out of a helicopter and immediately open your parachute, so your downward velocity satisfies the initial value problem

$$\frac{dv}{dt} = 32 - 1.6v, \qquad v(0) = 0$$

(with t in seconds and v in ft/s). Use the improved Euler method with a programmable calculator or computer to approximate the solution for $0 \leqq t \leqq 2$, first with step size $h = 0.01$ and then with $h = 0.005$, rounding off approximate v-values to three decimal places. What percentage of the limiting velocity 20 ft/s has been attained after 1 second? After 2 seconds?

26. As in Problem 26 of Section 2.4, suppose the deer population $P(t)$ in a small forest initially numbers 25 and satisfies the logistic equation

$$\frac{dP}{dt} = 0.0225P - 0.0003P^2$$

(with t in months). Use the improved Euler method with a programmable calculator or computer to approximate the solution for 10 years, first with step size $h = 1$ and then with $h = 0.5$, rounding off approximate P-values to three decimal places. What percentage of the limiting population of 75 deer has been attained after 5 years? After 10 years?

Use the improved Euler method with a computer system to find the desired solution values in Problems 27 and 28. Start with step size $h = 0.1$*, and then use successively smaller step sizes until successive approximate solution values at* $x = 2$ *agree rounded off to four decimal places.*

27. $y' = x^2 + y^2 - 1$, $y(0) = 0$; $y(2) = ?$
28. $y' = x + \frac{1}{2}y^2$, $y(-2) = 0$; $y(2) = ?$

29. Consider the crossbow bolt of Example 2 in Section 2.3, shot straight upward from the ground with an initial velocity of 49 m/s. Because of linear air resistance, its velocity function $v(t)$ satisfies the initial value problem

$$\frac{dv}{dt} = -(0.04)v - 9.8, \qquad v(0) = 49$$

with exact solution $v(t) = 294e^{-t/25} - 245$. Use a calculator or computer implementation of the improved Euler method to approximate $v(t)$ for $0 \leqq t \leqq 10$ using both $n = 50$ and $n = 100$ subintervals. Display the results at intervals of 1 second. Do the two approximations—each rounded to two decimal places—agree both with each other and with the exact solution? If the exact solution were unavailable, explain how you could use the improved Euler method to approximate closely **(a)** the bolt's time of ascent to its apex (given in Section 2.3 as 4.56 s) and **(b)** its impact velocity after 9.41 s in the air.

30. Consider now the crossbow bolt of Example 3 in Section 2.3. It still is shot straight upward from the ground with an initial velocity of 49 m/s, but because of air resistance proportional to the square of its velocity, its velocity function $v(t)$ satisfies the initial value problem

$$\frac{dv}{dt} = -(0.0011)v|v| - 9.8, \quad v(0) = 49.$$

The symbolic solution discussed in Section 2.3 required separate investigations of the bolt's ascent and its descent, with $v(t)$ given by a tangent function during ascent and by a hyperbolic tangent function during descent. But the improved Euler method requires no such distinction. Use a calculator or computer implementation of the improved Euler method to approximate $v(t)$ for $0 \leqq t \leqq 10$ using both $n = 100$ and $n = 200$ subintervals. Display the results at intervals of 1 second. Do the two approximations—each rounded to two decimal places—agree with each other? If an exact solution were unavailable, explain how you could use the improved Euler method to approximate closely **(a)** the bolt's time of ascent to its apex (given in Section 2.3 as 4.61 s) and **(b)** its impact velocity after 9.41 s in the air.

2.5 Application Improved Euler Implementation

Figure 2.5.10 lists TI-85 and BASIC programs implementing the improved Euler method to approximate the solution of the initial value problem

$$\frac{dy}{dx} = x + y, \quad y(0) = 1$$

considered in Example 2 of this section. The comments provided in the final column should make these programs intelligible even if you have little familiarity with the BASIC and TI programming languages.

TI-85	BASIC	Comment
`PROGRAM:IMPEULER`	`Program IMPEULER`	`Program title`
`:F=X+Y`	`DEF FN F(X,Y) = X + Y`	`Define function f`
`:10→N`	`N = 10`	`No. of steps`
`:0→X`	`X = 0`	`Initial x`
`:1→Y`	`Y = 1`	`Initial y`
`:1→X1`	`X1 = 1`	`Final x`
`:(X1-X)/N→H`	`H = (X1-X)/N`	`Step size`
`:For(I,1,N)`	`FOR I=1 TO N`	`Begin loop`
`:Y→Y0`	`Y0 = Y`	`Save previous y`
`:F→K1`	`K1 = FNF(X,Y)`	`First slope`
`:Y0+H*K1→Y`	`Y = Y0 + H*K1`	`Predictor`
`:X+H→X`	`X = X + H`	`New x`
`:F→K2`	`K2 = FNF(X,Y)`	`Second slope`
`:(K1+K2)/2→K`	`K = (K1 + K2)/2`	`Average slope`
`:Y0+H*K→Y`	`Y = Y0 + H*K`	`Corrector`
`:Disp X,Y`	`PRINT X,Y`	`Display results`
`:End`	`NEXT I`	`End loop`

FIGURE 2.5.10. TI-85 and BASIC improved Euler programs.

To apply the improved Euler method to a differential equation $dy/dx = f(x, y)$, one need only change the initial line of the program, in which the function f is defined. To increase the number of steps (and thereby decrease the step size), one need only change the value of **N** specified in the second line of the program.

Figure 2.5.11 exhibits one MATLAB implementation of the improved Euler method. The **impeuler** function takes as input the initial value **x**, the initial value **y**, the final value **x1** of x, and the desired number **n** of subintervals. As output it produces the resulting column vectors **X** and **Y** of x- and y-values. For instance, the MATLAB command

```
[X, Y] = impeuler(0, 1, 1, 10)
```

then generates the first and fourth columns of data shown in Fig. 2.5.4.

```
function yp = f(x,y)
yp = x + y;                          % yp = y'

function [X,Y] = impeuler(x,y,x1,n)
h = (x1 - x)/n;                      % step size
X = x;                               % initial x
Y = y;                               % initial y
for i = 1:n;                         % begin loop
    k1 = f(x,y);                     % first slope
    k2 = f(x+h,y+h*k1);              % second slope
    k = (k1 + k2)/2;;                % average slope
    x = x + h;                       % new x
    y = y + h*k;                     % new y
    X = [X;x];                       % update x-column
    Y = [Y;y];                       % update y-column
    end                              % end loop
```

FIGURE 2.5.11. MATLAB implementation of improved Euler method.

You should begin this project by implementing the improved Euler method with your own calculator or computer system. Test your program by applying it first to the initial value problem of Example 1, then to some of the problems for this section.

Famous Numbers Revisited

The following problems describe the numbers $e \approx 2.7182818$, $\ln 2 \approx 0.6931472$, and $\pi \approx 3.1415927$ as specific values of certain initial value problems. In each case, apply the improved Euler method with $n = 10, 20, 40, \dots$ subintervals (doubling n each time). How many subintervals are needed to obtain—twice in succession—the correct value of the target number rounded to five decimal places?

1. The number $e = y(1)$, where $y(x)$ is the solution of the initial value problem $dy/dx = y$, $y(0) = 1$.
2. The number $\ln 2 = y(2)$, where $y(x)$ is the solution of the initial value problem $dy/dx = 1/x$, $y(1) = 0$.
3. The number $\pi = y(1)$, where $y(x)$ is the solution of the initial value problem $dy/dx = 4/(1 + x^2)$, $y(0) = 0$.

Logistic Population Investigation

Apply your improved Euler program to the initial value problem $dy/dx = \frac{1}{3}y(8-y)$, $y(0) = 1$ of Example 3. In particular, verify (as claimed) that the approximate solution with step size $h = 1$ levels off at $y \approx 4.3542$ rather than at the limiting value $y = 8$ of the exact solution. Perhaps a table of values for $0 \leqq x \leqq 100$ will make this apparent.

For your own logistic population to investigate, consider the initial value problem

$$\frac{dy}{dx} = \frac{1}{n}y(m - y), \quad y(0) = 1,$$

where m and n are (for instance) the largest and smallest nonzero digits in your student ID number. Does the improved Euler approximation with step size $h = 1$ level off at the "correct" limiting value of the exact solution? If not, find a smaller value of h so that it does.

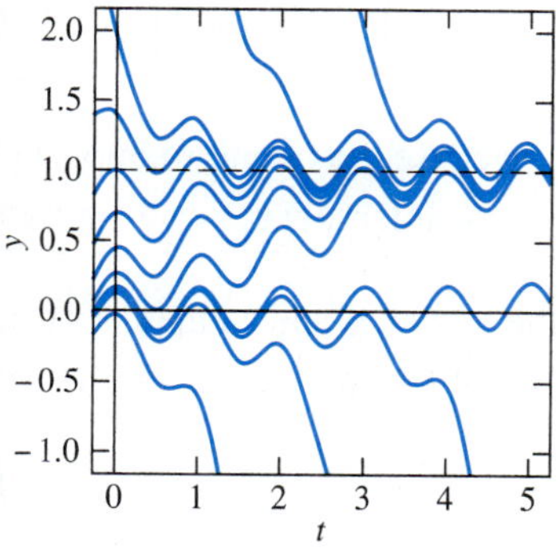

FIGURE 2.5.12. Solution curves of $dy/dt = y(1 - y) - \sin 2\pi t$.

Periodic Harvesting and Restocking

The differential equation

$$\frac{dy}{dt} = ky(M - y) - h \sin\left(\frac{2\pi t}{P}\right)$$

models a logistic population that is periodically harvested and restocked with period P and maximal harvesting/restocking rate h. A numerical approximation program was used to plot the typical solution curves for the case $k = M = h = P = 1$ that are shown in Fig. 2.5.12. This figure suggests—although it does not prove—the existence of a threshold initial population such that

- Beginning with an initial population above this threshold, the population oscillates (perhaps with period P?) around the (unharvested) stable limiting population $y(t) \equiv M$, whereas
- The population dies out if it begins with an initial population below this threshold.

Use an appropriate plotting utility to investigate your own logistic population with periodic harvesting and restocking (selecting typical values of the parameters k, M, h, and P). Do the observations indicated here appear to hold for your population?

2.6 The Runge–Kutta Method

We now discuss a method for approximating the solution $y = y(x)$ of the initial value problem

$$\frac{dy}{dx} = f(x, y), \quad y(x_0) = y_0 \tag{1}$$

that is considerably more accurate than the improved Euler method and is more widely used in practice than any of the numerical methods discussed in Sections 2.4 and 2.5. It is called the *Runge–Kutta method*, after the German mathematicians who developed it, Carl Runge (1856–1927) and Wilhelm Kutta (1867–1944).

With the usual notation, suppose that we have computed the approximations $y_1, y_2, y_3, \ldots, y_n$ to the actual values $y(x_1), y(x_2), y(x_3), \ldots, y(x_n)$ and now want to compute $y_{n+1} \approx y(x_{n+1})$. Then

$$y(x_{n+1}) - y(x_n) = \int_{x_n}^{x_{n+1}} y'(x)\,dx = \int_{x_n}^{x_n+h} y'(x)\,dx \tag{2}$$

by the fundamental theorem of calculus. Next, Simpson's rule for numerical integration yields

$$y(x_{n+1}) - y(x_n) \approx \frac{h}{6}\left[y'(x_n) + 4y'\left(x_n + \frac{h}{2}\right) + y'(x_{n+1})\right]. \tag{3}$$

Hence we want to define y_{n+1} so that

$$y_{n+1} \approx y_n + \frac{h}{6}\left[y'(x_n) + 2y'\left(x_n + \frac{h}{2}\right) + 2y'\left(x_n + \frac{h}{2}\right) + y'(x_{n+1})\right]; \tag{4}$$

we have split $4y'\left(x_n + \frac{1}{2}h\right)$ into a sum of two terms because we intend to approximate the slope $y'\left(x_n + \frac{1}{2}h\right)$ at the midpoint $x_n + \frac{1}{2}h$ of the interval $[x_n, x_{n+1}]$ in two different ways.

On the right-hand side in (4), we replace the [true] slope values $y'(x_n)$, $y'\left(x_n + \frac{1}{2}h\right)$, $y'\left(x_n + \frac{1}{2}h\right)$, and $y'(x_{n+1})$, respectively, with the following estimates.

$$k_1 = f(x_n, y_n) \tag{5a}$$

- This is the Euler method slope at x_n.

$$k_2 = f\left(x_n + \tfrac{1}{2}h,\ y_n + \tfrac{1}{2}hk_1\right) \tag{5b}$$

- This is an estimate of the slope at the midpoint of the interval $[x_n, x_{n+1}]$ using the Euler method to predict the ordinate there.

$$k_3 = f\left(x_n + \tfrac{1}{2}h,\ y_n + \tfrac{1}{2}hk_2\right) \tag{5c}$$

- This is an improved Euler value for the slope at the midpoint.

$$k_4 = f(x_{n+1},\ y_n + hk_3) \tag{5d}$$

- This is the Euler method slope at x_{n+1}, using the improved slope k_3 at the midpoint to step to x_{n+1}.

When these substitutions are made in (4), the result is the iterative formula

$$y_{n+1} = y_n + \frac{h}{6}(k_1 + 2k_2 + 2k_3 + k_4). \tag{6}$$

The use of this formula to compute the approximations $y_1, y_2, y_3, \ldots$ successively constitutes the **Runge–Kutta method.** Note that Eq. (6) takes the "Euler form"

$$y_{n+1} = y_n + h \cdot k$$

if we write

$$k = \frac{1}{6}(k_1 + 2k_2 + 2k_3 + k_4) \tag{7}$$

for the approximate *average slope* on the interval $[x_n, x_{n+1}]$.

The Runge–Kutta method is a *fourth-order* method—it can be proved that the cumulative error on a bounded interval $[a, b]$ with $a = x_0$ is of order h^4. (Thus the iteration in (6) is sometimes called the *fourth-order* Runge–Kutta method because it is possible to develop Runge–Kutta methods of other orders.) That is,

$$|y(x_n) - y_n| \leqq Ch^4, \tag{8}$$

where the constant C depends on the function $f(x, y)$ and the interval $[a, b]$, but does not depend on the step size h. The following example illustrates this high accuracy in comparison with the lower-order accuracy of our previous numerical methods.

Example 1 We first apply the Runge–Kutta method to the illustrative initial value problem

$$\frac{dy}{dx} = x + y, \quad y(0) = 1 \tag{9}$$

that we considered in Fig. 2.4.8 of Section 2.4 and again in Example 2 of Section 2.5. The exact solution of this problem is $y(x) = 2e^x - x - 1$. To make a point we use $h = 0.5$, a larger step size than in any previous example, so only two steps are required to go from $x = 0$ to $x = 1$.

In the first step we use the formulas in (5) and (6) to calculate

$$\begin{aligned}
k_1 &= 0 + 1 = 1,\\
k_2 &= (0 + 0.25) + [1 + (0.25) \cdot (1)] = 1.5,\\
k_3 &= (0 + 0.25) + [1 + (0.25) \cdot (1.5)] = 1.625,\\
k_4 &= (0.5) + [1 + (0.5) \cdot (1.625)] = 2.3125,
\end{aligned}$$

and then

$$y_1 = 1 + \frac{0.5}{6}[1 + 2 \cdot (1.5) + 2 \cdot (1.625) + 2.3125] \approx 1.7969.$$

Similarly, the second step yields $y_2 \approx 3.4347$.

Figure 2.6.1 presents these results together with the results (from Fig. 2.5.4) of applying the improved Euler method with step size $h = 0.1$. We see that even with the larger step size, the Runge–Kutta method gives (for this problem) four to five times the accuracy (in terms of relative percentage errors) of the improved Euler method. ■

	Improved Euler		**Runge–Kutta**		
x	**y with $h = 0.1$**	**Percent Error**	**y with $h = 0.5$**	**Percent Error**	**Actual y**
0.0	1.0000	0.00%	1.0000	0.00%	1.0000
0.5	1.7949	0.14%	1.7969	0.03%	1.7974
1.0	3.4282	0.24%	3.4347	0.05%	3.4366

FIGURE 2.6.1. Runge–Kutta and improved Euler results for the initial value problem $dy/dx = x + y$, $y(0) = 1$.

It is customary to measure the computational labor involved in solving $dy/dx = f(x, y)$ numerically by counting the number of evaluations of the function $f(x, y)$ that are required. In Example 1, the Runge–Kutta method required eight evaluations of $f(x, y) = x + y$ (four at each step), whereas the improved Euler method required 20 such evaluations (two for each of 10 steps). Thus the Runge–Kutta method gave over four times the accuracy with only 40% of the labor.

Computer programs implementing the Runge–Kutta method are listed in the project material for this section. Figure 2.6.2 shows the results obtained by applying the improved Euler and Runge–Kutta methods to the problem $dy/dx = x + y$, $y(0) = 1$ with the same step size $h = 0.1$. The relative error in the improved Euler value at $x = 1$ is about 0.24%, but for the Runge–Kutta value it is 0.00012%. In this comparison the Runge–Kutta method is about 2000 times as accurate, but requires only twice as many function evaluations, as the improved Euler method.

x	Improved Euler y	Runge–Kutta y	Actual y
0.1	1.1100	1.110342	1.110342
0.2	1.2421	1.242805	1.242806
0.3	1.3985	1.399717	1.399718
0.4	1.5818	1.583648	1.583649
0.5	1.7949	1.797441	1.797443
0.6	2.0409	2.044236	2.044238
0.7	2.3231	2.327503	2.327505
0.8	2.6456	2.651079	2.651082
0.9	3.0124	3.019203	3.019206
1.0	3.4282	3.436559	3.436564

FIGURE 2.6.2. Runge–Kutta and improved Euler results for the initial value problem $dy/dx = x + y$, $y(0) = 1$, with the same step size $h = 0.1$.

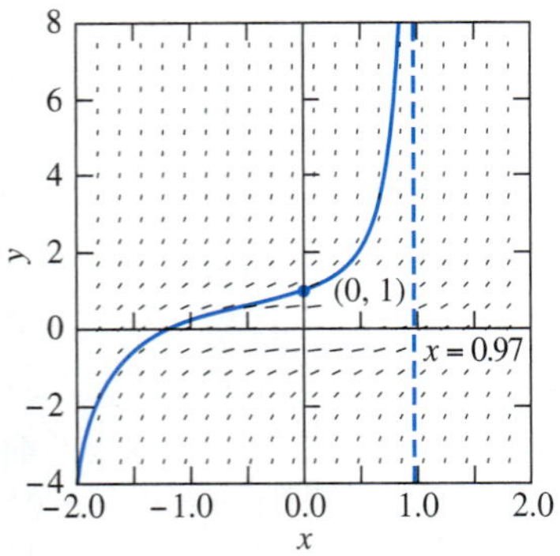

FIGURE 2.6.3. Solutions of $dy/dx = x^2 + y^2$, $y(0) = 1$.

The error bound

$$|y(x_n) - y_n| \leqq Ch^4 \tag{8}$$

for the Runge–Kutta method results in a rapid decrease in the magnitude of errors when the step size h is reduced (except for the possibility that very small step sizes may result in unacceptable roundoff errors). It follows from the inequality in (8) that (on a fixed bounded interval) halving the step size decreases the absolute error by a factor of $\left(\frac{1}{2}\right)^4 = \frac{1}{16}$. Consequently, the common practice of successively halving the step size until the computed results "stabilize" is particularly effective with the Runge–Kutta method.

Example 2 In Example 5 of Section 2.4 we saw that Euler's method is not adequate to approximate the solution $y(x)$ of the initial value problem

$$\frac{dy}{dx} = x^2 + y^2, \quad y(0) = 1 \tag{10}$$

as x approaches the infinite discontinuity near $x = 0.969811$ (see Fig. 2.6.3). Now we apply the Runge–Kutta method to this initial value problem.

Figure 2.6.4 shows Runge–Kutta results on the interval [0.0, 0.9], computed with step sizes $h = 0.1$, $h = 0.05$, and $h = 0.025$. There is still some difficulty near $x = 0.9$, but it seems safe to conclude from these data that $y(0.5) \approx 2.0670$.

x	y with $h = 0.1$	y with $h = 0.05$	y with $h = 0.025$
0.1	1.1115	1.1115	1.1115
0.3	1.4397	1.4397	1.4397
0.5	2.0670	2.0670	2.0670
0.7	3.6522	3.6529	3.6529
0.9	14.0218	14.2712	14.3021

FIGURE 2.6.4. Approximating the solution of the initial value problem in Eq. (10).

We therefore begin anew and apply the Runge–Kutta method to the initial value problem

$$\frac{dy}{dx} = x^2 + y^2, \quad y(0.5) = 2.0670. \tag{11}$$

Figure 2.6.5 shows results on the interval [0.5, 0.9], obtained with step sizes $h = 0.01$, $h = 0.005$, and $h = 0.0025$. We now conclude that $y(0.9) \approx 14.3049$.

x	y with $h = 0.01$	y with $h = 0.005$	y with $h = 0.0025$
0.5	2.0670	2.0670	2.0670
0.6	2.6440	2.6440	2.6440
0.7	3.6529	3.6529	3.6529
0.8	5.8486	5.8486	5.8486
0.9	14.3048	14.3049	14.3049

FIGURE 2.6.5. Approximating the solution of the initial value problem in Eq. (11).

Finally, Fig. 2.6.6 shows results on the interval [0.90, 0.95] for the initial value problem

$$\frac{dy}{dx} = x^2 + y^2, \quad y(0.9) = 14.3049, \tag{12}$$

obtained using step sizes $h = 0.002$, $h = 0.001$, and $h = 0.0005$. Our final approximate result is $y(0.95) \approx 50.4723$. The actual value of the solution at $x = 0.95$ is $y(0.95) \approx 50.471867$. Our slight overestimate results mainly from the fact that the four-place initial value in (12) is (in effect) the result of rounding *up* the actual value $y(0.9) \approx 14.304864$; such errors are magnified considerably as we approach the vertical asymptote. ■

x	y with $h = 0.002$	y with $h = 0.001$	y with $h = 0.0005$
0.90	14.3049	14.3049	14.3049
0.91	16.7024	16.7024	16.7024
0.92	20.0617	20.0617	20.0617
0.93	25.1073	25.1073	25.1073
0.94	33.5363	33.5363	33.5363
0.95	50.4722	50.4723	50.4723

FIGURE 2.6.6. Approximating the solution of the initial value problem in Eq. (12).

Example 3 A skydiver with a mass of 60 kg jumps from a helicopter hovering at an initial altitude of 5 kilometers. Assume that she falls vertically with initial velocity zero and experiences an upward force F_R of air resistance given in terms of her velocity v (in meters per second) by

$$F_R = (0.0096)(100v + 10v^2 + v^3)$$

(in newtons, and with the coordinate axis directed downward so that $v > 0$ during her descent to the ground). If she does not open her parachute, what will be her terminal velocity? How fast will she be falling after 5 s have elapsed? After 10 s? After 20 s?

Solution Newton's law $F = ma$ gives

$$m\frac{dv}{dt} = mg - F_R;$$

that is,

$$60\frac{dv}{dt} = (60)(9.8) - (0.0096)(100v + 10v^2 + v^3) \tag{13}$$

because $m = 60$ and $g = 9.8$. Thus the velocity function $v(t)$ satisfies the initial value problem

$$\frac{dv}{dt} = f(v), \quad v(0) = 0, \tag{14}$$

where

$$f(v) = 9.8 - (0.00016)(100v + 10v^2 + v^3). \tag{15}$$

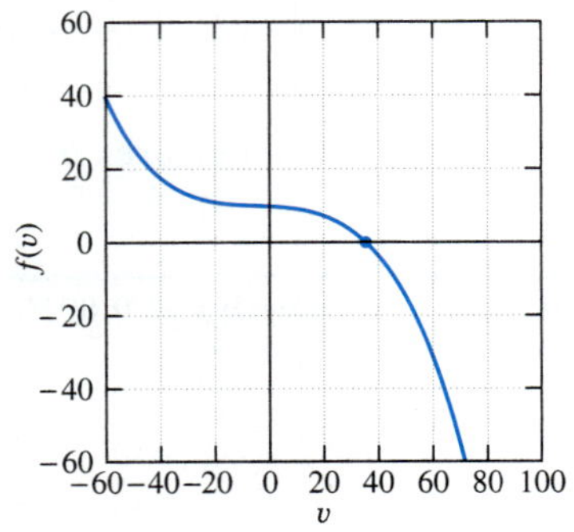

FIGURE 2.6.7. Graph of $f(v) = 9.8 - (0.00016)(100v + 10v^2 + v^3)$.

The skydiver reaches her terminal velocity when the forces of gravity and air resistance balance, so $f(v) = 0$. We can therefore calculate her terminal velocity immediately by solving the equation

$$f(v) = 9.8 - (0.00016)(100v + 10v^2 + v^3) = 0. \tag{16}$$

t (s)	v (m/s)	t (s)	v (m/s)
0	0	11	35.541
1	9.636	12	35.560
2	18.386	13	35.569
3	25.299	14	35.574
4	29.949	15	35.576
5	32.678	16	35.577
6	34.137	17	35.578
7	34.875	18	35.578
8	35.239	19	35.578
9	35.415	20	35.578
10	35.500		

FIGURE 2.6.8. The skydiver's velocity data.

Figure 2.6.7 shows the graph of the function $f(v)$ and exhibits the single real solution $v \approx 35.5780$ (found graphically or by using a calculator or computer **Solve** procedure). Thus the skydiver's terminal speed is approximately 35.578 m/s, about 128 km/h (almost 80 mi/h).

Figure 2.6.8 shows the results of Runge–Kutta approximations to the solution of the initial value problem in (14); the step sizes $h = 0.2$ and $h = 0.1$ yield the same results (to three decimal places). Observe that the terminal velocity is effectively attained in only 15 s. But the skydiver's velocity is 91.85% of her terminal velocity after only 5 s, and 99.78% after 10 s. ■

The final example of this section contains a *warning*: For certain types of initial value problems, the numerical methods we have discussed are not nearly so successful as in the previous examples.

Example 4 Consider the seemingly innocuous initial value problem

$$\frac{dy}{dx} = 5y - 6e^{-x}, \quad y(0) = 1 \tag{17}$$

whose exact solution is $y(x) = e^{-x}$. The table in Fig. 2.6.9 shows the results obtained by applying the Runge–Kutta method on the interval [0, 4] with step sizes

x	Runge–Kutta y with $h = 0.2$	Runge–Kutta y with $h = 0.1$	Runge–Kutta y with $h = 0.05$	Actual y
0.4	0.66880	0.67020	0.67031	0.67032
0.8	0.43713	0.44833	0.44926	0.44933
1.2	0.21099	0.29376	0.30067	0.30119
1.6	−0.46019	0.14697	0.19802	0.20190
2.0	−4.72142	−0.27026	0.10668	0.13534
2.4	−35.53415	−2.90419	−0.12102	0.09072
2.8	−261.25023	−22.05352	−1.50367	0.06081
3.2	−1,916.69395	−163.25077	−11.51868	0.04076
3.6	−14059.35494	−1205.71249	−85.38156	0.02732
4.0	−103,126.5270	−8903.12866	−631.03934	0.01832

FIGURE 2.6.9. Runge–Kutta attempts to solve numerically the initial value problem in Eq. (17).

$h = 0.2$, $h = 0.1$, and $h = 0.05$. Obviously these attempts are spectacularly unsuccessful. Although $y(x) = e^{-x} \to 0$ as $x \to +\infty$, it appears that our numerical approximations are headed toward $-\infty$ rather than zero.

The explanation lies in the fact that the general solution of the equation $dy/dx = 5y - 6e^{-x}$ is

$$y(x) = e^{-x} + Ce^{5x}. \tag{18}$$

The particular solution of (17) satisfying the initial condition $y(0) = 1$ is obtained with $C = 0$. But any departure, however small, from the exact solution $y(x) = e^{-x}$—even if due only to roundoff error—introduces [in effect] a nonzero value of C in Eq. (18). And as indicated in Fig. 2.6.10, all solution curves of the form in (18) with $C \neq 0$ diverge rapidly away from the one with $C = 0$, even if their initial values are close to 1. ■

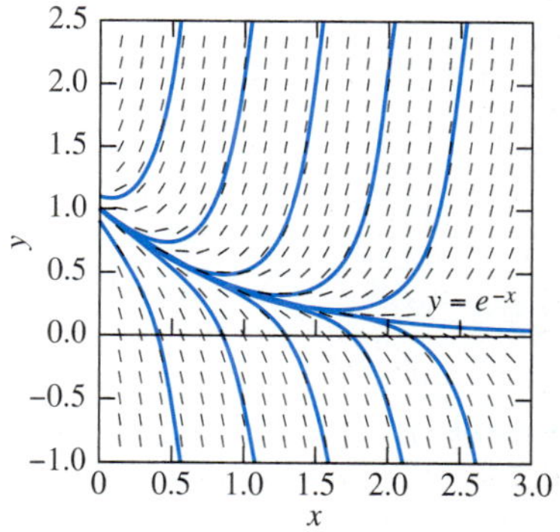

FIGURE 2.6.10. Direction field and solution curves for $dy/dx = 5y - 6e^{-x}$.

Difficulties of the sort illustrated by Example 4 sometimes are unavoidable, but one can at least hope to recognize such a problem when it appears. Approximate values whose order of magnitude varies with changing step size are a common indicator of such instability. These difficulties are discussed in numerical analysis textbooks and are the subject of current research in the field.

2.6 Problems

A hand-held calculator will suffice for Problems 1 through 10, where an initial value problem and its exact solution are given. Apply the Runge–Kutta method to approximate this solution on the interval $[0, 0.5]$ *with step size* $h = 0.25$. *Construct a table showing five-decimal-place values of the approximate solution and actual solution at the points* $x = 0.25$ *and* 0.5.

1. $y' = -y$, $y(0) = 2$; $y(x) = 2e^{-x}$
2. $y' = 2y$, $y(0) = \frac{1}{2}$; $y(x) = \frac{1}{2}e^{2x}$
3. $y' = y + 1$, $y(0) = 1$; $y(x) = 2e^x - 1$
4. $y' = x - y$, $y(0) = 1$; $y(x) = 2e^{-x} + x - 1$
5. $y' = y - x - 1$, $y(0) = 1$; $y(x) = 2 + x - e^x$
6. $y' = -2xy$, $y(0) = 2$; $y(x) = 2e^{-x^2}$
7. $y' = -3x^2y$, $y(0) = 3$; $y(x) = 3e^{-x^3}$
8. $y' = e^{-y}$, $y(0) = 0$; $y(x) = \ln(x + 1)$
9. $y' = \frac{1}{4}(1 + y^2)$, $y(0) = 1$; $y(x) = \tan\frac{1}{4}(x + \pi)$
10. $y' = 2xy^2$, $y(0) = 1$; $y(x) = \dfrac{1}{1 - x^2}$

Note: *The application following this problem set lists illustrative calculator/computer programs that can be used in the remaining problems.*

A programmable calculator or a computer will be useful for Problems 11 through 16. In each problem find the exact solution of the given initial value problem. Then apply the Runge–Kutta method twice to approximate (to five decimal places) this solution on the given interval, first with step size $h = 0.2$, *then with step size* $h = 0.1$. *Make a table showing the approximate values and the actual value, together with the percentage error in the more accurate approximation, for* x *an integral multiple of* 0.2. *Throughout, primes denote derivatives with respect to* x.

11. $y' = y - 2$, $y(0) = 1$; $0 \leqq x \leqq 1$
12. $y' = \frac{1}{2}(y - 1)^2$, $y(0) = 2$; $0 \leqq x \leqq 1$
13. $yy' = 2x^3$, $y(1) = 3$; $1 \leqq x \leqq 2$
14. $xy' = y^2$, $y(1) = 1$; $1 \leqq x \leqq 2$
15. $xy' = 3x - 2y$, $y(2) = 3$; $2 \leqq x \leqq 3$
16. $y^2y' = 2x^5$, $y(2) = 3$; $2 \leqq x \leqq 3$

A computer with a printer is required for Problems 17 through 24. In these initial value problems, use the Runge–Kutta method with step sizes $h = 0.2, 0.1, 0.05$, *and* 0.025 *to approximate to six decimal places the values of the solution at five equally spaced points of the given interval. Print the results in tabular form with appropriate headings to make it easy to gauge the effect of varying the step size* h. *Throughout, primes denote derivatives with respect to* x.

17. $y' = x^2 + y^2$, $y(0) = 0$; $0 \leqq x \leqq 1$
18. $y' = x^2 - y^2$, $y(0) = 1$; $0 \leqq x \leqq 2$
19. $y' = x + \sqrt{y}$, $y(0) = 1$; $0 \leqq x \leqq 2$
20. $y' = x + \sqrt[3]{y}$, $y(0) = -1$; $0 \leqq x \leqq 2$
21. $y' = \ln y$, $y(1) = 2$; $1 \leqq x \leqq 2$
22. $y' = x^{2/3} + y^{2/3}$, $y(0) = 1$; $0 \leqq x \leqq 2$
23. $y' = \sin x + \cos y$, $y(0) = 0$; $0 \leqq x \leqq 1$
24. $y' = \dfrac{x}{1 + y^2}$, $y(-1) = 1$; $-1 \leqq x \leqq 1$
25. As in Problem 25 of Section 2.5, you bail out of a helicopter and immediately open your parachute, so your downward velocity satisfies the initial value problem

$$\frac{dv}{dt} = 32 - 1.6v, \quad v(0) = 0$$

(with t in seconds and v in ft/s). Use the Runge–Kutta method with a programmable calculator or computer to approximate the solution for $0 \leqq t \leqq 2$, first with step size $h = 0.1$ and then with $h = 0.05$, rounding off approximate v-values to three decimal places. What percentage of the limiting velocity 20 ft/s has been attained after 1 second? After 2 seconds?

26. As in Problem 26 of Section 2.5, suppose the deer population $P(t)$ in a small forest initially numbers 25 and satisfies the logistic equation

$$\frac{dP}{dt} = 0.0225P - 0.0003P^2$$

(with t in months). Use the Runge–Kutta method with a programmable calculator or computer to approximate the solution for 10 years, first with step size $h = 6$ and then with $h = 3$, rounding off approximate P-values to four decimal places. What percentage of the limiting population of 75 deer has been attained after 5 years? After 10 years?

Use the Runge–Kutta method with a computer system to find the desired solution values in Problems 27 and 28. Start with step size $h = 1$, *and then use successively smaller step sizes until successive approximate solution values at* $x = 2$ *agree rounded off to five decimal places.*

27. $y' = x^2 + y^2 - 1$, $y(0) = 0$; $y(2) = ?$
28. $y' = x + \frac{1}{2}y^2$, $y(-2) = 0$; $y(2) = ?$

Velocity–Acceleration Problems

In Problems 29 and 30, the linear acceleration $a = dv/dt$ *of a moving particle is given by a formula* $dv/dt = f(t, v)$, *where the velocity* $v = dy/dt$ *is the derivative of the function* $y = y(t)$ *giving the position of the particle at time* t. *Suppose that the velocity* $v(t)$ *is approximated using the Runge–Kutta method to solve numerically the initial value problem*

$$\frac{dv}{dt} = f(t, v), \quad v(0) = v_0. \tag{19}$$

That is, starting with $t_0 = 0$ *and* v_0, *the formulas in Eqs. (5) and (6) are applied—with* t *and* v *in place of* x *and* y*—to calculate the successive approximate velocity values* v_1, v_2,

$v_3, \ldots, v_m$ at the successive times $t_1, t_2, t_3, \ldots, t_m$ (with $t_{n+1} = t_n + h$). Now suppose that we also want to approximate the distance *$y(t)$ traveled by the particle. We can do this by beginning with the initial position $y(0) = y_0$ and calculating*

$$y_{n+1} = y_n + v_n h + \tfrac{1}{2} a_n h^2 \qquad (20)$$

($n = 1, 2, 3, \ldots$), where $a_n = f(t_n, v_n) \approx v'(t_n)$ is the particle's approximate acceleration at time t_n. The formula in (20) would give the correct increment (from y_n to y_{n+1}) if the acceleration a_n remained constant during the time interval $[t_n, t_{n+1}]$.

Thus, once a table of approximate velocities has been calculated, Eq. (20) provides a simple way to calculate a table of corresponding successive positions. This process is illustrated in the project for this section, by beginning with the velocity data in Fig. 2.6.8 (Example 3) and proceeding to follow the skydiver's position during her descent to the ground.

29. Consider again the crossbow bolt of Example 2 in Section 2.3, shot straight upward from the ground with an initial velocity of 49 m/s. Because of linear air resistance, its velocity function $v = dy/dt$ satisfies the initial value problem

$$\frac{dv}{dt} = -(0.04)v - 9.8, \quad v(0) = 49$$

with exact solution $v(t) = 294e^{-t/25} - 245$. **(a)** Use a calculator or computer implementation of the Runge–Kutta method to approximate $v(t)$ for $0 \leqq t \leqq 10$ using both $n = 100$ and $n = 200$ subintervals. Display the results at intervals of 1 second. Do the two approximations—each rounded to four decimal places—agree both with each other and with the exact solution? **(b)** Now use the velocity data from part (a) to approximate $y(t)$ for $0 \leqq t \leqq 10$ using $n = 200$ subintervals. Display the results at intervals of 1 second. Do these approximate position values—each rounded to two decimal places—agree with the exact solution

$$y(t) = 7350\left(1 - e^{-t/25}\right) - 245t?$$

(c) If the exact solution were unavailable, explain how you could use the Runge–Kutta method to approximate closely the bolt's times of ascent and descent and the maximum height it attains.

30. Now consider again the crossbow bolt of Example 3 in Section 2.3. It still is shot straight upward from the ground with an initial velocity of 49 m/s, but because of air resistance proportional to the square of its velocity, its velocity function $v(t)$ satisfies the initial value problem

$$\frac{dv}{dt} = -(0.0011)v|v| - 9.8, \quad v(0) = 49.$$

Beginning with this initial value problem, repeat parts (a) through (c) of Problem 25 (except that you may need $n = 200$ subintervals to get four-place accuracy in part (a) and $n = 400$ subintervals for two-place accuracy in part (b)). According to the results of Problems 17 and 18 in Section 2.3, the bolt's velocity and position functions during ascent and descent are given by the following formulas.

Ascent:
$$v(t) = (94.388)\tan(0.478837 - [0.103827]t),$$
$$y(t) = 108.465 + (909.091)\ln\left(\cos(0.478837 - [0.103827]t)\right);$$

Descent:
$$v(t) = -(94.388)\tanh(0.103827[t - 4.6119]),$$
$$y(t) = 108.465 - (909.091)\ln\left(\cosh(0.103827[t - 4.6119])\right).$$

2.6 Application Runge–Kutta Implementation

Figure 2.6.11 lists TI-85 and BASIC programs implementing the Runge–Kutta method to approximate the solution of the initial value problem

$$\frac{dy}{dx} = x + y, \quad y(0) = 1$$

considered in Example 1 of this section. The comments provided in the final column should make these programs intelligible even if you have little familiarity with the BASIC and TI programming languages.

To apply the Runge–Kutta method to a different equation $dy/dx = f(x, y)$, one need only change the initial line of the program, in which the function f is defined. To increase the number of steps (and thereby decrease the step size), one need only change the value of **N** specified in the second line of the program.

TI-85	BASIC	Comment
`PROGRAM:RK`	`Program RK`	`Program title`
`:F=X+Y`	`DEF FN F(X,Y) = X + Y`	`Define function f`
`:10→N`	`N = 10`	`No. of steps`
`:0→X`	`X = 0`	`Initial x`
`:1→Y`	`Y = 1`	`Initial y`
`:1→X1`	`X1 = 1`	`Final x`
`:(X1-X)/N→H`	`H = (X1-X)/N`	`Step size`
`:For(I,1,N)`	`FOR I=1 TO N`	`Begin loop`
`:X→X0`	`X0 = X`	`Save previous x`
`:Y→Y0`	`Y0 = Y`	`Save previous y`
`:F→K1`	`K1 = FNF(X,Y)`	`First slope`
`:X0+H/2→X`	`X = X0 + H/2`	`Midpoint`
`:Y0+H*K1/2→Y`	`Y = Y0 + H*K1/2`	`Midpt predictor`
`:F→K2`	`K2 = FNF(X,Y)`	`Second slope`
`:Y0+H*K2/2→Y`	`Y = Y0 + H*K2/2`	`Midpt predictor`
`:F→K3`	`K3 = FNF(X,Y)`	`Third slope`
`:X0+H→X`	`X = X0 + H`	`New x`
`:Y0+H*K3→Y`	`Y = Y0 + H*K3`	`Endpt predictor`
`:F→K4`	`K4 = FNF(X,Y)`	`Fourth slope`
`:(K1+2*K2+2*K3 +K4)/6→K`	`K = (K1+2*K2+2*K3 +K4)/6`	`Average slope`
`:Y0+H*K→Y`	`Y = Y0 + H*K`	`Corrector`
`:Disp X,Y`	`PRINT X,Y`	`Display results`
`:End`	`NEXT I`	`End loop`

FIGURE 2.6.11. TI-85 and BASIC Runge–Kutta programs.

Figure 2.6.12 exhibits a MATLAB implementation of the Runge–Kutta method. Suppose that the function f describing the differential equation $y' = f(x, y)$ has been defined. Then the **rk** function takes as input the initial value **x**, the initial value **y**, the final value **x1** of x, and the desired number **n** of subintervals. As output it produces the resulting column vectors **X** and **Y** of x- and y-values. For instance, the MATLAB command

```
[X, Y] = rk(0, 1, 1, 10)
```

then generates the first and third columns of data shown in the table in Fig. 2.6.2.

You should begin this project by implementing the Runge–Kutta method with your own calculator or computer system. Test your program by applying it first to the initial value problem in Example 1, then to some of the problems for this section.

Famous Numbers Revisited, One Last Time

The following problems describe the numbers

$$e \approx 2.71828182846, \quad \ln 2 \approx 0.69314718056, \quad \text{and} \quad \pi \approx 3.14159265359$$

as specific values of certain initial value problems. In each case, apply the Runge–Kutta method with $n = 10, 20, 40, \ldots$ subintervals (doubling n each time). How

```
function yp = f(x,y)
yp = x + y;                          % yp = y'

function [X,Y] = rk(x,y,x1,n)
h = (x1 - x)/n;                      % step size
X = x;                               % initial x
Y = y;                               % initial y
for i = 1:n                          % begin loop
    k1 = f(x,y);                     % first slope
    k2 = f(x+h/2,y+h*k1/2);          % second slope
    k3 = f(x+h/2,y+h*k2/2);          % third slope
    k4 = f(x+h,y+h*k3);              % fourth slope
    k  = (k1+2*k2+2*k3+k4)/6;        % average slope
    x  = x + h;                      % new x
    y  = y + h*k;                    % new y
    X = [X;x];                       % update x-column
    Y = [Y;y];                       % update y-column
    end                              % end loop
```

FIGURE 2.6.12. MATLAB implementation of the Runge–Kutta method.

many subintervals are needed to obtain—twice in succession—the correct value of the target number rounded to nine decimal places?

1. The number $e = y(1)$, where $y(x)$ is the solution of the initial value problem $dy/dx = y$, $y(0) = 1$.
2. The number $\ln 2 = y(2)$, where $y(x)$ is the solution of the initial value problem $dy/dx = 1/x$, $y(1) = 0$.
3. The number $\pi = y(1)$, where $y(x)$ is the solution of the initial value problem $dy/dx = 4/(1 + x^2)$, $y(0) = 0$.

The Skydiver's Descent

The following MATLAB function describes the skydiver's acceleration function in Example 3.

```
function    vp = f(t,v)
vp = 9.8 - 0.00016*(100*v + 10*v^2 + v^3);
```

Then the commands

```
k = 200                          % 200 subintervals
[t,v] = rk(0, 0, 20, k);         % Runge-Kutta approximation
[t(1:10:k+1); v(1:10:k+1)]       % Display every 10th entry
```

produce the table of approximate velocities shown in Fig. 2.6.8. Finally, the commands

```
y = zeros(k+1,1):                             % initialize y
h = 0.1;                                      % step size
for n = 1:k                                   % for n = 1 to k
  a = f(t(n),v(n)):                           % acceleration
  y(n+1) = y(n) + v(n)*h + 0.5*a*h^2;         % Equation (20)
end                                           % end loop
[t(1:20:k+1),v(1:20:k+1),y(1:20:k+1)]         % each 20th entry
```

carry out the position function calculations described in Eq. (20) in the instructions for Problems 29 and 30. The results of these calculations are shown in the table in Fig. 2.6.13. It appears that the skydiver falls 629.866 m during her first 20 s of descent, and then free falls the remaining 4370.134 meters to the ground at her terminal speed of 35.578 m/s. Hence her total time of descent is $20 + (4370.134/35.578) \approx 142.833$ s, or about 2 min 23 s.

t (s)	v (m/s)	y (m)
0	0	0
2	18.386	18.984
4	29.949	68.825
6	34.137	133.763
8	35.239	203.392
10	35.500	274.192
12	35.560	345.266
14	35.574	416.403
16	35.577	487.555
18	35.578	558.710
20	35.578	629.866

FIGURE 2.6.13. The skydiver's velocity and position data.

For an individual problem to solve after implementing these methods using an available computer system, analyze your own skydive (perhaps from a different height), using your own mass m and a plausible air-resistance force of the form $F_R = av + bv^2 + cv^3$.

3 Linear Systems and Matrices

3.1 Introduction to Linear Systems

The subject of **linear algebra** centers around the problem of solving systems of (algebraic) *linear equations*. In high school algebra the *method of elimination* is used to solve systems of two and three (simultaneous) linear equations, with attention ordinarily confined to systems that have one and only one solution, and with applications to "word problems" that have one and only one "answer." In this section we introduce some of the basic terminology of linear algebra by reviewing the elementary technique of elimination from a slightly more general viewpoint. In subsequent sections we apply this same technique to the solution of systems involving many linear equations in many unknowns. The applications of the method of elimination are diverse and important, because so many mathematical problems involve the solution of systems of linear equations.

Recall that if a, b, and c are constants with a and b not both zero, then the graph of the equation

$$ax + by = c \tag{1}$$

is a (straight) *line* in the xy-plane. For this reason, an equation of the form in (1) is called a **linear equation** in the **variables** x and y. Similarly, an equation that can be written in the form

$$ax + by + cz = d \tag{2}$$

is called **linear** in the three variables x, y, and z (even though its graph in xyz-space is a plane rather than a line). Thus the equations

$$3x - 2y = 5 \quad \text{and} \quad x - 7y + 5z = -11$$

are linear, because they involve only the first powers of the variables. By contrast, the equations

$$x + y + xy = 5 \quad \text{and} \quad x^2 + y^3 + \sqrt{z} = 1$$

are *not* linear, because they cannot be rewritten to eliminate the higher powers, roots, and products of the variables.

Two Equations in Two Unknowns

A **system** of linear equations (also called a **linear system**) is simply a finite collection of linear equations involving certain variables. Sometimes we refer to the variables as the "unknowns" in the system. Thus a system of two linear equations in two unknowns x and y may be written in the form

$$\begin{aligned} a_1x + b_1y &= c_1 \\ a_2x + b_2y &= c_2. \end{aligned} \tag{3}$$

By a **solution** of the system in (3) is meant a pair (x, y) of values—normally real numbers—that satisfy *both* equations *simultaneously*.

Example 1 The values $x = 2$, $y = -1$ constitute a solution of the system

$$\begin{aligned} 2x - y &= 5 \\ x + 2y &= 0 \end{aligned} \tag{4}$$

because both $2(2) - (-1) = 5$ and $(2) + 2(-1) = 0$. The values $x = 3$, $y = 1$ satisfy the first equation in (4) but do not satisfy the second. Hence $(3, 1)$ is not a solution of the system in (4). ■

Example 2 The linear system

$$\begin{aligned} x + y &= 1 \\ 2x + 2y &= 3 \end{aligned} \tag{5}$$

has no solution at all, because if $x + y = 1$, then $2x + 2y = 2$, and so $2x + 2y \neq 3$. Thus, two numbers that satisfy the first equation in (5) cannot simultaneously satisfy the second. ■

A linear system is said to be **consistent** if it has at least one solution and **inconsistent** if it has none. Thus the system in Example 1 is consistent, whereas the system in Example 2 is inconsistent.

Three Possibilities

Given a system of linear equations, we ask: What is the set of all solutions of the system? In brief, what is the **solution set** of the system?

In the case of a linear system

$$\begin{aligned} a_1x + b_1y &= c_1 \\ a_2x + b_2y &= c_2 \end{aligned} \tag{6}$$

of two equations in two unknowns, we can use our knowledge of elementary geometry to sort out the possibilities for its solution set. If neither equation in (6) has both left-hand coefficients zero, then their graphs in the xy-plane are two straight lines L_1 and L_2. Then exactly one of the following situations must hold.

- The lines L_1 and L_2 intersect at a single point (as in Figure 3.1.1).
- The lines L_1 and L_2 are parallel nonintersecting lines (as in Figure 3.1.2).
- The lines L_1 and L_2 coincide—they actually are the same line (see Figure 3.1.3).

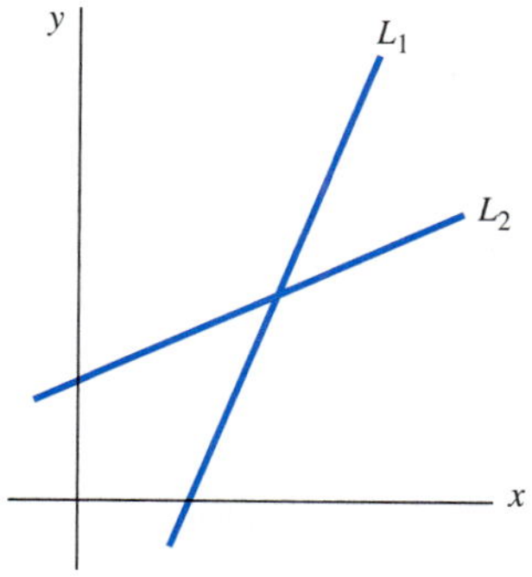

FIGURE 3.1.1. Two intersecting lines: a unique solution.

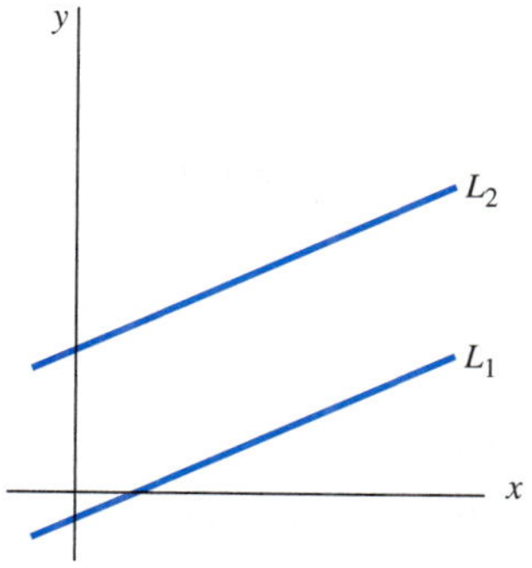

FIGURE 3.1.2. Two parallel lines: no solution.

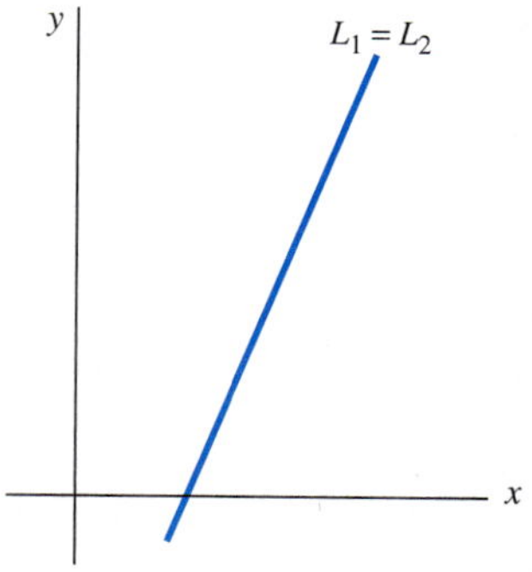

FIGURE 3.1.3. Two coincident lines: infinitely many solutions.

A pair (x, y) of real numbers constitutes a solution of the system in (6) if and only if the point (x, y) in the coordinate plane lies on *both* the two lines L_1 and L_2. In the case shown in Figure 3.1.1, there is exactly one such point. In the case shown in Figure 3.1.2, there is no such point, and in the case in Figure 3.1.3, there are infinitely many such points—every point on the line $L_1 = L_2$ is such a point. We therefore see that there are just *three possibilities* for a linear system of two equations in two unknowns: It has either

- exactly one solution;
- no solution; or
- infinitely many solutions.

It is a fundamental fact of linear algebra (which we will establish in Section 3.3) that, however many equations and variables may appear in a linear system, precisely these same three possibilities occur: A system of m linear equations in n variables either has a **unique** solution (that is, exactly one solution), or it has no solution, or it has infinitely many solutions. Thus it is impossible, for instance, for a linear system to have exactly 2 solutions or to have exactly 17 solutions.

The Method of Elimination

The next three examples illustrate how we can use the elementary method of elimination to solve a system of two equations in two unknowns. To **solve** a system means to determine what its solution set is. The basic idea of the method is this:

- First, we add an appropriate constant multiple of the first equation to the second equation. The idea is to choose the constant in such a way as to eliminate the variable x from the second equation.
- Next, the new second equation contains only the variable y, so we readily solve it for the value of y.
- Finally, we determine the value of x by "back substitution" of this value of y in the first equation.

Examples 3 through 5 illustrate this method in the three cases corresponding to Figures 3.1.1 through 3.1.3 (respectively).

Example 3 In order to solve the system

$$\begin{aligned} 5x + 3y &= 1 \\ x - 2y &= 8, \end{aligned} \tag{7}$$

we first interchange the two equations:

$$\begin{aligned} x - 2y &= 8 \\ 5x + 3y &= 1. \end{aligned}$$

Then we multiply the first equation by -5 and add the resulting terms to the second (without changing the first equation). The result is

$$\begin{aligned} x - 2y &= 8 \\ 13y &= -39. \end{aligned} \tag{8}$$

Now the second equation immediately yields the value $y = -3$, and *back substitution* of this value in the first equation yields

$$x = 2y + 8 = 2(-3) + 8 = 2.$$

Taking it as obvious (for the moment) that the systems in (7) and (8) have the same solution set, we conclude that the original system in (7) has the *unique solution* $x = 2$, $y = -3$. ■

Example 4 To solve the system

$$\begin{aligned} 2x + 6y &= 4 \\ 3x + 9y &= 11, \end{aligned} \tag{9}$$

we first multiply the first equation by $\frac{1}{2}$ and get

$$\begin{aligned} x + 3y &= 2 \\ 3x + 9y &= 11. \end{aligned}$$

We next multiply the first equation by -3 and add each term to the corresponding term in the second equation. The result is

$$\begin{aligned} x + 3y &= 2 \\ 0 &= 5. \end{aligned} \tag{10}$$

What, however, are we to make of the new second equation, $0 = 5$? The system in (10) actually is

$$\begin{aligned} x + 3y &= 2 \\ 0x + 0y &= 5. \end{aligned}$$

Because 0 is simply not equal to 5, there are *no* values of x and y that satisfy the second equation. Hence there certainly can be no values that satisfy both simultaneously. We conclude that the original system in (9) has *no solution*. ■

Example 5 If, instead of the system in (9), we had begun in Example 4 with the system

$$\begin{aligned} 2x + 6y &= 4 \\ 3x + 9y &= 6 \end{aligned} \tag{11}$$

and performed the same operations, we would have obtained, instead of (10), the system

$$\begin{aligned} x + 3y &= 2 \\ 0 &= 0. \end{aligned} \tag{12}$$

Here, $0 = 0$ is shorthand for the equation

$$0x + 0y = 0,$$

which is satisfied by *all* values of x and y. In terms of restrictions or conditions on x and y, one of our original two equations has in effect disappeared, leaving us with the single equation

$$x + 3y = 2. \tag{13}$$

Of course this is hardly surprising, because each equation in (11) is a multiple of the one in (13); in some sense we really had only one equation to begin with. At any rate, we can substitute any value of y we please in (13) and then solve for x. Thus our system in (11) has *infinitely many solutions*. To describe them explicitly, let us write $y = t$, where the **parameter** t is a new independent variable that we will use to generate solution pairs (x, y). Then Equation (13) yields $x = 2 - 3t$, so our infinite solution set of the system in (11) may be described as follows:

$$x = 2 - 3t, \quad y = t$$

as the arbitrary parameter t ranges over the set of all real numbers. For instance, $t = 2$ yields the solution $(-4, 2)$, and $t = -3$ yields the solution $(11, -3)$. ■

Remark: There is some latitude in how we parameterize the solutions of a system having infinitely many solutions, including the names of any parameters we introduce for this purpose. For example, we could parameterize the solutions of the system in (11) alternatively by writing $x = s$ in (13), thereby getting the *different* parameterization

$$x = s, \quad y = \tfrac{1}{3}(2 - s)$$

that yields the *same* solutions. For instance, the parameter values $s = -4$ and $s = 11$ here give the previously noted solutions $(-4, 2)$ and $(11, -3)$ of the original system in (13). ■

COMMENT These three examples illustrate the basic features of the method of elimination, which involves "transforming" a given linear system by means of a sequence of successive steps that do not change the solutions of the system. Each of these steps consists of performing one of the following three **elementary operations**:

1. Multiply one equation by a nonzero constant.
2. Interchange two equations.

3. Add a constant multiple of (the terms of) one equation to (corresponding terms of) another equation.

In subsequent sections of this chapter we discuss the systematic use of these elementary operations to eliminate variables successively in any linear system, whatever the number of equations and variables. In this way we will see that every linear system corresponds to precisely one of the three situations illustrated in Examples 3–5. That is, either

- we discover (as in Example 3) a unique solution of the system; or
- we eventually arrive at an inconsistent equation (as in Example 4), so that the system has no solution; or
- we find that the original system has infinitely many solutions.

Three Equations in Three Unknowns

In the case of a system of three linear equations in three variables x, y, and z, we may proceed as follows: Assuming that the first equation involves x, use the first equation to eliminate x from the second and third equations by adding appropriate multiples of the first equation. Then, assuming that the new second equation involves y, use the new second equation to eliminate y from the new third equation. Solve the new third equation for z, back-substitute in the second equation to determine y, and, finally, back-substitute y and z in the first equation to find x. The following two examples illustrate this procedure.

Example 6 Solve the linear system

$$\begin{aligned} x + 2y + z &= 4 \\ 3x + 8y + 7z &= 20 \\ 2x + 7y + 9z &= 23. \end{aligned} \tag{14}$$

Solution First, we add -3 times the first equation to the second equation; the result is

$$\begin{aligned} x + 2y + z &= 4 \\ 2y + 4z &= 8 \\ 2x + 7y + 9z &= 23. \end{aligned}$$

Then addition of -2 times the first equation to the third equation yields

$$\begin{aligned} x + 2y + z &= 4 \\ 2y + 4z &= 8 \\ 3y + 7z &= 15. \end{aligned}$$

We have now eliminated x from the second and third equations. To simplify the process of eliminating y from the third equation, we multiply the second equation by $\frac{1}{2}$, to obtain

$$\begin{aligned} x + 2y + z &= 4 \\ y + 2z &= 4 \\ 3y + 7z &= 15. \end{aligned}$$

Finally, addition of -3 times the second equation to the third equation gives

$$\begin{aligned} x + 2y + z &= 4 \\ y + 2z &= 4 \\ z &= 3. \end{aligned} \tag{15}$$

This system has a *triangular* form that makes its solution easy. By back substitution of $z = 3$ in the second equation in (15), we find that

$$y = 4 - 2z = 4 - 2(3) = -2;$$

then the first equation yields

$$\begin{aligned} x &= 4 - 2y - z \\ &= 4 - 2(-2) - (3) = 5. \end{aligned}$$

Thus our original system in (14) has the unique solution $x = 5$, $y = -2$, $z = 3$. ■

COMMENT The steps by which we transformed (14) into (15) show that every solution of the system in (14) is a solution of the system in (15). But these steps can be reversed to show similarly that every solution of the system in (15) is also a solution of the system in (14). Thus the two systems are equivalent in the sense that they have the same solution set. The computation at the end of Example 6 shows that (15) has the unique solution $(x, y, z) = (5, -2, 3)$, and it follows that this also is the unique solution of the original system in (14).

Example 7 Solve the system

$$\begin{aligned} 3x - 8y + 10z &= 22 \\ x - 3y + 2z &= 5 \\ 2x - 9y - 8z &= -11. \end{aligned} \tag{16}$$

Solution In order to avoid fractions in the elimination of x, we first interchange the first two equations to get

$$\begin{aligned} x - 3y + 2z &= 5 \\ 3x - 8y + 10z &= 22 \\ 2x - 9y - 8z &= -11. \end{aligned}$$

Addition of -3 times the first equation to the second gives

$$\begin{aligned} x - 3y + 2z &= 5 \\ y + 4z &= 7 \\ 2x - 9y - 8z &= -11, \end{aligned}$$

and then addition of -2 times the first equation to the third equation gives

$$\begin{aligned} x - 3y + 2z &= 5 \\ y + 4z &= 7 \\ -3y - 12z &= -21. \end{aligned}$$

Finally, addition of 3 times the second equation to the third equation yields

$$\begin{aligned} x - 3y + 2z &= 5 \\ y + 4z &= 7 \\ 0 &= 0. \end{aligned} \tag{17}$$

Because the third equation has disappeared, we can choose $z = t$ arbitrarily and then solve for x and y:

$$\begin{aligned} y &= 7 - 4z = 7 - 4t, \\ x &= 5 + 3y - 2z \\ &= 5 + 3(7 - 4t) - 2t = 26 - 14t. \end{aligned}$$

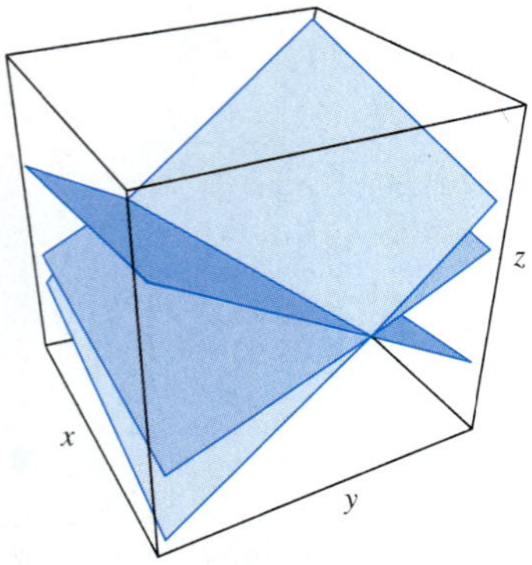

FIGURE 3.1.4. The three planes of Example 7.

Thus our original system in (16) has infinitely many solutions. Moreover, one convenient way of describing them is this:

$$x = 26 - 14t, \quad y = 7 - 4t, \quad z = t. \tag{18}$$

The arbitrary parameter t can take on all real number values, and in so doing generates all the (infinitely many) solutions of the original system in (16). Figure 3.1.4 shows the three planes corresponding to the three equations in (16). These planes intersect in a straight line that is parameterized by the equations in (18). ■

A Differential Equations Application

In Chapter 1 we saw that a general solution of a first-order differential equation involves an arbitrary constant that must be determined in order to satisfy a given initial condition. The next example illustrates the fact that a general solution of a second-order differential equation typically involves *two* arbitrary constants. An initial value problem for such a differential equation ordinarily involves *two* initial conditions, whose imposition leads to a system of two equations with the two arbitrary constants as unknowns.

Example 8 If A and B are constants and

$$y(x) = Ae^{3x} + Be^{-3x}, \tag{19}$$

then differentiation yields

$$y'(x) = 3Ae^{3x} - 3Be^{-3x} \tag{20}$$

and

$$y''(x) = 9Ae^{3x} + 9Be^{-3x} = 9y(x).$$

Thus the function $y(x)$ defined by (19) satisfies the second-order differential equation

$$y'' - 9y = 0. \tag{21}$$

Now, suppose that we want to solve the initial value problem consisting of this equation and the two initial conditions

$$y(0) = 7, \quad y'(0) = 9. \tag{22}$$

Then substitution of $x = 0$ in (19) and (20) yields the linear system

$$\begin{aligned} A + B &= 7 \\ 3A - 3B &= 9 \end{aligned}$$

that we readily solve for $A = 5$, $B = 2$. It follows that the particular solution

$$y(x) = 5e^{3x} + 2e^{-3x}$$

satisfies both the differential equation (21) and the initial conditions in (22). ■

In Chapter 5 we will study higher-order differential equations and initial value problems in detail, after developing the necessary methods of linear algebra in the intervening chapters.

3.1 Problems

In each of Problems 1–22, use the method of elimination to determine whether the given linear system is consistent or inconsistent. For each consistent system, find the solution if it is unique; otherwise, describe the infinite solution set in terms of an arbitrary parameter t (as in Examples 5 and 7).

1. $\begin{aligned} x + 3y &= 9 \\ 2x + y &= 8 \end{aligned}$

2. $\begin{aligned} 3x + 2y &= 9 \\ x - y &= 8 \end{aligned}$

3. $\begin{aligned} 2x + 3y &= 1 \\ 3x + 5y &= 3 \end{aligned}$

4. $\begin{aligned} 5x - 6y &= 1 \\ 6x - 5y &= 10 \end{aligned}$

5. $\begin{aligned} x + 2y &= 4 \\ 2x + 4y &= 9 \end{aligned}$

6. $\begin{aligned} 4x - 2y &= 4 \\ 6x - 3y &= 7 \end{aligned}$

7. $\begin{aligned} x - 4y &= -10 \\ -2x + 8y &= 20 \end{aligned}$

8. $\begin{aligned} 3x - 6y &= 12 \\ 2x - 4y &= 8 \end{aligned}$

9. $\begin{aligned} x + 5y + z &= 2 \\ 2x + y - 2z &= 1 \\ x + 7y + 2z &= 3 \end{aligned}$

10. $\begin{aligned} x + 3y + 2z &= 2 \\ 2x + 7y + 7z &= -1 \\ 2x + 5y + 2z &= 7 \end{aligned}$

11. $\begin{aligned} 2x + 7y + 3z &= 11 \\ x + 3y + 2z &= 2 \\ 3x + 7y + 9z &= -12 \end{aligned}$

12. $\begin{aligned} 3x + 5y - z &= 13 \\ 2x + 7y + z &= 28 \\ x + 7y + 2z &= 32 \end{aligned}$

13. $\begin{aligned} 3x + 9y + 7z &= 0 \\ 2x + 7y + 4z &= 0 \\ 2x + 6y + 5z &= 0 \end{aligned}$

14. $\begin{aligned} 4x + 9y + 12z &= -1 \\ 3x + y + 16z &= -46 \\ 2x + 7y + 3z &= 19 \end{aligned}$

15. $\begin{aligned} x + 3y + 2z &= 5 \\ x - y + 3z &= 3 \\ 3x + y + 8z &= 10 \end{aligned}$

16. $\begin{aligned} x - 3y + 2z &= 6 \\ x + 4y - z &= 4 \\ 5x + 6y + z &= 20 \end{aligned}$

17. $\begin{aligned} 2x - y + 4z &= 7 \\ 3x + 2y - 2z &= 3 \\ 5x + y + 2z &= 15 \end{aligned}$

18. $\begin{aligned} x + 5y + 6z &= 3 \\ 5x + 2y - 10z &= 1 \\ 8x + 17y + 8z &= 5 \end{aligned}$

19. $\begin{aligned} x - 2y + z &= 2 \\ 2x - y - 4z &= 13 \\ x - y - z &= 5 \end{aligned}$

20. $\begin{aligned} 2x + 3y + 7z &= 15 \\ x + 4y + z &= 20 \\ x + 2y + 3z &= 10 \end{aligned}$

21. $\begin{aligned} x + y - z &= 5 \\ 3x + y + 3z &= 11 \\ 4x + y + 5z &= 14 \end{aligned}$

22. $\begin{aligned} 4x - 2y + 6z &= 0 \\ x - y - z &= 0 \\ 2x - y + 3z &= 0 \end{aligned}$

In each of Problems 23–30, a second-order differential equation and its general solution y(x) are given. Determine the constants A and B so as to find a solution of the differential equation that satisfies the given initial conditions involving y(0) and y′(0).

23. $y'' + 4y = 0$, $y(x) = A\cos 2x + B\sin 2x$, $y(0) = 3$, $y'(0) = 8$

24. $y'' - 9y = 0$, $y(x) = A\cosh 3x + B\sinh 3x$, $y(0) = 5$, $y'(0) = 12$

25. $y'' - 25y = 0$, $y(x) = Ae^{5x} + Be^{-5x}$, $y(0) = 10$, $y'(0) = 20$

26. $y'' - 121y = 0$, $y(x) = Ae^{11x} + Be^{-11x}$, $y(0) = 44$, $y'(0) = 22$

27. $y'' + 2y' - 15y = 0$, $y(x) = Ae^{3x} + Be^{-5x}$, $y(0) = 40$, $y'(0) = -16$

28. $y'' - 10y' + 21y = 0$, $y(x) = Ae^{3x} + Be^{7x}$, $y(0) = 15$, $y'(0) = 13$

29. $6y'' - 5y' + y = 0$, $y(x) = Ae^{x/2} + Be^{x/3}$, $y(0) = 7$, $y'(0) = 11$

30. $15y'' + y' - 28y = 0$, $y(x) = Ae^{4x/3} + Be^{-7x/5}$, $y(0) = 41$, $y'(0) = 164$

31. A system of the form

$$\begin{aligned} a_1x + b_1y &= 0 \\ a_2x + b_2y &= 0, \end{aligned}$$

in which the constants on the right-hand side are all zero, is said to be **homogeneous**. Explain by geometric reasoning why such a system has either a unique solution or infinitely many solutions. In the former case, what is the unique solution?

32. Consider the system

$$\begin{aligned} a_1x + b_1y + c_1z &= d_1 \\ a_2x + b_2y + c_2z &= d_2 \end{aligned}$$

of two equations in three unknowns.

(a) Use the fact that the graph of each such equation is a plane in xyz-space to explain why such a system always has either no solution or infinitely many solutions.

(b) Explain why the system must have infinitely many solutions if $d_1 = 0 = d_2$.

33. The linear system

$$\begin{aligned} a_1x + b_1y &= c_1 \\ a_2x + b_2y &= c_2 \\ a_3x + b_3y &= c_3 \end{aligned}$$

of three equations in two unknowns represents three lines L_1, L_2, and L_3 in the xy-plane. Figure 3.1.5 shows six possible configurations of these three lines. In each case describe the solution set of the system.

34. Consider the linear system

$$\begin{aligned} a_1x + b_1y + c_1z &= d_1 \\ a_2x + b_2y + c_2z &= d_2 \\ a_3x + b_3y + c_3z &= d_3 \end{aligned}$$

of three equations in three unknowns to represent three planes P_1, P_2, and P_3 in xyz-space. Describe the solution set of the system in each of the following cases.

(a) The three planes are parallel and distinct.

(b) The three planes coincide—$P_1 = P_2 = P_3$.

(c) P_1 and P_2 coincide and are parallel to P_3.

(d) P_1 and P_2 intersect in a line L that is parallel to P_3.

(e) P_1 and P_2 intersect in a line L that lies in P_3.

(f) P_1 and P_2 intersect in a line L that intersects P_3 in a single point.

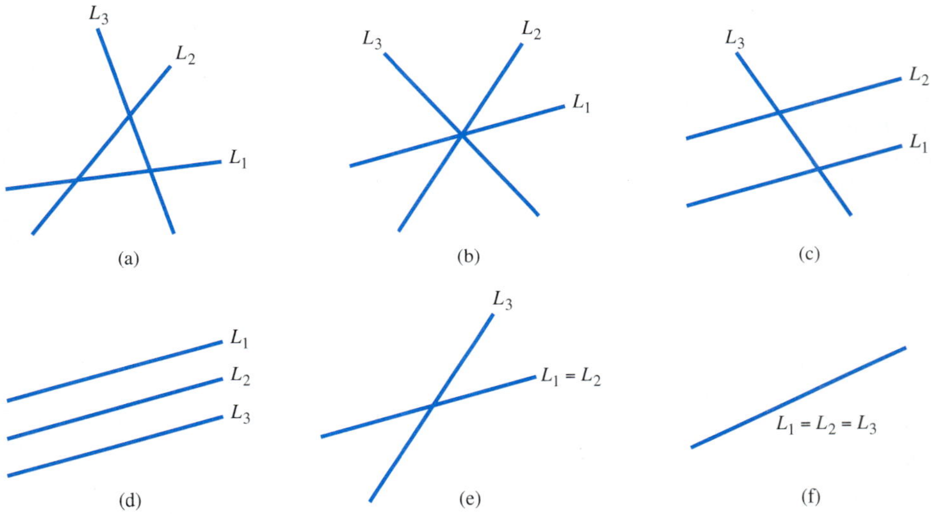

FIGURE 3.1.5. Three lines in the plane (Problem 33).

3.2 Matrices and Gaussian Elimination

In Example 6 of Section 3.1 we applied the method of elimination to solve the linear system

$$\begin{aligned} 1x + 2y + 1z &= 4 \\ 3x + 8y + 7z &= 20 \\ 2x + 7y + 9z &= 23. \end{aligned} \tag{1}$$

There we employed elementary operations to transform this system into the equivalent system

$$\begin{aligned} 1x + 2y + 1z &= 4 \\ 0x + 1y + 2z &= 4 \\ 0x + 0y + 1z &= 3, \end{aligned} \tag{2}$$

which we found easy to solve by back substitution. Here we have printed in color the coefficients and constants (including the 0s and 1s that would normally be omitted) because everything else—the symbols x, y, and z for the variables and the $+$ and $=$ signs—is excess baggage that means only extra writing, for we can keep track of these symbols mentally. In effect, in Example 6 we used an appropriate sequence of operations to transform the array

$$\begin{bmatrix} 1 & 2 & 1 & 4 \\ 3 & 8 & 7 & 20 \\ 2 & 7 & 9 & 23 \end{bmatrix} \tag{3}$$

of coefficients and constants in (1) into the array

$$\begin{bmatrix} 1 & 2 & 1 & 4 \\ 0 & 1 & 2 & 4 \\ 0 & 0 & 1 & 3 \end{bmatrix} \tag{4}$$

of constants and coefficients in (2).

Rectangular arrays of numbers like those in (3) and (4) are called *matrices.* Thus, a **matrix** is simply a rectangular array of numbers, which are called the **entries** or **elements** of the matrix. The **size**, or **shape**, of a matrix is specified by telling how many horizontal **rows** and vertical **columns** it has. Each matrix in (3) and (4) has three rows and four columns; therefore, each is a 3×4 (read "three by four") matrix. We *always* specify *first* the number of rows and *then* the number of columns.

Example 1 The matrices

$$\begin{bmatrix} 3 & -7 \\ -2 & 5 \end{bmatrix} \qquad \begin{bmatrix} 3 & 0 & -1 & 5 \end{bmatrix} \qquad \begin{bmatrix} 2 \\ 0 \\ -3 \end{bmatrix}$$

have sizes 2×2, 1×4, and 3×1, respectively. ■

Coefficient Matrices

A general system of m linear equations in the n variables $x_1, x_2, \ldots, x_n$ may be written in the form

$$\begin{aligned} a_{11}x_1 + a_{12}x_2 + a_{13}x_3 + \cdots + a_{1n}x_n &= b_1 \\ a_{21}x_1 + a_{22}x_2 + a_{23}x_3 + \cdots + a_{2n}x_n &= b_2 \\ &\;\;\vdots \\ a_{m1}x_1 + a_{m2}x_2 + a_{m3}x_3 + \cdots + a_{mn}x_n &= b_m. \end{aligned} \tag{5}$$

Observe that a_{ij} denotes the (constant) coefficient in the ith equation of the jth variable x_j and that b_i denotes the constant on the right-hand side in the ith equation. Thus the first subscript i refers to the equation and the second subscript j to the variable. For instance, a_{32} denotes the coefficient of x_2 in the 3rd equation. This scheme of systematic double subscripts enables us to specify readily the location of each coefficient in a system with many equations and variables. It also helps us get rid of the excess baggage in (5) by focusing our attention on the coefficients.

The **coefficient matrix** of the linear system in (5) is the $m \times n$ matrix

$$\mathbf{A} = \begin{bmatrix} a_{11} & a_{12} & a_{13} & \cdots & a_{1n} \\ a_{21} & a_{22} & a_{23} & \cdots & a_{2n} \\ \vdots & \vdots & \vdots & \ddots & \vdots \\ a_{m1} & a_{m2} & a_{m3} & \cdots & a_{mn} \end{bmatrix}. \tag{6}$$

We will use **boldface** capital letters to denote matrices and lowercase letters to denote numbers. When we want to refer more briefly to the matrix **A** and its entries, we can write

$$\mathbf{A} = \left[\, a_{ij} \,\right] = \begin{bmatrix} & \vdots & \\ \cdots\cdots & a_{ij} & \cdots\cdots \\ & \vdots & \end{bmatrix} \begin{matrix} \\ (i\text{th row}) \\ \\ \end{matrix} \tag{7}$$
$$(j\text{th column})$$

The first subscript i specifies the row and the second subscript j the column of **A** in which the element a_{ij} appears:

First subscript	Second subscript
Row	Column

Although double subscripts may seem tedious when first encountered, their usefulness should not be underestimated. For instance, they are consistent with the notation used in most programming languages. Many practical applications lead to linear systems with hundreds or even thousands of variables and equations. In a typical computer system, the coefficient matrix **A** of such a system would be represented by a two-dimensional array in which $A(i, j)$ denotes a_{ij}. With this approach, the computer can store a 100×100 matrix in the same way that it stores a 3×3 matrix.

The matrices in (3) and (4) include not only the coefficients but also the constants on the right-hand sides in the corresponding linear systems in (1) and (2). Let us write

$$\mathbf{b} = \begin{bmatrix} b_1 \\ b_2 \\ \vdots \\ b_m \end{bmatrix} \tag{8}$$

for the column of constants in the general system in (5). An $m \times 1$ matrix—that is, one with a single column—is often called a (**column**) **vector** and is denoted by a boldface letter. When we adjoin the constant vector **b** to the coefficient matrix **A** (as a final column), we get the matrix

$$\left[\, \mathbf{A} \quad \mathbf{b} \,\right] = \begin{bmatrix} a_{11} & a_{12} & \cdots & a_{1n} & b_1 \\ a_{21} & a_{22} & \cdots & a_{2n} & b_2 \\ \vdots & \vdots & \ddots & \vdots & \vdots \\ a_{m1} & a_{m2} & \cdots & a_{mn} & b_m \end{bmatrix}. \tag{9}$$

This $m \times (n + 1)$ matrix is called the **augmented coefficient matrix**, or simply the augmented matrix, of the $m \times n$ system in (5).

Although it takes a good deal of notation to describe the augmented matrix of a *general* linear system, it is a very simple matter to write the augmented matrix of a *particular* system. If the system is written in chalk on a blackboard with the variables in the same order in each equation, we merely erase all the x_j's and the plus and equal signs (retaining a minus sign for each negative coefficient) and insert a zero in each spot where a variable is missing in an equation.

Example 2 The augmented coefficient matrix of the system

$$\begin{aligned} 2x_1 + 3x_2 - 7x_3 + 4x_4 &= 6 \\ x_2 + 3x_3 - 5x_4 &= 0 \\ -x_1 + 2x_2 \qquad\quad - 9x_4 &= 17 \end{aligned}$$

of three equations in four variables is the 3×5 matrix

$$\begin{bmatrix} 2 & 3 & -7 & 4 & 6 \\ 0 & 1 & 3 & -5 & 0 \\ -1 & 2 & 0 & -9 & 17 \end{bmatrix}.$$

■

Elementary Row Operations

In Section 3.1 we described the three elementary operations that are used in the method of elimination. To each of these corresponds an *elementary row operation* on the augmented matrix of the system. For instance, when we interchange two equations in the system, we interchange the corresponding two rows of its augmented coefficient matrix.

DEFINITION Elementary Row Operations

The following are the three types of **elementary row operations** on the matrix **A**:

1. Multiply any (single) row of **A** by a nonzero constant.
2. Interchange two rows of **A**.
3. Add a constant multiple of one row of **A** to another row.

Figure 3.2.1 shows notation that we use to describe elementary row operations briefly. For instance,

$$\begin{bmatrix} 1 & 2 \\ 3 & 4 \end{bmatrix} \xrightarrow{(3)R_1+R_2} \begin{bmatrix} 1 & 2 \\ 6 & 10 \end{bmatrix} \tag{10}$$

shows the result of adding 3 times row 1 to row 2. Note that, when we perform the operation $(c)R_p + R_q$ on a matrix—that is, when we add c times row p to row q—row p itself remains unchanged.

Type	Row Operation	Notation
1	Multiply row p by c	cR_p
2	Interchange row p and row q	$\text{SWAP}(R_p, R_q)$
3	Add c times row p to row q	$(c)R_p + R_q$

FIGURE 3.2.1. Notation for elementary row operations.

Example 3 To solve the system

$$\begin{aligned} x_1 + 2x_2 + x_3 &= 4 \\ 3x_1 + 8x_2 + 7x_3 &= 20 \\ 2x_1 + 7x_2 + 9x_3 &= 23, \end{aligned} \tag{11}$$

whose augmented coefficient matrix is exhibited in (3), we carry out the following sequence of elementary row operations, corresponding to the steps in the solution of Example 6 in Section 3.1:

$$\begin{bmatrix} 1 & 2 & 1 & 4 \\ 3 & 8 & 7 & 20 \\ 2 & 7 & 9 & 23 \end{bmatrix} \tag{12}$$

$$\xrightarrow{(-3)R_1 + R_2} \begin{bmatrix} 1 & 2 & 1 & 4 \\ 0 & 2 & 4 & 8 \\ 2 & 7 & 9 & 23 \end{bmatrix}$$

$$\xrightarrow{(-2)R_1 + R_3} \begin{bmatrix} 1 & 2 & 1 & 4 \\ 0 & 2 & 4 & 8 \\ 0 & 3 & 7 & 15 \end{bmatrix}$$

$$\xrightarrow{\left(\frac{1}{2}\right)R_2} \begin{bmatrix} 1 & 2 & 1 & 4 \\ 0 & 1 & 2 & 4 \\ 0 & 3 & 7 & 15 \end{bmatrix}$$

$$\xrightarrow{(-3)R_2 + R_3} \begin{bmatrix} 1 & 2 & 1 & 4 \\ 0 & 1 & 2 & 4 \\ 0 & 0 & 1 & 3 \end{bmatrix}. \tag{13}$$

The final matrix here is the augmented coefficient matrix of the system

$$\begin{aligned} x_1 + 2x_2 + x_3 &= 4 \\ x_2 + 2x_3 &= 4. \\ x_3 &= 3 \end{aligned} \tag{14}$$

whose unique solution (readily found by back substitution) is $x_1 = 5$, $x_2 = -2$, $x_3 = 3$. ■

It is not quite self-evident that a sequence of elementary row operations produces a linear system having the same solution set as the original system. To state the pertinent result concisely, we need the following definition.

> **DEFINITION Row-Equivalent Matrices**
>
> Two matrices are called **row equivalent** if one can be obtained from the other by a (finite) sequence of elementary row operations.

Thus the two matrices in (10) are row equivalent, as are the two matrices in (12) and (13). In Problem 29 we ask you to show that if **B** can be obtained from **A** by elementary row operations, then **A** can be obtained from **B** by elementary row operations. This follows from the observation that row operations are "reversible." In Problem 30 we suggest a proof of the following theorem.

THEOREM 1 Equivalent Systems and Equivalent Matrices

If the augmented coefficient matrices of two linear systems are row equivalent, then the two systems have the same solution set.

Thus, the linear systems in (11) and (14) have the same solution set, because their augmented matrices in (12) and (13) are row equivalent.

Echelon Matrices

Up to this point we have been somewhat informal in our description of the method of elimination. Its objective is to transform a given linear system, by elementary row operations, into one for which back substitution leads easily and routinely to a solution. The following definition tells what should be the appearance of the augmented coefficient matrix of the transformed system.

DEFINITION Echelon Matrix

The matrix **E** is called an **echelon matrix** provided it has the following two properties:

1. Every row of **E** that consists entirely of zeros (if any) lies *beneath* every row that contains a nonzero element.
2. In each row of **E** that contains a nonzero element, the *first* nonzero element lies strictly to the *right* of the first nonzero element in the preceding row (if there is a preceding row).

Echelon matrices are sometimes called *row-echelon matrices*. Property 1 says that if **E** has any all-zero rows, then they are grouped together at the bottom of the matrix. The first (from the left) *nonzero* element in each of the other rows is called its **leading entry**. Property 2 says that the leading entries form a "descending staircase" pattern from upper left to lower right, as in the following echelon matrix.

$$\mathbf{E} = \begin{bmatrix} 2 & -1 & 0 & 4 & 7 \\ 0 & 1 & 2 & 0 & -5 \\ 0 & 0 & 0 & 3 & 0 \\ 0 & 0 & 0 & 0 & 0 \end{bmatrix}$$

Here we have highlighted the leading entries and indicated the descending staircase pattern.

The following list exhibits *all* possibilities for the form of a 3 × 3 echelon matrix, with asterisks denoting the (nonzero) leading entries; a # denotes an element that may be either zero or nonzero.

$$\begin{bmatrix} * & \# & \# \\ 0 & * & \# \\ 0 & 0 & * \end{bmatrix} \quad \begin{bmatrix} 0 & 0 & 0 \\ 0 & 0 & 0 \\ 0 & 0 & 0 \end{bmatrix}$$

$$\begin{bmatrix} * & \# & \# \\ 0 & * & \# \\ 0 & 0 & 0 \end{bmatrix} \quad \begin{bmatrix} * & \# & \# \\ 0 & 0 & * \\ 0 & 0 & 0 \end{bmatrix} \quad \begin{bmatrix} 0 & * & \# \\ 0 & 0 & * \\ 0 & 0 & 0 \end{bmatrix}$$

$$\begin{bmatrix} * & \# & \# \\ 0 & 0 & 0 \\ 0 & 0 & 0 \end{bmatrix} \quad \begin{bmatrix} 0 & * & \# \\ 0 & 0 & 0 \\ 0 & 0 & 0 \end{bmatrix} \quad \begin{bmatrix} 0 & 0 & * \\ 0 & 0 & 0 \\ 0 & 0 & 0 \end{bmatrix}$$

It follows from Properties 1 and 2 that the elements *beneath* any leading entry in the same column are all zero. The matrices

$$\mathbf{A} = \begin{bmatrix} 1 & 3 & -2 \\ 0 & 0 & 0 \\ 0 & 1 & 5 \end{bmatrix} \quad \text{and} \quad \mathbf{B} = \begin{bmatrix} 0 & 1 & 5 \\ 1 & 3 & -2 \\ 0 & 0 & 0 \end{bmatrix}$$

are not echelon matrices because **A** does not have Property 1 and **B** does not have Property 2.

Suppose that a linear system is in **echelon form**—its augmented matrix is an echelon matrix. Then those variables that correspond to *columns* containing leading entries are called **leading variables**; all the other variables are called **free variables**. The following algorithm describes the process of **back substitution** to solve such a system.

ALGORITHM Back Substitution

To solve a consistent linear system in echelon form by back substitution, carry out the following steps.

1. Set each free variable equal to an arbitrary parameter (a different parameter for each free variable).
2. Solve the *final* (nonzero) equation for its leading variable.
3. Substitute the result in the next-to-last equation and then solve for its leading variable.
4. Continuing in this fashion, work upward through the system of equations until all variables have been determined.

Example 4 The augmented coefficient matrix of the system

$$\begin{aligned} x_1 - 2x_2 + 3x_3 + 2x_4 + x_5 &= 10 \\ x_3 + 2x_5 &= -3 \\ x_4 - 4x_5 &= 7 \end{aligned} \tag{15}$$

is the echelon matrix

$$\begin{bmatrix} 1 & -2 & 3 & 2 & 1 & 10 \\ 0 & 0 & 1 & 0 & 2 & -3 \\ 0 & 0 & 0 & 1 & -4 & 7 \end{bmatrix}. \tag{16}$$

The leading entries are in the first, third, and fourth columns. Hence x_1, x_3, and x_4 are the leading variables and x_2 and x_5 are the free variables. To solve the system by back substitution, we first write

$$x_2 = s, \quad x_5 = t, \tag{17a}$$

where s and t are arbitrary parameters. Then the third equation in (15) gives

$$x_4 = 7 + 4x_5 = 7 + 4t; \tag{17b}$$

the second equation gives

$$x_3 = -3 - 2x_5 = -3 - 2t; \tag{17c}$$

finally, the first equation in (15) yields

$$\begin{aligned} x_1 &= 10 + 2x_2 - 3x_3 - 2x_4 - x_5 \\ &= 10 + 2s - 3(-3 - 2t) - 2(7 + 4t) - t. \end{aligned}$$

Therefore,

$$x_1 = 5 + 2s - 3t. \tag{17d}$$

Thus the system in (15) has an infinite solution set consisting of all $(x_1, x_2, x_3, x_4, x_5)$ given in terms of the two parameters s and t as follows:

$$\begin{aligned} x_1 &= 5 + 2s - 3t \\ x_2 &= s \\ x_3 &= -3 - 2t \\ x_4 &= 7 + 4t \\ x_5 &= t. \end{aligned}$$

For instance, with $s = 2$ and $t = 1$ we get the solution $x_1 = 6$, $x_2 = 2$, $x_3 = -5$, $x_4 = 11$, and $x_5 = 1$. ■

Gaussian Elimination

Because we can use back substitution to solve any linear system already in echelon form, it remains only to establish that we can transform any matrix (using elementary row operations) into an echelon matrix. The procedure that we have already illustrated in several examples is known as **Gaussian elimination**. The following systematic description of the procedure makes it clear that it succeeds with *any* given matrix **A**.

ALGORITHM Gaussian Elimination

1. Locate the first column of **A** that contains a nonzero element.
2. If the first (top) entry in this column is zero, interchange the first row of **A** with a row in which the corresponding entry is nonzero.
3. Now the first entry in our column is nonzero. Replace the entries below it in the same column with zeros by adding appropriate multiples of the first row of **A** to lower rows.
4. After Steps 1–3, the matrix looks like the matrix below, although there may be several initial columns of zeros or even none at all. Perform Steps 1–3 on the indicated lower right matrix $\mathbf{A}_1$.

ALGORITHM Gaussian Elimination (continued)

5. Repeat this cycle of steps until an echelon matrix is obtained.

$$\mathbf{A} = \left[\begin{array}{cc|cccc} 0 & * & \# & \# & \cdots & \# \\ \hline 0 & 0 & & & & \\ 0 & 0 & & \mathbf{A}_1 & & \\ \vdots & \vdots & & & & \\ 0 & 0 & & & & \end{array}\right] \tag{17}$$

In brief, we work on the matrix **A** one column at a time, from left to right. In each column containing a leading entry (perhaps after a row interchange), we "clear out" the nonzero elements below it and then move on to the next column.

Example 5 To solve the system

$$\begin{aligned} x_1 - 2x_2 + 3x_3 + 2x_4 + x_5 &= 10 \\ 2x_1 - 4x_2 + 8x_3 + 3x_4 + 10x_5 &= 7 \\ 3x_1 - 6x_2 + 10x_3 + 6x_4 + 5x_5 &= 27, \end{aligned} \tag{18}$$

we reduce its augmented coefficient matrix to echelon form as follows.

$$\begin{bmatrix} 1 & -2 & 3 & 2 & 1 & 10 \\ 2 & -4 & 8 & 3 & 10 & 7 \\ 3 & -6 & 10 & 6 & 5 & 27 \end{bmatrix}$$

$$\xrightarrow{(-2)R_1 + R_2} \begin{bmatrix} 1 & -2 & 3 & 2 & 1 & 10 \\ 0 & 0 & 2 & -1 & 8 & -13 \\ 3 & -6 & 10 & 6 & 5 & 27 \end{bmatrix}$$

$$\xrightarrow{(-3)R_1 + R_3} \begin{bmatrix} 1 & -2 & 3 & 2 & 1 & 10 \\ 0 & 0 & 2 & -1 & 8 & -13 \\ 0 & 0 & 1 & 0 & 2 & -3 \end{bmatrix}$$

$$\xrightarrow{\text{SWAP}(R_2, R_3)} \begin{bmatrix} 1 & -2 & 3 & 2 & 1 & 10 \\ 0 & 0 & 1 & 0 & 2 & -3 \\ 0 & 0 & 2 & -1 & 8 & -13 \end{bmatrix}$$

$$\xrightarrow{(-2)R_2 + R_3} \begin{bmatrix} 1 & -2 & 3 & 2 & 1 & 10 \\ 0 & 0 & 1 & 0 & 2 & -3 \\ 0 & 0 & 0 & -1 & 4 & -7 \end{bmatrix}$$

$$\xrightarrow{(-1)R_3} \begin{bmatrix} 1 & -2 & 3 & 2 & 1 & 10 \\ 0 & 0 & 1 & 0 & 2 & -3 \\ 0 & 0 & 0 & 1 & -4 & 7 \end{bmatrix}$$

Our final result is the echelon matrix in (16), so by Eqs. (17a)–(17d) in Example 4, the infinite solution set of the system in (18) is described in terms of arbitrary parameters s and t as follows:

$$\begin{aligned} x_1 &= 5 + 2s - 3t \\ x_2 &= s \\ x_3 &= -3 - 2t \\ x_4 &= 7 + 4t \\ x_5 &= t. \end{aligned} \tag{19}$$

Thus, the substitution of any two specific values for s and t in (19) yields a particular solution $(x_1, x_2, x_3, x_4, x_5)$ of the system, and each of the system's infinitely many different solutions is the result of some such substitution. ■

Examples 3 and 5 illustrate the ways in which Gaussian elimination can result in either a unique solution or infinitely many solutions. On the other hand, if the reduction of the augmented matrix to echelon form leads to a row of the form

$$0 \quad 0 \quad \cdots \quad 0 \quad 0 \quad *,$$

where the asterisk denotes a nonzero entry in the last column, then we have an inconsistent equation,

$$0x_1 + 0x_2 + \cdots + 0x_n = *,$$

and therefore the system has no solution.

Remark: We use algorithms such as the back substitution and Gaussian elimination algorithms of this section to outline the basic computational procedures of linear algebra. In modern numerical work, these procedures often are implemented on a computer. For instance, linear systems of more than four equations are usually solved in practice by using a computer to carry out the process of Gaussian elimination. ■

3.2 Problems

The linear systems in Problems 1–10 are in echelon form. Solve each by back substitution.

1. $\begin{aligned} x_1 + x_2 + 2x_3 &= 5 \\ x_2 + 3x_3 &= 6 \\ x_3 &= 2 \end{aligned}$

2. $\begin{aligned} 2x_1 - 5x_2 + x_3 &= 2 \\ 3x_2 - 2x_3 &= 9 \\ x_3 &= -3 \end{aligned}$

3. $\begin{aligned} x_1 - 3x_2 + 4x_3 &= 7 \\ x_2 - 5x_3 &= 2 \end{aligned}$

4. $\begin{aligned} x_1 - 5x_2 + 2x_3 &= 10 \\ x_2 - 7x_3 &= 5 \end{aligned}$

5. $\begin{aligned} x_1 + x_2 - 2x_3 + x_4 &= 9 \\ x_2 - x_3 + 2x_4 &= 1 \\ x_3 - 3x_4 &= 5 \end{aligned}$

6. $\begin{aligned} x_1 - 2x_2 + 5x_3 - 3x_4 &= 7 \\ x_2 - 3x_3 + 2x_4 &= 3 \\ x_4 &= -4 \end{aligned}$

7. $\begin{aligned} x_1 + 2x_2 + 4x_3 - 5x_4 &= 17 \\ x_2 - 2x_3 + 7x_4 &= 7 \end{aligned}$

8. $\begin{aligned} x_1 - 10x_2 + 3x_3 - 13x_4 &= 5 \\ x_3 + 3x_4 &= 10 \end{aligned}$

9. $\begin{aligned} 2x_1 + x_2 + x_3 + x_4 &= 6 \\ 3x_2 - x_3 - 2x_4 &= 2 \\ 3x_3 + 4x_4 &= 9 \\ x_4 &= 6 \end{aligned}$

10. $\begin{aligned} x_1 - 5x_2 + 2x_3 - 7x_4 + 11x_5 &= 0 \\ x_2 - 13x_3 + 3x_4 - 7x_5 &= 0 \\ x_4 - 5x_5 &= 0 \end{aligned}$

In Problems 11–22, use elementary row operations to transform each augmented coefficient matrix to echelon form. Then solve the system by back substitution.

11. $\begin{aligned} 2x_1 + 8x_2 + 3x_3 &= 2 \\ x_1 + 3x_2 + 2x_3 &= 5 \\ 2x_1 + 7x_2 + 4x_3 &= 8 \end{aligned}$

12. $\begin{aligned} 3x_1 + x_2 - 3x_3 &= 6 \\ 2x_1 + 7x_2 + x_3 &= -9 \\ 2x_1 + 5x_2 &= -5 \end{aligned}$

13. $\begin{aligned} x_1 + 3x_2 + 3x_3 &= 13 \\ 2x_1 + 5x_2 + 4x_3 &= 23 \\ 2x_1 + 7x_2 + 8x_3 &= 29 \end{aligned}$

14. $\begin{aligned} 3x_1 - 6x_2 - 2x_3 &= 1 \\ 2x_1 - 4x_2 + x_3 &= 17 \\ x_1 - 2x_2 - 2x_3 &= -9 \end{aligned}$

15. $\begin{aligned} 3x_1 + x_2 - 3x_3 &= -4 \\ x_1 + x_2 + x_3 &= 1 \\ 5x_1 + 6x_2 + 8x_3 &= 8 \end{aligned}$

16. $\begin{aligned} 2x_1 + 5x_2 + 12x_3 &= 6 \\ 3x_1 + x_2 + 5x_3 &= 12 \\ 5x_1 + 8x_2 + 21x_3 &= 17 \end{aligned}$

17. $$\begin{aligned} x_1 - 4x_2 - 3x_3 - 3x_4 &= 4 \\ 2x_1 - 6x_2 - 5x_3 - 5x_4 &= 5 \\ 3x_1 - x_2 - 4x_3 - 5x_4 &= -7 \end{aligned}$$

18. $$\begin{aligned} 3x_1 - 6x_2 + x_3 + 13x_4 &= 15 \\ 3x_1 - 6x_2 + 3x_3 + 21x_4 &= 21 \\ 2x_1 - 4x_2 + 5x_3 + 26x_4 &= 23 \end{aligned}$$

19. $$\begin{aligned} 3x_1 + x_2 + x_3 + 6x_4 &= 14 \\ x_1 - 2x_2 + 5x_3 - 5x_4 &= -7 \\ 4x_1 + x_2 + 2x_3 + 7x_4 &= 17 \end{aligned}$$

20. $$\begin{aligned} 2x_1 + 4x_2 - x_3 - 2x_4 + 2x_5 &= 6 \\ x_1 + 3x_2 + 2x_3 - 7x_4 + 3x_5 &= 9 \\ 5x_1 + 8x_2 - 7x_3 + 6x_4 + x_5 &= 4 \end{aligned}$$

21. $$\begin{aligned} x_1 + x_2 + x_3 \qquad &= 6 \\ 2x_1 - 2x_2 - 5x_3 \qquad &= -13 \\ 3x_1 \qquad + x_3 + x_4 &= 13 \\ 4x_1 - 2x_2 - 3x_3 + x_4 &= 1 \end{aligned}$$

22. $$\begin{aligned} 4x_1 - 2x_2 - 3x_3 + x_4 &= 3 \\ 2x_1 - 2x_2 - 5x_3 \qquad &= -10 \\ 4x_1 + x_2 + 2x_3 + x_4 &= 17 \\ 3x_1 \qquad + x_3 + x_4 &= 12 \end{aligned}$$

In Problems 23–27, determine for what values of k each system has (a) a unique solution; (b) no solution; (c) infinitely many solutions.

23. $$\begin{aligned} 3x + 2y &= 1 \\ 6x + 4y &= k \end{aligned}$$

24. $$\begin{aligned} 3x + 2y &= 0 \\ 6x + ky &= 0 \end{aligned}$$

25. $$\begin{aligned} 3x + 2y &= 11 \\ 6x + ky &= 21 \end{aligned}$$

26. $$\begin{aligned} 3x + 2y &= 1 \\ 7x + 5y &= k \end{aligned}$$

27. $$\begin{aligned} x + 2y + z &= 3 \\ 2x - y - 3z &= 5 \\ 4x + 3y - z &= k \end{aligned}$$

28. Under what condition on the constants a, b, and c does the system

$$\begin{aligned} 2x - y + 3z &= a \\ x + 2y + z &= b \\ 7x + 4y + 9z &= c \end{aligned}$$

have a unique solution? No solution? Infinitely many solutions?

29. This problem deals with the reversibility of elementary row operations.

(a) If the elementary row operation cR_p changes the matrix **A** to the matrix **B**, show that $(1/c)R_p$ changes **B** to **A**.

(b) If SWAP(R_p, R_q) changes **A** to **B**, show that SWAP(R_p, R_q) also changes **B** to **A**.

(c) If $cR_p + R_q$ changes **A** to **B**, show that $(-c)R_p + R_q$ changes **B** to **A**.

(d) Conclude that if **A** can be transformed into **B** by a finite sequence of elementary row operations, then **B** can similarly be transformed into **A**.

30. This problem outlines a proof that two linear systems LS_1 and LS_2 are equivalent (that is, have the same solution set) if their augmented coefficient matrices $\mathbf{A}_1$ and $\mathbf{A}_2$ are row equivalent.

(a) If a *single* elementary row operation transforms $\mathbf{A}_1$ to $\mathbf{A}_2$, show directly—considering separately the three cases—that every solution of LS_1 is also a solution of LS_2.

(b) Explain why it now follows from Problem 29 that every solution of either system is also a solution of the other system; thus the two systems have the same solution set.

3.2 Application Automated Row Operations

Computer algebra systems are often used to ease the labor of matrix computations, including elementary row operations. The 3×4 augmented coefficient matrix of Example 3 can be entered with the *Maple* command

```
with(linalg):
A := array( [[1, 2, 1,  4],
             [3, 8, 7, 20],
             [2, 7, 9, 23]] );
```

or the *Mathematica* command

```
A = {{1, 2, 1,  4},
     {3, 8, 7, 20},
     {2, 7, 9, 23}}
```

or the MATLAB command

```
A = [1 2 1  4
     3 8 7 20
     2 7 9 23]
```

The *Maple* **linalg** package has built-in elementary row operations that can be used to carry out the reduction of **A** exhibited in Example 3, as follows:

```
A := addrow(A,1,2,-3);                # (-3)R1 + R2
A := addrow(A,1,3,-2);                # (-2)R1 + R3
A := mulrow(A,2,1/2);                 # (1/2)R2
A := addrow(A,2,3,-3);                # (-3)R2 + R3
```

The inserted remarks, which follow the notation of Fig. 3.2.1, should make clear the structure of these row operations in *Maple*.

In *Mathematica* the ith row of the matrix **A** is denoted by **A[[i]]**, and we can operate directly on rows to carry out the reduction of Example 3, as follows:

```
A[[2]] = (-3)A[[1]] + A[[2]];      (*  (-3)R1 + R2  *)
A[[3]] = (-2)A[[1]] + A[[3]];      (*  (-2)R1 + R3  *)
A[[2]] = (1/2)A[[2]];              (*   (1/2)R2     *)
A[[3]] = (-3)A[[2]] + A[[3]];      (*  (-3)R2 + R3  *)
```

In MATLAB, the ith row of the matrix **A** is denoted by **A(i,:)**, and we can operate similarly on rows to carry out the reduction to echelon form, as follows:

```
A(2,:) = (-3)*A(1,:) + A(2,:)     %  (-3)R1 + R2
A(3,:) = (-2)*A(1,:) + A(3,:)     %  (-2)R1 + R3
A(2,:) = (1/2)*A(2,:)             %  (1/2)R2
A(3,:) = (-3)*A(2,:) + A(3,:)     %  (-3)R2 + R3
```

You should verify (by using the computer algebra system of your choice) that these operations yield the echelon matrix

$$\begin{bmatrix} 1 & 2 & 1 & 4 \\ 0 & 1 & 2 & 4 \\ 0 & 0 & 1 & 3 \end{bmatrix}.$$

If you wanted to interchange the 1st and 3rd rows—and proceed to solve the corresponding system by "forward substitution" rather than back substitution—this could be done as follows:

```
A := swaprow(A,1,3);              (Maple)
A[[{1,3}]] = A[[{3,1}]];          (Mathematica)
A([1 3],:) = A([3 1],:)           (MATLAB)
```

These "automated" elementary row operations can be used in Problems 11–22 of this section.

3.3 Reduced Row-Echelon Matrices

The result of the process of Gaussian elimination described in Section 3.2 is not uniquely determined. That is, two different sequences of elementary row operations, both starting with the same matrix **A**, may yield two different echelon matrices, each of course still row equivalent to **A**. For instance, the two echelon matrices

$$\begin{bmatrix} 1 & 2 & 3 \\ 0 & 4 & 5 \\ 0 & 0 & 6 \end{bmatrix} \quad \text{and} \quad \begin{bmatrix} 1 & 1 & 1 \\ 0 & 2 & 2 \\ 0 & 0 & 3 \end{bmatrix} \tag{1}$$

are readily seen (as in Problem 31) to be row equivalent. Hence it is possible to begin with an appropriate 3×3 matrix and derive by Gaussian elimination either of the two different echelon matrices in (1).

A full understanding of the structure of systems of linear equations depends on the definition of a special class of echelon matrices, a class having the property that every matrix in it is row equivalent to one and only one of these special echelon matrices. Recall that an echelon matrix **E** is one that has the following two properties:

1. Every all-zero row of **E** lies beneath every row that contains a nonzero element.
2. The leading nonzero entry in each row lies to the right of the leading nonzero entry in the preceding row.

A *reduced* echelon matrix (sometimes called a reduced row-echelon matrix) has two additional properties.

DEFINITION Reduced Echelon Matrix

A **reduced echelon matrix E** is an echelon matrix that has—in addition to Properties 1 and 2—the following properties:

3. Each leading entry of **E** is 1.
4. Each leading entry of **E** is the only nonzero element in its column.

A matrix is said to be in **reduced echelon form** if it is a reduced echelon matrix. Similarly, a linear system is in **reduced echelon form** if its augmented coefficient matrix is a reduced echelon matrix.

Example 1 The following matrices are reduced echelon matrices.

$$\begin{bmatrix} 1 & 0 \\ 0 & 1 \end{bmatrix} \qquad \begin{bmatrix} 1 & 0 & -3 & 0 \\ 0 & 1 & 4 & 0 \\ 0 & 0 & 0 & 1 \end{bmatrix}$$

$$\begin{bmatrix} 1 & -2 & 0 \\ 0 & 0 & 1 \\ 0 & 0 & 0 \end{bmatrix} \qquad \begin{bmatrix} 1 & 0 & -7 \\ 0 & 1 & 5 \\ 0 & 0 & 0 \end{bmatrix}$$

The echelon matrices

$$\mathbf{A} = \begin{bmatrix} 1 & 0 & 0 \\ 0 & 2 & 0 \\ 0 & 0 & 0 \end{bmatrix} \quad \text{and} \quad \mathbf{B} = \begin{bmatrix} 1 & 0 & 2 \\ 0 & 1 & 0 \\ 0 & 0 & 1 \end{bmatrix}$$

are not in *reduced* echelon form, because **A** does not have Property 3 and **B** does not have Property 4. ■

The process of transforming a matrix **A** into reduced echelon form is called **Gauss-Jordan elimination**.

ALGORITHM Gauss-Jordan Elimination

1. First transform **A** into echelon form by Gaussian elimination.
2. Then divide each element of each nonzero *row* by its leading entry (to satisfy Property 3).
3. Finally, use each leading 1 to "clear out" any remaining nonzero elements in its *column* (to satisfy Property 4).

The reduced echelon form of a matrix *is* unique. The special class of matrices that we mentioned at the beginning of this discussion is simply the class of all reduced echelon matrices. A proof of the following theorem may be found in Section 4.2 of B. Noble and J. W. Daniel, *Applied Linear Algebra*, 3rd ed. (Upper Saddle River, N.J.: Prentice Hall, 1988).

THEOREM 1 Unique Reduced Echelon Form

Every matrix is row equivalent to one and only one reduced echelon matrix.

Example 2 Find the reduced echelon form of the matrix

$$\mathbf{A} = \begin{bmatrix} 1 & 2 & 1 & 4 \\ 3 & 8 & 7 & 20 \\ 2 & 7 & 9 & 23 \end{bmatrix}.$$

Solution In Example 3 of Section 3.2 we found the echelon form

$$\begin{bmatrix} 1 & 2 & 1 & 4 \\ 0 & 1 & 2 & 4 \\ 0 & 0 & 1 & 3 \end{bmatrix},$$

which already satisfies Property 3. To clear out columns 2 and 3 (in order to satisfy Property 4), we continue the reduction as follows.

$$\begin{bmatrix} 1 & 2 & 1 & 4 \\ 0 & 1 & 2 & 4 \\ 0 & 0 & 1 & 3 \end{bmatrix} \xrightarrow{(-2)R_2 + R_1} \begin{bmatrix} 1 & 0 & -3 & -4 \\ 0 & 1 & 2 & 4 \\ 0 & 0 & 1 & 3 \end{bmatrix}$$

$$\xrightarrow{(-2)R_3 + R_2} \begin{bmatrix} 1 & 0 & -3 & -4 \\ 0 & 1 & 0 & -2 \\ 0 & 0 & 1 & 3 \end{bmatrix}$$

$$\xrightarrow{(3)R_3 + R_1} \begin{bmatrix} 1 & 0 & 0 & 5 \\ 0 & 1 & 0 & -2 \\ 0 & 0 & 1 & 3 \end{bmatrix}$$

For instance, we see immediately from this reduced echelon form that the linear system

$$\begin{aligned} x + 2y + z &= 4 \\ 3x + 8y + 7z &= 20 \\ 2x + 7y + 9z &= 23 \end{aligned}$$

with augmented coefficient matrix **A** has the unique solution $x = 5$, $y = -2$, $z = 3$. ■

Example 3 To use Gauss-Jordan elimination to solve the linear system

$$\begin{aligned} x_1 + x_2 + x_3 + x_4 &= 12 \\ x_1 + 2x_2 \quad\quad + 5x_4 &= 17 \\ 3x_1 + 2x_2 + 4x_3 - x_4 &= 31, \end{aligned} \tag{2}$$

we transform its augmented coefficient matrix into *reduced* echelon form, as follows:

$$\begin{bmatrix} 1 & 1 & 1 & 1 & 12 \\ 1 & 2 & 0 & 5 & 17 \\ 3 & 2 & 4 & -1 & 31 \end{bmatrix} \xrightarrow{(-1)R_1 + R_2} \begin{bmatrix} 1 & 1 & 1 & 1 & 12 \\ 0 & 1 & -1 & 4 & 5 \\ 3 & 2 & 4 & -1 & 31 \end{bmatrix}$$

$$\xrightarrow{(-3)R_1 + R_3} \begin{bmatrix} 1 & 1 & 1 & 1 & 12 \\ 0 & 1 & -1 & 4 & 5 \\ 0 & -1 & 1 & -4 & -5 \end{bmatrix}$$

$$\xrightarrow{(1)R_2 + R_3} \begin{bmatrix} 1 & 1 & 1 & 1 & 12 \\ 0 & 1 & -1 & 4 & 5 \\ 0 & 0 & 0 & 0 & 0 \end{bmatrix}$$

$$\xrightarrow{(-1)R_2 + R_1} \begin{bmatrix} 1 & 0 & 2 & -3 & 7 \\ 0 & 1 & -1 & 4 & 5 \\ 0 & 0 & 0 & 0 & 0 \end{bmatrix}.$$

Thus the reduced echelon form of the system in (2) is

$$\begin{aligned} x_1 \quad\quad + 2x_3 - 3x_4 &= 7 \\ x_2 - x_3 + 4x_4 &= 5 \\ 0 &= 0. \end{aligned} \tag{3}$$

The leading variables are x_1 and x_2; the free variables are x_3 and x_4. If we set

$$x_3 = s \quad \text{and} \quad x_4 = t,$$

then (3) immediately yields

$$\begin{aligned} x_1 &= 7 - 2s + 3t, \\ x_2 &= 5 + s - 4t. \end{aligned}$$

■

As a practical matter, Gauss-Jordan elimination generally offers no significant computational advantage over Gaussian elimination (transformation to nonreduced echelon form) followed by back substitution. Therefore, Gaussian elimination is commonly employed in practical procedures and in computer programs used to solve linear systems numerically.

The Three Possibilities

The chief importance of Gauss-Jordan elimination stems from the fact that the reduced echelon form of a general linear system

$$\begin{aligned}
a_{11}x_1 + a_{12}x_2 + a_{13}x_3 + \cdots + a_{1n}x_n &= b_1 \\
a_{21}x_1 + a_{22}x_2 + a_{23}x_3 + \cdots + a_{2n}x_n &= b_2 \\
&\vdots \\
a_{m1}x_1 + a_{m2}x_2 + a_{m3}x_3 + \cdots + a_{mn}x_n &= b_m
\end{aligned} \tag{4}$$

most clearly exhibits its underlying structure and enables us most readily to answer questions about the number and type of its solutions. If the leading variables are $x_{j_1}, x_{j_2}, \ldots, x_{j_r}$, then the reduced echelon form of the system in (4) looks like

$$\begin{aligned}
x_{j_1} \qquad\qquad\qquad + \sum_{k=r+1}^{n} c_{1k}x_{j_k} &= d_1 \\
x_{j_2} \qquad\qquad + \sum_{k=r+1}^{n} c_{2k}x_{j_k} &= d_2 \\
\ddots \qquad\qquad\qquad\qquad & \vdots \\
x_{j_r} + \sum_{k=r+1}^{n} c_{rk}x_{j_k} &= d_r \\
0 &= d_{r+1} \\
&\vdots \\
0 &= d_m,
\end{aligned} \tag{5}$$

where the summations involve only the (non-leading) free variables $x_{j_{r+1}}, \ldots, x_{j_n}$.

Now if any of the constants $d_{r+1}, d_{r+2}, \ldots, d_m$ in (5) is nonzero, then clearly the system has *no* real solution. On the other hand, suppose that $d_{r+1} = d_{r+2} = \cdots = d_m = 0$. If $r < n$, then there are free variables that we can set equal to arbitrary parameters, and so the system has *infinitely many* solutions. If instead $r = n$, then the summations in (5) disappear and we have the *unique* solution $x_1 = d_1$, $x_2 = d_2, \ldots, x_n = d_n$. These observations establish the following result, to which we have already alluded.

THEOREM 2 The Three Possibilities

A linear system of equations has either

- a unique solution, or
- no solution, or
- infinitely many solutions.

Homogeneous Systems

The linear system in (4) is called **homogeneous** provided that the constants $b_1, b_2, \ldots, b_m$ on the right-hand side are all zero. Thus a homogeneous system of m equations in n variables has the form

$$\begin{aligned} a_{11}x_1 + a_{12}x_2 + a_{13}x_3 + \cdots + a_{1n}x_n &= 0 \\ a_{21}x_1 + a_{22}x_2 + a_{23}x_3 + \cdots + a_{2n}x_n &= 0 \\ &\vdots \\ a_{m1}x_1 + a_{m2}x_2 + a_{m3}x_3 + \cdots + a_{mn}x_n &= 0. \end{aligned} \tag{6}$$

Every homogeneous system obviously has at least the **trivial solution**

$$x_1 = 0, \quad x_2 = 0, \ldots, x_n = 0. \tag{7}$$

Thus (by Theorem 2) we know from the outset that *every homogeneous linear system either has only the trivial solution or has infinitely many solutions.* If it has a *nontrivial* solution—one with not every x_i equal to zero—then it must have infinitely many solutions.

An important special case, in which a nontrivial solution is guaranteed, is that of a homogeneous system with *more variables than equations*: $m < n$. To see why there must be a solution, consider the reduced echelon system in (5) with the constants on the right-hand side all zero (because the original system is homogeneous). The number r of leading variables is at most the number m of equations (because there is at most one leading variable per equation). If $m < n$, it then follows that $r < n$, so there is at least one free variable that can be set equal to an arbitrary parameter, thereby yielding infinitely many solutions. This argument establishes the following key result.

THEOREM 3 Homogeneous Systems with More Variables than Equations

Every homogeneous linear system with more variables than equations has infinitely many solutions.

Example 4 The homogeneous linear system

$$\begin{aligned} 47x_1 - 73x_2 + 56x_3 + 21x_4 &= 0 \\ 19x_1 + 81x_2 - 17x_3 - 99x_4 &= 0 \\ 53x_1 + 62x_2 + 39x_3 + 25x_4 &= 0 \end{aligned}$$

of three equations in four unknowns necessarily has infinitely many solutions. The only question (which we could answer by reducing the system to echelon form) is whether the system has one, two, or three free variables. ■

The situation is different for a *nonhomogeneous* system with more variables than equations. The simple example

$$\begin{aligned} x_1 + x_2 + x_3 &= 0 \\ x_1 + x_2 + x_3 &= 1 \end{aligned}$$

shows that such a system may be inconsistent. But if it is consistent—meaning that $d_{r+1} = \cdots = d_m = 0$ in the reduced echelon form in (5)—then the fact that $m < n$ implies (just as in the proof of Theorem 3) that there is at least one free variable, and that the system therefore has infinitely many solutions. Hence *every nonhomogeneous system with more variables than equations either has no solution or has infinitely many solutions.*

Equal Numbers of Equations and Variables

An especially important case in the theory of linear systems is that of a homogeneous system

$$\begin{aligned} a_{11}x_1 + a_{12}x_2 + a_{13}x_3 + \cdots + a_{1n}x_n &= 0 \\ a_{21}x_1 + a_{22}x_2 + a_{23}x_3 + \cdots + a_{2n}x_n &= 0 \\ &\vdots \\ a_{n1}x_1 + a_{n2}x_2 + a_{n3}x_3 + \cdots + a_{nn}x_n &= 0 \end{aligned} \tag{8}$$

with the *same* number n of variables and equations. The coefficient matrix $\mathbf{A} = [a_{ij}]$ then has the same number of rows and columns and thus is an $n \times n$ **square matrix.**

Here we are most interested in the situation when (8) has *only* the trivial solution $x_1 = x_2 = \cdots = x_n = 0$. This occurs if and only if the reduced echelon system contains *no* free variables. That is, all n of the variables $x_1, x_2, \ldots, x_n$ must be leading variables. Because the system consists of exactly n equations, we conclude that the reduced echelon system is simply

$$\begin{aligned} x_1 \qquad\qquad\qquad &= 0 \\ x_2 \qquad\qquad &= 0 \\ x_3 \qquad &= 0 \\ \ddots \quad &\vdots \\ x_n &= 0, \end{aligned}$$

and, therefore, that the reduced echelon form of the coefficient matrix $\mathbf{A}$ is the matrix

$$\begin{bmatrix} 1 & 0 & 0 & \cdots & 0 \\ 0 & 1 & 0 & \cdots & 0 \\ 0 & 0 & 1 & \cdots & 0 \\ \vdots & \vdots & \vdots & \ddots & \vdots \\ 0 & 0 & 0 & \cdots & 1 \end{bmatrix}. \tag{9}$$

Such a (square) matrix, with ones on its **principal diagonal** (the one from upper left to lower right) and zeros elsewhere, is called an **identity matrix** (for reasons given in Section 3.4). For instance, the 2×2 and 3×3 identity matrices are

$$\begin{bmatrix} 1 & 0 \\ 0 & 1 \end{bmatrix} \quad \text{and} \quad \begin{bmatrix} 1 & 0 & 0 \\ 0 & 1 & 0 \\ 0 & 0 & 1 \end{bmatrix}.$$

The matrix in (9) is the $n \times n$ identity matrix. With this terminology, the preceding argument establishes the following theorem.

THEOREM 4 Homogeneous Systems with Unique Solutions

Let $\mathbf{A}$ be an $n \times n$ matrix. Then the homogeneous system with coefficient matrix $\mathbf{A}$ has only the trivial solution if and only if $\mathbf{A}$ is row equivalent to the $n \times n$ identity matrix.

Example 5 The computation in Example 2 (disregarding the fourth column in each matrix there) shows that the matrix

$$\mathbf{A} = \begin{bmatrix} 1 & 2 & 1 \\ 3 & 8 & 7 \\ 2 & 7 & 9 \end{bmatrix}$$

is row equivalent to the 3×3 identity matrix. Hence Theorem 4 implies that the homogeneous system

$$\begin{aligned} x_1 + 2x_2 + x_3 &= 0 \\ 3x_1 + 8x_2 + 7x_3 &= 0 \\ 2x_1 + 7x_2 + 9x_3 &= 0 \end{aligned}$$

with coefficient matrix $\mathbf{A}$ has only the trivial solution $x_1 = x_2 = x_3 = 0$. ■

3.3 Problems

Find the reduced echelon form of each of the matrices given in Problems 1–20.

1. $\begin{bmatrix} 1 & 2 \\ 3 & 7 \end{bmatrix}$

2. $\begin{bmatrix} 3 & 7 \\ 2 & 5 \end{bmatrix}$

3. $\begin{bmatrix} 3 & 7 & 15 \\ 2 & 5 & 11 \end{bmatrix}$

4. $\begin{bmatrix} 3 & 7 & -1 \\ 5 & 2 & 8 \end{bmatrix}$

5. $\begin{bmatrix} 1 & 2 & -11 \\ 2 & 3 & -19 \end{bmatrix}$

6. $\begin{bmatrix} 1 & -2 & 19 \\ 4 & -7 & 70 \end{bmatrix}$

7. $\begin{bmatrix} 1 & 2 & 3 \\ 1 & 4 & 1 \\ 2 & 1 & 9 \end{bmatrix}$

8. $\begin{bmatrix} 1 & -4 & -5 \\ 3 & -9 & 3 \\ 1 & -2 & 3 \end{bmatrix}$

9. $\begin{bmatrix} 5 & 2 & 18 \\ 0 & 1 & 4 \\ 4 & 1 & 12 \end{bmatrix}$

10. $\begin{bmatrix} 5 & 2 & -5 \\ 9 & 4 & -7 \\ 4 & 1 & -7 \end{bmatrix}$

11. $\begin{bmatrix} 3 & 9 & 1 \\ 2 & 6 & 7 \\ 1 & 3 & -6 \end{bmatrix}$

12. $\begin{bmatrix} 1 & -4 & -2 \\ 3 & -12 & 1 \\ 2 & -8 & 5 \end{bmatrix}$

13. $\begin{bmatrix} 2 & 7 & 4 & 0 \\ 1 & 3 & 2 & 1 \\ 2 & 6 & 5 & 4 \end{bmatrix}$

14. $\begin{bmatrix} 1 & 3 & 2 & 5 \\ 2 & 5 & 2 & 3 \\ 2 & 7 & 7 & 22 \end{bmatrix}$

15. $\begin{bmatrix} 2 & 2 & 4 & 2 \\ 1 & -1 & -4 & 3 \\ 2 & 7 & 19 & -3 \end{bmatrix}$

16. $\begin{bmatrix} 1 & 3 & 15 & 7 \\ 2 & 4 & 22 & 8 \\ 2 & 7 & 34 & 17 \end{bmatrix}$

17. $\begin{bmatrix} 1 & 1 & 1 & -1 & -4 \\ 1 & -2 & -2 & 8 & -1 \\ 2 & 3 & -1 & 3 & 11 \end{bmatrix}$

18. $\begin{bmatrix} 1 & -2 & -5 & -12 & 1 \\ 2 & 3 & 18 & 11 & 9 \\ 2 & 5 & 26 & 21 & 11 \end{bmatrix}$

19. $\begin{bmatrix} 2 & 7 & -10 & -19 & 13 \\ 1 & 3 & -4 & -8 & 6 \\ 1 & 0 & 2 & 1 & 3 \end{bmatrix}$

20. $\begin{bmatrix} 3 & 6 & 1 & 7 & 13 \\ 5 & 10 & 8 & 18 & 47 \\ 2 & 4 & 5 & 9 & 26 \end{bmatrix}$

21–30. Use the method of Gauss-Jordan elimination (transforming the augmented matrix into reduced echelon form) to solve Problems 11–20 in Section 3.2.

31. Show that the two matrices in (1) are both row equivalent to the 3×3 identity matrix (and hence, by Theorem 1, to each other).

32. Show that the 2×2 matrix

$$\mathbf{A} = \begin{bmatrix} a & b \\ c & d \end{bmatrix}$$

is row equivalent to the 2×2 identity matrix provided that $ad - bc \neq 0$.

33. List all possible reduced row-echelon forms of a 2×2 matrix, using asterisks to indicate elements that may be either zero or nonzero.

34. List all possible reduced row-echelon forms of a 3×3 matrix, using asterisks to indicate elements that may be either zero or nonzero.

35. Consider the homogeneous system

$$\begin{aligned} ax + by &= 0 \\ cx + dy &= 0. \end{aligned}$$

(a) If $x = x_0$ and $y = y_0$ is a solution and k is a real number, then show that $x = kx_0$ and $y = ky_0$ is also a solution.

(b) If $x = x_1$, $y = y_1$ and $x = x_2$, $y = y_2$ are both solutions, then show that $x = x_1 + x_2$, $y = y_1 + y_2$ is a solution.

36. Suppose that $ad - bc \neq 0$ in the homogeneous system of Problem 35. Use Problem 32 to show that its only solution is the trivial solution.

37. Show that the homogeneous system in Problem 35 has a nontrivial solution if and only if $ad - bc = 0$.

38. Use the result of Problem 37 to find all values of c for which the homogeneous system

$$\begin{aligned} (c+2)x + 3y &= 0 \\ 2x + (c-3)y &= 0 \end{aligned}$$

has a nontrivial solution.

39. Consider a *homogeneous* system of three equations in three unknowns. Suppose that the third equation is the sum of some multiple of the first equation and some multiple of the second equation. Show that the system has a nontrivial solution.

40. Let **E** be an echelon matrix that is row equivalent to the matrix **A**. Show that **E** has the same number of nonzero rows as does the reduced echelon form $\mathbf{E}^*$ of **A**. Thus the number of nonzero rows in an echelon form of **A** is an "invariant" of the matrix **A**. *Suggestion*: Consider reducing **E** to $\mathbf{E}^*$.

3.3 Application Automated Row Reduction

Most computer algebra systems include commands for the immediate reduction of matrices to reduced echelon form. For instance, if the matrix

$$\mathbf{A} = \begin{bmatrix} 1 & 2 & 1 & 4 \\ 3 & 8 & 7 & 20 \\ 2 & 7 & 9 & 23 \end{bmatrix}$$

of Example 2 has been entered—as illustrated in the 3.2 Application—then the *Maple* command

```
with(linalg): R := rref(A);
```

or the *Mathematica* command

```
R = RowReduce[A] // MatrixForm
```

or the MATLAB command

```
R = rref(A)
```

produces the reduced echelon matrix

$$\mathbf{R} = \begin{bmatrix} 1 & 0 & 0 & 5 \\ 0 & 1 & 0 & -2 \\ 0 & 0 & 1 & 3 \end{bmatrix}$$

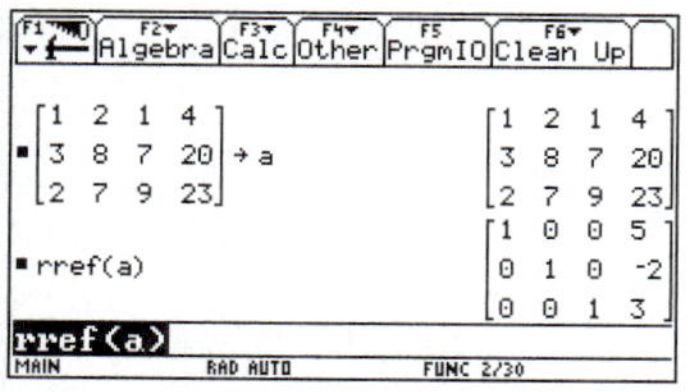

FIGURE 3.3.1. Finding a reduced row-echelon matrix with a TI-89 calculator.

that exhibits the solution of the linear system having augmented coefficient matrix **A**. The same calculation is illustrated in the calculator screen of Fig. 3.3.1. Solve similarly the systems in the following problems (which apparently would be somewhat tedious to solve by manual row reduction):

1. $\begin{aligned} 17x + 42y - 36z &= 213 \\ 13x + 45y - 34z &= 226 \\ 12x + 47y - 35z &= 197 \end{aligned}$

2. $\begin{aligned} 32x + 57y - 41z &= 713 \\ 23x + 43y - 37z &= 130 \\ 42x - 61y + 39z &= 221 \end{aligned}$

3. $\begin{aligned} 231x + 157y - 241z &= 420 \\ 323x + 181y - 375z &= 412 \\ 542x + 161y - 759z &= 419 \end{aligned}$

4. $\begin{aligned} 837x + 667y - 729z &= 1659 \\ 152x - 179y - 975z &= 1630 \\ 542x + 328y - 759z &= 1645 \end{aligned}$

5. $\begin{aligned} 49w - 57x + 37y - 59z &= 97 \\ 73w - 15x - 19y - 22z &= 99 \\ 52w - 51x + 14y - 29z &= 89 \\ 13w - 27x + 27y - 25z &= 73 \end{aligned}$

6. $\begin{aligned} 64w - 57x + 97y - 67z &= 485 \\ 92w + 77x - 34y - 37z &= 486 \\ 44w - 34x + 53y - 34z &= 465 \\ 27w + 57x - 69y + 29z &= 464 \end{aligned}$

3.4 Matrix Operations

As yet, we have used matrices only to simplify our record keeping in the solution of linear systems. But it turns out that matrices can be added and multiplied in ways similar to the ways in which numbers are added and multiplied and that these operations with matrices have far-reaching applications.

At the level of this text everyone "knows" that $2+3=5$, and we do not dwell on the underlying meaning of this equation. But in the case of matrices we must begin with precise definitions of what the familiar language of algebra is to mean when it is applied to matrices rather than to numbers.

Two matrices **A** and **B** *of the same size*—the same number of rows and the same number of columns—are called **equal** provided that each element of **A** is equal to the corresponding element of **B**. Thus two matrices of the same size are equal provided they are *elementwise equal*, and we write $\mathbf{A} = \mathbf{B}$ to denote equality of the two matrices **A** and **B**.

Example 1 If

$$\mathbf{A} = \begin{bmatrix} 3 & 4 \\ 5 & 6 \end{bmatrix}, \quad \mathbf{B} = \begin{bmatrix} 3 & 4 \\ 5 & 7 \end{bmatrix}, \quad \text{and} \quad \mathbf{C} = \begin{bmatrix} 3 & 4 & 7 \\ 5 & 6 & 8 \end{bmatrix},$$

then $\mathbf{A} \neq \mathbf{B}$ because $a_{22} = 6$, whereas $b_{22} = 7$, and $\mathbf{A} \neq \mathbf{C}$ because the matrices **A** and **C** are not of the same size. ■

The next two definitions are further examples of "doing it elementwise."

DEFINITION Addition of Matrices

If $\mathbf{A} = [\, a_{ij} \,]$ and $\mathbf{B} = [\, b_{ij} \,]$ are matrices *of the same size*, then their **sum** $\mathbf{A}+\mathbf{B}$ is the matrix obtained by adding corresponding elements of the matrices **A** and **B**. That is,

$$\mathbf{A} + \mathbf{B} = [\, a_{ij} + b_{ij} \,], \tag{1}$$

where the notation on the right signifies that the element in the ith row and jth column of the matrix $\mathbf{A}+\mathbf{B}$ is $a_{ij} + b_{ij}$.

Example 2 If

$$\mathbf{A} = \begin{bmatrix} 3 & 0 & -1 \\ 2 & -7 & 5 \end{bmatrix}, \quad \mathbf{B} = \begin{bmatrix} 4 & -3 & 6 \\ 9 & 0 & -2 \end{bmatrix}, \quad \text{and} \quad \mathbf{C} = \begin{bmatrix} 3 & -2 \\ -1 & 6 \end{bmatrix},$$

then

$$\mathbf{A} + \mathbf{B} = \begin{bmatrix} 7 & -3 & 5 \\ 11 & -7 & 3 \end{bmatrix},$$

but the sum $\mathbf{A} + \mathbf{C}$ is not defined because the matrices **A** and **C** are not of the same size. ■

DEFINITION Multiplication of a Matrix by a Number

If $\mathbf{A} = [\, a_{ij} \,]$ is a matrix and c is a number, then $c\mathbf{A}$ is the matrix obtained by multiplying each element of $\mathbf{A}$ by c. That is,

$$c\mathbf{A} = [\, ca_{ij} \,]. \tag{2}$$

Using multiplication of a matrix by a scalar, we define the **negative** $-\mathbf{A}$ of the matrix $\mathbf{A}$ and the **difference** $\mathbf{A} - \mathbf{B}$ of the two matrices $\mathbf{A}$ and $\mathbf{B}$ by writing

$$-\mathbf{A} = (-1)\mathbf{A} \quad \text{and} \quad \mathbf{A} - \mathbf{B} = \mathbf{A} + (-\mathbf{B}).$$

Example 3 If $\mathbf{A}$ and $\mathbf{B}$ are the 2×3 matrices of Example 2, then

$$3\mathbf{A} = \begin{bmatrix} 9 & 0 & -3 \\ 6 & -21 & 15 \end{bmatrix}, \quad -\mathbf{B} = \begin{bmatrix} -4 & 3 & -6 \\ -9 & 0 & 2 \end{bmatrix},$$

and

$$3\mathbf{A} - \mathbf{B} = \begin{bmatrix} 5 & 3 & -9 \\ -3 & -21 & 17 \end{bmatrix}.$$

■

Vectors

Our first application of these matrix operations is to vectors. As mentioned in Section 3.2, a **column vector** (or simply **vector**) is merely an $n \times 1$ matrix, one having a single column. We normally use boldface lowercase letters, rather than lightface uppercase letters, to denote vectors. If

$$\mathbf{a} = \begin{bmatrix} 6 \\ -2 \\ 5 \end{bmatrix} \quad \text{and} \quad \mathbf{b} = \begin{bmatrix} -2 \\ 3 \\ -4 \end{bmatrix},$$

then we can form such combinations as

$$3\mathbf{a} + 2\mathbf{b} = \begin{bmatrix} 18 \\ -6 \\ 15 \end{bmatrix} + \begin{bmatrix} -4 \\ 6 \\ -8 \end{bmatrix} = \begin{bmatrix} 14 \\ 0 \\ 7 \end{bmatrix}.$$

Largely for typographical reasons, we sometimes write

$$\mathbf{a} = \begin{bmatrix} a_1 \\ a_2 \\ \vdots \\ a_n \end{bmatrix} = (a_1, a_2, \ldots, a_n). \tag{3}$$

That is, $(a_1, a_2, \ldots, a_n)$ is simply another notation for the *column vector* with elements $a_1, a_2, \ldots, a_n$. It should not be confused with the *row vector*

$$\begin{bmatrix} a_1 & a_2 & \cdots & a_n \end{bmatrix}. \tag{4}$$

A **row vector** is a $1 \times n$ (rather than $n \times 1$) matrix having a single row, and

$$(3, 2, 1) = \begin{bmatrix} 3 \\ 2 \\ 1 \end{bmatrix} \neq \begin{bmatrix} 3 & 2 & 1 \end{bmatrix}$$

because the two matrices here have different sizes (even though they have the same elements).

Now consider the linear system

$$\begin{aligned} a_{11}x_1 + a_{12}x_2 + a_{13}x_3 + \cdots + a_{1n}x_n &= b_1 \\ a_{21}x_1 + a_{22}x_2 + a_{23}x_3 + \cdots + a_{2n}x_n &= b_2 \\ &\vdots \\ a_{m1}x_1 + a_{m2}x_2 + a_{m3}x_3 + \cdots + a_{mn}x_n &= b_m \end{aligned} \tag{5}$$

of m equations in n variables. We may regard a solution of this system as a *vector*

$$\mathbf{x} = \begin{bmatrix} x_1 \\ x_2 \\ x_3 \\ \vdots \\ x_n \end{bmatrix} = (x_1, x_2, x_3, \ldots, x_n) \tag{6}$$

whose elements satisfy each of the equations in (5). If we want to refer explicitly to the number of elements, we may call $\mathbf{x}$ an ***n*-vector**.

Example 4 Consider the homogeneous system

$$\begin{aligned} x_1 + 3x_2 - 15x_3 + 7x_4 &= 0 \\ x_1 + 4x_2 - 19x_3 + 10x_4 &= 0 \\ 2x_1 + 5x_2 - 26x_3 + 11x_4 &= 0. \end{aligned} \tag{7}$$

We find readily that the reduced echelon form of the augmented coefficient matrix of this system is

$$\begin{bmatrix} 1 & 0 & -3 & -2 & 0 \\ 0 & 1 & -4 & 3 & 0 \\ 0 & 0 & 0 & 0 & 0 \end{bmatrix}.$$

Hence x_1 and x_2 are leading variables and x_3 and x_4 are free variables. In the manner of Sections 3.2 and 3.3, we therefore see that the infinite solution set of the system in (7) is described by the equations

$$\begin{aligned} x_4 &= t, \\ x_3 &= s, \\ x_2 &= 4s - 3t, \\ x_1 &= 3s + 2t \end{aligned} \tag{8}$$

in terms of the arbitrary parameters s and t.

Now let us write the solution $\mathbf{x} = (x_1, x_2, x_3, x_4)$ in vector notation. The equations in (8) yield

$$\mathbf{x} = \begin{bmatrix} x_1 \\ x_2 \\ x_3 \\ x_4 \end{bmatrix} = \begin{bmatrix} 3s + 2t \\ 4s - 3t \\ s \\ t \end{bmatrix},$$

and "separating" the s and t parts gives

$$\mathbf{x} = \begin{bmatrix} 3s \\ 4s \\ s \\ 0 \end{bmatrix} + \begin{bmatrix} 2t \\ -3t \\ 0 \\ t \end{bmatrix} = s\begin{bmatrix} 3 \\ 4 \\ 1 \\ 0 \end{bmatrix} + t\begin{bmatrix} 2 \\ -3 \\ 0 \\ 1 \end{bmatrix}$$

—that is,

$$\mathbf{x} = s(3, 4, 1, 0) + t(2, -3, 0, 1) = s\mathbf{x}_1 + t\mathbf{x}_2. \tag{9}$$

Equation (9) expresses in **vector form** the general solution of the linear system in (7). It says that the vector $\mathbf{x}$ is a solution if and only if $\mathbf{x}$ is a *linear combination*—a sum of multiples—of the particular solutions $\mathbf{x}_1 = (3, 4, 1, 0)$ and $\mathbf{x}_2 = (2, -3, 0, 1)$. The parameters s and t are simply the coefficients in this "sum of multiples." ■

In the same manner as that in which we derived Eq. (9) from the equations in (8), the general solution of every homogeneous linear system can be expressed as a linear combination of particular solution vectors. For this reason (as well as others), linear combinations of vectors will play a central role in succeeding chapters.

Matrix Multiplication

The first surprise is that matrices are *not* multiplied elementwise. The initial purpose of matrix multiplication is to simplify the notation for systems of linear equations. If we write

$$\mathbf{A} = \begin{bmatrix} a_{ij} \end{bmatrix}, \quad \mathbf{x} = \begin{bmatrix} x_1 \\ x_2 \\ \vdots \\ x_n \end{bmatrix}, \quad \text{and} \quad \mathbf{b} = \begin{bmatrix} b_1 \\ b_2 \\ \vdots \\ b_m \end{bmatrix}, \tag{10}$$

then $\mathbf{A}$, $\mathbf{x}$, and $\mathbf{b}$ are, respectively, the coefficient matrix, the unknown vector, and the constant vector for the linear system in (5). We want to define the matrix product $\mathbf{Ax}$ in such a way that the entire system of linear equations reduces to the single matrix equation

$$\mathbf{Ax} = \mathbf{b}. \tag{11}$$

The first step is to define the product of a *row* vector $\mathbf{a}$ and a *column* vector $\mathbf{b}$,

$$\mathbf{a} = \begin{bmatrix} a_1 & a_2 & \cdots & a_n \end{bmatrix} \quad \text{and} \quad \mathbf{b} = \begin{bmatrix} b_1 \\ b_2 \\ \vdots \\ b_n \end{bmatrix},$$

each having n elements. In this case, the product **ab** is *defined* to be

$$\mathbf{ab} = a_1b_1 + a_2b_2 + \cdots + a_nb_n. \tag{12}$$

Thus **ab** is the *sum of products* of corresponding elements of **a** and **b**. For instance,

$$\begin{bmatrix} 2 & -3 \end{bmatrix}\begin{bmatrix} 3 \\ 5 \end{bmatrix} = (2)(3) + (-3)(5) = -9$$

and

$$\begin{bmatrix} 3 & 0 & -1 & 7 \end{bmatrix}\begin{bmatrix} 5 \\ 2 \\ -3 \\ 4 \end{bmatrix} = 3 \cdot 5 + 0 \cdot 2 + (-1)(-3) + 7 \cdot 4 = 46.$$

Note that if

$$\mathbf{a} = \begin{bmatrix} a_1 & a_2 & \cdots & a_n \end{bmatrix} \quad \text{and} \quad \mathbf{x} = \begin{bmatrix} x_1 \\ x_2 \\ \vdots \\ x_n \end{bmatrix},$$

then

$$\mathbf{ax} = a_1x_1 + a_2x_2 + \cdots + a_nx_n.$$

Hence the single equation

$$a_1x_1 + a_2x_2 + \cdots + a_nx_n = b \tag{13}$$

reduces to the equation

$$\mathbf{ax} = b, \tag{14}$$

which is a step toward the objective expressed in Eq. (11). This observation is the underlying motivation for the following definition.

DEFINITION Matrix Multiplication

Suppose that **A** is an $m \times p$ matrix and **B** is a $p \times n$ matrix. Then the **product AB** is the $m \times n$ matrix defined as follows: The element of **AB** in its ith row and jth column is the *sum of products* of corresponding elements in the ith row of **A** and the jth column of **B**.

That is, if the ith row of **A** is

$$\begin{bmatrix} a_{i1} & a_{i2} & a_{i3} & \cdots & a_{ip} \end{bmatrix}$$

and the jth column of **B** is

$$\begin{bmatrix} b_{1j} \\ b_{2j} \\ b_{3j} \\ \vdots \\ b_{pj} \end{bmatrix},$$

then the element in the ith row and jth column of the product **AB** is

$$a_{i1}b_{1j} + a_{i2}b_{2j} + a_{i3}b_{3j} + \cdots + a_{ip}b_{pj}.$$

Example 5 If

$$\mathbf{A} = \begin{bmatrix} 2 & -1 \\ -4 & 3 \end{bmatrix} \quad \text{and} \quad \mathbf{B} = \begin{bmatrix} 1 & 5 \\ 3 & 7 \end{bmatrix},$$

then $m = p = n = 2$, so **AB** will also be a 2×2 matrix. To find **AB**, we calculate sums of products as follows:

$$\begin{aligned} &\mathbf{AB}, \text{ row 1, column 1}: && (2)(1) + (-1)(3) = -1; \\ &\mathbf{AB}, \text{ row 1, column 2}: && (2)(5) + (-1)(7) = 3; \\ &\mathbf{AB}, \text{ row 2, column 1}: && (-4)(1) + (3)(3) = 5; \\ &\mathbf{AB}, \text{ row 2, column 2}: && (-4)(5) + (3)(7) = 1. \end{aligned}$$

Thus

$$\mathbf{AB} = \begin{bmatrix} -1 & 3 \\ 5 & 1 \end{bmatrix}.$$

■

For your first practice with matrix multiplication, you should compute

$$\mathbf{BA} = \begin{bmatrix} 1 & 5 \\ 3 & 7 \end{bmatrix} \begin{bmatrix} 2 & -1 \\ -4 & 3 \end{bmatrix} = \begin{bmatrix} -18 & 14 \\ -22 & 18 \end{bmatrix}.$$

Note that $\mathbf{AB} \neq \mathbf{BA}$. This shows that multiplication of matrices is *not* commutative! We must therefore be careful about the order in which we write the matrices in a matrix product.

The definition of the matrix product bears careful examination to see how it fits together. First, the fact that **A** is $m \times p$ and **B** is $p \times n$ implies that the number of *columns* of **A** is equal to the number of *rows* of **B**. If so, then the size of the product **AB** is obtained by a sort of cancellation of the "inside" dimensions:

$$\begin{array}{ccccc} \mathbf{A} & \text{times} & \mathbf{B} & = & \mathbf{AB} \\ m \times p & & p \times n & & m \times n. \end{array}$$

These "cancel."

If the inside dimensions are not equal, then the product **AB** is *not defined*.

Example 6 If **A** is a 3×2 matrix and **B** is a 2×3 matrix, then **AB** will be a 3×3 matrix, whereas **BA** will be a 2×2 matrix. If **C** is a 3×5 matrix and **D** is a 5×7 matrix, then **CD** will be a 3×7 matrix, but **DC** is undefined. ■

To emphasize the fact that the ijth element of **AB** is the product of the ith row of **A** and the jth column of **B**, we can write

$$\mathbf{AB} = \begin{bmatrix} \text{———} \; \mathbf{a}_1 \; \text{———} \\ \text{———} \; \mathbf{a}_2 \; \text{———} \\ \vdots \\ \text{———} \; \mathbf{a}_i \; \text{———} \\ \vdots \\ \text{———} \; \mathbf{a}_m \; \text{———} \end{bmatrix} \begin{bmatrix} | & | & & | & & | \\ \mathbf{b}_1 & \mathbf{b}_2 & \cdots & \mathbf{b}_j & \cdots & \mathbf{b}_n \\ | & | & & | & & | \end{bmatrix},$$

where $\mathbf{a}_1, \mathbf{a}_2, \dots, \mathbf{a}_m$ denote the m **row vectors** of $\mathbf{A}$ and $\mathbf{b}_1, \mathbf{b}_2, \dots, \mathbf{b}_n$ denote the n **column vectors** of $\mathbf{B}$. More briefly, if

$$\mathbf{A} = \begin{bmatrix} \mathbf{a}_1 \\ \mathbf{a}_2 \\ \vdots \\ \mathbf{a}_m \end{bmatrix} \quad \text{and} \quad \mathbf{B} = \begin{bmatrix} \mathbf{b}_1 & \mathbf{b}_2 & \cdots & \mathbf{b}_n \end{bmatrix}$$

in terms of the rows of $\mathbf{A}$ and the columns of $\mathbf{B}$, then

$$\mathbf{AB} = \begin{bmatrix} \mathbf{a}_i \mathbf{b}_j \end{bmatrix}. \tag{15}$$

Therefore, as mentioned earlier, the ijth element $\mathbf{a}_i \mathbf{b}_j$ of $\mathbf{AB}$ is given in terms of elements of $\mathbf{A}$ and $\mathbf{B}$ by

$$\mathbf{a}_i \mathbf{b}_j = \begin{bmatrix} a_{i1} & a_{i2} & \cdots & a_{ip} \end{bmatrix} \begin{bmatrix} b_{1j} \\ b_{2j} \\ \vdots \\ b_{pj} \end{bmatrix} = a_{i1}b_{1j} + a_{i2}b_{2j} + \cdots + a_{ip}b_{pj}.$$

That is,

$$\mathbf{a}_i \mathbf{b}_j = \sum_{k=1}^{p} a_{ik} b_{kj}. \tag{16}$$

One can visualize "pouring the ith row of $\mathbf{A}$ down the jth column of $\mathbf{B}$" until elements match in pairs, then forming the sum of the products of these pairs, to obtain the element c_{ij} of the matrix $\mathbf{C} = \mathbf{AB}$.

SUGGESTION The key to accuracy and confidence in computing matrix products lies in doing it systematically. Always perform your computations in the same order. *First* calculate the elements of the *first* row of $\mathbf{AB}$ by multiplying the *first* row of $\mathbf{A}$ by the successive columns of $\mathbf{B}$; *second*, calculate the elements of the *second* row of $\mathbf{AB}$ by multiplying the *second* row of $\mathbf{A}$ by the successive columns of $\mathbf{B}$; and so forth.

Computing systems often are used for the calculation of products of "large" matrices. If the matrices $\mathbf{A}$ and $\mathbf{B}$, with appropriate sizes, have been entered—as illustrated in the 3.2 Application—then the *Maple* command

```
with(linalg) : C := multiply(A,B),
```

or the *Mathematica* command

```
C = A.B,
```

or the MATLAB command

```
C = A*B
```

immediately yield the product matrix $\mathbf{C} = \mathbf{AB}$.

Matrix Equations

If $\mathbf{A} = \left[\, a_{ij} \,\right]$ is an $m \times n$ coefficient matrix and $\mathbf{x} = (x_1, x_2, \dots, x_n)$ is an $n \times 1$ variable (column) matrix, then the product $\mathbf{Ax}$ is the $m \times 1$ matrix

$$\mathbf{Ax} = \begin{bmatrix} a_{11} & a_{12} & \cdots & a_{1n} \\ a_{21} & a_{22} & \cdots & a_{2n} \\ \vdots & \vdots & \ddots & \vdots \\ a_{m1} & a_{m2} & \cdots & a_{mn} \end{bmatrix} \begin{bmatrix} x_1 \\ x_2 \\ \vdots \\ x_n \end{bmatrix}$$

$$= \begin{bmatrix} a_{11}x_1 + a_{12}x_2 + \cdots + a_{1n}x_n \\ a_{21}x_1 + a_{22}x_2 + \cdots + a_{2n}x_n \\ \vdots \\ a_{m1}x_1 + a_{m2}x_2 + \cdots + a_{mn}x_n \end{bmatrix} \overset{(?)}{=} \begin{bmatrix} b_1 \\ b_2 \\ \vdots \\ b_m \end{bmatrix} = \mathbf{b}.$$

We therefore see that

$$\mathbf{Ax} = \mathbf{b} \tag{17}$$

if and only if $\mathbf{x} = (x_1, x_2, \dots, x_n)$ is a solution of the linear system in (5). Thus, matrix multiplication enables us to "boil down" a system of m scalar equations in n unknowns to the single matrix equation in (17), which is analogous in notation to the single scalar equation $ax = b$ in a single variable x.

Example 7 The system

$$\begin{aligned} 3x_1 - 4x_2 + x_3 + 7x_4 &= 10 \\ 4x_1 \phantom{{}-4x_2} - 5x_3 + 2x_4 &= 0 \\ x_1 + 9x_2 + 2x_3 - 6x_4 &= 5 \end{aligned}$$

of three equations in four unknowns is equivalent to the single matrix equation

$$\begin{bmatrix} 3 & -4 & 1 & 7 \\ 4 & 0 & -5 & 2 \\ 1 & 9 & 2 & -6 \end{bmatrix} \begin{bmatrix} x_1 \\ x_2 \\ x_3 \\ x_4 \end{bmatrix} = \begin{bmatrix} 10 \\ 0 \\ 5 \end{bmatrix}.$$

■

Matrix Algebra

The definitions of matrix addition and multiplication can be used to establish the rules of matrix algebra listed in the following theorem.

THEOREM 1 Rules of Matrix Algebra

If **A**, **B**, and **C** are matrices of appropriate sizes to make the indicated operations possible, then the following identities hold.

Commutative law of addition: $\mathbf{A} + \mathbf{B} = \mathbf{B} + \mathbf{A}$

Associative law of addition: $\mathbf{A} + (\mathbf{B} + \mathbf{C}) = (\mathbf{A} + \mathbf{B}) + \mathbf{C}$

Associative law of multiplication: $\mathbf{A}(\mathbf{BC}) = (\mathbf{AB})\mathbf{C}$

Distributive laws: $\mathbf{A}(\mathbf{B} + \mathbf{C}) = \mathbf{AB} + \mathbf{AC}$

and

$(\mathbf{A} + \mathbf{B})\mathbf{C} = \mathbf{AC} + \mathbf{BC}$

The only verification that is not entirely routine is that of the associative law of multiplication; see Problem 44 for an outline. Each of the others follows quickly from the corresponding law for the ordinary arithmetic of real numbers. As an illustration, we prove the first distributive law. Suppose that $\mathbf{A} = \begin{bmatrix} a_{ij} \end{bmatrix}$ is an $m \times p$ matrix and that $\mathbf{B} = \begin{bmatrix} b_{ij} \end{bmatrix}$ and $\mathbf{C} = \begin{bmatrix} c_{ij} \end{bmatrix}$ are $p \times n$ matrices. Then

$$\mathbf{B} + \mathbf{C} = \begin{bmatrix} b_{ij} + c_{ij} \end{bmatrix},$$

so by (16) the ijth element of the $m \times n$ matrix $\mathbf{A}(\mathbf{B} + \mathbf{C})$ is

$$\sum_{k=1}^{p} a_{ik}(b_{kj} + c_{kj}). \tag{18}$$

The ijth element of the $m \times n$ matrix $\mathbf{AB} + \mathbf{AC}$ is

$$\sum_{k=1}^{p} a_{ik}b_{kj} + \sum_{k=1}^{p} a_{ik}c_{kj} = \sum_{k=1}^{p} (a_{ik}b_{kj} + a_{ik}c_{kj}). \tag{19}$$

But the distributive law for real numbers, $a(b + c) = ab + ac$, tells us that corresponding terms of the sums in (18) and (19) are equal. Hence, the ijth terms of the two $m \times n$ matrices $\mathbf{A}(\mathbf{B} + \mathbf{C})$ and $\mathbf{AB} + \mathbf{AC}$ are equal, and so these matrices are equal: $\mathbf{A}(\mathbf{B} + \mathbf{C}) = \mathbf{AB} + \mathbf{AC}$.

If a and b are real numbers, then rules such as

$$(a + b)\mathbf{C} = a\mathbf{C} + b\mathbf{C}, \quad (ab)\mathbf{C} = a(b\mathbf{C}), \quad a(\mathbf{BC}) = (a\mathbf{B})\mathbf{C}$$

are even easier to verify. What all these rules amount to is this: *In matrix manipulations, pairs of parentheses can be inserted or deleted in the same ways as in the ordinary algebra of real numbers.*

But not all of the rules of "ordinary" algebra carry over to matrix algebra. In Example 5 we saw that multiplication of matrices is *not* commutative—in general, $\mathbf{AB} \neq \mathbf{BA}$. Other exceptions are associated with zero matrices. A **zero matrix** is one whose elements are *all* zero, such as

$$\begin{bmatrix} 0 & 0 \\ 0 & 0 \end{bmatrix}, \quad \begin{bmatrix} 0 & 0 & 0 \\ 0 & 0 & 0 \end{bmatrix}, \quad \begin{bmatrix} 0 & 0 \\ 0 & 0 \\ 0 & 0 \end{bmatrix}, \quad \begin{bmatrix} 0 \\ 0 \end{bmatrix}.$$

We ordinarily denote a zero matrix (whatever its size) by $\mathbf{0}$. It should be clear that for any matrix $\mathbf{A}$,

$$\mathbf{0} + \mathbf{A} = \mathbf{A} = \mathbf{A} + \mathbf{0}, \quad \mathbf{A0} = \mathbf{0}, \quad \text{and} \quad \mathbf{0A} = \mathbf{0},$$

where in each case $\mathbf{0}$ is a zero matrix of appropriate size. Thus zero matrices appear to play a role in the arithmetic of matrices similar to the role of the real number 0 in ordinary arithmetic.

For real numbers, the following two rules are familiar:

- If $ab = ac$ and $a \neq 0$, then $b = c$
  (the "cancellation law").
- If $ad = 0$, then either $a = 0$ or $d = 0$.

The following example shows that matrices do *not* obey either of these rules.

Example 8 If

$$\mathbf{A} = \begin{bmatrix} 4 & 1 & -2 & 7 \\ 3 & 1 & -1 & 5 \end{bmatrix}, \quad \mathbf{B} = \begin{bmatrix} 1 & 5 \\ 3 & -1 \\ -2 & 4 \\ 2 & -3 \end{bmatrix}, \quad \text{and} \quad \mathbf{C} = \begin{bmatrix} 3 & 4 \\ 2 & 1 \\ -2 & 3 \\ 1 & -3 \end{bmatrix},$$

then $\mathbf{B} \neq \mathbf{C}$, but

$$\mathbf{AB} = \begin{bmatrix} 25 & -10 \\ 18 & -5 \end{bmatrix} = \mathbf{AC}. \qquad \text{(Check this!)}$$

Thus the cancellation law does not generally hold for matrices. If

$$\mathbf{D} = \mathbf{B} - \mathbf{C} = \begin{bmatrix} -2 & 1 \\ 1 & -2 \\ 0 & 1 \\ 1 & 0 \end{bmatrix},$$

then

$$\mathbf{AD} = \begin{bmatrix} 0 & 0 \\ 0 & 0 \end{bmatrix} = \mathbf{0},$$

despite the fact that neither **A** nor **D** is a zero matrix. See Problems 31–38 for additional ways in which the algebra of matrices differs significantly from the familiar algebra of real numbers. ■

Recall that an *identity matrix* is a square matrix **I** that has ones on its principal diagonal and zeros elsewhere. Identity matrices play a role in matrix arithmetic which is strongly analogous to that of the real number 1, for which $a \cdot 1 = 1 \cdot a = a$ for all values of the real number a. For instance, you can check that

$$\begin{bmatrix} a & b \\ c & d \end{bmatrix}\begin{bmatrix} 1 & 0 \\ 0 & 1 \end{bmatrix} = \begin{bmatrix} 1 & 0 \\ 0 & 1 \end{bmatrix}\begin{bmatrix} a & b \\ c & d \end{bmatrix} = \begin{bmatrix} a & b \\ c & d \end{bmatrix}.$$

Similarly, if

$$\mathbf{A} = \begin{bmatrix} a_{11} & a_{12} & a_{13} \\ a_{21} & a_{22} & a_{23} \\ a_{31} & a_{32} & a_{33} \end{bmatrix} \quad \text{and} \quad \mathbf{I} = \begin{bmatrix} 1 & 0 & 0 \\ 0 & 1 & 0 \\ 0 & 0 & 1 \end{bmatrix},$$

then $\mathbf{AI} = \mathbf{IA} = \mathbf{A}$. For instance, the element in the second row and third column of **AI** is

$$(a_{21})(0) + (a_{22})(0) + (a_{23})(1) = a_{23}.$$

If a is a nonzero real number and $b = a^{-1}$, then $ab = ba = 1$. Given a nonzero square matrix **A**, the question as to whether there exists an *inverse matrix* **B**, one such that $\mathbf{AB} = \mathbf{BA} = \mathbf{I}$, is more complicated and is investigated in Section 3.5.

3.4 Problems

In Problems 1–4, two matrices **A** *and* **B** *and two numbers c and d are given. Compute the matrix c***A** + *d***B**.

1. $\mathbf{A} = \begin{bmatrix} 3 & -5 \\ 2 & 7 \end{bmatrix}$, $\mathbf{B} = \begin{bmatrix} -1 & 0 \\ 3 & -4 \end{bmatrix}$, $c = 3$, $d = 4$

2. $\mathbf{A} = \begin{bmatrix} 2 & 0 & -3 \\ -1 & 5 & 6 \end{bmatrix}$, $\mathbf{B} = \begin{bmatrix} -2 & 3 & 1 \\ 7 & 1 & 5 \end{bmatrix}$, $c = 5$, $d = -3$

3. $\mathbf{A} = \begin{bmatrix} 5 & 0 \\ 0 & 7 \\ 3 & -1 \end{bmatrix}$, $\mathbf{B} = \begin{bmatrix} -4 & 5 \\ 3 & 2 \\ 7 & 4 \end{bmatrix}$, $c = -2$, $d = 4$

4. $\mathbf{A} = \begin{bmatrix} 2 & -1 & 0 \\ 4 & 0 & -3 \\ 5 & -2 & 7 \end{bmatrix}$, $\mathbf{B} = \begin{bmatrix} 6 & -3 & -4 \\ 5 & 2 & -1 \\ 0 & 7 & 9 \end{bmatrix}$, $c = 7$, $d = 5$

In Problems 5–12, two matrices **A** *and* **B** *are given. Calculate whichever of the matrices* **AB** *and* **BA** *is defined.*

5. $\mathbf{A} = \begin{bmatrix} 2 & -1 \\ 3 & 2 \end{bmatrix}$, $\mathbf{B} = \begin{bmatrix} -4 & 2 \\ 1 & 3 \end{bmatrix}$

6. $\mathbf{A} = \begin{bmatrix} 1 & 0 & -3 \\ 3 & 2 & 4 \\ 2 & -3 & 5 \end{bmatrix}$, $\mathbf{B} = \begin{bmatrix} 7 & -4 & 3 \\ 1 & 5 & -2 \\ 0 & 3 & 9 \end{bmatrix}$

7. $\mathbf{A} = \begin{bmatrix} 1 & 2 & 3 \end{bmatrix}$, $\mathbf{B} = \begin{bmatrix} 3 \\ 4 \\ 5 \end{bmatrix}$

8. $\mathbf{A} = \begin{bmatrix} 1 & 0 & 3 \\ 2 & -5 & 4 \end{bmatrix}$, $\mathbf{B} = \begin{bmatrix} 3 & 0 \\ -1 & 4 \\ 6 & 5 \end{bmatrix}$

9. $\mathbf{A} = \begin{bmatrix} 3 \\ -2 \end{bmatrix}$, $\mathbf{B} = \begin{bmatrix} 0 & -2 \\ 3 & 1 \\ -4 & 5 \end{bmatrix}$

10. $\mathbf{A} = \begin{bmatrix} 2 & 1 \\ 4 & 3 \end{bmatrix}$, $\mathbf{B} = \begin{bmatrix} -1 & 0 & 4 \\ 3 & -2 & 5 \end{bmatrix}$

11. $\mathbf{A} = \begin{bmatrix} 3 & -5 \end{bmatrix}$, $\mathbf{B} = \begin{bmatrix} 2 & 7 & 5 & 6 \\ -1 & 4 & 2 & 3 \end{bmatrix}$

12. $\mathbf{A} = \begin{bmatrix} 1 & 0 & 3 & -2 \end{bmatrix}$, $\mathbf{B} = \begin{bmatrix} 2 & -7 & 5 \\ 3 & 9 & 10 \end{bmatrix}$

In Problems 13–16, three matrices **A**, **B**, *and* **C** *are given. Verify by computation of both sides the associative law* **A**(**BC**) = (**AB**)**C**.

13. $\mathbf{A} = \begin{bmatrix} 3 & 1 \\ -1 & 4 \end{bmatrix}$, $\mathbf{B} = \begin{bmatrix} 2 & 5 \\ -3 & 1 \end{bmatrix}$, $\mathbf{C} = \begin{bmatrix} 0 & 1 \\ 2 & 3 \end{bmatrix}$

14. $\mathbf{A} = \begin{bmatrix} 2 & -1 \end{bmatrix}$, $\mathbf{B} = \begin{bmatrix} 2 & 5 \\ -3 & 1 \end{bmatrix}$, $\mathbf{C} = \begin{bmatrix} 6 \\ -5 \end{bmatrix}$

15. $\mathbf{A} = \begin{bmatrix} 3 \\ 2 \end{bmatrix}$, $\mathbf{B} = \begin{bmatrix} 1 & -1 & 2 \end{bmatrix}$, $\mathbf{C} = \begin{bmatrix} 2 & 0 \\ 0 & 3 \\ 1 & 4 \end{bmatrix}$

16. $\mathbf{A} = \begin{bmatrix} 2 & 0 \\ 0 & 3 \\ 1 & 4 \end{bmatrix}$, $\mathbf{B} = \begin{bmatrix} 1 & -1 \\ 3 & -2 \end{bmatrix}$,

$$\mathbf{C} = \begin{bmatrix} 1 & 0 & -1 & 2 \\ 3 & 2 & 0 & 1 \end{bmatrix}$$

In Problems 17–22, first write each given homogeneous system in the matrix form **Ax** = **0**. *Then find the solution in vector form, as in Eq. (9).*

17. $\begin{aligned} x_1 \qquad - 5x_3 + 4x_4 &= 0 \\ x_2 + 2x_3 - 7x_4 &= 0 \end{aligned}$

18. $\begin{aligned} x_1 - 3x_2 \qquad + 6x_4 &= 0 \\ x_3 + 9x_4 &= 0 \end{aligned}$

19. $\begin{aligned} x_1 \qquad + 3x_4 - x_5 &= 0 \\ x_2 \qquad - 2x_4 + 6x_5 &= 0 \\ x_3 + x_4 - 8x_5 &= 0 \end{aligned}$

20. $\begin{aligned} x_1 - 3x_2 \qquad + 7x_5 &= 0 \\ x_3 \qquad - 2x_5 &= 0 \\ x_4 - 10x_5 &= 0 \end{aligned}$

21. $\begin{aligned} x_1 \qquad - x_3 + 2x_4 + 7x_5 &= 0 \\ x_2 + 2x_3 - 3x_4 + 4x_5 &= 0 \end{aligned}$

22. $\begin{aligned} x_1 - x_2 \qquad + 7x_4 + 3x_5 &= 0 \\ x_3 - x_4 - 2x_5 &= 0 \end{aligned}$

23. Let

$$\mathbf{A} = \begin{bmatrix} 2 & 1 \\ 3 & 2 \end{bmatrix}, \quad \mathbf{B} = \begin{bmatrix} a & b \\ c & d \end{bmatrix},$$

and

$$\mathbf{I} = \begin{bmatrix} 1 & 0 \\ 0 & 1 \end{bmatrix}.$$

Find **B** so that **AB** = **I** = **BA** as follows: First equate entries on the two sides of the equation **AB** = **I**. Then solve the resulting four equations for a, b, c, and d. Finally verify that **BA** = **I** as well.

24. Repeat Problem 23, but with **A** replaced by the matrix

$$\mathbf{A} = \begin{bmatrix} 3 & 4 \\ 5 & 7 \end{bmatrix}.$$

25. Repeat Problem 23, but with **A** replaced by the matrix

$$\mathbf{A} = \begin{bmatrix} 5 & 7 \\ 2 & 3 \end{bmatrix}.$$

26. Use the technique of Problem 23 to show that if

$$\mathbf{A} = \begin{bmatrix} 1 & -2 \\ -2 & 4 \end{bmatrix},$$

then there does *not* exist a matrix **B** such that **AB** = **I**. *Suggestion*: Show that the system of four equations in a, b, c, and d is inconsistent.

27. A **diagonal matrix** is a square matrix of the form

$$\begin{bmatrix} a_1 & 0 & 0 & \cdots & 0 \\ 0 & a_2 & 0 & \cdots & 0 \\ 0 & 0 & a_3 & \cdots & 0 \\ \vdots & \vdots & \vdots & \ddots & \vdots \\ 0 & 0 & 0 & \cdots & a_n \end{bmatrix},$$

in which every element off the main diagonal is zero. Show that the product $\mathbf{AB}$ of two $n \times n$ diagonal matrices $\mathbf{A}$ and $\mathbf{B}$ is again a diagonal matrix. State a concise rule for quickly computing $\mathbf{AB}$. Is it clear that $\mathbf{AB} = \mathbf{BA}$? Explain.

28. The positive integral powers of a square matrix $\mathbf{A}$ are defined as follows:

$$\mathbf{A}^1 = \mathbf{A}, \quad \mathbf{A}^2 = \mathbf{AA}, \quad \mathbf{A}^3 = \mathbf{AA}^2,$$
$$\mathbf{A}^4 = \mathbf{AA}^3, \ldots, \quad \mathbf{A}^{n+1} = \mathbf{AA}^n, \ldots.$$

Suppose that r and s are positive integers. Prove that $\mathbf{A}^r\mathbf{A}^s = \mathbf{A}^{r+s}$ and that $(\mathbf{A}^r)^s = \mathbf{A}^{rs}$ (in close analogy with the laws of exponents for real numbers).

29. If $\mathbf{A} = \begin{bmatrix} a & b \\ c & d \end{bmatrix}$, then show that

$$\mathbf{A}^2 = (a+d)\mathbf{A} - (ad - bc)\mathbf{I},$$

where $\mathbf{I}$ denotes the 2×2 identity matrix. Thus every 2×2 matrix $\mathbf{A}$ satisfies the equation

$$\mathbf{A}^2 - (\text{trace}\,\mathbf{A})\mathbf{A} + (\det \mathbf{A})\mathbf{I} = \mathbf{0}$$

where $\det \mathbf{A} = ad - bc$ denotes the determinant of the matrix $\mathbf{A}$, and trace $\mathbf{A}$ denotes the sum of its diagonal elements. This result is the 2-dimensional case of the Cayley-Hamilton theorem of Section 6.3.

30. The formula in Problem 29 can be used to compute $\mathbf{A}^2$ without an explicit matrix multiplication. It follows that

$$\mathbf{A}^3 = (a+d)\mathbf{A}^2 - (ad - bc)\mathbf{A}$$

without an explicit matrix multiplication,

$$\mathbf{A}^4 = (a+d)\mathbf{A}^3 - (ad - bc)\mathbf{A}^2,$$

and so on. Use this method to compute $\mathbf{A}^2$, $\mathbf{A}^3$, $\mathbf{A}^4$, and $\mathbf{A}^5$ given

$$\mathbf{A} = \begin{bmatrix} 2 & 1 \\ 1 & 2 \end{bmatrix}.$$

Problems 31–38 illustrate ways in which the algebra of matrices is not analogous to the algebra of real numbers.

31. **(a)** Suppose that $\mathbf{A}$ and $\mathbf{B}$ are the matrices of Example 5. Show that $(\mathbf{A}+\mathbf{B})(\mathbf{A}-\mathbf{B}) \neq \mathbf{A}^2 - \mathbf{B}^2$.

(b) Suppose that $\mathbf{A}$ and $\mathbf{B}$ are square matrices with the property that $\mathbf{AB} = \mathbf{BA}$. Show that $(\mathbf{A}+\mathbf{B})(\mathbf{A}-\mathbf{B}) = \mathbf{A}^2 - \mathbf{B}^2$.

32. **(a)** Suppose that $\mathbf{A}$ and $\mathbf{B}$ are the matrices of Example 5. Show that $(\mathbf{A}+\mathbf{B})^2 \neq \mathbf{A}^2 + 2\mathbf{AB} + \mathbf{B}^2$.

(b) Suppose that $\mathbf{A}$ and $\mathbf{B}$ are square matrices such that $\mathbf{AB} = \mathbf{BA}$. Show that $(\mathbf{A}+\mathbf{B})^2 = \mathbf{A}^2 + 2\mathbf{AB} + \mathbf{B}^2$.

33. Find four different 2×2 matrices $\mathbf{A}$, with each main diagonal element either $+1$ or -1, such that $\mathbf{A}^2 = \mathbf{I}$.

34. Find a 2×2 matrix $\mathbf{A}$ with each element $+1$ or -1 such that $\mathbf{A}^2 = \mathbf{0}$. The formula of Problem 29 may be helpful.

35. Use the formula of Problem 29 to find a 2×2 matrix $\mathbf{A}$ such that $\mathbf{A} \neq \mathbf{0}$ and $\mathbf{A} \neq \mathbf{I}$ but such that $\mathbf{A}^2 = \mathbf{A}$.

36. Find a 2×2 matrix $\mathbf{A}$ with each main diagonal element zero such that $\mathbf{A}^2 = \mathbf{I}$.

37. Find a 2×2 matrix $\mathbf{A}$ with each main diagonal element zero such that $\mathbf{A}^2 = -\mathbf{I}$.

38. This is a continuation of the previous two problems. Find two nonzero 2×2 matrices $\mathbf{A}$ and $\mathbf{B}$ such that $\mathbf{A}^2 + \mathbf{B}^2 = \mathbf{0}$.

39. Use matrix multiplication to show that if $\mathbf{x}_1$ and $\mathbf{x}_2$ are two solutions of the homogeneous system $\mathbf{Ax} = \mathbf{0}$ and c_1 and c_2 are real numbers, then $c_1\mathbf{x}_1 + c_2\mathbf{x}_2$ is also a solution.

40. **(a)** Use matrix multiplication to show that if $\mathbf{x}_0$ is a solution of the homogeneous system $\mathbf{Ax} = \mathbf{0}$ and $\mathbf{x}_1$ is a solution of the nonhomogeneous system $\mathbf{Ax} = \mathbf{b}$, then $\mathbf{x}_0 + \mathbf{x}_1$ is also a solution of the nonhomogeneous system.

(b) Suppose that $\mathbf{x}_1$ and $\mathbf{x}_2$ are solutions of the nonhomogeneous system of part (a). Show that $\mathbf{x}_1 - \mathbf{x}_2$ is a solution of the homogeneous system $\mathbf{Ax} = \mathbf{0}$.

41. This is a continuation of Problem 32. Show that if $\mathbf{A}$ and $\mathbf{B}$ are square matrices such that $\mathbf{AB} = \mathbf{BA}$, then

$$(\mathbf{A}+\mathbf{B})^3 = \mathbf{A}^3 + 3\mathbf{A}^2\mathbf{B} + 3\mathbf{AB}^2 + \mathbf{B}^3$$

and

$$(\mathbf{A}+\mathbf{B})^4 = \mathbf{A}^4 + 4\mathbf{A}^3\mathbf{B} + 6\mathbf{A}^2\mathbf{B}^2 + 4\mathbf{AB}^3 + \mathbf{B}^4.$$

42. Let

$$\mathbf{A} = \begin{bmatrix} 1 & 2 & 0 \\ 0 & 1 & 2 \\ 0 & 0 & 1 \end{bmatrix} = \begin{bmatrix} 1 & 0 & 0 \\ 0 & 1 & 0 \\ 0 & 0 & 1 \end{bmatrix} + \begin{bmatrix} 0 & 2 & 0 \\ 0 & 0 & 2 \\ 0 & 0 & 0 \end{bmatrix} = \mathbf{I} + \mathbf{N}.$$

(a) Show that $\mathbf{N}^2 \neq \mathbf{0}$ but $\mathbf{N}^3 = \mathbf{0}$.

(b) Use the binomial formulas of Problem 41 to compute

$$\mathbf{A}^2 = (\mathbf{I}+\mathbf{N})^2 = \mathbf{I} + 2\mathbf{N} + \mathbf{N}^2,$$
$$\mathbf{A}^3 = (\mathbf{I}+\mathbf{N})^3 = \mathbf{I} + 3\mathbf{N} + 3\mathbf{N}^2,$$

and

$$\mathbf{A}^4 = (\mathbf{I}+\mathbf{N})^4 = \mathbf{I} + 4\mathbf{N} + 6\mathbf{N}^2.$$

43. Consider the 3×3 matrix

$$\mathbf{A} = \begin{bmatrix} 2 & -1 & -1 \\ -1 & 2 & -1 \\ -1 & -1 & 2 \end{bmatrix}.$$

First verify by direct computation that $\mathbf{A}^2 = 3\mathbf{A}$. Then conclude that $\mathbf{A}^{n+1} = 3^n\mathbf{A}$ for every positive integer n.

44. Let $\mathbf{A} = \begin{bmatrix} a_{hi} \end{bmatrix}$, $\mathbf{B} = \begin{bmatrix} b_{ij} \end{bmatrix}$, and $\mathbf{C} = \begin{bmatrix} c_{jk} \end{bmatrix}$ be matrices of sizes $m \times n$, $n \times p$, and $p \times q$, respectively. To establish the associative law $\mathbf{A}(\mathbf{BC}) = (\mathbf{AB})\mathbf{C}$, proceed as follows. By Equation (16) the hjth element of $\mathbf{AB}$ is

$$\sum_{i=1}^{n} a_{hi} b_{ij}.$$

By another application of Equation (16), the hkth element of $(\mathbf{AB})\mathbf{C}$ is

$$\sum_{j=1}^{p} \left(\sum_{i=1}^{n} a_{hi} b_{ij} \right) c_{jk} = \sum_{i=1}^{n} \sum_{j=1}^{p} a_{hi} b_{ij} c_{jk}.$$

Show similarly that the double sum on the right is also equal to the hkth element of $\mathbf{A}(\mathbf{BC})$. Hence the $m \times q$ matrices $(\mathbf{AB})\mathbf{C}$ and $\mathbf{A}(\mathbf{BC})$ are equal.

3.5 Inverses of Matrices

Recall that the $n \times n$ **identity matrix** is the diagonal matrix

$$\mathbf{I} = \begin{bmatrix} 1 & 0 & 0 & \cdots & 0 \\ 0 & 1 & 0 & \cdots & 0 \\ 0 & 0 & 1 & \cdots & 0 \\ \vdots & \vdots & \vdots & \ddots & \vdots \\ 0 & 0 & 0 & \cdots & 1 \end{bmatrix} \tag{1}$$

having ones on its main diagonal and zeros elsewhere. It is not difficult to deduce directly from the definition of the matrix product that $\mathbf{I}$ acts like an identity for matrix multiplication:

$$\mathbf{AI} = \mathbf{A} \quad \text{and} \quad \mathbf{IB} = \mathbf{B} \tag{2}$$

if the sizes of $\mathbf{A}$ and $\mathbf{B}$ are such that the products $\mathbf{AI}$ and $\mathbf{IB}$ are defined. It is, nevertheless, instructive to derive the identities in (2) formally from the two basic facts about matrix multiplication that we state below. First, recall that the notation

$$\mathbf{A} = \begin{bmatrix} \mathbf{a}_1 & \mathbf{a}_2 & \mathbf{a}_3 & \cdots & \mathbf{a}_n \end{bmatrix} \tag{3}$$

expresses the $m \times n$ matrix $\mathbf{A}$ in terms of its column vectors $\mathbf{a}_1, \mathbf{a}_2, \mathbf{a}_3, \ldots, \mathbf{a}_n$.

Fact 1 **Ax** *in terms of columns of* **A**

If $\mathbf{A} = \begin{bmatrix} \mathbf{a}_1 & \mathbf{a}_2 & \cdots & \mathbf{a}_n \end{bmatrix}$ and $\mathbf{x} = (x_1, x_2, \ldots, x_n)$ is an n-vector, then

$$\mathbf{Ax} = x_1\mathbf{a}_1 + x_2\mathbf{a}_2 + \cdots + x_n\mathbf{a}_n. \tag{4}$$

The reason is that when each row vector of $\mathbf{A}$ is multiplied by the column vector $\mathbf{x}$, its jth element is multiplied by x_j.

Fact 2 **AB** *in terms of columns of* **B**

If $\mathbf{A}$ is an $m \times n$ matrix and $\mathbf{B} = \begin{bmatrix} \mathbf{b}_1 & \mathbf{b}_2 & \cdots & \mathbf{b}_p \end{bmatrix}$ is an $n \times p$ matrix, then

$$\mathbf{AB} = \begin{bmatrix} \mathbf{Ab}_1 & \mathbf{Ab}_2 & \cdots & \mathbf{Ab}_p \end{bmatrix}. \tag{5}$$

That is, *the jth column of* **AB** *is the product of* **A** *and the jth column of* **B**. The reason is that the elements of the jth column of $\mathbf{AB}$ are obtained by multiplying the individual rows of $\mathbf{A}$ by the jth column of $\mathbf{B}$.

Example 1 The third column of the product **AB** of the matrices

$$\mathbf{A} = \begin{bmatrix} 2 & -1 & 0 \\ 4 & 0 & 3 \end{bmatrix} \quad \text{and} \quad \mathbf{B} = \begin{bmatrix} 3 & 7 & 5 & -4 \\ -2 & 6 & 3 & 6 \\ 5 & 1 & -2 & -1 \end{bmatrix}$$

is

$$\mathbf{Ab}_3 = \begin{bmatrix} 2 & -1 & 0 \\ 4 & 0 & 3 \end{bmatrix} \begin{bmatrix} 5 \\ 3 \\ -2 \end{bmatrix} = \begin{bmatrix} 7 \\ 14 \end{bmatrix}.$$

■

To prove that $\mathbf{AI} = \mathbf{A}$, note first that

$$\mathbf{I} = \begin{bmatrix} \mathbf{e}_1 & \mathbf{e}_2 & \cdots & \mathbf{e}_n \end{bmatrix}, \tag{6}$$

where the jth column vector of **I** is the jth **basic unit vector**

$$\mathbf{e}_j = \begin{bmatrix} 0 \\ \vdots \\ 1 \\ \vdots \\ 0 \end{bmatrix} \leftarrow j\text{th entry}. \tag{7}$$

If $\mathbf{A} = \begin{bmatrix} \mathbf{a}_1 & \mathbf{a}_2 & \cdots & \mathbf{a}_n \end{bmatrix}$, then Fact 1 yields

$$\mathbf{Ae}_j = 0 \cdot \mathbf{a}_1 + \cdots + 1 \cdot \mathbf{a}_j + \cdots + 0 \cdot \mathbf{a}_n = \mathbf{a}_j. \tag{8}$$

Hence Fact 2 gives

$$\begin{aligned} \mathbf{AI} &= \mathbf{A}\begin{bmatrix} \mathbf{e}_1 & \mathbf{e}_2 & \cdots & \mathbf{e}_n \end{bmatrix} \\ &= \begin{bmatrix} \mathbf{Ae}_1 & \mathbf{Ae}_2 & \cdots & \mathbf{Ae}_n \end{bmatrix} = \begin{bmatrix} \mathbf{a}_1 & \mathbf{a}_2 & \cdots & \mathbf{a}_n \end{bmatrix}; \end{aligned}$$

that is, $\mathbf{AI} = \mathbf{A}$. The proof that $\mathbf{IB} = \mathbf{B}$ is similar. (See Problems 41 and 42.)

The Inverse Matrix $\mathbf{A}^{-1}$

If $a \neq 0$, then there is a number $b = a^{-1} = 1/a$ such that $ab = ba = 1$. Given a nonzero matrix **A**, we therefore wonder whether there is a matrix **B** such that $\mathbf{AB} = \mathbf{BA} = \mathbf{I}$. The following two examples show that the answer to this question depends upon the particular matrix **A**.

Example 2 If

$$\mathbf{A} = \begin{bmatrix} 4 & 9 \\ 3 & 7 \end{bmatrix} \quad \text{and} \quad \mathbf{B} = \begin{bmatrix} 7 & -9 \\ -3 & 4 \end{bmatrix},$$

then

$$\mathbf{AB} = \begin{bmatrix} 4 & 9 \\ 3 & 7 \end{bmatrix} \begin{bmatrix} 7 & -9 \\ -3 & 4 \end{bmatrix} = \begin{bmatrix} 1 & 0 \\ 0 & 1 \end{bmatrix} = \mathbf{I};$$

$\mathbf{BA} = \mathbf{I}$ by a similar computation. ■

Example 3 Let

$$\mathbf{A} = \begin{bmatrix} 1 & -3 \\ -2 & 6 \end{bmatrix} \quad \text{and} \quad \mathbf{B} = \begin{bmatrix} a & b \\ c & d \end{bmatrix}.$$

If the matrix **B** had the property that $\mathbf{AB} = \mathbf{BA} = \mathbf{I}$, then

$$\mathbf{AB} = \begin{bmatrix} 1 & -3 \\ -2 & 6 \end{bmatrix}\begin{bmatrix} a & b \\ c & d \end{bmatrix}$$
$$= \begin{bmatrix} a - 3c & b - 3d \\ -2a + 6c & -2b + 6d \end{bmatrix} = \begin{bmatrix} 1 & 0 \\ 0 & 1 \end{bmatrix}.$$

But upon equating corresponding elements of **AB** and the 2×2 identity matrix in the last line, we find that

$$\begin{aligned} a - 3c &= 1 \\ -2a + 6c &= 0 \end{aligned} \quad \text{and} \quad \begin{aligned} b - 3d &= 0 \\ -2b + 6d &= 1. \end{aligned}$$

It is clear that these equations are inconsistent. Thus there can exist *no* 2×2 matrix **B** such that $\mathbf{AB} = \mathbf{I}$. ■

DEFINITION Invertible Matrix

The square matrix **A** is called **invertible** if there exists a matrix **B** such that

$$\mathbf{AB} = \mathbf{BA} = \mathbf{I}.$$

Thus the matrix **A** of Example 2 is invertible, whereas the matrix **A** of Example 3 is not invertible.

A matrix **B** such that $\mathbf{AB} = \mathbf{BA} = \mathbf{I}$ is called an **inverse matrix** of the matrix **A**. The following theorem says that no matrix can have two different inverse matrices.

THEOREM 1 Uniqueness of Inverse Matrices

If the matrix **A** is invertible, then there exists precisely one matrix **B** such that $\mathbf{AB} = \mathbf{BA} = \mathbf{I}$.

***Proof*:** If **C** is a (possibly different) matrix such that $\mathbf{AC} = \mathbf{CA} = \mathbf{I}$ as well, then the associative law of multiplication gives

$$\mathbf{C} = \mathbf{CI} = \mathbf{C}(\mathbf{AB}) = (\mathbf{CA})\mathbf{B} = \mathbf{IB} = \mathbf{B}.$$

Thus **C** is in fact the same matrix as **B**. ■

The unique inverse of an invertible matrix **A** is denoted by $\mathbf{A}^{-1}$. Thus we say in Example 2 that

$$\text{If} \quad \mathbf{A} = \begin{bmatrix} 4 & 9 \\ 3 & 7 \end{bmatrix} \quad \text{then} \quad \mathbf{A}^{-1} = \begin{bmatrix} 7 & -9 \\ -3 & 4 \end{bmatrix}.$$

In the case of a 2×2 matrix **A**, it is easy to determine whether or not **A** is invertible and to find $\mathbf{A}^{-1}$ if it exists. In Problems 36 and 37 we ask you to verify the following result.

THEOREM 2 Inverses of 2 x 2 Matrices

The 2×2 matrix

$$\mathbf{A} = \begin{bmatrix} a & b \\ c & d \end{bmatrix}$$

is invertible if and only if $ad - bc \neq 0$, in which case

$$\mathbf{A}^{-1} = \frac{1}{ad - bc} \begin{bmatrix} d & -b \\ -c & a \end{bmatrix}. \tag{9}$$

Equation (9) gives us the following prescription for writing the inverse of an invertible 2×2 matrix:

- First, interchange the two main diagonal entries.
- Then, change the signs of the two off-diagonal elements.
- Finally, divide each element of the resulting matrix by $ad - bc$.

You might check that this is how $\mathbf{B} = \mathbf{A}^{-1}$ is obtained from $\mathbf{A}$ in Example 2 (in which $ad - bc = 1$).

Example 4 If

$$\mathbf{A} = \begin{bmatrix} 4 & 6 \\ 5 & 9 \end{bmatrix},$$

then $ad - bc = 36 - 30 = 6 \neq 0$, so

$$\mathbf{A}^{-1} = \tfrac{1}{6} \begin{bmatrix} 9 & -6 \\ -5 & 4 \end{bmatrix} = \begin{bmatrix} \frac{3}{2} & -1 \\ -\frac{5}{6} & \frac{2}{3} \end{bmatrix}.$$ ■

Arbitrary integral powers of a square matrix $\mathbf{A}$ are defined as follows, though in the case of a negative exponent we must assume that $\mathbf{A}$ is also invertible. If n is a positive integer, we define

$$\begin{aligned} \mathbf{A}^0 &= \mathbf{I} \qquad \text{and} \qquad \mathbf{A}^1 = \mathbf{A}; \\ \mathbf{A}^{n+1} &= \mathbf{A}^n\mathbf{A} \quad \text{for } n \geq 1; \\ \mathbf{A}^{-n} &= (\mathbf{A}^{-1})^n. \end{aligned}$$

In Problem 28 of Section 3.4, we asked you to verify the laws of exponents

$$\mathbf{A}^r\mathbf{A}^s = \mathbf{A}^{r+s}, \quad (\mathbf{A}^r)^s = \mathbf{A}^{rs} \tag{10}$$

in the case of positive integral exponents, and Problem 31 of this section deals with the case of negative integral exponents. In Problem 29 we ask you to establish parts (a) and (b) of the following theorem.

THEOREM 3 Algebra of Inverse Matrices

If the matrices $\mathbf{A}$ and $\mathbf{B}$ of the same size are invertible, then

(a) $\mathbf{A}^{-1}$ is invertible and $(\mathbf{A}^{-1})^{-1} = \mathbf{A}$;

(b) If n is a nonnegative integer, then $\mathbf{A}^n$ is invertible and $(\mathbf{A}^n)^{-1} = (\mathbf{A}^{-1})^n$;

(c) The product $\mathbf{AB}$ is invertible and

$$(\mathbf{AB})^{-1} = \mathbf{B}^{-1}\mathbf{A}^{-1}. \tag{11}$$

***Proof of* (c)**

$$(\mathbf{AB})(\mathbf{B}^{-1}\mathbf{A}^{-1}) = \mathbf{A}(\mathbf{BB}^{-1})\mathbf{A}^{-1} = \mathbf{AIA}^{-1} = \mathbf{AA}^{-1} = \mathbf{I};$$
$$(\mathbf{B}^{-1}\mathbf{A}^{-1})(\mathbf{AB}) = \mathbf{B}^{-1}(\mathbf{A}^{-1}\mathbf{A})\mathbf{B} = \mathbf{B}^{-1}\mathbf{IB} = \mathbf{B}^{-1}\mathbf{B} = \mathbf{I}.$$

Thus we get $\mathbf{I}$ when we multiply $\mathbf{AB}$ *on either side* by $\mathbf{B}^{-1}\mathbf{A}^{-1}$. Because the inverse of the matrix $\mathbf{AB}$ is unique, this proves that $\mathbf{AB}$ is invertible and that its inverse matrix is $\mathbf{B}^{-1}\mathbf{A}^{-1}$. ■

In mathematics it is frequently important to note the surprises. The surprise in Eq. (11) is the reversal of the natural order of the factors in the right-hand side. You should now be able to show that

$$(\mathbf{ABC})^{-1} = \mathbf{C}^{-1}\mathbf{B}^{-1}\mathbf{A}^{-1}.$$

In general, any product of invertible matrices of the same size is again invertible, and the inverse of a product of invertible matrices is the product *in reverse order* of their inverses.

THEOREM 4 Inverse Matrix Solution of Ax = b

If the $n \times n$ matrix $\mathbf{A}$ is invertible, then for any n-vector $\mathbf{b}$ the system

$$\mathbf{Ax} = \mathbf{b} \tag{12}$$

has the unique solution

$$\mathbf{x} = \mathbf{A}^{-1}\mathbf{b} \tag{13}$$

that is obtained by multiplying both sides in (12) on the left by the matrix $\mathbf{A}^{-1}$.

***Proof*:** We must show that $\mathbf{x} = \mathbf{A}^{-1}\mathbf{b}$ is a solution and that it is the only solution of Eq. (12). First, the computation

$$\mathbf{A}(\mathbf{A}^{-1}\mathbf{b}) = (\mathbf{AA}^{-1})\mathbf{b} = \mathbf{Ib} = \mathbf{b}$$

shows that $\mathbf{x} = \mathbf{A}^{-1}\mathbf{b}$ *is* a solution. Second, if $\mathbf{x}_1$ is any (possibly different) solution, we observe that multiplication of each side of the equation $\mathbf{Ax}_1 = \mathbf{b}$ on the left by $\mathbf{A}^{-1}$ yields $\mathbf{x}_1 = \mathbf{A}^{-1}\mathbf{b}$, and hence $\mathbf{x}_1$ is the same solution as $\mathbf{x}$ after all. ■

Example 5 To solve the system

$$\begin{aligned} 4x_1 + 6x_2 &= 6 \\ 5x_1 + 9x_2 &= 18, \end{aligned}$$

we use the inverse of the coefficient matrix

$$\mathbf{A} = \begin{bmatrix} 4 & 6 \\ 5 & 9 \end{bmatrix}$$

that we found in Example 4. Then Eq. (13) yields

$$\begin{aligned} \mathbf{x} = \mathbf{A}^{-1}\mathbf{b} &= \begin{bmatrix} 4 & 6 \\ 5 & 9 \end{bmatrix}^{-1} \begin{bmatrix} 6 \\ 18 \end{bmatrix} \\ &= \begin{bmatrix} \frac{3}{2} & -1 \\ -\frac{5}{6} & \frac{2}{3} \end{bmatrix} \begin{bmatrix} 6 \\ 18 \end{bmatrix} = \begin{bmatrix} -9 \\ 7 \end{bmatrix}. \end{aligned}$$

Thus $x_1 = -9$, $x_2 = 7$ is the unique solution. ■

How to Find $\mathbf{A}^{-1}$

Theorem 2 tells us only how to invert 2×2 matrices. The development of a method for inverting larger matrices involves a special class of matrices, which we define next.

DEFINITION Elementary Matrix

The $n \times n$ matrix $\mathbf{E}$ is called an **elementary matrix** if it can be obtained by performing a *single* elementary row operation on the $n \times n$ identity matrix $\mathbf{I}$.

Example 6 We obtain some typical elementary matrices as follows.

$$\begin{bmatrix} 1 & 0 \\ 0 & 1 \end{bmatrix} \xrightarrow{(3)R_1} \begin{bmatrix} 3 & 0 \\ 0 & 1 \end{bmatrix} = \mathbf{E}_1$$

$$\begin{bmatrix} 1 & 0 & 0 \\ 0 & 1 & 0 \\ 0 & 0 & 1 \end{bmatrix} \xrightarrow{(2)R_1 + R_3} \begin{bmatrix} 1 & 0 & 0 \\ 0 & 1 & 0 \\ 2 & 0 & 1 \end{bmatrix} = \mathbf{E}_2$$

$$\begin{bmatrix} 1 & 0 & 0 \\ 0 & 1 & 0 \\ 0 & 0 & 1 \end{bmatrix} \xrightarrow{\text{SWAP}(R_1, R_2)} \begin{bmatrix} 0 & 1 & 0 \\ 1 & 0 & 0 \\ 0 & 0 & 1 \end{bmatrix} = \mathbf{E}_3$$

The three elementary matrices $\mathbf{E}_1$, $\mathbf{E}_2$, and $\mathbf{E}_3$ correspond to three typical elementary row operations. ■

Now, suppose that the $m \times m$ elementary matrix $\mathbf{E}$ corresponds to a certain elementary row operation. It turns out that if we perform this same operation on an arbitrary $m \times n$ matrix $\mathbf{A}$, we get the product matrix $\mathbf{EA}$ that results upon multiplying $\mathbf{A}$ on the *left* by the matrix $\mathbf{E}$. Thus we can carry out an elementary row operation by means of left multiplication by the corresponding elementary matrix. Problems 38–40 illustrate typical cases in the proof of the following theorem.

THEOREM 5 Elementary Matrices and Row Operations

If an elementary row operation is performed on the $m \times n$ matrix $\mathbf{A}$, then the result is the product matrix $\mathbf{EA}$, where $\mathbf{E}$ is the elementary matrix obtained by performing the same row operation on the $m \times m$ identity matrix.

Elementary row operations are reversible. That is, to every elementary row operation there corresponds an *inverse* elementary row operation that cancels its effects (see Figure 3.5.1). It follows that *every elementary matrix is invertible.* To see why, let $\mathbf{E}$ be a given elementary matrix and let $\mathbf{E}_1$ be the elementary matrix corresponding to the inverse of the row operation that transforms $\mathbf{I}$ into $\mathbf{E}$. Then the inverse operation transforms $\mathbf{E}$ to $\mathbf{I}$, so Theorem 5 implies that $\mathbf{E}_1\mathbf{E} = \mathbf{I}$. We see similarly that $\mathbf{E}\mathbf{E}_1 = \mathbf{I}$. Hence, the elementary matrix $\mathbf{E}$ is invertible with $\mathbf{E}^{-1} = \mathbf{E}_1$.

Elementary Row Operation	Inverse Operation
$(c)R_i$	$\dfrac{1}{c}R_i$
$\text{SWAP}(R_i, R_j)$	$\text{SWAP}(R_i, R_j)$
$(c)R_i + R_j$	$(-c)R_i + R_j$

FIGURE 3.5.1. Inverse elementary row operations.

Elementary matrices are not ordinarily used for computational purposes; it is simpler to carry out row operations directly than to multiply by elementary matrices. Instead, their principal role is in the proof of the following theorem, which leads in turn to a practical method for inverting matrices.

THEOREM 6 Invertible Matrices and Row Operations

The $n \times n$ matrix $\mathbf{A}$ is invertible if and only if it is row equivalent to the $n \times n$ identity matrix $\mathbf{I}$.

***Proof*:** Assume first that $\mathbf{A}$ is invertible. Then, by Theorem 4 (with $\mathbf{b} = \mathbf{0}$), it follows that $\mathbf{Ax} = \mathbf{0}$ has only the trivial solution $\mathbf{x} = \mathbf{0}$. But Theorem 4 in Section 3.3 implies that this is so (if and) only if $\mathbf{A}$ is row equivalent to $\mathbf{I}$.

Now assume, conversely, that $\mathbf{A}$ is row equivalent to $\mathbf{I}$. That is, there is a finite sequence of elementary row operations that transforms $\mathbf{A}$ into $\mathbf{I}$. According to Theorem 5, each of these operations can be performed by multiplying on the left by the corresponding elementary matrix. If $\mathbf{E}_1, \mathbf{E}_2, \ldots, \mathbf{E}_k$ are the elementary matrices corresponding to these row operations, it follows that

$$\mathbf{E}_k\mathbf{E}_{k-1}\cdots\mathbf{E}_2\mathbf{E}_1\mathbf{A} = \mathbf{I}. \tag{14}$$

If we now multiply each side in Eq. (14) by the inverse matrices $(\mathbf{E}_k)^{-1}$, $(\mathbf{E}_{k-1})^{-1}$, $\ldots$, $(\mathbf{E}_2)^{-1}$, $(\mathbf{E}_1)^{-1}$ in turn, we find that

$$\mathbf{A} = (\mathbf{E}_1)^{-1}(\mathbf{E}_2)^{-1}\cdots(\mathbf{E}_{k-1})^{-1}(\mathbf{E}_k)^{-1}. \tag{15}$$

Thus **A** is a product of invertible elementary matrices, and it follows from part (c) of Theorem 3 that **A** is invertible. ■

The proof of Theorem 6 actually tells us *how to find* the inverse matrix of **A**. If we invert each side in Eq. (15) (remembering to reverse the order on the right), we get

$$\mathbf{A}^{-1} = \mathbf{E}_k\mathbf{E}_{k-1}\cdots\mathbf{E}_2\mathbf{E}_1\mathbf{I}. \tag{16}$$

Because each left multiplication by an elementary matrix is equivalent to performing the corresponding row operation, we see by comparison of Eqs. (14) and (16) that *the same sequence of elementary row operations that transforms* **A** *into* **I** *also transforms* **I** *into* $\mathbf{A}^{-1}$.

ALGORITHM Finding $\mathbf{A}^{-1}$

To find the inverse $\mathbf{A}^{-1}$ of the invertible $n \times n$ matrix **A**, find a sequence of elementary row operations that reduces **A** to the $n \times n$ identity matrix **I**. Then apply the same sequence of operations in the same order to **I** to transform it into $\mathbf{A}^{-1}$.

As a practical matter, it generally is more convenient to carry out the two reductions—from **A** to **I** and from **I** to $\mathbf{A}^{-1}$—in parallel, as illustrated in our next example.

Example 7 Find the inverse of the 3×3 matrix

$$\mathbf{A} = \begin{bmatrix} 4 & 3 & 2 \\ 5 & 6 & 3 \\ 3 & 5 & 2 \end{bmatrix}.$$

Solution We want to reduce **A** to the 3×3 identity matrix **I** while simultaneously performing the same sequence of row operations on **I** to obtain $\mathbf{A}^{-1}$. In order to carry out this process efficiently, we adjoin **I** on the right of **A** to form the 3×6 matrix

$$\left[\begin{array}{ccc|ccc} 4 & 3 & 2 & 1 & 0 & 0 \\ 5 & 6 & 3 & 0 & 1 & 0 \\ 3 & 5 & 2 & 0 & 0 & 1 \end{array}\right].$$

We now apply the following sequence of elementary row operations to this 3×6 matrix (designed to transform its left half into the 3×3 identity matrix).

$$\xrightarrow{(-1)R_3 + R_1} \left[\begin{array}{ccc|ccc} 1 & -2 & 0 & 1 & 0 & -1 \\ 5 & 6 & 3 & 0 & 1 & 0 \\ 3 & 5 & 2 & 0 & 0 & 1 \end{array}\right]$$

$$\xrightarrow{(-1)R_3 + R_2} \left[\begin{array}{ccc|ccc} 1 & -2 & 0 & 1 & 0 & -1 \\ 2 & 1 & 1 & 0 & 1 & -1 \\ 3 & 5 & 2 & 0 & 0 & 1 \end{array}\right]$$

$$\xrightarrow{(-2)R_1 + R_2} \left[\begin{array}{ccc|ccc} 1 & -2 & 0 & 1 & 0 & -1 \\ 0 & 5 & 1 & -2 & 1 & 1 \\ 3 & 5 & 2 & 0 & 0 & 1 \end{array}\right]$$

$$\xrightarrow{(-3)R_1 + R_3} \left[\begin{array}{ccc|ccc} 1 & -2 & 0 & 1 & 0 & -1 \\ 0 & 5 & 1 & -2 & 1 & 1 \\ 0 & 11 & 2 & -3 & 0 & 4 \end{array}\right]$$

$$\xrightarrow{(-2)R_2 + R_3} \left[\begin{array}{ccc|ccc} 1 & -2 & 0 & 1 & 0 & -1 \\ 0 & 5 & 1 & -2 & 1 & 1 \\ 0 & 1 & 0 & 1 & -2 & 2 \end{array}\right]$$

$$\xrightarrow{\text{SWAP}(R_2,\, R_3)} \left[\begin{array}{ccc|ccc} 1 & -2 & 0 & 1 & 0 & -1 \\ 0 & 1 & 0 & 1 & -2 & 2 \\ 0 & 5 & 1 & -2 & 1 & 1 \end{array}\right]$$

$$\xrightarrow{(2)R_2 + R_1} \left[\begin{array}{ccc|ccc} 1 & 0 & 0 & 3 & -4 & 3 \\ 0 & 1 & 0 & 1 & -2 & 2 \\ 0 & 5 & 1 & -2 & 1 & 1 \end{array}\right]$$

$$\xrightarrow{(-5)R_2 + R_3} \left[\begin{array}{ccc|ccc} 1 & 0 & 0 & 3 & -4 & 3 \\ 0 & 1 & 0 & 1 & -2 & 2 \\ 0 & 0 & 1 & -7 & 11 & -9 \end{array}\right]$$

Now that we have reduced the left half of the 3×6 matrix to $\mathbf{I}$, we simply examine its right half to see that the inverse of $\mathbf{A}$ is

$$\mathbf{A}^{-1} = \begin{bmatrix} 3 & -4 & 3 \\ 1 & -2 & 2 \\ -7 & 11 & -9 \end{bmatrix}.$$

■

Remark: Ordinarily, we do not know in advance whether a given square matrix is invertible or not. To find out, we attempt to carry out the reduction process illustrated in Example 7. If we succeed in reducing $\mathbf{A}$ to $\mathbf{I}$, then $\mathbf{A}$ is invertible and thereby we find $\mathbf{A}^{-1}$. Otherwise—if, somewhere along the way, an all-zero row appears in the left half—we conclude that $\mathbf{A}$ is not row equivalent to $\mathbf{I}$, and therefore (by Theorem 6) $\mathbf{A}$ is not invertible. ■

Matrix Equations

In certain applications, one needs to solve a system $\mathbf{Ax} = \mathbf{b}$ of n equations in n unknowns several times in succession—with the same $n \times n$ coefficient matrix $\mathbf{A}$ each time, but with different constant vectors $\mathbf{b}_1, \mathbf{b}_2, \ldots, \mathbf{b}_k$ on the right. Thus we want to find solution vectors $\mathbf{x}_1, \mathbf{x}_2, \ldots, \mathbf{x}_k$ such that

$$\mathbf{Ax}_1 = \mathbf{b}_1, \quad \mathbf{Ax}_2 = \mathbf{b}_2, \quad \ldots, \quad \mathbf{Ax}_k = \mathbf{b}_k. \tag{17}$$

By Fact 2 at the beginning of this section,

$$\begin{bmatrix} \mathbf{Ax}_1 & \mathbf{Ax}_2 & \cdots & \mathbf{Ax}_k \end{bmatrix} = \mathbf{A}\begin{bmatrix} \mathbf{x}_1 & \mathbf{x}_2 & \cdots & \mathbf{x}_k \end{bmatrix}.$$

So the k equations in (17) are equivalent to the single matrix equation

$$\mathbf{AX} = \mathbf{B}, \tag{18}$$

where

$$\mathbf{X} = \begin{bmatrix} \mathbf{x}_1 & \mathbf{x}_2 & \cdots & \mathbf{x}_k \end{bmatrix} \quad \text{and} \quad \mathbf{B} = \begin{bmatrix} \mathbf{b}_1 & \mathbf{b}_2 & \cdots & \mathbf{b}_k \end{bmatrix}.$$

If $\mathbf{A}$ is invertible and we know $\mathbf{A}^{-1}$, we can find the $n \times k$ matrix of "unknowns" by multiplying each term in Equation (18) on the left by $\mathbf{A}^{-1}$:

$$\mathbf{X} = \mathbf{A}^{-1}\mathbf{B}. \tag{19}$$

Note that this equation is a generalization of Eq. (13) in Theorem 4. If $k = 1$, it usually is simplest to solve the system by Gaussian elimination, but when several different solutions are sought, it may be simpler to find $\mathbf{A}^{-1}$ first and then to apply (19).

Example 8 Find a 3×4 matrix $\mathbf{X}$ such that

$$\begin{bmatrix} 4 & 3 & 2 \\ 5 & 6 & 3 \\ 3 & 5 & 2 \end{bmatrix} \mathbf{X} = \begin{bmatrix} 3 & -1 & 2 & 6 \\ 7 & 4 & 1 & 5 \\ 5 & 2 & 4 & 1 \end{bmatrix}.$$

Solution The coefficient matrix is the matrix $\mathbf{A}$ whose inverse we found in Example 7, so Eq. (19) yields

$$\mathbf{X} = \mathbf{A}^{-1}\mathbf{B} = \begin{bmatrix} 3 & -4 & 3 \\ 1 & -2 & 2 \\ -7 & 11 & -9 \end{bmatrix} \begin{bmatrix} 3 & -1 & 2 & 6 \\ 7 & 4 & 1 & 5 \\ 5 & 2 & 4 & 1 \end{bmatrix},$$

and hence

$$\mathbf{X} = \begin{bmatrix} -4 & -13 & 14 & 1 \\ -1 & -5 & 8 & -2 \\ 11 & 33 & -39 & 4 \end{bmatrix}.$$

By looking at the third columns of $\mathbf{B}$ and $\mathbf{X}$, for instance, we see that the solution of

$$\begin{aligned} 4x_1 + 3x_2 + 2x_3 &= 2 \\ 5x_1 + 6x_2 + 3x_3 &= 1 \\ 3x_1 + 5x_2 + 2x_3 &= 4 \end{aligned}$$

is $x_1 = 14$, $x_2 = 8$, $x_3 = -39$. ■

Nonsingular Matrices

Theorem 6 tells us that the square matrix $\mathbf{A}$ is invertible if and only if it is row equivalent to the identity matrix $\mathbf{I}$, and Theorem 4 in Section 3.3 implies that the latter is true if and only if the system $\mathbf{Ax} = \mathbf{0}$ has only the trivial solution $\mathbf{x} = \mathbf{0}$. A square matrix having these equivalent properties is sometimes called a **nonsingular matrix**.

THEOREM 7 Properties of Nonsingular Matrices

The following properties of an $n \times n$ matrix $\mathbf{A}$ are equivalent.

(a) $\mathbf{A}$ is invertible.

(b) $\mathbf{A}$ is row equivalent to the $n \times n$ identity matrix $\mathbf{I}$.

(c) $\mathbf{Ax} = \mathbf{0}$ has only the trivial solution.

(d) For every n-vector $\mathbf{b}$, the system $\mathbf{Ax} = \mathbf{b}$ has a unique solution.

(e) For every n-vector $\mathbf{b}$, the system $\mathbf{Ax} = \mathbf{b}$ is consistent.

***Proof*:** By the remarks preceding the statement of Theorem 7, we already know that properties (a), (b), and (c) are equivalent—if $\mathbf{A}$ has any one of these properties, then it also has the other two. We can, therefore, complete the proof by establishing the chain of logical implications

$$(c) \Rightarrow (d) \Rightarrow (e) \Rightarrow (a).$$

That is, we need to show that if $\mathbf{A}$ has property (c), then it has property (d), and, similarly, that (d) implies (e) and that (e) implies (a).

(c) $\Rightarrow$ (d): We already know that (c) implies (a), and Theorem 4 says that (a) implies (d). Therefore, (c) implies (d).

(d) $\Rightarrow$ (e): This is obvious, because if the system $\mathbf{Ax} = \mathbf{b}$ has a unique solution, then it certainly has a solution, and thus is consistent.

(e) $\Rightarrow$ (a): Given the hypothesis that $\mathbf{Ax} = \mathbf{b}$ is consistent for every $\mathbf{b}$, we must prove that $\mathbf{A}$ is invertible. Let $\mathbf{b} = \mathbf{e}_j$, the jth column vector of the identity matrix $\mathbf{I}$. Then the consistency of $\mathbf{Ax} = \mathbf{e}_j$ yields an n-vector $\mathbf{x}_j$ such that

$$\mathbf{Ax}_j = \mathbf{e}_j. \tag{20}$$

Let the vectors $\mathbf{x}_1, \mathbf{x}_2, \dots, \mathbf{x}_n$ be obtained in this way for $j = 1, 2, \dots, n$, and let $\mathbf{B}$ be the $n \times n$ matrix with these vectors as its columns:

$$\mathbf{B} = \begin{bmatrix} \mathbf{x}_1 & \mathbf{x}_2 & \dots & \mathbf{x}_n \end{bmatrix}.$$

Then

$$\begin{aligned} \mathbf{AB} &= \mathbf{A}\begin{bmatrix} \mathbf{x}_1 & \mathbf{x}_2 & \cdots & \mathbf{x}_n \end{bmatrix} \\ &= \begin{bmatrix} \mathbf{Ax}_1 & \mathbf{Ax}_2 & \cdots & \mathbf{Ax}_n \end{bmatrix} \\ &= \begin{bmatrix} \mathbf{e}_1 & \mathbf{e}_2 & \cdots & \mathbf{e}_n \end{bmatrix} \qquad \text{[by (20)]}. \end{aligned}$$

Therefore, $\mathbf{AB} = \mathbf{I}$, and thus we have found a matrix $\mathbf{B}$ such that $\mathbf{AB} = \mathbf{I}$.

We next show that $\mathbf{B}$ is invertible by showing that $\mathbf{Bx} = \mathbf{0}$ has only the trivial solution [and by using the fact that property (c) implies property (a)]. But if $\mathbf{Bx} = \mathbf{0}$, then

$$\mathbf{A}(\mathbf{Bx}) = \mathbf{A0} = \mathbf{0},$$

which implies that $\mathbf{Ix} = \mathbf{0}$ and thus that $\mathbf{x} = \mathbf{0}$. So $\mathbf{B}$ is indeed invertible. We can therefore multiply each term in the equation $\mathbf{AB} = \mathbf{I}$ on the right by $\mathbf{B}^{-1}$ to get

$$\mathbf{ABB}^{-1} = \mathbf{IB}^{-1},$$

so that $\mathbf{A} = \mathbf{B}^{-1}$. Thus $\mathbf{A}$ is the inverse of an invertible matrix, and so is itself invertible. This establishes that property (e) implies property (a), and therefore we have completed the proof. ■

The proof of Theorem 7 is a bit long, but it summarizes most of the basic theory of Chapter 1 and is therefore well worth the effort. Indeed, this theorem is one of the central theorems of elementary linear algebra, and we will need to refer to it repeatedly in subsequent chapters.

3.5 Problems

In Problems 1–8, first apply the formulas in (9) to find $\mathbf{A}^{-1}$. *Then use* $\mathbf{A}^{-1}$ *(as in Example 5) to solve the system* $\mathbf{Ax} = \mathbf{b}$.

1. $\mathbf{A} = \begin{bmatrix} 3 & 2 \\ 4 & 3 \end{bmatrix}, \mathbf{b} = \begin{bmatrix} 5 \\ 6 \end{bmatrix}$

2. $\mathbf{A} = \begin{bmatrix} 3 & 7 \\ 2 & 5 \end{bmatrix}, \mathbf{b} = \begin{bmatrix} -1 \\ 3 \end{bmatrix}$

3. $\mathbf{A} = \begin{bmatrix} 6 & 7 \\ 5 & 6 \end{bmatrix}, \mathbf{b} = \begin{bmatrix} 2 \\ -3 \end{bmatrix}$

4. $\mathbf{A} = \begin{bmatrix} 5 & 12 \\ 7 & 17 \end{bmatrix}, \mathbf{b} = \begin{bmatrix} 5 \\ 5 \end{bmatrix}$

5. $\mathbf{A} = \begin{bmatrix} 3 & 2 \\ 5 & 4 \end{bmatrix}, \mathbf{b} = \begin{bmatrix} 5 \\ 6 \end{bmatrix}$

6. $\mathbf{A} = \begin{bmatrix} 4 & 7 \\ 3 & 6 \end{bmatrix}, \mathbf{b} = \begin{bmatrix} 10 \\ 5 \end{bmatrix}$

7. $\mathbf{A} = \begin{bmatrix} 7 & 9 \\ 5 & 7 \end{bmatrix}, \mathbf{b} = \begin{bmatrix} 3 \\ 2 \end{bmatrix}$

8. $\mathbf{A} = \begin{bmatrix} 8 & 15 \\ 5 & 10 \end{bmatrix}, \mathbf{b} = \begin{bmatrix} 7 \\ 3 \end{bmatrix}$

In Problems 9–22, use the method of Example 7 to find the inverse $\mathbf{A}^{-1}$ *of each given matrix* $\mathbf{A}$.

9. $\begin{bmatrix} 5 & 6 \\ 4 & 5 \end{bmatrix}$

10. $\begin{bmatrix} 5 & 7 \\ 4 & 6 \end{bmatrix}$

11. $\begin{bmatrix} 1 & 5 & 1 \\ 2 & 5 & 0 \\ 2 & 7 & 1 \end{bmatrix}$

12. $\begin{bmatrix} 1 & 3 & 2 \\ 2 & 8 & 3 \\ 3 & 10 & 6 \end{bmatrix}$

13. $\begin{bmatrix} 2 & 7 & 3 \\ 1 & 3 & 2 \\ 3 & 7 & 9 \end{bmatrix}$

14. $\begin{bmatrix} 3 & 5 & 6 \\ 2 & 4 & 3 \\ 2 & 3 & 5 \end{bmatrix}$

15. $\begin{bmatrix} 1 & 1 & 5 \\ 1 & 4 & 13 \\ 3 & 2 & 12 \end{bmatrix}$

16. $\begin{bmatrix} 1 & -3 & -3 \\ -1 & 1 & 2 \\ 2 & -3 & -3 \end{bmatrix}$

17. $\begin{bmatrix} 1 & -3 & 0 \\ -1 & 2 & -1 \\ 0 & -2 & 2 \end{bmatrix}$

18. $\begin{bmatrix} 1 & -2 & 2 \\ 3 & 0 & 1 \\ 1 & -1 & 2 \end{bmatrix}$

19. $\begin{bmatrix} 1 & 4 & 3 \\ 1 & 4 & 5 \\ 2 & 5 & 1 \end{bmatrix}$

20. $\begin{bmatrix} 2 & 0 & -1 \\ 1 & 0 & 3 \\ 1 & 1 & 1 \end{bmatrix}$

21. $\begin{bmatrix} 0 & 0 & 1 & 0 \\ 1 & 0 & 0 & 0 \\ 0 & 1 & 2 & 0 \\ 3 & 0 & 0 & 1 \end{bmatrix}$

22. $\begin{bmatrix} 4 & 0 & 1 & 1 \\ 3 & 1 & 3 & 1 \\ 0 & 1 & 2 & 0 \\ 3 & 2 & 4 & 1 \end{bmatrix}$

In Problems 23–28, use the method of Example 8 to find a matrix $\mathbf{X}$ *such that* $\mathbf{AX} = \mathbf{B}$.

23. $\mathbf{A} = \begin{bmatrix} 4 & 3 \\ 5 & 4 \end{bmatrix}, \mathbf{B} = \begin{bmatrix} 1 & 3 & -5 \\ -1 & -2 & 5 \end{bmatrix}$

24. $\mathbf{A} = \begin{bmatrix} 7 & 6 \\ 8 & 7 \end{bmatrix}, \mathbf{B} = \begin{bmatrix} 2 & 0 & 4 \\ 0 & 5 & -3 \end{bmatrix}$

25. $\mathbf{A} = \begin{bmatrix} 1 & 4 & 1 \\ 2 & 8 & 3 \\ 2 & 7 & 4 \end{bmatrix}, \mathbf{B} = \begin{bmatrix} 1 & 0 & 3 \\ 0 & 2 & 2 \\ -1 & 1 & 0 \end{bmatrix}$

26. $\mathbf{A} = \begin{bmatrix} 1 & 5 & 1 \\ 2 & 1 & -2 \\ 1 & 7 & 2 \end{bmatrix}, \mathbf{B} = \begin{bmatrix} 2 & 0 & 1 \\ 0 & 3 & 0 \\ 1 & 0 & 2 \end{bmatrix}$

27. $\mathbf{A} = \begin{bmatrix} 1 & -2 & 3 \\ 2 & 1 & 7 \\ 2 & 2 & 7 \end{bmatrix}, \mathbf{B} = \begin{bmatrix} 0 & 0 & 1 & 1 \\ 0 & 1 & 0 & 1 \\ 1 & 0 & 1 & 0 \end{bmatrix}$

28. $\mathbf{A} = \begin{bmatrix} 6 & 5 & 3 \\ 5 & 3 & 2 \\ 3 & 4 & 2 \end{bmatrix}, \mathbf{B} = \begin{bmatrix} 2 & 1 & 0 & 2 \\ -1 & 3 & 5 & 0 \\ 1 & 1 & 0 & 5 \end{bmatrix}$

29. Verify parts (a) and (b) of Theorem 3.

30. Suppose that $\mathbf{A}$, $\mathbf{B}$, and $\mathbf{C}$ are invertible matrices of the same size. Show that the product $\mathbf{ABC}$ is invertible and that $(\mathbf{ABC})^{-1} = \mathbf{C}^{-1}\mathbf{B}^{-1}\mathbf{A}^{-1}$.

31. Suppose that $\mathbf{A}$ is an invertible matrix and that r and s are negative integers. Verify that $\mathbf{A}^r\mathbf{A}^s = \mathbf{A}^{r+s}$ and that $(\mathbf{A}^r)^s = \mathbf{A}^{rs}$.

32. Prove that if $\mathbf{A}$ is an invertible matrix and $\mathbf{AB} = \mathbf{AC}$, then $\mathbf{B} = \mathbf{C}$. Thus *invertible* matrices can be canceled.

33. Let $\mathbf{A}$ be an $n \times n$ matrix such that $\mathbf{Ax} = \mathbf{x}$ for every n-vector $\mathbf{x}$. Show that $\mathbf{A} = \mathbf{I}$.

34. Show that a diagonal matrix is invertible if and only if each diagonal element is nonzero. In this case, state concisely how the inverse matrix is obtained.

35. Let $\mathbf{A}$ be an $n \times n$ matrix with either a row or a column consisting only of zeros. Show that $\mathbf{A}$ is not invertible.

36. Show that $\mathbf{A} = \begin{bmatrix} a & b \\ c & d \end{bmatrix}$ is not invertible if $ad - bc = 0$.

37. Suppose that $ad - bc \neq 0$ and $\mathbf{A}^{-1}$ is defined as in Equation (9). Verify directly that $\mathbf{AA}^{-1} = \mathbf{A}^{-1}\mathbf{A} = \mathbf{I}$.

38. Let $\mathbf{E}$ be the elementary matrix $\mathbf{E}_1$ of Example 6. If $\mathbf{A}$ is a 2×2 matrix, show that $\mathbf{EA}$ is the result of multiplying the first row of $\mathbf{A}$ by 3.

39. Let $\mathbf{E}$ be the elementary matrix $\mathbf{E}_2$ of Example 6 and suppose that $\mathbf{A}$ is a 3×3 matrix. Show that $\mathbf{EA}$ is the result upon adding twice the first row of $\mathbf{A}$ to its third row.

40. Let $\mathbf{E}$ be the elementary matrix $\mathbf{E}_3$ of Example 6. Show that $\mathbf{EA}$ is the result of interchanging the first two rows of the matrix $\mathbf{A}$.

41. Show that the ith row of the product $\mathbf{AB}$ is $\mathbf{A}_i\mathbf{B}$, where $\mathbf{A}_i$ is the ith row of the matrix $\mathbf{A}$.

42. Apply the result of Problem 41 to show that if $\mathbf{B}$ is an $m \times n$ matrix and $\mathbf{I}$ is the $m \times m$ identity matrix, then $\mathbf{IB} = \mathbf{B}$.

43. Suppose that the matrices $\mathbf{A}$ and $\mathbf{B}$ are row equivalent. Use Theorem 5 to prove that $\mathbf{B} = \mathbf{GA}$, where $\mathbf{G}$ is a product of elementary matrices.

44. Show that every invertible matrix is a product of elementary matrices.

45. Extract from the proof of Theorem 7 a self-contained proof of the following fact: If $\mathbf{A}$ and $\mathbf{B}$ are square matrices such that $\mathbf{AB} = \mathbf{I}$, then $\mathbf{A}$ and $\mathbf{B}$ are invertible.

46. Deduce from the result of Problem 45 that if $\mathbf{A}$ and $\mathbf{B}$ are square matrices whose product $\mathbf{AB}$ is invertible, then $\mathbf{A}$ and $\mathbf{B}$ are themselves invertible.

3.5 Application Automated Solution of Linear Systems

Linear systems with more than two or three equations are most frequently solved with the aid of calculators or computers. If an $n \times n$ linear system is written in the matrix form $\mathbf{Ax} = \mathbf{b}$, then we need to calculate first the inverse matrix $\mathbf{A}^{-1}$ and then the matrix product $\mathbf{x} = \mathbf{A}^{-1}\mathbf{b}$. Suppose the $n \times n$ matrix $\mathbf{A}$ and the column vector $\mathbf{b}$ have been entered (as illustrated in the 3.2 Application). If $\mathbf{A}$ is invertible, then the inverse matrix $\mathbf{A}^{-1}$ is calculated by the *Maple* command `with(linalg): inverse(A)`, the *Mathematica* command `Inverse[A]`, or the MATLAB command `inv(A)`. Consequently, the solution vector $\mathbf{x}$ is calculated by the *Maple* command

```
with(linalg): x := multiply(inverse(A),b);
```

or the *Mathematica* command

```
x = Inverse[A].b
```

or the MATLAB command

```
x = inv(A)*b
```

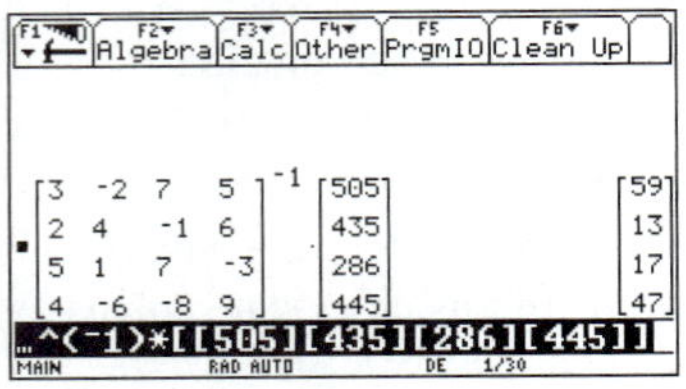

FIGURE 3.5.2. TI-89 solution of a linear system $\mathbf{Ax} = \mathbf{b}$.

Figure 3.5.2 illustrates a similar calculator solution of the linear system

$$\begin{aligned}
3x_1 - 2x_2 + 7x_3 + 5x_4 &= 505\\
2x_1 + 4x_2 - x_3 + 6x_4 &= 435\\
5x_1 + x_2 + 7x_3 - 3x_4 &= 286\\
4x_1 - 6x_2 - 8x_3 + 9x_4 &= 445
\end{aligned}$$

for the solution $x_1 = 59$, $x_2 = 13$, $x_3 = 17$, $x_4 = 47$.

Remark: Whereas the preceding commands illustrate the handy use of conveniently available inverse matrices to solve linear systems, it might be mentioned that modern computer systems employ direct methods—involving Gaussian elimination and still more sophisticated techniques—that are more efficient and numerically reliable to solve a linear system $\mathbf{Ax} = \mathbf{b}$ *without* first calculating the inverse matrix $\mathbf{A}^{-1}$. ■

Use an available calculator or computer system to solve the linear systems in Problems 1–6 of the 3.3 Application. The applied problems below are elementary in character—resembling the "word problems" of high school algebra—but might illustrate the practical advantages of automated solutions.

1. You are walking down the street minding your own business when you spot a small but heavy leather bag lying on the sidewalk. It turns out to contain U.S. Mint American Eagle gold coins of the following types:

- One-half ounce gold coins that sell for \$285 each,
- One-quarter ounce gold coins that sell for \$150 each, and
- One-tenth ounce gold coins that sell for \$70 each.

A bank receipt found in the bag certifies that it contains 258 such coins with a total weight of 67 ounces and a total value of exactly \$40,145. How many coins of each type are there?

2. Now you really strike it rich! You find a bag containing one-ounce U.S. American Eagle gold coins valued at \$550 each, together with half-ounce and quarter-ounce coins valued as in the preceding problem. If this bag contains a total of 365 coins with a total weight of exactly 11 pounds and a total value of \$100,130, how many gold coins of each type are there?

3. A commercial customer orders 81 gallons of paint that contains equal amounts of red paint, green paint, and blue paint—and, hence, could be prepared by mixing 27 gallons of each. However, the store wishes to prepare this order by mixing three types of paint that are already available in large quantity:

- a *reddish* paint that is a mixture of 50% red, 25% green, and 25% blue paint;
- a *greenish* paint that is 12.5% red, 75% green, and 12.5% blue paint; and
- a *bluish* paint that is 20% red, 20% green, and 60% blue paint.

How many gallons of each are needed to prepare the customer's order?

4. Now the paint store receives a really big order—for 244 gallons of paint that is 1/2 red paint, 1/4 green paint, and 1/4 blue paint. The store has three already-mixed types of paint available in large quantity—the greenish paint and the bluish paint of the preceding problem, plus a reddish paint that is 2/3 red paint, 1/6 green paint, and 1/6 blue paint. How many gallons of each must be mixed in order to fill this order?

5. A tour busload of 45 people attended two Florida theme parks on successive days. On Day 1 the entrance fee was \$15 per adult, \$8 per child, \$12 per senior citizen and the total charge was \$558. On Day 2 the entrance fee was \$20 per adult, \$12 per child, \$17 per senior citizen and the total charge was \$771. How many adults, children, and senior citizens were on this tour bus?

6. For some crazy reason, the lunches bought at the first theme park were totaled separately for the adults, children, and seniors. The adults ordered 34 hot dogs, 15 French fries, and 24 soft drinks for a total bill of \$70.85. The children ordered 20 hot dogs, 14 French fries, and 15 soft drinks for a total bill of \$46.65. The senior citizens ordered 11 hot dogs, 10 French fries, and 12 soft drinks for a total bill of \$30.05. What were the prices of a hot dog, an order of French fries, and a soft drink?

7. A fast-food restaurant sells four types of sandwiches—hamburgers, cheeseburgers, roast beef, and chicken—and has four cash registers. At the end of each day, each cash register tallies the number of each type of sandwich sold, and the total sandwich receipts for the day. The four cash register operators work at different speeds, and one day's totals were as follows:

	Hamburgers	**Cheeseburgers**	**Roast Beef**	**Chicken**	**Receipts**
Register 1	37	44	17	23	\$232.99
Register 2	28	35	13	17	\$178.97
Register 3	32	39	19	21	\$215.99
Register 4	47	51	25	29	\$294.38

What was the price of each of the four types of sandwiches?

8. The fast-food restaurant of the preceding problem adds a ham sandwich to its menu and, because of increased business, it also adds a fifth cash register and reduces prices. After this expansion, one day's totals were as follows:

	Hamburgers	Cheeseburgers	Roast Beef	Chicken	Ham	Receipts
Register 1	41	49	22	26	19	$292.79
Register 2	34	39	18	20	16	$236.73
Register 3	36	43	23	24	18	$270.70
Register 4	49	52	26	31	24	$340.19
Register 5	52	55	24	28	25	$341.64

What were the new prices of the five types of sandwiches?

3.6 Determinants

In Theorem 2 of Section 3.5 we saw that the 2×2 matrix

$$\mathbf{A} = \begin{bmatrix} a & b \\ c & d \end{bmatrix}$$

is invertible if and only if $ad - bc \neq 0$. The number $ad - bc$ is called the **determinant** of the 2×2 matrix $\mathbf{A}$. There are several common notations for determinants:

$$\det \mathbf{A} = \det \begin{bmatrix} a & b \\ c & d \end{bmatrix} = \begin{vmatrix} a & b \\ c & d \end{vmatrix} = ad - bc. \tag{1}$$

In particular, note the vertical bars that distinguish a determinant from a matrix. Sometimes we write $\det(\mathbf{A})$ to emphasize that (1) defines a function which associates a number $|\mathbf{A}|$ with each 2×2 matrix $\mathbf{A}$.

Example 1

$$\begin{vmatrix} 3 & 7 \\ 4 & -6 \end{vmatrix} = (3)(-6) - (4)(7) = -46$$ ■

Determinants are often used in elementary algebra to solve a 2×2 system of the form

$$\begin{aligned} a_{11}x + a_{12}y &= b_1 \\ a_{21}x + a_{22}y &= b_2. \end{aligned} \tag{2}$$

It follows from Theorems 2 and 7 in Section 3.5 that this system has a unique solution if and only if its **coefficient determinant**—the determinant of its coefficient matrix $\mathbf{A}$—is nonzero:

$$\det \mathbf{A} = \begin{vmatrix} a_{11} & a_{12} \\ a_{21} & a_{22} \end{vmatrix} \neq 0. \tag{3}$$

Then **Cramer's rule** says that this unique solution is given by

$$x = \frac{\begin{vmatrix} b_1 & a_{12} \\ b_2 & a_{22} \end{vmatrix}}{\begin{vmatrix} a_{11} & a_{12} \\ a_{21} & a_{22} \end{vmatrix}}, \qquad y = \frac{\begin{vmatrix} a_{11} & b_1 \\ a_{21} & b_2 \end{vmatrix}}{\begin{vmatrix} a_{11} & a_{12} \\ a_{21} & a_{22} \end{vmatrix}}. \tag{4}$$

Thus Cramer's rule gives each of x and y as a quotient of two determinants, the denominator in each case being the determinant of the coefficient matrix. In the numerator for x the coefficients a_{11} and a_{21} of x are replaced with the right-side coefficients b_1 and b_2, whereas in the numerator for y the coefficients of y are replaced with b_1 and b_2.

The proof of Cramer's rule is a routine calculation—to verify that evaluation of the determinant in (4) gives the same expressions for x and y as are obtained by solution of (2) by the method of elimination.

Example 2 To apply Cramer's rule to solve the system

$$\begin{aligned} 7x + 8y &= 5 \\ 6x + 9y &= 4 \end{aligned}$$

with coefficient determinant

$$\begin{vmatrix} 7 & 8 \\ 6 & 9 \end{vmatrix} = 15,$$

we simply substitute in the equations in (4) to get

$$x = \frac{\begin{vmatrix} 5 & 8 \\ 4 & 9 \end{vmatrix}}{15} = \frac{13}{15}, \qquad y = \frac{\begin{vmatrix} 7 & 5 \\ 6 & 4 \end{vmatrix}}{15} = -\frac{2}{15}.$$

■

Higher-Order Determinants

We define 3×3 determinants in terms of 2×2 determinants, 4×4 determinants in terms of 3×3 determinants, and so on. This type of definition—one dimension at a time, with the definition in each dimension depending on its meaning in lower dimensions—is called an *inductive definition.*

The determinant $\det \mathbf{A} = |a_{ij}|$ of a 3×3 matrix $\mathbf{A} = \begin{bmatrix} a_{ij} \end{bmatrix}$ is defined as follows:

$$\begin{vmatrix} a_{11} & a_{12} & a_{13} \\ a_{21} & a_{22} & a_{23} \\ a_{31} & a_{32} & a_{33} \end{vmatrix} = a_{11} \begin{vmatrix} a_{22} & a_{23} \\ a_{32} & a_{33} \end{vmatrix} - a_{12} \begin{vmatrix} a_{21} & a_{23} \\ a_{31} & a_{33} \end{vmatrix} + a_{13} \begin{vmatrix} a_{21} & a_{22} \\ a_{31} & a_{32} \end{vmatrix}. \tag{5}$$

Note the single minus sign on the right-hand side. The three 2×2 determinants in (5) are multiplied by the elements a_{11}, a_{12}, and a_{13} along the *first row* of the matrix $\mathbf{A}$. Each of these elements a_{1j} is multiplied by the determinant of the 2×2 submatrix of $\mathbf{A}$ that remains after the row and column containing a_{1j} are deleted.

Example 3

$$\begin{aligned} \begin{vmatrix} 5 & -2 & -3 \\ 4 & 0 & 1 \\ 3 & -1 & 2 \end{vmatrix} &= (5) \begin{vmatrix} 0 & 1 \\ -1 & 2 \end{vmatrix} - (-2) \begin{vmatrix} 4 & 1 \\ 3 & 2 \end{vmatrix} + (-3) \begin{vmatrix} 4 & 0 \\ 3 & -1 \end{vmatrix} \\ &= (5)(1) + (2)(5) - (3)(-4) = 27 \end{aligned}$$

■

The definition of higher-order determinants is simplified by the following notation and terminology.

DEFINITION Minors and Cofactors

Let $\mathbf{A} = \left[a_{ij} \right]$ be an $n \times n$ matrix. The ijth **minor** of $\mathbf{A}$ (also called the **minor** of a_{ij}) is the determinant M_{ij} of the $(n-1) \times (n-1)$ submatrix that remains after deleting the ith row and the jth column of $\mathbf{A}$. The ijth **cofactor** A_{ij} of $\mathbf{A}$ (or the **cofactor** of a_{ij}) is defined to be

$$A_{ij} = (-1)^{i+j} M_{ij}. \tag{6}$$

For example, the minor of a_{12} in a 3×3 matrix is

$$M_{12} = \begin{vmatrix} a_{11} & a_{12} & a_{13} \\ a_{21} & a_{22} & a_{23} \\ a_{31} & a_{32} & a_{33} \end{vmatrix} = \begin{vmatrix} a_{21} & a_{23} \\ a_{31} & a_{33} \end{vmatrix}.$$

The minor of a_{32} in a 4×4 matrix is

$$M_{32} = \begin{vmatrix} a_{11} & a_{12} & a_{13} & a_{14} \\ a_{21} & a_{22} & a_{23} & a_{24} \\ a_{31} & a_{32} & a_{33} & a_{34} \\ a_{41} & a_{42} & a_{43} & a_{44} \end{vmatrix} = \begin{vmatrix} a_{11} & a_{13} & a_{14} \\ a_{21} & a_{23} & a_{24} \\ a_{41} & a_{43} & a_{44} \end{vmatrix}.$$

According to Eq. (6), the cofactor A_{ij} is obtained by attaching the sign $(-1)^{i+j}$ to the minor M_{ij}. This sign is most easily remembered as the one that appears in the ijth position in checkerboard arrays such as

$$\begin{bmatrix} + & - & + \\ - & + & - \\ + & - & + \end{bmatrix} \quad \text{and} \quad \begin{bmatrix} + & - & + & - \\ - & + & - & + \\ + & - & + & - \\ - & + & - & + \end{bmatrix}.$$

Note that a plus sign always appears in the upper left corner and that the signs alternate horizontally and vertically. In the 4×4 case, for instance,

$$\begin{aligned} A_{11} &= +M_{11}, & A_{12} &= -M_{12}, & A_{13} &= +M_{13}, & A_{14} &= -M_{14}, \\ A_{21} &= -M_{21}, & A_{22} &= +M_{22}, & A_{23} &= -M_{23}, & A_{24} &= +M_{24}, \end{aligned}$$

and so forth.

With this notation, the definition of 3×3 determinants in (5) can be rewritten as

$$\begin{aligned} \det \mathbf{A} &= a_{11}M_{11} - a_{12}M_{12} + a_{13}M_{13} \\ &= a_{11}A_{11} + a_{12}A_{12} + a_{13}A_{13}. \end{aligned} \tag{7}$$

The last formula is the **cofactor expansion** of $\det \mathbf{A}$ along the *first row* of $\mathbf{A}$. Its natural generalization yields the definition of the determinant of an $n \times n$ matrix, under the inductive assumption that $(n-1) \times (n-1)$ determinants have already been defined.

DEFINITION ***n* × *n* Determinants**

The **determinant** $\det \mathbf{A} = |a_{ij}|$ of an $n \times n$ matrix $\mathbf{A} = \begin{bmatrix} a_{ij} \end{bmatrix}$ is defined as

$$\det \mathbf{A} = a_{11}A_{11} + a_{12}A_{12} + \cdots + a_{1n}A_{1n}. \tag{8}$$

Thus we multiply each element of the first row of $\mathbf{A}$ by its cofactor and then add these n products to get $\det \mathbf{A}$.

In numerical computations, it frequently is more convenient to work first with minors rather than with cofactors and then attach signs in accord with the checkerboard pattern illustrated previously. Note that determinants have been defined only for *square* matrices.

Example 4 To evaluate the determinant of

$$\mathbf{A} = \begin{bmatrix} 2 & 0 & 0 & -3 \\ 0 & -1 & 0 & 0 \\ 7 & 4 & 3 & 5 \\ -6 & 2 & 2 & 4 \end{bmatrix},$$

we observe that there are only two nonzero terms in the cofactor expansion along the first row. We need not compute the cofactors of zeros, because they will be multiplied by zero in computing the determinant; hence

$$\det \mathbf{A} = +(2)\begin{vmatrix} -1 & 0 & 0 \\ 4 & 3 & 5 \\ 2 & 2 & 4 \end{vmatrix} - (-3)\begin{vmatrix} 0 & -1 & 0 \\ 7 & 4 & 3 \\ -6 & 2 & 2 \end{vmatrix}.$$

Each of the 3×3 determinants on the right-hand side has only a single nonzero term in its cofactor expansion along the first row, so

$$\begin{aligned} \det \mathbf{A} &= (2)(-1)\begin{vmatrix} 3 & 5 \\ 2 & 4 \end{vmatrix} + (3)(+1)\begin{vmatrix} 7 & 3 \\ -6 & 2 \end{vmatrix} \\ &= (-2)(12 - 10) + (3)(14 + 18) = 92. \end{aligned}$$

■

Note that, if we could expand along the second row in Example 4, there would be only a single 3×3 determinant to evaluate. It is in fact true that a determinant can be evaluated by expansion along *any* row *or* column. The proof of the following theorem is included in Appendix B.

THEOREM 1 Cofactor Expansions of Determinants

The determinant of an $n \times n$ matrix $\mathbf{A} = \begin{bmatrix} a_{ij} \end{bmatrix}$ can be obtained by expansion along any row or column. The cofactor expansion along the ith row is

$$\det \mathbf{A} = a_{i1}A_{i1} + a_{i2}A_{i2} + \cdots + a_{in}A_{in}. \tag{9}$$

The cofactor expansion along the jth column is

$$\det \mathbf{A} = a_{1j}A_{1j} + a_{2j}A_{2j} + \cdots + a_{nj}A_{nj}. \tag{10}$$

The formulas in (9) and (10) provide $2n$ different cofactor expansions of an $n \times n$ determinant. For $n = 3$, for instance, we have

$$\begin{aligned}\det \mathbf{A} &= a_{11}A_{11} + a_{12}A_{12} + a_{13}A_{13} \\ &= a_{21}A_{21} + a_{22}A_{22} + a_{23}A_{23} \\ &= a_{31}A_{31} + a_{32}A_{32} + a_{33}A_{33} \end{aligned} \quad \text{row expansions}$$

$$\begin{aligned} &= a_{11}A_{11} + a_{21}A_{21} + a_{31}A_{31} \\ &= a_{12}A_{12} + a_{22}A_{22} + a_{32}A_{32} \\ &= a_{13}A_{13} + a_{23}A_{23} + a_{33}A_{33}. \end{aligned} \quad \text{column expansions}$$

In a specific example, we naturally attempt to choose the expansion that requires the least computational labor.

Example 5 To evaluate the determinant of

$$\mathbf{A} = \begin{bmatrix} 7 & 6 & 0 \\ 9 & -3 & 2 \\ 4 & 5 & 0 \end{bmatrix},$$

we expand along the third column, because it has only a single nonzero entry. Thus

$$\det \mathbf{A} = -(2)\begin{vmatrix} 7 & 6 \\ 4 & 5 \end{vmatrix} = (-2)(35 - 24) = -22. \quad \blacksquare$$

In addition to providing ways of evaluating determinants, the theorem on cofactor expansions is a valuable tool for investigating the general properties of determinants. For instance, it follows immediately from the formulas in (9) and (10) that, *if the square matrix* $\mathbf{A}$ *has either an all-zero row or an all-zero column, then* $\det \mathbf{A} = 0$. For example, by expanding along the second row we see immediately that

$$\begin{vmatrix} 17 & 33 & -24 \\ 0 & 0 & 0 \\ 80 & -62 & 41 \end{vmatrix} = 0.$$

Row and Column Properties

We now list seven properties of determinants that simplify their computation. Each of these properties is readily verified directly in the case of 2×2 determinants. Just as our definition of $n \times n$ determinants was inductive, the following discussion of Properties 1–7 of determinants is inductive. That is, we suppose that $n \geq 3$ and that these properties have already been verified for $(n-1) \times (n-1)$ determinants.

PROPERTY 1: If the $n \times n$ matrix $\mathbf{B}$ is obtained from $\mathbf{A}$ by multiplying a single row (or a column) of $\mathbf{A}$ by the constant k, then $\det \mathbf{B} = k \det \mathbf{A}$.

For instance, if the ith row of $\mathbf{A}$ is multiplied by k, then the elements *off* the ith row of $\mathbf{A}$ are unchanged. Hence for each $j = 1, 2, \ldots, n$, the ijth cofactors of $\mathbf{A}$ and $\mathbf{B}$ are equal: $A_{ij} = B_{ij}$. Therefore, expansion of $\mathbf{B}$ along the ith row gives

$$\begin{aligned}\det \mathbf{B} &= (ka_{i1})B_{i1} + (ka_{i2})B_{i2} + \cdots + (ka_{in})B_{in} \\ &= k(a_{i1}A_{i1} + a_{i2}A_{i2} + \cdots + a_{in}A_{in}),\end{aligned}$$

and thus $\det \mathbf{B} = k \det \mathbf{A}$.

Property 1 implies simply that a constant can be factored out of a single row or column of a determinant. Thus we see that

$$\begin{vmatrix} 7 & 15 & -17 \\ -2 & 9 & 6 \\ 5 & -12 & 10 \end{vmatrix} = (3) \begin{vmatrix} 7 & 5 & -17 \\ -2 & 3 & 6 \\ 5 & -4 & 10 \end{vmatrix}$$

by factoring 3 out of the second column.

PROPERTY 2: If the $n \times n$ matrix $\mathbf{B}$ is obtained from $\mathbf{A}$ by interchanging two rows (or two columns), then $\det \mathbf{B} = -\det \mathbf{A}$.

To see why this is so, suppose (for instance) that the first row is *not* one of the two that are interchanged (recall that $n \geq 3$). Then for each $j = 1, 2, \ldots, n$, the cofactor B_{1j} is obtained by interchanging two rows of the cofactor A_{1j}. Therefore, $B_{1j} = -A_{1j}$ by Property 2 for $(n-1) \times (n-1)$ determinants. Because $b_{1j} = a_{1j}$ for each j, it follows by expanding along the first row that

$$\begin{aligned} \det \mathbf{B} &= b_{11}B_{11} + b_{12}B_{12} + \cdots + b_{1n}B_{1n} \\ &= a_{11}(-A_{11}) + a_{12}(-A_{12}) + \cdots + a_{1n}(-A_{1n}) \\ &= -(a_{11}A_{11} + a_{12}A_{12} + \cdots + a_{1n}A_{1n}), \end{aligned}$$

and thus $\det \mathbf{B} = -\det \mathbf{A}$.

PROPERTY 3: If two rows (or two columns) of the $n \times n$ matrix $\mathbf{A}$ are identical, then $\det \mathbf{A} = 0$.

To see why, let $\mathbf{B}$ denote the matrix obtained by interchanging the two identical rows of $\mathbf{A}$. Then $\mathbf{B} = \mathbf{A}$, so $\det \mathbf{B} = \det \mathbf{A}$. But Property 2 implies that $\det \mathbf{B} = -\det \mathbf{A}$. Thus $\det \mathbf{A} = -\det \mathbf{A}$, and it follows immediately that $\det \mathbf{A} = 0$.

PROPERTY 4: Suppose that the $n \times n$ matrices $\mathbf{A}_1$, $\mathbf{A}_2$, and $\mathbf{B}$ are identical except for their ith rows—that is, the other $n-1$ rows of the three matrices are identical—and that the ith row of $\mathbf{B}$ is the sum of the ith rows of $\mathbf{A}_1$ and $\mathbf{A}_2$. Then

$$\det \mathbf{B} = \det \mathbf{A}_1 + \det \mathbf{A}_2.$$

This result also holds if columns are involved instead of rows.

Property 4 is readily established by expanding $\mathbf{B}$ along its ith row. In Problem 45 we ask you to supply the details for a typical case. The main importance (at this point) of Property 4 is that it implies the following property relating determinants and elementary row operations.

PROPERTY 5: If the $n \times n$ matrix $\mathbf{B}$ is obtained by adding a constant multiple of one row (or column) of $\mathbf{A}$ to another row (or column) of $\mathbf{A}$, then $\det \mathbf{B} = \det \mathbf{A}$.

Thus the value of a determinant is *not* changed either by the type of elementary row operation described or by the corresponding type of elementary column operation. The following computation with 3×3 matrices illustrates the general proof of Property 5. Let

$$\mathbf{A} = \begin{bmatrix} a_{11} & a_{12} & a_{13} \\ a_{21} & a_{22} & a_{23} \\ a_{31} & a_{32} & a_{33} \end{bmatrix} \quad \text{and} \quad \mathbf{B} = \begin{bmatrix} a_{11} & a_{12} & a_{13} + ka_{11} \\ a_{21} & a_{22} & a_{23} + ka_{21} \\ a_{31} & a_{32} & a_{33} + ka_{31} \end{bmatrix}.$$

So $\mathbf{B}$ is the result of adding k times the first column of $\mathbf{A}$ to its third column. Then

$$\det \mathbf{B} = \begin{vmatrix} a_{11} & a_{12} & a_{13} + ka_{11} \\ a_{21} & a_{22} & a_{23} + ka_{21} \\ a_{31} & a_{32} & a_{33} + ka_{31} \end{vmatrix} \tag{11}$$

$$= \begin{vmatrix} a_{11} & a_{12} & a_{13} \\ a_{21} & a_{22} & a_{23} \\ a_{31} & a_{32} & a_{33} \end{vmatrix} + k \begin{vmatrix} a_{11} & a_{12} & a_{11} \\ a_{21} & a_{22} & a_{21} \\ a_{31} & a_{32} & a_{31} \end{vmatrix}.$$

Here we first applied Property 4 and then factored k out of the second summand with the aid of Property 1. Now the first determinant on the right-hand side in (11) is simply $\det \mathbf{A}$, whereas the second determinant is zero (by Property 3—its first and third columns are identical). We therefore have shown that $\det \mathbf{B} = \det \mathbf{A}$, as desired.

Although we use only elementary row operations in reducing a matrix to echelon form, Properties 1, 2, and 5 imply that we may use both elementary row operations and the analogous elementary column operations in simplifying the evaluation of determinants.

Example 6 The matrix

$$\mathbf{A} = \begin{bmatrix} 2 & -3 & -4 \\ -1 & 4 & 2 \\ 3 & 10 & 1 \end{bmatrix}$$

has no zero elements to simplify the computation of its determinant *as it stands*, but we notice that we can "knock out" the first two elements of its third column by adding twice the first column to the third column. This gives

$$\det \mathbf{A} = \begin{vmatrix} 2 & -3 & -4 \\ -1 & 4 & 2 \\ 3 & 10 & 1 \end{vmatrix} = \begin{vmatrix} 2 & -3 & 0 \\ -1 & 4 & 0 \\ 3 & 10 & 7 \end{vmatrix}$$
$$= (+7) \begin{vmatrix} 2 & -3 \\ -1 & 4 \end{vmatrix} = 35.$$

The moral of the example is this: Evaluate determinants with your eyes open. ■

An **upper triangular matrix** is a square matrix having only zeros *below* its main diagonal. A **lower triangular matrix** is a square matrix having only zeros *above* its main diagonal. A **triangular matrix** is one that is either upper triangular or lower triangular, and thus looks like

$$\begin{bmatrix} 1 & 6 & 10 \\ 0 & 5 & 8 \\ 0 & 0 & 7 \end{bmatrix} \quad \text{or} \quad \begin{bmatrix} 1 & 0 & 0 \\ 3 & 7 & 0 \\ 4 & 6 & 5 \end{bmatrix}.$$

The next property tells us that determinants of triangular matrices are especially easy to evaluate.

PROPERTY 6: The determinant of a triangular matrix is equal to the product of its diagonal elements.

The reason is that the determinant of any triangular matrix can be evaluated as in the following computation:

$$\begin{vmatrix} 3 & 11 & 9 & 2 \\ 0 & -2 & 8 & -6 \\ 0 & 0 & 5 & 17 \\ 0 & 0 & 0 & -4 \end{vmatrix} = (3)\begin{vmatrix} -2 & 8 & -6 \\ 0 & 5 & 17 \\ 0 & 0 & -4 \end{vmatrix}$$

$$= (3)(-2)\begin{vmatrix} 5 & 17 \\ 0 & -4 \end{vmatrix} = (3)(-2)(5)(-4) = 120.$$

At each step we expand along the first column and pick up another diagonal element as a factor of the determinant.

Example 7 To evaluate

$$\begin{vmatrix} 2 & -1 & 3 \\ -2 & 1 & 5 \\ 4 & -2 & 10 \end{vmatrix},$$

we first add the first row to the second and then subtract twice the first row from the third. This yields

$$\begin{vmatrix} 2 & -1 & 3 \\ -2 & 1 & 5 \\ 4 & -2 & 10 \end{vmatrix} = \begin{vmatrix} 2 & -1 & 3 \\ 0 & 0 & 8 \\ 0 & 0 & 4 \end{vmatrix} = 0,$$

because we produced a triangular matrix having a zero on its main diagonal. (Can you see an even quicker way to do it by keeping your eyes open?) ■

The Transpose of a Matrix

The transpose $\mathbf{A}^T$ of a 2×2 matrix $\mathbf{A}$ is obtained by interchanging its off-diagonal elements:

$$\text{If} \quad \mathbf{A} = \begin{bmatrix} a & b \\ c & d \end{bmatrix} \quad \text{then} \quad \mathbf{A}^T = \begin{bmatrix} a & c \\ b & d \end{bmatrix}. \tag{12}$$

More generally, the **transpose** of the $m \times n$ matrix $\mathbf{A} = \begin{bmatrix} a_{ij} \end{bmatrix}$ is the $n \times m$ matrix $\mathbf{A}^T$ defined by

$$\mathbf{A}^T = \begin{bmatrix} a_{ji} \end{bmatrix}. \tag{13}$$

Note the reversal of subscripts; this means that the element of $\mathbf{A}^T$ in its jth row and ith column is the element of $\mathbf{A}$ in the ith row and the jth column of $\mathbf{A}$. Hence the rows of the transpose $\mathbf{A}^T$ are (in order) the columns of $\mathbf{A}$, and the columns of $\mathbf{A}^T$ are the rows of $\mathbf{A}$. Thus we obtain $\mathbf{A}^T$ by changing the rows of $\mathbf{A}$ into columns. For instance,

$$\begin{bmatrix} 2 & 0 & 5 \\ 4 & -1 & 7 \end{bmatrix}^T = \begin{bmatrix} 2 & 4 \\ 0 & -1 \\ 5 & 7 \end{bmatrix}$$

and

$$\begin{bmatrix} 7 & -2 & 6 \\ 1 & 2 & 3 \\ 5 & 0 & 4 \end{bmatrix}^T = \begin{bmatrix} 7 & 1 & 5 \\ -2 & 2 & 0 \\ 6 & 3 & 4 \end{bmatrix}.$$

If the matrix $\mathbf{A}$ is square, then we get $\mathbf{A}^T$ by interchanging elements of $\mathbf{A}$ that are located symmetrically with respect to its main diagonal. Thus $\mathbf{A}^T$ is the mirror reflection of $\mathbf{A}$ through its main diagonal.

In Problems 47 and 48 we ask you to verify the following properties of the transpose operation (under the assumption that $\mathbf{A}$ and $\mathbf{B}$ are matrices of appropriate sizes and c is a number):

(i) $(\mathbf{A}^T)^T = \mathbf{A}$;

(ii) $(\mathbf{A} + \mathbf{B})^T = \mathbf{A}^T + \mathbf{B}^T$;

(iii) $(c\mathbf{A})^T = c\mathbf{A}^T$;

(iv) $(\mathbf{AB})^T = \mathbf{B}^T\mathbf{A}^T$.

PROPERTY 7: If $\mathbf{A}$ is a square matrix, then $\det(\mathbf{A}^T) = \det \mathbf{A}$.

This property of determinants can be verified by writing the cofactor expansion of $\det \mathbf{A}$ along its first *row* and the cofactor expansion of $\det(\mathbf{A}^T)$ along its first *column*. When this is done, we see that the two expansions are identical, because the rows of $\mathbf{A}$ are the columns of $\mathbf{A}^T$ (as in Problem 49).

Determinants and Invertibility

We began this section with the remark that a 2×2 matrix $\mathbf{A}$ is invertible if and only if its determinant is nonzero: $|\mathbf{A}| \neq 0$. This result holds more generally for $n \times n$ matrices.

THEOREM 2 Determinants and Invertibility

The $n \times n$ matrix $\mathbf{A}$ is invertible if and only if $\det \mathbf{A} \neq 0$.

This theorem (along with the others in this section) is proved in Appendix B. So now we can add the statement that $\det \mathbf{A} \neq 0$ to the list of equivalent properties of nonsingular matrices stated in Theorem 7 of Section 3.5. Indeed, some texts *define* the square matrix $\mathbf{A}$ to be nonsingular if and only if $\det \mathbf{A} \neq 0$.

Example 8 According to Problem 61, the determinant of the matrix

$$\mathbf{V} = \begin{bmatrix} 1 & a & a^2 \\ 1 & b & b^2 \\ 1 & c & c^2 \end{bmatrix}$$

is

$$|\mathbf{V}| = (b - a)(c - a)(c - b).$$

Note that $|\mathbf{V}| \neq 0$ if and only if the three numbers a, b, and c are *distinct*. Hence it follows from Theorem 2 that $\mathbf{V}$ is invertible if and only if a, b, and c are distinct. For instance, with $a = -2$, $b = 3$, and $c = -4$ we see that the matrix

$$\begin{bmatrix} 1 & -2 & 4 \\ 1 & 3 & 9 \\ 1 & -4 & 16 \end{bmatrix}$$

is invertible. ■

The connection between nonzero determinants and matrix invertibility is closely related to the fact that the determinant function "respects" matrix multiplication in the sense of the following theorem.

THEOREM 3 Determinants and Matrix Products

If $\mathbf{A}$ and $\mathbf{B}$ are $n \times n$ matrices, then

$$|\mathbf{AB}| = |\mathbf{A}|\,|\mathbf{B}|. \tag{14}$$

The fact that $|\mathbf{AB}| = |\mathbf{A}|\,|\mathbf{B}|$ seems so natural that we might fail to note that it is also quite remarkable. Contrast the simplicity of the equation $|\mathbf{AB}| = |\mathbf{A}|\,|\mathbf{B}|$ with the complexity of the seemingly unrelated definitions of determinants and matrix products. For another contrast, we can mention that $|\mathbf{A} + \mathbf{B}|$ is generally *not* equal to $|\mathbf{A}| + |\mathbf{B}|$. (Pick a pair of 2×2 matrices and calculate their determinants and that of their sum.)

As a first application of Theorem 3, we can calculate the determinant of the *inverse* of an invertible matrix $\mathbf{A}$:

$$\mathbf{AA}^{-1} = \mathbf{I},$$

so

$$|\mathbf{A}|\,|\mathbf{A}^{-1}| = |\mathbf{AA}^{-1}| = |\mathbf{I}| = 1,$$

and therefore

$$|\mathbf{A}^{-1}| = \frac{1}{|\mathbf{A}|}. \tag{15}$$

Now $|\mathbf{A}| \neq 0$ because $\mathbf{A}$ is invertible. Thus $\det(\mathbf{A}^{-1})$ *is the reciprocal of* $\det \mathbf{A}$.

Example 9 The determinant of

$$\mathbf{A} = \begin{bmatrix} 2 & 2 & -2 & 0 \\ -4 & 2 & 3 & 1 \\ 0 & 4 & -1 & -4 \\ 3 & 1 & 3 & -1 \end{bmatrix}$$

happens to be $|\mathbf{A}| = 320 \neq 0$. Hence $\mathbf{A}$ is invertible, and *without* finding $\mathbf{A}^{-1}$ we know from Eq. (15) that $|\mathbf{A}^{-1}| = 1/320$. ■

Cramer's Rule for $n \times n$ Systems

Suppose that we need to solve the $n \times n$ linear system

$$\mathbf{Ax} = \mathbf{b}, \tag{16}$$

where

$$\mathbf{A} = \begin{bmatrix} a_{ij} \end{bmatrix}, \quad \mathbf{x} = \begin{bmatrix} x_1 \\ x_2 \\ \vdots \\ x_n \end{bmatrix}, \quad \text{and} \quad \mathbf{b} = \begin{bmatrix} b_1 \\ b_2 \\ \vdots \\ b_n \end{bmatrix}.$$

We assume that the coefficient matrix $\mathbf{A}$ is invertible, so we know in advance that a unique solution $\mathbf{x}$ of (16) exists. The question is how to write $\mathbf{x}$ explicitly in terms of the coefficients a_{ij} and the constants b_i. In the following discussion, we think of $\mathbf{x}$ as a fixed (though as yet unknown) vector.

If we denote by $\mathbf{a}_1, \mathbf{a}_2, \ldots, \mathbf{a}_n$ the column vectors of the $n \times n$ matrix $\mathbf{A}$, then

$$\mathbf{A} = \begin{bmatrix} \mathbf{a}_1 & \mathbf{a}_2 & \cdots & \mathbf{a}_n \end{bmatrix}.$$

The desired formula for the ith unknown x_i involves the determinant of the matrix $\begin{bmatrix} \mathbf{a}_1 & \cdots & \mathbf{b} & \cdots & \mathbf{a}_n \end{bmatrix}$ that is obtained upon replacing the ith column $\mathbf{a}_i$ of $\mathbf{A}$ with the constant vector $\mathbf{b}$.

THEOREM 4 Cramer's Rule

Consider the $n \times n$ linear system $\mathbf{Ax} = \mathbf{b}$, with

$$\mathbf{A} = \begin{bmatrix} \mathbf{a}_1 & \mathbf{a}_2 & \cdots & \mathbf{a}_n \end{bmatrix}.$$

If $|\mathbf{A}| \neq 0$, then the ith entry of the unique solution $\mathbf{x} = (x_1, x_2, \ldots, x_n)$ is given by

$$x_i = \frac{\begin{vmatrix} \mathbf{a}_1 & \cdots & \mathbf{b} & \cdots & \mathbf{a}_n \end{vmatrix}}{|\mathbf{A}|} = \frac{1}{|\mathbf{A}|}\begin{vmatrix} a_{11} & \cdots & b_1 & \cdots & a_{1n} \\ a_{21} & \cdots & b_2 & \cdots & a_{2n} \\ \vdots & \ddots & \vdots & \ddots & \vdots \\ a_{n1} & \cdots & b_n & \cdots & a_{nn} \end{vmatrix}, \tag{17}$$

where in the last expression the constant vector $\mathbf{b}$ replaces the ith column vector $\mathbf{a}_i$ of $\mathbf{A}$.

For instance, the unique solution (x_1, x_2, x_3) of the 3×3 system

$$\begin{aligned} a_{11}x_1 + a_{12}x_2 + a_{13}x_3 &= b_1 \\ a_{21}x_1 + a_{22}x_2 + a_{23}x_3 &= b_2 \\ a_{31}x_1 + a_{32}x_2 + a_{33}x_3 &= b_3 \end{aligned} \tag{18}$$

with $|\mathbf{A}| = |a_{ij}| \neq 0$ is given by the formulas

$$\begin{aligned} x_1 &= \frac{1}{|\mathbf{A}|}\begin{vmatrix} b_1 & a_{12} & a_{13} \\ b_2 & a_{22} & a_{23} \\ b_3 & a_{32} & a_{33} \end{vmatrix}, \\ x_2 &= \frac{1}{|\mathbf{A}|}\begin{vmatrix} a_{11} & b_1 & a_{13} \\ a_{21} & b_2 & a_{23} \\ a_{31} & b_3 & a_{33} \end{vmatrix}, \\ x_3 &= \frac{1}{|\mathbf{A}|}\begin{vmatrix} a_{11} & a_{12} & b_1 \\ a_{21} & a_{22} & b_2 \\ a_{31} & a_{32} & b_3 \end{vmatrix}. \end{aligned} \tag{19}$$

Example 10 Use Cramer's rule to solve the system

$$\begin{aligned} x_1 + 4x_2 + 5x_3 &= 2 \\ 4x_1 + 2x_2 + 5x_3 &= 3 \\ -3x_1 + 3x_2 - x_3 &= 1. \end{aligned}$$

Solution We find that

$$|\mathbf{A}| = \begin{vmatrix} 1 & 4 & 5 \\ 4 & 2 & 5 \\ -3 & 3 & -1 \end{vmatrix} = 29;$$

then the formulas in (19) yield

$$x_1 = \frac{1}{29}\begin{vmatrix} 2 & 4 & 5 \\ 3 & 2 & 5 \\ 1 & 3 & -1 \end{vmatrix} = \frac{33}{29},$$

$$x_2 = \frac{1}{29}\begin{vmatrix} 1 & 2 & 5 \\ 4 & 3 & 5 \\ -3 & 1 & -1 \end{vmatrix} = \frac{35}{29},$$

and

$$x_3 = \frac{1}{29}\begin{vmatrix} 1 & 4 & 2 \\ 4 & 2 & 3 \\ -3 & 3 & 1 \end{vmatrix} = -\frac{23}{29}.$$

■

Inverses and the Adjoint Matrix

The inverse $\mathbf{A}^{-1}$ of the invertible $n \times n$ matrix $\mathbf{A}$ can be found by solving the matrix equation

$$\mathbf{AX} = \mathbf{I}. \tag{20}$$

If we write the coefficient matrix $\mathbf{A} = \begin{bmatrix} \mathbf{a}_1 & \mathbf{a}_2 & \cdots & \mathbf{a}_n \end{bmatrix}$, the unknown matrix $\mathbf{X} = \begin{bmatrix} \mathbf{x}_1 & \mathbf{x}_2 & \cdots & \mathbf{x}_n \end{bmatrix}$, and the identity matrix $\mathbf{I} = \begin{bmatrix} \mathbf{e}_1 & \mathbf{e}_2 & \cdots & \mathbf{e}_n \end{bmatrix}$ in terms of their columns, then (20) says that

$$\mathbf{Ax}_j = \mathbf{e}_j \tag{21}$$

for $j = 1, 2, \ldots, n$. Each of these n equations can then be solved explicitly using Cramer's rule. In particular, Eq. (17) says that the ith element of the column matrix $\mathbf{x}_j$ is given by

$$x_{ij} = \frac{\begin{vmatrix} \mathbf{a}_1 & \cdots & \mathbf{a}_{i-1} & \mathbf{e}_j & \mathbf{a}_{i+1} & \cdots & \mathbf{a}_n \end{vmatrix}}{|\mathbf{A}|} \tag{22}$$

for $i, j = 1, 2, \ldots, n$. When these results are assembled (Appendix B), we get an explicit formula for the inverse matrix.

THEOREM 5 The Inverse Matrix

The inverse of the invertible matrix **A** is given by the formula

$$\mathbf{A}^{-1} = \frac{\left[\, A_{ij} \,\right]^T}{|\mathbf{A}|}, \tag{23}$$

where, as usual, A_{ij} denotes the *ij*th cofactor of **A**; that is, A_{ij} is the product of $(-1)^{i+j}$ and the *ij*th minor determinant of **A**.

We see in (23) the *transpose* of the **cofactor matrix** $\left[\, A_{ij} \,\right]$ of the $n \times n$ matrix **A**. This transposed cofactor matrix is called the **adjoint matrix** of **A** and is denoted by

$$\operatorname{adj} \mathbf{A} = \left[\, A_{ij} \,\right]^T = \left[\, A_{ji} \,\right]. \tag{24}$$

Example 11 Apply the formula in (23) to find the inverse of the matrix

$$\mathbf{A} = \begin{bmatrix} 1 & 4 & 5 \\ 4 & 2 & 5 \\ -3 & 3 & -1 \end{bmatrix}$$

of Example 10, in which we saw that $|\mathbf{A}| = 29$.

Solution First we calculate the cofactors of **A**, arranging our computations in a natural 3×3 array:

$$\begin{aligned}
A_{11} &= + \begin{vmatrix} 2 & 5 \\ 3 & -1 \end{vmatrix} = -17, & A_{12} &= - \begin{vmatrix} 4 & 5 \\ -3 & -1 \end{vmatrix} = -11, & A_{13} &= + \begin{vmatrix} 4 & 2 \\ -3 & 3 \end{vmatrix} = 18, \\
A_{21} &= - \begin{vmatrix} 4 & 5 \\ 3 & -1 \end{vmatrix} = 19, & A_{22} &= + \begin{vmatrix} 1 & 5 \\ -3 & -1 \end{vmatrix} = 14, & A_{23} &= - \begin{vmatrix} 1 & 4 \\ -3 & 3 \end{vmatrix} = -15, \\
A_{31} &= + \begin{vmatrix} 4 & 5 \\ 2 & 5 \end{vmatrix} = 10, & A_{32} &= - \begin{vmatrix} 1 & 5 \\ 4 & 5 \end{vmatrix} = 15, & A_{33} &= + \begin{vmatrix} 1 & 4 \\ 4 & 2 \end{vmatrix} = -14.
\end{aligned}$$

(Note the familiar checkerboard pattern of signs.) Thus the cofactor matrix of **A** is

$$\left[\, A_{ij} \,\right] = \begin{bmatrix} -17 & -11 & 18 \\ 19 & 14 & -15 \\ 10 & 15 & -14 \end{bmatrix}.$$

Next, we interchange rows and columns to obtain the adjoint matrix

$$\operatorname{adj} \mathbf{A} = \left[\, A_{ij} \,\right]^T = \begin{bmatrix} -17 & 19 & 10 \\ -11 & 14 & 15 \\ 18 & -15 & -14 \end{bmatrix}.$$

Finally, in accord with Eq. (23), we divide by $|\mathbf{A}| = 29$ to get the inverse matrix

$$\mathbf{A}^{-1} = \frac{1}{29} \begin{bmatrix} -17 & 19 & 10 \\ -11 & 14 & 15 \\ 18 & -15 & -14 \end{bmatrix}.$$

■

Computational Efficiency

The amount of labor required to complete a numerical calculation is measured by the number of arithmetical operations it involves. So let us count just the number of multiplications required to evaluate an $n \times n$ determinant using cofactor expansions. If $n = 5$, for instance, then the cofactor expansion along a row or column requires computations of *five* 4×4 determinants. A cofactor expansion of each of these five 4×4 determinants involves *four* 3×3 determinants. Each of these four 3×3 determinants leads to *three* 2×2 determinants, and finally each of these 2×2 determinants requires *two* multiplications for its evaluation. Hence the total number of multiplications required to evaluate the original 5×5 determinant is

$$5 \cdot 4 \cdot 3 \cdot 2 = 5! = 120.$$

In general, *the number of multiplications required to evaluate an $n \times n$ determinant completely by cofactor expansions is*

$$n! = n \cdot (n-1) \cdots 3 \cdot 2 \cdot 1 \qquad (n \text{ factorial}).$$

Thus, if we ignore the smaller number of additions involved, then our *operations count* for an $n \times n$ determinant is $n!$. For instance, the evaluation of a 25×25 determinant by cofactor expansion would require about $25! \approx 1.55 \times 10^{25}$ operations. If we used a supercomputer capable of a billion operations per second this task would require about $(1.55 \times 10^{16})/(365.25 \times 24 \times 3600) \approx 492$ million years! This same supercomputer would require about 9.64×10^{47} years—incomparably longer than the estimated age of the universe (perhaps 20 billion years)—to evaluate a 50×50 determinant by cofactor expansion. Contemporary scientific and engineering applications routinely involve matrices of size 1000×1000 (or larger) whose cofactor expansions would require incomprehensibly long times with any conceivable computer.

However, a typical 2000-era desktop computer, using MATLAB and performing "only" about 100 million operations per second, calculates the determinant or inverse of a randomly generated 100×100 matrix almost instantaneously. How is this possible? Obviously not by using cofactor expansions!

The answer is that determinants are much more efficiently calculated by Gaussian elimination—that is, by reduction to triangular form, so that only the product of the remaining diagonal elements need be calculated. Similarly, the inverse of the invertible matrix $\mathbf{A}$ is much more efficiently calculated by reduction of the augmented matrix $\begin{bmatrix} \mathbf{A} & \mathbf{I} \end{bmatrix}$ to reduced echelon form (as in Section 3.5) than by use of Cramer's rule (as in Theorem 5 here). A careful count reveals that the number of arithmetic operations required to reduce an $n \times n$ matrix to echelon form is of the order of n^3 rather than $n!$. As indicated in the table of Fig. 3.6.1, n^3 is dramatically smaller than $n!$ if n is fairly large. Consequently, cofactor expansions of determinants and Cramer's rule for the solution of systems are primarily of theoretical importance, and are seldom used in numerical problems where n is greater than 3 or 4.

n	5	10	25	50	100
$\frac{2}{3}n^3$	83	667	10,417	83,333	666,667
$n!$	120	3,628,800	1.55×10^{25}	3.04×10^{64}	9.33×10^{157}

FIGURE 3.6.1. Approximate operations counts for evaluating an $n \times n$ determinant by Gaussian elimination ($\frac{2}{3}n^3$) and by cofactor expansion ($n!$).

3.6 Problems

Use cofactor expansions to evaluate the determinants in Problems 1–6. Expand along the row or column that minimizes the amount of computation that is required.

1. $\begin{vmatrix} 0 & 0 & 3 \\ 4 & 0 & 0 \\ 0 & 5 & 0 \end{vmatrix}$

2. $\begin{vmatrix} 2 & 1 & 0 \\ 1 & 2 & 1 \\ 0 & 1 & 2 \end{vmatrix}$

3. $\begin{vmatrix} 1 & 0 & 0 & 0 \\ 2 & 0 & 5 & 0 \\ 3 & 6 & 9 & 8 \\ 4 & 0 & 10 & 7 \end{vmatrix}$

4. $\begin{vmatrix} 5 & 11 & 8 & 7 \\ 3 & -2 & 6 & 23 \\ 0 & 0 & 0 & -3 \\ 0 & 4 & 0 & 17 \end{vmatrix}$

5. $\begin{vmatrix} 0 & 0 & 1 & 0 & 0 \\ 2 & 0 & 0 & 0 & 0 \\ 0 & 0 & 0 & 3 & 0 \\ 0 & 0 & 0 & 0 & 4 \\ 0 & 5 & 0 & 0 & 0 \end{vmatrix}$

6. $\begin{vmatrix} 3 & 0 & 11 & -5 & 0 \\ -2 & 4 & 13 & 6 & 5 \\ 0 & 0 & 5 & 0 & 0 \\ 7 & 6 & -9 & 17 & 7 \\ 0 & 0 & 8 & 2 & 0 \end{vmatrix}$

In Problems 7–12, evaluate each given determinant after first simplifying the computation (as in Example 6) by adding an appropriate multiple of some row or column to another.

7. $\begin{vmatrix} 1 & 1 & 1 \\ 2 & 2 & 2 \\ 3 & 3 & 3 \end{vmatrix}$

8. $\begin{vmatrix} 2 & 3 & 4 \\ -2 & -3 & 1 \\ 3 & 2 & 7 \end{vmatrix}$

9. $\begin{vmatrix} 3 & -2 & 5 \\ 0 & 5 & 17 \\ 6 & -4 & 12 \end{vmatrix}$

10. $\begin{vmatrix} -3 & 6 & 5 \\ 1 & -2 & -4 \\ 2 & -5 & 12 \end{vmatrix}$

11. $\begin{vmatrix} 1 & 2 & 3 & 4 \\ 0 & 5 & 6 & 7 \\ 0 & 0 & 8 & 9 \\ 2 & 4 & 6 & 9 \end{vmatrix}$

12. $\begin{vmatrix} 2 & 0 & 0 & -3 \\ 0 & 1 & 11 & 12 \\ 0 & 0 & 5 & 13 \\ -4 & 0 & 0 & 7 \end{vmatrix}$

Use the method of elimination to evaluate the determinants in Problems 13–20.

13. $\begin{vmatrix} -4 & 4 & -1 \\ -1 & -2 & 2 \\ 1 & 4 & 3 \end{vmatrix}$

14. $\begin{vmatrix} 4 & 2 & -2 \\ 3 & 1 & -5 \\ -5 & -4 & 3 \end{vmatrix}$

15. $\begin{vmatrix} -2 & 5 & 4 \\ 5 & 3 & 1 \\ 1 & 4 & 5 \end{vmatrix}$

16. $\begin{vmatrix} 2 & 4 & -2 \\ -5 & -4 & -1 \\ -4 & 2 & 1 \end{vmatrix}$

17. $\begin{vmatrix} 2 & 3 & 3 & 1 \\ 0 & 4 & 3 & -3 \\ 2 & -1 & -1 & -3 \\ 0 & -4 & -3 & 2 \end{vmatrix}$

18. $\begin{vmatrix} 1 & 4 & 4 & 1 \\ 0 & 1 & -2 & 2 \\ 3 & 3 & 1 & 4 \\ 0 & 1 & -3 & -2 \end{vmatrix}$

19. $\begin{vmatrix} 1 & 0 & 0 & 3 \\ 0 & 1 & -2 & 0 \\ -2 & 3 & -2 & 3 \\ 0 & -3 & 3 & 3 \end{vmatrix}$

20. $\begin{vmatrix} 1 & 2 & 1 & -1 \\ 2 & 1 & 3 & 3 \\ 0 & 1 & -2 & 3 \\ -1 & 4 & -2 & 4 \end{vmatrix}$

Use Cramer's rule to solve the systems in Problems 21–32.

21. $\begin{aligned} 3x + 4y &= 2 \\ 5x + 7y &= 1 \end{aligned}$

22. $\begin{aligned} 5x + 8y &= 3 \\ 8x + 13y &= 5 \end{aligned}$

23. $\begin{aligned} 17x + 7y &= 6 \\ 12x + 5y &= 4 \end{aligned}$

24. $\begin{aligned} 11x + 15y &= 10 \\ 8x + 11y &= 7 \end{aligned}$

25. $\begin{aligned} 5x + 6y &= 12 \\ 3x + 4y &= 6 \end{aligned}$

26. $\begin{aligned} 6x + 7y &= 3 \\ 8x + 9y &= 4 \end{aligned}$

27. $\begin{aligned} 5x_1 + 2x_2 - 2x_3 &= 1 \\ x_1 + 5x_2 - 3x_3 &= -2 \\ 5x_1 - 3x_2 + 5x_3 &= 2 \end{aligned}$

28. $\begin{aligned} 5x_1 + 4x_2 - 2x_3 &= 4 \\ 2x_1 \qquad\quad + 3x_3 &= 2 \\ 2x_1 - x_2 + x_3 &= 1 \end{aligned}$

29. $\begin{aligned} 3x_1 - x_2 - 5x_3 &= 3 \\ 4x_1 - 4x_2 - 3x_3 &= -4 \\ x_1 \qquad\quad - 5x_3 &= 2 \end{aligned}$

30. $\begin{aligned} x_1 + 4x_2 + 2x_3 &= 3 \\ 4x_1 + 2x_2 + x_3 &= 1 \\ 2x_1 - 2x_2 - 5x_3 &= -3 \end{aligned}$

31. $\begin{aligned} 2x_1 \qquad\quad - 5x_3 &= -3 \\ 4x_1 - 5x_2 + 3x_3 &= 3 \\ -2x_1 + x_2 + x_3 &= 1 \end{aligned}$

32. $\begin{aligned} 3x_1 + 4x_2 - 3x_3 &= 5 \\ 3x_1 - 2x_2 + 4x_3 &= 7 \\ 3x_1 + 2x_2 - x_3 &= 3 \end{aligned}$

Apply Theorem 5 to find the inverse $\mathbf{A}^{-1}$ *of each matrix* $\mathbf{A}$ *given in Problems 33–40.*

33. $\begin{bmatrix} -5 & -2 & 2 \\ 1 & 5 & -3 \\ 5 & -3 & 1 \end{bmatrix}$

34. $\begin{bmatrix} 2 & 0 & 3 \\ -5 & -4 & 2 \\ 2 & -1 & 1 \end{bmatrix}$

35. $\begin{bmatrix} 3 & 5 & 2 \\ -2 & 3 & -4 \\ -5 & 0 & -5 \end{bmatrix}$

36. $\begin{bmatrix} -4 & 4 & 3 \\ 3 & -1 & -5 \\ 1 & 0 & -5 \end{bmatrix}$

37. $\begin{bmatrix} -4 & 1 & 5 \\ -2 & 4 & 5 \\ -3 & -3 & -1 \end{bmatrix}$

38. $\begin{bmatrix} 3 & 4 & -3 \\ 3 & 2 & -1 \\ -3 & 2 & -4 \end{bmatrix}$

39. $\begin{bmatrix} -3 & -2 & 3 \\ 0 & 3 & 2 \\ 2 & 3 & -5 \end{bmatrix}$

40. $\begin{bmatrix} 2 & 4 & -3 \\ 2 & -3 & -1 \\ -5 & 0 & -3 \end{bmatrix}$

41. Show that $(\mathbf{AB})^T = \mathbf{B}^T\mathbf{A}^T$ if $\mathbf{A}$ and $\mathbf{B}$ are arbitrary 2×2 matrices.

42. Consider the 2×2 matrices

$$\mathbf{A} = \begin{bmatrix} a & b \\ c & d \end{bmatrix} \quad \text{and} \quad \mathbf{B} = \begin{bmatrix} \mathbf{x} \\ \mathbf{y} \end{bmatrix},$$

where $\mathbf{x}$ and $\mathbf{y}$ denote the row vectors of $\mathbf{B}$. Then the product $\mathbf{AB}$ can be written in the form

$$\mathbf{AB} = \begin{bmatrix} a\mathbf{x} + b\mathbf{y} \\ c\mathbf{x} + d\mathbf{y} \end{bmatrix}.$$

Use this expression and the properties of determinants to show that

$$\det \mathbf{AB} = (ad - bc)\begin{vmatrix} \mathbf{x} \\ \mathbf{y} \end{vmatrix} = (\det \mathbf{A})(\det \mathbf{B}).$$

Thus the determinant of a product of 2×2 matrices is equal to the product of their determinants.

Each of Problems 43–46 lists a special case of one of Property 1 through Property 5. Verify it by expanding the determinant on the left-hand side along an appropriate row or column.

43. $$\begin{vmatrix} ka_{11} & a_{12} & a_{13} \\ ka_{21} & a_{22} & a_{23} \\ ka_{31} & a_{32} & a_{33} \end{vmatrix} = k \begin{vmatrix} a_{11} & a_{12} & a_{13} \\ a_{21} & a_{22} & a_{23} \\ a_{31} & a_{32} & a_{33} \end{vmatrix}$$

44. $$\begin{vmatrix} a_{21} & a_{22} & a_{23} \\ a_{11} & a_{12} & a_{13} \\ a_{31} & a_{32} & a_{33} \end{vmatrix} = - \begin{vmatrix} a_{11} & a_{12} & a_{13} \\ a_{21} & a_{22} & a_{23} \\ a_{31} & a_{32} & a_{33} \end{vmatrix}$$

45. $$\begin{vmatrix} a_1 & b_1 & c_1 + d_1 \\ a_2 & b_2 & c_2 + d_2 \\ a_3 & b_3 & c_3 + d_3 \end{vmatrix} = \begin{vmatrix} a_1 & b_1 & c_1 \\ a_2 & b_2 & c_2 \\ a_3 & b_3 & c_3 \end{vmatrix} + \begin{vmatrix} a_1 & b_1 & d_1 \\ a_2 & b_2 & d_2 \\ a_3 & b_3 & d_3 \end{vmatrix}$$

46. $$\begin{vmatrix} a_{11} + ka_{12} & a_{12} & a_{13} \\ a_{21} + ka_{22} & a_{22} & a_{23} \\ a_{31} + ka_{32} & a_{32} & a_{33} \end{vmatrix} = \begin{vmatrix} a_{11} & a_{12} & a_{13} \\ a_{21} & a_{22} & a_{23} \\ a_{31} & a_{32} & a_{33} \end{vmatrix}$$

47. Suppose that $\mathbf{A}$ and $\mathbf{B}$ are matrices of the same size. Show that: (a) $(\mathbf{A}^T)^T = \mathbf{A}$; (b) $(c\mathbf{A})^T = c\mathbf{A}^T$; and (c) $(\mathbf{A} + \mathbf{B})^T = \mathbf{A}^T + \mathbf{B}^T$.

48. Let $\mathbf{A}$ and $\mathbf{B}$ be matrices such that $\mathbf{AB}$ is defined. Show that $(\mathbf{AB})^T = \mathbf{B}^T\mathbf{A}^T$. Begin by recalling that the ijth element of $\mathbf{AB}$ is obtained by multiplying elements in the ith row of $\mathbf{A}$ with those in the jth column of $\mathbf{B}$. What is the ijth element of $\mathbf{B}^T\mathbf{A}^T$?

49. Let $\mathbf{A} = \begin{bmatrix} a_{ij} \end{bmatrix}$ be a 3×3 matrix. Show that $\det(\mathbf{A}^T) = \det \mathbf{A}$ by expanding $\det \mathbf{A}$ along its first row and $\det(\mathbf{A}^T)$ along its first column.

50. Suppose that $\mathbf{A}^2 = \mathbf{A}$. Prove that $|\mathbf{A}| = 0$ or $|\mathbf{A}| = 1$.

51. Suppose that $\mathbf{A}^n = \mathbf{0}$ (the zero matrix) for some positive integer n. Prove that $|\mathbf{A}| = 0$.

52. The square matrix $\mathbf{A}$ is called **orthogonal** provided that $\mathbf{A}^T = \mathbf{A}^{-1}$. Show that the determinant of such a matrix must be either $+1$ or -1.

53. The matrices $\mathbf{A}$ and $\mathbf{B}$ are said to be **similar** provided that $\mathbf{A} = \mathbf{P}^{-1}\mathbf{BP}$ for some invertible matrix $\mathbf{P}$. Show that if $\mathbf{A}$ and $\mathbf{B}$ are similar, then $|\mathbf{A}| = |\mathbf{B}|$.

54. Deduce from Theorems 2 and 3 that if $\mathbf{A}$ and $\mathbf{B}$ are $n \times n$ invertible matrices, then $\mathbf{AB}$ is invertible if and only if both $\mathbf{A}$ and $\mathbf{B}$ are invertible.

55. Let $\mathbf{A}$ and $\mathbf{B}$ be $n \times n$ matrices. Suppose it is known that *either* $\mathbf{AB} = \mathbf{I}$ *or* $\mathbf{BA} = \mathbf{I}$. Use the result of Problem 54 to conclude that $\mathbf{B} = \mathbf{A}^{-1}$.

56. Let $\mathbf{A}$ be an $n \times n$ matrix with $\det \mathbf{A} = 1$ and with all elements of $\mathbf{A}$ integers.

(a) Show that $\mathbf{A}^{-1}$ has only integer entries.

(b) Suppose that $\mathbf{b}$ is an n-vector with only integer entries. Show that the solution vector $\mathbf{x}$ of $\mathbf{Ax} = \mathbf{b}$ has only integer entries.

57. Let $\mathbf{A}$ be a 3×3 upper triangular matrix with nonzero determinant. Show by explicit computation that $\mathbf{A}^{-1}$ is also upper triangular.

58. Figure 3.6.2 shows an acute triangle with angles A, B, and C and opposite sides a, b, and c. By dropping a perpendicular from each vertex to the opposite side, derive the equations

$$c \cos B + b \cos C = a$$
$$c \cos A + a \cos C = b$$
$$a \cos B + b \cos A = c.$$

Regarding these as linear equations in the unknowns $\cos A$, $\cos B$, and $\cos C$, use Cramer's rule to derive the **law of cosines** by solving for

$$\cos A = \frac{b^2 + c^2 - a^2}{2bc}.$$

Thus

$$a^2 = b^2 + c^2 - 2bc \cos A.$$

Note that the case $A = \pi/2$ (90°) reduces to the Pythagorean theorem.

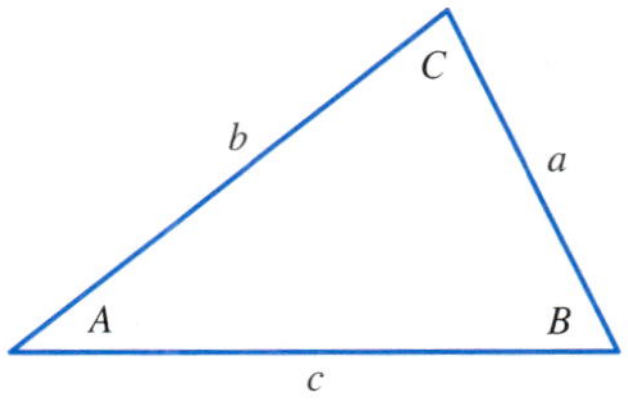

FIGURE 3.6.2. The triangle of Problem 58.

59. Show that

$$\begin{vmatrix} 2 & 1 \\ 1 & 2 \end{vmatrix} = 3 \quad \text{and} \quad \begin{vmatrix} 2 & 1 & 0 \\ 1 & 2 & 1 \\ 0 & 1 & 2 \end{vmatrix} = 4.$$

60. Consider the $n \times n$ determinant

$$B_n = \begin{vmatrix} 2 & 1 & 0 & 0 & \cdots & 0 & 0 \\ 1 & 2 & 1 & 0 & \cdots & 0 & 0 \\ 0 & 1 & 2 & 1 & \cdots & 0 & 0 \\ \vdots & \vdots & \vdots & \vdots & \ddots & \vdots & \vdots \\ 0 & 0 & 0 & 0 & \cdots & 2 & 1 \\ 0 & 0 & 0 & 0 & \cdots & 1 & 2 \end{vmatrix}$$

in which each entry on the main diagonal is a 2, each entry on the two adjacent diagonals is a 1, and every other entry is zero.

(a) Expand along the first row to show that

$$B_n = 2B_{n-1} - B_{n-2}.$$

(b) Prove by induction on n that $B_n = n + 1$ for $n \geq 2$.

Problems 61–64 deal with the ***Vandermonde determinant***

$$V(x_1, x_2, \ldots, x_n) = \begin{vmatrix} 1 & x_1 & x_1^2 & \cdots & x_1^{n-1} \\ 1 & x_2 & x_2^2 & \cdots & x_2^{n-1} \\ \vdots & \vdots & \vdots & \ddots & \vdots \\ 1 & x_n & x_n^2 & \cdots & x_n^{n-1} \end{vmatrix}$$

that will play an important role in Section 3.7.

61. Show by direct computation that $V(a, b) = b - a$ and that

$$V(a, b, c) = \begin{vmatrix} 1 & a & a^2 \\ 1 & b & b^2 \\ 1 & c & c^2 \end{vmatrix} = (b - a)(c - a)(c - b).$$

62. The formulas in Problem 61 are the cases $n = 2$ and $n = 3$ of the general formula

$$V(x_1, x_2, \ldots, x_n) = \prod_{\substack{i,j=1 \\ i>j}}^{n} (x_i - x_j). \tag{25}$$

The case $n = 4$ is

$$V(x_1, x_2, x_3, x_4) = (x_2 - x_1)(x_3 - x_1)(x_3 - x_2) \times (x_4 - x_1)(x_4 - x_2)(x_4 - x_3).$$

Prove this as follows. Given x_1, x_2, and x_3, define the cubic polynomial $P(y)$ to be

$$P(y) = \begin{vmatrix} 1 & x_1 & x_1^2 & x_1^3 \\ 1 & x_2 & x_2^2 & x_2^3 \\ 1 & x_3 & x_3^2 & x_3^3 \\ 1 & y & y^2 & y^3 \end{vmatrix}. \tag{26}$$

Because $P(x_1) = P(x_2) = P(x_3) = 0$ (why?), the roots of $P(y)$ are x_1, x_2, and x_3. It follows that

$$P(y) = k(y - x_1)(y - x_2)(y - x_3),$$

where k is the coefficient of y^3 in $P(y)$. Finally, observe that expansion of the 4×4 determinant in (26) along its last row gives $k = V(x_1, x_2, x_3)$ and that $V(x_1, x_2, x_3, x_4) = P(x_4)$.

63. Generalize the argument in Problem 62 to prove the formula in (25) by induction on n. Begin with the $(n-1)$st-degree polynomial

$$P(y) = \begin{vmatrix} 1 & x_1 & x_1^2 & \cdots & x_1^{n-1} \\ 1 & x_2 & x_2^2 & \cdots & x_2^{n-1} \\ \vdots & \vdots & \vdots & \ddots & \vdots \\ 1 & x_{n-1} & x_{n-1}^2 & \cdots & x_{n-1}^{n-1} \\ 1 & y & y^2 & \cdots & y^{n-1} \end{vmatrix}.$$

64. Use the formula in (25) to evaluate the two determinants given next.

(a) $\begin{vmatrix} 1 & 1 & 1 & 1 \\ 1 & 2 & 4 & 8 \\ 1 & 3 & 9 & 27 \\ 1 & 4 & 16 & 64 \end{vmatrix}$

(b) $\begin{vmatrix} 1 & -1 & 1 & -1 \\ 1 & 2 & 4 & 8 \\ 1 & -2 & 4 & -8 \\ 1 & 3 & 9 & 27 \end{vmatrix}$

3.7 Linear Equations and Curve Fitting

Linear algebra has important applications to the common scientific problem of representing empirical data by means of equations or functions of specified types. We give here only a brief introduction to this extensive subject.

Typically, we begin with a collection of given *data points* (x_0, y_0), (x_1, y_1), $\ldots$, (x_n, y_n) that are to be represented by a specific type of function $y = f(x)$. For instance, y might be the volume of a sample of gas when its temperature is x. Thus the given data points are the results of experiment or measurement, and we want to determine the curve $y = f(x)$ in the xy-plane so that it passes through each of these points; see Figure 3.7.1. Thus we speak of "fitting" the curve to the data points.

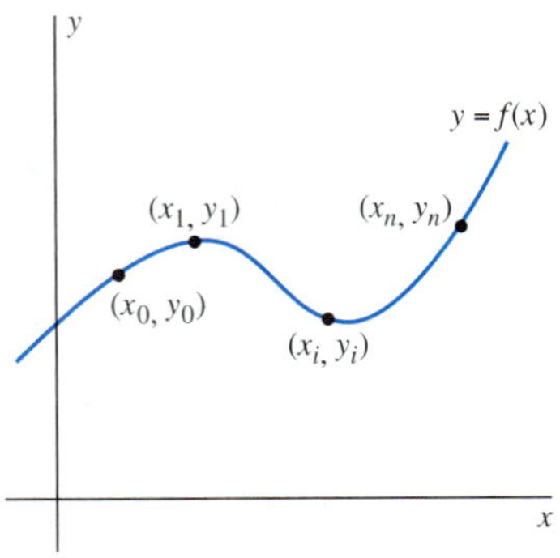

FIGURE 3.7.1. A curve $y = f(x)$ interpolating (that is, passing through) given data points.

We will confine our attention largely to polynomial curves. A **polynomial of degree** n is a function of the form

$$f(x) = a_0 + a_1 x + a_2 x^2 + \cdots + a_n x^n, \tag{1}$$

where the coefficients $a_0, a_1, a_2, \ldots, a_n$ are constants. The data point (x_i, y_i) lies on the curve $y = f(x)$ provided that $f(x_i) = y_i$. The condition that this be so for each $i = 0, 1, 2, \ldots, n$ yields the $n+1$ equations

$$\begin{aligned}
a_0 + a_1 x_0 + a_2(x_0)^2 + \cdots + a_n(x_0)^n &= y_0 \\
a_0 + a_1 x_1 + a_2(x_1)^2 + \cdots + a_n(x_1)^n &= y_1 \\
a_0 + a_1 x_2 + a_2(x_2)^2 + \cdots + a_n(x_2)^n &= y_2 \\
&\vdots \\
a_0 + a_1 x_n + a_2(x_n)^2 + \cdots + a_n(x_n)^n &= y_n.
\end{aligned} \tag{2}$$

Because the numbers x_i and y_i are given, this is a system of $n+1$ *linear* equations in the $n+1$ unknowns $a_0, a_1, a_2, \ldots, a_n$ (the coefficients that determine the polynomial in (1)).

The $(n+1) \times (n+1)$ coefficient matrix of the system in (1) is the Vandermonde matrix

$$\mathbf{A} = \begin{bmatrix} 1 & x_0 & (x_0)^2 & \cdots & (x_0)^n \\ 1 & x_1 & (x_1)^2 & \cdots & (x_1)^n \\ 1 & x_2 & (x_2)^2 & \cdots & (x_2)^n \\ \vdots & \vdots & \vdots & \ddots & \vdots \\ 1 & x_n & (x_n)^2 & \cdots & (x_n)^n \end{bmatrix}, \tag{3}$$

whose determinant is discussed in Problems 61–63 of Section 3.6. It follows from Eq. (25) there that, if the x-coordinates $x_0, x_1, x_2, \ldots, x_n$ are distinct, then the matrix $\mathbf{A}$ is nonsingular. Hence Theorem 7 in Section 3.5 implies that the system in (2) has a unique solution for the coefficients $a_0, a_1, a_2, \ldots, a_n$ in (1). Thus there is a unique nth degree polynomial that fits the $n+1$ given data points. We call it an **interpolating polynomial**, and say that it **interpolates** the given points.

Example 1 Find a cubic polynomial of the form

$$y = A + Bx + Cx^2 + Dx^3$$

that interpolates the data points $(-1, 4)$, $(1, 2)$, $(2, 1)$, and $(3, 16)$.

Solution In a particular problem, it generally is simpler to use distinct capital letters rather than subscripted symbols to denote the coefficients. Here we want to find the values of A, B, C, and D so that $y(-1) = 4$, $y(1) = 2$, $y(2) = 1$, and $y(3) = 16$. These conditions yield the four linear equations

$$\begin{aligned}
A - B + C - D &= 4 \\
A + B + C + D &= 2 \\
A + 2B + 4C + 8D &= 1 \\
A + 3B + 9C + 27D &= 16.
\end{aligned}$$

We readily reduce this system to the echelon form

$$\begin{aligned}
A - B + C - D &= 4 \\
B \qquad + D &= -1 \\
C + 2D &= 0 \\
D &= 2,
\end{aligned}$$

and then back substitution yields $A = 7$, $B = -3$, $C = -4$, and $D = 2$. Thus the desired cubic polynomial is

$$y = 7 - 3x - 4x^2 + 2x^3.$$

The graph of this cubic is shown in Fig. 3.7.2, along with the four original data points. ■

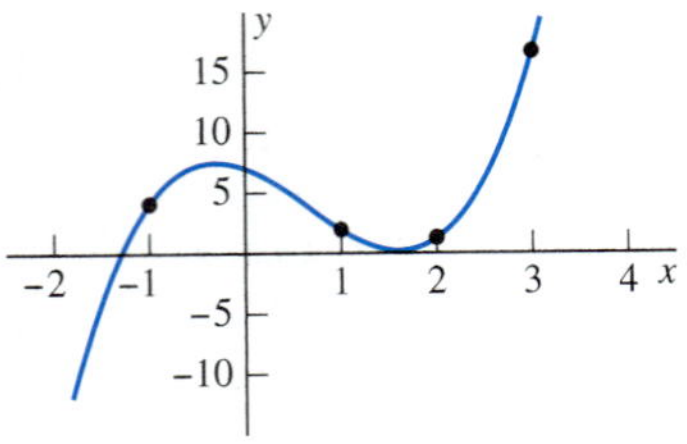

FIGURE 3.7.2. Cubic curve through the four data points of Example 1.

Modeling World Population Growth

As a concrete example of interpolation, we consider the growth of the world's human population. The table in Fig. 3.7.3 shows the total world population (in billions) at 5-year intervals. Actual populations are shown for the years 1960–1995. The figures listed for the years 2000–2025 are populations predicted by the United Nations on the basis of detailed demographic analysis of population trends on a country-by-country basis throughout the world. Each entry in the final column of this table gives the average annual percentage growth rate during the preceding 5-year period. For instance, $3.039(1.0194)^5 \approx 3.345$ for the growth during the 5-year period 1960–1965, so the average annual growth during this period is about 1.94%.

Year	World Population (billions)	Percent Growth
1960	3.039	
1965	3.345	1.94%
1970	3.707	2.08%
1975	4.086	1.97%
1980	4.454	1.74%
1985	4.851	1.72%
1990	5.279	1.71%
1995	5.688	1.50%
2000	6.083	1.35%
2005	6.468	1.23%
2010	6.849	1.15%
2015	7.227	1.08%
2020	7.585	0.97%
2025	7.923	0.88%

FIGURE 3.7.3. World population data.

We see that the world population grew at an annual rate of about 2% during the 1960s, but the rate of growth has slowed since then, and it is expected to slow even more during the early decades of the 21st century. In particular, the growth of the world population at the present time in history is *not* natural or exponential in character—that characterization would imply a constant percentage rate of growth. We explore the possibility of interpolating world population data with polynomial models that might be usable to predict future populations. It seems natural to expect better results with higher-degree interpolating polynomials. Let's see whether this is so.

Example 2 First, we fit a linear polynomial $P_1(t) = a + bt$ (with $t = 0$ in 1900) to the 1980 and 1990 world population values. We need only solve the equations

$$\begin{aligned} a + 80b &= 4.454 \\ a + 90b &= 5.279 \end{aligned}$$

for $a = -2.146$, $b = 0.0825$. Thus our linear interpolating polynomial is

$$P_1(t) = -2.146 + 0.0825t. \tag{4}$$

■

Example 3 Now let's fit a quadratic polynomial $P_2(t) = a + bt + ct^2$ (with $t = 0$ in 1900) to the 1980, 1985, and 1990 world population values. With the three data points (80, 4.454), (85, 4.851), and (90, 5.279), the system in (3) yields the equations

$$\begin{aligned} a + 80b + 80^2c &= 4.454 \\ a + 85b + 85^2c &= 4.851 \\ a + 90b + 90^2c &= 5.279 \end{aligned}$$

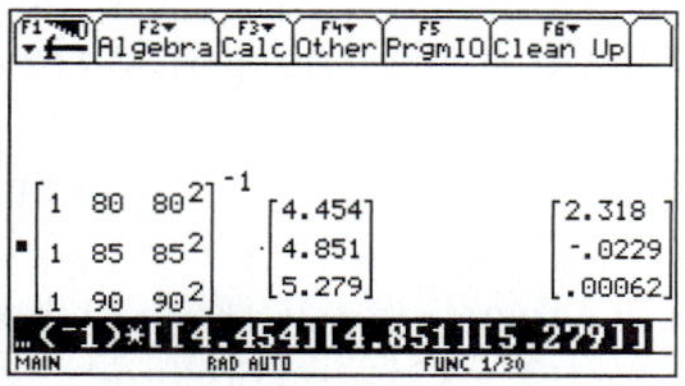

FIGURE 3.7.4. TI-89 calculator solution of the 3×3 system in Example 3.

having the calculator solution $a = 2.318$, $b = -0.0229$, $c = 0.00062$ (Fig. 3.7.4). Thus our quadratic interpolating polynomial is

$$P_2(t) = 2.318 - 0.0229t + 0.00062t^2. \tag{5}$$

■

Example 4 Next we fit a cubic polynomial $P_3(t) = a + bt + ct^2 + dt^3$ (with $t = 0$ in 1900) to the 1980, 1985, 1990, and 1995 world population values. With the four data points (80, 4.454), (85, 4.851), (90, 5.279), and (95, 5.688), the system in (3) yields the four equations

$$\begin{aligned} 4.454 &= a + 80b + 80^2c + 80^3d \\ 4.851 &= a + 85b + 85^2c + 85^3d \\ 5.279 &= a + 90b + 90^2c + 90^3d \\ 5.688 &= a + 95b + 95^2c + 95^3d. \end{aligned}$$

As in the 3.5 Application, a calculator or computer yields the solution

$$\begin{bmatrix} a \\ b \\ c \\ d \end{bmatrix} = \begin{bmatrix} 1 & 80 & 80^2 & 80^3 \\ 1 & 85 & 85^2 & 85^3 \\ 1 & 90 & 90^2 & 90^3 \\ 1 & 95 & 95^2 & 95^3 \end{bmatrix}^{-1} \begin{bmatrix} 4.454 \\ 4.851 \\ 5.279 \\ 5.688 \end{bmatrix} = \begin{bmatrix} 43.118 \\ -1.46623 \\ 0.01762 \\ -0.000066667 \end{bmatrix}.$$

Thus our cubic interpolating polynomial is

$$P_3(t) = 43.118 - 1.46623t + 0.01762t^2 - 0.000066667t^3. \tag{6}$$

■

Example 5 In order to fit a fourth-degree population model of the form

$$P_4(t) = a + bt + ct^2 + dt^3 + et^4$$

to the 1975-1980-1985-1990-1995 world population data, we need to solve the linear system

$$\begin{bmatrix} 1 & 75 & 75^2 & 75^3 & 75^4 \\ 1 & 80 & 80^2 & 80^3 & 80^4 \\ 1 & 85 & 85^2 & 85^3 & 85^4 \\ 1 & 90 & 90^2 & 90^3 & 90^4 \\ 1 & 95 & 95^2 & 95^3 & 95^4 \end{bmatrix} \begin{bmatrix} a \\ b \\ c \\ d \\ e \end{bmatrix} = \begin{bmatrix} 4.086 \\ 4.454 \\ 4.851 \\ 5.279 \\ 5.688 \end{bmatrix}$$

to find the values of the coefficients a, b, c, d, and e. The result is

$$P_4(t) = -158.434 + 7.78543t - 0.141413t^2 + 0.00114667t^3 - 0.00000346667t^4. \tag{7}$$

■

The table in Fig. 3.7.5 compares our linear, quadratic, cubic, and quartic predictions with the "correct" United Nations prediction for the year 2000. Each "error" in the third column of this table is the amount by which the corresponding prediction undershoots (positive error) or overshoots (negative error) the U.N. prediction. We see that the cubic prediction is better than the quartic but worse than the linear prediction. The quadratic prediction is even worse. Thus there apparently is no correlation between the degree of the polynomial model and the accuracy of its predictions.

Figure 3.7.6 shows the U.N. world population data points for the years 1960 through 2025, together with the plots of the linear, quadratic, cubic, and quartic population functions of Examples 2 through 5. It looks as though, the more work we do to find a polynomial fitting selected data points, the less we get for our effort. It certainly is true in this figure that—outside the interval from 1975 to 1995—the higher the degree of the polynomial is, the worse it appears to fit the given data points. The issue here is the difference between

- *interpolating* data points *within* the interval of given points being fitted, and
- *extrapolating* data points *outside* this interval.

	Year 2000 Prediction	Error
Linear	6.104	−0.021
Quadratic	6.228	−0.145
Cubic	6.028	+0.055
Quartic	5.976	+0.107
United Nations	6.083	

FIGURE 3.7.5. Predictions of the 2000 world population.

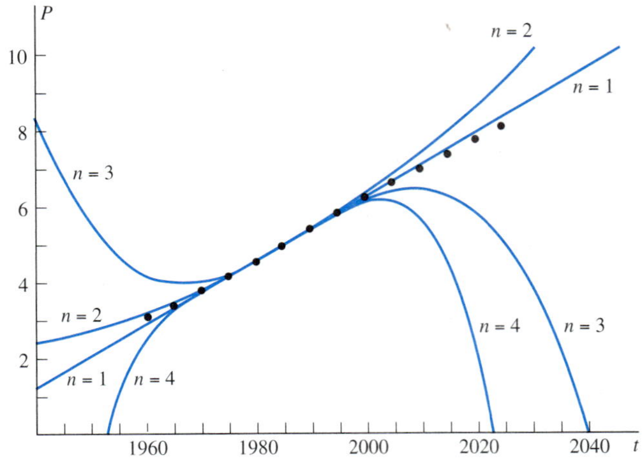

FIGURE 3.7.6. Comparison of world population data and the interpolating polynomials $P_n(t)$ for $t = 1, 2, 3, 4$.

All four of our polynomials appear to do a good job of interpolating but, somewhat paradoxically, the higher is the degree, the worse is the apparent accuracy of extrapolation. The highly questionable accuracy of data extrapolation outside the interval of interpolation has significant implications. For instance, consider a news report that when a certain alleged carcinogen was fed to mice in sufficient amounts to kill an elephant, the mice got cancer. It is then argued that moderate amounts of this carcinogen may cause cancer in humans; or that if 1 part per billion of this carcinogen in the environment kills 1 person, then 1 part per million (a thousand times as much) will kill 1000 people. Such arguments are common, but they may well be cases of extrapolation beyond the range of accuracy. The bottom line is that interpolation is fairly safe—though hardly fail-safe—but extrapolation is risky.

Geometric Applications

In contrast with population prediction, there are interesting situations where curve fitting is exact. For instance, the fact that "two points determine a line" in the plane means that, when we fit the linear function $y = a + bx$ to a given pair of points, we get precisely the one and only straight line in the plane that passes through these points. Similarly, "three points determine a circle," meaning that there is one and only one circle in the plane that passes through three given noncollinear points. In order to find this particular circle, we recall that the equation of a circle with center (h, k) and radius r is

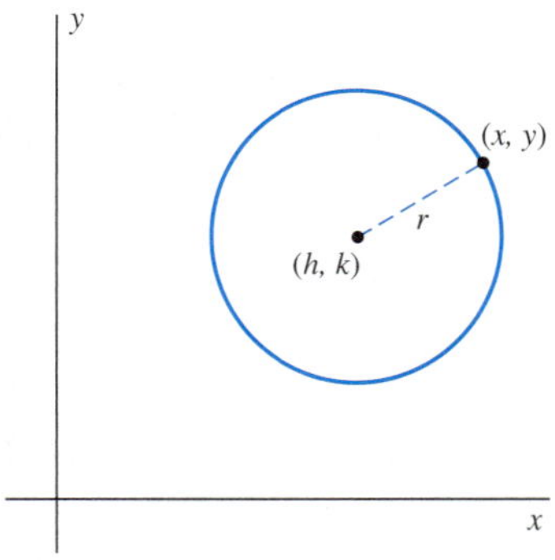

FIGURE 3.7.7. The circle with center (h, k) and radius r.

$$(x - h)^2 + (y - k)^2 = r^2 \quad \text{(Fig. 3.7.7).} \tag{8}$$

Simplification gives

$$(x^2 - 2hx + h^2) + (y^2 - 2ky + k^2) = r^2,$$

that is,

$$x^2 + y^2 + Ax + By + C = 0 \tag{9}$$

(where $A = -2h$, $B = -2k$, and $C = h^2 + k^2 - r^2$) as the general equation of a circle in the plane.

Example 6 Find the equation of the circle that is determined by the points $P(-1, 5)$, $Q(5, -3)$, and $R(6, 4)$.

Solution Substitution of the xy-coordinates of each of the three points P, Q, and R into (9) gives the three equations

$$\begin{aligned} -A + 5B + C &= -26 \\ 5A - 3B + C &= -34 \\ 6A + 4B + C &= -52. \end{aligned}$$

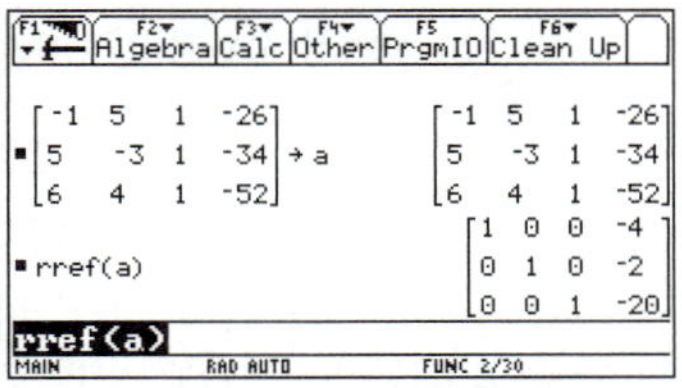

FIGURE 3.7.8. TI-89 calculator solution of the 3×3 system in Example 6.

Reduction of the corresponding augmented coefficient matrix to reduced row-echelon form (Fig. 3.7.8) yields $A = -4$, $B = -2$, and $C = -20$. Thus the equation of the desired circle is

$$x^2 + y^2 - 4x - 2y - 20 = 0.$$

To find its center and radius, we complete the squares in x and y and get

$$(x - 2)^2 + (y - 1)^2 = 25.$$

Thus the circle has center $(2, 1)$ and radius 5 (Fig. 3.7.9). ■

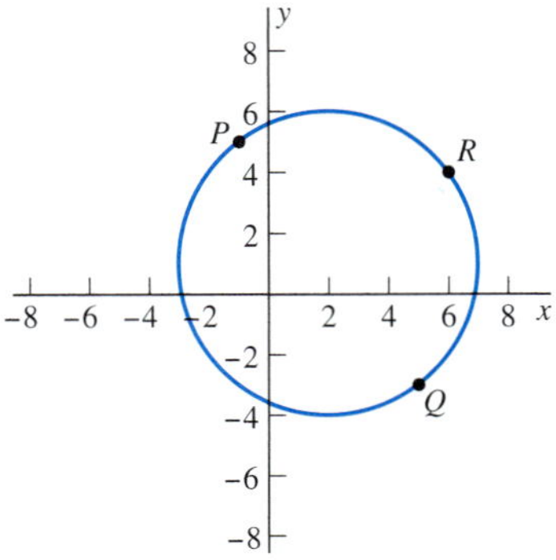

FIGURE 3.7.9. The circle of Example 6.

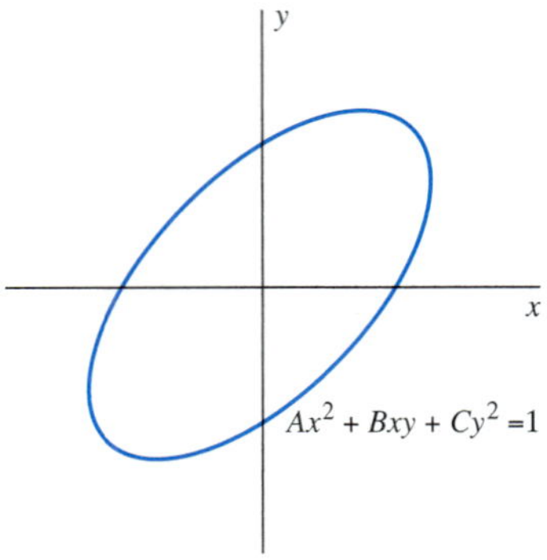

FIGURE 3.7.10. A rotated central ellipse.

Three appropriate points in the plane also determine a **central conic** with equation of the form

$$Ax^2 + Bxy + Cy^2 = 1. \tag{10}$$

This is a rotated conic section—an ellipse, parabola, or hyperbola—centered at the origin of the *xy*-coordinate system. Figure 3.7.10 shows a typical rotated ellipse in the plane.

Example 7 Find the equation of the central conic that passes through the same three points $P(-1, 5)$, $Q(5, -3)$, and $R(6, 4)$ of Example 6.

Solution Substitution of the *xy*-coordinates of each of the three points P, Q, and R into (10) gives the linear system of three equations

$$\begin{aligned} A - 5B + 25C &= 1 \\ 25A - 15B + 9C &= 1 \\ 36A + 24B + 16C &= 1 \end{aligned} \tag{11}$$

in the three unknowns A, B, and C. Reduction of the corresponding augmented coefficient matrix to reduced row-echelon form (Fig. 3.7.11) yields the values

$$A = \frac{277}{14212}, \quad B = -\frac{172}{14212}, \quad \text{and} \quad C = \frac{523}{14212}.$$

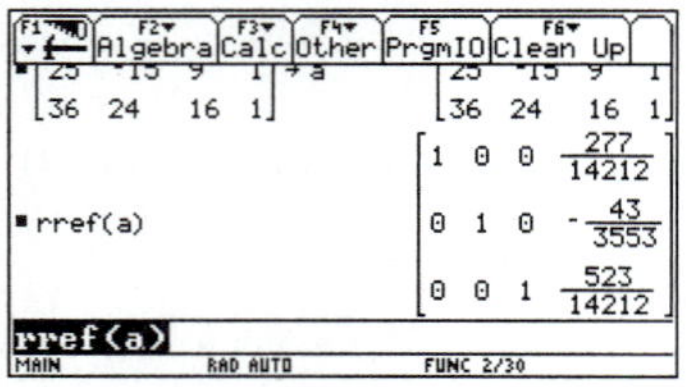

FIGURE 3.7.11. TI-89 calculator solution of the reduced row-echelon form of the augmented coefficient matrix in (11).

If we substitute these coefficient values in (10) and multiply the result by the common denominator 14212, we get the desired equation

$$277x^2 - 172xy + 523y^2 = 14212 \tag{12}$$

of our central conic. The computer plot in Fig. 3.7.12 verifies that this rotated ellipse does indeed pass through all three points P, Q, and R. ■

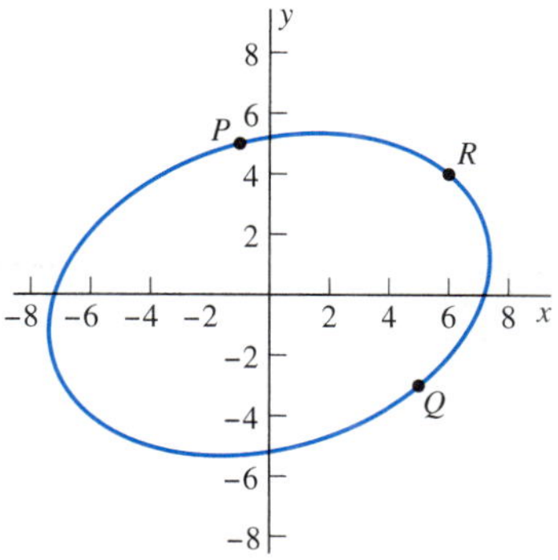

FIGURE 3.7.12. Central ellipse passing through the points P, Q, and R of Example 7.

3.7 Problems

In each of Problems 1–10, $n+1$ data points are given. Find the nth degree polynomial $y = f(x)$ that fits these points.

1. (1, 1) and (3, 7)
2. (−1, 11) and (2, −10)
3. (0, 3), (1, 1), and (2, −5)
4. (−1, 1), (1, 5), and (2, 16)
5. (1, 3), (2, 3), and (3, 5)
6. (−1, −1), (3, −13), and (5, 5)
7. (−1, 1), (0, 0), (1, 1), and (2, −4)
8. (−1, 3), (0, 5), (1, 7), and (2, 3)
9. (−2, −2), (−1, 2), (1, 10), and (2, 26)
10. (−1, 27), (1, 13), (2, 3), and (3, −25)

Three points are given in each of Problems 11–14. Find the equation of the circle determined by these points, as well as its center and radius.

11. (−1, −1), (6, 6), and (7, 5)
12. (3, −4), (5, 10), and (−9, 12)
13. (1, 0), (0, −5), and (−5, −4)
14. (0, 0), (10, 0), and (−7, 7)

In Problems 15–18, find an equation of the central ellipse that passes through the three given points.

15. (0, 5), (5, 0), and (5, 5)
16. (0, 5), (5, 0), and (10, 10)
17. (0, 1), (1, 0), and (10, 10)
18. (0, 4), (3, 0), and (5, 5)
19. Find a curve of the form $y = A + (B/x)$ that passes through the points (1, 5) and (2, 4).
20. Find a curve of the form $y = Ax + (B/x) + (C/x^2)$ that passes through the points (1, 2), (2, 20), and (4, 41).

A sphere in space with ***center*** *(h, k, l) and* ***radius*** *r has equation*

$$(x-h)^2 + (y-k)^2 + (z-l)^2 = r^2.$$

Four given points in space suffice to determine the values of h, k, l, and r. In Problems 21 and 22, find the center and radius of the sphere that passes through the four given points P, Q, R, and S. Hint: Substitute each given triple of coordinates into the sphere equation above to obtain four equations that h, k, l, and r must satisfy. To solve these equations, first subtract the first one from each of the other three. How many unknowns are left in the three equations that result?

21. $P(4, 6, 15)$, $Q(13, 5, 7)$, $R(5, 14, 6)$, $S(5, 5, -9)$
22. $P(11, 17, 17)$, $Q(29, 1, 15)$, $R(13, -1, 33)$, $S(-19, -13, 1)$

Problems 23–34 are intended as calculator or computer problems and are based on the U.S. census data in the table of Fig. 3.7.13, listed by national region in millions for the census years 1950–1990. See **www.census.gov/population/censusdata/table-16.pdf** *for further details.*

	1950	1960	1970	1980	1990
Northeast	39.478	44.678	49.061	49.137	50.809
Midwest	44.461	51.619	56.590	58.867	59.669
South	47.197	54.973	62.813	75.367	85.446
West	20.190	28.053	34.838	43.171	52.786
U.S.	151.326	179.323	203.302	226.542	248.710

FIGURE 3.7.13. Regional population data (in millions) for Problems 23–34.

In Problems 23–26, fit a quadratic function to the 1970, 1980, and 1990 population values for the indicated region.

23. The Northeast

24. The Midwest

25. The South

26. The West

27–30. The same as Problems 23–26, except fit a cubic polynomial to the 1960, 1970, 1980, and 1990 population data for the indicated region.

31–34. The same as Problems 23–26, except fit a quartic polynomial to the 1950, 1960, 1970, 1980, and 1990 population data for the indicated region.

35. Explain why the determinant equation

$$\begin{vmatrix} y & x^2 & x & 1 \\ y_1 & x_1^2 & x_1 & 1 \\ y_2 & x_2^2 & x_2 & 1 \\ y_3 & x_3^2 & x_3 & 1 \end{vmatrix} = 0$$

fits a quadratic polynomial of the form $y = Ax^2 + Bx + C$ to the three given points (x_1, y_1), (x_2, y_2), and (x_3, y_3).

36. Expand the determinant in Problem 35 to find a parabola that interpolates the points $(1, 3)$, $(2, 3)$, and $(3, 7)$.

37. Explain why the determinant equation

$$\begin{vmatrix} x^2 + y^2 & x & y & 1 \\ x_1^2 + y_1^2 & x_1 & y_1 & 1 \\ x_2^2 + y_2^2 & x_2 & y_2 & 1 \\ x_3^2 + y_3^2 & x_3 & y_3 & 1 \end{vmatrix} = 0$$

fits a circle of the form $x^2 + y^2 + Ax + By + C = 0$ to the three given points (x_1, y_1), (x_2, y_2), and (x_3, y_3).

38. Expand the determinant in Problem 37 to find the equation of a circle passing through the three points $(3, -4)$, $(5, 10)$, and $(-9, 12)$. Then find its center and radius.

39. Explain why the determinant equation

$$\begin{vmatrix} x^2 & xy & y^2 & 1 \\ x_1^2 & x_1y_1 & y_1^2 & 1 \\ x_2^2 & x_2y_2 & y_2^2 & 1 \\ x_3^2 & x_3y_3 & y_3^2 & 1 \end{vmatrix} = 0$$

fits a central conic equation of the form $Ax^2 + Bxy + Cy^2 = 1$ to the three given points (x_1, y_1), (x_2, y_2), and (x_3, y_3).

40. Expand the determinant in Problem 39 to find the equation of the ellipse passing through the three points $(0, 4)$, $(3, 0)$, and $(5, 5)$.

4 Vector Spaces

4.1 The Vector Space $\mathbf{R}^3$

Here we take a fresh look—from the viewpoint of linear algebra—at the familiar 3-dimensional space of the physical world around us (and of multivariable calculus). This review, combining old and new ideas, will provide an introduction to basic concepts that are explored further in subsequent sections of the chapter.

We define **three-dimensional coordinate space $\mathbf{R}^3$** to be the set of all ordered triples (a, b, c) of real numbers. In ordinary language, the elements of $\mathbf{R}^3$ are called **points**. The numbers a, b, and c are called the **coordinates** of the point $P(a, b, c)$ and can be regarded as specifying the location of P in a fixed xyz-coordinate system.

The location of the point P can also be specified by means of the *arrow* or directed line segment (Fig. 4.1.1) that points from the origin (its *initial point*) to P (its *terminal point*). Arrows are often used in physics to represent *vector quantities*, such as force and velocity, that possess both magnitude and direction. For instance, the *velocity vector* $\mathbf{v}$ of a point moving in space may be represented by an arrow that points in the direction of its motion, with the length of the arrow being equal to the speed of the moving point. If we locate this arrow with its initial point at the

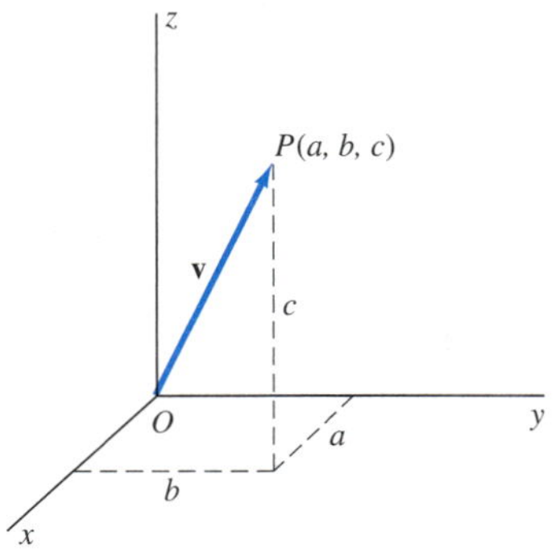

FIGURE 4.1.1. The arrow $\overrightarrow{OP}$ representing the vector $\mathbf{v} = (a, b, c)$.

origin O, then its direction and length are determined by its terminal point (a, b, c). But the arrow is merely a pictorial object; the mathematical object associated with the vector $\mathbf{v}$ is simply the point (a, b, c). For this reason, it is customary to use the words point and vector interchangeably for elements of 3-space $\mathbf{R}^3$ and thus to refer either to the *point* (a, b, c) or to the *vector* (a, b, c).

DEFINITION Vector

A **vector** $\mathbf{v}$ in 3-space $\mathbf{R}^3$ is simply an ordered triple (a, b, c) of real numbers. We write $\mathbf{v} = (a, b, c)$ and call the numbers a, b, and c the **components** of the vector $\mathbf{v}$.

Thus "a vector is a point is a vector," but vector terminology often aids us in visualizing geometric relationships between different points. The point $P(a, b, c)$ determines the vector $\mathbf{v} = (a, b, c)$ in $\mathbf{R}^3$, and $\mathbf{v}$ is represented geometrically (as in Fig. 4.1.1) by the *position vector* $\overrightarrow{OP}$ from the origin $O(0, 0, 0)$ to P—or equally well by any parallel translate of this arrow. What is important about an arrow usually is not where it is, but how long it is and which way it points.

As in Section 3.4, we adopt the convention that the vector $\mathbf{v}$ with components v_1, v_2, and v_3 may be written interchangeably as either

$$\mathbf{v} = (v_1, v_2, v_3) \quad \text{or} \quad \mathbf{v} = \begin{bmatrix} v_1 \\ v_2 \\ v_3 \end{bmatrix}$$

with the column matrix regarded as just another symbol representing one and the same ordered triple of real numbers. Then the following definitions of addition of vectors and of multiplication of vectors by scalars are consistent with the matrix operations defined in Section 3.4.

DEFINITION Addition of Vectors

The **sum** $\mathbf{u} + \mathbf{v}$ of the two vectors $\mathbf{u} = (u_1, u_2, u_3)$ and $\mathbf{v} = (v_1, v_2, v_3)$ is the vector

$$\mathbf{u} + \mathbf{v} = (u_1 + v_1, u_2 + v_2, u_3 + v_3) \tag{1}$$

that is obtained upon addition of respective components of $\mathbf{u}$ and $\mathbf{v}$.

Thus we add vectors by adding corresponding components—that is, by *componentwise addition* (just as we add matrices). For instance, the sum of the vectors $\mathbf{u} = (4, 3, -5)$ and $\mathbf{v} = (-5, 2, 15)$ is the vector

$$\mathbf{u} + \mathbf{v} = (4, 3, -5) + (-5, 2, 15) = (4 - 5, 3 + 2, -5 + 15) = (-1, 5, 10).$$

The geometric representation of vectors as arrows often converts an algebraic relation into a picture that is readily understood and remembered. Addition of vectors is defined algebraically by Eq. (1). The geometric interpretation of vector addition is the *triangle law of addition* illustrated in Fig. 4.1.2 (for the case of 2-dimensional vectors in the plane), where the labeled lengths indicate why this interpretation is valid. An equivalent interpretation is the *parallelogram law of addition*, illustrated in Fig. 4.1.3.

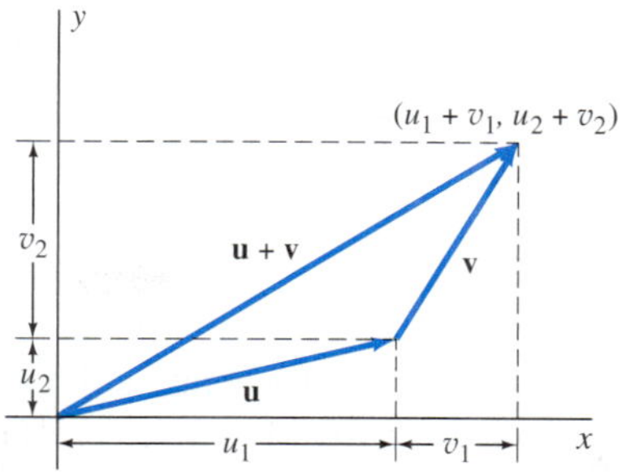

FIGURE 4.1.2. The triangle law of vector addition.

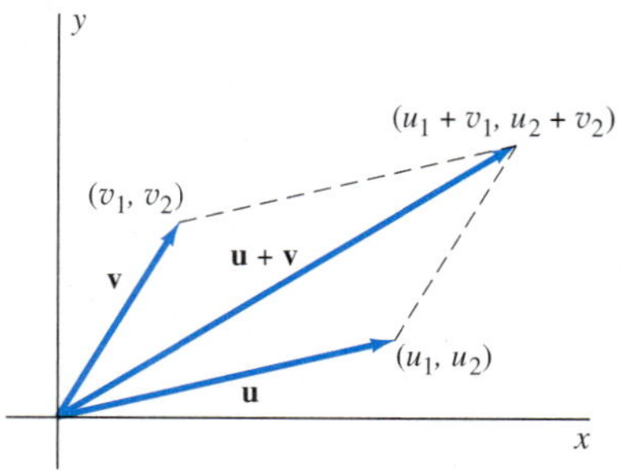

FIGURE 4.1.3. The parallelogram law of vector addition.

Multiplication of a vector by a scalar (a real number) is also defined in a componentwise manner.

> **DEFINITION Multiplication of a Vector by a Scalar**
>
> If $\mathbf{v} = (v_1, v_2, v_3)$ is a vector and c is a real number, then the **scalar multiple** $c\mathbf{v}$ is the vector
>
> $$c\mathbf{v} = (cv_1, cv_2, cv_3) \tag{2}$$
>
> that is obtained upon multiplying each component of $\mathbf{v}$ by c.

The **length** $|\mathbf{v}|$ of the vector $\mathbf{v} = (a, b, c)$ is defined to be the distance of the point $P(a, b, c)$ from the origin,

$$|\mathbf{v}| = \sqrt{a^2 + b^2 + c^2}. \tag{3}$$

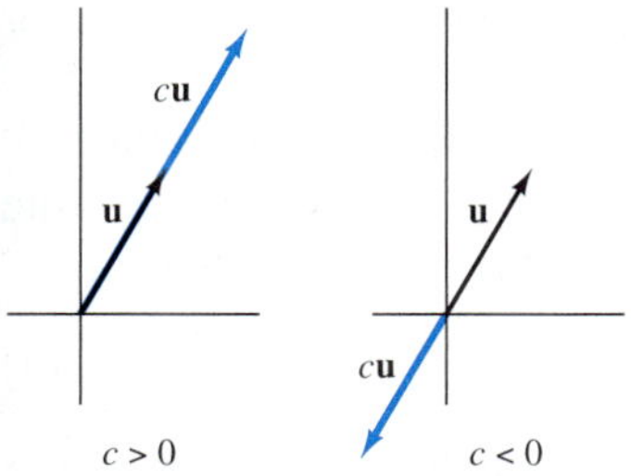

FIGURE 4.1.4. The vector $c\mathbf{u}$ may have the same direction as $\mathbf{u}$ or the opposite direction.

The length of $c\mathbf{v}$ is $|c|$ times the length of $\mathbf{v}$. For instance, if $\mathbf{v} = (4, 3, -12)$, then

$$\begin{aligned}
(-7)\mathbf{v} &= (-7 \cdot 4, -7 \cdot 3, -7 \cdot (-12)) = (-28, -21, 84), \\
|\mathbf{v}| &= \sqrt{(4)^2 + (3)^2 + (-12)^2} = \sqrt{169} = 13, \quad \text{and} \\
|-7\mathbf{v}| &= |-7| \cdot |\mathbf{v}| = 7 \cdot 13 = 91.
\end{aligned}$$

The geometric interpretation of scalar multiplication is that $c\mathbf{v}$ is a vector of length $|c| \cdot |\mathbf{v}|$, with the same direction as $\mathbf{v}$ if $c > 0$ but the opposite direction if $c < 0$ (Fig. 4.1.4).

With vector addition and multiplication by scalars defined as in (1) and (2), $\mathbf{R}^3$ is a **vector space**. That is, these operations satisfy the conditions in (a)–(h) of the following theorem.

> **THEOREM 1 $\mathbf{R}^3$ as a Vector Space**
>
> If $\mathbf{u}$, $\mathbf{v}$, and $\mathbf{w}$ are vectors in $\mathbf{R}^3$, and r and s are real numbers, then
>
> **(a)** $\mathbf{u} + \mathbf{v} = \mathbf{v} + \mathbf{u}$ (commutativity)
>
> **(b)** $\mathbf{u} + (\mathbf{v} + \mathbf{w}) = (\mathbf{u} + \mathbf{v}) + \mathbf{w}$ (associativity)
>
> **(c)** $\mathbf{u} + \mathbf{0} = \mathbf{0} + \mathbf{u} = \mathbf{u}$ (zero element)
>
> **(d)** $\mathbf{u} + (-\mathbf{u}) = (-\mathbf{u}) + \mathbf{u} = \mathbf{0}$ (additive inverse)

(e) $r(\mathbf{u}+\mathbf{v}) = r\mathbf{u}+r\mathbf{v}$	(distributivity)
(f) $(r+s)\mathbf{u} = r\mathbf{u}+s\mathbf{u}$	
(g) $r(s\mathbf{u}) = (rs)\mathbf{u}$	
(h) $1(\mathbf{u}) = \mathbf{u}$	(multiplicative identity).

Of course, $\mathbf{0} = (0, 0, 0)$ denotes the **zero vector** in (c) and (d), and $-\mathbf{u} = (-1)\mathbf{u}$ in (d). Each of properties (a)–(h) in Theorem 1 is readily verified in a componentwise manner. For instance, if $\mathbf{u} = (u_1, u_2, u_3)$ and $\mathbf{v} = (v_1, v_2, v_3)$, then use of the (ordinary) distributive law of real numbers gives

$$\begin{aligned} r(\mathbf{u}+\mathbf{v}) &= r(u_1+v_1, u_2+v_2, u_3+v_3) \\ &= (r(u_1+v_1), r(u_2+v_2), r(u_3+v_3)) \\ &= (ru_1+rv_1, ru_2+rv_2, ru_3+rv_3) \\ &= (ru_1, ru_2, ru_3) + (rv_1, rv_2, rv_3) \\ &= r(u_1, u_2, u_3) + r(v_1, v_2, v_3) = r\mathbf{u}+r\mathbf{v}, \end{aligned}$$

so we have verified property (e).

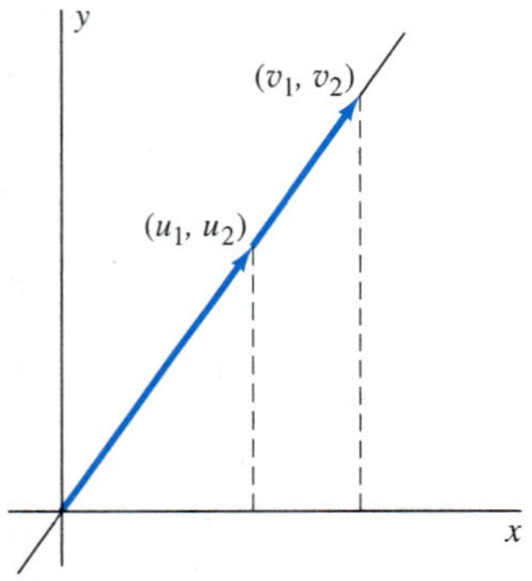

FIGURE 4.1.5. Two linearly dependent vectors **u** and **v**.

The Vector Space $\mathbf{R}^2$

The familiar coordinate plane $\mathbf{R}^2$ is the set of all ordered pairs (a, b) of real numbers. We may regard $\mathbf{R}^2$ as the xy-plane in $\mathbf{R}^3$ by identifying (a, b) with the vector $(a, b, 0)$ in $\mathbf{R}^3$. Then $\mathbf{R}^2$ is simply the set of all 3-dimensional vectors that have third component 0.

Clearly, the sum of any two vectors in $\mathbf{R}^2$ is again a vector in $\mathbf{R}^2$, as is any scalar multiple of a vector in $\mathbf{R}^2$. Indeed, vectors in $\mathbf{R}^2$ satisfy all the properties of a vector space enumerated in Theorem 1. Consequently, the plane $\mathbf{R}^2$ is a vector space in its own right.

The two vectors **u** and **v** are collinear—they lie on the same line through the origin and hence point either in the same direction (Fig. 4.1.5) or in opposite directions—if and only if one is a scalar multiple of the other; that is, either

$$\mathbf{u} = c\mathbf{v} \quad \text{or} \quad \mathbf{v} = c\mathbf{u}, \tag{4}$$

for some scalar c. The scalar c merely adjusts the length and direction of one vector to fit the other. If **u** and **v** are nonzero vectors, then $c = \pm|\mathbf{u}|/|\mathbf{v}|$ in the first relation, with c being positive if the two vectors point in the same direction, negative otherwise.

If one of the relations in (4) holds for some scalar c, then we say that the two vectors are **linearly dependent**. Note that if $\mathbf{u} = \mathbf{0}$ while $\mathbf{v} \neq \mathbf{0}$, then $\mathbf{u} = 0\mathbf{v}$ but **v** is not a scalar multiple of **u**. (Why?) Thus, if precisely one of the two vectors **u** and **v** is the zero vector, then **u** and **v** are linearly dependent, but only one of the two relations in (4) holds.

If **u** and **v** are linearly dependent vectors with $\mathbf{u} = c\mathbf{v}$ (for instance), then $1 \cdot \mathbf{u} + (-c) \cdot \mathbf{v} = \mathbf{0}$. Thus there exist scalars a and b *not both zero* such that

$$a\mathbf{u} + b\mathbf{v} = \mathbf{0}. \tag{5}$$

Conversely, suppose that Eq. (5) holds with a and b not both zero. If $a \neq 0$ (for instance) then we can solve for

$$\mathbf{u} = -\frac{b}{a}\mathbf{v} = c\mathbf{v}$$

with $c = -b/a$, so it follows that **u** and **v** are linearly dependent. Therefore, we have proved the following theorem.

THEOREM 2 Two Linearly Dependent Vectors

The two vectors **u** and **v** are linearly dependent if and only if there exist scalars a and b *not both zero* such that

$$a\mathbf{u} + b\mathbf{v} = \mathbf{0}. \quad (5)$$

The most interesting pairs of vectors are those that are not linearly dependent. The two vectors **u** and **v** are said to be **linearly independent** provided that they are *not* linearly dependent. Thus **u** and **v** are linearly independent if and only if neither is a scalar multiple of the other. By Theorem 2 this is equivalent to the following statement:

The two vectors **u** and **v** are linearly independent if and only if the relation

$$a\mathbf{u} + b\mathbf{v} = \mathbf{0} \quad (5)$$

implies that $a = b = 0$.

Thus the vectors **u** and **v** are linearly independent provided that no *nontrivial* linear combination of them is equal to the zero vector.

Example 1 If $\mathbf{u} = (3, -2)$, $\mathbf{v} = (-6, 4)$, and $\mathbf{w} = (5, -7)$, then **u** and **v** are linearly dependent, because $\mathbf{v} = -2\mathbf{u}$. On the other hand, **u** and **w** are linearly independent. Here is an argument to establish this fact: Suppose that there were scalars a and b such that

$$a\mathbf{u} + b\mathbf{w} = \mathbf{0}.$$

Then

$$a(3, -2) + b(5, -7) = \mathbf{0},$$

and thus we get the simultaneous equations

$$\begin{aligned} 3a + 5b &= 0 \\ -2a - 7b &= 0. \end{aligned}$$

It is now easy to show that $a = b = 0$ is the (unique) solution of this system. This shows that whenever

$$a\mathbf{u} + b\mathbf{w} = \mathbf{0},$$

it follows that $a = b = 0$. Therefore, **u** and **w** are linearly independent.

Alternatively, we could prove that **u** and **w** are linearly independent by showing that neither is a scalar multiple of the other (because $\frac{5}{3} \neq \frac{7}{2}$). ■

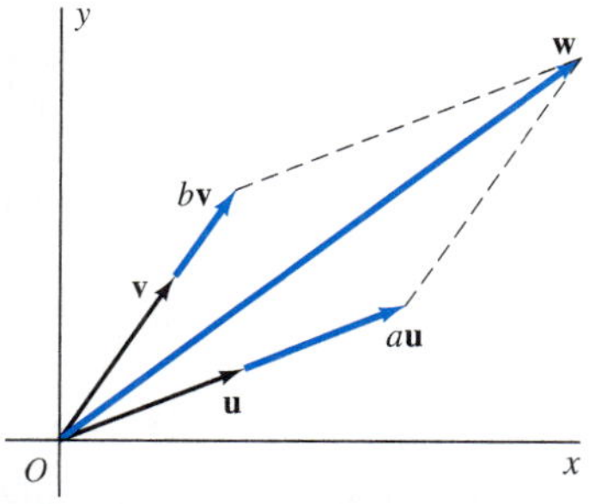

FIGURE 4.1.6. The vector **w** as a linear combination of the two linearly independent vectors **u** and **v**.

The most important property of linearly independent pairs of plane vectors is this: If **u** and **v** are linearly independent vectors in the plane, then any third vector **w** in $\mathbf{R}^2$ can be expressed as a *linear combination* of **u** and **v**. That is, there exist scalars a and b such that $\mathbf{w} = a\mathbf{u} + b\mathbf{v}$ (Fig. 4.1.6). This is a statement that the two linearly independent vectors **u** and **v** suffice (in an obvious sense) to "generate" the whole plane $\mathbf{R}^2$. This general fact—which Section 4.3 discusses in a broader context—is illustrated computationally by the following example.

Example 2 Express the vector $\mathbf{w} = (11, 4)$ as a linear combination of the vectors $\mathbf{u} = (3, -2)$ and $\mathbf{v} = (-2, 7)$.

Solution We want to find numbers a and b such that $a\mathbf{u} + b\mathbf{v} = \mathbf{w}$; that is,

$$a\begin{bmatrix} 3 \\ -2 \end{bmatrix} + b\begin{bmatrix} -2 \\ 7 \end{bmatrix} = \begin{bmatrix} 11 \\ 4 \end{bmatrix}.$$

This vector equation is equivalent to the 2×2 linear system

$$\begin{bmatrix} 3 & -2 \\ -2 & 7 \end{bmatrix}\begin{bmatrix} a \\ b \end{bmatrix} = \begin{bmatrix} 11 \\ 4 \end{bmatrix},$$

which (using Gaussian elimination or Cramer's rule) we readily solve for $a = 5$, $b = 2$. Thus $\mathbf{w} = 5\mathbf{u} + 2\mathbf{v}$. ■

Linear Independence in $\mathbf{R}^3$

We have said that the two vectors are linearly dependent provided that they lie on the same line through the origin. For *three* vectors $\mathbf{u} = (u_1, u_2, u_3)$, $\mathbf{v} = (v_1, v_2, v_3)$, and $\mathbf{w} = (w_1, w_2, w_3)$ in space, the analogous condition is that the three points (u_1, u_2, u_3), (v_1, v_2, v_3), and (w_1, w_2, w_3) lie in the same *plane* through the origin in $\mathbf{R}^3$. Given $\mathbf{u}$, $\mathbf{v}$, and $\mathbf{w}$, how can we determine whether the vectors $\mathbf{u}$, $\mathbf{v}$, and $\mathbf{w}$ are *coplanar*? The key to the answer is the following observation: If r and s are scalars, then the parallelogram law of addition implies that the vectors $\mathbf{u}$, $\mathbf{v}$, and $r\mathbf{u} + s\mathbf{v}$ are coplanar; specifically, they lie in the plane through the origin that is determined by the parallelogram with vertices $\mathbf{0}$, $r\mathbf{u}$, $s\mathbf{v}$, and $r\mathbf{u} + s\mathbf{v}$. Thus any linear combination of $\mathbf{u}$ and $\mathbf{v}$ is coplanar with $\mathbf{u}$ and $\mathbf{v}$. This is the motivation for our next definition.

DEFINITION Linearly Dependent Vectors in $\mathbf{R}^3$

The three vectors $\mathbf{u}$, $\mathbf{v}$, and $\mathbf{w}$ in $\mathbf{R}^3$ are said to be **linearly dependent** provided that one of them is a linear combination of the other two—that is, either

$$\begin{aligned} \mathbf{w} &= r\mathbf{u} + s\mathbf{v} \quad \text{or} \\ \mathbf{u} &= r\mathbf{v} + s\mathbf{w} \quad \text{or} \\ \mathbf{v} &= r\mathbf{u} + s\mathbf{w} \end{aligned} \tag{6}$$

for appropriate scalars r and s.

Note that each of the three equations in (6) implies that there exist three scalars a, b, and c *not all zero* such that

$$a\mathbf{u} + b\mathbf{v} + c\mathbf{w} = \mathbf{0}. \tag{7}$$

For if $\mathbf{w} = r\mathbf{u} + s\mathbf{v}$ (for instance), then

$$r\mathbf{u} + s\mathbf{v} + (-1)\mathbf{w} = \mathbf{0},$$

so we can take $a = r$, $b = s$, and $c = -1 \neq 0$. Conversely, suppose that (7) holds with a, b, and c not all zero. If $c \neq 0$ (for instance), then we can solve for

$$\mathbf{w} = -\frac{a}{c}\mathbf{u} - \frac{b}{c}\mathbf{v} = r\mathbf{u} + s\mathbf{v}$$

with $r = -a/c$ and $s = -b/c$, so it follows that the three vectors **u**, **v**, and **w** are linearly dependent. Therefore, we have proved the following theorem.

THEOREM 3 Three Linearly Dependent Vectors

The three vectors **u**, **v**, and **w** in $\mathbf{R}^3$ are linearly dependent if and only if there exist scalars a, b, and c *not all zero* such that

$$a\mathbf{u} + b\mathbf{v} + c\mathbf{w} = \mathbf{0}. \tag{7}$$

The three vectors **u**, **v**, and **w** are called **linearly independent** provided that they are *not* linearly dependent. Thus **u**, **v**, and **w** are linearly independent if and only if neither of them is a linear combination of the other two. As a consequence of Theorem 3, this is equivalent to the following statement:

The vectors **u**, **v**, and **w** are linearly independent if and only if the relation

$$a\mathbf{u} + b\mathbf{v} + c\mathbf{w} = \mathbf{0} \tag{7}$$

implies that $a = b = c = 0$.

Thus the three vectors **u**, **v**, and **w** are linearly independent provided that no *nontrivial* linear combination of them is equal to the zero vector.

Given two vectors, we can see at a glance whether either is a scalar multiple of the other. By contrast, it is not evident at a glance whether or not three given vectors in $\mathbf{R}^3$ are linearly independent. The following theorem provides one way to resolve this question.

THEOREM 4 Three Linearly Independent Vectors

The vectors $\mathbf{u} = (u_1, u_2, u_3)$, $\mathbf{v} = (v_1, v_2, v_3)$, and $\mathbf{w} = (w_1, w_2, w_3)$ are linearly independent if and only if

$$\begin{vmatrix} u_1 & v_1 & w_1 \\ u_2 & v_2 & w_2 \\ u_3 & v_3 & w_3 \end{vmatrix} \neq 0. \tag{8}$$

***Proof*:** We want to show that **u**, **v**, and **w** are linearly independent if and only if the matrix

$$\mathbf{A} = \begin{bmatrix} u_1 & v_1 & w_1 \\ u_2 & v_2 & w_2 \\ u_3 & v_3 & w_3 \end{bmatrix}$$

—with column vectors **u**, **v**, and **w**—has nonzero determinant. By Theorem 3, the vectors **u**, **v**, and **w** are linearly independent if and only if the equation

$$a\begin{bmatrix} u_1 \\ u_2 \\ u_3 \end{bmatrix} + b\begin{bmatrix} v_1 \\ v_2 \\ v_3 \end{bmatrix} + c\begin{bmatrix} w_1 \\ w_2 \\ w_3 \end{bmatrix} = \mathbf{0}$$

implies that $a = b = c = 0$—that is, if and only if the system

$$\begin{bmatrix} u_1 & v_1 & w_1 \\ u_2 & v_2 & w_2 \\ u_3 & v_3 & w_3 \end{bmatrix} \begin{bmatrix} a \\ b \\ c \end{bmatrix} = \begin{bmatrix} 0 \\ 0 \\ 0 \end{bmatrix} \tag{9}$$

with unknowns a, b, and c has only the trivial solution. But Theorem 7 in Section 3.5 implies that this is so if and only if the coefficient matrix $\mathbf{A}$ is invertible, and Theorem 2 in Section 3.6 implies that $\mathbf{A}$ is invertible if and only if $|\mathbf{A}| \neq 0$. Therefore, $\mathbf{u}$, $\mathbf{v}$, and $\mathbf{w}$ are linearly independent if and only if $|\mathbf{A}| \neq 0$. ■

Hence, in order to determine whether or not three given vectors $\mathbf{u}$, $\mathbf{v}$, and $\mathbf{w}$ are linearly independent, we can calculate the determinant in (8). In practice, however, it is usually more efficient to set up and solve the linear system in (9). If we obtain only the trivial solution $a = b = c = 0$, then the three given vectors are linearly independent. But if we find a nontrivial solution, we then can express one of the vectors as a linear combination of the other two and thus see *how* the three vectors are linearly dependent.

Example 3 To determine whether the three vectors $\mathbf{u} = (1, 2, -3)$, $\mathbf{v} = (3, 1, -2)$, and $\mathbf{w} = (5, -5, 6)$ are linearly independent or dependent, we need to solve the system

$$a\mathbf{u} + b\mathbf{v} + c\mathbf{w} = \begin{bmatrix} 1 & 3 & 5 \\ 2 & 1 & -5 \\ -3 & -2 & 6 \end{bmatrix} \begin{bmatrix} a \\ b \\ c \end{bmatrix} = \begin{bmatrix} 0 \\ 0 \\ 0 \end{bmatrix}.$$

By Gaussian elimination, we readily reduce this system to the echelon form

$$\begin{bmatrix} 1 & 3 & 5 \\ 0 & 1 & 3 \\ 0 & 0 & 0 \end{bmatrix} \begin{bmatrix} a \\ b \\ c \end{bmatrix} = \begin{bmatrix} 0 \\ 0 \\ 0 \end{bmatrix}.$$

Therefore, we can choose $c = 1$, and it follows that $b = -3c = -3$ and $a = -3b - 5c = 4$. Therefore,

$$4\mathbf{u} + (-3)\mathbf{v} + \mathbf{w} = \mathbf{0},$$

and hence $\mathbf{u}$, $\mathbf{v}$, and $\mathbf{w}$ are linearly dependent, with $\mathbf{w} = -4\mathbf{u} + 3\mathbf{v}$. ■

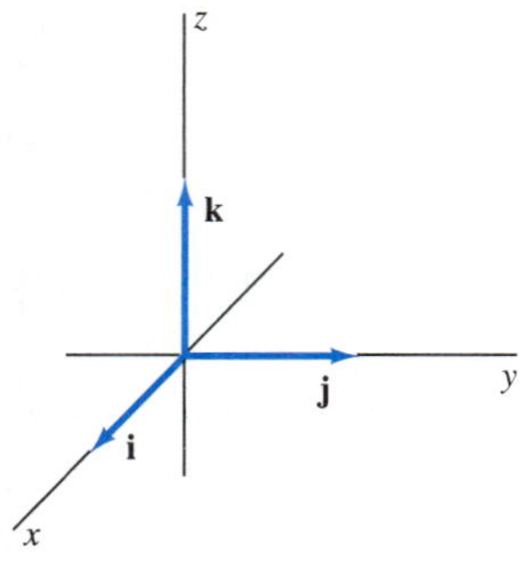

FIGURE 4.1.7. The basic unit vectors **i**, **j**, and **k**.

Basis Vectors in $\mathbf{R}^3$

Perhaps the most familiar triple of linearly independent vectors in $\mathbf{R}^3$ consists of the **basic unit vectors**

$$\mathbf{i} = (1, 0, 0), \quad \mathbf{j} = (0, 1, 0), \quad \text{and} \quad \mathbf{k} = (0, 0, 1). \tag{10}$$

When represented by arrows with the initial points at the origin, these three vectors point in the positive directions along the three coordinate axes (Fig. 4.1.7). The expression

$$\mathbf{v} = a\mathbf{i} + b\mathbf{j} + c\mathbf{k} = (a, b, c)$$

shows both that

- the three vectors $\mathbf{i}$, $\mathbf{j}$, and $\mathbf{k}$ are linearly independent (because $\mathbf{v} = \mathbf{0}$ immediately implies $a = b = c = 0$), and that
- any vector in $\mathbf{R}^3$ can be expressed as a linear combination of $\mathbf{i}$, $\mathbf{j}$, and $\mathbf{k}$.

A **basis** for $\mathbf{R}^3$ is a triple $\mathbf{u}$, $\mathbf{v}$, $\mathbf{w}$ of vectors such that every vector $\mathbf{t}$ in $\mathbf{R}^3$ can be expressed as a linear combination

$$\mathbf{t} = a\mathbf{u} + b\mathbf{v} + c\mathbf{w} \tag{11}$$

of them. That is, given any vector $\mathbf{t}$ in $\mathbf{R}^3$, there exist scalars a, b, c such that Eq. (11) holds. Thus the unit vectors $\mathbf{i}$, $\mathbf{j}$, and $\mathbf{k}$ constitute a basis for $\mathbf{R}^3$. The following theorem says that any three *linearly independent* vectors constitute a basis for $\mathbf{R}^3$.

THEOREM 5 Basis for $\mathbf{R}^3$

If the vectors $\mathbf{u}$, $\mathbf{v}$, and $\mathbf{w}$ in $\mathbf{R}^3$ are linearly independent, then they constitute a basis for $\mathbf{R}^3$.

***Proof*:** Let $\mathbf{u} = (u_1, u_2, u_3)$, $\mathbf{v} = (v_1, v_2, v_3)$, and $\mathbf{w} = (w_1, w_2, w_3)$. Given any fourth vector $\mathbf{t} = (t_1, t_2, t_3)$, we must show that there exist scalars a, b, and c such that

$$\mathbf{t} = a\mathbf{u} + b\mathbf{v} + c\mathbf{w} \tag{11}$$

—that is, such that the system

$$\begin{bmatrix} u_1 & v_1 & w_1 \\ u_2 & v_2 & w_2 \\ u_3 & v_3 & w_3 \end{bmatrix} \begin{bmatrix} a \\ b \\ c \end{bmatrix} = \begin{bmatrix} t_1 \\ t_2 \\ t_3 \end{bmatrix} \tag{12}$$

has a solution (a, b, c). But by Theorem 4 the fact that $\mathbf{u}$, $\mathbf{v}$, and $\mathbf{w}$ are linearly independent implies that the coefficient matrix $\mathbf{A}$ in (12) has nonzero determinant and is therefore invertible. Hence, there is a (unique) solution, which may be obtained by multiplying both sides in (12) by $\mathbf{A}^{-1}$. ■

Example 4 In order to express the vector $\mathbf{t} = (4, 20, 23)$ as a combination of the linearly independent vectors $\mathbf{u} = (1, 3, 2)$, $\mathbf{v} = (2, 8, 7)$, and $\mathbf{w} = (1, 7, 9)$, we need to solve the system

$$\begin{bmatrix} 1 & 2 & 1 \\ 3 & 8 & 7 \\ 2 & 7 & 9 \end{bmatrix} \begin{bmatrix} a \\ b \\ c \end{bmatrix} = \begin{bmatrix} 4 \\ 20 \\ 23 \end{bmatrix}$$

that we obtain by substitution in Eq. (12). The echelon form found by Gaussian elimination is

$$\begin{bmatrix} 1 & 2 & 1 \\ 0 & 1 & 2 \\ 0 & 0 & 1 \end{bmatrix} \begin{bmatrix} a \\ b \\ c \end{bmatrix} = \begin{bmatrix} 4 \\ 4 \\ 3 \end{bmatrix},$$

so $c = 3$, $b = 4 - 2c = -2$, and $a = 4 - 2b - c = 5$. Thus

$$\mathbf{t} = 5\mathbf{u} - 2\mathbf{v} + 3\mathbf{w}.$$ ■

Subspaces of $\mathbf{R}^3$

Up until this point, we have used the words *line* and *plane* only in an informal or intuitive way. It is now time for us to say precisely what is meant by a line or plane *through the origin* in $\mathbf{R}^3$. Each is an example of a *subspace* of $\mathbf{R}^3$.

The nonempty subset V of $\mathbf{R}^3$ is called a **subspace** of $\mathbf{R}^3$ provided that V itself is a vector space under the operations of vector addition and multiplication of vectors by scalars. Suppose that the nonempty subset V of $\mathbf{R}^3$ is **closed** under these operations—that is, that the sum of any two vectors in V is also in V and that every scalar multiple of a vector in V is also in V. Then the vectors in V automatically satisfy properties (a) through (h) of Theorem 1, because these properties are "inherited" from $\mathbf{R}^3$; they hold for *all* vectors in $\mathbf{R}^3$, including those in V. Consequently, we see that a nonempty subset V of $\mathbf{R}^3$ is a **subspace** of $\mathbf{R}^3$ if and only if it satisfies the following two conditions:

(i) If $\mathbf{u}$ and $\mathbf{v}$ are vectors in V, then $\mathbf{u} + \mathbf{v}$ is also in V (closure under addition).

(ii) If $\mathbf{u}$ is a vector in V and c is a scalar, then $c\mathbf{u}$ is in V (closure under multiplication by scalars).

It is immediate that $V = \mathbf{R}^3$ is a subspace: $\mathbf{R}^3$ is a subspace of itself. At the opposite extreme, the subset $V = \{\mathbf{0}\}$, containing only the zero vector, is also a subspace of $\mathbf{R}^3$, because $\mathbf{0} + \mathbf{0} = \mathbf{0}$ and $c\mathbf{0} = \mathbf{0}$ for every scalar c. Thus $V = \{\mathbf{0}\}$ satisfies conditions (i) and (ii). The subspaces $\{\mathbf{0}\}$ and $\mathbf{R}^3$ are sometimes called the *trivial* subspaces of $\mathbf{R}^3$ (because the verification that they *are* subspaces is quite trivial). All subspaces other than $\{\mathbf{0}\}$ and $\mathbf{R}^3$ itself are called **proper subspaces** of $\mathbf{R}^3$.

Now we want to show that the proper subspaces of $\mathbf{R}^3$ are what we customarily call lines and planes through the origin. Let V be a subspace of $\mathbf{R}^3$ that is neither $\{\mathbf{0}\}$ nor $\mathbf{R}^3$ itself. There are two cases to consider, depending on whether or not V contains two linearly independent vectors.

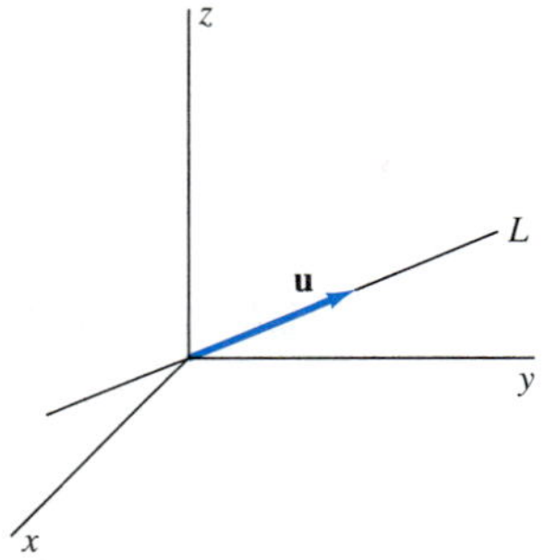

FIGURE 4.1.8. The line L spanned by the vector $\mathbf{u}$.

CASE 1: Suppose that V does *not* contain two linearly independent vectors. If $\mathbf{u}$ is a fixed nonzero vector in V, then, by condition (ii) above, every scalar multiple $c\mathbf{u}$ is also in V. Conversely, if $\mathbf{v}$ is any other vector in V, then $\mathbf{u}$ and $\mathbf{v}$ are linearly dependent, so it follows that $\mathbf{v} = c\mathbf{u}$ for some scalar c. Thus the subspace V is the set of all scalar multiples of the fixed nonzero vector $\mathbf{u}$ and is therefore what we call a **line** through the origin in $\mathbf{R}^3$. (See Fig. 4.1.8.)

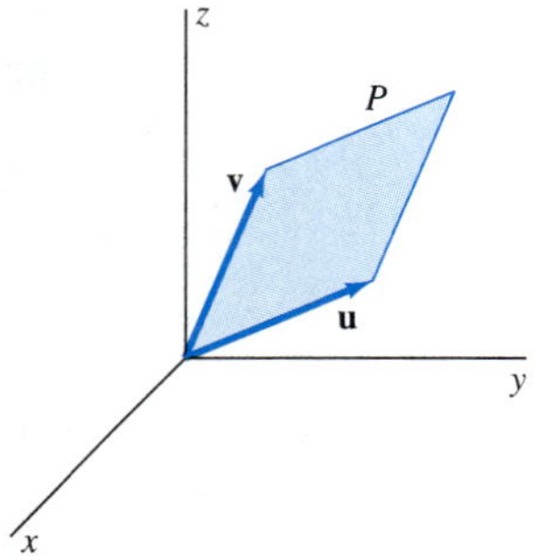

FIGURE 4.1.9. The plane P spanned by the vectors $\mathbf{u}$ and $\mathbf{v}$.

CASE 2: Suppose that V contains two linearly independent vectors $\mathbf{u}$ and $\mathbf{v}$. It then follows from conditions (i) and (ii) that V contains every linear combination $a\mathbf{u} + b\mathbf{v}$ of $\mathbf{u}$ and $\mathbf{v}$. (See Problem 38.) Conversely, let $\mathbf{w}$ be any other vector in V. If $\mathbf{u}$, $\mathbf{v}$, $\mathbf{w}$ were linearly independent, then, by Theorem 4, V would be all of $\mathbf{R}^3$. Therefore, $\mathbf{u}$, $\mathbf{v}$, $\mathbf{w}$ are linearly dependent, so it follows that there exist scalars a and b such that $\mathbf{w} = a\mathbf{u} + b\mathbf{v}$. (See Problem 40.) Thus the subspace V is the set of all linear combinations $a\mathbf{u} + b\mathbf{v}$ of the two linearly independent vectors $\mathbf{u}$ and $\mathbf{v}$ and is therefore what we call a **plane** through the origin in $\mathbf{R}^3$. (See Fig. 4.1.9.)

Subspaces of the coordinate plane $\mathbf{R}^2$ are defined similarly—they are the nonempty subsets of $\mathbf{R}^2$ that are closed under addition and multiplication by scalars. In Problem 39 we ask you to show that every proper subspace of $\mathbf{R}^2$ is a line through the origin.

Example 5 Let V be the set of all vectors (x, y) in $\mathbf{R}^2$ such that $y = x$. Given $\mathbf{u}$ and $\mathbf{v}$ in V, we may write $\mathbf{u} = (u, u)$ and $\mathbf{v} = (v, v)$. Then $\mathbf{u}+\mathbf{v} = (u+v, u+v)$ and $c\mathbf{u} = (cu, cu)$ are in V. It follows that V is a subspace of $\mathbf{R}^2$. ■

Example 6 Let V be the set of all vectors (x, y) in $\mathbf{R}^2$ such that $x+y=1$. Thus V is the straight line that passes through the unit points on the x- and y-axes. Then $\mathbf{u} = (1, 0)$ and $\mathbf{v} = (0, 1)$ are in V, but the vector $\mathbf{u} + \mathbf{v} = (1, 1)$ is not. It follows that V is *not* a subspace of $\mathbf{R}^2$. ■

Example 6 illustrates the fact that lines that do not pass through the origin are not subspaces of $\mathbf{R}^2$. Because every subspace must contain the zero vector (per Problem 37), only lines and planes that pass through the origin are subspaces of $\mathbf{R}^3$.

4.1 Problems

In Problems 1–4, find $|\mathbf{a} - \mathbf{b}|$, $2\mathbf{a} + \mathbf{b}$, *and* $3\mathbf{a} - 4\mathbf{b}$.

1. $\mathbf{a} = (2, 5, -4)$, $\mathbf{b} = (1, -2, -3)$
2. $\mathbf{a} = (-1, 0, 2)$, $\mathbf{b} = (3, 4, -5)$
3. $\mathbf{a} = 2\mathbf{i} - 3\mathbf{j} + 5\mathbf{k}$, $\mathbf{b} = 5\mathbf{i} + 3\mathbf{j} - 7\mathbf{k}$
4. $\mathbf{a} = 2\mathbf{i} - \mathbf{j}$, $\mathbf{b} = \mathbf{j} - 3\mathbf{k}$

In Problems 5–8, determine whether the given vectors $\mathbf{u}$ *and* $\mathbf{v}$ *are linearly dependent or linearly independent.*

5. $\mathbf{u} = (0, 2)$, $\mathbf{v} = (0, 3)$
6. $\mathbf{u} = (0, 2)$, $\mathbf{v} = (3, 0)$
7. $\mathbf{u} = (2, 2)$, $\mathbf{v} = (2, -2)$
8. $\mathbf{u} = (2, -2)$, $\mathbf{v} = (-2, 2)$

In Problems 9–14, express $\mathbf{w}$ *as a linear combination of* $\mathbf{u}$ *and* $\mathbf{v}$.

9. $\mathbf{u} = (1, -2)$, $\mathbf{v} = (-1, 3)$, $\mathbf{w} = (1, 0)$
10. $\mathbf{u} = (3, 4)$, $\mathbf{v} = (2, 3)$, $\mathbf{w} = (0, -1)$
11. $\mathbf{u} = (5, 7)$, $\mathbf{v} = (2, 3)$, $\mathbf{w} = (1, 1)$
12. $\mathbf{u} = (4, 1)$, $\mathbf{v} = (-2, -1)$, $\mathbf{w} = (2, -2)$
13. $\mathbf{u} = (7, 5)$, $\mathbf{v} = (3, 4)$, $\mathbf{w} = (5, -2)$
14. $\mathbf{u} = (5, -2)$, $\mathbf{v} = (-6, 4)$, $\mathbf{w} = (5, 6)$

In Problems 15–18, apply Theorem 4 (that is, calculate a determinant) to determine whether the given vectors $\mathbf{u}$, $\mathbf{v}$, *and* $\mathbf{w}$ *are linearly dependent or independent.*

15. $\mathbf{u} = (3, -1, 2)$, $\mathbf{v} = (5, 4, -6)$, $\mathbf{w} = (8, 3, -4)$
16. $\mathbf{u} = (5, -2, 4)$, $\mathbf{v} = (2, -3, 5)$, $\mathbf{w} = (4, 5, -7)$
17. $\mathbf{u} = (1, -1, 2)$, $\mathbf{v} = (3, 0, 1)$, $\mathbf{w} = (1, -2, 2)$
18. $\mathbf{u} = (1, 1, 0)$, $\mathbf{v} = (4, 3, 1)$, $\mathbf{w} = (3, -2, -4)$

In Problems 19–24, use the method of Example 3 to determine whether the given vectors $\mathbf{u}$, $\mathbf{v}$, *and* $\mathbf{w}$ *are linearly independent or dependent. If they are linearly dependent, find scalars* a, b, *and* c *not all zero such that* $a\mathbf{u} + b\mathbf{v} + c\mathbf{w} = \mathbf{0}$.

19. $\mathbf{u} = (2, 0, 1)$, $\mathbf{v} = (-3, 1, -1)$, $\mathbf{w} = (0, -2, -1)$
20. $\mathbf{u} = (5, 5, 4)$, $\mathbf{v} = (2, 3, 1)$, $\mathbf{w} = (4, 1, 5)$
21. $\mathbf{u} = (1, 1, -2)$, $\mathbf{v} = (-2, -1, 6)$, $\mathbf{w} = (3, 7, 2)$
22. $\mathbf{u} = (1, 1, 0)$, $\mathbf{v} = (5, 1, 3)$, $\mathbf{w} = (0, 1, 2)$
23. $\mathbf{u} = (2, 0, 3)$, $\mathbf{v} = (5, 4, -2)$, $\mathbf{w} = (2, -1, 1)$
24. $\mathbf{u} = (1, 4, 5)$, $\mathbf{v} = (4, 2, 5)$, $\mathbf{w} = (-3, 3, -1)$

In Problems 25–28, express the vector $\mathbf{t}$ *as a linear combination of the vectors* $\mathbf{u}$, $\mathbf{v}$, *and* $\mathbf{w}$.

25. $\mathbf{t} = (2, -7, 9)$, $\mathbf{u} = (1, -2, 2)$, $\mathbf{v} = (3, 0, 1)$, $\mathbf{w} = (1, -1, 2)$
26. $\mathbf{t} = (5, 30, -21)$, $\mathbf{u} = (5, 2, -2)$, $\mathbf{v} = (1, 5, -3)$, $\mathbf{w} = (5, -3, 4)$
27. $\mathbf{t} = (0, 0, 19)$, $\mathbf{u} = (1, 4, 3)$, $\mathbf{v} = (-1, -2, 2)$, $\mathbf{w} = (4, 4, 1)$
28. $\mathbf{t} = (7, 7, 7)$, $\mathbf{u} = (2, 5, 3)$, $\mathbf{v} = (4, 1, -1)$, $\mathbf{w} = (1, 1, 5)$

In Problems 29–32, show that the given set V is closed under addition and under multiplication by scalars and is therefore a subspace of $\mathbf{R}^3$.

29. V is the set of all (x, y, z) such that $x = 0$.
30. V is the set of all (x, y, z) such that $x + y + z = 0$.
31. V is the set of all (x, y, z) such that $2x = 3y$.
32. V is the set of all (x, y, z) such that $z = 2x + 3y$.

In Problems 33–36, show that the given set V is not a subspace of $\mathbf{R}^3$.

33. V is the set of all (x, y, z) such that $y = 1$.
34. V is the set of all (x, y, z) such that $x + y + z = 3$.
35. V is the set of all (x, y, z) such that $z \geq 0$.
36. V is the set of all (x, y, z) such that $xyz = 1$.
37. Show that every subspace V of $\mathbf{R}^3$ contains the zero vector $\mathbf{0}$.
38. Suppose that V is a subspace of $\mathbf{R}^3$. Show that V is closed under the operation of taking linear combinations of pairs of vectors. That is, show that if $\mathbf{u}$ and $\mathbf{v}$ are in V and a and b are scalars, then $a\mathbf{u} + b\mathbf{v}$ is in V.
39. Suppose that V is a proper subspace of $\mathbf{R}^2$ and that $\mathbf{u}$ is a nonzero vector in V. Show that V is the set of all scalar multiples of $\mathbf{u}$ and therefore that V is a line through the origin.
40. Suppose that $\mathbf{u}$, $\mathbf{v}$, and $\mathbf{w}$ are vectors in $\mathbf{R}^3$ such that $\mathbf{u}$ and $\mathbf{v}$ are linearly independent but $\mathbf{u}$, $\mathbf{v}$, and $\mathbf{w}$ are linearly dependent. Show that there exist scalars a and b such that $\mathbf{w} = a\mathbf{u} + b\mathbf{v}$.
41. Let V_1 and V_2 be subspaces of $\mathbf{R}^3$. Their *intersection* $V = V_1 \cap V_2$ is the set of all vectors that lie both in V_1 and in V_2. Show that V is a subspace of $\mathbf{R}^3$.

4.2 The Vector Space $\mathbf{R}^n$ and Subspaces

In Section 4.1 we defined 3-dimensional space $\mathbf{R}^3$ to be the set of all triples (x, y, z) of real numbers. This definition provides a *mathematical model* of the physical space in which we live, because geometric intuition and experience require that the location of every point be specified uniquely by *three* coordinates.

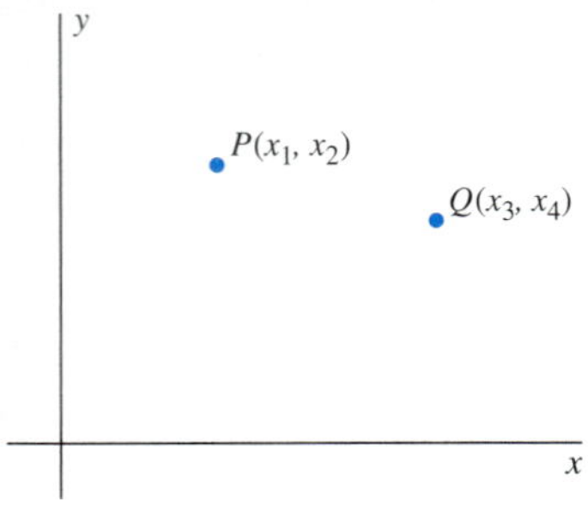

FIGURE 4.2.1. Two points $P(x_1, x_2)$ and $Q(x_3, x_4)$ in $\mathbf{R}^4$.

In science fiction, the fourth dimension often plays a rather exotic role. But there are common and ordinary situations where it is convenient to use four (or even more) coordinates rather than just two or three. For example, suppose we want to describe the motion of two points P and Q that are moving in the plane $\mathbf{R}^2$ under the action of some given physical law. (See Fig. 4.2.1.) In order to tell where P and Q are at a given instant, we need to give two coordinates for P and two coordinates for Q. So let us write $P(x_1, x_2)$ and $Q(x_3, x_4)$ to indicate these *four* coordinates. Then the two points P and Q determine a quadruple or 4-tuple (x_1, x_2, x_3, x_4) of real numbers, and any such 4-tuple determines a possible pair of locations of P and Q. In this way the set of all pairs of points P and Q in the plane corresponds to the set of all 4-tuples of real numbers. By analogy with our definition of $\mathbf{R}^3$, we may define *4-dimensional space* $\mathbf{R}^4$ to be the set of all such 4-tuples (x_1, x_2, x_3, x_4). Then we can specify a pair of points P and Q in $\mathbf{R}^2$ by specifying a single *point* (x_1, x_2, x_3, x_4) in $\mathbf{R}^4$, and this viewpoint may simplify our analysis of the motions of the original points P and Q. For instance, it may turn out that their coordinates satisfy some single equation such as

$$3x_1 - 4x_2 + 2x_3 - 5x_4 = 0$$

that is better understood in terms of a single point in $\mathbf{R}^4$ than in terms of the separate points P and Q in $\mathbf{R}^2$. Finally, note that in this example the fourth dimension is quite tangible—it refers simply to the second coordinate x_4 of the point Q.

But the number 4 is no more special in this context than the numbers 2 and 3. To describe similarly the motion of three points in $\mathbf{R}^3$, we would need 9 coordinates rather than 4. For instance, we might be studying the sun-earth-moon system in 3-dimensional space, with (x_1, x_2, x_3) denoting the coordinates of the sun, (x_4, x_5, x_6) the coordinates of the earth, and (x_7, x_8, x_9) the coordinates of the moon at a given instant. Then the single list $(x_1, x_2, x_3, x_4, x_5, x_6, x_7, x_8, x_9)$ of 9 coordinates could be used to specify simultaneously the locations of the sun, earth, and moon. Application of Newton's second law of motion to each of these three bodies would then lead to a system of differential equations involving the dependent variables $x_i(t)$, $i = 1, 2, \ldots, 9$ as functions of time t. (We will discuss this very situation in Section 7.6.)

An efficient description of another situation might require fifty, a hundred, or even a thousand coordinates. For instance, an investor's stock portfolio might contain 50 different high-tech stocks. If x_i denotes the value of the ith stock on a particular day, then the state of the whole portfolio is described by the single list $(x_1, x_2, x_3, \ldots, x_{49}, x_{50})$ of 50 separate numbers. And the evolution of this portfolio with time could then be described by the motion of a single point in 50-dimensional space!

An ***n*-tuple** of real numbers is an (ordered) list $(x_1, x_2, x_3, \ldots, x_n)$ of n real numbers. Thus $(1, 3, 4, 2)$ is a 4-tuple and $(0, -3, 7, 5, 2, -1)$ is a 6-tuple. A 2-tuple is an ordered pair and a 3-tuple is an ordered triple of real numbers, so the following definition generalizes our earlier definitions of the plane $\mathbf{R}^2$ and 3-space $\mathbf{R}^3$.

DEFINITION ***n*-Space $\mathbf{R}^n$**

The ***n*-dimensional space $\mathbf{R}^n$** is the set of all n-tuples $(x_1, x_2, x_3, \ldots, x_n)$ of real numbers.

The elements of n-space $\mathbf{R}^n$ are called **points** or **vectors**, and we ordinarily use boldface letters to denote vectors. The ith entry of the vector $\mathbf{x} = (x_1, x_2, x_3, \ldots, x_n)$ is called its ith **coordinate** or its ith **component**. For consistency with matrix operations, we agree that

$$\mathbf{x} = (x_1, x_2, x_3, \ldots, x_n) = \begin{bmatrix} x_1 \\ x_2 \\ \vdots \\ x_n \end{bmatrix},$$

so that the $n \times 1$ matrix, or *column* vector, is simply another notation for the same ordered list of n real numbers.

If $n > 3$, we cannot visualize vectors in $\mathbf{R}^n$ in the concrete way that we can "see" vectors in $\mathbf{R}^2$ and $\mathbf{R}^3$. Nevertheless, we saw in Section 4.1 that the geometric properties of $\mathbf{R}^2$ and $\mathbf{R}^3$ stem ultimately from algebraic operations with vectors, and these algebraic operations can be defined for vectors in $\mathbf{R}^n$ by analogy with their definitions in dimensions 2 and 3.

If $\mathbf{u} = (u_1, u_2, \ldots, u_n)$ and $\mathbf{v} = (v_1, v_2, \ldots, v_n)$ are vectors in $\mathbf{R}^n$, then their **sum** is the vector $\mathbf{u} + \mathbf{v}$ given by

$$\mathbf{u} + \mathbf{v} = (u_1 + v_1, u_2 + v_2, \ldots, u_n + v_n). \tag{1}$$

Thus addition of vectors in $\mathbf{R}^n$ is defined in a componentwise manner, just as in $\mathbf{R}^2$ and $\mathbf{R}^3$, and we can visualize $\mathbf{u} + \mathbf{v}$ as a diagonal vector in a parallelogram determined by $\mathbf{u}$ and $\mathbf{v}$, as in Fig. 4.2.2. If c is a scalar—a real number—then the **scalar multiple** $c\mathbf{u}$ is also defined in componentwise fashion:

$$c\mathbf{u} = (cu_1, cu_2, \ldots, cu_n). \tag{2}$$

We can visualize $c\mathbf{u}$ as a vector collinear with $\mathbf{u}$, with only its length (and possibly its direction along the same line) altered by multiplication by the scalar c. See Fig. 4.2.3.

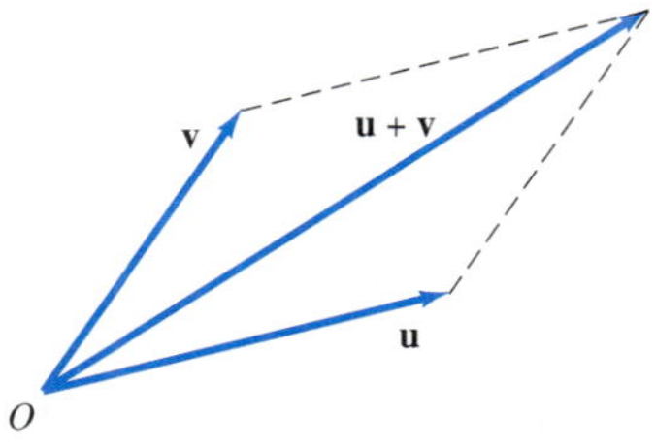

FIGURE 4.2.2. The parallelogram law for addition of vectors, in $\mathbf{R}^n$ just as in $\mathbf{R}^2$ or $\mathbf{R}^3$.

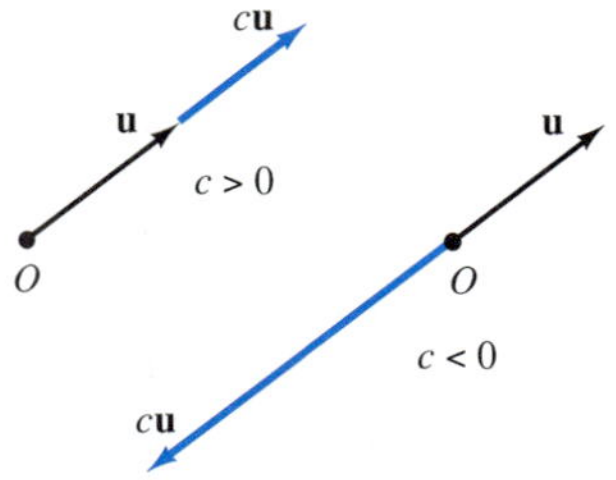

FIGURE 4.2.3. Multiplication of the vector $\mathbf{u}$ by the scalar c, in $\mathbf{R}^n$ just as in $\mathbf{R}^2$ or $\mathbf{R}^3$.

In Section 4.1 we saw that, for vectors in $\mathbf{R}^2$ and $\mathbf{R}^3$, the operations in (1) and (2) satisfy a list of properties that constitute the definition of a *vector space*; the same is true for vectors in $\mathbf{R}^n$.

Definition of a Vector Space

Let V be a set of elements called *vectors*, in which the operations of addition of vectors and multiplication of vectors by scalars are defined. That is, given vectors $\mathbf{u}$ and $\mathbf{v}$ in V and a scalar c, the vectors $\mathbf{u}+\mathbf{v}$ and $c\mathbf{u}$ are also in V (so that V is *closed* under vector addition and multiplication by scalars). Then, with these operations, V is called a **vector space** provided that—given any vectors $\mathbf{u}$, $\mathbf{v}$, and $\mathbf{w}$ in V and any scalars a and b—the following properties hold true:

(a) $\mathbf{u}+\mathbf{v}=\mathbf{v}+\mathbf{u}$ (commutativity)

(b) $\mathbf{u}+(\mathbf{v}+\mathbf{w})=(\mathbf{u}+\mathbf{v})+\mathbf{w}$ (associativity)

(c) $\mathbf{u}+\mathbf{0}=\mathbf{0}+\mathbf{u}=\mathbf{u}$ (zero element)

(d) $\mathbf{u}+(-\mathbf{u})=(-\mathbf{u})+\mathbf{u}=\mathbf{0}$ (additive inverse)

(e) $a(\mathbf{u}+\mathbf{v})=a\mathbf{u}+a\mathbf{v}$ (distributivity)

(f) $(a+b)\mathbf{u}=a\mathbf{u}+b\mathbf{u}$

(g) $a(b\mathbf{u})=(ab)\mathbf{u}$

(h) $(1)\mathbf{u}=\mathbf{u}$

In property (c), it is meant that there exists a **zero vector 0** in V such that $\mathbf{u}+\mathbf{0}=\mathbf{u}$. The zero vector in $\mathbf{R}^n$ is

$$\mathbf{0}=(0,0,\ldots,0).$$

Similarly, property (d) actually means that, given the vector $\mathbf{u}$ in V, there exists a vector $-\mathbf{u}$ in V such that $\mathbf{u}+(-\mathbf{u})=\mathbf{0}$. In $\mathbf{R}^n$, we clearly have

$$-\mathbf{u}=(-u_1,-u_2,\ldots,-u_n).$$

The fact that $\mathbf{R}=\mathbf{R}^1$ satisfies properties (a)–(h), with the real numbers playing the dual roles of scalars *and* vectors, means that the real line may be regarded as a vector space. If $n>1$, then each of properties (a)–(h) may be readily verified for $\mathbf{R}^n$ by working with components and applying the corresponding properties of real numbers. For example, to verify the commutativity of vector addition in (a), we begin with Eq. (1) and write

$$\begin{aligned}\mathbf{u}+\mathbf{v}&=(u_1+v_1,u_2+v_2,\ldots,u_n+v_n)\\&=(v_1+u_1,v_2+u_2,\ldots,v_n+u_n)\\&=\mathbf{v}+\mathbf{u}.\end{aligned}$$

Thus n-space $\mathbf{R}^n$ *is a vector space* with the operations defined in Eqs. (1) and (2). We will understand throughout that our scalars are the *real numbers*, though in more advanced treatments of linear algebra other scalars (such as the complex numbers) are sometimes used.

Observe that, in the definition of a vector space, nothing is said about the elements of V (the "vectors") being n-tuples. They can be any objects for which addition and multiplication by scalars are defined and which satisfy properties (a)–(h). Although many of the vectors you will see in this text actually are n-tuples of real numbers, an example in which they are not may help to clarify the definition of a vector space.

Example 1 Let $\mathcal{F}$ be the set of all real-valued functions defined on the real number line $\mathbf{R}$. Then each vector in $\mathcal{F}$ is a function $\mathbf{f}$ such that the real number $\mathbf{f}(x)$ is defined for all x in $\mathbf{R}$. Given $\mathbf{f}$ and $\mathbf{g}$ in $\mathcal{F}$ and a real number c, the functions $\mathbf{f}+\mathbf{g}$ and $c\mathbf{f}$ are defined in the natural way,

$$(\mathbf{f}+\mathbf{g})(x) = \mathbf{f}(x) + \mathbf{g}(x)$$

and

$$(c\mathbf{f})(x) = c(\mathbf{f}(x)).$$

Then each of properties (a)–(h) of a vector space follows readily from the corresponding property of the real numbers. For instance, if a is a scalar, then

$$\begin{aligned}[a(\mathbf{f}+\mathbf{g})](x) &= a[(\mathbf{f}+\mathbf{g})(x)] \\ &= a[\mathbf{f}(x) + \mathbf{g}(x)] \\ &= a\mathbf{f}(x) + a\mathbf{g}(x) \\ &= (a\mathbf{f} + a\mathbf{g})(x).\end{aligned}$$

Thus $a(\mathbf{f}+\mathbf{g}) = a\mathbf{f} + a\mathbf{g}$, so $\mathcal{F}$ enjoys property (e). ■

After verification of the other seven properties, we conclude that $\mathcal{F}$ is, indeed, a vector space—one that differs in a fundamental way from each of the vector spaces $\mathbf{R}^n$. We naturally call $\mathbf{R}^n$ an n-dimensional vector space, and in Section 4.4 we will define the word *dimension* in such a way that the dimension of $\mathbf{R}^n$ actually *is* n. But the vector space $\mathcal{F}$ of functions turns out not to have dimension n for any integer n; it is an example of an *infinite-dimensional* vector space. (See Section 4.4.)

Subspaces

Let $W = \{\mathbf{0}\}$ be the subset of $\mathbf{R}^n$ that contains only the zero vector $\mathbf{0}$. Then W satisfies properties (a)–(h) of a vector space, because each reduces trivially to the equation $\mathbf{0} = \mathbf{0}$. Thus W is a subset of the vector space $V = \mathbf{R}^n$ that is itself a vector space. According to the following definition, a subset of a vector space V that is itself a vector space is called a *subspace* of V.

DEFINITION Subspace

Let W be a nonempty subset of the vector space V. Then W is a **subspace** of V provided that W itself is a vector space with the operations of addition and multiplication by scalars as defined in V.

In order for the subset W to be a subspace of the vector space V, it first must be closed under the operations of vector addition and multiplication by scalars. Then it must satisfy properties (a)–(h) of a vector space. But W "inherits" all these properties from V, because the vectors in W all are vectors in V, and the vectors in V all satisfy properties (a)–(h). The following "subspace criterion" says that, in order to determine whether the subset W is a vector space, we need only check the two closure conditions.

THEOREM 1 Conditions for a Subspace

The nonempty subset W of the vector space V is a subspace of V if and only if it satisfies the following two conditions:

(i) If $\mathbf{u}$ and $\mathbf{v}$ are vectors in W, then $\mathbf{u}+\mathbf{v}$ is also in W.

(ii) If $\mathbf{u}$ is in W and c is a scalar, then the vector $c\mathbf{u}$ is also in W.

In Section 4.1 we saw that lines through the origin in $\mathbf{R}^2$ form subspaces of $\mathbf{R}^2$ and that lines and planes through the origin are subspaces of $\mathbf{R}^3$. The subspace W of the following example may be regarded as a higher-dimensional "plane" (or "hyperplane") through the origin in $\mathbf{R}^n$.

Example 2 Let W be the subset of $\mathbf{R}^n$ consisting of all those vectors $(x_1, x_2, \ldots, x_n)$ whose coordinates satisfy the single homogeneous linear equation

$$a_1x_1 + a_2x_2 + \cdots + a_nx_n = 0,$$

where the given coefficients $a_1, a_2, \ldots, a_n$ are not all zero. If $\mathbf{u} = (u_1, u_2, \ldots, u_n)$ and $\mathbf{v} = (v_1, v_2, \ldots, v_n)$ are vectors in W, then

$$\begin{aligned} &a_1(u_1 + v_1) + a_2(u_2 + v_2) + \cdots + a_n(u_n + v_n) \\ &\quad = (a_1u_1 + a_2u_2 + \cdots + a_nu_n) + (a_1v_1 + a_2v_2 + \cdots + a_nv_n) \\ &\quad = 0 + 0 = 0, \end{aligned}$$

so $\mathbf{u} + \mathbf{v} = (u_1 + v_1, \ldots, u_n + v_n)$ is also in W. If c is a scalar, then

$$\begin{aligned} a_1(cu_1) + a_2(cu_2) + \cdots + a_n(cu_n) &= c(a_1u_1 + a_2u_2 + \cdots + a_nu_n) \\ &= (c)(0) = 0, \end{aligned}$$

so $c\mathbf{u} = (cu_1, cu_2, \ldots, cu_n)$ is in W. Thus we have shown that W satisfies conditions (i) and (ii) of Theorem 1 and is therefore a subspace of $\mathbf{R}^n$. ■

In order to apply Theorem 1 to show that W *is* a subspace of the vector space V, we must show that W satisfies *both* conditions in the theorem. But to apply Theorem 1 to show that W is *not* a subspace of V, we need only show *either* that condition (i) fails *or* that condition (ii) fails.

Example 3 Let W be the set of all those vectors (x_1, x_2, x_3, x_4) in $\mathbf{R}^4$ whose four coordinates are all nonnegative: $x_i \geq 0$ for $i = 1, 2, 3, 4$. Then it should be clear that the sum of two vectors in W is also a vector in W, because the sum of two nonnegative numbers is nonnegative. Thus W satisfies condition (i) of Theorem 1. But if we take $\mathbf{u} = (1, 1, 1, 1)$ in W and $c = -1$, then we find that the scalar multiple

$$c\mathbf{u} = (-1)(1, 1, 1, 1) = (-1, -1, -1, -1)$$

is *not* in W. Thus W fails to satisfy condition (ii) and therefore is *not* a subspace of $\mathbf{R}^4$. ■

Example 4 Let W be the set of all those vectors (x_1, x_2, x_3, x_4) in $\mathbf{R}^4$ such that $x_1x_4 = 0$. Now W satisfies condition (ii) of Theorem 1, because $x_1x_4 = 0$ implies that $(cx_1)(cx_4) = 0$ for any scalar c. But if we take the vectors $\mathbf{u} = (1, 1, 0, 0)$ and $\mathbf{v} = (0, 0, 1, 1)$ in W, we see that their sum $\mathbf{u} + \mathbf{v} = (1, 1, 1, 1)$ is *not* in W. Thus W does not satisfy condition (i) and therefore is *not* a subspace of $\mathbf{R}^4$. ■

Example 2 implies that the solution set of a homogeneous linear equation

$$a_1x_1 + a_2x_2 + \cdots + a_nx_n = 0$$

in n variables is always a *subspace* of $\mathbf{R}^n$. Theorem 2 further implies that the same is true of a *homogeneous* system of linear equations. Recall that any such system of m equations in n unknowns can be written in the form $\mathbf{Ax} = \mathbf{0}$, where $\mathbf{A}$ is the $m \times n$ coefficient matrix and $\mathbf{x} = (x_1, x_2, \ldots, x_n)$ is regarded as a column vector. The solution set of $\mathbf{Ax} = \mathbf{0}$ is then the set of all vectors $\mathbf{x}$ in $\mathbf{R}^n$ that satisfy this equation—that is, the set of all its solution vectors.

THEOREM 2 Solution Subspaces

If $\mathbf{A}$ is a (constant) $m \times n$ matrix, then the solution set of the homogeneous linear system

$$\mathbf{Ax} = \mathbf{0} \tag{3}$$

is a subspace of $\mathbf{R}^n$.

***Proof*:** Let W denote the solution set of Eq. (3). If $\mathbf{u}$ and $\mathbf{v}$ are vectors in W, then $\mathbf{Au} = \mathbf{Av} = \mathbf{0}$. Hence

$$\mathbf{A}(\mathbf{u} + \mathbf{v}) = \mathbf{Au} + \mathbf{Av} = \mathbf{0} + \mathbf{0} = \mathbf{0}.$$

Thus the sum $\mathbf{u} + \mathbf{v}$ is also in W, and hence W satisfies condition (i) of Theorem 1. If c is a scalar, then

$$\mathbf{A}(c\mathbf{u}) = c(\mathbf{Au}) = c\mathbf{0} = \mathbf{0},$$

so $c\mathbf{u}$ is in W if $\mathbf{u}$ is in W. Thus W also satisfies condition (ii) of Theorem 1. It therefore follows that W is a subspace of $\mathbf{R}^n$. ■

Note that, in order to conclude by Theorem 2 that the solution set of a linear system is a subspace, it is necessary for the system to be *homogeneous*. Indeed, the solution set of a nonhomogeneous linear system

$$\mathbf{Ax} = \mathbf{b} \tag{4}$$

with $\mathbf{b} \neq \mathbf{0}$ is *never* a subspace. For if $\mathbf{u}$ were a solution vector of the system in (4), then

$$\mathbf{A}(2\mathbf{u}) = 2(\mathbf{Au}) = 2\mathbf{b} \neq \mathbf{b}$$

because $\mathbf{b} \neq \mathbf{0}$. Thus the scalar multiple $2\mathbf{u}$ of the solution vector $\mathbf{u}$ is *not* a solution vector. Therefore, Theorem 1 implies that the solution set of (4) is *not* a subspace.

Because the solution set of a *homogeneous* linear system is a subspace, we often call it the **solution space** of the system. The subspaces of $\mathbf{R}^n$ are the possible solution spaces of a homogeneous linear system with n unknowns, and this is one of the principal reasons for our interest in subspaces of $\mathbf{R}^n$.

At opposite extremes as subspaces of $\mathbf{R}^n$ lie the **zero subspace** $\{\mathbf{0}\}$ and $\mathbf{R}^n$ itself. Every other subspace of $\mathbf{R}^n$, each one that is neither $\{\mathbf{0}\}$ nor $\mathbf{R}^n$, is called a **proper subspace** of $\mathbf{R}^n$. The proper subspaces of $\mathbf{R}^n$ play the same role in $\mathbf{R}^n$ that lines and planes through the origin play in $\mathbf{R}^3$. In the following two sections we develop the tools that are needed to analyze the structure of a given proper subspace of $\mathbf{R}^n$. In particular, given the homogeneous system $\mathbf{Ax} = \mathbf{0}$, we ask how we can describe its solution space in a concise and illuminating way, beyond the mere statement that it is the set of all solution vectors of the system. Example 5 illustrates one possible way of doing this.

Example 5 In Example 4 of Section 3.4 we considered the homogeneous system

$$\begin{aligned} x_1 + 3x_2 - 15x_3 + 7x_4 &= 0 \\ x_1 + 4x_2 - 19x_3 + 10x_4 &= 0 \\ 2x_1 + 5x_2 - 26x_3 + 11x_4 &= 0. \end{aligned} \tag{5}$$

The reduced echelon form of the coefficient matrix of this system is

$$\begin{bmatrix} 1 & 0 & -3 & -2 \\ 0 & 1 & -4 & 3 \\ 0 & 0 & 0 & 0 \end{bmatrix}.$$

Hence x_1 and x_2 are the leading variables and x_3 and x_4 are free variables. Back substitution yields the general solution

$$x_3 = s, \quad x_4 = t, \quad x_2 = 4s - 3t, \quad x_1 = 3s + 2t$$

in terms of arbitrary parameters s and t. Thus a typical solution vector of the system in (5) has the form

$$\mathbf{x} = \begin{bmatrix} x_1 \\ x_2 \\ x_3 \\ x_4 \end{bmatrix} = \begin{bmatrix} 3s + 2t \\ 4s - 3t \\ s \\ t \end{bmatrix} = s \begin{bmatrix} 3 \\ 4 \\ 1 \\ 0 \end{bmatrix} + t \begin{bmatrix} 2 \\ -3 \\ 0 \\ 1 \end{bmatrix}.$$

It follows that the solution space of the system in (5) can be described as the set of all linear combinations of the form

$$\mathbf{x} = s\mathbf{u} + t\mathbf{v}, \tag{6}$$

where $\mathbf{u} = (3, 4, 1, 0)$ and $\mathbf{v} = (2, -3, 0, 1)$. Thus we have found two particular solution vectors $\mathbf{u}$ and $\mathbf{v}$ of our system that completely determine its solution space [by the formula in (6)]. ■

4.2 Problems

In Problems 1–14, a subset W of some n-space $\mathbf{R}^n$ is defined by means of a given condition imposed on the typical vector $(x_1, x_2, \ldots, x_n)$. Apply Theorem 1 to determine whether or not W is a subspace of $\mathbf{R}^n$.

1. W is the set of all vectors in $\mathbf{R}^3$ such that $x_3 = 0$.
2. W is the set of all vectors in $\mathbf{R}^3$ such that $x_1 = 5x_2$.
3. W is the set of all vectors in $\mathbf{R}^3$ such that $x_2 = 1$.
4. W is the set of all vectors in $\mathbf{R}^3$ such that $x_1 + x_2 + x_3 = 1$.
5. W is the set of all vectors in $\mathbf{R}^4$ such that $x_1 + 2x_2 + 3x_3 + 4x_4 = 0$.
6. W is the set of all vectors in $\mathbf{R}^4$ such that $x_1 = 3x_3$ and $x_2 = 4x_4$.
7. W is the set of all vectors in $\mathbf{R}^2$ such that $|x_1| = |x_2|$.
8. W is the set of all vectors in $\mathbf{R}^2$ such that $(x_1)^2 + (x_2)^2 = 0$.
9. W is the set of all vectors in $\mathbf{R}^2$ such that $(x_1)^2 + (x_2)^2 = 1$.
10. W is the set of all vectors in $\mathbf{R}^2$ such that $|x_1| + |x_2| = 1$.
11. W is the set of all vectors in $\mathbf{R}^4$ such that $x_1 + x_2 = x_3 + x_4$.
12. W is the set of all vectors in $\mathbf{R}^4$ such that $x_1 x_2 = x_3 x_4$.
13. W is the set of all vectors in $\mathbf{R}^4$ such that $x_1 x_2 x_3 x_4 = 0$.
14. W is the set of all those vectors in $\mathbf{R}^4$ whose components are all nonzero.

In Problems 15–18, apply the method of Example 5 to find two solution vectors $\mathbf{u}$ and $\mathbf{v}$ such that the solution space is the set of all linear combinations of the form $s\mathbf{u} + t\mathbf{v}$.

15. $$\begin{aligned} x_1 - 4x_2 + x_3 - 4x_4 &= 0 \\ x_1 + 2x_2 + x_3 + 8x_4 &= 0 \\ x_1 + x_2 + x_3 + 6x_4 &= 0 \end{aligned}$$

16. $$\begin{aligned} x_1 - 4x_2 - 3x_3 - 7x_4 &= 0 \\ 2x_1 - x_2 + x_3 + 7x_4 &= 0 \\ x_1 + 2x_2 + 3x_3 + 11x_4 &= 0 \end{aligned}$$

17.
$$\begin{aligned} x_1 + 3x_2 + 8x_3 - x_4 &= 0 \\ x_1 - 3x_2 - 10x_3 + 5x_5 &= 0 \\ x_1 + 4x_2 + 11x_3 - 2x_4 &= 0 \end{aligned}$$

18.
$$\begin{aligned} x_1 + 3x_2 + 2x_3 + 5x_4 - x_5 &= 0 \\ 2x_1 + 7x_2 + 4x_3 + 11x_4 + 2x_5 &= 0 \\ 2x_1 + 6x_2 + 5x_3 + 12x_4 - 7x_5 &= 0 \end{aligned}$$

In Problems 19–22, reduce the given system to echelon form to find a single solution vector **u** *such that the solution space is the set of all scalar multiples of* **u**.

19.
$$\begin{aligned} x_1 - 3x_2 - 5x_3 - 6x_4 &= 0 \\ 2x_1 + x_2 + 4x_3 - 4x_4 &= 0 \\ x_1 + 3x_2 + 7x_3 + x_4 &= 0 \end{aligned}$$

20.
$$\begin{aligned} x_1 + 5x_2 + x_3 - 8x_4 &= 0 \\ 2x_1 + 5x_2 \qquad - 5x_4 &= 0 \\ 2x_1 + 7x_2 + x_3 - 9x_4 &= 0 \end{aligned}$$

21.
$$\begin{aligned} x_1 + 7x_2 + 2x_3 - 3x_4 &= 0 \\ 2x_1 + 7x_2 + x_3 - 4x_4 &= 0 \\ 3x_1 + 5x_2 - x_3 - 5x_4 &= 0 \end{aligned}$$

22.
$$\begin{aligned} x_1 + 3x_2 + 3x_3 + 3x_4 &= 0 \\ 2x_1 + 7x_2 + 5x_3 - x_4 &= 0 \\ 2x_1 + 7x_2 + 4x_3 - 4x_4 &= 0 \end{aligned}$$

23. Show that every subspace W of a vector space V contains the zero vector **0**.

24. Apply the properties of a vector space V to show each of the following.

(a) $0\mathbf{u} = \mathbf{0}$ for every **u** in V.
(b) $c\mathbf{0} = \mathbf{0}$ for every scalar c.
(c) $(-1)\mathbf{u} = -\mathbf{u}$ for every **u** in V.

Do *not* assume that the vectors in V are n-tuples of real numbers.

25. Show that the nonempty subset W of a vector space V is a subspace of V if and only if for every pair of vectors **u** and **v** in W and every pair of scalars a and b, $a\mathbf{u} + b\mathbf{v}$ is also in W.

26. *Prove*: If **u** is a (fixed) vector in the vector space V, then the set W of all scalar multiples $c\mathbf{u}$ of **u** is a subspace of V.

27. Let **u** and **v** be (fixed) vectors in the vector space V. Show that the set W of all linear combinations $a\mathbf{u} + b\mathbf{v}$ of **u** and **v** is a subspace of V.

28. Suppose that **A** is an $n \times n$ matrix and that k is a (constant) scalar. Show that the set of all vectors **x** such that $\mathbf{Ax} = k\mathbf{x}$ is a subspace of $\mathbf{R}^n$.

29. Let **A** be an $n \times n$ matrix, **b** be a nonzero vector, and $\mathbf{x}_0$ be a solution vector of the system $\mathbf{Ax} = \mathbf{b}$. Show that **x** is a solution of the nonhomogeneous system $\mathbf{Ax} = \mathbf{b}$ if and only if $\mathbf{y} = \mathbf{x} - \mathbf{x}_0$ is a solution of the homogeneous system $\mathbf{Ay} = \mathbf{0}$.

30. Let U and V be subspaces of the vector space W. Their **intersection** $U \cap V$ is the set of all vectors that are both in U and in V. Show that $U \cap V$ is a subspace of W. If U and V are two planes through the origin in $\mathbf{R}^3$, what is $U \cap V$?

31. Let U and V be subspaces of the vector space W. Their **sum** $U + V$ is the set of all vectors **w** of the form

$$\mathbf{w} = \mathbf{u} + \mathbf{v},$$

where **u** is in U and **v** is in V. Show that $U + V$ is a subspace of W. If U and V are lines through the origin in $\mathbf{R}^3$, what is $U + V$?

4.3 Linear Combinations and Independence of Vectors

In Example 5 of Section 4.2 we solved the homogeneous linear system

$$\begin{aligned} x_1 + 3x_2 - 15x_3 + 7x_4 &= 0 \\ x_1 + 4x_2 - 19x_3 + 10x_4 &= 0 \\ 2x_1 + 5x_2 - 26x_3 + 11x_4 &= 0. \end{aligned} \tag{1}$$

We found that its solution space W consists of all those vectors **x** in $\mathbf{R}^4$ that have the form

$$\mathbf{x} = s(3, 4, 1, 0) + t(2, -3, 0, 1). \tag{2}$$

We therefore can visualize W as the plane in $\mathbf{R}^4$ determined by the vectors $\mathbf{v}_1 = (3, 4, 1, 0)$ and $\mathbf{v}_2 = (2, -3, 0, 1)$. The fact that every solution vector is a combination [as in (2)] of the particular solution vectors $\mathbf{v}_1$ and $\mathbf{v}_2$ gives us a tangible understanding of the solution space W of the system in (1).

More generally, we know from Theorem 2 in Section 4.2 that the solution set V of any $m \times n$ homogeneous linear system $\mathbf{Ax} = \mathbf{0}$ is a subspace of $\mathbf{R}^n$. In order to understand such a vector space V better, we would like to find a minimal set of vectors $\mathbf{v}_1, \mathbf{v}_2, \ldots, \mathbf{v}_k$ in V such that *every* vector in V is a sum of scalar multiples of these particular vectors.

The vector $\mathbf{w}$ is called a **linear combination** of the vectors $\mathbf{v}_1, \mathbf{v}_2, \ldots, \mathbf{v}_k$ provided that there exist scalars $c_1, c_2, \ldots, c_k$ such that

$$\mathbf{w} = c_1\mathbf{v}_1 + c_2\mathbf{v}_2 + \cdots + c_k\mathbf{v}_k. \tag{3}$$

Given a vector $\mathbf{w}$ in $\mathbf{R}^n$, the problem of determining whether or not $\mathbf{w}$ is a linear combination of the vectors $\mathbf{v}_1, \mathbf{v}_2, \ldots, \mathbf{v}_k$ amounts to solving a linear system to see whether we can find scalars $c_1, c_2, \ldots, c_k$ so that (3) holds.

Example 1 To determine whether the vector $\mathbf{w} = (2, -6, 3)$ in $\mathbf{R}^3$ is a linear combination of the vectors $\mathbf{v}_1 = (1, -2, -1)$ and $\mathbf{v}_2 = (3, -5, 4)$, we write the equation $c_1\mathbf{v}_1 + c_2\mathbf{v}_2 = \mathbf{w}$ in matrix form:

$$c_1\begin{bmatrix} 1 \\ -2 \\ -1 \end{bmatrix} + c_2\begin{bmatrix} 3 \\ -5 \\ 4 \end{bmatrix} = \begin{bmatrix} 2 \\ -6 \\ 3 \end{bmatrix}$$

—that is,

$$\begin{aligned} c_1 + 3c_2 &= 2 \\ -2c_1 - 5c_2 &= -6 \\ -c_1 + 4c_2 &= 3. \end{aligned}$$

The augmented coefficient matrix

$$\begin{bmatrix} 1 & 3 & 2 \\ -2 & -5 & -6 \\ -1 & 4 & 3 \end{bmatrix}$$

can be reduced by elementary row operations to echelon form:

$$\begin{bmatrix} 1 & 3 & 2 \\ 0 & 1 & -2 \\ 0 & 0 & 19 \end{bmatrix}.$$

We see now, from the third row, that our system is inconsistent, so the desired scalars c_1 and c_2 do not exist. Thus $\mathbf{w}$ is *not* a linear combination of $\mathbf{v}_1$ and $\mathbf{v}_2$. ■

Example 2 To express the vector $\mathbf{w} = (-7, 7, 11)$ as a linear combination of the vectors $\mathbf{v}_1 = (1, 2, 1)$, $\mathbf{v}_2 = (-4, -1, 2)$, and $\mathbf{v}_3 = (-3, 1, 3)$, we write the equation $c_1\mathbf{v}_1 + c_2\mathbf{v}_2 + c_3\mathbf{v}_3 = \mathbf{w}$ in the form

$$c_1\begin{bmatrix} 1 \\ 2 \\ 1 \end{bmatrix} + c_2\begin{bmatrix} -4 \\ -1 \\ 2 \end{bmatrix} + c_3\begin{bmatrix} -3 \\ 1 \\ 3 \end{bmatrix} = \begin{bmatrix} -7 \\ 7 \\ 11 \end{bmatrix}$$

—that is,

$$\begin{aligned} c_1 - 4c_2 - 3c_3 &= -7 \\ 2c_1 - c_2 + c_3 &= 7 \\ c_1 + 2c_2 + 3c_3 &= 11. \end{aligned}$$

The reduced echelon form of the augmented coefficient matrix of this system is

$$\begin{bmatrix} 1 & 0 & 1 & 5 \\ 0 & 1 & 1 & 3 \\ 0 & 0 & 0 & 0 \end{bmatrix}.$$

Thus c_3 is a free variable. With $c_3 = t$, back substitution yields $c_1 = 5 - t$ and $c_2 = 3 - t$. For instance, $t = 1$ gives $c_1 = 4$, $c_2 = 2$, and $c_3 = 1$, so

$$\mathbf{w} = 4\mathbf{v}_1 + 2\mathbf{v}_2 + \mathbf{v}_3.$$

But $t = -2$ yields $c_1 = 7$, $c_2 = 5$, and $c_3 = -2$, so $\mathbf{w}$ can also be expressed as

$$\mathbf{w} = 7\mathbf{v}_1 + 5\mathbf{v}_2 - 2\mathbf{v}_3.$$

We have found not only that $\mathbf{w}$ can be expressed as a linear combination of the vectors $\mathbf{v}_1, \mathbf{v}_2, \mathbf{v}_3$ but also that this can be done in many different ways (one for each choice of the parameter t). ■

We began this section with the observation that every solution vector of the linear system in (1) is a linear combination of the vectors $\mathbf{v}_1$ and $\mathbf{v}_2$ that appear in the right-hand side of Eq. (2). A brief way of saying this is that the vectors $\mathbf{v}_1$ and $\mathbf{v}_2$ *span* the solution space. More generally, suppose that $\mathbf{v}_1, \mathbf{v}_2, \ldots, \mathbf{v}_k$ are vectors in a vector space V. Then we say that the vectors $\mathbf{v}_1, \mathbf{v}_2, \ldots, \mathbf{v}_k$ **span** the vector space V provided that *every* vector in V is a linear combination of these k vectors. We may also say that the set $S = \{\mathbf{v}_1, \mathbf{v}_2, \ldots, \mathbf{v}_k\}$ of vectors is a **spanning set** for V.

Example 3 The familiar unit vectors $\mathbf{i} = (1, 0, 0)$, $\mathbf{j} = (0, 1, 0)$, and $\mathbf{k} = (0, 0, 1)$ span $\mathbf{R}^3$, because every vector $\mathbf{x} = (x_1, x_2, x_3)$ in $\mathbf{R}^3$ can be expressed as the linear combination

$$\mathbf{x} = x_1\mathbf{i} + x_2\mathbf{j} + x_3\mathbf{k}$$

of these three vectors **i**, **j**, and **k**. ■

If the vectors $\mathbf{v}_1, \mathbf{v}_2, \ldots, \mathbf{v}_k$ in the vector space V do not span V, we can ask about the subset of V consisting of all those vectors that are linear combinations of $\mathbf{v}_1, \mathbf{v}_2, \ldots, \mathbf{v}_k$. The following theorem implies that this subset is always a *subspace* of V.

THEOREM 1 The Span of a Set of Vectors

Let $\mathbf{v}_1, \mathbf{v}_2, \ldots, \mathbf{v}_k$ be vectors in the vector space V. Then the set W of all linear combinations of $\mathbf{v}_1, \mathbf{v}_2, \ldots, \mathbf{v}_k$ is a subspace of V.

***Proof*:** We must show that W is closed under addition of vectors and multiplication by scalars. If

$$\mathbf{u} = a_1\mathbf{v}_1 + a_2\mathbf{v}_2 + \cdots + a_k\mathbf{v}_k$$

and

$$\mathbf{v} = b_1\mathbf{v}_1 + b_2\mathbf{v}_2 + \cdots + b_k\mathbf{v}_k$$

are vectors in W, then

$$\mathbf{u} + \mathbf{v} = (a_1 + b_1)\mathbf{v}_1 + (a_2 + b_2)\mathbf{v}_2 + \cdots + (a_k + b_k)\mathbf{v}_k$$

and

$$c\mathbf{u} = (ca_1)\mathbf{v}_1 + (ca_2)\mathbf{v}_2 + \cdots + (ca_k)\mathbf{v}_k$$

for any scalar c. Thus $\mathbf{u} + \mathbf{v}$ and $c\mathbf{u}$ are linear combinations of $\mathbf{v}_1, \mathbf{v}_2, \ldots, \mathbf{v}_k$ and therefore are vectors in W. It is clear that W is nonempty, and hence W is a subspace of V. ■

We say that the subspace W of Theorem 1 is the space **spanned** by the vectors $\mathbf{v}_1, \mathbf{v}_2, \ldots, \mathbf{v}_k$ (or is the **span** of the set $S = \{\mathbf{v}_1, \mathbf{v}_2, \ldots, \mathbf{v}_k\}$ of vectors). We sometimes write

$$W = \text{span}(S) = \text{span}\{\mathbf{v}_1, \mathbf{v}_2, \ldots, \mathbf{v}_k\}.$$

Thus Example 3 implies that $\mathbf{R}^3 = \text{span}\{\mathbf{i}, \mathbf{j}, \mathbf{k}\}$. The question as to whether a given vector $\mathbf{w}$ in $\mathbf{R}^n$ lies in the subspace $\text{span}\{\mathbf{v}_1, \mathbf{v}_2, \ldots, \mathbf{v}_k\}$ reduces to solving a linear system, as illustrated by Examples 1 and 2.

It is easy to verify that the space $W = \text{span}\{\mathbf{v}_1, \mathbf{v}_2, \ldots, \mathbf{v}_k\}$ of Theorem 1 is the smallest subspace of V that contains all the vectors $\mathbf{v}_1, \mathbf{v}_2, \ldots, \mathbf{v}_k$—meaning that every other subspace of V that contains these k vectors must also contain W (Problem 30).

Linear Independence

Henceforth, when we solve a homogeneous system of linear equations, we generally will seek a set $\mathbf{v}_1, \mathbf{v}_2, \ldots, \mathbf{v}_k$ of solution vectors that span the solution space W of the system. Perhaps the most concrete way to describe a subspace W of a vector space V is to give explicitly a set $\mathbf{v}_1, \mathbf{v}_2, \ldots, \mathbf{v}_k$ of vectors that span W. And this type of representation is most useful and desirable (as well as most aesthetically pleasing) when each vector $\mathbf{w}$ in W is expressible in a *unique* way as a linear combination of $\mathbf{v}_1, \mathbf{v}_2, \ldots, \mathbf{v}_k$. (For instance, each vector in $\mathbf{R}^3$ is a unique linear combination of the vectors $\mathbf{i}$, $\mathbf{j}$, and $\mathbf{k}$ of Example 3.) But Example 2 demonstrates that a vector $\mathbf{w}$ may well be expressed in many different ways as a linear combination of given vectors $\mathbf{v}_1, \mathbf{v}_2, \ldots, \mathbf{v}_k$.

Thus not all spanning sets enjoy the uniqueness property that we desire. Two questions arise:

- Given a subspace W of a vector space V, does there necessarily exist a spanning set with the uniqueness property that we desire?
- If so, how do we find such a spanning set for W?

The following definition provides the key to both answers.

DEFINITION Linear Independence

The vectors $\mathbf{v}_1, \mathbf{v}_2, \ldots, \mathbf{v}_k$ in a vector space V are said to be **linearly independent** provided that the equation

$$c_1\mathbf{v}_1 + c_2\mathbf{v}_2 + \cdots + c_k\mathbf{v}_k = \mathbf{0} \tag{4}$$

has only the trivial solution $c_1 = c_2 = \cdots = c_k = 0$. That is, the only linear combination of $\mathbf{v}_1, \mathbf{v}_2, \ldots, \mathbf{v}_k$ that represents the zero vector $\mathbf{0}$ is the trivial combination $0\mathbf{v}_1 + 0\mathbf{v}_2 + \cdots + 0\mathbf{v}_k$.

Remark: We can immediately verify that any vector $\mathbf{w}$ in the subspace W spanned by the linearly independent vectors $\mathbf{v}_1, \mathbf{v}_2, \ldots, \mathbf{v}_k$ is *uniquely* expressible as a linear combination of these vectors. For

$$\mathbf{w} = \sum_{i=1}^{k} a_i \mathbf{v}_i = \sum_{i=1}^{k} b_i \mathbf{v}_i \quad \text{implies that} \quad \sum_{i=1}^{k} (a_i - b_i)\mathbf{v}_i = \mathbf{0}.$$

Hence, with $c_i = a_i - b_i$ for each $i = 1, \ldots, k$, the linear independence of $\mathbf{v}_1, \mathbf{v}_2, \ldots, \mathbf{v}_k$ implies that $c_1 = c_2 = \cdots = c_k = 0$. Thus

$$\sum_{i=1}^{k} a_i \mathbf{v}_i = \sum_{i=1}^{k} b_i \mathbf{v}_i \quad \text{implies that} \quad a_i = b_i \text{ for } i = 1, \ldots, k,$$

so we see that $\mathbf{w}$ can be expressed in only *one* way as a combination of the linearly independent vectors $\mathbf{v}_1, \mathbf{v}_2, \ldots, \mathbf{v}_k$. ■

Example 4 The **standard unit vectors**

$$\begin{aligned}
\mathbf{e}_1 &= (1, 0, 0, \ldots, 0), \\
\mathbf{e}_2 &= (0, 1, 0, \ldots, 0), \\
&\vdots \\
\mathbf{e}_n &= (0, 0, 0, \ldots, 1)
\end{aligned}$$

in $\mathbf{R}^n$ are linearly independent. The reason is that the equation

$$c_1\mathbf{e}_1 + c_2\mathbf{e}_2 + \cdots + c_n\mathbf{e}_n = \mathbf{0}$$

evidently reduces to

$$(c_1, c_2, \ldots, c_n) = (0, 0, \ldots, 0)$$

and thus has only the trivial solution $c_1 = c_2 = \cdots = c_n = 0$. ■

Example 5 To determine whether the vectors $\mathbf{v}_1 = (1, 2, 2, 1)$, $\mathbf{v}_2 = (2, 3, 4, 1)$, and $\mathbf{v}_3 = (3, 8, 7, 5)$ in $\mathbf{R}^4$ are linearly independent, we write the equation $c_1\mathbf{v}_1 + c_2\mathbf{v}_2 + c_3\mathbf{v}_3 = \mathbf{0}$ as the linear system

$$\begin{aligned}
c_1 + 2c_2 + 3c_3 &= 0 \\
2c_1 + 3c_2 + 8c_3 &= 0 \\
2c_1 + 4c_2 + 7c_3 &= 0 \\
c_1 + c_2 + 5c_3 &= 0
\end{aligned}$$

and then solve for c_1, c_2, and c_3. The augmented coefficient matrix of this system reduces to the echelon form

$$\begin{bmatrix} 1 & 2 & 3 & 0 \\ 0 & 1 & -2 & 0 \\ 0 & 0 & 1 & 0 \\ 0 & 0 & 0 & 0 \end{bmatrix},$$

so we see that the only solution is $c_1 = c_2 = c_3 = 0$. Thus the vectors $\mathbf{v}_1$, $\mathbf{v}_2$, and $\mathbf{v}_3$ are linearly independent. ■

Observe that linear independence of the vectors $\mathbf{v}_1, \mathbf{v}_2, \ldots, \mathbf{v}_k$ actually is a property of the *set* $S = \{\mathbf{v}_1, \mathbf{v}_2, \ldots, \mathbf{v}_k\}$ whose elements are these vectors. Occasionally the phraseology "the set $S = \{\mathbf{v}_1, \mathbf{v}_2, \ldots, \mathbf{v}_k\}$ is linearly independent" is more convenient. For instance, *any subset of a linearly independent set* $S = \{\mathbf{v}_1, \mathbf{v}_2, \ldots, \mathbf{v}_k\}$ *is a linearly independent set of vectors* (Problem 29).

Now we show that *the coefficients in a linear combination of the linearly independent vectors* $\mathbf{v}_1, \mathbf{v}_2, \ldots, \mathbf{v}_k$ *are unique*. If both

$$\mathbf{w} = a_1\mathbf{v}_1 + a_2\mathbf{v}_2 + \cdots + a_k\mathbf{v}_k \tag{5}$$

and

$$\mathbf{w} = b_1\mathbf{v}_1 + b_2\mathbf{v}_2 + \cdots + b_k\mathbf{v}_k, \tag{6}$$

then

$$a_1\mathbf{v}_1 + a_2\mathbf{v}_2 + \cdots + a_k\mathbf{v}_k = b_1\mathbf{v}_1 + b_2\mathbf{v}_2 + \cdots + b_k\mathbf{v}_k,$$

so it follows that

$$(a_1 - b_1)\mathbf{v}_1 + (a_2 - b_2)\mathbf{v}_2 + \cdots + (a_k - b_k)\mathbf{v}_k = \mathbf{0}. \tag{7}$$

Because the vectors $\mathbf{v}_1, \mathbf{v}_2, \ldots, \mathbf{v}_k$ are linearly independent, each of the coefficients in (7) must vanish. Therefore, $a_1 = b_1, a_2 = b_2, \ldots, a_k = b_k$, so we have shown that the linear combinations in (5) and (6) actually are identical. Hence, if a vector $\mathbf{w}$ is in the set span$\{\mathbf{v}_1, \mathbf{v}_2, \ldots, \mathbf{v}_k\}$, then it can be expressed in only one way as a linear combination of these linearly independent vectors.

A set of vectors is called **linearly dependent** provided it is not linearly independent. Hence the vectors $\mathbf{v}_1, \mathbf{v}_2, \ldots, \mathbf{v}_k$ are linearly dependent if and only if there exist scalars $c_1, c_2, \ldots, c_k$ *not all zero* such that

$$c_1\mathbf{v}_1 + c_2\mathbf{v}_2 + \cdots + c_k\mathbf{v}_k = \mathbf{0}. \tag{8}$$

In short, a (finite) set of vectors is linearly dependent provided that some *nontrivial* linear combination of them equals the zero vector.

Example 6 Let $\mathbf{v}_1 = (2, 1, 3)$, $\mathbf{v}_2 = (5, -2, 4)$, $\mathbf{v}_3 = (3, 8, -6)$, and $\mathbf{v}_4 = (2, 7, -4)$. Then the equation $c_1\mathbf{v}_1 + c_2\mathbf{v}_2 + c_3\mathbf{v}_3 + c_4\mathbf{v}_4 = \mathbf{0}$ is equivalent to the linear system

$$\begin{aligned} 2c_1 + 5c_2 + 3c_3 + 2c_4 &= 0 \\ c_1 - 2c_2 + 8c_3 + 7c_4 &= 0 \\ 3c_1 + 4c_2 - 6c_3 - 4c_4 &= 0 \end{aligned}$$

of three equations in four unknowns. Because this homogeneous system has more unknowns than equations, Theorem 3 in Section 3.3 implies that it has a nontrivial solution. Therefore we may conclude—without even solving explicitly for c_1, c_2, c_3, and c_4—that the vectors $\mathbf{v}_1$, $\mathbf{v}_2$, $\mathbf{v}_3$, and $\mathbf{v}_4$ are linearly dependent. (It happens that

$$2\mathbf{v}_1 - \mathbf{v}_2 + 3\mathbf{v}_3 - 4\mathbf{v}_4 = \mathbf{0},$$

as you can verify easily.) ■

The argument in Example 6 may be generalized in an obvious way to prove that *any set of more than n vectors in* $\mathbf{R}^n$ *is linearly dependent*. For if $k > n$, then Eq. (8) is equivalent to a homogeneous linear system with more unknowns (k) than equations (n), so Theorem 3 in Section 3.3 yields a nontrivial solution.

We now look at the way in which the elements of a linearly dependent set of vectors $\mathbf{v}_1, \mathbf{v}_2, \ldots, \mathbf{v}_k$ "depend" on one another. We know that there exist scalars $c_1, c_2, \ldots, c_k$ not all zero such that

$$c_1\mathbf{v}_1 + c_2\mathbf{v}_2 + \cdots + c_k\mathbf{v}_k = \mathbf{0}. \tag{9}$$

Suppose that the pth coefficient is nonzero: $c_p \neq 0$. Then we can solve Eq. (9) for $c_p\mathbf{v}_p$ and next divide by c_p to get

$$\mathbf{v}_p = a_1\mathbf{v}_1 + \cdots + a_{p-1}\mathbf{v}_{p-1} + a_{p+1}\mathbf{v}_{p+1} + \cdots + a_k\mathbf{v}_k, \tag{10}$$

where $a_i = -c_i/c_p$ for $i \neq p$. Thus at least one of the linearly dependent vectors is a linear combination of the other $k-1$. Conversely, suppose we are given a set of vectors $\mathbf{v}_1, \mathbf{v}_2, \ldots, \mathbf{v}_k$ with one of them dependent on the others as in Eq. (10). Then we can transpose all the terms to the left-hand side to get an equation of the form in (9) with $c_p = 1 \neq 0$. This shows that the vectors are linearly dependent. Therefore, we have proved that *the vectors* $\mathbf{v}_1, \mathbf{v}_2, \ldots, \mathbf{v}_k$ *are linearly dependent if and only if at least one of them is a linear combination of the others.*

For instance (as we saw in Section 4.1), two vectors are linearly dependent if and only if one of them is a scalar multiple of the other, in which case the two vectors are collinear. Three vectors are linearly dependent if and only if one of them is a linear combination of the other two, in which case the three vectors are coplanar.

In Theorem 4 of Section 4.1 we saw that the *determinant* provides a criterion for deciding whether three vectors in $\mathbf{R}^3$ are linearly independent: The vectors $\mathbf{v}_1$, $\mathbf{v}_2$, $\mathbf{v}_3$ in $\mathbf{R}^3$ are linearly independent if and only if the determinant of the 3×3 matrix

$$\mathbf{A} = \begin{bmatrix} \mathbf{v}_1 & \mathbf{v}_2 & \mathbf{v}_3 \end{bmatrix}$$

is nonzero. The proof given there in the three-dimensional case generalizes readily to the n-dimensional case. Given n vectors $\mathbf{v}_1, \mathbf{v}_2, \ldots, \mathbf{v}_n$ in $\mathbf{R}^n$, we consider the $n \times n$ matrix

$$\mathbf{A} = \begin{bmatrix} \mathbf{v}_1 & \mathbf{v}_2 & \cdots & \mathbf{v}_n \end{bmatrix}$$

having these vectors as its column vectors. Then, by Theorem 2 in Section 3.6, $\det \mathbf{A} \neq 0$ if and only if $\mathbf{A}$ is invertible, in which case the system $\mathbf{Ac} = \mathbf{0}$ has only the trivial solution $c_1 = c_2 = \cdots = c_n = 0$, so the vectors $\mathbf{v}_1, \mathbf{v}_2, \ldots, \mathbf{v}_n$ must be linearly independent.

THEOREM 2 Independence of n Vectors in $\mathbf{R}^n$

The n vectors $\mathbf{v}_1, \mathbf{v}_2, \ldots, \mathbf{v}_n$ in $\mathbf{R}^n$ are linearly independent if and only if the $n \times n$ matrix

$$\mathbf{A} = \begin{bmatrix} \mathbf{v}_1 & \mathbf{v}_2 & \cdots & \mathbf{v}_n \end{bmatrix}$$

having them as its column vectors has nonzero determinant.

We saw earlier that a set of *more* than n vectors in $\mathbf{R}^n$ is always linearly dependent. The following theorem shows us how the determinant provides a criterion in the case of *fewer* than n vectors in $\mathbf{R}^n$.

THEOREM 3 Independence of Fewer Than n Vectors in $\mathbf{R}^n$

Consider k vectors $\mathbf{v}_1, \mathbf{v}_2, \ldots, \mathbf{v}_k$ in $\mathbf{R}^n$, with $k < n$. Let

$$\mathbf{A} = \begin{bmatrix} \mathbf{v}_1 & \mathbf{v}_2 & \cdots & \mathbf{v}_k \end{bmatrix}$$

be the $n \times k$ matrix having them as its column vectors. Then the vectors $\mathbf{v}_1, \mathbf{v}_2, \ldots, \mathbf{v}_k$ are linearly independent if and only if some $k \times k$ submatrix of $\mathbf{A}$ has nonzero determinant.

Rather than including a complete proof, we will simply illustrate the "if" part of Theorem 3 in the case $n = 5$, $k = 3$. Let $\mathbf{v}_1 = (a_1, a_2, a_3, a_4, a_5)$, $\mathbf{v}_2 = (b_1, b_2, b_3, b_4, b_5)$, and $\mathbf{v}_3 = (c_1, c_2, c_3, c_4, c_5)$ be three vectors in $\mathbf{R}^5$ such that the 5×3 matrix

$$\mathbf{A} = \begin{bmatrix} a_1 & b_1 & c_1 \\ a_2 & b_2 & c_2 \\ a_3 & b_3 & c_3 \\ a_4 & b_4 & c_4 \\ a_5 & b_5 & c_5 \end{bmatrix}$$

has a 3×3 submatrix with nonzero determinant. Suppose, for instance, that

$$\begin{vmatrix} a_1 & b_1 & c_1 \\ a_3 & b_3 & c_3 \\ a_5 & b_5 & c_5 \end{vmatrix} \neq 0.$$

Then Theorem 2 implies that the three vectors $\mathbf{u}_1 = (a_1, a_3, a_5)$, $\mathbf{u}_2 = (b_1, b_3, b_5)$, and $\mathbf{u}_3 = (c_1, c_3, c_5)$ in $\mathbf{R}^3$ are linearly independent. Now suppose that $c_1\mathbf{v}_1 + c_2\mathbf{v}_2 + c_3\mathbf{v}_3 = \mathbf{0}$. Then by deleting the second and fourth components of each vector in this equation, we find that $c_1\mathbf{u}_1 + c_2\mathbf{u}_2 + c_3\mathbf{u}_3 = \mathbf{0}$. But the fact that $\mathbf{u}_1$, $\mathbf{u}_2$, $\mathbf{u}_3$ are linearly independent implies that $c_1 = c_2 = c_3 = 0$, and it now follows that $\mathbf{v}_1$, $\mathbf{v}_2$, $\mathbf{v}_3$ are linearly independent.

4.3 Problems

In Problems 1–8, determine whether the given vectors $\mathbf{v}_1, \mathbf{v}_2, \ldots, \mathbf{v}_k$ are linearly independent or linearly dependent. Do this essentially by inspection—that is, without solving a linear system of equations.

1. $\mathbf{v}_1 = (4, -2, 6, -4)$, $\mathbf{v}_2 = (6, -3, 9, -6)$
2. $\mathbf{v}_1 = (3, 9, -3, 6)$, $\mathbf{v}_2 = (2, 6, -2, 4)$
3. $\mathbf{v}_1 = (3, 4)$, $\mathbf{v}_2 = (6, -1)$, $\mathbf{v}_3 = (7, 5)$
4. $\mathbf{v}_1 = (4, -2, 2)$, $\mathbf{v}_2 = (5, 4, -3)$, $\mathbf{v}_3 = (4, 6, 5)$, $\mathbf{v}_4 = (-7, 9, 3)$
5. $\mathbf{v}_1 = (1, 0, 0)$, $\mathbf{v}_2 = (0, -2, 0)$, $\mathbf{v}_3 = (0, 0, 3)$
6. $\mathbf{v}_1 = (1, 0, 0)$, $\mathbf{v}_2 = (1, 1, 0)$, $\mathbf{v}_3 = (1, 1, 1)$
7. $\mathbf{v}_1 = (2, 1, 0, 0)$, $\mathbf{v}_2 = (3, 0, 1, 0)$, $\mathbf{v}_3 = (4, 0, 0, 1)$
8. $\mathbf{v}_1 = (1, 0, 3, 0)$, $\mathbf{v}_2 = (0, 2, 0, 4)$, $\mathbf{v}_3 = (1, 2, 3, 4)$

In Problems 9–16, express the indicated vector $\mathbf{w}$ as a linear combination of the given vectors $\mathbf{v}_1, \mathbf{v}_2, \ldots, \mathbf{v}_k$ if this is possible. If not, show that it is impossible.

9. $\mathbf{w} = (1, 0, -7)$; $\mathbf{v}_1 = (5, 3, 4)$, $\mathbf{v}_2 = (3, 2, 5)$
10. $\mathbf{w} = (3, -1, -2)$; $\mathbf{v}_1 = (-3, 1, -2)$, $\mathbf{v}_2 = (6, -2, 3)$
11. $\mathbf{w} = (1, 0, 0, -1)$; $\mathbf{v}_1 = (7, -6, 4, 5)$, $\mathbf{v}_2 = (3, -3, 2, 3)$
12. $\mathbf{w} = (4, -4, 3, 3)$; $\mathbf{v}_1 = (7, 3, -1, 9)$, $\mathbf{v}_2 = (-2, -2, 1, -3)$
13. $\mathbf{w} = (5, 2, -2)$; $\mathbf{v}_1 = (1, 5, -3)$, $\mathbf{v}_2 = (5, -3, 4)$
14. $\mathbf{w} = (2, -3, 2, -3)$; $\mathbf{v}_1 = (1, 0, 0, 3)$, $\mathbf{v}_2 = (0, 1, -2, 0)$, $\mathbf{v}_3 = (0, -1, 1, 1)$
15. $\mathbf{w} = (4, 5, 6)$; $\mathbf{v}_1 = (2, -1, 4)$, $\mathbf{v}_2 = (3, 0, 1)$, $\mathbf{v}_3 = (1, 2, -1)$
16. $\mathbf{w} = (7, 7, 9, 11)$; $\mathbf{v}_1 = (2, 0, 3, 1)$, $\mathbf{v}_2 = (4, 1, 3, 2)$, $\mathbf{v}_3 = (1, 3, -1, 3)$

In Problems 17–22, three vectors $\mathbf{v}_1$, $\mathbf{v}_2$, and $\mathbf{v}_3$ are given. If they are linearly independent, show this; otherwise find a nontrivial linear combination of them that is equal to the zero vector.

17. $\mathbf{v}_1 = (1, 0, 1)$, $\mathbf{v}_2 = (2, -3, 4)$, $\mathbf{v}_3 = (3, 5, 2)$
18. $\mathbf{v}_1 = (2, 0, -3)$, $\mathbf{v}_2 = (4, -5, -6)$, $\mathbf{v}_3 = (-2, 1, 3)$

19. $\mathbf{v}_1 = (2, 0, 3, 0)$, $\mathbf{v}_2 = (5, 4, -2, 1)$, $\mathbf{v}_3 = (2, -1, 1, -1)$
20. $\mathbf{v}_1 = (1, 1, -1, 1)$, $\mathbf{v}_2 = (2, 1, 1, 1)$, $\mathbf{v}_3 = (3, 1, 4, 1)$
21. $\mathbf{v}_1 = (3, 0, 1, 2)$, $\mathbf{v}_2 = (1, -1, 0, 1)$, $\mathbf{v}_3 = (1, 2, 1, 0)$
22. $\mathbf{v}_1 = (3, 9, 0, 5)$, $\mathbf{v}_2 = (3, 0, 9, -7)$, $\mathbf{v}_3 = (4, 7, 5, 0)$

In Problems 23–26, the vectors $\{\mathbf{v}_i\}$ are known to be linearly independent. Apply the definition of linear independence to show that the vectors $\{\mathbf{u}_i\}$ are also linearly independent.

23. $\mathbf{u}_1 = \mathbf{v}_1 + \mathbf{v}_2$, $\mathbf{u}_2 = \mathbf{v}_1 - \mathbf{v}_2$
24. $\mathbf{u}_1 = \mathbf{v}_1 + \mathbf{v}_2$, $\mathbf{u}_2 = 2\mathbf{v}_1 + 3\mathbf{v}_2$
25. $\mathbf{u}_1 = \mathbf{v}_1$, $\mathbf{u}_2 = \mathbf{v}_1 + 2\mathbf{v}_2$, $\mathbf{u}_3 = \mathbf{v}_1 + 2\mathbf{v}_2 + 3\mathbf{v}_3$
26. $\mathbf{u}_1 = \mathbf{v}_2 + \mathbf{v}_3$, $\mathbf{u}_2 = \mathbf{v}_1 + \mathbf{v}_3$, $\mathbf{u}_3 = \mathbf{v}_1 + \mathbf{v}_2$
27. *Prove*: If the (finite) set S of vectors contains the zero vector, then S is linearly dependent.
28. *Prove*: If the set S of vectors is linearly dependent and the (finite) set T contains S, then T is also linearly dependent. You may assume that $S = \{\mathbf{v}_1, \mathbf{v}_2, \ldots, \mathbf{v}_k\}$ and that $T = \{\mathbf{v}_1, \mathbf{v}_2, \ldots, \mathbf{v}_m\}$ with $m > k$.
29. Show that if the (finite) set S of vectors is linearly independent, then any subset T of S is also linearly independent.
30. Suppose that the subspace U of the vector space V contains the vectors $\mathbf{v}_1, \mathbf{v}_2, \ldots, \mathbf{v}_k$. Show that U contains the subspace spanned by these vectors.
31. Let S and T be sets of vectors in a vector space such that S is a subset of span(T). Show that span(S) is also a subset of span(T).
32. Let $\mathbf{v}_1, \mathbf{v}_2, \ldots, \mathbf{v}_k$ be linearly independent vectors in the set S of vectors. *Prove*: If no set of more than k vectors in S is linearly independent, then every vector in S is a linear combination of $\mathbf{v}_1, \mathbf{v}_2, \ldots, \mathbf{v}_k$.

In Problems 33–35, let $\mathbf{v}_1, \mathbf{v}_2, \ldots, \mathbf{v}_k$ be vectors in $\mathbf{R}^n$ and let

$$\mathbf{A} = \begin{bmatrix} \mathbf{v}_1 & \mathbf{v}_2 & \cdots & \mathbf{v}_k \end{bmatrix}$$

be the $n \times k$ matrix with these vectors as its column vectors.

33. *Prove*: If some $k \times k$ submatrix of $\mathbf{A}$ is the $k \times k$ identity matrix, then $\mathbf{v}_1, \mathbf{v}_2, \ldots, \mathbf{v}_k$ are linearly independent.
34. Suppose that $k = n$, that the vectors $\mathbf{v}_1, \mathbf{v}_2, \ldots, \mathbf{v}_k$ are linearly independent, and that $\mathbf{B}$ is a nonsingular $n \times n$ matrix. Prove that the column vectors of the matrix $\mathbf{AB}$ are linearly independent.
35. Suppose that $k < n$, that the vectors $\mathbf{v}_1, \mathbf{v}_2, \ldots, \mathbf{v}_k$ are linearly independent, and that $\mathbf{B}$ is a nonsingular $k \times k$ matrix. Use Theorem 3 to show that the column vectors of $\mathbf{AB}$ are linearly independent.

4.4 Bases and Dimension for Vector Spaces

An especially useful way of describing the solution space of a homogeneous linear system is to list explicitly a set S of solution vectors such that *every* solution vector is a unique linear combination of these particular ones. The following definition specifies the properties of such a set S of "basic" solution vectors, and the concept is equally important for vector spaces other than solution spaces.

DEFINITION Basis

A finite set S of vectors in a vector space V is called a **basis** for V provided that
(a) the vectors in S are linearly independent, and
(b) the vectors in S span V.

In short, a basis for the vector space V is a linearly independent spanning set of vectors in V. Thus, if $S = \{\mathbf{v}_1, \mathbf{v}_2, \ldots, \mathbf{v}_n\}$ is a basis for V, then any vector $\mathbf{w}$ in V can be expressed as a linear combination

$$\mathbf{w} = c_1\mathbf{v}_1 + c_2\mathbf{v}_2 + \cdots + c_n\mathbf{v}_n \tag{1}$$

of the vectors in S, and we saw in Section 4.3 that the linear independence of S implies that the coefficients $c_1, c_2, \ldots, c_n$ in (1) are *unique*. That is, $\mathbf{w}$ cannot be expressed differently as a linear combination of the basis vectors $\mathbf{v}_1, \mathbf{v}_2, \ldots, \mathbf{v}_n$.

Example 1 The **standard basis** for $\mathbf{R}^n$ consists of the unit vectors

$$\mathbf{e}_1 = (1, 0, 0, \ldots, 0),\ \ \mathbf{e}_2 = (0, 1, 0, \ldots, 0),\ \ \ldots,\ \ \mathbf{e}_n = (0, 0, 0, \ldots, 1).$$

If $\mathbf{x} = (x_1, x_2, \ldots, x_n)$ is a vector in $\mathbf{R}^n$, then

$$\mathbf{x} = x_1\mathbf{e}_1 + x_2\mathbf{e}_2 + \cdots + x_n\mathbf{e}_n.$$

Thus the column vectors

$$\mathbf{e}_1 = \begin{bmatrix} 1 \\ 0 \\ \vdots \\ 0 \end{bmatrix}, \quad \mathbf{e}_2 = \begin{bmatrix} 0 \\ 1 \\ \vdots \\ 0 \end{bmatrix}, \quad \ldots, \quad \mathbf{e}_n = \begin{bmatrix} 0 \\ 0 \\ \vdots \\ 1 \end{bmatrix}$$

of the $n \times n$ identity matrix span $\mathbf{R}^n$, and we noted in Example 4 of Section 4.3 that these *standard unit vectors* are linearly independent. ■

Example 2 Let $\mathbf{v}_1, \mathbf{v}_2, \ldots, \mathbf{v}_n$ be n linearly independent vectors in $\mathbf{R}^n$. We saw in Section 4.3 that any set of more than n vectors in $\mathbf{R}^n$ is linearly dependent. Hence, given a vector $\mathbf{w}$ in $\mathbf{R}^n$, there exist scalars $c, c_1, c_2, \ldots, c_n$ not all zero such that

$$c\mathbf{w} + c_1\mathbf{v}_1 + c_2\mathbf{v}_2 + \cdots + c_n\mathbf{v}_n = \mathbf{0}. \tag{2}$$

If c were zero, then (2) would imply that the vectors $\mathbf{v}_1, \mathbf{v}_2, \ldots, \mathbf{v}_n$ are linearly dependent. Hence $c \neq 0$, so Eq. (2) can be solved for $\mathbf{w}$ as a linear combination of $\mathbf{v}_1, \mathbf{v}_2, \ldots, \mathbf{v}_n$. Thus the linearly independent vectors $\mathbf{v}_1, \mathbf{v}_2, \ldots, \mathbf{v}_n$ also *span* $\mathbf{R}^n$ and therefore constitute a basis for $\mathbf{R}^n$. ■

Example 2 shows that *any set of n linearly independent vectors in $\mathbf{R}^n$ is a basis for $\mathbf{R}^n$*. By Theorem 2 in Section 4.3, we can therefore determine whether n given vectors $\mathbf{v}_1, \mathbf{v}_2, \ldots, \mathbf{v}_n$ form a basis for $\mathbf{R}^n$ by calculating the determinant of the $n \times n$ matrix

$$\mathbf{A} = \begin{bmatrix} \mathbf{v}_1 & \mathbf{v}_2 & \cdots & \mathbf{v}_n \end{bmatrix}$$

with these vectors as its column vectors. They constitute a basis for $\mathbf{R}^n$ if and only if $\det \mathbf{A} \neq 0$.

Example 3 Let $\mathbf{v}_1 = (1, -1, -2, -3)$, $\mathbf{v}_2 = (1, -1, 2, 3)$, $\mathbf{v}_3 = (1, -1, -3, -2)$, and $\mathbf{v}_4 = (0, 3, -1, 2)$. Then we find that

$$\begin{vmatrix} 1 & 1 & 1 & 0 \\ -1 & -1 & -1 & 3 \\ -2 & 2 & -3 & -1 \\ -3 & 3 & -2 & 2 \end{vmatrix} = 30 \neq 0,$$

so it follows that $\{\mathbf{v}_1, \mathbf{v}_2, \mathbf{v}_3, \mathbf{v}_4\}$ is a basis for $\mathbf{R}^4$. ■

Theorem 1 has the following import: Just as in $\mathbf{R}^n$, a basis for *any* vector space V contains the largest possible number of linearly independent vectors in V.

THEOREM 1 Bases as Maximal Linearly Independent Sets

Let $S = \{\mathbf{v}_1, \mathbf{v}_2, \ldots, \mathbf{v}_n\}$ be a basis for the vector space V. Then any set of more than n vectors in V is linearly dependent.

***Proof*:** We need to show that if $m > n$, then any set $T = \{\mathbf{w}_1, \mathbf{w}_2, \ldots, \mathbf{w}_m\}$ of m vectors in V is linearly dependent. Because the basis S spans V, each vector $\mathbf{w}_j$ in T can be expressed as a linear combination of $\mathbf{v}_1, \mathbf{v}_2, \ldots, \mathbf{v}_n$:

$$\begin{aligned}
\mathbf{w}_1 &= a_{11}\mathbf{v}_1 + a_{21}\mathbf{v}_2 + \cdots + a_{n1}\mathbf{v}_n \\
\mathbf{w}_2 &= a_{12}\mathbf{v}_1 + a_{22}\mathbf{v}_2 + \cdots + a_{n2}\mathbf{v}_n \\
&\vdots \\
\mathbf{w}_m &= a_{1m}\mathbf{v}_1 + a_{2m}\mathbf{v}_2 + \cdots + a_{nm}\mathbf{v}_n.
\end{aligned} \tag{3}$$

Now we need to find scalars $c_1, c_2, \ldots, c_m$ not all zero such that

$$c_1\mathbf{w}_1 + c_2\mathbf{w}_2 + \cdots + c_m\mathbf{w}_m = \mathbf{0}. \tag{4}$$

Substituting in (4) the expressions in (3) for each $\mathbf{w}_j$, we get

$$\begin{aligned}
&c_1(a_{11}\mathbf{v}_1 + a_{21}\mathbf{v}_2 + \cdots + a_{n1}\mathbf{v}_n) \\
&+ c_2(a_{12}\mathbf{v}_1 + a_{22}\mathbf{v}_2 + \cdots + a_{n2}\mathbf{v}_n) + \cdots \\
&+ c_m(a_{1m}\mathbf{v}_1 + a_{2m}\mathbf{v}_2 + \cdots + a_{nm}\mathbf{v}_n) = \mathbf{0}.
\end{aligned} \tag{5}$$

Because the vectors $\mathbf{v}_1, \mathbf{v}_2, \ldots, \mathbf{v}_n$ are linearly independent, the coefficient in (5) of each $\mathbf{v}_i$ must vanish:

$$\begin{aligned}
a_{11}c_1 + a_{12}c_2 + \cdots + a_{1m}c_m &= 0 \\
a_{21}c_1 + a_{22}c_2 + \cdots + a_{2m}c_m &= 0 \\
&\vdots \\
a_{n1}c_1 + a_{n2}c_2 + \cdots + a_{nm}c_m &= 0.
\end{aligned} \tag{6}$$

But because $m > n$, this is a homogeneous linear system with more unknowns than equations, and therefore has a nontrivial solution $(c_1, c_2, \ldots, c_m)$. Then $c_1, c_2, \ldots, c_m$ are scalars not all zero such that (4) holds, so we have shown that the set T of more than n vectors is linearly dependent. ■

Now let $S = \{\mathbf{v}_1, \mathbf{v}_2, \ldots, \mathbf{v}_n\}$ and $T = \{\mathbf{w}_1, \mathbf{w}_2, \ldots, \mathbf{w}_m\}$ be two different bases for the same vector space V. Because S is a basis and T is linearly independent, Theorem 1 implies that $m \leq n$. Next, reverse the roles of S and T; the fact that T is a basis and S is linearly independent implies similarly that $n \leq m$. Hence $m = n$, so we have proved the following theorem for any vector space with a finite basis.

THEOREM 2 The Dimension of a Vector Space

Any two bases for a vector space consist of the same number of vectors.

A nonzero vector space V is called **finite dimensional** provided that there exists a basis for V consisting of a finite number of vectors from V. In this case the number n of vectors in each basis for V is called the **dimension** of V, denoted by $n = \dim V$. Then V is an *n-dimensional* vector space. The standard basis of Example 1 shows that $\mathbf{R}^n$ is, indeed, an n-dimensional vector space.

Note that the zero vector space $\{\mathbf{0}\}$ has no basis because it contains *no* linearly independent set of vectors. (Sometimes it is convenient to adopt the convention that the null set is a basis for $\{\mathbf{0}\}$.) Here we define $\dim\{\mathbf{0}\}$ to be zero. A nonzero vector space that has no finite basis is called **infinite dimensional**. Infinite dimensional vector spaces are discussed in Section 4.5, but we include an illustrative example of one here.

Example 4 Let $\mathcal{P}$ be the set of all polynomials of the form

$$p(x) = a_0 + a_1 x + a_2 x^2 + \cdots + a_n x^n,$$

where the largest exponent $n \geq 0$ that appears is the *degree* of the polynomial $p(x)$, and the coefficients $a_0, a_1, a_2, \ldots, a_n$ are real numbers. We add polynomials in $\mathcal{P}$ and multiply them by scalars in the usual way—that is, by collecting coefficients of like powers of x. For instance, if

$$p(x) = 3 + 2x + 5x^3 \quad \text{and} \quad q(x) = 7 + 4x + 3x^2 + 9x^4,$$

then

$$\begin{aligned}(p+q)(x) &= (3+7) + (2+4)x + (0+3)x^2 + (5+0)x^3 + (0+9)x^4 \\ &= 10 + 6x + 3x^2 + 5x^3 + 9x^4\end{aligned}$$

and

$$(7p)(x) = 7(3 + 2x + 5x^3) = 21 + 14x + 35x^3.$$

It is readily verified that, with these operations, $\mathcal{P}$ is a vector space. But $\mathcal{P}$ has no finite basis. For if $p_1, p_2, \ldots, p_n$ are elements of $\mathcal{P}$, then the degree of any linear combination of them is at most the maximum of *their* degrees. Hence no polynomial in $\mathcal{P}$ of higher degree lies in $\operatorname{span}\{p_1, p_2, \ldots, p_n\}$. Thus no finite subset of $\mathcal{P}$ spans $\mathcal{P}$, and therefore $\mathcal{P}$ is an infinite dimensional vector space. ■

Here, our concern is with finite-dimensional vector spaces, and we note first that any proper subspace W of a finite-dimensional vector space V is itself finite dimensional, with $\dim W < \dim V$. For if $\dim V = n$, let $k \leq n$ be the largest integer such that W contains k linearly independent vectors $\mathbf{v}_1, \mathbf{v}_2, \ldots, \mathbf{v}_k$. Then, by the same argument as in Example 2, we see that $\{\mathbf{v}_1, \mathbf{v}_2, \ldots, \mathbf{v}_k\}$ is a basis for W, so W is finite dimensional with $\dim W = k$, and $k < n$ because W is a proper subspace.

Moreover, an n-dimensional vector space V contains proper subspaces of each dimension $k = 1, 2, \ldots, n-1$. For instance, if $\{\mathbf{v}_1, \mathbf{v}_2, \ldots, \mathbf{v}_n\}$ is a basis for V and $k < n$, then $W = \operatorname{span}\{\mathbf{v}_1, \mathbf{v}_2, \ldots, \mathbf{v}_k\}$ is a subspace of dimension k. Thus $\mathbf{R}^4$ contains proper subspaces of dimensions 1, 2, and 3; $\mathbf{R}^5$ contains proper subspaces of dimensions 1, 2, 3, and 4; and so on. The proper subspaces of $\mathbf{R}^n$ of dimensions $1, 2, \ldots, n-1$ are the higher-dimensional analogues of lines and planes through the origin in $\mathbf{R}^3$.

Suppose that V is an n-dimensional vector space and that $S = \{\mathbf{v}_1, \mathbf{v}_2, \ldots, \mathbf{v}_n\}$ is a set of n vectors in V. Then, in order to show that S is a basis for V, it is not necessary to prove *both* that S is linearly independent *and* that S spans V, because it turns out (for n vectors in an n-dimensional vector space) that either property of S implies the other. For instance, if $S = \{\mathbf{v}_1, \mathbf{v}_2, \ldots, \mathbf{v}_n\}$ is a set of n linearly independent vectors in V, then Theorem 1 and the argument of Example 2 (once again) imply that S spans V and, hence, is a basis for V. This proves part (a) of Theorem 3. The remaining parts are left to the problems. (See Problems 27–32.)

THEOREM 3 Independent Sets, Spanning Sets, and Bases

Let V be an n-dimensional vector space and let S be a subset of V. Then
(a) If S is linearly independent and consists of n vectors, then S is a basis for V;
(b) If S spans V and consists of n vectors, then S is a basis for V;
(c) If S is linearly independent, then S is contained in a basis for V;
(d) If S spans V, then S contains a basis for V.

Part (c) of Theorem 3 is often applied in the following form: If W is a k-dimensional subspace of the n-dimensional vector space V, then any basis $\{\mathbf{v}_1, \mathbf{v}_2, \ldots, \mathbf{v}_k\}$ for W can be "extended" to a basis $\{\mathbf{v}_1, \mathbf{v}_2, \ldots, \mathbf{v}_n\}$ for V, consisting of the original basis vectors for W together with $n-k$ additional vectors $\mathbf{v}_{k+1}, \mathbf{v}_{k+2}, \ldots, \mathbf{v}_n$.

Bases for Solution Spaces

We consider now the homogeneous linear system

$$\mathbf{Ax} = \mathbf{0}, \tag{7}$$

in which $\mathbf{A}$ is an $m \times n$ matrix, so the system consists of m equations in the n variables $x_1, x_2, \ldots, x_n$. Its solution space W is then a subspace of $\mathbf{R}^n$. We want to determine the dimension of W and, moreover, to find an explicit basis for W. Thus we seek a maximal set of linearly independent solution vectors of (7).

Recall the Gaussian elimination method of Section 3.2. We use elementary row operations to reduce the coefficient matrix $\mathbf{A}$ to an echelon matrix $\mathbf{E}$ and note the (nonzero) *leading entries* in the rows of $\mathbf{E}$. The *leading variables* are those that correspond to the columns of $\mathbf{E}$ containing the leading entries. The remaining variables (if any) are the *free variables*. If there are no free variables, then our system has only the trivial solution, so $W = \{\mathbf{0}\}$. If there are free variables, we set each of them (separately) equal to a parameter and solve (by back substitution) for the leading variables as linear combinations of these parameters. The solution space W is then the set of all solution vectors obtained in this manner (for all possible values of the parameters).

To illustrate the general situation, let us suppose that the leading variables are the first r variables $x_1, x_2, \ldots, x_r$, so the $k = n - r$ variables $x_{r+1}, x_{r+2}, \ldots, x_n$ are free variables. The reduced system $\mathbf{Ex} = \mathbf{0}$ then takes the form

$$\begin{aligned}
b_{11}x_1 + b_{12}x_2 + \cdots + b_{1r}x_r + \cdots + b_{1n}x_n &= 0 \\
b_{22}x_2 + \cdots + b_{2r}x_r + \cdots + b_{2n}x_n &= 0 \\
&\vdots \\
b_{rr}x_r + \cdots + b_{rn}x_n &= 0 \\
0 &= 0 \\
&\vdots \\
0 &= 0
\end{aligned} \tag{8}$$

with the last $m - r$ equations being "trivial." We set

$$x_{r+1} = t_1, \quad x_{r+2} = t_2, \ldots, \quad x_n = t_k \tag{9}$$

and then solve (by back substitution) the equations in (8) for the leading variables

$$\begin{aligned} x_1 &= c_{11}t_1 + c_{12}t_2 + \cdots + c_{1k}t_k \\ x_2 &= c_{21}t_1 + c_{22}t_2 + \cdots + c_{2k}t_k \\ &\vdots \\ x_r &= c_{r1}t_1 + c_{r2}t_2 + \cdots + c_{rk}t_k. \end{aligned} \tag{10}$$

The typical solution vector $(x_1, x_2, \ldots, x_n)$ is given in terms of the k parameters $t_1, t_2, \ldots, t_k$ by the equations in (9) and (10).

We now choose k particular solution vectors $\mathbf{v}_1, \mathbf{v}_2, \ldots, \mathbf{v}_k$ as follows: To get $\mathbf{v}_j$ we set the jth parameter t_j equal to 1 and set all other parameters equal to zero. Then

$$\mathbf{v}_j = (c_{1j}, c_{2j}, \ldots, c_{rj}, 0, \ldots, 1, \ldots, 0), \tag{11}$$

with the 1 appearing as the $(r+j)$th entry. The vectors $\mathbf{v}_1, \mathbf{v}_2, \ldots, \mathbf{v}_k$ are the column vectors of the $n \times k$ matrix

$$\begin{bmatrix} c_{11} & c_{12} & \cdots & c_{1k} \\ c_{21} & c_{22} & \cdots & c_{2k} \\ \vdots & \vdots & \ddots & \vdots \\ c_{r1} & c_{r2} & \cdots & c_{rk} \\ 1 & 0 & \cdots & 0 \\ 0 & 1 & \cdots & 0 \\ \vdots & \vdots & \ddots & \vdots \\ 0 & 0 & \cdots & 1 \end{bmatrix}. \tag{12}$$

Because of the presence of the lower $k \times k$ identity matrix, it is clear that the vectors $\mathbf{v}_1, \mathbf{v}_2, \ldots, \mathbf{v}_k$ are linearly independent. (See Problem 36.) But Eqs. (9) and (10) show that the typical solution vector $\mathbf{x}$ is a linear combination

$$\mathbf{x} = t_1\mathbf{v}_1 + t_2\mathbf{v}_2 + \cdots + t_k\mathbf{v}_k. \tag{13}$$

Therefore, the vectors $\mathbf{v}_1, \mathbf{v}_2, \ldots, \mathbf{v}_k$ defined in (11) form a basis for the solution space W of the original system in (7).

The following algorithm summarizes the steps in this procedure.

ALGORITHM A Basis for the Solution Space

To find a basis for the solution space W of the homogeneous linear system $\mathbf{Ax} = \mathbf{0}$, carry out the following steps.

1. Reduce the coefficient matrix $\mathbf{A}$ to echelon form.
2. Identify the r leading variables and the $k = n - r$ free variables. If $k = 0$, then $W = \{\mathbf{0}\}$.
3. Set the free variables equal to parameters $t_1, t_2, \ldots, t_k$, and then solve by back substitution for the leading variables in terms of these parameters.
4. Let $\mathbf{v}_j$ be the solution vector obtained by setting t_j equal to 1 and the other parameters equal to zero. Then $\{\mathbf{v}_1, \mathbf{v}_2, \ldots, \mathbf{v}_k\}$ is a basis for W.

Example 5 Find a basis for the solution space of the homogeneous linear system

$$\begin{aligned} 3x_1 + 6x_2 - x_3 - 5x_4 + 5x_5 &= 0 \\ 2x_1 + 4x_2 - x_3 - 3x_4 + 2x_5 &= 0 \\ 3x_1 + 6x_2 - 2x_3 - 4x_4 + x_5 &= 0. \end{aligned} \tag{14}$$

Solution We readily reduce the coefficient matrix **A** to the echelon form

$$\mathbf{E} = \begin{bmatrix} 1 & 2 & 0 & -2 & 3 \\ 0 & 0 & 1 & -1 & 4 \\ 0 & 0 & 0 & 0 & 0 \end{bmatrix}.$$

The leading entries are in the first and third columns, so the leading variables are x_1 and x_3; the free variables are x_2, x_4, and x_5. To avoid subscripts, we use r, s, and t rather than t_1, t_2, and t_3 to denote the three parameters. Thus we set

$$x_2 = r, \quad x_4 = s, \quad \text{and} \quad x_5 = t. \tag{15}$$

Then back substitution in the reduced system

$$\begin{aligned} x_1 + 2x_2 \qquad - 2x_4 + 3x_5 &= 0 \\ x_3 - x_4 + 4x_5 &= 0 \end{aligned}$$

yields

$$x_1 = -2r + 2s - 3t \quad \text{and} \quad x_3 = s - 4t. \tag{16}$$

The equations in (15) and (16) give the typical solution vector $(x_1, x_2, x_3, x_4, x_5)$ in terms of the parameters r, s, and t.

With $r = 1$ and $s = t = 0$,

$$\text{we obtain} \quad \mathbf{v}_1 = (-2, 1, 0, 0, 0).$$

With $s = 1$ and $r = t = 0$,

$$\text{we obtain} \quad \mathbf{v}_2 = (2, 0, 1, 1, 0).$$

With $t = 1$ and $r = s = 0$,

$$\text{we obtain} \quad \mathbf{v}_3 = (-3, 0, -4, 0, 1).$$

Thus the solution space of the system in (14) is a 3-dimensional subspace of $\mathbf{R}^5$ with basis $\{\mathbf{v}_1, \mathbf{v}_2, \mathbf{v}_3\}$. ■

4.4 Problems

In Problems 1–8, determine whether or not the given vectors in $\mathbf{R}^n$ form a basis for $\mathbf{R}^n$.

1. $\mathbf{v}_1 = (4, 7)$, $\mathbf{v}_2 = (5, 6)$

2. $\mathbf{v}_1 = (3, -1, 2)$, $\mathbf{v}_2 = (6, -2, 4)$, $\mathbf{v}_3 = (5, 3, -1)$

3. $\mathbf{v}_1 = (1, 7, -3)$, $\mathbf{v}_2 = (2, 1, 4)$, $\mathbf{v}_3 = (6, 5, 1)$, $\mathbf{v}_4 = (0, 7, 13)$

4. $\mathbf{v}_1 = (3, -7, 5, 2)$, $\mathbf{v}_2 = (1, -1, 3, 4)$, $\mathbf{v}_3 = (7, 11, 3, 13)$

5. $\mathbf{v}_1 = (0, 7, -3)$, $\mathbf{v}_2 = (0, 5, 4)$, $\mathbf{v}_3 = (0, 5, 10)$

6. $\mathbf{v}_1 = (0, 0, 1)$, $\mathbf{v}_2 = (0, 1, 2)$, $\mathbf{v}_3 = (1, 2, 3)$

7. $\mathbf{v}_1 = (0, 0, 1)$, $\mathbf{v}_2 = (7, 4, 11)$, $\mathbf{v}_3 = (5, 3, 13)$

8. $\mathbf{v}_1 = (2, 0, 0, 0)$, $\mathbf{v}_2 = (0, 3, 0, 0)$, $\mathbf{v}_3 = (0, 0, 7, 6)$, $\mathbf{v}_4 = (0, 0, 4, 5)$

In Problems 9–11, find a basis for the indicated subspace of $\mathbf{R}^3$.

9. The plane with equation $x - 2y + 5z = 0$.

10. The plane with equation $y = z$.

11. The line of intersection of the planes described in Problems 9 and 10.

In Problems 12–14, find a basis for the indicated subspace of $\mathbf{R}^4$.

12. The set of all vectors of the form (a, b, c, d) for which $a = b + c + d$.

13. The set of all vectors of the form (a, b, c, d) such that $a = 3c$ and $b = 4d$.

14. The set of all vectors of the form (a, b, c, d) for which $a + 2b = c + 3d = 0$.

In Problems 15–26, find a basis for the solution space of the given homogeneous linear system.

15. $$\begin{aligned} x_1 - 2x_2 + 3x_3 &= 0 \\ 2x_1 - 3x_2 - x_3 &= 0 \end{aligned}$$

16. $$\begin{aligned} x_1 + 3x_2 + 4x_3 &= 0 \\ 3x_1 + 8x_2 + 7x_3 &= 0 \end{aligned}$$

17. $$\begin{aligned} x_1 - 3x_2 + 2x_3 - 4x_4 &= 0 \\ 2x_1 - 5x_2 + 7x_3 - 3x_4 &= 0 \end{aligned}$$

18. $$\begin{aligned} x_1 + 3x_2 + 4x_3 + 5x_4 &= 0 \\ 2x_1 + 6x_2 + 9x_3 + 5x_4 &= 0 \end{aligned}$$

19. $$\begin{aligned} x_1 - 3x_2 - 9x_3 - 5x_4 &= 0 \\ 2x_1 + x_2 - 4x_3 + 11x_4 &= 0 \\ x_1 + 3x_2 + 3x_3 + 13x_4 &= 0 \end{aligned}$$

20. $$\begin{aligned} x_1 - 3x_2 - 10x_3 + 5x_4 &= 0 \\ x_1 + 4x_2 + 11x_3 - 2x_4 &= 0 \\ x_1 + 3x_2 + 8x_3 - x_4 &= 0 \end{aligned}$$

21. $$\begin{aligned} x_1 - 4x_2 - 3x_3 - 7x_4 &= 0 \\ 2x_1 - x_2 + x_3 + 7x_4 &= 0 \\ x_1 + 2x_2 + 3x_3 + 11x_4 &= 0 \end{aligned}$$

22. $$\begin{aligned} x_1 - 2x_2 - 3x_3 - 16x_4 &= 0 \\ 2x_1 - 4x_2 + x_3 + 17x_4 &= 0 \\ x_1 - 2x_2 + 3x_3 + 26x_4 &= 0 \end{aligned}$$

23. $$\begin{aligned} x_1 + 5x_2 + 13x_3 + 14x_4 &= 0 \\ 2x_1 + 5x_2 + 11x_3 + 12x_4 &= 0 \\ 2x_1 + 7x_2 + 17x_3 + 19x_4 &= 0 \end{aligned}$$

24. $$\begin{aligned} x_1 + 3x_2 - 4x_3 - 8x_4 + 6x_5 &= 0 \\ x_1 + 2x_3 + x_4 + 3x_5 &= 0 \\ 2x_1 + 7x_2 - 10x_3 - 19x_4 + 13x_5 &= 0 \end{aligned}$$

25. $$\begin{aligned} x_1 + 2x_2 + 7x_3 - 9x_4 + 31x_5 &= 0 \\ 2x_1 + 4x_2 + 7x_3 - 11x_4 + 34x_5 &= 0 \\ 3x_1 + 6x_2 + 5x_3 - 11x_4 + 29x_5 &= 0 \end{aligned}$$

26. $$\begin{aligned} 3x_1 + x_2 - 3x_3 + 11x_4 + 10x_5 &= 0 \\ 5x_1 + 8x_2 + 2x_3 - 2x_4 + 7x_5 &= 0 \\ 2x_1 + 5x_2 - x_4 + 14x_5 &= 0 \end{aligned}$$

27. Suppose that S is a set of n linearly independent vectors in the n-dimensional vector space V. Prove that S is a basis for V.

28. Suppose that S is a set of n vectors that span the n-dimensional vector space V. Prove that S is a basis for V.

29. Let $\{\mathbf{v}_1, \mathbf{v}_2, \ldots, \mathbf{v}_k\}$ be a basis for the proper subspace W of the vector space V, and suppose that the vector $\mathbf{v}$ of V is not in W. Show that the vectors $\mathbf{v}_1, \mathbf{v}_2, \ldots, \mathbf{v}_k, \mathbf{v}$ are linearly independent.

30. Use the result of Problem 29 to prove that every linearly independent set of vectors in a finite-dimensional vector space V is contained in a basis for V.

31. Suppose that the vectors $\mathbf{v}_1, \mathbf{v}_2, \ldots, \mathbf{v}_k, \mathbf{v}_{k+1}$ span the vector space V and that $\mathbf{v}_{k+1}$ is a linear combination of $\mathbf{v}_1, \mathbf{v}_2, \ldots, \mathbf{v}_k$. Show that the vectors $\mathbf{v}_1, \mathbf{v}_2, \ldots, \mathbf{v}_k$ span V.

32. Use the result of Problem 31 to prove that every spanning set for a finite-dimensional vector space V contains a basis for V.

33. Let S be a linearly independent set of vectors in the finite-dimensional vector space V. Then S is called a **maximal linearly independent set** provided that if any other vector is adjoined to S, then the resulting set is linearly dependent. Prove that every maximal linearly independent set in V is a basis for V.

34. Let S be a finite set of vectors that span the vector space V. Then S is called a **minimal spanning set** provided that no proper subset of S spans V. Prove that every minimal spanning set in V is a basis for V.

35. Let S be a finite set of vectors that span the vector space V. Then S is called a **uniquely spanning set** provided that each vector in V can be expressed in one and only one way as a linear combination of the vectors in S. Prove that every uniquely spanning set in V is a basis for V.

36. Apply the definition of linear independence to show directly that the column vectors of the matrix in (12) are linearly independent.

4.5 Row and Column Spaces

In numerous examples we have observed the phenomenon of "disappearing equations" that sometimes occurs when we solve a linear system using the method of Gaussian elimination. The appearance in this process of a trivial equation $0 = 0$

means that one of the original equations was redundant. For instance, in the system

$$\begin{aligned} x - 2y + 2z &= 0 \\ x + 4y + 3z &= 0 \\ 2x + 2y + 5z &= 0, \end{aligned}$$

the third equation provides no additional information about a solution (x, y, z) because it is merely the sum of the first two equations.

Given a homogeneous linear system, it is natural to ask how many of the equations are "irredundant," and which ones they are. We will see that an answer to this question leads to a natural and simple relation between the number of irredundant equations, the number of unknowns, and the number of linearly independent solutions.

Row Space and Row Rank

The individual equations of the homogeneous linear system $\mathbf{Ax} = \mathbf{0}$ correspond to the "row matrices"

$$\begin{gathered} \begin{bmatrix} a_{11} & a_{12} & \cdots & a_{1n} \end{bmatrix} \\ \begin{bmatrix} a_{21} & a_{22} & \cdots & a_{2n} \end{bmatrix} \\ \vdots \\ \begin{bmatrix} a_{m1} & a_{m2} & \cdots & a_{mn} \end{bmatrix} \end{gathered}$$

of the $m \times n$ matrix $\mathbf{A} = \begin{bmatrix} a_{ij} \end{bmatrix}$. The **row vectors** of $\mathbf{A}$ are the m vectors

$$\begin{aligned} \mathbf{r}_1 &= (a_{11}, a_{12}, \ldots, a_{1n}) \\ \mathbf{r}_2 &= (a_{21}, a_{22}, \ldots, a_{2n}) \\ &\vdots \\ \mathbf{r}_m &= (a_{m1}, a_{m2}, \ldots, a_{mn}) \end{aligned} \tag{1}$$

in $\mathbf{R}^n$. Recalling from Section 3.4 the convection that n-tuples denote column-vector elements of $\mathbf{R}^n$, we see that the row vectors of $\mathbf{A}$ are the transposes of its row matrices; that is,

$$\mathbf{r}_i = \begin{bmatrix} a_{i1} \\ a_{i2} \\ \vdots \\ a_{in} \end{bmatrix} = \begin{bmatrix} a_{i1} & a_{i2} & \cdots & a_{in} \end{bmatrix}^T$$

for $i = 1, 2, \ldots, m$, and thus actually are column vectors (despite being called "row vectors"). The subspace of $\mathbf{R}^n$ spanned by the m row vectors $\mathbf{r}_1, \mathbf{r}_2, \ldots, \mathbf{r}_m$ is called the **row space** Row($\mathbf{A}$) of the matrix $\mathbf{A}$. The *dimension* of the row space Row($\mathbf{A}$) is called the **row rank** of the matrix $\mathbf{A}$.

Example 1 Consider the 4×5 echelon matrix

$$\mathbf{A} = \begin{bmatrix} 1 & -3 & 2 & 5 & 3 \\ 0 & 0 & 1 & -4 & 2 \\ 0 & 0 & 0 & 1 & 7 \\ 0 & 0 & 0 & 0 & 0 \end{bmatrix}.$$

Its row vectors are $\mathbf{r}_1 = (1, -3, 2, 5, 3)$, $\mathbf{r}_2 = (0, 0, 1, -4, 2)$, $\mathbf{r}_3 = (0, 0, 0, 1, 7)$, and $\mathbf{r}_4 = (0, 0, 0, 0, 0)$. We want to show that the row vectors that are not zero are linearly independent. To this end we calculate the linear combination

$$c_1\mathbf{r}_1 + c_2\mathbf{r}_2 + c_3\mathbf{r}_3 = (c_1, -3c_1, 2c_1 + c_2, 5c_1 - 4c_2 + c_3, 3c_1 + 2c_2 + 7c_3).$$

If $c_1\mathbf{r}_1 + c_2\mathbf{r}_2 + c_3\mathbf{r}_3 = \mathbf{0}$, then the first component gives $c_1 = 0$, next the third component $2c_1 + c_2 = 0$ yields $c_2 = 0$, and finally the fourth component $5c_1 - 4c_2 + c_3 = 0$ yields $c_3 = 0$. Thus the vectors $\mathbf{r}_1$, $\mathbf{r}_2$, and $\mathbf{r}_3$ are linearly independent and therefore form a basis for the row space $\text{Row}(\mathbf{A})$. Hence $\text{Row}(\mathbf{A})$ is a 3-dimensional subspace of $\mathbf{R}^5$, and the row rank of $\mathbf{A}$ is 3. ■

It should be apparent that the method used in Example 1 applies to any *echelon* matrix $\mathbf{E} = [e_{ij}]$. If the first column of $\mathbf{E}$ is not all zero, then its *nonzero* row vectors $\mathbf{r}_1, \mathbf{r}_2, \ldots, \mathbf{r}_k$ have the forms

$$\begin{aligned}
\mathbf{r}_1 &= (e_{11}, \ldots, \; e_{1p}, \ldots, \; e_{1q}, \ldots), \\
\mathbf{r}_2 &= (0, \ldots, 0, \; e_{2p}, \ldots, \; e_{2q}, \ldots), \\
\mathbf{r}_3 &= (0, \ldots, \quad 0, \ldots, \quad e_{3q}, \ldots),
\end{aligned}$$

and so forth (where e_{11}, e_{2p}, and e_{3q} denote nonzero leading entries). Then the equation

$$c_1\mathbf{r}_1 + c_2\mathbf{r}_2 + \cdots + c_k\mathbf{r}_k = \mathbf{0}$$

implies that

$$c_1e_{11} = 0, \quad c_1e_{1p} + c_2e_{2p} = 0, \quad c_1e_{1q} + c_2e_{2q} + c_3e_{3q} = 0,$$

and so forth. We can therefore conclude in turn that $c_1 = 0$, $c_2 = 0$, $\ldots$, $c_k = 0$. Thus the row vectors $\mathbf{r}_1, \mathbf{r}_2, \ldots, \mathbf{r}_k$ are linearly independent, and hence we have the following result.

THEOREM 1 Row Space of an Echelon Matrix

The nonzero row vectors of an echelon matrix are linearly independent and therefore form a basis for its row space.

We investigate the row space of an arbitrary matrix $\mathbf{A}$ by reducing $\mathbf{A}$ to an echelon matrix $\mathbf{E}$. The following theorem then guarantees that $\mathbf{A}$ and $\mathbf{E}$ have the same row space. Recall that two matrices are (row) *equivalent* provided that each can be transformed to the other by elementary row operations.

THEOREM 2 Row Spaces of Equivalent Matrices

If two matrices $\mathbf{A}$ and $\mathbf{B}$ are equivalent, then they have the same row space.

***Proof*:** Because $\mathbf{A}$ can be transformed to $\mathbf{B}$ by elementary row operations, it follows that each row vector of $\mathbf{B}$ is a linear combination of the row vectors of $\mathbf{A}$; the fact that this is obviously true if $\mathbf{A}$ is transformed into $\mathbf{B}$ by a single row operation implies its truth for any (finite) number of row operations. But then each vector in $\text{Row}(\mathbf{B})$ is a linear combination of linear combinations of row vectors of

A and hence is a linear combination of row vectors of **A**, which means that it is also a vector in Row(**A**). Therefore, the fact that **A** can be transformed to **B** implies that Row(**A**) contains Row(**B**).

But elementary row operations are "reversible," so **B** can also be transformed to **A** by row operations. Hence the same argument shows that Row(**B**) contains Row(**A**). Finally, the fact that each of the two row spaces Row(**A**) and Row(**B**) contains the other means that they are, indeed, identical. ■

Theorems 1 and 2 together provide a routine procedure for finding a basis for the row space of a given matrix, thereby determining its row rank.

> **ALGORITHM 1 A Basis for the Row Space**
>
> To find a basis for the row space of a matrix **A**, use elementary row operations to reduce **A** to an echelon matrix **E**. Then the nonzero row vectors of **E** form a basis for Row(**A**).

Example 2 To find a basis for the row space of the matrix

$$\mathbf{A} = \begin{bmatrix} 1 & 2 & 1 & 3 & 2 \\ 3 & 4 & 9 & 0 & 7 \\ 2 & 3 & 5 & 1 & 8 \\ 2 & 2 & 8 & -3 & 5 \end{bmatrix}, \tag{2}$$

we reduce it to its echelon form

$$\mathbf{E} = \begin{bmatrix} 1 & 2 & 1 & 3 & 2 \\ 0 & 1 & -3 & 5 & -4 \\ 0 & 0 & 0 & 1 & -7 \\ 0 & 0 & 0 & 0 & 0 \end{bmatrix}. \tag{3}$$

Then the nonzero row vectors $\mathbf{v}_1 = (1, 2, 1, 3, 2)$, $\mathbf{v}_2 = (0, 1, -3, 5, -4)$, and $\mathbf{v}_3 = (0, 0, 0, 1, -7)$ form a basis for the row space of **A**. Thus Row(**A**) is a 3-dimensional subspace of $\mathbf{R}^5$ and the row rank of **A** is 3. ■

In Example 2, note that Algorithm 1 does *not* tell us that the first three row vectors of the matrix **A** itself are linearly independent. We know that *some* three row vectors of **A** span Row(**A**), but without further investigation we do not know *which* three.

Column Space and Column Rank

Now we turn our attention from row vectors to column vectors. Given an $m \times n$ matrix $\mathbf{A} = [a_{ij}]$, the **column vectors** of **A** are the n vectors

$$\mathbf{c}_1 = \begin{bmatrix} a_{11} \\ a_{21} \\ \vdots \\ a_{m1} \end{bmatrix}, \quad \mathbf{c}_2 = \begin{bmatrix} a_{12} \\ a_{22} \\ \vdots \\ a_{m2} \end{bmatrix}, \quad \ldots, \quad \mathbf{c}_n = \begin{bmatrix} a_{1n} \\ a_{2n} \\ \vdots \\ a_{mn} \end{bmatrix} \tag{4}$$

in $\mathbf{R}^m$. The subspace of $\mathbf{R}^m$ spanned by the n column vectors $\mathbf{c}_1, \mathbf{c}_2, \ldots, \mathbf{c}_n$ is called the **column space** Col(**A**) of the matrix **A**. The *dimension* of the space Col(**A**) is called the **column rank** of the matrix **A**.

Example 3 Consider the 4×5 echelon matrix

$$\mathbf{E} = \begin{bmatrix} 1 & 2 & 1 & 3 & 2 \\ 0 & 1 & -3 & 5 & -4 \\ 0 & 0 & 0 & 1 & -7 \\ 0 & 0 & 0 & 0 & 0 \end{bmatrix}. \tag{5}$$

Its five column vectors $\mathbf{c}_1, \mathbf{c}_2, \mathbf{c}_3, \mathbf{c}_4, \mathbf{c}_5$ all lie in the subspace $\mathbf{R}^3 = \{(x_1, x_2, x_3, 0)\}$ of $\mathbf{R}^4$. The column vectors that contain the leading entries in the nonzero rows of $\mathbf{E}$ are

$$\mathbf{c}_1 = \begin{bmatrix} 1 \\ 0 \\ 0 \\ 0 \end{bmatrix}, \quad \mathbf{c}_2 = \begin{bmatrix} 2 \\ 1 \\ 0 \\ 0 \end{bmatrix}, \quad \text{and} \quad \mathbf{c}_4 = \begin{bmatrix} 3 \\ 5 \\ 1 \\ 0 \end{bmatrix}. \tag{6}$$

Note that

$$a_1\mathbf{c}_1 + a_2\mathbf{c}_2 + a_4\mathbf{c}_4 = (a_1 + 2a_2 + 3a_4, a_2 + 5a_4, a_4, 0).$$

Hence $a_1\mathbf{c}_1 + a_2\mathbf{c}_2 + a_4\mathbf{c}_4 = \mathbf{0}$ readily implies that $a_1 = a_2 = a_4 = 0$. Thus the vectors $\mathbf{c}_1$, $\mathbf{c}_2$, and $\mathbf{c}_4$ are linearly independent and therefore form a basis for $\mathbf{R}^3 = \{(x_1, x_2, x_3, 0)\}$. Hence $\text{Col}(\mathbf{E})$ is a 3-dimensional subspace of $\mathbf{R}^4$ and the column rank of $\mathbf{E}$ is 3. ■

To find the column rank of an arbitrary $m \times n$ matrix $\mathbf{A}$, we begin in the same way as when finding the row rank. First we use elementary row operations to reduce $\mathbf{A}$ to an echelon matrix $\mathbf{E}$. But the relationship between the column spaces of $\mathbf{A}$ and $\mathbf{E}$ is far more subtle than between their row spaces; the reason is that elementary row operations generally do *not* preserve column spaces. (See Problem 36.)

To analyze this situation, let us denote by $\mathbf{c}_1, \mathbf{c}_2, \ldots, \mathbf{c}_n$ the column vectors of the original matrix $\mathbf{A}$, and by $\mathbf{c}_1^*, \mathbf{c}_2^*, \ldots, \mathbf{c}_n^*$ the column vectors of the echelon matrix to which we have reduced $\mathbf{A}$. Suppose that $\mathbf{E}$ has k nonzero rows. Then each column vector of $\mathbf{E}$ has the form $\mathbf{c}_j^* = (*, \ldots, *, 0, \ldots, 0)$ with $m - k$ final zeros. Hence the column space $\text{Col}(\mathbf{E})$ is contained in $\mathbf{R}^k$ (considered as the subspace of $\mathbf{R}^m$ for which $x_{k+1} = \cdots = x_m = 0$).

We are particularly interested in those columns of $\mathbf{E}$ that contain the nonzero leading entries in the nonzero rows of $\mathbf{E}$. These k columns are called the **pivot columns** of $\mathbf{E}$. The echelon matrix $\mathbf{E}$ has the general form

$$\mathbf{E} = \begin{bmatrix} d_1 & * & * & * & \cdots & * & * & * \\ 0 & d_2 & * & * & \cdots & * & * & * \\ 0 & 0 & 0 & d_3 & \cdots & * & * & * \\ \vdots & \vdots & \vdots & \vdots & \ddots & \vdots & \vdots & \vdots \\ 0 & 0 & 0 & 0 & \cdots & d_k & * & * \\ 0 & 0 & 0 & 0 & \cdots & 0 & 0 & 0 \\ \vdots & \vdots & \vdots & \vdots & \ddots & \vdots & \vdots & \vdots \\ 0 & 0 & 0 & 0 & \cdots & 0 & 0 & 0 \end{bmatrix}, \tag{7}$$

where $d_1, d_2, \ldots, d_k$ are the nonzero leading entries. Hence the k pivot column vectors of **E** look like

$$\begin{bmatrix} d_1 \\ 0 \\ 0 \\ \vdots \\ 0 \\ 0 \\ \vdots \\ 0 \end{bmatrix}, \quad \begin{bmatrix} * \\ d_2 \\ 0 \\ \vdots \\ 0 \\ 0 \\ \vdots \\ 0 \end{bmatrix}, \quad \begin{bmatrix} * \\ * \\ d_3 \\ \vdots \\ 0 \\ 0 \\ \vdots \\ 0 \end{bmatrix}, \ldots, \begin{bmatrix} * \\ * \\ * \\ \vdots \\ d_k \\ 0 \\ \vdots \\ 0 \end{bmatrix}. \tag{8}$$

Because of the upper triangular pattern visible here, it should be apparent that the argument in Example 3 can be used to show that these k pivot vectors form a basis for Col(**E**). Thus *the column rank of the echelon matrix* **E** *is equal to the number k of its nonzero rows, and hence is equal to its row rank.*

Now comes the subtle part. We want to show that the column rank of the original matrix **A** also is k. Note first that the homogeneous linear systems

$$\mathbf{Ax} = \mathbf{0} \quad \text{and} \quad \mathbf{Ex} = \mathbf{0} \tag{9}$$

have the same solution set because the matrices **A** and **E** are equivalent. If $\mathbf{x} = (x_1, x_2, \ldots, x_n)$, then the left-hand sides in (9) are linear combinations of column vectors:

$$\begin{aligned} \mathbf{Ax} &= x_1\mathbf{c}_1 + x_2\mathbf{c}_2 + \cdots + x_n\mathbf{c}_n, \\ \mathbf{Ex} &= x_1\mathbf{c}_1^* + x_2\mathbf{c}_2^* + \cdots + x_n\mathbf{c}_n^*. \end{aligned} \tag{10}$$

Because **x** satisfies *either* both *or* neither of the systems in (9), it follows that

$$x_1\mathbf{c}_1 + x_2\mathbf{c}_2 + \cdots + x_n\mathbf{c}_n = \mathbf{0}$$

if and only if

$$x_1\mathbf{c}_1^* + x_2\mathbf{c}_2^* + \cdots + x_n\mathbf{c}_n^* = \mathbf{0}.$$

Hence every linear dependence between column vectors of **E** is mirrored in a linear dependence *with the same coefficients* between the corresponding column vectors of **A**. In particular, because the k pivot column vectors of **E** are linearly independent, it follows that the corresponding k column vectors of **A** are linearly independent. And because the k pivot column vectors of **E** span Col(**E**), it follows that the corresponding k column vectors of **A** span Col(**A**). Thus the latter k vectors form a basis for the column space of **A**, and the column rank of **A** is k. Consequently we have established the following method for finding a basis for the column space of a given matrix.

ALGORITHM 2 A Basis for the Column Space

To find a basis for the column space of a matrix **A**, use elementary row operations to reduce **A** to an echelon matrix **E**. Then the column vectors of **A** that correspond to the pivot columns of **E** form a basis for Col(**A**).

Remark: Since the pivot column vectors of the matrix **A**—that is, the column vectors corresponding to the pivot column vectors of the echelon matrix **E**—form a basis for the column space of **A**, it follows that every non-pivot column vector of **A** is a linear combination of its pivot column vectors. ■

Example 4 Consider again the 4×5 matrix

$$\mathbf{A} = \begin{bmatrix} 1 & 2 & 1 & 3 & 2 \\ 3 & 4 & 9 & 0 & 7 \\ 2 & 3 & 5 & 1 & 8 \\ 2 & 2 & 8 & -3 & 5 \end{bmatrix}$$

of Example 2, in which we reduced it to the echelon matrix

$$\mathbf{E} = \begin{bmatrix} 1 & 2 & 1 & 3 & 2 \\ 0 & 1 & -3 & 5 & -4 \\ 0 & 0 & 0 & 1 & -7 \\ 0 & 0 & 0 & 0 & 0 \end{bmatrix}.$$

The pivot columns of **E** are its first, second, and fourth columns, so the 3-dimensional column space of **A** has a basis consisting of its first, second, and fourth column vectors $\mathbf{c}_1 = (1, 3, 2, 2)$, $\mathbf{c}_2 = (2, 4, 3, 2)$, and $\mathbf{c}_4 = (3, 0, 1, -3)$. ■

The fact that Algorithm 2 provides a basis for Col(**A**) consisting of column vectors of **A** itself (rather than the echelon matrix **E**) enables us to apply it to the problem of extracting a maximal linearly independent subset from a given set of vectors in $\mathbf{R}^n$.

Example 5 Find a subset of the vectors $\mathbf{v}_1 = (1, -1, 2, 2)$, $\mathbf{v}_2 = (-3, 4, 1, -2)$, $\mathbf{v}_3 = (0, 1, 7, 4)$, and $\mathbf{v}_4 = (-5, 7, 4, -2)$ that forms a basis for the subspace W of $\mathbf{R}^4$ spanned by these four vectors.

Solution It is not clear at the outset whether W is 2-dimensional, 3-dimensional, or 4-dimensional. To apply Algorithm 2, we arrange the given vectors as the *column* vectors of the matrix

$$\mathbf{A} = \begin{bmatrix} 1 & -3 & 0 & -5 \\ -1 & 4 & 1 & 7 \\ 2 & 1 & 7 & 4 \\ 2 & -2 & 4 & -2 \end{bmatrix},$$

which reduces readily to the echelon matrix

$$\mathbf{E} = \begin{bmatrix} 1 & -3 & 0 & -5 \\ 0 & 1 & 1 & 2 \\ 0 & 0 & 0 & 0 \\ 0 & 0 & 0 & 0 \end{bmatrix}.$$

The pivot columns of **E** are its first two columns, so the first two column vectors $\mathbf{v}_1 = (1, -1, 2, 2)$ and $\mathbf{v}_2 = (-3, 4, 1, -2)$ of **A** form a basis for the column space W. In particular, we see that W is 2-dimensional. (You can confirm that $\mathbf{v}_3 = 3\mathbf{v}_1 + \mathbf{v}_2$ and $\mathbf{v}_4 = \mathbf{v}_1 + 2\mathbf{v}_2$.) ■

Rank and Dimension

We have seen that the row rank *and* the column rank of an echelon matrix **E** *both* equal the number of rows of **E** that are not all zeros. But if **A** is any matrix that reduces by row operations to the echelon matrix **E**, then—according to Algorithm 1—the row rank of **A** is equal to the row rank of **E**, while Algorithm 2 implies that the column rank of **A** is equal to the column rank of **E**. We therefore have the following fundamental result.

THEOREM 3 Equality of Row Rank and Column Rank

The row rank and column rank of any matrix are equal.

For instance, if **A** is a 5×7 matrix and Row(**A**) is a 3-dimensional subspace of $\mathbf{R}^7$, then Theorem 3 tells us that Col(**A**) must be a 3-dimensional subspace of $\mathbf{R}^5$.

To see what an extraordinary theorem this is, think of a random 11×17 matrix **A**, possibly printed by a computer that uses a random number generator to produce independently the 187 entries in **A**. If it turns out that precisely 9 (no more) of the 11 row vectors of **A** are linearly independent, then Theorem 3 implies that precisely 9 (no more) of the 17 column vectors of **A** are linearly independent.

The common value of the row rank and the column rank of the matrix **A** is simply called the **rank** of **A** and is denoted by rank(**A**). The rank of **A** provides a meaning for the number of irredundant equations in the homogeneous linear system

$$\mathbf{Ax} = \mathbf{0}. \tag{11}$$

Indeed, the individual scalar equations in (11) correspond to the column vectors of the transpose matrix $\mathbf{A}^T$, whose rank r equals that of **A** (Problem 25). Then the r columns vectors of $\mathbf{A}^T$ that form a basis for $\mathrm{Col}(\mathbf{A}^T)$ correspond to the r equations in (11) that are irredundant; the remaining equations are linear combinations of these irredundant ones and hence are redundant.

The solution space of the homogeneous system $\mathbf{Ax} = \mathbf{0}$ is sometimes called the **null space** of **A**, denoted by Null(**A**). Note that if **A** is an $m \times n$ matrix, then Null(**A**) and Row(**A**) are subspaces of $\mathbf{R}^n$, whereas Col(**A**) is a subspace of $\mathbf{R}^m$. If $r = \mathrm{rank}(\mathbf{A})$, then the r column vectors of **A** that form a basis for Col(**A**) correspond to the r leading variables in a solution of $\mathbf{Ax} = \mathbf{0}$ by Gaussian elimination. Moreover, we know from the algorithm in Section 4.4 that the dimension of the solution space Null(**A**) is equal to the number $n - r$ of free variables. We therefore obtain the important identity

$$\mathrm{rank}(\mathbf{A}) + \dim \mathrm{Null}(\mathbf{A}) = n \tag{12}$$

for any $m \times n$ matrix **A**. For instance, if **A** is the matrix of Example 4 with rank 3 and $n = 5$, then the dimension of the null space of **A** is $5 - 3 = 2$.

For another typical application of Equation (12), consider a homogeneous system of five linear equations in seven unknowns. If the 5×7 coefficient matrix of the system has rank 3, so only 3 of the 5 equations are irredundant, then (because $n = 7$) the system has $7 - 3 = 4$ linearly independent solutions.

If **A** is an $m \times n$ matrix with rank m, then Equation (12) implies that the system $\mathbf{Ax} = \mathbf{0}$ has $n - m$ linearly independent solutions. Thus $\mathrm{rank}(\mathbf{A}) = m$ is the condition under which the "conventional wisdom" about the relation between the numbers of equations, unknowns, and solutions is valid.

Nonhomogeneous Linear Systems

It follows immediately from Equation (10) and the definition of Col(**A**) that *the nonhomogeneous linear system* $\mathbf{Ax} = \mathbf{b}$ *is consistent if and only if the vector* **b** *is in the column space of* **A**. In the problems below we outline some of the applications to nonhomogeneous systems of the results in this section.

If we can find a single particular solution $\mathbf{x}_0$ of the nonhomogeneous system

$$\mathbf{Ax} = \mathbf{b}, \tag{13}$$

then the determination of its solution set reduces to solving the associated homogeneous system

$$\mathbf{Ax} = \mathbf{0}. \tag{14}$$

For then Problem 29 in Section 4.2 implies that the solution set of (13) is the set of all vectors **x** of the form

$$\mathbf{x} = \mathbf{x}_0 + \mathbf{x}_h,$$

where $\mathbf{x}_h$ is a solution of (14). Hence if $\mathbf{x}_1, \mathbf{x}_2, \dots, \mathbf{x}_r$ is a basis for the solution space of (14), then the solution set of the nonhomogeneous system is the set of all vectors of the form

$$\mathbf{x} = \mathbf{x}_0 + c_1\mathbf{x}_1 + c_2\mathbf{x}_2 + \cdots + c_r\mathbf{x}_r.$$

We may describe this solution set as the *translate* by the vector $\mathbf{x}_0$ of the r-dimensional solution space of the corresponding homogeneous system.

4.5 Problems

In Problems 1–12, find both a basis for the row space and a basis for the column space of the given matrix **A**.

1. $\begin{bmatrix} 1 & 2 & 3 \\ 1 & 5 & -9 \\ 2 & 5 & 2 \end{bmatrix}$

2. $\begin{bmatrix} 5 & 2 & 4 \\ 2 & 1 & 1 \\ 4 & 1 & 5 \end{bmatrix}$

3. $\begin{bmatrix} 1 & -4 & -3 & -7 \\ 2 & -1 & 1 & 7 \\ 1 & 2 & 3 & 11 \end{bmatrix}$

4. $\begin{bmatrix} 1 & -3 & -9 & -5 \\ 2 & 1 & 4 & 11 \\ 1 & 3 & 3 & 13 \end{bmatrix}$

5. $\begin{bmatrix} 1 & 1 & 1 & 1 \\ 3 & 1 & -3 & 4 \\ 2 & 5 & 11 & 12 \end{bmatrix}$

6. $\begin{bmatrix} 1 & 4 & 9 & 2 \\ 2 & 2 & 6 & -3 \\ 2 & 7 & 16 & 3 \end{bmatrix}$

7. $\begin{bmatrix} 1 & 1 & -1 & 7 \\ 1 & 4 & 5 & 16 \\ 1 & 3 & 3 & 13 \\ 2 & 5 & 4 & 23 \end{bmatrix}$

8. $\begin{bmatrix} 1 & -2 & -3 & -5 \\ 1 & 4 & 9 & 2 \\ 1 & 3 & 7 & 1 \\ 2 & 2 & 6 & -3 \end{bmatrix}$

9. $\begin{bmatrix} 1 & 3 & 3 & 9 \\ 2 & 7 & 4 & 8 \\ 2 & 7 & 5 & 12 \\ 2 & 8 & 3 & 2 \end{bmatrix}$

10. $\begin{bmatrix} 1 & 2 & 3 & 1 & 3 \\ 1 & 3 & 4 & 3 & 6 \\ 2 & 2 & 4 & 3 & 5 \\ 2 & 1 & 3 & 2 & 3 \end{bmatrix}$

11. $\begin{bmatrix} 1 & 1 & 3 & 3 & 1 \\ 2 & 3 & 7 & 8 & 2 \\ 2 & 3 & 7 & 8 & 3 \\ 3 & 1 & 7 & 5 & 4 \end{bmatrix}$

12. $\begin{bmatrix} 1 & 1 & 3 & 3 & 0 \\ -1 & 0 & -2 & -1 & 1 \\ 2 & 3 & 7 & 8 & 1 \\ -2 & 4 & 0 & 6 & 7 \end{bmatrix}$

In Problems 13–16, a set S of vectors in $\mathbf{R}^4$ *is given. Find a subset of S that forms a basis for the subspace of* $\mathbf{R}^4$ *spanned by S.*

13. $\mathbf{v}_1 = (1, 3, -2, 4)$, $\mathbf{v}_2 = (2, -1, 3, 2)$, $\mathbf{v}_3 = (5, 1, 4, 8)$

14. $\mathbf{v}_1 = (1, -1, 2, 3)$, $\mathbf{v}_2 = (2, 3, 4, 1)$, $\mathbf{v}_3 = (1, 1, 2, 1)$, $\mathbf{v}_4 = (4, 1, 8, 7)$

15. $\mathbf{v}_1 = (3, 2, 2, 2)$, $\mathbf{v}_2 = (2, 1, 2, 1)$, $\mathbf{v}_3 = (4, 3, 2, 3)$, $\mathbf{v}_4 = (1, 2, 3, 4)$

16. $\mathbf{v}_1 = (5, 4, 2, 2)$, $\mathbf{v}_2 = (3, 1, 2, 3)$, $\mathbf{v}_3 = (7, 7, 2, 1)$, $\mathbf{v}_4 = (1, -1, 2, 4)$, $\mathbf{v}_5 = (5, 4, 6, 7)$

Let $S = \{\mathbf{v}_1, \mathbf{v}_2, \dots, \mathbf{v}_k\}$ *be a basis for the subspace W of* $\mathbf{R}^n$. *Then a basis T for* $\mathbf{R}^n$ *that contains S can be found by applying the method of Example 5 to the vectors*

$$\mathbf{v}_1, \mathbf{v}_2, \ldots, \mathbf{v}_k, \mathbf{e}_1, \mathbf{e}_2, \ldots, \mathbf{e}_n.$$

Do this in Problems 17–20.

17. Find a basis T for $\mathbf{R}^3$ that contains the vectors $\mathbf{v}_1 = (1, 2, 2)$ and $\mathbf{v}_2 = (2, 3, 3)$.

18. Find a basis T for $\mathbf{R}^3$ that contains the vectors $\mathbf{v}_1 = (3, 2, -1)$ and $\mathbf{v}_2 = (2, -2, 1)$.

19. Find a basis T for $\mathbf{R}^4$ that contains the vectors $\mathbf{v}_1 = (1, 1, 1, 1)$ and $\mathbf{v}_2 = (2, 3, 3, 3)$.

20. Find a basis T for $\mathbf{R}^4$ that contains the vectors $\mathbf{v}_1 = (3, 2, 3, 3)$ and $\mathbf{v}_2 = (5, 4, 5, 5)$.

Given a homogeneous system $\mathbf{Ax} = \mathbf{0}$ *of (scalar) linear equations, we say that a subset of these equations is* irredundant *provided that the corresponding column vectors of the transpose* $\mathbf{A}^T$ *are linearly independent. In Problems 21–24, extract from each given system a maximal subset of irredundant equations.*

21.
$$\begin{aligned} x_1 - 3x_2 + 2x_3 &= 0 \\ 2x_1 + 3x_2 + 2x_3 &= 0 \\ 4x_1 - 3x_2 + 6x_3 &= 0 \end{aligned}$$

22.
$$\begin{aligned} 3x_1 + 2x_2 + 2x_3 + 2x_4 &= 0 \\ 2x_1 + 3x_2 + 3x_3 + 3x_4 &= 0 \\ 8x_1 + 7x_2 + 7x_3 + 7x_4 &= 0 \end{aligned}$$

23.
$$\begin{aligned} x_1 + 2x_2 - x_3 + 2x_4 &= 0 \\ 3x_1 - x_2 + 3x_3 + x_4 &= 0 \\ 5x_1 + 3x_2 + x_3 + 5x_4 &= 0 \\ 2x_1 \qquad + 5x_3 + 4x_4 &= 0 \end{aligned}$$

24.
$$\begin{aligned} 3x_1 + 2x_2 + 2x_3 &= 0 \\ 2x_1 + 3x_2 + 3x_3 &= 0 \\ 7x_1 + 8x_2 + 8x_3 &= 0 \\ 8x_1 + 7x_2 + 7x_3 &= 0 \\ 5x_1 + 6x_2 + 5x_3 &= 0 \end{aligned}$$

25. Explain why the rank of a matrix $\mathbf{A}$ is equal to the rank of its transpose $\mathbf{A}^T$.

26. Explain why the $n \times n$ matrix $\mathbf{A}$ is invertible if and only if its rank is n.

27. Let $\mathbf{A}$ be a 3×5 matrix whose three row vectors are linearly independent. Prove that, for each $\mathbf{b}$ in $\mathbf{R}^3$, the nonhomogeneous system $\mathbf{Ax} = \mathbf{b}$ has a solution.

28. Let $\mathbf{A}$ be a 5×3 matrix that has three linearly independent row vectors. Suppose that $\mathbf{b}$ is a vector in $\mathbf{R}^5$ such that the nonhomogeneous system $\mathbf{Ax} = \mathbf{b}$ has a solution. Prove that this solution is unique.

29. Let $\mathbf{A}$ be an $m \times n$ matrix with $m < n$. Given $\mathbf{b}$ in $\mathbf{R}^m$, note that any solution of $\mathbf{Ax} = \mathbf{b}$ expresses $\mathbf{b}$ as a linear combination of the column vectors of $\mathbf{A}$. Then prove that no such solution is unique.

30. Given the $m \times n$ matrix $\mathbf{A}$ with $m > n$, show that there exists a vector $\mathbf{b}$ in $\mathbf{R}^m$ such that the system $\mathbf{Ax} = \mathbf{b}$ has no solution.

31. (Existence of Solutions) Let $\mathbf{A}$ be an $m \times n$ matrix. Prove that the system $\mathbf{Ax} = \mathbf{b}$ is consistent for every $\mathbf{b}$ in $\mathbf{R}^m$ if and only if the rank of $\mathbf{A}$ is equal to m.

32. (Uniqueness of Solutions) Let $\mathbf{A}$ be an $m \times n$ matrix and suppose that the system $\mathbf{Ax} = \mathbf{b}$ is consistent. Prove that its solution is unique if and only if the rank of $\mathbf{A}$ is equal to n.

33. Prove that the pivot column vectors in (8) are linearly independent.

34. Let $\mathbf{A}$ be a matrix with rank r, and suppose that $\mathbf{A}$ can be reduced to echelon form *without row interchanges*. Show that the first r row vectors of $\mathbf{A}$ are linearly independent.

35. Deduce from Theorem 3 in Section 4.3 that the rank of the matrix $\mathbf{A}$ is the largest integer r such that $\mathbf{A}$ has a nonsingular $r \times r$ submatrix.

36. Let $\mathbf{A}$ be the 3×2 matrix whose columns are the unit basis vectors $\mathbf{i} = (1, 0, 0)$ and $\mathbf{j} = (0, 1, 0)$. If $\mathbf{B}$ is the row-equivalent matrix obtained by adding the first and second rows of $\mathbf{A}$ to its third row, show that $\mathbf{A}$ and $\mathbf{B}$ do *not* have the same column space.

4.6 Orthogonal Vectors in $\mathbf{R}^n$

In this section we show that the geometrical concepts of *distance* and *angle* in n-dimensional space can be based on the definition of the *dot product* of two vectors in $\mathbf{R}^n$. Recall from elementary calculus that the dot product $\mathbf{u} \cdot \mathbf{v}$ of two 3-dimensional vectors $\mathbf{u} = (u_1, u_2, u_3)$ and $\mathbf{v} = (v_1, v_2, v_3)$ is (by definition) the sum

$$\mathbf{u} \cdot \mathbf{v} = u_1 v_1 + u_2 v_2 + u_3 v_3$$

of the products of corresponding scalar components of the two vectors.

Similarly, the **dot product** $\mathbf{u} \cdot \mathbf{v}$ of two n-dimensional vectors $\mathbf{u} = (u_1, u_2, \ldots, u_n)$ and $\mathbf{v} = (v_1, v_2, \ldots, v_n)$ is defined by

$$\mathbf{u} \cdot \mathbf{v} = u_1 v_1 + u_2 v_2 + \cdots + u_n v_n \tag{1}$$

(with one additional scalar product term for each additional dimension). And just as in $\mathbf{R}^3$, it follows readily from the formula in (1) that if $\mathbf{u}$, $\mathbf{v}$, and $\mathbf{w}$ are vectors in

$\mathbf{R}^n$ and c is a scalar, then

$$\mathbf{u} \cdot \mathbf{v} = \mathbf{v} \cdot \mathbf{u} \qquad \text{(symmetry)} \tag{2}$$

$$\mathbf{u} \cdot (\mathbf{v} + \mathbf{w}) = \mathbf{u} \cdot \mathbf{v} + \mathbf{u} \cdot \mathbf{w} \qquad \text{(distributivity)} \tag{3}$$

$$(c\mathbf{u}) \cdot \mathbf{v} = c(\mathbf{u} \cdot \mathbf{v}) \qquad \text{(homogeneity)} \tag{4}$$

$$\left.\begin{array}{l} \mathbf{u} \cdot \mathbf{u} \geqq 0; \\ \mathbf{u} \cdot \mathbf{u} = 0 \quad \text{if and only if} \\ \mathbf{u} = \mathbf{0}. \end{array}\right\} \qquad \text{(positivity)} \tag{5}$$

Therefore, the dot product in $\mathbf{R}^n$ is an example of an *inner product.*

DEFINITION Inner Product

An **inner product** on a vector space V is a function that associates with each pair of vectors $\mathbf{u}$ and $\mathbf{v}$ in V a scalar $\langle \mathbf{u}, \mathbf{v} \rangle$ such that, if $\mathbf{u}$, $\mathbf{v}$, and $\mathbf{w}$ are vectors and c is a scalar, then

(i) $\langle \mathbf{u}, \mathbf{v} \rangle = \langle \mathbf{v}, \mathbf{u} \rangle$;

(ii) $\langle \mathbf{u}, \mathbf{v} + \mathbf{w} \rangle = \langle \mathbf{u}, \mathbf{v} \rangle + \langle \mathbf{u}, \mathbf{w} \rangle$;

(iii) $\langle c\mathbf{u}, \mathbf{v} \rangle = c\langle \mathbf{u}, \mathbf{v} \rangle$;

(iv) $\langle \mathbf{u}, \mathbf{u} \rangle \geqq 0$; $\langle \mathbf{u}, \mathbf{u} \rangle = 0$ if and only if $\mathbf{u} = \mathbf{0}$.

The dot product on $\mathbf{R}^n$ is sometimes called the **Euclidean inner product**, and with this inner product $\mathbf{R}^n$ is sometimes called **Euclidean n-dimensional space**. We can use any of the notations in

$$\mathbf{u} \cdot \mathbf{v} = \langle \mathbf{u}, \mathbf{v} \rangle = \mathbf{u}^T \mathbf{v}$$

for the dot product of the two $n \times 1$ column vectors $\mathbf{u}$ and $\mathbf{v}$. (Note in the last expression that $\mathbf{u}^T = (u_1, u_2, \dots, u_n)^T$ is the $1 \times n$ *row* vector with the indicated entries, so the 1×1 matrix product $\mathbf{u}^T \mathbf{v}$ is simply a scalar.) Here we will ordinarily use the notation $\mathbf{u} \cdot \mathbf{v}$.

The **length** $|\mathbf{u}|$ of the vector $\mathbf{u} = (u_1, u_2, \dots, u_n)$ is defined as follows:

$$|\mathbf{u}| = \sqrt{(\mathbf{u} \cdot \mathbf{u})} = \left(u_1^2 + u_2^2 + \cdots + u_n^2\right)^{1/2}. \tag{6}$$

Note that the case $n = 2$ is a consequence of the familiar Pythagorean formula in the plane.

Theorem 1 gives one of the most important inequalities in mathematics. Many proofs are known, but none of them seems direct and well motivated.

THEOREM 1 The Cauchy-Schwarz Inequality

If $\mathbf{u}$ and $\mathbf{v}$ are vectors in $\mathbf{R}^n$, then

$$|\mathbf{u} \cdot \mathbf{v}| \leqq |\mathbf{u}||\mathbf{v}|. \tag{7}$$

Proof: If $\mathbf{u} = \mathbf{0}$, then $|\mathbf{u} \cdot \mathbf{v}| = |\mathbf{u}| = 0$, so the inequality is satisfied trivially. If $\mathbf{u} \neq \mathbf{0}$, then we let $a = \mathbf{u} \cdot \mathbf{u}$, $b = 2\mathbf{u} \cdot \mathbf{v}$, and $c = \mathbf{v} \cdot \mathbf{v}$. For any real number x, the distributivity and positivity properties of the dot product then yield

$$\begin{aligned} 0 &\leqq (x\mathbf{u} + \mathbf{v}) \cdot (x\mathbf{u} + \mathbf{v}) \\ &= (\mathbf{u} \cdot \mathbf{u})x^2 + 2(\mathbf{u} \cdot \mathbf{v})x + (\mathbf{v} \cdot \mathbf{v}), \end{aligned}$$

so that

$$0 \leqq ax^2 + bx + c.$$

Thus the quadratic equation $ax^2 + bx + c = 0$ either has no real roots or has a repeated real root. Hence the quadratic formula

$$x = \frac{-b \pm \sqrt{b^2 - 4ac}}{2a}$$

implies that the discriminant $b^2 - 4ac$ cannot be positive; that is, $b^2 \leqq 4ac$, so

$$4(\mathbf{u} \cdot \mathbf{v})^2 \leqq 4(\mathbf{u} \cdot \mathbf{u})(\mathbf{v} \cdot \mathbf{v}).$$

We get the Cauchy-Schwarz inequality in (7) when we take square roots, remembering that the numbers $|\mathbf{u}| = (\mathbf{u} \cdot \mathbf{u})^{1/2}$ and $|\mathbf{v}| = (\mathbf{v} \cdot \mathbf{v})^{1/2}$ are nonnegative. ■

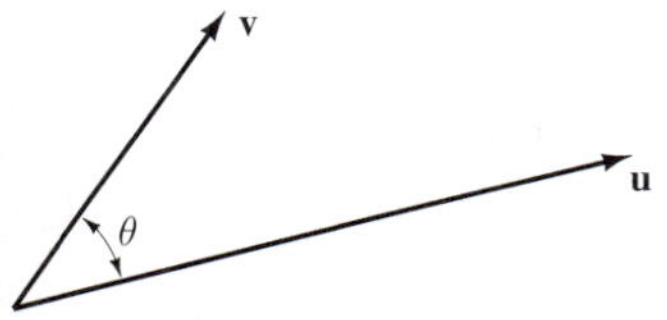

FIGURE 4.6.1. The angle θ between the vectors **u** and **v**.

The Cauchy-Schwarz inequality enables us to *define* the angle θ between the nonzero vectors **u** and **v**. (See Figure 4.6.1.) Division by the positive number $|\mathbf{u}||\mathbf{v}|$ in (7) yields $|\mathbf{u} \cdot \mathbf{v}|/(|\mathbf{u}||\mathbf{v}|) \leqq 1$, so

$$-1 \leqq \frac{\mathbf{u} \cdot \mathbf{v}}{|\mathbf{u}||\mathbf{v}|} \leqq +1. \tag{8}$$

Hence there is a unique angle θ between 0 and π radians, inclusive (that is, between 0° and 180°), such that

$$\cos\theta = \frac{\mathbf{u} \cdot \mathbf{v}}{|\mathbf{u}||\mathbf{v}|}. \tag{9}$$

Thus we obtain the same geometric interpretation

$$\mathbf{u} \cdot \mathbf{v} = |\mathbf{u}||\mathbf{v}| \cos\theta \tag{10}$$

of the dot product in $\mathbf{R}^n$ as one sees (for 3-dimensional vectors) in elementary calculus textbooks—for instance, see Section 11.2 of Edwards and Penney, *Calculus: Early Transcendentals*, 7th edition (Prentice Hall) 2008.

On the basis of (10) we call the vectors **u** and **v** **orthogonal** provided that

$$\mathbf{u} \cdot \mathbf{v} = 0. \tag{11}$$

If **u** and **v** are nonzero vectors this means that $\cos\theta = 0$, so $\theta = \pi/2$ (90°). Note that $\mathbf{u} = \mathbf{0}$ satisfies (11) for all **v**, so the zero vector is orthogonal to *every* vector.

Example 1 Find the angle θ_n in $\mathbf{R}^n$ between the x_1-axis and the line through the origin and the point $(1, 1, \ldots, 1)$.

Solution We take $\mathbf{u} = (1, 0, 0, \ldots, 0)$ on the x_1-axis and $\mathbf{v} = (1, 1, \ldots, 1)$. Then $|\mathbf{u}| = 1$, $|\mathbf{v}| = \sqrt{n}$, and $\mathbf{u} \cdot \mathbf{v} = 1$, so the formula in (9) gives

$$\cos\theta_n = \frac{\mathbf{u} \cdot \mathbf{v}}{|\mathbf{u}||\mathbf{v}|} = \frac{1}{\sqrt{n}}.$$

For instance, if

$$\begin{aligned}
n &= 3, \quad \text{then} \quad \theta_3 = \cos^{-1}\left(\frac{1}{\sqrt{3}}\right) \approx 0.9553 \quad (55°);\\
n &= 4, \quad \text{then} \quad \theta_4 = \cos^{-1}\left(\frac{1}{\sqrt{4}}\right) \approx 1.0472 \quad (60°);\\
n &= 5, \quad \text{then} \quad \theta_5 = \cos^{-1}\left(\frac{1}{\sqrt{5}}\right) \approx 1.1071 \quad (63°);\\
n &= 100, \quad \text{then} \quad \theta_{100} = \cos^{-1}\left(\frac{1}{10}\right) \approx 1.4706 \quad (84°).
\end{aligned}$$

It is interesting to note that θ_n increases as n increases. Indeed, θ_n approaches $\cos^{-1}(0) = \pi/2$ (90°) as n increases without bound (so that $1/\sqrt{n}$ approaches zero). ■

In addition to angles, the dot product provides a definition of distance in $\mathbf{R}^n$. The **distance** $d(\mathbf{u}, \mathbf{v})$ between the points (vectors) $\mathbf{u} = (u_1, u_2, \ldots, u_n)$ and $\mathbf{v} = (v_1, v_2, \ldots, v_n)$ is defined to be

$$\begin{aligned}
d(\mathbf{u}, \mathbf{v}) &= |\mathbf{u} - \mathbf{v}|\\
&= \left[(u_1 - v_1)^2 + (u_2 - v_2)^2 + \cdots + (u_n - v_n)^2\right]^{1/2}.
\end{aligned} \tag{12}$$

Example 2 The distance between the points $\mathbf{u} = (1, -1, -2, 3, 5)$ and $\mathbf{v} = (4, 3, 4, 5, 9)$ in $\mathbf{R}^5$ is

$$|\mathbf{u} - \mathbf{v}| = \sqrt{3^2 + 4^2 + 6^2 + 2^2 + 4^2} = \sqrt{81} = 9.$$ ■

The *triangle inequality* of Theorem 2 relates the three sides of the triangle shown in Figure 4.6.2.

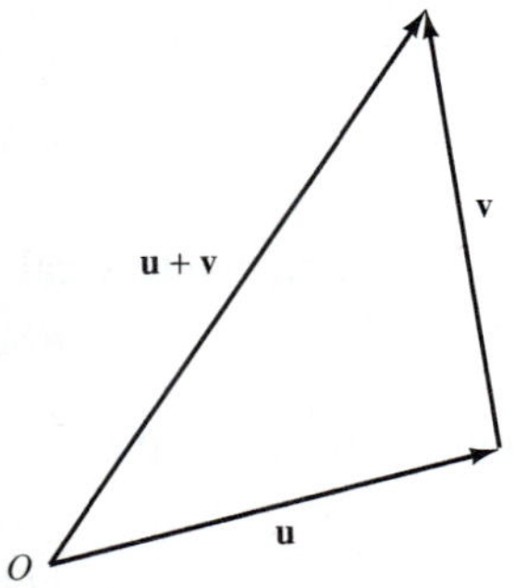

FIGURE 4.6.2. The "triangle" of the triangle inequality.

THEOREM 2 The Triangle Inequality

If $\mathbf{u}$ and $\mathbf{v}$ are vectors in $\mathbf{R}^n$, then

$$|\mathbf{u} + \mathbf{v}| \leqq |\mathbf{u}| + |\mathbf{v}|. \tag{13}$$

***Proof*:** We apply the Cauchy-Schwarz inequality to find that

$$\begin{aligned}
|\mathbf{u} + \mathbf{v}|^2 &= (\mathbf{u} + \mathbf{v}) \cdot (\mathbf{u} + \mathbf{v})\\
&= \mathbf{u} \cdot \mathbf{u} + 2\mathbf{u} \cdot \mathbf{v} + \mathbf{v} \cdot \mathbf{v}\\
&\leqq \mathbf{u} \cdot \mathbf{u} + 2|\mathbf{u}||\mathbf{v}| + \mathbf{v} \cdot \mathbf{v}\\
&= |\mathbf{u}|^2 + 2|\mathbf{u}||\mathbf{v}| + |\mathbf{v}|^2,
\end{aligned} \tag{14}$$

and therefore

$$|\mathbf{u}+\mathbf{v}|^2 \leqq (|\mathbf{u}|+|\mathbf{v}|)^2.$$

We now get (13) when we take square roots. ■

The vectors $\mathbf{u}$ and $\mathbf{v}$ are orthogonal if and only if $\mathbf{u}\cdot\mathbf{v}=0$, so line (14) in the proof of the triangle inequality yields the fact that the **Pythagorean formula**

$$|\mathbf{u}+\mathbf{v}|^2 = |\mathbf{u}|^2+|\mathbf{v}|^2 \tag{15}$$

holds if and only if the triangle with "adjacent side vectors" $\mathbf{u}$ and $\mathbf{v}$ is a right triangle with hypotenuse vector $\mathbf{u}+\mathbf{v}$ (see Figure 4.6.3).

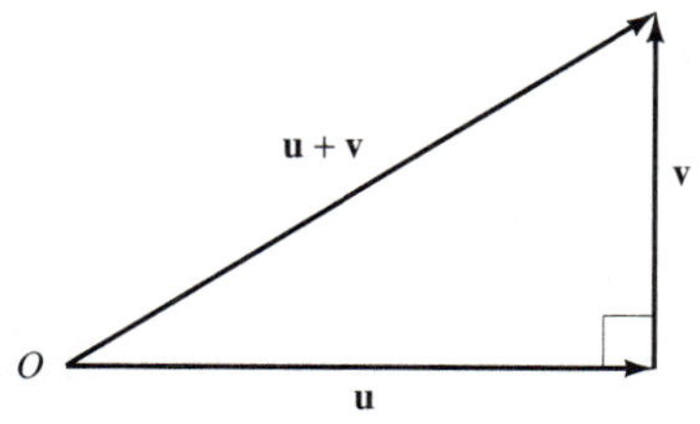

FIGURE 4.6.3. A right triangle in $\mathbf{R}^n$.

The following theorem states a simple relationship between orthogonality and linear independence.

THEOREM 3 Orthogonality and Linear Independence

If the nonzero vectors $\mathbf{v}_1, \mathbf{v}_2, \ldots, \mathbf{v}_k$ are mutually orthogonal—that is, each two of them are orthogonal—then they are linearly independent.

***Proof*:** Suppose that

$$c_1\mathbf{v}_1 + c_2\mathbf{v}_2 + \cdots + c_k\mathbf{v}_k = \mathbf{0},$$

where, as usual, $c_1, c_2, \ldots, c_k$ are scalars. When we take the dot product of each side of this equation with $\mathbf{v}_i$, we find that

$$c_i\mathbf{v}_i\cdot\mathbf{v}_i = c_i|\mathbf{v}_i|^2 = 0.$$

Now $|\mathbf{v}_i| \neq 0$ because $\mathbf{v}_i$ is a nonzero vector. It follows that $c_i = 0$. Thus $c_1 = c_2 = \cdots = c_k = 0$, and therefore the mutually orthogonal nonzero vectors $\mathbf{v}_1, \mathbf{v}_2, \ldots, \mathbf{v}_k$ are linearly independent. ■

In particular, any set of n mutually orthogonal nonzero vectors in $\mathbf{R}^n$ constitutes a basis for $\mathbf{R}^n$. Such a basis is called an **orthogonal basis**. For instance, the standard unit vectors $\mathbf{e}_1, \mathbf{e}_2, \ldots, \mathbf{e}_n$ form an orthogonal basis for $\mathbf{R}^n$.

Orthogonal Complements

Now we want to relate orthogonality to the solution of systems of linear equations. Consider the homogeneous linear system

$$\mathbf{A}\mathbf{x} = \mathbf{0} \tag{16}$$

of m equations in n unknowns. If $\mathbf{v}_1, \mathbf{v}_2, \ldots, \mathbf{v}_m$ are the row vectors of the $m\times n$ coefficient matrix $\mathbf{A}$, then the system looks like

$$\begin{bmatrix} \text{———} \mathbf{v}_1 \text{———} \\ \text{———} \mathbf{v}_2 \text{———} \\ \vdots \\ \text{———} \mathbf{v}_m \text{———} \end{bmatrix} \begin{bmatrix} | \\ \mathbf{x} \\ | \end{bmatrix} = \begin{bmatrix} \mathbf{v}_1\cdot\mathbf{x} \\ \mathbf{v}_2\cdot\mathbf{x} \\ \vdots \\ \mathbf{v}_m\cdot\mathbf{x} \end{bmatrix} = \begin{bmatrix} 0 \\ 0 \\ \vdots \\ 0 \end{bmatrix}.$$

Consequently, it is clear that $\mathbf{x}$ is a solution vector of $\mathbf{Ax} = \mathbf{0}$ if and only if $\mathbf{x}$ is orthogonal to each row vector of $\mathbf{A}$. But in the latter event $\mathbf{x}$ is orthogonal to every linear combination of row vectors of $\mathbf{A}$ because

$$\begin{aligned}\mathbf{x} \cdot (c_1\mathbf{v}_1 + c_2\mathbf{v}_2 + \cdots + c_m\mathbf{v}_m)\\ &= c_1\mathbf{x} \cdot \mathbf{v}_1 + c_2\mathbf{x} \cdot \mathbf{v}_2 + \cdots + c_m\mathbf{x} \cdot \mathbf{v}_m\\ &= (c_1)(0) + (c_2)(0) + \cdots + (c_m)(0) = 0.\end{aligned}$$

Thus we have shown that *the vector* $\mathbf{x}$ *in* $\mathbf{R}^n$ *is a solution vector of* $\mathbf{Ax} = \mathbf{0}$ *if and only if* $\mathbf{x}$ *is orthogonal to each vector in the row space* $\text{Row}(\mathbf{A})$ *of the matrix* $\mathbf{A}$. This situation motivates the following definition.

DEFINITION The Orthogonal Complement of a Subspace

The vector $\mathbf{u}$ is **orthogonal** to the subspace V of $\mathbf{R}^n$ provided that $\mathbf{u}$ is orthogonal to every vector in V. The **orthogonal complement** $V^\perp$ (read "V perp") of V is the set of all those vectors in $\mathbf{R}^n$ that are orthogonal to the subspace V.

If $\mathbf{u}_1$ and $\mathbf{u}_2$ are vectors in $V^\perp$, $\mathbf{v}$ is in V, and c_1 and c_2 are scalars, then

$$\begin{aligned}(c_1\mathbf{u}_1 + c_2\mathbf{u}_2) \cdot \mathbf{v} &= c_1\mathbf{u}_1 \cdot \mathbf{v} + c_2\mathbf{u}_2 \cdot \mathbf{v}\\ &= (c_1)(0) + (c_2)(0) = 0.\end{aligned}$$

Thus any linear combination of vectors in $V^\perp$ is orthogonal to every vector in V and hence is a vector in $V^\perp$. Therefore *the orthogonal complement* $V^\perp$ *of a subspace* V *is itself a subspace of* $\mathbf{R}^n$. The standard picture of two complementary subspaces V and $V^\perp$ consists of an orthogonal line and plane through the origin in $\mathbf{R}^3$ (see Fig. 4.6.4). The proofs of the remaining parts of Theorem 4 are left to the problems.

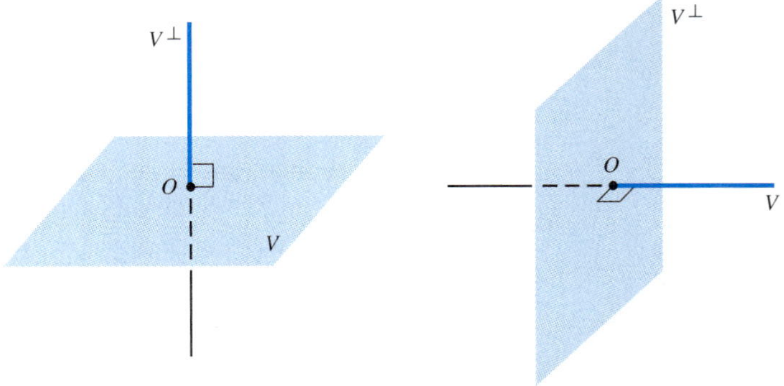

FIGURE 4.6.4. Orthogonal complements.

THEOREM 4 Properties of Orthogonal Complements

Let V be a subspace of $\mathbf{R}^n$. Then

1. Its orthogonal complement $V^\perp$ is also a subspace of $\mathbf{R}^n$;
2. The only vector that lies in both V and $V^\perp$ is the zero vector;
3. The orthogonal complement of $V^\perp$ is V—that is, $(V^\perp)^\perp = V$;
4. If S is a spanning set for V, then the vector $\mathbf{u}$ is in $V^\perp$ if and only if $\mathbf{u}$ is orthogonal to every vector in S.

In our discussion of the homogeneous linear system $\mathbf{Ax} = \mathbf{0}$ in (16), we showed that a vector space lies in the **null space** Null(**A**) of **A**—that is, in the solution space of $\mathbf{Ax} = \mathbf{0}$—if and only if it is orthogonal to each vector in the row space of **A**. In the language of orthogonal complements, this proves Theorem 5.

THEOREM 5 The Row Space and the Null Space

Let **A** be an $m \times n$ matrix. Then the row space Row(**A**) and the null space Null(**A**) are orthogonal complements in $\mathbf{R}^n$. That is,

$$\text{If} \quad V = \text{Row}(\mathbf{A}), \quad \text{then} \quad V^{\perp} = \text{Null}(\mathbf{A}). \tag{17}$$

Now suppose that a subspace V of $\mathbf{R}^n$ is given, with $\mathbf{v}_1, \mathbf{v}_2, \ldots, \mathbf{v}_m$ a set of vectors that span V. For instance, these vectors may form a given basis for V. Then the implication in (17) provides the following algorithm for finding a basis for the orthogonal complement $V^{\perp}$ of V.

1. Let **A** be the $m \times n$ matrix with row vectors $\mathbf{v}_1, \mathbf{v}_2, \ldots, \mathbf{v}_m$.
2. Reduce **A** to echelon form and use the algorithm of Section 4.4 to find a basis $\{\mathbf{u}_1, \mathbf{u}_2, \ldots, \mathbf{u}_k\}$ for the solution space Null(**A**) of $\mathbf{Ax} = \mathbf{0}$. Because $V^{\perp} =$ Null(**A**), this will be a basis for the orthogonal complement of V.

Example 3 Let V be the 1-dimensional subspace of $\mathbf{R}^3$ spanned by the vector $\mathbf{v}_1 = (1, -3, 5)$. Then

$$\mathbf{A} = \begin{bmatrix} 1 & -3 & 5 \end{bmatrix}$$

and our linear system $\mathbf{Ax} = \mathbf{0}$ consists of the single equation

$$x_1 - 3x_2 + 5x_3 = 0.$$

If $x_2 = s$ and $x_3 = t$, then $x_1 = 3s - 5t$. With $s = 1$ and $t = 0$ we get the solution vector $\mathbf{u}_1 = (3, 1, 0)$, whereas with $s = 0$ and $t = 1$ we get the solution vector $\mathbf{u}_2 = (-5, 0, 1)$. Thus the orthogonal complement $V^{\perp}$ is the 2-dimensional subspace of $\mathbf{R}^3$ having $\mathbf{u}_1 = (3, 1, 0)$ and $\mathbf{u}_2 = (-5, 0, 1)$ as basis vectors. ■

Example 4 Let V be the 2-dimensional subspace of $\mathbf{R}^5$ that has $\mathbf{v}_1 = (1, 2, 1, -3, -3)$ and $\mathbf{v}_2 = (2, 5, 6, -10, -12)$ as basis vectors. The matrix

$$\mathbf{A} = \begin{bmatrix} 1 & 2 & 1 & -3 & -3 \\ 2 & 5 & 6 & -10 & -12 \end{bmatrix}$$

with row vectors $\mathbf{v}_1$ and $\mathbf{v}_2$ has reduced echelon form

$$\mathbf{E} = \begin{bmatrix} 1 & 0 & -7 & 5 & 9 \\ 0 & 1 & 4 & -4 & -6 \end{bmatrix}.$$

Hence the solution space of $\mathbf{Ax} = \mathbf{0}$ is described parametrically by

$$x_3 = r, \quad x_4 = s, \quad x_5 = t,$$
$$x_2 = -4r + 4s + 6t$$
$$x_1 = 7r - 5s - 9t.$$

Then the choice

$$\begin{aligned} r = 1, \quad s = 0, \quad t = 0 \quad &\text{yields} \quad \mathbf{u}_1 = (7, -4, 1, 0, 0); \\ r = 0, \quad s = 1, \quad t = 0 \quad &\text{yields} \quad \mathbf{u}_2 = (-5, 4, 0, 1, 0); \\ r = 0, \quad s = 0, \quad t = 1 \quad &\text{yields} \quad \mathbf{u}_3 = (-9, 6, 0, 0, 1). \end{aligned}$$

Thus the orthogonal complement $V^\perp$ is the 3-dimensional subspace of $\mathbf{R}^5$ with basis $\{\mathbf{u}_1, \mathbf{u}_2, \mathbf{u}_3\}$. ■

Observe that $\dim V + \dim V^\perp = 3$ in Example 3, but $\dim V + \dim V^\perp = 5$ in Example 4. It is no coincidence that in each case the dimensions of V and $V^\perp$ add up to the dimension n of the Euclidean space containing them. To see why, suppose that V is a subspace of $\mathbf{R}^n$ and let $\mathbf{A}$ be an $m \times n$ matrix whose row vectors span V. Then Equation (12) in Section 4.5 implies that

$$\operatorname{rank}(\mathbf{A}) + \dim \operatorname{Null}(\mathbf{A}) = n.$$

But

$$\dim V = \dim \operatorname{Row}(\mathbf{A}) = \operatorname{rank}(\mathbf{A})$$

and

$$\dim V^\perp = \dim \operatorname{Null}(\mathbf{A})$$

by Theorem 5, so it follows that

$$\dim V + \dim V^\perp = n. \tag{18}$$

Moreover, it should be apparent intuitively that if

$$\{\mathbf{v}_1, \mathbf{v}_2, \ldots, \mathbf{v}_m\} \text{ is a basis for } V$$

and

$$\{\mathbf{u}_1, \mathbf{u}_2, \ldots, \mathbf{u}_k\} \text{ is a basis for } V^\perp,$$

then

$$\{\mathbf{v}_1, \mathbf{v}_2, \ldots, \mathbf{v}_m, \mathbf{u}_1, \mathbf{u}_2, \ldots, \mathbf{u}_k\} \text{ is a basis for } \mathbf{R}^n.$$

That is, *the union of a basis for V and a basis for $V^\perp$ is a basis for $\mathbf{R}^n$*. In Problem 34 of this section we ask you to prove that this is so.

4.6 Problems

In Problems 1–4, determine whether the given vectors are mutually orthogonal.

1. $\mathbf{v}_1 = (2, 1, 2, 1)$, $\mathbf{v}_2 = (3, -6, 1, -2)$, $\mathbf{v}_3 = (3, -1, -5, 5)$

2. $\mathbf{v}_1 = (3, -2, 3, -4)$, $\mathbf{v}_2 = (6, 3, 4, 6)$, $\mathbf{v}_3 = (17, -12, -21, 3)$

3. $\mathbf{v}_1 = (5, 2, -4, -1)$, $\mathbf{v}_2 = (3, -5, 1, 1)$, $\mathbf{v}_3 = (3, 0, 8, -17)$

4. $\mathbf{v}_1 = (1, 2, 3, -2, 1)$, $\mathbf{v}_2 = (3, 2, 3, 6, -4)$, $\mathbf{v}_3 = (6, 2, -4, 1, 4)$

In Problems 5–8, the three vertices A, B, and C of a triangle are given. Prove that each triangle is a right triangle by showing that its sides a, b, and c satisfy the Pythagorean relation $a^2 + b^2 = c^2$.

5. $A(6, 6, 5, 8)$, $B(6, 8, 6, 5)$, $C(5, 7, 4, 6)$

6. $A(3, 5, 1, 3)$, $B(4, 2, 6, 4)$, $C(1, 3, 4, 2)$

7. $A(4, 5, 3, 5, -1)$, $B(3, 4, -1, 4, 4)$, $C(1, 3, 1, 3, 1)$

8. $A(2, 8, -3, -1, 2)$, $B(-2, 5, 6, 2, 12)$, $C(-5, 3, 2, -3, 5)$

9–12. Find the acute angles (in degrees) of each of the right triangles of Problems 5–8, respectively.

In Problems 13–22, the given vectors span a subspace V of the indicated Euclidean space. Find a basis for the orthogonal complement $V^\perp$ of V.

13. $\mathbf{v}_1 = (1, -2, 3)$ **14.** $\mathbf{v}_1 = (1, 5, -3)$

15. $\mathbf{v}_1 = (1, -2, -3, 5)$ **16.** $\mathbf{v}_1 = (1, 7, -6, -9)$

17. $\mathbf{v}_1 = (1, 3, 2, 4)$, $\mathbf{v}_2 = (2, 7, 7, 3)$

18. $\mathbf{v}_1 = (1, -3, 3, 5)$, $\mathbf{v}_2 = (2, -5, 9, 3)$

19. $\mathbf{v}_1 = (1, 2, 5, 2, 3)$, $\mathbf{v}_2 = (3, 7, 11, 9, 5)$

20. $\mathbf{v}_1 = (2, 5, 5, 4, 3)$, $\mathbf{v}_2 = (3, 7, 8, 8, 8)$

21. $\mathbf{v}_1 = (1, 2, 3, 1, 3)$, $\mathbf{v}_2 = (1, 3, 4, 3, 6)$, $\mathbf{v}_3 = (2, 2, 4, 3, 5)$

22. $\mathbf{v}_1 = (1, 1, 1, 1, 3)$, $\mathbf{v}_2 = (2, 3, 1, 4, 7)$, $\mathbf{v}_3 = (5, 3, 7, 1, 5)$

23. Prove: For arbitrary vectors $\mathbf{u}$ and $\mathbf{v}$,

(a) $|\mathbf{u}+\mathbf{v}|^2 + |\mathbf{u}-\mathbf{v}|^2 = 2|\mathbf{u}|^2 + 2|\mathbf{v}|^2$;

(b) $|\mathbf{u}+\mathbf{v}|^2 - |\mathbf{u}-\mathbf{v}|^2 = 4\mathbf{u}\cdot\mathbf{v}$.

24. Given mutually orthogonal vectors $\mathbf{v}_1, \mathbf{v}_2, \dots, \mathbf{v}_k$, show that

$$|\mathbf{v}_1 + \mathbf{v}_2 + \cdots + \mathbf{v}_k|^2 = |\mathbf{v}_1|^2 + |\mathbf{v}_2|^2 + \cdots + |\mathbf{v}_k|^2.$$

25. Suppose that A, B, and C are the unit points on distinct coordinate axes in $\mathbf{R}^n$ (for example, perhaps $A = \mathbf{e}_1$, $B = \mathbf{e}_2$, and $C = \mathbf{e}_3$). Show that ABC is an equilateral triangle.

26. Prove that $|\mathbf{u}\cdot\mathbf{v}| = |\mathbf{u}||\mathbf{v}|$ if and only if the vectors $\mathbf{u}$ and $\mathbf{v}$ are linearly dependent.

In Problems 27–29, V denotes a subspace of $\mathbf{R}^n$.

27. Show that the only vector that lies in both V and $V^\perp$ is the zero vector.

28. Prove that if $W = V^\perp$ then $W^\perp = V$.

29. Let S be a spanning set for V. Show that the vector $\mathbf{u}$ is in $V^\perp$ if and only if $\mathbf{u}$ is orthogonal to every vector in S.

30. Suppose that the vectors $\mathbf{u}$ and $\mathbf{v}$ are orthogonal and that $\mathbf{u}+\mathbf{v} = \mathbf{0}$. Show that $\mathbf{u} = \mathbf{v} = \mathbf{0}$.

31. Let S and T be two sets of vectors in $\mathbf{R}^n$ such that each vector in S is orthogonal to every vector in T. Prove that each vector in span(S) is orthogonal to every vector in span(T).

32. Let $S = \{\mathbf{u}_1, \mathbf{u}_2\}$ and $T = \{\mathbf{v}_1, \mathbf{v}_2\}$ be linearly independent sets of vectors such that each $\mathbf{u}_i$ in S is orthogonal to every vector $\mathbf{v}_j$ in T. Then use the results of Problems 30 and 31 to show that the four vectors $\mathbf{u}_1, \mathbf{u}_2, \mathbf{v}_1, \mathbf{v}_2$ are linearly independent.

33. Let $S = \{\mathbf{u}_1, \mathbf{u}_2, \dots, \mathbf{u}_k\}$ and $T = \{\mathbf{v}_1, \mathbf{v}_2, \dots, \mathbf{v}_m\}$ be linearly independent sets of vectors such that each $\mathbf{u}_i$ in S is orthogonal to every vector $\mathbf{v}_j$ in T. Then generalize Problem 32 to show that the $k+m$ vectors $\mathbf{u}_1, \mathbf{u}_2, \dots, \mathbf{u}_k, \mathbf{v}_1, \mathbf{v}_2, \dots, \mathbf{v}_m$ are linearly independent.

34. Deduce from the result of Problem 33 that if V is a subspace of $\mathbf{R}^n$, then the union of a basis for V and a basis for $V^\perp$ is a basis for $\mathbf{R}^n$.

35. Let $\mathbf{A}$ be an $m \times n$ matrix and $\mathbf{b}$ a vector in $\mathbf{R}^m$. Combine the observations listed below to prove that *the nonhomogeneous linear system* $\mathbf{Ax} = \mathbf{b}$ *is consistent if and only if* $\mathbf{b}$ *is orthogonal to the solution space of the homogeneous system* $A^T\mathbf{y} = \mathbf{0}$.

(a) $\mathbf{Ax} = \mathbf{b}$ is consistent if and only if $\mathbf{b}$ is in Col($\mathbf{A}$).

(b) $\mathbf{b}$ is in Col($\mathbf{A}$) if and only if $\mathbf{b}$ is orthogonal to $\{\text{Col}(\mathbf{A})\}^\perp = \{\text{Row}(\mathbf{A}^T)\}^\perp$.

(c) $\{\text{Row}(\mathbf{A}^T)\}^\perp = \text{Null}(\mathbf{A}^T)$.

4.7 General Vector Spaces

In the previous six sections of this chapter, almost all the specific vector spaces appearing in our examples and problems have been vector spaces of n-tuples of real numbers. We have thus confined our attention largely to Euclidean spaces and their subspaces. In this section we discuss examples of some other types of vector spaces that play important roles in various branches of mathematics and their applications.

Example 1 Given fixed positive integers m and n, let M_{mn} denote the set of all $m \times n$ matrices with real number entries. Then M_{mn} is a vector space with the usual operations of addition of matrices and multiplication of matrices by scalars (real numbers). That is, these operations satisfy properties (a)–(h) in the definition of a vector space (Section 4.2). In particular, the zero element in M_{mn} is the $m \times n$ matrix $\mathbf{0}$ matrix whose elements are all zeros, and the negative $-\mathbf{A}$ of the matrix $\mathbf{A}$ is (as usual) the matrix whose elements are the negatives of the corresponding entries of $\mathbf{A}$.

Given positive integers i and j with $1 \le i \le m$ and $1 \le j \le n$, let $\mathbf{E}_{ij}$ denote the $m \times n$ matrix whose only nonzero entry is the number 1 in the ith row and the jth column. Then it should be clear that the mn elements $\{\mathbf{E}_{ij}\}$ of M_{mn} form a basis for M_{mn}, and so M_{mn} is a finite-dimensional vector space of dimension mn.

For instance, in the case of the vector space M_{22} of all 2×2 matrices, these alleged basis elements are the four matrices

$$\mathbf{E}_{11} = \begin{bmatrix} 1 & 0 \\ 0 & 0 \end{bmatrix}, \quad \mathbf{E}_{12} = \begin{bmatrix} 0 & 1 \\ 0 & 0 \end{bmatrix},$$
$$\mathbf{E}_{21} = \begin{bmatrix} 0 & 0 \\ 1 & 0 \end{bmatrix}, \quad \text{and} \quad \mathbf{E}_{22} = \begin{bmatrix} 0 & 0 \\ 0 & 1 \end{bmatrix}.$$

Then any matrix $\mathbf{A}$ in M_{22} can be expressed as

$$\mathbf{A} = \begin{bmatrix} a & b \\ c & d \end{bmatrix} = a\mathbf{E}_{11} + b\mathbf{E}_{12} + c\mathbf{E}_{21} + d\mathbf{E}_{22},$$

so the set $\{\mathbf{E}_{11}, \mathbf{E}_{12}, \mathbf{E}_{21}, \mathbf{E}_{22}\}$ spans M_{22}. Moreover,

$$a\mathbf{E}_{11} + b\mathbf{E}_{12} + c\mathbf{E}_{21} + d\mathbf{E}_{22} = \begin{bmatrix} a & b \\ c & d \end{bmatrix} = \begin{bmatrix} 0 & 0 \\ 0 & 0 \end{bmatrix}$$

implies immediately that $a = b = c = d = 0$, so the set $\{\mathbf{E}_{11}, \mathbf{E}_{12}, \mathbf{E}_{21}, \mathbf{E}_{22}\}$ is also linearly independent and therefore forms a basis for M_{22}. Thus, we have shown that M_{22} is a 4-dimensional vector space. ■

Example 2 Let C denote the subset of M_{22} consisting of all 2×2 matrices of the form

$$\begin{bmatrix} a & -b \\ b & a \end{bmatrix}. \tag{1}$$

Obviously the sum of any two such matrices is such a matrix, as is any scalar multiple of such a matrix. Thus C is closed under the operations of addition of matrices and multiplication by scalars and is therefore a *subspace* of M_{22}.

To determine the *dimension* of the subspace C, we consider the two special matrices

$$\mathbf{Re} = \begin{bmatrix} 1 & 0 \\ 0 & 1 \end{bmatrix} \quad \text{and} \quad \mathbf{Im} = \begin{bmatrix} 0 & -1 \\ 1 & 0 \end{bmatrix} \tag{2}$$

in C. Here $\mathbf{Re}$ is another notation for the 2×2 identity matrix $\mathbf{I}$, whereas the matrix $\mathbf{Im}$ is not so familiar. The matrices $\mathbf{Re}$ and $\mathbf{Im}$ are linearly independent—obviously neither is a scalar multiple of the other—and

$$\begin{bmatrix} a & -b \\ b & a \end{bmatrix} = a\mathbf{Re} + b\mathbf{Im}, \tag{3}$$

so these two matrices span C. Thus $\{\mathbf{Re}, \mathbf{Im}\}$ is a basis for C, and so C is a 2-dimensional subspace of the 4-dimensional vector space M_{22} of 2×2 matrices.

Now observe that

$$(\mathbf{Im})^2 = \begin{bmatrix} 0 & -1 \\ 1 & 0 \end{bmatrix}\begin{bmatrix} 0 & -1 \\ 1 & 0 \end{bmatrix} = \begin{bmatrix} -1 & 0 \\ 0 & -1 \end{bmatrix} = -\mathbf{Re}. \tag{4}$$

It follows that the matrix product of any two elements $a\mathbf{Re} + b\mathbf{Im}$ and $c\mathbf{Re} + d\mathbf{Im}$ of C is given by

$$\begin{aligned}(a\mathbf{Re} + b\mathbf{Im})(c\mathbf{Re} + d\mathbf{Im}) &= ac(\mathbf{Re})^2 + bc\mathbf{Im}\,\mathbf{Re} + ad\mathbf{Re}\,\mathbf{Im} + bd(\mathbf{Im})^2 \\ &= (ac - bd)\mathbf{Re} + (ad + bc)\mathbf{Im}. \qquad (5)\end{aligned}$$

This explicit formula for the product of two matrices in C shows that, in addition to being closed under matrix addition and multiplication by scalars, the subspace C of M_{22} also is *closed under matrix multiplication.* ■

If you feel that the idea of a space of matrices being closed under matrix multiplication is a rather complex one, you are correct! Indeed, the vector space C of Example 2 can be taken as a "model" for the set of all *complex numbers* of the form $a + bi$, where a and b are real numbers and $i = \sqrt{-1}$ denotes the "imaginary" square root of -1 (so that $i^2 = -1$). Because $(\mathbf{Im})^2 = -\mathbf{Re}$, the matrix $\mathbf{Im}$ plays the role of i and the matrix

$$\begin{bmatrix} a & -b \\ b & a \end{bmatrix} = a\mathbf{Re} + b\mathbf{Im}$$

corresponds to the complex number $a + bi$. And the matrix multiplication formula in (5) is then analogous to the formula

$$(a + bi)(c + di) = (ac - bd) + (ad + bc)i$$

for the multiplication of complex numbers. In addition to providing a concrete interpretation of complex numbers in terms of real numbers, the matrix model C provides a convenient way to manipulate complex numbers in computer programming languages in which matrix operations are "built in" but complex operations are not.

Function Spaces

In Example 1 of Section 4.2 we introduced the vector space $\mathcal{F}$ of all real-valued functions defined on the real line $\mathbf{R}$. If $\mathbf{f}$ and $\mathbf{g}$ are elements of $\mathcal{F}$ and c is a scalar, then

$$(\mathbf{f} + \mathbf{g})(x) = \mathbf{f}(x) + \mathbf{g}(x)$$

and (6)

$$(c\mathbf{f})(x) = c(\mathbf{f}(x))$$

for all x in $\mathbf{R}$. The zero element in $\mathcal{F}$ is the function $\mathbf{0}$ such that $\mathbf{0}(x) = 0$ for all x. We frequently refer to a function in $\mathcal{F}$ by simply specifying its formula. For instance, by "the function x^2" we mean that function whose value at each x in $\mathbf{R}$ is x^2.

The functions $\mathbf{f}_1, \mathbf{f}_2, \ldots, \mathbf{f}_k$ in $\mathcal{F}$ are linearly dependent provided that there exist scalars $c_1, c_2, \ldots, c_k$, not all zero, such that

$$c_1\mathbf{f}_1(x) + c_2\mathbf{f}_2(x) + \cdots + c_k\mathbf{f}_k(x) = 0$$

for all x in $\mathbf{R}$. We often can determine whether two given functions are linearly dependent simply by observing whether one is a scalar multiple of the other.

Example 3 The functions e^x and e^{-2x} are linearly independent because either of the equations $e^x = ae^{-2x}$ or $e^{-2x} = be^x$ would imply that e^{3x} is a constant, which obviously is not so. By contrast, the functions $\sin 2x$ and $\sin x \cos x$ are linearly dependent, because of the trigonometric identity $\sin 2x = 2 \sin x \cos x$. The three functions 1, $\cos^2 x$, and $\sin^2 x$ are linearly dependent, because the fundamental identity $\cos^2 x + \sin^2 x = 1$ can be written in the form

$$(-1)(1) + (1)(\cos^2 x) + (1)(\sin^2 x) = 0.$$ ■

A subspace of $\mathcal{F}$ is called a **function space**. An example of a function space is the vector space $\mathcal{P}$ of all polynomials in $\mathcal{F}$ (Example 4 in Section 4.4). Recall that a function $\mathbf{p}(x)$ in $\mathcal{F}$ is called a **polynomial of degree** $n \geqq 0$ if it can be expressed in the form

$$\mathbf{p}(x) = a_0 + a_1 x + a_2 x^2 + \cdots + a_n x^n \tag{7}$$

with $a_n \neq 0$. Polynomials are added and multiplied by scalars in the usual manner—by collecting coefficients of like powers of x. Clearly any linear combination of polynomials is a polynomial, so $\mathcal{P}$ is, indeed, a subspace of $\mathcal{F}$. (Recall the subspace criterion of Theorem 1 in Section 4.2.)

Example 4 Given $n \geq 0$, denote by $\mathcal{P}_n$ the set of all polynomials of degree at most n. That is, $\mathcal{P}_n$ consists of all polynomials of degrees $0, 1, 2, \ldots, n$. Any linear combination of two polynomials of degree at most n is again a polynomial of degree at most n, so $\mathcal{P}_n$ is a subspace of $\mathcal{P}$ (and of $\mathcal{F}$). The formula in (7) shows that the $n + 1$ polynomials

$$1, x, x^2, x^3, \ldots, x^n \tag{8}$$

span $\mathcal{P}_n$. To show that these "monomials" are linearly independent, suppose that

$$c_0 + c_1 x + c_2 x^2 + \cdots + c_n x^n = 0 \tag{9}$$

for all x in **R**. If $a_0, a_1, \ldots, a_n$ are fixed *distinct* real numbers, then each a_i satisfies Eq. (9). In matrix notation this means that

$$\begin{bmatrix} 1 & a_0 & a_0^2 & \cdots & a_0^n \\ 1 & a_1 & a_1^2 & \cdots & a_1^n \\ 1 & a_2 & a_2^2 & \cdots & a_2^n \\ \vdots & \vdots & \vdots & \ddots & \vdots \\ 1 & a_n & a_n^2 & \cdots & a_n^n \end{bmatrix} \begin{bmatrix} c_0 \\ c_1 \\ c_2 \\ \vdots \\ c_n \end{bmatrix} = \begin{bmatrix} 0 \\ 0 \\ 0 \\ \vdots \\ 0 \end{bmatrix}. \tag{10}$$

The coefficient matrix in (10) is a Vandermonde matrix **V**, and Problems 61–63 in Section 3.6 imply that any Vandermonde matrix is nonsingular. Therefore the system $\mathbf{Vc} = \mathbf{0}$ has only the trivial solution $\mathbf{c} = \mathbf{0}$, so it follows from (10) that $c_0 = c_1 = c_2 = \cdots = c_n = 0$. Thus we have proved that the $n + 1$ monomials $1, x, x^2, \ldots, x^n$ are linearly independent and hence constitute a basis for $\mathcal{P}_n$. Consequently $\mathcal{P}_n$ is an $(n + 1)$-dimensional vector space. ■

The fact that the monomials in (8) are linearly independent implies that the coefficients in a polynomial are unique. That is, if

$$a_0 + a_1 x + \cdots + a_n x^n = b_0 + b_1 x + \cdots + b_n x^n$$

for all x, then $a_0 = b_0$, $a_1 = b_1$, ..., and $a_n = b_n$. (See the remark that follows the definition of linear independence in Section 4.3.) This *identity principle* for polynomials is often used without proof in elementary algebra. A typical application is the method of *partial-fraction decomposition* illustrated in Example 5.

Example 5 Find constants A, B, and C such that

$$\frac{6x}{(x-1)(x+1)(x+2)} = \frac{A}{x-1} + \frac{B}{x+1} + \frac{C}{x+2} \tag{11}$$

for all x (other than $x = 1, -1$, or -2).

Solution Multiplication of each side of the equation in (11) by the denominator on the left-hand side yields

$$6x = A(x+1)(x+2) + B(x-1)(x+2) + C(x-1)(x+1);$$
$$6x = (2A - 2B - C) + (3A + B)x + (A + B + C)x^2.$$

Then the identity principle for polynomials yields the linear equations

$$\begin{aligned} 2A - 2B - C &= 0 \\ 3A + B \phantom{{}- C} &= 6 \\ A + B + C &= 0, \end{aligned}$$

which we readily solve for $A = 1$, $B = 3$, and $C = -4$. Therefore

$$\frac{6x}{(x-1)(x+1)(x+2)} = \frac{1}{x-1} + \frac{3}{x+1} - \frac{4}{x+2}$$

if $x \neq 1, -1, -2$. ■

Example 6 Show that the four polynomials

$$1, \quad x, \quad 3x^2 - 1, \quad \text{and} \quad 5x^3 - 3x \tag{12}$$

form a basis for $\mathcal{P}_3$.

Solution In order to show simultaneously that these four polynomials span $\mathcal{P}_3$ and are linearly independent, it suffices to see that every polynomial

$$p(x) = b_0 + b_1 x + b_2 x^2 + b_3 x^3 \tag{13}$$

in $\mathcal{P}_3$ can be expressed uniquely as a linear combination

$$\begin{aligned} &c_0(1) + c_1(x) + c_2(3x^2 - 1) + c_3(5x^3 - 3x) \\ &\quad = (c_0 - c_2) + (c_1 - 3c_3)x + (3c_2)x^2 + (5c_3)x^3 \end{aligned} \tag{14}$$

of the polynomials in (12). But upon comparing the coefficients in (13) and (14), we see that we need only observe that the linear system

$$\begin{bmatrix} 1 & 0 & -1 & 0 \\ 0 & 1 & 0 & -3 \\ 0 & 0 & 3 & 0 \\ 0 & 0 & 0 & 5 \end{bmatrix} \begin{bmatrix} c_0 \\ c_1 \\ c_2 \\ c_3 \end{bmatrix} = \begin{bmatrix} b_0 \\ b_1 \\ b_2 \\ b_3 \end{bmatrix}$$

obviously has a unique solution for c_0, c_1, c_2, c_3 in terms of b_0, b_1, b_2, b_3. ■

Because the vector space $\mathcal{P}$ of all polynomials contains $n+1$ linearly independent functions for every integer $n \geq 0$, it follows that $\mathcal{P}$ is an infinite-dimensional vector space. So too is any function space, such as $\mathcal{F}$ itself, that contains $\mathcal{P}$. Another important function space is the set $\mathcal{C}^{(0)}$ of all continuous functions on $\mathbf{R}$; $\mathcal{C}^{(0)}$ is a subspace of $\mathcal{F}$ because every linear combination of continuous functions is again a continuous function. Every polynomial is a continuous function, so $\mathcal{C}^{(0)}$ contains $\mathcal{P}$ and is therefore an infinite-dimensional vector space. Similarly, the set $\mathcal{C}^{(k)}$ of all functions in $\mathcal{F}$ that have continuous kth-order derivatives is an infinite-dimensional function space.

Solution Spaces of Differential Equations

In Section 1.5 we saw that the (now familiar) first-order differential equation

$$y' = ky \qquad (k \text{ constant}) \tag{15}$$

has the general solution $y(x) = Ce^{kx}$. Thus the "solution space" of (15) is the set of all constant multiples of the single function e^{kx}, and it is therefore the 1-dimensional function space with basis $\{e^{kx}\}$.

In Section 5.1 we will see that the set of all solutions $y(x)$ of a linear second-order differential equation of the form

$$ay'' + by' + cy = 0 \tag{16}$$

(with constant coefficients a, b, and c) is a 2-dimensional function space $\mathcal{S}$, called the **solution space** of the differential equation. Here we will illustrate this general fact with three simple examples of second-order equations that can be solved in an elementary manner.

Example 7 With $a = 1$ and $b = c = 0$ in (16), we get the differential equation

$$y'' = 0. \tag{17}$$

If $y(x)$ is any solution of this equation, so $y''(x) = 0$, then two integrations give first

$$y'(x) = \int y''(x)\,dx = \int (0)\,dx = A$$

and then

$$y(x) = \int y'(x)\,dx = \int A\,dx = Ax + B,$$

where A and B are arbitrary constants of integration. Thus every solution of (17) is of the form $y(x) = Ax + B$, and clearly every such function is a solution of the differential equation. Thus the solution space of (17) is simply the 2-dimensional space $\mathcal{P}_1$ of linear polynomials generated by the basis $\{1, x\}$. ■

Example 8 With $a = 1$, $b = -2$, and $c = 0$ in (16), we get the differential equation

$$y'' - 2y' = 0. \tag{18}$$

If $y(x)$ is any solution of this equation and if $v(x) = y'(x)$, then (18) says that

$$v'(x) = 2v(x).$$

But this is a familiar first-order differential equation with general solution $v(x) = Ce^{2x}$. Then integration gives

$$y(x) = \int y'(x)\,dx = \int v(x)\,dx = \int Ce^{2x}\,dx = \tfrac{1}{2}Ce^{2x} + A.$$

Thus every solution of (18) is of the form $y(x) = A + Be^{2x}$ (where $B = \frac{1}{2}C$), and you should show by substitution in (18) that every function of this form is a solution of the differential equation. Therefore, the solution space of (18) is the 2-dimensional function space that is generated by the basis $\{1, e^{2x}\}$. ■

Example 9 With $a = 1$, $b = 0$, and $c = -1$ in (16), we get the differential equation

$$y'' - y = 0. \tag{19}$$

If $y(x)$ is any solution of this equation and if $v(x) = y'(x)$, then (19) says that

$$y'' = \frac{dv}{dx} = \frac{dv}{dy}\frac{dy}{dx} = v\frac{dv}{dy} = y;$$

so $v\,dv = y\,dy$. Integration therefore gives

$$\tfrac{1}{2}v^2 = \tfrac{1}{2}y^2 + C, \quad \text{so} \quad v^2 = y^2 \pm a^2, \tag{20}$$

where the final sign depends on the sign of C. The final result is independent of this choice (Problem 27); we'll take the negative sign for the sake of illustration. Since $v = dy/dx$, it then follows from (20) that

$$x = \pm \int \frac{dy}{\sqrt{y^2 - a^2}}. \tag{21}$$

Here again the final result is independent of the choice of sign; this time we'll take the positive sign. Then the substitution $y = au$ and the standard integral

$$\int \frac{du}{\sqrt{u^2 - 1}} = \cosh^{-1} u + C$$

(see Section 7.6 of Edwards and Penney, *Calculus: Early Transcendentals*, 7th edition, Prentice Hall, 2008) yield

$$x = \cosh^{-1}\frac{y}{a} + b,$$

so finally,

$$y = a\cosh(x - b) = a(\cosh x \cosh b - \sinh x \sinh b),$$

which can be written as

$$y(x) = A\cosh x + B\sinh x, \tag{22}$$

with $A = a\cosh b$ and $B = -a\sinh b$. Thus every solution of the differential equation $y'' - y = 0$ in (19) is of the form $y(x) = A\cosh x + B\sinh x$, and you should show by substitution in (19) that every function of this form is a solution of the differential equation. Therefore, the solution space $\mathcal{S}$ of (19) is the 2-dimensional function space that is generated by the basis $\{\cosh x, \sinh x\}$. If we substitute into (22) the relations

$$\cosh x = \frac{e^x + e^{-x}}{2}, \quad \sinh x = \frac{e^x - e^{-x}}{2}$$

that define the hyperbolic cosine and sine functions in terms of exponentials, we see that every solution of $y'' - y = 0$ can also be written in the form of the linear combination

$$y(x) = Ce^x + De^{-x}.$$

Consequently, $\{e^x, e^{-x}\}$ is a second basis for the 2-dimensional solution space $\mathcal{S}$. ■

In Section 5.3 we will learn a much quicker way to solve linear second-order differential equations like the equation $y'' - y = 0$ of Example 9 and the equation $y'' + y = 0$ of Problem 28.

4.7 Problems

In Problems 1–4, determine whether or not the indicated set of 3×3 matrices is a subspace of M_{33}.

1. The set of all diagonal 3×3 matrices.
2. The set of all symmetric 3×3 matrices (that is, matrices $\mathbf{A} = [\,a_{ij}\,]$ such that $a_{ij} = a_{ji}$ for $1 \le i \le 3$, $1 \le j \le 3$).
3. The set of all nonsingular 3×3 matrices.
4. The set of all singular 3×3 matrices.

*In Problems 5–8, determine whether or not each indicated set of functions is a subspace of the space $\mathcal{F}$ of all real-valued functions on **R**.*

5. The set of all f such that $f(0) = 0$.
6. The set of all f such that $f(x) \neq 0$ for all x.
7. The set of all f such that $f(0) = 0$ and $f(1) = 1$.
8. The set of all f such that $f(-x) = -f(x)$ for all x.

In Problems 9–12, a condition on the coefficients of a polynomial $a_0 + a_1x + a_2x^2 + a_3x^3$ is given. Determine whether or not the set of all such polynomials satisfying this condition is a subspace of the space $\mathcal{P}$ of all polynomials.

9. $a_3 \neq 0$
10. $a_0 = a_1 = 0$
11. $a_0 + a_1 + a_2 + a_3 = 0$
12. a_0, a_1, a_2, and a_3 are all integers

In Problems 13–18, determine whether the given functions are linearly independent.

13. $\sin x$ and $\cos x$
14. e^x and xe^x
15. $1 + x$, $1 - x$, and $1 - x^2$
16. $1 + x$, $x + x^2$, and $1 - x^2$
17. $\cos 2x$, $\sin^2 x$, and $\cos^2 x$
18. $2\cos x + 3\sin x$ and $4\cos x + 5\sin x$

In Problems 19–22, use the method of Example 5 to find the constants A, B, and C in the indicated partial-fraction decompositions.

19. $\dfrac{x-5}{(x-2)(x-3)} = \dfrac{A}{x-2} + \dfrac{B}{x-3}$

20. $\dfrac{2}{x(x^2-1)} = \dfrac{A}{x} + \dfrac{B}{x-1} + \dfrac{C}{x+1}$

21. $\dfrac{8}{x(x^2+4)} = \dfrac{A}{x} + \dfrac{Bx+C}{x^2+4}$

22. $\dfrac{2x}{(x+1)(x+2)(x+3)} = \dfrac{A}{x+1} + \dfrac{B}{x+2} + \dfrac{C}{x+3}$

In Problems 23 and 24, use the method of Example 7 to find a basis for the solution space of the given differential equation. (It's 3-dimensional in Problem 23 and 4-dimensional in Problem 24.)

23. $y''' = 0$
24. $y^{(4)} = 0$

In Problems 25 and 26, use the method of Example 8 to find a basis for the 2-dimensional solution space of the given differential equation.

25. $y'' - 5y' = 0$
26. $y'' + 10y' = 0$

27. Take the positive sign in Eq. (20), and then use the standard integral

$$\int \frac{du}{\sqrt{u^2+1}} = \sinh^{-1} u + C$$

to derive the same general solution $y(x) = A\cosh x + B\sinh x$ given in Eq. (22).

28. Use the method of Example 9 and the standard integral

$$\int \frac{du}{\sqrt{1-u^2}} = \sin^{-1} u + C$$

to derive the general solution $y(x) = A\cos x + B\sin x$ of the second-order differential equation $y'' + y = 0$. Thus its solution space has basis $\{\cos x, \sin x\}$.

29. Let V be the set of all infinite sequences $\{x_n\} = \{x_1, x_2, x_3, \dots\}$ of real numbers. Let addition of elements of V and multiplication by scalars be defined as follows:

$$\{x_n\} + \{y_n\} = \{x_n + y_n\}$$

and

$$c\{x_n\} = \{cx_n\}.$$

(a) Show that V is a vector space with these operations.
(b) Prove that V is infinite dimensional.

30. Let V be the vector space of Problem 29 and let the subset W consist of those elements $\{x_n\}$ of V such that $x_n = x_{n-1} + x_{n-2}$ for $n \geq 2$. Thus a typical element of W is the *Fibonacci sequence*

$$\{1, 1, 2, 3, 5, 8, 13, 21, 34, 55, \dots\}.$$

(a) Show that W is a subspace of V.
(b) Prove that W is 2-dimensional.

31. Motivated by Example 2, let us define a function or transformation T from the set of all complex numbers to the space of all 2×2 matrices of the form in (1) as follows. Given a complex number $z = a + bi$, let

$$T(z) = \begin{bmatrix} a & -b \\ b & a \end{bmatrix}.$$

(a) Suppose that z_1 and z_2 are complex numbers and c_1 and c_2 are real numbers. Show that

$$T(c_1 z_1 + c_2 z_2) = c_1 T(z_1) + c_2 T(z_2).$$

(b) Show that

$$T(z_1 z_2) = T(z_1)T(z_2)$$

for all complex numbers z_1 and z_2. Note the complex multiplication on the left in contrast with the matrix multiplication on the right.

(c) Prove that if z is an arbitrary nonzero complex number, then

$$T(z^{-1}) = \{T(z)\}^{-1},$$

where $z^{-1} = 1/z$ and $\{T(z)\}^{-1}$ is the inverse of the matrix $T(z)$.

32. Let **A** and **B** be 4×4 (real) matrices partitioned into 2×2 submatrices or "blocks":

$$\mathbf{A} = \begin{bmatrix} \mathbf{A}_{11} & \mathbf{A}_{12} \\ \mathbf{A}_{21} & \mathbf{A}_{22} \end{bmatrix}, \quad \mathbf{B} = \begin{bmatrix} \mathbf{B}_{11} & \mathbf{B}_{12} \\ \mathbf{B}_{21} & \mathbf{B}_{22} \end{bmatrix}.$$

Then verify that **AB** can be calculated in "blockwise" fashion:

$$\mathbf{AB} = \begin{bmatrix} \mathbf{A}_{11}\mathbf{B}_{11} + \mathbf{A}_{12}\mathbf{B}_{21} & \mathbf{A}_{11}\mathbf{B}_{12} + \mathbf{A}_{12}\mathbf{B}_{22} \\ \mathbf{A}_{21}\mathbf{B}_{11} + \mathbf{A}_{22}\mathbf{B}_{21} & \mathbf{A}_{21}\mathbf{B}_{12} + \mathbf{A}_{22}\mathbf{B}_{22} \end{bmatrix}.$$

33. Given a 2×2 matrix $\mathbf{M} = \begin{bmatrix} z_{ij} \end{bmatrix}$ whose entries are complex numbers, let $T(\mathbf{M})$ denote the 4×4 matrix of real numbers given in block form by

$$T(\mathbf{M}) = \begin{bmatrix} T(z_{11}) & T(z_{12}) \\ T(z_{21}) & T(z_{22}) \end{bmatrix}.$$

The transformation T on the right-hand side is the one we defined in Problem 31. Suppose that **M** and **N** are 2×2 complex matrices. Use Problems 31(b) and 32 to show that

$$T(\mathbf{MN}) = T(\mathbf{M})T(\mathbf{N}).$$

Hence one can find the product of the 2×2 complex matrices **M** and **N** by calculating the product of the 4×4 real matrices $T(\mathbf{M})$ and $T(\mathbf{N})$. For instance, if

$$\mathbf{M} = \begin{bmatrix} 1+i & 2+i \\ 1+2i & 1+3i \end{bmatrix}$$

and

$$\mathbf{N} = \begin{bmatrix} 3-i & 1-2i \\ 2-i & 3-2i \end{bmatrix},$$

then

$$T(\mathbf{M})T(\mathbf{N}) = \left[\begin{array}{rr|rr} 1 & -1 & 2 & -1 \\ 1 & 1 & 1 & 2 \\ \hline 1 & -2 & 1 & -3 \\ 2 & 1 & 3 & 1 \end{array}\right] \times \left[\begin{array}{rr|rr} 3 & 1 & 1 & 2 \\ -1 & 3 & -2 & 1 \\ \hline 2 & 1 & 3 & 2 \\ -1 & 2 & -2 & 3 \end{array}\right]$$

$$= \left[\begin{array}{rr|rr} 9 & -2 & 11 & 2 \\ 2 & 9 & -2 & 11 \\ \hline 10 & -10 & 14 & -7 \\ 10 & 10 & 7 & 14 \end{array}\right].$$

Therefore,

$$\mathbf{MN} = \begin{bmatrix} 9+2i & 11-2i \\ 10+10i & 14+7i \end{bmatrix}.$$

If **M** is a nonsingular 2×2 complex matrix, it can be shown that $T(\mathbf{M}^{-1}) = \{T(\mathbf{M})\}^{-1}$. The $n \times n$ versions of these results are sometimes used to carry out complex matrix operations in computer languages that support only real arithmetic.

5 Higher-Order Linear Differential Equations

5.1 Introduction: Second-Order Linear Equations

In Chapters 1 and 2 we investigated first-order differential equations. We now turn to equations of higher order $n \geqq 2$, beginning in this chapter with equations that are linear. The general theory of linear differential equations parallels the second-order case ($n = 2$), which we outline in this initial section.

Recall that a second-order differential equation in the (unknown) function $y(x)$ is one of the form

$$G(x, y, y', y'') = 0. \tag{1}$$

This differential equation is said to be **linear** provided that G is linear in the dependent variable y and its derivatives y' and y''. Thus a linear second-order equation takes (or can be written in) the form

$$A(x)y'' + B(x)y' + C(x)y = F(x). \tag{2}$$

Unless otherwise noted, we will always assume that the (known) coefficient functions $A(x)$, $B(x)$, $C(x)$, and $F(x)$ are continuous on some open interval I (perhaps unbounded) on which we wish to solve this differential equation, but we do *not* require that they be linear functions of x. Thus the differential equation

$$e^x y'' + (\cos x)y' + (1 + \sqrt{x})y = \tan^{-1} x$$

is linear because the dependent variable y and its derivatives y' and y'' appear linearly. By contrast, the equations

$$y'' = yy' \quad \text{and} \quad y'' + 3(y')^2 + 4y^3 = 0$$

are not linear because products and powers of y or its derivatives appear.

If the function $F(x)$ on the right-hand side of Eq. (2) vanishes identically on I, then we call Eq. (2) a **homogeneous** linear equation; otherwise, it is **nonhomogeneous**. For example, the second-order equation

$$x^2y'' + 2xy' + 3y = \cos x$$

is nonhomogeneous; its *associated* homogeneous equation is

$$x^2y'' + 2xy' + 3y = 0.$$

In general, the homogeneous linear equation **associated** with Eq. (2) is

$$A(x)y'' + B(x)y' + C(x)y = 0. \tag{3}$$

In case the differential equation in (2) models a physical system, the nonhomogeneous term $F(x)$ frequently corresponds to some *external* influence on the system.

Remark: Note that the meaning of the term "homogeneous" for a second-order linear differential equation is quite different from its meaning for a first-order differential equation (as in Section 1.6). Of course, it is not unusual—either in mathematics or in the English language more generally—for the same word to have different meanings in different contexts.

A Typical Application

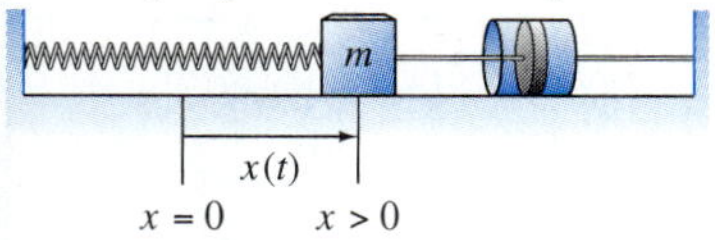

FIGURE 5.1.1. A mass-spring-dashpot system.

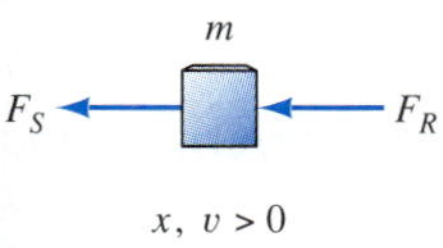

FIGURE 5.1.2. Directions of the forces acting on m.

Linear differential equations frequently appear as mathematical models of mechanical systems and electrical circuits. For example, suppose that a mass m is attached both to a spring that exerts on it a force F_S and to a dashpot (shock absorber) that exerts a force F_R on the mass (Fig. 5.1.1). Assume that the restoring force F_S of the spring is proportional to the *displacement* x of the mass from its equilibrium position and acts opposite to the direction of displacement (Fig. 5.1.2). Then

$$F_S = -kx \qquad (\text{with } k > 0)$$

so $F_S < 0$ if $x > 0$ (spring stretched) while $F_S > 0$ if $x < 0$ (spring compressed). We assume that the dashpot force F_R is proportional to the *velocity* $v = dx/dt$ of the mass and acts opposite to the direction of motion. Then

$$F_R = -cv = -c\frac{dx}{dt} \qquad (\text{with } c > 0)$$

so $F_R < 0$ if $v > 0$ (motion to the right) while $F_R > 0$ if $v < 0$ (motion to the left).

If F_R and F_S are the only forces acting on the mass m and its resulting acceleration is $a = dv/dt$, then Newton's law $F = ma$ gives

$$mx'' = F_S + F_R; \tag{4}$$

that is,

$$m\frac{d^2x}{dt^2} + c\frac{dx}{dt} + kx = 0. \tag{5}$$

Thus we have a differential equation satisfied by the position function $x(t)$ of the mass m. This homogeneous second-order linear equation governs the *free vibrations* of the mass; we will return to this problem in detail in Section 5.4.

If, in addition to F_S and F_R, the mass m is acted on by an external force $F(t)$—which must then be added to the right-hand side in Eq. (4)—the resulting equation is

$$m\frac{d^2x}{dt^2} + c\frac{dx}{dt} + kx = F(t). \tag{6}$$

This nonhomogeneous linear differential equation governs the *forced vibrations* of the mass under the influence of the external force $F(t)$.

Homogeneous Second-Order Linear Equations

Consider the general second-order linear equation

$$A(x)y'' + B(x)y' + C(x)y = F(x), \tag{7}$$

where the coefficient functions A, B, C, and F are continuous on the open interval I. Here we assume in addition that $A(x) \neq 0$ at each point of I, so we can divide each term in Eq. (7) by $A(x)$ and write it in the form

$$y'' + p(x)y' + q(x)y = f(x). \tag{8}$$

We will discuss first the associated homogeneous equation

$$y'' + p(x)y' + q(x)y = 0. \tag{9}$$

A particularly useful property of this *homogeneous* linear equation is the fact that the sum of any two solutions of Eq. (9) is again a solution, as is any constant multiple of a solution. This is the central idea of the following theorem.

THEOREM 1 Principle of Superposition for Homogeneous Equations

Let y_1 and y_2 be two solutions of the homogeneous linear equation in (9) on the interval I. If c_1 and c_2 are constants, then the linear combination

$$y = c_1y_1 + c_2y_2 \tag{10}$$

is also a solution of Eq. (9) on I.

***Proof*:** The conclusion follows almost immediately from the linearity of the operation of differentiation, which gives

$$y' = c_1y_1' + c_2y_2' \quad \text{and} \quad y'' = c_1y_1'' + c_2y_2''.$$

Then

$$\begin{aligned} y'' + py' + qy &= (c_1y_1 + c_2y_2)'' + p(c_1y_1 + c_2y_2)' + q(c_1y_1 + c_2y_2) \\ &= (c_1y_1'' + c_2y_2'') + p(c_1y_1' + c_2y_2') + q(c_1y_1 + c_2y_2) \\ &= c_1(y_1'' + py_1' + qy_1) + c_2(y_2'' + py_2' + qy_2) \\ &= c_1 \cdot 0 + c_2 \cdot 0 = 0 \end{aligned}$$

because y_1 and y_2 are solutions. Thus $y = c_1y_1 + c_2y_2$ is also a solution. ■

Remark: According to Theorem 1 in Section 4.2, Theorem 1 here implies that the set $\mathcal{S}$ of all solutions of the homogeneous linear second-order equation (9) is a subspace of the vector space of all functions on the real line **R**—and hence is itself a vector space. We therefore call S the **solution space** of the differential equation. ■

Example 1 We can see by inspection that

$$y_1(x) = \cos x \quad \text{and} \quad y_2(x) = \sin x$$

are two solutions of the equation

$$y'' + y = 0.$$

For instance,

$$y_1'' + y_1 = (\cos x)'' + \cos x = (-\cos x) + \cos x = 0.$$

Theorem 1 tells us that any linear combination of these solutions, such as

$$y(x) = 3y_1(x) - 2y_2(x) = 3\cos x - 2\sin x,$$

is also a solution. Conversely, Problem 28 in Section 4.5 says that *every* solution of $y'' + y = 0$ is a linear combination of these two particular solutions y_1 and y_2. Thus, a general solution of $y'' + y = 0$ is given by

$$y(x) = c_1 \cos x + c_2 \sin x.$$

It is important to understand that this single formula for the general solution encompasses a "twofold infinity" of particular solutions. Indeed, the pair $\{\cos x, \sin x\}$ constitutes a basis of the 2-dimensional solution space $\mathcal{S}$ of the differential equation $y'' + y = 0$. Figures 5.1.3 through 5.1.5 illustrate some of the solutions obtained either with one of the coefficients c_1 and c_2 set equal to zero, or with both nonzero. ■

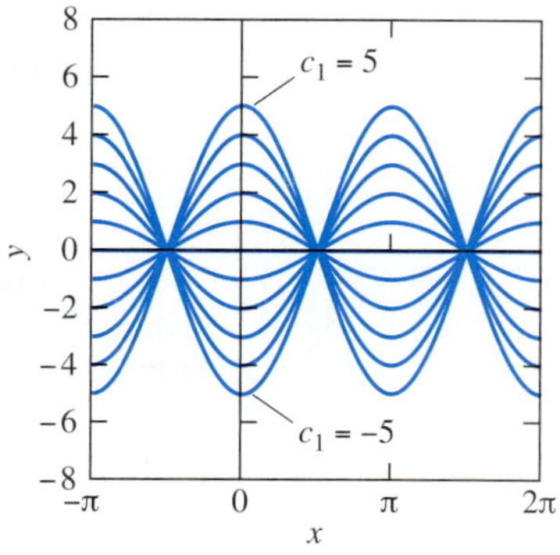

FIGURE 5.1.3. Solutions $y(x) = c_1 \cos x$ of $y'' + y = 0$.

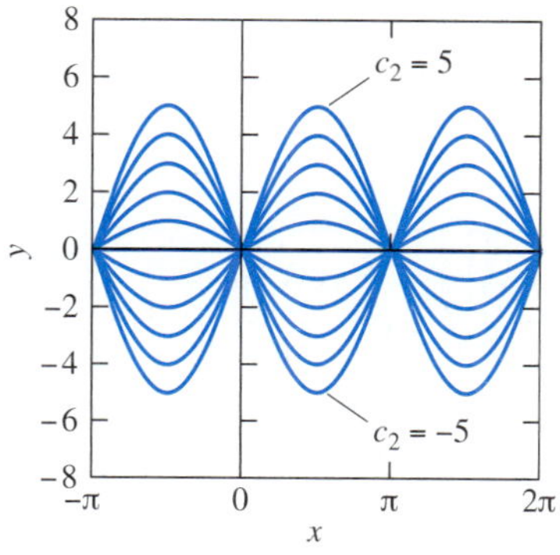

FIGURE 5.1.4. Solutions $y(x) = c_2 \sin x$ of $y'' + y = 0$.

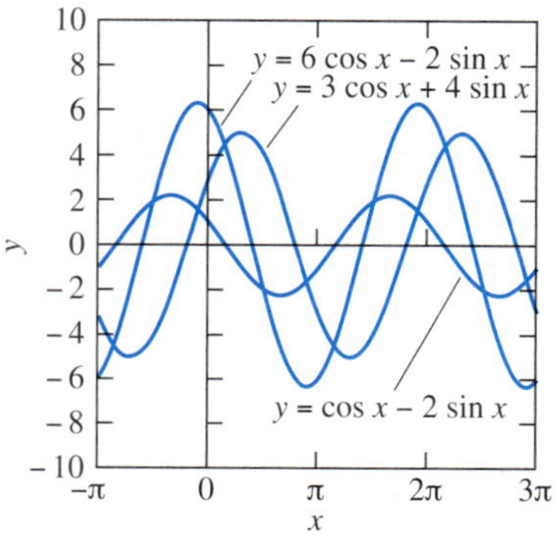

FIGURE 5.1.5. Solutions of $y'' + y = 0$ with c_1 and c_2 both nonzero.

Earlier in this section we gave the linear equation $mx'' + cx' + kx = F(t)$ as a mathematical model of the motion of the mass shown in Fig. 5.1.1. Physical considerations suggest that the motion of the mass should be determined by its initial position and initial velocity. Hence, given any preassigned values of $x(0)$ and $x'(0)$, Eq. (6) ought to have a *unique* solution satisfying these initial conditions. More generally, in order to be a "good" mathematical model of a deterministic physical situation, a differential equation must have unique solutions satisfying any appropriate initial conditions. The following existence and uniqueness theorem (proved in Appendix A) gives us this assurance for the general second-order equation.

THEOREM 2 Existence and Uniqueness for Linear Equations

Suppose that the functions p, q, and f are continuous on the open interval I containing the point a. Then, given any two numbers b_0 and b_1, the equation

$$y'' + p(x)y' + q(x)y = f(x) \tag{8}$$

has a unique (that is, one and only one) solution on the entire interval I that satisfies the initial conditions

$$y(a) = b_0, \quad y'(a) = b_1. \tag{11}$$

Remark 1: Equation (8) and the conditions in (11) constitute a second-order linear **initial value problem.** Theorem 2 tells us that any such initial value problem has a unique solution on the *whole* interval I, where the coefficient functions in (8) are continuous. Recall from Section 1.3 that a *nonlinear* differential equation generally has a unique solution on only a smaller interval.

Remark 2: Whereas a *first-order* differential equation $dy/dx = F(x, y)$ generally admits only a single solution curve $y = y(x)$ passing through a given initial point (a, b), Theorem 2 implies that the *second-order* equation in (8) has infinitely many solution curves passing through the point (a, b_0)—namely, one for each (real number) value of the initial slope $y'(a) = b_1$. That is, instead of there being only one line through (a, b_0) tangent to a solution curve, *every* nonvertical straight line through (a, b_0) is tangent to some solution curve of Eq. (8). Figure 5.1.6 shows a number of solution curves of the equation $y'' + 3y' + 2y = 0$ all having the same initial value $y(0) = 1$, while Fig. 5.1.7 shows a number of solution curves all having the same initial slope $y'(0) = 1$. The application at the end of this section suggests how to construct such families of solution curves for a given homogeneous second-order linear differential equation. ■

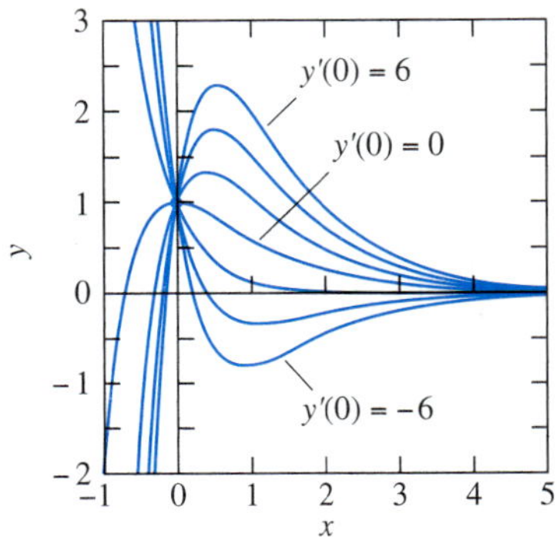

FIGURE 5.1.6. Solutions of $y'' + 3y' + 2y = 0$ with the same initial value $y(0) = 1$ but different initial slopes.

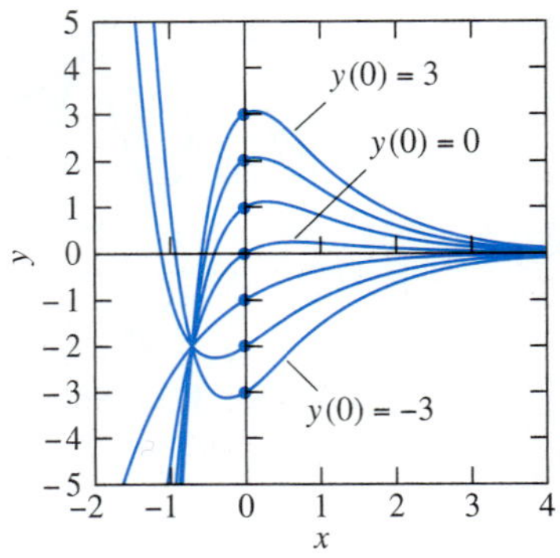

FIGURE 5.1.7. Solutions of $y'' + 3y' + 2y = 0$ with the same initial slope $y'(0) = 1$ but different initial values.

Example 1 *Continued* We saw in the first part of Example 1 that $y(x) = 3\cos x - 2\sin x$ is a solution (on the entire real line) of $y'' + y = 0$. It has the initial values $y(0) = 3$, $y'(0) = -2$. Theorem 2 tells us that this is the *only* solution with these initial values. More generally, the solution

$$y(x) = b_0 \cos x + b_1 \sin x$$

satisfies the *arbitrary* initial conditions $y(0) = b_0$, $y'(0) = b_1$; this illustrates the *existence* of such a solution, also as guaranteed by Theorem 2. ■

Example 1 suggests how, given a *homogeneous* second-order linear equation, we might actually find the solution $y(x)$ whose existence is assured by Theorem 2. First, we find two "essentially different" solutions y_1 and y_2; second, we attempt to impose on the general solution

$$y = c_1 y_1 + c_2 y_2 \tag{12}$$

the initial conditions $y(a) = b_0$, $y'(a) = b_1$. That is, we attempt to solve the simultaneous equations

$$\begin{aligned} c_1 y_1(a) + c_2 y_2(a) &= b_0, \\ c_1 y_1'(a) + c_2 y_2'(a) &= b_1 \end{aligned} \tag{13}$$

for the coefficients c_1 and c_2.

Example 2 Verify that the functions

$$y_1(x) = e^x \quad \text{and} \quad y_2(x) = xe^x$$

are solutions of the differential equation

$$y'' - 2y' + y = 0,$$

and then find a solution satisfying the initial conditions $y(0) = 3$, $y'(0) = 1$.

Solution The verification is routine; we omit it. We impose the given initial conditions on the general solution

$$y(x) = c_1 e^x + c_2 x e^x,$$

for which

$$y'(x) = (c_1 + c_2)e^x + c_2 x e^x,$$

to obtain the simultaneous equations

$$\begin{aligned} y(0) &= c_1 &= 3, \\ y'(0) &= c_1 + c_2 &= 1. \end{aligned}$$

The resulting solution is $c_1 = 3$, $c_2 = -2$. Hence the solution of the original initial value problem is

$$y(x) = 3e^x - 2xe^x.$$

Figure 5.1.8 shows several additional solutions of $y'' - 2y' + y = 0$, all having the same initial value $y(0) = 3$. ■

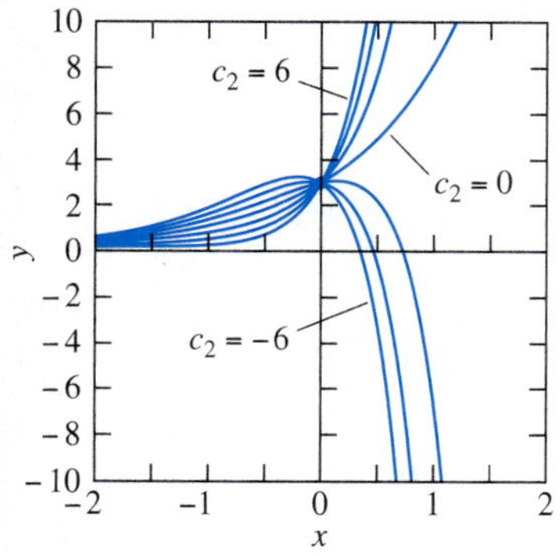

FIGURE 5.1.8. Different solutions $y(x) = 3e^x + c_2 x e^x$ of $y'' - 2y' + y = 0$ with the same initial value $y(0) = 3$.

Linearly Independent Solutions

In order for the procedure of Example 2 to succeed, the two solutions y_1 and y_2 must have the property that the equations in (13) can always be solved for c_1 and c_2, no matter what the initial values $y(a) = b_0$ and $y'(a) = b_1$ might be. This is so, provided that the two functions y_1 and y_2 are linearly independent. According to the following definition, linear independence of functions defined on an interval I is analogous to linear independence of ordinary vectors.

DEFINITION Linear Independence of Two Functions

Two functions defined on an open interval I are said to be **linearly independent** on I provided that neither is a constant multiple of the other.

Two functions are said to be **linearly dependent** on an open interval provided that they are not linearly independent there; that is, one of them is a constant multiple of the other. We can always determine whether two given functions f and g are linearly dependent on an interval I by noting at a glance whether either of the two quotients f/g or g/f is a constant-valued function on I.

Example 3 Thus it is clear that the following pairs of functions are linearly independent on the entire real line:

$$\begin{aligned} \sin x \quad &\text{and} \quad \cos x; \\ e^x \quad &\text{and} \quad e^{-2x}; \\ e^x \quad &\text{and} \quad xe^x; \\ x+1 \quad &\text{and} \quad x^2; \\ x \quad &\text{and} \quad |x|. \end{aligned}$$

That is, neither $\sin x/\cos x = \tan x$ nor $\cos x/\sin x = \cot x$ is a constant-valued function; neither $e^x/e^{-2x} = e^{3x}$ nor e^{-2x}/e^x is a constant-valued function; and so forth. But the identically zero function $f(x) \equiv 0$ and any other function g are linearly dependent on every interval, because $0 \cdot g(x) = 0 = f(x)$. Also, the functions

$$f(x) = \sin 2x \quad \text{and} \quad g(x) = \sin x \cos x$$

are linearly dependent on any interval because $f(x) = 2g(x)$ is the familiar trigonometric identity $\sin 2x = 2 \sin x \cos x$. ■

General Solutions

But does the homogeneous equation $y'' + py' + qy = 0$ always have two linearly independent solutions? Theorem 2 says yes! We need only choose y_1 and y_2 so that

$$y_1(a) = 1, \;\; y_1'(a) = 0 \quad \text{and} \quad y_2(a) = 0, \;\; y_2'(a) = 1.$$

It is then impossible that either $y_1 = ky_2$ or $y_2 = ky_1$ because $k \cdot 0 \neq 1$ for any constant k. Theorem 2 tells us that two such linearly independent solutions *exist*; actually finding them is a crucial matter that we will discuss briefly at the end of this section, and in greater detail beginning in Section 5.3.

We want to show, finally, that given *any* two linearly independent solutions y_1 and y_2 of the homogeneous equation

$$y''(x) + p(x)y'(x) + q(x)y(x) = 0, \tag{9}$$

every solution y of Eq. (9) can be expressed as a linear combination

$$y = c_1y_1 + c_2y_2 \tag{12}$$

of y_1 and y_2. This means that the function in (12) is a *general solution* of Eq. (9)—it provides all possible solutions of the differential equation.

As suggested by the equations in (13), the determination of the constants c_1 and c_2 in (12) depends on a certain 2×2 determinant of values of y_1, y_2, and their derivatives. Given two functions f and g, the **Wronskian** of f and g is the determinant

$$W = \begin{vmatrix} f & g \\ f' & g' \end{vmatrix} = fg' - f'g.$$

We write either $W(f, g)$ or $W(x)$, depending on whether we wish to emphasize the two functions or the point x at which the Wronskian is to be evaluated. For example,

$$W(\cos x, \sin x) = \begin{vmatrix} \cos x & \sin x \\ -\sin x & \cos x \end{vmatrix} = \cos^2 x + \sin^2 x = 1$$

and

$$W(e^x, xe^x) = \begin{vmatrix} e^x & xe^x \\ e^x & e^x + xe^x \end{vmatrix} = e^{2x}.$$

These are examples of linearly *independent* pairs of solutions of differential equations (see Examples 1 and 2). Note that in both cases the Wronskian is everywhere *nonzero*.

On the other hand, if the functions f and g are linearly dependent, with $f = kg$ (for example), then

$$W(f, g) = \begin{vmatrix} kg & g \\ kg' & g' \end{vmatrix} = kgg' - kg'g \equiv 0.$$

Thus the Wronskian of two linearly *dependent* functions is identically zero. In Section 5.2 we will prove that, if the two functions y_1 and y_2 are solutions of a homogeneous second-order linear equation, then the strong converse stated in part (b) of Theorem 3 holds.

THEOREM 3 Wronskians of Solutions

Suppose that y_1 and y_2 are two solutions of the homogeneous second-order linear equation [Eq. (9)]

$$y'' + p(x)y' + q(x)y = 0$$

on an open interval I on which p and q are continuous.

(a) If y_1 and y_2 are linearly dependent, then $W(y_1, y_2) \equiv 0$ on I.

(b) If y_1 and y_2 are linearly independent, then $W(y_1, y_2) \neq 0$ at each point of I.

Thus, given two solutions of Eq. (9), there are just two possibilities: The Wronskian W is identically zero if the solutions are linearly dependent; the Wronskian is never zero if the solutions are linearly independent. The latter fact is what we need to show that $y = c_1y_1 + c_2y_2$ is the general solution of Eq. (9) if y_1 and y_2 are linearly independent solutions.

THEOREM 4 General Solutions of Homogeneous Equations

Let y_1 and y_2 be two linearly independent solutions of the homogeneous equation [Eq. (9)]

$$y'' + p(x)y' + q(x)y = 0$$

with p and q continuous on the open interval I. If Y is any solution whatsoever of Eq. (9) on I, then there exist numbers c_1 and c_2 such that

$$Y(x) = c_1y_1(x) + c_2y_2(x)$$

for all x in I.

Remark: In essence, Theorem 4 tells us that, when we have found *two* linearly independent solutions y_1 and y_2 of the homogeneous equation in (9), we have then found *all* of its solutions. Specifically, $\{y_1, y_2\}$ is then a basis for the solution space of the differential equation. We therefore call the linear combination $Y = c_1y_1 + c_2y_2$ a *general solution* of the differential equation. ■

***Proof of Theorem 4*:** Choose a point a of I, and consider the simultaneous equations

$$\begin{aligned} c_1y_1(a) + c_2y_2(a) &= Y(a), \\ c_1y_1'(a) + c_2y_2'(a) &= Y'(a). \end{aligned} \tag{14}$$

The determinant of the coefficients in this system of linear equations in the unknowns c_1 and c_2 is simply the Wronskian $W(y_1, y_2)$ evaluated at $x = a$. By Theorem 3, this determinant is nonzero, so by elementary algebra it follows that the equations in (14) can be solved for c_1 and c_2. With these values of c_1 and c_2, we define the solution

$$G(x) = c_1y_1(x) + c_2y_2(x)$$

of Eq. (9); then

$$G(a) = c_1y_1(a) + c_2y_2(a) = Y(a)$$

and

$$G'(a) = c_1y_1'(a) + c_2y_2'(a) = Y'(a).$$

Thus the two solutions Y and G have the same initial values at a; likewise, so do Y' and G'. By the uniqueness of a solution determined by such initial values (Theorem 2), it follows that Y and G agree on I. Thus we see that

$$Y(x) \equiv G(x) = c_1 y_1(x) + c_2 y_2(x),$$

as desired. ■

Example 4 If $y_1(x) = e^{2x}$ and $y_2(x) = e^{-2x}$, then

$$y_1'' = (2)(2)e^{2x} = 4e^{2x} = 4y_1 \quad \text{and} \quad y_2'' = (-2)(-2)e^{-2x} = 4e^{-2x} = 4y_2.$$

Therefore, y_1 and y_2 are linearly independent solutions of

$$y'' - 4y = 0. \tag{15}$$

But $y_3(x) = \cosh 2x$ and $y_4(x) = \sinh 2x$ are also solutions of Eq. (15), because

$$\frac{d^2}{dx^2}(\cosh 2x) = \frac{d}{dx}(2\sinh 2x) = 4\cosh 2x$$

and, similarly, $(\sinh 2x)'' = 4\sinh 2x$. It therefore follows from Theorem 4 that the functions $\cosh 2x$ and $\sinh 2x$ can be expressed as linear combinations of $y_1(x) = e^{2x}$ and $y_2(x) = e^{-2x}$. Of course, this is no surprise, because

$$\cosh 2x = \tfrac{1}{2}e^{2x} + \tfrac{1}{2}e^{-2x} \quad \text{and} \quad \sinh 2x = \tfrac{1}{2}e^{2x} - \tfrac{1}{2}e^{-2x}$$

by the definitions of the hyperbolic cosine and hyperbolic sine functions. Thus the solution space of the differential equation $y'' - 4y = 0$ has the two different bases $\{e^{2x}, e^{-2x}\}$ and $\{\cosh 2x, \sinh 2x\}$. ■

Remark: Because e^{2x}, e^{-2x} and $\cosh 2x$, $\sinh 2x$ are two different pairs of linearly independent solutions of the equation $y'' - 4y = 0$ in (15) , Theorem 4 implies that every particular solution $Y(x)$ of this equation can be written both in the form

$$Y(x) = c_1 e^{2x} + c_2 e^{-2x}$$

and in the form

$$Y(x) = a\cosh 2x + b\sinh 2x.$$

Thus these two different linear combinations (with arbitrary constant coefficients) provide two different descriptions of the set of all solutions of the same differential equation $y'' - 4y = 0$. Hence each of these two linear combinations is a general solution of the equation. Indeed, this is why it is accurate to refer to a specific such linear combination as "*a* general solution" rather than as "*the* general solution." ■

Linear Second-Order Equations with Constant Coefficients

As an illustration of the general theory introduced in this section, we discuss the homogeneous second-order linear differential equation

$$ay'' + by' + cy = 0 \tag{16}$$

with *constant* coefficients a, b, and c. We first look for a *single* solution of Eq. (16) and begin with the observation that

$$(e^{rx})' = re^{rx} \quad \text{and} \quad (e^{rx})'' = r^2e^{rx}, \tag{17}$$

so any derivative of e^{rx} is a constant multiple of e^{rx}. Hence, if we substituted $y = e^{rx}$ in Eq. (16), then each term would be a constant multiple of e^{rx}, with the constant coefficients dependent on r and the coefficients a, b, and c. This suggests that we try to find a value of r so that these multiples of e^{rx} will have sum zero. If we succeed, then $y = e^{rx}$ will be a solution of Eq. (16).

For example, if we substitute $y = e^{rx}$ in the equation

$$y'' - 5y' + 6y = 0,$$

we obtain

$$r^2e^{rx} - 5re^{rx} + 6e^{rx} = 0.$$

Thus

$$(r^2 - 5r + 6)e^{rx} = 0; \quad (r - 2)(r - 3)e^{rx} = 0.$$

Hence $y = e^{rx}$ will be a solution if either $r = 2$ or $r = 3$. So, in searching for a single solution, we actually have found two solutions: $y_1(x) = e^{2x}$ and $y_2(x) = e^{3x}$.

To carry out this procedure in the general case, we substitute $y = e^{rx}$ in Eq. (16). With the aid of the equations in (17), we find the result to be

$$ar^2e^{rx} + bre^{rx} + ce^{rx} = 0.$$

Because e^{rx} is never zero, we conclude that $y(x) = e^{rx}$ will satisfy the differential equation in (16) precisely when r is a root of the *algebraic* equation

$$ar^2 + br + c = 0. \tag{18}$$

This quadratic equation is called the **characteristic equation** of the homogeneous linear differential equation

$$ay'' + by' + cy = 0. \tag{16}$$

If Eq. (18) has two *distinct* (unequal) real roots r_1 and r_2, then the corresponding solutions $y_1(x) = e^{r_1x}$ and $y_2(x) = e^{r_2x}$ of (16) are linearly independent. (Why?) This gives the following result.

THEOREM 5 Distinct Real Roots

If the roots r_1 and r_2 of the characteristic equation in (18) are real and distinct, then

$$y(x) = c_1e^{r_1x} + c_2e^{r_2x} \tag{19}$$

is a general solution of Eq. (16). Thus the solution space of the equation $ay'' + by' + cy = 0$ has basis $\{e^{r_1x}, e^{r_2x}\}$.

Example 5 Find the general solution of

$$2y'' - 7y' + 3y = 0.$$

Solution We can solve the characteristic equation

$$2r^2 - 7r + 3 = 0$$

by factoring:

$$(2r - 1)(r - 3) = 0.$$

The roots $r_1 = \frac{1}{2}$ and $r_2 = 3$ are real and distinct, so Theorem 5 yields the general solution

$$y(x) = c_1 e^{x/2} + c_2 e^{3x}.$$

■

Example 6 The differential equation $y'' + 2y' = 0$ has characteristic equation

$$r^2 + 2r = r(r + 2) = 0$$

with distinct real roots $r_1 = 0$ and $r_2 = -2$. Because $e^{0 \cdot x} \equiv 1$, we get the general solution

$$y(x) = c_1 + c_2 e^{-2x}.$$

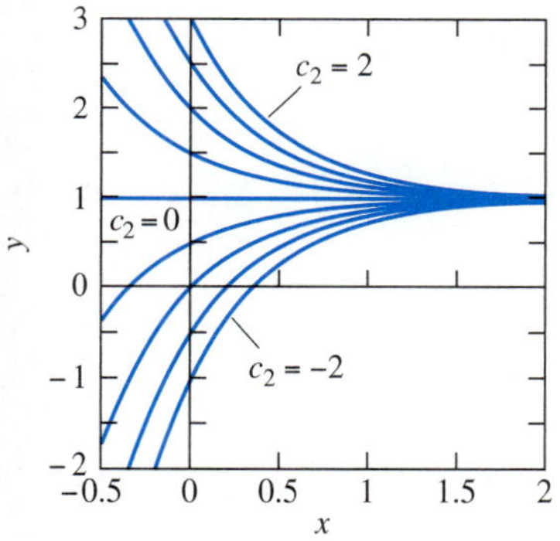

FIGURE 5.1.9. Solutions $y(x) = 1 + c_2 e^{-2x}$ of $y'' + 2y' = 0$ with different values of c_2.

Figure 5.1.9 shows several different solution curves with $c_1 = 1$, all appearing to approach the solution curve $y(x) \equiv 1$ (with $c_2 = 0$) as $x \to +\infty$. ■

Remark: Note that Theorem 5 changes a problem involving a differential equation into one involving only the solution of an *algebraic* equation. ■

If the characteristic equation in (18) has equal roots $r_1 = r_2$, we get (at first) only the single solution $y_1(x) = e^{r_1 x}$ of Eq. (16). The problem in this case is to produce the "missing" second solution of the differential equation.

A double root $r = r_1$ will occur precisely when the characteristic equation is a constant multiple of the equation

$$(r - r_1)^2 = r^2 - 2r_1 r + r_1^2 = 0.$$

Any differential equation with this characteristic equation is equivalent to

$$y'' - 2r_1 y' + r_1^2 y = 0. \tag{20}$$

But it is easy to verify by direct substitution that $y = xe^{r_1 x}$ is a second solution of Eq. (20). It is clear (but you should verify) that

$$y_1(x) = e^{r_1 x} \quad \text{and} \quad y_2(x) = xe^{r_1 x}$$

are linearly independent functions, so the general solution of the differential equation in (20) is

$$y(x) = c_1 e^{r_1 x} + c_2 x e^{r_1 x}.$$

THEOREM 6 Repeated Roots

If the characteristic equation in (18) has equal (necessarily real) roots $r_1 = r_2$, then

$$y(x) = (c_1 + c_2 x)e^{r_1 x} \tag{21}$$

is a general solution of Eq. (16). In this case the solution space of the equation $ay'' + by' + cy = 0$ has basis $\{e^{r_1 x}, xe^{r_1 x}\}$.

Example 7 To solve the initial value problem

$$\begin{aligned} &y'' + 2y' + y = 0; \\ &y(0) = 5, \quad y'(0) = -3, \end{aligned}$$

we note first that the characteristic equation

$$r^2 + 2r + 1 = (r+1)^2 = 0$$

has equal roots $r_1 = r_2 = -1$. Hence the general solution provided by Theorem 6 is

$$y(x) = c_1 e^{-x} + c_2 x e^{-x}.$$

Differentiation yields

$$y'(x) = -c_1 e^{-x} + c_2 e^{-x} - c_2 x e^{-x},$$

so the initial conditions yield the equations

$$\begin{aligned} y(0) &= c_1 &&= 5, \\ y'(0) &= -c_1 + c_2 &&= -3, \end{aligned}$$

which imply that $c_1 = 5$ and $c_2 = 2$. Thus the desired particular solution of the initial value problem is

$$y(x) = 5e^{-x} + 2xe^{-x}.$$

This particular solution, together with several others of the form $y(x) = c_1 e^{-x} + 2xe^{-x}$, is illustrated in Fig. 5.1.10. ■

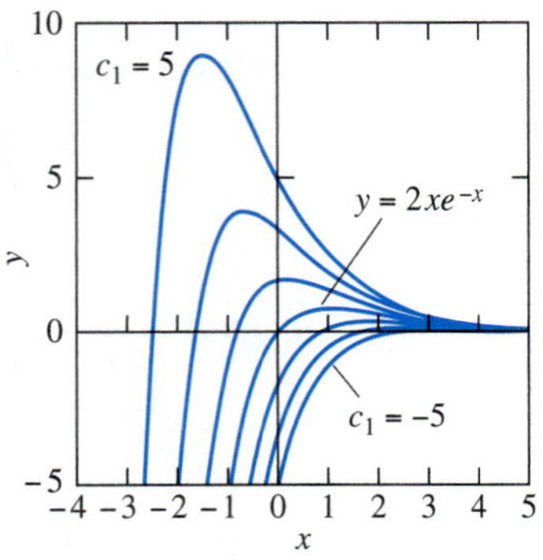

FIGURE 5.1.10. Solutions $y(x) = c_1 e^{-x} + 2xe^{-x}$ of $y'' + 2y' + y = 0$ with different values of c_1.

The characteristic equation in (18) may have either real or complex roots. The case of complex roots will be discussed in Section 5.3.

5.1 Problems

In Problems 1 through 16, a homogeneous second-order linear differential equation, two functions y_1 and y_2, and a pair of initial conditions are given. First verify that y_1 and y_2 are solutions of the differential equation. Then find a particular solution of the form $y = c_1y_1 + c_2y_2$ that satisfies the given initial conditions. Primes denote derivatives with respect to x.

1. $y'' - y = 0$; $y_1 = e^x$, $y_2 = e^{-x}$; $y(0) = 0$, $y'(0) = 5$
2. $y'' - 9y = 0$; $y_1 = e^{3x}$, $y_2 = e^{-3x}$; $y(0) = -1$, $y'(0) = 15$
3. $y'' + 4y = 0$; $y_1 = \cos 2x$, $y_2 = \sin 2x$; $y(0) = 3$, $y'(0) = 8$
4. $y'' + 25y = 0$; $y_1 = \cos 5x$, $y_2 = \sin 5x$; $y(0) = 10$, $y'(0) = -10$
5. $y'' - 3y' + 2y = 0$; $y_1 = e^x$, $y_2 = e^{2x}$; $y(0) = 1$, $y'(0) = 0$
6. $y'' + y' - 6y = 0$; $y_1 = e^{2x}$, $y_2 = e^{-3x}$; $y(0) = 7$, $y'(0) = -1$
7. $y'' + y' = 0$; $y_1 = 1$, $y_2 = e^{-x}$; $y(0) = -2$, $y'(0) = 8$
8. $y'' - 3y' = 0$; $y_1 = 1$, $y_2 = e^{3x}$; $y(0) = 4$, $y'(0) = -2$
9. $y'' + 2y' + y = 0$; $y_1 = e^{-x}$, $y_2 = xe^{-x}$; $y(0) = 2$, $y'(0) = -1$
10. $y'' - 10y' + 25y = 0$; $y_1 = e^{5x}$, $y_2 = xe^{5x}$; $y(0) = 3$, $y'(0) = 13$
11. $y'' - 2y' + 2y = 0$; $y_1 = e^x \cos x$, $y_2 = e^x \sin x$; $y(0) = 0$, $y'(0) = 5$
12. $y'' + 6y' + 13y = 0$; $y_1 = e^{-3x} \cos 2x$, $y_2 = e^{-3x} \sin 2x$; $y(0) = 2$, $y'(0) = 0$
13. $x^2y'' - 2xy' + 2y = 0$; $y_1 = x$, $y_2 = x^2$; $y(1) = 3$, $y'(1) = 1$
14. $x^2y'' + 2xy' - 6y = 0$; $y_1 = x^2$, $y_2 = x^{-3}$; $y(2) = 10$, $y'(2) = 15$
15. $x^2y'' - xy' + y = 0$; $y_1 = x$, $y_2 = x \ln x$; $y(1) = 7$, $y'(1) = 2$
16. $x^2y'' + xy' + y = 0$; $y_1 = \cos(\ln x)$, $y_2 = \sin(\ln x)$; $y(1) = 2$, $y'(1) = 3$

The following three problems illustrate the fact that the superposition principle does not generally hold for nonlinear equations.

17. Show that $y = 1/x$ is a solution of $y' + y^2 = 0$, but that if $c \neq 0$ and $c \neq 1$, then $y = c/x$ is not a solution.
18. Show that $y = x^3$ is a solution of $yy'' = 6x^4$, but that if $c^2 \neq 1$, then $y = cx^3$ is not a solution.
19. Show that $y_1 \equiv 1$ and $y_2 = \sqrt{x}$ are solutions of $yy'' + (y')^2 = 0$, but that their sum $y = y_1 + y_2$ is not a solution.

Determine whether the pairs of functions in Problems 20 through 26 are linearly independent or linearly dependent on the real line.

20. $f(x) = \pi$, $g(x) = \cos^2 x + \sin^2 x$
21. $f(x) = x^3$, $g(x) = x^2|x|$
22. $f(x) = 1 + x$, $g(x) = 1 + |x|$
23. $f(x) = xe^x$, $g(x) = |x|e^x$
24. $f(x) = \sin^2 x$, $g(x) = 1 - \cos 2x$
25. $f(x) = e^x \sin x$, $g(x) = e^x \cos x$
26. $f(x) = 2\cos x + 3\sin x$, $g(x) = 3\cos x - 2\sin x$

27. Let y_p be a particular solution of the nonhomogeneous equation $y'' + py' + qy = f(x)$ and let y_c be a solution of its associated homogeneous equation. Show that $y = y_c + y_p$ is a solution of the given nonhomogeneous equation.
28. With $y_p = 1$ and $y_c = c_1 \cos x + c_2 \sin x$ in the notation of Problem 27, find a solution of $y'' + y = 1$ satisfying the initial conditions $y(0) = -1 = y'(0)$.
29. Show that $y_1 = x^2$ and $y_2 = x^3$ are two different solutions of $x^2y'' - 4xy' + 6y = 0$, both satisfying the initial conditions $y(0) = 0 = y'(0)$. Explain why these facts do not contradict Theorem 2 (with respect to the guaranteed uniqueness).
30. **(a)** Show that $y_1 = x^3$ and $y_2 = |x^3|$ are linearly independent solutions on the real line of the equation $x^2y'' - 3xy' + 3y = 0$. **(b)** Verify that $W(y_1, y_2)$ is identically zero. Why do these facts not contradict Theorem 3?
31. Show that $y_1 = \sin x^2$ and $y_2 = \cos x^2$ are linearly independent functions, but that their Wronskian vanishes at $x = 0$. Why does this imply that there is *no* differential equation of the form $y'' + p(x)y' + q(x)y = 0$, with both p and q continuous everywhere, having both y_1 and y_2 as solutions?
32. Let y_1 and y_2 be two solutions of $A(x)y'' + B(x)y' + C(x)y = 0$ on an open interval I where A, B, and C are continuous and $A(x)$ is never zero. **(a)** Let $W = W(y_1, y_2)$. Show that

$$A(x)\frac{dW}{dx} = (y_1)(Ay_2'') - (y_2)(Ay_1'').$$

Then substitute for Ay_2'' and Ay_1'' from the original differential equation to show that

$$A(x)\frac{dW}{dx} = -B(x)W(x).$$

(b) Solve this first-order equation to deduce **Abel's formula**

$$W(x) = K \exp\left(-\int \frac{B(x)}{A(x)}\,dx\right),$$

where K is a constant. **(c)** Why does Abel's formula imply that the Wronskian $W(y_1, y_2)$ is either zero everywhere or nonzero everywhere (as stated in Theorem 3)?

Apply Theorems 5 and 6 to find general solutions of the differential equations given in Problems 33 through 42. Primes denote derivatives with respect to x.

33. $y'' - 3y' + 2y = 0$
34. $y'' + 2y' - 15y = 0$
35. $y'' + 5y' = 0$
36. $2y'' + 3y' = 0$
37. $2y'' - y' - y = 0$
38. $4y'' + 8y' + 3y = 0$
39. $4y'' + 4y' + y = 0$
40. $9y'' - 12y' + 4y = 0$
41. $6y'' - 7y' - 20y = 0$
42. $35y'' - y' - 12y = 0$

Each of Problems 43 through 48 gives a general solution $y(x)$ of a homogeneous second-order differential equation $ay'' + by' + cy = 0$ with constant coefficients. Find such an equation.

43. $y(x) = c_1 + c_2 e^{-10x}$ **44.** $y(x) = c_1 e^{10x} + c_2 e^{-10x}$

45. $y(x) = c_1 e^{-10x} + c_2 x e^{-10x}$

46. $y(x) = c_1 e^{10x} + c_2 e^{100x}$ **47.** $y(x) = c_1 + c_2 x$

48. $y(x) = e^x \left(c_1 e^{x\sqrt{2}} + c_2 e^{-x\sqrt{2}}\right)$

Problems 49 and 50 deal with the solution curves of $y'' + 3y' + 2y = 0$ shown in Figs. 5.1.6 and 5.1.7.

49. Find the highest point on the solution curve with $y(0) = 1$ and $y'(0) = 6$ in Fig. 5.1.6.

50. Find the third-quadrant point of intersection of the solution curves shown in Fig. 5.1.7.

51. A second-order **Euler equation** is one of the form

$$ax^2y'' + bxy' + cy = 0, \tag{22}$$

where a, b, c are constants. **(a)** Show that if $x > 0$, then the substitution $v = \ln x$ transforms Eq. (22) into the constant-coefficient linear equation

$$a\frac{d^2y}{dv^2} + (b - a)\frac{dy}{dv} + cy = 0 \tag{23}$$

with independent variable v. **(b)** If the roots r_1 and r_2 of the characteristic equation of Eq. (23) are real and distinct, conclude that a general solution of the Euler equation in (22) is $y(x) = c_1 x^{r_1} + c_2 x^{r_2}$.

Make the substitution $v = \ln x$ of Problem 51 to find general solutions (for $x > 0$) of the Euler equations in Problems 52 through 56.

52. $x^2y'' + xy' - y = 0$ **53.** $x^2y'' + 2xy' - 12y = 0$

54. $4x^2y'' + 8xy' - 3y = 0$ **55.** $x^2y'' + xy' = 0$

56. $x^2y'' - 3xy' + 4y = 0$

5.1 Application Plotting Second-Order Solution Families

This application deals with the plotting by computer of families of solutions such as those illustrated in Figs. 5.1.6 and 5.1.7. Show first that the general solution of the differential equation

$$y'' + 3y' + 2y = 0 \tag{1}$$

is

$$y(x) = c_1 e^{-x} + c_2 e^{-2x}. \tag{2}$$

For Fig. 5.1.6, show that the particular solution of Eq. (1) satisfying the initial conditions $y(0) = 1$, $y'(0) = b$ is given by

$$y(x) = (b + 2)e^{-x} - (b + 1)e^{-2x}. \tag{3}$$

The MATLAB loop

```
x = - 1 : 0.02 : 5      % x-vector from x = - 1 to x = 5
for b = -6 : 2 : 6      % for b = -6 to 6 with db = 2 do
   y = (b + 2)*exp(-x) - (b + 1)*exp(-2*x);
   plot(x,y)
end
```

was used to generate Fig. 5.1.6.

For Fig. 5.1.7, show that the particular solution of Eq. (1) satisfying the initial conditions $y(0) = a$, $y'(0) = 1$ is given by

$$y(x) = (2a + 1)e^{-x} - (a + 1)e^{-2x}. \tag{4}$$

The MATLAB loop

```
x = -2 : 0.02 : 4       % x-vector from x = -2 to x = 4
for a = -3 : 1 : 3      % for a = -3 to 3 with da = 1 do
```

```
        y = (2*a + 1)*exp(-x) - (a + 1)*exp(-2*x);
        plot(x,y)
    end
```

was used to generate Fig. 5.1.7.

Computer algebra systems such as *Maple* and *Mathematica*, as well as graphing calculators, have commands to carry out for-loops such as the two shown here. Begin by reproducing Figs. 5.1.6 and 5.1.7. Then, for each of the following differential equations, construct both a family of solution curves satisfying $y(0) = 1$ and a family of solution curves satisfying the initial condition $y'(0) = 1$.

1. $y'' - y = 0$

2. $y'' - 3y' + 2y = 0$

3. $2y'' + 3y' + y = 0$

4. $y'' + y = 0$ (see Example 1)

5. $y'' + 2y' + 2y = 0$, which has general solution $y(x) = e^{-x}(c_1 \cos x + c_2 \sin x)$.

5.2 General Solutions of Linear Equations

We now show that our discussion in Section 5.1 of second-order linear equations generalizes in a very natural way to the general ***n*th-order linear** differential equation of the form

$$P_0(x)y^{(n)} + P_1(x)y^{(n-1)} + \cdots + P_{n-1}(x)y' + P_n(x)y = F(x). \tag{1}$$

Unless otherwise noted, we will always assume that the coefficient functions $P_i(x)$ and $F(x)$ are continuous on some open interval I (perhaps unbounded), where we wish to solve the equation. Under the additional assumption that $P_0(x) \neq 0$ at each point of I, we can divide each term in Eq. (1) by $P_0(x)$ to obtain an equation with leading coefficient 1, of the form

$$y^{(n)} + p_1(x)y^{(n-1)} + \cdots + p_{n-1}(x)y' + p_n(x)y = f(x). \tag{2}$$

The **homogeneous** linear equation **associated with** Eq. (2) is

$$y^{(n)} + p_1(x)y^{(n-1)} + \cdots + p_{n-1}(x)y' + p_n(x)y = 0. \tag{3}$$

Just as in the second-order case, a *homogeneous* nth-order linear differential equation has the valuable property that any superposition, or *linear combination,* of solutions of the equation is again a solution. The proof of the following theorem is essentially the same—a routine verification—as that of Theorem 1 of Section 5.1.

THEOREM 1 Principle of Superposition for Homogeneous Equations

Let $y_1, y_2, \ldots, y_n$ be n solutions of the homogeneous linear equation in (3) on the interval I. If $c_1, c_2, \ldots, c_n$ are constants, then the linear combination

$$y = c_1 y_1 + c_2 y_2 + \cdots + c_n y_n \tag{4}$$

is also a solution of Eq. (3) on I.

Example 1 It is easy to verify that the three functions

$$y_1(x) = e^{-3x}, \quad y_2(x) = \cos 2x, \quad \text{and} \quad y_3(x) = \sin 2x$$

are all solutions of the homogeneous third-order equation

$$y^{(3)} + 3y'' + 4y' + 12y = 0$$

on the entire real line. Theorem 1 tells us that any linear combination of these solutions, such as

$$y(x) = -3y_1(x) + 3y_2(x) - 2y_3(x) = -3e^{-3x} + 3\cos 2x - 2\sin 2x,$$

is also a solution on the entire real line. We will see that, conversely, every solution of the differential equation of this example is a linear combination of the three particular solutions y_1, y_2, and y_3. Thus a general solution is given by

$$y(x) = c_1 e^{-3x} + c_2 \cos 2x + c_3 \sin 2x.$$ ■

Existence and Uniqueness of Solutions

We saw in Section 5.1 that a particular solution of a *second-order* linear differential equation is determined by *two* initial conditions. Similarly, a particular solution of an nth-order linear differential equation is determined by n initial conditions. The following theorem, proved in the Appendix, is the natural generalization of Theorem 2 of Section 5.1.

THEOREM 2 Existence and Uniqueness for Linear Equations

Suppose that the functions $p_1, p_2, \ldots, p_n$, and f are continuous on the open interval I containing the point a. Then, given n numbers $b_0, b_1, \ldots, b_{n-1}$, the nth-order linear equation [Eq. (2)]

$$y^{(n)} + p_1(x)y^{(n-1)} + \cdots + p_{n-1}(x)y' + p_n(x)y = f(x)$$

has a unique (that is, one and only one) solution on the entire interval I that satisfies the n initial conditions

$$y(a) = b_0, \quad y'(a) = b_1, \quad \ldots, \quad y^{(n-1)}(a) = b_{n-1}. \tag{5}$$

Equation (2) and the conditions in (5) constitute an nth-order **initial value problem.** Theorem 2 tells us that any such initial value problem has a unique solution on the *whole* interval I where the coefficient functions in (2) are continuous. It tells us nothing, however, about how to find this solution. In Section 5.3 we will see how to construct explicit solutions of initial value problems in the *constant* coefficient case that occurs often in applications.

Example 1 *Continued*

We saw earlier that

$$y(x) = -3e^{-3x} + 3\cos 2x - 2\sin 2x$$

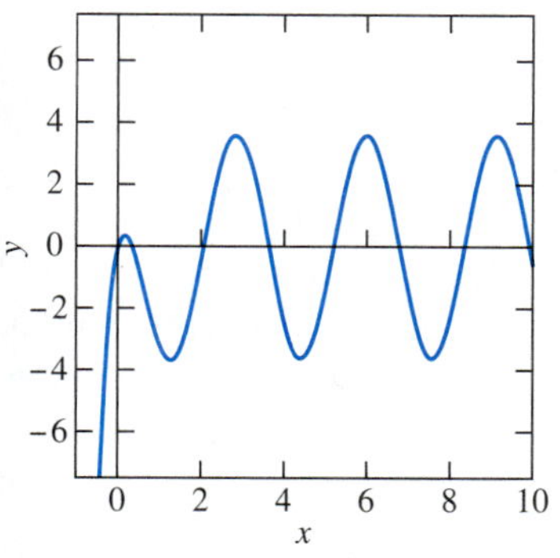

FIGURE 5.2.1. The particular solution $y(x) = -3e^{-3x} + 3\cos 2x - 2\sin 2x$.

is a solution of

$$y^{(3)} + 3y'' + 4y' + 12y = 0$$

on the entire real line. This particular solution has initial values $y(0) = 0$, $y'(0) = 5$, and $y''(0) = -39$, and Theorem 2 implies that there is no other solution with these same initial values. Note that its graph (in Fig. 5.2.1) looks periodic on the right. Indeed, because of the negative exponent, we see that $y(x) \approx 3\cos 2x - 2\sin 2x$ for large positive x. ■

Remark: Because its general solution involves the three arbitrary constants c_1, c_2, and c_3, the third-order equation in Example 1 has a "threefold infinity" of solutions, including three families of especially simple solutions:

- $y(x) = c_1 e^{-3x}$ (obtained from the general solution with $c_2 = c_3 = 0$),
- $y(x) = c_2 \cos 2x$ (with $c_1 = c_3 = 0$), and
- $y(x) = c_3 \sin 2x$ (with $c_1 = c_2 = 0$).

Alternatively, Theorem 2 suggests a threefold infinity of particular solutions corresponding to independent choices of the three initial values $y(0) = b_0$, $y'(0) = b_1$, and $y''(0) = b_2$. Figures 5.2.2 through 5.2.4 illustrate three corresponding families of solutions—for each of which, two of these three initial values are zero. ■

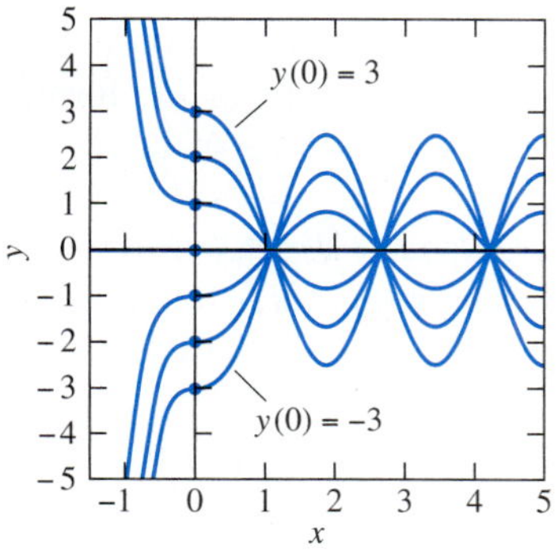

FIGURE 5.2.2. Solutions of $y^{(3)} + 3y'' + 4y' + 12y = 0$ with $y'(0) = y''(0) = 0$ but with different values for $y(0)$.

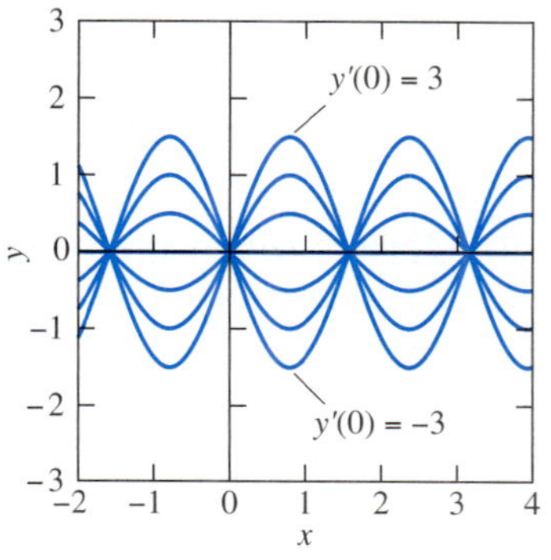

FIGURE 5.2.3. Solutions of $y^{(3)} + 3y'' + 4y' + 12y = 0$ with $y(0) = y''(0) = 0$ but with different values for $y'(0)$.

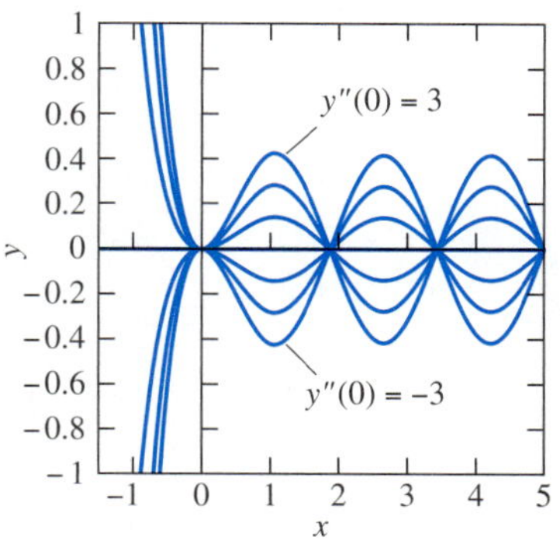

FIGURE 5.2.4. Solutions of $y^{(3)} + 3y'' + 4y' + 12y = 0$ with $y(0) = y'(0) = 0$ but with different values for $y''(0)$.

Note that Theorem 2 implies that the *trivial* solution $y(x) \equiv 0$ is the only solution of the *homogeneous* equation

$$y^{(n)} + p_1(x)y^{(n-1)} + \cdots + p_{n-1}(x)y' + p_n(x)y = 0 \tag{3}$$

that satisfies the *trivial* initial conditions

$$y(a) = y'(a) = \cdots = y^{(n-1)}(a) = 0.$$

Example 2 It is easy to verify that

$$y_1(x) = x^2 \quad \text{and} \quad y_2(x) = x^3$$

are two different solutions of

$$x^2y'' - 4xy' + 6y = 0,$$

and that both satisfy the initial conditions $y(0) = y'(0) = 0$. Why does this not contradict the uniqueness part of Theorem 2? It is because the leading coefficient in this differential equation vanishes at $x = 0$, so this equation cannot be written in the form of Eq. (3) with coefficient functions *continuous* on an open interval containing the point $x = 0$. ■

Linearly Independent Solutions

On the basis of our knowledge of general solutions of second-order linear equations, we anticipate that a general solution of the *homogeneous* nth-order linear equation

$$y^{(n)} + p_1(x)y^{(n-1)} + \cdots + p_{n-1}(x)y' + p_n(x)y = 0 \tag{3}$$

will be a linear combination

$$y = c_1y_1 + c_2y_2 + \cdots + c_ny_n, \tag{4}$$

where $y_1, y_2, \ldots, y_n$ are particular solutions of Eq. (3). But these n particular solutions must be "sufficiently independent" that we can always choose the coefficients $c_1, c_2, \ldots, c_n$ in (4) to satisfy arbitrary initial conditions of the form in (5). The question is this: What should be meant by *independence* of three or more functions?

Recall that *two* functions f_1 and f_2 are linearly *dependent* if one is a constant multiple of the other; that is, if either $f_1 = kf_2$ or $f_2 = kf_1$ for some constant k. If we write these equations as

$$(1)f_1 + (-k)f_2 = 0 \quad \text{or} \quad (k)f_1 + (-1)f_2 = 0,$$

we see that the linear dependence of f_1 and f_2 implies that there exist two constants c_1 and c_2 *not both zero* such that

$$c_1f_1 + c_2f_2 = 0. \tag{6}$$

Conversely, if c_1 and c_2 are not both zero, then Eq. (6) certainly implies that f_1 and f_2 are linearly dependent.

In analogy with Eq. (6), we say that n functions $f_1, f_2, \ldots, f_n$ are *linearly dependent* provided that some *nontrivial* linear combination

$$c_1f_1 + c_2f_2 + \cdots + c_nf_n$$

of them vanishes identically; *nontrivial* means that *not all* of the coefficients $c_1, c_2, \ldots, c_n$ are zero (although some of them may be zero).

DEFINITION Linear Dependence of Functions

The n functions $f_1, f_2, \ldots, f_n$ are said to be **linearly dependent** on the interval I provided that there exist constants $c_1, c_2, \ldots, c_n$ not all zero such that

$$c_1 f_1 + c_2 f_2 + \cdots + c_n f_n = 0 \tag{7}$$

on I; that is,

$$c_1 f_1(x) + c_2 f_2(x) + \cdots + c_n f_n(x) = 0$$

for all x in I.

Thus linear dependence of functions on an interval I is precisely analogous to linear dependence of ordinary vectors (Section 4.3).

If not all the coefficients in Eq. (7) are zero, then, clearly, we can solve for at least one of the functions as a linear combination of the others, and conversely. Thus the functions $f_1, f_2, \ldots, f_n$ are linearly dependent if and only if at least one of them is a linear combination of the others.

Example 3 The functions

$$f_1(x) = \sin 2x, \quad f_2(x) = \sin x \cos x, \quad \text{and} \quad f_3(x) = e^x$$

are linearly dependent on the real line because

$$(1) f_1 + (-2) f_2 + (0) f_3 = 0$$

(by the familiar trigonometric identity $\sin 2x = 2 \sin x \cos x$). ■

The n functions $f_1, f_2, \ldots, f_n$ are called **linearly independent** on the interval I provided that they are not linearly dependent there. Equivalently, they are linearly independent on I provided that the identity

$$c_1 f_1 + c_2 f_2 + \cdots + c_n f_n = 0 \tag{7}$$

holds on I only in the trivial case

$$c_1 = c_2 = \cdots = c_n = 0;$$

that is, *no* nontrivial linear combination of these functions vanishes on I. Put yet another way, the functions $f_1, f_2, \ldots, f_n$ are linearly independent if no one of them is a linear combination of the others. (Why?)

Sometimes one can show that n given functions are linearly dependent by finding, as in Example 3, nontrivial values of the coefficients so that Eq. (7) holds. But in order to show that n given functions are linearly independent, we must prove that nontrivial values of the coefficients *cannot* be found, and this is seldom easy to do in any direct or obvious manner.

Fortunately, in the case of n solutions of a homogeneous nth-order linear equation, there is a tool that makes the determination of their linear dependence or independence a routine matter in many examples. This tool is the Wronskian determinant, which we introduced (for the case $n = 2$) in Section 5.1. Suppose that the n

functions $f_1, f_2, \ldots, f_n$ are each $n-1$ times differentiable. Then their **Wronskian** is the $n \times n$ determinant

$$W = \begin{vmatrix} f_1 & f_2 & \cdots & f_n \\ f_1' & f_2' & \cdots & f_n' \\ \vdots & \vdots & & \vdots \\ f_1^{(n-1)} & f_2^{(n-1)} & \cdots & f_n^{(n-1)} \end{vmatrix}. \tag{8}$$

We write $W(f_1, f_2, \ldots, f_n)$ or $W(x)$, depending on whether we wish to emphasize the functions or the point x at which their Wronskian is to be evaluated. The Wronskian is named after the Polish mathematician J. M. H. Wronski (1778–1853).

We saw in Section 5.1 that the Wronskian of two linearly dependent functions vanishes identically. More generally, *the Wronskian of n linearly dependent functions $f_1, f_2, \ldots, f_n$ is identically zero.* To prove this, assume that Eq. (7) holds on the interval I for some choice of the constants $c_1, c_2, \ldots, c_n$ not all zero. We then differentiate this equation $n-1$ times in succession, obtaining the n equations

$$\begin{aligned} c_1 f_1(x) + c_2 f_2(x) \quad &+ \cdots + \quad c_n f_n(x) = 0, \\ c_1 f_1'(x) + c_2 f_2'(x) \quad &+ \cdots + \quad c_n f_n'(x) = 0, \\ &\vdots \\ c_1 f_1^{(n-1)}(x) + c_2 f_2^{(n-1)}(x) &+ \cdots + c_n f_n^{(n-1)}(x) = 0, \end{aligned} \tag{9}$$

which hold for all x in I. We recall from Theorem 7 in Section 3.5 that a *homogeneous $n \times n$* linear system of equations has a nontrivial solution if and only if its coefficient matrix is *not* invertible, which by Theorem 2 in Section 3.6 is so if and only if the coefficient determinant vanishes. In Eq. (9) the unknowns are the constants $c_1, c_2, \ldots, c_n$ and the determinant of coefficients is simply the Wronskian $W(f_1, f_2, \ldots, f_n)$ evaluated at the typical point x of I. Because we know that the c_i are not all zero, it follows that $W(x) \equiv 0$, as we wanted to prove.

Therefore, to show that the functions $f_1, f_2, \ldots, f_n$ are *linearly independent* on the interval I, it suffices to show that their Wronskian is nonzero at just one point of I.

Example 4 Show that the functions $y_1(x) = e^{-3x}$, $y_2(x) = \cos 2x$, and $y_3(x) = \sin 2x$ (of Example 1) are linearly independent.

Solution Their Wronskian is

$$W = \begin{vmatrix} e^{-3x} & \cos 2x & \sin 2x \\ -3e^{-3x} & -2\sin 2x & 2\cos 2x \\ 9e^{-3x} & -4\cos 2x & -4\sin 2x \end{vmatrix}$$

$$= e^{-3x} \begin{vmatrix} -2\sin 2x & 2\cos 2x \\ -4\cos 2x & -4\sin 2x \end{vmatrix} + 3e^{-3x} \begin{vmatrix} \cos 2x & \sin 2x \\ -4\cos 2x & -4\sin 2x \end{vmatrix}$$

$$+ 9e^{-3x} \begin{vmatrix} \cos 2x & \sin 2x \\ -2\sin 2x & 2\cos 2x \end{vmatrix} = 26e^{-3x} \neq 0.$$

Because $W \neq 0$ everywhere, it follows that y_1, y_2, and y_3 are linearly independent on any open interval (including the entire real line). ■

Example 5 Show first that the three solutions

$$y_1(x) = x, \quad y_2(x) = x \ln x, \quad \text{and} \quad y_3(x) = x^2$$

of the third-order equation

$$x^3 y^{(3)} - x^2 y'' + 2xy' - 2y = 0 \tag{10}$$

are linearly independent on the open interval $x > 0$. Then find a particular solution of Eq. (10) that satisfies the initial conditions

$$y(1) = 3, \quad y'(1) = 2, \quad y''(1) = 1. \tag{11}$$

Solution Note that for $x > 0$, we could divide each term in (10) by x^3 to obtain a homogeneous linear equation of the standard form in (3). When we compute the Wronskian of the three given solutions, we find that

$$W = \begin{vmatrix} x & x \ln x & x^2 \\ 1 & 1 + \ln x & 2x \\ 0 & \dfrac{1}{x} & 2 \end{vmatrix} = x.$$

Thus $W(x) \neq 0$ for $x > 0$, so y_1, y_2, and y_3 are linearly independent on the interval $x > 0$. To find the desired particular solution, we impose the initial conditions in (11) on

$$\begin{aligned} y(x) &= c_1 x + c_2 x \ln x + c_3 x^2, \\ y'(x) &= c_1 + c_2(1 + \ln x) + 2c_3 x, \\ y''(x) &= 0 + \frac{c_2}{x} + 2c_3. \end{aligned}$$

This yields the simultaneous equations

$$\begin{aligned} y(1) &= c_1 + c_3 = 3, \\ y'(1) &= c_1 + c_2 + 2c_3 = 2, \\ y''(1) &= c_2 + 2c_3 = 1; \end{aligned}$$

we solve to find $c_1 = 1$, $c_2 = -3$, and $c_3 = 2$. Thus the particular solution in question is

$$y(x) = x - 3x \ln x + 2x^2.$$ ■

Provided that $W(y_1, y_2, \dots, y_n) \neq 0$, it turns out (Theorem 4) that we can always find values of the coefficients in the linear combination

$$y = c_1 y_1 + c_2 y_2 + \cdots + c_n y_n$$

that satisfy any given initial conditions of the form in (5). Theorem 3 provides the necessary nonvanishing of W in the case of linearly independent solutions.

THEOREM 3 Wronskians of Solutions

Suppose that $y_1, y_2, \ldots, y_n$ are n solutions of the homogeneous nth-order linear equation

$$y^{(n)} + p_1(x)y^{(n-1)} + \cdots + p_{n-1}(x)y' + p_n(x)y = 0 \tag{3}$$

on an open interval I where each p_i is continuous. Let

$$W = W(y_1, y_2, \ldots, y_n).$$

(a) If $y_1, y_2, \ldots, y_n$ are linearly dependent, then $W \equiv 0$ on I.

(b) If $y_1, y_2, \ldots, y_n$ are linearly independent, then $W \neq 0$ at each point of I.
Thus there are just two possibilities: Either $W = 0$ everywhere on I, or $W \neq 0$ everywhere on I.

***Proof*:** We have already proven part (a). To prove part (b), it is sufficient to assume that $W(a) = 0$ at some point of I and show this implies that the solutions $y_1, y_2, \ldots, y_n$ are linearly dependent. But $W(a)$ is simply the determinant of coefficients of the system of n homogeneous linear equations

$$\begin{aligned}
c_1y_1(a) + \quad & c_2y_2(a) + \cdots + \quad c_ny_n(a) = 0, \\
c_1y_1'(a) + \quad & c_2y_2'(a) + \cdots + \quad c_ny_n'(a) = 0, \\
& \vdots \\
c_1y_1^{(n-1)}(a) + & c_2y_2^{(n-1)}(a) + \cdots + c_ny_n^{(n-1)}(a) = 0
\end{aligned} \tag{12}$$

in the n unknowns $c_1, c_2, \ldots, c_n$. Because $W(a) = 0$, the basic fact from linear algebra quoted just after (9) implies that the equations in (12) have a nontrivial solution. That is, the numbers $c_1, c_2, \ldots, c_n$ are not all zero.

We now use these values to define the particular solution

$$Y(x) = c_1y_1(x) + c_2y_2(x) + \cdots + c_ny_n(x) \tag{13}$$

of Eq. (3). The equations in (12) then imply that Y satisfies the trivial initial conditions

$$Y(a) = Y'(a) = \cdots = Y^{(n-1)}(a) = 0.$$

Theorem 2 (uniqueness) therefore implies that $Y(x) \equiv 0$ on I. In view of (13) and the fact that $c_1, c_2, \ldots, c_n$ are not all zero, this is the desired conclusion that the solutions $y_1, y_2, \ldots, y_n$ are linearly dependent. This completes the proof of Theorem 3. ■

Remark: According to Problem 35, the Wronskian of Theorem 3 satisfies the first-order equation $W' = -p_1(x)W$. Solution of this equation yields *Abel's formula*

$$W(x) = K \exp\left(-\int p_1(x)\,dx\right)$$

for the homogeneous linear equation in (3). According as the constant $K = 0$ or $K \neq 0$, it follows immediately from Abel's formula that either W vanishes everywhere or W is nonzero everywhere. ■

General Solutions

We can now show that, given any fixed set of n linearly independent solutions of a *homogeneous* nth-order equation, *every* (other) solution of the equation can be expressed as a linear combination of those n particular solutions. Using the fact from Theorem 3 that the Wronskian of n linearly independent solutions is nonzero, the proof of the following theorem is essentially the same as the proof of Theorem 4 of Section 5.1 (the case $n = 2$).

THEOREM 4 General Solutions of Homogeneous Equations

Let $y_1, y_2, \ldots, y_n$ be n linearly independent solutions of the homogeneous equation

$$y^{(n)} + p_1(x)y^{(n-1)} + \cdots + p_{n-1}(x)y' + p_n(x)y = 0 \tag{3}$$

on an open interval I, where the p_i are continuous. If Y is any solution whatsoever of Eq. (3), then there exist numbers $c_1, c_2, \ldots, c_n$ such that

$$Y(x) = c_1y_1(x) + c_2y_2(x) + \cdots + c_ny_n(x)$$

for all x in I.

Remark: Theorem 4 tells us that, once we have found n linearly independent solutions $y_1, y_2, \ldots, y_n$ of the nth-order homogeneous linear equation in (3), we really have found *all* of its solutions. For then the solution space $\mathcal{S}$ of the equation is an n-dimensional vector space with basis $\{y_1, y_2, \ldots, y_n\}$. Because *every* solution of (3) can be expressed as a linear combination of the form

$$y = c_1y_1 + c_2y_2 + \cdots + c_ny_n,$$

we call such a linear combination of n linearly independent particular solutions a **general solution** of the homogeneous linear differential equation. ■

Example 6 According to Example 4, the particular solutions $y_1(x) = e^{-3x}$, $y_2(x) = \cos 2x$, and $y_3(x) = \sin 2x$ of the linear differential equation $y^{(3)} + 3y'' + 4y' + 12y = 0$ are linearly independent. Now Theorem 2 says that—given b_0, b_1, and b_2—there exists a particular solution $y(x)$ satisfying the initial conditions $y(0) = b_0$, $y'(0) = b_1$, and $y''(0) = b_2$. Hence Theorem 4 implies that *this* particular solution is a linear combination of y_1, y_2, and y_3. That is, there exist coefficients c_1, c_2, and c_3 such that

$$y(x) = c_1e^{-3x} + c_2\cos 2x + c_3\sin 2x.$$

Upon successive differentiation and substitution of $x = 0$, we discover that to find these coefficients, we need only solve the three linear equations

$$\begin{aligned} c_1 + c_2 \qquad\quad &= b_0, \\ -3c_1 \qquad + 2c_3 &= b_1, \\ 9c_1 - 4c_2 \qquad\quad &= b_2. \end{aligned}$$

(See the application for this section.) ■

Nonhomogeneous Equations

We now consider the *nonhomogeneous* nth-order linear differential equation

$$y^{(n)} + p_1(x)y^{(n-1)} + \cdots + p_{n-1}(x)y' + p_n(x)y = f(x) \tag{2}$$

with associated homogeneous equation

$$y^{(n)} + p_1(x)y^{(n-1)} + \cdots + p_{n-1}(x)y' + p_n(x)y = 0. \tag{3}$$

Suppose that a single fixed particular solution y_p of the nonhomogeneous equation in (2) is known, and that Y is any other solution of Eq. (2). If $y_c = Y - y_p$, then substitution of y_c in the differential equation gives (using the linearity of differentiation)

$$\begin{aligned} y_c^{(n)} + p_1 y_c^{(n-1)} + \cdots + p_{n-1}y_c' + p_n y_c \\ &= (Y^{(n)} + p_1 Y^{(n-1)} + \cdots + p_{n-1}Y' + p_n Y) \\ &\quad - (y_p^{(n)} + p_1 y_p^{(n-1)} + \cdots + p_{n-1}y_p' + p_n y_p) \\ &= f(x) - f(x) = 0. \end{aligned}$$

Thus $y_c = Y - y_p$ is a solution of the associated homogeneous equation in (3). Then

$$Y = y_c + y_p, \tag{14}$$

and it follows from Theorem 4 that

$$y_c = c_1 y_1 + c_2 y_2 + \cdots + c_n y_n, \tag{15}$$

where $y_1, y_2, \ldots, y_n$ are linearly independent solutions of the associated *homogeneous* equation. We call y_c a **complementary function** of the nonhomogeneous equation and have thus proved that a *general solution* of the nonhomogeneous equation in (2) is the sum of its complementary function y_c and a single particular solution y_p of Eq. (2).

THEOREM 5 Solutions of Nonhomogeneous Equations

Let y_p be a particular solution of the nonhomogeneous equation in (2) on an open interval I where the functions p_i and f are continuous. Let $y_1, y_2, \ldots, y_n$ be linearly independent solutions of the associated homogeneous equation in (3). If Y is any solution whatsoever of Eq. (2) on I, then there exist numbers $c_1, c_2, \ldots, c_n$ such that

$$Y(x) = c_1 y_1(x) + c_2 y_2(x) + \cdots + c_n y_n(x) + y_p(x) \tag{16}$$

for all x in I.

Example 7 It is evident that $y_p = 3x$ is a particular solution of the equation

$$y'' + 4y = 12x \tag{17}$$

and that $y_c(x) = c_1 \cos 2x + c_2 \sin 2x$ is its complementary solution. Find a solution of Eq. (17) that satisfies the initial conditions $y(0) = 5$, $y'(0) = 7$.

Solution The general solution of Eq. (17) is

$$y(x) = c_1 \cos 2x + c_2 \sin 2x + 3x.$$

Now

$$y'(x) = -2c_1 \sin 2x + 2c_2 \cos 2x + 3.$$

Hence the initial conditions give

$$y(0) = c_1 = 5,$$

$$y'(0) = 2c_2 + 3 = 7.$$

We find that $c_1 = 5$ and $c_2 = 2$. Thus the desired solution is

$$y(x) = 5 \cos 2x + 2 \sin 2x + 3x.$$ ■

5.2 Problems

In Problems 1 through 6, show directly that the given functions are linearly dependent on the real line. That is, find a nontrivial linear combination of the given functions that vanishes identically.

1. $f(x) = 2x,\ g(x) = 3x^2,\ h(x) = 5x - 8x^2$
2. $f(x) = 5,\ g(x) = 2 - 3x^2,\ h(x) = 10 + 15x^2$
3. $f(x) = 0,\ g(x) = \sin x,\ h(x) = e^x$
4. $f(x) = 17,\ g(x) = 2\sin^2 x,\ h(x) = 3\cos^2 x$
5. $f(x) = 17,\ g(x) = \cos^2 x,\ h(x) = \cos 2x$
6. $f(x) = e^x,\ g(x) = \cosh x,\ h(x) = \sinh x$

In Problems 7 through 12, use the Wronskian to prove that the given functions are linearly independent on the indicated interval.

7. $f(x) = 1,\ g(x) = x,\ h(x) = x^2$; the real line
8. $f(x) = e^x,\ g(x) = e^{2x},\ h(x) = e^{3x}$; the real line
9. $f(x) = e^x,\ g(x) = \cos x,\ h(x) = \sin x$; the real line
10. $f(x) = e^x,\ g(x) = x^{-2},\ h(x) = x^{-2}\ln x$; $x > 0$
11. $f(x) = x,\ g(x) = xe^x,\ h(x) = x^2e^x$; the real line
12. $f(x) = x,\ g(x) = \cos(\ln x),\ h(x) = \sin(\ln x)$; $x > 0$

In Problems 13 through 20, a third-order homogeneous linear equation and three linearly independent solutions are given. Find a particular solution satisfying the given initial conditions.

13. $y^{(3)} + 2y'' - y' - 2y = 0$; $y(0) = 1,\ y'(0) = 2,\ y''(0) = 0$; $y_1 = e^x,\ y_2 = e^{-x},\ y_3 = e^{-2x}$
14. $y^{(3)} - 6y'' + 11y' - 6y = 0$; $y(0) = 0,\ y'(0) = 0,\ y''(0) = 3$; $y_1 = e^x,\ y_2 = e^{2x},\ y_3 = e^{3x}$
15. $y^{(3)} - 3y'' + 3y' - y = 0$; $y(0) = 2,\ y'(0) = 0,\ y''(0) = 0$; $y_1 = e^x,\ y_2 = xe^x,\ y_3 = x^2e^x$
16. $y^{(3)} - 5y'' + 8y' - 4y = 0$; $y(0) = 1,\ y'(0) = 4,\ y''(0) = 0$; $y_1 = e^x,\ y_2 = e^{2x},\ y_3 = xe^{2x}$
17. $y^{(3)} + 9y' = 0$; $y(0) = 3,\ y'(0) = -1,\ y''(0) = 2$; $y_1 = 1,\ y_2 = \cos 3x,\ y_3 = \sin 3x$
18. $y^{(3)} - 3y'' + 4y' - 2y = 0$; $y(0) = 1,\ y'(0) = 0,\ y''(0) = 0$; $y_1 = e^x,\ y_2 = e^x\cos x,\ y_3 = e^x\sin x$
19. $x^3y^{(3)} - 3x^2y'' + 6xy' - 6y = 0$; $y(1) = 6,\ y'(1) = 14,\ y''(1) = 22$; $y_1 = x,\ y_2 = x^2,\ y_3 = x^3$
20. $x^3y^{(3)} + 6x^2y'' + 4xy' - 4y = 0$; $y(1) = 1,\ y'(1) = 5,\ y''(1) = -11$; $y_1 = x,\ y_2 = x^{-2},\ y_3 = x^{-2}\ln x$

In Problems 21 through 24, a nonhomogeneous differential equation, a complementary solution y_c, and a particular solution y_p are given. Find a solution satisfying the given initial conditions.

21. $y'' + y = 3x$; $y(0) = 2,\ y'(0) = -2$; $y_c = c_1\cos x + c_2\sin x$; $y_p = 3x$
22. $y'' - 4y = 12$; $y(0) = 0,\ y'(0) = 10$; $y_c = c_1e^{2x} + c_2e^{-2x}$; $y_p = -3$
23. $y'' - 2y' - 3y = 6$; $y(0) = 3,\ y'(0) = 11$; $y_c = c_1e^{-x} + c_2e^{3x}$; $y_p = -2$
24. $y'' - 2y' + 2y = 2x$; $y(0) = 4,\ y'(0) = 8$; $y_c = c_1e^x\cos x + c_2e^x\sin x$; $y_p = x + 1$
25. Let $Ly = y'' + py' + qy$. Suppose that y_1 and y_2 are two functions such that

$$Ly_1 = f(x) \quad \text{and} \quad Ly_2 = g(x).$$

Show that their sum $y = y_1 + y_2$ satisfies the nonhomogeneous equation $Ly = f(x) + g(x)$.

26. **(a)** Find by inspection particular solutions of the two nonhomogeneous equations

$$y'' + 2y = 4 \quad \text{and} \quad y'' + 2y = 6x.$$

(b) Use the method of Problem 25 to find a particular solution of the differential equation $y'' + 2y = 6x + 4$.

27. Prove directly that the functions

$$f_1(x) \equiv 1, \quad f_2(x) = x, \quad \text{and} \quad f_3(x) = x^2$$

are linearly independent on the whole real line. (*Suggestion*: Assume that $c_1 + c_2x + c_3x^2 = 0$. Differentiate this equation twice, and conclude from the equations you get that $c_1 = c_2 = c_3 = 0$.)

28. Generalize the method of Problem 27 to prove directly that the functions

$$f_0(x) \equiv 1, \; f_1(x) = x, \; f_2(x) = x^2, \ldots, \; f_n(x) = x^n$$

are linearly independent on the real line.

29. Use the result of Problem 28 and the definition of linear independence to prove directly that, for any constant r, the functions

$$f_0(x) = e^{rx}, \quad f_1(x) = xe^{rx}, \ldots, \quad f_n(x) = x^n e^{rx}$$

are linearly independent on the whole real line.

30. Verify that $y_1 = x$ and $y_2 = x^2$ are linearly independent solutions on the entire real line of the equation

$$x^2y'' - 2xy' + 2y = 0,$$

but that $W(x, x^2)$ vanishes at $x = 0$. Why do these observations not contradict part (b) of Theorem 3?

31. This problem indicates why we can impose *only* n initial conditions on a solution of an nth-order linear differential equation. **(a)** Given the equation

$$y'' + py' + qy = 0,$$

explain why the value of $y''(a)$ is determined by the values of $y(a)$ and $y'(a)$. **(b)** Prove that the equation

$$y'' - 2y' - 5y = 0$$

has a solution satisfying the conditions

$$y(0) = 1, \quad y'(0) = 0, \quad \text{and} \quad y''(0) = C$$

if and only if $C = 5$.

32. Prove that an nth-order homogeneous linear differential equation satisfying the hypotheses of Theorem 2 has n linearly independent solutions $y_1, y_2, \ldots, y_n$. [*Suggestion*: Let y_i be the unique solution such that

$$y_i^{(i-1)}(a) = 1 \quad \text{and} \quad y_i^{(k)}(a) = 0 \quad \text{if } k \neq i - 1.]$$

33. Suppose that the three numbers r_1, r_2, and r_3 are distinct. Show that the three functions $\exp(r_1x)$, $\exp(r_2x)$, and $\exp(r_3x)$ are linearly independent by showing that their Wronskian

$$W = \exp[(r_1 + r_2 + r_3)x] \cdot \begin{vmatrix} 1 & 1 & 1 \\ r_1 & r_2 & r_3 \\ r_1^2 & r_2^2 & r_3^2 \end{vmatrix}$$

is nonzero for all x.

34. Assume as known that the **Vandermonde determinant**

$$V = \begin{vmatrix} 1 & 1 & \cdots & 1 \\ r_1 & r_2 & \cdots & r_n \\ r_1^2 & r_2^2 & \cdots & r_n^2 \\ \vdots & \vdots & & \vdots \\ r_1^{n-1} & r_2^{n-1} & \cdots & r_n^{n-1} \end{vmatrix}$$

is nonzero if the numbers $r_1, r_2, \ldots, r_n$ are distinct. Prove by the method of Problem 33 that the functions

$$f_i(x) = \exp(r_ix), \quad 1 \leqq i \leqq n$$

are linearly independent.

35. According to Problem 32 of Section 5.1, the Wronskian $W(y_1, y_2)$ of two solutions of the second-order equation

$$y'' + p_1(x)y' + p_2(x)y = 0$$

is given by Abel's formula

$$W(x) = K \exp\left(-\int p_1(x)\,dx\right)$$

for some constant K. It can be shown that the Wronskian of n solutions $y_1, y_2, \ldots, y_n$ of the nth-order equation

$$y^{(n)} + p_1(x)y^{(n-1)} + \cdots + p_{n-1}(x)y' + p_n(x)y = 0$$

satisfies the same identity. Prove this for the case $n = 3$ as follows: **(a)** The derivative of a determinant of functions is the sum of the determinants obtained by separately differentiating the rows of the original determinant. Conclude that

$$W' = \begin{vmatrix} y_1 & y_2 & y_3 \\ y_1' & y_2' & y_3' \\ y_1^{(3)} & y_2^{(3)} & y_3^{(3)} \end{vmatrix}.$$

(b) Substitute for $y_1^{(3)}$, $y_2^{(3)}$, and $y_3^{(3)}$ from the equation

$$y^{(3)} + p_1y'' + p_2y' + p_3y = 0,$$

and then show that $W' = -p_1W$. Integration now gives Abel's formula.

36. Suppose that one solution $y_1(x)$ of the homogeneous second-order linear differential equation

$$y'' + p(x)y' + q(x)y = 0 \tag{18}$$

is known (on an interval I where p and q are continuous functions). The method of **reduction of order** consists of substituting $y_2(x) = v(x)y_1(x)$ in (18) and attempting to determine the function $v(x)$ so that $y_2(x)$ is a second linearly independent solution of (18). After substituting $y = v(x)y_1(x)$ in Eq. (18), use the fact that $y_1(x)$ is a solution to deduce that

$$y_1v'' + (2y_1' + py_1)v' = 0. \tag{19}$$

If $y_1(x)$ is known, then (19) is a separable equation that is readily solved for the derivative $v'(x)$ of $v(x)$. Integration of $v'(x)$ then gives the desired (nonconstant) function $v(x)$.

37. Before applying Eq. (19) with a given homogeneous second-order linear differential equation and a known solution $y_1(x)$, the equation must first be written in the form of (18) with leading coefficient 1 in order to correctly determine the coefficient function $p(x)$. Frequently it is more convenient to simply substitute $y = v(x)y_1(x)$ in the given differential equation and then proceed directly to find $v(x)$. Thus, starting with the readily verified solution $y_1(x) = x^3$ of the equation

$$x^2y'' - 5xy' + 9y = 0 \quad (x > 0),$$

substitute $y = vx^3$ and deduce that $xv'' + v' = 0$. Then solve for $v(x) = C \ln x$, and thereby obtain (with $C = 1$) the second solution $y_2(x) = x^3 \ln x$.

In each of Problems 38 through 42, a differential equation and one solution y_1 are given. Use the method of reduction of order as in Problem 37 to find a second linearly independent solution y_2.

38. $x^2y'' + xy' - 9y = 0$ $(x > 0)$; $y_1(x) = x^3$

39. $4y'' - 4y' + y = 0$; $y_1(x) = e^{x/2}$

40. $x^2y'' - x(x+2)y' + (x+2)y = 0$ $(x > 0)$; $y_1(x) = x$

41. $(x+1)y'' - (x+2)y' + y = 0$ $(x > -1)$; $y_1(x) = e^x$

42. $(1-x^2)y'' + 2xy' - 2y = 0$ $(-1 < x < 1)$; $y_1(x) = x$

43. First note that $y_1(x) = x$ is one solution of Legendre's equation of order 1,

$$(1-x^2)y'' - 2xy' + 2y = 0.$$

Then use the method of reduction of order to derive the second solution

$$y_2(x) = 1 - \frac{x}{2}\ln\frac{1+x}{1-x} \quad (\text{for } -1 < x < 1).$$

44. First verify by substitution that $y_1(x) = x^{-1/2}\cos x$ is one solution (for $x > 0$) of Bessel's equation of order $\frac{1}{2}$,

$$x^2y'' + xy' + (x^2 - \tfrac{1}{4})y = 0.$$

Then derive by reduction of order the second solution $y_2(x) = x^{-1/2}\sin x$.

5.2 Application Plotting Third-Order Solution Families

This application deals with the plotting by computer of families of solutions such as those illustrated in Figs. 5.2.2 through 5.2.4. We know from Example 6 that the general solution of

$$y^{(3)} + 3y'' + 4y' + 12y = 0 \tag{1}$$

is

$$y(x) = c_1e^{-3x} + c_2\cos 2x + c_3\sin 2x. \tag{2}$$

For Fig. 5.2.2, use the method of Example 6 to show that the particular solution of Eq. (1) satisfying the initial conditions $y(0) = a$, $y'(0) = 0$, and $y''(0) = 0$ is given by

$$y(x) = \frac{a}{13}\left(4e^{-3x} + 9\cos 2x + 6\sin 2x\right). \tag{3}$$

The MATLAB loop

```
x = -1.5 : 0.02 : 5      % x-vector from x = -1.5 to x = 5
for a = -3 : 1 : 3       % for a = -3 to 3 with da = 1 do
   c1 = 4*a/13;
   c2 = 9*a/13;
   c3 = 6*a/13;
   y = c1*exp(-3*x) + c2*cos(2*x) + c3*sin(2*x);
   plot(x,y)
end
```

was used to generate Fig. 5.2.2.

For Fig. 5.2.3, show that the particular solution of Eq. (1) satisfying the initial conditions $y(0) = 0$, $y'(0) = b$, and $y''(0) = 0$ is given by

$$y(x) = \frac{b}{2}\sin 2x, \tag{4}$$

and alter the preceding for-loop accordingly.

For Fig. 5.2.4, show that the particular solution of Eq. (1) satisfying the initial conditions $y(0) = 0$, $y'(0) = 0$, and $y''(0) = c$ is given by

$$y(x) = \frac{c}{26}\left(2e^{-3x} - 2\cos 2x + 3\sin 2x\right). \tag{5}$$

Computer algebra systems such as *Maple* and *Mathematica,* as well as graphing calculators, have commands to carry out for-loops such as the one shown here. Begin by reproducing Figs. 5.2.2 through 5.2.4. Then plot similar families of solution curves for the differential equations in Problems 13 through 20.

5.3 Homogeneous Equations with Constant Coefficients

In Section 5.2 we saw that a general solution of an nth-order homogeneous linear equation is a linear combination of n linearly independent particular solutions, but we said little about how actually to find even a single solution. The solution of a linear differential equation with *variable* coefficients ordinarily requires numerical methods (Chapter 2) or infinite series methods (Chapter 11). But we can now show how to find, explicitly and in a rather straightforward way, n linearly independent solutions of a given nth-order linear equation if it has *constant* coefficients. The general such equation may be written in the form

$$a_n y^{(n)} + a_{n-1}y^{(n-1)} + \cdots + a_2 y'' + a_1 y' + a_0 y = 0, \tag{1}$$

where the coefficients $a_0, a_1, a_2, \ldots, a_n$ are real constants with $a_n \neq 0$.

The Characteristic Equation

We first look for a *single* solution of Eq. (1) and begin with the observation that

$$\frac{d^k}{dx^k}(e^{rx}) = r^k e^{rx}, \tag{2}$$

so any derivative of e^{rx} is a constant multiple of e^{rx}. Hence, if we substituted $y = e^{rx}$ in Eq. (1), each term would be a constant multiple of e^{rx}, with the constant coefficients depending on r and the coefficients a_k. This suggests that we try to find r so that all these multiples of e^{rx} will have sum zero, in which case $y = e^{rx}$ will be a solution of Eq. (1).

For example, in Section 5.1 we substituted $y = e^{rx}$ in the second-order equation

$$ay'' + by' + cy = 0$$

to derive the characteristic equation

$$ar^2 + br + c = 0$$

that r must satisfy.

To carry out this technique in the general case, we substitute $y = e^{rx}$ in Eq. (1), and with the aid of Eq. (2) we find the result to be

$$a_n r^n e^{rx} + a_{n-1} r^{n-1} e^{rx} + \cdots + a_2 r^2 e^{rx} + a_1 r e^{rx} + a_0 e^{rx} = 0;$$

that is,

$$e^{rx}\left(a_n r^n + a_{n-1} r^{n-1} + \cdots + a_2 r^2 + a_1 r + a_0\right) = 0.$$

Because e^{rx} is never zero, we see that $y = e^{rx}$ will be a solution of Eq. (1) precisely when r is a root of the equation

$$a_n r^n + a_{n-1} r^{n-1} + \cdots + a_2 r^2 + a_1 r + a_0 = 0. \tag{3}$$

This equation is called the **characteristic equation** or **auxiliary equation** of the differential equation in (1). Our problem, then, is reduced to the solution of this purely algebraic equation.

According to the fundamental theorem of algebra, every nth-degree polynomial—such as the one in Eq. (3)—has n zeros, though not necessarily distinct and not necessarily real. Finding the exact values of these zeros may be difficult or even impossible; the quadratic formula is sufficient for second-degree equations, but for equations of higher degree we may need either to spot a fortuitous factorization or to apply a numerical technique such as Newton's method (or use a calculator/computer `solve` command).

Distinct Real Roots

Whatever the method we use, let us suppose that we have solved the characteristic equation. Then we can always write a general solution of the differential equation. The situation is slightly more complicated in the case of repeated roots or complex roots of Eq. (3), so let us first examine the simplest case—in which the characteristic equation has n distinct (no two equal) *real* roots $r_1, r_2, \ldots, r_n$. Then the functions

$$e^{r_1 x}, e^{r_2 x} \ldots, e^{r_n x}$$

are all solutions of Eq. (1), and (by Problem 34 of Section 5.2) these n solutions are linearly independent on the entire real line. In summary, we have proved Theorem 1.

THEOREM 1 Distinct Real Roots

If the roots $r_1, r_2, \ldots, r_n$ of the characteristic equation in (3) are real and distinct, then

$$y(x) = c_1 e^{r_1 x} + c_2 e^{r_2 x} + \cdots + c_n e^{r_n x} \tag{4}$$

is a general solution of Eq. (1). Thus the n linearly independent functions $\{e^{r_1 x}, e^{r_2 x}, \ldots, e^{r_n x}\}$ constitute a basis for the n-dimensional solution space of Eq. (1).

Example 1 Solve the initial value problem

$$y^{(3)} + 3y'' - 10y' = 0;$$
$$y(0) = 7, \quad y'(0) = 0, \quad y''(0) = 70.$$

Solution The characteristic equation of the given differential equation is

$$r^3 + 3r^2 - 10r = 0.$$

We solve by factoring:

$$r(r^2 + 3r - 10) = r(r + 5)(r - 2) = 0,$$

and so the characteristic equation has the three distinct real roots $r = 0$, $r = -5$, and $r = 2$. Because $e^0 = 1$, Theorem 1 gives the general solution

$$y(x) = c_1 + c_2 e^{-5x} + c_3 e^{2x}.$$

Then the given initial conditions yield the linear equations

$$\begin{aligned} y(0) &= c_1 + \quad c_2 + \quad c_3 = 7, \\ y'(0) &= \quad\; - \; 5c_2 + 2c_3 = 0, \\ y''(0) &= \quad\;\; 25c_2 + 4c_3 = 70 \end{aligned}$$

in the coefficients c_1, c_2, and c_3. The last two equations give $y''(0) - 2y'(0) = 35c_2 = 70$, so $c_2 = 2$. Then the second equation gives $c_3 = 5$, and finally the first equation gives $c_1 = 0$. Thus the desired particular solution is

$$y(x) = 2e^{-5x} + 5e^{2x}.$$

■

Polynomial Differential Operators

If the roots of the characteristic equation in (3) are *not* distinct—there are repeated roots—then we cannot produce n linearly independent solutions of Eq. (1) by the method of Theorem 1. For example, if the roots are 1, 2, 2, and 2, we obtain only the *two* functions e^x and e^{2x}. The problem, then, is to produce the missing linearly independent solutions. For this purpose, it is convenient to adopt "operator notation" and write Eq. (1) in the form $Ly = 0$, where the **operator**

$$L = a_n \frac{d^n}{dx^n} + a_{n-1} \frac{d^{n-1}}{dx^{n-1}} + \cdots + a_2 \frac{d^2}{dx^2} + a_1 \frac{d}{dx} + a_0. \tag{5}$$

operates on the n times differentiable function $y(x)$ to produce the linear combination

$$Ly = a_n y^{(n)} + a_{n-1} y^{(n-1)} + \cdots + a_2 y^{(2)} + a_1 y' + a_0 y$$

of y and its first n derivatives. We also denote by $D = d/dx$ the operation of differentiation with respect to x, so that

$$Dy = y', \quad D^2 y = y'', \quad D^3 y = y^{(3)},$$

and so on. In terms of D, the operator L in (5) may be written

$$L = a_n D^n + a_{n-1} D^{n-1} + \cdots + a_2 D^2 + a_1 D + a_0, \tag{6}$$

and we will find it useful to think of the right-hand side in Eq. (6) as a (formal) nth-degree polynomial in the "variable" D; it is a **polynomial differential operator.**

A first-degree polynomial operator with leading coefficient 1 has the form $D - a$, where a is a real number. It operates on a function $y = y(x)$ to produce

$$(D - a)y = Dy - ay = y' - ay.$$

The important fact about such operators is that any two of them *commute*:

$$(D - a)(D - b)y = (D - b)(D - a)y \tag{7}$$

for any twice differentiable function $y = y(x)$. The proof of the formula in (7) is the following computation:

$$\begin{aligned}(D - a)(D - b)y &= (D - a)(y' - by)\\ &= D(y' - by) - a(y' - by)\\ &= y'' - (b + a)y' + aby = y'' - (a + b)y' + bay\\ &= D(y' - ay) - b(y' - ay)\\ &= (D - b)(y' - ay) = (D - b)(D - a)y.\end{aligned}$$

We see here also that $(D - a)(D - b) = D^2 - (a + b)D + ab$. Similarly, it can be shown by induction on the number of factors that an operator product of the form $(D - a_1)(D - a_2) \cdot \cdots \cdot (D - a_n)$ expands—by multiplying out and collecting coefficients—in the same way as does an ordinary product $(x - a_1)(x - a_2) \cdot \cdots \cdot (x - a_n)$ of linear factors with x denoting a real variable. Consequently, the algebra of polynomial differential operators closely resembles the algebra of ordinary real polynomials.

Repeated Real Roots

Let us now consider the possibility that the characteristic equation

$$a_n r^n + a_{n-1} r^{n-1} + \cdots + a_2 r^2 + a_1 r + a_0 = 0 \tag{3}$$

has *repeated* roots. For example, suppose that Eq. (3) has only two distinct roots, r_0 of multiplicity 1 and r_1 of multiplicity $k = n - 1 > 1$. Then (after dividing by a_n) Eq. (3) can be rewritten in the form

$$(r - r_1)^k (r - r_0) = (r - r_0)(r - r_1)^k = 0. \tag{8}$$

Similarly, the corresponding operator L in (6) can be written as

$$L = (D - r_1)^k (D - r_0) = (D - r_0)(D - r_1)^k, \tag{9}$$

the order of the factors making no difference because of the formula in (7).

Two solutions of the differential equation $Ly = 0$ are certainly $y_0 = e^{r_0 x}$ and $y_1 = e^{r_1 x}$. This is, however, not sufficient; we need $k + 1$ linearly independent solutions in order to construct a general solution, because the equation is of order $k + 1$. To find the missing $k - 1$ solutions, we note that

$$Ly = (D - r_0)[(D - r_1)^k y] = 0.$$

Consequently, *every* solution of the kth-order equation

$$(D - r_1)^k y = 0 \tag{10}$$

will also be a solution of the original equation $Ly = 0$. Hence our problem is reduced to that of finding a general solution of the differential equation in (10).

The fact that $y_1 = e^{r_1 x}$ is one solution of Eq. (10) suggests that we try the substitution

$$y(x) = u(x)y_1(x) = u(x)e^{r_1 x}, \tag{11}$$

where $u(x)$ is a function yet to be determined. Observe that

$$(D - r_1)\left[ue^{r_1 x}\right] = (Du)e^{r_1 x} + u(r_1 e^{r_1 x}) - r_1(ue^{r_1 x}) = (Du)e^{r_1 x}. \tag{12}$$

Upon k applications of this fact, it follows that

$$(D - r_1)^k\left[ue^{r_1 x}\right] = (D^k u)e^{r_1 x} \tag{13}$$

for any sufficiently differentiable function $u(x)$. Hence $y = ue^{r_1 x}$ will be a solution of Eq. (10) if and only if $D^k u = u^{(k)} = 0$. But this is so if and only if

$$u(x) = c_1 + c_2 x + c_3 x^2 + \cdots + c_k x^{k-1},$$

a polynomial of degree at most $k - 1$. Hence our desired solution of Eq. (10) is

$$y(x) = ue^{r_1 x} = (c_1 + c_2 x + c_3 x^2 + \cdots + c_k x^{k-1})e^{r_1 x}.$$

In particular, we see here the additional solutions $xe^{r_1 x}$, $x^2 e^{r_1 x}$, $\dots$, $x^{k-1}e^{r_1 x}$ of the original differential equation $Ly = 0$.

The preceding analysis can be carried out with the operator $D - r_1$ replaced with an arbitrary polynomial operator. When this is done, the result is a proof of the following theorem.

THEOREM 2 Repeated Roots

If the characteristic equation in (3) has a repeated root r of multiplicity k, then the part of a general solution of the differential equation in (1) corresponding to r is of the form

$$(c_1 + c_2 x + c_3 x^2 + \cdots + c_k x^{k-1})e^{rx}. \tag{14}$$

Remark: According to Problem 29 of Section 5.2, the k functions e^{rx}, xe^{rx}, $x^2 e^{rx}$, $\dots$, $x^{k-1}e^{rx}$ involved in (14) are linearly independent. Thus a root of multiplicity $k \geqq 1$ corresponds to k linearly independent "basis solutions" that generate a k-dimensional subspace of the n-dimensional solution space $\mathcal{S}$ of the nth-order homogeneous linear differential equation. Because the sum of the multiplicities of the separate roots of the nth-degree characteristic equation is n, we get a basis $\{y_1, y_2, \dots, y_n\}$ for $\mathcal{S}$ when we assemble all the basis solutions corresponding to these separate roots. We can then write a general solution $y = c_1 y_1 + c_2 y_2 + \cdots + c_n y_n$ of the differential equation. In effect, we thereby construct a general solution by combining different "parts" corresponding to the separate roots of the characteristic equation. ■

Example 2 Find a general solution of the fifth-order differential equation

$$9y^{(5)} - 6y^{(4)} + y^{(3)} = 0.$$

Solution The characteristic equation is

$$9r^5 - 6r^4 + r^3 = r^3(9r^2 - 6r + 1) = r^3(3r-1)^2 = 0.$$

It has the triple root $r = 0$ and the double root $r = \frac{1}{3}$. The triple root $r = 0$ contributes

$$c_1 e^{0 \cdot x} + c_2 x e^{0 \cdot x} + c_3 x^2 e^{0 \cdot x} = c_1 + c_2 x + c_3 x^2$$

to the solution, while the double root $r = \frac{1}{3}$ contributes $c_4 e^{x/3} + c_5 x e^{x/3}$. Hence a general solution of the given differential equation is

$$y(x) = c_1 + c_2 x + c_3 x^2 + c_4 e^{x/3} + c_5 x e^{x/3}.$$

■

Complex-Valued Functions and Euler's Formula

Because we have assumed that the coefficients of the differential equation and its characteristic equation are real, any complex (nonreal) roots will occur in complex conjugate pairs $a \pm bi$ where a and b are real and $i = \sqrt{-1}$. This raises the question as to what might be meant by an exponential such as $e^{(a+bi)x}$.

To answer this question, we recall from elementary calculus the Taylor (or Maclaurin) series for the exponential function

$$e^t = \sum_{n=0}^{\infty} \frac{t^n}{n!} = 1 + t + \frac{t^2}{2!} + \frac{t^3}{3!} + \frac{t^4}{4!} + \cdots .$$

If we substitute $t = i\theta$ in this series and recall that $i^2 = -1$, $i^3 = -i$, $i^4 = 1$, and so on, we get

$$\begin{aligned} e^{i\theta} &= \sum_{n=0}^{\infty} \frac{(i\theta)^n}{n!} \\ &= 1 + i\theta - \frac{\theta^2}{2!} - \frac{i\theta^3}{3!} + \frac{\theta^4}{4!} + \frac{i\theta^5}{5!} - \cdots \\ &= \left(1 - \frac{\theta^2}{2!} + \frac{\theta^4}{4!} - \cdots\right) + i\left(\theta - \frac{\theta^3}{3!} + \frac{\theta^5}{5!} - \cdots\right). \end{aligned}$$

Because the two real series in the last line are the Taylor series for $\cos\theta$ and $\sin\theta$, respectively, this implies that

$$e^{i\theta} = \cos\theta + i \sin\theta. \tag{15}$$

This result is known as **Euler's formula**. Because of it, we *define* the exponential function e^z, for $z = x + iy$ an arbitrary complex number, to be

$$e^z = e^{x+iy} = e^x e^{iy} = e^x(\cos y + i \sin y). \tag{16}$$

Thus it appears that complex roots of the characteristic equation will lead to complex-valued solutions of the differential equation. A **complex-valued function** F of the real variable x associates with each real number x (in its domain of definition) the complex number

$$F(x) = f(x) + ig(x). \tag{17}$$

The real-valued functions f and g are called the **real** and **imaginary** parts, respectively, of F. If they are differentiable, we define the **derivative** F' of F by

$$F'(x) = f'(x) + ig'(x). \tag{18}$$

Thus we simply differentiate the real and imaginary parts of F separately.

We say that the complex-valued function $F(x)$ satisfies the homogeneous linear differential equation $L[F(x)] = 0$ provided that its real and imaginary parts in (17) separately satisfy this equation—so $L[F(x)] = L[f(x)] + iL[g(x)] = 0$.

The particular complex-valued functions of interest here are of the form $F(x) = e^{rx}$, where $r = a \pm bi$. We note from Euler's formula that

$$e^{(a+bi)x} = e^{ax}e^{ibx} = e^{ax}(\cos bx + i \sin bx) \tag{19a}$$

and

$$e^{(a-bi)x} = e^{ax}e^{-ibx} = e^{ax}(\cos bx - i \sin bx). \tag{19b}$$

The most important property of e^{rx} is that

$$D_x(e^{rx}) = re^{rx}, \tag{20}$$

if r is a complex number. The proof of this assertion is a straightforward computation based on the definitions and formulas given earlier:

$$\begin{aligned} D_x(e^{rx}) &= D_x(e^{ax}\cos bx) + iD_x(e^{ax}\sin bx) \\ &= \left(ae^{ax}\cos bx - be^{ax}\sin bx\right) + i\left(ae^{ax}\sin bx + be^{ax}\cos bx\right) \\ &= (a + bi)(e^{ax}\cos bx + ie^{ax}\sin bx) = re^{rx}. \end{aligned}$$

Complex Roots

It follows from Eq. (20) that when r is complex (just as when r is real), e^{rx} will be a solution of the differential equation in (1) if and only if r is a root of its characteristic equation. If the complex conjugate pair of roots $r_1 = a + bi$ and $r_2 = a - bi$ are simple (nonrepeated), then the corresponding part of a general solution of Eq. (1) is

$$\begin{aligned} y(x) &= C_1e^{r_1x} + C_2e^{r_2x} = C_1e^{(a+bi)x} + C_2e^{(a-bi)x} \\ &= C_1e^{ax}(\cos bx + i \sin bx) + C_2e^{ax}(\cos bx - i \sin bx) \\ y(x) &= (C_1 + C_2)e^{ax}\cos bx + i(C_1 - C_2)e^{ax}\sin bx, \end{aligned}$$

where the arbitrary constants C_1 and C_2 can be complex. For instance, the choice $C_1 = C_2 = \frac{1}{2}$ gives the real-valued solution $y_1(x) = e^{ax}\cos bx$, while the choice $C_1 = -\frac{1}{2}i$, $C_2 = \frac{1}{2}i$ gives the independent real-valued solution $y_2(x) = e^{ax}\sin bx$. This yields the following result.

THEOREM 3 Complex Roots

If the characteristic equation in (3) has an unrepeated pair of complex conjugate roots $a \pm bi$ (with $b \neq 0$), then the corresponding part of a general solution of Eq. (1) has the form

$$e^{ax}(c_1 \cos bx + c_2 \sin bx). \tag{21}$$

Thus the linearly independent solutions $e^{ax}\cos bx$ and $e^{ax}\sin bx$ (corresponding to the complex conjugate characteristic roots $a \pm bi$) generate a 2-dimensional subspace of the solution space of the differential equation.

Example 3 The characteristic equation of

$$y'' + b^2 y = 0 \quad (b > 0)$$

is $r^2 + b^2 = 0$, with roots $r = \pm bi$. So Theorem 3 (with $a = 0$) gives the general solution

$$y(x) = c_1 \cos bx + c_2 \sin bx.$$ ■

Example 4 Find the particular solution of

$$y'' - 4y' + 5y = 0$$

for which $y(0) = 1$ and $y'(0) = 5$.

Solution Completion of the square in the characteristic equation yields

$$r^2 - 4r + 5 = (r-2)^2 + 1 = 0,$$

so $r - 2 = \pm\sqrt{-1} = \pm i$. Thus we obtain the complex conjugate roots $2 \pm i$ (which could also be found directly using the quadratic formula). Hence Theorem 3 with $a = 2$ and $b = 1$ gives the general solution

$$y(x) = e^{2x}(c_1 \cos x + c_2 \sin x).$$

Then

$$y'(x) = 2e^{2x}(c_1 \cos x + c_2 \sin x) + e^{2x}(-c_1 \sin x + c_2 \cos x),$$

so the initial conditions give

$$y(0) = c_1 = 1 \quad \text{and} \quad y'(0) = 2c_1 + c_2 = 5.$$

It follows that $c_2 = 3$, and so the desired particular solution is

$$y(x) = e^{2x}(\cos x + 3 \sin x).$$ ■

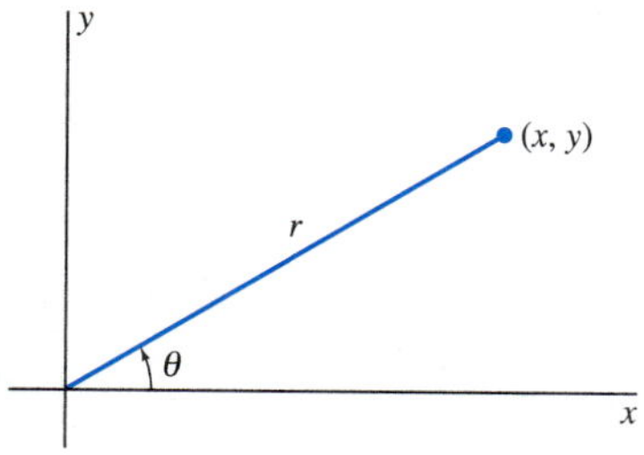

FIGURE 5.3.1. Modulus and argument of the complex number $x + iy$.

In Example 5 below, we employ the **polar form**

$$z = x + iy = re^{i\theta} \tag{22}$$

of the complex number z. This form follows from Euler's formula upon writing

$$z = r\left(\frac{x}{r} + i\frac{y}{r}\right) = r(\cos\theta + i\sin\theta) = re^{i\theta}$$

in terms of the **modulus** $r = \sqrt{x^2 + y^2} > 0$ of the number z and its **argument** θ indicated in Fig. 5.3.1. For instance, the imaginary number i has modulus 1 and argument $\frac{\pi}{2}$, so $i = e^{i\pi/2}$. Similarly, $-i = e^{i3\pi/2}$. Another consequence is the fact that the nonzero complex number $z = re^{i\theta}$ has the two square roots

$$\sqrt{z} = \pm(re^{i\theta})^{1/2} = \pm\sqrt{r}\, e^{i\theta/2}, \tag{23}$$

where $\sqrt{r}$ denotes (as usual for a positive real number) the positive square root of the modulus of z.

Example 5 Find a general solution of the differential equation $y^{(4)} + 4y = 0$.

Solution The characteristic equation is

$$\begin{aligned} r^4 + 4 &= (r^2)^2 - (2i)^2 \\ &= (r^2 + 2i)(r^2 - 2i) = 0, \end{aligned}$$

and its four roots are $\pm\sqrt{\pm 2i}$. Since $i = e^{i\pi/2}$ and $-i = e^{i3\pi/2}$, we find that

$$\sqrt{2i} = \left(2e^{i\pi/2}\right)^{1/2} = \sqrt{2}\, e^{i\pi/4} = \sqrt{2}\left(\cos\frac{\pi}{4} + i\sin\frac{\pi}{4}\right) = 1 + i$$

and

$$\sqrt{-2i} = \left(2e^{i3\pi/2}\right)^{1/2} = \sqrt{2}\, e^{i3\pi/4} = \sqrt{2}\left(\cos\frac{3\pi}{4} + i\sin\frac{3\pi}{4}\right) = -1 + i.$$

Thus the four (distinct) roots of the characteristic equation are $r = \pm(1 \pm i)$. These two pairs of complex conjugate roots, $1 \pm i$ and $-1 \pm i$, give a general solution

$$y(x) = e^x(c_1\cos x + c_2\sin x) + e^{-x}(c_3\cos x + c_4\sin x)$$

of the differential equation $y^{(4)} + 4y = 0$. ■

Repeated Complex Roots

Theorem 2 holds for repeated complex roots. If the conjugate pair $a \pm bi$ has multiplicity k, then the corresponding part of the general solution has the form

$$\begin{aligned} (A_1 + A_2x + \cdots + A_kx^{k-1})e^{(a+bi)x} &+ (B_1 + B_2x + \cdots + B_kx^{k-1})e^{(a-bi)x} \\ &= \sum_{p=0}^{k-1} x^p e^{ax}(c_p\cos bx + d_p\sin bx). \end{aligned} \tag{24}$$

It can be shown that the $2k$ functions

$$x^p e^{ax}\cos bx, \quad x^p e^{ax}\sin bx, \quad 0 \leqq p \leqq k - 1$$

that appear in Eq. (24) are linearly independent, and therefore generate a $2k$-dimensional subspace of the solution space of the differential equation.

Example 6 Find a general solution of $(D^2 + 6D + 13)^2 y = 0$.

Solution By completing the square, we see that the characteristic equation

$$(r^2 + 6r + 13)^2 = [(r + 3)^2 + 4]^2 = 0$$

has as its roots the conjugate pair $-3 \pm 2i$ of multiplicity $k = 2$. Hence Eq. (24) gives the general solution

$$y(x) = e^{-3x}(c_1\cos 2x + d_1\sin 2x) + xe^{-3x}(c_2\cos 2x + d_2\sin 2x).$$ ■

In applications we are seldom presented in advance with a factorization as convenient as the one in Example 6. Often the most difficult part of solving a homogeneous linear equation is finding the roots of its characteristic equation. Example 7 illustrates an approach that may succeed when a root of the characteristic equation can be found by inspection. The application material for this section illustrates other possibilities.

Example 7 The characteristic equation of the differential equation

$$y^{(3)} + y' - 10y = 0$$

is the cubic equation

$$r^3 + r - 10 = 0.$$

By a standard theorem of elementary algebra, the only possible rational roots are the factors ± 1, ± 2, ± 5, and ± 10 of the constant term 10. By trial and error (if not by inspection) we discover the root 2. The factor theorem of elementary algebra implies that $r - 2$ is a factor of $r^3 + r - 10$, and division of the former into the latter produces as quotient the quadratic polynomial

$$r^2 + 2r + 5 = (r + 1)^2 + 4.$$

The roots of this quotient are the complex conjugates $-1 \pm 2i$. The three roots we have found now yield the general solution

$$y(x) = c_1 e^{2x} + e^{-x}(c_2 \cos 2x + c_3 \sin 2x). \quad \blacksquare$$

Example 8 The roots of the characteristic equation of a certain differential equation are 3, -5, 0, 0, 0, 0, -5, $2 \pm 3i$, and $2 \pm 3i$. Write a general solution of this homogeneous differential equation.

Solution The solution can be read directly from the list of roots. It is

$$\begin{aligned} y(x) = c_1 + c_2 x + c_3 x^2 + c_4 x^3 + c_5 e^{3x} + c_6 e^{-5x} + c_7 x e^{-5x} \\ + e^{2x}(c_8 \cos 3x + c_9 \sin 3x) + x e^{2x}(c_{10} \cos 3x + c_{11} \sin 3x). \end{aligned} \quad \blacksquare$$

5.3 Problems

Find the general solutions of the differential equations in Problems 1 through 20.

1. $y'' - 4y = 0$
2. $2y'' - 3y' = 0$
3. $y'' + 3y' - 10y = 0$
4. $2y'' - 7y' + 3y = 0$
5. $y'' + 6y' + 9y = 0$
6. $y'' + 5y' + 5y = 0$
7. $4y'' - 12y' + 9y = 0$
8. $y'' - 6y' + 13y = 0$
9. $y'' + 8y' + 25y = 0$
10. $5y^{(4)} + 3y^{(3)} = 0$
11. $y^{(4)} - 8y^{(3)} + 16y'' = 0$
12. $y^{(4)} - 3y^{(3)} + 3y'' - y' = 0$
13. $9y^{(3)} + 12y'' + 4y' = 0$
14. $y^{(4)} + 3y'' - 4y = 0$
15. $y^{(4)} - 8y'' + 16y = 0$
16. $y^{(4)} + 18y'' + 81y = 0$
17. $6y^{(4)} + 11y'' + 4y = 0$
18. $y^{(4)} = 16y$
19. $y^{(3)} + y'' - y' - y = 0$
20. $y^{(4)} + 2y^{(3)} + 3y'' + 2y' + y = 0$ [*Suggestion*: Expand $(r^2 + r + 1)^2$.]

Solve the initial value problems given in Problems 21 through 26.

21. $y'' - 4y' + 3y = 0$; $y(0) = 7$, $y'(0) = 11$
22. $9y'' + 6y' + 4y = 0$; $y(0) = 3$, $y'(0) = 4$
23. $y'' - 6y' + 25y = 0$; $y(0) = 3$, $y'(0) = 1$
24. $2y^{(3)} - 3y'' - 2y' = 0$; $y(0) = 1$, $y'(0) = -1$, $y''(0) = 3$
25. $3y^{(3)} + 2y'' = 0$; $y(0) = -1$, $y'(0) = 0$, $y''(0) = 1$
26. $y^{(3)} + 10y'' + 25y' = 0$; $y(0) = 3$, $y'(0) = 4$, $y''(0) = 5$

Find general solutions of the equations in Problems 27 through 32. First find a small integral root of the characteristic equation by inspection; then factor by division.

27. $y^{(3)} + 3y'' - 4y = 0$ **28.** $2y^{(3)} - y'' - 5y' - 2y = 0$
29. $y^{(3)} + 27y = 0$ **30.** $y^{(4)} - y^{(3)} + y'' - 3y' - 6y = 0$
31. $y^{(3)} + 3y'' + 4y' - 8y = 0$
32. $y^{(4)} + y^{(3)} - 3y'' - 5y' - 2y = 0$

In Problems 33 through 36, one solution of the differential equation is given. Find the general solution.

33. $y^{(3)} + 3y'' - 54y = 0$; $y = e^{3x}$
34. $3y^{(3)} - 2y'' + 12y' - 8y = 0$; $y = e^{2x/3}$
35. $6y^{(4)} + 5y^{(3)} + 25y'' + 20y' + 4y = 0$; $y = \cos 2x$
36. $9y^{(3)} + 11y'' + 4y' - 14y = 0$; $y = e^{-x}\sin x$
37. Find a function $y(x)$ such that $y^{(4)}(x) = y^{(3)}(x)$ for all x and $y(0) = 18$, $y'(0) = 12$, $y''(0) = 13$, and $y^{(3)}(0) = 7$.
38. Solve the initial value problem

$$y^{(3)} - 5y'' + 100y' - 500y = 0;$$
$$y(0) = 0, \quad y'(0) = 10, \quad y''(0) = 250$$

given that $y_1(x) = e^{5x}$ is one particular solution of the differential equation.

In Problems 39 through 42, find a linear homogeneous constant-coefficient equation with the given general solution.

39. $y(x) = (A + Bx + Cx^2)e^{2x}$
40. $y(x) = Ae^{2x} + B\cos 2x + C\sin 2x$
41. $y(x) = A\cos 2x + B\sin 2x + C\cosh 2x + D\sinh 2x$
42. $y(x) = (A + Bx + Cx^2)\cos 2x + (D + Ex + Fx^2)\sin 2x$

Problems 43 through 47 pertain to the solution of differential equations with complex coefficients.

43. **(a)** Use Euler's formula to show that every complex number can be written in the form $re^{i\theta}$, where $r \geqq 0$ and $-\pi < \theta \leqq \pi$. **(b)** Express the numbers 4, -2, $3i$, $1+i$, and $-1+i\sqrt{3}$ in the form $re^{i\theta}$. **(c)** The two square roots of $re^{i\theta}$ are $\pm\sqrt{r}e^{i\theta/2}$. Find the square roots of the numbers $2 - 2i\sqrt{3}$ and $-2 + 2i\sqrt{3}$.
44. Use the quadratic formula to solve the following equations. Note in each case that the roots are not complex conjugates.

(a) $x^2 + ix + 2 = 0$ **(b)** $x^2 - 2ix + 3 = 0$

45. Find a general solution of $y'' - 2iy' + 3y = 0$.
46. Find a general solution of $y'' - iy' + 6y = 0$.
47. Find a general solution of $y'' = \left(-2 + 2i\sqrt{3}\right)y$.
48. Solve the initial value problem

$$y^{(3)} = y; \quad y(0) = 1, \quad y'(0) = y''(0) = 0.$$

[*Suggestion*: Impose the given initial conditions on the general solution $y(x) = Ae^x + Be^{\alpha x} + Ce^{\beta x}$, where α and β are the complex conjugate roots of $r^3 - 1 = 0$, to discover that

$$y(x) = \frac{1}{3}\left(e^x + 2e^{-x/2}\cos\frac{x\sqrt{3}}{2}\right)$$

is a solution.]

49. Solve the initial value problem

$$y^{(4)} = y^{(3)} + y'' + y' + 2y;$$
$$y(0) = y'(0) = y''(0) = 0,\ y^{(3)}(0) = 30.$$

50. The differential equation

$$y'' + (\operatorname{sgn} x)y = 0 \tag{25}$$

has the discontinuous coefficient function

$$\operatorname{sgn} x = \begin{cases} +1 & \text{if } x > 0, \\ -1 & \text{if } x < 0. \end{cases}$$

Show that Eq. (25) nevertheless has two linearly independent solutions $y_1(x)$ and $y_2(x)$ defined for all x such that

- Each satisfies Eq. (25) at each point $x \neq 0$;
- Each has a continuous derivative at $x = 0$;
- $y_1(0) = y_2'(0) = 1$ and $y_2(0) = y_1'(0) = 0$.

[*Suggestion*: Each $y_i(x)$ will be defined by one formula for $x < 0$ and by another for $x \geqq 0$.] The graphs of these two solutions are shown in Fig. 5.3.2.

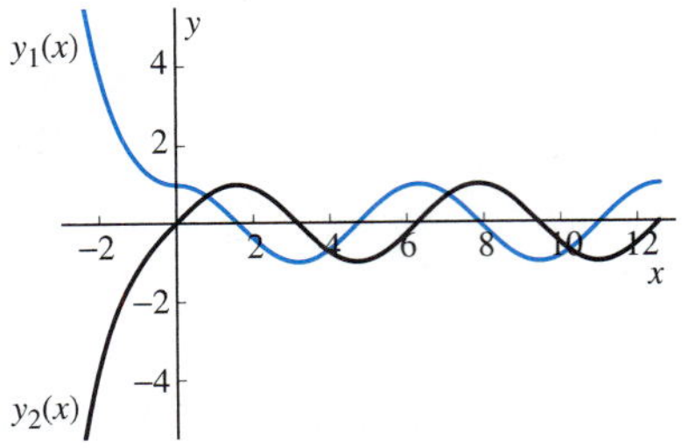

FIGURE 5.3.2. Graphs of $y_1(x)$ and $y_2(x)$ in Problem 50.

51. According to Problem 51 in Section 5.1, the substitution $v = \ln x$ $(x > 0)$ transforms the second-order Euler equation $ax^2y'' + bxy' + cy = 0$ to a constant-coefficient homogeneous linear equation. Show similarly that this same substitution transforms the third-order Euler equation

$$ax^3y''' + bx^2y'' + cxy' + dy = 0$$

(where a, b, c, d are constants) into the constant-coefficient equation

$$a\frac{d^3y}{dv^3} + (b - 3a)\frac{d^2y}{dv^2} + (c - b + 2a)\frac{dy}{dv} + dy = 0.$$

Make the substitution $v = \ln x$ of Problem 51 to find general solutions (for $x > 0$) of the Euler equations in Problems 52 through 58.

52. $x^2y'' + xy' + 9y = 0$
53. $x^2y'' + 7xy' + 25y = 0$
54. $x^3y''' + 6x^2y'' + 4xy' = 0$
55. $x^3y''' - x^2y'' + xy' = 0$
56. $x^3y''' + 3x^2y'' + xy' = 0$
57. $x^3y''' - 3x^2y'' + xy' = 0$
58. $x^3y''' + 6x^2y'' + 7xy' + y = 0$

5.3 Application Approximate Solutions of Linear Equations

To meet the needs of applications such as those of this section, polynomial-solving utilities are now a common feature of calculator and computer systems and can be used to solve a characteristic equation numerically even when no simple factorization is evident or even possible. For instance, suppose that we want to solve the homogeneous linear differential equation

$$y^{(3)} - 3y'' + y = 0 \tag{1}$$

with characteristic equation

$$r^3 - 3r^2 + 1 = 0. \tag{2}$$

A typical graphing calculator has a **`solve`** command that can be used to find the approximate roots of a polynomial equation. As indicated in Figs. 5.3.3 and 5.3.4, we find that the roots of Eq. (2) are given by $r \approx -0.5321$, 0.6527, and 2.8794. Some analogous computer algebra system commands are

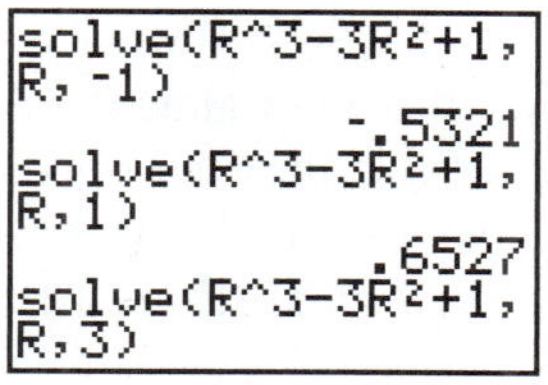

FIGURE 5.3.3. Solving the equation $r^3 - 3r^2 + 1 = 0$ with a TI-83 calculator that requires an estimate of each root.

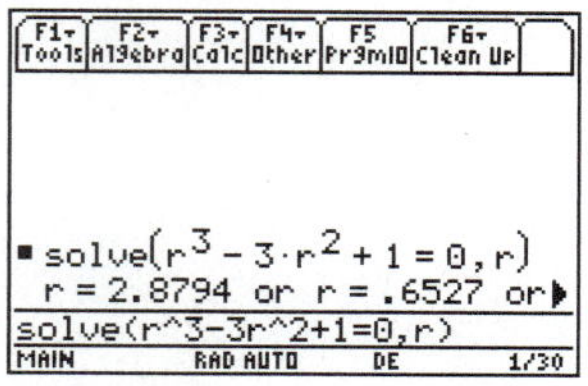

FIGURE 5.3.4. Solving the characteristic equation with a TI-89 calculator having a more sophisticated **`solve`** facility.

```
fsolve(r^3 - 3*r^2 + 1 = 0, r);      (Maple)
NSolve[r^3 - 3*r^2 + 1 == 0, r]      (Mathematica)
roots([1  -3  0  1])                 (MATLAB)
```

(In the MATLAB command, one enters the polynomial's vector **`[1 -3 0 1]`** of coefficients, listed in descending order.) However we find these approximate roots, it follows that a general solution of the differential equation in (1) is given (approximately) by

$$y(x) = c_1 e^{-(0.5321)x} + c_2 e^{(0.6527)x} + c_3 e^{(2.8794)x}. \tag{3}$$

Use calculator or computer methods like those indicated here to find general solutions (in approximate numerical form) of the following differential equations.

1. $y^{(3)} - 3y' + y = 0$
2. $y^{(3)} + 3y'' - 3y = 0$
3. $y^{(3)} + y' + y = 0$
4. $y^{(3)} + 3y' + 5y = 0$
5. $y^{(4)} + 2y^{(3)} - 3y = 0$
6. $y^{(4)} + 3y' - 4y = 0$

5.4 Mechanical Vibrations

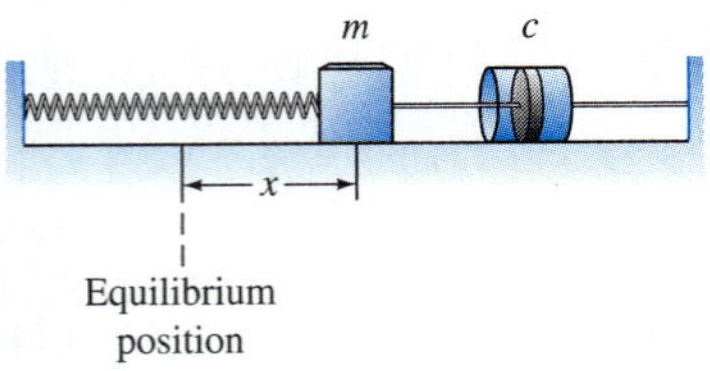

FIGURE 5.4.1. A mass-spring-dashpot system.

The motion of a mass attached to a spring serves as a relatively simple example of the vibrations that occur in more complex mechanical systems. For many such systems, the analysis of these vibrations is a problem in the solution of linear differential equations with constant coefficients.

We consider a body of mass m attached to one end of an ordinary spring that resists compression as well as stretching; the other end of the spring is attached to a fixed wall, as shown in Fig. 5.4.1. Assume that the body rests on a frictionless horizontal plane, so that it can move only back and forth as the spring compresses and stretches. Denote by x the distance of the body from its **equilibrium position**—its position when the spring is unstretched. We take $x > 0$ when the spring is stretched, and thus $x < 0$ when it is compressed.

According to Hooke's law, the restorative force F_S that the spring exerts on the mass is proportional to the distance x that the spring has been stretched or compressed. Because this is the same as the displacement x of the mass m from its equilibrium position, it follows that

$$F_S = -kx. \tag{1}$$

The positive constant of proportionality k is called the **spring constant.** Note that F_S and x have opposite signs: $F_S < 0$ when $x > 0$, $F_S > 0$ when $x < 0$.

Figure 5.4.1 shows the mass attached to a dashpot—a device, like a shock absorber, that provides a force directed opposite to the instantaneous direction of motion of the mass m. We assume the dashpot is so designed that this force F_R is proportional to the velocity $v = dx/dt$ of the mass; that is,

$$F_R = -cv = -c\frac{dx}{dt}. \tag{2}$$

The positive constant c is the **damping constant** of the dashpot. More generally, we may regard Eq. (2) as specifying frictional forces in our system (including air resistance to the motion of m).

If, in addition to the forces F_S and F_R, the mass is subjected to a given **external force** $F_E = F(t)$, then the total force acting on the mass is $F = F_S + F_R + F_E$. Using Newton's law,

$$F = ma = m\frac{d^2x}{dt^2} = mx'',$$

we obtain the second-order linear differential equation

$$mx'' + cx' + kx = F(t) \tag{3}$$

that governs the motion of the mass.

If there is no dashpot (and we ignore all frictional forces), then we set $c = 0$ in Eq. (3) and call the motion **undamped**; it is **damped** motion if $c > 0$. If there is no external force, we replace $F(t)$ with 0 in Eq. (3). We refer to the motion as **free** in this case and **forced** in the case $F(t) \neq 0$. Thus the homogeneous equation

$$mx'' + cx' + kx = 0 \tag{4}$$

describes free motion of a mass on a spring with dashpot but with no external forces applied. We will defer discussion of forced motion until Section 5.6.

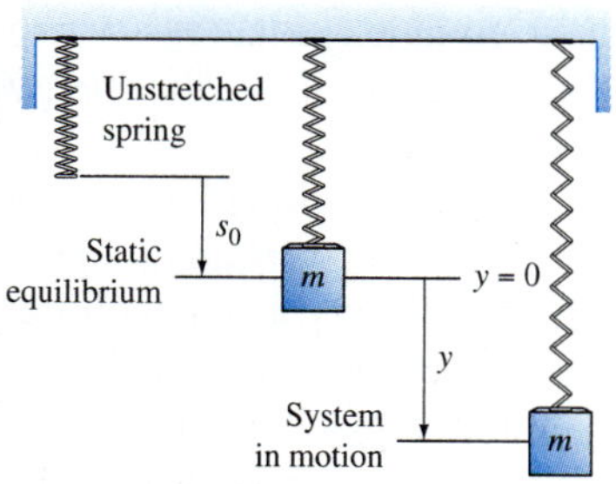

FIGURE 5.4.2. A mass suspended vertically from a spring.

For an alternative example, we might attach the mass to the lower end of a spring that is suspended vertically from a fixed support, as in Fig. 5.4.2. In this case the weight $W = mg$ of the mass would stretch the spring a distance s_0 determined by Eq. (1) with $F_S = -W$ and $x = s_0$. That is, $mg = ks_0$, so that $s_0 = mg/k$. This gives the **static** equilibrium position of the mass. If y denotes the displacement of the mass in motion, measured downward from its static equilibrium position, then we ask you to show in Problem 9 that y satisfies Eq. (3); specifically, that

$$my'' + cy' + ky = F(t) \tag{5}$$

if we include damping and external forces (meaning those other than gravity).

The Simple Pendulum

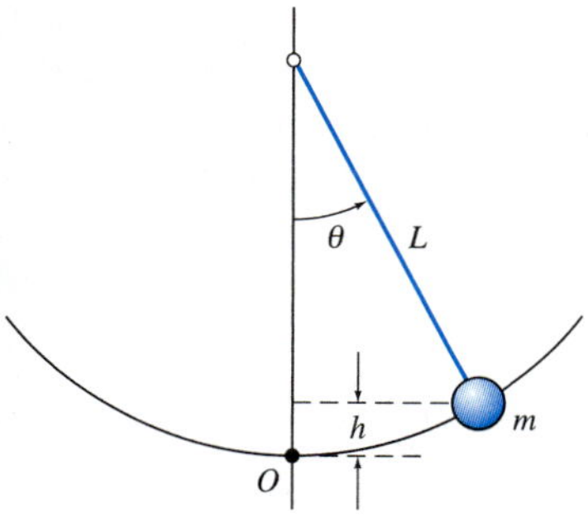

FIGURE 5.4.3. The simple pendulum.

The importance of the differential equation that appears in Eqs. (3) and (5) stems from the fact that it describes the motion of many other simple mechanical systems. For example, a **simple pendulum** consists of a mass m swinging back and forth on the end of a string (or better, a *massless rod*) of length L, as shown in Fig. 5.4.3. We may specify the position of the mass at time t by giving the counterclockwise angle $\theta = \theta(t)$ that the string or rod makes with the vertical at time t. To analyze the motion of the mass m, we will apply the law of the *conservation of mechanical energy*, according to which the sum of the kinetic energy and the potential energy of m remains constant.

The distance along the circular arc from 0 to m is $s = L\theta$, so the velocity of the mass is $v = ds/dt = L(d\theta/dt)$, and therefore its kinetic energy is

$$T = \frac{1}{2}mv^2 = \frac{1}{2}m\left(\frac{ds}{dt}\right)^2 = \frac{1}{2}mL^2\left(\frac{d\theta}{dt}\right)^2.$$

We next choose as reference point the lowest point O reached by the mass (see Fig. 5.4.3). Then its potential energy V is the product of its weight mg and its vertical height $h = L(1 - \cos\theta)$ above O, so

$$V = mgL(1 - \cos\theta).$$

The fact that the sum of T and V is a constant C therefore gives

$$\frac{1}{2}mL^2\left(\frac{d\theta}{dt}\right)^2 + mgL(1 - \cos\theta) = C.$$

We differentiate both sides of this identity with respect to t to obtain

$$mL^2\left(\frac{d\theta}{dt}\right)\left(\frac{d^2\theta}{dt^2}\right) + mgL(\sin\theta)\frac{d\theta}{dt} = 0,$$

so

$$\frac{d^2\theta}{dt^2} + \frac{g}{L}\sin\theta = 0 \tag{6}$$

after removal of the common factor $mL^2(d\theta/dt)$. This differential equation can be derived in a seemingly more elementary manner using the familiar second law $F = ma$ of Newton (applied to tangential components of the acceleration of the mass and the force acting on it). However, derivations of differential equations based on conservation of energy are often seen in more complex situations where Newton's law is not so directly applicable, and it may be instructive to see the energy method in a simpler application like the pendulum.

Now recall that if θ is small, then $\sin\theta \approx \theta$ (this approximation obtained by retaining just the first term in the Taylor series for $\sin\theta$). In fact, $\sin\theta$ and θ agree to two decimal places when $|\theta|$ is at most $\pi/12$ (that is, 15°). In a typical pendulum clock, for example, θ would never exceed 15°. It therefore seems reasonable to simplify our mathematical model of the simple pendulum by replacing $\sin\theta$ with θ in Eq. (6). If we also insert a term $c\theta'$ to account for the frictional resistance of the surrounding medium, the result is an equation in the form of Eq. (4):

$$\theta'' + c\theta' + k\theta = 0, \tag{7}$$

where $k = g/L$. Note that this equation is independent of the mass m on the end of the rod. We might, however, expect the effects of the discrepancy between θ and $\sin\theta$ to accumulate over a period of time, so that Eq. (7) will probably not describe accurately the actual motion of the pendulum over a long period of time.

In the remainder of this section, we first analyze free undamped motion and then free damped motion.

Free Undamped Motion

If we have only a mass on a spring, with neither damping nor external force, then Eq. (3) takes the simpler form

$$mx'' + kx = 0. \tag{8}$$

It is convenient to define

$$\omega_0 = \sqrt{\frac{k}{m}} \tag{9}$$

and rewrite Eq. (8) as

$$x'' + \omega_0^2 x = 0. \tag{8$'$}$$

The general solution of Eq. (8′) is

$$x(t) = A\cos\omega_0 t + B\sin\omega_0 t. \tag{10}$$

To analyze the motion described by this solution, we choose constants C and α so that

$$C = \sqrt{A^2 + B^2}, \quad \cos\alpha = \frac{A}{C}, \quad \text{and} \quad \sin\alpha = \frac{B}{C}, \tag{11}$$

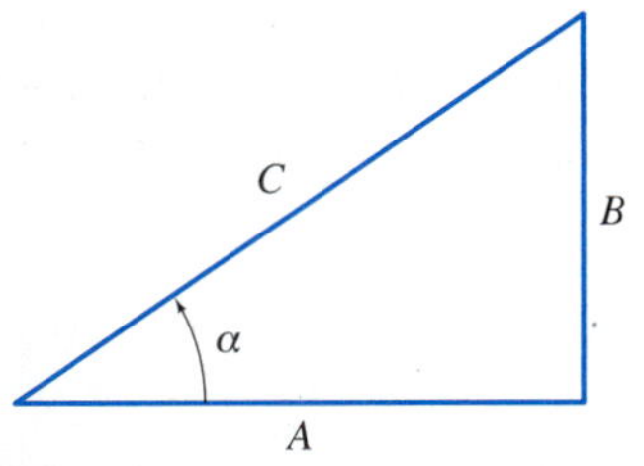

FIGURE 5.4.4. The angle α.

as indicated in Fig. 5.4.4. Note that, although $\tan\alpha = B/A$, the angle α is *not* given by the principal branch of the inverse tangent function (which gives values only in the interval $-\pi/2 < x < \pi/2$). Instead, α is the angle between 0 and 2π whose cosine and sine have the signs given in (11), where either A or B or both may be negative. Thus

$$\alpha = \begin{cases} \tan^{-1}(B/A) & \text{if } A > 0,\ B > 0 \text{ (first quadrant)}, \\ \pi + \tan^{-1}(B/A) & \text{if } A < 0 \text{ (second or third quadrant)}, \\ 2\pi + \tan^{-1}(B/A) & \text{if } A > 0,\ B < 0 \text{ (fourth quadrant)}, \end{cases}$$

where $\tan^{-1}(B/A)$ is the angle in $(-\pi/2, \pi/2)$ given by a calculator or computer.

In any event, from (10) and (11) we get

$$x(t) = C\left(\frac{A}{C}\cos\omega_0 t + \frac{B}{C}\sin\omega_0 t\right) = C(\cos\alpha\cos\omega_0 t + \sin\alpha\sin\omega_0 t).$$

With the aid of the cosine addition formula, we find that

$$x(t) = C\cos(\omega_0 t - \alpha). \tag{12}$$

Thus the mass oscillates to and fro about its equilibrium position with

1. **Amplitude** C,
2. **Circular frequency** ω_0, and
3. **Phase angle** α.

Such motion is called **simple harmonic motion.**

If time t is measured in seconds, the circular frequency ω_0 has dimensions of radians per second (rad/s). The **period** of the motion is the time required for the system to complete one full oscillation, so is given by

$$T = \frac{2\pi}{\omega_0} \tag{13}$$

seconds; its **frequency** is

$$\nu = \frac{1}{T} = \frac{\omega_0}{2\pi} \tag{14}$$

in hertz (Hz), which measures the number of complete cycles per second. Note that frequency is measured in cycles per second, whereas circular frequency has the dimensions of radians per second.

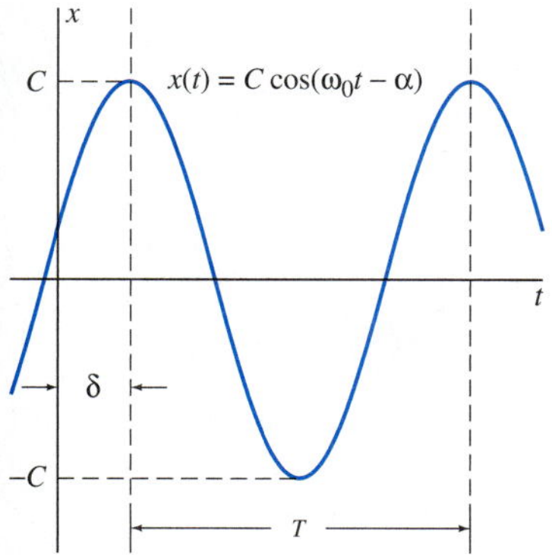

FIGURE 5.4.5. Simple harmonic motion.

A typical graph of a simple harmonic position function

$$x(t) = C\cos(\omega_0 t - \alpha) = C\cos\left(\omega_0\left(t - \frac{\alpha}{\omega_0}\right)\right) = C\cos(\omega_0(t - \delta))$$

is shown in Fig. 5.4.5, where the geometric significance of the amplitude C, the period T, and the **time lag**

$$\delta = \frac{\alpha}{\omega_0}$$

are indicated.

If the initial position $x(0) = x_0$ and initial velocity $x'(0) = v_0$ of the mass are given, we first determine the values of the coefficients A and B in Eq. (10), then find the amplitude C and phase angle α by carrying out the transformation of $x(t)$ to the form in Eq. (12), as indicated previously.

Example 1 A body with mass $m = \frac{1}{2}$ kilogram (kg) is attached to the end of a spring that is stretched 2 meters (m) by a force of 100 newtons (N). It is set in motion with initial position $x_0 = 1$ (m) and initial velocity $v_0 = -5$ (m/s). (Note that these initial conditions indicate that the body is displaced to the right and is moving to the left at time $t = 0$.) Find the position function of the body as well as the amplitude, frequency, period of oscillation, and time lag of its motion.

Solution The spring constant is $k = (100\text{ N})/(2\text{ m}) = 50$ (N/m), so Eq. (8) yields $\frac{1}{2}x'' + 50x = 0$; that is,

$$x'' + 100x = 0.$$

Consequently, the circular frequency of the resulting simple harmonic motion of the body will be $\omega_0 = \sqrt{100} = 10$ (rad/s). Hence it will oscillate with period

$$T = \frac{2\pi}{\omega_0} = \frac{2\pi}{10} \approx 0.6283\text{ s}$$

and with frequency

$$\nu = \frac{1}{T} = \frac{\omega_0}{2\pi} = \frac{10}{2\pi} \approx 1.5915\text{ Hz}.$$

We now impose the initial conditions $x(0) = 1$ and $x'(0) = -5$ on the position function

$$x(t) = A\cos 10t + B\sin 10t \quad \text{with} \quad x'(t) = -10A\sin 10t + 10B\cos 10t.$$

It follows readily that $A = 1$ and $B = -\frac{1}{2}$, so the position function of the body is

$$x(t) = \cos 10t - \frac{1}{2}\sin 10t.$$

Hence its amplitude of motion is

$$C = \sqrt{(1)^2 + (-\tfrac{1}{2})^2} = \tfrac{1}{2}\sqrt{5}\text{ m}.$$

To find the time lag, we write

$$x(t) = \frac{\sqrt{5}}{2}\left(\frac{2}{\sqrt{5}}\cos 10t - \frac{1}{\sqrt{5}}\sin 10t\right) = \frac{\sqrt{5}}{2}\cos(10t - \alpha),$$

where the phase angle α satisfies

$$\cos\alpha = \frac{2}{\sqrt{5}} > 0 \quad \text{and} \quad \sin\alpha = -\frac{1}{\sqrt{5}} < 0.$$

Hence α is the fourth-quadrant angle

$$\alpha = 2\pi + \tan^{-1}\left(\frac{-1/\sqrt{5}}{2/\sqrt{5}}\right) = 2\pi - \tan^{-1}(\tfrac{1}{2}) \approx 5.8195,$$

and the time lag of the motion is

$$\delta = \frac{\alpha}{\omega_0} \approx 0.5820\text{ s}.$$

With the amplitude and approximate phase angle shown explicitly, the position function of the body takes the form

$$x(t) \approx \tfrac{1}{2}\sqrt{5}\cos(10t - 5.8195),$$

and its graph is shown in Fig. 5.4.6. ■

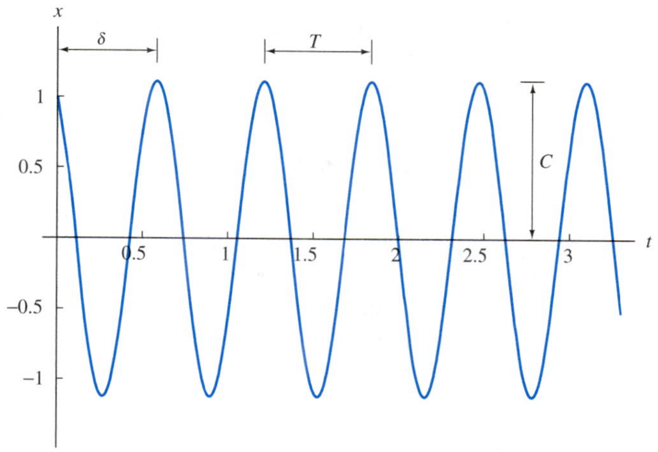

FIGURE 5.4.6. Graph of the position function $x(t) = C\cos(\omega_0 t - \alpha)$ in Example 1, with amplitude $C \approx 1.118$, period $T \approx 0.628$, and time lag $\delta \approx 0.582$.

Free Damped Motion

With damping but no external force, the differential equation we have been studying takes the form $mx'' + cx' + kx = 0$; alternatively,

$$x'' + 2px' + \omega_0^2 x = 0, \tag{15}$$

where $\omega_0 = \sqrt{k/m}$ is the corresponding *undamped* circular frequency and

$$p = \frac{c}{2m} > 0. \tag{16}$$

The characteristic equation $r^2 + 2pr + \omega_0^2 = 0$ of Eq. (15) has roots

$$r_1,\ r_2 = -p \pm (p^2 - \omega_0^2)^{1/2} \tag{17}$$

that depend on the sign of

$$p^2 - \omega_0^2 = \frac{c^2}{4m^2} - \frac{k}{m} = \frac{c^2 - 4km}{4m^2}. \tag{18}$$

The **critical damping** c_{cr} is given by $c_{cr} = \sqrt{4km}$, and we distinguish three cases, according as $c > c_{cr}$, $c = c_{cr}$, or $c < c_{cr}$.

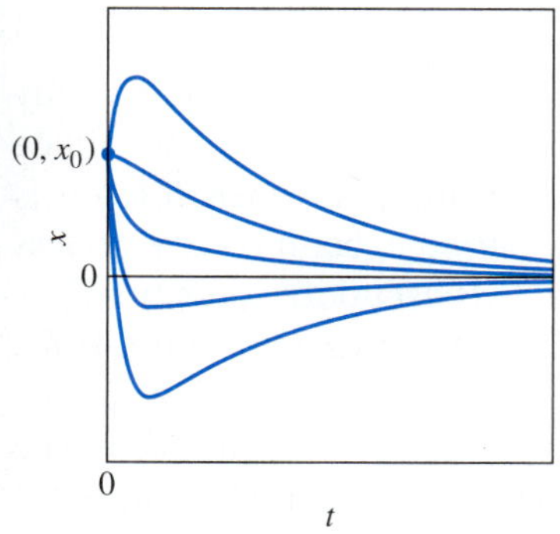

FIGURE 5.4.7. Overdamped motion: $x(t) = c_1 e^{r_1 t} + c_2 e^{r_2 t}$ with $r_1 < 0$ and $r_2 < 0$. Solution curves are graphed with the same initial position x_0 and different initial velocities.

OVERDAMPED CASE: $c > c_{cr}$ $(c^2 > 4km)$. Because c is relatively large in this case, we are dealing with a strong resistance in comparison with a relatively weak spring or a small mass. Then (17) gives distinct real roots r_1 and r_2, both of which are negative. The position function has the form

$$x(t) = c_1 e^{r_1 t} + c_2 e^{r_2 t}. \tag{19}$$

It is easy to see that $x(t) \to 0$ as $t \to +\infty$ and that the body settles to its equilibrium position without any oscillations (Problem 29). Figure 5.4.7 shows some typical graphs of the position function for the overdamped case; we chose x_0 a fixed positive number and illustrated the effects of changing the initial velocity v_0. In every case the would-be oscillations are damped out.

CRITICALLY DAMPED CASE: $c = c_{cr}(c^2 = 4km)$. In this case, (17) gives equal roots $r_1 = r_2 = -p$ of the characteristic equation, so the general solution is

$$x(t) = e^{-pt}(c_1 + c_2 t). \tag{20}$$

Because $e^{-pt} > 0$ and $c_1 + c_2 t$ has at most one positive zero, the body passes through its equilibrium position at most once, and it is clear that $x(t) \to 0$ as $t \to +\infty$. Some graphs of the motion in the critically damped case appear in Fig. 5.4.8, and they resemble those of the overdamped case (Fig. 5.4.7). In the critically damped case, the resistance of the dashpot is just large enough to damp out any oscillations, but even a slight reduction in resistance will bring us to the remaining case, the one that shows the most dramatic behavior.

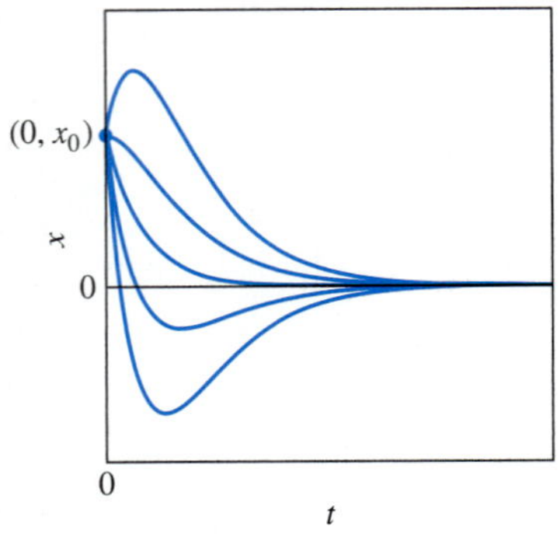

FIGURE 5.4.8. Critically damped motion: $x(t) = (c_1 + c_2 t)e^{-pt}$ with $p > 0$. Solution curves are graphed with the same initial position x_0 and different initial velocities.

UNDERDAMPED CASE: $c < c_{cr}(c^2 < 4km)$. The characteristic equation now has two complex conjugate roots $-p \pm i\sqrt{\omega_0^2 - p^2}$, and the general solution is

$$x(t) = e^{-pt}(A\cos\omega_1 t + B\sin\omega_1 t), \tag{21}$$

where

$$\omega_1 = \sqrt{\omega_0^2 - p^2} = \frac{\sqrt{4km - c^2}}{2m}. \tag{22}$$

Using the cosine addition formula as in the derivation of Eq. (12), we may rewrite Eq. (20) as

$$x(t) = Ce^{-pt}\left(\frac{A}{C}\cos\omega_1 t + \frac{B}{C}\sin\omega_1 t\right),$$

so

$$x(t) = Ce^{-pt}\cos(\omega_1 t - \alpha), \tag{23}$$

where

$$C = \sqrt{A^2 + B^2}, \quad \cos\alpha = \frac{A}{C}, \quad \text{and} \quad \sin\alpha = \frac{B}{C}.$$

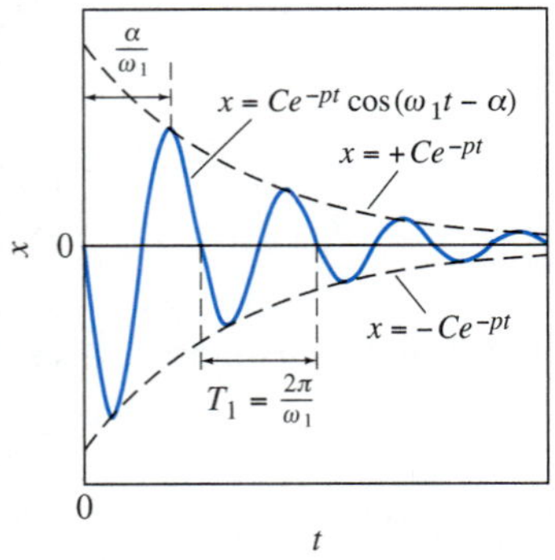

FIGURE 5.4.9. Underdamped oscillations: $x(t) = Ce^{-pt}\cos(\omega_1 t - \alpha)$.

The solution in (21) represents exponentially damped oscillations of the body around its equilibrium position. The graph of $x(t)$ lies between the "amplitude envelope" curves $x = -Ce^{-pt}$ and $x = Ce^{-pt}$ and touches them when $\omega_1 t - \alpha$ is an integral multiple of π. The motion is not actually periodic, but it is nevertheless useful to call ω_1 its **circular frequency** (more properly, its **pseudofrequency**), $T_1 = 2\pi/\omega_1$ its **pseudoperiod** of oscillation, and Ce^{-pt} its **time-varying amplitude.** Most of these quantities are shown in the typical graph of underdamped motion shown in Fig. 5.4.9. Note from Eq. (22) that in this case ω_1 is less than the undamped circular frequency ω_0, so T_1 is larger than the period T of oscillation of the same mass without damping on the same spring. Thus the action of the dashpot has at least two effects:

1. It exponentially damps the oscillations, in accord with the time-varying amplitude.
2. It slows the motion; that is, the dashpot decreases the frequency of the motion.

As the following example illustrates, damping typically also delays the motion further—that is, increases the time lag—as compared with undamped motion with the same initial conditions.

Example 2 The mass and spring of Example 1 are now attached also to a dashpot that provides 1 N of resistance for each meter per second of velocity. The mass is set in motion with the same initial position $x(0) = 1$ and initial velocity $x'(0) = -5$ as in Example 1. Now find the position function of the mass, its new frequency and pseudoperiod of motion, its new time lag, and the times of its first four passages through the initial position $x = 0$.

Solution Rather than memorizing the various formulas given in the preceding discussion, it is better practice in a particular case to set up the differential equation and then solve it directly. Recall that $m = \frac{1}{2}$ and $k = 50$; we are now given $c = 1$ in mks units. Hence Eq. (4) is $\frac{1}{2}x'' + x' + 50x = 0$; that is,

$$x'' + 2x' + 100x = 0.$$

The characteristic equation $r^2 + 2r + 100 = (r+1)^2 + 99 = 0$ has roots r_1, $r_2 = -1 \pm \sqrt{99}\,i$, so the general solution is

$$x(t) = e^{-t}(A\cos\sqrt{99}\,t + B\sin\sqrt{99}\,t). \tag{24}$$

Consequently, the new circular (pseudo)frequency is $\omega_1 = \sqrt{99} \approx 9.9499$ (as compared with $\omega_0 = 10$ in Example 1). The new (pseudo)period and frequency are

$$T_1 = \frac{2\pi}{\omega_1} = \frac{2\pi}{\sqrt{99}} \approx 0.6315 \text{ s}$$

and

$$\nu_1 = \frac{1}{T_1} = \frac{\omega_1}{2\pi} = \frac{\sqrt{99}}{2\pi} \approx 1.5836 \text{ Hz}$$

(as compared with $T \approx 0.6283 < T_1$ and $\nu \approx 1.5915 > \nu_1$ in Example 1).

We now impose the initial conditions $x(0) = 1$ and $x'(0) = -5$ on the position function in (23) and the resulting velocity function

$$x'(t) = -e^{-t}(A\cos\sqrt{99}\,t + B\sin\sqrt{99}\,t) + \sqrt{99}\,e^{-t}(-A\sin\sqrt{99}\,t + B\cos\sqrt{99}\,t).$$

It follows that

$$x(0) = A = 1 \quad \text{and} \quad x'(0) = -A + B\sqrt{99} = -5,$$

whence we find that $A = 1$ and $B = -4/\sqrt{99}$. Thus the new position function of the body is

$$x(t) = e^{-t}\left(\cos\sqrt{99}\,t - \frac{4}{\sqrt{99}}\sin\sqrt{99}\,t\right).$$

Hence its time-varying amplitude of motion is

$$C_1 e^{-t} = \sqrt{(1)^2 + \left(-\frac{4}{\sqrt{99}}\right)^2}\, e^{-t} = \sqrt{\frac{115}{99}}\, e^{-t}.$$

We therefore write

$$x(t) = \frac{\sqrt{115}}{\sqrt{99}} e^{-t} \left(\frac{\sqrt{99}}{\sqrt{115}} \cos \sqrt{99}\, t - \frac{4}{\sqrt{115}} \sin \sqrt{99}\, t \right)$$
$$= \sqrt{\frac{115}{99}} e^{-t} \cos(\sqrt{99}\, t - \alpha_1),$$

where the phase angle α_1 satisfies

$$\cos \alpha_1 = \frac{\sqrt{99}}{\sqrt{115}} > 0 \quad \text{and} \quad \sin \alpha_1 = -\frac{4}{\sqrt{115}} < 0.$$

Hence α_1 is the fourth-quadrant angle

$$\alpha_1 = 2\pi + \tan^{-1}\left(\frac{-4/\sqrt{115}}{\sqrt{99}/\sqrt{115}} \right) = 2\pi - \tan^{-1}\left(\frac{4}{\sqrt{99}} \right) \approx 5.9009,$$

and the time lag of the motion is

$$\delta_1 = \frac{\alpha_1}{\omega_1} \approx 0.5931 \text{ s}$$

(as compared with $\delta \approx 0.5820 < \delta_1$ in Example 1). With the time-varying amplitude and approximate phase angle shown explicitly, the position function of the mass takes the form

$$x(t) \approx \sqrt{\frac{115}{99}}\, e^{-t} \cos(\sqrt{99}\, t - 5.9009), \tag{25}$$

and its graph is the damped exponential that is shown in Fig. 5.4.10 (in comparison with the undamped oscillations of Example 1.

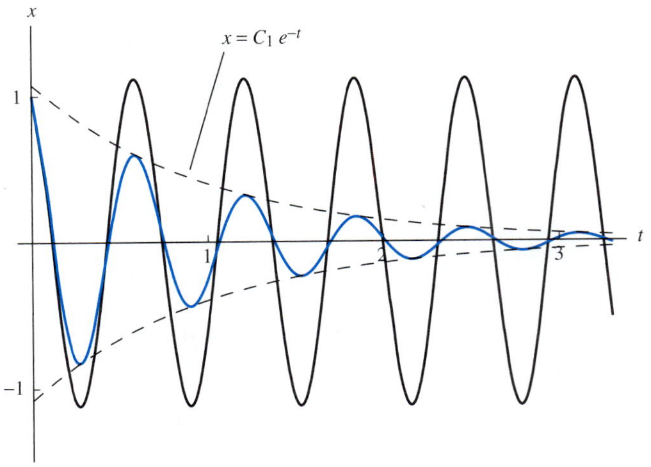

FIGURE 5.4.10. Graphs of the position function $x(t) = C_1 e^{-t} \cos(\omega_1 t - \alpha_1)$ of Example 2 (damped oscillations), the position function $x(t) = C \cos(\omega_0 t - \alpha)$ of Example 1 (undamped oscillations), and the envelope curves $x(t) = \pm C_1 e^{-t}$.

From (24) we see that the mass passes through its equilibrium position $x = 0$ when $\cos(\omega_1 t - \alpha_1) = 0$, and thus when

$$\omega_1 t - \alpha_1 = -\frac{3\pi}{2}, -\frac{\pi}{2}, \frac{\pi}{2}, \frac{3\pi}{2}, \dots, \quad \text{that is, when}$$

$$t = \delta_1 - \frac{3\pi}{2\omega_1}, \delta_1 - \frac{\pi}{2\omega_1}, \delta_1 + \frac{\pi}{2\omega_1}, \delta_1 + \frac{3\pi}{2\omega_1}, \dots .$$

We see similarly that the undamped mass of Example 1 passes through equilibrium when

$$t = \delta_0 - \frac{3\pi}{2\omega_0}, \delta_0 - \frac{\pi}{2\omega_0}, \delta_0 + \frac{\pi}{2\omega_0}, \delta_0 + \frac{3\pi}{2\omega_0}, \dots .$$

The following table compares the first four values t_1, t_2, t_3, t_4 we calculate for the undamped and damped cases, respectively.

n	1	2	3	4
t_n (undamped)	0.1107	0.4249	0.7390	1.0532
t_n (damped)	0.1195	0.4352	0.7509	1.0667

Accordingly, in Fig. 5.4.11 (where only the first three equilibrium passages are shown) we see the damped oscillations lagging slightly behind the undamped ones. ■

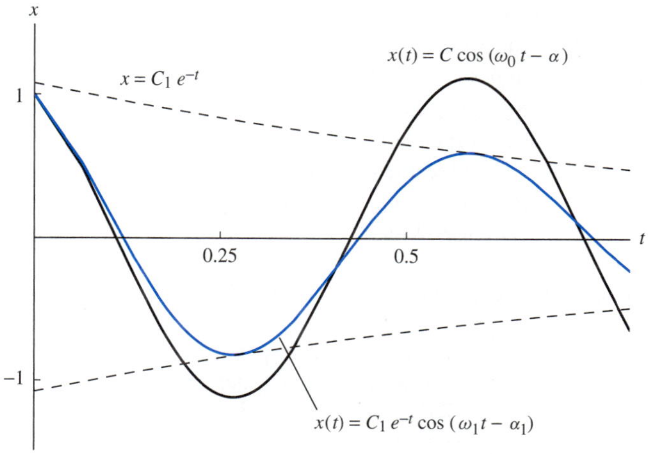

FIGURE 5.4.11. Graphs on the interval $0 \le t \le 0.8$ illustrating the additional delay associated with damping.

5.4 Problems

1. Determine the period and frequency of the simple harmonic motion of a 4-kg mass on the end of a spring with spring constant 16 N/m.
2. Determine the period and frequency of the simple harmonic motion of a body of mass 0.75 kg on the end of a spring with spring constant 48 N/m.
3. A mass of 3 kg is attached to the end of a spring that is stretched 20 cm by a force of 15 N. It is set in motion with initial position $x_0 = 0$ and initial velocity $v_0 = -10$ m/s. Find the amplitude, period, and frequency of the resulting motion.
4. A body with mass 250 g is attached to the end of a spring that is stretched 25 cm by a force of 9 N. At time $t = 0$ the body is pulled 1 m to the right, stretching the spring,

and set in motion with an initial velocity of 5 m/s to the left. **(a)** Find $x(t)$ in the form $C\cos(\omega_0 t - \alpha)$. **(b)** Find the amplitude and period of motion of the body.

In Problems 5 through 8, assume that the differential equation of a simple pendulum of length L is $L\theta'' + g\theta = 0$, *where* $g = GM/R^2$ *is the gravitational acceleration at the location of the pendulum (at distance R from the center of the earth; M denotes the mass of the earth).*

5. Two pendulums are of lengths L_1 and L_2 and—when located at the respective distances R_1 and R_2 from the center of the earth—have periods p_1 and p_2. Show that

$$\frac{p_1}{p_2} = \frac{R_1\sqrt{L_1}}{R_2\sqrt{L_2}}.$$

6. A certain pendulum keeps perfect time in Paris, where the radius of the earth is $R = 3956$ (mi). But this clock loses 2 min 40 s per day at a location on the equator. Use the result of Problem 5 to find the amount of the equatorial bulge of the earth.

7. A pendulum of length 100.10 in., located at a point at sea level where the radius of the earth is $R = 3960$ (mi), has the same period as does a pendulum of length 100.00 in. atop a nearby mountain. Use the result of Problem 5 to find the height of the mountain.

8. Most grandfather clocks have pendulums with adjustable lengths. One such clock loses 10 min per day when the length of its pendulum is 30 in. With what length pendulum will this clock keep perfect time?

9. Derive Eq. (5) describing the motion of a mass attached to the bottom of a vertically suspended spring. [*Suggestion*: First denote by $x(t)$ the displacement of the mass below the unstretched position of the spring; set up the differential equation for x. Then substitute $y = x - x_0$ in this differential equation.]

10. Consider a floating cylindrical buoy with radius r, height h, and uniform density $\rho \leqq 0.5$ (recall that the density of water is 1 g/cm^3). The buoy is initially suspended at rest with its bottom at the top surface of the water and is released at time $t = 0$. Thereafter it is acted on by two forces: a downward gravitational force equal to its weight $mg = \pi r^2 hg$ and (by Archimedes' principle of buoyancy) an upward force equal to the weight $\pi r^2 xg$ of water displaced, where $x = x(t)$ is the depth of the bottom of the buoy beneath the surface at time t (Fig. 5.4.12). Conclude that the buoy undergoes simple harmonic motion around its equilibrium position $x_e = \rho h$ with period $p = 2\pi\sqrt{\rho h/g}$. Compute p and the amplitude of the motion if $\rho = 0.5$ g/cm^3, $h = 200$ cm, and $g = 980$ cm/s^2.

11. A cylindrical buoy weighing 100 lb (thus of mass $m = 3.125$ slugs in ft-lb-s (fps) units) floats in water with its axis vertical (as in Problem 10). When depressed slightly and released, it oscillates up and down four times every 10 s. Assume that friction is negligible. Find the radius of the buoy.

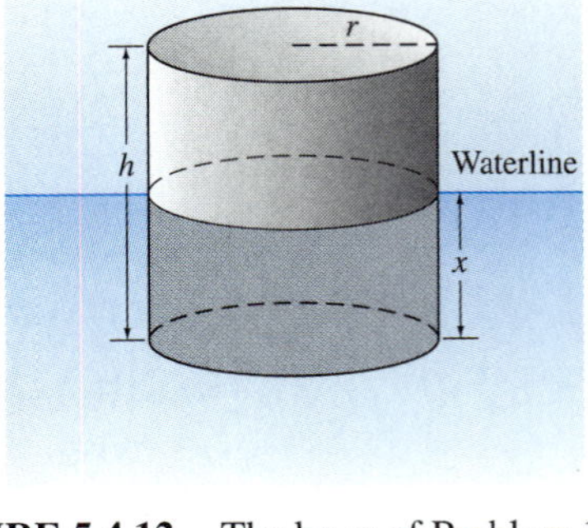

FIGURE 5.4.12. The buoy of Problem 10.

12. Assume that the earth is a solid sphere of uniform density, with mass M and radius $R = 3960$ (mi). For a particle of mass m *within* the earth at distance r from the center of the earth, the gravitational force attracting m toward the center is $F_r = -GM_r m/r^2$, where M_r is the mass of the part of the earth within a sphere of radius r. **(a)** Show that $F_r = -GMmr/R^3$. **(b)** Now suppose that a small hole is drilled straight through the center of the earth, thus connecting two antipodal points on its surface. Let a particle of mass m be dropped at time $t = 0$ into this hole with initial speed zero, and let $r(t)$ be its distance from the center of the earth at time t (Fig. 5.4.13). Conclude from Newton's second law and part (a) that $r''(t) = -k^2 r(t)$, where $k^2 = GM/R^3 = g/R$.

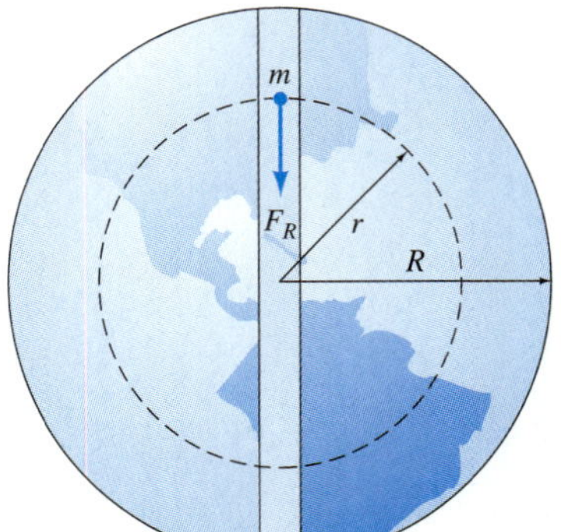

FIGURE 5.4.13. A mass m falling down a hole through the center of the earth (Problem 12).

(c) Take $g = 32.2$ ft/s^2, and conclude from part (b) that the particle undergoes simple harmonic motion back and forth between the ends of the hole, with a period of about 84 min. **(d)** Look up (or derive) the period of a satellite that just skims the surface of the earth; compare with the result in part (c). How do you explain the coincidence? Or *is* it a coincidence? **(e)** With what speed (in miles per hour) does the particle pass through the center of the earth? **(f)** Look up (or derive) the orbital velocity of a satellite that just skims the surface of the earth; compare with the result in part (e). How do you explain the coincidence? Or *is* it a coincidence?

13. Suppose that the mass in a mass-spring-dashpot system with $m = 10$, $c = 9$, and $k = 2$ is set in motion with $x(0) = 0$ and $x'(0) = 5$. **(a)** Find the position function $x(t)$ and show that its graph looks as indicated in Fig. 5.4.14. **(b)** Find how far the mass moves to the right before starting back toward the origin.

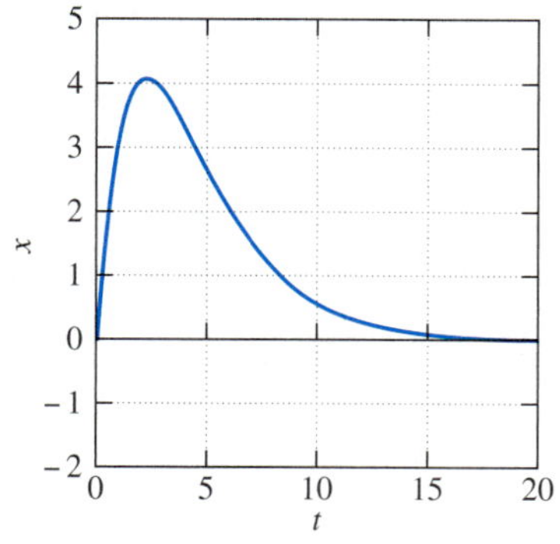

FIGURE 5.4.14. The position function $x(t)$ of Problem 13.

14. Suppose that the mass in a mass-spring-dashpot system with $m = 25$, $c = 10$, and $k = 226$ is set in motion with $x(0) = 20$ and $x'(0) = 41$. **(a)** Find the position function $x(t)$ and show that its graph looks as indicated in Fig. 5.4.15. **(b)** Find the pseudoperiod of the oscillations and the equations of the "envelope curves" that are dashed in the figure.

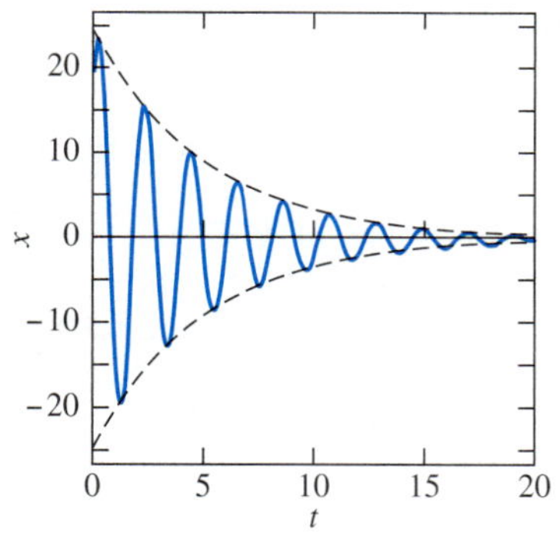

FIGURE 5.4.15. The position function $x(t)$ of Problem 14.

The remaining problems in this section deal with free damped motion. In Problems 15 through 21, a mass m is attached to both a spring (with given spring constant k) and a dashpot (with given damping constant c). The mass is set in motion with initial position x_0 and initial velocity v_0. Find the position function $x(t)$ and determine whether the motion is overdamped, critically damped, or underdamped. If it is underdamped, write the position function in the form $x(t) = C_1 e^{-pt}\cos(\omega_1 t - \alpha_1)$. Also, find the undamped position function $u(t) = C_0\cos(\omega_0 t - \alpha_0)$ that would result if the mass on the spring were set in motion with the same initial position and velocity, but with the dashpot disconnected (so $c = 0$). Finally, construct a figure that illustrates the effect of damping by comparing the graphs of $x(t)$ and $u(t)$.

15. $m = \frac{1}{2}$, $c = 3$, $k = 4$; $x_0 = 2$, $v_0 = 0$

16. $m = 3$, $c = 30$, $k = 63$; $x_0 = 2$, $v_0 = 2$

17. $m = 1$, $c = 8$, $k = 16$; $x_0 = 5$, $v_0 = -10$

18. $m = 2$, $c = 12$, $k = 50$; $x_0 = 0$, $v_0 = -8$

19. $m = 4$, $c = 20$, $k = 169$; $x_0 = 4$, $v_0 = 16$

20. $m = 2$, $c = 16$, $k = 40$; $x_0 = 5$, $v_0 = 4$

21. $m = 1$, $c = 10$, $k = 125$; $x_0 = 6$, $v_0 = 50$

22. A 12-lb weight (mass $m = 0.375$ slugs in fps units) is attached both to a vertically suspended spring that it stretches 6 in. and to a dashpot that provides 3 lb of resistance for every foot per second of velocity. **(a)** If the weight is pulled down 1 ft below its static equilibrium position and then released from rest at time $t = 0$, find its position function $x(t)$. **(b)** Find the frequency, time-varying amplitude, and phase angle of the motion.

23. This problem deals with a highly simplified model of a car of weight 3200 lb (mass $m = 100$ slugs in fps units). Assume that the suspension system acts like a single spring and its shock absorbers like a single dashpot, so that its vertical vibrations satisfy Eq. (4) with appropriate values of the coefficients. **(a)** Find the stiffness coefficient k of the spring if the car undergoes free vibrations at 80 cycles per minute (cycles/min) when its shock absorbers are disconnected. **(b)** With the shock absorbers connected the car is set into vibration by driving it over a bump, and the resulting damped vibrations have a frequency of 78 cycles/min. After how long will the time-varying amplitude be 1% of its initial value?

Problems 24 through 34 deal with a mass-spring-dashpot system having position function $x(t)$ satisfying Eq. (4). We write $x_0 = x(0)$ and $v_0 = x'(0)$ and recall that $p = c/(2m)$, $\omega_0^2 = k/m$, and $\omega_1^2 = \omega_0^2 - p^2$. The system is critically damped, overdamped, or underdamped, as specified in each problem.

24. (Critically damped) Show in this case that

$$x(t) = (x_0 + v_0 t + p x_0 t)e^{-pt}.$$

25. (Critically damped) Deduce from Problem 24 that the mass passes through $x = 0$ at some instant $t > 0$ if and only if x_0 and $v_0 + px_0$ have opposite signs.

26. (Critically damped) Deduce from Problem 24 that $x(t)$ has a local maximum or minimum at some instant $t > 0$ if and only if v_0 and $v_0 + px_0$ have the same sign.

27. (Overdamped) Show in this case that

$$x(t) = \frac{1}{2\gamma}\left[(v_0 - r_2 x_0)e^{r_1 t} - (v_0 - r_1 x_0)e^{r_2 t}\right],$$

where $r_1, r_2 = -p \pm \sqrt{p^2 - \omega_0^2}$ and $\gamma = (r_1 - r_2)/2 > 0$.

28. (Overdamped) If $x_0 = 0$, deduce from Problem 27 that

$$x(t) = \frac{v_0}{\gamma}e^{-pt}\sinh \gamma t.$$

29. (Overdamped) Prove that in this case the mass can pass through its equilibrium position $x = 0$ at most once.

30. (Underdamped) Show that in this case

$$x(t) = e^{-pt}\left(x_0 \cos\omega_1 t + \frac{v_0 + px_0}{\omega_1}\sin\omega_1 t\right).$$

31. (Underdamped) If the damping constant c is small in comparison with $\sqrt{8mk}$, apply the binomial series to show that

$$\omega_1 \approx \omega_0\left(1 - \frac{c^2}{8mk}\right).$$

32. (Underdamped) Show that the local maxima and minima of

$$x(t) = Ce^{-pt}\cos(\omega_1 t - \alpha)$$

occur where

$$\tan(\omega_1 t - \alpha) = -\frac{p}{\omega_1}.$$

Conclude that $t_2 - t_1 = 2\pi/\omega_1$ if two consecutive maxima occur at times t_1 and t_2.

33. (Underdamped) Let x_1 and x_2 be two consecutive local maximum values of $x(t)$. Deduce from the result of Problem 32 that

$$\ln\frac{x_1}{x_2} = \frac{2\pi p}{\omega_1}.$$

The constant $\Delta = 2\pi p/\omega_1$ is called the **logarithmic decrement** of the oscillation. Note also that $c = m\omega_1\Delta/\pi$ because $p = c/(2m)$.

Note: *The result of Problem 33 provides an accurate method for measuring the* viscosity *of a fluid, which is an important parameter in fluid dynamics but which is not easy to measure directly. According to Stokes's drag law, a spherical body of radius a moving at a (relatively slow) speed through a fluid of viscosity μ experiences a resistive force $F_R = 6\pi\mu a v$. Thus if a spherical mass on a spring is immersed in the fluid and set in motion, this drag resistance damps its oscillations with damping constant $c = 6\pi a\mu$. The frequency ω_1 and logarithmic decrement Δ of the oscillations can be measured by direct observation. The final formula in Problem 33 then gives c and hence the viscosity of the fluid.*

34. (Underdamped) A body weighing 100 lb (mass m = 3.125 slugs in fps units) is oscillating attached to a spring and a dashpot. Its first two maximum displacements of 6.73 in. and 1.46 in. are observed to occur at times 0.34 s and 1.17 s, respectively. Compute the damping constant (in pound-seconds per foot) and spring constant (in pounds per foot).

Differential Equations and Determinism

Given a mass m, a dashpot constant c, and a spring constant k, Theorem 2 of Section 5.1 implies that the equation

$$mx'' + cx' + kx = 0 \tag{26}$$

has a unique *solution for $t \geqq 0$ satisfying given initial conditions $x(0) = x_0$, $x'(0) = v_0$. Thus the future motion of an ideal mass-spring-dashpot system is completely determined by the differential equation and the initial conditions. Of course in a real physical system it is impossible to measure the parameters m, c, and k* precisely. *Problems 35 through 38 explore the resulting uncertainty in predicting the future behavior of a physical system.*

35. Suppose that $m = 1$, $c = 2$, and $k = 1$ in Eq. (26). Show that the solution with $x(0) = 0$ and $x'(0) = 1$ is

$$x_1(t) = te^{-t}.$$

36. Suppose that $m = 1$ and $c = 2$ but $k = 1 - 10^{-2n}$. Show that the solution of Eq. (26) with $x(0) = 0$ and $x'(0) = 1$ is

$$x_2(t) = 10^n e^{-t}\sinh 10^{-n}t.$$

37. Suppose that $m = 1$ and $c = 2$ but that $k = 1 + 10^{-2n}$. Show that the solution of Eq. (26) with $x(0) = 0$ and $x'(0) = 1$ is

$$x_3(t) = 10^n e^{-t}\sin 10^{-n}t.$$

38. Whereas the graphs of $x_1(t)$ and $x_2(t)$ resemble those shown in Figs. 5.4.7 and 5.4.8, the graph of $x_3(t)$ exhibits damped oscillations like those illustrated in Fig. 5.4.9, but with a very long pseudoperiod. Nevertheless, show that for each fixed $t > 0$ it is true that

$$\lim_{n\to\infty} x_2(t) = \lim_{n\to\infty} x_3(t) = x_1(t).$$

Conclude that *on a given finite time interval* the three solutions are in "practical" agreement if n is sufficiently large.

5.5 Nonhomogeneous Equations and Undetermined Coefficients

We learned in Section 5.3 how to solve homogeneous linear equations with constant coefficients, but we saw in Section 5.4 that an external force in a simple mechanical system contributes a nonhomogeneous term to its differential equation. The general nonhomogeneous nth-order linear equation with constant coefficients has the form

$$a_n y^{(n)} + a_{n-1}y^{(n-1)} + \cdots + a_1 y' + a_0 y = f(x). \tag{1}$$

By Theorem 5 of Section 5.2, a general solution of Eq. (1) has the form

$$y = y_c + y_p, \tag{2}$$

where the complementary function $y_c(x)$ is a general solution of the associated homogeneous equation

$$a_n y^{(n)} + a_{n-1} y^{(n-1)} + \cdots + a_1 y' + a_0 y = 0, \tag{3}$$

and $y_p(x)$ is a particular solution of Eq. (1). Thus our remaining task is to find y_p.

The **method of undetermined coefficients** is a straightforward way of doing this when the given function $f(x)$ in Eq. (1) is sufficiently simple that we can make an intelligent guess as to the general form of y_p. For example, suppose that $f(x)$ is a polynomial of degree m. Then, because the derivatives of a polynomial are themselves polynomials of lower degree, it is reasonable to suspect a particular solution

$$y_p(x) = A_m x^m + A_{m-1} x^{m-1} + \cdots + A_1 x + A_0$$

that is also a polynomial of degree m, but with as yet undetermined coefficients. We may, therefore, substitute this expression for y_p into Eq. (1), and then—by equating coefficients of like powers of x on the two sides of the resulting equation—attempt to determine the coefficients $A_0, A_1, \ldots, A_m$ so that y_p will, indeed, be a particular solution of Eq. (1).

Similarly, suppose that

$$f(x) = a \cos kx + b \sin kx.$$

Then it is reasonable to expect a particular solution of the same form:

$$y_p(x) = A \cos kx + B \sin kx,$$

a linear combination with undetermined coefficients A and B. The reason is that any derivative of such a linear combination of $\cos kx$ and $\sin kx$ has the same form. We may therefore substitute this form of y_p in Eq. (1), and then—by equating coefficients of $\cos kx$ and $\sin kx$ on both sides of the resulting equation—attempt to determine the coefficients A and B so that y_p will, indeed, be a particular solution.

It turns out that this approach does succeed whenever all the derivatives of $f(x)$ have the same form as $f(x)$ itself. Before describing the method in full generality, we illustrate it with several preliminary examples.

Example 1 Find a particular solution of $y'' + 3y' + 4y = 3x + 2$.

Solution Here $f(x) = 3x + 2$ is a polynomial of degree 1, so our guess is that

$$y_p(x) = Ax + B.$$

Then $y_p' = A$ and $y_p'' = 0$, so y_p will satisfy the differential equation provided that

$$(0) + 3(A) + 4(Ax + B) = 3x + 2,$$

that is,

$$(4A)x + (3A + 4B) = 3x + 2$$

for all x. This will be true if the x-terms and constant terms on the two sides of this equation agree. It therefore suffices for A and B to satisfy the two linear equations $4A = 3$ and $3A + 4B = 2$ that we readily solve for $A = \frac{3}{4}$ and $B = -\frac{1}{16}$. Thus we have found the particular solution

$$y_p(x) = \tfrac{3}{4}x - \tfrac{1}{16}.$$

■

Example 2 Find a particular solution of $y'' - 4y = 2e^{3x}$.

Solution Any derivative of e^{3x} is a constant multiple of e^{3x}, so it is reasonable to try

$$y_p(x) = Ae^{3x}.$$

Then $y_p'' = 9Ae^{3x}$, so the given differential equation will be satisfied provided that

$$9Ae^{3x} - 4(Ae^{3x}) = 2e^{3x};$$

that is, $5A = 2$, so that $A = \frac{2}{5}$. Thus our particular solution is $y_p(x) = \frac{2}{5}e^{3x}$. ■

Example 3 Find a particular solution of $3y'' + y' - 2y = 2\cos x$.

Solution A first guess might be $y_p(x) = A\cos x$, but the presence of y' on the left-hand side signals that we probably need a term involving $\sin x$ as well. So we try

$$\begin{aligned} y_p(x) &= A\cos x + B\sin x; \\ y_p'(x) &= -A\sin x + B\cos x, \\ y_p''(x) &= -A\cos x - B\sin x. \end{aligned}$$

Then substitution of y_p and its derivatives into the given differential equation yields

$$3(-A\cos x - B\sin x) + (-A\sin x + B\cos x) - 2(A\cos x + B\sin x) = 2\cos x,$$

that is (collecting coefficients on the left),

$$(-5A + B)\cos x + (-A - 5B)\sin x = 2\cos x.$$

This will be true for all x provided that the cosine and sine terms on the two sides of this equation agree. It therefore suffices for A and B to satisfy the two linear equations

$$\begin{aligned} -5A + B &= 2, \\ -A - 5B &= 0 \end{aligned}$$

with readily found solution $A = -\frac{5}{13}$, $B = \frac{1}{13}$. Hence a particular solution is

$$y_p(x) = -\tfrac{5}{13}\cos x + \tfrac{1}{13}\sin x.$$ ■

The following example, which superficially resembles Example 2, indicates that the method of undetermined coefficients is not always quite so simple as we have made it appear.

Example 4 Find a particular solution of $y'' - 4y = 2e^{2x}$.

Solution If we try $y_p(x) = Ae^{2x}$, we find that

$$y_p'' - 4y_p = 4Ae^{2x} - 4Ae^{2x} = 0 \neq 2e^{2x}.$$

Thus, no matter how A is chosen, Ae^{2x} cannot satisfy the given nonhomogeneous equation. In fact, the preceding computation shows that Ae^{2x} satisfies instead the associated *homogeneous* equation. Therefore, we should begin with a trial function $y_p(x)$ whose derivative involves both e^{2x} *and something else* that can cancel upon substitution into the differential equation to leave the e^{2x} term that we need. A reasonable guess is

$$y_p(x) = Axe^{2x},$$

for which

$$y_p'(x) = Ae^{2x} + 2Axe^{2x} \quad \text{and} \quad y_p''(x) = 4Ae^{2x} + 4Axe^{2x}.$$

Substitution into the original differential equation yields

$$(4Ae^{2x} + 4Axe^{2x}) - 4(Axe^{2x}) = 2e^{2x}.$$

The terms involving xe^{2x} obligingly cancel, leaving only $4Ae^{2x} = 2e^{2x}$, so that $A = \frac{1}{2}$. Consequently, a particular solution is

$$y_p(x) = \tfrac{1}{2}xe^{2x}. \qquad ■$$

The General Approach

Our initial difficulty in Example 4 resulted from the fact that $f(x) = 2e^{2x}$ satisfies the associated homogeneous equation. Rule 1, given shortly, tells what to do when we do not have this difficulty, and Rule 2 tells what to do when we do have it.

The method of undetermined coefficients applies whenever the function $f(x)$ in Eq. (1) is a linear combination of (finite) products of functions of the following three types:

1. A polynomial in x;

2. An exponential function e^{rx}; (4)

3. $\cos kx$ or $\sin kx$.

Any such function, for example,

$$f(x) = (3 - 4x^2)e^{5x} - 4x^3 \cos 10x,$$

has the crucial property that only *finitely* many linearly independent functions appear as terms (summands) in $f(x)$ and its derivatives of all orders. In Rules 1 and 2 we assume that $Ly = f(x)$ is a nonhomogeneous linear equation with constant coefficients and that $f(x)$ is a function of this kind.

RULE 1 Method of Undetermined Coefficients

Suppose that no term appearing either in $f(x)$ or in any of its derivatives satisfies the associated homogeneous equation $Ly = 0$. Then take as a trial solution for y_p a linear combination of all linearly independent such terms and their derivatives. Then determine the coefficients by substitution of this trial solution into the nonhomogeneous equation $Ly = f(x)$.

Note that this rule is not a theorem requiring proof; it is merely a procedure to be followed in searching for a particular solution y_p. If we succeed in finding y_p, then nothing more need be said. (It can be proved, however, that this procedure will always succeed under the conditions specified here.)

In practice we check the supposition made in Rule 1 by first using the characteristic equation to find the complementary function y_c, and then write a list of all the terms appearing in $f(x)$ and its successive derivatives. If none of the terms in this list duplicates a term in y_c, then we proceed with Rule 1.

Example 5 Find a particular solution of

$$y'' + 4y = 3x^3. \tag{5}$$

Solution The (familiar) complementary solution of Eq. (5) is

$$y_c(x) = c_1 \cos 2x + c_2 \sin 2x.$$

The function $f(x) = 3x^3$ and its derivatives are constant multiples of the linearly independent functions x^3, x^2, x, and 1. Because none of these appears in y_c, we try

$$\begin{aligned} y_p &= Ax^3 + Bx^2 + Cx + D, \\ y_p' &= 3Ax^2 + 2Bx + C, \\ y_p'' &= 6Ax + 2B. \end{aligned}$$

Substitution in Eq. (5) gives

$$\begin{aligned} y_p'' + 4y_p &= (6Ax + 2B) + 4(Ax^3 + Bx^2 + Cx + D) \\ &= 4Ax^3 + 4Bx^2 + (6A + 4C)x + (2B + D) = 3x^3. \end{aligned}$$

We equate coefficients of like powers of x in the last equation to get

$$\begin{aligned} 4A &= 3, & 4B &= 0, \\ 6A + 4C &= 0, & 2B + D &= 0 \end{aligned}$$

with solution $A = \frac{3}{4}$, $B = 0$, $C = -\frac{9}{8}$, and $D = 0$. Hence a particular solution of Eq. (5) is

$$y_p(x) = \tfrac{3}{4}x^3 - \tfrac{9}{8}x.$$ ■

Example 6 Solve the initial value problem

$$\begin{aligned} &y'' - 3y' + 2y = 3e^{-x} - 10\cos 3x; \\ &y(0) = 1, \quad y'(0) = 2. \end{aligned} \tag{6}$$

Solution The characteristic equation $r^2 - 3r + 2 = 0$ has roots $r = 1$ and $r = 2$, so the complementary function is

$$y_c(x) = c_1 e^x + c_2 e^{2x}.$$

The terms involved in $f(x) = 3e^{-x} - 10\cos 3x$ and its derivatives are e^{-x}, $\cos 3x$, and $\sin 3x$. Because none of these appears in y_c, we try

$$\begin{aligned} y_p &= Ae^{-x} + B\cos 3x + C\sin 3x, \\ y_p' &= -Ae^{-x} - 3B\sin 3x + 3C\cos 3x, \\ y_p'' &= Ae^{-x} - 9B\cos 3x - 9C\sin 3x. \end{aligned}$$

After we substitute these expressions into the differential equation in (6) and collect coefficients, we get

$$\begin{aligned} y_p'' - 3y_p' + 2y_p &= 6Ae^{-x} + (-7B - 9C)\cos 3x + (9B - 7C)\sin 3x \\ &= 3e^{-x} - 10\cos 3x. \end{aligned}$$

We equate the coefficients of the terms involving e^{-x}, those involving $\cos 3x$, and those involving $\sin 3x$. The result is the system

$$\begin{aligned} 6A &= 3, \\ -7B - 9C &= -10, \\ 9B - 7C &= 0 \end{aligned}$$

with solution $A = \frac{1}{2}$, $B = \frac{7}{13}$, and $C = \frac{9}{13}$. This gives the particular solution

$$y_p(x) = \tfrac{1}{2}e^{-x} + \tfrac{7}{13}\cos 3x + \tfrac{9}{13}\sin 3x,$$

which, however, does not have the required initial values in (6).

To satisfy those initial conditions, we begin with the *general* solution

$$\begin{aligned} y(x) &= y_c(x) + y_p(x) \\ &= c_1e^x + c_2e^{2x} + \tfrac{1}{2}e^{-x} + \tfrac{7}{13}\cos 3x + \tfrac{9}{13}\sin 3x, \end{aligned}$$

with derivative

$$y'(x) = c_1e^x + 2c_2e^{2x} - \tfrac{1}{2}e^{-x} - \tfrac{21}{13}\sin 3x + \tfrac{27}{13}\cos 3x.$$

The initial conditions in (6) lead to the equations

$$\begin{aligned} y(0) &= c_1 + c_2 + \tfrac{1}{2} + \tfrac{7}{13} = 1, \\ y'(0) &= c_1 + 2c_2 - \tfrac{1}{2} + \tfrac{27}{13} = 2 \end{aligned}$$

with solution $c_1 = -\frac{1}{2}$, $c_2 = \frac{6}{13}$. The desired particular solution is therefore

$$y(x) = -\tfrac{1}{2}e^x + \tfrac{6}{13}e^{2x} + \tfrac{1}{2}e^{-x} + \tfrac{7}{13}\cos 3x + \tfrac{9}{13}\sin 3x.$$ ■

Example 7 Find the general form of a particular solution of

$$y^{(3)} + 9y' = x\sin x + x^2e^{2x}. \tag{7}$$

Solution The characteristic equation $r^3 + 9r = 0$ has roots $r = 0$, $r = -3i$, and $r = 3i$. So the complementary function is

$$y_c(x) = c_1 + c_2 \cos 3x + c_3 \sin 3x.$$

The derivatives of the right-hand side in Eq. (7) involve the terms

$$\cos x, \quad \sin x, \quad x \cos x, \quad x \sin x,$$
$$e^{2x}, \quad xe^{2x}, \quad \text{and} \quad x^2 e^{2x}.$$

Because there is no duplication with the terms of the complementary function, the trial solution takes the form

$$y_p(x) = A \cos x + B \sin x + Cx \cos x + Dx \sin x + Ee^{2x} + Fxe^{2x} + Gx^2 e^{2x}.$$

Upon substituting y_p in Eq. (7) and equating coefficients of like terms, we get seven equations determining the seven coefficients A, B, C, D, E, F, and G. ■

The Case of Duplication

Now we turn our attention to the situation in which Rule 1 does not apply: Some of the terms involved in $f(x)$ and its derivatives satisfy the associated homogeneous equation. For instance, suppose that we want to find a particular solution of the differential equation

$$(D - r)^3 y = (2x - 3)e^{rx}. \tag{8}$$

Proceeding as in Rule 1, our first guess would be

$$y_p(x) = Ae^{rx} + Bxe^{rx}. \tag{9}$$

This form of $y_p(x)$ will not be adequate because the complementary function of Eq. (8) is

$$y_c(x) = c_1 e^{rx} + c_2 x e^{rx} + c_3 x^2 e^{rx}, \tag{10}$$

so substitution of (9) in the left-hand side of (8) would yield zero rather than $(2x - 3)e^{rx}$.

To see how to amend our first guess, we observe that

$$(D - r)^2[(2x - 3)e^{rx}] = [D^2(2x - 3)]e^{rx} = 0$$

by Eq. (13) of Section 5.3. If $y(x)$ is *any* solution of Eq. (8) and we apply the operator $(D - r)^2$ to both sides, we see that $y(x)$ is also a solution of the equation $(D - r)^5 y = 0$. The general solution of this *homogeneous* equation can be written as

$$y(x) = \underbrace{c_1 e^{rx} + c_2 x e^{rx} + c_3 x^2 e^{rx}}_{y_c} + \underbrace{Ax^3 e^{rx} + Bx^4 e^{rx}}_{y_p}.$$

Thus *every* solution of our original equation in (8) is the sum of a complementary function and a *particular solution* of the form

$$y_p(x) = Ax^3 e^{rx} + Bx^4 e^{rx}. \tag{11}$$

Note that the right-hand side in Eq. (11) can be obtained by multiplying each term of our first guess in (9) by the least positive integral power of x (in this case, x^3) that suffices to eliminate duplication between the terms of the resulting trial solution $y_p(x)$ and the complementary function $y_c(x)$ given in (10). This procedure succeeds in the general case.

To simplify the general statement of Rule 2, we observe that to find a particular solution of the nonhomogeneous linear differential equation

$$Ly = f_1(x) + f_2(x), \tag{12}$$

it suffices to find *separately* particular solutions $Y_1(x)$ and $Y_2(x)$ of the two equations

$$Ly = f_1(x) \quad \text{and} \quad Ly = f_2(x), \tag{13}$$

respectively. For linearity then gives

$$L[Y_1 + Y_2] = LY_1 + LY_2 = f_1(x) + f_2(x),$$

and therefore $y_p = Y_1 + Y_2$ is a particular solution of Eq. (12). (This is a type of "superposition principle" for nonhomogeneous linear equations.)

Now our problem is to find a particular solution of the equation $Ly = f(x)$, where $f(x)$ is a linear combination of products of the elementary functions listed in (4). Thus $f(x)$ can be written as a sum of terms each of the form

$$P_m(x)e^{rx}\cos kx \quad \text{or} \quad P_m(x)e^{rx}\sin kx, \tag{14}$$

where $P_m(x)$ is a polynomial in x of degree m. Note that any derivative of such a term is of the same form but with *both* sines and cosines appearing. The procedure by which we arrived earlier at the particular solution in (11) of Eq. (8) can be generalized to show that the following procedure is always successful.

RULE 2 Method of Undetermined Coefficients

If the function $f(x)$ is of either form in (14), take as the trial solution

$$\begin{aligned} y_p(x) = x^s[&(A_0 + A_1x + A_2x^2 + \cdots + A_mx^m)e^{rx}\cos kx \\ &+ (B_0 + B_1x + B_2x^2 + \cdots + B_mx^m)e^{rx}\sin kx], \end{aligned} \tag{15}$$

where s is the smallest nonnegative integer such that no term in y_p duplicates a term in the complementary function y_c. Then determine the coefficients in Eq. (15) by substituting y_p into the nonhomogeneous equation.

In practice we seldom need to deal with a function $f(x)$ exhibiting the full generality in (14). The table in Fig. 5.5.1 lists the form of y_p in various common cases, corresponding to the possibilities $m = 0$, $r = 0$, and $k = 0$.

On the other hand, it is common to have

$$f(x) = f_1(x) + f_2(x),$$

where $f_1(x)$ and $f_2(x)$ are different functions of the sort listed in the table in Fig. 5.5.1. In this event we take as y_p the sum of the trial solutions for $f_1(x)$ and $f_2(x)$, choosing s *separately* for each part to eliminate duplication with the complementary function. This procedure is illustrated in Examples 8 through 10.

$f(x)$	y_p
$P_m = b_0 + b_1 x + b_2 x^2 + \cdots + b_m x^m$	$x^s(A_0 + A_1 x + A_2 x^2 + \cdots + A_m x^m)$
$a \cos kx + b \sin kx$	$x^s(A \cos kx + B \sin kx)$
$e^{rx}(a \cos kx + b \sin kx)$	$x^s e^{rx}(A \cos kx + B \sin kx)$
$P_m(x)e^{rx}$	$x^s(A_0 + A_1 x + A_2 x^2 + \cdots + A_m x^m)e^{rx}$
$P_m(x)(a \cos kx + b \sin kx)$	$x^s[(A_0 + A_1 x + \cdots + A_m x^m) \cos kx$ $+ (B_0 + B_1 x + \cdots + B_m x^m) \sin kx]$

FIGURE 5.5.1. Substitutions in the method of undetermined coefficients.

Example 8 Find a particular solution of

$$y^{(3)} + y'' = 3e^x + 4x^2. \tag{16}$$

Solution The characteristic equation $r^3 + r^2 = 0$ has roots $r_1 = r_2 = 0$ and $r_3 = -1$, so the complementary function is

$$y_c(x) = c_1 + c_2 x + c_3 e^{-x}.$$

As a first step toward our particular solution, we form the sum

$$(Ae^x) + (B + Cx + Dx^2).$$

The part Ae^x corresponding to $3e^x$ does not duplicate any part of the complementary function, but the part $B + Cx + Dx^2$ must be multiplied by x^2 to eliminate duplication. Hence we take

$$\begin{aligned}
y_p &= Ae^x + Bx^2 + Cx^3 + Dx^4, \\
y_p' &= Ae^x + 2Bx + 3Cx^2 + 4Dx^3, \\
y_p'' &= Ae^x + 2B + 6Cx + 12Dx^2, \quad \text{and} \\
y_p^{(3)} &= Ae^x + 6C + 24Dx.
\end{aligned}$$

Substitution of these derivatives in Eq. (16) yields

$$2Ae^x + (2B + 6C) + (6C + 24D)x + 12Dx^2 = 3e^x + 4x^2.$$

The system of equations

$$\begin{aligned}
2A &= 3, \qquad & 2B + 6C &= 0, \\
6C + 24D &= 0, \qquad & 12D &= 4
\end{aligned}$$

has the solution $A = \frac{3}{2}$, $B = 4$, $C = -\frac{4}{3}$, and $D = \frac{1}{3}$. Hence the desired particular solution is

$$y_p(x) = \tfrac{3}{2}e^x + 4x^2 - \tfrac{4}{3}x^3 + \tfrac{1}{3}x^4.$$

■

Example 9 Determine the appropriate form for a particular solution of

$$y'' + 6y' + 13y = e^{-3x} \cos 2x.$$

Solution The characteristic equation $r^2+6r+13 = 0$ has roots $-3\pm 2i$, so the complementary function is

$$y_c(x) = e^{-3x}(c_1 \cos 2x + c_2 \sin 2x).$$

This is the same form as a first attempt $e^{-3x}(A \cos 2x + B \sin 2x)$ at a particular solution, so we must multiply by x to eliminate duplication. Hence we would take

$$y_p(x) = e^{-3x}(Ax \cos 2x + Bx \sin 2x).$$ ■

Example 10 Determine the appropriate form for a particular solution of the fifth-order equation

$$(D-2)^3(D^2+9)y = x^2e^{2x} + x \sin 3x.$$

Solution The characteristic equation $(r-2)^3(r^2+9) = 0$ has roots $r = 2, 2, 2, 3i$, and $-3i$, so the complementary function is

$$y_c(x) = c_1e^{2x} + c_2xe^{2x} + c_3x^2e^{2x} + c_4 \cos 3x + c_5 \sin 3x.$$

As a first step toward the form of a particular solution, we examine the sum

$$[(A + Bx + Cx^2)e^{2x}] + [(D + Ex) \cos 3x + (F + Gx) \sin 3x].$$

To eliminate duplication with terms of $y_c(x)$, the first part—corresponding to x^2e^{2x}—must be multiplied by x^3, and the second part—corresponding to $x \sin 3x$—must be multiplied by x. Hence we would take

$$y_p(x) = (Ax^3 + Bx^4 + Cx^5)e^{2x} + (Dx + Ex^2) \cos 3x + (Fx + Gx^2) \sin 3x.$$ ■

Variation of Parameters

Finally, let us point out the kind of situation in which the method of undetermined coefficients cannot be used. Consider, for example, the equation

$$y'' + y = \tan x, \tag{17}$$

which at first glance may appear similar to those considered in the preceding examples. Not so; the function $f(x) = \tan x$ has *infinitely many* linearly independent derivatives

$$\sec^2 x, \quad 2\sec^2 x \tan x, \quad 4\sec^2 x \tan^2 x + 2\sec^4 x, \quad \ldots.$$

Therefore, we do not have available a *finite* linear combination to use as a trial solution.

We discuss here the method of **variation of parameters**, which—in principle (that is, if the integrals that appear can be evaluated)—can always be used to find a particular solution of the nonhomogeneous linear differential equation

$$y^{(n)} + p_{n-1}(x)y^{(n-1)} + \cdots + p_1(x)y' + p_0(x)y = f(x), \tag{18}$$

provided that we already know the general solution

$$y_c = c_1 y_1 + c_2 y_2 + \cdots + c_n y_n \tag{19}$$

of the associated homogeneous equation

$$y^{(n)} + p_{n-1}(x)y^{(n-1)} + \cdots + p_1(x)y' + p_0(x)y = 0. \tag{20}$$

Here, in brief, is the basic idea of the method of variation of parameters. Suppose that we replace the constants, or *parameters*, $c_1, c_2, \ldots, c_n$ in the complementary function in Eq. (19) with *variables*: functions $u_1, u_2, \ldots, u_n$ of x. We ask whether it is possible to choose these functions in such a way that the combination

$$y_p(x) = u_1(x)y_1(x) + u_2(x)y_2(x) + \cdots + u_n(x)y_n(x) \tag{21}$$

is a particular solution of the nonhomogeneous equation in (18). It turns out that this *is* always possible.

The method is essentially the same for all orders $n \geqq 2$, but we will describe it in detail only for the case $n = 2$. So we begin with the second-order nonhomogeneous equation

$$L[y] = y'' + P(x)y' + Q(x)y = f(x) \tag{22}$$

with complementary function

$$y_c(x) = c_1 y_1(x) + c_2 y_2(x) \tag{23}$$

on some open interval I where the functions P and Q are continuous. We want to find functions u_1 and u_2 such that

$$y_p(x) = u_1(x)y_1(x) + u_2(x)y_2(x) \tag{24}$$

is a particular solution of Eq. (22).

One condition on the two functions u_1 and u_2 is that $L[y_p] = f(x)$. Because two conditions are required to determine two functions, we are free to impose an additional condition of our choice. We will do so in a way that simplifies the computations as much as possible. But first, to impose the condition $L[y_p] = f(x)$, we must compute the derivatives y_p' and y_p''. The product rule gives

$$y_p' = (u_1 y_1' + u_2 y_2') + (u_1' y_1 + u_2' y_2).$$

To avoid the appearance of the second derivatives u_1'' and u_2'', the additional condition that we now impose is that the second sum here must vanish:

$$u_1' y_1 + u_2' y_2 = 0. \tag{25}$$

Then

$$y_p' = u_1 y_1' + u_2 y_2', \tag{26}$$

and the product rule gives

$$y_p'' = (u_1 y_1'' + u_2 y_2'') + (u_1' y_1' + u_2' y_2'). \tag{27}$$

But both y_1 and y_2 satisfy the homogeneous equation

$$y'' + Py' + Qy = 0$$

associated with the nonhomogeneous equation in (22), so

$$y_i'' = -Py_i' - Qy_i \tag{28}$$

for $i = 1, 2$. It therefore follows from Eq. (27) that

$$y_p'' = (u_1' y_1' + u_2' y_2') - P \cdot (u_1 y_1' + u_2 y_2') - Q \cdot (u_1 y_1 + u_2 y_2).$$

In view of Eqs. (24) and (26), this means that

$$y_p'' = (u_1' y_1' + u_2' y_2') - Py_p' - Qy_p;$$

hence

$$L[y_p] = u_1' y_1' + u_2' y_2'. \tag{29}$$

The requirement that y_p satisfy the nonhomogeneous equation in (22)—that is, that $L[y_p] = f(x)$—therefore implies that

$$u_1' y_1' + u_2' y_2' = f(x). \tag{30}$$

Finally, Eqs. (25) and (30) determine the functions u_1 and u_2 that we need. Collecting these equations, we obtain a system

$$\begin{aligned} u_1' y_1 + u_2' y_2 &= 0, \\ u_1' y_1' + u_2' y_2' &= f(x) \end{aligned} \tag{31}$$

of two linear equations in the two *derivatives* u_1' and u_2'. Note that the determinant of coefficients in (31) is simply the Wronskian $W(y_1, y_2)$. Once we have solved the equations in (31) for the derivatives u_1' and u_2', we integrate each to obtain the functions u_1 and u_2 such that

$$y_p = u_1 y_1 + u_2 y_2 \tag{32}$$

is the desired particular solution of Eq. (22). In Problem 63 we ask you to carry out this process explicitly and thereby verify the formula for $y_p(x)$ in the following theorem.

THEOREM 1 Variation of Parameters

If the nonhomogeneous equation $y'' + P(x)y' + Q(x)y = f(x)$ has complementary function $y_c(x) = c_1 y_1(x) + c_2 y_2(x)$, then a particular solution is given by

$$y_p(x) = \left(-\int \frac{y_2(x)f(x)}{W(x)}\,dx\right) y_1(x) + \left(\int \frac{y_1(x)f(x)}{W(x)}\,dx\right) y_2(x), \qquad (33)$$

where $W = W(y_1, y_2)$ is the Wronskian of the two independent solutions y_1 and y_2 of the associated homogeneous equation.

Example 11 Find a particular solution of the equation $y'' + y = \tan x$.

Solution The complementary function is $y_c(x) = c_1 \cos x + c_2 \sin x$, and we could simply substitute directly in Eq. (33). But it is more instructive to set up the equations in (31) and solve for u_1' and u_2', so we begin with

$$\begin{aligned} y_1 &= \cos x, & y_2 &= \sin x, \\ y_1' &= -\sin x, & y_2' &= \cos x. \end{aligned}$$

Hence the equations in (31) are

$$\begin{aligned} (u_1')(\cos x) + (u_2')(\sin x) &= 0, \\ (u_1')(-\sin x) + (u_2')(\cos x) &= \tan x. \end{aligned}$$

We easily solve these equations for

$$\begin{aligned} u_1' &= -\sin x \tan x = -\frac{\sin^2 x}{\cos x} = \cos x - \sec x, \\ u_2' &= \cos x \tan x = \sin x. \end{aligned}$$

Hence we take

$$u_1 = \int (\cos x - \sec x)\,dx = \sin x - \ln|\sec x + \tan x|$$

and

$$u_2 = \int \sin x\,dx = -\cos x.$$

(Do you see why we choose the constants of integration to be zero?) Thus our particular solution is

$$\begin{aligned} y_p(x) &= u_1(x)y_1(x) + u_2(x)y_2(x) \\ &= (\sin x - \ln|\sec x + \tan x|)\cos x + (-\cos x)(\sin x); \end{aligned}$$

that is,

$$y_p(x) = -(\cos x)\ln|\sec x + \tan x|. \qquad \blacksquare$$

5.5 Problems

In Problems 1 through 20, find a particular solution y_p of the given equation. In all these problems, primes denote derivatives with respect to x.

1. $y'' + 16y = e^{3x}$
2. $y'' - y' - 2y = 3x + 4$
3. $y'' - y' - 6y = 2\sin 3x$
4. $4y'' + 4y' + y = 3xe^x$
5. $y'' + y' + y = \sin^2 x$
6. $2y'' + 4y' + 7y = x^2$
7. $y'' - 4y = \sinh x$
8. $y'' - 4y = \cosh 2x$
9. $y'' + 2y' - 3y = 1 + xe^x$
10. $y'' + 9y = 2\cos 3x + 3\sin 3x$
11. $y^{(3)} + 4y' = 3x - 1$
12. $y^{(3)} + y' = 2 - \sin x$
13. $y'' + 2y' + 5y = e^x \sin x$
14. $y^{(4)} - 2y'' + y = xe^x$
15. $y^{(5)} + 5y^{(4)} - y = 17$
16. $y'' + 9y = 2x^2 e^{3x} + 5$
17. $y'' + y = \sin x + x\cos x$
18. $y^{(4)} - 5y'' + 4y = e^x - xe^{2x}$
19. $y^{(5)} + 2y^{(3)} + 2y'' = 3x^2 - 1$
20. $y^{(3)} - y = e^x + 7$

In Problems 21 through 30, set up the appropriate form of a particular solution y_p, but do not determine the values of the coefficients.

21. $y'' - 2y' + 2y = e^x \sin x$
22. $y^{(5)} - y^{(3)} = e^x + 2x^2 - 5$
23. $y'' + 4y = 3x\cos 2x$
24. $y^{(3)} - y'' - 12y' = x - 2xe^{-3x}$
25. $y'' + 3y' + 2y = x(e^{-x} - e^{-2x})$
26. $y'' - 6y' + 13y = xe^{3x}\sin 2x$
27. $y^{(4)} + 5y'' + 4y = \sin x + \cos 2x$
28. $y^{(4)} + 9y'' = (x^2 + 1)\sin 3x$
29. $(D-1)^3(D^2-4)y = xe^x + e^{2x} + e^{-2x}$
30. $y^{(4)} - 2y'' + y = x^2\cos x$

Solve the initial value problems in Problems 31 through 40.

31. $y'' + 4y = 2x$; $y(0) = 1$, $y'(0) = 2$
32. $y'' + 3y' + 2y = e^x$; $y(0) = 0$, $y'(0) = 3$
33. $y'' + 9y = \sin 2x$; $y(0) = 1$, $y'(0) = 0$
34. $y'' + y = \cos x$; $y(0) = 1$, $y'(0) = -1$
35. $y'' - 2y' + 2y = x + 1$; $y(0) = 3$, $y'(0) = 0$
36. $y^{(4)} - 4y'' = x^2$; $y(0) = y'(0) = 1$, $y''(0) = y^{(3)}(0) = -1$
37. $y^{(3)} - 2y'' + y' = 1 + xe^x$; $y(0) = y'(0) = 0$, $y''(0) = 1$
38. $y'' + 2y' + 2y = \sin 3x$; $y(0) = 2$, $y'(0) = 0$
39. $y^{(3)} + y'' = x + e^{-x}$; $y(0) = 1$, $y'(0) = 0$, $y''(0) = 1$
40. $y^{(4)} - y = 5$; $y(0) = y'(0) = y''(0) = y^{(3)}(0) = 0$

41. Find a particular solution of the equation

$$y^{(4)} - y^{(3)} - y'' - y' - 2y = 8x^5.$$

42. Find the solution of the initial value problem consisting of the differential equation of Problem 41 and the initial conditions

$$y(0) = y'(0) = y''(0) = y^{(3)}(0) = 0.$$

43. **(a)** Write

$$\cos 3x + i\sin 3x = e^{3ix} = (\cos x + i\sin x)^3$$

by Euler's formula, expand, and equate real and imaginary parts to derive the identities

$$\cos^3 x = \tfrac{3}{4}\cos x + \tfrac{1}{4}\cos 3x,$$
$$\sin^3 x = \tfrac{3}{4}\sin x - \tfrac{1}{4}\sin 3x.$$

(b) Use the result of part (a) to find a general solution of

$$y'' + 4y = \cos^3 x.$$

Use trigonometric identities to find general solutions of the equations in Problems 44 through 46.

44. $y'' + y' + y = \sin x \sin 3x$
45. $y'' + 9y = \sin^4 x$
46. $y'' + y = x\cos^3 x$

In Problems 47 through 56, use the method of variation of parameters to find a particular solution of the given differential equation.

47. $y'' + 3y' + 2y = 4e^x$
48. $y'' - 2y' - 8y = 3e^{-2x}$
49. $y'' - 4y' + 4y = 2e^{2x}$
50. $y'' - 4y = \sinh 2x$
51. $y'' + 4y = \cos 3x$
52. $y'' + 9y = \sin 3x$
53. $y'' + 9y = 2\sec 3x$
54. $y'' + y = \csc^2 x$
55. $y'' + 4y = \sin^2 x$
56. $y'' - 4y = xe^x$

57. You can verify by substitution that $y_c = c_1 x + c_2 x^{-1}$ is a complementary function for the nonhomogeneous second-order equation

$$x^2 y'' + xy' - y = 72x^5.$$

But before applying the method of variation of parameters, you must first divide this equation by its leading coefficient x^2 to rewrite it in the standard form

$$y'' + \frac{1}{x}y' - \frac{1}{x^2}y = 72x^3.$$

Thus $f(x) = 72x^3$ in Eq. (22). Now proceed to solve the equations in (31) and thereby derive the particular solution $y_p = 3x^5$.

In Problems 58 through 62, a nonhomogeneous second-order linear equation and a complementary function y_c are given. Apply the method of Problem 57 to find a particular solution of the equation.

58. $x^2y'' - 4xy' + 6y = x^3$; $y_c = c_1x^2 + c_2x^3$
59. $x^2y'' - 3xy' + 4y = x^4$; $y_c = x^2(c_1 + c_2\ln x)$
60. $4x^2y'' - 4xy' + 3y = 8x^{4/3}$; $y_c = c_1x^{1/2} + c_2x^{3/2}$
61. $x^2y'' + xy' + y = \ln x$; $y_c = c_1\cos(\ln x) + c_2\sin(\ln x)$
62. $(x^2-1)y'' - 2xy' + 2y = x^2 - 1$; $y_c = c_1x + c_2(1 + x^2)$

63. Carry out the solution process indicated in the text to derive the variation of parameters formula in (33) from Eqs. (31) and (32).

64. Apply the variation of parameters formula in (33) to find the particular solution $y_p(x) = -x\cos x$ of the nonhomogeneous equation $y'' + y = 2\sin x$.

5.5 Application Automated Variation of Parameters

The variation of parameters formula in (33) is especially amenable to implementation in a computer algebra system when the indicated integrals would be too tedious or inconvenient for manual evaluation. For example, suppose that we want to find a particular solution of the nonhomogeneous equation

$$y'' + y = \tan x$$

of Example 11, with complementary function $y_c(x) = c_1 \cos x + c_2 \sin x$. Then the *Maple* commands

```
y1 := cos(x):
y2 := sin(x):
f := tan(x):
W := y1*diff(y2,x) - y2*diff(y1,x):
W := simplify(W):
yp := -y1*int(y2*f/W,x) + y2*int(y1*f/W,x):
simplify(yp);
```

implement (33) and produce the result

$$y_p(x) = -(\cos x) \ln \left(\frac{1 + \sin x}{\cos x} \right)$$

equivalent to the result $y_p(x) = -(\cos x) \ln(\sec x + \tan x)$ found in Example 11. The analogous *Mathematica* commands

```
y1 = Cos[x];
y2 = Sin[x];
f = Tan[x];
W = y1*D[y2,x] - y2*D[y1,x]   //  Simplify
yp = -y1*Integrate[y2*f/W,x] + y2*Integrate[y1*f/W,x];
Simplify[yp]
```

produce the result

$$y_p(x) = -(\cos x) \ln \left(\frac{\cos(x/2) + \sin(x/2)}{\cos(x/2) - \sin(x/2)} \right),$$

which (by the usual difference-of-squares technique) also is equivalent to the result found in Example 11.

To solve similarly a second-order linear equation $y'' + P(x)y' + Q(x)y = f(x)$ whose complementary function $y_c(x) = c_1y_1(x) + c_2y_2(x)$ is known, we need only insert the corresponding definitions of $y_1(x)$, $y_2(x)$, and $f(x)$ in the initial lines shown here. Find in this way the indicated particular solution $y_p(x)$ of the nonhomogeneous equations in Problems 1 through 6.

1. $y'' + y = 2 \sin x$ $\quad y_p(x) = -x \cos x$

2. $y'' + y = 4x \sin x$ $\quad y_p(x) = x \sin x - x^2 \cos x$

3. $y'' + y = 12x^2 \sin x$ $\quad y_p(x) = 3x^2 \sin x + (3x - 2x^3) \cos x$

4. $y'' - 2y' + 2y = 2e^x \sin x$ $\quad y_p(x) = -xe^x \cos x$

5. $y'' - 2y' + 2y = 4xe^x \sin x$ $\quad y_p(x) = e^x(x \sin x - x^2 \cos x)$

6. $y'' - 2y' + 2y = 12x^2e^x \sin x$ $\quad y_p(x) = e^x \left[3x^2 \sin x + (3x - 2x^3) \cos x\right]$

5.6 Forced Oscillations and Resonance

In Section 5.4 we derived the differential equation

$$mx'' + cx' + kx = F(t) \tag{1}$$

that governs the 1-dimensional motion of a mass m that is attached to a spring (with constant k) and a dashpot (with constant c) and is also acted on by an external force $F(t)$. Machines with rotating components commonly involve mass-spring systems (or their equivalents) in which the external force is simple harmonic:

$$F(t) = F_0 \cos \omega t \quad \text{or} \quad F(t) = F_0 \sin \omega t, \tag{2}$$

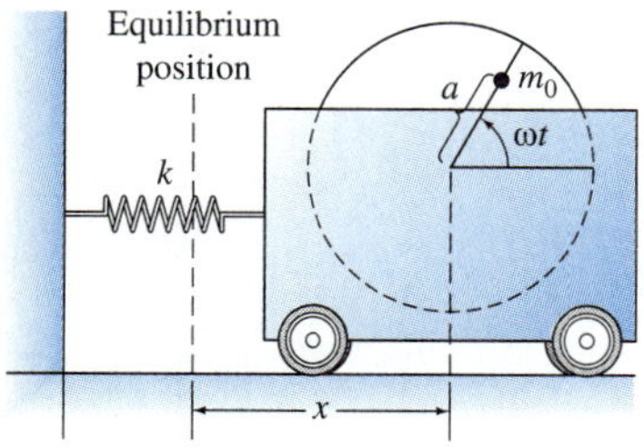

FIGURE 5.6.1. The cart-with-flywheel system.

where the constant F_0 is the amplitude of the periodic force and ω is its circular frequency.

For an example of how a rotating machine component can provide a simple harmonic force, consider the cart with a rotating vertical flywheel shown in Fig. 5.6.1. The cart has mass $m - m_0$, not including the flywheel of mass m_0. The centroid of the flywheel is off center at a distance a from its center, and its angular speed is ω radians per second. The cart is attached to a spring (with constant k) as shown. Assume that the centroid of the cart itself is directly beneath the center of the flywheel, and denote by $x(t)$ its displacement from its equilibrium position (where the spring is unstretched). Figure 5.6.1 helps us to see that the displacement $\overline{x}$ of the centroid of the combined cart plus flywheel is given by

$$\overline{x} = \frac{(m - m_0)x + m_0(x + a\cos\omega t)}{m} = x + \frac{m_0 a}{m}\cos\omega t.$$

Let us ignore friction and apply Newton's second law $m\overline{x}'' = -kx$, because the force exerted by the spring is $-kx$. We substitute for $\overline{x}$ in the last equation to obtain

$$mx'' - m_0 a\omega^2 \cos\omega t = -kx;$$

that is,

$$mx'' + kx = m_0 a\omega^2 \cos\omega t. \tag{3}$$

Thus the cart with its rotating flywheel acts like a mass on a spring under the influence of a simple harmonic external force with amplitude $F_0 = m_0 a\omega^2$. Such a system is a reasonable model of a front-loading washing machine with the clothes being washed loaded off center. This illustrates the practical importance of analyzing solutions of Eq. (1) with external forces as in (2).

Undamped Forced Oscillations

To study undamped oscillations under the influence of the external force $F(t) = F_0 \cos\omega t$, we set $c = 0$ in Eq. (1) and thereby begin with the equation

$$mx'' + kx = F_0 \cos\omega t \tag{4}$$

whose complementary function is $x_c = c_1\cos\omega_0 t + c_2\sin\omega_0 t$. Here

$$\omega_0 = \sqrt{\frac{k}{m}}$$

(as in Eq. (9) of Section 5.4) is the (circular) **natural frequency** of the mass-spring system. The fact that the angle $\omega_0 t$ is measured in (dimensionless) radians reminds us that if t is measured in seconds (s), then ω_0 is measured in radians per second—that is, in inverse seconds (s^{-1}). Also recall from Eq. (14) in Section 5.4 that division of a circular frequency ω by the number 2π of radians in a cycle gives the corresponding (ordinary) **frequency** $\nu = \omega/2\pi$ in Hz (hertz = cycles per second).

Let us assume initially that the external and natural frequencies are *unequal*: $\omega \neq \omega_0$. We substitute $x_p = A\cos\omega t$ in Eq. (4) to find a particular solution. [No sine term is needed in x_p because there is no term involving x' on the left-hand side in Eq. (4).] This gives

$$-m\omega^2 A\cos\omega t + kA\cos\omega t = F_0\cos\omega t,$$

so

$$A = \frac{F_0}{k - m\omega^2} = \frac{F_0/m}{\omega_0^2 - \omega^2}, \tag{5}$$

and thus

$$x_p(t) = \frac{F_0/m}{\omega_0^2 - \omega^2}\cos\omega t. \tag{6}$$

Therefore, the general solution $x = x_c + x_p$ is given by

$$x(t) = c_1\cos\omega_0 t + c_2\sin\omega_0 t + \frac{F_0/m}{\omega_0^2 - \omega^2}\cos\omega t, \tag{7}$$

where the constants c_1 and c_2 are determined by the initial values $x(0)$ and $x'(0)$. Equivalently, as in Eq. (12) of Section 5.4, we can rewrite Eq. (7) as

$$x(t) = C\cos(\omega_0 t - \alpha) + \frac{F_0/m}{\omega_0^2 - \omega^2}\cos\omega t, \tag{8}$$

so we see that the resulting motion is a superposition of two oscillations, one with natural circular frequency ω_0, the other with the frequency ω of the external force.

Example 1 Suppose that $m = 1$, $k = 9$, $F_0 = 80$, and $\omega = 5$, so the differential equation in (4) is

$$x'' + 9x = 80\cos 5t.$$

Find $x(t)$ if $x(0) = x'(0) = 0$.

Solution Here the natural frequency $\omega_0 = 3$ and the frequency $\omega = 5$ of the external force are unequal, as in the preceding discussion. First we substitute $x_p = A\cos 5t$ in the differential equation and find that $-25A + 9A = 80$, so that $A = -5$. Thus a particular solution is

$$x_p(t) = -5\cos 5t.$$

The complementary function is $x_c = c_1\cos 3t + c_2\sin 3t$, so the general solution of the given nonhomogeneous equation is

$$x(t) = c_1\cos 3t + c_2\sin 3t - 5\cos 5t,$$

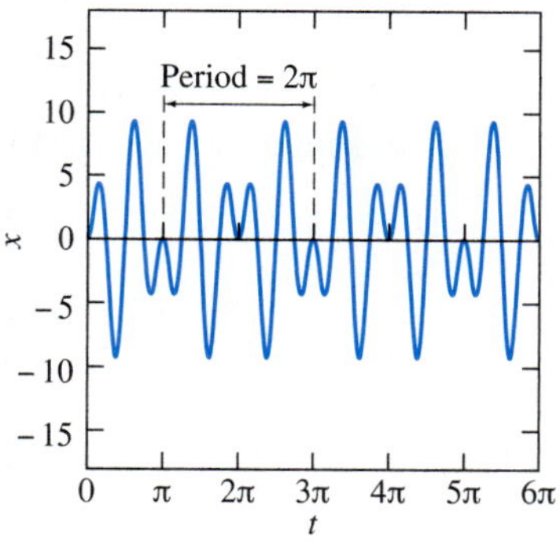

FIGURE 5.6.2. The response $x(t) = 5\cos 3t - 5\cos 5t$ in Example 1.

with derivative

$$x'(t) = -3c_1 \sin 3t + 3c_2 \cos 3t + 25 \sin 5t.$$

The initial conditions $x(0) = 0$ and $x'(0) = 0$ now yield $c_1 = 5$ and $c_2 = 0$, so the desired particular solution is

$$x(t) = 5\cos 3t - 5\cos 5t.$$

As indicated in Fig. 5.6.2, the period of $x(t)$ is the least common multiple 2π of the periods $2\pi/3$ and $2\pi/5$ of the two cosine terms. ■

Beats

If we impose the initial conditions $x(0) = x'(0) = 0$ on the solution in (7), we find that

$$c_1 = -\frac{F_0}{m(\omega_0^2 - \omega^2)} \quad \text{and} \quad c_2 = 0,$$

so the particular solution is

$$x(t) = \frac{F_0}{m(\omega_0^2 - \omega^2)}(\cos \omega t - \cos \omega_0 t). \tag{9}$$

The trigonometric identity $2 \sin A \sin B = \cos(A - B) - \cos(A + B)$, applied with $A = \frac{1}{2}(\omega_0 + \omega)t$ and $B = \frac{1}{2}(\omega_0 - \omega)t$, enables us to rewrite Eq. (9) in the form

$$x(t) = \frac{2F_0}{m(\omega_0^2 - \omega^2)} \sin \tfrac{1}{2}(\omega_0 - \omega)t \sin \tfrac{1}{2}(\omega_0 + \omega)t. \tag{10}$$

Suppose now that $\omega \approx \omega_0$, so that $\omega_0 + \omega$ is very large in comparison with $|\omega_0 - \omega|$. Then $\sin \frac{1}{2}(\omega_0 + \omega)t$ is a *rapidly* varying function, whereas $\sin \frac{1}{2}(\omega_0 - \omega)t$ is a *slowly* varying function. We may therefore interpret Eq. (10) as a rapid oscillation with circular frequency $\frac{1}{2}(\omega_0 + \omega)$,

$$x(t) = A(t) \sin \tfrac{1}{2}(\omega_0 + \omega)t,$$

but with a slowly varying amplitude

$$A(t) = \frac{2F_0}{m(\omega_0^2 - \omega^2)} \sin \tfrac{1}{2}(\omega_0 - \omega)t.$$

Example 2 With $m = 0.1$, $F_0 = 50$, $\omega_0 = 55$, and $\omega = 45$, Eq. (10) gives

$$x(t) = \sin 5t \sin 50t.$$

Figure 5.6.3 shows the corresponding oscillation of frequency $\frac{1}{2}(\omega_0 + \omega) = 50$ that is "modulated" by the amplitude function $A(t) = \sin 5t$ of frequency $\frac{1}{2}(\omega_0 - \omega) = 5$. ■

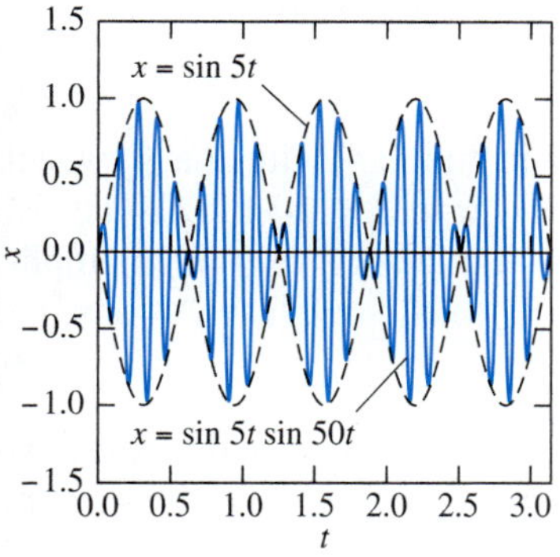

FIGURE 5.6.3. The phenomenon of beats.

A rapid oscillation with a (comparatively) slowly varying periodic amplitude exhibits the phenomenon of *beats*. For example, if two horns not exactly attuned to one another simultaneously play their middle C, one at $\omega_0/(2\pi) = 258$ Hz and the other at $\omega/(2\pi) = 254$ Hz, then one hears a beat—an audible variation in the *amplitude* of the combined sound—with a frequency of

$$\frac{(\omega_0 - \omega)/2}{2\pi} = \frac{258 - 254}{2} = 2 \text{ (Hz)}.$$

Resonance

Looking at Eq. (6), we see that the amplitude A of x_p is large when the natural and external frequencies ω_0 and ω are approximately equal. It is sometimes useful to rewrite Eq. (5) in the form

$$A = \frac{F_0}{k - m\omega^2} = \frac{F_0/k}{1-(\omega/\omega_0)^2} = \pm\frac{\rho F_0}{k}, \tag{11}$$

where F_0/k is the **static displacement** of a spring with constant k due to a *constant* force F_0, and the **amplification factor** ρ is defined to be

$$\rho = \frac{1}{|1-(\omega/\omega_0)^2|}. \tag{12}$$

It is clear that $\rho \to +\infty$ as $\omega \to \omega_0$. This is the phenomenon of **resonance**—the increase without bound (as $\omega \to \omega_0$) in the amplitude of oscillations of an undamped system with natural frequency ω_0 in response to an external force with frequency $\omega \approx \omega_0$.

We have been assuming that $\omega \neq \omega_0$. What sort of catastrophe should one expect if ω and ω_0 are precisely equal? Then Eq. (4), upon division of each term by m, becomes

$$x'' + \omega_0^2 x = \frac{F_0}{m}\cos\omega_0 t. \tag{13}$$

Because $\cos\omega_0 t$ is a term of the complementary function, the method of undetermined coefficients calls for us to try

$$x_p(t) = t(A\cos\omega_0 t + B\sin\omega_0 t).$$

We substitute this in Eq. (13) and thereby find that $A = 0$ and $B = F_0/(2m\omega_0)$. Hence the particular solution is

$$x_p(t) = \frac{F_0}{2m\omega_0} t\sin\omega_0 t. \tag{14}$$

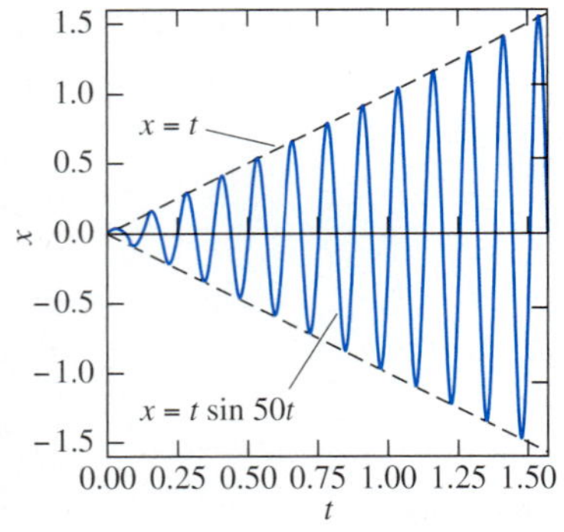

FIGURE 5.6.4. The phenomenon of resonance.

The graph of $x_p(t)$ in Fig. 5.6.4 (in which $m = 1$, $F_0 = 100$, and $\omega_0 = 50$) shows vividly how the amplitude of the oscillation theoretically would increase without bound in this case of *resonance*, $\omega = \omega_0$. We may interpret this phenomenon as reinforcement of the natural vibrations of the system by externally impressed vibrations at the same frequency.

Example 3 Suppose that $m = 5$ kg and that $k = 500$ N/m in the cart with the flywheel of Fig. 5.6.1. Then the natural frequency is $\omega_0 = \sqrt{k/m} = 10$ rad/s; that is, $10/(2\pi) \approx 1.59$ Hz. We would therefore expect oscillations of very large amplitude to occur if the flywheel revolves at about $(1.59)(60) \approx 95$ revolutions per minute (rpm). ■

In practice, a mechanical system with very little damping can be destroyed by resonance vibrations. A spectacular example can occur when a column of soldiers marches in step over a bridge. Any complicated structure such as a bridge has many natural frequencies of vibration. If the frequency of the soldiers' cadence is approximately equal to one of the natural frequencies of the structure, then—just as in our simple example of a mass on a spring—resonance will occur. Indeed, the resulting resonance vibrations can be of such large amplitude that the bridge will collapse. This has actually happened—for example, the collapse of Broughton Bridge near Manchester, England, in 1831—and it is the reason for the now-standard practice of breaking cadence when crossing a bridge. Resonance may have been involved in the 1981 Kansas City disaster in which a hotel balcony (called a *skywalk*) collapsed with dancers on it. The collapse of a building in an earthquake is sometimes due to resonance vibrations caused by the ground oscillating at one of the natural frequencies of the structure; this happened to many buildings in the Mexico City earthquake of September 19, 1985. On occasion an airplane has crashed because of resonant wing oscillations caused by vibrations of the engines. It is reported that for some of the first commercial jet aircraft, the natural frequency of the vertical vibrations of the airplane during turbulence was almost exactly that of the mass-spring system consisting of the pilot's head (mass) and spine (spring). Resonance occurred, causing pilots to have difficulty in reading the instruments. Large modern commercial jets have different natural frequencies, so that this resonance problem no longer occurs.

Modeling Mechanical Systems

The avoidance of destructive resonance vibrations is an ever-present consideration in the design of mechanical structures and systems of all types. Often the most important step in determining the natural frequency of vibration of a system is the formulation of its differential equation. In addition to Newton's law $F = ma$, the principle of conservation of energy is sometimes useful for this purpose (as in the derivation of the pendulum equation in Section 5.4). The following kinetic and potential energy formulas are often useful.

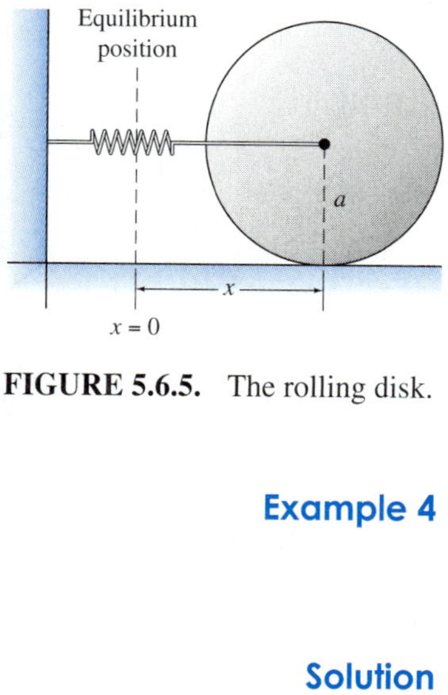

FIGURE 5.6.5. The rolling disk.

1. *Kinetic energy*: $T = \frac{1}{2}mv^2$ for translation of a mass m with velocity v;
2. *Kinetic energy*: $T = \frac{1}{2}I\omega^2$ for rotation of a body of a moment of inertia I with angular velocity ω;
3. *Potential energy*: $V = \frac{1}{2}kx^2$ for a spring with constant k stretched or compressed a distance x;
4. *Potential energy*: $V = mgh$ for the gravitational potential energy of a mass m at height h above the reference level (the level at which $V = 0$), provided that g may be regarded as essentially constant.

Example 4 Find the natural frequency of a mass m on a spring with constant k if, instead of sliding without friction, it is a uniform disk of radius a that rolls without slipping, as shown in Fig 5.6.5.

Solution With the preceding notation, the principle of conservation of energy gives

$$\frac{1}{2}mv^2 + \frac{1}{2}I\omega^2 + \frac{1}{2}kx^2 = E,$$

where E is a constant (the total mechanical energy of the system). We note that $v = a\omega$ and recall that $I = ma^2/2$ for a uniform circular disk. Then we may simplify the last equation to

$$\frac{3}{4}mv^2 + \frac{1}{2}kx^2 = E.$$

Because the right-hand side of this equation is constant, differentiation with respect to t (with $v = x'$ and $v' = x''$) now gives

$$\tfrac{3}{2}mx'x'' + kxx' = 0.$$

We divide each term by $\frac{3}{2}mx'$ to obtain

$$x'' + \frac{2k}{3m}x = 0.$$

Thus the natural frequency of horizontal back-and-forth oscillation of our rolling disk is $\sqrt{2k/3m}$, which is $\sqrt{2/3} \approx 0.8165$ times the familiar natural frequency $\sqrt{k/m}$ of a mass on a spring that is sliding without friction rather than rolling without sliding. It is interesting (and perhaps surprising) that this natural frequency does not depend on the *radius* of the disk. It could be either a dime or a large disk with radius of one meter (but of the same mass). ■

Example 5 Suppose that a car oscillates vertically as if it were a mass $m = 800$ kg on a single spring (with constant $k = 7 \times 10^4$ N/m), attached to a single dashpot (with constant $c = 3000$ N·s/m). Suppose that this car with the dashpot *disconnected* is driven along a washboard road surface with an amplitude of 5 cm and a wavelength of $L = 10$ m (Fig. 5.6.6). At what car speed will resonance vibrations occur?

Solution We think of the car as a unicycle, as pictured in Fig. 5.6.7. Let $x(t)$ denote the upward displacement of the mass m from its equilibrium position; we ignore the force of gravity, because it merely displaces the equilibrium position as in Problem 9 of Section 5.4. We write the equation of the road surface as

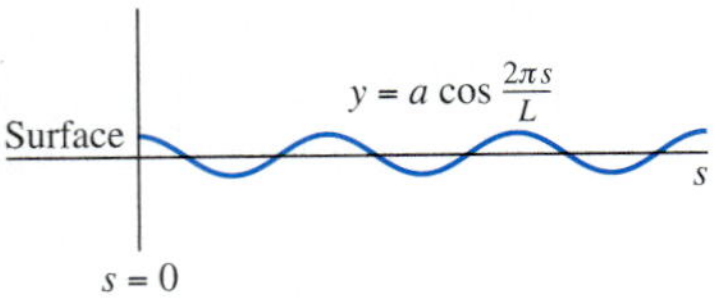

FIGURE 5.6.6. The washboard road surface of Example 5.

$$y = a\cos\frac{2\pi s}{L} \quad (a = 0.05 \text{ m}, L = 10 \text{ m}). \tag{15}$$

When the car is in motion, the spring is stretched by the amount $x - y$, so Newton's second law, $F = ma$, gives

$$mx'' = -k(x - y);$$

that is,

$$mx'' + kx = ky. \tag{16}$$

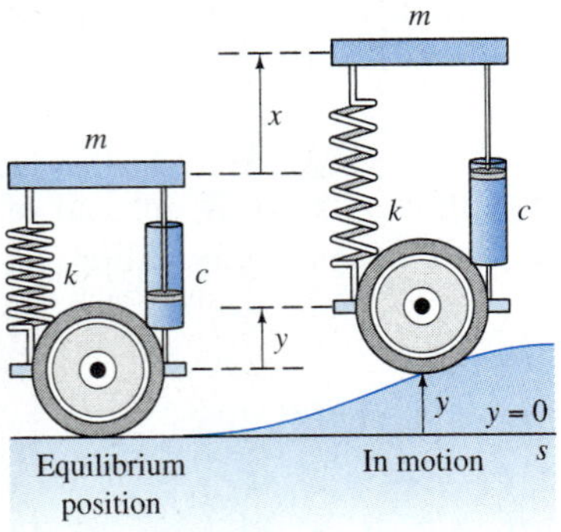

FIGURE 5.6.7. The "unicycle model" of a car.

If the velocity of the car is v, then $s = vt$ in Eq. (15), so Eq. (16) takes the form

$$mx'' + kx = ka\cos\frac{2\pi vt}{L}. \tag{16'}$$

This is the differential equation that governs the vertical oscillations of the car. In comparing it with Eq. (4), we see that we have forced oscillations with circular frequency $\omega = 2\pi v/L$. Resonance will occur when $\omega = \omega_0 = \sqrt{k/m}$. We use our numerical data to find the speed of the car at resonance:

$$v = \frac{L}{2\pi}\sqrt{\frac{k}{m}} = \frac{10}{2\pi}\sqrt{\frac{7 \times 10^4}{800}} \approx 14.89 \text{ (m/s)};$$

that is, about 33.3 mi/h (using the conversion factor of 2.237 mi/h per m/s). ■

Damped Forced Oscillations

In real physical systems there is always some damping, from frictional effects if nothing else. The complementary function x_c of the equation

$$mx'' + cx' + kx = F_0 \cos \omega t \tag{17}$$

is given by Eq. (19), (20), or (21) of Section 5.4, depending on whether $c > c_{cr} = \sqrt{4km}$, $c = c_{cr}$, or $c < c_{cr}$. The specific form is not important here. What is important is that, in any case, these formulas show that $x_c(t) \to 0$ as $t \to +\infty$. Thus x_c is a **transient solution** of Eq. (17)—one that dies out with the passage of time, leaving only the particular solution x_p.

The method of undetermined coefficients indicates that we should substitute

$$x(t) = A \cos \omega t + B \sin \omega t$$

in Eq. (17). Then collecting terms and equating coefficients of $\cos \omega t$ and $\sin \omega t$ in the result yields the two equations

$$(k - m\omega^2)A + c\omega B = F_0, \qquad -c\omega A + (k - m\omega^2)B = 0 \tag{18}$$

which we solve without difficulty for

$$A = \frac{(k - m\omega^2)F_0}{(k - m\omega^2)^2 + (c\omega)^2}, \qquad B = \frac{c\omega F_0}{(k - m\omega^2)^2 + (c\omega)^2}. \tag{19}$$

If we write

$$A \cos \omega t + B \sin \omega t = C(\cos \omega t \cos \alpha + \sin \omega t \sin \alpha) = C \cos(\omega t - \alpha)$$

as usual, we see that the resulting steady periodic oscillation

$$x_p(t) = C \cos(\omega t - \alpha) \tag{20}$$

has amplitude

$$C = \sqrt{A^2 + B^2} = \frac{F_0}{\sqrt{(k - m\omega^2)^2 + (c\omega)^2}}. \tag{21}$$

Now (19) implies that $\sin \alpha = B/C > 0$, so it follows that the phase angle α lies in the first or second quadrant. Thus

$$\tan \alpha = \frac{B}{A} = \frac{c\omega}{k - m\omega^2} \quad \text{with} \quad 0 < \alpha < \pi, \tag{22}$$

so

$$\alpha = \begin{cases} \tan^{-1} \dfrac{c\omega}{k - m\omega^2} & \text{if } k > m\omega^2, \\[2ex] \pi + \tan^{-1} \dfrac{c\omega}{k - m\omega^2} & \text{if } k < m\omega^2 \end{cases}$$

(whereas $\alpha = \pi/2$ if $k = m\omega^2$).

Note that if $c > 0$, then the "forced amplitude"—defined as a function $C(\omega)$ by (21)—always remains finite, in contrast with the case of resonance in the undamped case when the forcing frequency ω equals the critical frequency $\omega_0 =$

$\sqrt{k/m}$. But the forced amplitude may attain a maximum for some value of ω, in which case we speak of *practical resonance*. To see if and when practical resonance occurs, we need only graph C as a function of ω and look for a global maximum. It can be shown (Problem 27) that C is a steadily decreasing function of ω if $c \geqq \sqrt{2km}$. But if $c < \sqrt{2km}$, then the amplitude of C attains a maximum value—and so practical resonance occurs—at some value of ω less than ω_0, and then approaches zero as $\omega \to +\infty$. It follows that an underdamped system typically will undergo forced oscillations whose amplitude is

- Large if ω is close to the critical resonance frequency;
- Close to F_0/k if ω is very small;
- Very small if ω is very large.

Example 6 Find the transient motion and steady periodic oscillations of a damped mass-and-spring system with $m = 1$, $c = 2$, and $k = 26$ under the influence of an external force $F(t) = 82\cos 4t$ with $x(0) = 6$ and $x'(0) = 0$. Also investigate the possibility of practical resonance for this system.

Solution The resulting motion $x(t) = x_{\text{tr}}(t) + x_{\text{sp}}(t)$ of the mass satisfies the initial value problem

$$x'' + 2x' + 26x = 82\cos 4t; \qquad x(0) = 6,\ x'(0) = 0. \tag{23}$$

Instead of applying the general formulas derived earlier in this section, it is better in a concrete problem to work it directly. The roots of the characteristic equation

$$r^2 + 2r + 26 = (r+1)^2 + 25 = 0$$

are $r = -1 \pm 5i$, so the complementary function is

$$x_c(t) = e^{-t}(c_1 \cos 5t + c_2 \sin 5t).$$

When we substitute the trial solution

$$x(t) = A\cos 4t + B\sin 4t$$

in the given equation, collect like terms, and equate coefficients of $\cos 4t$ and $\sin 4t$, we get the equations

$$\begin{aligned} 10A + \ \ 8B &= 82, \\ -8A + 10B &= \ \ 0 \end{aligned}$$

with solution $A = 5$, $B = 4$. Hence the general solution of the equation in (23) is

$$x(t) = e^{-t}(c_1 \cos 5t + c_2 \sin 5t) + 5\cos 4t + 4\sin 4t.$$

At this point we impose the initial conditions $x(0) = 6$, $x'(0) = 0$ and find that $c_1 = 1$ and $c_2 = -3$. Therefore, the transient motion and the steady periodic oscillation of the mass are given by

$$x_{\text{tr}}(t) = e^{-t}(\cos 5t - 3\sin 5t)$$

and

$$x_{sp}(t) = 5\cos 4t + 4\sin 4t = \sqrt{41}\left(\frac{5}{\sqrt{41}}\cos 4t + \frac{4}{\sqrt{41}}\sin 4t\right)$$
$$= \sqrt{41}\cos(4t - \alpha),$$

where $\alpha = \tan^{-1}\left(\frac{4}{5}\right) \approx 0.6747$.

Figure 5.6.8 shows graphs of the solution $x(t) = x_{tr}(t) + x_{sp}(t)$ of the initial value problem

$$x'' + 2x' + 26x = 82\cos 4t, \quad x(0) = x_0,\ x'(0) = 0 \tag{24}$$

for the different values $x_0 = -20, -10, 0, 10,$ and 20 of the initial position. Here we see clearly what it means for the transient solution $x_{tr}(t)$ to "die out with the passage of time," leaving only the steady periodic motion $x_{sp}(t)$. Indeed, because $x_{tr}(t) \to 0$ exponentially, within a very few cycles the full solution $x(t)$ and the steady periodic solution $x_{sp}(t)$ are virtually indistinguishable (whatever the initial position x_0).

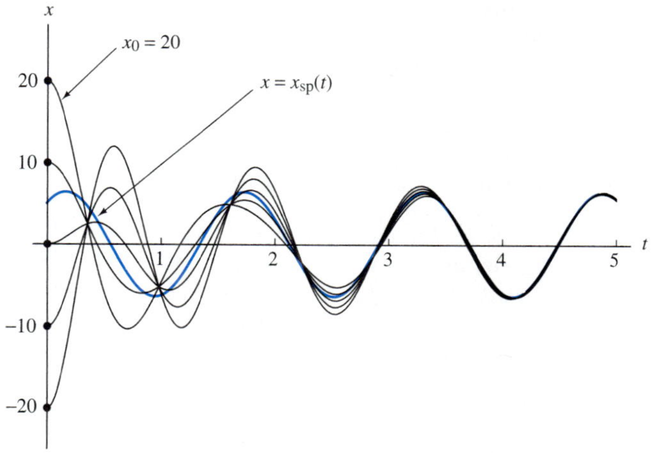

FIGURE 5.6.8. Solutions of the initial value problem in (24) with $x_0 = -20, -10, 0, 10,$ and 20.

To investigate the possibility of practical resonance in the given system, we substitute the values $m = 1$, $c = 2$, and $k = 26$ in (21) and find that the forced amplitude at frequency ω is

$$C(\omega) = \frac{82}{\sqrt{676 - 48\omega^2 + \omega^4}}.$$

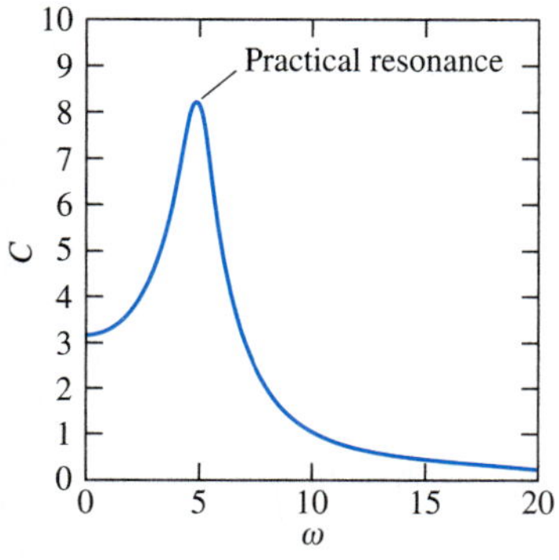

FIGURE 5.6.9. Plot of amplitude C versus external frequency ω.

The graph of $C(\omega)$ is shown in Fig. 5.6.9. The maximum amplitude occurs when

$$C'(\omega) = \frac{-41(4\omega^3 - 96\omega)}{(676 - 48\omega^2 + \omega^4)^{3/2}} = \frac{-164\omega(\omega^2 - 24)}{(676 - 48\omega^2 + \omega^4)^{3/2}} = 0.$$

Thus practical resonance occurs when the external frequency is $\omega = \sqrt{24}$ (a bit less than the mass-and-spring's undamped critical frequency of $\omega_0 = \sqrt{k/m} = \sqrt{26}$). ■

Remark: Note that the resonance frequency $\omega = \sqrt{24}$ found in Example 6 agrees with the formula

$$\omega = \sqrt{\frac{k}{m} - \frac{c^2}{2m^2}} = \sqrt{\frac{2mk - c^2}{2m^2}}$$

of Problem 27 for the resonance frequency of a system $mx'' + cx' + kx = F_0 \cos \omega t$ with $c < \sqrt{2mk}$. ■

5.6 Problems

In Problems 1 through 6, express the solution of the given initial value problem as a sum of two oscillations as in Eq. (8). Throughout, primes denote derivatives with respect to time t. In Problems 1 through 4, graph the solution function x(t) in such a way that you can identify and label (as in Fig. 5.6.2) its period.

1. $x'' + 9x = 10 \cos 2t$; $x(0) = x'(0) = 0$
2. $x'' + 4x = 5 \sin 3t$; $x(0) = x'(0) = 0$
3. $x'' + 100x = 225 \cos 5t + 300 \sin 5t$; $x(0) = 375$, $x'(0) = 0$
4. $x'' + 25x = 90 \cos 4t$; $x(0) = 0$, $x'(0) = 90$
5. $mx'' + kx = F_0 \cos \omega t$ with $\omega \neq \omega_0$; $x(0) = x_0$, $x'(0) = 0$
6. $mx'' + kx = F_0 \cos \omega t$ with $\omega = \omega_0$; $x(0) = 0$, $x'(0) = v_0$

In each of Problems 7 through 10, find the steady periodic solution $x_{sp}(t) = C\cos(\omega t - \alpha)$ of the given equation $mx'' + cx' + kx = F(t)$ with periodic forcing function $F(t)$ of frequency ω. Then graph $x_{sp}(t)$ together with (for comparison) the adjusted forcing function $F_1(t) = F(t)/m\omega$.

7. $x'' + 4x' + 4x = 10 \cos 3t$
8. $x'' + 3x' + 5x = -4 \cos 5t$
9. $2x'' + 2x' + x = 3 \sin 10t$
10. $x'' + 3x' + 3x = 8 \cos 10t + 6 \sin 10t$

In each of Problems 11 through 14, find and plot both the steady periodic solution $x_{sp}(t) = C\cos(\omega t - \alpha)$ of the given differential equation and the actual solution $x(t) = x_{sp}(t) + x_{tr}(t)$ that satisfies the given initial conditions.

11. $x'' + 4x' + 5x = 10 \cos 3t$; $x(0) = x'(0) = 0$
12. $x'' + 6x' + 13x = 10 \sin 5t$; $x(0) = x'(0) = 0$
13. $x'' + 2x' + 26x = 600 \cos 10t$; $x(0) = 10$, $x'(0) = 0$
14. $x'' + 8x' + 25x = 200 \cos t + 520 \sin t$; $x(0) = -30$, $x'(0) = -10$

Each of Problems 15 through 18 gives the parameters for a forced mass-spring-dashpot system with equation $mx'' + cx' + kx = F_0 \cos \omega t$. Investigate the possibility of practical resonance of this system. In particular, find the amplitude $C(\omega)$ of steady periodic forced oscillations with frequency ω. Sketch the graph of $C(\omega)$ and find the practical resonance frequency ω (if any).

15. $m = 1$, $c = 2$, $k = 2$, $F_0 = 2$
16. $m = 1$, $c = 4$, $k = 5$, $F_0 = 10$
17. $m = 1$, $c = 6$, $k = 45$, $F_0 = 50$
18. $m = 1$, $c = 10$, $k = 650$, $F_0 = 100$

19. A mass weighing 100 lb (mass $m = 3.125$ slugs in fps units) is attached to the end of a spring that is stretched 1 in. by a force of 100 lb. A force $F_0 \cos \omega t$ acts on the mass. At what frequency (in hertz) will resonance oscillations occur? Neglect damping.

20. A front-loading washing machine is mounted on a thick rubber pad that acts like a spring; the weight $W = mg$ (with $g = 9.8$ m/s^2) of the machine depresses the pad exactly 0.5 cm. When its rotor spins at ω radians per second, the rotor exerts a vertical force $F_0 \cos \omega t$ newtons on the machine. At what speed (in revolutions per minute) will resonance vibrations occur? Neglect friction.

21. Figure 5.6.10 shows a mass m on the end of a pendulum (of length L) also attached to a horizontal spring (with constant k). Assume small oscillations of m so that the spring remains essentially horizontal and neglect damping. Find the natural circular frequency ω_0 of motion of the mass in terms of L, k, m, and the gravitational constant g.

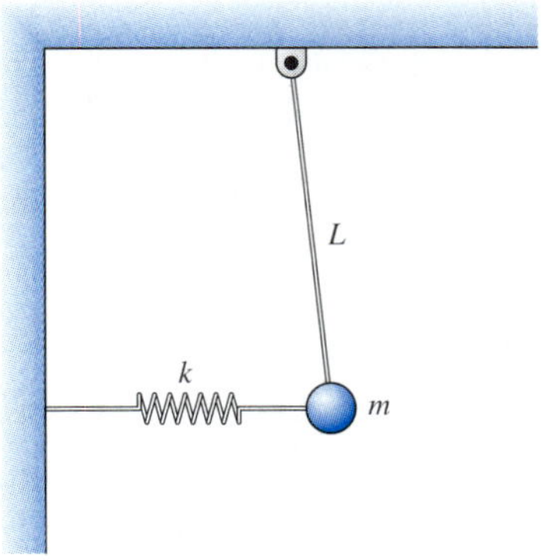

FIGURE 5.6.10. The pendulum-and-spring system of Problem 21.

22. A mass m hangs on the end of a cord around a pulley of radius a and moment of inertia I, as shown in Fig. 5.6.11. The rim of the pulley is attached to a spring (with constant k). Assume small oscillations so that the spring remains essentially horizontal and neglect friction. Find the natural circular frequency of the system in terms of m, a, k, I, and g.

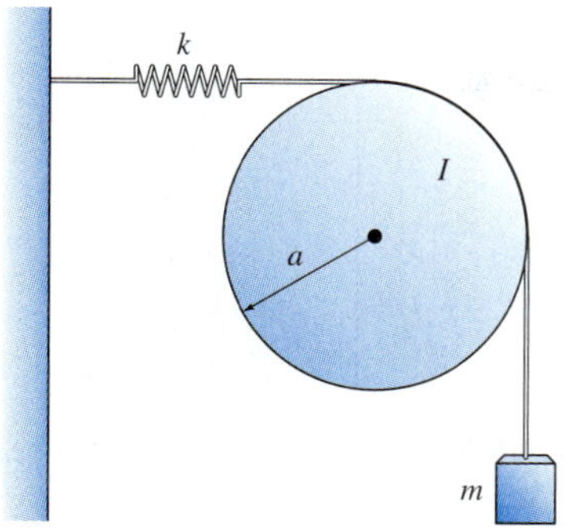

FIGURE 5.6.11. The mass-spring-pulley system of Problem 22.

23. A building consists of two floors. The first floor is attached rigidly to the ground, and the second floor is of mass $m = 1000$ slugs (fps units) and weighs 16 tons (32,000 lb). The elastic frame of the building behaves as a spring that resists horizontal displacements of the second floor; it requires a horizontal force of 5 tons to displace the second floor a distance of 1 ft. Assume that in an earthquake the ground oscillates horizontally with amplitude A_0 and circular frequency ω, resulting in an external horizontal force $F(t) = mA_0\omega^2 \sin \omega t$ on the second floor. **(a)** What is the natural frequency (in hertz) of oscillations of the second floor? **(b)** If the ground undergoes one oscillation every 2.25 s with an amplitude of 3 in., what is the amplitude of the resulting forced oscillations of the second floor?

24. A mass on a spring without damping is acted on by the external force $F(t) = F_0 \cos^3 \omega t$. Show that there are *two* values of ω for which resonance occurs, and find both.

25. Derive the steady periodic solution of

$$mx'' + cx' + kx = F_0 \sin \omega t.$$

In particular, show that it is what one would expect—the same as the formula in (20) with the same values of C and ω, except with $\sin(\omega t - \alpha)$ in place of $\cos(\omega t - \alpha)$.

26. Given the differential equation

$$mx'' + cx' + kx = E_0 \cos \omega t + F_0 \sin \omega t$$

—with both cosine and sine forcing terms—derive the steady periodic solution

$$x_{sp}(t) = \frac{\sqrt{E_0^2 + F_0^2}}{\sqrt{(k - m\omega^2)^2 + (c\omega)^2}} \cos(\omega t - \alpha - \beta),$$

where α is defined in Eq. (22) and $\beta = \tan^{-1}(F_0/E_0)$. [*Suggestion*: Add the steady periodic solutions separately corresponding to $E_0 \cos \omega t$ and $F_0 \sin \omega t$ (see Problem 25).]

27. According to Eq. (21), the amplitude of forced steady periodic oscillations for the system $mx'' + cx' + kx = F_0 \cos \omega t$ is given by

$$C(\omega) = \frac{F_0}{\sqrt{(k - m\omega^2)^2 + (c\omega)^2}}.$$

(a) If $c \geqq c_{cr}/\sqrt{2}$, where $c_{cr} = \sqrt{4km}$, show that C steadily decreases as ω increases. **(b)** If $c < c_{cr}/\sqrt{2}$, show that C attains a maximum value (practical resonance) when

$$\omega = \omega_m = \sqrt{\frac{k}{m} - \frac{c^2}{2m^2}} < \omega_0 = \sqrt{\frac{k}{m}}.$$

28. As indicated by the cart-with-flywheel example discussed in this section, an unbalanced rotating machine part typically results in a force having amplitude proportional to the *square* of the frequency ω. **(a)** Show that the amplitude of the steady periodic solution of the differential equation

$$mx'' + cx' + kx = mA\omega^2 \cos \omega t$$

[with a forcing term similar to that in Eq. (17)] is given by

$$C(\omega) = \frac{mA\omega^2}{\sqrt{(k - m\omega^2)^2 + (c\omega)^2}}.$$

(b) Suppose that $c^2 < 2mk$. Show that the maximum amplitude occurs at the frequency ω_m given by

$$\omega_m = \sqrt{\frac{k}{m}\left(\frac{2mk}{2mk - c^2}\right)}.$$

Thus the resonance frequency in this case is *larger* (in contrast with the result of Problem 27) than the natural frequency $\omega_0 = \sqrt{k/m}$. (*Suggestion*: Maximize the *square* of C.)

Automobile Vibrations

Problems 29 and 30 deal further with the car of Example 5. Its upward displacement function satisfies the equation $mx'' + cx' + kx = cy' + ky$ when the shock absorber is connected (so that $c > 0$). With $y = a \sin \omega t$ for the road surface, this differential equation becomes

$$mx'' + cx' + kx = E_0 \cos \omega t + F_0 \sin \omega t,$$

where $E_0 = c\omega a$ and $F_0 = ka$.

29. Apply the result of Problem 26 to show that the amplitude C of the resulting steady periodic oscillation for the car is given by

$$C = \frac{a\sqrt{k^2 + (c\omega)^2}}{\sqrt{(k - m\omega^2)^2 + (c\omega)^2}}.$$

Because $\omega = 2\pi v/L$ when the car is moving with velocity v, this gives C as a function of v.

30. Figure 5.6.12 shows the graph of the amplitude function $C(\omega)$ using the numerical data given in Example 5 (including $c = 3000$ N·s/m). It indicates that, as the car accelerates gradually from rest, it initially oscillates with amplitude slightly over 5 cm. Maximum resonance vibrations with amplitude about 14 cm occur around 32 mi/h, but then subside to more tolerable levels at high speeds. Verify these graphically based conclusions by analyzing the function $C(\omega)$. In particular, find the practical resonance frequency and the corresponding amplitude.

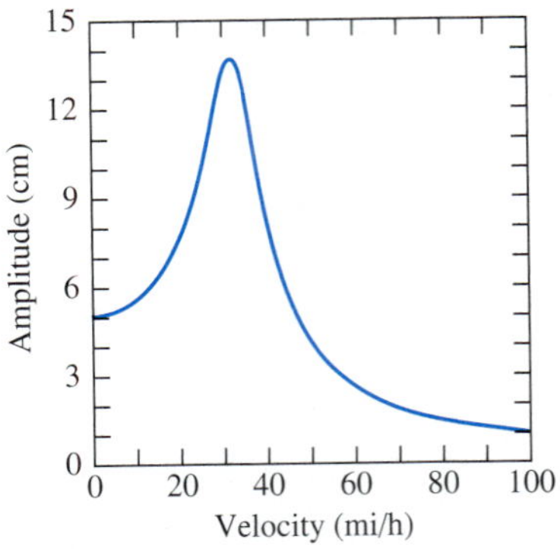

FIGURE 5.6.12. Amplitude of vibrations of the car on a washboard surface.

5.6 Application Forced Vibrations and Resonance

Here we investigate forced vibrations of the mass-spring-dashpot system with equation

$$mx'' + cx' + kx = F(t). \tag{1}$$

To simplify the notation, let's take $m = p^2$, $c = 2p$, and $k = p^2q^2 + 1$, where $p > 0$ and $q > 0$. Then the complementary function of Eq. (1) is

$$x_c(t) = e^{-t/p}(c_1 \cos qt + c_2 \sin qt). \tag{2}$$

We will take $p = 5$, $q = 3$, and thus investigate the transient and steady periodic solutions corresponding to

$$25x'' + 10x' + 226x = F(t), \quad x(0) = 0, \quad x'(0) = 0 \tag{3}$$

with several illustrative possibilities for the external force $F(t)$. For your personal investigations to carry out similarly, you might select integers p and q with $6 \leqq p \leqq 9$ and $2 \leqq q \leqq 5$.

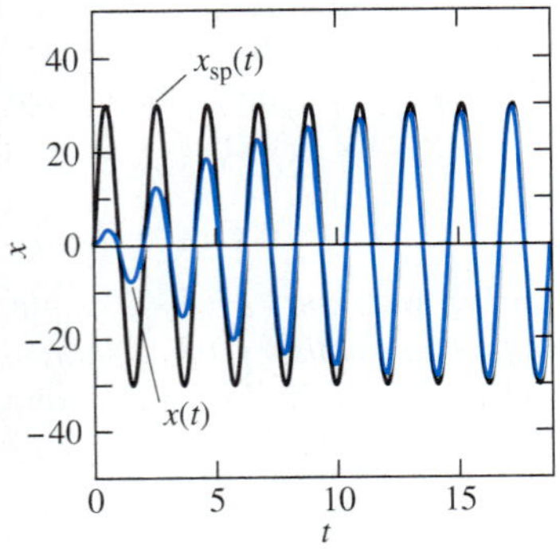

FIGURE 5.6.13. The solution $x(t) = x_{tr}(t) + x_{sp}(t)$ and the steady periodic solution $x(t) = x_{sp}(t)$ with periodic external force $F(t) = 901 \cos 3t$.

INVESTIGATION 1: With periodic external force $F(t) = 901 \cos 3t$, the MATLAB commands

```
x = dsolve('25*D2x+10*Dx+226*x=901*cos(3*t)',
      'x(0)=0, Dx(0)=0');
x = simple(x);
syms t, xsp = cos(3*t) + 30*sin(3*t);
ezplot(x, [0  6*pi]),hold on
ezplot(xsp, [0  6*pi])
```

produce the plot shown in Fig. 5.6.13. We see the (transient plus steady periodic) solution

$$x(t) = \cos 3t + 30 \sin 3t + e^{-t/5}\left(-\cos 3t - \tfrac{451}{15} \sin 3t\right)$$

rapidly "building up" to the steady periodic oscillation $x_{sp}(t) = \cos 3t + 30 \sin 3t$.

INVESTIGATION 2: With damped oscillatory external force

$$F(t) = 900e^{-t/5}\cos 3t,$$

we have duplication with the complementary function in (2). The *Maple* commands

```
de2 := 25*diff(x(t),t,t)+10*diff(x(t),t)+226*x(t) =
       900*exp(-t/5)*cos(3*t);
dsolve({de2,x(0)=0,D(x)(0)=0}, x(t));
x := simplify(combine(rhs(%),trig));
C := 6*t*exp(-t/5);
plot({x,C,-C},t=0..8*Pi);
```

produce the plot shown in Fig. 5.6.14. We see the solution

$$x(t) = 6te^{-t/5}\sin 3t$$

oscillating up-and-down between the envelope curves $x = \pm 6te^{-t/5}$. (Note the factor of t that signals a resonance situation.)

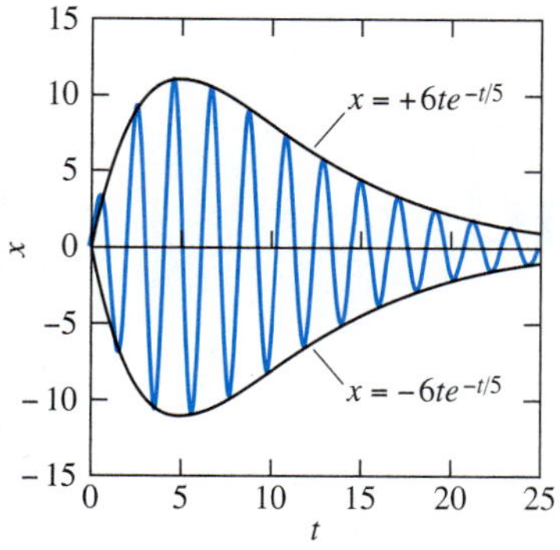

FIGURE 5.6.14. The solution $x(t) = 6te^{-t/5}\sin 3t$ and the envelope curves $x(t) = \pm 6te^{-t/5}$ with damped oscillatory force $F(t) = 900e^{-t/5}\cos 3t$.

INVESTIGATION 3: With damped oscillatory external force

$$F(t) = 2700te^{-t/5}\cos 3t,$$

we have a still more complicated resonance situation. The *Mathematica* commands

```
de3 = 25 x''[t] + 10 x'[t] + 226 x[t] ==
          2700 t Exp[-t/5] Cos[3t]
soln = DSolve[{de3, x[0] == 0, x'[0] == 0}, x[t], t]
x = First[x[t]  /.  soln]
amp = Exp[-t/5] Sqrt[(3t)^2 + (9t^2 - 1)^2]
Plot[{x, amp, -amp}, {t, 0, 10 Pi}];
```

produce the plot shown in Fig. 5.6.15. We see the solution

$$x(t) = e^{-t/5}\left[3t\cos 3t + (9t^2 - 1)\sin 3t\right]$$

oscillating up-and-down between the envelope curves

$$x = \pm e^{-t/5}\sqrt{(3t)^2 + (9t^2 - 1)^2}.$$

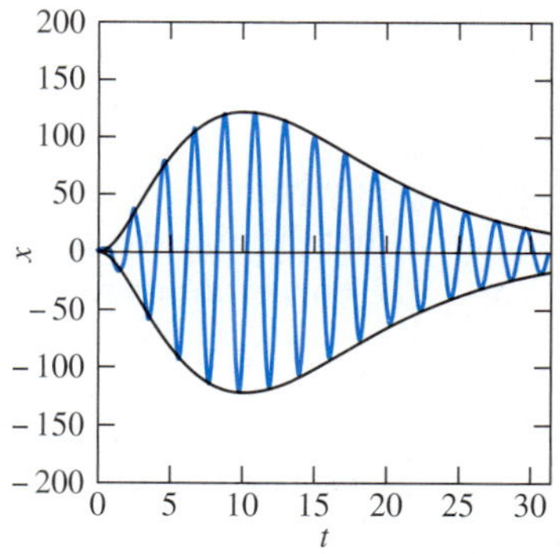

FIGURE 5.6.15. The solution $x(t) = e^{-t/5}[3t\cos 3t + (9t^2 - 1)\sin 3t]$ and the envelope curves $x(t) = \pm e^{-t/5}\sqrt{(3t)^2 + (9t^2 - 1)^2}$ with external force $F(t) = 2700te^{-t/5}\cos 3t$.

6 Eigenvalues and Eigenvectors

6.1 Introduction to Eigenvalues

Given a square matrix $\mathbf{A}$, let us pose the following question: Does there exist a nonzero vector $\mathbf{v}$ such that the result $\mathbf{Av}$ of multiplying $\mathbf{v}$ by the matrix $\mathbf{A}$ is simply a scalar multiple of $\mathbf{v}$? Thus we ask whether or not there exist a nonzero vector $\mathbf{v}$ *and* a scalar λ such that

$$\mathbf{Av} = \lambda\mathbf{v}. \tag{1}$$

Section 6.3 and subsequent chapters include interesting applications in which this question arises. The following definition provides the terminology that we use in discussing Eq. (1).

DEFINITION Eigenvalues and Eigenvectors

The number λ is said to be an **eigenvalue** of the $n \times n$ matrix $\mathbf{A}$ provided there exists a *nonzero* vector $\mathbf{v}$ such that

$$\mathbf{Av} = \lambda\mathbf{v}, \tag{1}$$

in which case the vector $\mathbf{v}$ is called an **eigenvector** of the matrix $\mathbf{A}$. We also say that the eigenvector $\mathbf{v}$ is **associated** with the eigenvalue λ, or that the eigenvalue λ **corresponds** to the eigenvector $\mathbf{v}$.

The prefix *eigen* is a German word that (in some contexts) may be translated as *proper*, so eigenvalues are sometimes called *proper values*. Eigenvalues and eigenvectors are also called *characteristic values* and *characteristic vectors* in some books.

Example 1 Consider the 2×2 matrix

$$\mathbf{A} = \begin{bmatrix} 5 & -6 \\ 2 & -2 \end{bmatrix}.$$

If $\mathbf{v} = (2, 1) = \begin{bmatrix} 2 & 1 \end{bmatrix}^T$, then

$$\mathbf{Av} = \begin{bmatrix} 5 & -6 \\ 2 & -2 \end{bmatrix} \begin{bmatrix} 2 \\ 1 \end{bmatrix} = \begin{bmatrix} 4 \\ 2 \end{bmatrix} = 2\mathbf{v}.$$

Thus, $\mathbf{v} = \begin{bmatrix} 2 & 1 \end{bmatrix}^T$ is an eigenvector of $\mathbf{A}$ that is associated with the eigenvalue $\lambda = 2$. If $\mathbf{v} = \begin{bmatrix} 3 & 2 \end{bmatrix}^T$, then

$$\mathbf{Av} = \begin{bmatrix} 5 & -6 \\ 2 & -2 \end{bmatrix} \begin{bmatrix} 3 \\ 2 \end{bmatrix} = \begin{bmatrix} 3 \\ 2 \end{bmatrix} = 1\mathbf{v},$$

so $\mathbf{v} = \begin{bmatrix} 3 & 2 \end{bmatrix}^T$ is an eigenvector of $\mathbf{A}$ associated with the eigenvalue $\lambda = 1$. In summary, we see that the scalars $\lambda_1 = 2$ and $\lambda_2 = 1$ are both eigenvalues of the matrix $\mathbf{A}$; they correspond to the eigenvectors $\mathbf{v}_1 = \begin{bmatrix} 2 & 1 \end{bmatrix}^T$ and $\mathbf{v}_2 = \begin{bmatrix} 3 & 2 \end{bmatrix}^T$, respectively. ■

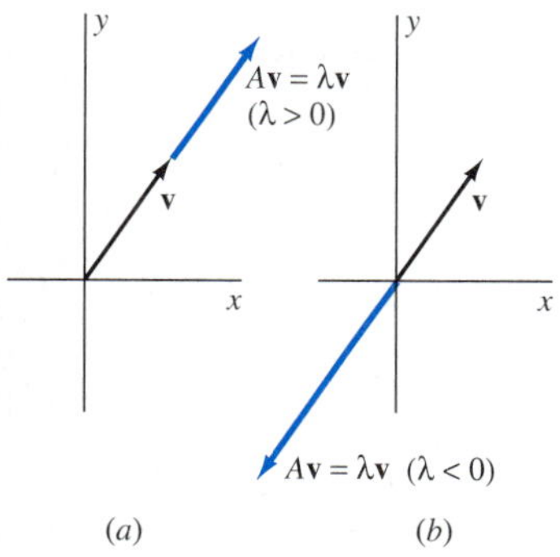

FIGURE 6.1.1. (a) A positive eigenvalue; (b) a negative eigenvalue.

Eigenvalues and eigenvectors have a simple geometric interpretation. Suppose that λ is an eigenvalue of the matrix $\mathbf{A}$ and has the associated eigenvector $\mathbf{v}$, so that $\mathbf{Av} = \lambda\mathbf{v}$. Then the length $|\mathbf{Av}|$ of the vector $\mathbf{Av}$ is $\pm\lambda|\mathbf{v}|$, depending on the sign of λ. Thus, if $\lambda > 0$, then multiplication of $\mathbf{v}$ by the matrix $\mathbf{A}$ expands or contracts the vector $\mathbf{v}$ while preserving its direction; if $\lambda < 0$, then multiplication of $\mathbf{v}$ by $\mathbf{A}$ reverses the direction of $\mathbf{v}$ (see Fig. 6.1.1).

Remark 1 If $\mathbf{v} = \mathbf{0}$, then the equation $\mathbf{Av} = \lambda\mathbf{v}$ holds for every scalar λ and hence is of no significance. This is why only *nonzero* vectors qualify as eigenvectors in the definition.

Remark 2 Let λ and $\mathbf{v}$ be an eigenvalue and associated eigenvector of the matrix $\mathbf{A}$. If k is any nonzero scalar and $\mathbf{u} = k\mathbf{v}$, then

$$\begin{aligned} \mathbf{Au} &= \mathbf{A}(k\mathbf{v}) = k(\mathbf{Av}) \\ &= k(\lambda\mathbf{v}) = \lambda(k\mathbf{v}) = \lambda\mathbf{u}, \end{aligned}$$

so $\mathbf{u} = k\mathbf{v}$ is also an eigenvector associated with λ. Thus, *any nonzero scalar multiple of an eigenvector is also an eigenvector and is associated with the same eigenvalue*. In Example 1, for instance, $\mathbf{u}_1 = -3\mathbf{v}_1 = \begin{bmatrix} -6 & -3 \end{bmatrix}^T$ is an eigenvector associated with $\lambda_1 = 2$, and $\mathbf{u}_2 = 4\mathbf{v}_2 = \begin{bmatrix} 12 & 8 \end{bmatrix}^T$ is an eigenvector associated with $\lambda_2 = 1$. ■

The Characteristic Equation

We now attack the problem of finding the eigenvalues and eigenvectors of a given $n \times n$ square matrix $\mathbf{A}$. According to the definition, the nonzero vector $\mathbf{v}$ is an eigenvector of $\mathbf{A}$ associated with the eigenvalue λ exactly when

$$\mathbf{Av} = \lambda\mathbf{v} = \lambda\mathbf{Iv};$$

that is, when

$$(\mathbf{A} - \lambda\mathbf{I})\mathbf{v} = \mathbf{0}. \tag{2}$$

For a fixed value of λ, Eq. (2) is a homogeneous linear system of n equations in the n components of $\mathbf{v}$. By Theorem 2 in Section 3.6 and Theorem 7 in Section 3.5, this system has a nontrivial solution $\mathbf{v} \neq \mathbf{0}$ if and only if the determinant

$$\det(\mathbf{A} - \lambda\mathbf{I}) = |\mathbf{A} - \lambda\mathbf{I}|$$

of its coefficient matrix is *zero*. The equation $|\mathbf{A} - \lambda\mathbf{I}| = 0$ is called the **characteristic equation** of the square matrix $\mathbf{A}$, and we have proved that there exists an eigenvector $\mathbf{v}$ associated with λ if and only if λ satisfies this equation.

THEOREM 1 The Characteristic Equation

The number λ is an eigenvalue of the $n \times n$ matrix $\mathbf{A}$ if and only if λ satisfies the characteristic equation

$$|\mathbf{A} - \lambda\mathbf{I}| = 0. \tag{3}$$

Now let us see just what sort of equation the characteristic equation in (3) is. Note that the matrix

$$\mathbf{A} - \lambda\mathbf{I} = \begin{bmatrix} a_{11} - \lambda & a_{12} & \cdots & a_{1n} \\ a_{21} & a_{22} - \lambda & \cdots & a_{2n} \\ \vdots & \vdots & \ddots & \vdots \\ a_{n1} & a_{n2} & \cdots & a_{nn} - \lambda \end{bmatrix} \tag{4}$$

is obtained simply by subtracting λ from each diagonal element of $\mathbf{A}$. If we think of expanding the determinant by cofactors, we see that $|\mathbf{A} - \lambda\mathbf{I}|$ is a *polynomial* in the variable λ, and that the highest power of λ comes from the product of the diagonal elements of the matrix in (4). Therefore, the characteristic equation of the $n \times n$ matrix $\mathbf{A}$ takes the form

$$(-1)^n\lambda^n + c_{n-1}\lambda^{n-1} + \cdots + c_1\lambda + c_0 = 0, \tag{5}$$

an nth-degree polynomial equation in λ.

According to the fundamental theorem of algebra, every nth-degree polynomial equation in one variable has n solutions (counting multiple solutions), but some of them can be complex. Hence, we can say that *an $n \times n$ matrix $\mathbf{A}$ always has n eigenvalues*, though they might not be distinct and might not all be real. In this chapter we confine our attention mainly to real eigenvalues, but in Chapter 7 we will see important applications of complex eigenvalues to the solution of linear systems of differential equations.

ALGORITHM Eigenvalues and Eigenvectors

To find the eigenvalues and associated eigenvectors of the $n \times n$ matrix $\mathbf{A}$:

1. First solve the characteristic equation

$$|\mathbf{A} - \lambda\mathbf{I}| = 0.$$

2. Then, for *each* eigenvalue λ thereby found, solve the linear system

$$(\mathbf{A} - \lambda\mathbf{I})\mathbf{v} = \mathbf{0}$$

to find the eigenvectors associated with λ.

Solving the characteristic equation is almost always easier said than done. In the examples that follow and in the problems, we have chosen matrices for which the characteristic polynomial $|\mathbf{A} - \lambda\mathbf{I}|$ readily factors to reveal the eigenvalues.

Example 2 Find the eigenvalues and associated eigenvectors of the matrix

$$\mathbf{A} = \begin{bmatrix} 5 & 7 \\ -2 & -4 \end{bmatrix}.$$

Solution Here

$$\mathbf{A} - \lambda\mathbf{I} = \begin{bmatrix} 5-\lambda & 7 \\ -2 & -4-\lambda \end{bmatrix}, \tag{6}$$

so the characteristic equation of $\mathbf{A}$ is

$$\begin{aligned} 0 &= |\mathbf{A} - \lambda\mathbf{I}| \\ &= \begin{vmatrix} 5-\lambda & 7 \\ -2 & -4-\lambda \end{vmatrix} \\ &= (5-\lambda)(-4-\lambda) - (-2)(7) \\ &= (\lambda-5)(\lambda+4) + 14 = \lambda^2 - \lambda - 6; \end{aligned}$$

that is, $(\lambda + 2)(\lambda - 3) = 0$. Thus, the matrix $\mathbf{A}$ has the two eigenvalues -2 and 3. To distinguish them, we write $\lambda_1 = -2$ and $\lambda_2 = 3$. To find the associated eigenvectors, we must *separately* substitute each eigenvalue in (6) and then solve the resulting system $(\mathbf{A} - \lambda\mathbf{I})\mathbf{v} = \mathbf{0}$.

CASE 1: $\lambda_1 = -2$. With $\mathbf{v} = \begin{bmatrix} x & y \end{bmatrix}^T$, the system $(\mathbf{A} - \lambda\mathbf{I})\mathbf{v} = \mathbf{0}$ is

$$\begin{bmatrix} 7 & 7 \\ -2 & -2 \end{bmatrix}\begin{bmatrix} x \\ y \end{bmatrix} = \begin{bmatrix} 0 \\ 0 \end{bmatrix}.$$

Each of the two scalar equations here is a multiple of the equation $x + y = 0$, and any nontrivial solution $\mathbf{v} = \begin{bmatrix} x & y \end{bmatrix}^T$ of this equation is a nonzero multiple of $\begin{bmatrix} 1 & -1 \end{bmatrix}^T$. Hence, to within a constant multiple, the only eigenvector associated with $\lambda_1 = -2$ is $\mathbf{v}_1 = \begin{bmatrix} 1 & -1 \end{bmatrix}^T$.

CASE 2: $\lambda_2 = 3$. With $\mathbf{v} = \begin{bmatrix} x & y \end{bmatrix}^T$, the system $(\mathbf{A} - \lambda\mathbf{I})\mathbf{v} = \mathbf{0}$ is

$$\begin{bmatrix} 2 & 7 \\ -2 & -7 \end{bmatrix}\begin{bmatrix} x \\ y \end{bmatrix} = \begin{bmatrix} 0 \\ 0 \end{bmatrix}.$$

Again, we have only a single equation, $2x + 7y = 0$, and any nontrivial solution of this equation will suffice. The choice $y = -2$ yields $x = 7$, so (to within a constant multiple) the only eigenvector associated with $\lambda_2 = 3$ is $\mathbf{v}_2 = \begin{bmatrix} 7 & -2 \end{bmatrix}^T$.

Finally, note that it is not enough to say simply that the given matrix **A** has eigenvalues -2 and 3 and has eigenvectors $\begin{bmatrix} 1 & -1 \end{bmatrix}^T$ and $\begin{bmatrix} 7 & -2 \end{bmatrix}^T$. To give complete information, we must say which eigenvector is associated with each eigenvalue. ■

If the elements of the matrix **A** are all real, then so are the coefficients of its characteristic equation $|\mathbf{A} - \lambda\mathbf{I}| = 0$. In this event, it follows that the complex eigenvalues (if any) of **A** occur only in conjugate pairs. The next example illustrates the possibility of imaginary or complex eigenvalues and eigenvectors.

Example 3 The characteristic equation of the matrix

$$\mathbf{A} = \begin{bmatrix} 0 & 8 \\ -2 & 0 \end{bmatrix}$$

is

$$|\mathbf{A} - \lambda\mathbf{I}| = \begin{bmatrix} -\lambda & 8 \\ -2 & -\lambda \end{bmatrix} = \lambda^2 + 16 = 0.$$

Hence the matrix **A** has the complex conjugate eigenvalues $\lambda = \pm 4i$.

CASE 1: $\lambda_1 = -4i$. With $\mathbf{v} = \begin{bmatrix} x & y \end{bmatrix}^T$, the system $(\mathbf{A} - \lambda\mathbf{I})\mathbf{v} = \mathbf{0}$ is

$$\begin{bmatrix} 4i & 8 \\ -2 & 4i \end{bmatrix}\begin{bmatrix} x \\ y \end{bmatrix} = \begin{bmatrix} 0 \\ 0 \end{bmatrix}.$$

The first equation here, $4ix + 8y = 0$, is $-2i$ times the second one and, obviously, is satisfied by $x = 2i$, $y = 1$. Thus $\mathbf{v}_1 = \begin{bmatrix} 2i & 1 \end{bmatrix}^T$ is a complex eigenvector associated with the complex eigenvalue $\lambda_1 = -4i$.

CASE 2: $\lambda_2 = +4i$. In this case, you may verify similarly that the conjugate $\mathbf{v}_2 = \begin{bmatrix} -2i & 1 \end{bmatrix}^T$ of $\mathbf{v}_1$ is an eigenvector associated with the conjugate λ_2 of the eigenvalue λ_1. Thus the complex conjugate eigenvalues $\mp 4i$ of the matrix **A** correspond to the complex conjugate eigenvectors $\begin{bmatrix} \pm 2i & 1 \end{bmatrix}^T$ (taking either both upper signs or both lower signs). ■

Example 4 The 2×2 identity matrix **I** has characteristic equation

$$\begin{vmatrix} 1-\lambda & 0 \\ 0 & 1-\lambda \end{vmatrix} = (1-\lambda)^2 = 0,$$

so **I** has the single eigenvalue $\lambda = 1$. The equation $(\mathbf{I} - 1\lambda)\mathbf{v} = \mathbf{0}$ is

$$\begin{bmatrix} 0 & 0 \\ 0 & 0 \end{bmatrix}\begin{bmatrix} x \\ y \end{bmatrix} = \begin{bmatrix} 0 \\ 0 \end{bmatrix},$$

so *every* nonzero vector $\mathbf{v} = \begin{bmatrix} x & y \end{bmatrix}^T$ is an eigenvector of $\mathbf{I}$. In particular, the single eigenvalue $\lambda = 1$ corresponds to the *two* linearly independent eigenvectors $\mathbf{v}_1 = \begin{bmatrix} 1 & 0 \end{bmatrix}^T$ and $\mathbf{v}_2 = \begin{bmatrix} 0 & 1 \end{bmatrix}^T$. ■

Example 5 The characteristic equation of the matrix

$$\mathbf{A} = \begin{bmatrix} 2 & 3 \\ 0 & 2 \end{bmatrix}$$

is $(2 - \lambda)^2 = 0$, so $\mathbf{A}$ has the single eigenvalue $\lambda = 2$. The equation $(\mathbf{A} - 2\mathbf{I})\mathbf{v} = \mathbf{0}$ is

$$\begin{bmatrix} 0 & 3 \\ 0 & 0 \end{bmatrix} \begin{bmatrix} x \\ y \end{bmatrix} = \begin{bmatrix} 0 \\ 0 \end{bmatrix}.$$

Thus x is arbitrary, but $y = 0$, so the eigenvalue $\lambda = 2$ corresponds (to within a constant multiple) to the single eigenvector $\mathbf{v} = \begin{bmatrix} 1 & 0 \end{bmatrix}^T$. ■

Examples 2–5 illustrate the four possibilities for a 2×2 matrix $\mathbf{A}$. It can have either

- two distinct real eigenvalues, each corresponding to a single eigenvector;
- a single real eigenvalue corresponding to a single eigenvector;
- a single real eigenvalue corresponding to two linearly independent eigenvectors; or
- two complex conjugate eigenvalues corresponding to complex conjugate eigenvectors.

The characteristic equation of a 3×3 matrix is, by Eq. (5), of the form

$$-\lambda^3 + c_2\lambda^2 + c_1\lambda + c_0 = 0. \tag{7}$$

Now the value of the continuous function $-\lambda^3 + \cdots + c_0$ on the left-hand side here approaches $+\infty$ as $\lambda \to -\infty$ and $-\infty$ as $\lambda \to +\infty$. Hence every such cubic equation has at least one real solution, so every 3×3 matrix has (in contrast with 2×2 matrices) at least one real eigenvalue. A 3×3 matrix may have one, two, or three distinct eigenvalues, and a single eigenvalue of a 3×3 matrix may correspond to one, two, or three linearly independent eigenvectors. The remaining two examples of this section illustrate some of the possibilities. The next example shows also that, whereas the zero vector $\mathbf{0}$ cannot be an eigenvector of a matrix, there is nothing to prevent $\lambda = 0$ from being an eigenvalue.

Example 6 Find the eigenvalues and associated eigenvectors of the matrix

$$\mathbf{A} = \begin{bmatrix} 3 & 0 & 0 \\ -4 & 6 & 2 \\ 16 & -15 & -5 \end{bmatrix}.$$

Solution The matrix $\mathbf{A} - \lambda\mathbf{I}$ is

$$\mathbf{A} - \lambda\mathbf{I} = \begin{bmatrix} 3 - \lambda & 0 & 0 \\ -4 & 6 - \lambda & 2 \\ 16 & -15 & -5 - \lambda \end{bmatrix}. \tag{8}$$

Upon expansion of the determinant along its first row, we find that

$$\begin{aligned}|\mathbf{A} - \lambda\mathbf{I}| &= (3-\lambda)[(6-\lambda)(-5-\lambda)+30] \\ &= (3-\lambda)(\lambda^2-\lambda) = \lambda(\lambda-1)(3-\lambda).\end{aligned}$$

Hence the characteristic equation $|\mathbf{A} - \lambda\mathbf{I}| = 0$ yields the three eigenvalues $\lambda_1 = 0$, $\lambda_2 = 1$, and $\lambda_3 = 3$. To find the associated eigenvectors, we must solve the system $(\mathbf{A} - \lambda\mathbf{I})\mathbf{v} = \mathbf{0}$ separately for each of these three eigenvalues.

CASE 1: $\lambda_1 = 0$. We write $\mathbf{v} = \begin{bmatrix} x & y & z \end{bmatrix}^T$ and substitute $\lambda = 0$ in the coefficient matrix in (8) to obtain the system

$$\begin{bmatrix} 3 & 0 & 0 \\ -4 & 6 & 2 \\ 16 & -15 & -5 \end{bmatrix}\begin{bmatrix} x \\ y \\ z \end{bmatrix} = \begin{bmatrix} 0 \\ 0 \\ 0 \end{bmatrix}.$$

From the first of the three equations here, $3x = 0$, we see that $x = 0$. Then each of the remaining two equations is a multiple of the equation $3y + z = 0$. The choice $y = 1$ yields $z = -3$. Thus the eigenvector $\mathbf{v}_1 = \begin{bmatrix} 0 & 1 & -3 \end{bmatrix}^T$ is associated with $\lambda_1 = 0$.

CASE 2: $\lambda_2 = 1$. Substitution of $\lambda = 1$ in the coefficient matrix in (8) yields the system

$$\begin{bmatrix} 2 & 0 & 0 \\ -4 & 5 & 2 \\ 16 & -15 & -6 \end{bmatrix}\begin{bmatrix} x \\ y \\ z \end{bmatrix} = \begin{bmatrix} 0 \\ 0 \\ 0 \end{bmatrix}$$

for $\mathbf{v} = \begin{bmatrix} x & y & z \end{bmatrix}^T$. The first equation $2x = 0$ implies that $x = 0$. Then the third equation is a multiple of the second equation, $5y + 2z = 0$. The choice of $y = 2$ yields $z = -5$, so the eigenvector $\mathbf{v}_2 = \begin{bmatrix} 0 & 2 & -5 \end{bmatrix}^T$ is associated with $\lambda_2 = 1$.

CASE 3: $\lambda_3 = 3$. Substitution of $\lambda = 3$ in the coefficient matrix in (8) yields the system

$$\begin{bmatrix} 0 & 0 & 0 \\ -4 & 3 & 2 \\ 16 & -15 & -8 \end{bmatrix}\begin{bmatrix} x \\ y \\ z \end{bmatrix} = \begin{bmatrix} 0 \\ 0 \\ 0 \end{bmatrix}.$$

In this case, the first equation yields no information, but the result of adding 4 times the second equation to the third equation is $-3y = 0$, so $y = 0$. Consequently, the second and third equations are both multiples of the equation $2x - z = 0$. The choice $x = 1$ yields $z = 2$, so the eigenvector $\mathbf{v}_3 = \begin{bmatrix} 1 & 0 & 2 \end{bmatrix}^T$ is associated with $\lambda_3 = 3$.

In summary, we have found the eigenvectors $\mathbf{v}_1 = \begin{bmatrix} 0 & 1 & -3 \end{bmatrix}^T$, $\mathbf{v}_2 = \begin{bmatrix} 0 & 2 & -5 \end{bmatrix}^T$, and $\mathbf{v}_3 = \begin{bmatrix} 1 & 0 & 2 \end{bmatrix}^T$ associated with the distinct eigenvalues $\lambda_1 = 0$, $\lambda_2 = 1$, and $\lambda_3 = 3$, respectively. ■

Remark: Substitution of $\lambda = 0$ in the characteristic equation $|\mathbf{A} - \lambda\mathbf{I}| = 0$ yields $|\mathbf{A}| = 0$. Therefore, $\lambda = 0$ is an eigenvalue of the matrix $\mathbf{A}$ if and only if $\mathbf{A}$ is singular: $|\mathbf{A}| = 0$. ■

Eigenspaces

Let λ be a fixed eigenvalue of the $n \times n$ matrix $\mathbf{A}$. Then the set of all eigenvectors associated with $\mathbf{A}$ is the set of all nonzero solution vectors of the system

$$(\mathbf{A} - \lambda\mathbf{I})\mathbf{v} = \mathbf{0}. \tag{9}$$

The solution space of this system is called the **eigenspace** of $\mathbf{A}$ associated with the eigenvalue λ. This subspace of $\mathbf{R}^n$ consists of all eigenvectors associated with λ *together with the zero vector*. In Example 6 we found (to within a constant multiple) only a single eigenvector associated with each eigenvalue λ; in this case, the eigenspace of λ is 1-dimensional. In the case of an eigenspace of higher dimension, we generally want to find a basis for the solution space of Eq. (9).

Example 7 Find bases for the eigenspaces of the matrix

$$\mathbf{A} = \begin{bmatrix} 4 & -2 & 1 \\ 2 & 0 & 1 \\ 2 & -2 & 3 \end{bmatrix}.$$

Solution Here we have

$$\mathbf{A} - \lambda\mathbf{I} = \begin{bmatrix} 4-\lambda & -2 & 1 \\ 2 & -\lambda & 1 \\ 2 & -2 & 3-\lambda \end{bmatrix}. \tag{10}$$

We expand along the first row to obtain

$$\begin{aligned} |\mathbf{A} - \lambda\mathbf{I}| &= (4-\lambda)(\lambda^2 - 3\lambda + 2) - (-2)(4 - 2\lambda) + (1)(-4 + 2\lambda) \\ &= -\lambda^3 + 7\lambda^2 - 16\lambda + 12. \end{aligned}$$

Thus, to find the eigenvalues we need to solve the cubic equation

$$\lambda^3 - 7\lambda^2 + 16\lambda - 12 = 0. \tag{11}$$

We look (hopefully) for integer solutions. The factor theorem of algebra implies that if the polynomial equation

$$\lambda^n + c_{n-1}\lambda^{n-1} + \cdots + c_1\lambda + c_0 = 0$$

with integer coefficients and leading coefficient 1 has an integer solution, then that integer is a divisor of the constant c_0. In the case of the cubic equation in (11), the possibilities for such a solution are ± 1, ± 2, ± 3, ± 4, ± 6, and ± 12. We substitute these numbers successively in (11) and thereby find that $+1$ and -1 are not solutions but that $\lambda = +2$ is a solution. Hence $\lambda - 2$ is a factor of the cubic polynomial in (11). Next, the long division

$$\begin{array}{r|l} & \lambda^2 - 5\lambda \;\; + 6 \\ \hline \lambda - 2 & \lambda^3 - 7\lambda^2 + 16\lambda - 12 \\ & \lambda^3 - 2\lambda^2 \\ \hline & \quad - 5\lambda^2 + 16\lambda - 12 \\ & \quad - 5\lambda^2 + 10\lambda \\ \hline & \qquad\qquad 6\lambda - 12 \\ & \qquad\qquad 6\lambda - 12 \\ \hline & \qquad\qquad\qquad 0 \end{array}$$

shows that

$$\begin{aligned}\lambda^3 - 7\lambda^2 + 16\lambda - 12 &= (\lambda - 2)(\lambda^2 - 5\lambda + 6)\\ &= (\lambda - 2)(\lambda - 2)(\lambda - 3)\\ &= (\lambda - 2)^2(\lambda - 3).\end{aligned}$$

Thus we see finally that the given matrix **A** has the repeated eigenvalue $\lambda = 2$ and the eigenvalue $\lambda = 3$.

CASE 1: $\lambda = 2$. The system $(\mathbf{A} - \lambda\mathbf{I})\mathbf{v} = \mathbf{0}$ is

$$\begin{bmatrix} 2 & -2 & 1 \\ 2 & -2 & 1 \\ 2 & -2 & 1 \end{bmatrix}\begin{bmatrix} x \\ y \\ z \end{bmatrix} = \begin{bmatrix} 0 \\ 0 \\ 0 \end{bmatrix},$$

which reduces to the single equation $2x - 2y + z = 0$. This equation obviously has a 2-dimensional solution space. With $y = 1$ and $z = 0$, we get $x = 1$ and, hence, obtain the basis eigenvector $\mathbf{v}_1 = \begin{bmatrix} 1 & 1 & 0 \end{bmatrix}^T$. With $y = 0$ and $z = 2$, we get $x = -1$ and, hence, the basis eigenvector $\mathbf{v}_2 = \begin{bmatrix} -1 & 0 & 2 \end{bmatrix}^T$. The 2-dimensional eigenspace of **A** associated with the repeated eigenvalue $\lambda = 2$ has basis $\{\mathbf{v}_1, \mathbf{v}_2\}$.

CASE 2: $\lambda = 3$. The system $(\mathbf{A} - \lambda\mathbf{I})\mathbf{v} = \mathbf{0}$ is

$$\begin{bmatrix} 1 & -2 & 1 \\ 2 & -3 & 1 \\ 2 & -2 & 0 \end{bmatrix}\begin{bmatrix} x \\ y \\ z \end{bmatrix} = \begin{bmatrix} 0 \\ 0 \\ 0 \end{bmatrix}.$$

The last equation here implies that $x = y$, and then each of the first two equations yields $x = y = z$. It follows that the eigenspace of **A** associated with $\lambda = 3$ is 1-dimensional and has $\mathbf{v}_3 = \begin{bmatrix} 1 & 1 & 1 \end{bmatrix}^T$ as a basis eigenvector. ■

Remark: The typical higher-degree polynomial is not so easy to factor as the one in Example 7. Hence a numerical technique such as Newton's method is often needed to solve the characteristic equation. Moreover, for an $n \times n$ matrix with n greater than about 4, the amount of labor required to find the characteristic equation by expanding the determinant $|\mathbf{A} - \lambda\mathbf{I}|$ is generally prohibitive; because of the presence of the variable λ, row and column elimination methods do not work as they do with numerical determinants. Consequently, specialized techniques, beyond the scope of the present discussion, are often required to find the eigenvalues and eigenvectors of the large matrices that occur in many applications. Problems 40 and 41 at the end of this section outline a numerical technique that sometimes is useful with matrices of moderate size. ■

6.1 Problems

In Problems 1 through 26, find the (real) eigenvalues and associated eigenvectors of the given matrix **A**. *Find a basis for each eigenspace of dimension 2 or larger.*

1. $\begin{bmatrix} 4 & -2 \\ 1 & 1 \end{bmatrix}$ **2.** $\begin{bmatrix} 5 & -6 \\ 3 & -4 \end{bmatrix}$ **3.** $\begin{bmatrix} 8 & -6 \\ 3 & -1 \end{bmatrix}$

4. $\begin{bmatrix} 4 & -3 \\ 2 & -1 \end{bmatrix}$ **5.** $\begin{bmatrix} 10 & -9 \\ 6 & -5 \end{bmatrix}$ **6.** $\begin{bmatrix} 6 & -4 \\ 3 & -1 \end{bmatrix}$

7. $\begin{bmatrix} 10 & -8 \\ 6 & -4 \end{bmatrix}$ **8.** $\begin{bmatrix} 7 & -6 \\ 12 & -10 \end{bmatrix}$ **9.** $\begin{bmatrix} 8 & -10 \\ 2 & -1 \end{bmatrix}$

10. $\begin{bmatrix} 9 & -10 \\ 2 & 0 \end{bmatrix}$ **11.** $\begin{bmatrix} 19 & -10 \\ 21 & -10 \end{bmatrix}$ **12.** $\begin{bmatrix} 13 & -15 \\ 6 & -6 \end{bmatrix}$

13. $\begin{bmatrix} 2 & 0 & 0 \\ 2 & -2 & -1 \\ -2 & 6 & 3 \end{bmatrix}$ **14.** $\begin{bmatrix} 5 & 0 & 0 \\ 4 & -4 & -2 \\ -2 & 12 & 6 \end{bmatrix}$

15. $\begin{bmatrix} 2 & -2 & 0 \\ 2 & -2 & -1 \\ -2 & 2 & 3 \end{bmatrix}$ **16.** $\begin{bmatrix} 1 & 0 & -1 \\ -2 & 3 & -1 \\ -6 & 6 & 0 \end{bmatrix}$

17. $\begin{bmatrix} 3 & 5 & -2 \\ 0 & 2 & 0 \\ 0 & 2 & 1 \end{bmatrix}$ **18.** $\begin{bmatrix} 1 & 0 & 0 \\ -6 & 8 & 2 \\ 12 & -15 & -3 \end{bmatrix}$

19. $\begin{bmatrix} 3 & 6 & -2 \\ 0 & 1 & 0 \\ 0 & 0 & 1 \end{bmatrix}$ **20.** $\begin{bmatrix} 1 & 0 & 0 \\ -4 & 7 & 2 \\ 10 & -15 & -4 \end{bmatrix}$

21. $\begin{bmatrix} 4 & -3 & 1 \\ 2 & -1 & 1 \\ 0 & 0 & 2 \end{bmatrix}$ **22.** $\begin{bmatrix} 5 & -6 & 3 \\ 6 & -7 & 3 \\ 6 & -6 & 2 \end{bmatrix}$

23. $\begin{bmatrix} 1 & 2 & 2 & 2 \\ 0 & 2 & 2 & 2 \\ 0 & 0 & 3 & 2 \\ 0 & 0 & 0 & 4 \end{bmatrix}$ **24.** $\begin{bmatrix} 1 & 0 & 4 & 0 \\ 0 & 1 & 4 & 0 \\ 0 & 0 & 3 & 0 \\ 0 & 0 & 0 & 3 \end{bmatrix}$

25. $\begin{bmatrix} 1 & 0 & 1 & 0 \\ 0 & 1 & 1 & 0 \\ 0 & 0 & 2 & 0 \\ 0 & 0 & 0 & 2 \end{bmatrix}$ **26.** $\begin{bmatrix} 4 & 0 & 0 & -3 \\ 0 & 2 & 0 & 0 \\ 0 & 0 & -1 & 0 \\ 6 & 0 & 0 & -5 \end{bmatrix}$

Find the complex conjugate eigenvalues and corresponding eigenvectors of the matrices given in Problems 27 through 32.

27. $\mathbf{A} = \begin{bmatrix} 0 & 1 \\ -1 & 0 \end{bmatrix}$ **28.** $\mathbf{A} = \begin{bmatrix} 0 & -6 \\ 6 & 0 \end{bmatrix}$

29. $\mathbf{A} = \begin{bmatrix} 0 & -3 \\ 12 & 0 \end{bmatrix}$ **30.** $\mathbf{A} = \begin{bmatrix} 0 & -12 \\ 12 & 0 \end{bmatrix}$

31. $\mathbf{A} = \begin{bmatrix} 0 & 24 \\ -6 & 0 \end{bmatrix}$ **32.** $\mathbf{A} = \begin{bmatrix} 0 & -4 \\ 36 & 0 \end{bmatrix}$

33. Suppose that λ is an eigenvalue of the matrix $\mathbf{A}$ with associated eigenvector $\mathbf{v}$ and that n is a positive integer. Show that λ^n is an eigenvalue of $\mathbf{A}^n$ with associated eigenvector $\mathbf{v}$.

34. Show that λ is an eigenvalue of the invertible matrix $\mathbf{A}$ if and only if λ^{-1} is an eigenvalue of $\mathbf{A}^{-1}$. Are the associated eigenvectors the same?

35. **(a)** Suppose that $\mathbf{A}$ is a square matrix. Use the characteristic equation to show that $\mathbf{A}$ and $\mathbf{A}^T$ have the same eigenvalues.
(b) Give an example of a 2×2 matrix $\mathbf{A}$ such that $\mathbf{A}$ and $\mathbf{A}^T$ do not have the same eigenvectors.

36. Show that the eigenvalues of a triangular $n \times n$ matrix are its diagonal elements.

37. Suppose that the characteristic equation $|\mathbf{A} - \lambda\mathbf{I}| = 0$ is written as a polynomial equation [Eq. (5)]. Show that the constant term is $c_0 = \det \mathbf{A}$. *Suggestion*: Substitute an appropriate value for λ.

38. If $\mathbf{A} = \begin{bmatrix} a_{ij} \end{bmatrix}$ is an $n \times n$ matrix, then the **trace** $\operatorname{Tr}\mathbf{A}$ of $\mathbf{A}$ is defined to be

$$\operatorname{Tr}\mathbf{A} = a_{11} + a_{22} + \cdots + a_{nn},$$

the sum of the diagonal elements of $\mathbf{A}$. It can be proved that the coefficient of λ^{n-1} in Eq. (5) is $c_{n-1} = (-1)^{n-1}(\operatorname{Tr}\mathbf{A})$. Show explicitly that this is true in the case of a 2×2 matrix.

39. Suppose that the $n \times n$ matrix $\mathbf{A}$ has n (real) eigenvalues $\lambda_1, \lambda_2, \ldots, \lambda_n$. Assuming the general result stated in Problem 38, prove that

$$\begin{aligned} \lambda_1 + \lambda_2 + \cdots + \lambda_n &= \operatorname{Tr}\mathbf{A} \\ &= a_{11} + a_{22} + \cdots + a_{nn}. \end{aligned}$$

40. According to the results stated in Problems 37 and 38, the characteristic polynomial

$$p(\lambda) = |\mathbf{A} - \lambda\mathbf{I}|$$

of a 3×3 matrix $\mathbf{A}$ is given by

$$p(\lambda) = -\lambda^3 + (\operatorname{Tr}\mathbf{A})\lambda^2 + c_1\lambda + (\det \mathbf{A}).$$

The remaining coefficient c_1 can be found by substituting $\lambda = 1$ and then calculating the two determinants $|\mathbf{A}|$ and $p(1) = |\mathbf{A} - \mathbf{I}|$. Use this method to find the characteristic equation, eigenvalues, and associated eigenvectors of the matrix

$$\mathbf{A} = \begin{bmatrix} 32 & -67 & 47 \\ 7 & -14 & 13 \\ -7 & 15 & -6 \end{bmatrix}.$$

41. According to the results stated in Problems 37 and 38, the characteristic polynomial

$$p(\lambda) = |\mathbf{A} - \lambda\mathbf{I}|$$

of a 4×4 matrix $\mathbf{A}$ is given by

$$p(\lambda) = \lambda^4 - (\operatorname{Tr}\mathbf{A})\lambda^3 + c_2\lambda^2 + c_1\lambda + (\det \mathbf{A}).$$

The remaining coefficients c_1 and c_2 can be found by substituting $\lambda = \pm 1$ and calculating the three determinants $|\mathbf{A}|$, $p(1) = |\mathbf{A} - \mathbf{I}|$, and $p(-1) = |\mathbf{A} + \mathbf{I}|$. Use this method to find the characteristic equation, eigenvalues, and associated eigenvectors of the matrix

$$\mathbf{A} = \begin{bmatrix} 22 & -9 & -8 & -8 \\ 10 & -7 & -14 & 2 \\ 10 & 0 & 8 & -10 \\ 29 & -9 & -3 & -15 \end{bmatrix}.$$

42. Combine ideas from Problems 37–40 to show explicitly that

(a) If $\mathbf{A}$ is a 2×2 matrix then $\det(\mathbf{A} - \lambda\mathbf{I}) = (-\lambda)^2 + \operatorname{Tr}(\mathbf{A})(-\lambda) + \det(\mathbf{A})$.

(b) If $\mathbf{A}$ is a 3×3 matrix then $\det(\mathbf{A} - \lambda\mathbf{I}) = (-\lambda)^3 + \operatorname{Tr}(\mathbf{A})(-\lambda)^2 + c_1(-\lambda) + \det(\mathbf{A})$, where c_1 is the sum of the minors of the diagonal elements of $\mathbf{A}$. (Recall from Section 3.6 the distinction between minors and cofactors.)

6.2 Diagonalization of Matrices

Given an $n \times n$ matrix **A**, we may ask *how many* linearly independent eigenvectors the matrix **A** has. In Section 6.1, we saw several examples (with $n = 2$ and $n = 3$) in which the $n \times n$ matrix **A** has n linearly independent eigenvectors—the largest possible number. By contrast, in Example 5 of Section 6.1, we saw that the 2×2 matrix

$$\mathbf{A} = \begin{bmatrix} 2 & 3 \\ 0 & 2 \end{bmatrix}$$

has the single eigenvalue $\lambda = 2$ corresponding to the single eigenvector $\mathbf{v} = \begin{bmatrix} 1 & 0 \end{bmatrix}^T$.

Something very nice happens when the $n \times n$ matrix **A** does have n linearly independent eigenvectors. Suppose that the eigenvalues $\lambda_1, \lambda_2, \ldots, \lambda_n$ (not necessarily distinct) of **A** correspond to the n linearly independent eigenvectors $\mathbf{v}_1, \mathbf{v}_2, \ldots, \mathbf{v}_n$, respectively. Let

$$\mathbf{P} = \begin{bmatrix} | & | & & | \\ \mathbf{v}_1 & \mathbf{v}_2 & \cdots & \mathbf{v}_n \\ | & | & & | \end{bmatrix} \tag{1}$$

be the $n \times n$ matrix having these eigenvectors as its *column* vectors. Then

$$\mathbf{AP} = \mathbf{A} \begin{bmatrix} | & | & & | \\ \mathbf{v}_1 & \mathbf{v}_2 & \cdots & \mathbf{v}_n \\ | & | & & | \end{bmatrix}$$

$$= \begin{bmatrix} | & | & & | \\ \mathbf{Av}_1 & \mathbf{Av}_2 & \cdots & \mathbf{Av}_n \\ | & | & & | \end{bmatrix}$$

and hence

$$\mathbf{AP} = \begin{bmatrix} | & | & & | \\ \lambda_1\mathbf{v}_1 & \lambda_2\mathbf{v}_2 & \cdots & \lambda_n\mathbf{v}_n \\ | & | & & | \end{bmatrix}, \tag{2}$$

because $\mathbf{Av}_j = \lambda_j \mathbf{v}_j$ for each $j = 1, 2, \ldots, n$. Thus the product matrix $\mathbf{AP}$ has column vectors $\lambda_1\mathbf{v}_1, \lambda_2\mathbf{v}_2, \ldots, \lambda_n\mathbf{v}_n$.

Now consider the diagonal matrix

$$\mathbf{D} = \begin{bmatrix} \lambda_1 & 0 & \cdots & 0 \\ 0 & \lambda_2 & \cdots & 0 \\ \vdots & \vdots & \ddots & \vdots \\ 0 & 0 & \cdots & \lambda_n \end{bmatrix}, \tag{3}$$

whose diagonal elements are the eigenvalues corresponding (in the same order) to the eigenvectors forming the columns of $\mathbf{P}$. Then

$$\mathbf{PD} = \begin{bmatrix} | & | & & | \\ \mathbf{v}_1 & \mathbf{v}_2 & \cdots & \mathbf{v}_n \\ | & | & & | \end{bmatrix} \begin{bmatrix} \lambda_1 & 0 & \cdots & 0 \\ 0 & \lambda_2 & \cdots & 0 \\ \vdots & \vdots & \ddots & \vdots \\ 0 & 0 & \cdots & \lambda_n \end{bmatrix}$$

$$= \begin{bmatrix} | & | & & | \\ \lambda_1\mathbf{v}_1 & \lambda_2\mathbf{v}_2 & \cdots & \lambda_n\mathbf{v}_n \\ | & | & & | \end{bmatrix}, \tag{4}$$

because the product of the ith row of $\mathbf{P}$ and the jth column of $\mathbf{D}$ is simply the product of λ_j and the ith component of $\mathbf{v}_j$.

Finally, upon comparing the results in (2) and (4), we see that

$$\mathbf{AP} = \mathbf{PD}. \tag{5}$$

But the matrix $\mathbf{P}$ is invertible, because its n column vectors are linearly independent. So we may multiply on the right by $\mathbf{P}^{-1}$ to obtain

$$\mathbf{A} = \mathbf{PDP}^{-1}. \tag{6}$$

Equation (6) expresses the $n \times n$ matrix $\mathbf{A}$ having n linearly independent eigenvectors in terms of the *eigenvector matrix* $\mathbf{P}$ and the diagonal *eigenvalue matrix* $\mathbf{D}$. It can be rewritten as $\mathbf{D} = \mathbf{P}^{-1}\mathbf{AP}$, but the form in (6) is the one that should be memorized.

Example 1 In Example 1 of Section 6.1 we saw that the matrix

$$\mathbf{A} = \begin{bmatrix} 5 & -6 \\ 2 & -2 \end{bmatrix}$$

has eigenvalues $\lambda_1 = 2$ and $\lambda_2 = 1$ corresponding to the linearly independent eigenvectors $\mathbf{v}_1 = \begin{bmatrix} 2 & 1 \end{bmatrix}^T$ and $\mathbf{v}_2 = \begin{bmatrix} 3 & 2 \end{bmatrix}^T$, respectively. Then

$$\mathbf{P} = \begin{bmatrix} 2 & 3 \\ 1 & 2 \end{bmatrix}, \quad \mathbf{D} = \begin{bmatrix} 2 & 0 \\ 0 & 1 \end{bmatrix}, \quad \text{and} \quad \mathbf{P}^{-1} = \begin{bmatrix} 2 & -3 \\ -1 & 2 \end{bmatrix}.$$

So

$$\mathbf{PDP}^{-1} = \begin{bmatrix} 2 & 3 \\ 1 & 2 \end{bmatrix}\begin{bmatrix} 2 & 0 \\ 0 & 1 \end{bmatrix}\begin{bmatrix} 2 & -3 \\ -1 & 2 \end{bmatrix}$$

$$= \begin{bmatrix} 4 & 3 \\ 2 & 2 \end{bmatrix}\begin{bmatrix} 2 & -3 \\ -1 & 2 \end{bmatrix} = \begin{bmatrix} 5 & -6 \\ 2 & -2 \end{bmatrix} = \mathbf{A},$$

in accord with Eq. (6). ■

Similarity and Diagonalization

The following definition embodies the precise relationship in (6) between the original matrix **A** and the diagonal matrix **D**.

DEFINITION Similar Matrices

The $n \times n$ matrices **A** and **B** are called **similar** provided that there exists an invertible matrix **P** such that

$$\mathbf{B} = \mathbf{P}^{-1}\mathbf{AP}. \tag{7}$$

Note that this relationship between **A** and **B** is symmetric, for if $\mathbf{B} = \mathbf{P}^{-1}\mathbf{AP}$, then $\mathbf{A} = \mathbf{Q}^{-1}\mathbf{BQ}$ for some invertible matrix **Q**—just take $\mathbf{Q} = \mathbf{P}^{-1}$.

An $n \times n$ matrix **A** is called **diagonalizable** if it is similar to a diagonal matrix **D**; that is, there exist a diagonal matrix **D** and an invertible matrix **P** such that $\mathbf{A} = \mathbf{PDP}^{-1}$, and so

$$\mathbf{P}^{-1}\mathbf{AP} = \mathbf{D}. \tag{8}$$

The process of finding the diagonalizing matrix **P** and the diagonal matrix **D** in (8) is called **diagonalization** of the matrix **A**. In Example 1 we showed that the matrices

$$\mathbf{A} = \begin{bmatrix} 5 & -6 \\ 2 & -2 \end{bmatrix} \quad \text{and} \quad \mathbf{D} = \begin{bmatrix} 2 & 0 \\ 0 & 1 \end{bmatrix}$$

are similar, and hence that the 2×2 matrix **A** is diagonalizable.

Now we ask under what conditions a given square matrix is diagonalizable. In deriving Eq. (6), we showed that *if* the $n \times n$ matrix **A** has n linearly independent eigenvectors, *then* **A** is diagonalizable. The converse of this statement is also true.

THEOREM 1 Criterion for Diagonalizability

The $n \times n$ matrix **A** is diagonalizable if and only if it has n linearly independent eigenvectors.

***Proof*:** It remains only to show that, if the $n \times n$ matrix **A** is diagonalizable, then it has n linearly independent eigenvectors. Suppose that **A** is similar to the diagonal matrix **D** with diagonal elements $d_1, d_2, \ldots, d_n$, and let

$$\mathbf{P} = \begin{bmatrix} \mathbf{v}_1 & \mathbf{v}_2 & \cdots & \mathbf{v}_n \end{bmatrix}$$

be an invertible matrix such that $\mathbf{D} = \mathbf{P}^{-1}\mathbf{A}\mathbf{P}$. Then

$$\begin{aligned}\mathbf{AP} &= \mathbf{A}\begin{bmatrix} \mathbf{v}_1 & \mathbf{v}_2 & \cdots & \mathbf{v}_n \end{bmatrix} \\ &= \begin{bmatrix} \mathbf{Av}_1 & \mathbf{Av}_2 & \cdots & \mathbf{Av}_n \end{bmatrix}\end{aligned}$$

and

$$\mathbf{PD} = \begin{bmatrix} d_1\mathbf{v}_1 & d_2\mathbf{v}_2 & \cdots & d_n\mathbf{v}_n \end{bmatrix},$$

by essentially the same computation as in Eq. (4). But $\mathbf{AP} = \mathbf{PD}$ because $\mathbf{D} = \mathbf{P}^{-1}\mathbf{AP}$, so it follows that

$$\mathbf{Av}_j = d_j\mathbf{v}_j$$

for each $j = 1, 2, \ldots, n$. Thus the vectors $\mathbf{v}_1, \mathbf{v}_2, \ldots, \mathbf{v}_n$ are eigenvectors of $\mathbf{A}$ associated with the eigenvalues $d_1, d_2, \ldots, d_n$, respectively. And it follows from Theorem 2 in Section 3.6 and Theorem 2 in Section 4.3 that these n eigenvectors of the matrix $\mathbf{A}$ are linearly independent, because they are the column vectors of the invertible matrix $\mathbf{P}$. ■

Remark: It is important to remember not only the fact that an $n \times n$ matrix $\mathbf{A}$ having n linearly independent eigenvectors is diagonalizable, but also the specific diagonalization $\mathbf{A} = \mathbf{PDP}^{-1}$ in Eq. (6), where the matrix $\mathbf{P}$ has the n eigenvectors as its columns, and the corresponding eigenvalues are the diagonal elements of the diagonal matrix $\mathbf{D}$. ■

Example 2 In Example 5 of Section 6.1 we saw that the matrix

$$\mathbf{A} = \begin{bmatrix} 2 & 3 \\ 0 & 2 \end{bmatrix}$$

has only one eigenvalue, $\lambda = 2$, and that (to within a constant multiple) only the single eigenvector $\mathbf{v} = \begin{bmatrix} 1 & 0 \end{bmatrix}^T$ is associated with this eigenvalue. Thus the 2×2 matrix $\mathbf{A}$ does *not* have $n = 2$ linearly independent eigenvectors. Hence Theorem 1 implies that $\mathbf{A}$ is not diagonalizable. ■

Example 3 In Example 6 of Section 6.1 we saw that the matrix

$$\mathbf{A} = \begin{bmatrix} 3 & 0 & 0 \\ -4 & 6 & 2 \\ 16 & -15 & -5 \end{bmatrix}$$

has the following eigenvalues and associated eigenvectors:

$$\begin{aligned}\lambda_1 = 3: &\quad \mathbf{v}_1 = \begin{bmatrix} 1 & 0 & 2 \end{bmatrix}^T \\ \lambda_2 = 1: &\quad \mathbf{v}_2 = \begin{bmatrix} 0 & 2 & -5 \end{bmatrix}^T \\ \lambda_3 = 0: &\quad \mathbf{v}_3 = \begin{bmatrix} 0 & 1 & -3 \end{bmatrix}^T.\end{aligned}$$

It is obvious (why?) that the three eigenvectors $\mathbf{v}_1$, $\mathbf{v}_2$, $\mathbf{v}_3$ are linearly independent, so Theorem 1 implies that the 3×3 matrix $\mathbf{A}$ is diagonalizable. In particular, the inverse of the eigenvector matrix

$$\mathbf{P} = \begin{bmatrix} \mathbf{v}_1 & \mathbf{v}_2 & \mathbf{v}_3 \end{bmatrix} = \begin{bmatrix} 1 & 0 & 0 \\ 0 & 2 & 1 \\ 2 & -5 & -3 \end{bmatrix}$$

is

$$\mathbf{P}^{-1} = \begin{bmatrix} 1 & 0 & 0 \\ -2 & 3 & 1 \\ 4 & -5 & -2 \end{bmatrix},$$

and the diagonal eigenvalue matrix is

$$\mathbf{D} = \begin{bmatrix} \lambda_1 & 0 & 0 \\ 0 & \lambda_2 & 0 \\ 0 & 0 & \lambda_3 \end{bmatrix} = \begin{bmatrix} 3 & 0 & 0 \\ 0 & 1 & 0 \\ 0 & 0 & 0 \end{bmatrix}.$$

Therefore, Eq. (6) in the form $\mathbf{P}^{-1}\mathbf{AP} = \mathbf{D}$ yields the diagonalization

$$\mathbf{P}^{-1}\mathbf{AP} = \begin{bmatrix} 1 & 0 & 0 \\ -2 & 3 & 1 \\ 4 & -5 & -2 \end{bmatrix} \begin{bmatrix} 3 & 0 & 0 \\ -4 & 6 & 2 \\ 16 & -15 & -5 \end{bmatrix} \begin{bmatrix} 1 & 0 & 0 \\ 0 & 2 & 1 \\ 2 & -5 & -3 \end{bmatrix}$$
$$= \begin{bmatrix} 3 & 0 & 0 \\ 0 & 1 & 0 \\ 0 & 0 & 0 \end{bmatrix}$$

of the matrix **A**. ■

The following theorem tells us that any set of eigenvectors associated with *distinct* eigenvalues (as in Example 3) is automatically linearly independent.

THEOREM 2 Eigenvectors Associated with Distinct Eigenvalues

Suppose that the eigenvectors $\mathbf{v}_1, \mathbf{v}_2, \ldots, \mathbf{v}_k$ are associated with the distinct eigenvalues $\lambda_1, \lambda_2, \ldots, \lambda_k$ of the matrix **A**. Then these k eigenvectors are linearly independent.

***Proof*:** Our proof will be by induction on k. The theorem is certainly true in the case $k = 1$, because any single (nonzero) eigenvector constitutes a linearly independent set. Now assume inductively that any set of $k - 1$ eigenvectors associated with distinct eigenvalues is linearly independent. Supposing that

$$c_1\mathbf{v}_1 + c_2\mathbf{v}_2 + \cdots + c_k\mathbf{v}_k = \mathbf{0}, \tag{9}$$

we need to show that $c_1 = c_2 = \cdots = c_k = 0$. To do this, we will multiply in Eq. (9) by the matrix $\mathbf{A} - \lambda_1\mathbf{I}$. First note that

$$(\mathbf{A} - \lambda_1\mathbf{I})\mathbf{v}_j = \mathbf{A}\mathbf{v}_j - \lambda_1\mathbf{v}_j = \begin{cases} \mathbf{0} & \text{if } j = 1, \\ (\lambda_j - \lambda_1)\mathbf{v}_j & \text{if } j > 1, \end{cases}$$

because $\mathbf{A}\mathbf{v}_j = \lambda_j\mathbf{v}_j$ for each j. Therefore, the result of multiplying Eq. (9) by $\mathbf{A} - \lambda_1\mathbf{I}$ is

$$c_2(\lambda_2 - \lambda_1)\mathbf{v}_2 + \cdots + c_k(\lambda_k - \lambda_1)\mathbf{v}_k = \mathbf{0}. \tag{10}$$

But the $k - 1$ eigenvectors $\mathbf{v}_2, \mathbf{v}_3, \ldots, \mathbf{v}_k$ are linearly independent by the inductive assumption, so each of the scalar coefficients $c_j(\lambda_j - \lambda_1)$ here must be zero. Now our hypothesis that the eigenvalues of **A** are distinct implies that $\lambda_j - \lambda_1 \neq 0$ for each $j > 1$. It therefore follows from Eq. (10) that $c_2 = c_3 = \cdots = c_k = 0$. But then Eq. (9) reduces to $c_1\mathbf{v}_1 = \mathbf{0}$, so it now follows (because $\mathbf{v}_1 \neq \mathbf{0}$) that $c_1 = 0$ as well. Thus we have shown that all the coefficients in Eq. (9) must vanish, and hence that the k eigenvectors $\mathbf{v}_1, \mathbf{v}_2, \ldots, \mathbf{v}_k$ are linearly independent. Theorem 2 now follows by induction. ■

If the $n \times n$ matrix $\mathbf{A}$ has n distinct eigenvalues, then by Theorem 2 the n associated eigenvectors are linearly independent, so Theorem 1 implies that the matrix $\mathbf{A}$ is diagonalizable. Thus we have the sufficient condition for diagonalizability given in Theorem 3.

THEOREM 3 An $n \times n$ Matrix with n Distinct Eigenvalues

If the $n \times n$ matrix $\mathbf{A}$ has n distinct eigenvalues, then it is diagonalizable.

In general, however, an $n \times n$ matrix $\mathbf{A}$ can be expected to have fewer than n distinct eigenvalues $\lambda_1, \lambda_2, \ldots, \lambda_k$. If $k < n$, then we may *attempt* to diagonalize $\mathbf{A}$ by carrying out the following procedure.

Step 1. Find a basis S_i for the eigenspace associated with each eigenvalue λ_i.

Step 2. Form the union S of the bases $S_1, S_2, \ldots, S_k$. According to Theorem 4 in this section, the set S of eigenvectors of $\mathbf{A}$ is linearly independent.

Step 3. If S contains n eigenvectors $\mathbf{v}_1, \mathbf{v}_2, \ldots, \mathbf{v}_n$, then the matrix

$$\mathbf{P} = \begin{bmatrix} \mathbf{v}_1 & \mathbf{v}_2 & \cdots & \mathbf{v}_n \end{bmatrix}$$

diagonalizes $\mathbf{A}$: that is, $\mathbf{P}^{-1}\mathbf{AP} = \mathbf{D}$, where the diagonal elements of $\mathbf{D}$ are the eigenvalues (repeated as necessary) corresponding to the n eigenvectors $\mathbf{v}_1, \mathbf{v}_2, \ldots, \mathbf{v}_n$.

If the set S—obtained by "merging" the bases for all the eigenspaces of $\mathbf{A}$—contains fewer than n eigenvectors, then it can be proved that the matrix $\mathbf{A}$ is not diagonalizable.

Example 4 In Example 7 of Section 6.1, we saw that the matrix

$$\mathbf{A} = \begin{bmatrix} 4 & -2 & 1 \\ 2 & 0 & 1 \\ 2 & -2 & 3 \end{bmatrix}$$

has only two distinct eigenvalues, $\lambda_1 = 2$ and $\lambda_2 = 3$. We found that the eigenvalue $\lambda_1 = 2$ corresponds to a 2-dimensional eigenspace with basis vectors $\mathbf{v}_1 = \begin{bmatrix} 1 & 1 & 0 \end{bmatrix}^T$ and $\mathbf{v}_2 = \begin{bmatrix} -1 & 0 & 2 \end{bmatrix}^T$, and that $\lambda_2 = 3$ corresponds to a 1-dimensional eigenspace with basis vector $\mathbf{v}_3 = \begin{bmatrix} 1 & 1 & 1 \end{bmatrix}^T$. By Theorem 4 (or by explicit verification), these three eigenvectors are linearly independent, so Theorem 1 implies that the 3×3 matrix $\mathbf{A}$ is diagonalizable. The diagonalizing matrix

$$\mathbf{P} = \begin{bmatrix} \mathbf{v}_1 & \mathbf{v}_2 & \mathbf{v}_3 \end{bmatrix} = \begin{bmatrix} 1 & -1 & 1 \\ 1 & 0 & 1 \\ 0 & 2 & 1 \end{bmatrix}$$

has inverse matrix

$$\mathbf{P}^{-1} = \begin{bmatrix} -2 & 3 & -1 \\ -1 & 1 & 0 \\ 2 & -2 & 1 \end{bmatrix},$$

so we obtain the diagonalization

$$\mathbf{P}^{-1}\mathbf{AP} = \begin{bmatrix} -2 & 3 & -1 \\ -1 & 1 & 0 \\ 2 & -2 & 1 \end{bmatrix} \begin{bmatrix} 4 & -2 & 1 \\ 2 & 0 & 1 \\ 2 & -2 & 3 \end{bmatrix} \begin{bmatrix} 1 & -1 & 1 \\ 1 & 0 & 1 \\ 0 & 2 & 1 \end{bmatrix}$$
$$= \begin{bmatrix} 2 & 0 & 0 \\ 0 & 2 & 0 \\ 0 & 0 & 3 \end{bmatrix} = \mathbf{D}$$

of the matrix **A**. ■

THEOREM 4 Complete Independence of Eigenvectors

Let $\lambda_1, \lambda_2, \ldots, \lambda_k$ be the distinct eigenvalues of the $n \times n$ matrix **A**. For each $i = 1, 2, \ldots, k$, let S_i be a basis for the eigenspace associated with λ_i. Then the union S of the bases $S_1, S_2, \ldots, S_k$ is a linearly independent set of eigenvectors of **A**.

***Proof*:** To simplify the notation, we will illustrate the proof with the typical case $k = 3$, with **A** having three distinct eigenvalues λ_1, λ_2, and λ_3. Let

$$S_1 = \{\mathbf{u}_1, \mathbf{u}_2, \ldots, \mathbf{u}_p\},$$
$$S_2 = \{\mathbf{v}_1, \mathbf{v}_2, \ldots, \mathbf{v}_q\}, \quad \text{and}$$
$$S_3 = \{\mathbf{w}_1, \mathbf{w}_2, \ldots, \mathbf{w}_r\}$$

be bases for the eigenspaces associated with the eigenvalues λ_1, λ_2, and λ_3. Assuming that a linear combination of the vectors in $S = S_1 \cup S_2 \cup S_3$ vanishes—

$$\begin{aligned} a_1\mathbf{u}_1 + a_2\mathbf{u}_2 + \cdots + a_p\mathbf{u}_p \\ + b_1\mathbf{v}_1 + b_2\mathbf{v}_2 + \cdots + b_q\mathbf{v}_q \\ + c_1\mathbf{w}_1 + c_2\mathbf{w}_2 + \cdots + c_r\mathbf{w}_r = \mathbf{0} \end{aligned} \tag{11}$$

—we need to show that the coefficients are all zero. If we write

$$\mathbf{u} = a_1\mathbf{u}_1 + a_2\mathbf{u}_2 + \cdots + a_p\mathbf{u}_p,$$
$$\mathbf{v} = b_1\mathbf{v}_1 + b_2\mathbf{v}_2 + \cdots + b_q\mathbf{v}_q, \quad \text{and}$$
$$\mathbf{w} = c_1\mathbf{w}_1 + c_2\mathbf{w}_2 + \cdots + c_r\mathbf{w}_r,$$

then Eq. (11) takes the simple form

$$\mathbf{u} + \mathbf{v} + \mathbf{w} = \mathbf{0}. \tag{12}$$

But the vectors **u**, **v**, and **w** either are zero vectors or are eigenvectors associated with the distinct eigenvalues λ_1, λ_2, and λ_3. In the latter event, Theorem 2 would imply that **u**, **v**, and **w** are linearly independent. Therefore, Eq. (12) implies that $\mathbf{u} = \mathbf{v} = \mathbf{w} = \mathbf{0}$. Finally, the fact that the $\{\mathbf{u}_i\}$ are linearly independent implies that $a_1 = a_2 = \cdots = a_p = 0$; the fact that the $\{\mathbf{v}_i\}$ are linearly independent implies that $b_1 = b_2 = \cdots = b_q = 0$; similarly, $c_1 = c_2 = \cdots = c_r = 0$. Thus we have shown that the coefficients in (11) all vanish, and hence that the vectors in $S = S_1 \cup S_2 \cup S_3$ are linearly independent. ■

6.2 Problems

In Problems 1 through 28, determine whether or not the given matrix **A** *is diagonalizable. If it is, find a diagonalizing matrix* **P** *and a diagonal matrix* **D** *such that* $\mathbf{P}^{-1}\mathbf{AP} = \mathbf{D}$.

1. $\begin{bmatrix} 5 & -4 \\ 2 & -1 \end{bmatrix}$ **2.** $\begin{bmatrix} 6 & -6 \\ 4 & -4 \end{bmatrix}$ **3.** $\begin{bmatrix} 5 & -3 \\ 2 & 0 \end{bmatrix}$

4. $\begin{bmatrix} 5 & -4 \\ 3 & -2 \end{bmatrix}$ **5.** $\begin{bmatrix} 9 & -8 \\ 6 & -5 \end{bmatrix}$ **6.** $\begin{bmatrix} 10 & -6 \\ 12 & -7 \end{bmatrix}$

7. $\begin{bmatrix} 6 & -10 \\ 2 & -3 \end{bmatrix}$ **8.** $\begin{bmatrix} 11 & -15 \\ 6 & -8 \end{bmatrix}$ **9.** $\begin{bmatrix} -1 & 4 \\ -1 & 3 \end{bmatrix}$

10. $\begin{bmatrix} 3 & -1 \\ 1 & 1 \end{bmatrix}$ **11.** $\begin{bmatrix} 5 & 1 \\ -9 & -1 \end{bmatrix}$ **12.** $\begin{bmatrix} 11 & 9 \\ -16 & -13 \end{bmatrix}$

13. $\begin{bmatrix} 1 & 3 & 0 \\ 0 & 2 & 0 \\ 0 & 0 & 2 \end{bmatrix}$ **14.** $\begin{bmatrix} 2 & -2 & 1 \\ 2 & -2 & 1 \\ 2 & -2 & 1 \end{bmatrix}$

15. $\begin{bmatrix} 3 & -3 & 1 \\ 2 & -2 & 1 \\ 0 & 0 & 1 \end{bmatrix}$ **16.** $\begin{bmatrix} 3 & -2 & 0 \\ 0 & 1 & 0 \\ -4 & 4 & 1 \end{bmatrix}$

17. $\begin{bmatrix} 7 & -8 & 3 \\ 6 & -7 & 3 \\ 2 & -2 & 2 \end{bmatrix}$ **18.** $\begin{bmatrix} 6 & -5 & 2 \\ 4 & -3 & 2 \\ 2 & -2 & 3 \end{bmatrix}$

19. $\begin{bmatrix} 1 & 1 & -1 \\ -2 & 4 & -1 \\ -4 & 4 & 1 \end{bmatrix}$ **20.** $\begin{bmatrix} 2 & 0 & 0 \\ -6 & 11 & 2 \\ 6 & -15 & 0 \end{bmatrix}$

21. $\begin{bmatrix} 0 & 1 & 0 \\ -1 & 2 & 0 \\ -1 & 1 & 1 \end{bmatrix}$ **22.** $\begin{bmatrix} 2 & -2 & 1 \\ -1 & 2 & 0 \\ -5 & 7 & -1 \end{bmatrix}$

23. $\begin{bmatrix} -2 & 4 & -1 \\ -3 & 5 & -1 \\ -1 & 1 & 1 \end{bmatrix}$ **24.** $\begin{bmatrix} 3 & -2 & 1 \\ 1 & 0 & 1 \\ -1 & 1 & 2 \end{bmatrix}$

25. $\begin{bmatrix} 1 & 0 & -2 & 0 \\ 0 & 1 & -2 & 0 \\ 0 & 0 & -1 & 0 \\ 0 & 0 & 0 & -1 \end{bmatrix}$ **26.** $\begin{bmatrix} 1 & 0 & 0 & 1 \\ 0 & 1 & 0 & 1 \\ 0 & 0 & 1 & 1 \\ 0 & 0 & 0 & 2 \end{bmatrix}$

27. $\begin{bmatrix} 1 & 1 & 0 & 0 \\ 0 & 1 & 1 & 0 \\ 0 & 0 & 1 & 1 \\ 0 & 0 & 0 & 2 \end{bmatrix}$ **28.** $\begin{bmatrix} 1 & 1 & 0 & 1 \\ 0 & 1 & 1 & 1 \\ 0 & 0 & 2 & 1 \\ 0 & 0 & 0 & 2 \end{bmatrix}$

29. Prove: If the matrices **A** and **B** are similar and the matrices **B** and **C** are similar, then the matrices **A** and **C** are similar.

30. Suppose that the matrices **A** and **B** are similar and that n is a positive integer. Prove that the matrices $\mathbf{A}^n$ and $\mathbf{B}^n$ are similar.

31. Suppose that the invertible matrices **A** and **B** are similar. Prove that their inverses $\mathbf{A}^{-1}$ and $\mathbf{B}^{-1}$ are also similar.

32. Show that if the $n \times n$ matrices **A** and **B** are similar, then they have the same characteristic equation and therefore have the same eigenvalues.

33. Suppose that the $n \times n$ matrices **A** and **B** are similar and that each has n real eigenvalues. Show that $\det \mathbf{A} = \det \mathbf{B}$ and that $\operatorname{Tr} \mathbf{A} = \operatorname{Tr} \mathbf{B}$. See Problems 38 and 39 in Section 6.1.

34. Consider the 2×2 matrix

$$\mathbf{A} = \begin{bmatrix} a & b \\ c & d \end{bmatrix}$$

and let $\Delta = (a - d)^2 + 4bc$. Then show that

(a) **A** is diagonalizable if $\Delta > 0$;

(b) **A** is not diagonalizable if $\Delta < 0$;

(c) If $\Delta = 0$, then **A** may be diagonalizable or it may not be.

35. Let **A** be a 3×3 matrix with three distinct eigenvalues. Tell how to construct six different invertible matrices $\mathbf{P}_1, \mathbf{P}_2, \dots, \mathbf{P}_6$ and six different diagonal matrices $\mathbf{D}_1, \mathbf{D}_2, \dots, \mathbf{D}_6$ such that $\mathbf{P}_i\mathbf{D}_i(\mathbf{P}_i)^{-1} = \mathbf{A}$ for each $i = 1, 2, \dots, 6$.

36. Prove: If the diagonalizable matrices **A** and **B** have the same eigenvalues (with the same multiplicities), then **A** and **B** are similar.

37. Given: The diagonalizable matrix **A**. Show that the eigenvalues of $\mathbf{A}^2$ are the squares of the eigenvalues of **A** but that **A** and $\mathbf{A}^2$ have the same eigenvectors.

38. Suppose that the $n \times n$ matrix **A** has n linearly independent eigenvectors associated with a single eigenvalue λ. Show that **A** is a diagonal matrix.

39. Let λ_i be an eigenvalue of the $n \times n$ matrix **A**, and assume that the characteristic equation of **A** has only real solutions. The **algebraic multiplicity** of λ_i is the largest positive integer $p(i)$ such that $(\lambda - \lambda_i)^{p(i)}$ is a factor of the characteristic polynomial $|\mathbf{A} - \lambda\mathbf{I}|$. The **geometric multiplicity** of λ_i is the dimension $q(i)$ of the eigenspace associated with λ_i. It can be shown that $p(i) \geq q(i)$ for every eigenvalue λ_i. Taking this as already established, prove that the given matrix **A** is diagonalizable if and only if the geometric multiplicity of each eigenvalue is equal to its algebraic multiplicity.

6.3 Applications Involving Powers of Matrices

In this section we discuss applications that depend on an ability to compute the matrix $\mathbf{A}^k$ for large values of k, given the $n \times n$ matrix **A**. If **A** is diagonalizable, then $\mathbf{A}^k$ can be found directly by a method that avoids the labor of calculating the powers $\mathbf{A}^2, \mathbf{A}^3, \mathbf{A}^4, \dots$ by successive matrix multiplications.

Recall from Section 6.2 that, if the $n \times n$ matrix $\mathbf{A}$ has n linearly independent eigenvectors $\mathbf{v}_1, \mathbf{v}_2, \ldots, \mathbf{v}_n$ associated with the eigenvalues $\lambda_1, \lambda_2, \ldots, \lambda_n$ (not necessarily distinct), then

$$\mathbf{A} = \mathbf{PDP}^{-1}, \tag{1}$$

where

$$\mathbf{P} = \begin{bmatrix} | & | & & | \\ \mathbf{v}_1 & \mathbf{v}_2 & \cdots & \mathbf{v}_n \\ | & | & & | \end{bmatrix}$$

and

$$\mathbf{D} = \begin{bmatrix} \lambda_1 & 0 & \cdots & 0 \\ 0 & \lambda_2 & \cdots & 0 \\ \vdots & \vdots & \ddots & \vdots \\ 0 & 0 & \cdots & \lambda_n \end{bmatrix}.$$

Note that (1) yields

$$\mathbf{A}^2 = (\mathbf{PDP}^{-1})(\mathbf{PDP}^{-1}) = \mathbf{PD}(\mathbf{P}^{-1}\mathbf{P})\mathbf{DP}^{-1} = \mathbf{PD}^2\mathbf{P}^{-1}$$

because $\mathbf{P}^{-1}\mathbf{P} = \mathbf{I}$. More generally, for each positive integer k,

$$\begin{aligned} \mathbf{A}^k &= (\mathbf{PDP}^{-1})^k \\ &= (\mathbf{PDP}^{-1})(\mathbf{PDP}^{-1}) \cdots (\mathbf{PDP}^{-1})(\mathbf{PDP}^{-1}) \\ &= \mathbf{PD}(\mathbf{P}^{-1}\mathbf{P})\mathbf{D} \cdots (\mathbf{P}^{-1}\mathbf{P})\mathbf{DP}^{-1}; \\ \mathbf{A}^k &= \mathbf{PD}^k\mathbf{P}^{-1}. \end{aligned} \tag{2}$$

But the kth power $\mathbf{D}^k$ of the diagonal matrix $\mathbf{D}$ is easily computed:

$$\mathbf{D}^k = \begin{bmatrix} \lambda_1^k & 0 & \cdots & 0 \\ 0 & \lambda_2^k & \cdots & 0 \\ \vdots & \vdots & \ddots & \vdots \\ 0 & 0 & \cdots & \lambda_n^k \end{bmatrix}. \tag{3}$$

Consequently the formula in (2) gives a quick and effective way to calculate any desired power of the diagonalizable matrix $\mathbf{A}$, once its eigenvalues and associated eigenvectors—and hence the matrices $\mathbf{P}$ and $\mathbf{D}$—have been found.

Example 1 Find $\mathbf{A}^5$ if

$$\mathbf{A} = \begin{bmatrix} 4 & -2 & 1 \\ 2 & 0 & 1 \\ 2 & -2 & 3 \end{bmatrix}.$$

Solution In Example 7 of Section 6.1, we found that the 3×3 matrix $\mathbf{A}$ has the eigenvalue $\lambda_1 = 3$ with associated eigenvector $\mathbf{v}_1 = \begin{bmatrix} 1 & 1 & 1 \end{bmatrix}^T$, and the repeated eigenvalue $\lambda_2 = 2$ with associated eigenvectors $\mathbf{v}_2 = \begin{bmatrix} 1 & 1 & 0 \end{bmatrix}^T$ and $\mathbf{v}_3 = \begin{bmatrix} -1 & 0 & 2 \end{bmatrix}^T$. Therefore, $\mathbf{A} = \mathbf{PDP}^{-1}$ with

$$\mathbf{P} = \begin{bmatrix} 1 & 1 & -1 \\ 1 & 1 & 0 \\ 1 & 0 & 2 \end{bmatrix} \quad \text{and} \quad \mathbf{D} = \begin{bmatrix} 3 & 0 & 0 \\ 0 & 2 & 0 \\ 0 & 0 & 2 \end{bmatrix}.$$

We first calculate

$$\mathbf{P}^{-1} = \begin{bmatrix} 2 & -2 & 1 \\ -2 & 3 & -1 \\ -1 & 1 & 0 \end{bmatrix}$$

and

$$\mathbf{D}^5 = \begin{bmatrix} 243 & 0 & 0 \\ 0 & 32 & 0 \\ 0 & 0 & 32 \end{bmatrix}.$$

Then the formula in (3) yields

$$\begin{aligned} \mathbf{A}^5 &= \begin{bmatrix} 1 & 1 & -1 \\ 1 & 1 & 0 \\ 1 & 0 & 2 \end{bmatrix} \begin{bmatrix} 243 & 0 & 0 \\ 0 & 32 & 0 \\ 0 & 0 & 32 \end{bmatrix} \begin{bmatrix} 2 & -2 & 1 \\ -2 & 3 & -1 \\ -1 & 1 & 0 \end{bmatrix} \\ &= \begin{bmatrix} 243 & 32 & -32 \\ 243 & 32 & 0 \\ 243 & 0 & 64 \end{bmatrix} \begin{bmatrix} 2 & -2 & 1 \\ -2 & 3 & -1 \\ -1 & 1 & 0 \end{bmatrix} \\ &= \begin{bmatrix} 454 & -422 & 211 \\ 422 & -390 & 211 \\ 422 & -422 & 243 \end{bmatrix}. \end{aligned}$$

■

Transition Matrices

We want to apply this method of computing $\mathbf{A}^k$ to the analysis of a certain type of physical system that can be described by means of the following kind of mathematical model. Suppose that the sequence

$$\mathbf{x}_0, \mathbf{x}_1, \mathbf{x}_2, \ldots, \mathbf{x}_k, \ldots \tag{4}$$

of n-vectors is defined by its **initial vector** $\mathbf{x}_0$ and an $n \times n$ **transition matrix A** in the following manner:

$$\mathbf{x}_{k+1} = \mathbf{A}\mathbf{x}_k \qquad \text{for each } k \geq 0. \tag{5}$$

We envision a physical system—such as a population with n specified subpopulations—that evolve through a sequence of successive *states* described by the vectors in (4). Then our goal is to calculate the kth *state vector* $\mathbf{x}_k$. But using (5) repeatedly, we find that

$$\mathbf{x}_1 = \mathbf{A}\mathbf{x}_0, \quad \mathbf{x}_2 = \mathbf{A}\mathbf{x}_1 = \mathbf{A}^2\mathbf{x}_0, \quad \mathbf{x}_3 = \mathbf{A}\mathbf{x}_2 = \mathbf{A}^3\mathbf{x}_0,$$

and in general that

$$\mathbf{x}_k = \mathbf{A}^k \mathbf{x}_0. \tag{6}$$

Thus our task amounts to calculating the kth power $\mathbf{A}^k$ of the transition matrix $\mathbf{A}$.

Example 2 Consider a metropolitan area with a *constant* total population of 1 million individuals. This area consists of a city and its suburbs, and we want to analyze the changing urban and suburban populations. Let C_k denote the city population and S_k the suburban population after k years. Suppose that each year 15% of the people in the city move to the suburbs, whereas 10% of the people in the suburbs move to the city. Then it follows that

$$\begin{aligned} C_{k+1} &= 0.85C_k + 0.10S_k \\ S_{k+1} &= 0.15C_k + 0.90S_k \end{aligned} \tag{7}$$

for each $k \geq 0$. (For instance, next year's city population C_{k+1} will equal 85% of this year's city population C_k plus 10% of this year's suburban population S_k.) Thus the metropolitan area's *population vector* $\mathbf{x}_k = \begin{bmatrix} C_k & S_k \end{bmatrix}^T$ satisfies

$$\mathbf{x}_{k+1} = \mathbf{A}\mathbf{x}_k \quad \text{and hence} \quad \mathbf{x}_k = \mathbf{A}^k \mathbf{x}_0 \tag{8}$$

(for each $k \geq 0$) with *transition matrix*

$$\mathbf{A} = \begin{bmatrix} 0.85 & 0.10 \\ 0.15 & 0.90 \end{bmatrix}.$$

The characteristic equation of the transition matrix $\mathbf{A}$ is

$$\begin{aligned} \left(\frac{17}{20} - \lambda\right)\left(\frac{9}{10} - \lambda\right) - \left(\frac{3}{20}\right)\left(\frac{1}{10}\right) &= 0; \\ (17 - 20\lambda)(9 - 10\lambda) - 3 &= 0; \\ 200\lambda^2 - 350\lambda + 150 &= 0; \\ 4\lambda^2 - 7\lambda + 3 &= 0; \\ (\lambda - 1)(4\lambda - 3) &= 0. \end{aligned}$$

Thus the eigenvalues of $\mathbf{A}$ are $\lambda_1 = 1$ and $\lambda_2 = 0.75$. For $\lambda_1 = 1$, the system $(\mathbf{A} - \lambda\mathbf{I})\mathbf{v} = \mathbf{0}$ is

$$\begin{bmatrix} -0.15 & 0.10 \\ 0.15 & -0.10 \end{bmatrix}\begin{bmatrix} x \\ y \end{bmatrix} = \begin{bmatrix} 0 \\ 0 \end{bmatrix},$$

so an associated eigenvector is $\mathbf{v}_1 = \begin{bmatrix} 2 & 3 \end{bmatrix}^T$. For $\lambda_2 = 0.75$, the system $(\mathbf{A} - \lambda\mathbf{I})\mathbf{v} = \mathbf{0}$ is

$$\begin{bmatrix} 0.10 & 0.10 \\ 0.15 & 0.15 \end{bmatrix}\begin{bmatrix} x \\ y \end{bmatrix} = \begin{bmatrix} 0 \\ 0 \end{bmatrix},$$

so an associated eigenvector is $\mathbf{v}_2 = \begin{bmatrix} -1 & 1 \end{bmatrix}^T$. It follows that $\mathbf{A} = \mathbf{P}\mathbf{D}\mathbf{P}^{-1}$, where

$$\mathbf{P} = \begin{bmatrix} 2 & -1 \\ 3 & 1 \end{bmatrix} \quad \text{and} \quad \mathbf{D} = \begin{bmatrix} 1 & 0 \\ 0 & \frac{3}{4} \end{bmatrix},$$

and

$$\mathbf{P}^{-1} = \frac{1}{5}\begin{bmatrix} 1 & 1 \\ -3 & 2 \end{bmatrix}.$$

Now suppose that our goal is to determine the long-term distribution of population between the city and its suburbs. Note first that $(\frac{3}{4})^k$ is "negligible" when k is sufficiently large; for instance, $(\frac{3}{4})^{40} \approx 0.00001$. It follows that if $k \geq 40$, then the formula $\mathbf{A}^k = \mathbf{P}\mathbf{D}^k\mathbf{P}^{-1}$ yields

$$\begin{aligned} \mathbf{A}^k &= \begin{bmatrix} 2 & -1 \\ 3 & 1 \end{bmatrix}\begin{bmatrix} 1 & 0 \\ 0 & (\frac{3}{4})^k \end{bmatrix}\left(\tfrac{1}{5}\right)\begin{bmatrix} 1 & 1 \\ -3 & 2 \end{bmatrix} \\ &\approx \frac{1}{5}\begin{bmatrix} 2 & -1 \\ 3 & 1 \end{bmatrix}\begin{bmatrix} 1 & 0 \\ 0 & 0 \end{bmatrix}\begin{bmatrix} 1 & 1 \\ -3 & 2 \end{bmatrix} \\ &= \frac{1}{5}\begin{bmatrix} 2 & 0 \\ 3 & 0 \end{bmatrix}\begin{bmatrix} 1 & 1 \\ -3 & 2 \end{bmatrix} = \frac{1}{5}\begin{bmatrix} 2 & 2 \\ 3 & 3 \end{bmatrix}. \end{aligned} \tag{9}$$

Hence it follows that, if k is sufficiently large, then

$$\begin{aligned} \mathbf{x}_k = \mathbf{A}^k\mathbf{x}_0 &\approx \frac{1}{5}\begin{bmatrix} 2 & 2 \\ 3 & 3 \end{bmatrix}\begin{bmatrix} C_0 \\ S_0 \end{bmatrix} \\ &= (C_0 + S_0)\begin{bmatrix} 0.4 \\ 0.6 \end{bmatrix} = \begin{bmatrix} 0.4 \\ 0.6 \end{bmatrix}, \end{aligned}$$

because $C_0 + S_0 = 1$ (million), the constant total population of the metropolitan area. Thus, our analysis shows that, irrespective of the initial distribution of population between the city and its suburbs, the long-term distribution consists of 40% in the city and 60% in the suburbs. ■

Remark: The result in Example 2—that the long-term situation is independent of the initial situation—is characteristic of a general class of common problems. Note that the transition matrix **A** in (8) has the property that *the sum of the elements in each column is* 1. An $n \times n$ matrix with nonnegative entries having this property is called a **stochastic matrix**. It can be proved that, if **A** is a stochastic matrix having only positive entries, then $\lambda_1 = 1$ is one eigenvalue of **A** and $|\lambda_i| < 1$ for the others. (See Problems 39 and 40.) Moreover, as $k \to \infty$, the matrix $\mathbf{A}^k$ approaches the constant matrix

$$\begin{bmatrix} \mathbf{v}_1 & \mathbf{v}_1 & \cdots & \mathbf{v}_1 \end{bmatrix},$$

each of whose identical columns is the eigenvector of **A** associated with $\lambda_1 = 1$ that has the sum of its elements equal to 1. The 2×2 stochastic matrix **A** of Example 1 illustrates this general result, with $\lambda_1 = 1$, $\lambda_2 = \frac{3}{4}$, and $\mathbf{v}_1 = (\frac{2}{5}, \frac{3}{5})$. ■

Predator-Prey Models

Next, we consider a predator-prey population consisting of the foxes and rabbits living in a certain forest. Initially, there are F_0 foxes and R_0 rabbits; after k months, there are F_k foxes and R_k rabbits. We assume that the transition from each month to the next is described by the equations

$$\begin{aligned} F_{k+1} &= 0.4F_k + 0.3R_k \\ R_{k+1} &= -rF_k + 1.2R_k, \end{aligned} \tag{10}$$

where the constant $r > 0$ is the "capture rate" representing the average number of rabbits consumed monthly by each fox. Thus

$$\mathbf{x}_{k+1} = \mathbf{A}\mathbf{x}_k \quad \text{and hence} \quad \mathbf{x}_k = \mathbf{A}^k\mathbf{x}_0, \tag{11}$$

where

$$\mathbf{x}_k = \begin{bmatrix} F_k \\ R_k \end{bmatrix} \quad \text{and} \quad \mathbf{A} = \begin{bmatrix} 0.4 & 0.3 \\ -r & 1.2 \end{bmatrix}. \tag{12}$$

The term $0.4F_k$ in the first equation in (10) indicates that, without rabbits to eat, only 40% of the foxes would survive each month, while the term $0.3R_k$ represents the growth in the fox population due to the available food supply of rabbits. The term $1.2R_k$ in the second equation indicates that, in the absence of any foxes, the rabbit population would increase by 20% each month. We want to investigate the long-term behavior of the fox and rabbit populations for different values of the capture rate r of rabbits by foxes.

The characteristic equation of the transition matrix $\mathbf{A}$ in (12) is

$$\begin{aligned} (0.4 - \lambda)(1.2 - \lambda) + (0.3)r &= 0; \\ (4 - 10\lambda)(12 - 10\lambda) + 30r &= 0; \\ 100\lambda^2 - 160\lambda + (48 + 30r) &= 0. \end{aligned}$$

The quadratic formula then yields the equation

$$\begin{aligned} \lambda &= \frac{1}{200}\left[160 \pm \sqrt{(160)^2 - (400)(48 + 30r)}\right] \\ &= \frac{1}{10}\left(8 \pm \sqrt{16 - 30r}\right), \end{aligned} \tag{13}$$

which gives the eigenvalues of $\mathbf{A}$ in terms of the capture rate r. Examples 3, 4, and 5 illustrate three possibilities (for different values of r) for what may happen to the fox and rabbit populations as k increases:

- F_k and R_k may approach constant nonzero values. This is the case of *stable limiting populations* that coexist in equilibrium with one another.
- F_k and R_k may both approach zero. This is the case of *mutual extinction* of the two species.
- F_k and R_k may both increase without bound. This is the case of a *population explosion*.

Example 3 *Stable Limiting Population*

If $r = 0.4$, then Eq. (13) gives the eigenvalues $\lambda_1 = 1$ and $\lambda_2 = 0.6$. For $\lambda_1 = 1$, the system $(\mathbf{A} - \lambda\mathbf{I})\mathbf{v} = \mathbf{0}$ is

$$\begin{bmatrix} -0.6 & 0.3 \\ -0.4 & 0.2 \end{bmatrix}\begin{bmatrix} x \\ y \end{bmatrix} = \begin{bmatrix} 0 \\ 0 \end{bmatrix},$$

so an associated eigenvector is $\mathbf{v}_1 = \begin{bmatrix} 1 & 2 \end{bmatrix}^T$. For $\lambda_2 = 0.6$, the system $(\mathbf{A} - \lambda\mathbf{I})\mathbf{v} = \mathbf{0}$ is

$$\begin{bmatrix} -0.2 & 0.3 \\ -0.4 & 0.6 \end{bmatrix}\begin{bmatrix} x \\ y \end{bmatrix} = \begin{bmatrix} 0 \\ 0 \end{bmatrix},$$

so an associated eigenvector is $\mathbf{v}_2 = \begin{bmatrix} 3 & 2 \end{bmatrix}^T$. It follows that $\mathbf{A} = \mathbf{PDP}^{-1}$, where

$$\mathbf{P} = \begin{bmatrix} 1 & 3 \\ 2 & 2 \end{bmatrix}, \quad \mathbf{D} = \begin{bmatrix} 1 & 0 \\ 0 & 0.6 \end{bmatrix}, \quad \text{and} \quad \mathbf{P}^{-1} = -\tfrac{1}{4}\begin{bmatrix} 2 & -3 \\ -2 & 1 \end{bmatrix}.$$

We are now ready to calculate $\mathbf{A}^k$. If k is sufficiently large that $(0.6)^k \approx 0$—for instance, $(0.6)^{25} \approx 0.000003$—then the formula $\mathbf{A}^k = \mathbf{PD}^k\mathbf{P}^{-1}$ yields

$$\begin{aligned}
\mathbf{A}^k &= \begin{bmatrix} 1 & 3 \\ 2 & 2 \end{bmatrix}\begin{bmatrix} 1 & 0 \\ 0 & (0.6)^k \end{bmatrix}\left(-\tfrac{1}{4}\right)\begin{bmatrix} 2 & -3 \\ -2 & 1 \end{bmatrix} \\
&\approx -\tfrac{1}{4}\begin{bmatrix} 1 & 3 \\ 2 & 2 \end{bmatrix}\begin{bmatrix} 1 & 0 \\ 0 & 0 \end{bmatrix}\begin{bmatrix} 2 & -3 \\ -2 & 1 \end{bmatrix} \\
&= -\tfrac{1}{4}\begin{bmatrix} 1 & 0 \\ 2 & 0 \end{bmatrix}\begin{bmatrix} 2 & -3 \\ -2 & 1 \end{bmatrix} \\
&= \tfrac{1}{4}\begin{bmatrix} -2 & 3 \\ -4 & 6 \end{bmatrix}.
\end{aligned}$$

Hence it follows that if k is sufficiently large, then

$$\mathbf{x}_k = \mathbf{A}^k\mathbf{x}_0 = \tfrac{1}{4}\begin{bmatrix} -2 & 3 \\ -4 & 6 \end{bmatrix}\begin{bmatrix} F_0 \\ R_0 \end{bmatrix} = \tfrac{1}{4}\begin{bmatrix} 3R_0 - 2F_0 \\ 6R_0 - 4F_0 \end{bmatrix}$$

—that is,

$$\begin{bmatrix} F_k \\ R_k \end{bmatrix} = \alpha\begin{bmatrix} 1 \\ 2 \end{bmatrix}, \quad \text{where} \quad \alpha = \tfrac{1}{4}(3R_0 - 2F_0). \tag{14}$$

Assuming that the initial populations are such that $\alpha > 0$ (that is, $3R_0 > 2F_0$), (14) implies that, as k increases, the fox and rabbit populations approach a stable situation in which there are twice as many rabbits as foxes. For instance, if $F_0 = R_0 = 100$, then when k is sufficiently large, there will be 25 foxes and 50 rabbits. ■

Example 4 *Mutual Extinction*

If $r = 0.5$, then Eq. (13) gives the eigenvalues $\lambda_1 = 0.9$ and $\lambda_2 = 0.7$. For $\lambda_1 = 0.9$, the system $(\mathbf{A} - \lambda\mathbf{I})\mathbf{v} = \mathbf{0}$ is

$$\begin{bmatrix} -0.5 & 0.3 \\ -0.5 & 0.3 \end{bmatrix}\begin{bmatrix} x \\ y \end{bmatrix} = \begin{bmatrix} 0 \\ 0 \end{bmatrix},$$

so an associated eigenvector is $\mathbf{v}_1 = \begin{bmatrix} 3 & 5 \end{bmatrix}^T$. For $\lambda_2 = 0.7$, the system $(\mathbf{A} - \lambda\mathbf{I})\mathbf{v} = \mathbf{0}$ is

$$\begin{bmatrix} -0.3 & 0.3 \\ -0.5 & 0.5 \end{bmatrix}\begin{bmatrix} x \\ y \end{bmatrix} = \begin{bmatrix} 0 \\ 0 \end{bmatrix},$$

so an associated eigenvector is $\mathbf{v}_2 = \begin{bmatrix} 1 & 1 \end{bmatrix}^T$. It follows that $\mathbf{A} = \mathbf{PDP}^{-1}$, with

$$\mathbf{P} = \begin{bmatrix} 3 & 1 \\ 5 & 1 \end{bmatrix}, \quad \mathbf{D} = \begin{bmatrix} 0.9 & 0 \\ 0 & 0.7 \end{bmatrix}, \quad \text{and} \quad \mathbf{P}^{-1} = -\tfrac{1}{2}\begin{bmatrix} 1 & -1 \\ -5 & 3 \end{bmatrix}.$$

Now both $(0.9)^k$ and $(0.7)^k$ approach 0 as k increases without bound ($k \to +\infty$). Hence if k is sufficiently large, then the formula $\mathbf{A}^k = \mathbf{P}\mathbf{D}^k\mathbf{P}^{-1}$ yields

$$\mathbf{A}^k = \begin{bmatrix} 3 & 1 \\ 5 & 1 \end{bmatrix} \begin{bmatrix} (0.9)^k & 0 \\ 0 & (0.7)^k \end{bmatrix} \left(-\tfrac{1}{2}\right) \begin{bmatrix} 1 & -1 \\ -5 & 3 \end{bmatrix}$$
$$\approx -\tfrac{1}{2} \begin{bmatrix} 3 & 1 \\ 5 & 1 \end{bmatrix} \begin{bmatrix} 0 & 0 \\ 0 & 0 \end{bmatrix} \begin{bmatrix} 1 & -1 \\ -5 & 3 \end{bmatrix} = \begin{bmatrix} 0 & 0 \\ 0 & 0 \end{bmatrix},$$

so

$$\begin{bmatrix} F_k \\ R_k \end{bmatrix} = \mathbf{A}^k \begin{bmatrix} F_0 \\ R_0 \end{bmatrix} \approx \begin{bmatrix} 0 & 0 \\ 0 & 0 \end{bmatrix} \begin{bmatrix} F_0 \\ R_0 \end{bmatrix} = \begin{bmatrix} 0 \\ 0 \end{bmatrix}. \tag{15}$$

Thus F_k and R_k both approach zero as $k \to +\infty$, so both the foxes and the rabbits die out—mutual extinction occurs. ■

Example 5 *Population Explosion*

If $r = 0.325$, then Eq. (13) gives the eigenvalues $\lambda_1 = 1.05$ and $\lambda_2 = 0.55$. For $\lambda_1 = 1.05$, the system $(\mathbf{A} - \lambda\mathbf{I})\mathbf{v} = \mathbf{0}$ is

$$\begin{bmatrix} -0.650 & 0.30 \\ -0.325 & 0.15 \end{bmatrix} \begin{bmatrix} x \\ y \end{bmatrix} = \begin{bmatrix} 0 \\ 0 \end{bmatrix}.$$

Each equation is a multiple of $-13x + 6y = 0$, so an associated eigenvector is $\mathbf{v}_1 = \begin{bmatrix} 6 & 13 \end{bmatrix}^T$. For $\lambda_2 = 0.55$, the system $(\mathbf{A} - \lambda\mathbf{I})\mathbf{v} = \mathbf{0}$ is

$$\begin{bmatrix} -0.150 & 0.30 \\ -0.325 & 0.65 \end{bmatrix} \begin{bmatrix} x \\ y \end{bmatrix} = \begin{bmatrix} 0 \\ 0 \end{bmatrix},$$

so an associated eigenvector is $\mathbf{v}_2 = \begin{bmatrix} 2 & 1 \end{bmatrix}^T$. It follows that $\mathbf{A} = \mathbf{P}\mathbf{D}\mathbf{P}^{-1}$ with

$$\mathbf{P} = \begin{bmatrix} 6 & 2 \\ 13 & 1 \end{bmatrix}, \quad \mathbf{D} = \begin{bmatrix} 1.05 & 0 \\ 0 & 0.55 \end{bmatrix}, \quad \text{and} \quad \mathbf{P}^{-1} = -\tfrac{1}{20} \begin{bmatrix} 1 & -2 \\ -13 & 6 \end{bmatrix}.$$

Note that $(0.55)^k$ approaches zero but that $(1.05)^k$ increases without bound as $k \to +\infty$. It follows that if k is sufficiently large, then the formula $\mathbf{A}^k = \mathbf{P}\mathbf{D}^k\mathbf{P}^{-1}$ yields

$$\mathbf{A}^k = \begin{bmatrix} 6 & 2 \\ 13 & 1 \end{bmatrix} \begin{bmatrix} (1.05)^k & 0 \\ 0 & (0.55)^k \end{bmatrix} \left(-\tfrac{1}{20}\right) \begin{bmatrix} 1 & -2 \\ -13 & 6 \end{bmatrix}$$
$$\approx -\tfrac{1}{20} \begin{bmatrix} 6 & 2 \\ 13 & 1 \end{bmatrix} \begin{bmatrix} (1.05)^k & 0 \\ 0 & 0 \end{bmatrix} \begin{bmatrix} 1 & -2 \\ -13 & 6 \end{bmatrix}$$
$$= -\tfrac{1}{20} \begin{bmatrix} (6)(1.05)^k & 0 \\ (13)(1.05)^k & 0 \end{bmatrix} \begin{bmatrix} 1 & -2 \\ -13 & 6 \end{bmatrix},$$

and therefore

$$\mathbf{A}^k \approx -\tfrac{1}{20}(1.05)^k \begin{bmatrix} 6 & -12 \\ 13 & -26 \end{bmatrix}. \tag{16}$$

Hence, if k is sufficiently large, then

$$\mathbf{x}_k = \mathbf{A}^k\mathbf{x}_0 \approx -\tfrac{1}{20}(1.05)^k \begin{bmatrix} 6 & -12 \\ 13 & -26 \end{bmatrix}\begin{bmatrix} F_0 \\ R_0 \end{bmatrix}$$

$$= -\tfrac{1}{20}(1.05)^k \begin{bmatrix} 6F_0 - 12R_0 \\ 13F_0 - 26R_0 \end{bmatrix},$$

and so

$$\begin{bmatrix} F_k \\ R_k \end{bmatrix} \approx (1.05)^k\gamma \begin{bmatrix} 6 \\ 13 \end{bmatrix}, \quad \text{where} \quad \gamma = \tfrac{1}{20}(2R_0 - F_0). \tag{17}$$

If $\gamma > 0$ (that is, if $2R_0 > F_0$), then the factor $(1.05)^k$ in (17) implies that the fox and rabbit populations both increase at a rate of 5% per month, and thus each increases without bound as $k \to +\infty$. Moreover, when k is sufficiently large, the two populations maintain a constant ratio of 6 foxes for every 13 rabbits. It is also of interest to note that the monthly "population multiplier" is the larger eigenvalue $\lambda_1 = 1.05$ and that the limiting *ratio* of populations is determined by the associated eigenvector $\mathbf{v}_1 = \begin{bmatrix} 6 & 13 \end{bmatrix}^T$. ■

In summary, let us compare the results in Examples 3, 4, and 5. The critical capture rate $r = 0.4$ of Example 3 represents a monthly consumption of 0.4 rabbits per fox, resulting in stable limiting populations of both species. But if the foxes are greedier and consume more than 0.4 rabbits per fox monthly, then the result is extinction of both species (as in Example 4). If the rabbits become more skilled at evading foxes, so that less than 0.4 rabbits per fox are consumed each month, then both populations grow without bound, as in Example 5.

The Cayley-Hamilton Theorem

One of the more notable and important theorems in advanced linear algebra says that *every matrix satisfies its own characteristic equation* (as proved for the 2×2 case in Problem 29 of Section 3.4).

THEOREM 1 Cayley-Hamilton

If the $n \times n$ matrix $\mathbf{A}$ has the characteristic polynomial

$$p(\lambda) = (-1)^n\lambda^n + c_{n-1}\lambda^{n-1} + \cdots + c_2\lambda^2 + c_1\lambda + c_0,$$

then

$$p(\mathbf{A}) = (-1)^n\mathbf{A}^n + c_{n-1}\mathbf{A}^{n-1} + \cdots + c_2\mathbf{A}^2 + c_1\mathbf{A} + c_0\mathbf{I} = \mathbf{0}. \tag{18}$$

***Proof*:** We establish (18) only for the special case in which the matrix $\mathbf{A}$ is *diagonalizable*; the general case will be discussed in Section 7.5. If $\lambda_1, \lambda_2, \ldots, \lambda_n$ are the eigenvalues of $\mathbf{A}$, then

$$\mathbf{A}^k = \mathbf{P}\mathbf{D}^k\mathbf{P}^{-1}, \quad \text{where} \quad \mathbf{D}^k = \begin{bmatrix} \lambda_1^k & 0 & \cdots & 0 \\ 0 & \lambda_2^k & \cdots & 0 \\ \vdots & \vdots & \ddots & \vdots \\ 0 & 0 & \cdots & \lambda_n^k \end{bmatrix}$$

as in Eqs. (2) and (3) at the beginning of this section. First, we note that

$$
\begin{aligned}
p(\mathbf{D}) &= (-1)^n\mathbf{D}^n + c_{n-1}\mathbf{D}^{n-1} + \cdots + c_2\mathbf{D}^2 + c_1\mathbf{D} + c_0\mathbf{I} \\
&= (-1)^n \begin{bmatrix} \lambda_1^n & 0 & \cdots & 0 \\ 0 & \lambda_2^n & \cdots & 0 \\ \vdots & \vdots & \ddots & \vdots \\ 0 & 0 & \cdots & \lambda_n^n \end{bmatrix} + c_{n-1} \begin{bmatrix} \lambda_1^{n-1} & 0 & \cdots & 0 \\ 0 & \lambda_2^{n-1} & \cdots & 0 \\ \vdots & \vdots & \ddots & \vdots \\ 0 & 0 & \cdots & \lambda_n^{n-1} \end{bmatrix} \\
&\quad + \cdots + c_2 \begin{bmatrix} \lambda_1^2 & 0 & \cdots & 0 \\ 0 & \lambda_2^2 & \cdots & 0 \\ \vdots & \vdots & \ddots & \vdots \\ 0 & 0 & \cdots & \lambda_n^2 \end{bmatrix} + c_1 \begin{bmatrix} \lambda_1 & 0 & \cdots & 0 \\ 0 & \lambda_2 & \cdots & 0 \\ \vdots & \vdots & \ddots & \vdots \\ 0 & 0 & \cdots & \lambda_n \end{bmatrix} \\
&\quad + c_0 \begin{bmatrix} 1 & 0 & \cdots & 0 \\ 0 & 1 & \cdots & 0 \\ \vdots & \vdots & \ddots & \vdots \\ 0 & 0 & \cdots & 1 \end{bmatrix} \\
&= \begin{bmatrix} p(\lambda_1) & 0 & \cdots & 0 \\ 0 & p(\lambda_2) & \cdots & 0 \\ \vdots & \vdots & \ddots & \vdots \\ 0 & 0 & \cdots & p(\lambda_n) \end{bmatrix} = \mathbf{0},
\end{aligned}
$$

because $p(\lambda_k) = 0$ for $k = 1, 2, \ldots, n$. Thus the diagonal form $\mathbf{D}$ of the matrix $\mathbf{A}$ satisfies the characteristic equation of $\mathbf{A}$. It therefore follows that

$$
\begin{aligned}
p(\mathbf{A}) &= (-1)^n\mathbf{P}\mathbf{D}^n\mathbf{P}^{-1} + c_{n-1}\mathbf{P}\mathbf{D}^{n-1}\mathbf{P}^{-1} + \cdots \\
&\quad + c_2\mathbf{P}\mathbf{D}^2\mathbf{P}^{-1} + c_1\mathbf{P}\mathbf{D}\mathbf{P}^{-1} + c_0\mathbf{P}\mathbf{I}\mathbf{P}^{-1} \\
&= \mathbf{P}p(\mathbf{D})\mathbf{P}^{-1} \\
&= \mathbf{P}\,\mathbf{0}\,\mathbf{P}^{-1} = \mathbf{0},
\end{aligned}
$$

as desired. ■

Example 6 In Example 7 of Section 6.1, we found that the matrix

$$
\mathbf{A} = \begin{bmatrix} 4 & -2 & 1 \\ 2 & 0 & 1 \\ 2 & -2 & 3 \end{bmatrix}
$$

has the characteristic polynomial

$$
p(\lambda) = -\lambda^3 + 7\lambda^2 - 16\lambda + 12,
$$

so

$$
-\mathbf{A}^3 + 7\mathbf{A}^2 - 16\mathbf{A} + 12\mathbf{I} = \mathbf{0}, \tag{19}
$$

by the Cayley-Hamilton theorem. Because

$$\mathbf{A}^2 = \begin{bmatrix} 14 & -10 & 5 \\ 10 & -6 & 5 \\ 10 & -10 & 9 \end{bmatrix},$$

it follows from (19) that

$$\begin{aligned}
\mathbf{A}^3 &= 7\mathbf{A}^2 - 16\mathbf{A} + 12\mathbf{I} \\
&= 7\begin{bmatrix} 14 & -10 & 5 \\ 10 & -6 & 5 \\ 10 & -10 & 9 \end{bmatrix} - 16\begin{bmatrix} 4 & -2 & 1 \\ 2 & 0 & 1 \\ 2 & -2 & 3 \end{bmatrix} + 12\begin{bmatrix} 1 & 0 & 0 \\ 0 & 1 & 0 \\ 0 & 0 & 1 \end{bmatrix} \\
&= \begin{bmatrix} 46 & -38 & 19 \\ 38 & -30 & 19 \\ 38 & -38 & 27 \end{bmatrix}.
\end{aligned}$$

Multiplication by $\mathbf{A}$ now gives

$$\begin{aligned}
\mathbf{A}^4 &= 7\mathbf{A}^3 - 16\mathbf{A}^2 + 12\mathbf{A} \\
&= 7(7\mathbf{A}^2 - 16\mathbf{A} + 12\mathbf{I}) - 16\mathbf{A}^2 + 12\mathbf{A} \\
&= 33\mathbf{A}^2 - 100\mathbf{A} + 84\mathbf{I} \\
&= 33\begin{bmatrix} 14 & -10 & 5 \\ 10 & -6 & 5 \\ 10 & -10 & 9 \end{bmatrix} - 100\begin{bmatrix} 4 & -2 & 1 \\ 2 & 0 & 1 \\ 2 & -2 & 3 \end{bmatrix} + 84\begin{bmatrix} 1 & 0 & 0 \\ 0 & 1 & 0 \\ 0 & 0 & 1 \end{bmatrix} \\
&= \begin{bmatrix} 146 & -130 & 65 \\ 130 & -114 & 65 \\ 130 & -130 & 81 \end{bmatrix}.
\end{aligned}$$

Thus we can use the Cayley-Hamilton theorem to express (positive integral) powers of the 3×3 matrix $\mathbf{A}$ in terms of $\mathbf{A}$ and $\mathbf{A}^2$. We also can use the Cayley-Hamilton theorem to find the inverse matrix $\mathbf{A}^{-1}$. If we multiply Eq. (19) by $\mathbf{A}^{-1}$, we can solve the resulting equation $-\mathbf{A}^2 + 7\mathbf{A} - 16\mathbf{I} + 12\mathbf{A}^{-1} = \mathbf{0}$:

$$\begin{aligned}
\mathbf{A}^{-1} &= \frac{1}{12}(\mathbf{A}^2 - 7\mathbf{A} + 16\mathbf{I}) \\
&= \frac{1}{12}\begin{bmatrix} 14 & -10 & 5 \\ 10 & -6 & 5 \\ 10 & -10 & 9 \end{bmatrix} - \frac{7}{12}\begin{bmatrix} 4 & -2 & 1 \\ 2 & 0 & 1 \\ 2 & -2 & 3 \end{bmatrix} + \frac{16}{12}\begin{bmatrix} 1 & 0 & 0 \\ 0 & 1 & 0 \\ 0 & 0 & 1 \end{bmatrix} \\
&= \frac{1}{6}\begin{bmatrix} 1 & 2 & -1 \\ -2 & 5 & -1 \\ -2 & 2 & 2 \end{bmatrix}.
\end{aligned}$$

■

6.3 Problems

In Problems 1 through 10, a matrix **A** *is given. Use the method of Example 1 to compute* $\mathbf{A}^5$.

1. $\begin{bmatrix} 3 & -2 \\ 1 & 0 \end{bmatrix}$ **2.** $\begin{bmatrix} 5 & -6 \\ 3 & -4 \end{bmatrix}$

3. $\begin{bmatrix} 6 & -6 \\ 4 & -4 \end{bmatrix}$ **4.** $\begin{bmatrix} 4 & -3 \\ 2 & -1 \end{bmatrix}$

5. $\begin{bmatrix} 5 & -4 \\ 3 & -2 \end{bmatrix}$ **6.** $\begin{bmatrix} 6 & -10 \\ 2 & -3 \end{bmatrix}$

7. $\begin{bmatrix} 1 & 3 & 0 \\ 0 & 2 & 0 \\ 0 & 0 & 2 \end{bmatrix}$ **8.** $\begin{bmatrix} 1 & -2 & 1 \\ 0 & 1 & 0 \\ 0 & -2 & 2 \end{bmatrix}$

9. $\begin{bmatrix} 1 & -3 & 1 \\ 0 & 2 & 0 \\ 0 & 0 & 2 \end{bmatrix}$ **10.** $\begin{bmatrix} 4 & -3 & 1 \\ 2 & -1 & 1 \\ 0 & 0 & 2 \end{bmatrix}$

Find $\mathbf{A}^{10}$ *for each matrix* **A** *given in Problems 11 through 14.*

11. $\begin{bmatrix} 1 & 0 & 0 \\ 6 & 5 & 2 \\ 21 & -15 & -6 \end{bmatrix}$ **12.** $\begin{bmatrix} 11 & -6 & -2 \\ 20 & -11 & -4 \\ 0 & 0 & 1 \end{bmatrix}$

13. $\begin{bmatrix} 1 & -1 & 1 \\ 2 & -2 & 1 \\ 4 & -4 & 1 \end{bmatrix}$ **14.** $\begin{bmatrix} 5 & -5 & -3 \\ 2 & -2 & -1 \\ 4 & -4 & -3 \end{bmatrix}$

15–24. Use the Cayley-Hamilton theorem (as in Example 6) to find $\mathbf{A}^{-1}$, $\mathbf{A}^3$, and $\mathbf{A}^4$ for each matrix **A** given in Problems 5–14 (respectively).

In Problems 25 through 30, a city-suburban population transition matrix **A** *(as in Example 2) is given. Find the resulting long-term distribution of a constant total population between the city and its suburbs.*

25. $\mathbf{A} = \begin{bmatrix} 0.9 & 0.1 \\ 0.1 & 0.9 \end{bmatrix}$ **26.** $\mathbf{A} = \begin{bmatrix} 0.85 & 0.05 \\ 0.15 & 0.95 \end{bmatrix}$

27. $\mathbf{A} = \begin{bmatrix} 0.75 & 0.15 \\ 0.25 & 0.85 \end{bmatrix}$ **28.** $\mathbf{A} = \begin{bmatrix} 0.8 & 0.1 \\ 0.2 & 0.9 \end{bmatrix}$

29. $\mathbf{A} = \begin{bmatrix} 0.9 & 0.05 \\ 0.1 & 0.95 \end{bmatrix}$ **30.** $\mathbf{A} = \begin{bmatrix} 0.8 & 0.15 \\ 0.2 & 0.85 \end{bmatrix}$

Problems 31 through 33 deal with a fox-rabbit population as in Examples 3 through 5, except with the transition matrix

$$\mathbf{A} = \begin{bmatrix} 0.6 & 0.5 \\ -r & 1.2 \end{bmatrix}$$

in place of the one used in the text.

31. If $r = 0.16$, show that in the long term the populations of foxes and rabbits are stable, with 5 foxes for each 4 rabbits.

32. If $r = 0.175$, show that in the long term the populations of foxes and rabbits both die out.

33. If $r = 0.135$, show that in the long term the fox and rabbit populations both increase at the rate of 5% per month, maintaining a constant ratio of 10 foxes for each 9 rabbits.

34. Suppose that the 2×2 matrix **A** has eigenvalues $\lambda_1 = 1$ and $\lambda_2 = -1$ with eigenvectors $\mathbf{v}_1 = (3, 4)$ and $\mathbf{v}_2 = (5, 7)$, respectively. Find the matrix **A** and the powers $\mathbf{A}^{99}$ and $\mathbf{A}^{100}$.

35. Suppose that $|\lambda| = 1$ for each eigenvalue λ of the diagonalizable matrix **A**. Show that $\mathbf{A}^n = \mathbf{I}$ for every even positive integer n.

36. Suppose that

$$\mathbf{A} = \begin{bmatrix} 0 & 1 \\ 1 & 0 \end{bmatrix}.$$

Show that $\mathbf{A}^{2n} = \mathbf{I}$ and that $\mathbf{A}^{2n+1} = \mathbf{A}$ for every positive integer n.

37. Suppose that

$$\mathbf{A} = \begin{bmatrix} 0 & 1 \\ -1 & 0 \end{bmatrix}.$$

Show that $\mathbf{A}^{4n} = \mathbf{I}$, $\mathbf{A}^{4n+1} = \mathbf{A}$, $\mathbf{A}^{4n+2} = -\mathbf{I}$, and $\mathbf{A}^{4n+3} = -\mathbf{A}$ for every positive integer n.

38. The matrix

$$\mathbf{A} = \begin{bmatrix} 1 & 1 \\ 0 & 1 \end{bmatrix}$$

is not diagonalizable. (Why not?) Write $\mathbf{A} = \mathbf{I} + \mathbf{B}$. Show that $\mathbf{B}^2 = \mathbf{0}$ and thence that

$$\mathbf{A}^n = \begin{bmatrix} 1 & n \\ 0 & 1 \end{bmatrix}.$$

39. Consider the stochastic matrix

$$\mathbf{A} = \begin{bmatrix} p & 1-q \\ 1-p & q \end{bmatrix},$$

where $0 < p < 1$ and $0 < q < 1$. Show that the eigenvalues of **A** are $\lambda_1 = 1$ and $\lambda_2 = p + q - 1$, so that $|\lambda_2| < 1$.

40. Suppose that **A** is an $n \times n$ stochastic matrix—the sum of the elements of each column vector is 1. If $\mathbf{v} = (1, 1, \dots, 1)$, show that $\mathbf{A}^T\mathbf{v} = \mathbf{v}$. Why does it follow that $\lambda = 1$ is an eigenvalue of **A**?

41. In his *Liber abaci* (Book of Calculation) published in 1202, Leonardo Fibonacci* asked the following question: How many pairs of rabbits are produced from a single original pair in one year, if every month each pair begets a new pair, which is similarly productive beginning in the second succeeding month? The answer is provided by the **Fibonacci sequence**

$$1, 1, 2, 3, 5, 8, 13, 21, 34, 55, \dots$$

in which each term is the sum of its two immediate predecessors. That is, the sequence is defined recursively as follows:

$$s_0 = 1 = s_1, \quad s_{n+1} = s_n + s_{n-1} \quad \text{for } n \geq 1.$$

* "Leonardo, son of Bonacci," 1175–1250?, one of the outstanding mathematicians of the middle ages. He is also known as Leonardo of Pisa (Leonardo Pisano).

Then s_n is the number of rabbit pairs present after n months. Note that if

$$\mathbf{x}_n = \begin{bmatrix} s_{n+1} \\ s_n \end{bmatrix} \quad \text{and} \quad \mathbf{A} = \begin{bmatrix} 1 & 1 \\ 1 & 0 \end{bmatrix},$$

then

$$\mathbf{x}_n = \mathbf{A}\mathbf{x}_{n-1} \quad \text{and so} \quad \mathbf{x}_n = \mathbf{A}^n\mathbf{x}_0,$$

where $\mathbf{x}_0 = (1, 1)$.

(a) Show that $\mathbf{A}$ has eigenvalues $\lambda_1 = \frac{1}{2}(1 + \sqrt{5})$ and $\lambda_2 = \frac{1}{2}(1 - \sqrt{5})$ with associated eigenvectors $\mathbf{v}_1 = (1 + \sqrt{5}, 2)$ and $\mathbf{v}_2 = (1 - \sqrt{5}, 2)$, respectively.

(b) Compute

$$\begin{bmatrix} s_{n+1} \\ s_n \end{bmatrix} = \mathbf{A}^n\mathbf{x}_0 = \mathbf{P}\mathbf{D}^n\mathbf{P}^{-1}\begin{bmatrix} 1 \\ 1 \end{bmatrix}$$

to derive the amazing formula

$$s_n = \frac{1}{\sqrt{5}}\left[\left(\frac{1+\sqrt{5}}{2}\right)^{n+1} - \left(\frac{1-\sqrt{5}}{2}\right)^{n+1}\right].$$

Thus after 1 year the number of rabbit pairs is

$$s_{12} = \frac{1}{\sqrt{5}}\left[\left(\frac{1+\sqrt{5}}{2}\right)^{13} - \left(\frac{1-\sqrt{5}}{2}\right)^{13}\right] = 233.$$

Show that there are 75,025 rabbit pairs after 2 years and over 2.5 trillion rabbit pairs after 5 years.

7 Linear Systems of Differential Equations

7.1 First-Order Systems and Applications

In Chapters 1 and 5 we discussed methods for solving an ordinary differential equation that involves only one dependent variable. Many applications, however, require the use of two or more dependent variables, each a function of a single independent variable (typically time). Such a problem leads naturally to a *system* of simultaneous ordinary differential equations. We will usually denote the independent variable by t and the dependent variables (the unknown functions of t) by x_1, $x_2, x_3, \ldots$, or by $x, y, z, \ldots$. Primes will indicate derivatives with respect to t.

We will restrict our attention to systems in which the number of equations is the same as the number of dependent variables (unknown functions). For instance, a system of two first-order equations in the dependent variables x and y has the general form

$$\begin{aligned} f(t, x, y, x', y') &= 0, \\ g(t, x, y, x', y') &= 0, \end{aligned} \tag{1}$$

where the functions f and g are given. A **solution** of this system is a pair $x(t)$, $y(t)$ of functions of t that satisfy both equations identically over some interval of values of t.

For an example of a second-order system, consider a particle of mass m that moves in space under the influence of a force field $\mathbf{F}$ that depends on time t, the position $(x(t), y(t), z(t))$ of the particle, and its velocity $(x'(t), y'(t), z'(t))$. Writing Newton's law $m\mathbf{a} = \mathbf{F}$ componentwise, we get the system

$$\begin{aligned} mx'' &= F_1(t, x, y, z, x', y', z'), \\ my'' &= F_2(t, x, y, z, x', y', z'), \\ mz'' &= F_3(t, x, y, z, x', y', z') \end{aligned} \tag{2}$$

of three second-order equations with independent variable t and dependent variables x, y, z; the three right-hand side functions F_1, F_2, F_3 are the components of the vector-valued function $\mathbf{F}$.

Initial Applications

Examples 1 and 2 further illustrate how systems of differential equations arise naturally in scientific problems.

Example 1

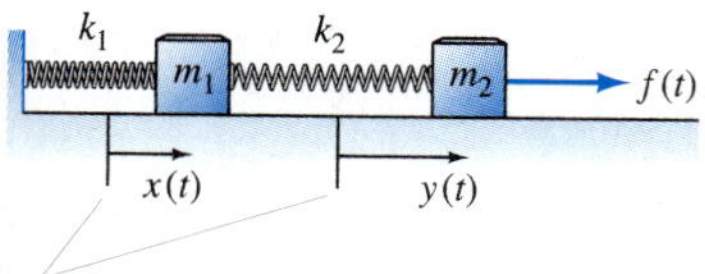

FIGURE 7.1.1. The mass-and-spring system of Example 1.

Consider the system of two masses and two springs shown in Fig. 7.1.1, with a given external force $f(t)$ acting on the right-hand mass m_2. We denote by $x(t)$ the displacement (to the right) of the mass m_1 from its static equilibrium position [when the system is motionless and in equilibrium and $f(t) = 0$] and by $y(t)$ the displacement of the mass m_2 from its static position. Thus the two springs are neither stretched nor compressed when x and y are zero.

In the configuration in Fig. 7.1.1, the first spring is stretched x units and the second by $y - x$ units. We apply Newton's law of motion to the two "free body diagrams" shown in Fig. 7.1.2; we thereby obtain the system

$$\begin{aligned} m_1 x'' &= -k_1 x + k_2(y - x), \\ m_2 y'' &= -k_2(y - x) + f(t) \end{aligned} \tag{3}$$

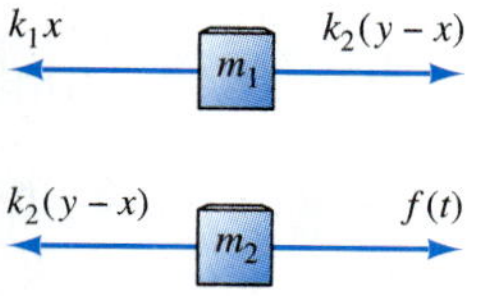

FIGURE 7.1.2. The free body diagrams for the system of Example 1.

of differential equations that the position functions $x(t)$ and $y(t)$ must satisfy. For instance, if $m_1 = 2$, $m_2 = 1$, $k_1 = 4$, $k_2 = 2$, and $f(t) = 40 \sin 3t$ in appropriate physical units, then the system in (3) reduces to

$$\begin{aligned} 2x'' &= -6x + 2y, \\ y'' &= 2x - 2y + 40 \sin 3t. \end{aligned} \tag{4}$$

■

Example 2

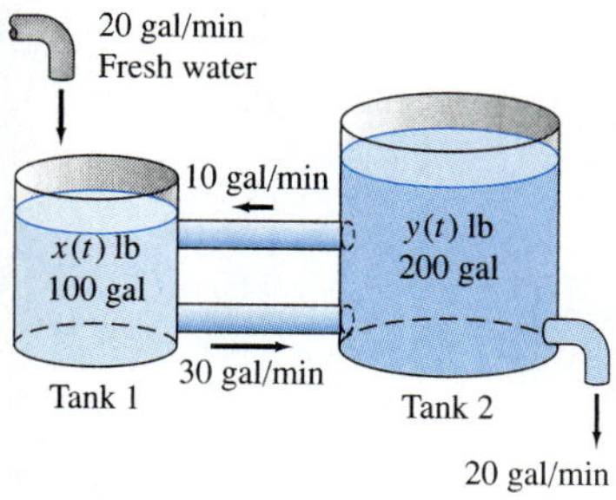

FIGURE 7.1.3. The two brine tanks of Example 2.

Consider two brine tanks connected as shown in Fig. 7.1.3. Tank 1 contains $x(t)$ pounds of salt in 100 gal of brine and tank 2 contains $y(t)$ pounds of salt in 200 gal of brine. The brine in each tank is kept uniform by stirring, and brine is pumped from each tank to the other at the rates indicated in Fig. 7.1.3. In addition, fresh water flows into tank 1 at 20 gal/min, and the brine in tank 2 flows out at 20 gal/min (so the total volume of brine in the two tanks remains constant). The salt concentrations in the two tanks are $x/100$ pounds per gallon and $y/200$ pounds per gallon, respectively. When we compute the rates of change of the amount of salt in the two tanks, we therefore get the system of differential equations that $x(t)$ and $y(t)$ must satisfy:

$$\begin{aligned} x' &= -30 \cdot \frac{x}{100} + 10 \cdot \frac{y}{200} = -\frac{3}{10}x + \frac{1}{20}y, \\ y' &= 30 \cdot \frac{x}{100} - 10 \cdot \frac{y}{200} - 20 \cdot \frac{y}{200} = \frac{3}{10}x - \frac{3}{20}y \end{aligned}$$

—that is,

$$\begin{aligned} 20x' &= -6x + y, \\ 20y' &= 6x - 3y. \end{aligned} \tag{5}$$

■

First-Order Systems

Consider a system of differential equations that can be solved for the highest-order derivatives of the dependent variables that appear, as explicit functions of t and lower-order derivatives of the dependent variables. For instance, in the case of a system of two second-order equations, our assumption is that it can be written in the form

$$\begin{aligned} x_1'' &= f_1(t, x_1, x_2, x_1', x_2'), \\ x_2'' &= f_2(t, x_1, x_2, x_1', x_2'). \end{aligned} \tag{6}$$

It is of both practical and theoretical importance that any such higher-order system can be transformed into an equivalent system of *first-order* equations.

To describe how such a transformation is accomplished, we consider first the "system" consisting of the single nth-order equation

$$x^{(n)} = f(t, x, x', \ldots, x^{(n-1)}). \tag{7}$$

We introduce the dependent variables $x_1, x_2, \ldots, x_n$ defined as follows:

$$x_1 = x, \quad x_2 = x', \quad x_3 = x'', \quad \ldots, \quad x_n = x^{(n-1)}. \tag{8}$$

Note that $x_1' = x' = x_2$, $x_2' = x'' = x_3$, and so on. Hence the substitution of (8) in Eq. (7) yields the system

$$\begin{aligned} x_1' &= x_2, \\ x_2' &= x_3, \\ &\vdots \\ x_{n-1}' &= x_n, \\ x_n' &= f(t, x_1, x_2, \ldots, x_n) \end{aligned} \tag{9}$$

of n *first-order* equations. Evidently, this system is equivalent to the original nth-order equation in (7), in the sense that $x(t)$ is a solution of Eq. (7) if and only if the functions $x_1(t), x_2(t), \ldots, x_n(t)$ defined in (8) satisfy the system of equations in (9).

Example 3 The third-order equation

$$x^{(3)} + 3x'' + 2x' - 5x = \sin 2t$$

is of the form in (7) with

$$f(t, x, x', x'') = 5x - 2x' - 3x'' + \sin 2t.$$

Hence the substitutions

$$x_1 = x, \quad x_2 = x' = x_1', \quad x_3 = x'' = x_2'$$

yield the system

$$\begin{aligned} x_1' &= x_2, \\ x_2' &= x_3, \\ x_3' &= 5x_1 - 2x_2 - 3x_3 + \sin 2t \end{aligned}$$

of three first-order equations. ■

It may appear that the first-order system obtained in Example 3 offers little advantage, because we could use the methods of Chapter 5 to solve the original (linear) third-order equation. But suppose that we were confronted with the nonlinear equation

$$x'' = x^3 + (x')^3,$$

to which none of our earlier methods can be applied. The corresponding first-order system is

$$\begin{aligned} x_1' &= x_2, \\ x_2' &= (x_1)^3 + (x_2)^3, \end{aligned} \tag{10}$$

and we will see in Section 7.6 that there exist effective numerical techniques for approximating the solution of essentially any first-order system. So, in this case, the transformation to a first-order system *is* advantageous. From a practical viewpoint, large systems of higher-order differential equations typically are solved numerically with the aid of the computer, and the first step is to transform such a system into a first-order system for which a standard computer program is available.

Example 4 The system

$$\begin{aligned} 2x'' &= -6x + 2y, \\ y'' &= \quad 2x - 2y + 40\sin 3t \end{aligned} \tag{4}$$

of second-order equations was derived in Example 1. Transform this system into an equivalent first-order system.

Solution Motivated by the equations in (8), we define

$$x_1 = x, \quad x_2 = x' = x_1', \quad y_1 = y, \quad y_2 = y' = y_1'.$$

Then the system in (4) yields the system

$$\begin{aligned} x_1' &= x_2, \\ 2x_2' &= -6x_1 + 2y_1, \\ y_1' &= y_2, \\ y_2' &= 2x_1 - 2y_1 + 40\sin 3t \end{aligned} \tag{11}$$

of four first-order equations in the dependent variables x_1, x_2, y_1, and y_2. ■

Simple Two-Dimensional Systems

The linear second-order differential equation

$$x'' + px' + qx = 0 \tag{12}$$

(with constant coefficients and independent variable t) transforms via the substitutions $x' = y$, $x'' = y'$ into the two-dimensional linear system

$$\begin{aligned} x' &= y, \\ y' &= -qx - py. \end{aligned} \tag{13}$$

Conversely, we can solve this system in (13) by solving the familiar single equation in (12).

Example 5 To solve the two-dimensional system

$$\begin{aligned} x' &= -2y, \\ y' &= \tfrac{1}{2}x, \end{aligned} \tag{14}$$

we begin with the observation that

$$x'' = -2y' = -2\left(\tfrac{1}{2}x\right) = -x.$$

This gives the single second-order equation $x'' + x = 0$ having general solution

$$x(t) = A\cos t + B\sin t = C\cos(t - \alpha),$$

where $A = C\cos\alpha$ and $B = C\sin\alpha$. Then

$$\begin{aligned} y(t) &= -\tfrac{1}{2}x'(t) = -\tfrac{1}{2}(-A\sin t + B\cos t) \\ &= \tfrac{1}{2}C\sin(t - \alpha). \end{aligned}$$

The identity $\cos^2\theta + \sin^2\theta = 1$ therefore implies that, for each value of t, the point $(x(t), y(t))$ lies on the ellipse

$$\frac{x^2}{C^2} + \frac{y^2}{(C/2)^2} = 1$$

with semiaxes C and $C/2$. Figure 7.1.4 shows several such ellipses in the xy-plane. ■

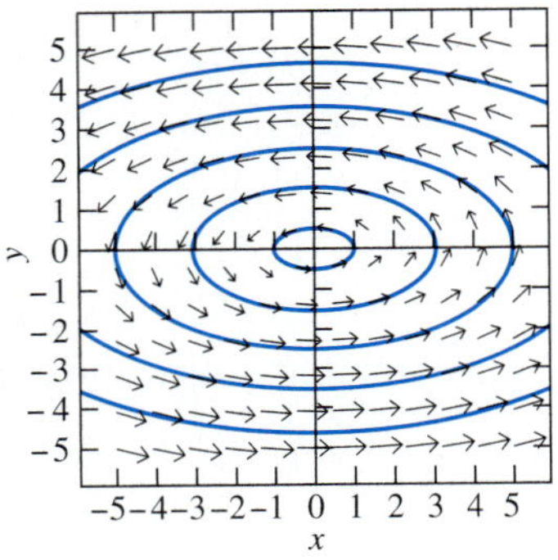

FIGURE 7.1.4. Direction field and solution curves for the system $x' = -2y$, $y' = \frac{1}{2}x$ of Example 5.

A solution $(x(t), y(t))$ of a two-dimensional system

$$\begin{aligned} x' &= f(t, x, y), \\ y' &= g(t, x, y) \end{aligned}$$

can be regarded as a parametrization of a **solution curve** or **trajectory** of the system in the xy-plane. Thus the trajectories of the system in (14) are the ellipses of Fig. 7.1.4. The choice of an initial point $(x(0), y(0))$ determines which one of these trajectories a particular solution parametrizes.

The picture showing a system's trajectories in the xy-plane—its so-called *phase plane portrait*—fails to reveal precisely how the point $(x(t), y(t))$ moves along its trajectory. If the functions f and g do not involve the independent variable t, then a direction field—showing typical arrows representing vectors with components (proportional to) the derivatives $x' = f(x, y)$ and $y' = g(x, y)$—can be plotted. Because the moving point $(x(t), y(t))$ has velocity vector $(x'(t), y'(t))$, this direction field indicates the point's direction of motion along its trajectory. For instance, the direction field plotted in Fig. 7.1.4 indicates that each such point moves counterclockwise around its elliptical trajectory. Additional information can be shown in the separate graphs of $x(t)$ and $y(t)$ as functions of t.

Example 5 *Continued*

With initial values $x(0) = 2$, $y(0) = 0$, the general solution in Example 5 yields

$$x(0) = A = 2, \quad y(0) = -\tfrac{1}{2}B = 0.$$

The resulting particular solution is given by

$$x(t) = 2\cos t, \quad y(t) = \sin t.$$

The graphs of the two functions are shown in Fig. 7.1.5. We see that $x(t)$ initially decreases while $y(t)$ increases. It follows that, as t increases, the solution point $(x(t), y(t))$ traverses the trajectory $\frac{1}{4}x^2 + y^2 = 1$ in the counterclockwise direction, as indicated by the direction field vectors in Fig. 7.1.4. ■

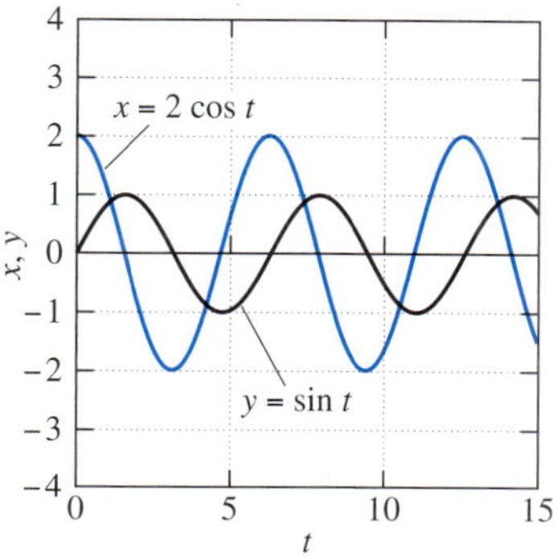

FIGURE 7.1.5. x- and y-solution curves for the initial value problem $x' = -2y$, $y' = \frac{1}{2}x$, $x(0) = 2$, $y(0) = 0$.

Example 6

To find a general solution of the system

$$\begin{aligned} x' &= y, \\ y' &= 2x + y, \end{aligned} \tag{15}$$

we begin with the observation that

$$x'' = y' = 2x + y = x' + 2x.$$

This gives the single linear second-order equation

$$x'' - x' - 2x = 0$$

having the characteristic equation

$$r^2 - r - 2 = (r + 1)(r - 2) = 0$$

and the general solution

$$x(t) = Ae^{-t} + Be^{2t}. \tag{16}$$

Therefore,

$$y(t) = x'(t) = -Ae^{-t} + 2Be^{2t}. \tag{17}$$

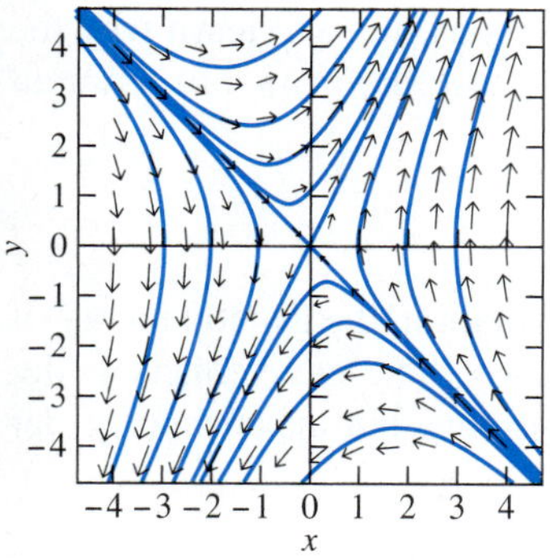

FIGURE 7.1.6. Direction field and solution curves for the system $x' = y$, $y' = 2x + y$ of Example 6.

Typical phase plane trajectories of the system in (15) parametrized by Eqs. (16) and (17) are shown in Fig. 7.1.6. These trajectories may resemble hyperbolas sharing common asymptotes, but Problem 23 shows that their actual form is somewhat more complicated. ■

Example 7 To solve the initial value problem

$$\begin{aligned} x' &= -y, \\ y' &= (1.01)x - (0.2)y, \\ x(0) &= 0, \quad y(0) = 1, \end{aligned} \tag{18}$$

we begin with the observation that

$$x'' = -y' = -[(1.01)x - (0.2)y] = (-1.01)x - (0.2)x'.$$

This gives the single linear second-order equation

$$x'' + (0.2)x' + (1.01)x = 0$$

having the characteristic equation

$$r^2 + (0.2)r + 1.01 = (r + 0.1)^2 + 1 = 0,$$

characteristic roots $-0.1 \pm i$, and the general solution

$$x(t) = e^{-t/10}(A \cos t + B \sin t).$$

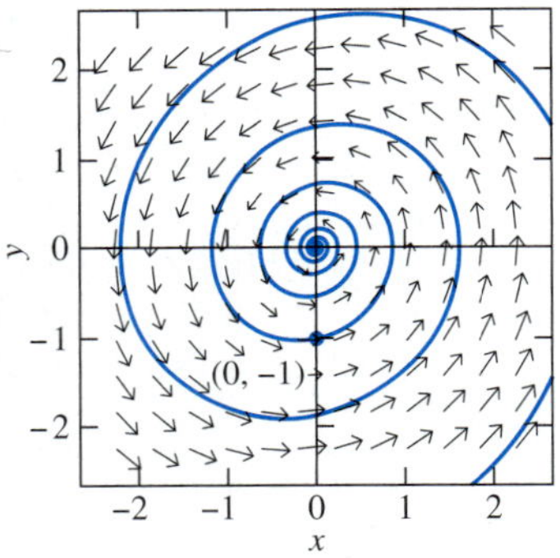

FIGURE 7.1.7. Direction field and solution curve for the system $x' = -y$, $y' = (1.01)x - (0.2)y$ of Example 7.

Then $x(0) = A = 0$, so

$$\begin{aligned} x(t) &= Be^{-t/10} \sin t, \\ y(t) &= -x'(t) = \tfrac{1}{10}Be^{-t/10} \sin t - Be^{-t/10} \cos t. \end{aligned}$$

Finally, $y(0) = -B = 1$, so the desired solution of the system in (18) is

$$\begin{aligned} x(t) &= e^{-t/10} \sin t, \\ y(t) &= \tfrac{1}{10}e^{-t/10}(\sin t + 10 \cos t). \end{aligned} \tag{19}$$

These equations parametrize the spiral trajectory in Fig. 7.1.7; the trajectory approaches the origin as $t \to +\infty$. Figure 7.1.8 shows the x- and y-solution curves given in (19). ■

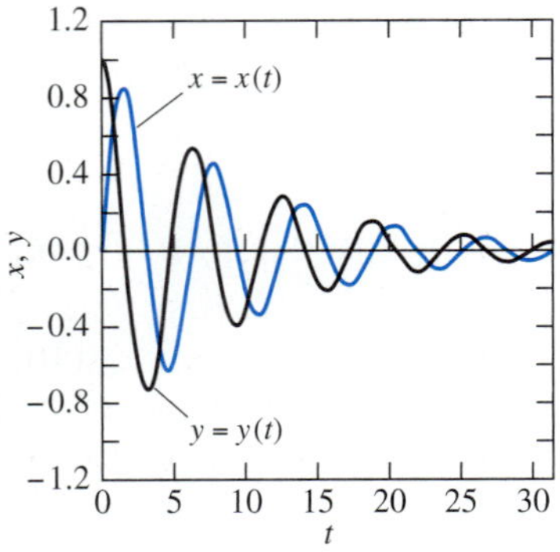

FIGURE 7.1.8. x- and y-solution curves for the initial value problem of Example 7.

When we study linear systems in subsequent sections, we will learn why the superficially similar systems in Examples 5 through 7 have the markedly different trajectories shown in Figs. 7.1.4, 7.1.6, and 7.1.7.

Linear Systems

In addition to practical advantages for numerical computation, the general theory of systems and systematic solution techniques are more easily and more concisely described for first-order systems than for higher-order systems. For instance, consider a *linear* first-order system of the form

$$\begin{aligned} x_1' &= p_{11}(t)x_1 + p_{12}(t)x_2 + \cdots + p_{1n}(t)x_n + f_1(t), \\ x_2' &= p_{21}(t)x_1 + p_{22}(t)x_2 + \cdots + p_{2n}(t)x_n + f_2(t), \\ &\vdots \\ x_n' &= p_{n1}(t)x_1 + p_{n2}(t)x_2 + \cdots + p_{nn}(t)x_n + f_n(t). \end{aligned} \tag{20}$$

We say that this system is **homogeneous** if the functions $f_1, f_2, \ldots, f_n$ are all identically zero; otherwise, it is **nonhomogeneous.** Thus the linear system in (5) is homogeneous, whereas the linear system in (11) is nonhomogeneous. The system in (10) is nonlinear because the right-hand side of the second equation is not a linear function of the dependent variables x_1 and x_2.

A **solution** of the system in (20) is an n-tuple of functions $x_1(t), x_2(t), \ldots, x_n(t)$ that (on some interval) identically satisfy each of the equations in (20). We will see that the general theory of a system of n linear first-order equations shares many similarities with the general theory of a single nth-order linear differential equation. Theorem 1 (proved in Appendix A) is analogous to Theorem 2 of Section 5.2. It tells us that if the coefficient functions p_{ij} and f_j in (20) are continuous, then the system has a unique solution satisfying given initial conditions.

THEOREM 1 Existence and Uniqueness for Linear Systems

Suppose that the functions $p_{11}, p_{12}, \ldots, p_{nn}$ and the functions $f_1, f_2, \ldots, f_n$ are continuous on the open interval I containing the point a. Then, given the n numbers $b_1, b_2, \ldots, b_n$, the system in (20) has a unique solution on the entire interval I that satisfies the n initial conditions

$$x_1(a) = b_1, \quad x_2(a) = b_2, \quad \ldots, \quad x_n(a) = b_n. \tag{21}$$

Thus n initial conditions are needed to determine a solution of a system of n linear first-order equations, and we therefore expect a general solution of such a system to involve n arbitrary constants. For instance, we saw in Example 4 that the second-order linear system

$$\begin{aligned} 2x'' &= -6x + 2y, \\ y'' &= 2x - 2y + 40\sin 3t, \end{aligned}$$

which describes the position functions $x(t)$ and $y(t)$ of Example 1, is equivalent to the system of *four* first-order linear equations in (11). Hence four initial conditions would be needed to determine the subsequent motions of the two masses in Example 1. Typical initial values would be the initial positions $x(0)$ and $y(0)$ and the initial velocities $x'(0)$ and $y'(0)$. On the other hand, we found that the amounts $x(t)$ and $y(t)$ of salt in the two tanks of Example 2 are described by the system

$$\begin{aligned} 20x' &= -6x + y, \\ 20y' &= 6x - 3y \end{aligned}$$

of *two* first-order linear equations. Hence the two initial values $x(0)$ and $y(0)$ should suffice to determine the solution. Given a higher-order system, we often must transform it into an equivalent first-order system to discover how many initial conditions are needed to determine a unique solution. Theorem 1 tells us that the number of such conditions is precisely the same as the number of equations in the equivalent first-order system.

7.1 Problems

In Problems 1 through 10, transform the given differential equation or system into an equivalent system of first-order differential equations.

1. $x'' + 3x' + 7x = t^2$

2. $x^{(4)} + 6x'' - 3x' + x = \cos 3t$

3. $t^2x'' + tx' + (t^2 - 1)x = 0$

4. $t^3x^{(3)} - 2t^2x'' + 3tx' + 5x = \ln t$

5. $x^{(3)} = (x')^2 + \cos x$

6. $x'' - 5x + 4y = 0,\ y'' + 4x - 5y = 0$

7. $x'' = -\dfrac{kx}{(x^2 + y^2)^{3/2}},\ y'' = -\dfrac{ky}{(x^2 + y^2)^{3/2}}$

8. $x'' + 3x' + 4x - 2y = 0,\ y'' + 2y' - 3x + y = \cos t$

9. $x'' = 3x - y + 2z,\ y'' = x + y - 4z,\ z'' = 5x - y - z$

10. $x'' = (1 - y)x,\ y'' = (1 - x)y$

Use the method of Examples 5, 6, and 7 to find general solutions of the systems in Problems 11 through 20. If initial conditions are given, find the corresponding particular solution. For each problem, use a computer system or graphing calculator to construct a direction field and typical solution curves for the given system.

11. $x' = y,\ y' = -x$

12. $x' = y,\ y' = x$

13. $x' = -2y,\ y' = 2x;\ x(0) = 1,\ y(0) = 0$

14. $x' = 10y,\ y' = -10x;\ x(0) = 3,\ y(0) = 4$

15. $x' = \frac{1}{2}y,\ y' = -8x$

16. $x' = 8y,\ y' = -2x$

17. $x' = y,\ y' = 6x - y;\ x(0) = 1,\ y(0) = 2$

18. $x' = -y,\ y' = 10x - 7y;\ x(0) = 2,\ y(0) = -7$

19. $x' = -y,\ y' = 13x + 4y;\ x(0) = 0,\ y(0) = 3$

20. $x' = y,\ y' = -9x + 6y$

21. (a) Calculate $[x(t)]^2 + [y(t)]^2$ to show that the trajectories of the system $x' = y$, $y' = -x$ of Problem 11 are circles. (b) Calculate $[x(t)]^2 - [y(t)]^2$ to show that the trajectories of the system $x' = y$, $y' = x$ of Problem 12 are hyperbolas.

22. (a) Beginning with the general solution of the system $x' = -2y$, $y' = 2x$ of Problem 13, calculate $x^2 + y^2$ to show that the trajectories are circles. (b) Show similarly that the trajectories of the system $x' = \frac{1}{2}y$, $y' = -8x$ of Problem 15 are ellipses with equations of the form $16x^2 + y^2 = C^2$.

23. First solve Eqs. (16) and (17) for e^{-t} and e^{2t} in terms of $x(t)$, $y(t)$, and the constants A and B. Then substitute the results in $(e^{2t})(e^{-t})^2 = 1$ to show that the trajectories of the system $x' = y$, $y' = 2x + y$ in Example 6 satisfy an equation of the form

$$4x^3 - 3xy^2 + y^3 = C \quad \text{(constant)}.$$

Then show that $C = 0$ yields the straight lines $y = -x$ and $y = 2x$ that are visible in Fig. 7.1.6.

24. Derive the equations

$$\begin{aligned} m_1x_1'' &= -(k_1 + k_2)x_1 + k_2x_2, \\ m_2x_2'' &= k_2x_1 - (k_2 + k_3)x_2 \end{aligned}$$

for the displacements (from equilibrium) of the two masses shown in Fig. 7.1.9.

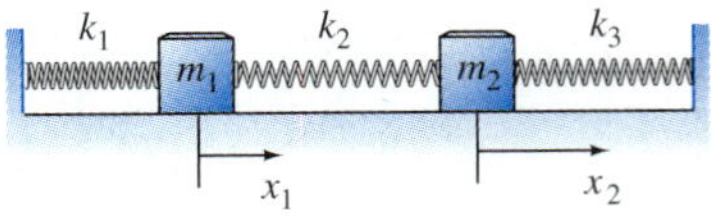

FIGURE 7.1.9. The system of Problem 24.

25. Two particles each of mass m are attached to a string under (constant) tension T, as indicated in Fig. 7.1.10. Assume that the particles oscillate vertically (that is, parallel to the y-axis) with amplitudes so small that the sines of the angles shown are accurately approximated by their tangents. Show that the displacements y_1 and y_2 satisfy the equations

$$ky_1'' = -2y_1 + y_2, \quad ky_2'' = y_1 - 2y_2,$$

where $k = mL/T$.

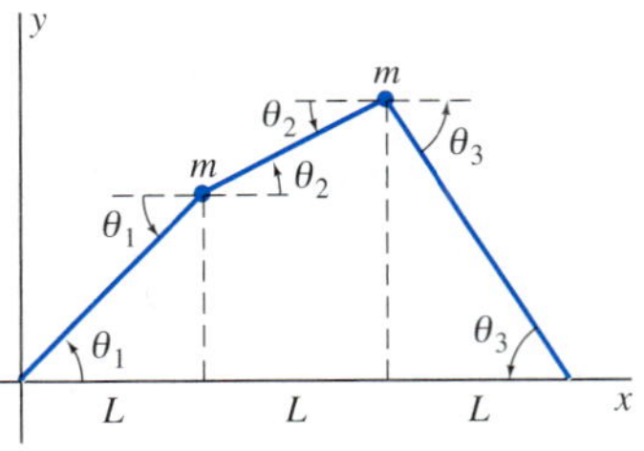

FIGURE 7.1.10. The mechanical system of Problem 25.

26. Three 100-gal fermentation vats are connected as indicated in Fig. 7.1.11, and the mixtures in each tank are kept uniform by stirring. Denote by $x_i(t)$ the amount (in pounds) of alcohol in tank T_i at time t ($i = 1, 2, 3$). Suppose that the mixture circulates between the tanks at the rate of 10 gal/min. Derive the equations

$$\begin{aligned} 10x_1' &= -x_1 + x_3 \\ 10x_2' &= x_1 - x_2 \\ 10x_3' &= x_2 - x_3. \end{aligned}$$

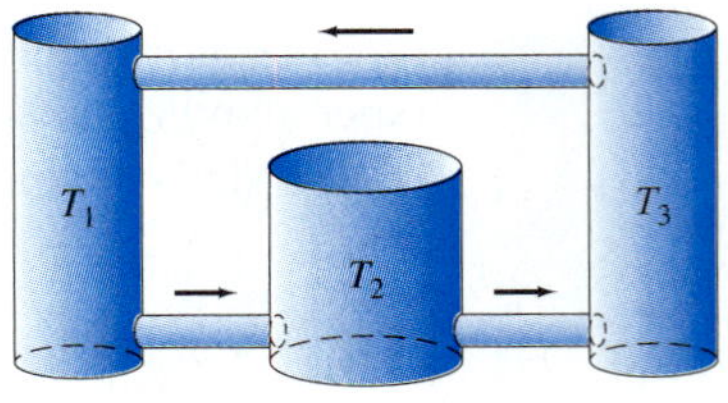

FIGURE 7.1.11. The fermentation tanks of Problem 26.

27. A particle of mass m moves in the plane with coordinates $(x(t), y(t))$ under the influence of a force that is directed toward the origin and has magnitude $k/(x^2 + y^2)$—an inverse-square central force field. Show that

$$mx'' = -\frac{kx}{r^3} \quad \text{and} \quad my'' = -\frac{ky}{r^3},$$

where $r = \sqrt{x^2 + y^2}$.

28. Suppose that a projectile of mass m moves in a vertical plane in the atmosphere near the surface of the earth under the influence of two forces: a downward gravitational force of magnitude mg, and a resistive force $\mathbf{F}_R$ that is directed opposite to the velocity vector $\mathbf{v}$ and has magnitude kv^2 (where $v = |\mathbf{v}|$ is the speed of the projectile; see Fig. 7.1.12). Show that the equations of motion of the projectile are

$$mx'' = -kvx', \quad my'' = -kvy' - mg,$$

where $v = \sqrt{(x')^2 + (y')^2}$.

29. Suppose that a particle with mass m and electrical charge q moves in the xy-plane under the influence of the magnetic field $\mathbf{B} = B\mathbf{k}$ (thus a uniform field parallel to the z-axis), so the force on the particle is $\mathbf{F} = q\mathbf{v} \times \mathbf{B}$ if its velocity is $\mathbf{v}$. Show that the equations of motion of the particle are

$$mx'' = +qBy', \quad my'' = -qBx'.$$

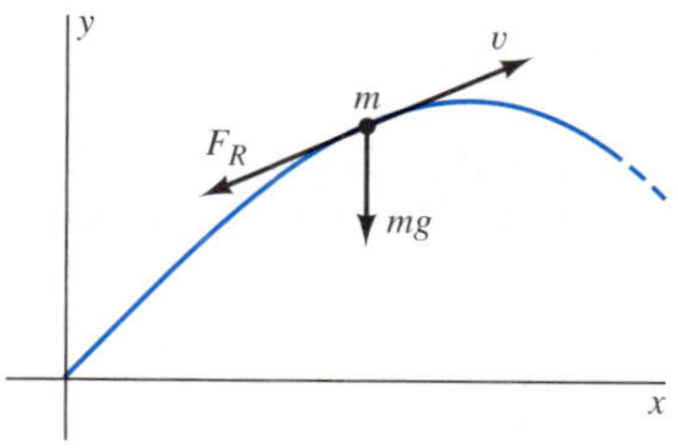

FIGURE 7.1.12. The trajectory of the projectile of Problem 28.

7.1 Application Gravitation and Kepler's Laws of Planetary Motion

Around the turn of the seventeenth century, Johannes Kepler analyzed a lifetime of planetary observations by the astronomer Tycho Brahe. Kepler concluded that the motion of the planets around the sun is described by the following three propositions, now known as **Kepler's laws of planetary motion:**

1. The orbit of each planet is an ellipse with the sun at one focus.
2. The radius vector from the sun to each planet sweeps out area at a constant rate.
3. The square of the planet's period of revolution is proportional to the cube of the major semiaxis of its elliptical orbit.

In his *Principia Mathematica* (1687) Isaac Newton deduced the inverse square law of gravitation from Kepler's laws. In this application we lead you (in the opposite direction) through a derivation of Kepler's first two laws from Newton's law of gravitation. Newton's explanation of planetary motion in terms of gravitation is a landmark of scientific and intellectual history.

Assume that the sun is located at the origin in the plane of motion of a planet, and write the position vector of the planet in the form

$$\mathbf{r}(t) = (x(t), y(t)) = x\mathbf{i} + y\mathbf{j}, \tag{1}$$

where $\mathbf{i} = (1, 0)$ and $\mathbf{j} = (0, 1)$ denote the unit vectors in the positive x- and y-directions. Then the inverse-square law of gravitation implies (Problem 27) that the acceleration vector $\mathbf{r}''(t)$ of the planet is given by

$$\mathbf{r}'' = -\frac{k\mathbf{r}}{r^3}, \tag{2}$$

where $r = \sqrt{x^2 + y^2}$ is the distance from the sun to the planet. If the polar coordinates of the planet at time t are $(r(t), \theta(t))$, then the radial and transverse unit vectors shown in Fig. 7.1.13 are given by

$$\mathbf{u}_r = \mathbf{i}\cos\theta + \mathbf{j}\sin\theta \quad \text{and} \quad \mathbf{u}_\theta = -\mathbf{i}\sin\theta + \mathbf{j}\cos\theta. \tag{3}$$

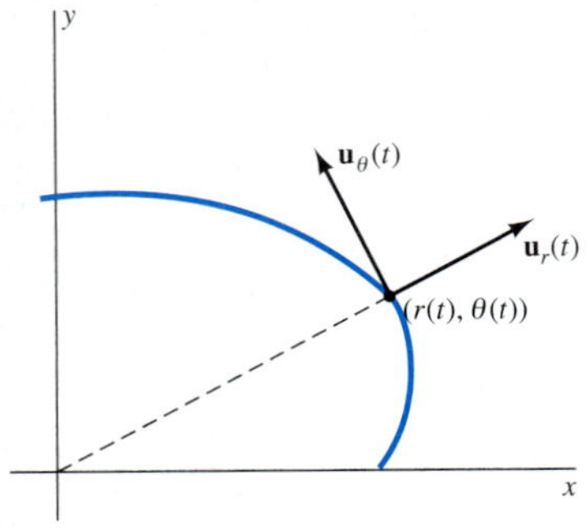

FIGURE 7.1.13. The radial and transverse unit vectors $\mathbf{u}_r$ and $\mathbf{u}_\theta$.

The radial unit vector $\mathbf{u}_r$ (when located at the planet's position) always points directly away from the origin, so $\mathbf{u}_r = \mathbf{r}/r$, and the transverse unit vector $\mathbf{u}_\theta$ is obtained from $\mathbf{u}_r$ by a 90° counterclockwise rotation.

STEP 1: Differentiate the equations in (3) componentwise to show that

$$\frac{d\mathbf{u}_r}{dt} = \mathbf{u}_\theta \frac{d\theta}{dt} \quad \text{and} \quad \frac{d\mathbf{u}_\theta}{dt} = -\mathbf{u}_r \frac{d\theta}{dt}. \tag{4}$$

STEP 2: Use the equations in (4) to differentiate the planet's position vector $\mathbf{r} = r\mathbf{u}_r$ and thereby show that its velocity vector is given by

$$\mathbf{v} = \frac{d\mathbf{r}}{dt} = \mathbf{u}_r \frac{dr}{dt} + r\frac{d\theta}{dt}\mathbf{u}_\theta. \tag{5}$$

STEP 3: Differentiate again to show that the planet's acceleration vector $\mathbf{a} = d\mathbf{v}/dt$ is given by

$$\mathbf{a} = \left[\frac{d^2r}{dt^2} - r\left(\frac{d\theta}{dt}\right)^2\right]\mathbf{u}_r + \left[\frac{1}{r}\frac{d}{dt}\left(r^2\frac{d\theta}{dt}\right)\right]\mathbf{u}_\theta. \tag{6}$$

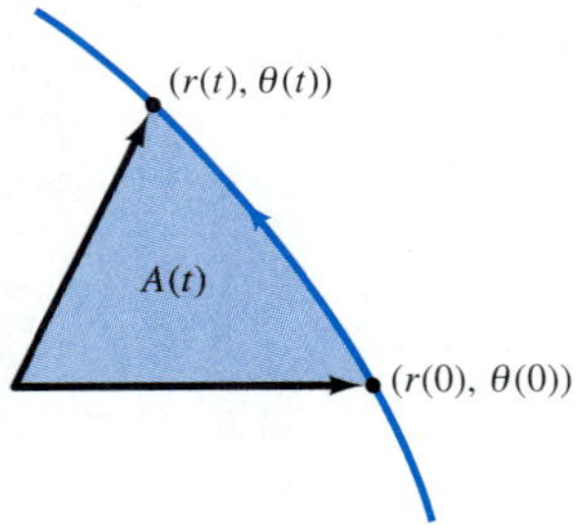

FIGURE 7.1.14. Area swept out by the radius vector.

STEP 4: The radial and transverse components on the right-hand sides in Eqs. (2) and (6) must agree. Equating the transverse components—that is, the coefficients of $\mathbf{u}_\theta$—we get

$$\frac{1}{r}\frac{d}{dt}\left(r^2\frac{d\theta}{dt}\right) = 0, \tag{7}$$

so it follows that

$$r^2\frac{d\theta}{dt} = h, \tag{8}$$

where h is a constant. Because the polar-coordinate area element—for computation of the area $A(t)$ in Fig. 7.1.14—is given by $dA = \frac{1}{2}r^2d\theta$, Eq. (8) implies that the derivative $A'(t)$ is *constant*, which is a statement of Kepler's second law.

STEP 5: Equate radial components in (2) and (6), and then use the result in (8) to show that the planet's radial coordinate function $r(t)$ satisfies the second-order differential equation

$$\frac{d^2r}{dt^2} - \frac{h^2}{r^3} = -\frac{k}{r^2}. \tag{9}$$

STEP 6: Although the differential equation in (9) is nonlinear, it can be transformed to a linear equation by means of a simple substitution. For this purpose, assume that the orbit can be written in the polar-coordinate form $r = r(\theta)$, and first use the chain rule and Eq. (8) to show that if $r = 1/z$, then

$$\frac{dr}{dt} = -h\frac{dz}{d\theta}.$$

Differentiate again, to deduce from Eq. (9) that the function $z(\theta) = 1/r(\theta)$ satisfies the second-order equation

$$\frac{d^2z}{d\theta^2} + z = \frac{k}{h^2}. \tag{10}$$

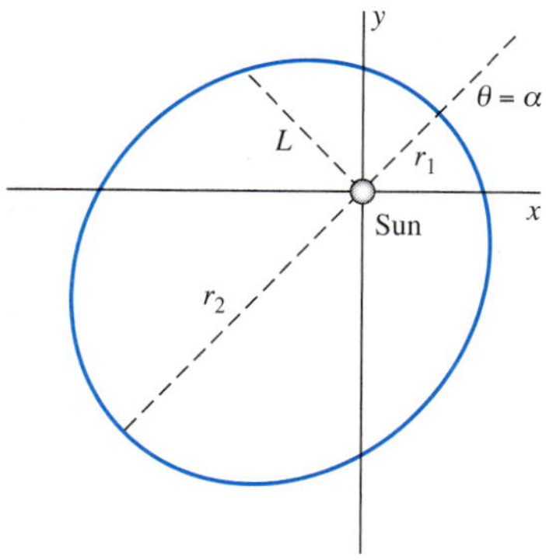

FIGURE 7.1.15. The elliptical orbit

$$r = \frac{L}{1 + e\cos(\theta - \alpha)}$$

with perihelion distance $r_1 = L/(1+e)$ and aphelion distance $r_2 = L/(1-e)$.

STEP 7: Show that the general solution of Eq. (10) is

$$z(\theta) = A\cos\theta + B\sin\theta + \frac{k}{h^2}. \tag{11}$$

STEP 8: Finally, deduce from Eq. (11) that $r(\theta) = 1/z(\theta)$ is given by

$$r(\theta) = \frac{L}{1 + e\cos(\theta - \alpha)}, \tag{12}$$

with $e = Ch^2/k$, $C\cos\alpha = A$, $C\sin\alpha = B$, and $L = h^2/k$. The polar-coordinate graph of Eq. (12) is a conic section of eccentricity e—an ellipse if $0 \leqq e < 1$, a parabola if $e = 1$, and a hyperbola if $e > 1$—with focus at the origin. Planetary orbits are bounded and therefore are ellipses, with eccentricity $e < 1$. As indicated in Fig. 7.1.15, the major axis of the ellipse lies along the radial line $\theta = \alpha$.

STEP 9: Plot some typical elliptical orbits—as described by (12)—having different eccentricities, sizes, and orientations. In rectangular coordinates you can write

$$x(t) = r(t)\cos t, \quad y(t) = r(t)\sin t, \quad 0 \leqq t \leqq 2\pi$$

to plot an elliptical orbit with eccentricity e, semilatus rectum L (Fig. 7.1.15), and rotation angle α. The eccentricity of the earth's orbit is $e \approx 0.0167$, so close to zero that the orbit looks nearly circular (though with the sun off center), and the eccentricities of the other planetary orbits range from 0.0068 for Venus and 0.0933 for Mars to 0.2056 for Mercury (and 0.2486 for Pluto, no longer classified a planet). But many comets have highly eccentric orbits–Halley's comet has $e \approx 0.97$ (Fig.7.1.16).

FIGURE 7.1.16. The shape of the orbit of Halley's comet.

7.2 Matrices and Linear Systems

A system of differential equations often can be simplified by expressing it as a single differential equation involving a matrix-valued function. A **matrix-valued function**, or simply **matrix function**, is a matrix such as

$$\mathbf{x}(t) = \begin{bmatrix} x_1(t) \\ x_2(t) \\ \vdots \\ x_n(t) \end{bmatrix} \tag{1}$$

or

$$\mathbf{A}(t) = \begin{bmatrix} a_{11}(t) & a_{12}(t) & \cdots & a_{1n}(t) \\ a_{21}(t) & a_{22}(t) & \cdots & a_{2n}(t) \\ \vdots & \vdots & & \vdots \\ a_{m1}(t) & a_{m2}(t) & \cdots & a_{mn}(t) \end{bmatrix}, \tag{2}$$

in which each entry is a function of t. We say that the matrix function $\mathbf{A}(t)$ is **continuous** (or **differentiable**) at a point (or on an interval) if each of its elements has the same property. The **derivative** of a differentiable matrix function is defined by elementwise differentiation; that is,

$$\mathbf{A}'(t) = \frac{d\mathbf{A}}{dt} = \left[\frac{da_{ij}}{dt} \right]. \tag{3}$$

Example 1 If

$$\mathbf{x}(t) = \begin{bmatrix} t \\ t^2 \\ e^{-t} \end{bmatrix} \quad \text{and} \quad \mathbf{A}(t) = \begin{bmatrix} \sin t & 1 \\ t & \cos t \end{bmatrix},$$

then

$$\frac{d\mathbf{x}}{dt} = \begin{bmatrix} 1 \\ 2t \\ -e^{-t} \end{bmatrix} \quad \text{and} \quad \mathbf{A}'(t) = \begin{bmatrix} \cos t & 0 \\ 1 & -\sin t \end{bmatrix}.$$ ■

The differentiation rules

$$\frac{d}{dt}(\mathbf{A} + \mathbf{B}) = \frac{d\mathbf{A}}{dt} + \frac{d\mathbf{B}}{dt} \tag{4}$$

and

$$\frac{d}{dt}(\mathbf{AB}) = \mathbf{A}\frac{d\mathbf{B}}{dt} + \frac{d\mathbf{A}}{dt}\mathbf{B} \tag{5}$$

follow readily by elementwise application of the analogous differentiation rules of elementary calculus for real-valued functions. If c is a (constant) real number and $\mathbf{C}$ is a constant matrix, then

$$\frac{d}{dt}(c\mathbf{A}) = c\frac{d\mathbf{A}}{dt}, \quad \frac{d}{dt}(\mathbf{CA}) = \mathbf{C}\frac{dA}{dt}, \quad \text{and} \quad \frac{d}{dt}(\mathbf{AC}) = \frac{d\mathbf{A}}{dt}\mathbf{C}. \tag{6}$$

Because of the noncommutativity of matrix multiplication, it is important not to reverse the order of the factors in Eqs. (5) and (6).

First-Order Linear Systems

The notation and terminology of matrices and vectors may seem rather elaborate when first encountered, but it is readily assimilated with practice. Our main use for matrix notation will be the simplification of computations with systems of differential equations, especially those computations that would be burdensome in scalar notation.

We discuss here the general system of n first-order linear equations

$$\begin{aligned}
x_1' &= p_{11}(t)x_1 + p_{12}(t)x_2 + \cdots + p_{1n}(t)x_n + f_1(t), \\
x_2' &= p_{21}(t)x_1 + p_{22}(t)x_2 + \cdots + p_{2n}(t)x_n + f_2(t), \\
x_3' &= p_{31}(t)x_1 + p_{32}(t)x_2 + \cdots + p_{3n}(t)x_n + f_3(t), \\
&\;\;\vdots \\
x_n' &= p_{n1}(t)x_1 + p_{n2}(t)x_2 + \cdots + p_{nn}(t)x_n + f_n(t).
\end{aligned} \tag{7}$$

If we introduce the *coefficient matrix*

$$\mathbf{P}(t) = \left[\, p_{ij}(t) \,\right]$$

and the column vectors

$$\mathbf{x} = \left[\, x_i \,\right] \quad \text{and} \quad \mathbf{f}(t) = \left[\, f_i(t) \,\right],$$

then—as illustrated in Example 2 below—the system in (7) takes the form of a single matrix equation

$$\frac{d\mathbf{x}}{dt} = \mathbf{P}(t)\mathbf{x} + \mathbf{f}(t). \tag{8}$$

We will see that the general theory of the linear system in (7) closely parallels that of a single nth-order equation. The matrix notation used in Eq. (8) not only emphasizes this analogy, but also saves a great deal of space.

A **solution** of Eq. (8) on the open interval I is a column vector function $\mathbf{x}(t) = \left[x_i(t)\right]$ such that the component functions of $\mathbf{x}$ satisfy the system in (7) identically on I. If the functions $p_{ij}(t)$ and $f_i(t)$ are all continuous on I, then Theorem 1 of Section 7.1 guarantees the existence on I of a unique solution $\mathbf{x}(t)$ satisfying preassigned initial conditions $\mathbf{x}(a) = \mathbf{b}$.

Example 2 The first-order system

$$\begin{aligned} x_1' &= 4x_1 - 3x_2, \\ x_2' &= 6x_1 - 7x_2 \end{aligned}$$

can be recast as a matrix equation by writing

$$\mathbf{x}' = \begin{bmatrix} x_1' \\ x_2' \end{bmatrix} = \begin{bmatrix} 4x_1 - 3x_2 \\ 6x_1 - 7x_2 \end{bmatrix} = \begin{bmatrix} 4 & -3 \\ 6 & -7 \end{bmatrix}\begin{bmatrix} x_1 \\ x_2 \end{bmatrix} = \begin{bmatrix} 4 & -3 \\ 6 & -7 \end{bmatrix}\mathbf{x}.$$

Thus we obtain a matrix equation of the form in (8),

$$\frac{d\mathbf{x}}{dt} = \mathbf{P}(t)\mathbf{x} + \mathbf{f}(t) \quad \text{with} \quad \mathbf{P}(t) = \begin{bmatrix} 4 & -3 \\ 6 & -7 \end{bmatrix} \quad \text{and} \quad \mathbf{f}(t) = \begin{bmatrix} 0 \\ 0 \end{bmatrix} = \mathbf{0}.$$

To verify that the vector functions

$$\mathbf{x}_1(t) = \begin{bmatrix} 3e^{2t} \\ 2e^{2t} \end{bmatrix} \quad \text{and} \quad \mathbf{x}_2(t) = \begin{bmatrix} e^{-5t} \\ 3e^{-5t} \end{bmatrix}$$

are both solutions of the matrix differential equation with coefficient matrix $\mathbf{P}$, we need only calculate

$$\mathbf{P}\mathbf{x}_1 = \begin{bmatrix} 4 & -3 \\ 6 & -7 \end{bmatrix}\begin{bmatrix} 3e^{2t} \\ 2e^{2t} \end{bmatrix} = \begin{bmatrix} 6e^{2t} \\ 4e^{2t} \end{bmatrix} = \mathbf{x}_1'$$

and

$$\mathbf{P}\mathbf{x}_2 = \begin{bmatrix} 4 & -3 \\ 6 & -7 \end{bmatrix}\begin{bmatrix} e^{-5t} \\ 3e^{-5t} \end{bmatrix} = \begin{bmatrix} -5e^{-5t} \\ -15e^{-5t} \end{bmatrix} = \mathbf{x}_2'.$$

■

To investigate the general nature of the solutions of Eq. (8), we consider first the **associated homogeneous equation**

$$\frac{d\mathbf{x}}{dt} = \mathbf{P}(t)\mathbf{x}, \tag{9}$$

which has the form shown in Eq. (8), but with $\mathbf{f}(t) \equiv \mathbf{0}$. We expect it to have n solutions $\mathbf{x}_1, \mathbf{x}_2, \ldots, \mathbf{x}_n$ that are independent in some appropriate sense, and that are such that every solution of Eq. (9) is a linear combination of these n particular solutions. Given n solutions $\mathbf{x}_1, \mathbf{x}_2, \ldots, \mathbf{x}_n$ of Eq. (9), let us write

$$\mathbf{x}_j(t) = \begin{bmatrix} x_{1j}(t) \\ \vdots \\ x_{ij}(t) \\ \vdots \\ x_{nj}(t) \end{bmatrix}. \tag{10}$$

Thus $x_{ij}(t)$ denotes the ith component of the vector $\mathbf{x}_j(t)$, so the second subscript refers to the vector function $\mathbf{x}_j(t)$, whereas the first subscript refers to a component of this function. Theorem 1 is analogous to Theorem 1 of Section 5.2.

THEOREM 1 Principle of Superposition

Let $\mathbf{x}_1, \mathbf{x}_2, \ldots, \mathbf{x}_n$ be n solutions of the homogeneous linear equation in (9) on the open interval I. If $c_1, c_2, \ldots, c_n$ are constants, then the linear combination

$$\mathbf{x}(t) = c_1\mathbf{x}_1(t) + c_2\mathbf{x}_2(t) + \cdots + c_n\mathbf{x}_n(t) \tag{11}$$

is also a solution of Eq. (9) on I.

***Proof*:** We know that $\mathbf{x}_i' = \mathbf{P}(t)\mathbf{x}_i$ for each i $(1 \leqq i \leqq n)$, so it follows immediately that

$$\begin{aligned} \mathbf{x}' &= c_1\mathbf{x}_1' + c_2\mathbf{x}_2' + \cdots + c_n\mathbf{x}_n' \\ &= c_1\mathbf{P}(t)\mathbf{x}_1 + c_2\mathbf{P}(t)\mathbf{x}_2 + \cdots + c_n\mathbf{P}(t)\mathbf{x}_n \\ &= \mathbf{P}(t)(c_1\mathbf{x}_1 + c_2\mathbf{x}_2 + \cdots + c_n\mathbf{x}_n). \end{aligned}$$

That is, $\mathbf{x}' = \mathbf{P}(t)\mathbf{x}$, as desired. The remarkable simplicity of this proof demonstrates clearly one advantage of matrix notation. ■

Example 2 *Continued*

If $\mathbf{x}_1$ and $\mathbf{x}_2$ are the two solutions of

$$\frac{d\mathbf{x}}{dt} = \begin{bmatrix} 4 & -3 \\ 6 & -7 \end{bmatrix}\mathbf{x}$$

discussed in Example 2, then the linear combination

$$\mathbf{x}(t) = c_1\mathbf{x}_1(t) + c_2\mathbf{x}_2(t) = c_1\begin{bmatrix} 3e^{2t} \\ 2e^{2t} \end{bmatrix} + c_2\begin{bmatrix} e^{-5t} \\ 3e^{-5t} \end{bmatrix}$$

is also a solution. In scalar form with $\mathbf{x} = [x_1 \quad x_2]^T$, this gives the solution

$$\begin{aligned} x_1(t) &= 3c_1e^{2t} + c_2e^{-5t}, \\ x_2(t) &= 2c_1e^{2t} + 3c_2e^{-5t}. \end{aligned}$$

■

Independence and General Solutions

Linear independence is defined in the same way for vector-valued functions as for real-valued functions (Section 5.2). The vector-valued functions $\mathbf{x}_1, \mathbf{x}_2, \ldots, \mathbf{x}_n$ are **linearly dependent** on the interval I provided that there exist constants $c_1, c_2, \ldots, c_n$, *not all zero*, such that

$$c_1\mathbf{x}_1(t) + c_2\mathbf{x}_2(t) + \cdots + c_n\mathbf{x}_n(t) = \mathbf{0} \tag{12}$$

for all t in I. Otherwise, they are **linearly independent.** Equivalently, they are linearly independent provided that no one of them is a linear combination of the others. For instance, the two solutions $\mathbf{x}_1$ and $\mathbf{x}_2$ of Example 2 are linearly independent because, clearly, neither is a scalar multiple of the other.

Just as in the case of a single nth-order equation, there is a Wronskian determinant that tells us whether or not n given solutions of the homogeneous equation in (9) are linearly dependent. If $\mathbf{x}_1, \mathbf{x}_2, \ldots, \mathbf{x}_n$ are such solutions, then their **Wronskian** is the $n \times n$ determinant

$$W(t) = \begin{vmatrix} x_{11}(t) & x_{12}(t) & \cdots & x_{1n}(t) \\ x_{21}(t) & x_{22}(t) & \cdots & x_{2n}(t) \\ \vdots & \vdots & & \vdots \\ x_{n1}(t) & x_{n2}(t) & \cdots & x_{nn}(t) \end{vmatrix}, \tag{13}$$

using the notation in (10) for the components of the solutions. We may write either $W(t)$ or $W(\mathbf{x}_1, \mathbf{x}_2, \ldots, \mathbf{x}_n)$. Note that W is the determinant of the matrix that has as its *column* vectors the solutions $\mathbf{x}_1, \mathbf{x}_2, \ldots, \mathbf{x}_n$. Theorem 2 is analogous to Theorem 3 of Section 5.2. Moreover, its proof is essentially the same, with the definition of $W(\mathbf{x}_1, \mathbf{x}_2, \ldots, \mathbf{x}_n)$ in Eq. (13) substituted for the definition of the Wronskian of n solutions of a single nth-order equation. (See Problems 34 through 36.)

THEOREM 2 Wronskians of Solutions

Suppose that $\mathbf{x}_1, \mathbf{x}_2, \ldots, \mathbf{x}_n$ are n solutions of the homogeneous linear equation $\mathbf{x}' = \mathbf{P}(t)\mathbf{x}$ on an open interval I. Suppose also that $\mathbf{P}(t)$ is continuous on I. Let

$$W = W(\mathbf{x}_1, \mathbf{x}_2, \ldots, \mathbf{x}_n).$$

Then

- If $\mathbf{x}_1, \mathbf{x}_2, \ldots, \mathbf{x}_n$ are linearly dependent on I, then $W = 0$ at every point of I.
- If $\mathbf{x}_1, \mathbf{x}_2, \ldots, \mathbf{x}_n$ are linearly independent on I, then $W \neq 0$ at each point of I.

Thus there are only two possibilities for solutions of homogeneous systems: Either $W = 0$ at *every* point of I, or $W = 0$ at *no* point of I.

Example 3 It is readily verified (as in Example 2) that

$$\mathbf{x}_1(t) = \begin{bmatrix} 2e^t \\ 2e^t \\ e^t \end{bmatrix}, \quad \mathbf{x}_2(t) = \begin{bmatrix} 2e^{3t} \\ 0 \\ -e^{3t} \end{bmatrix}, \quad \text{and} \quad \mathbf{x}_3(t) = \begin{bmatrix} 2e^{5t} \\ -2e^{5t} \\ e^{5t} \end{bmatrix}$$

are solutions of the equation

$$\frac{d\mathbf{x}}{dt} = \begin{bmatrix} 3 & -2 & 0 \\ -1 & 3 & -2 \\ 0 & -1 & 3 \end{bmatrix} \mathbf{x}. \tag{14}$$

The Wronskian of these solutions is

$$W = \begin{vmatrix} 2e^t & 2e^{3t} & 2e^{5t} \\ 2e^t & 0 & -2e^{5t} \\ e^t & -e^{3t} & e^{5t} \end{vmatrix} = e^{9t} \begin{vmatrix} 2 & 2 & 2 \\ 2 & 0 & -2 \\ 1 & -1 & 1 \end{vmatrix} = -16e^{9t},$$

which is never zero. Hence Theorem 2 implies that the solutions $\mathbf{x}_1$, $\mathbf{x}_2$, and $\mathbf{x}_3$ are linearly independent (on any open interval). ■

Theorem 3 is analogous to Theorem 4 of Section 5.2. It says that the **general solution** of the *homogeneous* $n \times n$ system $\mathbf{x}' = \mathbf{P}(t)\mathbf{x}$ is a linear combination

$$\mathbf{x} = c_1\mathbf{x}_1 + c_2\mathbf{x}_2 + \cdots + c_n\mathbf{x}_n \tag{15}$$

of any n given linearly independent solutions $\mathbf{x}_1, \mathbf{x}_2, \ldots, \mathbf{x}_n$.

THEOREM 3 General Solutions of Homogeneous Systems

Let $\mathbf{x}_1, \mathbf{x}_2, \ldots, \mathbf{x}_n$ be n linearly independent solutions of the homogeneous linear equation $\mathbf{x}' = \mathbf{P}(t)\mathbf{x}$ on an open interval I where $\mathbf{P}(t)$ is continuous. If $\mathbf{x}(t)$ is any solution whatsoever of the equation $\mathbf{x}' = \mathbf{P}(t)\mathbf{x}$ on I, then there exist numbers $c_1, c_2, \ldots, c_n$ such that

$$\mathbf{x}(t) = c_1\mathbf{x}_1(t) + c_2\mathbf{x}_2(t) + \cdots + c_n\mathbf{x}_n(t) \tag{15}$$

for all t in I.

***Proof*:** Let a be a fixed point of I. We first show that there exist numbers c_1, $c_2, \ldots, c_n$ such that the solution

$$\mathbf{y}(t) = c_1\mathbf{x}_1(t) + c_2\mathbf{x}_2(t) + \cdots + c_n\mathbf{x}_n(t) \tag{16}$$

has the same initial values at $t = a$ as does the given solution $\mathbf{x}(t)$; that is, such that

$$c_1\mathbf{x}_1(a) + c_2\mathbf{x}_2(a) + \cdots + c_n\mathbf{x}_n(a) = \mathbf{x}(a). \tag{17}$$

Let $\mathbf{X}(t)$ be the $n \times n$ matrix with column vectors $\mathbf{x}_1, \mathbf{x}_2, \ldots, \mathbf{x}_n$, and let $\mathbf{c}$ be the column vector with components $c_1, c_2, \ldots, c_n$. Then Eq. (17) may be written in the form

$$\mathbf{X}(a)\mathbf{c} = \mathbf{x}(a). \tag{18}$$

The Wronskian determinant $W(a) = |\mathbf{X}(a)|$ is nonzero because the solutions $\mathbf{x}_1$, $\mathbf{x}_2, \ldots, \mathbf{x}_n$ are linearly independent. Hence the matrix $\mathbf{X}(a)$ has an inverse matrix $\mathbf{X}(a)^{-1}$. Therefore the vector $\mathbf{c} = \mathbf{X}(a)^{-1}\mathbf{x}(a)$ satisfies Eq. (18), as desired.

Finally, note that the given solution $\mathbf{x}(t)$ and the solution $\mathbf{y}(t)$ of Eq. (16)—with the values of c_i determined by the equation $\mathbf{c} = \mathbf{X}(a)^{-1}\mathbf{x}(a)$—have the same initial values (at $t = a$). It follows from the existence-uniqueness theorem of Section 7.1 that $\mathbf{x}(t) = \mathbf{y}(t)$ for all t in I. This establishes Eq. (15). ■

Remark: Every $n \times n$ system $\mathbf{x}' = \mathbf{P}(t)\mathbf{x}$ with continuous coefficient matrix does have a set of n linearly independent solutions $\mathbf{x}_1, \mathbf{x}_2, \ldots, \mathbf{x}_n$ as in the hypotheses of Theorem 3. It suffices to choose for $\mathbf{x}_j(t)$ the unique solution such that

$$\mathbf{x}_j(a) = \begin{bmatrix} 0 \\ 0 \\ 0 \\ \vdots \\ 0 \\ 1 \\ 0 \\ \vdots \\ 0 \end{bmatrix} \begin{matrix} \\ \\ \\ \\ \\ \longleftarrow \text{position } j \\ \\ \\ \\ \end{matrix}$$

—that is, the column vector with all elements zero except for a 1 in row j. (In other words, $\mathbf{x}_j(a)$ is merely the jth column of the identity matrix.) Then

$$W(\mathbf{x}_1, \mathbf{x}_2, \ldots, \mathbf{x}_n)\Big|_{t=a} = |\mathbf{I}| \neq 0,$$

so the solutions $\mathbf{x}_1, \mathbf{x}_2, \ldots, \mathbf{x}_n$ are linearly independent by Theorem 2. How to actually find these solutions explicitly is another matter—one that we address in Section 7.3 (for the case of constant-coefficient matrices). ■

Initial Value Problems and Elementary Row Operations

The general solution in Eq. (15) of the homogeneous linear system $\mathbf{x}' = \mathbf{P}(t)\mathbf{x}$ can be written in the form

$$\mathbf{x}(t) = \mathbf{X}(t)\mathbf{c}, \tag{19}$$

where

$$\mathbf{X}(t) = \begin{bmatrix} \mathbf{x}_1(t) & \mathbf{x}_2(t) & \cdots & \mathbf{x}_n(t) \end{bmatrix} \tag{20}$$

is the $n \times n$ matrix whose *column vectors* are the linearly independent solutions $\mathbf{x}_1$, $\mathbf{x}_2, \ldots, \mathbf{x}_n$, and where $\mathbf{c} = \begin{bmatrix} c_1 & c_2 & \cdots & c_n \end{bmatrix}^T$ is the vector of coefficients in the linear combination

$$\mathbf{x}(t) = c_1\mathbf{x}_1(t) + c_2\mathbf{x}_2(t) + \cdots + c_n\mathbf{x}_n(t). \tag{15}$$

Suppose now that we wish to solve the *initial value problem*

$$\frac{d\mathbf{x}}{dt} = \mathbf{P}\mathbf{x}, \quad \mathbf{x}(a) = \mathbf{b}, \tag{21}$$

where the initial vector $\mathbf{b} = \begin{bmatrix} b_1 & b_2 & \cdots & b_n \end{bmatrix}^T$ is given. Then, according to Eq. (19), it suffices to solve the system

$$\mathbf{X}(a)\mathbf{c} = \mathbf{b} \tag{22}$$

to find the coefficients $c_1, c_2, \ldots, c_n$ in Eq. (15). For instance, we can use the row-reduction techniques of Sections 3.2 and 3.3.

Example 4 Use the solution vectors given in Example 3 to solve the initial value problem

$$\frac{d\mathbf{x}}{dt} = \begin{bmatrix} 3 & -2 & 0 \\ -1 & 3 & -2 \\ 0 & -1 & 3 \end{bmatrix} \mathbf{x}, \quad \mathbf{x}(0) = \begin{bmatrix} 0 \\ 2 \\ 6 \end{bmatrix}. \tag{23}$$

Solution It follows from Theorem 3 that the linear combination

$$\begin{aligned} \mathbf{x}(t) &= c_1\mathbf{x}_1(t) + c_2\mathbf{x}_2(t) + c_3\mathbf{x}_3(t) \\ &= c_1 \begin{bmatrix} 2e^t \\ 2e^t \\ e^t \end{bmatrix} + c_2 \begin{bmatrix} 2e^{3t} \\ 0 \\ -e^{3t} \end{bmatrix} + c_3 \begin{bmatrix} 2e^{5t} \\ -2e^{5t} \\ e^{5t} \end{bmatrix} \end{aligned}$$

is a general solution of the 3×3 linear system in (23). In scalar form, this gives the general solution

$$\begin{aligned} x_1(t) &= 2c_1e^t + 2c_2e^{3t} + 2c_3e^{5t}, \\ x_2(t) &= 2c_1e^t \qquad\qquad - 2c_3e^{5t}, \\ x_3(t) &= c_1e^t - c_2e^{3t} + c_3e^{5t}. \end{aligned}$$

We seek the particular solution satisfying the initial conditions

$$x_1(0) = 0, \quad x_2(0) = 2, \quad x_3(0) = 6.$$

When we substitute these values in the preceding three scalar equations, we get the algebraic linear system

$$\begin{aligned} 2c_1 + 2c_2 + 2c_3 &= 0, \\ 2c_1 \qquad - 2c_3 &= 2, \\ c_1 - c_2 + c_3 &= 6 \end{aligned}$$

having the augmented coefficient matrix

$$\left[\begin{array}{ccc|c} 2 & 2 & 2 & 0 \\ 2 & 0 & -2 & 2 \\ 1 & -1 & 1 & 6 \end{array}\right].$$

Multiplication of each of the first two rows by $\frac{1}{2}$ gives

$$\left[\begin{array}{ccc|c} 1 & 1 & 1 & 0 \\ 1 & 0 & -1 & 1 \\ 1 & -1 & 1 & 6 \end{array}\right];$$

then subtraction of the first row both from the second row and from the third row gives the matrix

$$\left[\begin{array}{ccc|c} 1 & 1 & 1 & 0 \\ 0 & -1 & -2 & 1 \\ 0 & -2 & 0 & 6 \end{array}\right].$$

The first column of this matrix now has the desired form.

Now, we multiply the second row by -1, then add twice the result to the third row. Thereby, we get the upper triangular augmented coefficient matrix

$$\left[\begin{array}{ccc|c} 1 & 1 & 1 & 0 \\ 0 & 1 & 2 & -1 \\ 0 & 0 & 4 & 4 \end{array}\right]$$

that corresponds to the transformed system

$$\begin{aligned} c_1 + c_2 + c_3 &= 0, \\ c_2 + 2c_3 &= -1, \\ 4c_3 &= 4. \end{aligned}$$

We finally solve in turn for $c_3 = 1$, $c_2 = -3$, and $c_1 = 2$. Thus the desired particular solution is given by

$$\mathbf{x}(t) = 2\mathbf{x}_1(t) - 3\mathbf{x}_2(t) + \mathbf{x}_3(t) = \begin{bmatrix} 4e^t - 6e^{3t} + 2e^{5t} \\ 4e^t \phantom{- 6e^{3t}} - 2e^{5t} \\ 2e^t + 3e^{3t} + e^{5t} \end{bmatrix}.$$ ■

Nonhomogeneous Solutions

We finally turn our attention to a *nonhomogeneous* linear system of the form

$$\frac{d\mathbf{x}}{dt} = \mathbf{P}(t)\mathbf{x} + \mathbf{f}(t). \tag{24}$$

The following theorem is analogous to Theorem 5 of Section 5.2 and is proved in precisely the same way, substituting the preceding theorems in this section for the analogous theorems of Section 5.2. In brief, Theorem 4 means that the general solution of Eq. (24) has the form

$$\mathbf{x}(t) = \mathbf{x}_c(t) + \mathbf{x}_p(t), \tag{25}$$

where $\mathbf{x}_p(t)$ is a single particular solution of Eq. (24) and the **complementary function** $\mathbf{x}_c(t)$ is a general solution of the associated homogeneous equation $\mathbf{x}' = \mathbf{P}(t)\mathbf{x}$.

THEOREM 4 Solutions of Nonhomogeneous Systems

Let $\mathbf{x}_p$ be a particular solution of the nonhomogeneous linear equation in (24) on an open interval I on which the functions $\mathbf{P}(t)$ and $\mathbf{f}(t)$ are continuous. Let $\mathbf{x}_1, \mathbf{x}_2, \ldots, \mathbf{x}_n$ be linearly independent solutions of the associated homogeneous equation on I. If $\mathbf{x}(t)$ is any solution whatsoever of Eq. (24) on I, then there exist numbers $c_1, c_2, \ldots, c_n$ such that

$$\mathbf{x}(t) = c_1\mathbf{x}_1(t) + c_2\mathbf{x}_2(t) + \cdots + c_n\mathbf{x}_n(t) + \mathbf{x}_p(t) \tag{26}$$

for all t in I.

Thus, finding a general solution of a homogeneous linear system involves two separate steps:

1. Finding the general solution $\mathbf{x}_c(t)$ of the associated homogeneous system;
2. Finding a single particular solution $\mathbf{x}_p(t)$ of the nonhomogeneous system.

The sum $\mathbf{x}(t) = \mathbf{x}_c(t) + \mathbf{x}_p(t)$ will then be a general solution of the nonhomogeneous system.

Example 5 The nonhomogeneous linear system

$$\begin{aligned} x_1' &= 3x_1 - 2x_2 \qquad\qquad - 9t + 13, \\ x_2' &= -x_1 + 3x_2 - 2x_3 + 7t - 15, \\ x_3' &= \qquad - x_2 + 3x_3 - 6t + 7 \end{aligned}$$

can be recast as a matrix equation by writing

$$\begin{aligned} \frac{d\mathbf{x}}{dt} = \begin{bmatrix} x_1' \\ x_2' \\ x_3' \end{bmatrix} &= \begin{bmatrix} 3x_1 - 2x_2 \qquad - 9t + 13 \\ -x_1 + 3x_2 - 2x_3 + 7t - 15 \\ - x_2 + 3x_3 - 6t + 7 \end{bmatrix} \\ &= \begin{bmatrix} 3x_1 - 2x_2 \\ -x_1 + 3x_2 - 2x_3 \\ - x_2 + 3x_3 \end{bmatrix} + \begin{bmatrix} -9t + 13 \\ 7t - 15 \\ -6t + 7 \end{bmatrix} \\ &= \begin{bmatrix} 3 & -2 & 0 \\ -1 & 3 & -2 \\ 0 & -1 & 3 \end{bmatrix} \begin{bmatrix} x_1 \\ x_2 \\ x_3 \end{bmatrix} + \begin{bmatrix} -9t + 13 \\ 7t - 15 \\ -6t + 7 \end{bmatrix}. \end{aligned}$$

Thus we get the form in (24) with

$$\mathbf{P}(t) = \begin{bmatrix} 3 & -2 & 0 \\ -1 & 3 & -2 \\ 0 & -1 & 3 \end{bmatrix}, \quad \mathbf{f}(t) = \begin{bmatrix} -9t + 13 \\ 7t - 15 \\ -6t + 7 \end{bmatrix}.$$

In Example 3 we saw that a general solution of the associated homogeneous linear system

$$\frac{d\mathbf{x}}{dt} = \begin{bmatrix} 3 & -2 & 0 \\ -1 & 3 & -2 \\ 0 & -1 & 3 \end{bmatrix} \mathbf{x}$$

is given by

$$\mathbf{x}_c(t) = \begin{bmatrix} 2c_1e^t + 2c_2e^{3t} + 2c_3e^{5t} \\ 2c_1e^t \qquad\qquad - 2c_3e^{5t} \\ c_1e^t - c_2e^{3t} + c_3e^{5t} \end{bmatrix},$$

and we can verify, by substitution, that the function

$$\mathbf{x}_p(t) = \begin{bmatrix} 3t \\ 5 \\ 2t \end{bmatrix}$$

(found by using a computer algebra system, or perhaps by a human being using a method discussed in Section 8.2) is a particular solution of the original nonhomogeneous system. Consequently, Theorem 4 implies that a general solution of the nonhomogeneous system is given by

$$\mathbf{x}(t) = \mathbf{x}_c(t) + \mathbf{x}_p(t)$$

—that is, by

$$\begin{aligned} x_1(t) &= 2c_1e^t + 2c_2e^{3t} + 2c_3e^{5t} + 3t, \\ x_2(t) &= 2c_1e^t \qquad\qquad - 2c_3e^{5t} + 5, \\ x_3(t) &= c_1e^t - c_2e^{3t} + c_3e^{5t} + 2t. \end{aligned}$$

■

7.2 Problems

In Problems 1 and 2, verify the product law for differentiation, $(\mathbf{AB})' = \mathbf{A}'\mathbf{B} + \mathbf{AB}'$.

1. $\mathbf{A}(t) = \begin{bmatrix} t & 2t-1 \\ t^3 & \frac{1}{t} \end{bmatrix}$ and $\mathbf{B}(t) = \begin{bmatrix} 1-t & 1+t \\ 3t^2 & 4t^3 \end{bmatrix}$.

2. $\mathbf{A}(t) = \begin{bmatrix} e^t & t & t^2 \\ -t & 0 & 2 \\ 8t & -1 & t^3 \end{bmatrix}$ and $\mathbf{B}(t) = \begin{bmatrix} 3 \\ 2e^{-t} \\ 3t \end{bmatrix}$.

In Problems 3 through 12, write the given system in the form $\mathbf{x}' = \mathbf{P}(t)\mathbf{x} + \mathbf{f}(t)$.

3. $x' = -3y,\ y' = 3x$

4. $x' = 3x - 2y,\ y' = 2x + y$

5. $x' = 2x + 4y + 3e^t,\ y' = 5x - y - t^2$

6. $x' = tx - e^t y + \cos t,\ y' = e^{-t}x + t^2 y - \sin t$

7. $x' = y + z,\ y' = z + x,\ z' = x + y$

8. $x' = 2x - 3y,\ y' = x + y + 2z,\ z' = 5y - 7z$

9. $x' = 3x - 4y + z + t,\ y' = x - 3z + t^2,\ z' = 6y - 7z + t^3$

10. $x' = tx - y + e^t z,\ y' = 2x + t^2 y - z,\ z' = e^{-t}x + 3ty + t^3 z$

11. $x_1' = x_2,\ x_2' = 2x_3,\ x_3' = 3x_4,\ x_4' = 4x_1$

12. $x_1' = x_2 + x_3 + 1,\ x_2' = x_3 + x_4 + t,$
$x_3' = x_1 + x_4 + t^2,\ x_4' = x_1 + x_2 + t^3$

In Problems 13 through 22, first verify that the given vectors are solutions of the given system. Then use the Wronskian to show that they are linearly independent. Finally, write the general solution of the system.

13. $\mathbf{x}' = \begin{bmatrix} 4 & 2 \\ -3 & -1 \end{bmatrix}\mathbf{x};\ \mathbf{x}_1 = \begin{bmatrix} 2e^t \\ -3e^t \end{bmatrix},\ \mathbf{x}_2 = \begin{bmatrix} e^{2t} \\ -e^{2t} \end{bmatrix}$

14. $\mathbf{x}' = \begin{bmatrix} -3 & 2 \\ -3 & 4 \end{bmatrix}\mathbf{x};\ \mathbf{x}_1 = \begin{bmatrix} e^{3t} \\ 3e^{3t} \end{bmatrix},\ \mathbf{x}_2 = \begin{bmatrix} 2e^{-2t} \\ e^{-2t} \end{bmatrix}$

15. $\mathbf{x}' = \begin{bmatrix} 3 & -1 \\ 5 & -3 \end{bmatrix}\mathbf{x};\ \mathbf{x}_1 = e^{2t}\begin{bmatrix} 1 \\ 1 \end{bmatrix},\ \mathbf{x}_2 = e^{-2t}\begin{bmatrix} 1 \\ 5 \end{bmatrix}$

16. $\mathbf{x}' = \begin{bmatrix} 4 & 1 \\ -2 & 1 \end{bmatrix}\mathbf{x};\ \mathbf{x}_1 = e^{3t}\begin{bmatrix} 1 \\ -1 \end{bmatrix},\ \mathbf{x}_2 = e^{2t}\begin{bmatrix} 1 \\ -2 \end{bmatrix}$

17. $\mathbf{x}' = \begin{bmatrix} 4 & -3 \\ 6 & -7 \end{bmatrix}\mathbf{x};\ \mathbf{x}_1 = \begin{bmatrix} 3e^{2t} \\ 2e^{2t} \end{bmatrix},\ \mathbf{x}_2 = \begin{bmatrix} e^{-5t} \\ 3e^{-5t} \end{bmatrix}$

18. $\mathbf{x}' = \begin{bmatrix} 3 & -2 & 0 \\ -1 & 3 & -2 \\ 0 & -1 & 3 \end{bmatrix}\mathbf{x};\ \mathbf{x}_1 = e^t\begin{bmatrix} 2 \\ 2 \\ 1 \end{bmatrix},$
$\mathbf{x}_2 = e^{3t}\begin{bmatrix} -2 \\ 0 \\ 1 \end{bmatrix},\ \mathbf{x}_3 = e^{5t}\begin{bmatrix} 2 \\ -2 \\ 1 \end{bmatrix}$

19. $\mathbf{x}' = \begin{bmatrix} 0 & 1 & 1 \\ 1 & 0 & 1 \\ 1 & 1 & 0 \end{bmatrix}\mathbf{x};\ \mathbf{x}_1 = e^{2t}\begin{bmatrix} 1 \\ 1 \\ 1 \end{bmatrix},$
$\mathbf{x}_2 = e^{-t}\begin{bmatrix} 1 \\ 0 \\ -1 \end{bmatrix},\ \mathbf{x}_3 = e^{-t}\begin{bmatrix} 0 \\ 1 \\ -1 \end{bmatrix}$

20. $\mathbf{x}' = \begin{bmatrix} 1 & 2 & 1 \\ 6 & -1 & 0 \\ -1 & -2 & -1 \end{bmatrix}\mathbf{x};\ \mathbf{x}_1 = \begin{bmatrix} 1 \\ 6 \\ -13 \end{bmatrix},$
$\mathbf{x}_2 = e^{3t}\begin{bmatrix} 2 \\ 3 \\ -2 \end{bmatrix},\ \mathbf{x}_3 = e^{-4t}\begin{bmatrix} -1 \\ 2 \\ 1 \end{bmatrix}$

21. $\mathbf{x}' = \begin{bmatrix} -8 & -11 & -2 \\ 6 & 9 & 2 \\ -6 & -6 & 1 \end{bmatrix}\mathbf{x};\ \mathbf{x}_1 = e^{-2t}\begin{bmatrix} 3 \\ -2 \\ 2 \end{bmatrix},$
$\mathbf{x}_2 = e^t\begin{bmatrix} 1 \\ -1 \\ 1 \end{bmatrix},\ \mathbf{x}_3 = e^{3t}\begin{bmatrix} 1 \\ -1 \\ 0 \end{bmatrix}$

22. $\mathbf{x}' = \begin{bmatrix} 1 & -4 & 0 & -2 \\ 0 & 1 & 0 & 0 \\ 6 & -12 & -1 & -6 \\ 0 & -4 & 0 & -1 \end{bmatrix}\mathbf{x};\ \mathbf{x}_1 = e^{-t}\begin{bmatrix} 1 \\ 0 \\ 0 \\ 1 \end{bmatrix},$
$\mathbf{x}_2 = e^{-t}\begin{bmatrix} 0 \\ 0 \\ 1 \\ 0 \end{bmatrix},\ \mathbf{x}_3 = e^t\begin{bmatrix} 0 \\ 1 \\ 0 \\ -2 \end{bmatrix},\ \mathbf{x}_4 = e^t\begin{bmatrix} 1 \\ 0 \\ 3 \\ 0 \end{bmatrix}$

In Problems 23 through 32, find a particular solution of the indicated linear system that satisfies the given initial conditions.

23. The system of Problem 14: $x_1(0) = 0,\ x_2(0) = 5$

24. The system of Problem 15: $x_1(0) = 5,\ x_2(0) = -3$

25. The system of Problem 16: $x_1(0) = 11,\ x_2(0) = -7$

26. The system of Problem 17: $x_1(0) = 8,\ x_2(0) = 0$

27. The system of Problem 18: $x_1(0) = 0,\ x_2(0) = 0,\ x_3(0) = 4$

28. The system of Problem 19: $x_1(0) = 10,\ x_2(0) = 12,\ x_3(0) = -1$

29. The system of Problem 21: $x_1(0) = 1,\ x_2(0) = 2,\ x_3(0) = 3$

30. The system of Problem 21: $x_1(0) = 5,\ x_2(0) = -7,\ x_3(0) = 11$

31. The system of Problem 22: $x_1(0) = x_2(0) = x_3(0) = x_4(0) = 1$

32. The system of Problem 22: $x_1(0) = 1,\ x_2(0) = 3,\ x_3(0) = 4,\ x_4(0) = 7$

33. (a) Show that the vector functions

$$\mathbf{x}_1(t) = \begin{bmatrix} t \\ t^2 \end{bmatrix} \quad \text{and} \quad \mathbf{x}_2 = \begin{bmatrix} t^2 \\ t^3 \end{bmatrix}$$

are linearly independent on the real line. (b) Why does it follow from Theorem 2 that there is *no* continuous matrix $\mathbf{P}(t)$ such that $\mathbf{x}_1$ and $\mathbf{x}_2$ are both solutions of $\mathbf{x}' = \mathbf{P}(t)\mathbf{x}$?

34. Suppose that one of the vector functions

$$\mathbf{x}_1(t) = \begin{bmatrix} x_{11}(t) \\ x_{21}(t) \end{bmatrix} \quad \text{and} \quad \mathbf{x}_2(t) = \begin{bmatrix} x_{12}(t) \\ x_{22}(t) \end{bmatrix}$$

is a constant multiple of the other on the open interval I. Show that their Wronskian $W(t) = |[x_{ij}(t)]|$ must vanish identically on I. This proves part (a) of Theorem 2 in the case $n = 2$.

35. Suppose that the vectors $\mathbf{x}_1(t)$ and $\mathbf{x}_2(t)$ of Problem 34 are solutions of the equation $\mathbf{x}' = \mathbf{P}(t)\mathbf{x}$, where the 2×2 matrix $\mathbf{P}(t)$ is continuous on the open interval I. Show that if there exists a point a of I at which their Wronskian $W(a)$ is zero, then there exist numbers c_1 and c_2 not both zero such that $c_1\mathbf{x}_1(a) + c_2\mathbf{x}_2(a) = \mathbf{0}$. Then conclude from the uniqueness of solutions of the equation $\mathbf{x}' = \mathbf{P}(t)\mathbf{x}$ that

$$c_1\mathbf{x}_1(t) + c_2\mathbf{x}_2(t) = \mathbf{0}$$

for all t in I; that is, that $\mathbf{x}_1$ and $\mathbf{x}_2$ are linearly dependent. This proves part (b) of Theorem 2 in the case $n = 2$.

36. Generalize Problems 34 and 35 to prove Theorem 2 for n an arbitrary positive integer.

37. Let $\mathbf{x}_1(t), \mathbf{x}_2(t), \ldots, \mathbf{x}_n(t)$ be vector functions whose ith components (for some fixed i) $x_{i1}(t), x_{i2}(t), \ldots, x_{in}(t)$ are linearly independent real-valued functions. Conclude that the vector functions are themselves linearly independent.

7.3 The Eigenvalue Method for Linear Systems

We now introduce a powerful method for constructing the general solution of a *homogeneous* first-order system with *constant* coefficients,

$$\begin{aligned} x_1' &= a_{11}x_1 + a_{12}x_2 + \cdots + a_{1n}x_n, \\ x_2' &= a_{21}x_1 + a_{22}x_2 + \cdots + a_{2n}x_n, \\ &\ \vdots \\ x_n' &= a_{n1}x_1 + a_{n2}x_2 + \cdots + a_{nn}x_n. \end{aligned} \tag{1}$$

By Theorem 3 of Section 7.2, we know that it suffices to find n linearly independent solution vectors $\mathbf{x}_1, \mathbf{x}_2, \ldots, \mathbf{x}_n$; the linear combination

$$\mathbf{x}(t) = c_1\mathbf{x}_1 + c_2\mathbf{x}_2 + \cdots + c_n\mathbf{x}_n \tag{2}$$

with arbitrary coefficients will then be a general solution of the system in (1).

To search for the n needed linearly independent solution vectors, we proceed by analogy with the characteristic root method for solving a single homogeneous equation with constant coefficients (Section 5.3). It is reasonable to anticipate solution vectors of the form

$$\mathbf{x}(t) = \begin{bmatrix} x_1 \\ x_2 \\ x_3 \\ \vdots \\ x_n \end{bmatrix} = \begin{bmatrix} v_1e^{\lambda t} \\ v_2e^{\lambda t} \\ v_3e^{\lambda t} \\ \vdots \\ v_ne^{\lambda t} \end{bmatrix} = \begin{bmatrix} v_1 \\ v_2 \\ v_3 \\ \vdots \\ v_n \end{bmatrix} e^{\lambda t} = \mathbf{v}e^{\lambda t}, \tag{3}$$

where $\lambda, v_1, v_2, v_3, \ldots, v_n$ are appropriate scalar constants. For if we substitute

$$x_i = v_ie^{\lambda t}, \qquad x_i' = \lambda v_ie^{\lambda t} \qquad (i = 1, 2, \ldots, n)$$

in (1), then each term in the resulting equations will have the factor $e^{\lambda t}$, so we can cancel it throughout. This will leave us with n linear equations that—for appropriate values of λ—we can hope to solve for values of the coefficients $v_1, v_2, \ldots, v_n$ in Eq. (3), so that $\mathbf{x}(t) = \mathbf{v}e^{\lambda t}$ is, indeed, a solution of the system in (1).

To investigate this possibility, it is more efficient to write the system in (1) in the matrix form

$$\mathbf{x}' = \mathbf{A}\mathbf{x}, \tag{4}$$

where $\mathbf{A} = \left[a_{ij} \right]$. When we substitute the trial solution $\mathbf{x} = \mathbf{v}e^{\lambda t}$ (having derivative $\mathbf{x}' = \lambda\mathbf{v}e^{\lambda t}$) in Eq. (4), the result is

$$\lambda\mathbf{v}e^{\lambda t} = \mathbf{A}\mathbf{v}e^{\lambda t}.$$

We cancel the nonzero scalar factor $e^{\lambda t}$ to get

$$\mathbf{A}\mathbf{v} = \lambda\mathbf{v}. \tag{5}$$

Recall, from Section 6.1, that Eq. (5) means that $\mathbf{v} \neq \mathbf{0}$ is an *eigenvector* of the matrix $\mathbf{A}$ associated with the *eigenvalue* λ. Our discussion of Eqs. (3)–(5) therefore provides a proof of the following theorem, which is the basis for the **eigenvalue method** of solving a first-order system with constant coefficients.

THEOREM 1 Eigenvalue Solutions of $\mathbf{x}' = \mathbf{A}\mathbf{x}$

Let λ be an eigenvalue of the [constant] coefficient matrix $\mathbf{A}$ of the first-order linear system

$$\frac{d\mathbf{x}}{dt} = \mathbf{A}\mathbf{x}.$$

If $\mathbf{v}$ is an eigenvector associated with λ, then

$$\mathbf{x}(t) = \mathbf{v}e^{\lambda t}$$

is a nontrivial solution of the system.

Recall that an eigenvalue λ of the matrix $\mathbf{A}$ is a solution of the *characteristic equation*

$$|\mathbf{A} - \lambda\mathbf{I}| = \det(\mathbf{A} - \lambda\mathbf{I}) = 0 \tag{6}$$

and that an eigenvector $\mathbf{v}$ associated with λ is then a solution of the *eigenvector equation*

$$(\mathbf{A} - \lambda\mathbf{I})\mathbf{v} = \mathbf{0}. \tag{7}$$

The Eigenvalue Method

In outline, this method for solving the $n \times n$ homogeneous constant-coefficient system $\mathbf{x}' = \mathbf{A}\mathbf{x}$ proceeds as follows:

1. First, we solve the characteristic equation in (6) for the eigenvalues $\lambda_1, \lambda_2, \ldots, \lambda_n$ of the matrix $\mathbf{A}$.
2. Next, we attempt to find n *linearly independent* eigenvectors $\mathbf{v}_1, \mathbf{v}_2, \ldots, \mathbf{v}_n$ associated with these eigenvalues.
3. Step 2 is not always possible, but, when it is, we get n linearly independent solutions

$$\mathbf{x}_1(t) = \mathbf{v}_1 e^{\lambda_1 t}, \quad \mathbf{x}_2(t) = \mathbf{v}_2 e^{\lambda_2 t}, \quad \ldots, \quad \mathbf{x}_n(t) = \mathbf{v}_n e^{\lambda_n t}. \tag{8}$$

In this case the general solution of $\mathbf{x}' = \mathbf{A}\mathbf{x}$ is a linear combination

$$\mathbf{x}(t) = c_1\mathbf{x}_1(t) + c_2\mathbf{x}_2(t) + \cdots + c_n\mathbf{x}_n(t)$$

of these n solutions.

We will discuss separately the various cases that can occur, depending on whether the eigenvalues are distinct or repeated, real or complex. The case of repeated eigenvalues—multiple roots of the characteristic equation—will be deferred to Section 7.5.

Distinct Real Eigenvalues

If the eigenvalues $\lambda_1, \lambda_2, \ldots, \lambda_n$ are real and distinct, then we substitute each of them in turn in Eq. (7) and solve for the associated eigenvectors $\mathbf{v}_1, \mathbf{v}_2, \ldots, \mathbf{v}_n$. It then follows from Theorem 2 in Section 6.2 that the particular solutions given in (8) are linearly independent. (In any particular example, such linear independence can always be verified by using the Wronskian determinant of Section 7.2.) The following example illustrates the eigenvalue method.

Example 1 Find a general solution of the system

$$\begin{aligned} x_1' &= 4x_1 + 2x_2, \\ x_2' &= 3x_1 - x_2. \end{aligned} \tag{9}$$

Solution The matrix form of the system in (9) is

$$\mathbf{x}' = \begin{bmatrix} 4 & 2 \\ 3 & -1 \end{bmatrix} \mathbf{x}. \tag{10}$$

The characteristic equation of the coefficient matrix is

$$\begin{aligned} \begin{vmatrix} 4-\lambda & 2 \\ 3 & -1-\lambda \end{vmatrix} &= (4-\lambda)(-1-\lambda) - 6 \\ &= \lambda^2 - 3\lambda - 10 = (\lambda+2)(\lambda-5) = 0, \end{aligned}$$

so we have the distinct real eigenvalues $\lambda_1 = -2$ and $\lambda_2 = 5$.

For the coefficient matrix $\mathbf{A}$ in Eq. (10), the eigenvector equation $(\mathbf{A} - \lambda\mathbf{I})\mathbf{v} = \mathbf{0}$ takes the form

$$\begin{bmatrix} 4-\lambda & 2 \\ 3 & -1-\lambda \end{bmatrix} \begin{bmatrix} a \\ b \end{bmatrix} = \begin{bmatrix} 0 \\ 0 \end{bmatrix} \tag{11}$$

for the associated eigenvector $\mathbf{v} = \begin{bmatrix} a & b \end{bmatrix}^T$.

CASE 1: $\lambda_1 = -2$. Substitution of the first eigenvalue $\lambda_1 = -2$ in Eq. (11) yields the system

$$\begin{bmatrix} 6 & 2 \\ 3 & 1 \end{bmatrix} \begin{bmatrix} a \\ b \end{bmatrix} = \begin{bmatrix} 0 \\ 0 \end{bmatrix}$$

—that is, the two scalar equations

$$\begin{aligned} 6a + 2b &= 0, \\ 3a + b &= 0. \end{aligned} \tag{12}$$

In contrast with the typical nonsingular (algebraic) linear system that has a unique solution, the homogeneous linear system in (12) is *singular*—the two scalar equations obviously are equivalent (each being a multiple of the other). Therefore,

Eq. (12) has infinitely many nonzero solutions—we can choose a arbitrarily (but nonzero) and then solve for b.

Substitution of an eigenvalue λ in the eigenvector equation $(\mathbf{A} - \lambda\mathbf{I})\mathbf{v} = \mathbf{0}$ always yields a singular homogeneous linear system, and among its infinity of solutions we generally seek a "simple" solution with small integer values (if possible). Looking at the second equation in (12), the choice $a = 1$ yields $b = -3$, and thus

$$\mathbf{v}_1 = \begin{bmatrix} 1 \\ -3 \end{bmatrix}$$

is an eigenvector associated with $\lambda_1 = -2$ (as is any nonzero constant multiple of $\mathbf{v}_1$).

Remark: If, instead of the "simplest" choice $a = 1$, $b = -3$, we had made another choice $a = c$, $b = -3c$, we would have obtained the eigenvector

$$\mathbf{v}_1 = \begin{bmatrix} c \\ -3c \end{bmatrix} = c\begin{bmatrix} 1 \\ -3 \end{bmatrix}.$$

Because this is a constant multiple of our previous result, any choice we make leads to (a constant multiple of) the same solution

$$\mathbf{x}_1(t) = \begin{bmatrix} 1 \\ -3 \end{bmatrix} e^{-2t}.$$

■

CASE 2: $\lambda_2 = 5$. Substitution of the second eigenvalue $\lambda = 5$ in (11) yields the pair

$$\begin{aligned} -a + 2b &= 0, \\ 3a - 6b &= 0 \end{aligned} \tag{13}$$

of equivalent scalar equations. With $b = 1$ in the first equation we get $a = 2$, so

$$\mathbf{v}_2 = \begin{bmatrix} 2 \\ 1 \end{bmatrix}$$

is an eigenvector associated with $\lambda_2 = 5$. A different choice—$a = 2c$, $b = c$—would merely give a [constant] multiple of $\mathbf{v}_2$.

These two eigenvalues and their associated eigenvectors yield the two solutions

$$\mathbf{x}_1(t) = \begin{bmatrix} 1 \\ -3 \end{bmatrix} e^{-2t} \quad \text{and} \quad \mathbf{x}_2(t) = \begin{bmatrix} 2 \\ 1 \end{bmatrix} e^{5t}.$$

They are linearly independent, because their Wronskian

$$\begin{vmatrix} e^{-2t} & 2e^{5t} \\ -3e^{-2t} & e^{5t} \end{vmatrix} = 7e^{3t}$$

is nonzero. Hence a general solution of the system in (10) is

$$\mathbf{x}(t) = c_1\mathbf{x}_1(t) + c_2\mathbf{x}_2(t) = c_1\begin{bmatrix} 1 \\ -3 \end{bmatrix} e^{-2t} + c_2\begin{bmatrix} 2 \\ 1 \end{bmatrix} e^{5t};$$

in scalar form,

$$\begin{aligned} x_1(t) &= c_1e^{-2t} + 2c_2e^{5t}, \\ x_2(t) &= -3c_1e^{-2t} + c_2e^{5t}. \end{aligned}$$

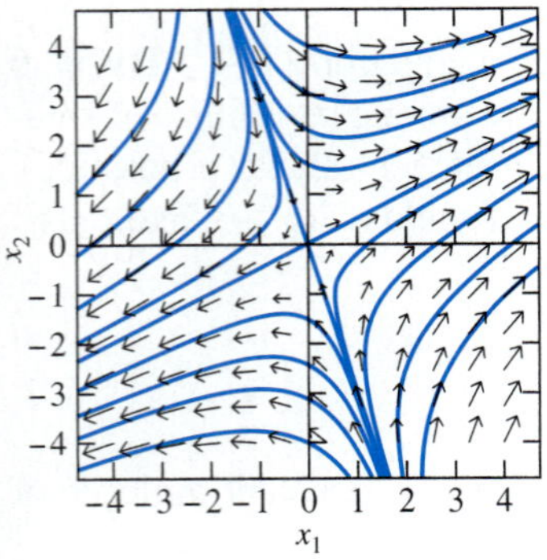

FIGURE 7.3.1. Direction field and solution curves for the linear system $x_1' = 4x_1 + 2x_2$, $x_2' = 3x_1 - x_2$ of Example 1.

Figure 7.3.1 shows some typical solution curves of the system in (10). We see two families of hyperbolas sharing the same pair of asymptotes: the line $x_1 = 2x_2$ obtained from the general solution with $c_1 = 0$, and the line $x_2 = -3x_1$ obtained with $c_2 = 0$. Given initial values $x_1(0) = b_1$, $x_2(0) = b_2$, it is apparent from the figure that

- If (b_1, b_2) lies to the right of the line $x_2 = -3x_1$, then $x_1(t)$ and $x_2(t)$ both tend to $+\infty$ as $t \to +\infty$;
- If (b_1, b_2) lies to the left of the line $x_2 = -3x_1$, then $x_1(t)$ and $x_2(t)$ both tend to $-\infty$ as $t \to +\infty$. ■

Remark: As in Example 1, it is convenient when discussing a linear system $\mathbf{x}' = \mathbf{A}\mathbf{x}$ to use vectors $\mathbf{x}_1, \mathbf{x}_2, \ldots, \mathbf{x}_n$ to denote different vector-valued solutions of the system, whereas the *scalars* $x_1, x_2, \ldots, x_n$ denote the components of a single vector-valued solution $\mathbf{x}$. ■

Compartmental Analysis

Frequently, a complex process or system can be broken down into simpler subsystems or "compartments" that can be analyzed separately. The whole system can then be modeled by describing the interactions between the various compartments. Thus a chemical plant may consist of a succession of separate stages (or even physical compartments) in which various reactants and products combine or are mixed. It may happen that a single differential equation describes each compartment of the system; then the whole physical system is modeled by a system of differential equations.

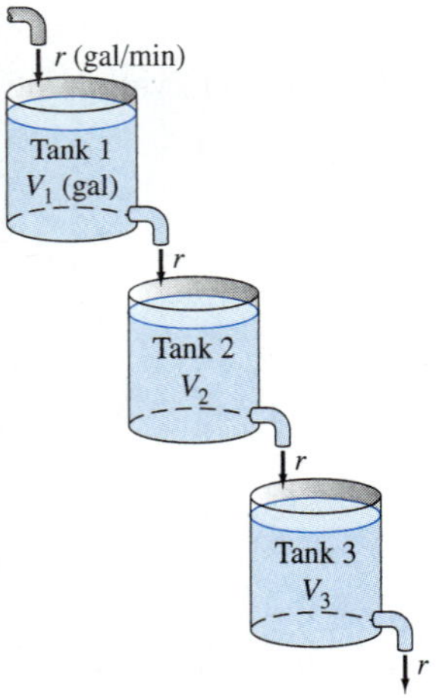

FIGURE 7.3.2. The three brine tanks of Example 2.

As a simple example of a three-stage system, Fig. 7.3.2 shows three brine tanks containing V_1, V_2, and V_3 gallons of brine, respectively. Fresh water flows into tank 1, while mixed brine flows from tank 1 into tank 2, from tank 2 into tank 3, and out of tank 3. Let $x_i(t)$ denote the amount (in pounds) of salt in tank i at time t, for $i = 1, 2$, and 3. If each flow rate is r gallons per minute, then a simple accounting of salt concentrations, as in Example 2 of Section 7.1, yields the first-order system

$$\begin{aligned} x_1' &= -k_1x_1, \\ x_2' &= k_1x_1 - k_2x_2, \\ x_3' &= k_2x_2 - k_3x_3, \end{aligned} \tag{14}$$

where

$$k_i = \frac{r}{V_i}, \quad i = 1, 2, 3. \tag{15}$$

Example 2 If $V_1 = 20$, $V_2 = 40$, $V_3 = 50$, $r = 10$ (gal/min), and the initial amounts of salt in the three brine tanks, in pounds, are

$$x_1(0) = 15, \quad x_2(0) = x_3(0) = 0,$$

find the amount of salt in each tank at time $t \geqq 0$.

Solution Substituting the given numerical values in (14) and (15), we get the initial value problem

$$\mathbf{x}'(t) = \begin{bmatrix} -0.5 & 0.0 & 0.0 \\ 0.5 & -0.25 & 0.0 \\ 0.0 & 0.25 & -0.2 \end{bmatrix} \mathbf{x}, \quad \mathbf{x}(0) = \begin{bmatrix} 15 \\ 0 \\ 0 \end{bmatrix} \tag{16}$$

for the vector $\mathbf{x}(t) = \begin{bmatrix} x_1(t) & x_2(t) & x_3(t) \end{bmatrix}^T$. The simple form of the matrix

$$\mathbf{A} - \lambda\mathbf{I} = \begin{bmatrix} -0.5 - \lambda & 0.0 & 0.0 \\ 0.5 & -0.25 - \lambda & 0.0 \\ 0.0 & 0.25 & -0.2 - \lambda \end{bmatrix} \tag{17}$$

leads readily to the characteristic equation

$$|\mathbf{A} - \lambda\mathbf{I}| = (-0.5 - \lambda)(-0.25 - \lambda)(-0.2 - \lambda) = 0.$$

Thus, the coefficient matrix $\mathbf{A}$ in (16) has the distinct eigenvalues $\lambda_1 = -0.5$, $\lambda_2 = -0.25$, and $\lambda_3 = -0.2$ and therefore has three linearly independent eigenvectors.

CASE 1: $\lambda_1 = -0.5$. Substituting $\lambda = -0.5$ in (17), we get the equation

$$\begin{bmatrix} \mathbf{A} + (0.5)\cdot\mathbf{I} \end{bmatrix}\mathbf{v} = \begin{bmatrix} 0.0 & 0.0 & 0.0 \\ 0.5 & 0.25 & 0.0 \\ 0.0 & 0.25 & 0.3 \end{bmatrix} \begin{bmatrix} a \\ b \\ c \end{bmatrix} = \begin{bmatrix} 0 \\ 0 \\ 0 \end{bmatrix}$$

for the associated eigenvector $\mathbf{v} = \begin{bmatrix} a & b & c \end{bmatrix}^T$. The last two rows, after division by 0.25 and 0.05, respectively, yield the scalar equations

$$\begin{aligned} 2a + b \qquad &= 0, \\ 5b + 6c &= 0. \end{aligned}$$

The second equation is satisfied by $b = -6$ and $c = 5$, and then the first equation gives $a = 3$. Thus the eigenvector

$$\mathbf{v}_1 = \begin{bmatrix} 3 & -6 & 5 \end{bmatrix}^T$$

is associated with the eigenvalue $\lambda_1 = -0.5$.

CASE 2: $\lambda_2 = -0.25$. Substituting $\lambda = -0.25$ in (17), we get the equation

$$\begin{bmatrix} \mathbf{A} + (0.25)\cdot\mathbf{I} \end{bmatrix}\mathbf{v} = \begin{bmatrix} -0.25 & 0 & 0 \\ 0.5 & 0 & 0 \\ 0 & 0.25 & 0.05 \end{bmatrix} \begin{bmatrix} a \\ b \\ c \end{bmatrix} = \begin{bmatrix} 0 \\ 0 \\ 0 \end{bmatrix}$$

for the associated eigenvector $\mathbf{v} = \begin{bmatrix} a & b & c \end{bmatrix}^T$. Each of the first two rows implies that $a = 0$, and division of the third row by 0.05 gives the equation

$$5b + c = 0,$$

which is satisfied by $b = 1$, $c = -5$. Thus the eigenvector

$$\mathbf{v}_2 = \begin{bmatrix} 0 & 1 & -5 \end{bmatrix}^T$$

is associated with the eigenvalue $\lambda_2 = -0.25$.

Case 3: $\lambda_3 = -0.2$. Substituting $\lambda = -0.2$ in (17), we get the equation

$$\left[\mathbf{A} + (0.2)\cdot\mathbf{I}\right]\mathbf{v} = \begin{bmatrix} -0.3 & 0.0 & 0.0 \\ 0.5 & -0.05 & 0.0 \\ 0.0 & 0.25 & 0.0 \end{bmatrix}\begin{bmatrix} a \\ b \\ c \end{bmatrix} = \begin{bmatrix} 0 \\ 0 \\ 0 \end{bmatrix}$$

for the eigenvector $\mathbf{v}$. The first and third rows imply that $a = 0$, and $b = 0$, respectively, but the all-zero third column leaves c arbitrary (but nonzero). Thus

$$\mathbf{v}_3 = \begin{bmatrix} 0 & 0 & 1 \end{bmatrix}^T$$

is an eigenvector associated with $\lambda_3 = -0.2$.

The general solution

$$\mathbf{x}(t) = c_1\mathbf{v}_1 e^{\lambda_1 t} + c_2\mathbf{v}_2 e^{\lambda_2 t} + c_3\mathbf{v}_3 e^{\lambda_3 t}$$

therefore takes the form

$$\mathbf{x}(t) = c_1\begin{bmatrix} 3 \\ -6 \\ 5 \end{bmatrix} e^{(-0.5)t} + c_2\begin{bmatrix} 0 \\ 1 \\ -5 \end{bmatrix} e^{(-0.25)t} + c_3\begin{bmatrix} 0 \\ 0 \\ 1 \end{bmatrix} e^{(-0.2)t}.$$

The resulting scalar equations are

$$\begin{aligned} x_1(t) &= 3c_1 e^{(-0.5)t}, \\ x_2(t) &= -6c_1 e^{(-0.5)t} + c_2 e^{(-0.25)t}, \\ x_3(t) &= 5c_1 e^{(-0.5)t} - 5c_2 e^{(-0.25)t} + c_3 e^{(-0.2)t}. \end{aligned}$$

When we impose the initial conditions $x_1(0) = 15$, $x_2(0) = x_3(0) = 0$, we get the equations

$$\begin{aligned} 3c_1 \phantom{{}-5c_2+c_3} &= 15, \\ -6c_1 + c_2 \phantom{{}+c_3} &= 0, \\ 5c_1 - 5c_2 + c_3 &= 0 \end{aligned}$$

that are readily solved (in turn) for $c_1 = 5$, $c_2 = 30$, and $c_3 = 125$. Thus, finally, the amounts of salt at time t in the three brine tanks are given by

$$\begin{aligned} x_1(t) &= 15e^{(-0.5)t}, \\ x_2(t) &= -30e^{(-0.5)t} + 30e^{(-0.25)t}, \\ x_3(t) &= 25e^{(-0.5)t} - 150e^{(-0.25)t} + 125e^{(-0.2)t}. \end{aligned}$$

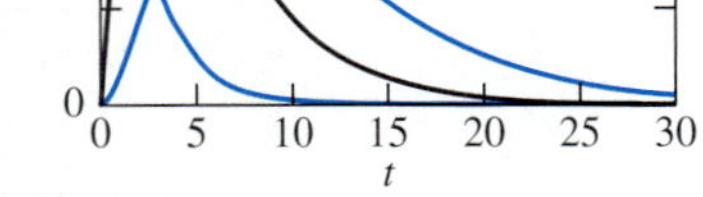

FIGURE 7.3.3. The salt content functions of Example 2.

Figure 7.3.3 shows the graphs of $x_1(t)$, $x_2(t)$, and $x_3(t)$. As we would expect, tank 1 is rapidly "flushed" by the incoming fresh water, and $x_1(t) \to 0$ as $t \to +\infty$. The amounts $x_2(t)$ and $x_3(t)$ of salt in tanks 2 and 3 peak in turn and then approach zero as the whole three-tank system is purged of salt as $t \to +\infty$. ■

Complex Eigenvalues

Even if some of the eigenvalues are complex, so long as they are distinct the method described previously still yields n linearly independent solutions. The only complication is that the eigenvectors associated with complex eigenvalues are ordinarily complex-valued, so we will have complex-valued solutions.

To obtain real-valued solutions, we note that—because we are assuming that the matrix $\mathbf{A}$ has only real entries—the coefficients in the characteristic equation will all be real. Consequently any complex eigenvalues must appear in complex-conjugate pairs. Suppose, then, that $\lambda = p + qi$ and $\overline{\lambda} = p - qi$ are such a pair of eigenvalues. If $\mathbf{v}$ is an eigenvector associated with λ, so that

$$(\mathbf{A} - \lambda\mathbf{I})\mathbf{v} = \mathbf{0},$$

then taking complex conjugates in this equation yields

$$(\mathbf{A} - \overline{\lambda}\mathbf{I})\overline{\mathbf{v}} = \mathbf{0}$$

since $\overline{\mathbf{A}} = \mathbf{A}$ and $\overline{\mathbf{I}} = \mathbf{I}$ (these matrices being real) and the conjugate of a complex product is the product of the conjugates of the factors. Thus the conjugate $\overline{\mathbf{v}}$ of $\mathbf{v}$ is an eigenvector associated with $\overline{\lambda}$. Of course, the conjugate of a vector is defined componentwise; if

$$\mathbf{v} = \begin{bmatrix} a_1 + b_1 i \\ a_2 + b_2 i \\ \vdots \\ a_n + b_n i \end{bmatrix} = \begin{bmatrix} a_1 \\ a_2 \\ \vdots \\ a_n \end{bmatrix} + \begin{bmatrix} b_1 \\ b_2 \\ \vdots \\ b_n \end{bmatrix} i = \mathbf{a} + \mathbf{b}i, \tag{18}$$

then $\overline{\mathbf{v}} = \mathbf{a} - \mathbf{b}i$. The complex-valued solution associated with λ and $\mathbf{v}$ is then

$$\mathbf{x}(t) = \mathbf{v}e^{\lambda t} = \mathbf{v}e^{(p+qi)t} = (\mathbf{a} + \mathbf{b}i)e^{pt}(\cos qt + i \sin qt)$$

—that is,

$$\mathbf{x}(t) = e^{pt}(\mathbf{a}\cos qt - \mathbf{b}\sin qt) + ie^{pt}(\mathbf{b}\cos qt + \mathbf{a}\sin qt). \tag{19}$$

Because the real and imaginary parts of a complex-valued solution are also solutions, we thus get the two *real-valued* solutions

$$\begin{aligned} \mathbf{x}_1(t) &= \operatorname{Re}[\mathbf{x}(t)] = e^{pt}(\mathbf{a}\cos qt - \mathbf{b}\sin qt), \\ \mathbf{x}_2(t) &= \operatorname{Im}[\mathbf{x}(t)] = e^{pt}(\mathbf{b}\cos qt + \mathbf{a}\sin qt) \end{aligned} \tag{20}$$

associated with the complex conjugate eigenvalues $p \pm qi$. It is easy to check that the same two real-valued solutions result from taking real and imaginary parts of $\overline{\mathbf{v}}e^{\overline{\lambda}t}$. Rather than memorizing the formulas in (20), it is preferable in a specific example to proceed as follows:

- First, find explicitly a single complex-valued solution $\mathbf{x}(t)$ associated with the complex eigenvalue λ;
- Then, find the real and imaginary parts $\mathbf{x}_1(t)$ and $\mathbf{x}_2(t)$, to get two independent real-valued solutions corresponding to the two complex conjugate eigenvalues λ and $\overline{\lambda}$.

Example 3 Find a general solution of the system

$$\begin{aligned}\frac{dx_1}{dt} &= 4x_1 - 3x_2,\\ \frac{dx_2}{dt} &= 3x_1 + 4x_2.\end{aligned} \tag{21}$$

Solution The coefficient matrix

$$\mathbf{A} = \begin{bmatrix} 4 & -3 \\ 3 & 4 \end{bmatrix}$$

has the characteristic equation

$$|\mathbf{A} - \lambda\mathbf{I}| = \begin{vmatrix} 4-\lambda & -3 \\ 3 & 4-\lambda \end{vmatrix} = (4-\lambda)^2 + 9 = 0$$

and hence has the complex conjugate eigenvalues $\lambda = 4 - 3i$ and $\overline{\lambda} = 4 + 3i$.

Substituting $\lambda = 4 - 3i$ in the eigenvector equation $(\mathbf{A} - \lambda\mathbf{I})\mathbf{v} = \mathbf{0}$, we get the equation

$$[\mathbf{A} - (4-3i)\cdot\mathbf{I}]\mathbf{v} = \begin{bmatrix} 3i & -3 \\ 3 & 3i \end{bmatrix}\begin{bmatrix} a \\ b \end{bmatrix} = \begin{bmatrix} 0 \\ 0 \end{bmatrix}$$

for an associated eigenvalue $\mathbf{v} = \begin{bmatrix} a & b \end{bmatrix}^T$. Division of each row by 3 yields the two scalar equations

$$\begin{aligned} ia - b &= 0,\\ a + ib &= 0,\end{aligned}$$

each of which is satisfied by $a = 1$ and $b = i$. Thus, $\mathbf{v} = \begin{bmatrix} 1 & i \end{bmatrix}^T$ is a complex eigenvector associated with the complex eigenvalue $\lambda = 4 - 3i$.

The corresponding complex-valued solution $\mathbf{x}(t) = \mathbf{v}e^{\lambda t}$ of $\mathbf{x}' = \mathbf{A}\mathbf{x}$ is then

$$\mathbf{x}(t) = \begin{bmatrix} 1 \\ i \end{bmatrix} e^{(4-3i)t} = \begin{bmatrix} 1 \\ i \end{bmatrix} e^{4t}(\cos 3t - i\sin 3t) = e^{4t}\begin{bmatrix} \cos 3t - i\sin 3t \\ i\cos 3t + \sin 3t \end{bmatrix}.$$

The real and imaginary parts of $\mathbf{x}(t)$ are the real-valued solutions

$$\mathbf{x}_1(t) = e^{4t}\begin{bmatrix} \cos 3t \\ \sin 3t \end{bmatrix} \quad \text{and} \quad \mathbf{x}_2(t) = e^{4t}\begin{bmatrix} -\sin 3t \\ \cos 3t \end{bmatrix}.$$

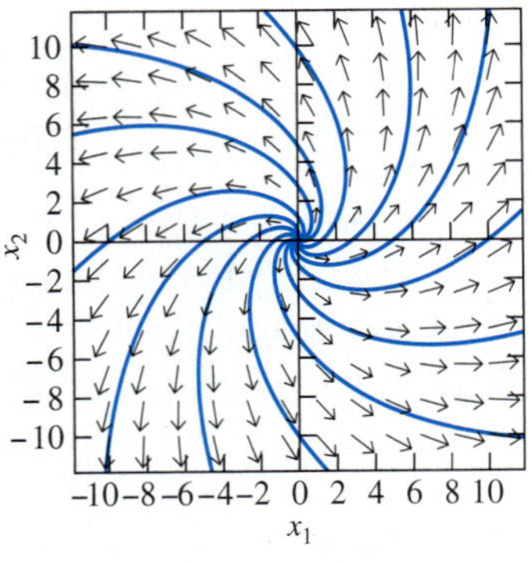

FIGURE 7.3.4. Direction field and solution curves for the linear system $x_1' = 4x_1 - 3x_2$, $x_2' = 3x_1 + 4x_2$ of Example 3.

A real-valued general solution of $\mathbf{x}' = \mathbf{A}\mathbf{x}$ is then given by

$$\mathbf{x}(t) = c_1\mathbf{x}_1(t) + c_2\mathbf{x}_2(t) = e^{4t}\begin{bmatrix} c_1\cos 3t - c_2\sin 3t \\ c_1\sin 3t + c_2\cos 3t \end{bmatrix}.$$

Finally, a general solution of the system in (21) in scalar form is

$$\begin{aligned} x_1(t) &= e^{4t}(c_1\cos 3t - c_2\sin 3t),\\ x_2(t) &= e^{4t}(c_1\sin 3t + c_2\cos 3t).\end{aligned}$$

Figure 7.3.4 shows some typical solution curves of the system in (21). Each appears to spiral counterclockwise as it emanates from the origin in the x_1x_2-plane. Actually, because of the factor e^{4t} in the general solution, we see that

- Along each solution curve, the point $(x_1(t), x_2(t))$ approaches the origin as $t \to -\infty$, whereas
- The absolute values of $x_1(t)$ and $x_2(t)$ both increase without bound as $t \to +\infty$. ■

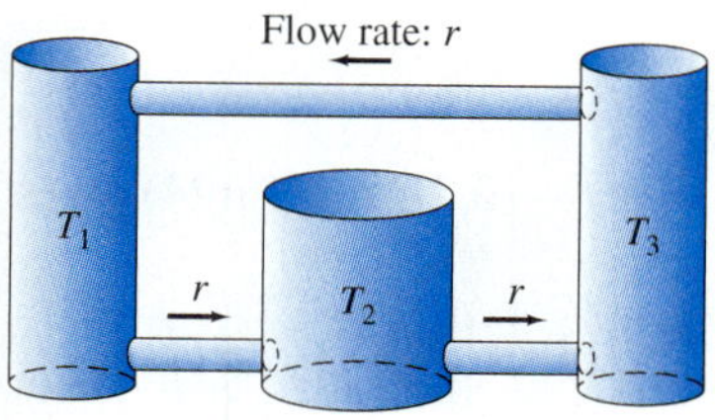

FIGURE 7.3.5. The three brine tanks of Example 4.

Figure 7.3.5 shows a "closed" system of three brine tanks with volumes V_1, V_2, and V_3. The difference between this system and the "open" system of Fig. 7.3.2 is that now the inflow to tank 1 is the outflow from tank 3. With the same notation as in Example 2, the appropriate modification of Eq. (14) is

$$\begin{aligned} \frac{dx_1}{dt} &= -k_1x_1 \qquad\qquad + k_3x_3, \\ \frac{dx_2}{dt} &= \quad k_1x_1 - k_2x_2, \\ \frac{dx_3}{dt} &= \qquad\qquad k_2x_2 - k_3x_3, \end{aligned} \tag{22}$$

where $k_i = r/V_i$ as in (15).

Example 4 Find the amounts $x_1(t)$, $x_2(t)$, and $x_3(t)$ of salt at time t in the three brine tanks of Fig. 7.3.5 if $V_1 = 50$ gal, $V_2 = 25$ gal, $V_3 = 50$ gal, and $r = 10$ gal/min.

Solution With the given numerical values, (22) takes the form

$$\frac{d\mathbf{x}}{dt} = \begin{bmatrix} -0.2 & 0 & 0.2 \\ 0.2 & -0.4 & 0 \\ 0 & 0.4 & -0.2 \end{bmatrix} \mathbf{x}, \tag{23}$$

where $\mathbf{x} = \begin{bmatrix} x_1 & x_2 & x_3 \end{bmatrix}^T$ as usual. When we expand the determinant of the matrix

$$\mathbf{A} - \lambda \cdot \mathbf{I} = \begin{bmatrix} -0.2-\lambda & 0.0 & 0.2 \\ 0.2 & -0.4-\lambda & 0.0 \\ 0.0 & 0.4 & -0.2-\lambda \end{bmatrix} \tag{24}$$

along its first row, we find that the characteristic equation of $\mathbf{A}$ is

$$\begin{aligned} (-0.2-\lambda)(-0.4-\lambda)(-0.2-\lambda) &+ (0.2)(0.2)(0.4) \\ &= -\lambda^3 - (0.8)\cdot\lambda^2 - (0.2)\cdot\lambda \\ &= -\lambda\left[(\lambda+0.4)^2 + (0.2)^2\right] = 0. \end{aligned}$$

Thus $\mathbf{A}$ has the zero eigenvalue $\lambda_0 = 0$ and the complex conjugate eigenvalues λ, $\overline{\lambda} = -0.4 \pm (0.2)i$. We anticipate one solution corresponding to the zero eigenvalue and two additional linearly independent solutions corresponding to the complex conjugate eigenvalues.

CASE 1: $\lambda_0 = 0$. Substitution of $\lambda = 0$ in Eq. (24) gives the eigenvector equation

$$(\mathbf{A} - 0\cdot\mathbf{I})\mathbf{v} = \begin{bmatrix} -0.2 & 0.0 & 0.2 \\ 0.2 & -0.4 & 0.0 \\ 0.0 & 0.4 & -0.2 \end{bmatrix} \begin{bmatrix} a \\ b \\ c \end{bmatrix} = \begin{bmatrix} 0 \\ 0 \\ 0 \end{bmatrix} \tag{25}$$

for $\mathbf{v} = \begin{bmatrix} a & b & c \end{bmatrix}^T$. The first row gives $a = c$ and the second row gives $a = 2b$, so $\mathbf{v}_0 = \begin{bmatrix} 2 & 1 & 2 \end{bmatrix}^T$ is an eigenvector associated with the eigenvalue $\lambda_0 = 0$. The corresponding solution $\mathbf{x}_0(t) = \mathbf{v}_0 e^{\lambda_0 t}$ of Eq. (23) is the constant solution

$$\mathbf{x}_0(t) = \begin{bmatrix} 2 \\ 1 \\ 2 \end{bmatrix}. \tag{26}$$

CASE 2: $\lambda = -0.4 - (0.2)i$. Substitution of $\lambda = -0.4 - (0.2)i$ in Eq. (24) gives the eigenvector equation

$$[\mathbf{A} - (-0.4 - (0.2)i)\mathbf{I}]\,\mathbf{v} = \begin{bmatrix} 0.2 + (0.2)i & 0.0 & 0.2 \\ 0.2 & (0.2)i & 0.0 \\ 0.0 & 0.4 & 0.2 + (0.2)i \end{bmatrix} \begin{bmatrix} a \\ b \\ c \end{bmatrix} = \begin{bmatrix} 0 \\ 0 \\ 0 \end{bmatrix}.$$

The second equation $(0.2)a + (0.2)ib = 0$ is satisfied by $a = 1$ and $b = i$. Then the first equation

$$[0.2 + (0.2)i]a + (0.2)c = 0$$

gives $c = -1 - i$. Thus, $\mathbf{v} = \begin{bmatrix} 1 & i & (-1 - i) \end{bmatrix}^T$ is a complex eigenvector associated with the complex eigenvalue $\lambda = -0.4 - (0.2)i$.

The corresponding complex-valued solution $\mathbf{x}(t) = \mathbf{v}e^{\lambda t}$ of (23) is

$$\begin{aligned} \mathbf{x}(t) &= \begin{bmatrix} 1 & i & -1 - i \end{bmatrix}^T e^{(-0.4 - 0.2i)t} \\ &= \begin{bmatrix} 1 & i & -1 - i \end{bmatrix}^T e^{(-0.4)t}(\cos 0.2t - i \sin 0.2t) \\ &= e^{(-0.4)t} \begin{bmatrix} \cos 0.2t - i \sin 0.2t \\ \sin 0.2t + i \cos 0.2t \\ -\cos 0.2t - \sin 0.2t - i \cos 0.2t + i \sin 0.2t \end{bmatrix}. \end{aligned}$$

The real and imaginary parts of $\mathbf{x}(t)$ are the real-valued solutions

$$\begin{aligned} \mathbf{x}_1(t) &= e^{(-0.4)t} \begin{bmatrix} \cos 0.2t \\ \sin 0.2t \\ -\cos 0.2t - \sin 0.2t \end{bmatrix}, \\ \mathbf{x}_2(t) &= e^{(-0.4)t} \begin{bmatrix} -\sin 0.2t \\ \cos 0.2t \\ -\cos 0.2t + \sin 0.2t \end{bmatrix}. \end{aligned} \tag{27}$$

The general solution

$$\mathbf{x}(t) = c_0\mathbf{x}_0(t) + c_1\mathbf{x}_1(t) + c_2\mathbf{x}_2(t)$$

has scalar components

$$\begin{aligned} x_1(t) &= 2c_0 + e^{(-0.4)t}(c_1 \cos 0.2t - c_2 \sin 0.2t), \\ x_2(t) &= c_0 + e^{(-0.4)t}(c_1 \sin 0.2t + c_2 \cos 0.2t), \\ x_3(t) &= 2c_0 + e^{(-0.4)t}[(-c_1 - c_2)\cos 0.2t + (-c_1 + c_2)\sin 0.2t] \end{aligned} \tag{28}$$

giving the amounts of salt in the three tanks at time t.

Observe that

$$x_1(t) + x_2(t) + x_3(t) \equiv 5c_0. \tag{29}$$

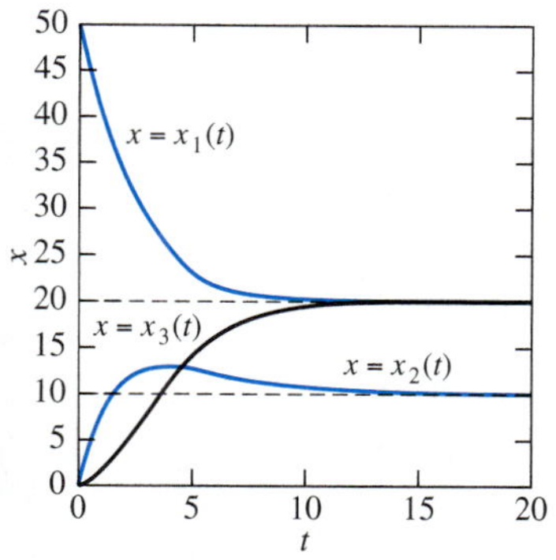

FIGURE 7.3.6. The salt content functions of Example 4.

Of course, the total amount of salt in the closed system is constant; the constant c_0 in (29) is one-fifth the total amount of salt. Because of the factors of $e^{(-0.4)t}$ in (28), we see that

$$\lim_{t\to\infty} x_1(t) = 2c_0, \quad \lim_{t\to\infty} x_2(t) = c_0, \quad \text{and} \quad \lim_{t\to\infty} x_3(t) = 2c_0.$$

Thus, as $t \to +\infty$ the salt in the system approaches a *steady-state* distribution with 40% of the salt in each of the two 50-gallon tanks and 20% in the 25-gallon tank. So whatever the initial distribution of salt among the three tanks, the limiting distribution is one of uniform concentration throughout the system. Figure 7.3.6 shows the graphs of the three solution functions with $c_0 = 10$, $c_1 = 30$, and $c_2 = -10$, in which case

$$x_1(0) = 50 \quad \text{and} \quad x_2(0) = x_3(0) = 0. \qquad \blacksquare$$

About the Cover

The graphic image on the cover of this book illustrates the space trajectories of solutions of the 3-dimensional linear system $\mathbf{x}' = \mathbf{Ax}$ with constant coefficient matrix

$$\mathbf{A} = \begin{bmatrix} 4 & 10 & 0 \\ -5 & -6 & 0 \\ 0 & 0 & 1 \end{bmatrix}.$$

To portray the motion in space of a point $\mathbf{x}(t)$ moving on a trajectory of this system, we can regard this trajectory as a necklace string on which colored beads are placed to mark its successive positions at fixed increments of time (so the point is moving fastest where the spacing between beads is greatest). In order to aid the eye in following the moving point's progress, the size of the beads decreases continuously with the passage of time and motion along the trajectory.

The matrix $\mathbf{A}$ has the single real eigenvalue -1 with the single (real) eigenvector $(0, 0, 1)$ and the complex conjugate eigenvalues $-1 \pm 5i$. The negative real eigenvalue corresponds to trajectories that lie on the z-axis and approach the origin as $t \to 0$ (as illustrated by the beads on the vertical axis of the figure). Thus the origin $(0, 0, 0)$ is a sink that "attracts" all the trajectories of the system.

The complex conjugate eigenvalues with negative real part correspond to trajectories in the horizontal xy-plane that spiral around the origin while approaching it. Any other trajectory—one which starts at a point lying neither on the z-axis nor in the xy-plane—combines the preceding behaviors by spiraling around the surface of a cone while approaching the origin at its vertex.

7.3 Problems

In Problems 1 through 16, apply the eigenvalue method of this section to find a general solution of the given system. If initial values are given, find also the corresponding particular solution. For each problem, use a computer system or graphing calculator to construct a direction field and typical solution curves for the given system.

1. $x_1' = x_1 + 2x_2, \; x_2' = 2x_1 + x_2$

2. $x_1' = 2x_1 + 3x_2, \; x_2' = 2x_1 + x_2$

3. $x_1' = 3x_1 + 4x_2, \; x_2' = 3x_1 + 2x_2; \; x_1(0) = x_2(0) = 1$

4. $x_1' = 4x_1 + x_2, \; x_2' = 6x_1 - x_2$

5. $x_1' = 6x_1 - 7x_2, \; x_2' = x_1 - 2x_2$

6. $x_1' = 9x_1 + 5x_2, \; x_2' = -6x_1 - 2x_2; \; x_1(0) = 1, \; x_2(0) = 0$

7. $x_1' = -3x_1 + 4x_2,\ x_2' = 6x_1 - 5x_2$

8. $x_1' = x_1 - 5x_2,\ x_2' = x_1 - x_2$

9. $x_1' = 2x_1 - 5x_2,\ x_2' = 4x_1 - 2x_2;\ x_1(0) = 2,\ x_2(0) = 3$

10. $x_1' = -3x_1 - 2x_2,\ x_2' = 9x_1 + 3x_2$

11. $x_1' = x_1 - 2x_2,\ x_2' = 2x_1 + x_2;\ x_1(0) = 0,\ x_2(0) = 4$

12. $x_1' = x_1 - 5x_2,\ x_2' = x_1 + 3x_2$

13. $x_1' = 5x_1 - 9x_2,\ x_2' = 2x_1 - x_2$

14. $x_1' = 3x_1 - 4x_2,\ x_2' = 4x_1 + 3x_2$

15. $x_1' = 7x_1 - 5x_2,\ x_2' = 4x_1 + 3x_2$

16. $x_1' = -50x_1 + 20x_2,\ x_2' = 100x_1 - 60x_2$

In Problems 17 through 25, the eigenvalues of the coefficient matrix can be found by inspection and factoring. Apply the eigenvalue method to find a general solution of each system.

17. $x_1' = 4x_1 + x_2 + 4x_3,\ x_2' = x_1 + 7x_2 + x_3,$
$x_3' = 4x_1 + x_2 + 4x_3$

18. $x_1' = x_1 + 2x_2 + 2x_3,\ x_2' = 2x_1 + 7x_2 + x_3,$
$x_3' = 2x_1 + x_2 + 7x_3$

19. $x_1' = 4x_1+x_2+x_3,\ x_2' = x_1+4x_2+x_3,\ x_3' = x_1+x_2+4x_3$

20. $x_1' = 5x_1 + x_2 + 3x_3,\ x_2' = x_1 + 7x_2 + x_3,$
$x_3' = 3x_1 + x_2 + 5x_3$

21. $x_1' = 5x_1-6x_3,\ x_2' = 2x_1-x_2-2x_3,\ x_3' = 4x_1-2x_2-4x_3$

22. $x_1' = 3x_1 + 2x_2 + 2x_3,\ x_2' = -5x_1 - 4x_2 - 2x_3,$
$x_3' = 5x_1 + 5x_2 + 3x_3$

23. $x_1' = 3x_1 + x_2 + x_3,\ x_2' = -5x_1 - 3x_2 - x_3,$
$x_3' = 5x_1 + 5x_2 + 3x_3$

24. $x_1' = 2x_1 + x_2 - x_3,\ x_2' = -4x_1 - 3x_2 - x_3,$
$x_3' = 4x_1 + 4x_2 + 2x_3$

25. $x_1' = 5x_1 + 5x_2 + 2x_3,\ x_2' = -6x_1 - 6x_2 - 5x_3,$
$x_3' = 6x_1 + 6x_2 + 5x_3$

26. Find the particular solution of the system

$$\begin{aligned}\frac{dx_1}{dt} &= 3x_1 \qquad\quad + x_3,\\ \frac{dx_2}{dt} &= 9x_1 - x_2 + 2x_3,\\ \frac{dx_3}{dt} &= -9x_1 + 4x_2 - x_3\end{aligned}$$

that satisfies the initial conditions $x_1(0) = 0$, $x_2(0) = 0$, $x_3(0) = 17$.

The amounts $x_1(t)$ and $x_2(t)$ of salt in the two brine tanks of Fig. 7.3.7 satisfy the differential equations

$$\frac{dx_1}{dt} = -k_1x_1, \qquad \frac{dx_2}{dt} = k_1x_1 - k_2x_2,$$

where $k_i = r/V_i$ for $i = 1, 2$. In Problems 27 and 28 the volumes V_1 and V_2 are given. First solve for $x_1(t)$ and $x_2(t)$, assuming that $r = 10$ (gal/min), $x_1(0) = 15$ (lb), and $x_2(0) = 0$. Then find the maximum amount of salt ever in tank 2. Finally, construct a figure showing the graphs of $x_1(t)$ and $x_2(t)$.

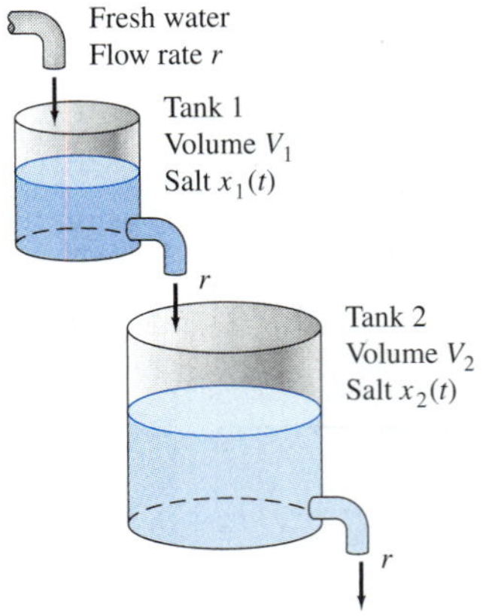

FIGURE 7.3.7. The two brine tanks of Problems 27 and 28.

27. $V_1 = 50$ (gal), $V_2 = 25$ (gal)

28. $V_1 = 25$ (gal), $V_2 = 40$ (gal)

The amounts $x_1(t)$ and $x_2(t)$ of salt in the two brine tanks of Fig. 7.3.8 satisfy the differential equations

$$\frac{dx_1}{dt} = -k_1x_1 + k_2x_2, \qquad \frac{dx_2}{dt} = k_1x_1 - k_2x_2,$$

where $k_i = r/V_i$ as usual. In Problems 29 and 30, solve for $x_1(t)$ and $x_2(t)$, assuming that $r = 10$ (gal/min), $x_1(0) = 15$ (lb), and $x_2(0) = 0$. Then construct a figure showing the graphs of $x_1(t)$ and $x_2(t)$.

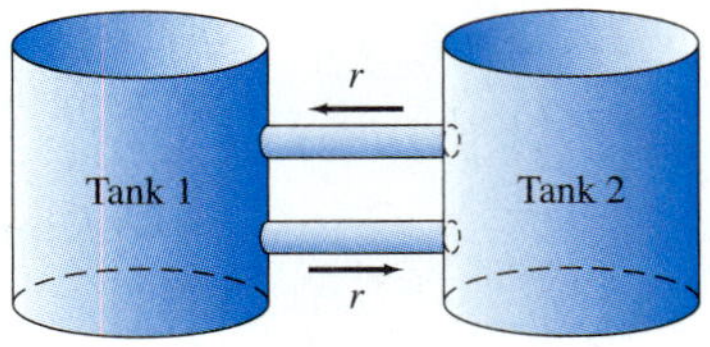

FIGURE 7.3.8. The two brine tanks of Problems 29 and 30.

29. $V_1 = 50$ (gal), $V_2 = 25$ (gal)

30. $V_1 = 25$ (gal), $V_2 = 40$ (gal)

Problems 31 through 34 deal with the open three-tank system of Fig. 7.3.2. Fresh water flows into tank 1; mixed brine flows from tank 1 into tank 2, from tank 2 into tank 3, and out of tank 3; all at the given flow rate r gallons per minute. The initial amounts $x_1(0) = x_0$ (lb), $x_2(0) = 0$, and $x_3(0) = 0$ of salt in the three tanks are given, as are their volumes V_1, V_2, and V_3 (in gallons). First solve for the amounts of salt in the three tanks at time t, then determine the maximal amount of salt that tank 3 ever contains. Finally, construct a figure showing the graphs of $x_1(t)$, $x_2(t)$, and $x_3(t)$.

31. $r = 30,\ x_0 = 27,\ V_1 = 30,\ V_2 = 15,\ V_3 = 10$

32. $r = 60,\ x_0 = 45,\ V_1 = 20,\ V_2 = 30,\ V_3 = 60$

33. $r = 60,\ x_0 = 45,\ V_1 = 15,\ V_2 = 10,\ V_3 = 30$

34. $r = 60,\ x_0 = 40,\ V_1 = 20,\ V_2 = 12,\ V_3 = 60$

Problems 35 through 37 deal with the closed three-tank system of Fig. 7.3.5, which is described by the equations in (24). Mixed brine flows from tank 1 into tank 2, from tank 2 into tank 3, and from tank 3 into tank 1, all at the given flow rate r gallons per minute. The initial amounts $x_1(0) = x_0$ *(pounds),* $x_2(0) = 0$, *and* $x_3(0) = 0$ *of salt in the three tanks are given, as are their volumes* V_1, V_2, *and* V_3 *(in gallons). First solve for the amounts of salt in the three tanks at time t, then determine the limiting amount (as* $t \to +\infty$*) of salt in each tank. Finally, construct a figure showing the graphs of* $x_1(t)$, $x_2(t)$, *and* $x_3(t)$.

35. $r = 120$, $x_0 = 33$, $V_1 = 20$, $V_2 = 6$, $V_3 = 40$
36. $r = 10$, $x_0 = 18$, $V_1 = 20$, $V_2 = 50$, $V_3 = 20$
37. $r = 60$, $x_0 = 55$, $V_1 = 60$, $V_2 = 20$, $V_3 = 30$

For each matrix **A** *given in Problems 38 through 40, the zeros in the matrix make its characteristic polynomial easy to calculate. Find the general solution of* $\mathbf{x}' = \mathbf{A}\mathbf{x}$.

38. $\mathbf{A} = \begin{bmatrix} 1 & 0 & 0 & 0 \\ 2 & 2 & 0 & 0 \\ 0 & 3 & 3 & 0 \\ 0 & 0 & 4 & 4 \end{bmatrix}$

39. $\mathbf{A} = \begin{bmatrix} -2 & 0 & 0 & 9 \\ 4 & 2 & 0 & -10 \\ 0 & 0 & -1 & 8 \\ 0 & 0 & 0 & 1 \end{bmatrix}$

40. $\mathbf{A} = \begin{bmatrix} 2 & 0 & 0 & 0 \\ -21 & -5 & -27 & -9 \\ 0 & 0 & 5 & 0 \\ 0 & 0 & -21 & -2 \end{bmatrix}$

41. The coefficient matrix **A** of the 4×4 system

$$\begin{aligned} x_1' &= 4x_1 + x_2 + x_3 + 7x_4, \\ x_2' &= x_1 + 4x_2 + 10x_3 + x_4, \\ x_3' &= x_1 + 10x_2 + 4x_3 + x_4, \\ x_4' &= 7x_1 + x_2 + x_3 + 4x_4 \end{aligned}$$

has eigenvalues $\lambda_1 = -3$, $\lambda_2 = -6$, $\lambda_3 = 10$, and $\lambda_4 = 15$. Find the particular solution of this system that satisfies the initial conditions

$$x_1(0) = 3, \quad x_2(0) = x_3(0) = 1, \quad x_4(0) = 3.$$

In Problems 42 through 50, use a calculator or computer system to calculate the eigenvalues and eigenvectors (as illustrated in Application 7.3) in order to find a general solution of the linear system $\mathbf{x}' = \mathbf{A}\mathbf{x}$ *with the given coefficient matrix* **A**.

42. $\mathbf{A} = \begin{bmatrix} -40 & -12 & 54 \\ 35 & 13 & -46 \\ -25 & -7 & 34 \end{bmatrix}$

43. $\mathbf{A} = \begin{bmatrix} -20 & 11 & 13 \\ 12 & -1 & -7 \\ -48 & 21 & 31 \end{bmatrix}$

44. $\mathbf{A} = \begin{bmatrix} 147 & 23 & -202 \\ -90 & -9 & 129 \\ 90 & 15 & -123 \end{bmatrix}$

45. $\mathbf{A} = \begin{bmatrix} 9 & -7 & -5 & 0 \\ -12 & 7 & 11 & 9 \\ 24 & -17 & -19 & -9 \\ -18 & 13 & 17 & 9 \end{bmatrix}$

46. $\mathbf{A} = \begin{bmatrix} 13 & -42 & 106 & 139 \\ 2 & -16 & 52 & 70 \\ 1 & 6 & -20 & -31 \\ -1 & -6 & 22 & 33 \end{bmatrix}$

47. $\mathbf{A} = \begin{bmatrix} 23 & -18 & -16 & 0 \\ -8 & 6 & 7 & 9 \\ 34 & -27 & -26 & -9 \\ -26 & 21 & 25 & 12 \end{bmatrix}$

48. $\mathbf{A} = \begin{bmatrix} 47 & -8 & 5 & -5 \\ -10 & 32 & 18 & -2 \\ 139 & -40 & -167 & -121 \\ -232 & 64 & 360 & 248 \end{bmatrix}$

49. $\mathbf{A} = \begin{bmatrix} 139 & -14 & -52 & -14 & 28 \\ -22 & 5 & 7 & 8 & -7 \\ 370 & -38 & -139 & -38 & 76 \\ 152 & -16 & -59 & -13 & 35 \\ 95 & -10 & -38 & -7 & 23 \end{bmatrix}$

50. $\mathbf{A} = \begin{bmatrix} 9 & 13 & 0 & 0 & 0 & -13 \\ -14 & 19 & -10 & -20 & 10 & 4 \\ -30 & 12 & -7 & -30 & 12 & 18 \\ -12 & 10 & -10 & -9 & 10 & 2 \\ 6 & 9 & 0 & 6 & 5 & -15 \\ -14 & 23 & -10 & -20 & 10 & 0 \end{bmatrix}$

7.3 Application Automatic Calculation of Eigenvalues and Eigenvectors

Most computational systems offer the capability to find eigenvalues and eigenvectors readily. For instance, Fig. 7.3.9 shows a graphing calculator computation of the eigenvalues and eigenvectors of the matrix

$$\mathbf{A} = \begin{bmatrix} -0.5 & 0.0 & 0.0 \\ 0.5 & -0.25 & 0.0 \\ 0.0 & 0.25 & -0.2 \end{bmatrix}$$

of Example 2. We see each eigenvector displayed as a column vector beneath its eigenvalue. Note that with results presented in decimal form, it is up to us to guess (and verify by matrix multiplication) that the exact eigenvector associated with the third eigenvalue $\lambda = -\frac{1}{2}$ is $\mathbf{v} = \begin{bmatrix} 1 & -2 & \frac{5}{3} \end{bmatrix}^T$. The *Maple* commands

```
with(linalg)
A := matrix(3,3,[-0.5,0,0,0.5,-0.25,0,0,0.25,-0.2]);
eigenvects(A);
```

the *Mathematica* commands

```
A = {{-0.5,0,0},{0.5,-0.25,0},{0,0.25,-0.2}}
Eigensystem[A]
```

and the MATLAB commands

```
A = [-0.5,0,0; 0.5,-0.25,0; 0,0.25,-0.2]
[V,D] = eig(A)
```

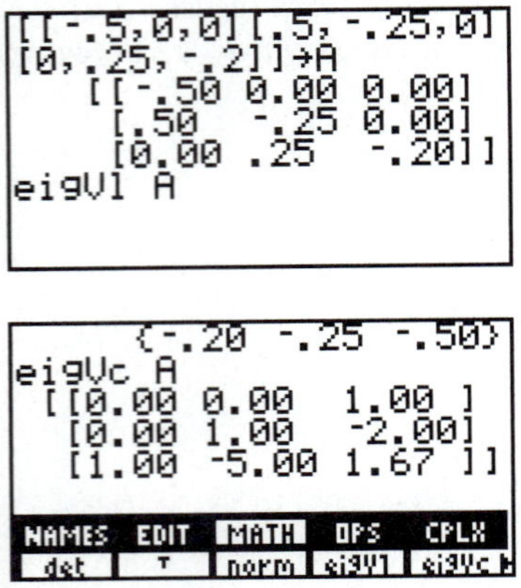

FIGURE 7.3.9. TI-86 calculation of the eigenvalues and eigenvectors of the matrix **A**.

(where `D` will be a diagonal matrix displaying the eigenvalues of `A` and the column vectors of `V` are the corresponding eigenvectors) produce similar results. You can use these commands to find the eigenvalues and eigenvectors needed for any of the problems in this section.

For a more substantial investigation, choose a positive integer $n < 10$ ($n = 5$, for instance) and let $q_1, q_2, \ldots, q_n$ denote the first n nonzero digits in your student ID number. Now consider an open system of brine tanks as in Fig. 7.3.2, except with n rather than three successive tanks having volumes $V_i = 10q_i$ ($i = 1, 2, \ldots, n$) in gallons. If each flow rate is $r = 10$ gallons per minute, then the salt amounts $x_1(t), x_2(t), \ldots, x_n(t)$ satisfy the linear system

$$\begin{aligned} x_1' &= -k_1 x_1, \\ x_i' &= k_{i-1}x_{i-1} - k_i x_i \quad (i = 2, 3, \ldots, n), \end{aligned}$$

where $k_i = r/V_i$. Apply the eigenvalue method to solve this system with initial conditions

$$x_1(0) = 10, \quad x_2(0) = x_3(0) = \cdots = x_n(0) = 0.$$

Graph the solution functions and estimate graphically the maximum amount of salt that each tank ever contains.

For an alternative investigation, suppose that the system of n tanks is closed as in Fig. 7.3.5, so that tank 1 receives as inflow the outflow from tank n (rather than fresh water). Then the first equation should be replaced with $x_1' = k_n x_n - k_1 x_1$. Now show that, in this *closed* system, as $t \to +\infty$ the salt originally in tank 1 distributes itself with constant density throughout the various tanks. A plot like Fig. 7.3.6 should make this fairly obvious.

7.4 Second-Order Systems and Mechanical Applications*

In this section we apply the matrix methods of Sections 7.2 and 7.3 to investigate the oscillations of typical mass-and-spring systems having two or more degrees of freedom. Our examples are chosen to illustrate phenomena that are generally characteristic of complex mechanical systems.

*This optional section may be omitted without loss of continuity. It provides a sample of the more technical applications of eigenvalues to physics and engineering problems.

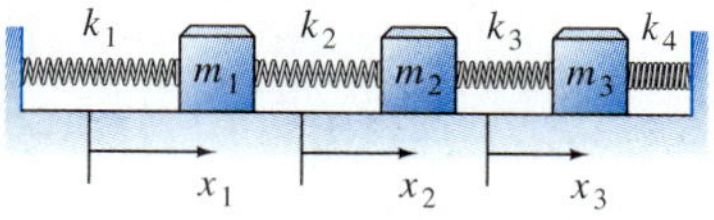

FIGURE 7.4.1. Three spring-coupled masses.

Figure 7.4.1 shows three masses connected to each other and to two walls by the four indicated springs. We assume that the masses slide without friction and that each spring obeys Hooke's law—its extension or compression x and force F of reaction are related by the formula $F = -kx$. If the rightward displacements x_1, x_2, and x_3 of the three masses (from their respective equilibrium positions) are all positive, then

- The first spring is stretched the distance x_1;
- The second spring is stretched the distance $x_2 - x_1$;
- The third spring is stretched the distance $x_3 - x_2$;
- The fourth spring is compressed the distance x_3.

Therefore, application of Newton's law $F = ma$ to the three masses (as in Example 1 of Section 7.1) yields their equations of motion:

$$\begin{aligned} m_1 x_1'' &= -k_1 x_1 \qquad\qquad\quad + k_2(x_2 - x_1), \\ m_2 x_2'' &= -k_2(x_2 - x_1) + k_3(x_3 - x_2), \\ m_3 x_3'' &= -k_3(x_3 - x_2) - k_4 x_3. \end{aligned} \tag{1}$$

Although we assumed in writing these equations that the displacements of the masses are all positive, they actually follow similarly from Hooke's and Newton's laws, whatever the signs of these displacements.

In terms of the **displacement vector** $\mathbf{x} = \begin{bmatrix} x_1 & x_2 & x_3 \end{bmatrix}^T$, the **mass matrix**

$$\mathbf{M} = \begin{bmatrix} m_1 & 0 & 0 \\ 0 & m_2 & 0 \\ 0 & 0 & m_3 \end{bmatrix}, \tag{2}$$

and the **stiffness matrix**

$$\mathbf{K} = \begin{bmatrix} -(k_1 + k_2) & k_2 & 0 \\ k_2 & -(k_2 + k_3) & k_3 \\ 0 & k_3 & -(k_3 + k_4) \end{bmatrix}, \tag{3}$$

the system in (1) takes the matrix form

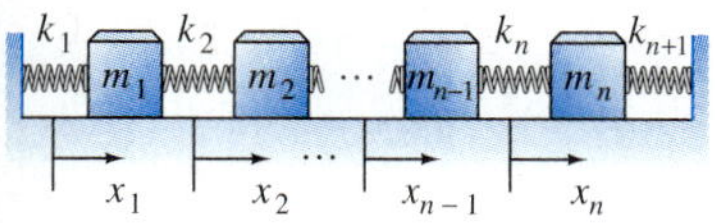

FIGURE 7.4.2. A system of n spring-coupled masses.

$$\mathbf{M}\mathbf{x}'' = \mathbf{K}\mathbf{x}. \tag{4}$$

The notation in Eqs. (1) through (4) generalizes in a natural way to the system of n spring-coupled masses shown in Fig. 7.4.2. We need only write

$$\mathbf{M} = \begin{bmatrix} m_1 & 0 & \cdots & 0 \\ 0 & m_2 & \cdot & 0 \\ \vdots & \vdots & & \vdots \\ 0 & 0 & \cdots & m_n \end{bmatrix} \tag{5}$$

and

$$\mathbf{K} = \begin{bmatrix} -(k_1+k_2) & k_2 & 0 & \cdots & 0 \\ k_2 & -(k_2+k_3) & k_3 & \cdots & 0 \\ 0 & k_3 & -(k_3+k_4) & \cdots & 0 \\ 0 & 0 & k_4 & \cdots & 0 \\ \vdots & \vdots & & & \vdots \\ 0 & 0 & \cdots & -(k_{n-1}+k_n) & k_n \\ 0 & 0 & \cdots & k_n & -(k_n+k_{n+1}) \end{bmatrix} \tag{6}$$

for the mass and stiffness matrices in Eq. (4).

The diagonal matrix $\mathbf{M}$ is obviously nonsingular; to get its inverse $\mathbf{M}^{-1}$ we need only replace each diagonal element with its reciprocal. Hence multiplication of each side in Eq. (4) by $\mathbf{M}^{-1}$ yields the homogeneous **second-order system**

$$\mathbf{x}'' = \mathbf{A}\mathbf{x}, \tag{7}$$

where $\mathbf{A} = \mathbf{M}^{-1}\mathbf{K}$. There is a wide variety of *frictionless* mechanical systems for which a displacement or position vector $\mathbf{x}$, a nonsingular mass matrix $\mathbf{M}$, and a stiffness matrix $\mathbf{K}$ satisfying Eq. (4) can be defined.

Solution of Second-Order Systems

To seek a solution of Eq. (7), we substitute (as in Section 7.3 for a first-order system) a trial solution of the form

$$\mathbf{x}(t) = \mathbf{v}e^{\alpha t}, \tag{8}$$

where $\mathbf{v}$ is a constant vector. Then $\mathbf{x}'' = \alpha^2\mathbf{v}e^{\alpha t}$, so substitution of Eq. (8) in (7) gives

$$\alpha^2\mathbf{v}e^{\alpha t} = \mathbf{A}\mathbf{v}e^{\alpha t},$$

which implies that

$$\mathbf{A}\mathbf{v} = \alpha^2\mathbf{v}. \tag{9}$$

Therefore, $\mathbf{x}(t) = \mathbf{v}e^{\alpha t}$ is a solution of $\mathbf{x}'' = \mathbf{A}\mathbf{x}$ if and only if $\alpha^2 = \lambda$, an eigenvalue of the matrix $\mathbf{A}$, and $\mathbf{v}$ is an associated eigenvector.

If $\mathbf{x}'' = \mathbf{A}\mathbf{x}$ models a mechanical system, then it is typical that the eigenvalues of $\mathbf{A}$ are *negative* real numbers. If

$$\alpha^2 = \lambda = -\omega^2 < 0,$$

then $\alpha = \pm\omega i$. In this case the solution given by Eq. (8) is

$$\mathbf{x}(t) = \mathbf{v}e^{i\omega t} = \mathbf{v}(\cos\omega t + i\sin\omega t).$$

The real and imaginary parts

$$\mathbf{x}_1(t) = \mathbf{v}\cos\omega t \quad \text{and} \quad \mathbf{x}_2(t) = \mathbf{v}\sin\omega t \tag{10}$$

of $\mathbf{x}(t)$ are then linearly independent *real-valued* solutions of the system. This analysis leads to the following theorem.

THEOREM 1 Second-Order Homogeneous Linear Systems

If the $n \times n$ matrix $\mathbf{A}$ has distinct negative eigenvalues $-\omega_1^2, -\omega_2^2, \ldots, -\omega_n^2$ with associated [real] eigenvectors $\mathbf{v}_1, \mathbf{v}_2, \ldots, \mathbf{v}_n$, then a general solution of

$$\mathbf{x}'' = \mathbf{A}\mathbf{x}$$

is given by

$$\mathbf{x}(t) = \sum_{i=1}^{n} (a_i \cos \omega_i t + b_i \sin \omega_i t)\mathbf{v}_i \tag{11}$$

with a_i and b_i arbitrary constants. In the special case of a nonrepeated zero eigenvalue λ_0 with associated eigenvector $\mathbf{v}_0$,

$$\mathbf{x}_0(t) = (a_0 + b_0 t)\mathbf{v}_0 \tag{12}$$

is the corresponding part of the general solution.

Remark: The nonzero vector $\mathbf{v}_0$ is an eigenvector corresponding to $\lambda_0 = 0$ provided that $\mathbf{A}\mathbf{v}_0 = \mathbf{0}$. If $\mathbf{x}(t) = (a_0 + b_0 t)\mathbf{v}_0$, then

$$\mathbf{x}'' = 0 \cdot \mathbf{v}_0 = (a_0 + b_0 t) \cdot \mathbf{0} = (a_0 + b_0 t) \cdot (\mathbf{A}\mathbf{v}_0) = \mathbf{A}\mathbf{x},$$

thus verifying the form in Eq. (12). ■

Example 1 Consider the mass-and-spring system with $n = 2$ shown in Fig. 7.4.3. Because there is no third spring connected to a right-hand wall, we set $k_3 = 0$. If $m_1 = 2$, $m_2 = 1$, $k_1 = 100$, and $k_2 = 50$, then the equation $\mathbf{M}\mathbf{x}'' = \mathbf{K}\mathbf{x}$ is

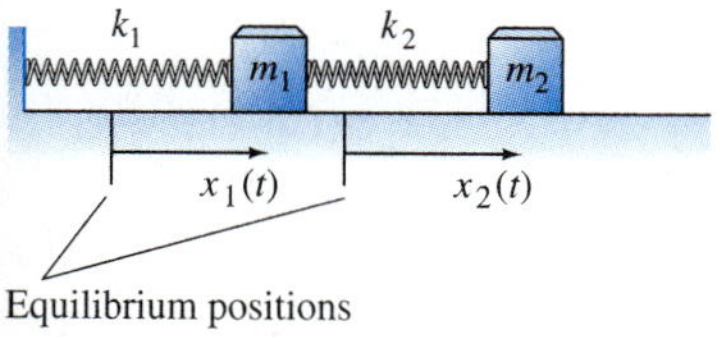

FIGURE 7.4.3. The mass-and-spring system of Example 1.

$$\begin{bmatrix} 2 & 0 \\ 0 & 1 \end{bmatrix} \mathbf{x}'' = \begin{bmatrix} -150 & 50 \\ 50 & -50 \end{bmatrix} \mathbf{x}, \tag{13}$$

which reduces to $\mathbf{x}'' = \mathbf{A}\mathbf{x}$ with

$$\mathbf{A} = \begin{bmatrix} -75 & 25 \\ 50 & -50 \end{bmatrix}.$$

The characteristic equation of $\mathbf{A}$ is

$$\begin{aligned} (-75 - \lambda)(-50 - \lambda) - 50 \cdot 25 &= \lambda^2 + 125\lambda + 2500 \\ &= (\lambda + 25)(\lambda + 100) = 0, \end{aligned}$$

so $\mathbf{A}$ has the negative eigenvalues $\lambda_1 = -25$ and $\lambda_2 = -100$. By Theorem 1, the system in (13) therefore has solutions with [circular] frequencies $\omega_1 = 5$ and $\omega_2 = 10$.

CASE 1: $\lambda_1 = -25$. The eigenvector equation $(\mathbf{A} - \lambda\mathbf{I})\mathbf{v} = \mathbf{0}$ is

$$\begin{bmatrix} -50 & 25 \\ 50 & -25 \end{bmatrix} \begin{bmatrix} a \\ b \end{bmatrix} = \begin{bmatrix} 0 \\ 0 \end{bmatrix},$$

so an eigenvector associated with $\lambda_1 = -25$ is $\mathbf{v}_1 = \begin{bmatrix} 1 & 2 \end{bmatrix}^T$.

CASE 2: $\lambda_2 = -100$. The eigenvector equation $(\mathbf{A} - \lambda\mathbf{I})\mathbf{v} = \mathbf{0}$ is

$$\begin{bmatrix} 25 & 25 \\ 50 & 50 \end{bmatrix}\begin{bmatrix} a \\ b \end{bmatrix} = \begin{bmatrix} 0 \\ 0 \end{bmatrix},$$

so an eigenvector associated with $\lambda_2 = -100$ is $\mathbf{v}_2 = \begin{bmatrix} 1 & -1 \end{bmatrix}^T$.

By Eq. (11) it follows that a general solution of the system in (13) is given by

$$\mathbf{x}(t) = (a_1 \cos 5t + b_1 \sin 5t)\mathbf{v}_1 + (a_2 \cos 10t + b_2 \sin 10t)\mathbf{v}_2. \tag{14}$$

The two terms on the right in Eq. (14) represent **free oscillations** of the mass-and-spring system. They describe the physical system's two **natural modes of oscillation** at its two [circular] **natural frequencies** $\omega_1 = 5$ and $\omega_2 = 10$. The natural mode

$$\mathbf{x}_1(t) = (a_1 \cos 5t + b_1 \sin 5t)\mathbf{v}_1 = c_1 \cos(5t - \alpha_1)\begin{bmatrix} 1 \\ 2 \end{bmatrix}$$

(with $c_1 = \sqrt{a_1^2 + b_1^2}$, $\cos\alpha_1 = a_1/c_1$, and $\sin\alpha_1 = b_1/c_1$) has the scalar component equations

$$\begin{aligned} x_1(t) &= \ \ c_1 \cos(5t - \alpha_1), \\ x_2(t) &= 2c_1 \cos(5t - \alpha_1), \end{aligned} \tag{15}$$

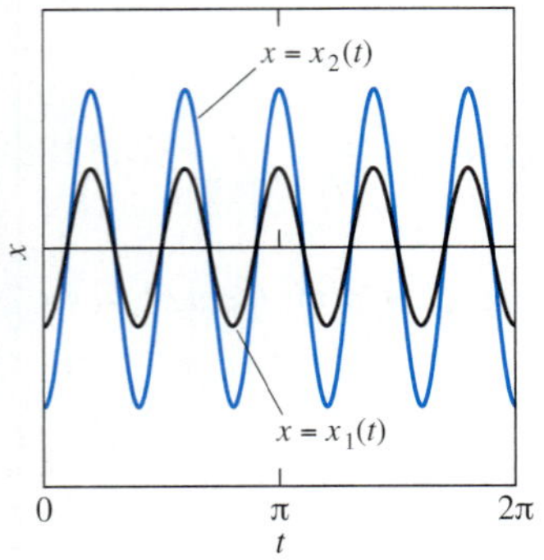

FIGURE 7.4.4. Oscillations in the same direction with frequency $\omega_1 = 5$; the amplitude of motion of mass 2 is twice that of mass 1.

and therefore describes a free oscillation in which the two masses move in synchrony in the same direction and with the same frequency $\omega_1 = 5$, but with the amplitude of motion of m_2 twice that of m_1 (see Fig. 7.4.4). The natural mode

$$\mathbf{x}_2(t) = (a_2 \cos 10t + b_2 \sin 10t)\mathbf{v}_2 = c_2 \cos(10t - \alpha_2)\begin{bmatrix} 1 \\ -1 \end{bmatrix}$$

has the scalar component equations

$$\begin{aligned} x_1(t) &= \ \ \ c_2 \cos(10t - \alpha_2), \\ x_2(t) &= -c_2 \cos(10t - \alpha_2), \end{aligned} \tag{16}$$

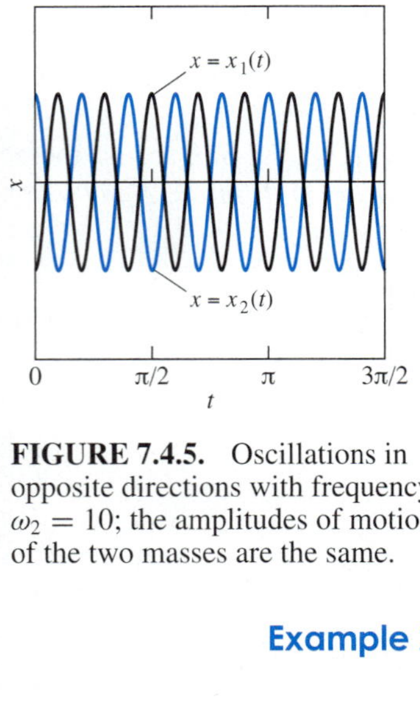

FIGURE 7.4.5. Oscillations in opposite directions with frequency $\omega_2 = 10$; the amplitudes of motion of the two masses are the same.

and therefore describes a free oscillation in which the two masses move in synchrony in opposite directions with the same frequency $\omega_2 = 10$ and with equal amplitudes of oscillation (see Fig. 7.4.5). ■

Example 2 Figure 7.4.6 shows three railway cars connected by buffer springs that react when compressed, but disengage instead of stretching. With $n = 3$, $k_2 = k_3 = k$, and $k_1 = k_4 = 0$ in Eqs. (2) through (4), we get the system

$$\begin{bmatrix} m_1 & 0 & 0 \\ 0 & m_2 & 0 \\ 0 & 0 & m_3 \end{bmatrix}\mathbf{x}'' = \begin{bmatrix} -k & k & 0 \\ k & -2k & k \\ 0 & k & -k \end{bmatrix}\mathbf{x}, \tag{17}$$

FIGURE 7.4.6. The three railway cars of Example 2.

which is equivalent to

$$\mathbf{x}'' = \begin{bmatrix} -c_1 & c_1 & 0 \\ c_2 & -2c_2 & c_2 \\ 0 & c_3 & -c_3 \end{bmatrix}\mathbf{x} \tag{18}$$

with

$$c_i = \frac{k}{m_i} \quad (i = 1, 2, 3). \tag{19}$$

If we assume further that $m_1 = m_3$, so that $c_1 = c_3$, then a brief computation gives

$$-\lambda(\lambda + c_1)(\lambda + c_1 + 2c_2) = 0 \tag{20}$$

for the characteristic equation of the coefficient matrix **A** in Eq. (18). Hence the matrix **A** has eigenvalues

$$\lambda_1 = 0, \quad \lambda_2 = -c_1, \quad \lambda_3 = -c_1 - 2c_2 \tag{21a}$$

corresponding to the natural frequencies

$$\omega_1 = 0, \quad \omega_2 = \sqrt{c_1}, \quad \omega_3 = \sqrt{c_1 + 2c_2} \tag{21b}$$

of the physical system.

For a numerical example, suppose that the first and third railway cars weigh 12 tons each, that the middle car weighs 8 tons, and that the spring constant is $k = 1.5$ tons/ft; i.e., $k = 3000$ lb/ft. Then, using fps units with mass measured in slugs (a weight of 32 pounds has a mass of 1 slug), we have

$$m_1 = m_3 = 750, \quad m_2 = 500,$$

and

$$c_1 = \frac{3000}{750} = 4, \quad c_2 = \frac{3000}{500} = 6.$$

Hence the coefficient matrix **A** is

$$\mathbf{A} = \begin{bmatrix} -4 & 4 & 0 \\ 6 & -12 & 6 \\ 0 & 4 & -4 \end{bmatrix}, \tag{22}$$

and the eigenvalue-frequency pairs given by (21a) and (21b) are $\lambda_1 = 0$, $\omega_1 = 0$; $\lambda_2 = -4$, $\omega_2 = 2$; and $\lambda_3 = -16$, $\omega_3 = 4$.

CASE 1: $\lambda_1 = 0, \omega_1 = 0.$ The eigenvector equation $(\mathbf{A} - \lambda\mathbf{I})\mathbf{v} = \mathbf{0}$ is

$$\mathbf{A}\mathbf{v} = \begin{bmatrix} -4 & 4 & 0 \\ 6 & -12 & 6 \\ 0 & 4 & -4 \end{bmatrix} \begin{bmatrix} a \\ b \\ c \end{bmatrix} = \begin{bmatrix} 0 \\ 0 \\ 0 \end{bmatrix},$$

so it is clear that $\mathbf{v}_1 = \begin{bmatrix} 1 & 1 & 1 \end{bmatrix}^T$ is an eigenvector associated with $\lambda_1 = 0$. According to Theorem 1, the corresponding part of a general solution of $\mathbf{x}'' = \mathbf{A}\mathbf{x}$ is

$$\mathbf{x}_1(t) = (a_1 + b_1 t)\mathbf{v}_1.$$

CASE 2: $\lambda_2 = -4$, $\omega_2 = 2$. The eigenvector equation $(\mathbf{A} - \lambda\mathbf{I})\mathbf{v} = \mathbf{0}$ is

$$(\mathbf{A} + 4\mathbf{I})\mathbf{v} = \begin{bmatrix} 0 & 4 & 0 \\ 6 & -8 & 6 \\ 0 & 4 & 0 \end{bmatrix} \begin{bmatrix} a \\ b \\ c \end{bmatrix} = \begin{bmatrix} 0 \\ 0 \\ 0 \end{bmatrix}$$

so it is clear that $\mathbf{v}_2 = \begin{bmatrix} 1 & 0 & -1 \end{bmatrix}^T$ is an eigenvector associated with $\lambda_2 = -4$. According to Theorem 1, the corresponding part of a general solution of $\mathbf{x}'' = \mathbf{A}\mathbf{x}$ is

$$\mathbf{x}_2(t) = (a_2 \cos 2t + b_2 \sin 2t)\mathbf{v}_2.$$

CASE 3: $\lambda_3 = -16$, $\omega_3 = 4$. The eigenvector equation $(\mathbf{A} - \lambda\mathbf{I})\mathbf{v} = \mathbf{0}$ is

$$(\mathbf{A} + 16\mathbf{I})\mathbf{v} = \begin{bmatrix} 12 & 4 & 0 \\ 6 & 4 & 6 \\ 0 & 4 & 12 \end{bmatrix} \begin{bmatrix} a \\ b \\ c \end{bmatrix} = \begin{bmatrix} 0 \\ 0 \\ 0 \end{bmatrix},$$

so it is clear that $\mathbf{v}_3 = \begin{bmatrix} 1 & -3 & 1 \end{bmatrix}^T$ is an eigenvector associated with $\lambda_3 = -16$. According to Theorem 1, the corresponding part of a general solution of $\mathbf{x}'' = \mathbf{A}\mathbf{x}$ is

$$\mathbf{x}_3(t) = (a_3 \cos 4t + b_3 \sin 4t)\mathbf{v}_3.$$

The general solution $\mathbf{x} = \mathbf{x}_1 + \mathbf{x}_2 + \mathbf{x}_3$ of $\mathbf{x}'' = \mathbf{A}\mathbf{x}$ is therefore given by

$$\begin{aligned} \mathbf{x}(t) = a_1 \begin{bmatrix} 1 \\ 1 \\ 1 \end{bmatrix} + b_1 t \begin{bmatrix} 1 \\ 1 \\ 1 \end{bmatrix} + a_2 \begin{bmatrix} 1 \\ 0 \\ -1 \end{bmatrix} \cos 2t \\ + b_2 \begin{bmatrix} 1 \\ 0 \\ -1 \end{bmatrix} \sin 2t + a_3 \begin{bmatrix} 1 \\ -3 \\ 1 \end{bmatrix} \cos 4t + b_3 \begin{bmatrix} 1 \\ -3 \\ 1 \end{bmatrix} \sin 4t. \end{aligned} \tag{23}$$

To determine a particular solution, let us suppose that the leftmost car is moving to the right with velocity v_0 and at time $t = 0$ strikes the other two cars, which are together but at rest. The corresponding initial conditions are

$$x_1(0) = x_2(0) = x_3(0) = 0, \tag{24a}$$

$$x_1'(0) = v_0, \quad x_2'(0) = x_3'(0) = 0. \tag{24b}$$

Then substitution of (24a) in (23) gives the scalar equations

$$\begin{aligned} a_1 + a_2 + a_3 &= 0, \\ a_1 \qquad - 3a_3 &= 0, \\ a_1 - a_2 + a_3 &= 0, \end{aligned}$$

which readily yield $a_1 = a_2 = a_3 = 0$. Hence the position functions of the three cars are

$$\begin{aligned} x_1(t) &= b_1 t + b_2 \sin 2t + b_3 \sin 4t, \\ x_2(t) &= b_1 t \qquad\qquad\quad - 3b_3 \sin 4t, \\ x_3(t) &= b_1 t - b_2 \sin 2t + b_3 \sin 4t, \end{aligned} \tag{25}$$

and their velocity functions are

$$\begin{aligned} x_1'(t) &= b_1 + 2b_2 \cos 2t + 4b_3 \cos 4t, \\ x_2'(t) &= b_1 \qquad\qquad\quad - 12b_3 \cos 4t, \\ x_3'(t) &= b_1 - 2b_2 \cos 2t + 4b_3 \cos 4t. \end{aligned} \tag{26}$$

Substitution of (24b) in (26) gives the equations

$$\begin{aligned} b_1 + 2b_2 + 4b_3 &= v_0, \\ b_1 \qquad - 12b_3 &= 0, \\ b_1 - 2b_2 + 4b_3 &= 0 \end{aligned}$$

that readily yield $b_1 = \frac{3}{8}v_0$, $b_2 = \frac{1}{4}v_0$, and $b_3 = \frac{1}{32}v_0$. Finally, the position functions in (25) are

$$\begin{aligned} x_1(t) &= \tfrac{1}{32}v_0(12t + 8\sin 2t + \sin 4t), \\ x_2(t) &= \tfrac{1}{32}v_0(12t \qquad\qquad - 3\sin 4t), \\ x_3(t) &= \tfrac{1}{32}v_0(12t - 8\sin 2t + \sin 4t). \end{aligned} \tag{27}$$

But these equations hold only so long as the two buffer springs remain compressed; that is, while both

$$x_2 - x_1 < 0 \quad \text{and} \quad x_3 - x_2 < 0.$$

To discover what this implies about t, we compute

$$\begin{aligned} x_2(t) - x_1(t) &= \tfrac{1}{32}v_0(-8\sin 2t - 4\sin 4t) \\ &= -\tfrac{1}{32}v_0(8\sin 2t + 8\sin 2t\cos 2t) \\ &= -\tfrac{1}{4}v_0(\sin 2t)(1 + \cos 2t) \end{aligned}$$

and, similarly,

$$x_3(t) - x_2(t) = -\tfrac{1}{4}v_0(\sin 2t)(1 - \cos 2t).$$

It follows that $x_2 - x_1 < 0$ and $x_3 - x_2 < 0$ until $t = \pi/2 \approx 1.57$ (seconds), at which time the equations in (26) and (27) give the values

$$x_1\left(\frac{\pi}{2}\right) = x_2\left(\frac{\pi}{2}\right) = x_3\left(\frac{\pi}{2}\right) = \frac{3\pi v_0}{16},$$

$$x_1'\left(\frac{\pi}{2}\right) = x_2'\left(\frac{\pi}{2}\right) = 0, \quad x_3'\left(\frac{\pi}{2}\right) = v_0.$$

We conclude that the three railway cars remain engaged and moving to the right until disengagement occurs at time $t = \pi/2$. Thereafter, cars 1 and 2 remain at rest (!), while car 3 continues to the right with speed v_0. If, for instance, $v_0 = 48$ feet per second (about 33 miles per hour), then the three cars travel a distance of $9\pi \approx 28.27$ (ft) during their 1.57 seconds of engagement, and

$$x_1(t) = x_2(t) = 9\pi, \quad x_3(t) = 48t - 15\pi \tag{27'}$$

for $t > \pi/2$. Figure 7.4.7 illustrates the "before" and "after" situations, and Fig. 7.4.8 shows the graphs of the functions $x_1(t)$, $x_2(t)$, and $x_3(t)$ in Eqs. (27) and (27′). ■

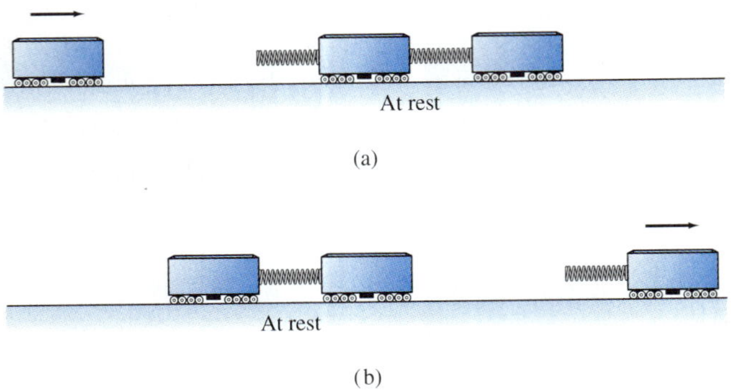

FIGURE 7.4.7. (a) Before; (b) after.

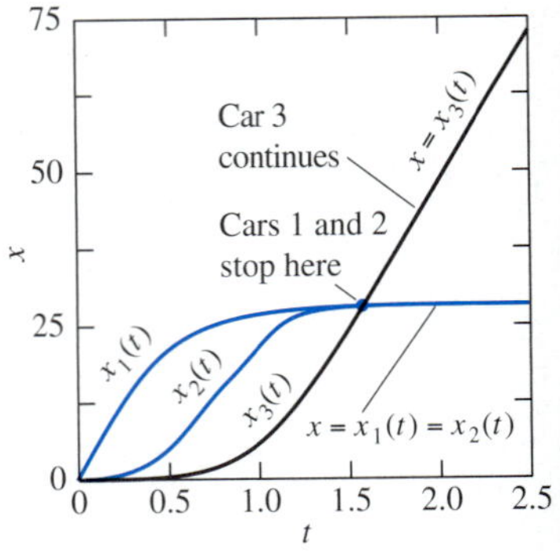

FIGURE 7.4.8. Position functions of the three railway cars of Example 2.

Forced Oscillations and Resonance

Suppose now that the ith mass of the mass-and-spring system in Fig. 7.4.2 is subject to an external force F_i ($i = 1, 2, \dots, n$) in addition to the forces exerted by the springs attached to it. Then the homogeneous equation $\mathbf{Mx}'' = \mathbf{Kx}$ is replaced with the nonhomogeneous equation

$$\mathbf{Mx}'' = \mathbf{Kx} + \mathbf{F}, \tag{28}$$

where $\mathbf{F} = \begin{bmatrix} F_1 & F_2 & \dots & F_n \end{bmatrix}^T$ is the **external force vector** for the system. Multiplication by $\mathbf{M}^{-1}$ yields

$$\mathbf{x}'' = \mathbf{Ax} + \mathbf{f}, \tag{29}$$

where $\mathbf{f}$ is the external force vector *per unit mass*. We are especially interested in the case of a *periodic external force*

$$\mathbf{f}(t) = \mathbf{F}_0 \cos \omega t \tag{30}$$

where $\mathbf{F}_0$ is a constant vector—as in the forced mass-spring system of Example 3 below and in the investigation of earthquake-induced building vibrations in the Application that follows this section. Thinking of the method of undetermined coefficients, we may anticipate a periodic particular solution of Eq. (29) of the form

$$\mathbf{x}_p(t) = \mathbf{c} \cos \omega t \tag{31}$$

with the known external frequency ω and with a coefficient vector $\mathbf{c}$ yet to be determined. Because $\mathbf{x}_p'' = -\omega^2 \mathbf{c} \cos \omega t$, substitution of (30) and (31) in (29), followed by cancellation of the common factor $\cos \omega t$, gives the linear system

$$(\mathbf{A} + \omega^2 \mathbf{I})\mathbf{c} = -\mathbf{F}_0 \tag{32}$$

to be solved for $\mathbf{c}$.

Observe that the matrix $\mathbf{A} + \omega^2 \mathbf{I}$ is nonsingular—in which case Eq. (32) can be solved for $\mathbf{c}$—unless $-\omega^2 = \lambda$, an eigenvalue of $\mathbf{A}$. Thus a periodic particular solution of the form in Eq. (31) exists provided that the external forcing frequency does *not* equal one of the natural frequencies $\omega_1, \omega_2, \dots, \omega_n$ of the system. The case in which ω *is* a natural frequency corresponds to the phenomenon of **resonance** discussed in Section 5.6.

Example 3

Suppose that the second mass in Example 1 is subjected to the external periodic force $50 \cos \omega t$. Then with $m_1 = 2$, $m_2 = 1$, $k_1 = 100$, $k_2 = 50$, and $F_0 = 50$ in Fig. 7.4.9, Eq. (29) takes the form

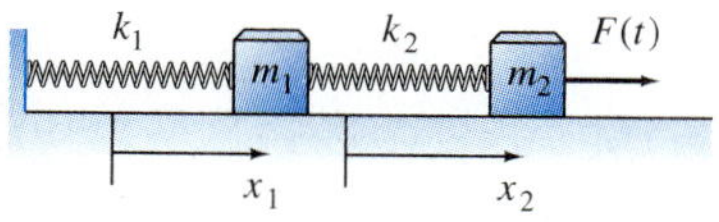

FIGURE 7.4.9. The forced mass-and-spring system of Example 3.

$$\mathbf{x}'' = \begin{bmatrix} -75 & 25 \\ 50 & -50 \end{bmatrix} \mathbf{x} + \begin{bmatrix} 0 \\ 50 \end{bmatrix} \cos \omega t, \tag{33}$$

and the substitution $\mathbf{x} = \mathbf{c} \cos \omega t$ leads to the equation

$$\begin{bmatrix} \omega^2 - 75 & 25 \\ 50 & \omega^2 - 50 \end{bmatrix} \mathbf{c} = \begin{bmatrix} 0 \\ -50 \end{bmatrix} \tag{34}$$

for the coefficient vector $\mathbf{c} = \begin{bmatrix} c_1 & c_2 \end{bmatrix}^T$. This system is readily solved for

$$c_1 = \frac{1250}{(\omega^2 - 25)(\omega^2 - 100)}, \quad c_2 = -\frac{50(\omega^2 - 75)}{(\omega^2 - 25)(\omega^2 - 100)}. \tag{35}$$

For instance, if the external squared frequency is $\omega^2 = 50$, then (35) yields $c_1 = -1$, $c_2 = -1$. The resulting forced periodic oscillation is described by

$$x_1(t) = -\cos \omega t, \quad x_2(t) = -\cos \omega t.$$

Thus the two masses oscillate in synchrony with equal amplitudes and in the same direction.

If the external squared frequency is $\omega^2 = 125$, then (35) yields $c_1 = \frac{1}{2}$, $c_2 = -1$. The resulting forced periodic oscillation is described by

$$x_1(t) = \tfrac{1}{2} \cos \omega t, \quad x_2(t) = -\cos \omega t,$$

and now the two masses oscillate in synchrony in opposite directions, but with the amplitude of motion of m_2 twice that of m_1.

It is evident from the denominators in (35) that c_1 and c_2 approach $+\infty$ as ω approaches either of the two natural frequencies $\omega_1 = 5$ and $\omega_2 = 10$ (found in Example 1). Figure 7.4.10 shows a plot of the amplitude $\sqrt{c_1^2 + c_2^2}$ of the forced periodic solution $\mathbf{x}(t) = \mathbf{c} \cos \omega t$ as a function of the forced frequency ω. The peaks at $\omega_2 = 5$ and $\omega_2 = 10$ exhibit visually the phenomenon of resonance. ■

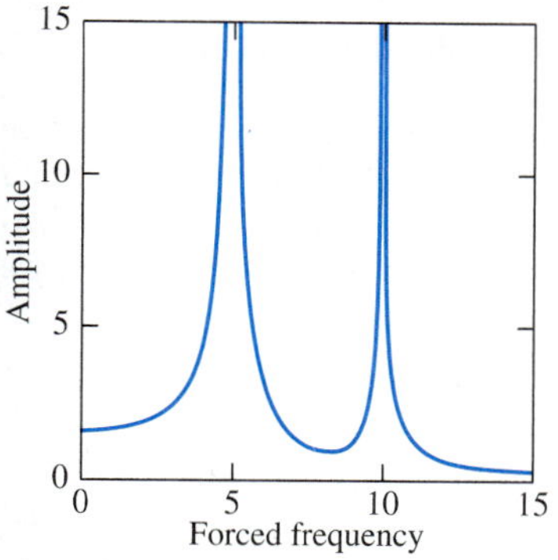

FIGURE 7.4.10. Frequency-amplitude plot for Example 3.

Periodic and Transient Solutions

It follows from Theorem 4 of Section 7.2 that a particular solution of the forced system

$$\mathbf{x}'' = \mathbf{A}\mathbf{x} + \mathbf{F}_0 \cos \omega t \tag{36}$$

will be of the form

$$\mathbf{x}(t) = \mathbf{x}_c(t) + \mathbf{x}_p(t), \tag{37}$$

where $\mathbf{x}_p(t)$ is a particular solution of the nonhomogeneous system and $\mathbf{x}_c(t)$ is a solution of the corresponding homogeneous system. It is typical for the effects of frictional resistance in mechanical systems to damp out the complementary function solution $\mathbf{x}_c(t)$, so that

$$\mathbf{x}_c(t) \to \mathbf{0} \quad \text{as} \quad t \to +\infty. \tag{38}$$

Hence $\mathbf{x}_c(t)$ is a **transient solution** that depends only on the initial conditions; it dies out with time, leaving the **steady periodic solution** $\mathbf{x}_p(t)$ resulting from the external driving force:

$$\mathbf{x}(t) \to \mathbf{x}_p(t) \quad \text{as} \quad t \to +\infty. \tag{39}$$

As a practical matter, every physical system includes frictional resistance (however small) that damps out transient solutions in this manner.

7.4 Problems

Problems 1 through 7 deal with the mass-and-spring system shown in Fig. 7.4.11 with stiffness matrix

$$\mathbf{K} = \begin{bmatrix} -(k_1 + k_2) & k_2 \\ k_2 & -(k_2 + k_3) \end{bmatrix}$$

and with the given mks values for the masses and spring constants. Find the two natural frequencies of the system and describe its two natural modes of oscillation.

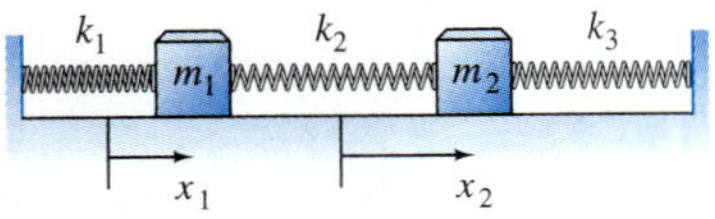

FIGURE 7.4.11. The mass-and-spring system for Problems 1 through 6.

1. $m_1 = m_2 = 1$; $k_1 = 0, k_2 = 2, k_3 = 0$ (no walls)
2. $m_1 = m_2 = 1$; $k_1 = 1, k_2 = 4, k_3 = 1$
3. $m_1 = 1, m_2 = 2$; $k_1 = 1, k_2 = k_3 = 2$
4. $m_1 = m_2 = 1$; $k_1 = 1, k_2 = 2, k_3 = 1$
5. $m_1 = m_2 = 1$; $k_1 = 2, k_2 = 1, k_3 = 2$
6. $m_1 = 1, m_2 = 2$; $k_1 = 2, k_2 = k_3 = 4$
7. $m_1 = m_2 = 1$; $k_1 = 4, k_2 = 6, k_3 = 4$

In Problems 8 through 10, the indicated mass-and-spring system is set in motion from rest ($x_1'(0) = x_2'(0) = 0$) in its equilibrium position ($x_1(0) = x_2(0) = 0$) with the given external forces $F_1(t)$ and $F_2(t)$ acting on the masses m_1 and m_2, respectively. Find the resulting motion of the system and describe it as a superposition of oscillations at three different frequencies.

8. The mass-and-spring system of Problem 2, with $F_1(t) = 96 \cos 5t$, $F_2(t) \equiv 0$

9. The mass-and-spring system of Problem 3, with $F_1(t) \equiv 0$, $F_2(t) = 120 \cos 3t$

10. The mass-and-spring system of Problem 7, with $F_1(t) = 30 \cos t$, $F_2(t) = 60 \cos t$

11. Consider a mass-and-spring system containing two masses $m_1 = 1$ and $m_2 = 1$ whose displacement functions $x(t)$ and $y(t)$ satisfy the differential equations

$$\begin{aligned} x'' &= -40x + 8y, \\ y'' &= 12x - 60y. \end{aligned}$$

(a) Describe the two fundamental modes of free oscillation of the system. **(b)** Assume that the two masses start in motion with the initial conditions

$$x(0) = 19, \quad x'(0) = 12$$

and

$$y(0) = 3, \quad y'(0) = 6$$

and are acted on by the same force, $F_1(t) = F_2(t) = -195 \cos 7t$. Describe the resulting motion as a superposition of oscillations at three different frequencies.

In Problems 12 and 13, find the natural frequencies of the three-mass system of Fig. 7.4.1, using the given masses and spring constants. For each natural frequency ω, give the ratio $a_1{:}a_2{:}a_3$ of amplitudes for a corresponding natural mode $x_1 = a_1 \cos \omega t$, $x_2 = a_2 \cos \omega t$, $x_3 = a_3 \cos \omega t$.

12. $m_1 = m_2 = m_3 = 1$; $k_1 = k_2 = k_3 = k_4 = 1$

13. $m_1 = m_2 = m_3 = 1$; $k_1 = k_2 = k_3 = k_4 = 2$ (*Hint*: One eigenvalue is $\lambda = -4$.)

14. In the system of Fig. 7.4.12, assume that $m_1 = 1$, $k_1 = 50$, $k_2 = 10$, and $F_0 = 5$ in mks units, and that $\omega = 10$. Then find m_2 so that in the resulting steady periodic oscillations, the mass m_1 will remain at rest(!). Thus the effect of the second mass-and-spring pair will be to neutralize the effect of the force on the first mass. This is an example of a *dynamic damper*. It has an electrical analogy that some cable companies use to prevent your reception of certain cable channels.

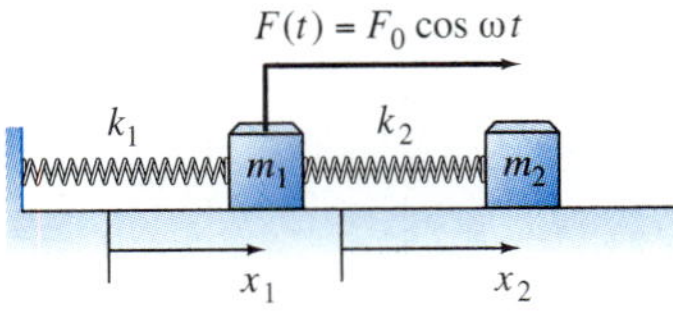

FIGURE 7.4.12. The mechanical system of Problem 14.

15. Suppose that $m_1 = 2$, $m_2 = \frac{1}{2}$, $k_1 = 75$, $k_2 = 25$, $F_0 = 100$, and $\omega = 10$ (all in mks units) in the forced mass-and-spring system of Fig. 7.4.9. Find the solution of the system $\mathbf{M}\mathbf{x}'' = \mathbf{K}\mathbf{x} + \mathbf{F}$ that satisfies the initial conditions $\mathbf{x}(0) = \mathbf{x}'(0) = \mathbf{0}$.

16. Figure 7.4.13 shows two railway cars with a buffer spring. We want to investigate the transfer of momentum that occurs after car 1 with initial velocity v_0 impacts car 2 at rest. The analog of Eq. (18) in the text is

$$\mathbf{x}'' = \begin{bmatrix} -c_1 & c_1 \\ c_2 & -c_2 \end{bmatrix} \mathbf{x}$$

with $c_i = k/m_i$ for $i = 1, 2$. Show that the eigenvalues of the coefficient matrix $\mathbf{A}$ are $\lambda_1 = 0$ and $\lambda_2 = -c_1 - c_2$, with associated eigenvectors $\mathbf{v}_1 = \begin{bmatrix} 1 & 1 \end{bmatrix}^T$ and $\mathbf{v}_2 = \begin{bmatrix} c_1 & -c_2 \end{bmatrix}^T$.

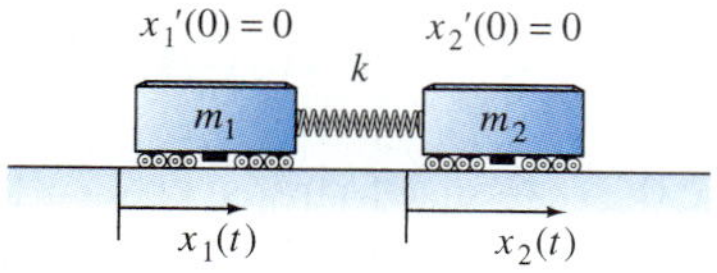

FIGURE 7.4.13. The two railway cars of Problems 16 through 19.

17. If the two cars of Problem 16 both weigh 16 tons (so that $m_1 = m_2 = 1000$ (slugs)) and $k = 1$ ton/ft (that is, 2000 lb/ft), show that the cars separate after $\pi/2$ seconds, and that $x_1'(t) = 0$ and $x_2'(t) = v_0$ thereafter. Thus the original momentum of car 1 is completely transferred to car 2.

18. If cars 1 and 2 weigh 8 and 16 tons, respectively, and $k = 3000$ lb/ft, show that the two cars separate after $\pi/3$ seconds, and that

$$x_1'(t) = -\tfrac{1}{3}v_0 \quad \text{and} \quad x_2'(t) = +\tfrac{2}{3}v_0$$

thereafter. Thus the two cars rebound in opposite directions.

19. If cars 1 and 2 weigh 24 and 8 tons, respectively, and $k = 1500$ lb/ft, show that the cars separate after $\pi/2$ seconds, and that

$$x_1'(t) = +\tfrac{1}{2}v_0 \quad \text{and} \quad x_2'(t) = +\tfrac{3}{2}v_0$$

thereafter. Thus both cars continue in the original direction of motion, but with different velocities.

Problems 20 through 23 deal with the same system of three railway cars (same masses) and two buffer springs (same spring constants) as shown in Fig. 7.4.6 and discussed in Example 2. The cars engage at time $t = 0$ with $x_1(0) = x_2(0) = x_3(0) = 0$ and with the given initial velocities (where $v_0 = 48$ ft/s). Show that the railway cars remain engaged until $t = \pi/2$ (s), after which time they proceed in their respective ways with constant velocities. Determine the values of these constant final velocities $x_1'(t)$, $x_2'(t)$, and $x_3'(t)$ of the three cars for $t > \pi/2$. In each problem you should find (as in Example 2) that the first and third railway cars exchange behaviors in some appropriate sense.

20. $x_1'(0) = v_0$, $x_2'(0) = 0$, $x_3'(0) = -v_0$

21. $x_1'(0) = 2v_0$, $x_2'(0) = 0$, $x_3'(0) = -v_0$

22. $x_1'(0) = v_0$, $x_2'(0) = v_0$, $x_3'(0) = -2v_0$

23. $x_1'(0) = 3v_0$, $x_2'(0) = 2v_0$, $x_3'(0) = 2v_0$

24. In the three-railway-car system of Fig. 7.4.6, suppose that cars 1 and 3 each weigh 32 tons, that car 2 weighs 8 tons, and that each spring constant is 4 tons/ft. If $x_1'(0) = v_0$ and $x_2'(0) = x_3'(0) = 0$, show that the two springs are compressed until $t = \pi/2$ and that

$$x_1'(t) = -\tfrac{1}{9}v_0 \quad \text{and} \quad x_2'(t) = x_3'(t) = +\tfrac{8}{9}v_0$$

thereafter. Thus car 1 rebounds, but cars 2 and 3 continue with the same velocity.

The Two-Axle Automobile

In Example 4 of Section 5.6 we investigated the vertical oscillations of a one-axle car—actually a unicycle. Now we can analyze a more realistic model: a car with two axles and with separate front and rear suspension systems. Figure 7.4.14 represents the suspension system of such a car. We assume that the car body acts as would a solid bar of mass m and length $L = L_1 + L_2$. It has moment of inertia I about its center of mass C, which is at distance L_1 from the front of the car. The car has front and back suspension springs with Hooke's constants k_1 and k_2, respectively. When the car is in motion, let $x(t)$ denote the vertical displacement of the center of mass of the car from equilibrium; let $\theta(t)$ denote its angular displacement (in radians) from the horizontal. Then Newton's laws of motion for linear and angular acceleration can be used to derive the equations

$$\begin{aligned} mx'' &= -(k_1 + k_2)x + (k_1L_1 - k_2L_2)\theta, \\ I\theta'' &= (k_1L_1 - k_2L_2)x - (k_1L_1^2 + k_2L_2^2)\theta. \end{aligned} \tag{40}$$

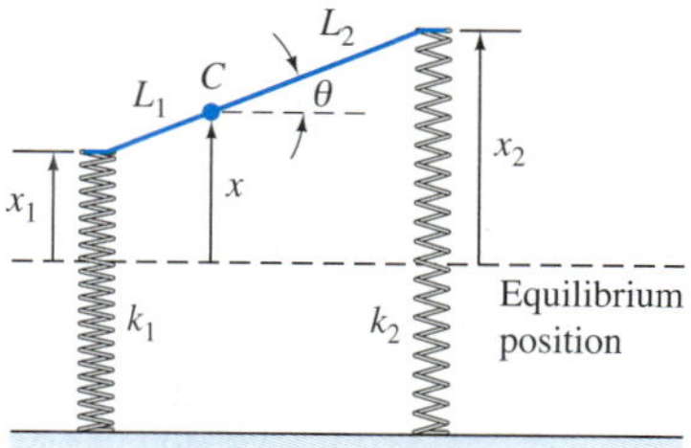

FIGURE 7.4.14. Model of the two-axle automobile.

25. Suppose that $m = 75$ slugs (the car weighs 2400 lb), $L_1 = 7$ ft, $L_2 = 3$ ft (it's a rear-engine car), $k_1 = k_2 = 2000$ lb/ft, and $I = 1000$ ft·lb·s^2. Then the equations in (40) take the form

$$\begin{aligned} 75x'' + 4000x - 8000\theta &= 0, \\ 1000\theta'' - 8000x + 116{,}000\theta &= 0. \end{aligned}$$

(a) Find the two natural frequencies ω_1 and ω_2 of the car. **(b)** Now suppose that the car is driven at a speed of v feet per second along a washboard surface shaped like a sine curve with a wavelength of 40 ft. The result is a periodic force on the car with frequency $\omega = 2\pi v/40 = \pi v/20$. Resonance occurs when $\omega = \omega_1$ or $\omega = \omega_2$. Find the corresponding two critical speeds of the car (in feet per second and in miles per hour).

26. Suppose that $k_1 = k_2 = k$ and $L_1 = L_2 = \frac{1}{2}L$ in Fig. 7.4.14 (the symmetric situation). Then show that every free oscillation is a combination of a vertical oscillation with frequency

$$\omega_1 = \sqrt{2k/m}$$

and an angular oscillation with frequency

$$\omega_2 = \sqrt{kL^2/(2I)}.$$

In Problems 27 through 29, the system of Fig. 7.4.14 is taken as a model for an undamped car with the given parameters in fps units. ***(a)*** *Find the two natural frequencies of oscillation (in hertz).* ***(b)*** *Assume that this car is driven along a sinusoidal washboard surface with a wavelength of 40 ft. Find the two critical speeds.*

27. $m = 100$, $I = 800$, $L_1 = L_2 = 5$, $k_1 = k_2 = 2000$

28. $m = 100$, $I = 1000$, $L_1 = 6$, $L_2 = 4$, $k_1 = k_2 = 2000$

29. $m = 100$, $I = 800$, $L_1 = L_2 = 5$, $k_1 = 1000$, $k_2 = 2000$

7.4 Application Earthquake-Induced Vibrations of Multistory Buildings

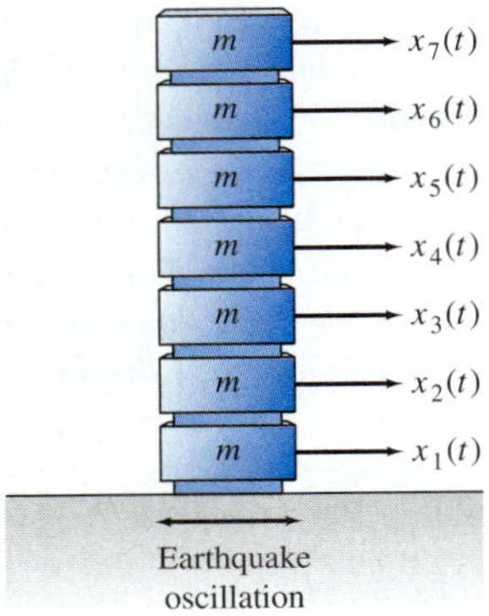

FIGURE 7.4.15. The seven-story building.

FIGURE 7.4.16. Forces on the ith floor.

In this application you are to investigate the response to transverse earthquake ground oscillations of the seven-story building illustrated in Fig. 7.4.15. Suppose that each of the seven (above-ground) floors weighs 16 tons, so the mass of each is $m = 1000$ (slugs). Also assume a horizontal restoring force of $k = 5$ (tons per foot) between adjacent floors. That is, the internal forces in response to horizontal displacements of the individual floors are those shown in Fig. 7.4.16. It follows that the free transverse oscillations indicated in Fig. 7.4.15 satisfy the equation $\mathbf{Mx}'' = \mathbf{Kx}$ with $n = 7$, $m_i = 1000$ (for each i), and $k_i = 10000$ (lb/ft) for $1 \leqq i \leqq 7$. The system then reduces to the form $\mathbf{x}'' = \mathbf{Ax}$ with

$$\mathbf{A} = \begin{bmatrix} -20 & 10 & 0 & 0 & 0 & 0 & 0 \\ 10 & -20 & 10 & 0 & 0 & 0 & 0 \\ 0 & 10 & -20 & 10 & 0 & 0 & 0 \\ 0 & 0 & 10 & -20 & 10 & 0 & 0 \\ 0 & 0 & 0 & 10 & -20 & 10 & 0 \\ 0 & 0 & 0 & 0 & 10 & -20 & 10 \\ 0 & 0 & 0 & 0 & 0 & 10 & -10 \end{bmatrix}. \tag{1}$$

Once the matrix **A** has been entered, the TI-86 command **`eigVl A`** takes about 15 seconds to calculate the seven eigenvalues shown in the λ-column of the table in Fig. 7.4.17. Alternatively, you can use the *Maple* command **`eigenvals(A)`**, the MATLAB command **`eig(A)`**, or the *Mathematica* command **`Eigenvalues[A]`**.

i	Eigenvalue λ	Frequency $\omega = \sqrt{-\lambda}$	Period $P = \frac{2\pi}{\omega}$ (sec)
1	−38.2709	6.1863	1.0157
2	−33.3826	5.7778	1.0875
3	−26.1803	5.1167	1.2280
4	−17.9094	4.2320	1.4847
5	−10.0000	3.1623	1.9869
6	−3.8197	1.9544	3.2149
7	−0.4370	0.6611	9.5042

FIGURE 7.4.17. Frequencies and periods of natural oscillations of the seven-story building.

Then calculate the entries in the remaining columns of the table showing the natural frequencies and periods of oscillation of the seven-story building. Note that a typical earthquake producing ground oscillations with a period of 2 seconds is uncomfortably close to the fifth natural frequency (with period 1.9869 seconds) of the building.

A horizontal earthquake oscillation $E\cos\omega t$ of the ground, with amplitude E and acceleration $a = -E\omega^2\cos\omega t$, produces an opposite inertial force $F = ma = mE\omega^2\cos\omega t$ on each floor of the building. The resulting nonhomogeneous system is

$$\mathbf{x}'' = \mathbf{A}\mathbf{x} + (E\omega^2\cos\omega t)\mathbf{b}, \tag{2}$$

where $\mathbf{b} = \begin{bmatrix} 1 & 1 & 1 & 1 & 1 & 1 & 1 \end{bmatrix}^T$ and $\mathbf{A}$ is the matrix of Eq. (1). Figure 7.4.18 shows a plot of *maximal amplitude* (for the forced oscillations of any single floor) versus the *period* of the earthquake vibrations. The spikes correspond to the first six of the seven resonant frequencies. We see, for instance, that whereas an earthquake with period 2 (s) likely would produce destructive resonance vibrations in the building, it probably would be unharmed by an earthquake with period 2.5 (s). Different buildings have different natural frequencies of vibration, and so a given earthquake may demolish one building but leave untouched the one next door. This seeming anomaly was observed in Mexico City after the devastating earthquake of September 19, 1985.

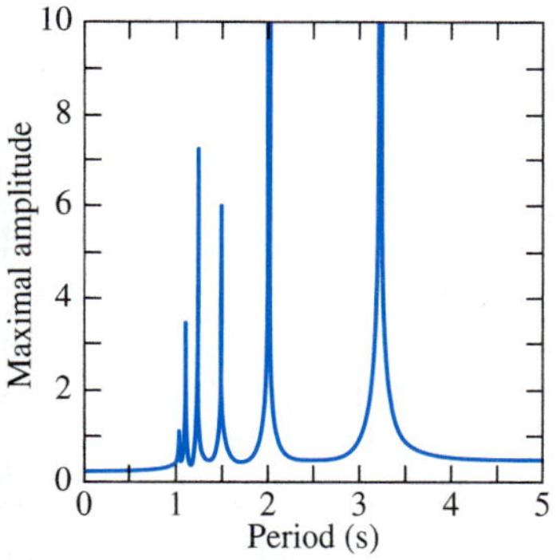

FIGURE 7.4.18. Resonance vibrations of a seven-story building—maximal amplitude as a function of period.

For your personal seven-story building to investigate, let the weight (in tons) of each story equal the largest digit of your student ID number and let k (in tons/ft) equal the smallest nonzero digit. Produce numerical and graphical results like those illustrated in Figs. 7.4.17 and 7.4.18. Is your building susceptible to likely damage from an earthquake with period in the 2- to 3-second range?

You might like to begin by working manually the following warm-up problems.

1. Find the periods of the natural vibrations of a building with two above-ground floors, each weighing 16 tons and with each restoring force being $k = 5$ tons/ft.
2. Find the periods of the natural vibrations of a building with three above-ground floors, with each weighing 16 tons and with each restoring force being $k = 5$ tons/ft.
3. Find the natural frequencies and natural modes of vibration of a building with three above-ground floors as in Problem 2, except that the upper two floors weigh 8 tons each instead of 16. Give the ratios of the amplitudes A, B, and C of oscillations of the three floors in the form $A : B : C$ with $A = 1$.
4. Suppose that the building of Problem 3 is subject to an earthquake in which the ground undergoes horizontal sinusoidal oscillations with a period of 3 s and an amplitude of 3 in. Find the amplitudes of the resulting steady periodic oscillations of the three above-ground floors. Assume the fact that a motion $E\sin\omega t$ of the ground, with acceleration $a = -E\omega^2\sin\omega t$, produces an opposite inertial force $F = -ma = mE\omega^2\sin\omega t$ on a floor of mass m.

7.5 Multiple Eigenvalue Solutions

In Section 7.3 we saw that if the $n \times n$ matrix $\mathbf{A}$ has n *distinct* (real or complex) eigenvalues $\lambda_1, \lambda_2, \dots, \lambda_n$ with respective associated eigenvectors $\mathbf{v}_1, \mathbf{v}_2, \dots, \mathbf{v}_n$, then a general solution of the system

$$\frac{d\mathbf{x}}{dt} = \mathbf{A}\mathbf{x} \tag{1}$$

is given by

$$\mathbf{x}(t) = c_1\mathbf{v}_1 e^{\lambda_1 t} + c_2\mathbf{v}_2 e^{\lambda_2 t} + \cdots + c_n\mathbf{v}_n e^{\lambda_n t} \tag{2}$$

with arbitrary constants $c_1, c_2, \dots, c_n$. In this section we discuss the situation when the characteristic equation

$$|\mathbf{A} - \lambda\mathbf{I}| = 0 \tag{3}$$

does *not* have n distinct roots, and thus has at least one repeated root.

An eigenvalue is of **multiplicity** k if it is a k-fold root of Eq. (3). For each eigenvalue λ, the eigenvector equation

$$(\mathbf{A} - \lambda\mathbf{I})\mathbf{v} = \mathbf{0} \tag{4}$$

has at least one nonzero solution $\mathbf{v}$, so there is at least one eigenvector associated with λ. But an eigenvalue of multiplicity $k > 1$ may have *fewer* than k linearly independent associated eigenvectors. In this case we are unable to find a "complete set" of n linearly independent eigenvectors of $\mathbf{A}$, as needed to form the general solution in (2).

Let us call an eigenvalue of multiplicity k **complete** if it has k linearly independent associated eigenvectors. If every eigenvalue of the matrix $\mathbf{A}$ is complete, then—because eigenvectors associated with different eigenvalues are linearly independent—it follows that $\mathbf{A}$ does have a complete set of n linearly independent eigenvectors $\mathbf{v}_1, \mathbf{v}_2, \dots, \mathbf{v}_n$ associated with the eigenvalues $\lambda_1, \lambda_2, \dots, \lambda_n$ (each repeated with its multiplicity). In this case a general solution of $\mathbf{x}' = \mathbf{A}\mathbf{x}$ is still given by the usual combination in (2).

Example 1 Find a general solution of the system

$$\mathbf{x}' = \begin{bmatrix} 9 & 4 & 0 \\ -6 & -1 & 0 \\ 6 & 4 & 3 \end{bmatrix} \mathbf{x}. \tag{5}$$

Solution The characteristic equation of the coefficient matrix $\mathbf{A}$ in Eq. (5) is

$$\begin{aligned} |\mathbf{A} - \lambda\mathbf{I}| &= \begin{vmatrix} 9-\lambda & 4 & 0 \\ -6 & -1-\lambda & 0 \\ 6 & 4 & 3-\lambda \end{vmatrix} \\ &= (3-\lambda)[(9-\lambda)(-1-\lambda) + 24] \\ &= (3-\lambda)(15 - 8\lambda + \lambda^2) \\ &= (5-\lambda)(3-\lambda)^2 = 0. \end{aligned}$$

Thus $\mathbf{A}$ has the distinct eigenvalue $\lambda_1 = 5$ and the repeated eigenvalue $\lambda_2 = 3$ of multiplicity $k = 2$.

CASE 1: $\lambda_1 = 5$. The eigenvector equation $(\mathbf{A}-\lambda\mathbf{I})\mathbf{v} = \mathbf{0}$, where $\mathbf{v} = [a \quad b \quad c]^T$, is

$$(\mathbf{A} - 5\mathbf{I})\mathbf{v} = \begin{bmatrix} 4 & 4 & 0 \\ -6 & -6 & 0 \\ 6 & 4 & -2 \end{bmatrix} \begin{bmatrix} a \\ b \\ c \end{bmatrix} = \begin{bmatrix} 0 \\ 0 \\ 0 \end{bmatrix}.$$

Each of the first two equations, $4a + 4b = 0$ and $-6a - 6b = 0$, yields $b = -a$. Then the third equation reduces to $2a - 2c = 0$, so that $c = a$. The choice $a = 1$ then yields the eigenvector

$$\mathbf{v}_1 = \begin{bmatrix} 1 & -1 & 1 \end{bmatrix}^T$$

associated with the eigenvalue $\lambda_1 = 5$.

CASE 2: $\lambda_2 = 3$. Now the eigenvector equation is

$$(\mathbf{A} - 3\mathbf{I})\mathbf{v} = \begin{bmatrix} 6 & 4 & 0 \\ -6 & -4 & 0 \\ 6 & 4 & 0 \end{bmatrix} \begin{bmatrix} a \\ b \\ c \end{bmatrix} = \begin{bmatrix} 0 \\ 0 \\ 0 \end{bmatrix},$$

so the nonzero vector $\mathbf{v} = \begin{bmatrix} a & b & c \end{bmatrix}^T$ is an eigenvector if and only if

$$6a + 4b = 0; \tag{6}$$

that is, $b = -\frac{3}{2}a$. The fact that Eq. (6) does not involve c means that c is arbitrary, subject to the condition $\mathbf{v} \neq \mathbf{0}$. If $c = 1$, then we may choose $a = b = 0$; this gives the eigenvector

$$\mathbf{v}_2 = \begin{bmatrix} 0 & 0 & 1 \end{bmatrix}^T$$

associated with $\lambda_2 = 3$. If $c = 0$, then we must choose a to be nonzero. For instance, if $a = 2$ (to avoid fractions), then $b = -3$, so

$$\mathbf{v}_3 = \begin{bmatrix} 2 & -3 & 0 \end{bmatrix}^T$$

is a second linearly independent eigenvector associated with the multiplicity 2 eigenvalue $\lambda_2 = 3$.

Thus we have found a complete set $\mathbf{v}_1$, $\mathbf{v}_2$, $\mathbf{v}_3$ of three eigenvectors associated with the eigenvalues 5, 3, 3. The corresponding general solution of Eq. (5) is

$$\begin{aligned} \mathbf{x}(t) &= c_1\mathbf{v}_1e^{5t} + c_2\mathbf{v}_2e^{3t} + c_3\mathbf{v}_3e^{3t} \\ &= c_1\begin{bmatrix} 1 \\ -1 \\ 1 \end{bmatrix} e^{5t} + c_2\begin{bmatrix} 0 \\ 0 \\ 1 \end{bmatrix} e^{3t} + c_3\begin{bmatrix} 2 \\ -3 \\ 0 \end{bmatrix} e^{3t}, \end{aligned} \tag{7}$$

with scalar component functions given by

$$\begin{aligned} x_1(t) &= c_1e^{5t} + 2c_3e^{3t}, \\ x_2(t) &= -c_1e^{5t} - 3c_3e^{3t}, \\ x_3(t) &= c_1e^{5t} + c_2e^{3t}. \end{aligned}$$

■

Remark: Our choice in Example 1 of the two eigenvectors

$$\mathbf{v}_2 = \begin{bmatrix} 0 & 0 & 1 \end{bmatrix}^T \quad \text{and} \quad \mathbf{v}_3 = \begin{bmatrix} 2 & -3 & 0 \end{bmatrix}^T$$

associated with the repeated eigenvalue $\lambda_2 = 3$ bears comment. The fact that $b = -\frac{3}{2}a$ for any eigenvector associated with $\lambda_2 = 3$ means that any such eigenvector can be written as

$$\mathbf{v} = \begin{bmatrix} a \\ -\frac{3}{2}a \\ c \end{bmatrix} = c\begin{bmatrix} 0 \\ 0 \\ 1 \end{bmatrix} + \frac{1}{2}a\begin{bmatrix} 2 \\ -3 \\ 0 \end{bmatrix} = c\mathbf{v}_2 + \tfrac{1}{2}a\mathbf{v}_3,$$

and thus is a linear combination of $\mathbf{v}_2$ and $\mathbf{v}_3$. Therefore, given a and c not both zero, we could choose $\mathbf{v}$ rather than $\mathbf{v}_3$ as our third eigenvector, and the new general solution

$$\mathbf{x}(t) = c_1\mathbf{v}_1e^{5t} + c_2\mathbf{v}_2e^{3t} + c_3\mathbf{v}e^{3t}$$

would be equivalent to the one in Eq. (7). Thus we need not worry about making the "right" choice of independent eigenvectors associated with a multiple eigenvalue. Any choice will do; we generally make the simplest one we can. ■

Defective Eigenvalues

The following example shows that—unfortunately—not all multiple eigenvalues are complete.

Example 2 The matrix

$$\mathbf{A} = \begin{bmatrix} 1 & -3 \\ 3 & 7 \end{bmatrix} \tag{8}$$

has characteristic equation

$$\begin{aligned} |\mathbf{A} - \lambda\mathbf{I}| &= \begin{vmatrix} 1-\lambda & -3 \\ 3 & 7-\lambda \end{vmatrix} \\ &= (1-\lambda)(7-\lambda) + 9 \\ &= \lambda^2 - 8\lambda + 16 = (\lambda - 4)^2 = 0. \end{aligned}$$

Thus $\mathbf{A}$ has the single eigenvalue $\lambda_1 = 4$ of multiplicity 2. The eigenvector equation

$$(\mathbf{A} - 4\mathbf{I})\mathbf{v} = \begin{bmatrix} -3 & -3 \\ 3 & 3 \end{bmatrix}\begin{bmatrix} a \\ b \end{bmatrix} = \begin{bmatrix} 0 \\ 0 \end{bmatrix}$$

then amounts to the equivalent scalar equations

$$-3a - 3b = 0, \quad 3a + 3b = 0.$$

Hence $b = -a$ if $\mathbf{v} = \begin{bmatrix} a & b \end{bmatrix}^T$ is to be an eigenvector of $\mathbf{A}$. Therefore, any eigenvector associated with $\lambda_1 = 4$ is a nonzero multiple of $\mathbf{v} = \begin{bmatrix} 1 & -1 \end{bmatrix}^T$. Thus the multiplicity 2 eigenvalue $\lambda_1 = 4$ has only one independent eigenvector, and hence is incomplete. ■

An eigenvalue λ of multiplicity $k > 1$ is called **defective** if it is not complete. If λ has only $p < k$ linearly independent eigenvectors, then the number

$$d = k - p \tag{9}$$

of "missing" eigenvectors is called the **defect** of the defective eigenvalue λ. Thus the defective eigenvalue $\lambda_1 = 4$ in Example 2 has multiplicity $k = 2$ and defect $d = 1$, because we saw that it has only $p = 1$ associated eigenvector.

If the eigenvalues of the $n \times n$ matrix $\mathbf{A}$ are not all complete, then the eigenvalue method as yet described will produce *fewer* than the needed n linearly independent solutions of the system $\mathbf{x}' = \mathbf{A}\mathbf{x}$. We therefore need to discover how to find the "missing solutions" corresponding to a defective eigenvalue λ of multiplicity $k > 1$.

The Case of Multiplicity $k = 2$

Let us begin with the case $k = 2$, and suppose that we have found (as in Example 2) that there is only a single eigenvector $\mathbf{v}_1$ associated with the defective eigenvalue λ. Then at this point we have found only the single solution

$$\mathbf{x}_1(t) = \mathbf{v}_1 e^{\lambda t} \tag{10}$$

of $\mathbf{x}' = \mathbf{A}\mathbf{x}$. By analogy with the case of a repeated characteristic root for a single linear differential equation (Section 5.3), we might hope to find a second solution of the form

$$\mathbf{x}_2(t) = (\mathbf{v}_2 t)e^{\lambda t} = \mathbf{v}_2 t e^{\lambda t}. \tag{11}$$

When we substitute $\mathbf{x} = \mathbf{v}_2 t e^{\lambda t}$ in $\mathbf{x}' = \mathbf{A}\mathbf{x}$, we get the equation

$$\mathbf{v}_2 e^{\lambda t} + \lambda \mathbf{v}_2 t e^{\lambda t} = \mathbf{A}\mathbf{v}_2 t e^{\lambda t}.$$

But because the coefficients of both $e^{\lambda t}$ and $te^{\lambda t}$ must balance, it follows that $\mathbf{v}_2 = \mathbf{0}$, and hence that $\mathbf{x}_2(t) \equiv \mathbf{0}$. This means that—contrary to our hope—the system $\mathbf{x}' = \mathbf{A}\mathbf{x}$ does not have a *nontrivial* solution of the form assumed in (11).

Instead of simply giving up on the idea behind Eq. (11), let us extend it slightly and replace $\mathbf{v}_2 t$ with $\mathbf{v}_1 t + \mathbf{v}_2$. Thus we explore the possibility of a second solution of the form

$$\mathbf{x}_2(t) = (\mathbf{v}_1 t + \mathbf{v}_2)e^{\lambda t} = \mathbf{v}_1 t e^{\lambda t} + \mathbf{v}_2 e^{\lambda t}, \tag{12}$$

where $\mathbf{v}_1$ and $\mathbf{v}_2$ are nonzero constant vectors. When we substitute $\mathbf{x} = \mathbf{v}_1 t e^{\lambda t} + \mathbf{v}_2 e^{\lambda t}$ in $\mathbf{x}' = \mathbf{A}\mathbf{x}$, we get the equation

$$\mathbf{v}_1 e^{\lambda t} + \lambda \mathbf{v}_1 t e^{\lambda t} + \lambda \mathbf{v}_2 e^{\lambda t} = \mathbf{A}\mathbf{v}_1 t e^{\lambda t} + \mathbf{A}\mathbf{v}_2 e^{\lambda t}. \tag{13}$$

We equate coefficients of $e^{\lambda t}$ and $te^{\lambda t}$ here, and thereby obtain the two equations

$$(\mathbf{A} - \lambda\mathbf{I})\mathbf{v}_1 = \mathbf{0} \tag{14}$$

and

$$(\mathbf{A} - \lambda\mathbf{I})\mathbf{v}_2 = \mathbf{v}_1 \tag{15}$$

that the vectors $\mathbf{v}_1$ and $\mathbf{v}_2$ must satisfy in order for (12) to give a solution of $\mathbf{x}' = \mathbf{A}\mathbf{x}$.

Note that Eq. (14) merely confirms that $\mathbf{v}_1$ is an eigenvector of $\mathbf{A}$ associated with the eigenvalue λ. Then Eq. (15) says that the vector $\mathbf{v}_2$ satisfies the equation

$$(\mathbf{A}-\lambda\mathbf{I})^2\mathbf{v}_2 = (\mathbf{A}-\lambda\mathbf{I})[(\mathbf{A}-\lambda\mathbf{I})\mathbf{v}_2] = (\mathbf{A}-\lambda\mathbf{I})\mathbf{v}_1 = \mathbf{0}.$$

It follows that, in order to solve simultaneously the two equations in (14) and (15), it suffices to find a solution $\mathbf{v}_2$ of the single equation $(\mathbf{A}-\lambda\mathbf{I})^2\mathbf{v}_2 = \mathbf{0}$ such that the resulting vector $\mathbf{v}_1 = (\mathbf{A}-\lambda\mathbf{I})\mathbf{v}_2$ is *nonzero*. It turns out that this is always possible if the defective eigenvalue λ of $\mathbf{A}$ is of multiplicity 2. Consequently, the procedure described in the following algorithm always succeeds in finding two independent solutions associated with such an eigenvalue.

ALGORITHM Defective Multiplicity 2 Eigenvalues

1. First find a nonzero solution $\mathbf{v}_2$ of the equation

$$(\mathbf{A}-\lambda\mathbf{I})^2\mathbf{v}_2 = \mathbf{0} \tag{16}$$

such that

$$(\mathbf{A}-\lambda\mathbf{I})\mathbf{v}_2 = \mathbf{v}_1 \tag{17}$$

is nonzero, and therefore is an eigenvector $\mathbf{v}_1$ associated with λ.

2. Then form the two independent solutions

$$\mathbf{x}_1(t) = \mathbf{v}_1 e^{\lambda t} \tag{18}$$

and

$$\mathbf{x}_2(t) = (\mathbf{v}_1 t + \mathbf{v}_2)e^{\lambda t} \tag{19}$$

of $\mathbf{x}' = \mathbf{A}\mathbf{x}$ corresponding to λ.

Example 3 Find a general solution of the system

$$\mathbf{x}' = \begin{bmatrix} 1 & -3 \\ 3 & 7 \end{bmatrix}\mathbf{x}. \tag{20}$$

Solution In Example 2 we found that the coefficient matrix $\mathbf{A}$ in Eq. (20) has the defective eigenvalue $\lambda = 4$ of multiplicity 2. We therefore begin by calculating

$$(\mathbf{A}-4\mathbf{I})^2 = \begin{bmatrix} -3 & -3 \\ 3 & 3 \end{bmatrix}\begin{bmatrix} -3 & -3 \\ 3 & 3 \end{bmatrix} = \begin{bmatrix} 0 & 0 \\ 0 & 0 \end{bmatrix}.$$

Hence Eq. (16) is

$$\begin{bmatrix} 0 & 0 \\ 0 & 0 \end{bmatrix}\mathbf{v}_2 = \mathbf{0},$$

and therefore is satisfied by *any* choice of $\mathbf{v}_2$. In principle, it could happen that $(\mathbf{A}-4\mathbf{I})\mathbf{v}_2$ is nonzero (as desired) for some choices of $\mathbf{v}_2$ though not for others. If we try $\mathbf{v}_2 = \begin{bmatrix} 1 & 0 \end{bmatrix}^T$ we find that

$$(\mathbf{A} - 4\mathbf{I})\mathbf{v}_2 = \begin{bmatrix} -3 & -3 \\ 3 & 3 \end{bmatrix} \begin{bmatrix} 1 \\ 0 \end{bmatrix} = \begin{bmatrix} -3 \\ 3 \end{bmatrix} = \mathbf{v}_1$$

is nonzero, and therefore is an eigenvector associated with $\lambda = 4$. (It is -3 times the eigenvector found in Example 2.) Therefore, the two solutions of Eq. (20) given by Eqs. (18) and (19) are

$$\mathbf{x}_1(t) = \mathbf{v}_1 e^{4t} = \begin{bmatrix} -3 \\ 3 \end{bmatrix} e^{4t},$$

$$\mathbf{x}_2(t) = (\mathbf{v}_1 t + \mathbf{v}_2)e^{4t} = \begin{bmatrix} -3t + 1 \\ 3t \end{bmatrix} e^{4t}.$$

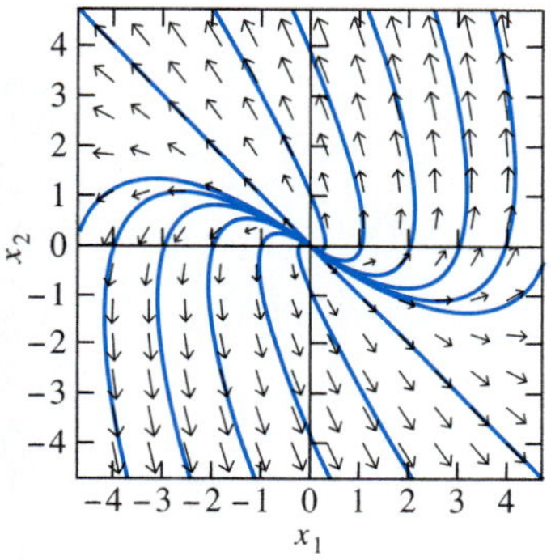

FIGURE 7.5.1. Direction field and solution curves for the linear system $x_1' = x_1 - 3x_2$, $x_2' = 3x_1 + 7x_2$ of Example 3.

The resulting general solution

$$\mathbf{x}(t) = c_1\mathbf{x}_1(t) + c_2\mathbf{x}_2(t)$$

has scalar component functions

$$x_1(t) = (-3c_2 t + c_2 - 3c_1)e^{4t},$$
$$x_2(t) = (3c_2 t + 3c_1)e^{4t}.$$

With $c_2 = 0$ these solution equations reduce to the equations $x_1(t) = -3c_1e^{4t}$, $x_2(t) = 3c_1e^{4t}$, which parametrize the line $x_1 = -x_2$ in the x_1x_2-plane. The point $(x_1(t), x_2(t))$ then recedes along this line away from the origin as $t \to +\infty$, to the northwest if $c_1 > 0$ and to the southeast if $c_1 < 0$. As indicated in Fig. 7.5.1, each solution curve with $c_2 \neq 0$ is tangent to the line $x_1 = -x_2$ at the origin; the point $(x_1(t), x_2(t))$ approaches the origin as $t \to -\infty$ and approaches $+\infty$ along the solution curve as $t \to +\infty$. ■

Generalized Eigenvectors

The vector $\mathbf{v}_2$ in Eq. (16) is an example of a generalized eigenvector. If λ is an eigenvalue of the matrix $\mathbf{A}$, then a **rank** r **generalized eigenvector** associated with λ is a vector $\mathbf{v}$ such that

$$(\mathbf{A} - \lambda\mathbf{I})^r\mathbf{v} = \mathbf{0} \quad \text{but} \quad (\mathbf{A} - \lambda\mathbf{I})^{r-1}\mathbf{v} \neq \mathbf{0}. \tag{21}$$

If $r = 1$, then (21) simply means that $\mathbf{v}$ is an eigenvector associated with λ (recalling the convention that the 0th power of a square matrix is the identity matrix). Thus a rank 1 generalized eigenvector is an ordinary eigenvector. The vector $\mathbf{v}_2$ in (16) is a rank 2 generalized eigenvector (and not an ordinary eigenvector).

The multiplicity 2 method described earlier boils down to finding a pair $\{\mathbf{v}_1, \mathbf{v}_2\}$ of generalized eigenvectors, one of rank 1 and one of rank 2, such that $(\mathbf{A} - \lambda\mathbf{I})\mathbf{v}_2 = \mathbf{v}_1$. Higher multiplicity methods involve longer "chains" of generalized eigenvectors. A **length** k **chain of generalized eigenvectors based on the eigenvector** $\mathbf{v}_1$ is a set $\{\mathbf{v}_1, \mathbf{v}_2, \ldots, \mathbf{v}_k\}$ of k generalized eigenvectors such that

$$\begin{aligned} (\mathbf{A} - \lambda\mathbf{I})\mathbf{v}_k &= \mathbf{v}_{k-1}, \\ (\mathbf{A} - \lambda\mathbf{I})\mathbf{v}_{k-1} &= \mathbf{v}_{k-2}, \\ &\vdots \\ (\mathbf{A} - \lambda\mathbf{I})\mathbf{v}_2 &= \mathbf{v}_1. \end{aligned} \tag{22}$$

Because $\mathbf{v}_1$ is an ordinary eigenvector, $(\mathbf{A} - \lambda\mathbf{I})\mathbf{v}_1 = \mathbf{0}$. Therefore, it follows from (22) that

$$(\mathbf{A} - \lambda\mathbf{I})^k\mathbf{v}_k = \mathbf{0}. \tag{23}$$

If $\{\mathbf{v}_1, \mathbf{v}_2, \mathbf{v}_3\}$ is a length 3 chain of generalized eigenvectors associated with the multiple eigenvalue λ of the matrix $\mathbf{A}$, then it is easy to verify that three linearly independent solutions of $\mathbf{x}' = \mathbf{A}\mathbf{x}$ are given by

$$\begin{aligned}
\mathbf{x}_1(t) &= \mathbf{v}_1 e^{\lambda t}, \\
\mathbf{x}_2(t) &= (\mathbf{v}_1 t + \mathbf{v}_2)e^{\lambda t}, \\
\mathbf{x}_3(t) &= \left(\tfrac{1}{2}\mathbf{v}_1 t^2 + \mathbf{v}_2 t + \mathbf{v}_3\right) e^{\lambda t}.
\end{aligned} \tag{24}$$

For instance, the equations in (22) give

$$\mathbf{A}\mathbf{v}_3 = \mathbf{v}_2 + \lambda\mathbf{v}_3, \quad \mathbf{A}\mathbf{v}_2 = \mathbf{v}_1 + \lambda\mathbf{v}_2, \quad \mathbf{A}\mathbf{v}_1 = \lambda\mathbf{v}_1,$$

so

$$\begin{aligned}
\mathbf{A}\mathbf{x}_3 &= \left[\tfrac{1}{2}\mathbf{A}\mathbf{v}_1 t^2 + \mathbf{A}\mathbf{v}_2 t + \mathbf{A}\mathbf{v}_3\right] e^{\lambda t} \\
&= \left[\tfrac{1}{2}\lambda\mathbf{v}_1 t^2 + (\mathbf{v}_1 + \lambda\mathbf{v}_2)t + (\mathbf{v}_2 + \lambda\mathbf{v}_3)\right] e^{\lambda t} \\
&= (\mathbf{v}_1 t + \mathbf{v}_2)e^{\lambda t} + \lambda\left(\tfrac{1}{2}\mathbf{v}_1 t^2 + \mathbf{v}_2 t + \mathbf{v}_3\right) e^{\lambda t} \\
&= \mathbf{x}_3'.
\end{aligned}$$

Therefore, $\mathbf{x}_3(t)$ in (24) does, indeed, define a solution of $\mathbf{x}' = \mathbf{A}\mathbf{x}$.

Consequently, in order to "handle" a multiplicity 3 eigenvalue λ, it suffices to find a length 3 chain $\{\mathbf{v}_1, \mathbf{v}_2, \mathbf{v}_3\}$ of generalized eigenvalues associated with λ. Looking at Eq. (23), we see that we need only find a solution $\mathbf{v}_3$ of

$$(\mathbf{A} - \lambda\mathbf{I})^3\mathbf{v}_3 = \mathbf{0}$$

such that the vectors

$$\mathbf{v}_2 = (\mathbf{A} - \lambda\mathbf{I})\mathbf{v}_3 \quad \text{and} \quad \mathbf{v}_1 = (\mathbf{A} - \lambda\mathbf{I})\mathbf{v}_2$$

are both nonzero (although, as we will see, this is not always possible).

Example 4 Find three linearly independent solutions of the system

$$\mathbf{x}' = \begin{bmatrix} 0 & 1 & 2 \\ -5 & -3 & -7 \\ 1 & 0 & 0 \end{bmatrix} \mathbf{x}. \tag{25}$$

Solution The characteristic equation of the coefficient matrix in Eq. (25) is

$$\begin{aligned} |\mathbf{A} - \lambda\mathbf{I}| &= \begin{bmatrix} -\lambda & 1 & 2 \\ -5 & -3-\lambda & -7 \\ 1 & 0 & -\lambda \end{bmatrix} \\ &= 1 \cdot [-7 - 2 \cdot (-3 - \lambda)] + (-\lambda)[(-\lambda)(-3-\lambda) + 5] \\ &= -\lambda^3 - 3\lambda^2 - 3\lambda - 1 = -(\lambda + 1)^3 = 0, \end{aligned}$$

and thus $\mathbf{A}$ has the eigenvalue $\lambda = -1$ of multiplicity 3. The eigenvector equation $(\mathbf{A} - \lambda\mathbf{I})\mathbf{v} = \mathbf{0}$ for an eigenvector $\mathbf{v} = \begin{bmatrix} a & b & c \end{bmatrix}^T$ is

$$(\mathbf{A} + \mathbf{I})\mathbf{v} = \begin{bmatrix} 1 & 1 & 2 \\ -5 & -2 & -7 \\ 1 & 0 & 1 \end{bmatrix} \begin{bmatrix} a \\ b \\ c \end{bmatrix} = \begin{bmatrix} 0 \\ 0 \\ 0 \end{bmatrix}.$$

The third row $a + c = 0$ gives $c = -a$; then the first row $a + b + 2c = 0$ gives $b = a$. Thus, to within a constant multiple, the eigenvalue $\lambda = -1$ has only the single associated eigenvector $\mathbf{v} = \begin{bmatrix} a & a & -a \end{bmatrix}^T$ with $a \neq 0$, and so the defect of $\lambda = -1$ is 2.

To apply the method described here for triple eigenvalues, we first calculate

$$(\mathbf{A} + \mathbf{I})^2 = \begin{bmatrix} 1 & 1 & 2 \\ -5 & -2 & -7 \\ 1 & 0 & 1 \end{bmatrix} \begin{bmatrix} 1 & 1 & 2 \\ -5 & -2 & -7 \\ 1 & 0 & 1 \end{bmatrix} = \begin{bmatrix} -2 & -1 & -3 \\ -2 & -1 & -3 \\ 2 & 1 & 3 \end{bmatrix}$$

and

$$(\mathbf{A} + \mathbf{I})^3 = \begin{bmatrix} 1 & 1 & 2 \\ -5 & -2 & -7 \\ 1 & 0 & 1 \end{bmatrix} \begin{bmatrix} -2 & -1 & -3 \\ -2 & -1 & -3 \\ 2 & 1 & 3 \end{bmatrix} = \begin{bmatrix} 0 & 0 & 0 \\ 0 & 0 & 0 \\ 0 & 0 & 0 \end{bmatrix}.$$

Thus *any* nonzero vector $\mathbf{v}_3$ will be a solution of the equation $(\mathbf{A} + \mathbf{I})^3\mathbf{v}_3 = \mathbf{0}$. Beginning with $\mathbf{v}_3 = \begin{bmatrix} 1 & 0 & 0 \end{bmatrix}^T$, for instance, we calculate

$$\mathbf{v}_2 = (\mathbf{A} + \mathbf{I})\mathbf{v}_3 = \begin{bmatrix} 1 & 1 & 2 \\ -5 & -2 & -7 \\ 1 & 0 & 1 \end{bmatrix} \begin{bmatrix} 1 \\ 0 \\ 0 \end{bmatrix} = \begin{bmatrix} 1 \\ -5 \\ 1 \end{bmatrix},$$

$$\mathbf{v}_1 = (\mathbf{A}+\mathbf{I})\mathbf{v}_2 = \begin{bmatrix} 1 & 1 & 2 \\ -5 & -2 & -7 \\ 1 & 0 & 1 \end{bmatrix} \begin{bmatrix} 1 \\ -5 \\ 1 \end{bmatrix} = \begin{bmatrix} -2 \\ -2 \\ 2 \end{bmatrix}.$$

Note that $\mathbf{v}_1$ is the previously found eigenvector $\mathbf{v}$ with $a = -2$; this agreement serves as a check of the accuracy of our matrix computations.

Thus we have found a length 3 chain $\{\mathbf{v}_1, \mathbf{v}_2, \mathbf{v}_3\}$ of generalized eigenvectors associated with the triple eigenvalue $\lambda = -1$. Substitution in (24) now yields the linearly independent solutions

$$\mathbf{x}_1(t) = \mathbf{v}_1 e^{-t} = \begin{bmatrix} -2 \\ -2 \\ 2 \end{bmatrix} e^{-t},$$

$$\mathbf{x}_2(t) = (\mathbf{v}_1 t + \mathbf{v}_2)e^{-t} = \begin{bmatrix} -2t+1 \\ -2t-5 \\ 2t+1 \end{bmatrix} e^{-t},$$

$$\mathbf{x}_3(t) = \left(\tfrac{1}{2}\mathbf{v}_1 t^2 + \mathbf{v}_2 t + \mathbf{v}_3\right) e^{-t} = \begin{bmatrix} -t^2+t+1 \\ -t^2-5t \\ t^2+t \end{bmatrix} e^{-t}$$

of the system $\mathbf{x}' = \mathbf{A}\mathbf{x}$. ■

The General Case

A fundamental theorem of linear algebra states that every $n \times n$ matrix $\mathbf{A}$ has n linearly independent generalized eigenvectors. These n generalized eigenvectors may be arranged in chains, with the sum of the lengths of the chains associated with a given eigenvalue λ equal to the multiplicity of λ. But the structure of these chains depends on the defect of λ, and can be quite complicated. For instance, a multiplicity 4 eigenvalue can correspond to

- Four length 1 chains (defect 0);
- Two length 1 chains and a length 2 chain (defect 1);
- Two length 2 chains (defect 2);
- A length 1 chain and a length 3 chain (defect 2); or
- A length 4 chain (defect 3).

Note that, in each of these cases, the length of the longest chain is at most $d+1$, where d is the defect of the eigenvalue. Consequently, once we have found all the ordinary eigenvectors associated with a multiple eigenvalue λ, and therefore know the defect d of λ, we can begin with the equation

$$(\mathbf{A} - \lambda\mathbf{I})^{d+1}\mathbf{u} = \mathbf{0} \tag{26}$$

to start building the chains of generalized eigenvectors associated with λ.

ALGORITHM Chains of Generalized Eigenvectors

Begin with a nonzero solution $\mathbf{u}_1$ of Eq. (26) and successively multiply by the matrix $\mathbf{A} - \lambda\mathbf{I}$ until the zero vector is obtained. If

$$(\mathbf{A} - \lambda\mathbf{I})\mathbf{u}_1 = \mathbf{u}_2 \neq \mathbf{0},$$
$$\vdots$$
$$(\mathbf{A} - \lambda\mathbf{I})\mathbf{u}_{k-1} = \mathbf{u}_k \neq \mathbf{0},$$

but $(\mathbf{A} - \lambda\mathbf{I})\mathbf{u}_k = \mathbf{0}$, then the vectors

$$\{\mathbf{v}_1, \mathbf{v}_2, \dots, \mathbf{v}_k\} = \{\mathbf{u}_k, \mathbf{u}_{k-1}, \dots, \mathbf{u}_2, \mathbf{u}_1\}$$

(listed in reverse order of their appearance) form a length k chain of generalized eigenvectors based on the (ordinary) eigenvector $\mathbf{v}_1$.

Each length k chain $\{\mathbf{v}_1, \mathbf{v}_2, \dots, \mathbf{v}_k\}$ of generalized eigenvectors (with $\mathbf{v}_1$ an ordinary eigenvector associated with λ) determines a set of k independent solutions of $\mathbf{x}' = \mathbf{A}\mathbf{x}$ corresponding to the eigenvalue λ:

$$\begin{aligned}
\mathbf{x}_1(t) &= \mathbf{v}_1 e^{\lambda t},\\
\mathbf{x}_2(t) &= (\mathbf{v}_1 t + \mathbf{v}_2)e^{\lambda t},\\
\mathbf{x}_3(t) &= \left(\tfrac{1}{2}\mathbf{v}_1 t^2 + \mathbf{v}_2 t + \mathbf{v}_3\right) e^{\lambda t},\\
&\ \vdots\\
\mathbf{x}_k(t) &= \left(\frac{\mathbf{v}_1 t^{k-1}}{(k-1)!} + \cdots + \frac{\mathbf{v}_{k-2} t^2}{2!} + \mathbf{v}_{k-1} t + \mathbf{v}_k\right) e^{\lambda t}.
\end{aligned} \tag{27}$$

Note that (27) reduces to Eqs. (18) through (19) and (24) in the cases $k = 2$ and $k = 3$, respectively.

To ensure that we obtain n generalized eigenvectors of the $n \times n$ matrix $\mathbf{A}$ that are actually linearly independent, and therefore produce a complete set of n linearly independent solutions of $\mathbf{x}' = \mathbf{A}\mathbf{x}$ when we amalgamate all the "chains of solutions" corresponding to different chains of generalized eigenvectors, we may rely on the following two facts:

- Any chain of generalized eigenvectors constitutes a linearly independent set of vectors.
- If two chains of generalized eigenvectors are based on linearly independent eigenvectors, then the union of these two chains is a linearly independent set of vectors (whether the two base eigenvectors are associated with different eigenvalues or with the same eigenvalue).

Example 5 Suppose that the 6×6 matrix $\mathbf{A}$ has two multiplicity 3 eigenvalues $\lambda_1 = -2$ and $\lambda_2 = 3$ with defects 1 and 2, respectively. Then λ_1 must have an associated eigenvector $\mathbf{u}_1$ and a length 2 chain $\{\mathbf{v}_1, \mathbf{v}_2\}$ of generalized eigenvectors (with the eigenvectors $\mathbf{u}_1$ and $\mathbf{v}_1$ being linearly independent), whereas λ_2 must have a length 3 chain $\{\mathbf{w}_1, \mathbf{w}_2, \mathbf{w}_3\}$ of generalized eigenvectors based on its single eigenvector $\mathbf{w}_1$. The six generalized eigenvectors $\mathbf{u}_1$, $\mathbf{v}_1$, $\mathbf{v}_2$, $\mathbf{w}_1$, $\mathbf{w}_2$, and $\mathbf{w}_3$ are then linearly independent and yield the following six independent solutions of $\mathbf{x}' = \mathbf{A}\mathbf{x}$:

$$\begin{aligned}
\mathbf{x}_1(t) &= \mathbf{u}_1 e^{-2t},\\
\mathbf{x}_2(t) &= \mathbf{v}_1 e^{-2t},\\
\mathbf{x}_3(t) &= (\mathbf{v}_1 t + \mathbf{v}_2)e^{-2t},\\
\mathbf{x}_4(t) &= \mathbf{w}_1 e^{3t},\\
\mathbf{x}_5(t) &= (\mathbf{w}_1 t + \mathbf{w}_2)e^{3t},\\
\mathbf{x}_6(t) &= \left(\tfrac{1}{2}\mathbf{w}_1 t^2 + \mathbf{w}_2 t + \mathbf{w}_3\right)e^{3t}.
\end{aligned}$$

■

As Example 5 illustrates, the computation of independent solutions corresponding to different eigenvalues and chains of generalized eigenvalues is a routine matter. The determination of the chain structure associated with a given multiple eigenvalue can be more interesting (as in Example 6).

An Application

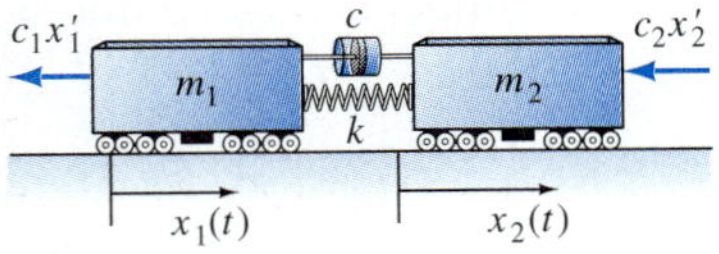

FIGURE 7.5.2. The railway cars of Example 6.

Figure 7.5.2 shows two railway cars that are connected with a spring (permanently attached to both cars) and with a damper that exerts opposite forces on the two cars, of magnitude $c(x_1' - x_2')$ proportional to their relative velocity. The two cars are also subject to frictional resistance forces c_1x_1' and c_2x_2' proportional to their respective velocities. An application of Newton's law $ma = F$ (as in Example 1 of Section 7.1) yields the equations of motion

$$\begin{aligned}
m_1x_1'' &= k(x_2 - x_1) - c_1x_1' - c(x_1' - x_2'),\\
m_2x_2'' &= k(x_1 - x_2) - c_2x_2' - c(x_2' - x_1').
\end{aligned} \tag{28}$$

In terms of the position vector $\mathbf{x}(t) = \begin{bmatrix} x_1(t) & x_2(t) \end{bmatrix}^T$, these equations can be written in the matrix form

$$\mathbf{M}\mathbf{x}'' = \mathbf{K}\mathbf{x} + \mathbf{R}\mathbf{x}', \tag{29}$$

where $\mathbf{M}$ and $\mathbf{K}$ are mass and stiffness matrices (as in Eqs. (2) and (3) of Section 7.4), and

$$\mathbf{R} = \begin{bmatrix} -(c + c_1) & c \\ c & -(c + c_2) \end{bmatrix}$$

is the *resistance matrix*. Unfortunately, because of the presence of the term involving $\mathbf{x}'$, the methods of Section 7.4 cannot be used.

Instead, we write (28) as a first-order system in the four unknown functions $x_1(t)$, $x_2(t)$, $x_3(t) = x_1'(t)$, and $x_4(t) = x_2'(t)$. If $m_1 = m_2 = 1$ we get

$$\mathbf{x}' = \mathbf{A}\mathbf{x}, \tag{30}$$

where now $\mathbf{x} = \begin{bmatrix} x_1 & x_2 & x_3 & x_4 \end{bmatrix}^T$ and

$$\mathbf{A} = \begin{bmatrix} 0 & 0 & 1 & 0 \\ 0 & 0 & 0 & 1 \\ -k & k & -(c + c_1) & c \\ k & -k & c & -(c + c_2) \end{bmatrix}. \tag{31}$$

Example 6 With $m_1 = m_2 = c = 1$ and $k = c_1 = c_2 = 2$, the system in Eq. (30) is

$$\mathbf{x}' = \begin{bmatrix} 0 & 0 & 1 & 0 \\ 0 & 0 & 0 & 1 \\ -2 & 2 & -3 & 1 \\ 2 & -2 & 1 & -3 \end{bmatrix} \mathbf{x}. \tag{32}$$

It is not too tedious to calculate manually—although a computer algebra system such as *Maple, Mathematica,* or MATLAB is useful here—the characteristic equation

$$\lambda^4 + 6\lambda^3 + 12\lambda^2 + 8\lambda = \lambda(\lambda + 2)^3 = 0$$

of the coefficient matrix $\mathbf{A}$ in Eq. (32). Thus $\mathbf{A}$ has the distinct eigenvalue $\lambda_0 = 0$ and the triple eigenvalue $\lambda_1 = -2$.

CASE 1: $\lambda_0 = 0$. The eigenvalue equation $(\mathbf{A} - \lambda\mathbf{I})\mathbf{v} = \mathbf{0}$ for the eigenvector $\mathbf{v} = \begin{bmatrix} a & b & c & d \end{bmatrix}^T$ is

$$\mathbf{A}\mathbf{v} = \begin{bmatrix} 0 & 0 & 1 & 0 \\ 0 & 0 & 0 & 1 \\ -2 & 2 & -3 & 1 \\ 2 & -2 & 1 & -3 \end{bmatrix} \begin{bmatrix} a \\ b \\ c \\ d \end{bmatrix} = \begin{bmatrix} 0 \\ 0 \\ 0 \\ 0 \end{bmatrix}.$$

The first two rows give $c = d = 0$; then the last two rows yield $a = b$. Thus

$$\mathbf{v}_0 = \begin{bmatrix} 1 & 1 & 0 & 0 \end{bmatrix}^T$$

is an eigenvector associated with $\lambda_0 = 0$.

CASE 2: $\lambda_1 = -2$. The eigenvalue equation $(\mathbf{A} - \lambda\mathbf{I})\mathbf{v} = \mathbf{0}$ is

$$(\mathbf{A} + 2\mathbf{I})\mathbf{v} = \begin{bmatrix} 2 & 0 & 1 & 0 \\ 0 & 2 & 0 & 1 \\ -2 & 2 & -1 & 1 \\ 2 & -2 & 1 & -1 \end{bmatrix} \begin{bmatrix} a \\ b \\ c \\ d \end{bmatrix} = \begin{bmatrix} 0 \\ 0 \\ 0 \\ 0 \end{bmatrix}.$$

The third and fourth scalar equations here are the differences of the first and second equations, and therefore are redundant. Hence $\mathbf{v}$ is determined by the first two equations,

$$2a + c = 0 \quad \text{and} \quad 2b + d = 0.$$

We can choose a and b independently, then solve for c and d. Thereby we obtain *two* eigenvectors associated with the triple eigenvalue $\lambda_1 = -2$. The choice $a = 1$, $b = 0$ yields $c = -2$, $d = 0$ and thereby the eigenvector

$$\mathbf{u}_1 = \begin{bmatrix} 1 & 0 & -2 & 0 \end{bmatrix}^T.$$

The choice $a = 0$, $b = 1$ yields $c = 0$, $d = -2$ and thereby the eigenvector

$$\mathbf{u}_2 = \begin{bmatrix} 0 & 1 & 0 & -2 \end{bmatrix}^T.$$

Because $\lambda_1 = -2$ has defect 1, we need a generalized eigenvector of rank 2, and hence a nonzero solution $\mathbf{v}_2$ of the equation

$$(\mathbf{A} + 2\mathbf{I})^2\mathbf{v}_2 = \begin{bmatrix} 2 & 2 & 1 & 1 \\ 2 & 2 & 1 & 1 \\ 0 & 0 & 0 & 0 \\ 0 & 0 & 0 & 0 \end{bmatrix} \mathbf{v}_2 = \mathbf{0}.$$

Obviously,

$$\mathbf{v}_2 = \begin{bmatrix} 0 & 0 & 1 & -1 \end{bmatrix}^T$$

is such a vector, and we find that

$$(\mathbf{A} + 2\mathbf{I})\mathbf{v}_2 = \begin{bmatrix} 2 & 0 & 1 & 0 \\ 0 & 2 & 0 & 1 \\ -2 & 2 & -1 & 1 \\ 2 & -2 & 1 & -1 \end{bmatrix} \begin{bmatrix} 0 \\ 0 \\ 1 \\ -1 \end{bmatrix} = \begin{bmatrix} 1 \\ -1 \\ -2 \\ 2 \end{bmatrix} = \mathbf{v}_1$$

is nonzero, and therefore is an eigenvector associated with $\lambda_1 = -2$. Then $\{\mathbf{v}_1, \mathbf{v}_2\}$ is the length 2 chain we need.

The eigenvector $\mathbf{v}_1$ just found is neither of the two eigenvectors $\mathbf{u}_1$ and $\mathbf{u}_2$ found previously, but we observe that $\mathbf{v}_1 = \mathbf{u}_1 - \mathbf{u}_2$. For a length 1 chain $\mathbf{w}_1$ to complete the picture, we can choose any linear combination of $\mathbf{u}_1$ and $\mathbf{u}_2$ that is independent of $\mathbf{v}_1$. For instance, we could choose either $\mathbf{w}_1 = \mathbf{u}_1$ or $\mathbf{w}_1 = \mathbf{u}_2$. However, we will see momentarily that the particular choice

$$\mathbf{w}_1 = \mathbf{u}_1 + \mathbf{u}_2 = \begin{bmatrix} 1 & 1 & -2 & -2 \end{bmatrix}^T$$

yields a solution of the system that is of physical interest.

Finally, the chains $\{\mathbf{v}_0\}$, $\{\mathbf{w}_1\}$, and $\{\mathbf{v}_1, \mathbf{v}_2\}$ yield the four independent solutions

$$\begin{aligned} \mathbf{x}_1(t) &= \mathbf{v}_0 e^{0 \cdot t} = \begin{bmatrix} 1 & 1 & 0 & 0 \end{bmatrix}^T, \\ \mathbf{x}_2(t) &= \mathbf{w}_1 e^{-2t} = \begin{bmatrix} 1 & 1 & -2 & -2 \end{bmatrix}^T e^{-2t}, \\ \mathbf{x}_3(t) &= \mathbf{v}_1 e^{-2t} = \begin{bmatrix} 1 & -1 & -2 & 2 \end{bmatrix}^T e^{-2t}, \\ \mathbf{x}_4(t) &= (\mathbf{v}_1 t + \mathbf{v}_2)e^{-2t} \\ &= \begin{bmatrix} t & -t & -2t+1 & 2t-1 \end{bmatrix}^T e^{-2t} \end{aligned} \tag{33}$$

of the system $\mathbf{x}' = \mathbf{A}\mathbf{x}$ in (32).

The four scalar components of the general solution

$$\mathbf{x}(t) = c_1\mathbf{x}_1(t) + c_2\mathbf{x}_2(t) + c_3\mathbf{x}_3(t) + c_4\mathbf{x}_4(t)$$

are described by the equations

$$\begin{aligned} x_1(t) &= c_1 + e^{-2t}(c_2 + c_3 + c_4 t), \\ x_2(t) &= c_1 + e^{-2t}(c_2 - c_3 - c_4 t), \\ x_3(t) &= e^{-2t}(-2c_2 - 2c_3 + c_4 - 2c_4 t), \\ x_4(t) &= e^{-2t}(-2c_2 + 2c_3 - c_4 + 2c_4 t). \end{aligned} \tag{34}$$

Recall that $x_1(t)$ and $x_2(t)$ are the position functions of the two masses, whereas $x_3(t) = x_1'(t)$ and $x_4(t) = x_2'(t)$ are their respective velocity functions.

For instance, suppose that $x_1(0) = x_2(0) = 0$ and that $x_1'(0) = x_2'(0) = v_0$. Then the equations

$$\begin{aligned} x_1(0) &= c_1 + c_2 + c_3 &&= 0, \\ x_2(0) &= c_1 + c_2 - c_3 &&= 0, \\ x_1'(0) &= \quad - 2c_2 - 2c_3 + c_4 &&= v_0, \\ x_2'(0) &= \quad - 2c_2 + 2c_3 - c_4 &&= v_0 \end{aligned}$$

are readily solved for $c_1 = \frac{1}{2}v_0$, $c_2 = -\frac{1}{2}v_0$, and $c_3 = c_4 = 0$, so

$$\begin{aligned} x_1(t) &= x_2(t) = \tfrac{1}{2}v_0\left(1 - e^{-2t}\right), \\ x_1'(t) &= x_2'(t) = v_0 e^{-2t}. \end{aligned}$$

In this case the two railway cars continue in the same direction with equal but exponentially damped velocities, approaching the displacements $x_1 = x_2 = \frac{1}{2}v_0$ as $t \to +\infty$.

It is of interest to interpret physically the individual generalized eigenvector solutions given in (33). The degenerate ($\lambda_0 = 0$) solution

$$\mathbf{x}_1(t) = \begin{bmatrix} 1 & 1 & 0 & 0 \end{bmatrix}^T$$

describes the two masses at rest with position functions $x_1(t) \equiv 1$ and $x_2(t) \equiv 1$. The solution

$$\mathbf{x}_2(t) = \begin{bmatrix} 1 & 1 & -2 & -2 \end{bmatrix}^T e^{-2t}$$

corresponding to the carefully chosen eigenvector $\mathbf{w}_1$ describes damped motions $x_1(t) = e^{-2t}$ and $x_2(t) = e^{-2t}$ of the two masses, with equal velocities in the same direction. Finally, the solutions $\mathbf{x}_3(t)$ and $\mathbf{x}_4(t)$ resulting from the length 2 chain $\{\mathbf{v}_1, \mathbf{v}_2\}$ both describe damped motion with the two masses moving in opposite directions. ■

The methods of this section apply to complex multiple eigenvalues just as to real multiple eigenvalues (although the necessary computations tend to be somewhat lengthy). Given a complex conjugate pair $\alpha \pm \beta i$ of eigenvalues of multiplicity k, we work with one of them (say, $\alpha - \beta i$) as if it were real to find k independent complex-valued solutions. The real and imaginary parts of these complex-valued solutions then provide $2k$ real-valued solutions associated with the two eigenvalues $\lambda = \alpha - \beta i$ and $\overline{\lambda} = \alpha + \beta i$ each of multiplicity k. See Problems 33 and 34.

The Jordan Normal Form

In Section 6.2 we saw that if the $n \times n$ matrix $\mathbf{A}$ has a complete set $\mathbf{v}_1, \mathbf{v}_2, \ldots, \mathbf{v}_n$ of n linearly independent eigenvectors, then it is similar to a diagonal matrix. In particular,

$$\mathbf{P}^{-1}\mathbf{A}\mathbf{P} = \mathbf{D}, \tag{35}$$

where $\mathbf{P} = \begin{bmatrix} \mathbf{v}_1 & \mathbf{v}_2 & \cdots & \mathbf{v}_n \end{bmatrix}$ and the diagonal elements of the diagonal matrix $\mathbf{D}$ are the corresponding eigenvalues $\lambda_1, \lambda_2, \ldots, \lambda_n$ (not necessarily distinct). This result is a special case of the following general theorem, a proof of which can be found in Appendix B of Gilbert Strang, *Linear Algebra and Its Applications* (4th edition, Brooks Cole, 2006).

THEOREM 1 Jordan Normal Form

If the $n \times n$ matrix $\mathbf{A}$ has s linearly independent eigenvectors $\mathbf{v}_1, \mathbf{v}_2, \ldots, \mathbf{v}_s$, then it is similar to a block-diagonal matrix of the *Jordan normal form*

$$\mathbf{J} = \begin{bmatrix} \mathbf{J}_1 & \mathbf{0} & \cdots & \mathbf{0} \\ \mathbf{0} & \mathbf{J}_2 & \cdots & \mathbf{0} \\ \vdots & \vdots & \ddots & \vdots \\ \mathbf{0} & \mathbf{0} & \cdots & \mathbf{J}_s \end{bmatrix}, \tag{36}$$

where each submatrix $\mathbf{J}_i$ is a $k \times k$ *Jordan block* of the form

$$\mathbf{J}_i = \begin{bmatrix} \lambda_i & 1 & 0 & \cdots & 0 \\ 0 & \lambda_i & 1 & \cdots & 0 \\ \vdots & \vdots & \ddots & \ddots & \vdots \\ 0 & 0 & \cdots & \lambda_i & 1 \\ 0 & 0 & 0 & \cdots & \lambda_i \end{bmatrix}, \tag{37}$$

with λ_i being the eigenvalue of $\mathbf{A}$ corresponding to the eigenvector $\mathbf{v}_i$; if $k = 1$, then $\mathbf{J}_i = \begin{bmatrix} \lambda_i \end{bmatrix}$.

If the Jordan block $\mathbf{J}_i$ in (37) is of size $k \times k$, then it corresponds to a length k chain of generalized eigenvectors based on the (ordinary) eigenvector $\mathbf{v}_i$. If all these generalized eigenvectors are arranged as column vectors in proper order corresponding to the appearance of the Jordan blocks in (36), the result is a nonsingular $n \times n$ matrix $\mathbf{Q}$ such that

$$\mathbf{Q}^{-1}\mathbf{A}\mathbf{Q} = \mathbf{J}. \tag{38}$$

The block-diagonal matrix $\mathbf{J}$ such that $\mathbf{A} = \mathbf{Q}\mathbf{J}\mathbf{Q}^{-1}$ is called the **Jordan normal form** of the matrix $\mathbf{A}$ and is unique [except for the order of appearance of the Jordan blocks in (36)].

Example 7 In Example 3 we saw that the matrix

$$\mathbf{A} = \begin{bmatrix} 1 & -3 \\ 3 & 7 \end{bmatrix}$$

has the single eigenvalue $\lambda = 4$ and the associated length 2 chain of generalized eigenvectors $\{\mathbf{v}_1, \mathbf{v}_2\}$, where $\mathbf{v}_1 = \begin{bmatrix} -3 & 3 \end{bmatrix}^T$ and $\mathbf{v}_2 = \begin{bmatrix} 1 & 0 \end{bmatrix}^T$, based on the eigenvector $\mathbf{v}_1 = \mathbf{A}\mathbf{v}_2$. If we define

$$\mathbf{Q} = \begin{bmatrix} \mathbf{v}_1 & \mathbf{v}_2 \end{bmatrix} = \begin{bmatrix} -3 & 1 \\ 3 & 0 \end{bmatrix},$$

then we find that the Jordan normal form of $\mathbf{A}$ is

$$\mathbf{J} = \mathbf{Q}^{-1}\mathbf{A}\mathbf{Q} = \begin{bmatrix} 4 & 1 \\ 0 & 4 \end{bmatrix}.$$

Here $\mathbf{J} = \mathbf{J}_1$ is a single 2×2 Jordan block corresponding to the single eigenvalue $\lambda = 4$ of $\mathbf{A}$. ■

Example 8 In Example 6 we saw that the matrix

$$\mathbf{A} = \begin{bmatrix} 0 & 0 & 1 & 0 \\ 0 & 0 & 0 & 1 \\ -2 & 2 & -3 & 1 \\ 2 & -2 & 1 & -3 \end{bmatrix}$$

has the distinct eigenvalue $\lambda_1 = 0$ corresponding to the eigenvector

$$\mathbf{v}_1 = \begin{bmatrix} 1 & 1 & 0 & 0 \end{bmatrix}^T$$

and the triple eigenvalue $\lambda_2 = -2$ corresponding both to the eigenvector $\mathbf{v}_2 = \begin{bmatrix} 1 & 0 & -2 & 0 \end{bmatrix}^T$ and to the length 2 chain of generalized eigenvectors $\{\mathbf{v}_3, \mathbf{v}_4\}$ with $\mathbf{v}_3 = \begin{bmatrix} 1 & -1 & -2 & 2 \end{bmatrix}^T$ and $\mathbf{v}_4 = \begin{bmatrix} 0 & 0 & 1 & -1 \end{bmatrix}^T$, such that $\mathbf{v}_3 = (\mathbf{A} + 2\mathbf{I})\mathbf{v}_4$. If we define

$$\mathbf{Q} = \begin{bmatrix} \mathbf{v}_1 & \mathbf{v}_2 & \mathbf{v}_3 & \mathbf{v}_4 \end{bmatrix} = \begin{bmatrix} 1 & 1 & 1 & 0 \\ 1 & 0 & -1 & 0 \\ 0 & -2 & -2 & 1 \\ 0 & 0 & 2 & -1 \end{bmatrix},$$

then (using a computer algebra system to ease the labor of calculation) we find that the Jordan normal form of $\mathbf{A}$ is

$$\mathbf{J} = \mathbf{Q}^{-1}\mathbf{A}\mathbf{Q} = \begin{bmatrix} 0 & 0 & 0 & 0 \\ 0 & -2 & 0 & 0 \\ 0 & 0 & -2 & 1 \\ 0 & 0 & 0 & -2 \end{bmatrix}.$$

Here we see the 1×1 Jordan block $\mathbf{J}_1 = \begin{bmatrix} 0 \end{bmatrix}$ corresponding to the eigenvalue $\lambda_1 = 0$ of $\mathbf{A}$, as well as the two Jordan blocks $\mathbf{J}_2 = \begin{bmatrix} -2 \end{bmatrix}$ and $\mathbf{J}_3 = \begin{bmatrix} -2 & 1 \\ 0 & -2 \end{bmatrix}$ corresponding to the two linearly independent eigenvectors $\mathbf{v}_2$ and $\mathbf{v}_3$ associated with the eigenvalue $\lambda_2 = -2$. ■

The General Cayley-Hamilton Theorem

In Section 6.3, we showed that every diagonalizable matrix $\mathbf{A}$ satisfies its characteristic equation $p(\lambda) = |\mathbf{A} - \lambda\mathbf{I}| = 0$, that is, $p(\mathbf{A}) = \mathbf{0}$. We can now use the Jordan normal form to show that this is true whether or not $\mathbf{A}$ is diagonalizable.

If $\mathbf{J} = \mathbf{Q}^{-1}\mathbf{A}\mathbf{Q}$ is the Jordan normal form of the matrix $\mathbf{A}$, then $p(\mathbf{A}) = \mathbf{Q}p(\mathbf{J})\mathbf{Q}^{-1}$, just as in the conclusion to the proof of Theorem 1 in Section 6.3. It therefore suffices to show that $p(\mathbf{J}) = \mathbf{0}$. If the Jordan blocks $\mathbf{J}_1, \mathbf{J}_2, \dots, \mathbf{J}_s$ in (36) have sizes $k_1, k_2, \dots, k_s$—that is, $\mathbf{J}_i$ is a $k_i \times k_i$ matrix—and corresponding eigenvalues $\lambda_1, \lambda_2, \dots, \lambda_s$ (respectively), then

$$p(\lambda) = (\lambda_1 - \lambda)^{k_1}(\lambda_2 - \lambda)^{k_2} \cdots (\lambda_s - \lambda)^{k_s},$$

and so

$$p(\mathbf{J}) = (\lambda_1\mathbf{I} - \mathbf{J})^{k_1}(\lambda_2\mathbf{I} - \mathbf{J})^{k_2} \cdots (\lambda_s\mathbf{I} - \mathbf{J})^{k_s}. \tag{39}$$

Now $p(\mathbf{J})$ has the same block-diagonal structure as $\mathbf{J}$ itself, and we see from (39) that the ith block of $p(\mathbf{J})$ involves the factor

$$(\lambda_i\mathbf{I} - \mathbf{J}_i)^{k_i} = \begin{bmatrix} 0 & -1 & 0 & \cdots & 0 \\ 0 & 0 & -1 & \cdots & 0 \\ \vdots & \vdots & \ddots & \ddots & \vdots \\ 0 & 0 & \cdots & 0 & -1 \\ 0 & 0 & 0 & \cdots & 0 \end{bmatrix}^{k_i},$$

where the $k_i \times k_i$ matrix $\lambda_i\mathbf{I} - \mathbf{J}_i$ has all elements zero except on its first superdiagonal. It is easily seen that $(\lambda_i\mathbf{I} - \mathbf{J}_i)^2$ then has nonzero elements only on its second superdiagonal; $(\lambda_i\mathbf{I} - \mathbf{J}_i)^3$ has nonzero elements only on its third superdiagonal; and so on, in turn, until we see that $(\lambda_i\mathbf{I} - \mathbf{J}_i)^{k_i} = \mathbf{0}$. Hence it follows from (39) that the ith block of $p(\mathbf{J})$ is the $k_i \times k_i$ zero matrix. This being true for each $i = 1, 2, \ldots, s$, we conclude that $p(\mathbf{J}) = \mathbf{0}$, as desired. This completes the proof of the general case of the Cayley-Hamilton theorem.

7.5 Problems

Find general solutions of the systems in Problems 1 through 22. In Problems 1 through 6, use a computer system or graphing calculator to construct a direction field and typical solution curves for the given system.

1. $\mathbf{x}' = \begin{bmatrix} -2 & 1 \\ -1 & -4 \end{bmatrix}\mathbf{x}$

2. $\mathbf{x}' = \begin{bmatrix} 3 & -1 \\ 1 & 1 \end{bmatrix}\mathbf{x}$

3. $\mathbf{x}' = \begin{bmatrix} 1 & -2 \\ 2 & 5 \end{bmatrix}\mathbf{x}$

4. $\mathbf{x}' = \begin{bmatrix} 3 & -1 \\ 1 & 5 \end{bmatrix}\mathbf{x}$

5. $\mathbf{x}' = \begin{bmatrix} 7 & 1 \\ -4 & 3 \end{bmatrix}\mathbf{x}$

6. $\mathbf{x}' = \begin{bmatrix} 1 & -4 \\ 4 & 9 \end{bmatrix}\mathbf{x}$

7. $\mathbf{x}' = \begin{bmatrix} 2 & 0 & 0 \\ -7 & 9 & 7 \\ 0 & 0 & 2 \end{bmatrix}\mathbf{x}$

8. $\mathbf{x}' = \begin{bmatrix} 25 & 12 & 0 \\ -18 & -5 & 0 \\ 6 & 6 & 13 \end{bmatrix}\mathbf{x}$

9. $\mathbf{x}' = \begin{bmatrix} -19 & 12 & 84 \\ 0 & 5 & 0 \\ -8 & 4 & 33 \end{bmatrix}\mathbf{x}$

10. $\mathbf{x}' = \begin{bmatrix} -13 & 40 & -48 \\ -8 & 23 & -24 \\ 0 & 0 & 3 \end{bmatrix}\mathbf{x}$

11. $\mathbf{x}' = \begin{bmatrix} -3 & 0 & -4 \\ -1 & -1 & -1 \\ 1 & 0 & 1 \end{bmatrix}\mathbf{x}$

12. $\mathbf{x}' = \begin{bmatrix} -1 & 0 & 1 \\ 0 & -1 & 1 \\ 1 & -1 & -1 \end{bmatrix}\mathbf{x}$

13. $\mathbf{x}' = \begin{bmatrix} -1 & 0 & 1 \\ 0 & 1 & -4 \\ 0 & 1 & -3 \end{bmatrix}\mathbf{x}$

14. $\mathbf{x}' = \begin{bmatrix} 0 & 0 & 1 \\ -5 & -1 & -5 \\ 4 & 1 & -2 \end{bmatrix}\mathbf{x}$

15. $\mathbf{x}' = \begin{bmatrix} -2 & -9 & 0 \\ 1 & 4 & 0 \\ 1 & 3 & 1 \end{bmatrix}\mathbf{x}$

16. $\mathbf{x}' = \begin{bmatrix} 1 & 0 & 0 \\ -2 & -2 & -3 \\ 2 & 3 & 4 \end{bmatrix}\mathbf{x}$

17. $\mathbf{x}' = \begin{bmatrix} 1 & 0 & 0 \\ 18 & 7 & 4 \\ -27 & -9 & -5 \end{bmatrix}\mathbf{x}$

18. $\mathbf{x}' = \begin{bmatrix} 1 & 0 & 0 \\ 1 & 3 & 1 \\ -2 & -4 & -1 \end{bmatrix}\mathbf{x}$

19. $\mathbf{x}' = \begin{bmatrix} 1 & -4 & 0 & -2 \\ 0 & 1 & 0 & 0 \\ 6 & -12 & -1 & -6 \\ 0 & -4 & 0 & -1 \end{bmatrix}\mathbf{x}$

20. $\mathbf{x}' = \begin{bmatrix} 2 & 1 & 0 & 1 \\ 0 & 2 & 1 & 0 \\ 0 & 0 & 2 & 1 \\ 0 & 0 & 0 & 2 \end{bmatrix} \mathbf{x}$

21. $\mathbf{x}' = \begin{bmatrix} -1 & -4 & 0 & 0 \\ 1 & 3 & 0 & 0 \\ 1 & 2 & 1 & 0 \\ 0 & 1 & 0 & 1 \end{bmatrix} \mathbf{x}$

22. $\mathbf{x}' = \begin{bmatrix} 1 & 3 & 7 & 0 \\ 0 & -1 & -4 & 0 \\ 0 & 1 & 3 & 0 \\ 0 & -6 & -14 & 1 \end{bmatrix} \mathbf{x}$

In Problems 23 through 32 the eigenvalues of the coefficient matrix $\mathbf{A}$ *are given. Find a general solution of the indicated system* $\mathbf{x}' = \mathbf{A}\mathbf{x}$. *Especially in Problems 29 through 32, use of a computer algebra system (as in the application material for this section) may be useful.*

23. $\mathbf{x}' = \begin{bmatrix} 39 & 8 & -16 \\ -36 & -5 & 16 \\ 72 & 16 & -29 \end{bmatrix} \mathbf{x}; \quad \lambda = -1, 3, 3$

24. $\mathbf{x}' = \begin{bmatrix} 28 & 50 & 100 \\ 15 & 33 & 60 \\ -15 & -30 & -57 \end{bmatrix} \mathbf{x}; \quad \lambda = -2, 3, 3$

25. $\mathbf{x}' = \begin{bmatrix} -2 & 17 & 4 \\ -1 & 6 & 1 \\ 0 & 1 & 2 \end{bmatrix} \mathbf{x}; \quad \lambda = 2, 2, 2$

26. $\mathbf{x}' = \begin{bmatrix} 5 & -1 & 1 \\ 1 & 3 & 0 \\ -3 & 2 & 1 \end{bmatrix} \mathbf{x}; \quad \lambda = 3, 3, 3$

27. $\mathbf{x}' = \begin{bmatrix} -3 & 5 & -5 \\ 3 & -1 & 3 \\ 8 & -8 & 10 \end{bmatrix} \mathbf{x}; \quad \lambda = 2, 2, 2$

28. $\mathbf{x}' = \begin{bmatrix} -15 & -7 & 4 \\ 34 & 16 & -11 \\ 17 & 7 & 5 \end{bmatrix} \mathbf{x}; \quad \lambda = 2, 2, 2$

29. $\mathbf{x}' = \begin{bmatrix} -1 & 1 & 1 & -2 \\ 7 & -4 & -6 & 11 \\ 5 & -1 & 1 & 3 \\ 6 & -2 & -2 & 6 \end{bmatrix} \mathbf{x}; \quad \lambda = -1, -1, 2, 2$

30. $\mathbf{x}' = \begin{bmatrix} 2 & 1 & -2 & 1 \\ 0 & 3 & -5 & 3 \\ 0 & -13 & 22 & -12 \\ 0 & -27 & 45 & -25 \end{bmatrix} \mathbf{x}; \quad \lambda = -1, -1, 2, 2$

31. $\mathbf{x}' = \begin{bmatrix} 35 & -12 & 4 & 30 \\ 22 & -8 & 3 & 19 \\ -10 & 3 & 0 & -9 \\ -27 & 9 & -3 & -23 \end{bmatrix} \mathbf{x}; \quad \lambda = 1, 1, 1, 1$

32. $\mathbf{x}' = \begin{bmatrix} 11 & -1 & 26 & 6 & -3 \\ 0 & 3 & 0 & 0 & 0 \\ -9 & 0 & -24 & -6 & 3 \\ 3 & 0 & 9 & 5 & -1 \\ -48 & -3 & -138 & -30 & 18 \end{bmatrix} \mathbf{x};$

$\lambda = 2, 2, 3, 3, 3$

33. The characteristic equation of the coefficient matrix $\mathbf{A}$ of the system

$$\mathbf{x}' = \begin{bmatrix} 3 & -4 & 1 & 0 \\ 4 & 3 & 0 & 1 \\ 0 & 0 & 3 & -4 \\ 0 & 0 & 4 & 3 \end{bmatrix} \mathbf{x}$$

is

$$\phi(\lambda) = (\lambda^2 - 6\lambda + 25)^2 = 0.$$

Therefore, $\mathbf{A}$ has the repeated complex conjugate pair $3 \pm 4i$ of eigenvalues. First show that the complex vectors

$$\mathbf{v}_1 = \begin{bmatrix} 1 & i & 0 & 0 \end{bmatrix}^T \quad \text{and} \quad \mathbf{v}_2 = \begin{bmatrix} 0 & 0 & 1 & i \end{bmatrix}^T$$

form a length 2 chain $\{\mathbf{v}_1, \mathbf{v}_2\}$ associated with the eigenvalue $\lambda = 3 - 4i$. Then calculate the real and imaginary parts of the complex-valued solutions

$$\mathbf{v}_1 e^{\lambda t} \quad \text{and} \quad (\mathbf{v}_1 t + \mathbf{v}_2) e^{\lambda t}$$

to find four independent real-valued solutions of $\mathbf{x}' = \mathbf{A}\mathbf{x}$.

34. The characteristic equation of the coefficient matrix $\mathbf{A}$ of the system

$$\mathbf{x}' = \begin{bmatrix} 2 & 0 & -8 & -3 \\ -18 & -1 & 0 & 0 \\ -9 & -3 & -25 & -9 \\ 33 & 10 & 90 & 32 \end{bmatrix} \mathbf{x}$$

is

$$\phi(\lambda) = (\lambda^2 - 4\lambda + 13)^2 = 0.$$

Therefore, $\mathbf{A}$ has the repeated complex conjugate pair $2 \pm 3i$ of eigenvalues. First show that the complex vectors

$$\mathbf{v}_1 = \begin{bmatrix} -i & 3+3i & 0 & -1 \end{bmatrix}^T,$$
$$\mathbf{v}_2 = \begin{bmatrix} 3 & -10+9i & -i & 0 \end{bmatrix}^T$$

form a length 2 chain $\{\mathbf{v}_1, \mathbf{v}_2\}$ associated with the eigenvalue $\lambda = 2 + 3i$. Then calculate (as in Problem 33) four independent real-valued solutions of $\mathbf{x}' = \mathbf{A}\mathbf{x}$.

35. Find the position functions $x_1(t)$ and $x_2(t)$ of the railway cars of Fig. 7.5.2 if the physical parameters are given by

$$m_1 = m_2 = c_1 = c_2 = c = k = 1$$

and the initial conditions are

$$x_1(0) = x_2(0) = 0, \quad x_1'(0) = x_2'(0) = v_0.$$

How far do the cars travel before stopping?

36. Repeat Problem 35 under the assumption that car 1 is shielded from air resistance by car 2, so now $c_1 = 0$. Show that, before stopping, the cars travel twice as far as those of Problem 35.

37–46. Use the method of Examples 7 and 8 to find the Jordan normal form $\mathbf{J}$ of each coefficient matrix $\mathbf{A}$ given in Problems 23 through 32 (respectively).

7.5 Application Defective Eigenvalues and Generalized Eigenvectors

A typical computer algebra system can calculate both the eigenvalues of a given matrix $\mathbf{A}$ and the linearly independent (ordinary) eigenvectors associated with each eigenvalue. For instance, consider the 4×4 matrix

$$\mathbf{A} = \begin{bmatrix} 35 & -12 & 4 & 30 \\ 22 & -8 & 3 & 19 \\ -10 & 3 & 0 & -9 \\ -27 & 9 & -3 & -23 \end{bmatrix} \tag{1}$$

of Problem 31 in this section. When the matrix $\mathbf{A}$ has been entered, the *Maple* calculation

```
with(linalg): eigenvectors(A);
[1, 4, {[-1, 0, 1, 1], [0, 1, 3, 0]}]
```

or the *Mathematica* calculation

```
Eigensystem[A]
{{1,1,1,1},
{{-3,-1,0,3}, {0,1,3,0}, {0,0,0,0}, {0,0,0,0}}}
```

reveals that the matrix $\mathbf{A}$ in Eq. (1) has the single eigenvalue $\lambda = 1$ of multiplicity 4 with only two independent associated eigenvectors $\mathbf{v}_1$ and $\mathbf{v}_2$. The MATLAB command

```
[V, D] = eig(sym(A))
```

provides the same information. The eigenvalue $\lambda = 1$ therefore has defect $d = 2$. If $\mathbf{B} = \mathbf{A} - (1)\mathbf{I}$, you should find that $\mathbf{B}^2 \neq \mathbf{0}$ but $\mathbf{B}^3 = \mathbf{0}$. If

$$\mathbf{u}_1 = \begin{bmatrix} 1 & 0 & 0 & 0 \end{bmatrix}^T, \quad \mathbf{u}_2 = \mathbf{B}\mathbf{u}_1, \quad \mathbf{u}_3 = \mathbf{B}\mathbf{u}_2,$$

then $\{\mathbf{u}_1, \mathbf{u}_2, \mathbf{u}_3\}$ should be a length 3 chain of generalized eigenvectors based on the ordinary eigenvector $\mathbf{u}_3$ (which should be a linear combination of the original eigenvectors $\mathbf{v}_1$ and $\mathbf{v}_2$). Use your computer algebra system to carry out this construction, and finally write four linearly independent solutions of the linear system $\mathbf{x}' = \mathbf{A}\mathbf{x}$.

For a more exotic matrix to investigate, consider the MATLAB's **gallery(5)** example matrix

$$\mathbf{A} = \begin{bmatrix} -9 & 11 & -21 & 63 & -252 \\ 70 & -69 & 141 & -421 & 1684 \\ -575 & 575 & -1149 & 3451 & -13801 \\ 3891 & -3891 & 7782 & -23345 & 93365 \\ 1024 & -1024 & 2048 & -6144 & 24572 \end{bmatrix}. \tag{2}$$

Use appropriate commands like those illustrated here to show that $\mathbf{A}$ has a single eigenvalue $\lambda = 0$ of multiplicity 5 and defect 4. Noting that $\mathbf{A} - (0)\mathbf{I} = \mathbf{A}$, you should find that $\mathbf{A}^4 \neq \mathbf{0}$ but that $\mathbf{A}^5 = \mathbf{0}$. Hence calculate the vectors

$$\mathbf{u}_1 = \begin{bmatrix} 1 & 0 & 0 & 0 & 0 \end{bmatrix}^T, \quad \mathbf{u}_2 = \mathbf{A}\mathbf{u}_1, \quad \mathbf{u}_3 = \mathbf{A}\mathbf{u}_2, \quad \mathbf{u}_4 = \mathbf{A}\mathbf{u}_3, \quad \mathbf{u}_5 = \mathbf{A}\mathbf{u}_4.$$

You should find that $\mathbf{u}_5$ is a nonzero vector such that $\mathbf{A}\mathbf{u}_5 = \mathbf{0}$, and is therefore an (ordinary) eigenvector of $\mathbf{A}$ associated with the eigenvalue $\lambda = 0$. Thus

$\{\mathbf{u}_1, \mathbf{u}_2, \mathbf{u}_3, \mathbf{u}_4, \mathbf{u}_5\}$ is a length 5 chain of generalized eigenvectors of the matrix $\mathbf{A}$ in Eq. (2), and you can finally write five linearly independent solutions of the linear system $\mathbf{x}' = \mathbf{A}\mathbf{x}$.

7.6 Numerical Methods for Systems

We now discuss the numerical approximation of solutions of systems of differential equations. Our goal is to apply the methods of Sections 2.4 through 2.6 to the initial value problem

$$\mathbf{x}' = \mathbf{f}(t, \mathbf{x}), \quad \mathbf{x}(t_0) = \mathbf{x}_0 \tag{1}$$

for a system of m first-order differential equations. In (1) the independent variable is the scalar t, and

$$\mathbf{x} = (x_1, x_2, \ldots, x_m) \quad \text{and} \quad \mathbf{f} = (f_1, f_2, \ldots, f_m)$$

are vector-valued functions. If the component functions of $\mathbf{f}$ and their first-order partial derivatives are all continuous in a neighborhood of the point $(t_0, \mathbf{x}_0)$, then Theorems 3 and 4 of Appendix A guarantee the existence and uniqueness of a solution $\mathbf{x} = \mathbf{x}(t)$ of (1) on some subinterval (of the t-axis) containing t_0. With this assurance we can proceed to discuss the numerical approximation of this solution.

Beginning with step size h, we want to approximate the values of $\mathbf{x}(t)$ at the points $t_1, t_2, t_3, \ldots$, where $t_{n+1} = t_n + h$ for $n \geqq 0$. Suppose that we have already computed the *approximations*

$$\mathbf{x}_1, \quad \mathbf{x}_2, \quad \mathbf{x}_3, \quad \ldots, \quad \mathbf{x}_n$$

to the *actual values*

$$\mathbf{x}(t_1), \quad \mathbf{x}(t_2), \quad \mathbf{x}(t_3), \quad \ldots, \quad \mathbf{x}(t_n)$$

of the exact solution of the system in (1). We can then make the step from $\mathbf{x}_n$ to the next approximation $\mathbf{x}_{n+1} \approx \mathbf{x}(t_{n+1})$ by any one of the methods of Sections 2.4 through 2.6. Essentially all that is required is to write the iterative formula of the selected method in the vector notation of the present discussion.

Euler Methods for Systems

For example, the iterative formula of Euler's method for systems is

$$\mathbf{x}_{n+1} = \mathbf{x}_n + h\mathbf{f}(t, \mathbf{x}_n). \tag{2}$$

To examine the case $m = 2$ of a pair of first-order differential equations, let us write

$$\mathbf{x} = \begin{bmatrix} x \\ y \end{bmatrix} \quad \text{and} \quad \mathbf{f} = \begin{bmatrix} f \\ g \end{bmatrix}.$$

Then the initial value problem in (1) is

$$\begin{aligned} x' &= f(t, x, y), \quad & x(t_0) &= x_0, \\ y' &= g(t, x, y), \quad & y(t_0) &= y_0, \end{aligned} \tag{3}$$

and the scalar components of the vector formula in (2) are

$$\begin{aligned} x_{n+1} &= x_n + hf(t_n, x_n, y_n), \\ y_{n+1} &= y_n + hg(t_n, x_n, y_n). \end{aligned} \tag{4}$$

Note that each iterative formula in (4) has the form of a single Euler iteration, but with y_n inserted like a parameter in the first formula (for x_{n+1}) and with x_n inserted like a parameter in the second formula (for y_{n+1}). The generalization to the system in (3) of each of the other methods in Sections 2.4 through 2.6 follows a similar pattern.

The improved Euler method for systems consists at each step of calculating first the predictor

$$\mathbf{u}_{n+1} = \mathbf{x}_n + h\mathbf{f}(t_n, \mathbf{x}_n) \tag{5}$$

and then the corrector

$$\mathbf{x}_{n+1} = \mathbf{x}_n + \frac{h}{2}[\mathbf{f}(t_n, \mathbf{x}_n) + \mathbf{f}(t_{n+1}, \mathbf{u}_{n+1})]. \tag{6}$$

For the case of the two-dimensional initial value problem in (3), the scalar components of the formulas in (5) and (6) are

$$\begin{aligned} u_{n+1} &= x_n + hf(t_n, x_n, y_n), \\ v_{n+1} &= y_n + hg(t_n, x_n, y_n) \end{aligned} \tag{7}$$

and

$$\begin{aligned} x_{n+1} &= x_n + \frac{h}{2}[f(t_n, x_n, y_n) + f(t_{n+1}, u_{n+1}, v_{n+1})], \\ y_{n+1} &= y_n + \frac{h}{2}[g(t_n, x_n, y_n) + g(t_{n+1}, u_{n+1}, v_{n+1})]. \end{aligned} \tag{8}$$

Example 1 Consider the initial value problem

$$\begin{aligned} x' &= 3x - 2y, \qquad & x(0) &= 3; \\ y' &= 5x - 4y, \qquad & y(0) &= 6. \end{aligned} \tag{9}$$

The exact solution of the system in (9) is

$$x(t) = 2e^{-2t} + e^t, \qquad y(t) = 5e^{-2t} + e^t. \tag{10}$$

Here we have $f(x, y) = 3x - 2y$ and $g(x, y) = 5x - 4y$ in (3), so the Euler iterative formulas in (4) are

$$x_{n+1} = x_n + h \cdot (3x_n - 2y_n), \qquad y_{n+1} = y_n + h \cdot (5x_n - 4y_n).$$

With step size $h = 0.1$ we calculate

$$\begin{aligned} x_1 &= 3 + (0.1) \cdot [3 \cdot 3 - 2 \cdot 6] = 2.7, \\ y_1 &= 6 + (0.1) \cdot [5 \cdot 3 - 4 \cdot 6] = 5.1 \end{aligned}$$

and

$$x_2 = 2.7 + (0.1) \cdot [3 \cdot (2.7) - 2 \cdot (5.1)] = 2.49,$$
$$y_2 = 5.1 + (0.1) \cdot [5 \cdot (2.7) - 4 \cdot (5.1)] = 4.41.$$

The actual values at $t_2 = 0.2$ given by (10) are $x(0.2) \approx 2.562$ and $y(0.2) \approx 4.573$.

To compute the improved Euler approximations to $x(0.2)$ and $y(0.2)$ with a single step of size $h = 0.2$, we first calculate the predictors

$$u_1 = 3 + (0.2) \cdot [3 \cdot 3 - 2 \cdot 6] = 2.4,$$
$$v_1 = 6 + (0.2) \cdot [5 \cdot 3 - 4 \cdot 6] = 4.2.$$

Then the corrector formulas in (8) yield

$$x_1 = 3 + (0.1) \cdot ([3 \cdot 3 - 2 \cdot 6] + [3 \cdot (2.4) - 2 \cdot (4.2)]) = 2.58,$$
$$y_1 = 6 + (0.1) \cdot ([5 \cdot 3 - 4 \cdot 6] + [5 \cdot (2.4) - 4 \cdot (4.2)]) = 4.62.$$

As we would expect, a single improved Euler step gives better accuracy than two ordinary Euler steps. ■

The Runge–Kutta Method and Second-Order Equations

The vector version of the iterative formula for the Runge–Kutta method is

$$\mathbf{x}_{n+1} = \mathbf{x}_n + \frac{h}{6}(\mathbf{k}_1 + 2\mathbf{k}_2 + 2\mathbf{k}_3 + \mathbf{k}_4) \tag{11}$$

where the vectors $\mathbf{k}_1$, $\mathbf{k}_2$, $\mathbf{k}_3$, and $\mathbf{k}_4$ are defined (by analogy with Eqs. (5a)–(5d) of Section 2.6) as follows:

$$\begin{aligned}
\mathbf{k}_1 &= \mathbf{f}(t_n, \mathbf{x}_n), \\
\mathbf{k}_2 &= \mathbf{f}\left(t_n + \tfrac{1}{2}h, \mathbf{x}_n + \tfrac{1}{2}h\mathbf{k}_1\right), \\
\mathbf{k}_3 &= \mathbf{f}\left(t_n + \tfrac{1}{2}h, \mathbf{x}_n + \tfrac{1}{2}h\mathbf{k}_2\right), \\
\mathbf{k}_4 &= \mathbf{f}(t_n + h, \mathbf{x}_n + h\mathbf{k}_3).
\end{aligned} \tag{12}$$

To describe in scalar notation the Runge–Kutta method for the two-dimensional initial value problem

$$\begin{aligned}
x' &= f(t, x, y), \qquad & x(t_0) &= x_0, \\
y' &= g(t, x, y), \qquad & y(t_0) &= y_0,
\end{aligned} \tag{3}$$

let us write

$$\mathbf{x} = \begin{bmatrix} x \\ y \end{bmatrix}, \quad \mathbf{f} = \begin{bmatrix} f \\ g \end{bmatrix}, \quad \text{and} \quad \mathbf{k}_i = \begin{bmatrix} F_i \\ G_i \end{bmatrix}.$$

Then the Runge–Kutta iterative formulas for the step from (x_n, y_n) to the next approximation $(x_{n+1}, y_{n+1}) \approx (x(t_{n+1}), y(t_{n+1}))$ are

$$\begin{aligned}
x_{n+1} &= x_n + \frac{h}{6}(F_1 + 2F_2 + 2F_3 + F_4), \\
y_{n+1} &= y_n + \frac{h}{6}(G_1 + 2G_2 + 2G_3 + G_4),
\end{aligned} \tag{13}$$

where the values F_1, F_2, F_3, and F_4 of the function f are

$$\begin{aligned} F_1 &= f(t_n, x_n, y_n), \\ F_2 &= f\left(t_n + \tfrac{1}{2}h, x_n + \tfrac{1}{2}hF_1, y_n + \tfrac{1}{2}hG_1\right), \\ F_3 &= f\left(t_n + \tfrac{1}{2}h, x_n + \tfrac{1}{2}hF_2, y_n + \tfrac{1}{2}hG_2\right), \\ F_4 &= f(t_n + h, x_n + hF_3, y_n + hG_3); \end{aligned} \tag{14}$$

G_1, G_2, G_3, and G_4 are the similarly defined values of the function g.

Perhaps the most common application of the two-dimensional Runge–Kutta method is to the numerical solution of second-order initial value problems of the form

$$\begin{aligned} &x'' = g(t, x, x'), \\ &x(t_0) = x_0, \quad x'(t_0) = y_0. \end{aligned} \tag{15}$$

If we introduce the auxiliary variable $y = x'$, then the problem in (15) translates into the two-dimensional first-order problem

$$\begin{aligned} x' &= y, & x(t_0) &= x_0, \\ y' &= g(t, x, y), & y(t_0) &= y_0. \end{aligned} \tag{16}$$

This is a problem of the form in (3) with $f(t, x, y) = y$.

If the functions f and g are not too complicated, then it is feasible to carry out manually a reasonable number of steps of the two-dimensional Runge–Kutta method described here. But the first operating electronic computers were constructed (during World War II) specifically to implement methods similar to the Runge–Kutta method for the numerical computation of trajectories of artillery projectiles. The application material for this section lists TI-85 and BASIC versions of Program RK2DIM that can be used with two-dimensional systems.

t	$x = \sin t$	$y = \cos t$
0.5	+0.47943	+0.87758
1.0	+0.84147	+0.54030
1.5	+0.99749	+0.07074
2.0	+0.90930	−0.41615
2.5	+0.59847	−0.80114
3.0	+0.14112	−0.98999
3.5	−0.35078	−0.93646
4.0	−0.75680	−0.65364
4.5	−0.97753	−0.21080
5.0	−0.95892	+0.28366

FIGURE 7.6.1. Runge–Kutta values (with $h = 0.05$) for the problem in Eq. (18).

Example 2 The exact solution of the initial value problem

$$x'' = -x; \quad x(0) = 0, \quad x'(0) = 1 \tag{17}$$

is $x(t) = \sin t$. The substitution $y = x'$ translates (17) into the two-dimensional problem

$$\begin{aligned} x' &= y, & x(0) &= 0; \\ y' &= -x, & y(0) &= 1, \end{aligned} \tag{18}$$

which has the form in (3) with $f(t, x, y) = y$ and $g(t, x, y) = -x$. The table in Fig. 7.6.1 shows the results produced for $0 \leqq t \leqq 5$ (radians) using Program RK2DIM with step size $h = 0.05$. The values shown for $x = \sin t$ and $y = \cos t$ are all accurate to five decimal places. ■

Example 3 In Example 4 of Section 2.3 we considered a lunar lander that initially is falling freely toward the surface of the moon. Its retrorockets, when fired, provide a deceleration of $T = 4\ \text{m/s}^2$. We found that a soft touchdown at the lunar surface is achieved by igniting these retrorockets when the lander is at a height of 41,870 meters (just over 26 miles) above the surface and is then descending at the rate of 450 m/s.

Now we want to compute the *descent time* of the lunar lander. Let the distance $x(t)$ of the lander from the center of the moon be measured in meters and measure time t in seconds. According to the analysis in Section 2.3 (where we used $r(t)$ instead of $x(t)$), $x(t)$ satisfies the initial value problem

$$\frac{d^2x}{dt^2} = T - \frac{GM}{x^2} = 4 - \frac{4.9044 \times 10^{12}}{x^2},$$
$$x(0) = R + 41870 = 1{,}781{,}870, \quad x'(0) = -450, \tag{19}$$

where $G \approx 6.6726 \times 10^{-11}$ N·(m/kg)^2 is the universal gravitational constant and $M = 7.35 \times 10^{22}$ kg and $R = 1.74 \times 10^6$ m are the mass and radius of the moon. We seek the value of t when $x(t) = R = 1{,}740{,}000$.

The problem in (19) is equivalent to the first-order system

$$\frac{dx}{dt} = y, \quad x(0) = 1{,}781{,}870;$$
$$\frac{dy}{dt} = 4 - \frac{4.9044 \times 10^{12}}{x^2}, \quad y(0) = -450. \tag{20}$$

The table in Fig. 7.6.2 shows the result of a Runge–Kutta approximation with step size $h = 1$ (the indicated data agreeing with those obtained with step size $h = 2$). Evidently, touchdown on the lunar surface ($x = 1{,}740{,}000$) occurs at some time between $t = 180$ and $t = 190$ seconds. The table in Fig. 7.6.3 shows a second Runge–Kutta approximation with $t(0) = 180$, $x(0) = 1{,}740{,}059$, $y(0) = -16.83$, and $h = 0.1$. Now it is apparent that the lander's time of descent to the lunar surface is very close to 187 seconds; that is, 3 min 7 s. (The final velocity terms in these two tables are positive because the lander would begin to ascend if its retrorockets were not turned off at touchdown.) ■

t (s)	x (m)	v (m/s)
0	1,781,870	−450.00
20	1,773,360	−401.04
40	1,765,826	−352.37
60	1,759,264	−303.95
80	1,753,667	−255.74
100	1,749,033	−207.73
120	1,745,357	−159.86
140	1,742,637	−112.11
160	1,740,872	−64.45
180	1,740,059	−16.83
200	1,740,199	30.77

FIGURE 7.6.2. The lander's descent to the lunar surface.

t (s)	x (m)	v (m/s)
180	1,740,059	−16.83
181	1,740,044	−14.45
182	1,740,030	−12.07
183	1,740,019	−9.69
184	1,740,011	−7.31
185	1,740,005	−4.93
186	1,740,001	−2.55
187	1,740,000	−0.17
188	1,740,001	2.21
189	1,740,004	4.59
190	1,740,010	6.97

FIGURE 7.6.3. Focusing on the lunar lander's soft touchdown.

Higher-Order Systems

As we saw in Section 7.1, any system of higher-order differential equations can be replaced with an equivalent system of first-order differential equations. For example, consider the system

$$\begin{aligned} x'' &= F(t, x, y, x', y'), \\ y'' &= G(t, x, y, x', y') \end{aligned} \tag{21}$$

of second-order equations. If we substitute

$$x = x_1, \quad y = x_2, \quad x' = x_3 = x_1', \quad y' = x_4 = x_2',$$

then we get the equivalent system

$$\begin{aligned} x_1' &= x_3, \\ x_2' &= x_4, \\ x_3' &= F(t, x_1, x_2, x_3, x_4), \\ x_4' &= G(t, x_1, x_2, x_3, x_4) \end{aligned} \tag{22}$$

of four first-order equations in the unknown functions $x_1(t) = x(t)$, $x_2(t) = y(t)$, $x_3(t)$, and $x_4(t)$. It would be a routine (if slightly tedious) matter to write a four-dimensional version of program RK2DIM for the purpose of solving such a system. But in a programming language that accommodates vectors, an n-dimensional Runge–Kutta program is scarcely more complicated than a one-dimensional program. For instance, the application material for this section lists the n-dimensional MATLAB program **rkn** that closely resembles the one-dimensional program **rk** of Fig. 2.6.11.

Example 4 Suppose that a batted ball starts at $x_0 = 0$, $y_0 = 0$ with initial velocity $v_0 = 160$ ft/s and with initial angle of inclination $\theta = 30°$. If air resistance is ignored, we find by the elementary methods of Section 1.2 that the baseball travels a [horizontal] distance of $400\sqrt{3}$ ft (approximately 693 ft) in 5 s before striking the ground. Now suppose that in addition to a downward gravitational acceleration ($g = 32$ ft/s^2), the baseball experiences an acceleration due to air resistance of $(0.0025)v^2$ feet per second per second, directed opposite to its instantaneous direction of motion. Determine how far the baseball will travel horizontally under these conditions.

Solution According to Problem 28 of Section 7.1, the equations of motion of the baseball are

$$\frac{d^2x}{dt^2} = -cv\frac{dx}{dt}, \quad \frac{d^2y}{dt^2} = -cv\frac{dy}{dt} - g, \tag{23}$$

where $v = \sqrt{(x')^2 + (y')^2}$ is the speed of the ball, and where $c = 0.0025$ and $g = 32$ in fps units. We convert to a first-order system as in (22) and thereby obtain the system

$$\begin{aligned} x_1' &= x_3, \\ x_2' &= x_4, \\ x_3' &= -cx_3\sqrt{x_3^2 + x_4^2}, \\ x_4' &= -cx_4\sqrt{x_3^2 + x_4^2} - g \end{aligned} \tag{24}$$

of four first-order differential equations with

$$\begin{aligned} x_1(0) &= x_2(0) = 0, \\ x_3(0) &= 80\sqrt{3}, \quad x_4(0) = 80. \end{aligned} \tag{25}$$

Note that $x_3(t)$ and $x_4(t)$ are simply the x- and y-components of the baseball's velocity vector, so $v = \sqrt{x_3^2 + x_4^2}$. We proceed to apply the Runge–Kutta method to investigate the motion of the batted baseball described by the initial value problem in (24) and (25), first taking $c = 0$ to ignore air resistance and then using $c = 0.0025$ to take air resistance into account.

WITHOUT AIR RESISTANCE: Figure 7.6.4 shows the numerical results obtained when a Runge–Kutta program such as **rkn** is applied with step size $h = 0.1$ and with $c = 0$ (no air resistance). For convenience in interpreting the results, the printed output at each selected step consists of the horizontal and vertical coordinates x and y of the baseball, its velocity v, and the angle of inclination α of its velocity vector (in degrees measured from the horizontal). These results agree with the exact solution when $c = 0$. The ball travels a horizontal distance of $400\sqrt{3} \approx 692.82$ ft in exactly 5 s, having reached a maximum height of 100 ft after 2.5 s. Note also that the ball strikes the ground at the same angle and with the same speed as its initial angle and speed.

t	x	y	v	α
0.0	0.00	0.00	160.00	+30
0.5	69.28	36.00	152.63	+25
1.0	138.56	64.00	146.64	+19
1.5	207.85	84.00	142.21	+13
2.0	277.13	96.00	139.48	+7
2.5	346.41	100.00	138.56	+0
3.0	415.69	96.00	139.48	−7
3.5	484.97	84.00	142.21	−13
4.0	554.26	64.00	146.64	−19
4.5	623.54	36.00	152.63	−25
5.0	692.82	0.00	160.00	−30

FIGURE 7.6.4. The batted baseball with no air resistance ($c = 0$).

t	x	y	v	α
0.0	0.00	0.00	160.00	+30
0.5	63.25	32.74	127.18	+24
1.0	117.11	53.20	104.86	+17
1.5	164.32	63.60	89.72	+8
2.0	206.48	65.30	80.17	−3
2.5	244.61	59.22	75.22	−15
3.0	279.29	46.05	73.99	−27
3.5	310.91	26.41	75.47	−37
4.0	339.67	0.91	78.66	−46

FIGURE 7.6.5. The batted baseball with air resistance ($c = 0.0025$).

WITH AIR RESISTANCE: Figure 7.6.5 shows the results obtained with the fairly realistic value of $c = 0.0025$ for the air resistance for a batted baseball. To within a hundredth of a foot in either direction, the same results are obtained with step sizes $h = 0.05$ and $h = 0.025$. We now see that with air resistance the ball travels a distance well under 400 ft in just over 4 s. The more refined data in Fig. 7.6.6 show that the ball travels horizontally only about 340 ft and that its maximum height is only about 66 ft. As illustrated in Fig. 7.6.7, air resistance has converted a massive home run into a routine fly ball (if hit straightaway to center field). Note also that when the ball strikes the ground, it has slightly under *half* its initial speed (only about 79 ft/s) and is falling at a steeper angle (about 46°). Every baseball fan has observed empirically these aspects of the trajectory of a fly ball.

t	x	y	v	α	
1.5	164.32	63.60	89.72	+8	
1.6	173.11	64.60	87.40	+5	
1.7	181.72	65.26	85.29	+3	
1.8	190.15	65.60	83.39	+1	← Apex
1.9	198.40	65.61	81.68	−1	
2.0	206.48	65.30	80.17	−3	
⋮	⋮	⋮	⋮	⋮	
3.8	328.50	11.77	77.24	−42	
3.9	334.14	6.45	77.93	−44	
4.0	339.67	0.91	78.66	−46	← Impact
4.1	345.10	−4.84	79.43	−47	
4.2	350.41	−10.79	80.22	−49	

FIGURE 7.6.6. The batted ball's apex and its impact with the ground.

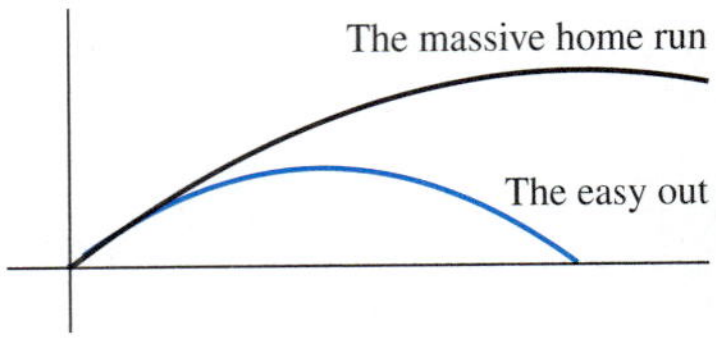

FIGURE 7.6.7. An "easy out" or a home run?

Variable Step Size Methods

The Runge–Kutta method for a large system requires an appreciable amount of computational labor, even when a computer is employed. Therefore, just as the step size h should not be so large that the resulting error in the solution is unacceptable, h ought not to be so small that too many steps are needed, hence requiring an unacceptable amount of computation. Thus the practical numerical solution of differential equations involves a tradeoff between accuracy and efficiency.

To facilitate this tradeoff, modern *variable step size methods* vary the step size h as the solution process proceeds. Large steps are taken in regions where the dependent variables are changing slowly; smaller steps are taken when these variables are changing rapidly, in order to prevent large errors.

An *adaptable* or *variable step size* Runge–Kutta method employs both a preassigned minimum error tolerance *MinTol* and a maximum error tolerance *MaxTol* to attempt to ensure that the error made in the typical step from $\mathbf{x}_n$ to $\mathbf{x}_{n+1}$ is neither too large (and hence inaccurate) nor too small (and hence inefficient). A fairly simple scheme for doing this may be outlined as follows:

- Having reached $\mathbf{x}_n$ with a Runge–Kutta step of length $t_n - t_{n-1} = h$, let $\mathbf{x}^{(1)}$ denote the result of a further Runge–Kutta step of length h and let $\mathbf{x}^{(2)}$ denote the result of *two* successive Runge–Kutta steps each of length $h/2$.
- On the grounds that $\mathbf{x}^{(2)}$ should be a more accurate approximation to $\mathbf{x}(t_n + h)$ than is $\mathbf{x}^{(1)}$, take
$$Err = |\mathbf{x}^{(1)} - \mathbf{x}^{(2)}|$$
as an estimate of the error in $\mathbf{x}^{(1)}$.
- If $MinTol \leqq Err \leqq MaxTol$, then let $\mathbf{x}_{n+1} = \mathbf{x}^{(1)}$, $t_{n+1} = t_n + h$, and proceed to the next step.
- If $Err < MinTol$, then the error is too small! Hence let $\mathbf{x}_{n+1} = \mathbf{x}^{(1)}$, $t_{n+1} = t_n + h$, but *double* the step size to $2h$ before making the next step.
- If $Err > MaxTol$, then the error is too large. Hence reject $\mathbf{x}^{(1)}$ and start afresh at $\mathbf{x}_n$ with the *halved* step size $h/2$.

The detailed implementation of such a scheme can be complicated. For a much more complete but readable discussion of adaptive Runge–Kutta methods, see Section 15.2 of William H. Press et al., *Numerical Recipes*: *The Art of Scientific Computing* (New York: Cambridge University Press, 1986).

Several widely available scientific computing packages (such as *Maple*, *Mathematica*, and MATLAB) include sophisticated variable step size programs that will accommodate an essentially arbitrary number of simultaneous differential equations. Such a general-purpose program might be used, for example, to model numerically the major components of the solar system: the sun and the nine (known) major planets. If m_i denotes the mass and $\mathbf{r}_i = (x_i, y_i, z_i)$ denotes the position vector of the ith one of these 10 bodies, then—by Newton's laws—the equation of motion of m_i is

$$m_i\mathbf{r}_i'' = \sum_{j\neq i} \frac{Gm_im_j}{(r_{ij})^3}(\mathbf{r}_j - \mathbf{r}_i), \tag{26}$$

where $r_{ij} = |\mathbf{r}_j - \mathbf{r}_i|$ denotes the distance between m_i and m_j. For each $i = 1, 2, \dots, 10$, the summation in Eq. (26) is over all values of $j \neq i$ from 1 to 10. The 10 vector equations in (26) constitute a system of 30 second-order scalar equations, and the equivalent first-order system consists of 60 differential equations in the coordinates and velocity components of the 10 major bodies in the solar system. Mathematical models that involve this many (or more) differential equations—and that require sophisticated software and hardware for their numerical analysis—are quite common in science, engineering, and applied technology.

Earth–Moon Satellite Orbits

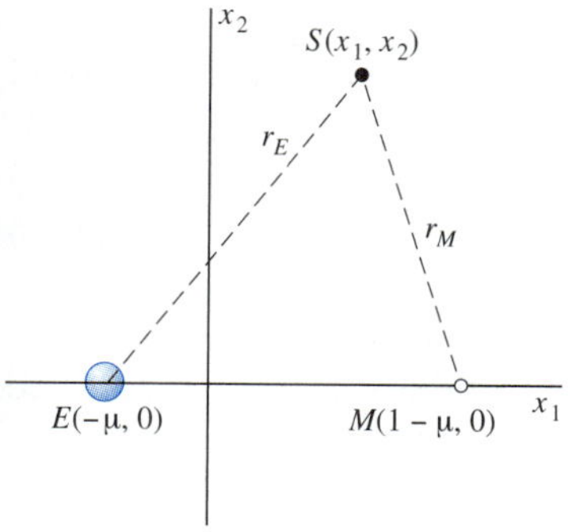

FIGURE 7.6.8. The Earth–Moon center-of-mass coordinate system.

For an example of a program whose efficient solution requires adaptive step size methods, we consider an Apollo satellite in orbit about the Earth E and Moon M. Figure 7.6.8 shows an x_1x_2-coordinate system whose origin lies at the center of mass of the Earth and the Moon and which rotates at the rate of one revolution per "moon month" of approximately $\tau = 27.32$ days, so the Earth and Moon remain fixed in their positions on the x_1-axis. If we take as unit distance the distance (about 384,000 kilometers, assumed constant) between the Earth and Moon centers, then their coordinates are $E(-\mu, 0)$ and $M(1-\mu, 0)$, where $\mu = m_M/(m_E + m_M)$ in terms of the Earth mass m_E and Moon mass m_M. If we take the total mass $m_E + m_M$ as the unit of mass and $\tau/(2\pi) \approx 4.35$ days as the unit of time, then the gravitational constant is $G = 1$ in Eq. (26), and the equations of motion of the satellite position $S(x_1, x_2)$ are

$$\begin{aligned} x_1'' &= x_1 + 2x_2' - \frac{(1-\mu)(x_1+\mu)}{(r_E)^3} - \frac{\mu(x_1 - 1 + \mu)}{(r_M)^3}, \\ x_2'' &= x_2 - 2x_1' - \frac{(1-\mu)x_2}{(r_E)^3} - \frac{\mu x_2}{(r_M)^3}, \end{aligned} \tag{27}$$

where r_E and r_M denote the satellite's distance to the Earth and Moon (indicated in Fig. 7.6.8). The initial two terms on the right-hand side of each equation result from the rotation of the coordinate system. In the system of units described here, the lunar mass is approximately $m_M = 0.012277471$. The second-order system in (27) can be converted to an equivalent first-order system (of four differential equations) by substituting

$$x_1' = x_3, \quad x_2' = x_4, \qquad \text{so that} \quad x_1'' = x_3', \quad x_2'' = x_4'.$$

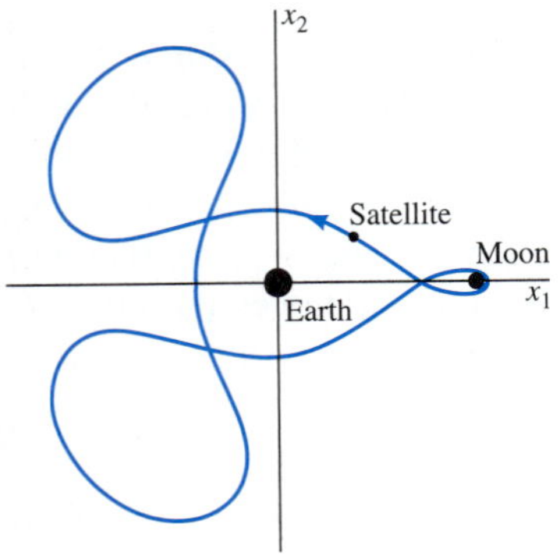

FIGURE 7.6.9. Apollo Moon–Earth bus orbit with insertion velocity $v_0 = 7476$ km/h.

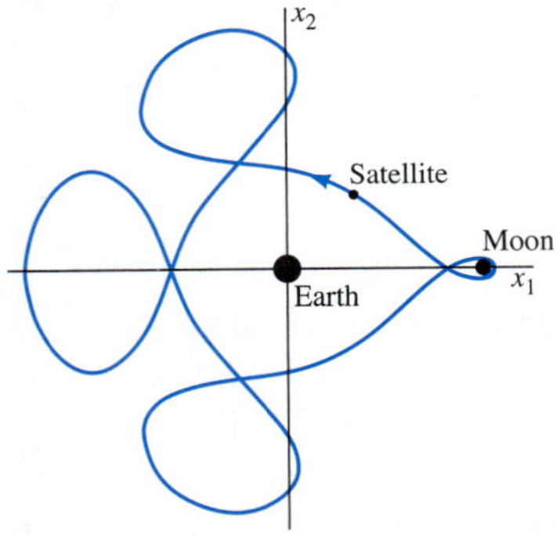

FIGURE 7.6.10. Apollo Moon–Earth bus orbit with insertion velocity $v_0 = 7365$ km/h.

Suppose that the satellite initially is in a clockwise circular orbit of radius about 2400 kilometers about the Moon. At its farthest point from the Earth ($x_1 = 0.994$) it is "launched" into Earth–Moon orbit with initial velocity v_0. The corresponding initial conditions are

$$x_1(0) = 0.994, \quad x_2(0) = 0, \quad x_3(0) = 0, \quad x_4(0) = -v_0.$$

An adaptive step size method (**`ode45`**) in the MATLAB software system was used to solve numerically the system in (27). The orbits in Figs. 7.6.9 and 7.6.10 were obtained with

$$v_0 = 2.031732629557 \quad \text{and} \quad v_0 = 2.001585106379,$$

respectively. [In the system of units used here, the unit of velocity is approximately 3680 km/h.] In each case a closed but multilooped periodic trajectory about the Earth and the Moon—a so-called bus orbit—is obtained, but a relatively small change in the initial velocity changes the number of loops! For more information, see NASA Contractor Report CR-61139, "Study of the Methods for the Numerical Solution of Ordinary Differential Equations," prepared by O. B. Francis, Jr. et al. for the NASA–George C. Marshall Space Flight Center, June 7, 1966.

So-called Moon–Earth "bus orbits" are periodic—that is, are closed trajectories traversed repeatedly by the satellite—only in a rotating x_1x_2-coordinate system as discussed above. The satellite of Fig. 7.6.9 traverses its closed orbit and returns to rendezvous with the Moon about 48.4 days after its insertion into orbit. Figures 7.6.11 and 7.6.12 illustrate the motion of the same satellite—but in an ordinary nonrotating xy-coordinate system centered at the Earth, in which the Moon encircles the Earth counterclockwise in a near-circular orbit, completing one revolution in about 27.3 days. The Moon starts at point S, and after 48.4 days it has completed a bit over 1.75 revolutions about the Earth and reaches the point R at which its rendezvous with the satellite occurs. Figure 7.6.11 shows the positions of Moon and satellite a day and a half after the satellite's insertion into its orbit, each traveling in a generally counterclockwise direction around the Earth. Figure 7.6.12 shows their positions a day and a half before their rendezvous at point R, the satellite meanwhile having encircled the Earth about 2.5 times in an orbit that (in the indicated xy-coordinate system) appears to resemble a slowly varying ellipse.

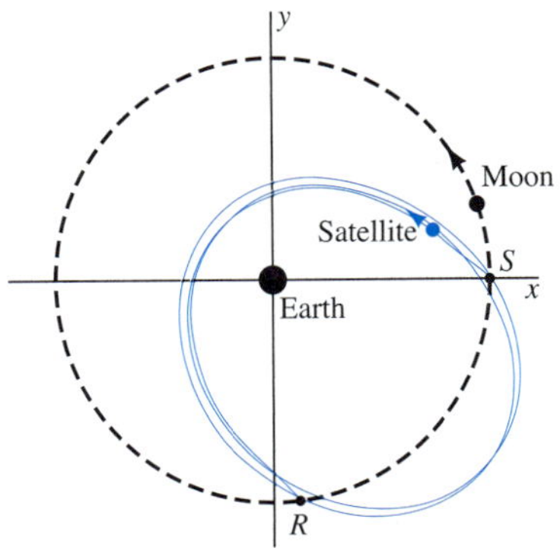

FIGURE 7.6.11. The moon and satellite in a nonrotating coordinate system, 1.5 days after orbital insertion of the satellite at starting point S.

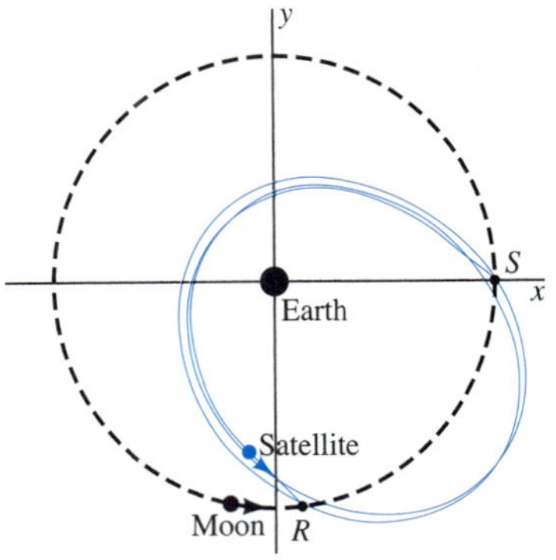

FIGURE 7.6.12. The moon and satellite in a nonrotating coordinate system, 1.5 days before their rendezvous at point R.

7.6 Problems

A hand-held calculator will suffice for Problems 1 through 8. In each problem an initial value problem and its exact solution are given. Approximate the values of $x(0.2)$ and $y(0.2)$ in three ways: **(a)** *by the Euler method with two steps of size $h = 0.1$;* **(b)** *by the improved Euler method with a single step of size $h = 0.2$; and* **(c)** *by the Runge–Kutta method with a single step of size $h = 0.2$. Compare the approximate values with the actual values $x(0.2)$ and $y(0.2)$.*

1. $x' = x + 2y$, $x(0) = 0$,
$y' = 2x + y$, $y(0) = 2$;
$x(t) = e^{3t} - e^{-t}$, $y(t) = e^{3t} + e^{-t}$

2. $x' = 2x + 3y$, $x(0) = 1$,
$y' = 2x + y$, $y(0) = -1$;
$x(t) = e^{-t}$, $y(t) = -e^{-t}$

3. $x' = 3x + 4y$, $x(0) = 1$,
$y' = 3x + 2y$, $y(0) = 1$;
$x(t) = \frac{1}{7}(8e^{6t} - e^{-t})$, $y(t) = \frac{1}{7}(6e^{6t} + e^{-t})$

4. $x' = 9x + 5y$, $x(0) = 1$,
$y' = -6x - 2y$, $y(0) = 0$;
$x(t) = -5e^{3t} + 6e^{4t}$, $y(t) = 6e^{3t} - 6e^{4t}$

5. $x' = 2x - 5y$, $x(0) = 2$,
$y' = 4x - 2y$, $y(0) = 3$;
$x(t) = 2\cos 4t - \frac{11}{4}\sin 4t$, $y(t) = 3\cos 4t + \frac{1}{2}\sin 4t$

6. $x' = x - 2y$, $x(0) = 0$,
$y' = 2x + y$, $y(0) = 4$;
$x(t) = -4e^{t}\sin 2t$, $y(t) = 4e^{t}\cos 2t$

7. $x' = 3x - y$, $x(0) = 2$,
$y' = x + y$, $y(0) = 1$;
$x(t) = (t + 2)e^{2t}$, $y(t) = (t + 1)e^{2t}$

8. $x' = 5x - 9y$, $x(0) = 0$,
$y' = 2x - y$, $y(0) = -1$;
$x(t) = 3e^{2t}\sin 3t$, $y(t) = e^{2t}(\sin 3t - \cos 3t)$

A computer will be required for the remaining problems in this section. In Problems 9 through 12, an initial value problem and its exact solution are given. In each of these four problems, use the Runge–Kutta method with step sizes $h = 0.1$ and $h = 0.05$ to approximate to five decimal places the values $x(1)$ and $y(1)$. Compare the approximations with the actual values.

9. $x' = 2x - y$, $\quad x(0) = 1$,
$y' = x + 2y$, $\quad y(0) = 0$;
$x(t) = e^{2t}\cos t$, $\quad y(t) = e^{2t}\sin t$

10. $x' = x + 2y$, $\quad x(0) = 0$,
$y' = x + e^{-t}$, $\quad y(0) = 0$;
$x(t) = \frac{1}{9}(2e^{2t} - 2e^{-t} + 6te^{-t})$,
$y(t) = \frac{1}{9}(e^{2t} - e^{-t} + 6te^{-t})$

11. $x' = -x + y - (1 + t^3)e^{-t}$, $\quad x(0) = 0$,
$y' = -x - y - (t - 3t^2)e^{-t}$, $\quad y(0) = 1$;
$x(t) = e^{-t}(\sin t - t)$, $\quad y(t) = e^{-t}(\cos t + t^3)$

12. $x'' + x = \sin t$, $\quad x(0) = 0$;
$x(t) = \frac{1}{2}(\sin t - t\cos t)$

13. Suppose that a crossbow bolt is shot straight upward with initial velocity 288 ft/s. If its deceleration due to air resistance is $(0.04)v$, then its height $x(t)$ satisfies the initial value problem

$$x'' = -32 - (0.04)x'; \quad x(0) = 0, \quad x'(0) = 288.$$

Find the maximum height that the bolt attains and the time required for it to reach this height.

14. Repeat Problem 13, but assume instead that the deceleration of the bolt due to air resistance is $(0.0002)v^2$.

15. Suppose that a projectile is fired straight upward with initial velocity v_0 from the surface of the earth. If air resistance is not a factor, then its height $x(t)$ at time t satisfies the initial value problem

$$\frac{d^2x}{dt^2} = -\frac{gR^2}{(x + R)^2}; \quad x(0) = 0, \quad x'(0) = v_0.$$

Use the values $g = 32.15$ ft/s$^2 \approx 0.006089$ mi/s^2 for the gravitational acceleration of the earth at its surface and $R = 3960$ mi as the radius of the earth. If $v_0 = 1$ mi/s, find the maximum height attained by the projectile and its time of ascent to this height.

Problems 16 through 18 deal with the batted baseball of Example 4, having initial velocity 160 ft/s and air resistance coefficient $c = 0.0025$.

16. Find the *range*—the horizontal distance the ball travels before it hits the ground—and its total time of flight with initial inclination angles 40°, 45°, and 50°.

17. Find (to the nearest degree) the initial inclination that maximizes the range. If there were no air resistance it would be exactly 45°, but your answer should be less than 45°.

18. Find (to the nearest half degree) the initial inclination angle greater than 45° for which the range is 300 ft.

19. Find the initial velocity of a baseball hit by Babe Ruth (with $c = 0.0025$ and initial inclination 40°) if it hit the bleachers at a point 50 ft high and 500 horizontal feet from home plate.

20. Consider the crossbow bolt of Problem 14, fired with the same initial velocity of 288 ft/s and with the air resistance deceleration $(0.0002)v^2$ directed opposite its direction of motion. Suppose that this bolt is fired from ground level at an initial angle of 45°. Find how high vertically and how far horizontally it goes, and how long it remains in the air.

21. Suppose that an artillery projectile is fired from ground level with initial velocity 3000 ft/s and initial inclination angle 40°. Assume that its air resistance deceleration is $(0.0001)v^2$. **(a)** What is the range of the projectile and what is its total time of flight? What is its speed at impact with the ground? **(b)** What is the maximum altitude of the projectile, and when is that altitude attained? **(c)** You will find that the projectile is still losing speed at the apex of its trajectory. What is the *minimum* speed that it attains during its descent?

7.6 Application Comets and Spacecraft

Figure 7.6.13 lists TI-85 and BASIC versions of the two-dimensional Runge–Kutta program RK2DIM. You should note that it closely parallels the one-dimensional Runge–Kutta program listed in Fig. 2.6.11, with a single line there replaced (where appropriate) with two lines here to calculate a *pair* of x- and y-values or slopes. Note also that the notation used is essentially that of Eqs. (13) and (14) in this section. The first several lines define the functions and initial data needed for Example 1.

Figure 7.6.14 exhibits an n-dimensional MATLAB implementation of the Runge–Kutta method. The MATLAB function **`f`** defines the vector of right-hand sides of the differential equations in the system $\mathbf{x}' = \mathbf{f}(t, \mathbf{x})$ to be solved. The **`rkn`** function then takes as input the initial t-value **`t`**, the column vector **`x`** of initial x-values, the final t-value **`t1`**, and the desired number **`n`** of subintervals. As output it produces the resulting column vector **`T`** of t-values and the matrix **`X`** whose rows give the corresponding x-values. For instance, with **f** as indicated in the figure, the MATLAB command

```
[T,X] = rkn(0, [0;1], 5, 50)
```

then generates the data shown in the table of Fig. 7.6.1 (which lists only every fifth value of each variable).

You can use Examples 1 through 3 in this section to test your own implementation of the Runge–Kutta method. Then investigate the comet and spacecraft problems described next. Additional application material at the Web site **`www.prenhall.com/edwards`** describes additional numerical ODE investigations ranging from batted baseballs to the Apollo orbits shown in Figs. 7.6.9 and 7.6.10.

Your Spacecraft Landing

Your spacecraft is traveling at constant velocity V, approaching a distant earthlike planet with mass M and radius R. When activated, your deceleration system provides a constant thrust T until impact with the surface of the planet. During the period of deceleration, your distance $x(t)$ from the center of the planet satisfies the differential equation

$$\frac{d^2x}{dt^2} = T - \frac{GM}{x^2}, \tag{1}$$

where $G \approx 6.6726 \times 10^{-11}$ N·(m/kg)^2 as in Example 3. Your question is this: At what altitude above the surface should your deceleration system be activated in order to achieve a soft touchdown? For a reasonable problem, you can take

$$M = 5.97 \times 10^{24} \text{ (kg)},$$
$$R = 6.38 \times 10^{6} \text{ (m)},$$
$$V = p \times 10^{4} \text{ (km/h)},$$
$$T = g + q \text{ (m/s}^2\text{)},$$

where $g = GM/R^2$ is the surface gravitational acceleration of the planet. Choose p to be the smallest nonzero digit and q the next-to-smallest nonzero digit in your ID number. Find the "ignition altitude" accurate to the nearest meter and the resulting "descent time" accurate to the nearest tenth of a second.

TI-85	BASIC	Comment
PROGRAM:RK2DIM	Program RK2DIM	Program title
:F=Y	DEF FN F(T,X,Y) = Y	Define function f
:G=-X	DEF FN G(T,X,Y) = -X	Define function g
:50→N	N = 50	No. of steps
:0→T	T = 0	Initial t
:0→X	X = 0	Initial x
:1→Y	Y = 1	Initial y
:5→T1	T1 = 5	Final t
:(T1-T)/N→H	H = (T1-T)/N	Step size
:For(I,1,N)	FOR I=1 TO N	Begin loop
:T→T0	T0 = T	Save previous t
:X→X0	X0 = X	Save previous x
:Y→Y0	Y0 = Y	Save previous y
:F→F1	F1 = FNF(T,X,Y)	First f-slope
:G→G1	G1 = FNG(T,X,Y)	First g-slope
:T0+H/2→T	T = T0 + H/2	Midpoint t
:X0+H*F1/2→X	X = X0 + H*F1/2	Midpt x-predictor
:Y0+H*G1/2→Y	Y = Y0 + H*G1/2	Midpt y-predictor
:F→F2	F2 = FNF(T,X,Y)	Second f-slope
:G→G2	G2 = FNG(T,X,Y)	Second g-slope
:X0+H*F2/2→X	X = X0 + H*F2/2	Midpt x-predictor
:Y0+H*G2/2→Y	Y = Y0 + H*G2/2	Midpt y-predictor
:F→F3	F3 = FNF(T,X,Y)	Third f-slope
:G→G3	G3 = FNG(T,X,Y)	Third g-slope
:T0+H→T	T = T0 + H	New t
:X0+H*F3→X	X = X0 + H*F3	Endpt x-predictor
:Y0+H*G3→Y	Y = Y0 + H*G3	Endpt y-predictor
:F→F4	F4 = FNF(T,X,Y)	Fourth f-slope
:G→G4	G4 = FNG(T,X,Y)	Fourth g-slope
:(F1+2*F2+2*F3 +F4)/6→FA	FA = (F1+2*F2+2*F3 +F4)/6	Average f-slope
:(G1+2*G2+2*G3 +G4)/6→GA	GA = (G1+2*G2+2*G3 +G4)/6	Average g-slope
:X0+H*FA→X	X = Y0 + H*FA	x-corrector
:Y0+H*GA→Y	Y = Y0 + H*GA	y-corrector
:Disp T,X,Y	PRINT T,X,Y	Display results
:End	NEXT I	End loop

FIGURE 7.6.13. TI-85 and BASIC two-dimensional Runge–Kutta programs.

```
function xp = f(t,x)
xp = x;
xp(1) = x(2);
xp(2) = -x(1);

function [T,Y] = rkn(t,x,t1,n)
h = (t1 - t)/n;                     % step size
T = t;                              % initial t
X = x';                             % initial x-vector
for i = 1:n                         % begin loop
    k1 = f(t,x);                    % first k-vector
    k2 = f(t+h/2,x+h*k1/2);         % second k-vector
    k3 = f(t+h/2,x+h*k2/2);         % third k-vector
    k4 = f(t+h    ,x+h*k3    );     % fourth k-vector
    k  = (k1+2*k2+2*k3+k4)/6;       % average k-vector
    t  = t + h;                     % new t
    x  = x + h*k;                   % new x
    T = [T;t];                      % update t-column
    X = [X;x'];                     % update x-matrix
    end                             % end loop
```

FIGURE 7.6.14. MATLAB implementation of the Runge–Kutta method.

Kepler's Law of Planetary (or Satellite) Motion

Consider a satellite in elliptical orbit around a planet of mass M, and suppose that physical units are so chosen that $GM = 1$ (where G is the gravitational constant). If the planet is located at the origin in the xy-plane, then the equations of motion of the satellite are

$$\frac{d^2x}{dt^2} = -\frac{x}{(x^2+y^2)^{3/2}}, \quad \frac{d^2y}{dt^2} = -\frac{y}{(x^2+y^2)^{3/2}}. \tag{2}$$

Let T denote the period of revolution of the satellite. Kepler's third law says that the *square* of T is proportional to the *cube* of the major semiaxis a of its elliptical orbit. In particular, if $GM = 1$, then

$$T^2 = 4\pi^2 a^3. \tag{3}$$

(For details, see Section 11.6 of Edwards and Penney, *Calculus: Early Transcendentals*, 7th ed. (Upper Saddle River, NJ: Prentice Hall, 2008).) If the satellite's x- and y-components of velocity, $x_3 = x' = x_1'$ and $x_4 = y' = x_2'$, are introduced, then the system in (2) translates into a system of four first-order differential equations having the form of those in Eq. (22) of this section.

(a) Solve this 4×4 system numerically with the initial conditions

$$x(0) = 1, \quad y(0) = 0, \quad x'(0) = 0, \quad y'(0) = 1$$

that correspond theoretically to a circular orbit of radius $a = 1$, so Eq. (3) gives $T = 2\pi$. Is this what you get?

(b) Now solve the system numerically with the initial conditions

$$x(0) = 1, \quad y(0) = 0, \quad x'(0) = 0, \quad y'(0) = \tfrac{1}{2}\sqrt{6}$$

that correspond theoretically to an elliptical orbit with major semiaxis $a = 2$, so Eq. (3) gives $T = 4\pi\sqrt{2}$. Is this what you get?

Halley's Comet

Halley's comet last reached perihelion (its point of closest approach to the sun at the origin) on February 9, 1986. Its position and velocity components at that time were

$$\mathbf{p}_0 = (0.325514, -0.459460, 0.166229) \quad \text{and}$$
$$\mathbf{v}_0 = (-9.096111, -6.916686, -1.305721)$$

(respectively), with position in AU (astronomical units, in which the unit of distance is the major semiaxis of the earth's orbit) and time in years. In this system, the three-dimensional equations of motion of the comet are

$$\frac{d^2x}{dt^2} = -\frac{\mu x}{r^3}, \quad \frac{d^2y}{dt^2} = -\frac{\mu y}{r^3}, \quad \frac{d^2z}{dt^2} = -\frac{\mu z}{r^3}, \tag{4}$$

where

$$\mu = 4\pi^2 \quad \text{and} \quad r = \sqrt{x^2 + y^2 + z^2}.$$

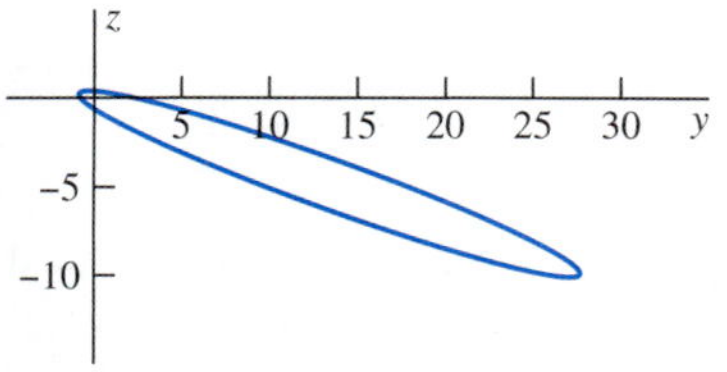

FIGURE 7.6.15. yz-projection of the orbit of Halley's comet.

Solve the equations in (4) numerically to verify the appearance of the yz-projection of the orbit of Halley's comet shown in Fig. 7.6.15. Plot the xy- and xz-projections as well.

Figure 7.6.16 shows the graph of the distance $r(t)$ of Halley's comet from the sun. Inspection of this graph indicates that Halley's comet reaches a maximum distance (at aphelion) of about 35 AU in a bit less than 40 years and returns to perihelion after about three-quarters of a century. The closer look in Fig. 7.6.17 indicates that the period of revolution of Halley's comet is about 76 years. Use your numerical solution to refine these observations. What is your best estimate of the calendar date of the comet's next perihelion passage?

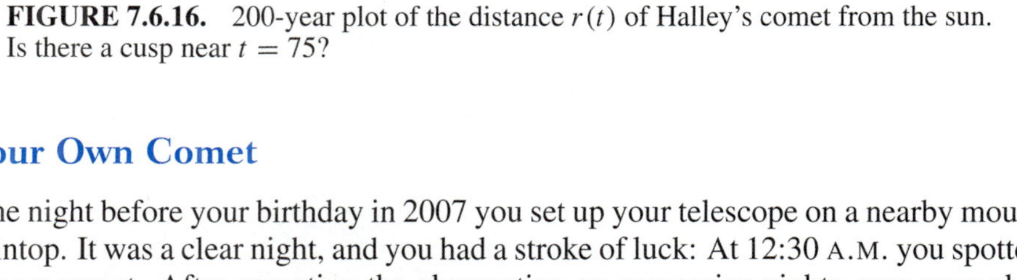

FIGURE 7.6.16. 200-year plot of the distance $r(t)$ of Halley's comet from the sun. Is there a cusp near $t = 75$?

Your Own Comet

The night before your birthday in 2007 you set up your telescope on a nearby mountaintop. It was a clear night, and you had a stroke of luck: At 12:30 A.M. you spotted a new comet. After repeating the observation on successive nights, you were able to calculate its solar system coordinates $\mathbf{p}_0 = (x_0, y_0, z_0)$ and its velocity vector $\mathbf{v}_0 = (v_{x0}, v_{y0}, v_{z0})$ on that first night. Using this information, determine the following:

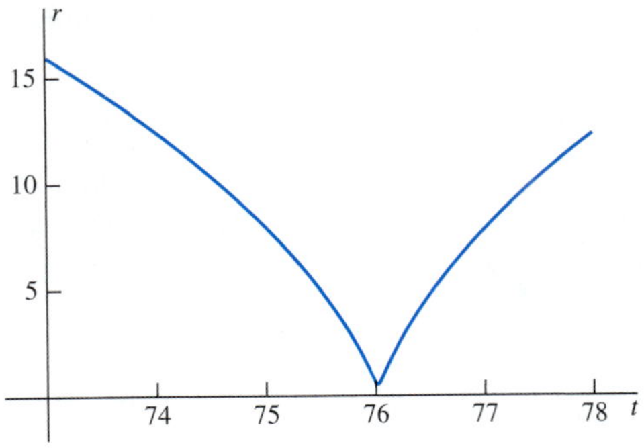

FIGURE 7.6.17. A closer look at Halley's perihelion passage after about 76 years.

- the comet's perihelion (point nearest the sun) and aphelion (point farthest from the sun),
- the comet's velocity at perihelion and at aphelion,
- the comet's period of revolution around the sun, and
- the comet's next two dates of perihelion passage.

Using units of length in AU and time in earth years, the equations of motion of your comet are given in (4). For your personal comet, begin with random initial position and velocity vectors with the same order of magnitude as those of Halley's comet. Repeat the random selection of initial position and velocity vectors, if necessary, until you get a plausible eccentric orbit that ranges well outside the earth's orbit (as most real comets do).

REFERENCES FOR FURTHER STUDY

The literature of the theory and applications of differential equations is vast. The following list includes a selection of books that might be useful to readers who wish to pursue further the topics introduced in this book.

1. ABRAMOWITZ, M. and I. A. STEGUN, *Handbook of Mathematical Functions*. New York: Dover, 1965. The comprehensive collection of tables to which frequent reference is made in the text.

2. BIRKHOFF, G. and G.-C. ROTA, *Ordinary Differential Equations* (4th ed.). New York: John Wiley, 1989. An intermediate-level text that includes a more complete treatment of existence and uniqueness theorems, Sturm-Liouville problems, and eigenfunction expansions.

3. BRAUN, M., *Differential Equations and Their Applications* (3rd ed.). New York: Springer-Verlag, 1983. An introductory text at a slightly higher level than this book; it includes several interesting "case study" applications.

4. CHURCHILL, R. V., *Operational Mathematics* (3rd ed.). New York: McGraw-Hill, 1972. The standard reference for theory and applications of Laplace transforms, starting at about the same level as Chapter 10 of this book.

5. CODDINGTON, E. A., *An Introduction to Ordinary Differential Equations*. Englewood Cliffs, N.J.: Prentice Hall, 1961. An intermediate-level introduction; Chapters 3 and 4 include proofs of the theorems on power series and Frobenius series solutions stated in Chapter 11 of this book.

6. CODDINGTON, E. A. and N. LEVINSON, *Theory of Ordinary Differential Equations*. New York: McGraw-Hill, 1955. An advanced theoretical text; Chapter 5 discusses solutions near an irregular singular point.

7. DORMAND, J. R., *Numerical Methods for Differential Equations*. Boca Raton: CRC Press, 1996. More complete coverage of modern computational methods for approximate solution of differential equations.

8. HUBBARD, J. H. and B. H. WEST, *Differential Equations: A Dynamical Systems Approach*. New York: Springer-Verlag, 1992 (part I) and 1995 (Higher-Dimensional Systems). Detailed treatment of qualitative phenomena, with a balanced combination of computational and theoretical viewpoints.

9. INCE, E. L., *Ordinary Differential Equations*. New York: Dover, 1956. First published in 1926, this is the classic older reference work on the subject.

10. MCLACHLAN, N. W., *Ordinary Non-Linear Differential Equations in Engineering and Physical Sciences*. London: Oxford University Press, 1956. A concrete introduction to the effects of nonlinearity in physical systems.

11. NOBLE, B. and J.W. DANIEL, *Applied Linear Algebra* (3rd ed.). Upper Saddle River, N.J.: Prentice Hall, 1988. An especially concrete introduction to linear algebra, with significant applications included.

12. POLKING, J. C. and D. ARNOLD, *Ordinary Differential Equations Using MATLAB* (3rd ed.). Upper Saddle River, N.J.: Prentice Hall, 2004. A manual for using MATLAB in an elementary differential equations course; based on the MATLAB programs **dfield** and **pplane** that are used and referenced in this text.

13. PRESS, W. H., B. P. FLANNERY, S. A. TEUKOLSKY, and W. T. VETTERLING, *Numerical Recipes: The Art of Scientific Computing*. Cambridge: Cambridge University Press, 1986. Chapter 15 discusses modern techniques for the numerical solution of differential equations. This edition include FORTRAN programs; Pascal and C editions have also been published.

14. SIMMONS, G. F., *Differential Equations* (2nd ed.). New York: McGraw-Hill, 1991. An introductory text with interesting historical notes and fascinating applications and with the most eloquent preface in any mathematics book currently in print.

15. STRANG, W. G., *Linear Algebra and Its Applications*. (4th ed.). New York: Brooks Cole, 2006. An introductory treatment of linear algebra with motivating applications. Appendix B contains a concrete derivation of the Jordan normal form for square matrices.

APPENDIX A

Existence and Uniqueness of Solutions

In Chapter 1 we saw that an initial value problem of the form

$$\frac{dy}{dx} = f(x, y), \quad y(a) = b \tag{1}$$

can fail (on a given interval containing the point $x = a$) to have a unique solution. For instance, in Example 4 of Section 1.3, we saw that the initial value problem

$$x^2\frac{dy}{dx} + y^2 = 0, \quad y(0) = b \tag{2}$$

has no solutions at all unless $b = 0$, in which case there are infinitely many solutions. According to Problem 31 of Section 1.3, the initial value problem

$$\frac{dy}{dx} = -\sqrt{1 - y^2}, \quad y(0) = 1 \tag{3}$$

has the two distinct solutions $y_1(x) \equiv 1$ and $y_2(x) = \cos x$ on the interval $0 \leqq x \leqq \pi$. In this appendix we investigate conditions on the function $f(x, y)$ that suffice to guarantee that the initial value problem in (1) has one and only one solution, and we then proceed to establish appropriate versions of the existence-uniqueness theorems that were stated without proof in Sections 1.3, 5.1, 5.2, and 7.1.

A.1 Existence of Solutions

The approach we employ is the **method of successive approximations**, which was developed by the French mathematician Emile Picard (1856–1941). This method is based on the fact that the function $y(x)$ satisfies the initial value problem in (1) on the open interval I containing $x = a$ if and only if it satisfies the integral equation

$$y(x) = b + \int_a^x f(t, y(t))\,dt \tag{4}$$

for all x in I. In particular, if $y(x)$ satisfies Eq. (4), then clearly $y(a) = b$, and differentiation of both sides in (4)—using the fundamental theorem of calculus—yields the differential equation $y'(x) = f(x, y(x))$.

To attempt to solve Eq. (4), we begin with the initial function

$$y_0(x) \equiv b \tag{5}$$

and then define iteratively a sequence $y_1, y_2, y_3, \ldots$ of functions that we hope will converge to the solution. Specifically, we let

$$y_1(x) = b + \int_a^x f(t, y_0(t))\,dt \quad \text{and} \quad y_2(x) = b + \int_a^x f(t, y_1(t))\,dt. \tag{6}$$

In general, y_{n+1} is obtained by substitution of y_n for y in the right-hand side in Eq. (4):

$$y_{n+1}(x) = b + \int_a^x f(t, y_n(t))\,dt. \tag{7}$$

Suppose we know that each of these functions $\{y_n(x)\}_0^\infty$ is defined on some open interval (the same for each n) containing $x = a$ and that the limit

$$y(x) = \lim_{n\to\infty} y_n(x) \tag{8}$$

exists at each point of this interval. Then it will follow that

$$\begin{aligned} y(x) = \lim_{n\to\infty} y_{n+1}(x) &= \lim_{n\to\infty}\left[b + \int_a^x f(t, y_n(t))\,dt\right] \\ &= b + \lim_{n\to\infty}\int_a^x f(t, y_n(t))\,dt & (9) \\ &= b + \int_a^x f\left(t, \lim_{n\to\infty} y_n(t)\right)dt & (10) \end{aligned}$$

and, hence, that

$$y(x) = b + \int_a^x f(t, y(t))\,dt,$$

provided that we can validate the interchange of limit operations involved in passing from (9) to (10). It is therefore reasonable to expect that, under favorable conditions, the sequence $\{y_n(x)\}$ defined iteratively in Eqs. (5) and (7) will converge to a solution $y(x)$ of the integral equation in (4), and hence to a solution of the original initial value problem in (1).

Example 1 To apply the method of successive approximations to the initial value problem

$$\frac{dy}{dx} = y, \quad y(0) = 1, \tag{11}$$

we write Eqs. (5) and (7), thereby obtaining

$$y_0(x) \equiv 1, \quad y_{n+1}(x) = 1 + \int_0^x y_n(t)\,dt. \tag{12}$$

The iteration formula in (12) yields

$$y_1(x) = 1 + \int_0^x 1\,dt = 1 + x,$$

$$y_2(x) = 1 + \int_0^x (1 + t)\,dt = 1 + x + \tfrac{1}{2}x^2,$$

$$y_3(x) = 1 + \int_0^x \left(1 + t + \tfrac{1}{2}t^2\right)dt = 1 + x + \tfrac{1}{2}x^2 + \tfrac{1}{6}x^3,$$

and

$$y_4(x) = 1 + \int_0^x \left(1 + t + \tfrac{1}{2}t^2 + \tfrac{1}{6}t^3\right) dt$$
$$= 1 + x + \tfrac{1}{2}x^2 + \tfrac{1}{6}x^3 + \tfrac{1}{24}x^4.$$

It is clear that we are generating the sequence of partial sums of a power series solution; indeed, we immediately recognize the series as that of $y(x) = e^x$. There is no difficulty in demonstrating that the exponential function is indeed the solution of the initial value problem in (11); moreover, a diligent student can verify (by using a proof by induction on n) that $y_n(x)$, obtained in the aforementioned manner, is indeed the nth partial sum for the Taylor series with center zero for $y(x) = e^x$. ■

Example 2 To apply the method of successive approximations to the initial value problem

$$\frac{dy}{dx} = 4xy, \qquad y(0) = 3, \tag{13}$$

we write Eqs. (5) and (7) as in Example 1. Now we obtain

$$y_0(x) \equiv 3, \quad y_{n+1}(x) = 3 + \int_0^x 4t y_n(t)\,dt. \tag{14}$$

The iteration formula in (14) yields

$$y_1(x) = 3 + \int_0^x (4t)(3)\,dt = 3 + 6x^2,$$

$$y_2(x) = 3 + \int_0^x (4t)(3 + 6t^2)\,dt = 3 + 6x^2 + 6x^4,$$

$$y_3(x) = 3 + \int_0^x (4t)(3 + 6t^2 + 6t^4)\,dt = 3 + 6x^2 + 6x^4 + 4x^6,$$

and

$$y_4(x) = 3 + \int_0^x (4t)(3 + 6t^2 + 6t^4 + 4t^6)\,dt$$
$$= 3 + 6x^2 + 6x^4 + 4x^6 + 2x^8.$$

It is again clear that we are generating partial sums of a power series solution. It is not quite so obvious what function has such a power series representation, but the initial value problem in (13) is readily solved by separation of variables:

$$y(x) = 3\exp\left(2x^2\right) = 3\sum_{n=0}^{\infty} \frac{(2x^2)^n}{n!}$$
$$= 3 + 6x^2 + 6x^4 + 4x^6 + 2x^8 + \tfrac{4}{5}x^{10} + \cdots.$$ ■

In some cases, it may be necessary to compute a much larger number of terms, either in order to identify the solution or to use a partial sum of its series with large subscript to approximate the solution accurately for x near its initial value. Fortunately, computer algebra systems such as *Maple* and *Mathematica* can perform the symbolic integrations (as opposed to numerical integrations) of the sort in Examples 1 and 2. If necessary, you could generate the first hundred terms in Example 2 in a matter of minutes.

In general, of course, we apply Picard's method because we cannot find a solution by elementary methods. Suppose that we have produced a large number of terms of what we believe to be the correct power series expansion of the solution. We *must* have conditions under which the sequence $\{y_n(x)\}$ provided by the method of successive approximations is guaranteed in advance to converge to a solution. It is just as convenient to discuss the initial value problem

$$\frac{d\mathbf{x}}{dt} = \mathbf{f}(\mathbf{x}, t), \quad \mathbf{x}(a) = \mathbf{b} \tag{15}$$

for a system of m first-order equations, where

$$\mathbf{x} = \begin{bmatrix} x_1 \\ x_2 \\ x_3 \\ \vdots \\ x_m \end{bmatrix}, \quad \mathbf{f} = \begin{bmatrix} f_1 \\ f_2 \\ f_3 \\ \vdots \\ f_m \end{bmatrix}, \quad \text{and} \quad \mathbf{b} = \begin{bmatrix} b_1 \\ b_2 \\ b_3 \\ \vdots \\ b_m \end{bmatrix}.$$

It turns out that with the aid of this vector notation (which we introduced in Section 7.1), most results concerning a single [scalar] equation $x' = f(x, t)$ can be generalized readily to analogous results for a system of m first-order equations, as abbreviated in (15). Consequently, the effort of using vector notation is amply justified by the generality it provides.

The method of successive approximations for the system in (15) calls for us to compute the sequence $\{\mathbf{x}_n(t)\}_0^\infty$ of vector-valued functions of t,

$$\mathbf{x}_n(t) = \begin{bmatrix} x_{1n}(t) \\ x_{2n}(t) \\ x_{3n}(t) \\ \vdots \\ x_{mn}(t) \end{bmatrix},$$

defined iteratively by

$$\mathbf{x}_0(a) \equiv \mathbf{b}, \quad \mathbf{x}_{n+1}(t) = \mathbf{b} + \int_a^t \mathbf{f}(\mathbf{x}_n(s), s)\, ds. \tag{16}$$

Recall that vector-valued functions are integrated componentwise.

Example 3 Consider the m-dimensional initial value problem

$$\frac{d\mathbf{x}}{dt} = \mathbf{A}\mathbf{x}, \quad \mathbf{x}(0) = \mathbf{b} \tag{17}$$

for a homogeneous linear system with $m \times m$ constant coefficient matrix $\mathbf{A}$. The equations in (16) take the form

$$\mathbf{x}_0(t) = \mathbf{b}, \quad \mathbf{x}_{n+1} = \mathbf{b} + \int_0^x \mathbf{A}\mathbf{x}_n(s)\, ds. \tag{18}$$

Thus

$$\mathbf{x}_1(t) = \mathbf{b} + \int_0^t \mathbf{Ab}\, ds = \mathbf{b} + \mathbf{Ab}t = (\mathbf{I} + \mathbf{A}t)\mathbf{b},$$

$$\mathbf{x}_2(t) = \mathbf{b} + \int_0^t \mathbf{A}(\mathbf{b} + \mathbf{Ab}s)\, ds = \mathbf{b} + \mathbf{Ab}t + \tfrac{1}{2}\mathbf{A}^2\mathbf{b}t^2 = (\mathbf{I} + \mathbf{A}t + \tfrac{1}{2}\mathbf{A}^2t^2)\mathbf{b},$$

and

$$\mathbf{x}_3(t) = \mathbf{b} + \int_0^t \mathbf{A}(\mathbf{b} + \mathbf{Ab}s + \tfrac{1}{2}\mathbf{A}^2\mathbf{b}s^2)\, ds = (\mathbf{I} + \mathbf{A}t + \tfrac{1}{2}\mathbf{A}^2t^2 + \tfrac{1}{6}\mathbf{A}^3t^3)\mathbf{b}.$$

We have therefore obtained the first several partial sums of the exponential series solution

$$\mathbf{x}(t) = e^{\mathbf{A}t}\mathbf{b} = \left(\sum_{n=0}^{\infty} \frac{(\mathbf{A}t)^n}{n!}\right)\mathbf{b} \tag{19}$$

of (17), which was derived earlier in Section 8.1. ■

The key to establishing convergence in the method of successive approximations is an appropriate condition on the rate at which $\mathbf{f}(\mathbf{x}, t)$ changes when $\mathbf{x}$ varies but t is held fixed. If R is a region in $(m+1)$-dimensional $(\mathbf{x}, t)$-space, then the function $\mathbf{f}(\mathbf{x}, t)$ is said to be **Lipschitz continuous** on R if there exists a constant $k > 0$ such that

$$|\mathbf{f}(\mathbf{x}_1, t) - \mathbf{f}(\mathbf{x}_2, t)| \leqq k|\mathbf{x}_1 - \mathbf{x}_2| \tag{20}$$

if $(\mathbf{x}_1, t)$ and $(\mathbf{x}_2, t)$ are points of R. Recall that the **norm** of an m-dimensional point or vector $\mathbf{x}$ is defined to be

$$|\mathbf{x}| = \sqrt{x_1^2 + x_2^2 + x_3^2 + \cdots + x_m^2}. \tag{21}$$

Then $|\mathbf{x}_1 - \mathbf{x}_2|$ is simply the Euclidean distance between the points $\mathbf{x}_1$ and $\mathbf{x}_2$.

Example 4 Let $f(x, t) = x^2 \exp(-t^2) \sin t$ and let R be the strip $0 \leqq x \leqq 2$ in the xy-plane. If (x_1, t) and (x_2, t) are both points of R, then

$$|f(x_1, t) - f(x_2, t)| = |\exp(-t^2) \sin t| \cdot |x_1 + x_2| \cdot |x_1 - x_2| \leqq 4|x_1 - x_2|,$$

because $\left|\exp(-t^2) \sin t\right| \leqq 1$ for all t and $|x_1 + x_2| \leqq 4$ if x_1 and x_2 are both in the interval $[0, 2]$. Thus f satisfies the Lipschitz condition in (20) with $k = 4$ and is therefore Lipschitz continuous in the strip R. ■

Example 5 Let $f(x, t) = t\sqrt{x}$ on the rectangle R consisting of the points (x, t) in the xt-plane for which $0 \leqq x \leqq 1$ and $0 \leqq t \leqq 1$. Then, taking $x_1 = x$, $x_2 = 0$, and $t = 1$, we find that

$$|f(x, 1) - f(0, 1)| = \sqrt{x} = \frac{1}{\sqrt{x}}|x - 0|.$$

Because $x^{-1/2} \to +\infty$ as $x \to 0^+$, we see that the Lipschitz condition in (20) cannot be satisfied by any (finite) constant $k > 0$. Thus the function f, though obviously continuous on R, is *not* Lipschitz continuous on R. ■

Suppose, however, that the function $f(x, t)$ has a continuous partial derivative $f_x(x, t)$ on the closed rectangle R in the xt-plane, and denote by k the maximum value of $|f_x(x, t)|$ on R. Then the mean-value theorem of differential calculus yields

$$|f(x_1, t) - f(x_2, t)| = |f_x(\overline{x}, t) \cdot (x_1 - x_2)|$$

for some $\overline{x}$ in (x_1, x_2), so it follows that

$$|f(x_1, t) - f(x_2, t)| \leqq k|x_1 - x_2|$$

because $|f_x(\overline{x}, t)| \leqq k$. Thus a continuously differentiable function $f(x, t)$ defined on a closed rectangle *is* Lipschitz continuous there. More generally, the multivariable mean value theorem of advanced calculus can be used similarly to prove that a *vector-valued function* $\mathbf{f}(\mathbf{x}, t)$ *with continuously differentiable component functions on a closed rectangular region* R *in* $(\mathbf{x}, t)$*-space is Lipschitz continuous on* R.

Example 6 The function $f(x, t) = x^2$ is Lipschitz continuous on any closed [bounded] region in the xt-plane. But consider this function on the infinite strip R consisting of the points (x, t) for which $0 \leqq t \leqq 1$ and x is arbitrary. Then

$$|f(x_1, t) - f(x_2, t)| = |x_1^2 - x_2^2| = |x_1 + x_2| \cdot |x_1 - x_2|.$$

Because $|x_1 + x_2|$ can be made arbitrarily large, it follows that f is *not* Lipschitz continuous on the infinite strip R. ■

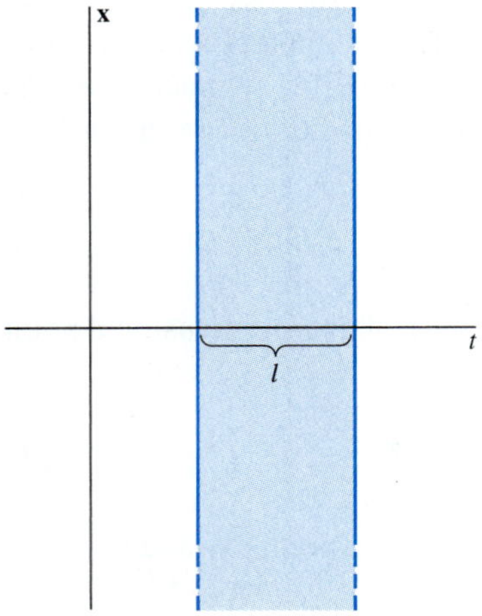

FIGURE A.1. An infinite slab in $(m + 1)$-space.

If I is an interval on the t-axis, then the set of all points $(\mathbf{x}, t)$ with t in I is an infinite strip or slab in $(m + 1)$-space (as indicated in Fig. A.1). Example 6 shows that Lipschitz continuity of $\mathbf{f}(\mathbf{x}, t)$ on such an infinite slab is a very strong condition. Nevertheless, the existence of a solution of the initial value problem

$$\frac{d\mathbf{x}}{dt} = \mathbf{f}(\mathbf{x}, t), \quad \mathbf{x}(a) = \mathbf{b} \tag{15}$$

under the hypothesis of Lipschitz continuity of $\mathbf{f}$ in such a slab is of considerable importance.

THEOREM 1 Global Existence of Solutions

Let $\mathbf{f}$ be a vector-valued function (with m components) of $m + 1$ real variables, and let I be a (bounded or unbounded) open interval containing $t = a$. If $\mathbf{f}(\mathbf{x}, t)$ is continuous and satisfies the Lipschitz condition in (20) for all t in I and for all $\mathbf{x}_1$ and $\mathbf{x}_2$, then the initial value problem in (15) has a solution on the entire interval I.

***Proof*:** We want to show that the sequence $\{\mathbf{x}_n(t)\}_0^\infty$ of successive approximations determined iteratively by

$$\mathbf{x}_0(a) = \mathbf{b}, \quad \mathbf{x}_{n+1}(t) = \mathbf{b} + \int_0^t \mathbf{f}(\mathbf{x}_n(s), s)\, ds \tag{16}$$

converges to a solution $\mathbf{x}(t)$ of (15). We see that each of these functions in turn is continuous on I, as each is an (indefinite) integral of a continuous function.

We may assume that $a = 0$, because the transformation $t \to t + a$ converts (15) into an equivalent problem with initial point $t = 0$. Also, we will consider only the portion $t \geqq 0$ of the interval I; the details for the case $t \leqq 0$ are very similar.

The main part of the proof consists in showing that if $[0, T]$ is a closed (and bounded) interval contained in I, then the sequence $\{\mathbf{x}_n(t)\}$ converges *uniformly* on $[0, T]$ to a limit function $\mathbf{x}(t)$. This means that, given $\epsilon > 0$, there exists an integer N such that

$$|\mathbf{x}_n(t) - \mathbf{x}(t)| < \epsilon \tag{22}$$

for all $n \geqq N$ and all t in $[0, T]$. For ordinary (perhaps nonuniform) convergence the integer N, for which (22) holds for all $n \geqq N$, might depend on t, with no single value of N working for all t in I. Once this uniform convergence of the sequence $\{\mathbf{x}_n(t)\}$ has been established, the following conclusions will follow from standard theorems of advanced calculus [see pages 620–622 of A. E. Taylor and W. R. Mann, *Advanced Calculus*, 3rd ed. (New York: John Wiley, 1983)].

1. The limit function $\mathbf{x}(t)$ is continuous on $[0, T]$.
2. If N is so chosen that the inequality in (22) holds for $n \geqq N$, then the Lipschitz continuity of $\mathbf{f}$ implies that

$$|\mathbf{f}(\mathbf{x}_n(t), t) - \mathbf{f}(\mathbf{x}(t), t)| \leqq k|\mathbf{x}_n(t) - \mathbf{x}(t)| < k\epsilon$$

for all t in $[0, T]$ and $n \geqq N$, so it follows that the sequence $\{\mathbf{f}(\mathbf{x}_n(t), t)\}_0^\infty$ converges uniformly to $\mathbf{f}(\mathbf{x}(t), t)$ on $[0, T]$.
3. But a uniformly convergent sequence or series can be integrated termwise, so it follows that, on taking limits in the iterative formula in (16),

$$\begin{aligned}\mathbf{x}(t) &= \lim_{n\to\infty} \mathbf{x}_{n+1}(t) = \mathbf{b} + \lim_{n\to\infty} \int_0^t \mathbf{f}(\mathbf{x}_n(s), s)\, ds \\ &= \mathbf{b} + \int_0^t \lim_{n\to\infty} \mathbf{f}(\mathbf{x}_n(s), s)\, ds;\end{aligned}$$

thus

$$\mathbf{x}(t) = \mathbf{b} + \int_0^t \mathbf{f}(\mathbf{x}(s), s)\, ds. \tag{23}$$

4. Because the function $\mathbf{x}(t)$ is continuous on $[0, T]$, the integral equation in (23) [analogous to the 1-dimensional case in (4)] implies that $\mathbf{x}'(t) = \mathbf{f}(\mathbf{x}(t), t)$ on $[0, T]$. If this is true on every closed subinterval of the open interval I, however, then it is true on the entire interval I as well.

It therefore remains only to prove that the sequence $\{\mathbf{x}_n(t)\}_0^\infty$ converges uniformly on the closed interval $[0, T]$. Let M be the maximum value of $|\mathbf{f}(\mathbf{b}, t)|$ for t in $[0, T]$. Then

$$|\mathbf{x}_1(t) - \mathbf{x}_0(t)| = \left|\int_0^t \mathbf{f}(\mathbf{x}_0(s), s)\, ds\right| \leqq \int_0^t |\mathbf{f}(\mathbf{b}, s)|\, ds \leqq Mt. \tag{24}$$

Next,

$$|\mathbf{x}_2(t) - \mathbf{x}_1(t)| = \left|\int_0^t [\mathbf{f}(\mathbf{x}_1(s), s) - \mathbf{f}(\mathbf{x}_0(s), s)]\, ds\right| \leqq k\int_0^t |\mathbf{x}_1(s) - \mathbf{x}_0(s)|\, ds,$$

and hence

$$|\mathbf{x}_2(t) - \mathbf{x}_1(t)| \leqq k\int_0^t Ms\, ds = \tfrac{1}{2}kMt^2. \tag{25}$$

We now proceed by induction. Assume that

$$|\mathbf{x}_n(t) - \mathbf{x}_{n-1}(t)| \leqq \frac{M}{k} \cdot \frac{(kt)^n}{n!}. \tag{26}$$

It then follows that

$$\begin{aligned}|\mathbf{x}_{n+1}(t) - \mathbf{x}_n(t)| &= \left| \int_0^t \mathbf{f}(\mathbf{x}_n(s), s) - \mathbf{f}(\mathbf{x}_{n-1}(s), s)]\, ds \right| \\ &\leqq k \int_0^t |\mathbf{x}_n(s) - \mathbf{x}_{n-1}(s)|\, ds;\end{aligned}$$

consequently,

$$|\mathbf{x}_{n+1}(t) - \mathbf{x}_n(t)| \leqq k \int_0^t \frac{M}{k} \cdot \frac{(ks)^n}{n!}\, ds.$$

It follows, upon evaluation of this integral, that

$$|\mathbf{x}_{n+1}(t) - \mathbf{x}_n(t)| \leqq \frac{M}{k} \cdot \frac{(kt)^{n+1}}{(n+1)!}.$$

Thus (26) holds on $[0, T]$ for all $n \geqq 1$.

Hence the terms of the infinite series

$$\mathbf{x}_0(t) + \sum_{n=1}^{\infty} \left[\mathbf{x}_n(t) - \mathbf{x}_{n-1}(t)\right] \tag{27}$$

are dominated (in magnitude on the interval $[0, T]$) by the terms of the convergent series

$$\sum_{n=1}^{\infty} \frac{M}{k} \cdot \frac{(kT)^{n+1}}{(n+1)!} = \frac{M}{k}(e^{kT} - 1), \tag{28}$$

which is a series of positive constants. It therefore follows (from the Weierstrass M-test on pages 618–619 of Taylor and Mann) that the series in (27) converges uniformly on $[0, T]$. But the sequence of partial sums of this series is, however, simply our original sequence $\{\mathbf{x}_n(t)\}_0^{\infty}$ of successive approximations, so the proof of Theorem 1 is finally complete. ■

A.2 Linear Systems

An important application of the global existence theorem just given is to the initial value problem

$$\frac{d\mathbf{x}}{dt} = \mathbf{A}(t)\mathbf{x} + \mathbf{g}(t), \quad \mathbf{x}(a) = \mathbf{b} \tag{29}$$

for a linear system, where the $m \times m$ matrix-valued function $\mathbf{A}(t)$ and the vector-valued function $\mathbf{g}(t)$ are continuous on a (bounded or unbounded) open interval I containing the point $t = a$. In order to apply Theorem 1 to the linear system in (29), we note first that the proof of Theorem 1 requires only that, for each closed and bounded subinterval J of I, there exists a Lipschitz constant k such that

$$|\mathbf{f}(\mathbf{x}_1, t) - \mathbf{f}(\mathbf{x}_2, t)| \leqq k|\mathbf{x}_1 - \mathbf{x}_2| \tag{20}$$

for all t in J (and all $\mathbf{x}_1$ and $\mathbf{x}_2$). Thus we do not need a single Lipschitz constant for the entire open interval I.

In (29) we have $\mathbf{f}(\mathbf{x}, t) = \mathbf{A}(t)\mathbf{x} + \mathbf{g}$, so

$$\mathbf{f}(\mathbf{x}_1, t) - \mathbf{f}(\mathbf{x}_2, t) = \mathbf{A}(t)(\mathbf{x}_1 - \mathbf{x}_2). \tag{30}$$

It therefore suffices to show that, if $\mathbf{A}(t)$ is continuous on the closed and bounded interval J, then there is a constant k such that

$$|\mathbf{A}(t)\mathbf{x}| \leqq k|\mathbf{x}| \tag{31}$$

for all t in J. But this follows from the fact (Problem 17) that

$$|\mathbf{A}\mathbf{x}| \leqq \|\mathbf{A}\| \cdot |\mathbf{x}|, \tag{32}$$

where the **norm** $\|\mathbf{A}\|$ of the matrix $\mathbf{A}$ is defined to be

$$\|\mathbf{A}\| = \left(\sum_{i,j=1}^{m} (a_{ij})^2 \right)^{1/2}. \tag{33}$$

Because $\mathbf{A}(t)$ is continuous on the closed and bounded interval J, the norm $\|\mathbf{A}\|$ is bounded on J, so Eq. (31) follows, as desired. Thus we have the following global existence theorem for the linear initial value problem in (29).

THEOREM 1 Existence for Linear Systems

Let the $m \times m$ matrix-valued function $\mathbf{A}(t)$ and the vector-valued function $\mathbf{g}(t)$ be continuous on the (bounded or unbounded) open interval I containing the point $t = a$. Then the initial value problem

$$\frac{d\mathbf{x}}{dt} = \mathbf{A}(t)\mathbf{x} + \mathbf{g}(t), \quad \mathbf{x}(a) = \mathbf{b} \tag{29}$$

has a solution on the (entire) interval I.

As we saw in Section 4.1, the mth-order initial value problem

$$\begin{aligned} &x^{(m)} + a_1(t)x^{(m-1)} + \cdots + a_{m-1}(t)x' + a_m(t)x = p(t), \\ &x(a) = b_0, \quad x'(a) = b_1, \quad \ldots, \quad x^{(m-1)}(a) = b_{m-1} \end{aligned} \tag{34}$$

is readily transformed into an equivalent $m \times m$ system of the form in (29). It therefore follows from Theorem 2 that if the functions $a_1(t), a_2(t), \ldots, a_m(t)$ and $p(t)$ in (34) are all continuous on the (bounded or unbounded) open interval I containing $t = a$, then the initial value problem in (34) has a solution on the (entire) interval I.

A.3 Local Existence

In the case of a *nonlinear* initial value problem

$$\frac{d\mathbf{x}}{dt} = \mathbf{f}(\mathbf{x}, t), \quad \mathbf{x}(a) = \mathbf{b}, \tag{35}$$

the hypothesis in Theorem 1 that $\mathbf{f}$ satisfies a Lipschitz condition on a slab $(\mathbf{x}, t)$ (t in I, all $\mathbf{x}$) is unrealistic and rarely satisfied. This is illustrated by the following simple example.

Example 1 Consider the initial value problem

$$\frac{dy}{dx} = x^2, \quad x(0) = b > 0. \tag{36}$$

As we saw in Example 6, the equation $x' = x^2$ does not satisfy a "strip Lipschitz condition." When we solve (36) by separation of variables, we get

$$x(t) = \frac{b}{1 - bt}. \tag{37}$$

Because the denominator vanishes for $t = 1/b$, Eq. (37) provides a solution of the initial value problem in (36) only for $t < 1/b$, despite the fact that the differential equation $x' = x^2$ "looks nice" on the entire real line—certainly the function appearing on the right-hand side of the equation is continuous everywhere. In particular, if b is large, then we have a solution only on a very small interval to the right of $t = 0$. ■

Although Theorem 2 assures us that *linear* equations have global solutions, Example 7 shows that, in general, even a "nice" nonlinear differential equation can be expected to have a solution only on a small interval around the initial point $t = a$, and it also shows that the length of this interval of existence can depend on the initial value $\mathbf{x}(a) = \mathbf{b}$, as well as on the differential equation itself. The reason is this: If $\mathbf{f}(\mathbf{x}, t)$ is continuously differentiable in a neighborhood of the point $(\mathbf{b}, a)$ in $(m+1)$-dimensional space, then—as indicated in the discussion preceding Example 6—we can conclude that $\mathbf{f}(\mathbf{x}, t)$ satisfies a Lipschitz condition on some rectangular region R centered at $(\mathbf{b}, a)$, of the form

$$|t - a| < A, \quad |x_i - b_i| < B_i \tag{38}$$

$(i = 1, 2, \ldots, m)$. In the proof of Theorem 1, we need to apply the Lipschitz condition on the function $\mathbf{f}$ in analyzing the iterative formula

$$\mathbf{x}_{n+1}(t) = \mathbf{b} + \int_a^t \mathbf{f}(\mathbf{x}_n(s), s)\, ds. \tag{39}$$

The potential difficulty is that, unless the values of t are suitably restricted, the point $(\mathbf{x}_n(t), t)$ appearing in the integrand in (39) may not lie in the region R where $\mathbf{f}$ is known to satisfy a Lipschitz condition. On the other hand, it can be shown that—on a sufficiently small open interval J containing the point $t = a$—the graphs of the functions $\{\mathbf{x}_n(t)\}$ given iteratively by the formula in (39) remain within the region R, so the proof of convergence can then be carried out as in the proof of Theorem 1. A proof of the following *local* existence theorem can be found in Chapter 6 of G. Birkhoff and G.-C. Rota, *Ordinary Differential Equations*, 4th ed. (New York: John Wiley, 1989).

THEOREM 1 Local Existence of Solutions

Let $\mathbf{f}$ be a vector-valued function (with m components) of the $m+1$ real variables $x_1, x_2, \ldots, x_m$, and t. If the first-order partial derivatives of $\mathbf{f}$ all exist and are continuous in some neighborhood of the point $\mathbf{x} = \mathbf{b}$, $t = a$, then the initial value problem

$$\frac{d\mathbf{x}}{dt} = \mathbf{f}(\mathbf{x}, t), \quad \mathbf{x}(a) = \mathbf{b}, \tag{35}$$

has a solution on some open interval containing the point $t = a$.

A.4 Uniqueness of Solutions

It is possible to establish the existence of solutions of the initial value problem in (35) under the much weaker hypothesis that $\mathbf{f}(\mathbf{x}, t)$ is merely continuous; techniques other than those used in this section are required. By contrast, the Lipschitz condition that we used in proving Theorem 1 is the key to *uniqueness* of solutions. In particular, the solution provided by Theorem 3 is unique near the point $t = a$.

THEOREM 1 Uniqueness of Solutions

Suppose that on some region R in $(m+1)$-space, the function $\mathbf{f}$ in (35) is continuous and satisfies the Lipschitz condition

$$|\mathbf{f}(\mathbf{x}_1, t) - \mathbf{f}(\mathbf{x}_2, t)| \leqq k \cdot |\mathbf{x}_1 - \mathbf{x}_2|. \tag{20}$$

If $\mathbf{x}_1(t)$ and $\mathbf{x}_2(t)$ are two solutions of the initial problem in (35) on some open interval I containing $x = a$, such that the solution curves $(\mathbf{x}_1(t), t)$ and $(\mathbf{x}_2(t), t)$ both lie in R for all t in I, then $\mathbf{x}_1(t) = \mathbf{x}_2(t)$ for all t in I.

We will outline the proof of Theorem 4 for the 1-dimensional case in which x is a real variable. A generalization of this proof to the multivariable case can be found in Chapter 6 of Birkhoff and Rota.

Let us consider the function

$$\phi(t) = [x_1(t) - x_2(t)]^2 \tag{40}$$

for which $\phi(a) = 0$, because $x_1(a) = x_2(a) = b$. We want to show that $\phi(t) \equiv 0$, so that $x_1(t) \equiv x_2(t)$. We will consider only the case $t \geqq a$; the details are similar for the case $t \leqq a$.

If we differentiate each side in Eq. (40), we find that

$$\begin{aligned} |\phi'(t)| &= |2[x_1(t) - x_2(t)] \cdot [x_1'(t) - x_2'(t)]| \\ &= |2[x_1(t) - x_2(t)] \cdot [f(x_1(t), t) - f(x_2(t), t)]| \\ &\leqq 2k|x_1(t) - x_2(t)|^2 = 2k\phi(t), \end{aligned}$$

by using the Lipschitz condition on f. Hence

$$\phi'(t) \leqq 2k\phi(t). \tag{41}$$

Now let us temporarily ignore the fact that $\phi(a) = 0$ and compare $\phi(t)$ with the solution of the differential equation

$$\Phi'(t) = 2k\Phi(t) \tag{42}$$

such that $\Phi(a) = \phi(a)$; clearly

$$\Phi(t) = \Phi(a)e^{2k(t-a)}. \tag{43}$$

In comparing (41) with (42), it seems inevitable that

$$\phi(t) \leqq \Phi(t) \qquad \text{for } t \geqq a, \tag{44}$$

and this is easily proved (Problem 18). Hence

$$0 \leqq [x_1(t) - x_2(t)]^2 \leqq [x_1(a) - x_2(a)]^2 e^{2k(t-a)}.$$

On taking square roots, we get

$$0 \leqq |x_1(t) - x_2(t)| \leqq |x_1(a) - x_2(a)|e^{k(t-a)}. \tag{45}$$

But $x_1(a) - x_2(a) = 0$, so (45) implies that $x_1(t) \equiv x_2(t)$.

Example 1 The initial value problem

$$\frac{dx}{dt} = 3x^{2/3}, \quad x(0) = 0 \tag{46}$$

has both the obvious solution $x_1(t) \equiv 0$ and the solution $x_2(t) = t^3$ (which is readily found by separation of variables). Hence the function $f(x,t)$ must *fail* to satisfy a Lipschitz condition near $(0, 0)$. Indeed, the mean value theorem yields

$$|f(x,0) - f(0,0)| = |f_x(\overline{x}, 0)| \cdot |x - 0|$$

for some $\overline{x}$ between 0 and x. But $f_x(x, 0) = 2x^{-1/3}$ is unbounded as $x \to 0$, so no Lipschitz condition can be satisfied. ■

A.5 Well-Posed Problems and Mathematical Models

In addition to uniqueness, another consequence of the inequality in (45) is the fact that solutions of the differential equation

$$\frac{dx}{dt} = f(x,t) \tag{47}$$

depend *continuously* on the initial value $x(a)$; that is, if $x_1(t)$ and $x_2(t)$ are two solutions of (47) on the interval $a \leqq t \leqq T$ such that the initial values $x_1(a)$ and $x_2(a)$ are sufficiently close to one another, then the values of $x_1(t)$ and $x_2(t)$ remain close to one another. In particular, if $|x_1(a) - x_2(a)| \leqq \delta$, then (45) implies that

$$|x_1(t) - x_2(t)| \leqq \delta e^{k(T-a)} = \epsilon \tag{48}$$

for all t with $a \leqq t \leqq T$. Obviously, we can make ϵ as small as we wish by choosing δ sufficiently close to zero.

This continuity of solutions of (47) with respect to initial values is important in practical applications where we are unlikely to know the initial value $x_0 = x(a)$ with absolute precision. For example, suppose that the initial value problem

$$\frac{dx}{dt} = f(x,t), \quad x(a) = x_0 \tag{49}$$

models a population for which we know only that the initial population is within $\delta > 0$ of the assumed value x_0. Then even if the function $f(x,t)$ is accurate, the solution $x(t)$ of (49) will be only an approximation to the actual population. But (45) implies that the actual population at time t will be within $\delta e^{k(T-a)}$ of the approximate population $x(t)$. Thus, on a given closed interval $[a, T]$, $x(t)$ will be a close approximation to the actual population provided that $\delta > 0$ is sufficiently small.

An initial value problem is usually considered *well posed* as a mathematical model for a real-world situation only if the differential equation has unique solutions that are continuous with respect to initial values. Otherwise it is unlikely that the initial value problem adequately mirrors the real-world situation.

An even stronger "continuous dependence" of solutions is often desirable. In addition to possible inaccuracy in the initial value, the function $f(x, t)$ may not model precisely the physical situation. For instance, it may involve physical parameters (such as resistance coefficients) whose values cannot be measured with absolute precision. Birkhoff and Rota generalize the proof of Theorem 4 to establish the following result.

THEOREM 1 Continuous Dependence of Solutions

Let $\mathbf{x}(t)$ and $\mathbf{y}(t)$ be solutions of the equations

$$\frac{d\mathbf{x}}{dt} = \mathbf{f}(\mathbf{x}, t) \quad \text{and} \quad \frac{d\mathbf{y}}{dt} = \mathbf{g}(\mathbf{y}, t) \tag{50}$$

on the closed interval $[a, T]$. Let $\mathbf{f}$ and $\mathbf{g}$ be continuous for $a \leqq t \leqq T$ and for $\mathbf{x}$ and $\mathbf{y}$ in a common region D of n-space and assume that $\mathbf{f}$ satisfies the Lipschitz condition in (20) on the region D. If

$$|\mathbf{f}(\mathbf{z}, t) - \mathbf{g}(\mathbf{z}, t)| \leqq \mu \tag{51}$$

for all t in the interval $[a, T]$ and all $\mathbf{z}$ in D, it then follows that

$$|\mathbf{x}(t) - \mathbf{y}(t)| \leqq |\mathbf{x}(a) - \mathbf{y}(a)| \cdot e^{k(t-a)} + \frac{\mu}{k}\left[e^{k(t-a)} - 1\right] \tag{52}$$

on the interval $[a, T]$.

If $\mu > 0$ is small, then (51) implies that the functions $\mathbf{f}$ and $\mathbf{g}$ appearing in the two differential equations, though different, are "close" to each other. If $\epsilon > 0$ is given, then it is apparent from (52) that

$$|\mathbf{x}(t) - \mathbf{y}(t)| \leqq \epsilon \tag{53}$$

for all t in $[a, T]$ if both $|\mathbf{x}(a) - \mathbf{y}(a)|$ and μ are sufficiently small. Thus Theorem 5 says (roughly) that if both the two initial values and the two differential equations in (50) are close to each other, then the two solutions remain close to each other for $a \leqq t \leqq T$.

For example, suppose that a falling body is subject both to constant gravitational acceleration g and to resistance proportional to some power of its velocity, so (with the positive axis directed downward) its velocity v satisfies the differential equation

$$\frac{dv}{dt} = g - cv^{\rho}. \tag{54}$$

Assume, however, that only an approximation $\overline{c}$ to the actual resistance c and an approximation $\overline{\rho}$ to the actual exponent ρ are known. Then our mathematical model is based on the differential equation

$$\frac{du}{dt} = g - \overline{c}u^{\overline{\rho}} \tag{55}$$

instead of the actual equation in (54). Thus if we solve Eq. (55), we obtain only an approximation $u(t)$ to the actual velocity $v(t)$. But if the parameters $\overline{c}$ and $\overline{\rho}$ are sufficiently close to the actual values c and ρ, then the right-hand sides in (54) and (55) will be close to each other. If so, then Theorem 5 implies that the actual and approximate velocity functions $v(t)$ and $u(t)$ are close to each other. In this case, the approximation in (55) will be a good model of the actual physical situation.

Problems

In Problems 1 through 8, apply the successive approximation formula to compute $y_n(x)$ for $n \leqq 4$. Then write the exponential series for which these approximations are partial sums (perhaps with the first term or two missing; for example,

$$e^x - 1 = x + \tfrac{1}{2}x^2 + \tfrac{1}{6}x^3 + \tfrac{1}{24}x^4 + \cdots).$$

1. $\dfrac{dy}{dx} = y,\ y(0) = 3$

2. $\dfrac{dy}{dx} = -2y,\ y(0) = 4$

3. $\dfrac{dy}{dx} = -2xy,\ y(0) = 1$

4. $\dfrac{dy}{dx} = 3x^2y,\ y(0) = 2$

5. $\dfrac{dy}{dx} = 2y + 2,\ y(0) = 0$

6. $\dfrac{dy}{dx} = x + y,\ y(0) = 0$

7. $\dfrac{dy}{dx} = 2x(1 + y),\ y(0) = 0$

8. $\dfrac{dy}{dx} = 4x(y + 2x^2),\ y(0) = 0$

In Problems 9 through 12, compute the successive approximations $y_n(x)$ for $n \leqq 3$; then compare them with the appropriate partial sums of the Taylor series of the exact solution.

9. $\dfrac{dy}{dx} = x + y,\ y(0) = 1$

10. $\dfrac{dy}{dx} = y + e^x,\ y(0) = 0$

11. $\dfrac{dy}{dx} = y^2,\ y(0) = 1$

12. $\dfrac{dy}{dx} = \tfrac{1}{2}y^3,\ y(0) = 1$

13. Apply the iterative formula in (16) to compute the first three successive approximations to the solution of the initial value problem

$$\frac{dx}{dt} = 2x - y, \qquad x(0) = 1;$$

$$\frac{dy}{dt} = 3x - 2y, \qquad y(0) = -1.$$

14. Apply the matrix exponential series in (19) to solve (in closed form) the initial value problem

$$\mathbf{x}'(t) = \begin{bmatrix} 1 & 1 \\ 0 & 1 \end{bmatrix}\mathbf{x}, \quad \mathbf{x}(0) = \begin{bmatrix} 1 \\ 1 \end{bmatrix}.$$

(*Suggestion*: Show first that

$$\begin{bmatrix} 1 & 1 \\ 0 & 1 \end{bmatrix}^n = \begin{bmatrix} 1 & n \\ 0 & 1 \end{bmatrix}$$

for each positive integer n.)

15. For the initial value problem $dy/dx = 1 + y^3$, $y(1) = 1$, show that the second Picard approximation is

$$y_2(x) = 1 + 2(x-1) + 3(x-1)^2 + 4(x-1)^3 + 2(x-1)^4.$$

Then compute $y_2(1.1)$ and $y_2(1.2)$. The fourth-order Runge–Kutta method with step size $h = 0.005$ yields $y(1.1) \approx 1.2391$ and $y(1.2) \approx 1.6269$.

16. For the initial value problem $dy/dx = x^2 + y^2$, $y(0) = 0$, show that the third Picard approximation is

$$y_3(x) = \frac{1}{3}x^3 + \frac{1}{63}x^7 + \frac{2}{2079}x^{11} + \frac{1}{59535}x^{15}.$$

Compute $y_3(1)$. The fourth-order Runge–Kutta method yields $y(1) \approx 0.350232$, both with step size $h = 0.05$ and with step size $h = 0.025$.

17. Prove as follows the inequality $|\mathbf{Ax}| \leqq \|\mathbf{A}\| \cdot |\mathbf{x}|$, where $\mathbf{A}$ is an $m \times m$ matrix with row vectors $\mathbf{a}_1, \mathbf{a}_2, \dots, \mathbf{a}_m$, and $\mathbf{x}$ is an m-dimensional vector. First note that the components of the vector $\mathbf{Ax}$ are $\mathbf{a}_1 \cdot \mathbf{x}, \mathbf{a}_2 \cdot \mathbf{x}, \dots, \mathbf{a}_m \cdot \mathbf{x}$, so

$$|\mathbf{Ax}| = \left[\sum_{n=1}^{m} (\mathbf{a}_i \cdot \mathbf{x})^2\right]^{1/2}.$$

Then use the Cauchy–Schwarz inequality $(\mathbf{a} \cdot \mathbf{x})^2 \leqq |\mathbf{a}|^2|\mathbf{x}|^2$ for the dot product.

18. Suppose that $\phi(t)$ is a differentiable function with

$$\phi'(t) \leqq k\phi(t) \qquad (k > 0)$$

for $t \geqq a$. Multiply both sides by e^{-kt}, then transpose to show that

$$\frac{d}{dt}\left[\phi(t)e^{-kt}\right] \leqq 0$$

for $t \geqq a$. Then apply the mean value theorem to conclude that

$$\phi(t) \leqq \phi(a)e^{k(t-a)}$$

for $t \geqq a$.

APPENDIX B

Theory of Determinants

The proof of the theorem on row and column cofactor expansions (Theorem 1 in Section 3.6) involves an alternative interpretation of determinants that in more advanced treatments is usually taken as the original definition. Consider the following elementary scheme for evaluating a 3×3 determinant:

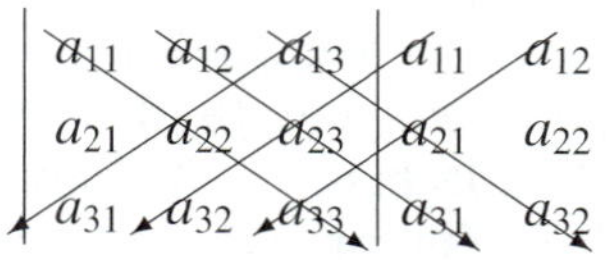

The first two columns of $\mathbf{A} = \left[a_{ij} \right]$ are duplicated to its right. It is readily verified that det $\mathbf{A}$ is equal to the sum of the three products along the indicated diagonals that point down and to the *right*, minus the sum of the products along the three diagonals that point down and to the *left*:

$$\begin{aligned} \det \mathbf{A} = &+ a_{11}a_{22}a_{33} + a_{12}a_{23}a_{31} + a_{13}a_{21}a_{32} \\ &- a_{13}a_{22}a_{31} - a_{11}a_{23}a_{32} - a_{12}a_{21}a_{33}. \end{aligned} \tag{1}$$

Remark: This scheme works **only** in dimensions 2 and 3. Neither the analogous scheme nor anything resembling it gives the correct value of an $n \times n$ determinant with $n \geq 4$.

Note that each of the six terms on the right in (1) is of the form $\pm a_{1i}a_{2j}a_{3k}$, where $(i \;\; j \;\; k)$ is a **permutation** of $(1 \;\; 2 \;\; 3)$. That is, the triple $(i \;\; j \;\; k)$ consists of the three distinct numbers 1, 2, and 3 written in some specific order. The six terms in (1) correspond to the six possible permutations of $(1 \;\; 2 \;\; 3)$:

$$\begin{matrix} (1 \;\; 2 \;\; 3) & (2 \;\; 1 \;\; 3) & (3 \;\; 1 \;\; 2) \\ (1 \;\; 3 \;\; 2) & (2 \;\; 3 \;\; 1) & (3 \;\; 2 \;\; 1). \end{matrix} \tag{2}$$

The $+$ and $-$ signs in (1) can also be explained in terms of these permutations. A **transposition** of an ordered sequence of objects (such as the numbers 1, 2, 3) is the operation of *interchanging* some single pair of them. For instance, the operation $(1 \;\; 2 \;\; 3) \to (1 \;\; 3 \;\; 2)$ is a transposition that consists of interchanging 2 and 3. But it requires two transpositions to change $(1 \;\; 2 \;\; 3)$ into the permutation $(3 \;\; 1 \;\; 2)$:

$$(1 \;\; 2 \;\; 3) \to (3 \;\; 2 \;\; 1) \to (3 \;\; 1 \;\; 2).$$

Given a permutation $P = (i \;\; j \;\; k)$ of $(1 \;\; 2 \;\; 3)$, we denote by $s(P)$ the *minimum* number of transpositions required to change $(1 \;\; 2 \;\; 3)$ into $(i \;\; j \;\; k)$. For the six permutations in (2) it is easy to verify that

$$\begin{matrix} s(1 \;\; 2 \;\; 3) = 0, & s(2 \;\; 1 \;\; 3) = 1, & s(3 \;\; 1 \;\; 2) = 2, \\ s(1 \;\; 3 \;\; 2) = 1, & s(2 \;\; 3 \;\; 1) = 2, & s(3 \;\; 2 \;\; 1) = 1. \end{matrix} \tag{3}$$

Now for the point to all this: If you check each of the six terms in (1) you will find that the *sign* of the term $a_{1i}a_{2j}a_{3k}$ is $(-1)^{s(P)}$, where $P = (i \;\; j \;\; k)$. Therefore, we can rewrite (1) more concisely as

$$\det \mathbf{A} = \sum_P (-1)^{s(P)} a_{1i}a_{2j}a_{3k}, \tag{4}$$

where there is one term on the right for each possible permutation $P = (i \;\; j \;\; k)$ of $(1 \;\; 2 \;\; 3)$.

The formula in (4) generalizes to determinants of higher order. If $\mathbf{A}$ is an $n \times n$ matrix, then

$$\det \mathbf{A} = \sum_P (-1)^{s(P)} a_{1i_1}a_{2i_2} \cdots a_{ni_n}, \tag{5}$$

where there is one term on the right for each possible permutation $P = (i_1 \;\; i_2 \;\; \cdots \;\; i_n)$ of $(1 \;\; 2 \;\; \cdots \;\; n)$, and $s(P)$ denotes the minimum number of transpositions required to change $(1 \;\; 2 \;\; \cdots \;\; n)$ into $(i_1 \;\; i_2 \;\; \cdots \;\; i_n)$. The formula in (5) can be established by induction on n. Assuming its validity for $(n-1) \times (n-1)$ determinants, the formula can be verified for an $n \times n$ matrix $\mathbf{A}$ by expanding along its first row. It is fairly easy to see that this gives terms of the form $\pm a_{1i_1}a_{2i_2} \cdots a_{ni_n}$, but somewhat more difficult to check the signs.

Finally, the proof of the cofactor expansion theorem consists of verifying similarly that the cofactor expansion of $\det \mathbf{A}$ along *any* row *or* column of $\mathbf{A}$ agrees with the formula in (5). The details are more lengthy than instructive, and therefore we omit them.

Determinants and Elementary Row Operations

As we described in Section 3.6, the determinant of any square matrix $\mathbf{A}$ can be evaluated directly by carrying out a cofactor expansion of $\det \mathbf{A}$ along any row or column of $\mathbf{A}$. Because this approach involves so much computational labor, it is more efficient instead to reduce $\mathbf{A}$ to an echelon matrix $\mathbf{R}$. Because any square echelon matrix is (upper) triangular, the determinant of the echelon matrix $\mathbf{R}$ is simply the product of its diagonal elements. But because we have altered the matrix $\mathbf{A}$ by transforming it into $\mathbf{R}$, the question is this: What effects do elementary row operations have on the determinant of $\mathbf{A}$?

The following theorem summarizes Properties 1, 2, and 5 of Section 3.6 and tells us how to keep track of the effect each elementary row operation will have on $\det \mathbf{A}$ as we reduce $\mathbf{A}$ to echelon form. We use here the concise notation

$$|\mathbf{A}| = \det \mathbf{A}$$

for the determinant of the matrix $\mathbf{A}$.

THEOREM 1 Effect of Elementary Row Operations

Let $\mathbf{A}$ be a square matrix and let $\mathbf{B}$ be the matrix obtained by performing one of the three types of elementary row operations on $\mathbf{A}$.

(a) Type 1: If $\mathbf{B}$ is obtained by *dividing* some row of $\mathbf{A}$ by k, then

$$|\mathbf{A}| = k|B|. \tag{6}$$

(b) Type 2: If $\mathbf{B}$ is obtained by interchanging two rows of $\mathbf{A}$, then

$$|\mathbf{A}| = -|\mathbf{B}|. \tag{7}$$

(c) Type 3: If **B** is obtained by adding a multiple of some row of **A** to another row, then

$$|\mathbf{A}| = |\mathbf{B}|. \tag{8}$$

The following example shows how to lay out the computation of $\det \mathbf{A}$ by using row operations to reduce **A** to echelon form.

Example 1

$$\begin{vmatrix} 2 & 2 & -2 & 0 \\ -4 & 2 & 3 & 1 \\ 0 & 4 & -1 & -4 \\ 3 & 1 & 3 & -1 \end{vmatrix} = 2\begin{vmatrix} 1 & 1 & -1 & 0 \\ -4 & 2 & 3 & 1 \\ 0 & 4 & -1 & -4 \\ 3 & 1 & 3 & -1 \end{vmatrix}$$

We have divided row 1 by 2.

$$= 2\begin{vmatrix} 1 & 1 & -1 & 0 \\ 0 & 6 & -1 & 1 \\ 0 & 4 & -1 & -4 \\ 0 & -2 & 6 & -1 \end{vmatrix}$$

We have added 4 times row 1 to row 2, and -3 times row 1 to row 4.

$$= +2\begin{vmatrix} 1 & 1 & -1 & 0 \\ 0 & 0 & 17 & -2 \\ 0 & 0 & 11 & -6 \\ 0 & -2 & 6 & -1 \end{vmatrix}$$

We have added 3 times row 4 to row 2, and 2 times row 4 to row 3.

$$= -2\begin{vmatrix} 1 & 1 & -1 & 0 \\ 0 & -2 & 6 & -1 \\ 0 & 0 & 11 & -6 \\ 0 & 0 & 17 & -2 \end{vmatrix}$$

We have interchanged rows 2 and 4.

$$= -22\begin{vmatrix} 1 & 1 & -1 & 0 \\ 0 & -2 & 6 & -1 \\ 0 & 0 & 1 & -\frac{6}{11} \\ 0 & 0 & 17 & -2 \end{vmatrix}$$

We have divided row 3 by 11.

$$= -22\begin{vmatrix} 1 & 1 & -1 & 0 \\ 0 & -2 & 6 & -1 \\ 0 & 0 & 1 & -\frac{6}{11} \\ 0 & 0 & 0 & \frac{80}{11} \end{vmatrix}$$

We have added -17 times row 3 to row 4.

$$=(-22)(1)(-2)(1)\left(\tfrac{80}{11}\right) = 320.$$

■

If the square matrix **A** has only *integer* entries, it is always possible to carry out the reduction to echelon form without introducing any fractions. In the computation of determinants (as opposed to the solution of linear systems), we have the additional flexibility of using elementary *column* operations; their effects are precisely analogous to those of elementary row operations, as described in Theorem 1. In Example 2, we illustrate this by finishing differently the evaluation in Example 1.

Example 2

$$\begin{vmatrix} 2 & 2 & -2 & 0 \\ -4 & 2 & 3 & 1 \\ 0 & 4 & -1 & -4 \\ 3 & 1 & 3 & -1 \end{vmatrix} = -2\begin{vmatrix} 1 & 1 & -1 & 0 \\ 0 & -2 & 6 & -1 \\ 0 & 0 & 11 & -6 \\ 0 & 0 & 17 & -2 \end{vmatrix} \qquad \text{(As in Example 1.)}$$

$$= -2\begin{vmatrix} 1 & 1 & -1 & 0 \\ 0 & -2 & 5 & -1 \\ 0 & 0 & 5 & -6 \\ 0 & 0 & 15 & -2 \end{vmatrix} \qquad \text{We have added column 4 to column 3.}$$

$$= -2\begin{vmatrix} 1 & 1 & -1 & 0 \\ 0 & -2 & 5 & -1 \\ 0 & 0 & 5 & -6 \\ 0 & 0 & 0 & 16 \end{vmatrix} \qquad \text{We have added } (-3) \text{ times row 3 to row 4.}$$

$$= (-2)(1)(-2)(5)(16) = 320.$$

Yet another alternative is to reduce only the first $n-2$ rows of the $n \times n$ matrix $\mathbf{A}$. We then directly calculate the 2×2 determinant that remains in the lower right-hand corner. We get $\det \mathbf{A}$ by multiplying this 2×2 determinant by the preceding factors and diagonal elements. Thus

$$\begin{vmatrix} 2 & 2 & -2 & 0 \\ -4 & 2 & 3 & 1 \\ 0 & 4 & -1 & -4 \\ 3 & 1 & 3 & -1 \end{vmatrix} = -2\begin{vmatrix} 1 & 1 & -1 & 0 \\ 0 & -2 & 6 & -1 \\ 0 & 0 & 11 & -6 \\ 0 & 0 & 17 & -2 \end{vmatrix}$$

$$= (-2)(1)(-2)\begin{vmatrix} 11 & -6 \\ 17 & -2 \end{vmatrix} = (4)(-22+102) = 320. \quad \blacksquare$$

Determinants and Invertibility

We began Section 3.6 with the remark that a 2×2 matrix $\mathbf{A}$ is invertible if and only if its determinant is nonzero: $|\mathbf{A}| \neq 0$. Now we want to show that this result also holds for $n \times n$ matrices. This connection between determinants and invertibility is closely related to the fact that the determinant function "respects" matrix multiplication in the sense that

$$|\mathbf{AB}| = |\mathbf{A}|\,|\mathbf{B}| \tag{9}$$

if $\mathbf{A}$ and $\mathbf{B}$ are $n \times n$ matrices. Our first step is to show that Eq. (9) holds if $\mathbf{A}$ is an *elementary matrix* obtained from the $n \times n$ identity matrix $\mathbf{I}$ by performing a single elementary row operation.

LEMMA Multiplication by an Elementary Matrix

If $\mathbf{B}$ is an $n \times n$ matrix and $\mathbf{E}$ is an $n \times n$ elementary matrix then

$$|\mathbf{EB}| = |\mathbf{E}|\,|\mathbf{B}|. \tag{10}$$

***Proof*:** This is really just a restatement of Theorem 1. For instance, suppose that $\mathbf{E}$ is obtained from $\mathbf{I}$ by multiplying the pth row by c, so that $|\mathbf{E}| = c$. Then

Theorem 5 in Section 3.5 tells us that the product **EB** is the result of multiplying the pth row of **B** by c. Therefore,

$$|\mathbf{EB}| = c|\mathbf{B}| = |\mathbf{E}|\,|\mathbf{B}|,$$

and so we have verified Eq. (10) for the first of the three types of elementary matrices. The verifications for the other two types are similar. ■

Now, let **A** be an $n \times n$ matrix whose invertibility we want to discuss, and let **R** be the *reduced* echelon form of **A**. If the elementary matrices $\mathbf{F}_1, \mathbf{F}_2, \ldots, \mathbf{F}_k$ correspond to the elementary row operations that reduce **A** to **R**, then

$$\mathbf{F}_k \cdots \mathbf{F}_2\mathbf{F}_1\mathbf{A} = \mathbf{R} \tag{11}$$

by Theorem 5 in Section 3.5. Recalling that every elementary matrix is invertible (Section 3.5), we can rewrite Eq. (11) as

$$\mathbf{A} = \mathbf{E}_1\mathbf{E}_2 \cdots \mathbf{E}_k\mathbf{R}, \tag{12}$$

where each $\mathbf{E}_i = (\mathbf{F}_i)^{-1}$ is an elementary matrix. It now follows, by k applications of the lemma, that

$$\mathbf{A} = |\mathbf{E}_1|\,|\mathbf{E}_2| \cdots |\mathbf{E}_k|\,|\mathbf{R}|. \tag{13}$$

This relation is the key both to the proof of (9) and to the proof of the following theorem.

THEOREM 2 Determinants and Invertibility

The $n \times n$ matrix **A** is invertible if and only if $\det \mathbf{A} \neq 0$.

***Proof*:** If (as previously) **R** is the reduced echelon form of **A**, then Theorem 6 in Section 3.5 implies that

$$\mathbf{A} \text{ is invertible} \quad \text{if and only if} \quad \mathbf{R} = \mathbf{I}. \tag{14}$$

Because **R** is a *square* reduced echelon matrix, we see that *either* **R** is the identity matrix **I** and $|\mathbf{R}| = 1$, *or* **R** has an all-zero row and, consequently, $|\mathbf{R}| = 0$. Therefore,

$$\mathbf{R} = \mathbf{I} \quad \text{if and only if} \quad |\mathbf{R}| \neq 0. \tag{15}$$

Finally, because $|\mathbf{E}| \neq 0$ if **E** is an elementary matrix, it follows immediately from Eq. (13) that

$$|\mathbf{R}| \neq 0 \quad \text{if and only if} \quad |\mathbf{A}| \neq 0. \tag{16}$$

Combining the statements in (14), (15), and (16), we see that **A** is invertible if and only if $|\mathbf{A}| \neq 0$. ■

So now we can add the statement $\det \mathbf{A} \neq 0$ to the list of equivalent properties of nonsingular matrices stated in Theorem 7 of Section 3.5. Indeed, some texts *define* the square matrix **A** to be nonsingular if and only if $\det \mathbf{A} \neq 0$.

THEOREM 3 Determinants and Matrix Products

If **A** and **B** are $n \times n$ matrices, then

$$|\mathbf{AB}| = |\mathbf{A}|\,|\mathbf{B}|. \tag{9}$$

***Proof*:** If **R** is the reduced echelon form of **A**, then we see that

$$\mathbf{A} = \mathbf{E}_1\mathbf{E}_2\cdots\mathbf{E}_k\mathbf{R}, \tag{12}$$

where $\mathbf{E}_1$, $\mathbf{E}_2$, ..., and $\mathbf{E}_k$ are elementary matrices. Hence

$$\mathbf{AB} = \mathbf{E}_1\mathbf{E}_2\cdots\mathbf{E}_k\mathbf{RB}.$$

We now take the determinant of both sides, using the lemma stated earlier to "split off" the elementary matrices:

$$\begin{aligned} |\mathbf{AB}| &= |\mathbf{E}_1|\,|\mathbf{E}_2\mathbf{E}_3\cdots\mathbf{E}_k\mathbf{RB}| && \text{(lemma once)} \\ &= |\mathbf{E}_1|\,|\mathbf{E}_2|\,|\mathbf{E}_3\cdots\mathbf{E}_k\mathbf{RB}| && \text{(lemma twice)} \\ &= |\mathbf{E}_1\mathbf{E}_2|\,|\mathbf{E}_3\cdots\mathbf{E}_k\mathbf{RB}|. && \text{(lemma thrice)} \end{aligned}$$

After $2k - 1$ steps, we get

$$|\mathbf{AB}| = |\mathbf{E}_1\mathbf{E}_2\cdots\mathbf{E}_k|\,|\mathbf{RB}|. \tag{17}$$

The remainder of the proof depends on whether or not **A** is invertible.

If **A** *is* invertible, then $\mathbf{R} = \mathbf{I}$, so Eq. (12) yields

$$\mathbf{A} = \mathbf{E}_1\mathbf{E}_2\cdots\mathbf{E}_k$$

and also that $\mathbf{RB} = \mathbf{IB} = \mathbf{B}$. In this case the meaning of Eq. (17) is precisely that $|\mathbf{AB}| = |\mathbf{A}|\,|\mathbf{B}|$.

If **A** is *not* invertible, then $|\mathbf{A}| = 0$ by Theorem 2. Also, as we noted previously, the reduced echelon form **R** of **A** has an all-zero row in this case. Hence it follows from the definition of matrix multiplication that the product **RB** has an all-zero row and, therefore, that $|\mathbf{RB}| = 0$. In this case Eq. (17) implies that $|\mathbf{AB}| = 0$. Because both $|\mathbf{A}| = 0$ and $|\mathbf{AB}| = 0$, the equation $|\mathbf{AB}| = |\mathbf{A}|\,|\mathbf{B}|$ holds, and the proof is complete. ■

Cramer's Rule and Inverse Matrices

Suppose that we need to solve the $n \times n$ linear system

$$\mathbf{Ax} = \mathbf{b}, \tag{18}$$

where

$$\mathbf{A} = \begin{bmatrix} a_{ij} \end{bmatrix}, \quad \mathbf{x} = \begin{bmatrix} x_1 \\ x_2 \\ \vdots \\ x_n \end{bmatrix} \quad \text{and} \quad \mathbf{b} = \begin{bmatrix} b_1 \\ b_2 \\ \vdots \\ b_n \end{bmatrix}. \tag{19}$$

We assume that the coefficient matrix **A** is invertible, so we know in advance that a unique solution **x** of (18) exists. The question is how to write **x** explicitly in terms of the coefficients a_{ij} and the constants b_i. In the following discussion, we think of **x** as a fixed (though as yet unknown) vector.

If we denote by $\mathbf{a}_1$, $\mathbf{a}_2$, ..., $\mathbf{a}_n$ the column vectors of the $n \times n$ matrix **A**, then

$$\mathbf{A} = \begin{bmatrix} \mathbf{a}_1 & \mathbf{a}_2 & \cdots & \mathbf{a}_n \end{bmatrix}. \tag{20}$$

By Fact 1 in Section 3.5, we can rewrite Eq. (18) as

$$x_1\mathbf{a}_1 + x_2\mathbf{a}_2 + \cdots + x_n\mathbf{a}_n = \mathbf{b}.$$

Thus the constant vector **b** is expressed in terms of the entries $x_1, x_2, \dots, x_n$ of the solution vector **x** and the column vectors of **A** by

$$\mathbf{b} = \sum_{j=1}^{n} x_j \mathbf{a}_j. \tag{21}$$

The trick for finding the ith unknown x_i is to compute the determinant of the matrix

$$\begin{bmatrix} \mathbf{a}_1 & \cdots & \mathbf{b} & \cdots & \mathbf{a}_n \end{bmatrix} = \begin{bmatrix} a_{11} & \cdots & b_1 & \cdots & a_{1n} \\ a_{21} & \cdots & b_2 & \cdots & a_{2n} \\ \vdots & \ddots & \vdots & \ddots & \vdots \\ a_{n1} & \cdots & b_n & \cdots & a_{nn} \end{bmatrix} \tag{22}$$

that we obtain by replacing the ith column $\mathbf{a}_i$ of **A** with the constant vector **b**. Using Eq. (21) to substitute for **b**, we find that

$$\begin{aligned} \begin{vmatrix} \mathbf{a}_1 & \cdots & \mathbf{b} & \cdots & \mathbf{a}_n \end{vmatrix} &= \begin{vmatrix} \mathbf{a}_1 & \cdots & \sum_{j=1}^{n} x_j \mathbf{a}_j & \cdots & \mathbf{a}_n \end{vmatrix} \\ &= \sum_{j=1}^{n} \begin{vmatrix} \mathbf{a}_1 & \cdots & x_j \mathbf{a}_j & \cdots & \mathbf{a}_n \end{vmatrix} && \text{(by Property 4 of determinants)} \\ &= \sum_{j=1}^{n} x_j \begin{vmatrix} \mathbf{a}_1 & \cdots & \mathbf{a}_j & \cdots & \mathbf{a}_n \end{vmatrix} && \text{(by Property 1 of determinants).} \end{aligned}$$

Note that, in the jth term of this summation, the vector $\mathbf{a}_j$ appears in the ith position. Thus we have found that

$$\begin{aligned} \begin{vmatrix} \mathbf{a}_1 & \cdots & \mathbf{b} & \cdots & \mathbf{a}_n \end{vmatrix} &= x_1 \begin{vmatrix} \mathbf{a}_1 & \mathbf{a}_2 & \cdots & \mathbf{a}_1 & \cdots & \mathbf{a}_n \end{vmatrix} \\ &+ x_2 \begin{vmatrix} \mathbf{a}_1 & \mathbf{a}_2 & \cdots & \mathbf{a}_2 & \cdots & \mathbf{a}_n \end{vmatrix} \\ &\;\;\vdots \\ &+ x_i \begin{vmatrix} \mathbf{a}_1 & \mathbf{a}_2 & \cdots & \mathbf{a}_i & \cdots & \mathbf{a}_n \end{vmatrix} \\ &\;\;\vdots \\ &+ x_n \begin{vmatrix} \mathbf{a}_1 & \mathbf{a}_2 & \cdots & \mathbf{a}_n & \cdots & \mathbf{a}_n \end{vmatrix}. \end{aligned}$$

Of the n determinants on the right-hand side here, all but the ith one have two identical columns and therefore are equal to zero. The coefficient of x_i in the ith term is simply

$$|\mathbf{A}| = \begin{vmatrix} \mathbf{a}_1 & \mathbf{a}_2 & \cdots & \mathbf{a}_i & \cdots & \mathbf{a}_n \end{vmatrix}.$$

Consequently, a result of our computation is that

$$\begin{vmatrix} \mathbf{a}_1 & \cdots & \mathbf{b} & \cdots & \mathbf{a}_n \end{vmatrix} = x_i |\mathbf{A}|. \tag{23}$$

We get the desired simple formula for x_i after we divide each side by $|\mathbf{A}| \neq 0$.

THEOREM 4 Cramer's Rule

Consider the $n \times n$ linear system $\mathbf{Ax} = \mathbf{b}$ with

$$\mathbf{A} = \begin{bmatrix} \mathbf{a}_1 & \mathbf{a}_2 & \cdots & \mathbf{a}_n \end{bmatrix}.$$

If $|\mathbf{A}| \neq 0$, then the ith entry of the unique solution $\mathbf{x} = (x_1, x_2, \ldots, x_n)$ is given by

$$\begin{aligned} x_i &= \frac{\begin{vmatrix} \mathbf{a}_1 & \cdots & \mathbf{b} & \cdots & \mathbf{a}_n \end{vmatrix}}{|\mathbf{A}|} \\ &= \frac{1}{|\mathbf{A}|}\begin{vmatrix} a_{11} & \cdots & b_1 & \cdots & a_{1n} \\ a_{21} & \cdots & b_2 & \cdots & a_{2n} \\ \vdots & \ddots & \vdots & \ddots & \vdots \\ a_{n1} & \cdots & b_n & \cdots & a_{nn} \end{vmatrix}, \end{aligned} \tag{24}$$

where (in the last expression) the constant vector $\mathbf{b}$ replaces the ith column vector $\mathbf{a}_i$ of $\mathbf{A}$.

Inverses and the Adjoint Matrix

We now use Cramer's rule to develop an explicit formula for the inverse $\mathbf{A}^{-1}$ of the invertible matrix $\mathbf{A}$. First, we need to rewrite Cramer's rule more concisely. Expansion of the determinant in the numerator in Eq. (24) along its ith column yields

$$x_i = \frac{1}{|\mathbf{A}|}(b_1 A_{1i} + b_2 A_{2i} + \cdots + b_n A_{ni}), \tag{25}$$

because the cofactor of b_p is simply the cofactor A_{pi} of a_{pi} in $|\mathbf{A}|$. The formula in Eq. (25) gives the solution vector

$$\mathbf{x} = \begin{bmatrix} x_1 \\ x_2 \\ \vdots \\ x_n \end{bmatrix} = \frac{1}{|\mathbf{A}|}\begin{bmatrix} b_1 A_{11} + b_2 A_{21} + \ldots + b_n A_{n1} \\ b_1 A_{12} + b_2 A_{22} + \ldots + b_n A_{n2} \\ \vdots \\ b_1 A_{1n} + b_2 A_{2n} + \ldots + b_n A_{nn} \end{bmatrix}.$$

Then the definition of matrix multiplication yields

$$\mathbf{x} = \frac{1}{|\mathbf{A}|}\begin{bmatrix} A_{11} & A_{21} & \cdots & A_{n1} \\ A_{12} & A_{22} & \cdots & A_{n2} \\ \vdots & \vdots & \ddots & \vdots \\ A_{1n} & A_{2n} & \cdots & A_{nn} \end{bmatrix}\begin{bmatrix} b_1 \\ b_2 \\ \vdots \\ b_n \end{bmatrix} \tag{26}$$

for the solution $\mathbf{x}$ of $\mathbf{Ax} = \mathbf{b}$.

Observe that the double subscripts in (26) are reversed from their usual order; the element in the ith row and jth column is A_{ji} (rather than A_{ij}). We therefore see in (26) the *transpose* of the **cofactor matrix** $\begin{bmatrix} A_{ij} \end{bmatrix}$ of the $n \times n$ matrix $\mathbf{A}$. The transpose of the cofactor matrix of $\mathbf{A}$ is called the **adjoint matrix** of $\mathbf{A}$ and is denoted by

$$\operatorname{adj}\mathbf{A} = \begin{bmatrix} A_{ij} \end{bmatrix}^T = \begin{bmatrix} A_{ji} \end{bmatrix}. \tag{27}$$

With the aid of this notation, Cramer's rule as expressed in Eq. (26) can be written in the especially simple form

$$\mathbf{x} = \frac{\left[A_{ji} \right] \mathbf{b}}{|\mathbf{A}|} = \frac{(\text{adj}\, \mathbf{A})\mathbf{b}}{|\mathbf{A}|}. \tag{28}$$

The fact that the formula in (28) gives the unique solution $\mathbf{x}$ of $\mathbf{Ax} = \mathbf{b}$ implies that

$$\mathbf{A}\frac{(\text{adj}\, \mathbf{A})\mathbf{b}}{|\mathbf{A}|} = \mathbf{b} \tag{29}$$

for every n-vector $\mathbf{b}$. If we write

$$\mathbf{C} = \frac{\text{adj}\, \mathbf{A}}{|\mathbf{A}|} \tag{30}$$

for brevity, then

$$\mathbf{ACb} = \mathbf{b} \tag{31}$$

for every n-vector $\mathbf{b}$. From this it follows (one column at a time, as we use Fact 2 in Section 3.5) that

$$\mathbf{ACB} = \mathbf{B} \tag{32}$$

for every matrix $\mathbf{B}$ having n rows. In particular, with $\mathbf{B} = \mathbf{I}$ (the $n \times n$ identity matrix), we see that

$$\mathbf{AC} = \mathbf{I}. \tag{33}$$

Therefore, we have discovered that the matrix $\mathbf{C}$ as defined in Eq. (30) is the inverse matrix $\mathbf{A}^{-1}$ of $\mathbf{A}$.

THEOREM 5 The Inverse Matrix

The inverse of the invertible matrix $\mathbf{A}$ is given by the formula

$$\mathbf{A}^{-1} = \frac{\text{adj}\, \mathbf{A}}{|\mathbf{A}|} = \frac{\left[a_{ij} \right]^T}{|\mathbf{A}|}, \tag{34}$$

where, as usual, A_{ij} is the ijth cofactor of $\mathbf{A}$—that is, A_{ij} is the product of $(-1)^{i+j}$ and the ijth minor of $\mathbf{A}$.

ANSWERS TO SELECTED PROBLEMS

Chapter 1

Section 1.1

11. If $y = y_1 = x^{-2}$, then $y'(x) = -2x^{-3}$ and $y''(x) = 6x^{-4}$, so $x^2y'' + 5xy' + 4y = x^2(6x^{-4}) + 5x(-2x^{-3}) + 4(x^{-2}) = 6x^{-2} - 10x^{-2} + 4x^{-2} = 0$. If $y = y_2 = x^{-2}\ln x$, then $y'(x) = x^{-3} - 2x^{-3}\ln x$ and $y''(x) = -5x^{-4} + 6x^{-4}\ln x$, so $x^2y'' + 5xy' + 4y = x^2(-5x^{-4} + 6x^{-4}\ln x) + 5x(x^{-3} - 2x^{-3}\ln x) + 4(x^{-2}\ln x) = 0$.

13. $r = \frac{2}{3}$ **14.** $r = \pm\frac{1}{2}$ **15.** $r = -2, 1$

16. $r = \frac{1}{6}(-3 \pm \sqrt{57})$

17. $C = 2$

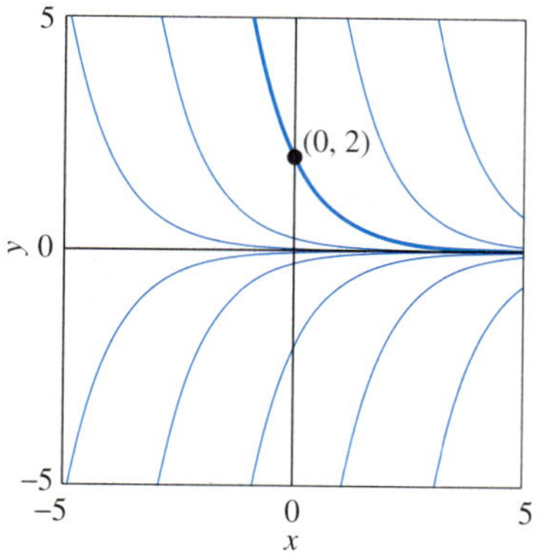

18. $C = 3$

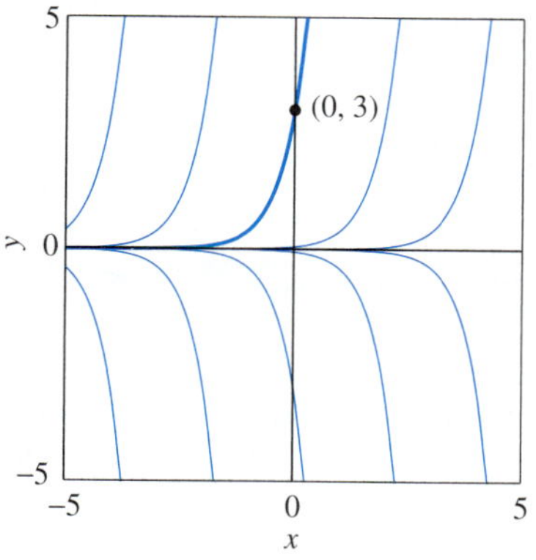

19. $C = 6$

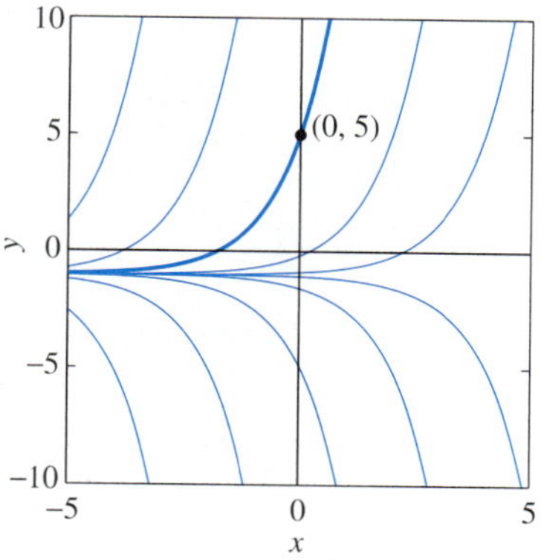

20. $C = 11$

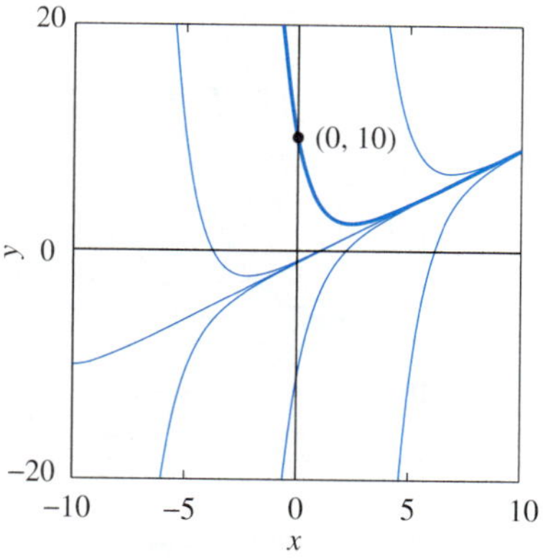

21. $C = 7$

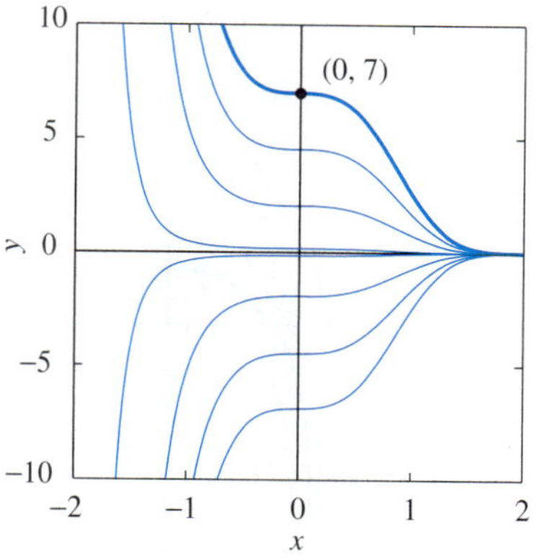

22. $C = 1$

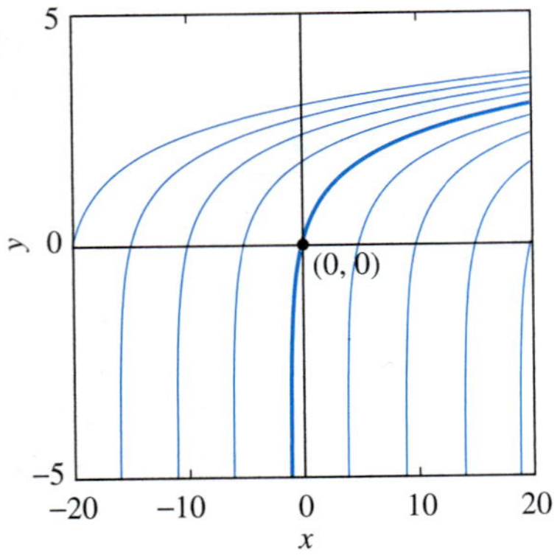

23. $C = -56$

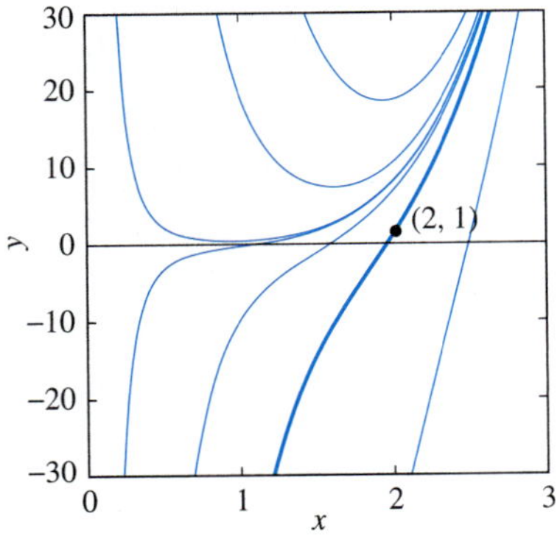

24. $C = 17$

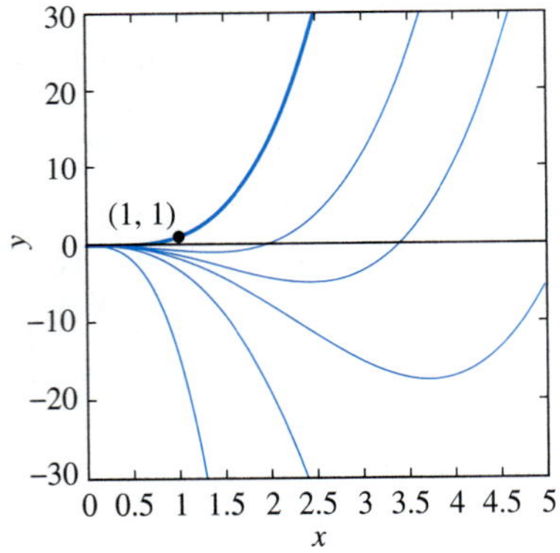

25. $C = \pi/4$

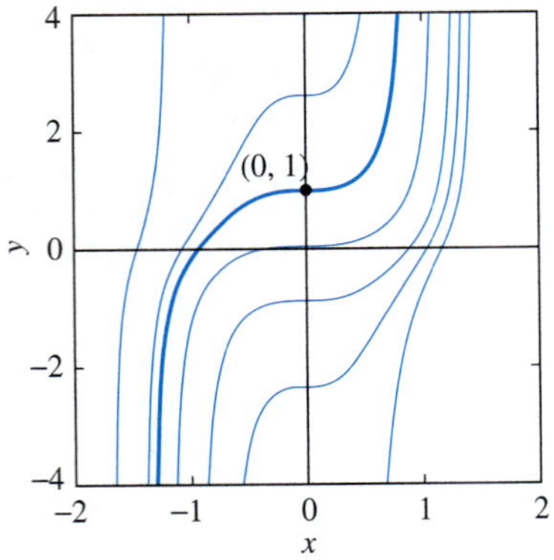

26. $C = -\pi$

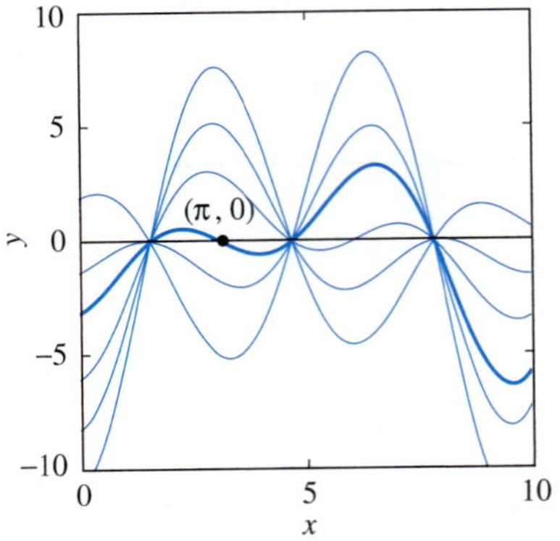

27. $y' = x + y$
28. $y' = 2y/x$
29. $y' = x/(1 - y)$
31. $y' = (y - x)/(y + x)$
32. $dP/dt = k\sqrt{P}$
33. $dv/dt = kv^2$
35. $dN/dt = k(P - N)$
37. $y \equiv 1$ or $y = x$
39. $y = x^2$
41. $y = \frac{1}{2}e^x$
42. $y = \cos x$ or $y = \sin x$
43. (b) The identically zero function $x(0) \equiv 0$
44. (a) The graphs (figure below) of typical solutions with $k = \frac{1}{2}$ suggest that (for each) the value $x(t)$ increases without bound as t increases.

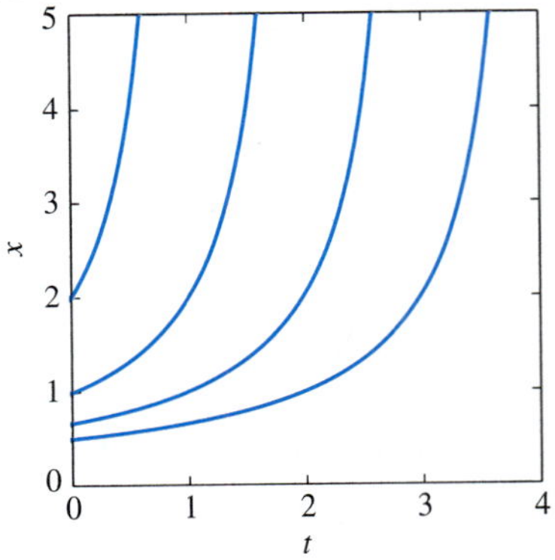

(b) The graphs (figure below) of typical solutions with $k = -\frac{1}{2}$ suggest that now the value $x(t)$ approaches 0 as t increases without bound.

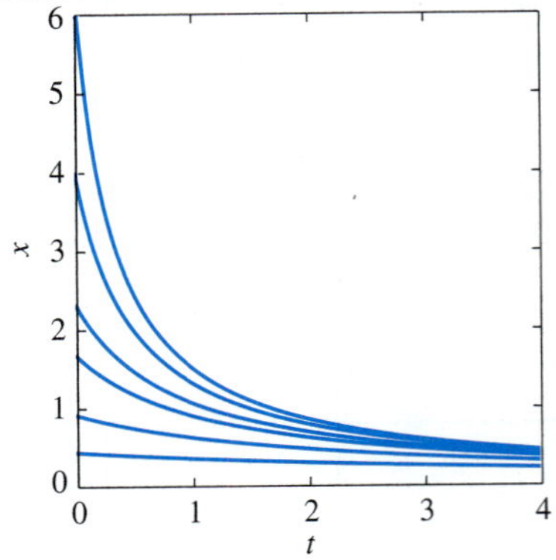

45. $P(t) = 100/(50 - t)$; $P = 100$ when $t = 49$, and $P = 1000$ when $t = 49.9$. Thus it appears that $P(t)$ grows without bound as t approaches 50.

46. $v(t) = 50/(5 + 2t)$; $v = 1$ when $t = 22.5$, and $v = \frac{1}{10}$ when $t = 247.5$. Thus it appears that $v(t)$ approaches 0 as t increases without bound.

47. (a) $C = 10.1$; (b) No such C, but the constant function $y(x) \equiv 0$ satisfies the conditions $y' = y^2$ and $y(0) = 0$.

Section 1.2

1. $y(x) = x^2 + x + 3$

2. $y(x) = \frac{1}{3}(x-2)^2 + 1$

3. $y(x) = \frac{1}{3}(2x^{3/2} - 16)$

4. $y(x) = -1/x + 6$

5. $y(x) = 2\sqrt{x+2} - 5$

6. $y(x) = \frac{1}{3}[(x^2+9)^{3/2} - 125]$

7. $y(x) = 10\tan^{-1} x$

8. $y(x) = \frac{1}{2}\sin 2x + 1$

9. $y(x) = \sin^{-1} x$

10. $y(x) = -(x+1)e^{-x} + 2$

11. $x(t) = 25t^2 + 10t + 20$

12. $x(t) = -10t^2 - 15t + 5$

13. $x(t) = \frac{1}{2}t^3 + 5t$

14. $x(t) = \frac{1}{3}t^3 + \frac{1}{2}t^2 - 7t + 4$

15. $x(t) = \frac{1}{3}(t+3)^4 - 37t - 26$

16. $x(t) = \frac{4}{3}(t+4)^{3/2} - 5t - \frac{29}{3}$

17. $x(t) = \frac{1}{2}\left[(t+1)^{-1} + t - 1\right]$

19. $x(t) = \begin{cases} 5t & \text{if } 0 \leq t \leq 5, \\ 10t - \frac{1}{2}t^2 - \frac{25}{2} & \text{if } 5 \leq t \leq 10. \end{cases}$

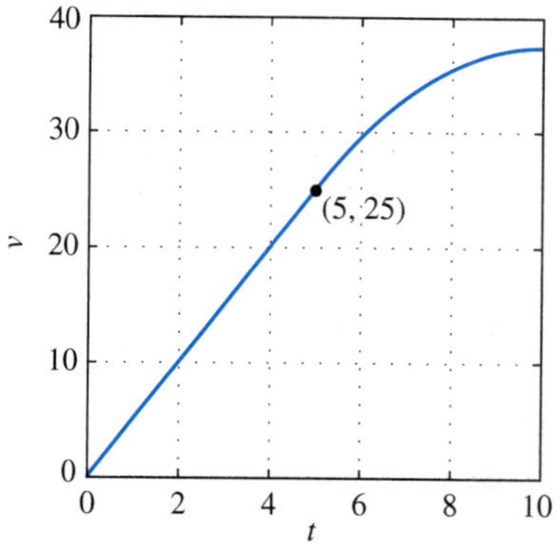

20. $x(t) = \begin{cases} \frac{1}{2}t^2 & \text{if } 0 \leq t \leq 5, \\ 5t - \frac{25}{2} & \text{if } 5 \leq t \leq 10. \end{cases}$

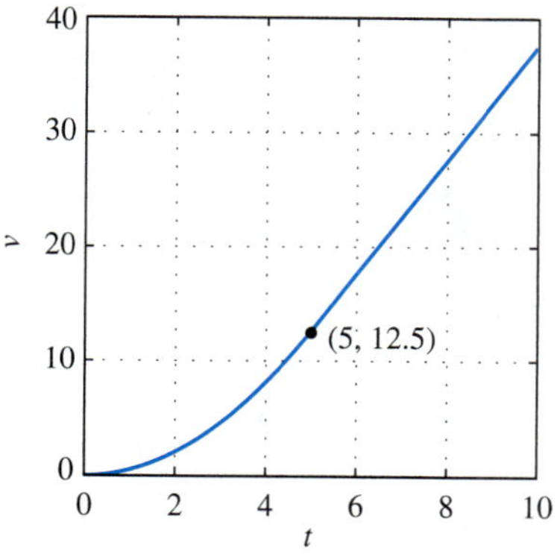

21. $x(t) = \begin{cases} \frac{1}{2}t^2 & \text{if } 0 \leq t \leq 5, \\ 10t - \frac{1}{2}t^2 - 25 & \text{if } 5 \leq t \leq 10. \end{cases}$

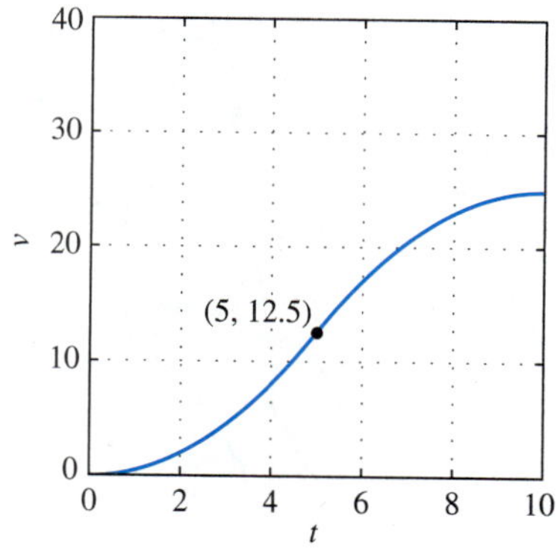

22. $x(t) = \begin{cases} \frac{5}{6}t^2 & \text{if } 0 \leq t \leq 3, \\ 5t - \frac{15}{2} & \text{if } 3 \leq t \leq 7, \\ \frac{1}{6}(-5t^2 + 100t - 290) & \text{if } 7 \leq t \leq 10. \end{cases}$

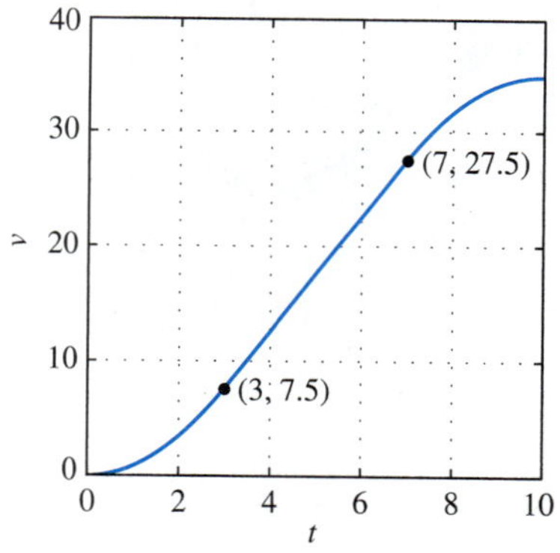

23. $v(t) = -(9.8)t + 49$, so the ball reaches its maximum height ($v = 0$) after $t = 5$ seconds. Its maximum height then is $y(5) = 122.5$ (m).

24. $v(5) = -160$ ft/s

25. The car stops when $t \approx 2.78$ (s), so the distance traveled before stopping is approximately $x(2.78) \approx 38.58$ (m).

26. (a) $y \approx 530$ m (b) $t \approx 20.41$ s (c) $t \approx 20.61$ s

27. $y_0 \approx 178.57$ (m)

28. $v(4.77) \approx -192.64$ ft/s

29. After 10 seconds the car has traveled 200 ft and is traveling at 70 ft/s.

30. $a = 22$ ft/s^2; it skids for 4 seconds.

31. $v_0 = 10\sqrt{30}$ (m/s), about 197.18 km/h

32. 60 m

33. $20\sqrt{10} \approx 63.25$ (ft/s)

34. 460.8 ft

36. About 13.6 ft

37. 25 (mi)

38. 1:10 P.M.

39. 6 mph

40. 2.4 mi

41. $\frac{544}{3} \approx 181.33$ ft/s

42. 25 mi

43. Time: 6.12245×10^9 s ≈ 194 years;
Distance: 1.8367×10^{17} m ≈ 19.4 light-years

44. About 54 mi/h

Section 1.3

1.

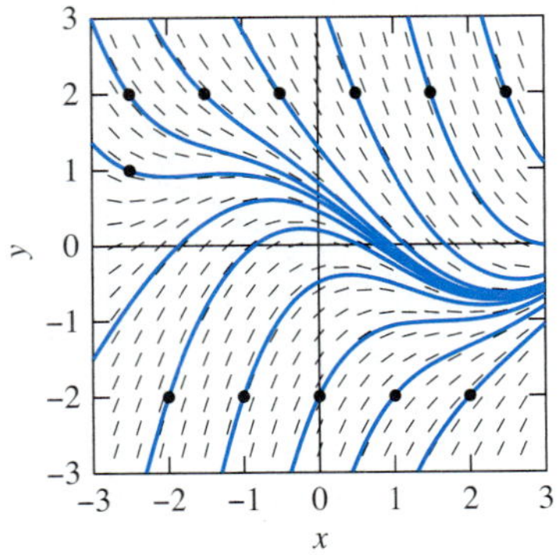

2.

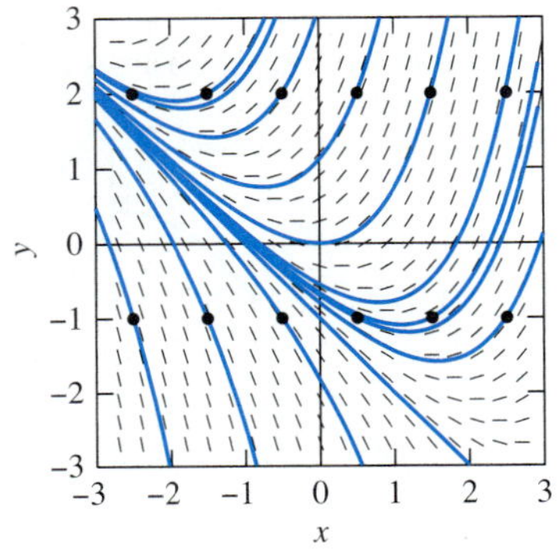

3.

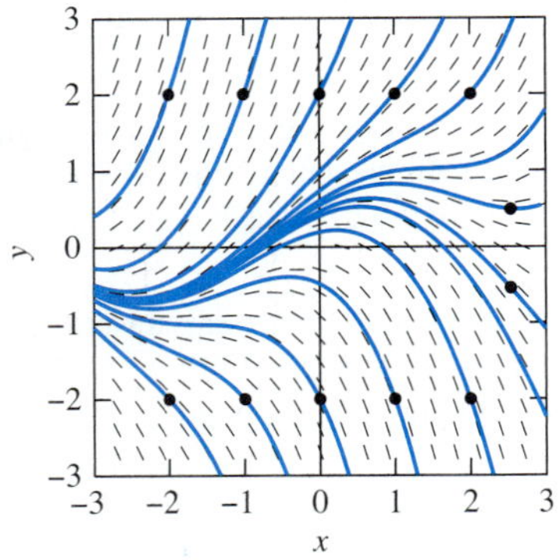

4.

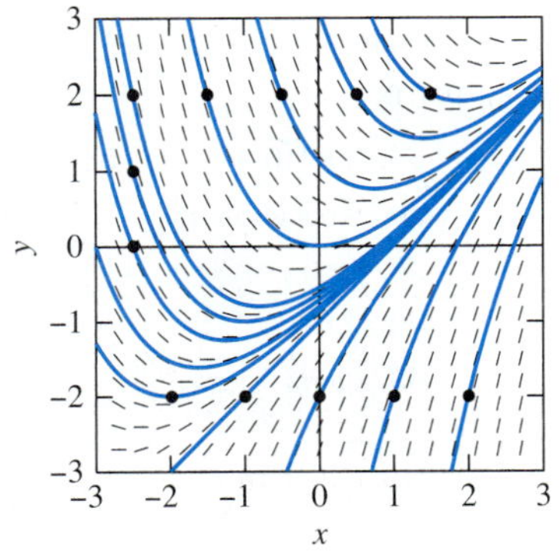

5.

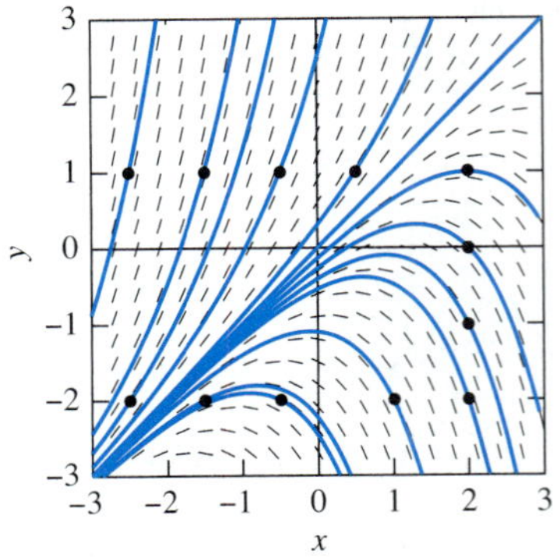

6.

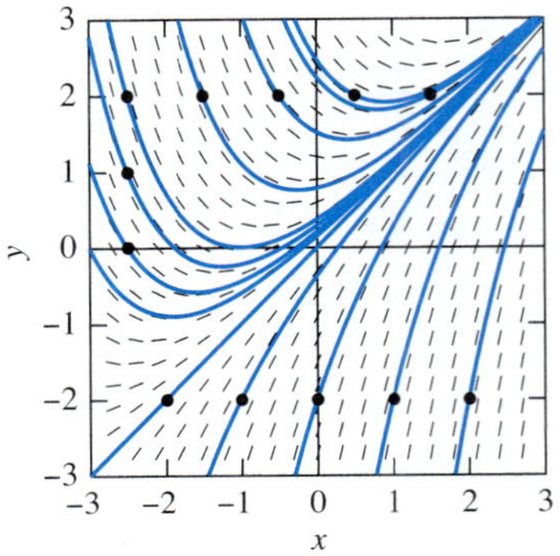

7.

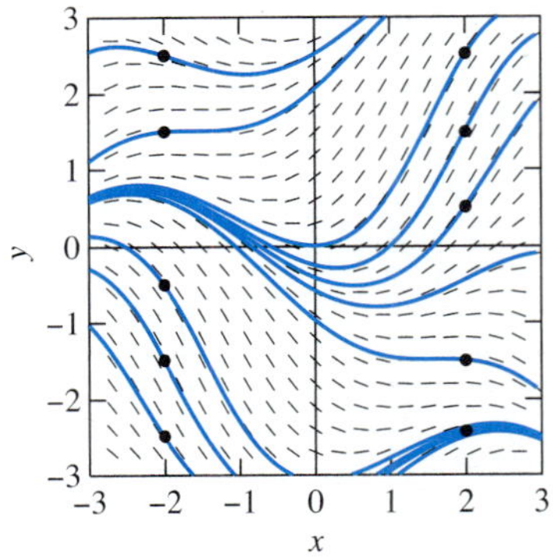

8.

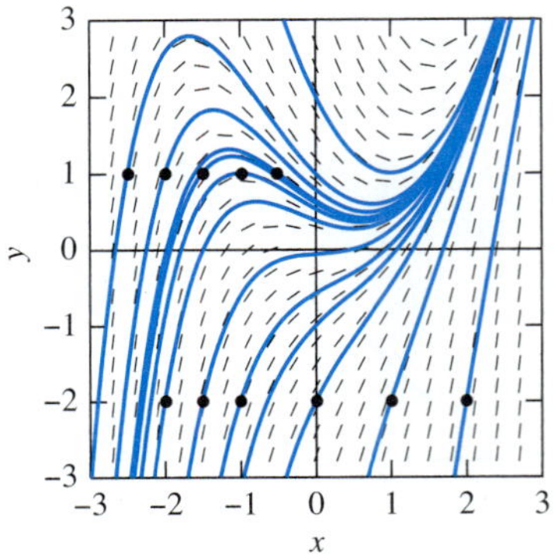

9.

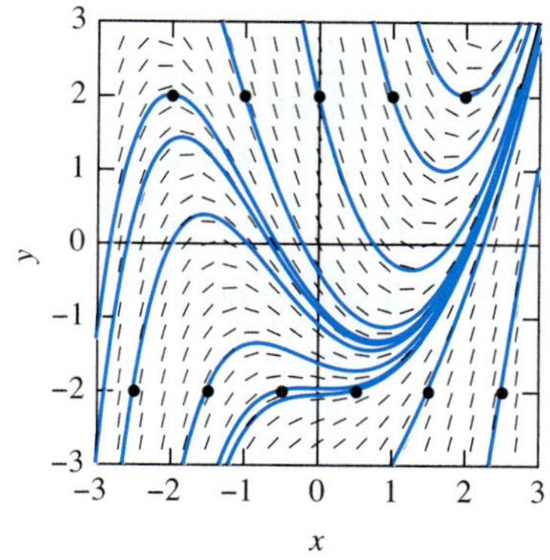

10.

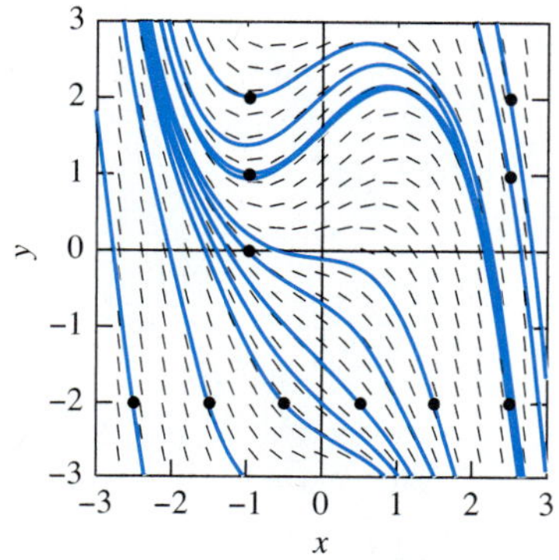

11. A unique solution exists in some neighborhood of $x = 1$.

12. A unique solution exists in some neighborhood of $x = 1$.

13. A unique solution exists in some neighborhood of $x = 0$.

14. Existence but not uniqueness is guaranteed in some neighborhood of $x = 0$.

15. Neither existence nor uniqueness is guaranteed in any neighborhood of $x = 2$.

16. A unique solution exists in some neighborhood of $x = 2$.

17. A unique solution exists in some neighborhood of $x = 0$.

18. Neither existence nor uniqueness is guaranteed.

19. A unique solution exists in some neighborhood of $x = 0$.

20. A unique solution exists in some neighborhood of $x = 0$.

21. Your figure should suggest that $y(-4) \approx 3$; an exact solution of the differential equation gives $y(-4) = 3 + e^{-4} \approx 3.0183$.

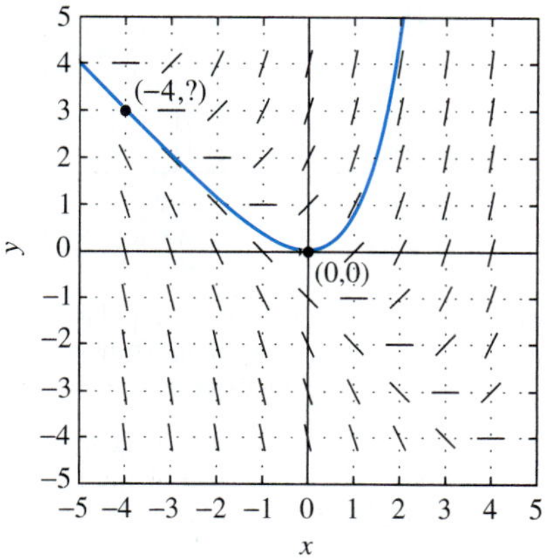

22. $y(-4) \approx -3$

23. Your figure should suggest that $y(2) \approx 1$; the actual value is closer to 1.004.

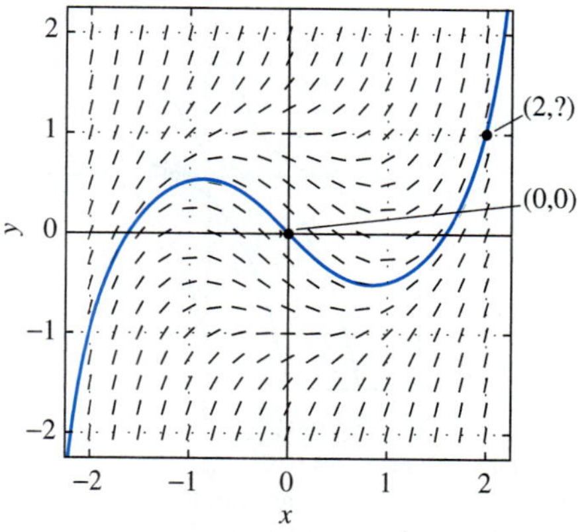

24. $y(2) \approx 1.5$

25. Your figure should suggest that the limiting velocity is about 20 ft/s (quite survivable) and that the time required to reach 19 ft/s is a little less than 2 seconds. An exact solution gives $v(t) = 19$ when $t = \frac{5}{8} \ln 20 \approx 1.8723$.

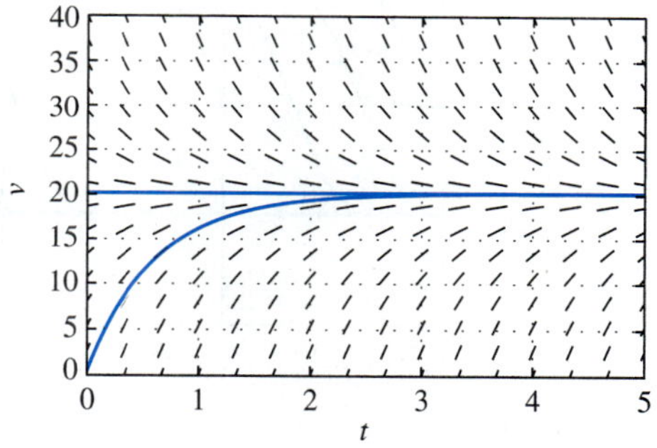

26. A figure suggests that there are 40 deer after about 60 months; a more accurate value is $t \approx 61.61$. The limiting population is 75 deer.

27. The initial value problem $y' = 2\sqrt{y}$, $y(0) = b$ has no solution if $b < 0$; a unique solution if $b > 0$; infinitely many solutions if $b = 0$.

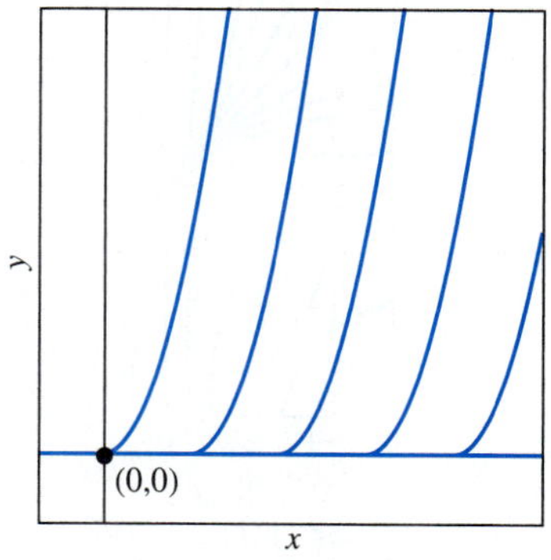

28. The initial value problem $xy' = y$, $y(a) = b$ has a unique solution if $a \neq 0$; infinitely many solutions if $a = b = 0$; no solution if $a = 0$ but $b \neq 0$.

29. If $b = 0$ then the initial value problem $y' = 3y^{2/3}$, $y(a) = b$ always has infinitely many solutions defined for all x. However, if $b \neq 0$ then it has a unique solution near $x = a$.

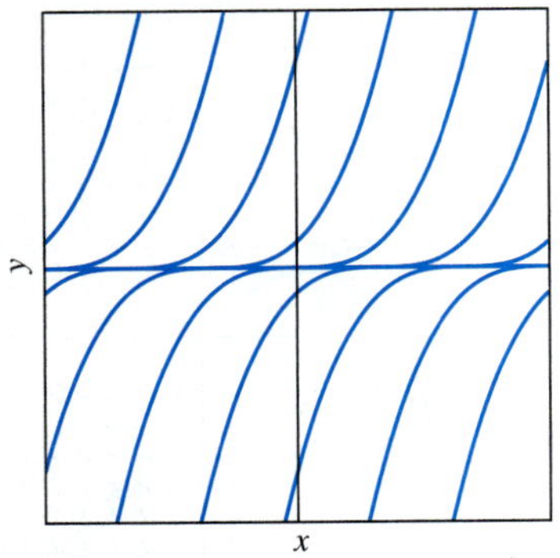

30. The initial value problem $y' = -\sqrt{1 - y^2}$, $y(a) = b$ has a unique solution if $|b| < 1$; no solution if $|b| > 1$, and infinitely many solutions (defined for all x) if $b = \pm 1$.

31. The initial value problem $y' = \sqrt{1 - y^2}$, $y(a) = b$ has a unique solution if $|b| < 1$; no solution if $|b| > 1$, and infinitely many solutions (defined for all x) if $b = \pm 1$.

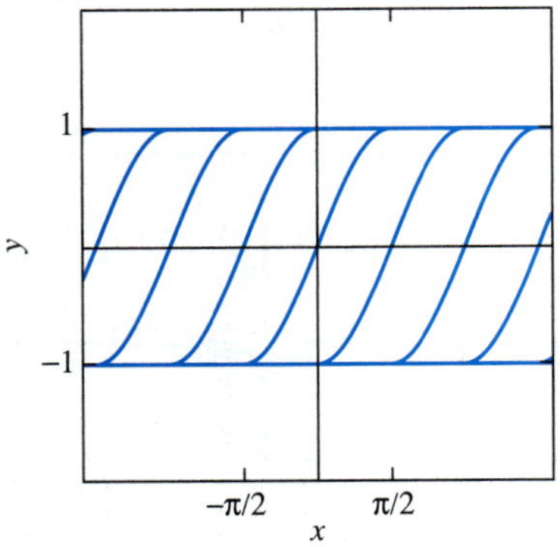

32. The initial value problem $y' = 4x\sqrt{y}$, $y(a) = b$ has infinitely many solutions (defined for all x) if $b \geqq 0$; no solutions if $b < 0$. However, if $b > 0$ then it has a unique solution near $x = a$.

33. The initial value problem $x^2 y' + y^2 = 0$, $y(a) = b$ has a unique solution with initial point (a, b) if $a \neq 0$, no solution if $a = 0$ but $b \neq 0$, infinitely many solutions if $a = b = 0$.

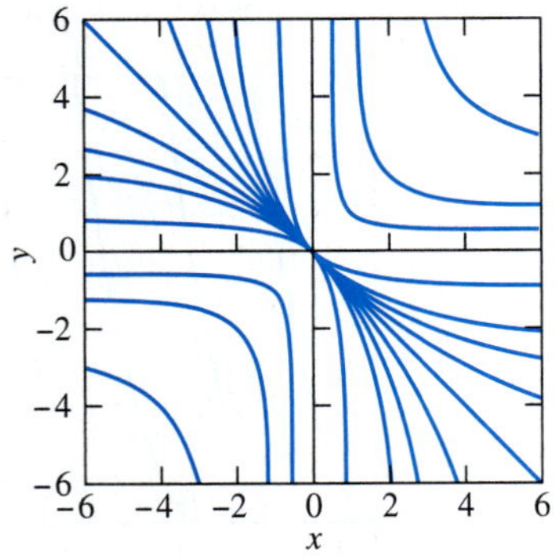

34. (a) If $y(-1) = -1.2$, then $y(1) \approx -0.48$. If $y(-1) = -0.8$, then $y(1) \approx 2.48$. (b) If $y(-3) = -3.01$, then $y(3) \approx -1.0343$. If $y(-3) = -2.99$, then $y(3) \approx 7.0343$.

35. (a) If $y(-3) = -0.2$, then $y(2) \approx 2.019$. If $y(-3) = +0.2$, then $y(2) \approx 2.022$. In either case, $y(2) \approx 2.02$.
(b) If $y(-3) \approx -0.5$, then $y(2) \approx 2.017$. If $y(-3) \approx +0.5$, then $y(2) \approx 2.024$. In either case, $y(2) \approx 2.02$.

Section 1.4

1. $y(x) = C\exp(-x^2)$

2. $y(x) = 1/(x^2 + C)$

3. $y(x) = C\exp(-\cos x)$

4. $y(x) = C(1 + x)^4$

5. $y(x) = \sin\left(C + \sqrt{x}\right)$

6. $y(x) = (x^{3/2} + C)^2$

7. $y(x) = (2x^{4/3} + C)^{3/2}$

8. $y(x) = \sin^{-1}(x^2 + C)$

9. $y(x) = C(1 + x)/(1 - x)$

10. $y(x) = (1 + x)/[1 + C(1 + x)] - 1$

11. $y(x) = (C - x^2)^{-1/2}$

12. $y^2 + 1 = Ce^{x^2}$

13. $\ln(y^4 + 1) = C + 4\sin x$

14. $3y + 2y^{3/2} = 3x + 2x^{3/2} + C$

15. $1/(3y^3) - 2/y = 1/x + \ln|x| + C$

16. $y(x) = \sec^{-1}(C\sqrt{1 + x^2})$

17. $\ln|1 + y| = x + \frac{1}{2}x^2 + C$

18. $y(x) = \tan\left(C - \dfrac{1}{x} - x\right)$

19. $y(x) = 2\exp(e^x)$

20. $y(x) = \tan(x^3 + \pi/4)$

21. $y^2 = 1 + \sqrt{x^2 - 16}$

22. $y(x) = -3\exp(x^4 - x)$

23. $y(x) = \frac{1}{2}(1 + e^{2x-2})$

24. $y(x) = \frac{\pi}{2}\sin x$

25. $y(x) = x\exp(x^2 - 1)$

26. $y(x) = 1/(1 - x^2 - x^3)$

27. $y = \ln(3e^{2x} - 2)$

28. $y(x) = \tan^{-1}(\sqrt{x} - 1)$

29. (a) General solution $y(x) = -1/(x - C)$; (b) The singular solution $y(x) \equiv 0$. (c) In the following figure we see that there is a unique solution through every point of the xy-plane.

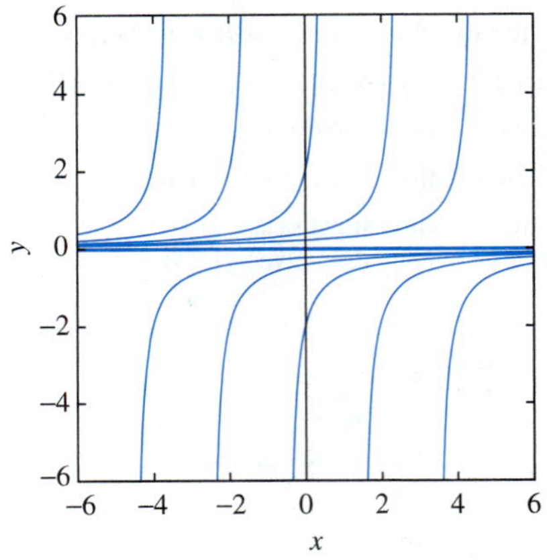

30. General solution $y(x) = (x - C)^2$; singular solution $y(x) \equiv 0$. (a) No solution if $b < 0$; (b) Infinitely many solutions (for all x) if $b \geqq 0$; (c) Two solutions near (a, b) if $b > 0$.

31. Separation of variables gives the same general solution $y = (x - C)^2$ as in Problem 30, but the restriction that $y' = 2\sqrt{y} \geqq 0$ implies that only the *right halves* of the parabolas qualify as solution curves. In the figure below we see that through the point (a, b) there passes (a) No solution curve if $b < 0$, (b) a unique solution curve if $b > 0$, (c) Infinitely many solution curves if $b = 0$.

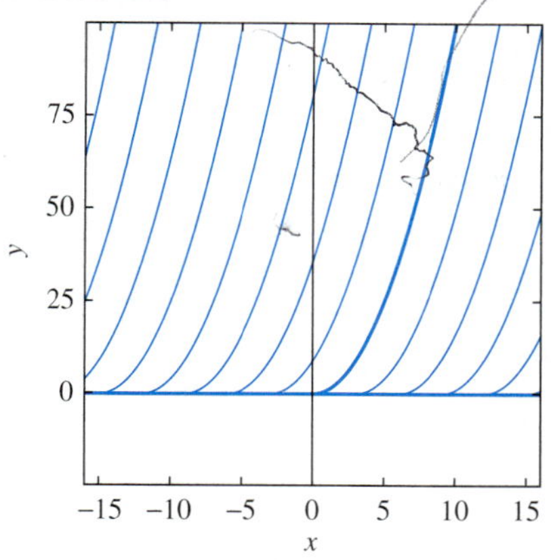

32. General solution $y(x) = \pm\sec(x - C)$; singular solutions $y(x) \equiv \pm 1$.
(a) No solution if $|b| < 1$; (b) A unique solution if $|b| > 1$; (c) Infinitely many solutions if $b = \pm 1$.

33. About 51840 persons
34. $t \approx 3.87$ hr
35. About 14735 years
36. Age about 686 years
37. \$21103.48
38. \$44.52
39. 2585 mg
40. About 35 years
41. About 4.86×10^9 years ago
42. About 1.25 billion years
43. After a total of about 63 min have elapsed
44. About 2.41 minutes
45. (a) 0.495 m; (b $(8.32 \times 10^{-7})I_0$; (c) 3.29 m
46. (a) About 9.60 inches; (b) About 18,200 ft
47. After about 46 days
48. About 6 billion years
49. After about 66 min 40 s
50. (a) $A(t) = 10 \cdot 3^{2t/15}$; (b) About 20.80 pu; (c) About 15.72 years
51. (a) $A(t) = 15 \cdot (\frac{2}{3})^{t/5}$; (b) approximately 7.84 su; (c) After about 33.4 months
52. About 120 thousand years ago
53. About 74 thousand years ago
54. 3 hours
55. 972 s
56. At time $t = 2048/1562 \approx 1.31$ (in hours)
58. 1:20 P.M.
59. (a) $y(t) = (8 - 7t)^{2/3}$; (b) at 1:08:34 P.M.; (c) $r = \frac{1}{60}\sqrt{\frac{7}{12}}$ (ft.) ≈ 0.15 (in.)
60. About 6 min 3 sec
61. Approximately 14 min 29 s
62. The tank is empty about 14 seconds after 2:00 P.M.
63. (a) 1:52:34 P.M.; (b) $r \approx 0.04442$ ft ≈ 0.53 in.
64. $r = \frac{1}{720}\sqrt{3}$ ft, about $\frac{1}{35}$ in.
65. At approximately 10:29 A.M.

Section 1.5

1. $y(x) = 2(1 - e^{-x})$
2. $y(x) = (3x + C)e^{2x}$
3. $y(x) = e^{-3x}(x^2 + C)$
4. $y(x) = (x + C)e^{x^2}$
5. $y(x) = x + 4x^{-2}$
6. $y(x) = x^2 + 32/x^5$
7. $y(x) = 5x^{1/2} + Cx^{-1/2}$
8. $y(x) = 3x + Cx^{-1/3}$
9. $y(x) = x(7 + \ln x)$
10. $y(x) = 3x^3 + Cx^{3/2}$
11. $y(x) \equiv 0$
12. $y(x) = \frac{1}{4}x^5 - 56x^{-3}$
13. $y(x) = (e^x + e^{-x})/2$
14. $y(x) = x^3 \ln x + 10x^3$
15. $y(x) = [1 - 5\exp(-x^2)]/2$
16. $y(x) = 1 + e^{-\sin x}$
17. $y(x) = (1 + \sin x)/(1 + x)$
18. $y(x) = x^2(\sin x + C)$
19. $y(x) = \frac{1}{2}\sin x + C\csc x$
20. $y(x) = -1 + \exp\left(x + \frac{1}{2}x^2\right)$
21. $y(x) = x^3 \sin x$
22. $y(x) = (x^3 + 5)e^{x^2}$
23. $y(x) = x^3(2 + Ce^{-2x})$
24. $y(x) = \frac{1}{3}[1 + 16(x^2 + 4)^{-3/2}]$
25. $y(x) = \left[\exp\left(-\frac{3}{2}x^2\right)\right]\left[3(x^2 + 1)^{3/2} - 2\right]$
26. $x(y) = 1/2y^2 + C/y^4$
27. $x(y) = e^y\left(C + \frac{1}{2}y^2\right)$
28. $x(y) = \frac{1}{2}[y + (1 + y^2)(\tan^{-1} y + C)]$
29. $y(x) = \left[\exp(x^2)\right]\left[C + \frac{1}{2}\sqrt{\pi}\,\mathrm{erf}(x)\right]$
30. $y(x) = x^{1/2}\int_1^x t^{-1/2}\cos t\,dt$
32. (a) $y(x) = \sin x - \cos x$; (b) $y(x) = Ce^{-x} + \sin x - \cos x$; (c) $y(x) = 2e^{-x} + \sin x - \cos x$
33. After about 7 min 41 s
34. About 22.2 days
35. About 5.5452 years
36. (a) $x(t) = (60 - t) - (60 - t)^3/3600$; (b) About 23.09 lb
37. 393.75 lb
38. (a) $x(t) = 50e^{-t/20}$; (b) $y(t) = 150e^{-t/40} - 100e^{-t/20}$; (c) 56.25 lb
39. (b) $y_{max} = 100e^{-1} \approx 36.79$ (gal)
41. (b) Approximately \$1,308,283
43. −50.0529, −28.0265, −6.0000, 16.0265, 38.0529
44. 3.99982, 4.00005, 4.00027, 4.00050, 4.00073
45. $x(t) = 20(1 - e^{-t/10})$; $x = 10$ after $t = 10\ln 2 \approx 6.93$ months.
46. $x(t) = \frac{20}{101}(101 - 102e^{-t/10} + \cos t + 10\sin t)$; $x = 10$ after $t = 6.47$ months.

Section 1.6

1. $x^2 - 2xy - y^2 = C$
2. $y^2 = x^2(\ln x + C)$
3. $y(x) = x(C + \ln|x|)^2$
4. $2\tan^{-1}(y/x) - \ln(y^2/x^2 + 1) = 2\ln x + C$
5. $\ln|xy| = C + xy^{-1}$
6. $2y\ln y = x + Cy$
7. $y^3 = 3x^3(C + \ln|x|)$
8. $y = -x\ln(C - \ln x)$
9. $y(x) = x/(C - \ln|x|)$
10. $x^2 + 2y^2 = Cx^6$
11. $y = C(x^2 + y^2)$
12. $4x^2 + y^2 = x^2(\ln x + C)^2$
13. $y + \sqrt{x^2 + y^2} = Cx^2$
14. $x - \sqrt{x^2 + y^2} = C$
15. $x^2(2xy + y^2) = C$
16. $x = 2\sqrt{x + y + 1} - 2\ln\left(1 + \sqrt{x + y + 1}\right) + C$
17. $y(x) = -4x + 2\tan(2x + C)$
18. $y = \ln(x + y + 1) + C$
19. $y^2 = x/(2 + Cx^5)$
20. $y^3 = 3 + Ce^{-3x^2}$
21. $y^2 = 1/(Ce^{-2x} - 1)$
22. $y^3 = 7x/(7Cx^7 + 15)$
23. $y(x) = (x + Cx^2)^{-3}$
24. $y^2 = e^{2x}/(C + \ln x)$
25. $2x^3y^3 = 3\sqrt{1 + x^4} + C$
26. $y^3 = e^{-x}(x + C)$
27. $y(x) = (x^4 + Cx)^{1/3}$
28. $y = \ln(Cx^2 + x^2e^{2x})$
29. $\sin^2 y = 4x^2 + Cx$
30. $x^2 - 2xe^y - e^{2y} = C$
31. $x^2 + 3xy + y^2 = C$
32. $2x^2 - xy + 3y^2 = C$
33. $x^3 + 2xy^2 + 2y^3 = C$
34. $x^3 + x^2y^2 + y^4 = C$
35. $3x^4 + 4y^3 + 12y\ln x = C$
36. $x + e^{xy} + y^2 = C$
37. $\sin x + x\ln y + e^y = C$

38. $x^2 + 2x\tan^{-1} y + \ln(1 + y^2) = C$
39. $5x^3y^3 + 5xy^4 + y^5 = C$
40. $e^x \sin y + x \tan y = C$
41. $x^2y^{-1} + y^2x^{-3} + 2y^{1/2} = C$
42. $xy^{-2/3} + x^{-3/2}y = C$
43. $y(x) = Ax^2 + B$
44. $x(y) = Ay^2 + B$
45. $y(x) = A\cos 2x + B\sin 2x$
46. $y(x) = x^2 + A\ln x + B$
47. $y(x) = A - \ln|x + B|$
48. $y(x) = \ln x + Ax^{-2} + B$
49. $y(x) = \pm(A + Be^x)^{1/2}$
50. $y(x) = \ln|\sec(x + A)| - \frac{1}{2}x^2 + B$
51. $x(y) = -\frac{1}{3}(y^3 + Ay + B)$
52. $Ay^2 - (Ax + B)^2 = 1$
53. $y(x) = A\tan(Ax + B)$
54. $Ay^2(B - x) = 1$
58. $y = \exp(x^2 + C/x^2)$
59. $x^2 - 2xy - y^2 - 2x - 6y = C$
60. $(x + 3y + 3)^5 = C(y - x - 5)$
61. $x = \tan(x - y) + \sec(x - y) + C$
64. $y(x) = x + e^{-x^2}\left[C + \frac{1}{2}\sqrt{\pi}\,\mathrm{erf}(x)\right]^{-1}$
65. $y(x) = x + (C - x)^{-1}$
69. Approximately 3.68 mi

Chapter 1 Review Problems

1. Linear: $y(x) = x^3(C + \ln x)$
2. Separable: $y(x) = x/(3 - Cx - x\ln x)$
3. Homogeneous: $y(x) = x/(C - \ln x)$
4. Exact: $x^2y^3 + e^x - \cos y = C$
5. Separable: $y(x) = C\exp(x^{-3} - x^{-2})$
6. Separable: $y(x) = x/(1 + Cx + 2x\ln x)$
7. Linear: $y(x) = x^{-2}(C + \ln x)$
8. Homogeneous: $y(x) = 3Cx/(C - x^3) = 3x/(1 + Kx^3)$
9. Bernoulli: $y(x) = (x^2 + Cx^{-1})^2$
10. Separable: $y(x) = \tan\left(C + x + \frac{1}{3}x^3\right)$
11. Homogeneous: $y(x) = x/(C - 3\ln x)$
12. Exact: $3x^2y^3 + 2xy^4 = C$
13. Separable: $y(x) = 1/(C + 2x^2 - x^5)$
14. Homogeneous: $y^2 = x^2/(C + 2\ln x)$
15. Linear: $y(x) = (x^3 + C)e^{-3x}$
16. Substitution: $v = y - x$; solution: $y - x - 1 = Ce^{2x}(y - x + 1)$
17. Exact: $e^x + e^y + e^{xy} = C$
18. Homogeneous: $y^2 = Cx^2(x^2 - y^2)$
19. Separable: $y(x) = x^2/(x^5 + Cx^2 + 1)$
20. Linear: $y(x) = 2x^{-3/2} + Cx^{-3}$
21. Linear: $y(x) = [C + \ln(x - 1)]/(x + 1)$
22. Bernoulli: $y(x) = (2x^4 + Cx^2)^3$
23. Exact: $xe^y + y\sin x = C$
24. Separable: $y(x) = x^{1/2}/(6x^2 + Cx^{1/2} + 2)$
25. Linear: $y(x) = (x + 1)^{-2}(x^3 + 3x^2 + 3x + C)$
$= x + 1 + K(x + 1)^{-2}$
26. Exact: $3x^{3/2}y^{4/3} - 5x^{6/5}y^{3/2} = C$
27. Bernoulli: $y(x) = x^{-1}(C + \ln x)^{-1/3}$
28. Linear: $y(x) = x^{-1}(C + e^{2x})$
29. Linear: $y(x) = (x^2 + x + C)(2x + 1)^{-1/2}$
30. Substitution: $v = x + y$; solution:
$x = 2(x + y)^{1/2} - 2\ln[1 + (x + y)^{1/2}] + C$
31. Separable and linear
32. Separable and Bernoulli
33. Exact and homogeneous
34. Exact and homogeneous
35. Separable and linear
36. Separable and Bernoulli

Chapter 2

Section 2.1

1. $x(t) = \dfrac{2}{2 - e^{-t}}$

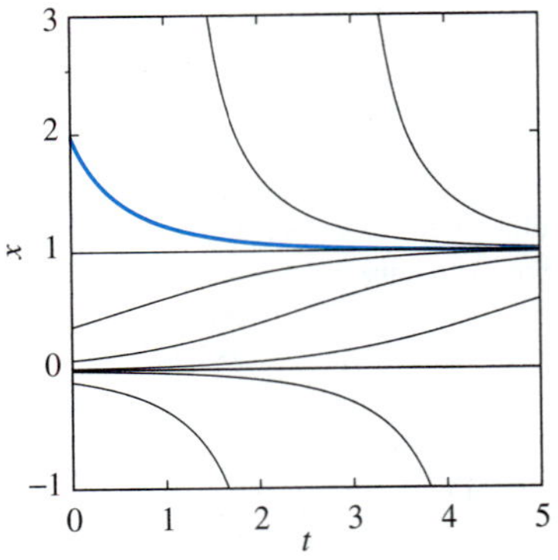

2. $x(t) = \dfrac{10}{1 + 9e^{-10t}}$

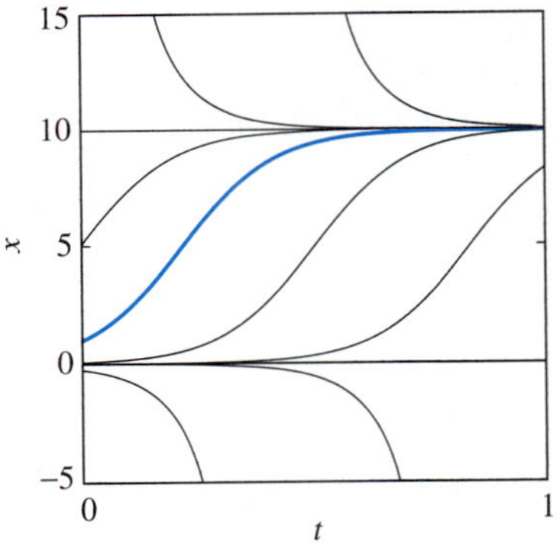

3. $x(t) = \dfrac{2 + e^{-2t}}{2 - e^{-2t}}$

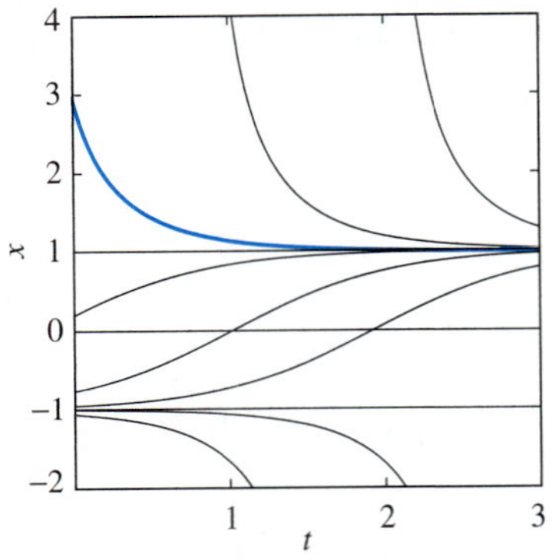

4. $x(t) = \dfrac{3(1 - e^{-12t})}{2(1 + e^{-12t})}$

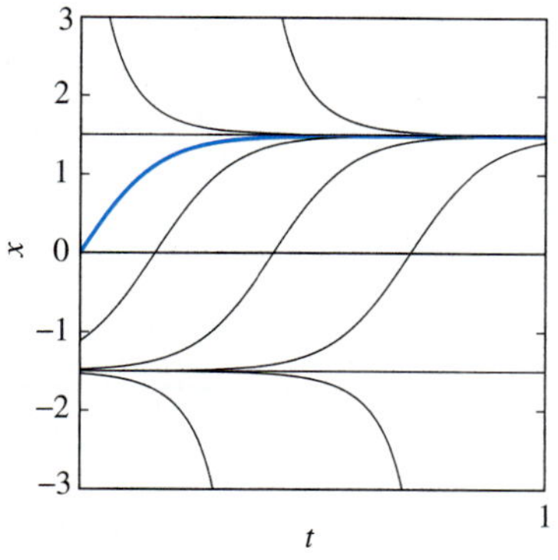

5. $x(t) = \dfrac{40}{8 - 3e^{-15t}}$

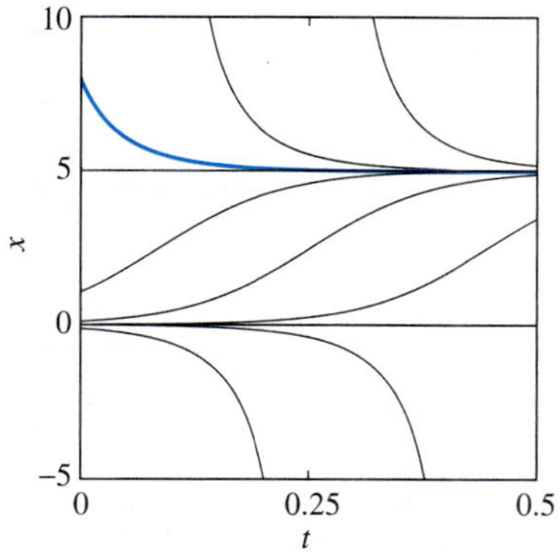

6. $x(t) = \dfrac{10}{2 + 3e^{15t}}$

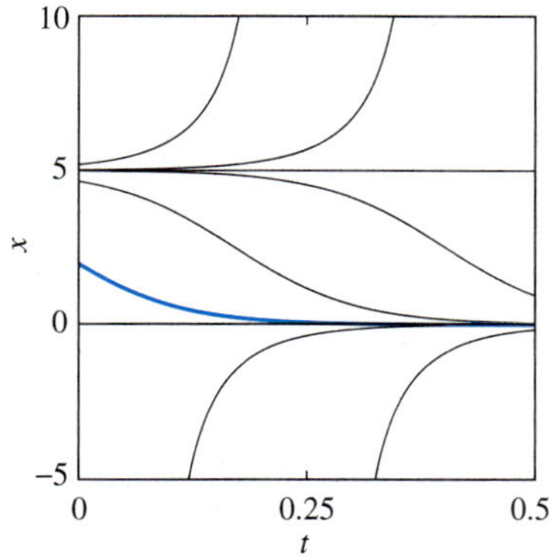

7. $x(t) = \dfrac{77}{11 - 4e^{-28t}}$

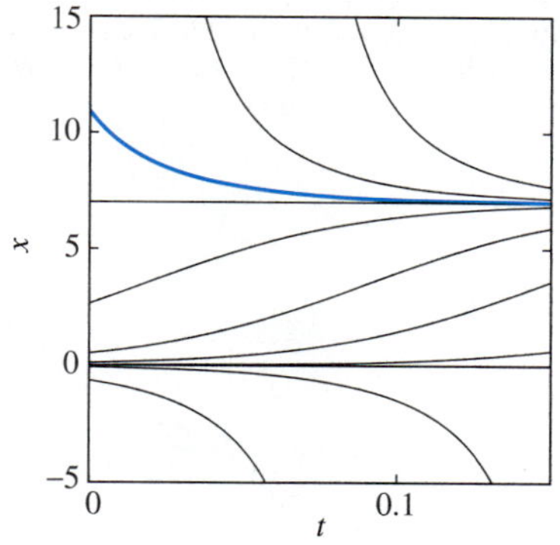

8. $x(t) = \dfrac{221}{17 - 4e^{91t}}$

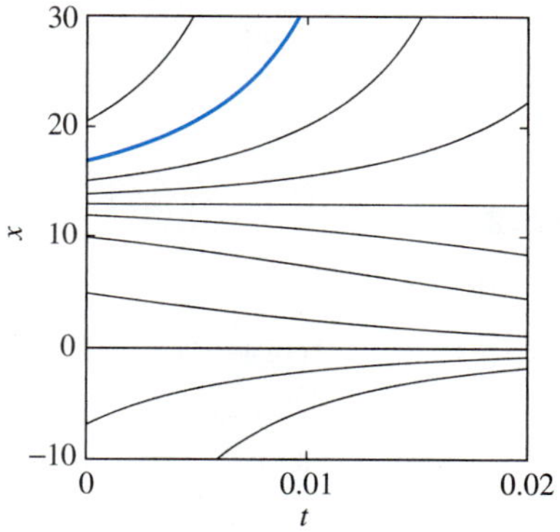

9. 484 **10.** 20 weeks

11. (b) $P(t) = \left(\frac{1}{2}t + 10\right)^2$

12. $P(t) = \dfrac{240}{20 - t}$

13. $P(t) = \dfrac{180}{30 - t}$

14. $P(t) = \dfrac{P_0}{1 + kP_0 t}$

16. About 27.69 months 17. About 44.22 months

19. About 24.41 months

20. About 42.12 months

21. $\dfrac{200}{1 + e^{-6/5}} \approx 153.7$ million

22. About 34.66 days

23. (a) $\lim_{t\to\infty} x(t) = 200$ grams (b) $\frac{5}{4}\ln 3 \approx 1.37$ seconds

24. About 9.24 days

25. (a) $M = 100$ and $k = 0.0002$; (b) In the year 2035

26. $50 \ln \frac{9}{8} \approx 5.89$ months

27. (a) $100 \ln \frac{9}{5} \approx 58.78$ months; (b) $100 \ln 2 \approx 69.31$ months

28. (a) The alligators eventually die out. (b) Doomsday occurs after about 9 years 2 months.

29. (a) $P(140) \approx 127.008$ million; (b) About 210.544 million; (c) In 2000 we get $P \approx 196.169$, whereas the actual 2000 population was about 281.422 million.

31. $\alpha \approx 0.3915$; 2.15×10^6 cells

37. $k \approx 0.000146679$, $M \approx 208.250$

38. $k \approx 0.0000668717$, $M \approx 338.027$

39. $P(t) = P_0 \exp\left(kt + \dfrac{b}{2\pi} \sin 2\pi t\right)$; the colored curve in the figure below shows the graph with $P_0 = 100$, $k = 0.03$, and $b = 0.06$. It oscillates about the black curve which represents natural growth with $P_0 = 100$ and $k = 0.03$. We see that the two agree at the end of each full year.

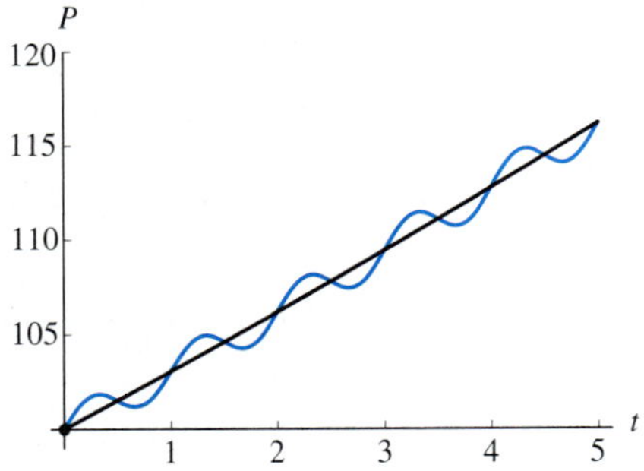

Section 2.2

1. Unstable critical point: $x = 4$;

$$x(t) = 4 + (x_0 - 4)e^t$$

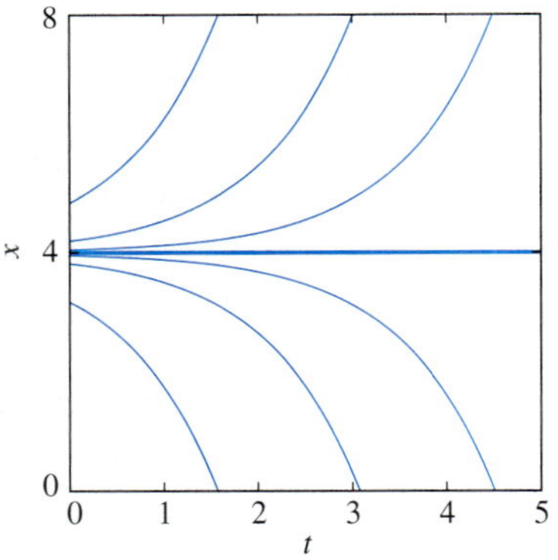

2. Stable critical point: $x = 3$;

$$x(t) = 3 + (x_0 - 3)e^{-t}$$

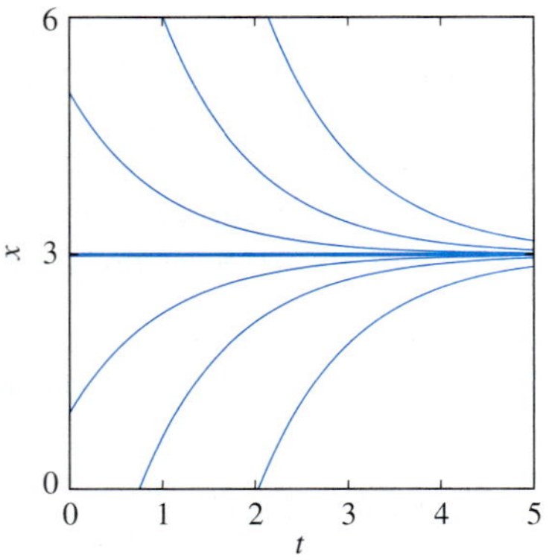

3. Stable critical point: $x = 0$; unstable critical point: $x = 4$;

$$x(t) = \frac{4x_0}{x_0 + (4 - x_0)e^{4t}}$$

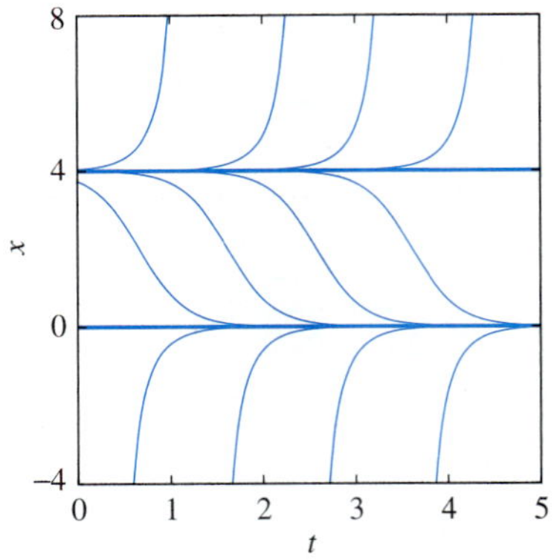

4. Stable critical point: $x = 3$; unstable critical point: $x = 0$;

$$x(t) = \frac{3x_0}{x_0 + (3 - x_0)e^{-3t}}$$

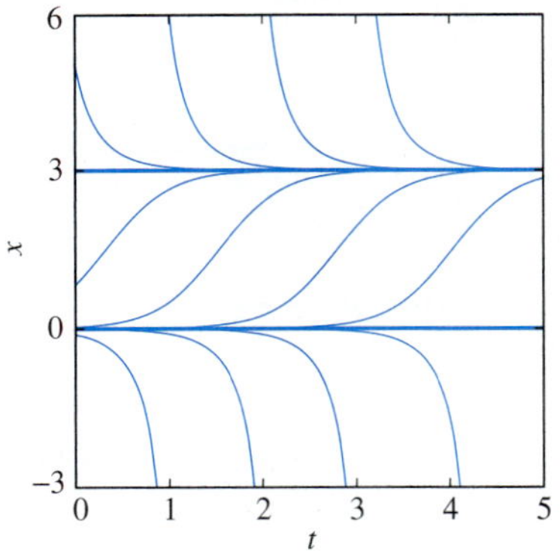

5. Stable critical point: $x = -2$; unstable critical point: $x = 2$;

$$x(t) = \frac{2[x_0 + 2 + (x_0 - 2)e^{4t}]}{x_0 + 2 - (x_0 - 2)e^{4t}}$$

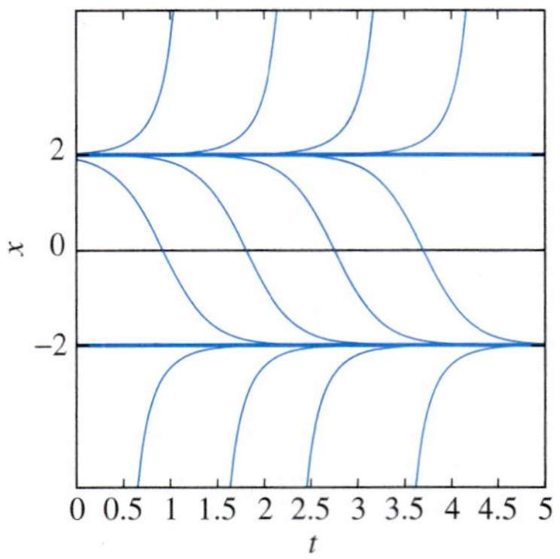

6. Stable critical point: $x = 3$; unstable critical point: $x = -3$;

$$x(t) = \frac{3[x_0 - 3 + (x_0 + 3)e^{6t}]}{3 - x_0 + (x_0 + 3)e^{6t}}$$

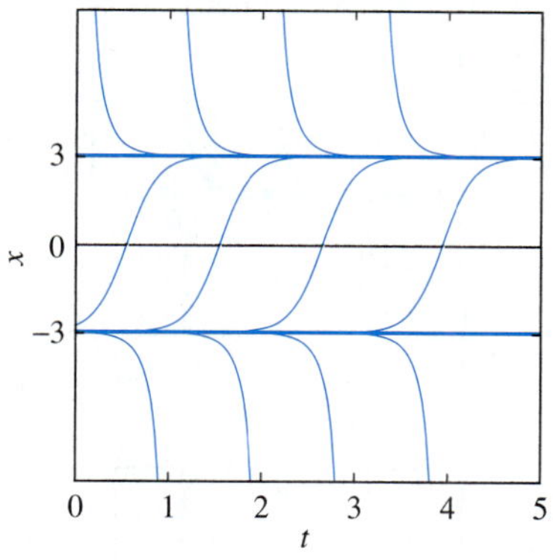

7. Semistable (see Problem 18) critical point: $x = 2$;

$$x(t) = \frac{(2t - 1)x_0 - 4t}{tx_0 - 2t - 1}$$

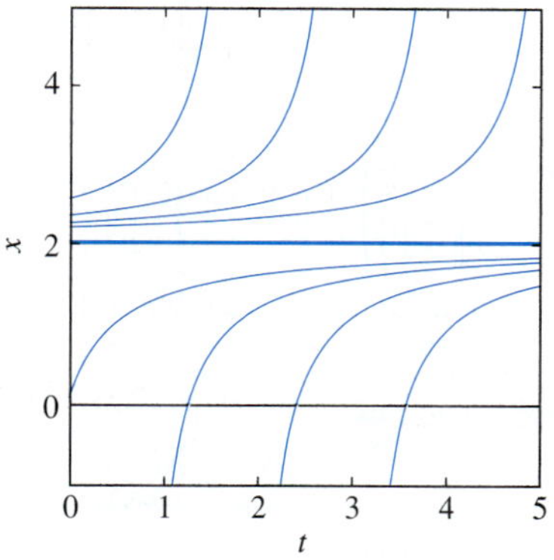

8. Semistable critical point: $x = 3$;

$$x(t) = \frac{(3t + 1)x_0 - 9t}{tx_0 - 3t + 1}$$

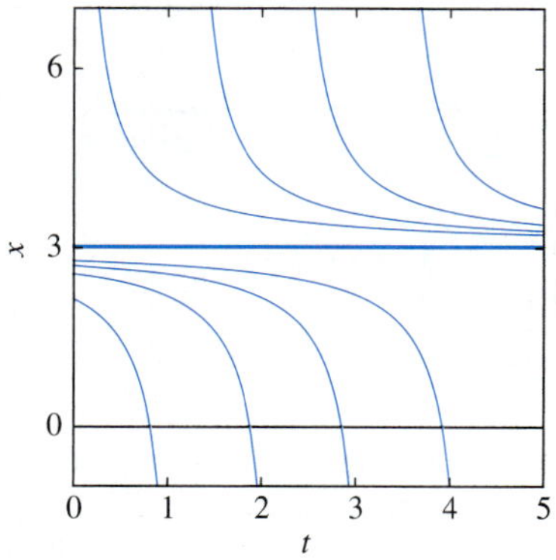

9. Stable critical point: $x = 1$; unstable critical point: $x = 4$;

$$x(t) = \frac{4(1 - x_0) + (x_0 - 4)e^{3t}}{1 - x_0 + (x_0 - 4)e^{3t}}$$

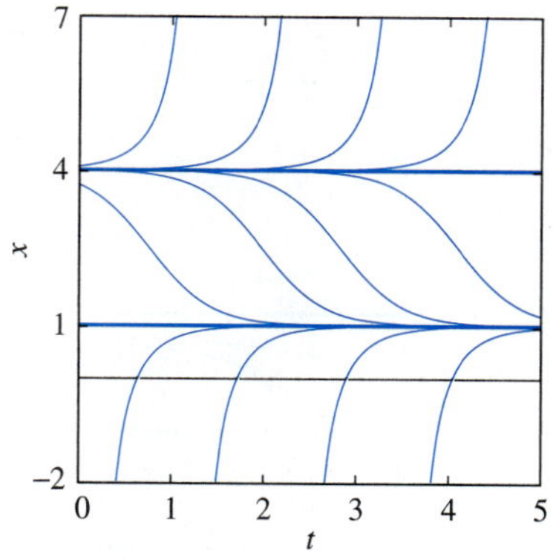

10. Stable critical point: $x = 5$; unstable critical point: $x = 2$;

$$x(t) = \frac{2(5 - x_0) + 5(x_0 - 2)e^{3t}}{5 - x_0 + (x_0 - 2)e^{3t}}$$

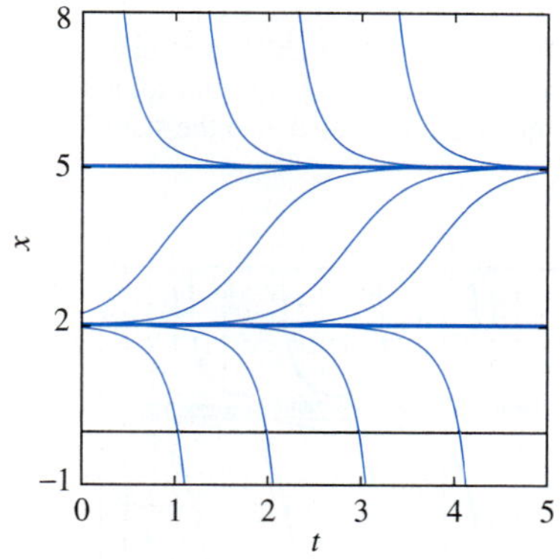

11. Unstable critical point: $x = 1$;

$$\frac{1}{(x(t) - 1)^2} = \frac{1}{(x_0 - 1)^2} - 2t$$

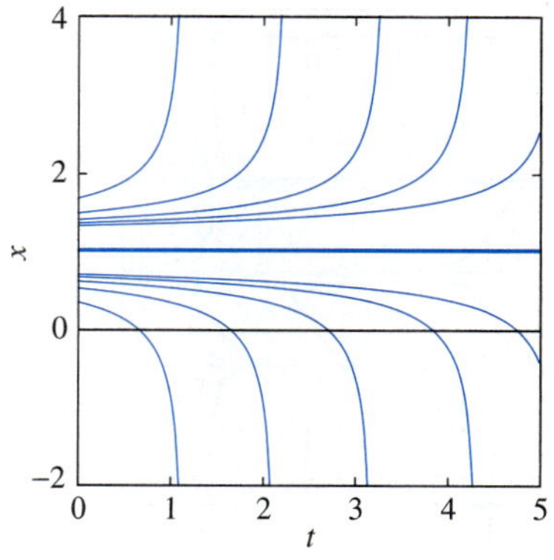

12. Stable critical point: $x = 2$;

$$\frac{1}{(2 - x(t))^2} = \frac{1}{(2 - x_0)^2} + 2t$$

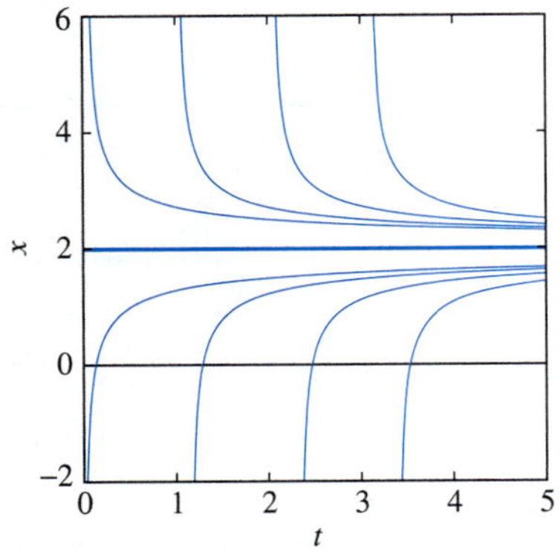

For each of Problems 13 through 18, we show a plot of slope field and typical solution curves. The equilibrium solutions of the given differential equation are labeled, and the stability or instability of each should be clear from the picture.

13.

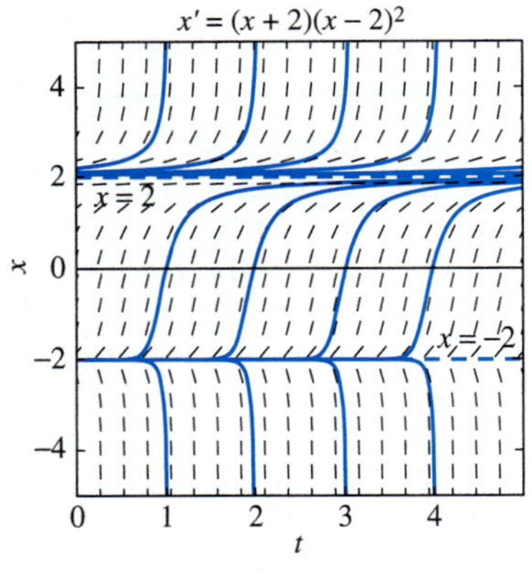

14.

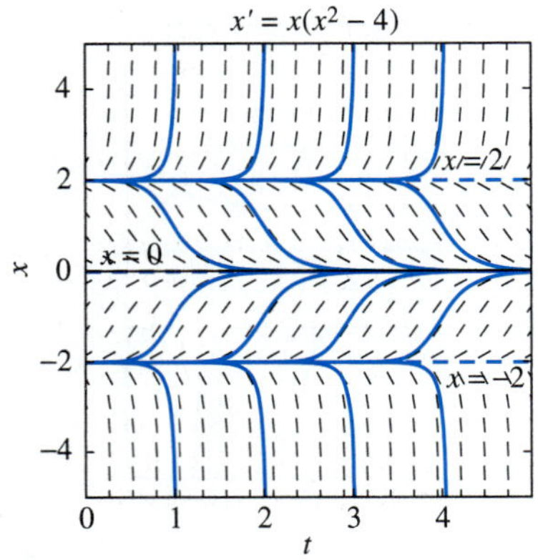

15.

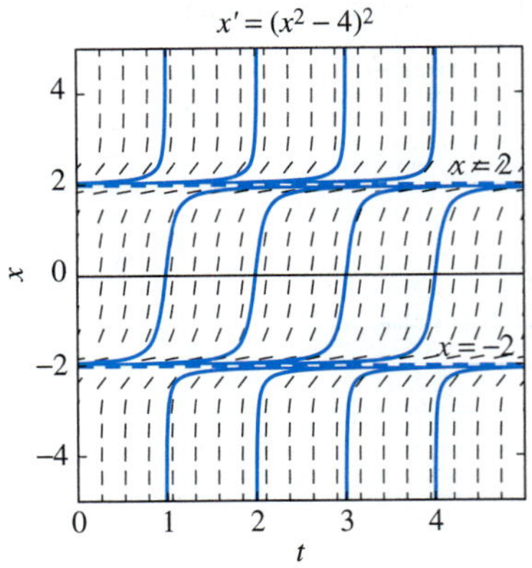

16.

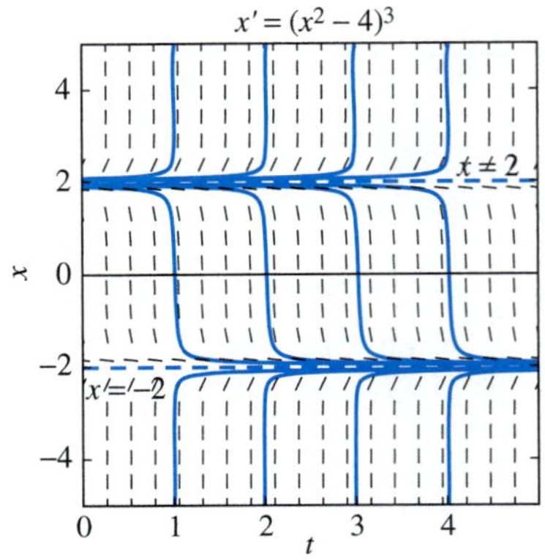

17.

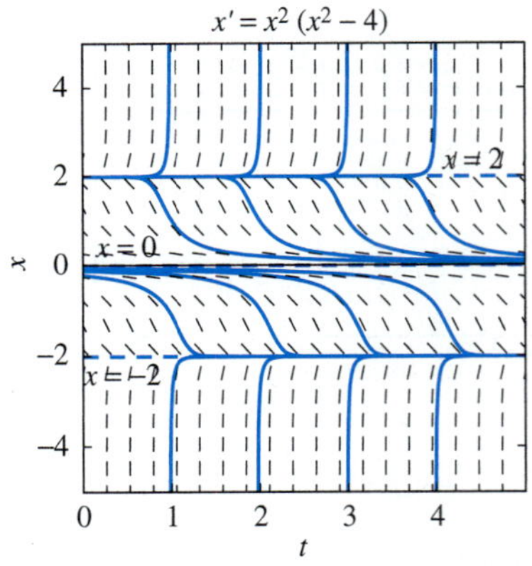

18.

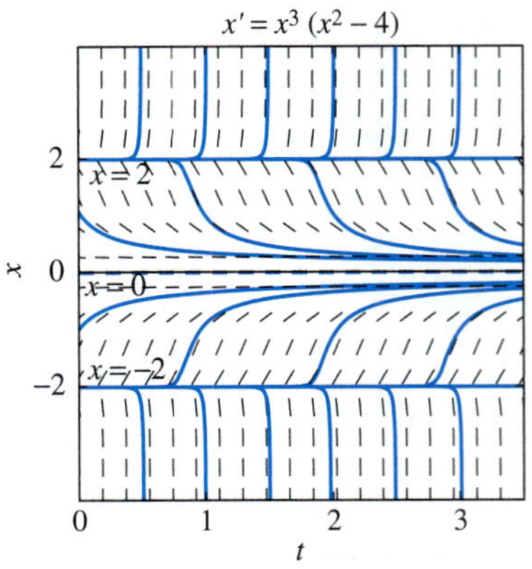

19. There are two critical points if $h < 2\frac{1}{2}$, one critical point if $h = 2\frac{1}{2}$, and no critical points if $h > 2\frac{1}{2}$. The bifurcation diagram is the parabola $(c - 5)^2 = 25 - 10h$ in the hc-plane.

20. There are two critical points if $s < \frac{1}{16}$, one critical point if $s = \frac{1}{16}$, and no critical points if $s > \frac{1}{16}$. The bifurcation diagram is the parabola $(2c - 5)^2 = 25(1 - 16s)$ in the sc-plane.

Section 2.3

1. Approximately 31.5 s
3. $400/(\ln 2) \approx 577$ ft
5. $400 \ln 7 \approx 778$ ft
7. (a) 100 ft/s; (b) about 23 s and 1403 ft to reach 90 ft/s
8. (a) 100 ft/s; (b) about 14.7 s and 830 ft to reach 90 ft/s
9. 50 ft/s
10. About 5 min 47 s
11. Time of fall: about 12.5 s
12. Approximately 648 ft
19. Approximately 30.46 ft/s; exactly 40 ft/s
20. Approximately 277.26 ft
22. Approximately 20.67 ft/s; about 484.57 s
23. Approximately 259.304 s
24. (a) About 0.88 cm; (b) about 2.91 km
25. (b) About 1.389 km/sec; (c) $r_{\max} = 100R/19 \approx 5.26R$
26. Yes
28. (b) After about $8\frac{1}{2}$ minutes it hits the surface at about 4.116 km/sec.
29. About 51.427 km
30. Approximately 11.11 km/s (as compared with the earth's escape velocity of about 11.18 km/sec).

Section 2.4

In Problems 1 through 10, we round off the indicated values to 3 decimal places.

1. Approximate values 1.125 and 1.181; true value 1.213
2. Approximate values 1.125 and 1.244; true value 1.359
3. Approximate values 2.125 and 2.221; true value 2.297
4. Approximate values 0.625 and 0.681; true value 0.713
5. Approximate values 0.938 and 0.889; true value 0.851
6. Approximate values 1.750 and 1.627; true value 1.558
7. Approximate values 2.859 and 2.737; true value 2.647
8. Approximate values 0.445 and 0.420; true value 0.405
9. Approximate values 1.267 and 1.278; true value 1.287
10. Approximate values 1.125 and 1.231; true value 1.333

Problems 11 through 24 call for tables of values that would occupy too much space for inclusion here. In Problems 11 through 16 we give first the final x-value, next the corresponding approximate y-values obtained with step sizes $h = 0.01$ and $h = 0.005$, and then the final true y-value. (All y-values rounded off accurate to 4 decimal places.)

11. 1.0, −0.7048, −0.7115, −0.7183
12. 1.0, 2.9864, 2.9931, 3.0000
13. 2.0, 4.8890, 4.8940, 4.8990
14. 2.0, 3.2031, 3.2304, 3.2589
15. 3.0, 3.4422, 3.4433, 3.4444
16. 3.0, 8.8440, 8.8445, 8.8451

In Problems 17 through 24, we give first the final x-value and then the corresponding approximate y-values obtained with step sizes $h = 0.1$, $h = 0.02$, $h = 0.004$, and $h = 0.0008$ respectively. (All y-values rounded off accurate to 4 decimal places.)

17. 1.0, 0.2925, 0.3379, 0.3477, 0.3497
18. 2.0, 1.6680, 1.6771, 1.6790, 1.6794
19. 2.0, 6.1831, 6.3653, 6.4022, 6.4096
20. 2.0, −1.3792, −1.2843, −1.2649, −1.2610
21. 2.0, 2.8508, 2.8681, 2.8716, 2.8723
22. 2.0, 6.9879, 7.2601, 7.3154, 7.3264
23. 1.0, 1.2262, 1.2300, 1.2306, 1.2307
24. 1.0, 0.9585, 0.9918, 0.9984, 0.9997

25. With both step sizes $h = 0.01$ and $h = 0.005$, the approximate velocity after 1 second is 16.0 ft/s (80% of the limiting velocity of 20 ft/sec); after 2 seconds it is 19.2 ft/s (96% of the limiting velocity).

26. With both step sizes $h = 1$ and $h = 0.5$, the approximate population after 5 years is 49 deer (65% of the limiting population of 75 deer); after 10 years it is 66 deer (88% of the limiting population).

27. With successive step sizes $h = 0.1, 0.01, 0.001, \ldots$ the first four approximations to $y(2)$ we obtain are 0.7772, 0.9777, 1.0017, and 1.0042. It therefore seems likely that $y(2) \approx 1.00$.

28. With successive step sizes $h = 0.1, 0.01, 0.001, \ldots$ the first four approximations to $y(2)$ we obtain are 1.2900, 1.4435, 1.4613, and 1.4631. It therefore seems likely that $y(2) \approx 1.46$.

29.

	$h = 0.15$	$h = 0.03$	$h = 0.006$
x	y	y	y
−1.0	1.0000	1.0000	1.0000
−0.7	1.0472	1.0512	1.0521
−0.4	1.1213	1.1358	1.1390
−0.1	1.2826	1.3612	1.3835
0.2	0.8900	1.4711	0.8210
0.5	0.7460	1.2808	0.7192

30.

	$h = 0.1$	$h = 0.01$
x	y	y
1.8	2.8200	4.3308
1.9	3.9393	7.9425
2.0	5.8521	28.3926

31.

	$h = 0.1$	$h = 0.01$
x	y	y
0.7	4.3460	6.4643
0.8	5.8670	11.8425
0.9	8.3349	39.5010

Section 2.5

1.

x	Improved Euler y	Actual y
0.1	1.8100	1.8097
0.2	1.6381	1.6375
0.3	1.4824	1.4816
0.4	1.3416	1.3406
0.5	1.2142	1.2131

Note: In Problems 2 through 10, we give the value of x, the corresponding improved Euler value of y, and the true value of y.

2. 0.5, 1.3514, 1.3191
3. 0.5, 2.2949, 2.2974
4. 0.5, 0.7142, 0.7131
5. 0.5, 0.8526, 0.8513
6. 0.5, 1.5575, 1.5576
7. 0.5, 2.6405, 2.6475
8. 0.5, 0.4053, 0.4055
9. 0.5, 1.2873, 1.2874
10. 0.5, 1.3309, 1.3333

In Problems 11 through 16, we give the final value of x, the corresponding values of y with $h = 0.01$ and with $h = 0.005$, and the true value of y.

11. 1.0, −0.71824, −0.71827, −0.71828
12. 1.0, 2.99995, 2.99999, 3.00000
13. 2.0, 4.89901, 4.89899, 4.89898
14. 2.0, 3.25847, 3.25878, 3.25889
15. 3.0, 3.44445, 3.44445, 3.44444
16. 3.0, 8.84511, 8.84509, 8.84509

In Problems 17 through 24, we give the final value of x and the corresponding values of y for $h = 0.1, 0.02, 0.004$, and 0.0008.

17. 1.0, 0.35183, 0.35030, 0.35023, 0.35023
18. 2.0, 1.68043, 1.67949, 1.67946, 1.67946
19. 2.0, 6.40834, 6.41134, 6.41147, 6.41147
20. 2.0, −1.26092, −1.26003, −1.25999, −1.25999
21. 2.0, 2.87204, 2.87245, 2.87247, 2.87247
22. 2.0, 7.31578, 7.32841, 7.32916, 7.32920
23. 1.0, 1.22967, 1.23069, 1.23073, 1.23073
24. 1.0, 1.00006, 1.00000, 1.00000, 1.00000
25. With both step sizes $h = 0.01$ and $h = 0.005$ the approximate velocity after 1 second is 15.962 ft/s (80% of the limiting velocity of 20 ft/sec); after 2 seconds it is 19.185 ft/s (96% of the limiting velocity).
26. With both step sizes $h = 1$ and $h = 0.5$ the approximate population after 5 years is 49.391 deer (65% of the limiting population of 75 deer); after 10 seconds it is 66.113 deer (88% of the limiting population).
27. With successive step sizes $h = 0.1, 0.01, 0.001, \ldots$ the first three approximations to $y(2)$ we obtain are 1.0109, 1.0045, and 1.0045. It therefore seems likely that $y(2) \approx 1.0045$.
28. With successive step sizes $h = 0.1, 0.01, 0.001, \ldots$ the first four approximations to $y(2)$ we obtain are 1.4662, 1.4634, 1.4633, and 1.4633. It therefore seems likely that $y(2) \approx 1.4633$.
29. Impact speed approximately 43.22 m/s
30. Impact speed approximately 43.48 m/s

Section 2.6

1. $y(0.25) \approx 1.55762$; $y(0.25) = 1.55760$.
$y(0.5) \approx 1.21309$; $y(0.5) = 1.21306$.
Solution: $y = 2e^{-x}$

In Problems 2 through 10, we give the approximation to $y(0.5)$, its true value, and the solution.

2. 1.35867, 1.35914; $y = \frac{1}{2}e^{2x}$
3. 2.29740, 2.29744; $y = 2e^x - 1$
4. 0.71309, 0.71306; $y = 2e^{-x} + x - 1$
5. 0.85130, 0.85128; $y = -e^x + x + 2$
6. 1.55759, 1.55760; $u = 2\exp\left(-x^2\right)$
7. 2.64745, 2.64749; $y = 3\exp\left(-x^3\right)$
8. 0.40547, 0.40547; $y = \ln(x+1)$
9. 1.28743, 1.28743; $y = \tan\frac{1}{4}(x+\pi)$
10. 1.33337, 1.33333; $y = (1-x^2)^{-1}$
11. *Solution*: $y(x) = 2 - e^x$.

x	$h = 0.2$ y	$h = 0.1$ y	Exact y
0.0	1.00000	1.00000	1.00000
0.2	0.77860	0.77860	0.77860
0.4	0.50818	0.50818	0.50818
0.6	0.17789	0.17788	0.17788
0.8	−0.22552	−0.22554	−0.22554
1.0	−0.71825	−0.71828	−0.71828

In Problems 12 through 16, we give the final value of x, the corresponding Runge–Kutta approximations with $h = 0.2$ and with $h = 0.1$, the exact value of y, and the solution.

12. 1.0, 2.99996, 3.00000, 3.00000; $y = 1 + 2/(2-x)$
13. 2.0, 4.89900, 4.89898, 4.89898; $y = \sqrt{8+x^4}$
14. 2.0, 3.25795, 3.25882, 3.25889; $y = 1/(1-\ln x)$
15. 3.0, 3.44445, 3.44444, 3.44444; $y = x + 4x^{-2}$
16. 3.0, 8.84515, 8.84509, 8.84509; $y = (x^6 - 37)^{1/3}$

In Problems 17 through 24, we give the final value of x and the corresponding values of y with $h = 0.2, 0.1, 0.05$, and 0.025.

17. 1.0, 0.350258, 0.350234, 0.350232, 0.350232
18. 2.0, 1.679513, 1.679461, 1.679459, 1.679459
19. 2.0, 6.411464, 6.411474, 6.411474, 6.411474
20. 2.0, −1.259990, −1.259992, −1.259993, −1.259993
21. 2.0, 2.872467, 2.872468, 2.872468, 2.872468
22. 2.0, 7.326761, 7.328452, 7.328971, 7.329134
23. 1.0, 1.230725, 1.230731, 1.230731, 1.230731
24. 1.0, 1.000000, 1.000000, 1.000000, 1.000000
25. With both step sizes $h = 0.1$ and $h = 0.05$, the approximate velocity after 1 second is 15.962 ft/s (80% of the limiting velocity of 20 ft/sec); after 2 seconds it is 19.185 ft/s (96% of the limiting velocity).

26. With both step sizes $h = 6$ and $h = 3$, the approximate population after 5 years is 49.3915 deer (65% of the limiting population of 75 deer); after 10 years it is 66.1136 deer (88% of the limiting population).

27. With successive step sizes $h = 1, 0.1, 0.01, \ldots$ the first four approximations to $y(2)$ we obtain are 1.05722, 1.00447, 1.00445, and 1.00445. Thus it seems likely that $y(2) \approx 1.00445$ accurate to 5 decimal places.

28. With successive step sizes $h = 1, 0.1, 0.01, \ldots$ the first four approximations to $y(2)$ we obtain are 1.48990, 1.46332, 1.46331, and 1.46331. Thus it seems likely that $y(2) \approx 1.4633$ accurate to 5 decimal places.

29. Time aloft: approximately 9.41 seconds

30. Time aloft: approximately 9.41 seconds

Chapter 3

Section 3.1

1. $x = 3, y = 2$

2. $x = 5, y = -3$

3. $x = -4, y = 3$

4. $x = 5, y = 4$

5. Inconsistent—no solution

6. Inconsistent—no solution

7. $x = -10 + 4t, y = t$ (infinitely many solutions)

8. $x = 4 + 2t, y = t$ (infinitely many solutions)

9. $x = 4, y = -1, z = 3$

10. $x = 3, y = 1, z = -2$

11. $x = 1, y = 3, z = -4$

12. $x = 1, y = 3, z = 5$

13. $x = y = z = 0$

14. $x = 5, y = 3, z = -4$

15. Inconsistent—no solution

16. Inconsistent—no solution

17. Inconsistent—no solution

18. Inconsistent—no solution

19. $x = 8 + 3t, y = 3 + 2t, z = t$ (infinitely many solutions)

20. $x = -5t, y = 5 + t, z = t$ (infinitely many solutions)

21. $x = 3 - 2t, y = 2 + 3t, z = t$ (infinitely many solutions)

22. $x = -4t, y = -5t, z = t$ (infinitely many solutions)

23. $y(x) = 3\cos 2x + 4\sin 2x$

24. $y(x) = 5\cosh 3x + 4\sinh 3x$

25. $y(x) = 7e^{5x} + 3e^{-5x}$

26. $y(x) = 23e^{11x} + 21e^{-11x}$

27. $y(x) = 23e^{3x} + 17e^{-5x}$

28. $y(x) = 23e^{3x} - 8e^{7x}$

29. $y(x) = 52e^{x/2} - 45e^{x/3}$

30. $y(x) = 81e^{4x/3} - 40e^{-7x/5}$

31. *Suggestion*: The two lines both pass through the origin.

32. *Suggestion*: Two distinct planes in space either are parallel or intersect in a straight line.

33. (a) No solution (b) A unique solution (c) No solution (d) No solution (e) A unique solution (f) Infinitely many solutions

34. (a) No solution (b) Infinitely many solutions (c) No solution (d) No solution (e) Infinitely many solutions (f) A unique solution

Section 3.2

1. $x_1 = 1, x_2 = 0, x_3 = 2$

2. $x_1 = 5, x_2 = 1, x_3 = -3$

3. $x_1 = 13 + 11t, x_2 = 2 + 5t, x_3 = t$

4. $x_1 = 35 + 33t, x_2 = 5 + 7t, x_3 = t$

5. $x_1 = 13 + 4t, x_2 = 6 + t, x_3 = 5 + 3t, x_4 = t$

6. $x_1 = 17 + t, x_2 = 11 + 3t, x_3 = t, x_4 = -4$

7. $x_1 = 3 - 8s + 19t, x_2 = 7 + 2s - 7t, x_3 = s, x_4 = t$

8. $x_1 = -25 + 10s + 22t, x_2 = s, x_3 = 10 - 3t, x_4 = t$

9. $x_1 = 1, x_2 = 3, x_3 = -5, x_4 = 6$

10. $x_1 = 63s - 16t, x_2 = 13s - 8t, x_3 = s, x_4 = 5t, x_5 = t$

11. $x_1 = 3, x_2 = -2, x_3 = 4$

12. $x_1 = 5, x_2 = -3, x_3 = 2$

13. $x_1 = 4 + 3t, x_2 = 3 - 2t, x_3 = t$

14. $x_1 = 5 + 2t, x_2 = t, x_3 = 7$

15. Inconsistent—no solution

16. Inconsistent—no solution

17. $x_1 = 3 - 2t, x_2 = -4 + t, x_3 = 5 - 3t, x_4 = t$

18. $x_1 = 4 + 2s - 3t, x_2 = s, x_3 = 3 - 4t, x_4 = t$

19. $x_1 = 3 - s - t, x_2 = 5 + 2s - 3t, x_3 = s, x_4 = t$

20. $x_1 = 2 + 3t, x_2 = 1 + s - 2t, x_3 = 2 + 2s, x_4 = s, x_5 = t$

21. $x_1 = 2, x_2 = 1, x_3 = 3, x_4 = 4$

22. $x_1 = 3, x_2 = -2, x_3 = 4, x_4 = -1$

23. (a) None (b) $k \neq 2$ (c) $k = 2$

24. (a) $k \neq 2$ (b) None (c) $k = 4$

25. (a) $k \neq 4$ (b) $k = 4$ (c) None

26. (a) All k (b) None (c) None

27. (a) None (b) $k \neq 11$ (c) $k = 11$

28. No solution unless $c = 2a + 3b$, in which case it has infinitely many solutions.

Section 3.3

1. $\begin{bmatrix} 1 & 0 \\ 0 & 1 \end{bmatrix}$

2. $\begin{bmatrix} 1 & 0 \\ 0 & 1 \end{bmatrix}$

3. $\begin{bmatrix} 1 & 0 & -2 \\ 0 & 1 & 3 \end{bmatrix}$

4. $\begin{bmatrix} 1 & 0 & 2 \\ 0 & 1 & -1 \end{bmatrix}$

5. $\begin{bmatrix} 1 & 0 & -5 \\ 0 & 1 & -3 \end{bmatrix}$

6. $\begin{bmatrix} 1 & 0 & 7 \\ 0 & 1 & -6 \end{bmatrix}$

7. $\begin{bmatrix} 1 & 0 & 5 \\ 0 & 1 & -1 \\ 0 & 0 & 0 \end{bmatrix}$

8. $\begin{bmatrix} 1 & 0 & 0 \\ 0 & 1 & 0 \\ 0 & 0 & 1 \end{bmatrix}$

9. $\begin{bmatrix} 1 & 0 & 2 \\ 0 & 1 & 4 \\ 0 & 0 & 0 \end{bmatrix}$

10. $\begin{bmatrix} 1 & 0 & -3 \\ 0 & 1 & 5 \\ 0 & 0 & 0 \end{bmatrix}$

11. $\begin{bmatrix} 1 & 3 & 0 \\ 0 & 0 & 1 \\ 0 & 0 & 0 \end{bmatrix}$

12. $\begin{bmatrix} 1 & -4 & 0 \\ 0 & 0 & 1 \\ 0 & 0 & 0 \end{bmatrix}$

13. $\begin{bmatrix} 1 & 0 & 0 & 3 \\ 0 & 1 & 0 & -2 \\ 0 & 0 & 1 & 2 \end{bmatrix}$

14. $\begin{bmatrix} 1 & 0 & 0 & 4 \\ 0 & 1 & 0 & -3 \\ 0 & 0 & 1 & 5 \end{bmatrix}$

15. $\begin{bmatrix} 1 & 0 & -1 & 2 \\ 0 & 1 & 3 & -1 \\ 0 & 0 & 0 & 0 \end{bmatrix}$

16. $\begin{bmatrix} 1 & 0 & 3 & -2 \\ 0 & 1 & 4 & 3 \\ 0 & 0 & 0 & 0 \end{bmatrix}$

17. $\begin{bmatrix} 1 & 0 & 0 & 2 & -3 \\ 0 & 1 & 0 & -1 & 4 \\ 0 & 0 & 1 & -2 & -5 \end{bmatrix}$

18. $\begin{bmatrix} 1 & 0 & 3 & -2 & 3 \\ 0 & 1 & 4 & 5 & 1 \\ 0 & 0 & 0 & 0 & 0 \end{bmatrix}$

19. $\begin{bmatrix} 1 & 0 & 2 & 1 & 3 \\ 0 & 1 & -2 & -3 & 1 \\ 0 & 0 & 0 & 0 & 0 \end{bmatrix}$

20. $\begin{bmatrix} 1 & 2 & 0 & 2 & 3 \\ 0 & 0 & 1 & 1 & 4 \\ 0 & 0 & 0 & 0 & 0 \end{bmatrix}$

21. $x_1 = 3, x_2 = -2, x_3 = 4$

22. $x_1 = 5, x_2 = -3, x_3 = 2$

23. $x_1 = 4 + 3t, x_2 = 3 - 2t, x_3 = t$

24. $x_1 = 5 + 2t, x_2 = t, x_3 = 7$

25. Inconsistent—no solution

26. Inconsistent—no solution

27. $x_1 = 3 - 2t, x_2 = -4 + t, x_3 = 5 - 3t, x_4 = t$

28. $x_1 = 4 + 2s - 3t, x_2 = s, x_3 = 3 - 4t, x_4 = t$

29. $x_1 = 3 - s - t, x_2 = 5 + 2s - 3t, x_3 = s, x_4 = t$

30. $x_1 = 2 + 3t, x_2 = 1 + s - 2t, x_3 = 2 + 2s, x_4 = s, x_5 = t$

31. The sequence $\frac{1}{6}R_3,\ R_2 - 5R_3,\ \frac{1}{4}R_2,\ R_1 - 2R_2,\ R_1 - 3R_3$ of row operations transforms the first matrix to the identity matrix.

33. $\begin{bmatrix} 1 & 0 \\ 0 & 1 \end{bmatrix}, \begin{bmatrix} 1 & * \\ 0 & 0 \end{bmatrix}, \begin{bmatrix} 0 & 1 \\ 0 & 0 \end{bmatrix}, \begin{bmatrix} 0 & 0 \\ 0 & 0 \end{bmatrix}$

Section 3.4

1. $\begin{bmatrix} 5 & -15 \\ 18 & 5 \end{bmatrix}$ **2.** $\begin{bmatrix} 16 & -9 & -18 \\ -26 & 22 & 15 \end{bmatrix}$

3. $\begin{bmatrix} -26 & 20 \\ 12 & -6 \\ 22 & 18 \end{bmatrix}$ **4.** $\begin{bmatrix} 44 & -22 & -20 \\ 53 & 10 & -26 \\ 35 & 21 & 94 \end{bmatrix}$

5. $\mathbf{AB} = \begin{bmatrix} -9 & 1 \\ -10 & 12 \end{bmatrix}, \mathbf{BA} = \begin{bmatrix} -2 & 8 \\ 11 & 5 \end{bmatrix}$

6. $\mathbf{AB} = \begin{bmatrix} 7 & -13 & -24 \\ 23 & 10 & 41 \\ 11 & -8 & 57 \end{bmatrix}, \mathbf{BA} = \begin{bmatrix} 1 & -17 & -22 \\ 12 & 16 & 7 \\ 27 & -21 & 57 \end{bmatrix}$

7. $\mathbf{AB} = \begin{bmatrix} 26 \end{bmatrix}, \mathbf{BA} = \begin{bmatrix} 3 & 6 & 9 \\ 4 & 8 & 12 \\ 5 & 10 & 15 \end{bmatrix}$

8. $\mathbf{AB} = \begin{bmatrix} 21 & 15 \\ 35 & 0 \end{bmatrix}, \mathbf{BA} = \begin{bmatrix} 3 & 0 & 9 \\ 7 & -20 & 13 \\ 16 & -25 & 38 \end{bmatrix}$

9. $\mathbf{BA} = \begin{bmatrix} 4 \\ 7 \\ -22 \end{bmatrix}$ but **AB** is not defined.

10. $\mathbf{AB} = \begin{bmatrix} 1 & -2 & 13 \\ 5 & -6 & 31 \end{bmatrix}$ but **BA** is not defined.

11. $\mathbf{AB} = \begin{bmatrix} 11 & 1 & 5 & 3 \end{bmatrix}$ but **BA** is not defined.

12. Neither product matrix **AB** nor **BA** is defined.

13. $\mathbf{A}(\mathbf{BC}) = (\mathbf{AB})\mathbf{C} = \begin{bmatrix} 32 & 51 \\ -2 & -17 \end{bmatrix}$

14. $\mathbf{A}(\mathbf{BC}) = (\mathbf{AB})\mathbf{C} = \begin{bmatrix} -3 \end{bmatrix}$

15. $\mathbf{A}(\mathbf{BC}) = (\mathbf{AB})\mathbf{C} = \begin{bmatrix} 12 & 15 \\ 8 & 10 \end{bmatrix}$

16. $\mathbf{A}(\mathbf{BC}) = (\mathbf{AB})\mathbf{C} = \begin{bmatrix} -4 & -4 & -2 & 2 \\ -9 & -12 & -9 & 12 \\ -14 & -18 & -13 & 17 \end{bmatrix}$

17. $\mathbf{x} = s(5, -2, 1, 0) + t(-4, 7, 0, 1)$

18. $\mathbf{x} = s(3, 1, 0, 0) + t(-6, 0, -9, 1)$

19. $\mathbf{x} = s(-3, 2, -1, 1, 0) + t(1, -6, 8, 0, 1)$

20. $\mathbf{x} = s(3, 1, 0, 0, 0) + t(-7, 0, 2, 10, 0)$

21. $\mathbf{x} = r(1, -2, 1, 0, 0) + s(-2, 3, 0, 1, 0) + t(-7, -4, 0, 0, 1)$

22. $\mathbf{x} = r(1, 1, 0, 0, 0) + s(-7, 0, 1, 1, 0) + t(-3, 0, 2, 0, 1)$

23. $\mathbf{B} = \begin{bmatrix} 2 & -1 \\ -3 & 2 \end{bmatrix}$ **24.** $\mathbf{B} = \begin{bmatrix} 7 & -4 \\ -5 & 3 \end{bmatrix}$

25. $\mathbf{B} = \begin{bmatrix} 3 & -7 \\ -2 & 5 \end{bmatrix}$

29. $\mathbf{A}^2 = \begin{bmatrix} a^2 + bc & ab + bd \\ ac + dc & bc + d^2 \end{bmatrix}$,

$\mathbf{A}^2 + (ad - bc)\mathbf{I} = \begin{bmatrix} a^2 + ad & ab + bd \\ ac + dc & d^2 + ad \end{bmatrix} =$

$\begin{bmatrix} a(a+d) & b(a+d) \\ c(a+d) & d(a+d) \end{bmatrix} = (a+d)\mathbf{A}$

31. (a) $(\mathbf{A} + \mathbf{B})(\mathbf{A} - \mathbf{B}) = \begin{bmatrix} -25 & -34 \\ -71 & -34 \end{bmatrix} \neq \begin{bmatrix} -8 & -45 \\ -44 & -51 \end{bmatrix}$

$= \mathbf{A}^2 - \mathbf{B}^2$

33. $\begin{bmatrix} 1 & 0 \\ 0 & 1 \end{bmatrix}, \begin{bmatrix} -1 & 0 \\ 0 & -1 \end{bmatrix}, \begin{bmatrix} +1 & 0 \\ 0 & -1 \end{bmatrix}, \begin{bmatrix} -1 & 0 \\ 0 & +1 \end{bmatrix}$

35. $\begin{bmatrix} 2 & -1 \\ 2 & -1 \end{bmatrix}$ **37.** $\begin{bmatrix} 0 & 1 \\ -1 & 0 \end{bmatrix}$

39. $\mathbf{A}(c_1\mathbf{x}_1 + c_2\mathbf{x}_2) = c_1\mathbf{A}\mathbf{x}_1 + c_2\mathbf{A}\mathbf{x}_2 = c_1 \cdot \mathbf{0} + c_2 \cdot \mathbf{0} = \mathbf{0}$

43. $\mathbf{A}^2 = \begin{bmatrix} 6 & -3 & -3 \\ -3 & 6 & -3 \\ -3 & -3 & 6 \end{bmatrix} = 3\mathbf{A}$

Section 3.5

1. $\mathbf{A}^{-1} = \begin{bmatrix} 3 & -2 \\ -4 & 3 \end{bmatrix}, \mathbf{x} = \begin{bmatrix} 3 \\ -2 \end{bmatrix}$

2. $\mathbf{A}^{-1} = \begin{bmatrix} 5 & -7 \\ -2 & 3 \end{bmatrix}, \mathbf{x} = \begin{bmatrix} -26 \\ 11 \end{bmatrix}$

3. $\mathbf{A}^{-1} = \begin{bmatrix} 6 & -7 \\ -5 & 6 \end{bmatrix}, \mathbf{x} = \begin{bmatrix} 33 \\ -28 \end{bmatrix}$

4. $\mathbf{A}^{-1} = \begin{bmatrix} 17 & -12 \\ -7 & 5 \end{bmatrix}, \mathbf{x} = \begin{bmatrix} 25 \\ -10 \end{bmatrix}$

5. $\mathbf{A}^{-1} = \frac{1}{2}\begin{bmatrix} 4 & -2 \\ -5 & 3 \end{bmatrix}, \mathbf{x} = \frac{1}{2}\begin{bmatrix} 8 \\ -7 \end{bmatrix}$

6. $\mathbf{A}^{-1} = \frac{1}{3}\begin{bmatrix} 6 & -7 \\ -3 & 4 \end{bmatrix}, \mathbf{x} = \frac{1}{3}\begin{bmatrix} 25 \\ -10 \end{bmatrix}$

7. $\mathbf{A}^{-1} = \frac{1}{4}\begin{bmatrix} 7 & -9 \\ -5 & 7 \end{bmatrix}, \mathbf{x} = \frac{1}{4}\begin{bmatrix} 3 \\ -1 \end{bmatrix}$

8. $\mathbf{A}^{-1} = \frac{1}{5}\begin{bmatrix} 10 & -15 \\ -5 & 8 \end{bmatrix}, \mathbf{x} = \frac{1}{5}\begin{bmatrix} 25 \\ -11 \end{bmatrix}$

9. $\mathbf{A}^{-1} = \begin{bmatrix} 5 & -6 \\ -4 & 5 \end{bmatrix}$ **10.** $\mathbf{A}^{-1} = \frac{1}{2}\begin{bmatrix} 6 & -7 \\ -4 & 5 \end{bmatrix}$

11. $\mathbf{A}^{-1} = \begin{bmatrix} -5 & -2 & 5 \\ 2 & 1 & -2 \\ -4 & -3 & 5 \end{bmatrix}$

12. $\mathbf{A}^{-1} = \begin{bmatrix} 18 & 2 & -7 \\ -3 & 0 & 1 \\ -4 & -1 & 2 \end{bmatrix}$

13. $\mathbf{A}^{-1} = \begin{bmatrix} -13 & 42 & -5 \\ 3 & -9 & 1 \\ 2 & -7 & 1 \end{bmatrix}$

14. $\mathbf{A}^{-1} = \begin{bmatrix} 11 & -7 & -9 \\ -4 & 3 & 3 \\ -2 & 1 & 2 \end{bmatrix}$

15. $\mathbf{A}^{-1} = \begin{bmatrix} -22 & 2 & 7 \\ -27 & 3 & 8 \\ 10 & -1 & -3 \end{bmatrix}$

16. $\mathbf{A}^{-1} = \frac{1}{3}\begin{bmatrix} -3 & 0 & 3 \\ -1 & -3 & -1 \\ -1 & 3 & 2 \end{bmatrix}$

17. $\mathbf{A}^{-1} = \frac{1}{4}\begin{bmatrix} -2 & -6 & -3 \\ -2 & -2 & -1 \\ -2 & -2 & 1 \end{bmatrix}$

18. $\mathbf{A}^{-1} = \frac{1}{5}\begin{bmatrix} 1 & 2 & -2 \\ -5 & 0 & 5 \\ -3 & -1 & 6 \end{bmatrix}$

19. $\mathbf{A}^{-1} = \frac{1}{6}\begin{bmatrix} -21 & 11 & 8 \\ 9 & -5 & -2 \\ -3 & 3 & 0 \end{bmatrix}$

20. $\mathbf{A}^{-1} = \frac{1}{7}\begin{bmatrix} 3 & 1 & 0 \\ -2 & -3 & 7 \\ -1 & 2 & 0 \end{bmatrix}$

21. $\mathbf{A}^{-1} = \begin{bmatrix} 0 & 1 & 0 & 0 \\ -2 & 0 & 1 & 0 \\ 1 & 0 & 0 & 0 \\ 0 & -3 & 0 & 1 \end{bmatrix}$

22. $\mathbf{A}^{-1} = \begin{bmatrix} 1 & -1 & 1 & 0 \\ 0 & -2 & -1 & 2 \\ 0 & 1 & 1 & -1 \\ -3 & 3 & -5 & 1 \end{bmatrix}$

23. $\mathbf{A}^{-1} = \begin{bmatrix} 4 & -3 \\ -5 & 4 \end{bmatrix}, \mathbf{X} = \begin{bmatrix} 7 & 18 & -35 \\ -9 & -23 & 45 \end{bmatrix}$

24. $\mathbf{A}^{-1} = \begin{bmatrix} 7 & -6 \\ -8 & 7 \end{bmatrix}, \mathbf{X} = \begin{bmatrix} 14 & -30 & 46 \\ -16 & 35 & -53 \end{bmatrix}$

25. $\mathbf{A}^{-1} = \begin{bmatrix} 11 & -9 & 4 \\ -2 & 2 & -1 \\ -2 & 1 & 0 \end{bmatrix}, \mathbf{X} = \begin{bmatrix} 7 & -14 & 15 \\ -1 & 3 & -2 \\ -2 & 2 & -4 \end{bmatrix}$

26. $\mathbf{A}^{-1} = \begin{bmatrix} -16 & 3 & 11 \\ 6 & -1 & -4 \\ -13 & 2 & 9 \end{bmatrix}, \mathbf{X} = \begin{bmatrix} -21 & 9 & 6 \\ 8 & -3 & -2 \\ -17 & 6 & 5 \end{bmatrix}$

27. $\mathbf{A}^{-1} = \begin{bmatrix} 7 & -20 & 17 \\ 0 & -1 & 1 \\ -2 & 6 & -5 \end{bmatrix}, \mathbf{X} = \begin{bmatrix} 17 & -20 & 24 & -13 \\ 1 & -1 & 1 & -1 \\ -5 & 6 & -7 & 4 \end{bmatrix}$

28. $\mathbf{A}^{-1} = \begin{bmatrix} -5 & 5 & 10 \\ -8 & 8 & 15 \\ 24 & -23 & -45 \end{bmatrix},$

$\mathbf{X} = \begin{bmatrix} -5 & 5 & 10 & 1 \\ -8 & 8 & 15 & 7 \\ 24 & -23 & -45 & -13 \end{bmatrix}$

Section 3.6

1. 60
2. 4
3. −210
4. −72
5. 120
6. 60
7. 0
8. 25
9. 30
10. 7
11. 40
12. 10
13. 78
14. −22
15. −74
16. 84
17. 8
18. 135
19. 39
20. 79
21. $\Delta = 1, x = 10, y = -7$
22. $\Delta = 1, x = -1, y = 1$
23. $\Delta = 1, x = 2, y = -4$
24. $\Delta = 1, x = 5, y = -3$
25. $\Delta = 2, x = 6, y = -3$
26. $\Delta = -2, x = \frac{1}{2}, y = 0$
27. $\Delta = 96, x_1 = \frac{1}{3}, x_2 = -\frac{2}{3}, x_3 = -\frac{1}{3}$
28. $\Delta = 35, x_1 = \frac{4}{7}, x_2 = \frac{3}{7}, x_3 = \frac{2}{7}$
29. $\Delta = 23, x_1 = 2, x_2 = 3, x_3 = 0$
30. $\Delta = 56, x_1 = -\frac{1}{7}, x_2 = \frac{9}{14}, x_3 = \frac{2}{7}$
31. $\Delta = 14, x_1 = -\frac{8}{7}, x_2 = -\frac{10}{7}, x_3 = \frac{1}{7}$
32. $\Delta = 6, x_1 = -\frac{7}{3}, x_2 = 9, x_3 = 8$

33. $\det \mathbf{A} = -4, \mathbf{A}^{-1} = \frac{1}{4}\begin{bmatrix} 4 & 4 & 4 \\ 16 & 15 & 13 \\ 28 & 25 & 23 \end{bmatrix}$

34. $\det \mathbf{A} = 35, \mathbf{A}^{-1} = \frac{1}{4}\begin{bmatrix} -2 & -3 & 12 \\ 9 & -4 & -19 \\ 13 & 2 & -8 \end{bmatrix}$

35. $\det \mathbf{A} = 35, \mathbf{A}^{-1} = \frac{1}{35}\begin{bmatrix} -15 & 25 & -26 \\ 10 & -5 & 8 \\ 15 & -25 & 19 \end{bmatrix}$

36. $\det \mathbf{A} = 23, \mathbf{A}^{-1} = \frac{1}{23}\begin{bmatrix} 5 & 20 & -17 \\ 10 & 17 & -11 \\ 1 & 4 & -8 \end{bmatrix}$

37. $\det \mathbf{A} = 29, \mathbf{A}^{-1} = \frac{1}{29}\begin{bmatrix} 11 & -14 & -15 \\ -17 & 19 & 10 \\ 18 & -15 & -14 \end{bmatrix}$

38. $\det \mathbf{A} = 6, \mathbf{A}^{-1} = \frac{1}{6}\begin{bmatrix} -6 & 10 & 2 \\ 15 & -21 & -6 \\ 12 & -18 & -6 \end{bmatrix}$

39. $\det \mathbf{A} = 37, \mathbf{A}^{-1} = \frac{1}{37}\begin{bmatrix} -21 & -1 & -13 \\ 4 & 9 & 6 \\ -6 & 5 & -9 \end{bmatrix}$

40. $\det \mathbf{A} = 107, \mathbf{A}^{-1} = \frac{1}{107}\begin{bmatrix} 9 & 12 & -13 \\ 11 & -21 & -4 \\ -15 & -20 & -14 \end{bmatrix}$

Section 3.7

1. $y(x) = -2 + 3x$
2. $y(x) = 4 - 7x$
3. $y(x) = 3 - 2x^2$
4. $y(x) = 2x + 3x^2$
5. $y(x) = 5 - 3x + x^2$
6. $y(x) = -10 - 7x + 2x^2$
7. $y(x) = \frac{1}{3}(4x + 3x^2 - 4x^3)$
8. $y(x) = 5 + 3x - x^3$
9. $y(x) = 4 + 3x + 2x^2 + x^3$
10. $y(x) = 17 - 5x + 3x^2 - 2x^3$
11. $x^2 + y^2 - 6x - 4y - 12 = 0$, center $(3, 2)$ and radius 5
12. $x^2 + y^2 + 6x - 8y - 75 = 0$, center $(-3, 4)$ and radius 10
13. $x^2 + y^2 + 4x + 4y - 5 = 0$, center $(-2, -2)$ and radius $\sqrt{13}$
14. $x^2 + y^2 - 10x - 24y = 0$, center $(5, 12)$ and radius 13
15. $x^2 - xy + y^2 = 25$
16. $4x^2 - 7xy + 4y^2 = 100$
17. $100x^2 - 199xy + 100y^2 = 100$
18. $400x^2 - 481xy + 225y^2 = 3600$
19. $y = 3 + \dfrac{2}{x}$
20. $y = 10x + \dfrac{8}{x} - \dfrac{16}{x^2}$
21. $x^2 + y^2 + z^2 - 2x - 4y - 6z - 155 = 0$, center $(1, 2, 3)$ and radius 13
22. $x^2 + y^2 + z^2 - 10x + 14y - 18z - 521 = 0$, center $(5, -7, 9)$ and radius 26
23. $P(t) = 49.061 - 0.0722t + 0.00798t^2$
24. $P(t) = 56.590 + 0.30145t - 0.007375t^2$
25. $P(t) = 62.813 + 1.37915t - 0.012375t^2$
26. $P(t) = 34.838 + 0.7692t + 0.0064t^2$
27. $P(t) = 44.678 + 0.850417t - 0.05105t^2 + 0.000983833t^3$
28. $P(t) = 51.619 + 0.672433t - 0.019565t^2 + 0.000203167t^3$
29. $P(t) = 54.973 + 0.308667t + 0.059515t^2 - 0.00119817t^3$
30. $P(t) = 28.053 + 0.592233t + 0.00907t^2 - 0.0000443333t^3$
31. $P(t) = 39.478 + 0.209692t + 0.0564163t^2 - 0.00292992t^3 + 0.0000391375t^4$

32. $P(t) = 44.461 + 0.7651t - 0.000489167t^2 - 0.000516t^3 + 7.19167 \times 10^{-6}t^4$

33. $P(t) = 47.197 + 1.22537t - 0.077192t^2 + 0.00373475t^3 - 0.0000493292t^4$

34. $P(t) = 20.190 + 1.00003t - 0.031775t^2 + 0.00116067t^3 - 0.000041205t^4$

36. $y = 2x^2 - 6x + 7$

38. $x^2 + y^2 + 6x - 8y - 75 = 0$, center $(-3, 4)$ and radius 10

40. $400x^2 - 481xy + 225y^2 - 3600 = 0$

Chapter 4

Section 4.1

1. $\sqrt{51}$, $(5, 8, -11)$, $(2, 23, 0)$

2. 9, $(1, 4, -1)$, $(-15, -16, 26)$

3. $3\sqrt{21}$, $9\mathbf{i} - 3\mathbf{j} + 3\mathbf{k}$, $-14\mathbf{i} - 21\mathbf{j} + 43\mathbf{k}$

4. $\sqrt{17}$, $4\mathbf{i} - \mathbf{j} - 3\mathbf{k}$, $6\mathbf{i} - 7\mathbf{j} + 12\mathbf{k}$

5. Linearly dependent

6. Linearly independent

7. Linearly independent

8. Linearly dependent

9. $\mathbf{w} = 3\mathbf{u} + 2\mathbf{v}$

10. $\mathbf{w} = 2\mathbf{u} - 3\mathbf{v}$

11. $\mathbf{w} = \mathbf{u} - 2\mathbf{v}$

12. $\mathbf{w} = 3\mathbf{u} + 5\mathbf{v}$

13. $\mathbf{w} = 2\mathbf{u} - 3\mathbf{v}$

14. $\mathbf{w} = 7\mathbf{u} + 5\mathbf{v}$

15. Linearly dependent

16. Linearly dependent

17. Linearly independent

18. Linearly independent

19. Linearly dependent; $a = 3, b = 2, c = 1$

20. Linearly dependent; $a = -2, b = 3, c = 1$

21. Linearly dependent; $a = 11, b = 4, c = -1$

22. Linearly independent

23. Linearly independent

24. Linearly independent

25. $\mathbf{t} = 2\mathbf{u} - \mathbf{v} + 3\mathbf{w}$

26. $\mathbf{t} = \mathbf{u} + 5\mathbf{v} - \mathbf{w}$

27. $\mathbf{t} = 2\mathbf{u} + 6\mathbf{v} + \mathbf{w}$

28. $\mathbf{t} = \mathbf{u} + \mathbf{v} + \mathbf{w}$

Section 4.2

1. W is a subspace of $\mathbf{R}^3$.

2. W is a subspace of $\mathbf{R}^3$.

3. W is not a subspace of $\mathbf{R}^3$.

4. W is not a subspace of $\mathbf{R}^3$.

5. W is a subspace of $\mathbf{R}^4$.

6. W is a subspace of $\mathbf{R}^4$.

7. W is not a subspace of $\mathbf{R}^2$.

8. W is a subspace of $\mathbf{R}^2$.

9. W is not a subspace of $\mathbf{R}^2$.

10. W is not a subspace of $\mathbf{R}^2$.

11. W is a subspace of $\mathbf{R}^4$.

12. W is not a subspace of $\mathbf{R}^4$.

13. W is not a subspace of $\mathbf{R}^4$.

14. W is not a subspace of $\mathbf{R}^4$.

15. $\mathbf{u} = (-1, 0, 1, 0)$ and $\mathbf{v} = (-4, -2, 0, 1)$

16. $\mathbf{u} = (-1, -1, 1, 0)$ and $\mathbf{v} = (-5, -3, 0, 1)$

17. $\mathbf{u} = (1, -3, 1, 0)$ and $\mathbf{v} = (-2, 1, 0, 1)$

18. $\mathbf{u} = (2, -1, -2, 1, 0)$ and $\mathbf{v} = (3, -4, 5, 0, 1)$

19. $\mathbf{u} = (1, 2, -1, 0)$

20. $\mathbf{u} = (-5, 3, -2, 1)$

21. $\mathbf{u} = (-3, 2, -4, 1)$

22. $\mathbf{u} = (-6, 4, -3, 1)$

Section 4.3

1. Linearly dependent

2. Linearly dependent

3. Linearly dependent

4. Linearly dependent

5. Linearly independent

6. Linearly independent

7. Linearly independent

8. Linearly dependent

9. $\mathbf{w} = 2\mathbf{v}_1 - 3\mathbf{v}_2$

10. $\mathbf{w} = 7\mathbf{v}_1 + 4\mathbf{v}_2$

11. $\mathbf{w} = \mathbf{v}_1 - 2\mathbf{v}_2$

12. $\mathbf{w} = 2\mathbf{v}_1 + 5\mathbf{v}_2$

13. $\mathbf{w}$ cannot be expressed as a linear combination of $\mathbf{v}_1$ and $\mathbf{v}_2$.

14. $\mathbf{w}$ cannot be expressed as a linear combination of $\mathbf{v}_1$, $\mathbf{v}_2$, and $\mathbf{v}_3$

15. $\mathbf{w} = 3\mathbf{v}_1 - 2\mathbf{v}_2 + 4\mathbf{v}_3$

16. $\mathbf{w} = 6\mathbf{v}_1 - 2\mathbf{v}_2 + 3\mathbf{v}_3$

17. The vectors $\mathbf{v}_1, \mathbf{v}_2, \mathbf{v}_3$ are linearly independent.

18. $3\mathbf{v}_1 + \mathbf{v}_2 + 5\mathbf{v}_3 = \mathbf{0}$

19. The vectors $\mathbf{v}_1, \mathbf{v}_2, \mathbf{v}_3$ are linearly independent.

20. The vectors $\mathbf{v}_1, \mathbf{v}_2, \mathbf{v}_3$ are linearly independent.

21. $\mathbf{v}_1 - 2\mathbf{v}_2 - \mathbf{v}_3 = \mathbf{0}$

22. $7\mathbf{v}_1 + 5\mathbf{v}_2 - 9\mathbf{v}_3 = \mathbf{0}$

Section 4.4

1. The vectors $\mathbf{v}_1$ and $\mathbf{v}_2$ form a basis for $\mathbf{R}^2$.

2. The vectors $\mathbf{v}_1, \mathbf{v}_2, \mathbf{v}_3$ do not form a basis for $\mathbf{R}^3$.

3. The given vectors do not form a basis for $\mathbf{R}^3$.

4. The given vectors do not form a basis for $\mathbf{R}^4$.

5. The three vectors $\mathbf{v}_1, \mathbf{v}_2, \mathbf{v}_3$ do not form a basis for $\mathbf{R}^3$.

6. The three given vectors form a basis for $\mathbf{R}^3$.

7. The three given vectors form a basis for $\mathbf{R}^3$.

8. The four given vectors form a basis for $\mathbf{R}^4$.

9. The plane $x - 2y + 5z = 0$ is a 2-dimensional subspace of $\mathbf{R}^3$ with basis consisting of the vectors $\mathbf{v}_1 = (2, 1, 0)$ and $\mathbf{v}_2 = (-5, 0, 1)$.

10. The plane $y - z = 0$ is a 2-dimensional subspace of $\mathbf{R}^3$ with basis consisting of the vectors $\mathbf{v}_1 = (1, 0, 0)$ and $\mathbf{v}_2 = (0, 1, 1)$.

11. The line is a 1-dimensional subspace of $\mathbf{R}^3$ with basis consisting of the vector $\mathbf{v} = (-3, 1, 1)$.

12. Hence the subspace consisting of all such vectors is 3-dimensional with basis consisting of the vectors $\mathbf{v}_1` = (1, 1, 0, 0)$, $\mathbf{v}_2 = (1, 0, 1, 0)$, and $\mathbf{v}_3 = (1, 0, 0, 1)$.

13. The subspace consisting of all such vectors is 2-dimensional with basis consisting of the vectors $\mathbf{v}_1 = (3, 0, 1, 0)$ and $\mathbf{v}_2 = (0, 4, 0, 1)$.

14. The subspace consisting of all such vectors is 2-dimensional with basis consisting of the vectors $\mathbf{v}_1 = (-2, 1, 0, 0)$ and $\mathbf{v}_2 = (0, 0, -3, 1)$.

15. The solution space of the given system is 1-dimensional with basis consisting of the vector $\mathbf{v}_1 = (11, 7, 1)$.

16. The solution space of the given system is 1-dimensional with basis consisting of the vector $\mathbf{v}_1 = (11, -5, 1)$.

17. The solution space of the given system is 2-dimensional with basis consisting of the vectors $\mathbf{v}_1 = (-11, -3, 1, 0)$ and $\mathbf{v}_2(-11, -5, 0, 1)$.

18. The solution space of the given system is 2-dimensional with basis consisting of the vectors $\mathbf{v}_1 = (-3, 1, 0, 0)$ and $\mathbf{v}_2(-25, 0, 5, 1)$.

19. The solution space of the given system is 2-dimensional with basis consisting of the vectors $\mathbf{v}_1 = (3, -2, 1, 0)$ and $\mathbf{v}_2 = (-4, -3, 0, 1)$.

20. The solution space of the given system is 2-dimensional with basis consisting of the vectors $\mathbf{v}_1 = (1, -3, 1, 0)$ and $\mathbf{v}_2 = (-2, 1, 0, 1)$.

21. The solution space of the given system is 2-dimensional with basis consisting of the vectors $\mathbf{v}_1 = (-1, -1, 1, 0)$ and $\mathbf{v}_2 = (-5, -3, 0, 1)$.

22. The solution space of the given system is 2-dimensional with basis consisting of the vectors $\mathbf{v}_1 = (2, 1, 0, 0)$ and $\mathbf{v}_2 = (-5, 0, -7, 1)$.

23. The solution space of the given system is 1-dimensional with basis consisting of the vector $\mathbf{v}_1 = (2, -3, 1, 0)$.

24. The solution space of the given system is 3-dimensional with basis consisting of the vectors $\mathbf{v}_1 = (-2, 2, 1, 0, 0)$, $\mathbf{v}_2 = (-1, 3, 0, 1, 0)$, and $\mathbf{v}_3 = (-3, -1, 0, 0, 1)$.

25. The solution space of the given system is 3-dimensional with basis consisting of the vectors $\mathbf{v}_1 = (-2, 1, 0, 0, 0)$, $\mathbf{v}_2 = (2, 0, 1, 1, 0)$, and $\mathbf{v}_3 = (-3, 0, -4, 0, 1)$.

26. The solution space of the given system is 2-dimensional with basis consisting of the vectors $\mathbf{v}_1 = (-2, 1, 2, 1, 0)$ and $\mathbf{v}_2 = (3, -4, 5, 0, 1)$.

Section 4.5

1. Row basis: The first and second row vectors of **E**.
Column basis: The first and second column vectors of **A**.

2. Row basis: The first and second row vectors of **E**.
Column basis: The first and second column vectors of **A**.

3. Row basis: The first and second row vectors of **E**.
Column basis: The first and second column vectors of **A**.

4. Row basis: The three row vectors of **E**.
Column basis: The first three column vectors of **A**.

5. Row basis: The three row vectors of **E**.
Column basis: The first, second, and fourth column vectors of **A**.

6. Row basis: The three row vectors of **E**.
Column basis: The first, second, and fourth column vectors of **A**.

7. Row basis: The first two row vectors of **E**.
Column basis: The first two column vectors of **A**.

8. Row basis: The first three row vectors of **E**.
Column basis: The first, second, and fourth column vectors of **A**.

9. Row basis: The first three row vectors of **E**.
Column basis: The first three column vectors of **A**.

10. Row basis: The first three row vectors of **E**.
Column basis: The first, second, and fourth column vectors of **A**.

11. Row basis: The first three row vectors of **E**.
Column basis: The first, second, and fifth column vectors of **A**.

12. Row basis: The first three row vectors of **E**.
Column basis: The first, second, and fifth column vectors of **A**.

13. Linearly independent: $\mathbf{v}_1$ and $\mathbf{v}_2$

14. Linearly independent: $\mathbf{v}_1$ and $\mathbf{v}_2$

15. Linearly independent: $\mathbf{v}_1$, $\mathbf{v}_2$, and $\mathbf{v}_4$

16. Linearly independent: $\mathbf{v}_1$, $\mathbf{v}_2$, $\mathbf{v}_4$, and $\mathbf{v}_5$

17. Basis vectors: $\mathbf{v}_1$, $\mathbf{v}_2$, $\mathbf{e}_2$

18. Basis vectors: $\mathbf{v}_1$, $\mathbf{v}_2$, $\mathbf{e}_2$

19. Basis vectors: $\mathbf{v}_1$, $\mathbf{v}_2$, $\mathbf{e}_2$, $\mathbf{e}_3$

20. Basis vectors: $\mathbf{v}_1$, $\mathbf{v}_2$, $\mathbf{e}_1$, $\mathbf{e}_3$

21. The first and second equations are irredundant.

22. The first and second equations are irredundant.

23. The first, second, and fourth equations are irredundant.

24. The first, second, and fifth equations are irredundant.

Section 4.6

1. Yes, the three vectors are mutually orthogonal.

2. Yes, the three vectors are mutually orthogonal.

3. Yes, the three vectors are mutually orthogonal.

4. Yes, the three vectors are mutually orthogonal.

5. $a^2 = 7, b^2 = 7, c^2 = 14$

6. $a^2 = 18, b^2 = 18, c^2 = 36$

7. $a^2 = 19, b^2 = 25, c^2 = 44$

8. $a^2 = 103, b^2 = 112, c^2 = 215$

9. $\angle A = \angle B = 45°$

10. $\angle A = \angle B = 45°$

11. $\angle A = 41.08°$, $\angle B = 48.92°$

12. $\angle A = 43.80°$, $\angle B = 46.20°$

13. $\mathbf{u}_1 = (2, 1, 0)$, $\mathbf{u}_2 = (-3, 0, 1)$

14. $\mathbf{u}_1 = (-5, 1, 0)$, $\mathbf{u}_2 = (3, 0, 1)$

15. $\mathbf{u}_1 = (2, 1, 0, 0)$, $\mathbf{u}_2 = (3, 0, 1, 0)$, $\mathbf{u}_3 = (-5, 0, 0, 1)$

16. $\mathbf{u}_1 = (-7, 1, 0, 0)$, $\mathbf{u}_2 = (6, 0, 1, 0)$, $\mathbf{u}_3 = (9, 0, 0, 1)$

17. $\mathbf{u}_1 = (7, -3, 1, 0)$, $\mathbf{u}_2 = (-19, 5, 0, 1)$

18. $\mathbf{u}_1 = (-12, -3, 1, 0)$, $\mathbf{u}_2 = (16, 7, 0, 1)$

19. $\mathbf{u}_1 = (-13, 4, 1, 0, 0)$, $\mathbf{u}_2 = (4, -3, 0, 1, 0)$, $\mathbf{u}_3 = (-11, 4, 0, 0, 1)$

20. $\mathbf{u}_1 = (-5, 1, 1, 0, 0)$, $\mathbf{u}_2 = (-12, 4, 0, 1, 0)$, $\mathbf{u}_3 = (-19, 7, 0, 0, 1)$

21. $\mathbf{u}_1 = (-1, -1, 1, 0, 0)$, $\mathbf{u}_2 = (0, -1, 0, -1, 1)$

22. $\mathbf{u}_1 = (-2, 1, 1, 0, 0)$, $\mathbf{u}_2 = (1, -2, 0, 1, 0)$

Section 4.7

1. It is a subspace.

2. It is a subspace.

3. It is not a subspace.

4. It is not a subspace.

5. It is a subspace.

6. It is not a subspace.

7. It is not a subspace.

8. It is a subspace.

9. It is not a subspace.

10. It is a subspace.

11. It is a subspace.

12. It is not a subspace.

13. The functions $\sin x$ and $\cos x$ are linearly independent.

14. The funtions e^x and xe^x are linearly independent.

15. The functions $(1 + x)$, $(1 - x)$, and $(1 - x^2)$ are linearly independent.

16. The three given polynomials are linearly dependent.

17. The three given trigonometric functions are linearly dependent.

18. The two given linear combinations of $\sin x$ and $\cos x$ are linearly independent.

19. $A = 3$ and $B = -2$.

20. $A = -2$ and $B = C = 1$.

21. $A = 2$, $B = -2$, and $C = 0$.

22. $A = -1$, $B = 4$, and $C = -3$.

23. The solution space is 3-dimensional with basis $\{1, x, x^2\}$.

24. The solution space is 4-dimensional with basis $\{1, x, x^2, x^3\}$.

25. The solution space is 2-dimensional with basis $\{1, e^{5x}\}$.

26. The solution space is 2-dimensional with basis $\{1, e^{-10x}\}$.

Chapter 5

Section 5.1

1. $y(x) = \frac{5}{2}e^x - \frac{5}{2}e^{-x}$

2. $y(x) = 2e^{3x} - 3e^{-3x}$

3. $y(x) = 3\cos 2x + 4\sin 2x$

4. $y(x) = 10\cos 5x - 2\sin 5x$

5. $y(x) = 2e^x - e^{2x}$

6. $y(x) = 4e^{2x} + 3e^{-3x}$

7. $y(x) = 6 - 8e^{-x}$

8. $y(x) = \frac{1}{3}(14 - 2e^{3x})$

9. $y(x) = 2e^{-x} + xe^{-x}$

10. $y(x) = 3e^{5x} - 2xe^{5x}$

11. $y(x) = 5e^x \sin x$

12. $y(x) = e^{-3x}(2\cos 2x + 3\sin 2x)$

13. $y(x) = 5x - 2x^2$

14. $y(x) = 3x^2 - 16/x^3$

15. $y(x) = 7x - 5x\ln x$

16. $y(x) = 2\cos(\ln x) + 3\sin(\ln x)$

21. Linearly independent

22. Linearly independent
23. Linearly independent
24. Linearly dependent
25. Linearly independent
26. Linearly independent
28. $y(x) = 1 - 2\cos x - \sin x$
29. There is no contradiction because if the given differential equation is divided by x^2 to get the form in Eq. (8), then the resulting coefficient functions $p(x) = -4/x$ and $q(x) = 6/x^2$ are not continuous at $x = 0$.
33. $y(x) = c_1e^x + c_2e^{2x}$
34. $y(x) = c_1e^{-5x} + c_2e^{3x}$
35. $y(x) = c_1 + c_2e^{-5x}$
36. $y(x) = c_1 + c_2e^{-3x/2}$
37. $y(x) = c_1e^{-x/2} + c_2e^x$
38. $y(x) = c_1e^{-x/2} + c_2e^{-3x/2}$
39. $y(x) = (c_1 + c_2x)e^{-x/2}$
40. $y(x) = (c_1 + c_2x)e^{2x/3}$
41. $y(x) = c_1e^{-4x/3} + c_2e^{5x/2}$
42. $y(x) = c_1e^{-4x/7} + c_2e^{3x/5}$
43. $y'' + 10y' = 0$
44. $y'' - 100y = 0$
45. $y'' + 20y' + 100y = 0$
46. $y'' - 110y' + 1000y = 0$
47. $y'' = 0$
48. $y'' - 2y' - y = 0$
49. The high point is $\left(\ln \frac{7}{4}, \frac{16}{7}\right)$.
50. $(-\ln 2, -2)$
52. $y(x) = c_1x + c_2/x$
53. $y(x) = c_1x^{-4} + c_2x^3$
54. $y(x) = c_1x^{-3/2} + c_2x^{1/2}$
55. $y(x) = c_1 + c_2\ln x$
56. $y(x) = x^2(c_1 + c_2\ln x)$

Section 5.2

1. $15\cdot(2x) - 16\cdot(3x^2) - 6\cdot(5x - 8x^2) \equiv 0$
2. $(-4)(5) + (5)(2 - 3x^2) + (1)(10 + 15x^2) \equiv 0$
3. $1\cdot 0 + 0\cdot\sin x + 0\cdot e^x \equiv 0$
4. $(6)(17) + (-51)(2\sin^2 x) + (-34)(3\cos^2 x) \equiv 0$
5. $1\cdot 17 - 34\cdot\cos^2 x + 17\cdot\cos 2x \equiv 0$
6. $(-1)(e^x) + (1)(\cosh x) + (1)(\sinh x) \equiv 0$
13. $y(x) = \frac{4}{3}e^x - \frac{1}{3}e^{-2x}$
14. $y(x) = \frac{1}{2}(3e^x - 6e^{2x} + 3e^{3x})$
15. $y(x) = (2 - 2x + x^2)e^x$
16. $y(x) = -12e^x + 13e^{2x} - 10xe^{2x}$
17. $y(x) = \frac{1}{9}(29 - 2\cos 3x - 3\sin 3x)$
18. $y(x) = e^x(2 - \cos x - \sin x)$
19. $y(x) = x + 2x^2 + 3x^3$
20. $y(x) = 2x - x^{-2} + x^{-2}\ln x$
21. $y(x) = 2\cos x - 5\sin x + 3x$
22. $y(x) = 4e^{2x} - e^{-2x} - 3$
23. $y(x) = e^{-x} + 4e^{3x} - 2$
24. $y(x) = e^x(3\cos x + 4\sin x) + x + 1$
38. $y_2(x) = \dfrac{1}{x^3}$
39. $y_2(x) = xe^{x/2}$
40. $y_2(x) = xe^x$
41. $y_2(x) = x + 2$
42. $y_2(x) = 1 + x^2$

Section 5.3

1. $y(x) = c_1e^{2x} + c_2e^{-2x}$
2. $y(x) = c_1 + c_2e^{3x/2}$
3. $y(x) = c_1e^{2x} + c_2e^{-5x}$
4. $y(x) = c_1e^{x/2} + c_2e^{3x}$
5. $y(x) = c_1e^{-3x} + c_2xe^{-3x}$
6. $y(x) = e^{-5x/2}\left[c_1\exp\left(\frac{1}{2}x\sqrt{5}\right) + c_2\exp\left(-\frac{1}{2}x\sqrt{5}\right)\right]$
7. $y(x) = c_1e^{3x/2} + c_2xe^{3x/2}$
8. $y(x) = e^{3x}(c_1\cos 2x + c_2\sin 2x)$
9. $y(x) = e^{-4x}(c_1\cos 3x + c_2\sin 3x)$
10. $y(x) = c_1 + c_2x + c_3x^2 + c_4e^{-3x/5}$
11. $y(x) = c_1 + c_2x + c_3e^{4x} + c_4xe^{4x}$
12. $y(x) = c_1 + c_2e^x + c_3xe^x + c_4x^2e^x$
13. $y(x) = c_1 + c_2e^{-2x/3} + c_3xe^{-2x/3}$
14. $y(x) = c_1e^x + c_2e^{-x} + c_3\cos 2x + c_4\sin 2x$
15. $y(x) = c_1e^{2x} + c_2xe^{2x} + c_3e^{-2x} + c_4xe^{-2x}$
16. $y(x) = (c_1 + c_2x)\cos 3x + (c_3 + c_4x)\sin 3x$
17. $y(x) = c_1\cos\left(x/\sqrt{2}\right) + c_2\sin\left(x/\sqrt{2}\right) + c_3\cos\left(2x/\sqrt{3}\right) + c_4\sin\left(2x/\sqrt{3}\right)$
18. $y(x) = c_1e^{2x} + c_2e^{-2x} + c_3\cos 2x + c_4\sin 2x$
19. $y(x) = c_1e^x + c_2e^{-x} + c_3xe^{-x}$
20. $y(x) = e^{-x/2}\left[(c_1 + c_2x)\cos\left(\frac{1}{2}x\sqrt{3}\right) + (c_3 + c_3x)\sin\left(\frac{1}{2}x\sqrt{3}\right)\right]$
21. $y(x) = 5e^x + 2e^{3x}$
22. $y(x) = e^{-x/3}\left[3\cos\left(x/\sqrt{3}\right) + 5\sqrt{3}\sin\left(x/\sqrt{3}\right)\right]$
23. $y(x) = e^{3x}(3\cos 4x - 2\sin 4x)$
24. $y(x) = \frac{1}{2}(-7 + e^{2x} + 8e^{-x/2})$
25. $y(x) = \frac{1}{4}(-13 + 6x + 9e^{-2x/3})$
26. $y(x) = \frac{1}{5}(24 - 9e^{-5x} - 25xe^{-5x})$
27. $y(x) = c_1e^x + c_2e^{-2x} + c_3xe^{-2x}$
28. $y(x) = c_1e^{2x} + c_2e^{-x} + c_3e^{-x/2}$
29. $y(x) = c_1e^{-3x} + e^{3x/2}\left[c_2\cos\left(\frac{3}{2}x\sqrt{3}\right) + c_3\sin\left(\frac{3}{2}x\sqrt{3}\right)\right]$
30. $y(x) = c_1e^{-x} + c_2e^{2x} + c_3\cos(x\sqrt{3}) + c_4\sin(x\sqrt{3})$
31. $y(x) = c_1e^x + e^{-2x}(c_2\cos 2x + c_3\sin 2x)$
32. $y(x) = c_1e^{2x} + (c_2 + c_3x + c_4x^2)e^{-x}$
33. $y(x) = c_1e^{3x} + e^{-3x}(c_2\cos 3x + c_3\sin 3x)$
34. $y(x) = c_1e^{2x/3} + c_2\cos 2x + c_3\sin 2x$
35. $y(x) = c_1e^{-x/2} + c_2e^{-x/3} + c_3\cos 2x + c_4\sin 2x$
36. $y(x) = c_1e^{7x/9} + e^{-x}(c_2\cos x + c_3\sin x)$
37. $y(x) = 11 + 5x + 3x^2 + 7e^x$
38. $y(x) = 2e^{5x} - 2\cos 10x$
39. $y^{(3)} - 6y'' + 12y' - 8y = 0$
40. $y^{(3)} - 2y'' + 4y' - 8y = 0$
41. $y^{(4)} - 16y = 0$
42. $y^{(6)} + 12y^{(4)} + 48y'' + 64y = 0$
44. (a) $x = i, -2i$ (b) $x = -i, 3i$
45. $y(x) = c_1e^{-ix} + c_2e^{3ix}$
46. $y(x) = c_1e^{3ix} + c_2e^{-2ix}$
47. $y(x) = c_1\exp\left(\left[1 + i\sqrt{3}\right]x\right) + c_2\exp\left(-\left[1 + i\sqrt{3}\right]x\right)$
48. $y(x) = \frac{1}{3}\left(e^x + \exp\left[\frac{1}{2}\left(-1 + i\sqrt{3}\right)x\right] + \exp\left[\frac{1}{2}\left(-1 - i\sqrt{3}\right)x\right]\right)$
49. $y(x) = 2e^{2x} - 5e^{-x} + 3\cos x - 9\sin x$
52. $y(x) = c_1\cos(3\ln x) + c_2\sin(3\ln x)$
53. $y(x) = x^{-3}[c_1\cos(4\ln x) + c_2\sin(4\ln x)]$
54. $y(x) = c_1 + c_2\ln x + c_3x^{-3}$
55. $y(x) = c_1 + x^2(c_2 + c_3\ln x)$
56. $y(x) = c_1 + c_2\ln x + c_3(\ln x)^2$
57. $y(x) = c_1 + x^3\left(c_2x^{-\sqrt{3}} + c_3x^{+\sqrt{3}}\right)$
58. $y(x) = x^{-1}[c_1 + c_2\ln x + c_3(\ln x)^2]$

Section 5.4

1. Frequency: 2 rad/s ($1/\pi$ Hz); period: π s

2. Frequency: 8 rad/s ($4/\pi$ Hz); period: $\pi/4$ s

3. Amplitude: 2 m; frequency: 5 rad/s; period: $2\pi/5$ s

4. (a) $x(t) = \frac{13}{12}\cos(12t - \alpha)$ with $\alpha = 2\pi - \tan^{-1}(5/12) \approx 5.8884$; (b) Amplitude: $\frac{13}{12}$ m; period: $\pi/6$ sec

6. About 7.33 mi

7. About 10450 ft

8. 29.59 in.

10. Amplitude: 100 cm; period: about 2.01 sec

11. About 3.8 in.

13. (a) $x(t) = 50(e^{-2t/5} - e^{-t/2})$; (b) 4.096 exactly

14. (a) $x(t) = 25e^{-t/5}\cos(3t - \alpha)$ with $\alpha = \tan^{-1}(3/4) \approx 0.6435$; (b) envelope curves $x = \pm 25e^{-t/5}$; pseudoperiod $2\pi/3$

15. $x(t) = 4e^{-2t} - 2e^{-4t}$, overdamped; $u(t) = 2\cos(2\sqrt{2}\,t)$

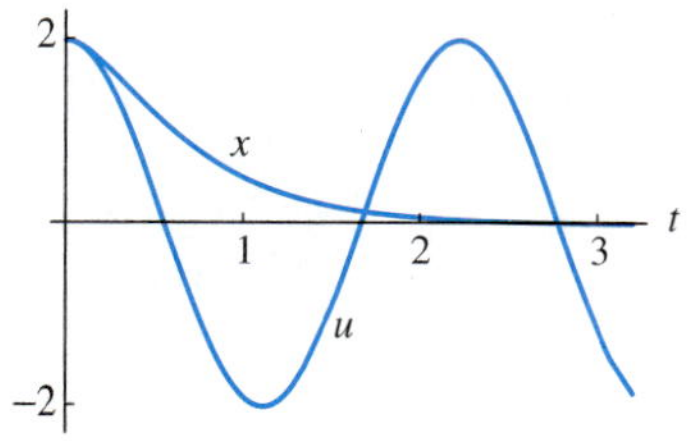

16. $x(t) = 4e^{-3t} - 2e^{-7t}$, overdamped; $u(t) \approx 2\sqrt{\frac{22}{21}}\cos(\sqrt{21}\,t - 0.2149)$

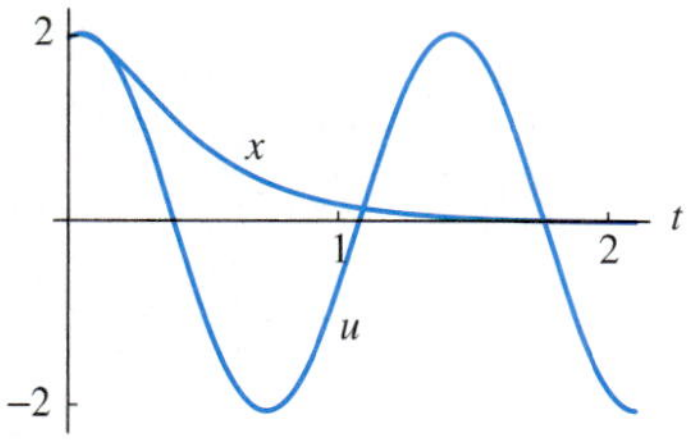

17. $x(t) = 5e^{-4t}(2t + 1)$, critically damped; $u(t) \approx \frac{5}{2}\sqrt{5}\cos(4t - 5.8195)$

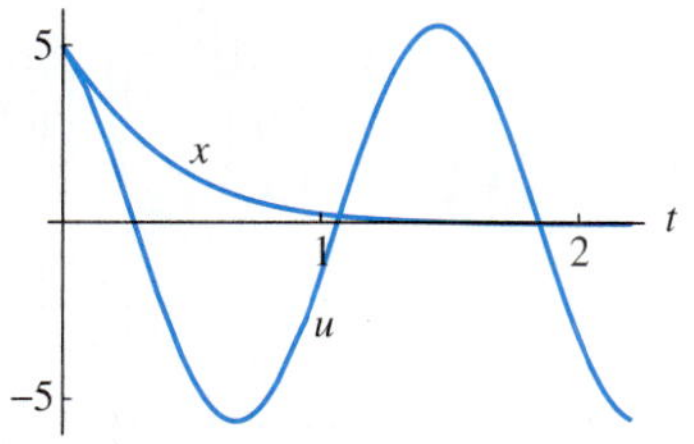

18. $x(t) = 2e^{-3t}\cos\left(4t - \frac{3\pi}{2}\right)$, underdamped; $u(t) = \frac{8}{5}\cos\left(5t - \frac{3\pi}{2}\right)$

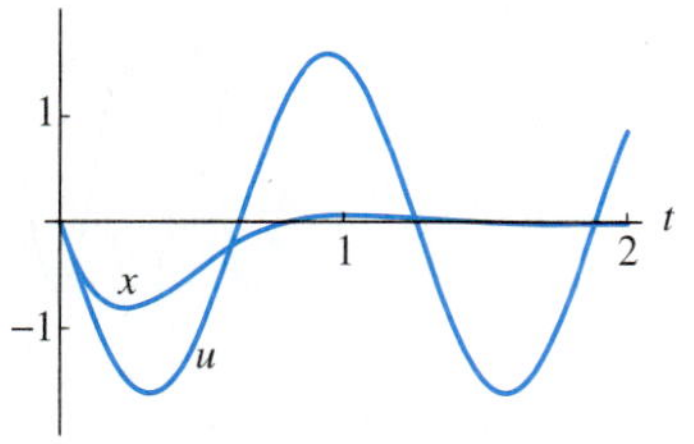

19. $x(t) \approx \frac{1}{3}\sqrt{313}\,e^{-5t/2}\cos(6t - 0.8254)$, underdamped; $u(t) \approx \frac{4}{13}\sqrt{233}\cos\left(\frac{13}{2}t - 0.5517\right)$

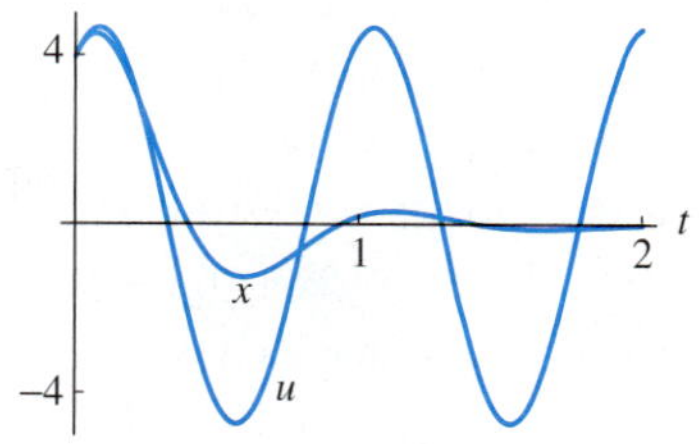

20. $x(t) \approx 13e^{-4t}\cos(2t - 1.1760)$, underdamped; $u(t) \approx \sqrt{\frac{129}{5}}\cos(2\sqrt{5}\,t - 0.1770)$

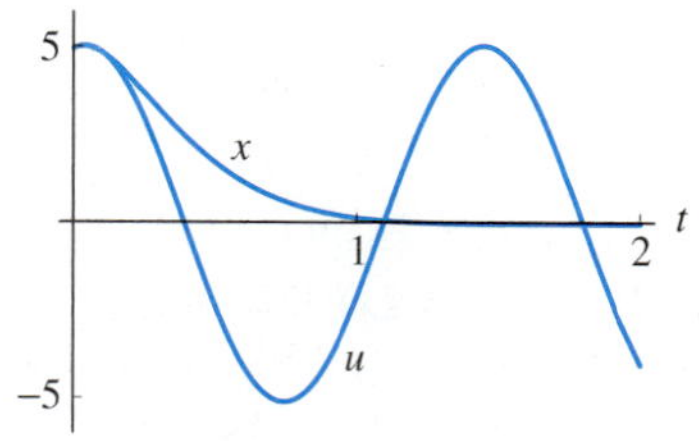

21. $x(t) \approx 10e^{-5t}\cos(10t - 0.9273)$, underdamped; $u(t) \approx 2\sqrt{14}\cos(5\sqrt{5}\,t - 0.6405)$

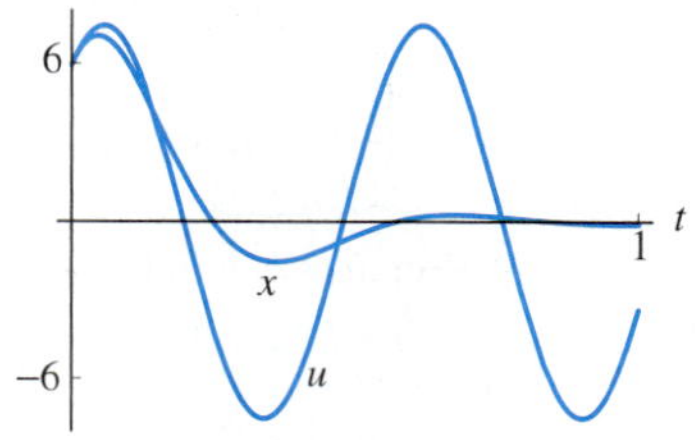

22. (b) The time-varying amplitude is $\frac{2}{3}\sqrt{3}$, the frequency is $4\sqrt{3}$ rad/s, and the phase angle is $\pi/6$.

23. (a) $k \approx 7018$ lb/ft; (b) After about 2.47 s

34. Damping constant: $c \approx 11.51$ lb/ft/s; spring constant: $k \approx 189.68$ lb/ft

Section 5.5

1. $y_p(x) = \frac{1}{25}e^{3x}$

2. $y_p(x) = -\frac{1}{4}(5 + 6x)$

3. $y_p(x) = \frac{1}{39}(\cos 3x - 5\sin 3x)$

4. $y_p(x) = \frac{1}{9}(-4e^x + 3xe^x)$

5. $y_p(x) = \frac{1}{26}(13 + 3\cos 2x - 2\sin 2x)$

6. $y_p(x) = \frac{1}{343}(4 - 56x + 49x^2)$

7. $y_p(x) = -\frac{1}{6}(e^x - e^{-x}) = -\frac{1}{3}\sinh x$

8. $y_p(x) = \frac{1}{4}x\sinh 2x$

9. $y_p(x) = -\frac{1}{3} + \frac{1}{16}(2x^2 - x)e^x$

10. $y_p(x) = \frac{1}{6}(2x\sin 3x - 3x\cos 3x)$

11. $y_p(x) = \frac{1}{8}(3x^2 - 2x)$

12. $y_p(x) = 2x + \frac{1}{2}x\sin x$

13. $y_p(x) = \frac{1}{65}e^x(7\sin x - 4\cos x)$

14. $y_p(x) = \frac{1}{24}(-3x^2e^x + x^3e^x)$

15. $y_p(x) \equiv -17$

16. $y_p(x) = \frac{1}{81}(45 + e^{3x} - 6xe^{3x} + 9x^2e^{3x})$

17. $y_p(x) = \frac{1}{4}(x^2\sin x - x\cos x)$

18. $y_p(x) = -\frac{1}{144}(24xe^x - 19xe^{2x} + 6x^2e^{2x})$

19. $y_p(x) = \frac{1}{8}(10x^2 - 4x^3 + x^4)$

20. $y_p(x) = -7 + \frac{1}{3}xe^x$

21. $y_p(x) = xe^x(A\cos x + B\sin x)$

22. $y_p(x) = Ax^3 + Bx^4 + Cx^5 + Dxe^x$

23. $y_p(x) = Ax\cos 2x + Bx\sin 2x + Cx^2\cos 2x + Dx^2\sin 2x$

24. $y_p(x) = Ax + Bx^2 + (Cx + Dx^2)e^{-3x}$

25. $y_p(x) = Axe^{-x} + Bx^2e^{-x} + Cxe^{-2x} + Dx^2e^{-2x}$

26. $y_p(x) = (Ax + Bx^2)e^{3x}\cos 2x + (Cx + Dx^2)e^{3x}\sin 2x$

27. $y_p(x) = Ax\cos x + Bx\sin x + Cx\cos 2x + Dx\sin 2x$

28. $y_p(x) = (Ax + Bx^2 + Cx^3)\cos 3x + (Dx + Ex^2 + Fx^3)\sin 3x$

29. $y_p(x) = Ax^3e^x + Bx^4e^x + Cxe^{2x} + Dxe^{-2x}$

30. $y_p(x) = (A + Bx + Cx^2)\cos x + (D + Ex + Fx^2)\sin x$

31. $y(x) = \cos 2x + \frac{3}{4}\sin 2x + \frac{1}{2}x$

32. $y(x) = \frac{1}{6}(15e^{-x} - 16e^{-2x} + e^x)$

33. $y(x) = \cos 3x - \frac{2}{15}\sin 3x + \frac{1}{5}\sin 2x$

34. $y(x) = \cos x - \sin x + \frac{1}{2}x\sin x$

35. $y(x) = e^x(2\cos x - \frac{5}{2}\sin x) + \frac{1}{2}x + 1$

36. $y(x) = \frac{1}{192}(234 + 240x - 9e^{-2x} - 33e^{2x} - 12x^2 - 4x^4)$

37. $y(x) = 4 - 4e^x + 3xe^x + x - \frac{1}{2}x^2e^x + \frac{1}{6}x^3e^x$

38. $y(x) = \frac{1}{85}[e^{-x}(176\cos x + 197\sin x) - (6\cos 3x + 7\sin 3x)]$

39. $y(x) = -3 + 3x - \frac{1}{2}x^2 + \frac{1}{6}x^3 + 4e^{-x} + xe^{-x}$

40. $y(x) = \frac{1}{4}(5e^{-x} + 5e^x + 10\cos x - 20)$

41. $y_p(x) = 255 - 450x + 30x^2 + 20x^3 + 10x^4 - 4x^5$

42. $y(x) = 10e^{-x} + 35e^{2x} + 210\cos x + 390\sin x + y_p(x)$ where $y_p(x)$ is the particular solution of Problem 41.

43. (b) $y(x) = c_1\cos 2x + c_2\sin 2x + \frac{1}{4}\cos x - \frac{1}{20}\cos 3x$

44. $y(x) = e^{-x/2}\left[c_1\cos\left(\frac{1}{2}x\sqrt{3}\right) + c_2\sin\left(\frac{1}{2}x\sqrt{3}\right)\right] + \frac{1}{26}(-3\cos 2x + 2\sin 2x) + \frac{1}{482}(15\cos 4x + 4\sin 4x)$

45. $y(x) = c_1\cos 3x + c_2\sin 3x + \frac{1}{24} - \frac{1}{10}\cos 2x - \frac{1}{56}\cos 4x$

46. $y(x) = c_1\cos x + c_2\sin x + \frac{1}{16}(3x\cos x + 3x^2\sin x) + \frac{1}{128}(3\sin 3x - 4x\cos 3x)$

47. $y_p(x) = \frac{2}{3}e^x$

48. $y_p(x) = -\frac{1}{12}(6x + 1)e^{-2x}$

49. $y_p(x) = x^2e^{2x}$

50. $y_p(x) = \frac{1}{16}(4x\cosh 2x - \sinh 2x)$

51. $y_p(x) = -\frac{1}{4}(\cos 2x\cos x - \sin 2x\sin x) + \frac{1}{20}(\cos 5x\cos 2x + \sin 5x\sin 2x) = -\frac{1}{5}\cos 3x$ (!)

52. $y_p(x) = -\frac{1}{6}x\cos 3x$

53. $y_p(x) = \frac{2}{3}x\sin 3x + \frac{2}{9}(\cos 3x)\ln|\cos 3x|$

54. $y_p(x) = -1 - (\cos x)\ln|\csc x - \cot x|$

55. $y_p(x) = \frac{1}{8}(1 - x\sin 2x)$

56. $y_p(x) = -\frac{1}{9}e^x(3x + 2)$

58. $y_p(x) = x^3(\ln x - 1)$

59. $y_p(x) = \frac{1}{4}x^4$

60. $y_p(x) = -\frac{72}{5}x^{4/3}$

61. $y_p(x) = \ln x$

62. $y_p(x) = -x^2 + x\ln\left|\dfrac{1+x}{1-x}\right| + \dfrac{1}{2}(1 + x^2)\ln|1 - x^2|$

Section 5.6

1. $x(t) = 2\cos 2t - 2\cos 3t$

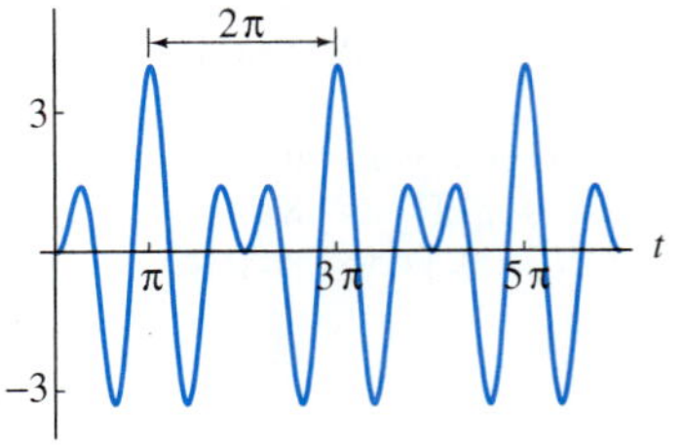

2. $x(t) = \frac{3}{2}\sin 2t - \sin 3t$

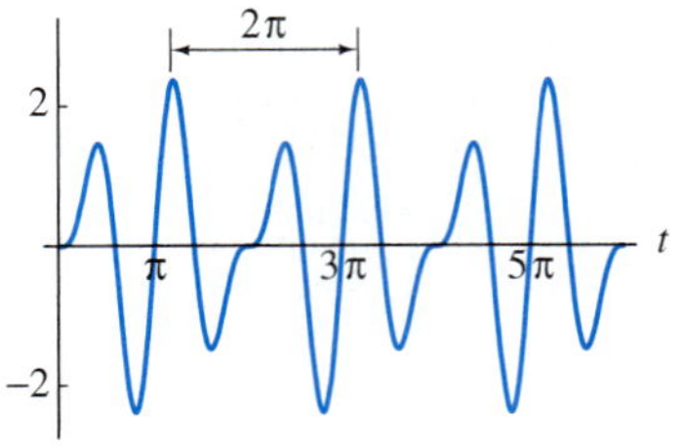

3. $x(t) = \sqrt{138388}\,\cos(10t - \alpha) + 5\cos(5t - \beta)$ with $\alpha = 2\pi - \tan^{-1}(1/186) \approx 6.2778$ and $\beta = \tan^{-1}(4/3) \approx 0.9273$.

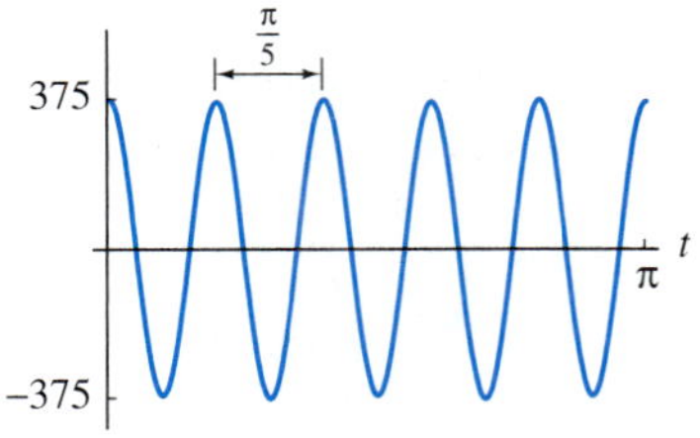

4. $x(t) = 2\sqrt{106}\,\cos(5t - \alpha) + 10\cos 4t$ with $\alpha = \pi - \tan^{-1}(9/5) \approx 2.0779$

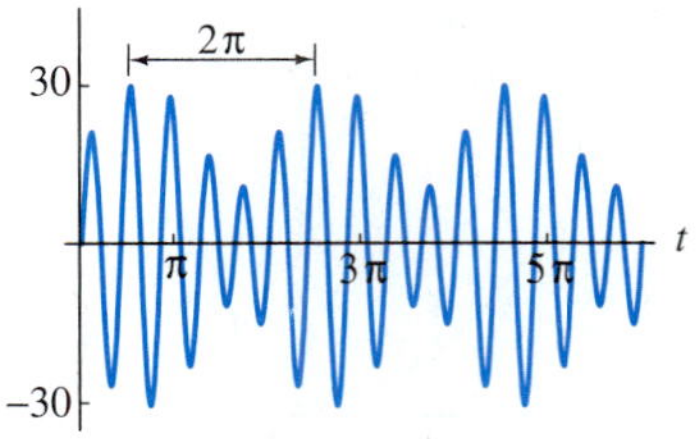

5. $x(t) = (x_0 - C)\cos\omega_0 t + C\cos\omega t$, where $C = F_0/(k - m\omega^2)$

7. $x_{sp}(t) = \frac{10}{13}\cos(3t - \alpha)$ with $\alpha = \pi - \tan^{-1}(12/5) \approx 1.9656$

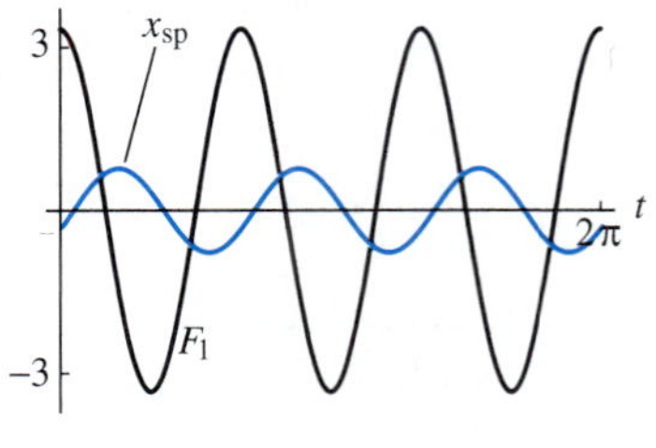

8. $x_{sp}(t) = \frac{4}{25}\cos(5t - \alpha)$ with $\alpha = 2\pi - \tan^{-1}(3/4) \approx 5.6397$

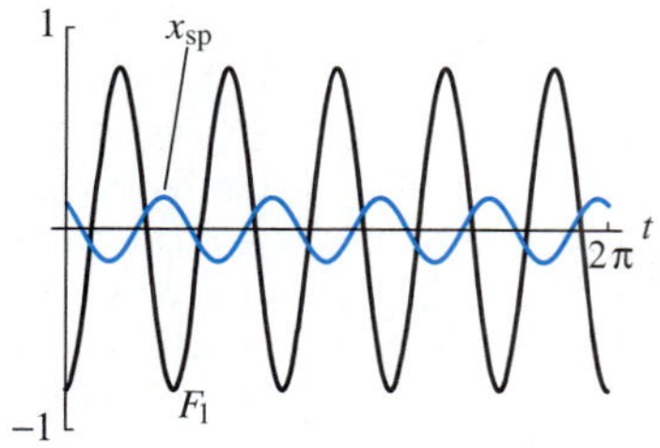

9. $x_{sp}(t) = \frac{3}{\sqrt{40001}} \cos(10t - \alpha)$ with
$\alpha = \pi + \tan^{-1}(199/20) \approx 4.6122$

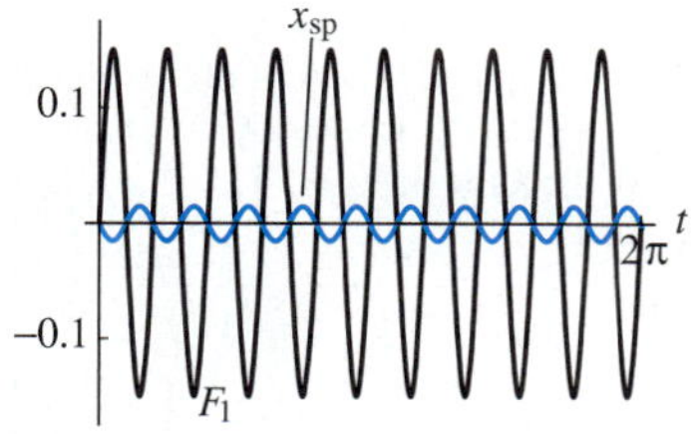

10. $x_{sp}(t) = \frac{10}{793}\sqrt{61} \cos(10t - \alpha)$ with
$\alpha = \pi + \tan^{-1}(171/478) \approx 3.4851$

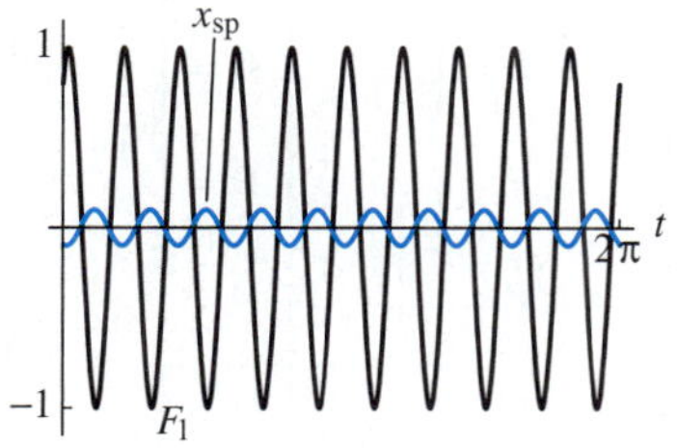

11. $x_{sp}(t) = \frac{\sqrt{10}}{4} \cos(3t - \alpha)$ with $\alpha = \pi - \tan^{-1}(3) \approx 1.8925$
$x_{tr}(t) = \frac{5}{4}\sqrt{2}\, e^{-2t} \cos(t - \beta)$ with $\beta = 2\pi - \tan^{-1}(7) \approx 4.8543$

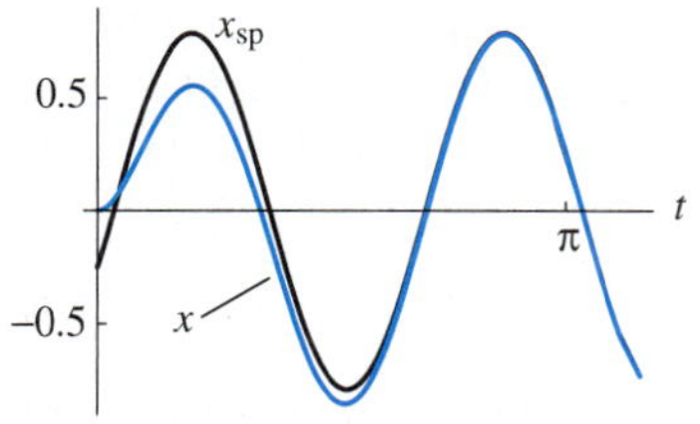

12. $x_{sp}(t) = \frac{5}{3\sqrt{29}} \cos(3t - \alpha)$ with $\alpha = \pi + \tan^{-1}(2/5) \approx 3.5221$
$x_{tr}(t) = \frac{25}{6\sqrt{29}} e^{-3t} \cos(2t - \beta)$ with $\beta = \tan^{-1}(5/2) \approx 1.1903$

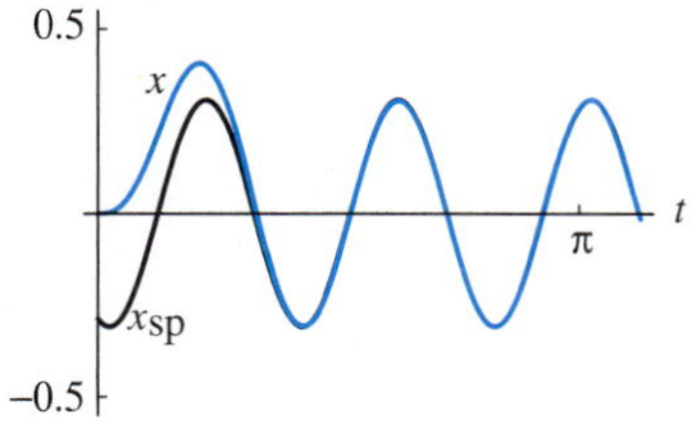

13. $x_{sp}(t) = \frac{300}{\sqrt{1469}} \cos(10t - \alpha)$ with
$\alpha = \pi - \tan^{-1}(10/37) \approx 2.8776$
$x_{tr}(t) = 2\sqrt{\frac{113314}{1469}}\, e^{-t} \cos(5t - \beta)$ with
$\beta = 2\pi - \tan^{-1}(421/12895) \approx 6.2505$

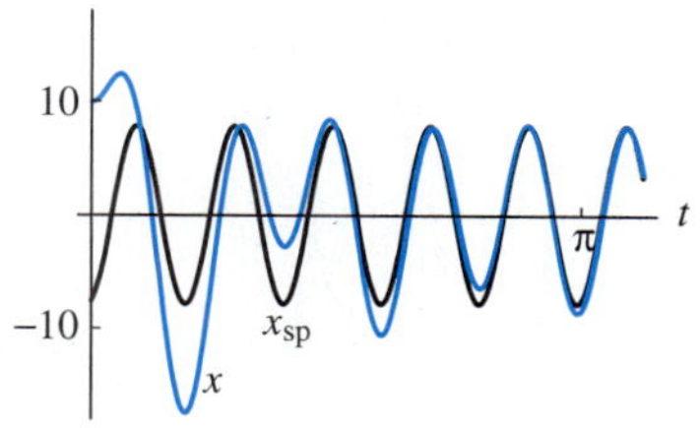

14. $x_{sp}(t) = \sqrt{485} \cos(t - \alpha)$ with $\alpha = \tan^{-1}(22) \approx 1.5254$
$x_{tr}(t) = \sqrt{3665}\, e^{-4t} \cos(3t - \beta)$ with
$\beta = \pi + \tan^{-1}(52/31) \approx 4.1748$

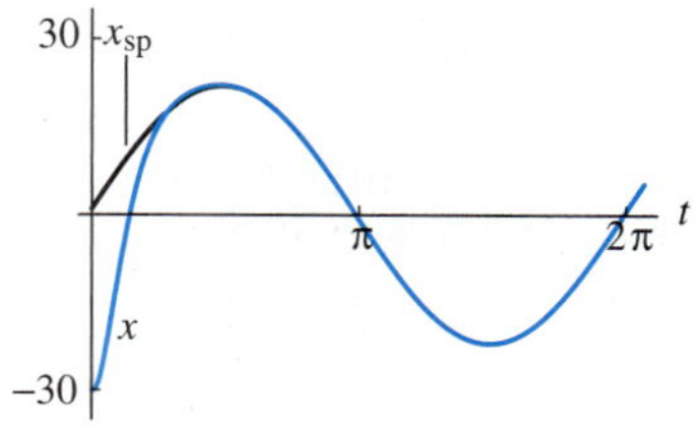

15. $C(\omega) = 2/\sqrt{4 + \omega^4}$; there is no practical resonance frequency.

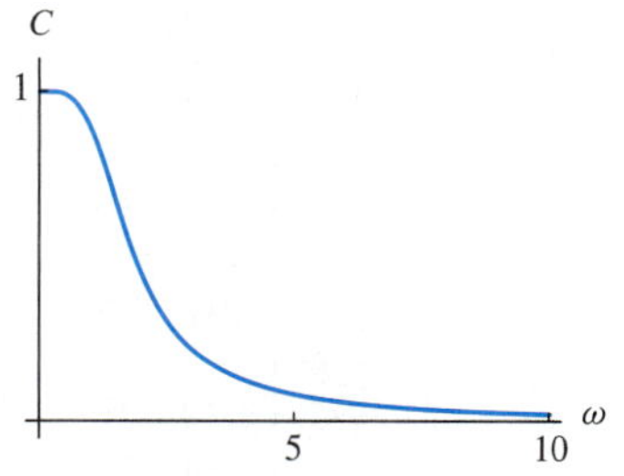

16. $C(\omega) = 10/\sqrt{25 + 6\omega^2 + \omega^4}$; there is no practical resonance frequency.

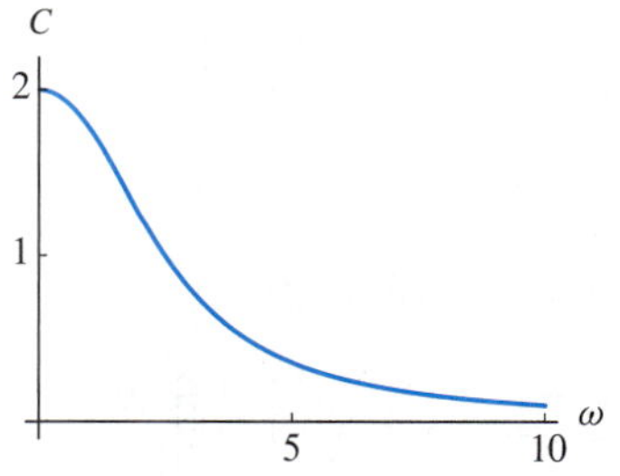

17. $C(\omega) = 50/\sqrt{2025 - 54\omega^2 + \omega^4}$; there is practical resonance at frequency $\omega = 3\sqrt{3}$.

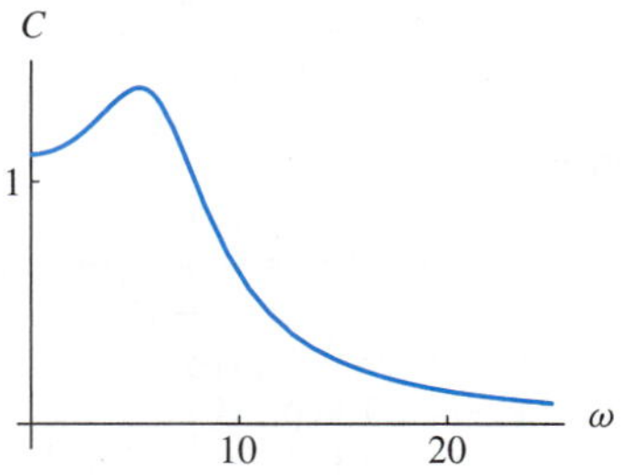

18. $C(\omega) = 100/\sqrt{422500 - 1200\omega^2 + \omega^4}$; there is practical resonance at frequency $\omega = 10\sqrt{6}$.

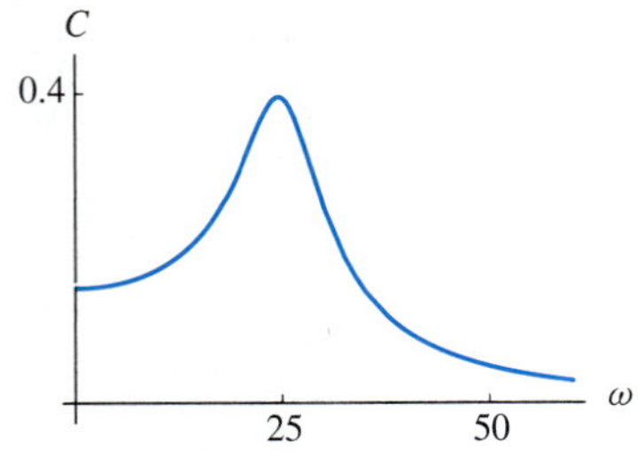

19. $\omega = \sqrt{384}$ rad/s (approximately 3.12 Hz)
20. $\omega \approx 44.27$ rad/s (approximately 7.05 Hz)
21. $\omega_0 = \sqrt{(g/L) + (k/m)}$
22. $\omega_0 = \sqrt{k/(m + I/a^2)}$
23. (a) Natural frequency: $\sqrt{10}$ rad/s (approximately 0.50 Hz); (b) amplitude: approximately 10.625 in.

Chapter 6

Section 6.1

1. $\lambda_1 = 2$: $\mathbf{v}_1 = (1, 1)$; $\lambda_2 = 3$: $\mathbf{v}_2 = (2, 1)$
2. $\lambda_1 = -1$: $\mathbf{v}_1 = (1, 1)$; $\lambda_2 = 2$: $\mathbf{v}_2 = (2, 1)$
3. $\lambda_1 = 2$: $\mathbf{v}_2 = (1, 1)$; $\lambda_2 = 5$: $\mathbf{v}_1 = (2, 1)$
4. $\lambda_1 = 1$: $\mathbf{v}_2 = (1, 1)$; $\lambda_2 = 2$: $\mathbf{v}_1 = (3, 2)$
5. $\lambda_1 = 1$: $\mathbf{v}_2 = (1, 1)$; $\lambda_2 = 4$: $\mathbf{v}_1 = (3, 2)$
6. $\lambda_1 = 2$: $\mathbf{v}_2 = (1, 1)$; $\lambda_2 = 3$: $\mathbf{v}_1 = (4, 3)$
7. $\lambda_1 = 2$: $\mathbf{v}_2 = (1, 1)$; $\lambda_2 = 4$: $\mathbf{v}_1 = (4, 3)$
8. $\lambda_1 = -2$: $\mathbf{v}_2 = (2, 3)$; $\lambda_2 = -1$: $\mathbf{v}_1 = (3, 4)$
9. $\lambda_1 = 3$: $\mathbf{v}_1 = (2, 1)$; $\lambda_2 = 4$: $\mathbf{v}_2 = (5, 2)$
10. $\lambda_1 = 4$: $\mathbf{v}_1 = (2, 1)$; $\lambda_2 = 5$: $\mathbf{v}_2 = (5, 2)$
11. $\lambda_1 = 4$: $\mathbf{v}_2 = (2, 3)$; $\lambda_2 = 5$: $\mathbf{v}_1 = (5, 7)$
12. $\lambda_1 = 3$: $\mathbf{v}_2 = (3, 2)$; $\lambda_2 = 4$: $\mathbf{v}_1 = (5, 3)$
13. $\lambda_1 = 0$: $\mathbf{v}_1 = (0, -1, 2)$; $\lambda_2 = 1$: $\mathbf{v}_2 = (0, -1, 3)$; $\lambda_3 = 2$: $\mathbf{v}_3 = (1, 0, 2)$
14. $\lambda_1 = 0$: $\mathbf{v}_1 = (0, -1, 2)$; $\lambda_2 = 2$: $\mathbf{v}_2 = (0, -1, 3)$; $\lambda_3 = 5$: $\mathbf{v}_3 = (1, 0, 2)$
15. $\lambda_1 = 0$: $\mathbf{v}_1 = (1, 1, 0)$; $\lambda_2 = 1$: $\mathbf{v}_2 = (2, 1, 1)$; $\lambda_3 = 2$: $\mathbf{v}_3 = (1, 0, 2)$
16. $\lambda_1 = 0$: $\mathbf{v}_1 = (1, 1, 1)$; $\lambda_2 = 1$: $\mathbf{v}_2 = (1, 1, 0)$; $\lambda_3 = 3$: $\mathbf{v}_3 = (-1, 0, 2)$
17. $\lambda_1 = 1$: $\mathbf{v}_1 = (1, 0, 1)$; $\lambda_2 = 2$: $\mathbf{v}_2 = (-1, 1, 2)$; $\lambda_3 = 3$: $\mathbf{v}_3 = (1, 0, 0)$
18. $\lambda_1 = 1$: $\mathbf{v}_1 = (1, 0, 3)$; $\lambda_2 = 2$: $\mathbf{v}_2 = (0, -1, 3)$; $\lambda_3 = 3$: $\mathbf{v}_3 = (0, -2, 5)$
19. $\lambda_1 = \lambda_2 = 1$: $\mathbf{v}_1 = (1, 0, 1)$, $\mathbf{v}_2 = (-3, 1, 0)$; $\lambda_3 = 3$, $\mathbf{v}_3 = (1, 0, 0)$
20. $\lambda_1 = \lambda_2 = 1$: $\mathbf{v}_1 = (1, 0, 2)$, $\mathbf{v}_2 = (3, 2, 0)$; $\lambda_3 = 2$, $\mathbf{v}_3 = (0, -2, 5)$
21. $\lambda_1 = 1$: $\mathbf{v}_1 = (1, 1, 0)$; $\lambda_2 = \lambda_3 = 2$: $\mathbf{v}_2 = (3, 2, 0)$, $\mathbf{v}_3 = (-1, 0, 2)$
22. $\lambda_1 = \lambda_2 = -1$: $\mathbf{v}_1 = (1, 1, 0)$, $\mathbf{v}_2 = (-1, 0, 2)$; $\lambda_3 = 2$, $\mathbf{v}_3 = (1, 1, 1)$
23. $\lambda_1 = 1$: $\mathbf{v}_1 = (1, 0, 0, 0)$; $\lambda_2 = 2$: $\mathbf{v}_2 = (2, 1, 0, 0)$; $\lambda_3 = 3$: $\mathbf{v}_3 = (3, 2, 1, 0)$; $\lambda_4 = 4$: $\mathbf{v}_4 = (4, 3, 2, 1)$
24. $\lambda_1 = \lambda_2 = 1$: $\mathbf{v}_1 = (1, 0, 0, 0)$, $\mathbf{v}_2 = (0, 1, 0, 0)$; $\lambda_3 = \lambda_4 = 3$: $\mathbf{v}_3 = (0, 0, 0, 1)$, $\mathbf{v}_4 = (2, 2, 1, 0)$
25. $\lambda_1 = \lambda_2 = 1$: $\mathbf{v}_1 = (1, 0, 0, 0)$, $\mathbf{v}_2 = (0, 1, 0, 0)$; $\lambda_3 = \lambda_4 = 2$: $\mathbf{v}_3 = (0, 0, 0, 1)$, $\mathbf{v}_4 = (1, 1, 1, 0)$
26. $\lambda_1 = -2$: $\mathbf{v}_1 = (1, 0, 0, 2)$; $\lambda_2 = -1$: $\mathbf{v}_2 = (0, 0, 1, 0)$; $\lambda_3 = 1$: $\mathbf{v}_3 = (1, 0, 0, 1)$; $\lambda_4 = 2$: $\mathbf{v}_4 = (0, 1, 0, 0)$
27. $\lambda_1 = -i$: $\mathbf{v}_1 = (+i, 1)$; $\lambda_2 = +i$: $\mathbf{v}_2 = (-i, 1)$
28. $\lambda_1 = -6i$: $\mathbf{v}_1 = (-i, 1)$; $\lambda_2 = +6i$: $\mathbf{v}_2 = (+i, 1)$
29. $\lambda_1 = -6i$: $\mathbf{v}_1 = (-i, 2)$; $\lambda_2 = +6i$: $\mathbf{v}_2 = (+i, 2)$
30. $\lambda_1 = -12i$: $\mathbf{v}_1 = (-i, 1)$; $\lambda_2 = +12i$: $\mathbf{v}_2 = (+i, 1)$
31. $\lambda_1 = -12i$: $\mathbf{v}_1 = (+2i, 1)$; $\lambda_2 = +12i$: $\mathbf{v}_2 = (-2i, 1)$
32. $\lambda_1 = -12i$: $\mathbf{v}_1 = (-i, 3)$; $\lambda_2 = +12i$: $\mathbf{v}_2 = (+i, 3)$
40. We find that Tr $\mathbf{A} = 12$ and det $\mathbf{A} = 60$, so the characteristic polynomial of the given matrix $\mathbf{A}$ is $p(\lambda) = -\lambda^3 + 12\lambda^2 + c_1\lambda + 60$. Eigenvalues and eigenvectors: $\lambda_1 = 3$: $\mathbf{v}_1 = (3, 2, 1)$; $\lambda_2 = 4$: $\mathbf{v}_2 = (5, 7, 7)$; $\lambda_3 = 5$: $\mathbf{v}_3 = (-1, 1, 2)$
41. We find that Tr $\mathbf{A} = 8$ and det $\mathbf{A} = -60$, so the characteristic polynomial of the given matrix $\mathbf{A}$ is $p(\lambda) = \lambda^4 - 8\lambda^3 + c_2\lambda^2 + c_1\lambda - 60$. Eigenvalues and eigenvectors: $\lambda_1 = -2$: $\mathbf{v}_1 = (1, 0, 1, 2)$; $\lambda_2 = 2$: $\mathbf{v}_2 = (3, 4, 0, 3)$; $\lambda_3 = 3$: $\mathbf{v}_3 = (3, 1, 2, 4)$; $\lambda_4 = 5$: $\mathbf{v}_4 = (1, 1, 0, 1)$

Section 6.2

1. $\lambda_1 = 1, \lambda_2 = 3$; $\mathbf{P} = \begin{bmatrix} 1 & 2 \\ 1 & 1 \end{bmatrix}$, $\mathbf{D} = \begin{bmatrix} 1 & 0 \\ 0 & 3 \end{bmatrix}$

2. $\lambda_1 = 0, \lambda_2 = 2$; $\mathbf{P} = \begin{bmatrix} 1 & 3 \\ 1 & 2 \end{bmatrix}$, $\mathbf{D} = \begin{bmatrix} 0 & 0 \\ 0 & 2 \end{bmatrix}$

3. $\lambda_1 = 2, \lambda_2 = 3$; $\mathbf{P} = \begin{bmatrix} 1 & 3 \\ 1 & 2 \end{bmatrix}$, $\mathbf{D} = \begin{bmatrix} 2 & 0 \\ 0 & 3 \end{bmatrix}$

4. $\lambda_1 = 1, \lambda_2 = 2$; $\mathbf{P} = \begin{bmatrix} 1 & 4 \\ 1 & 3 \end{bmatrix}$, $\mathbf{D} = \begin{bmatrix} 1 & 0 \\ 0 & 2 \end{bmatrix}$

5. $\lambda_1 = 1, \lambda_2 = 3$; $\mathbf{P} = \begin{bmatrix} 1 & 4 \\ 1 & 3 \end{bmatrix}$, $\mathbf{D} = \begin{bmatrix} 1 & 0 \\ 0 & 3 \end{bmatrix}$

6. $\lambda_1 = 1, \lambda_2 = 2$; $\mathbf{P} = \begin{bmatrix} 2 & 3 \\ 3 & 4 \end{bmatrix}$, $\mathbf{D} = \begin{bmatrix} 1 & 0 \\ 0 & 2 \end{bmatrix}$

7. $\lambda_1 = 1, \lambda_2 = 2$; $\mathbf{P} = \begin{bmatrix} 2 & 5 \\ 1 & 2 \end{bmatrix}$, $\mathbf{D} = \begin{bmatrix} 1 & 0 \\ 0 & 2 \end{bmatrix}$

8. $\lambda_1 = 1, \lambda_2 = 2$; $\mathbf{P} = \begin{bmatrix} 3 & 5 \\ 2 & 3 \end{bmatrix}$, $\mathbf{D} = \begin{bmatrix} 1 & 0 \\ 0 & 2 \end{bmatrix}$

9. The double eigenvalue $\lambda_1 = \lambda_2 = 1$ has only the single associated eigenvector $\mathbf{v}_1 = (2, 1)$, so the matrix $\mathbf{A}$ is not diagonalizable.
10. The double eigenvalue $\lambda_1 = \lambda_2 = 2$ has only the single associated eigenvector $\mathbf{v}_1 = (1, 1)$, so the matrix $\mathbf{A}$ is not diagonalizable.
11. The double eigenvalue $\lambda_1 = \lambda_2 = 2$ has only the single associated eigenvector $\mathbf{v}_1 = (-1, 3)$, so the matrix $\mathbf{A}$ is not diagonalizable.
12. The double eigenvalue $\lambda_1 = \lambda_2 = -1$ has only the single associated eigenvector $\mathbf{v}_1 = (-3, 4)$, so the matrix $\mathbf{A}$ is not diagonalizable.

13. $\lambda_1 = 1, \lambda_2 = \lambda_3 = 2$; $\mathbf{P} = \begin{bmatrix} 1 & 0 & 3 \\ 0 & 0 & 1 \\ 0 & 1 & 0 \end{bmatrix}$, $\mathbf{D} = \begin{bmatrix} 1 & 0 & 0 \\ 0 & 2 & 0 \\ 0 & 0 & 2 \end{bmatrix}$

14. $\lambda_1 = \lambda_2 = 0, \lambda_3 = 1$; $\mathbf{P} = \begin{bmatrix} -1 & 1 & 1 \\ 0 & 1 & 1 \\ 2 & 0 & 1 \end{bmatrix}$, $\mathbf{D} = \begin{bmatrix} 0 & 0 & 0 \\ 0 & 0 & 0 \\ 0 & 0 & 1 \end{bmatrix}$

15. $\lambda_1 = 0, \lambda_2 = \lambda_3 = 1$; $\mathbf{P} = \begin{bmatrix} 1 & -1 & 3 \\ 1 & 0 & 2 \\ 0 & 2 & 0 \end{bmatrix}$,

$$\mathbf{D} = \begin{bmatrix} 0 & 0 & 0 \\ 0 & 1 & 0 \\ 0 & 0 & 1 \end{bmatrix}$$

16. $\lambda_1 = \lambda_2 = 1, \lambda_3 = 3$; $\mathbf{P} = \begin{bmatrix} 0 & 1 & -1 \\ 0 & 1 & 0 \\ 1 & 0 & 2 \end{bmatrix}$,

$\mathbf{D} = \begin{bmatrix} 1 & 0 & 0 \\ 0 & 1 & 0 \\ 0 & 0 & 3 \end{bmatrix}$

17. $\lambda_1 = -1, \lambda_2 = 1, \lambda_3 = 2$; $\mathbf{P} = \begin{bmatrix} 1 & -1 & 1 \\ 1 & 0 & 1 \\ 0 & 2 & 1 \end{bmatrix}$,

$\mathbf{D} = \begin{bmatrix} -1 & 0 & 0 \\ 0 & 1 & 0 \\ 0 & 0 & 2 \end{bmatrix}$

18. $\lambda_1 = 1, \lambda_2 = 2, \lambda_3 = 3$; $\mathbf{P} = \begin{bmatrix} 1 & -1 & 1 \\ 1 & 0 & 1 \\ 0 & 2 & 1 \end{bmatrix}$,

$\mathbf{D} = \begin{bmatrix} 1 & 0 & 0 \\ 0 & 2 & 0 \\ 0 & 0 & 3 \end{bmatrix}$

19. $\lambda_1 = 1, \lambda_2 = 2, \lambda_3 = 3$; $\mathbf{P} = \begin{bmatrix} 1 & 1 & -1 \\ 1 & 1 & 0 \\ 1 & 0 & 2 \end{bmatrix}$,

$\mathbf{D} = \begin{bmatrix} 1 & 0 & 0 \\ 0 & 2 & 0 \\ 0 & 0 & 3 \end{bmatrix}$

20. $\lambda_1 = 2, \lambda_2 = 5, \lambda_3 = 6$; $\mathbf{P} = \begin{bmatrix} 1 & 0 & 0 \\ 0 & -1 & -2 \\ 3 & 3 & 5 \end{bmatrix}$,

$\mathbf{D} = \begin{bmatrix} 2 & 0 & 0 \\ 0 & 5 & 0 \\ 0 & 0 & 6 \end{bmatrix}$

21. The triple eigenvalue $\lambda_1 = \lambda_2 = \lambda_3 = 1$ has only the two associated eigenvectors $\mathbf{v}_1 = (0, 0, 1)$ and $\mathbf{v}_2 = (1, 1, 0)$, so the matrix **A** is not diagonalizable.

22. The triple eigenvalue $\lambda_1 = \lambda_2 = \lambda_3 = 1$ has only the single associated eigenvector $\mathbf{v}_1 = (1, 1, 1)$, so the matrix **A** is not diagonalizable.

23. The eigenvalues $\lambda_1 = \lambda_2 = 1$ and $\lambda_3 = 2$ have only the two associated eigenvectors $\mathbf{v}_1 = (1, 1, 1)$ and $\mathbf{v}_3 = (1, 1, 0)$, so the matrix **A** is not diagonalizable.

24. The eigenvalues $\lambda_1 = 1$ and $\lambda_2 = \lambda_3 = 2$ have only the two associated eigenvectors $\mathbf{v}_1 = (1, 1, 0)$ and $\mathbf{v}_2 = (1, 1, 1)$, so the matrix **A** is not diagonalizable.

25. $\lambda_1 = \lambda_2 = -1, \lambda_3 = \lambda_4 = 1$; $\mathbf{P} = \begin{bmatrix} 0 & 1 & 0 & 1 \\ 0 & 1 & 1 & 0 \\ 0 & 1 & 0 & 0 \\ 1 & 0 & 0 & 0 \end{bmatrix}$,

$\mathbf{D} = \begin{bmatrix} -1 & 0 & 0 & 0 \\ 0 & -1 & 0 & 0 \\ 0 & 0 & 1 & 0 \\ 0 & 0 & 0 & 1 \end{bmatrix}$

26. $\lambda_1 = \lambda_2 = \lambda_3 = 1, \lambda_4 = 2$; $\mathbf{P} = \begin{bmatrix} 0 & 0 & 1 & 1 \\ 0 & 1 & 0 & 1 \\ 1 & 0 & 0 & 1 \\ 0 & 0 & 0 & 1 \end{bmatrix}$,

$\mathbf{D} = \begin{bmatrix} 1 & 0 & 0 & 0 \\ 0 & 1 & 0 & 0 \\ 0 & 0 & 1 & 0 \\ 0 & 0 & 0 & 2 \end{bmatrix}$

27. The eigenvalues $\lambda_1 = \lambda_2 = \lambda_3 = 1$ and $\lambda_4 = 2$ have only the two associated eigenvectors $\mathbf{v}_1 = (1, 0, 0, 0)$ and $\mathbf{v}_4 = (1, 1, 1, 1)$, so the matrix **A** is not diagonalizable.

28. The eigenvalues $\lambda_1 = \lambda_2 = 1$ and $\lambda_3 = \lambda_4 = 2$ have only the two associated eigenvectors $\mathbf{v}_1 = (1, 0, 0, 0)$ and $\mathbf{v}_3 = (1, 1, 1, 0)$, so the matrix **A** is not diagonalizable.

Section 6.3

1. $\mathbf{P} = \begin{bmatrix} 1 & 2 \\ 1 & 1 \end{bmatrix}$, $\mathbf{D} = \begin{bmatrix} 1 & 0 \\ 0 & 2 \end{bmatrix}$, $A^5 = \begin{bmatrix} 63 & -62 \\ 31 & -30 \end{bmatrix}$

2. $\mathbf{P} = \begin{bmatrix} 1 & 2 \\ 1 & 1 \end{bmatrix}$, $\mathbf{D} = \begin{bmatrix} -1 & 0 \\ 0 & 2 \end{bmatrix}$, $A^5 = \begin{bmatrix} 65 & -66 \\ 33 & -34 \end{bmatrix}$

3. $\mathbf{P} = \begin{bmatrix} 1 & 3 \\ 1 & 2 \end{bmatrix}$, $\mathbf{D} = \begin{bmatrix} 0 & 0 \\ 0 & 2 \end{bmatrix}$, $A^5 = \begin{bmatrix} 96 & -96 \\ 64 & -64 \end{bmatrix}$

4. $\mathbf{P} = \begin{bmatrix} 1 & 3 \\ 1 & 2 \end{bmatrix}$, $\mathbf{D} = \begin{bmatrix} 1 & 0 \\ 0 & 2 \end{bmatrix}$, $A^5 = \begin{bmatrix} 94 & -93 \\ 62 & -61 \end{bmatrix}$

5. $\mathbf{P} = \begin{bmatrix} 1 & 4 \\ 1 & 3 \end{bmatrix}$, $\mathbf{D} = \begin{bmatrix} 1 & 0 \\ 0 & 2 \end{bmatrix}$, $A^5 = \begin{bmatrix} 125 & -124 \\ 93 & -92 \end{bmatrix}$

6. $\mathbf{P} = \begin{bmatrix} 2 & 5 \\ 1 & 2 \end{bmatrix}$, $\mathbf{D} = \begin{bmatrix} 1 & 0 \\ 0 & 2 \end{bmatrix}$, $A^5 = \begin{bmatrix} 156 & -310 \\ 62 & -123 \end{bmatrix}$

7. $\mathbf{P} = \begin{bmatrix} 1 & 0 & 3 \\ 0 & 0 & 1 \\ 0 & 1 & 0 \end{bmatrix}$, $\mathbf{D} = \begin{bmatrix} 1 & 0 & 0 \\ 0 & 2 & 0 \\ 0 & 0 & 2 \end{bmatrix}$,

$A^5 = \begin{bmatrix} 1 & 93 & 0 \\ 0 & 32 & 0 \\ 0 & 0 & 32 \end{bmatrix}$

8. $\mathbf{P} = \begin{bmatrix} 0 & 1 & 1 \\ 1 & 0 & 0 \\ 2 & 0 & 1 \end{bmatrix}$, $\mathbf{D} = \begin{bmatrix} 1 & 0 & 0 \\ 0 & 1 & 0 \\ 0 & 0 & 2 \end{bmatrix}$,

$A^5 = \begin{bmatrix} 1 & -62 & 31 \\ 0 & 1 & 0 \\ 0 & -62 & 32 \end{bmatrix}$

9. $\mathbf{P} = \begin{bmatrix} 1 & 1 & -3 \\ 0 & 0 & 1 \\ 0 & 1 & 0 \end{bmatrix}$, $\mathbf{D} = \begin{bmatrix} 1 & 0 & 0 \\ 0 & 2 & 0 \\ 0 & 0 & 2 \end{bmatrix}$,

$A^5 = \begin{bmatrix} 1 & -93 & 31 \\ 0 & 32 & 0 \\ 0 & 0 & 32 \end{bmatrix}$

10. $\mathbf{P} = \begin{bmatrix} 1 & -1 & 3 \\ 1 & 0 & 2 \\ 0 & 2 & 0 \end{bmatrix}$, $\mathbf{D} = \begin{bmatrix} 1 & 0 & 0 \\ 0 & 2 & 0 \\ 0 & 0 & 2 \end{bmatrix}$,

$A^5 = \begin{bmatrix} 94 & -93 & 31 \\ 62 & -61 & 31 \\ 0 & 0 & 32 \end{bmatrix}$

11. $\mathbf{P} = \begin{bmatrix} 0 & 0 & -1 \\ -1 & -2 & -42 \\ 3 & 5 & 87 \end{bmatrix}$, $\mathbf{D} = \begin{bmatrix} -1 & 0 & 0 \\ 0 & 0 & 0 \\ 0 & 0 & 1 \end{bmatrix}$,

$A^{10} = \begin{bmatrix} 1 & 0 & 0 \\ 78 & -5 & -2 \\ -195 & 15 & 6 \end{bmatrix}$

12. $\mathbf{P} = \begin{bmatrix} 1 & 1 & 3 \\ 2 & 0 & 5 \\ 0 & 5 & 0 \end{bmatrix}$, $\mathbf{D} = \begin{bmatrix} -1 & 0 & 0 \\ 0 & 1 & 0 \\ 0 & 0 & 1 \end{bmatrix}$,

$A^{10} = \begin{bmatrix} 1 & 0 & 0 \\ 0 & 1 & 0 \\ 0 & 0 & 1 \end{bmatrix}$

13. $\mathbf{P} = \begin{bmatrix} -1 & 1 & 1 \\ 0 & 1 & 1 \\ 2 & 0 & 1 \end{bmatrix}$, $\mathbf{D} = \begin{bmatrix} -1 & 0 & 0 \\ 0 & 0 & 0 \\ 0 & 0 & 1 \end{bmatrix}$,

$A^{10} = \begin{bmatrix} 3 & -3 & 1 \\ 2 & -2 & 1 \\ 0 & 0 & 1 \end{bmatrix}$

14. $\mathbf{P} = \begin{bmatrix} 1 & 1 & 2 \\ 0 & 1 & 1 \\ 2 & 0 & 1 \end{bmatrix}$, $\mathbf{D} = \begin{bmatrix} -1 & 0 & 0 \\ 0 & 0 & 0 \\ 0 & 0 & 1 \end{bmatrix}$,

$A^{10} = \begin{bmatrix} 3 & -3 & -1 \\ 2 & -2 & -1 \\ 0 & 0 & 1 \end{bmatrix}$

15. $\mathbf{A}^2 - 3\mathbf{A} + 2\mathbf{I} = \mathbf{0}$, $\mathbf{A}^3 = \begin{bmatrix} 29 & -28 \\ 21 & -20 \end{bmatrix}$, $\mathbf{A}^4 = \begin{bmatrix} 61 & -60 \\ 45 & -44 \end{bmatrix}$,

$\mathbf{A}^{-1} = \frac{1}{2}\begin{bmatrix} -2 & 4 \\ -3 & 5 \end{bmatrix}$

16. $\mathbf{A}^2 - 3\mathbf{A} + 2\mathbf{I} = \mathbf{0}$, $\mathbf{A}^3 = \begin{bmatrix} 36 & -70 \\ 14 & -27 \end{bmatrix}$, $\mathbf{A}^4 = \begin{bmatrix} 76 & -150 \\ 30 & -59 \end{bmatrix}$,

$\mathbf{A}^{-1} = \frac{1}{2}\begin{bmatrix} -3 & 10 \\ -2 & 6 \end{bmatrix}$

17. $-\mathbf{A}^3 + 5\mathbf{A}^2 - 8\mathbf{A} + 4\mathbf{I} = \mathbf{0}$, $\mathbf{A}^3 = \begin{bmatrix} 1 & 21 & 0 \\ 0 & 8 & 0 \\ 0 & 0 & 8 \end{bmatrix}$,

$\mathbf{A}^4 = \begin{bmatrix} 1 & 45 & 0 \\ 0 & 16 & 0 \\ 0 & 0 & 16 \end{bmatrix}$, $\mathbf{A}^{-1} = \frac{1}{2}\begin{bmatrix} 2 & -3 & 0 \\ 0 & 1 & 0 \\ 0 & 0 & 1 \end{bmatrix}$

18. $-\mathbf{A}^3 + 4\mathbf{A}^2 - 5\mathbf{A} + 2\mathbf{I} = \mathbf{0}$, $\mathbf{A}^3 = \begin{bmatrix} 1 & -14 & 7 \\ 0 & 1 & 0 \\ 0 & -14 & 8 \end{bmatrix}$,

$\mathbf{A}^4 = \begin{bmatrix} 1 & -30 & 15 \\ 0 & 1 & 0 \\ 0 & -30 & 16 \end{bmatrix}$ $\mathbf{A}^{-1} = \frac{1}{2}\begin{bmatrix} 2 & 2 & -1 \\ 0 & 2 & 0 \\ 0 & 2 & 1 \end{bmatrix}$

19. $-\mathbf{A}^3 + 5\mathbf{A}^2 - 8\mathbf{A} + 4\mathbf{I} = \mathbf{0}$, $\mathbf{A}^3 = \begin{bmatrix} 1 & -21 & 7 \\ 0 & 8 & 0 \\ 0 & 0 & 8 \end{bmatrix}$,

$\mathbf{A}^4 = \begin{bmatrix} 1 & -45 & 15 \\ 0 & 16 & 0 \\ 0 & 0 & 16 \end{bmatrix}$, $\mathbf{A}^{-1} = \frac{1}{2}\begin{bmatrix} 2 & 3 & -1 \\ 0 & 1 & 0 \\ 0 & 0 & 1 \end{bmatrix}$

20. $-\mathbf{A}^3 + 5\mathbf{A}^2 - 8\mathbf{A} + 4\mathbf{I} = \mathbf{0}$, $\mathbf{A}^3 = \begin{bmatrix} 22 & -21 & 7 \\ 14 & -13 & 7 \\ 0 & 0 & 8 \end{bmatrix}$,

$\mathbf{A}^4 = \begin{bmatrix} 46 & -45 & 15 \\ 30 & -29 & 15 \\ 0 & 0 & 16 \end{bmatrix}$, $\mathbf{A}^{-1} = \frac{1}{2}\begin{bmatrix} -1 & 3 & -1 \\ -2 & 4 & -1 \\ 0 & 0 & 1 \end{bmatrix}$

21. $-\mathbf{A}^3 + \mathbf{A} = \mathbf{0}$, $\mathbf{A}^3 = \mathbf{A} = \begin{bmatrix} 1 & 0 & 0 \\ 6 & 5 & 2 \\ 21 & -15 & -6 \end{bmatrix}$,

$\mathbf{A}^4 = \mathbf{A}^2 = \begin{bmatrix} 1 & 0 & 0 \\ 78 & -5 & -2 \\ -195 & 15 & 6 \end{bmatrix}$. Because $\lambda = 0$ is an eigenvalue, $\mathbf{A}$ is singular and $\mathbf{A}^{-1}$ does not exist.

22. $-\mathbf{A}^3 + \mathbf{A}^2 + \mathbf{A} - \mathbf{I} = \mathbf{0}$, $\mathbf{A}^3 = \mathbf{A} = \begin{bmatrix} 11 & -6 & -2 \\ 20 & -11 & -4 \\ 0 & 0 & 1 \end{bmatrix} = \mathbf{A}$,

$\mathbf{A}^4 = \begin{bmatrix} 1 & 0 & 0 \\ 0 & 1 & 0 \\ 0 & 0 & 1 \end{bmatrix} = \mathbf{I}$, $\mathbf{A}^{-1} = \mathbf{A}$

23. $-\mathbf{A}^3 + \mathbf{A} = \mathbf{0}$, $\mathbf{A}^3 = \mathbf{A} = \begin{bmatrix} 1 & -1 & 1 \\ 2 & -2 & 1 \\ 4 & -4 & 1 \end{bmatrix}$,

$\mathbf{A}^4 = \mathbf{A}^2 = \begin{bmatrix} 3 & -3 & 1 \\ 2 & -2 & 1 \\ 0 & 0 & 1 \end{bmatrix}$. Because $\lambda = 0$ is an eigenvalue, $\mathbf{A}$ is singular and $\mathbf{A}^{-1}$ does not exist.

24. $-\mathbf{A}^3 + \mathbf{A} = \mathbf{0}$, $\mathbf{A}^3 = \mathbf{A} = \begin{bmatrix} 5 & -5 & -3 \\ 2 & -2 & -1 \\ 4 & -4 & -3 \end{bmatrix}$,

$\mathbf{A}^4 = \mathbf{A}^2 = \begin{bmatrix} 3 & -3 & -1 \\ 2 & -2 & -1 \\ 0 & 0 & 1 \end{bmatrix}$. Because $\lambda = 0$ is an eigenvalue, $\mathbf{A}$ is singular and $\mathbf{A}^{-1}$ does not exist.

25. $\mathbf{x}_k = \mathbf{A}^k\mathbf{x}_0 = \begin{bmatrix} 1 & -1 \\ 1 & 1 \end{bmatrix}\begin{bmatrix} 1 & 0 \\ 0 & 4/5 \end{bmatrix}^k \frac{1}{2}\begin{bmatrix} 1 & 1 \\ -1 & 1 \end{bmatrix}\mathbf{x}_0 \to=$

$(C_0 + S_0)\begin{bmatrix} 1/2 \\ 1/2 \end{bmatrix}$ as $k \to \infty$. The long-term distribution of population is 50% city, 50% suburban.

26. $\mathbf{x}_k = \mathbf{A}^k\mathbf{x}_0 = \begin{bmatrix} 1 & -1 \\ 3 & 1 \end{bmatrix}\begin{bmatrix} 1 & 0 \\ 0 & 4/5 \end{bmatrix}^k \frac{1}{4}\begin{bmatrix} 1 & 1 \\ -3 & 1 \end{bmatrix}\mathbf{x}_0 \to=$

$(C_0 + S_0)\begin{bmatrix} 1/4 \\ 3/4 \end{bmatrix}$ as $k \to \infty$. The long-term distribution of population is 25% city, 75% suburban.

27. $\mathbf{x}_k = \mathbf{A}^k\mathbf{x}_0 = \begin{bmatrix} 3 & -1 \\ 5 & 1 \end{bmatrix}\begin{bmatrix} 1 & 0 \\ 0 & 3/5 \end{bmatrix}^k \frac{1}{8}\begin{bmatrix} 1 & 1 \\ -5 & 3 \end{bmatrix}\mathbf{x}_0 \to=$

$(C_0 + S_0)\begin{bmatrix} 3/8 \\ 5/8 \end{bmatrix}$ as $k \to \infty$. The long-term distribution of population is 3/8 city, 5/8 suburban.

28. $\mathbf{x}_k = \mathbf{A}^k\mathbf{x}_0 = \begin{bmatrix} 1 & -1 \\ 2 & 1 \end{bmatrix}\begin{bmatrix} 1 & 0 \\ 0 & 7/10 \end{bmatrix}^k \frac{1}{3}\begin{bmatrix} 1 & 1 \\ -2 & 1 \end{bmatrix}\mathbf{x}_0 \to=$

$(C_0 + S_0)\begin{bmatrix} 1/3 \\ 2/3 \end{bmatrix}$ as $k \to \infty$. The long-term distribution of population is 1/3 city, 2/3 suburban.

29. $\mathbf{x}_k = \mathbf{A}^k\mathbf{x}_0 = \begin{bmatrix} 1 & -1 \\ 2 & 1 \end{bmatrix}\begin{bmatrix} 1 & 0 \\ 0 & 17/20 \end{bmatrix}^k \frac{1}{3}\begin{bmatrix} 1 & 1 \\ -2 & 1 \end{bmatrix}\mathbf{x}_0 \to=$

$(C_0 + S_0)\begin{bmatrix} 1/3 \\ 2/3 \end{bmatrix}$ as $k \to \infty$. The long-term distribution of population is 1/3 city, 2/3 suburban.

30. $\mathbf{x}_k = \mathbf{A}^k\mathbf{x}_0 = \begin{bmatrix} 3 & -1 \\ 4 & 1 \end{bmatrix}\begin{bmatrix} 1 & 0 \\ 0 & 13/20 \end{bmatrix}^k \frac{1}{7}\begin{bmatrix} 1 & 1 \\ -4 & 3 \end{bmatrix}\mathbf{x}_0 \to=$

$(C_0 + S_0)\begin{bmatrix} 3/7 \\ 4/7 \end{bmatrix}$ as $k \to \infty$. The long-term distribution of population is 3/7 city, 4/7 suburban.

31. $\mathbf{x}_k = \mathbf{A}^k\mathbf{x}_0 = \begin{bmatrix} 5 & 5 \\ 4 & 2 \end{bmatrix}\begin{bmatrix} 1 & 0 \\ 0 & 4/5 \end{bmatrix}^k \frac{1}{10}\begin{bmatrix} -2 & 5 \\ 4 & -5 \end{bmatrix}\mathbf{x}_0 \to=$

$\begin{bmatrix} 2.5R_0 - F_0 \\ 2R_0 - 0.8F_0 \end{bmatrix}$ as $k \to \infty$. The fox–rabbit population approaches a stable situation with $2.5R_0 - F_0$ foxes and $2R_0 - 0.8F_0$ rabbits.

32. $\mathbf{x}_k = \mathbf{A}^k\mathbf{x}_0 =$

$\begin{bmatrix} 10 & 2 \\ 7 & 1 \end{bmatrix}\begin{bmatrix} 19/20 & 0 \\ 0 & 17/20 \end{bmatrix}^k \frac{1}{4}\begin{bmatrix} -1 & 2 \\ 7 & -10 \end{bmatrix}\mathbf{x}_0 \to= \begin{bmatrix} 0 \\ 0 \end{bmatrix}$

as $k \to \infty$. The fox and rabbit population both die out.

33. $\mathbf{x}_k = \mathbf{A}^k\mathbf{x}_0 =$

$\begin{bmatrix} 10 & 10 \\ 9 & 3 \end{bmatrix}\begin{bmatrix} 21/20 & 0 \\ 0 & 3/4 \end{bmatrix}^k \frac{1}{60}\begin{bmatrix} -3 & 10 \\ 9 & -10 \end{bmatrix}\mathbf{x}_0 \approx$

$\frac{1}{60}(1.05)^k(10R_0 - 3F_0)\begin{bmatrix} 10 \\ 9 \end{bmatrix}$ as when k is sufficiently large. The fox and rabbit populations are both increasing at 5% per year, with 10 foxes for each 9 rabbits.

34. $\mathbf{A} = \mathbf{PDP}^{-1} = \begin{bmatrix} 41 & -30 \\ 56 & -41 \end{bmatrix}$. If n is even, then $\mathbf{D}^n = \mathbf{I}$ so $\mathbf{A}^n = \mathbf{PD}^n\mathbf{P}^{-1} = \mathbf{PIP}^{-1} = \mathbf{I}$. If n is odd, then $\mathbf{A}^n = \mathbf{A}^{n-1}\mathbf{A} = \mathbf{IA} = \mathbf{A}$. Thus $\mathbf{A}^{99} = \mathbf{A}$ and $\mathbf{A}^{100} = \mathbf{I}$.

35. $\lambda = \pm 1$ implies that $\mathbf{D}^n = \mathbf{I}$ if n is even, in which case $\mathbf{A}^n = \mathbf{PD}^n\mathbf{P}^{-1} = \mathbf{I}$.

36. $\mathbf{A}^2 = \mathbf{I}$, so $\mathbf{A}^3 = \mathbf{A}^2\mathbf{A} = \mathbf{I}\mathbf{A} = \mathbf{A}$, $\mathbf{A}^4 = \mathbf{A}^3\mathbf{A} = \mathbf{A}^2 = \mathbf{I}$, and so forth.

37. $\mathbf{A}^2 = -\mathbf{I}$, so $\mathbf{A}^3 = \mathbf{A}^2\mathbf{A} = -\mathbf{I}\mathbf{A} = -\mathbf{A}$, $\mathbf{A}^4 = \mathbf{A}^3\mathbf{A} = -\mathbf{A}^2 = \mathbf{I}$, and so forth.

38. If $\mathbf{B} = \begin{bmatrix} 0 & 1 \\ 0 & 0 \end{bmatrix}$ so $\mathbf{B}^2 = \mathbf{0}$, and it follows that

$$\mathbf{A}^n = \mathbf{I} + n\mathbf{B} = \begin{bmatrix} 1 & n \\ 0 & 1 \end{bmatrix}.$$

Chapter 7

Section 7.1

1. $x_1' = x_2,\ x_2' = -7x_1 - 3x_2 + t^2$

2. $x_1' = x_2,\ x_2' = x_3,\ x_3' = x_4,\ x_4' = -x_1 + 3x_2 - 6x_3 + \cos 3t$

3. $x_1' = x_2,\ t^2x_2' = (1 - t^2)x_1 - tx_2$

4. $x_1' = x_2, x_2' = x_3, t^3x_3' = -5x_1 - 3tx_2 + 2t^2x_3 + \ln t$

5. $x_1' = x_2,\ x_2' = x_3,\ x_3' = x_2^2 + \cos x_1$

6. $x_1' = x_2, x_2' = 5x_1 - 4y_1,\ y_1' = y_2,\ y_2' = -4x_1 + 5y_1$

7. $x_1' = x_2,\ y_1' = y_2,\ x_2' = -kx_1 \cdot (x_1^2 + y_1^2)^{-3/2}$,
$y_2' = -ky_1 \cdot (x_1^2 + y_1^2)^{-3/2}$

8. $x_1' = x_2, x_2' = -4x_1 + 2y_1 - 3x_2$
$y_1' = y_2, y_2' = 3x_1 - y_1 - 2y_2 + \cos t$

9. $x_1' = x_2, y_1' = y_2,\ z_1' = z_2,\ x_2' = 3x_1 - y_1 + 2z_1$,
$y_2' = x_1 + y_1 - 4z_1,\ z_2' = 5x_1 - y_1 - z_1$

10. $x_1' = x_2, x_2' = x_1(1 - y_1)$
$y_1' = y_2, y_2' = y_1(1 - x_1)$

11. $x(t) = A\cos t + B\sin t,\ y(t) = B\cos t - A\sin t$

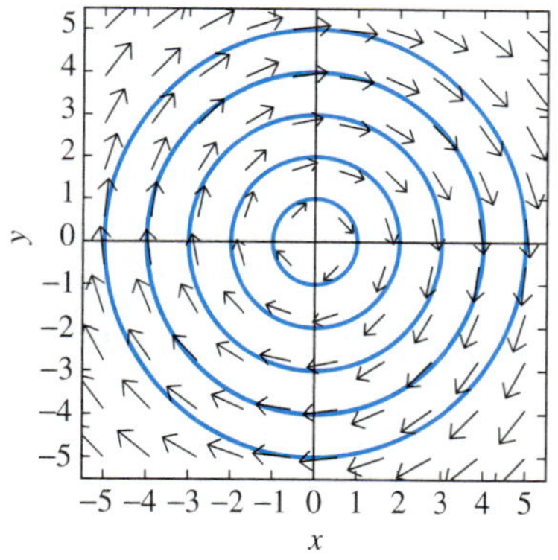

12. $x(t) = Ae^t + Be^{-t},\ y(t) = Ae^t - Be^{-t}$

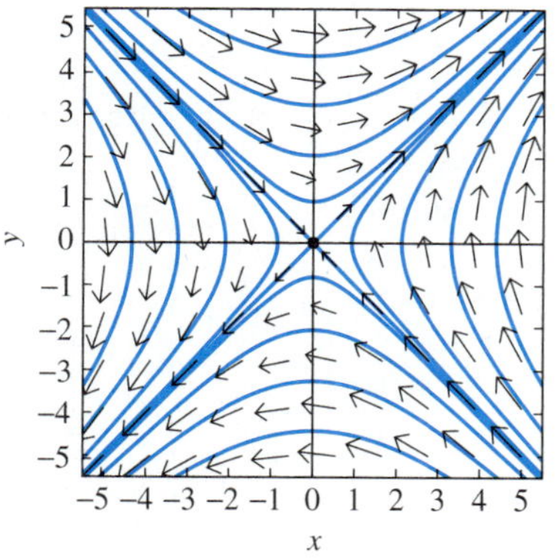

13. $x(t) = A\cos 2t + B\sin 2t,\ y(t) = -B\cos 2t + A\sin 2t$;
$x(t) = \cos 2t,\ y(t) = \sin 2t$

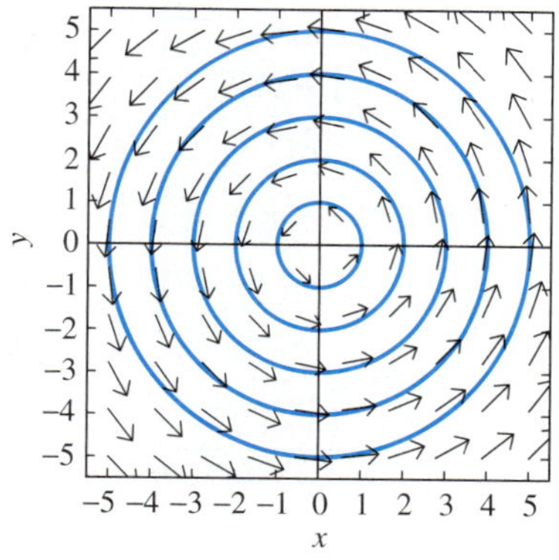

14. $x(t) = A\cos 10t + B\sin 10t,\ y(t) = B\cos 10t - A\sin 10t$;
$x(t) = 3\cos 10t + 4\sin 10t,\ y(t) = 4\cos 10t - 3\sin 10t$

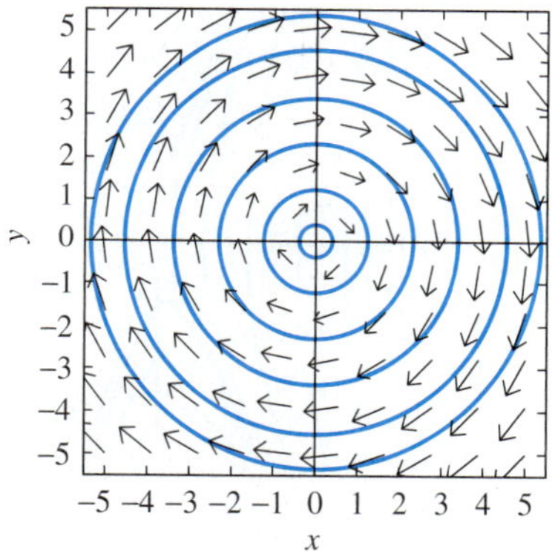

15. $x(t) = A\cos 2t + B\sin 2t,\ y(t) = 4B\cos 2t - 4A\sin 2t$

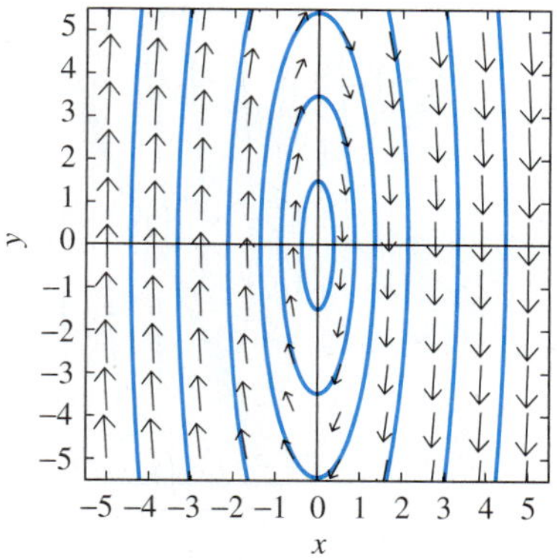

16. $x(t) = A\cos 4t + B\sin 4t,\ y(t) = \frac{1}{2}B\cos 4t - \frac{1}{2}A\sin 4t$

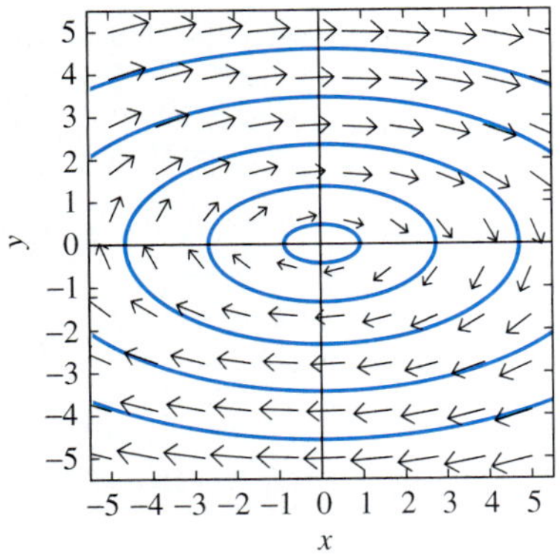

17. $x(t) = Ae^{-3t} + Be^{2t},\ y(t) = -3Ae^{-3t} + 2Be^{2t};\ x(t) = e^{2t},$
$y(t) = 2e^{2t}$

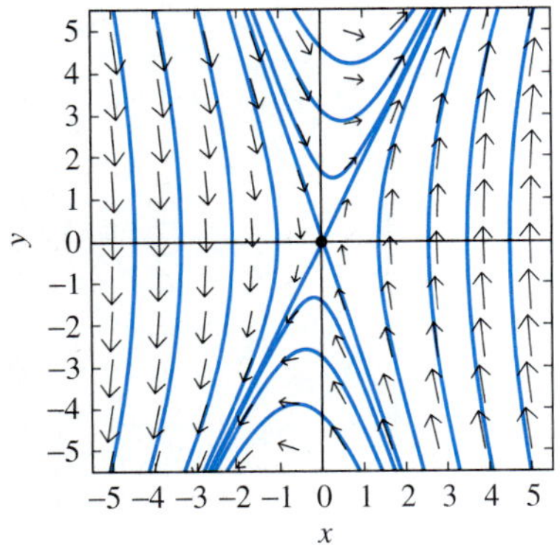

18. $x(t) = Ae^{-2t} + Be^{-5t},\ y(t) = 2Ae^{-2t} + 5Be^{-5t};\ A = \frac{17}{3}$ and $B = -\frac{11}{3}$ in the particular solution.

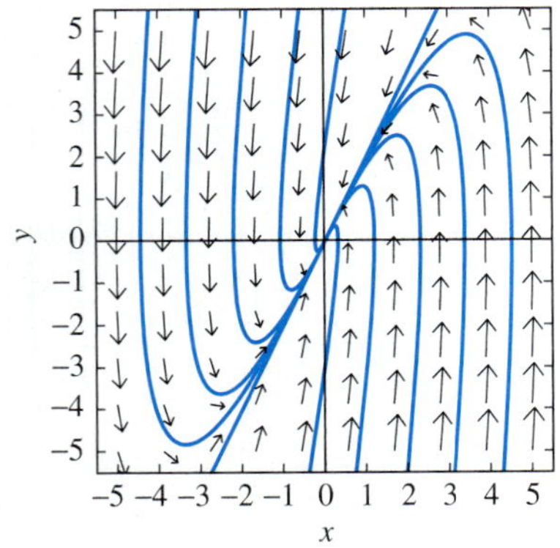

19. $x(t) = -e^{-2t}\sin 3t,\ y(t) = e^{-2t}(3\cos 3t + 2\sin 3t)$

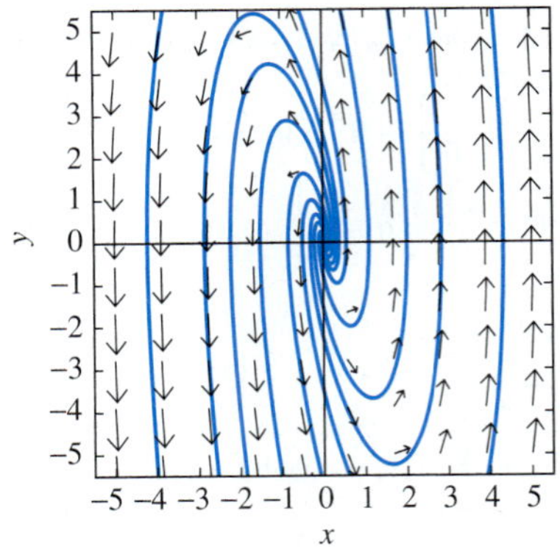

20. $x(t) = (A + Bt)e^{3t},\ \ y(t) = (3A + B + 3Bt)e^{3t}$

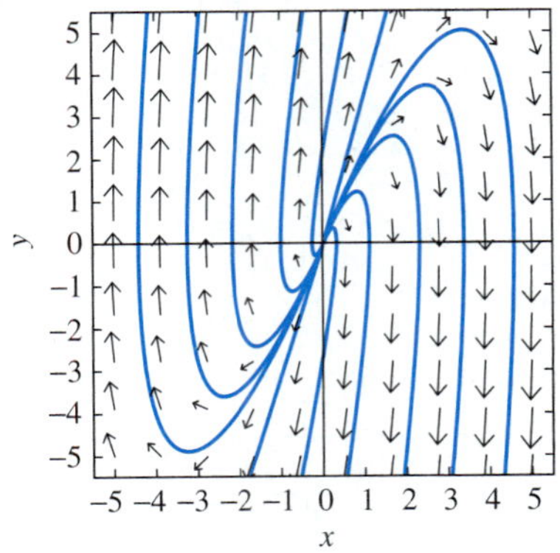

Section 7.2

1. $(\mathbf{AB})' = \begin{bmatrix} 1 - 8t + 18t^2 & 1 + 2t - 12t^2 + 32t^3 \\ 3 + 3t^2 - 4t^3 & 8t + 3t^2 + 4t^3 \end{bmatrix}$

3. $\mathbf{x} = \begin{bmatrix} x \\ y \end{bmatrix}, \mathbf{P}(t) = \begin{bmatrix} 0 & -3 \\ 3 & 0 \end{bmatrix}, \mathbf{f}(t) = \begin{bmatrix} 0 \\ 0 \end{bmatrix}$

5. $\mathbf{x} = \begin{bmatrix} x \\ y \end{bmatrix}, \mathbf{P}(t) = \begin{bmatrix} 2 & 4 \\ 5 & -1 \end{bmatrix}, \mathbf{f}(t) = \begin{bmatrix} 3e^t \\ -t^2 \end{bmatrix}$

7. $\mathbf{x} = \begin{bmatrix} x \\ y \\ z \end{bmatrix}, \mathbf{P}(t) = \begin{bmatrix} 0 & 1 & 1 \\ 1 & 0 & 1 \\ 1 & 1 & 0 \end{bmatrix}, \mathbf{f}(t) = \begin{bmatrix} 0 \\ 0 \\ 0 \end{bmatrix}$

9. $\mathbf{x} = \begin{bmatrix} x \\ y \\ z \end{bmatrix}, \mathbf{P}(t) = \begin{bmatrix} 3 & -4 & 1 \\ 1 & 0 & -3 \\ 0 & 6 & -7 \end{bmatrix}, \mathbf{f}(t) = \begin{bmatrix} t \\ t^2 \\ t^3 \end{bmatrix}$

11. $\mathbf{x} = \begin{bmatrix} x_1 \\ x_2 \\ x_3 \\ x_4 \end{bmatrix}, \mathbf{P}(t) = \begin{bmatrix} 0 & 1 & 0 & 0 \\ 0 & 0 & 2 & 0 \\ 0 & 0 & 0 & 3 \\ 4 & 0 & 0 & 0 \end{bmatrix}, \mathbf{f}(t) = \begin{bmatrix} 0 \\ 0 \\ 0 \\ 0 \end{bmatrix}$

13. $W(t) = e^{3t};\ \mathbf{x}(t) = \begin{bmatrix} 2c_1e^t + c_2e^{2t} \\ -3c_1e^t - c_2e^{2t} \end{bmatrix}$

15. $W(t) \equiv 4;\ \mathbf{x}(t) = \begin{bmatrix} c_1e^{2t} + c_2e^{-2t} \\ c_1e^{2t} + 5c_2e^{-2t} \end{bmatrix}$

17. $W(t) = 7e^{-3t};\ \mathbf{x}(t) = \begin{bmatrix} 3c_1e^{2t} + \ c_2e^{-5t} \\ 2c_1e^{2t} + 3c_2e^{-5t} \end{bmatrix}$

19. $W(t) \equiv 3;\ \mathbf{x}(t) = \begin{bmatrix} c_1e^{2t} + c_2e^{-t} \\ c_1e^{2t} + c_3e^{-t} \\ c_1e^{2t} - (c_2 + c_3)e^{-t} \end{bmatrix}$

21. $W(t) = e^{2t};\ \mathbf{x}(t) = \begin{bmatrix} 3c_1e^{-2t} + c_2e^t + c_3e^{3t} \\ -2c_1e^{-2t} - c_2e^t - c_3e^{3t} \\ 2c_1e^{-2t} + c_2e^t \end{bmatrix}$

23. $\mathbf{x} = 2\mathbf{x}_1 - \mathbf{x}_2$ **25.** $\mathbf{x} = 15\mathbf{x}_1 - 4\mathbf{x}_2$

27. $\mathbf{x} = \mathbf{x}_1 + 2\mathbf{x}_2 + \mathbf{x}_3$ **29.** $\mathbf{x} = 3\mathbf{x}_1 - 3\mathbf{x}_2 - 5\mathbf{x}_3$

31. $\mathbf{x} = 3\mathbf{x}_1 + 7\mathbf{x}_2 + \mathbf{x}_3 - 2\mathbf{x}_4$ **32.** $\mathbf{x} = 13\mathbf{x}_1 + 41\mathbf{x}_2 + 3\mathbf{x}_3 - 12\mathbf{x}_4$

33. (a) $\mathbf{x}_2 = t\mathbf{x}_1$, so neither is a constant multiple of the other.
(b) $W(\mathbf{x}_1, \mathbf{x}_2) \equiv 0$, whereas Theorem 2 implies that $W \neq 0$ if $\mathbf{x}_1$ and $\mathbf{x}_2$ were independent solutions of a system of the indicated form.

Section 7.3

1. $x_1(t) = c_1e^{-t} + c_2e^{3t}$, $x_2(t) = -c_1e^{-t} + c_2e^{3t}$

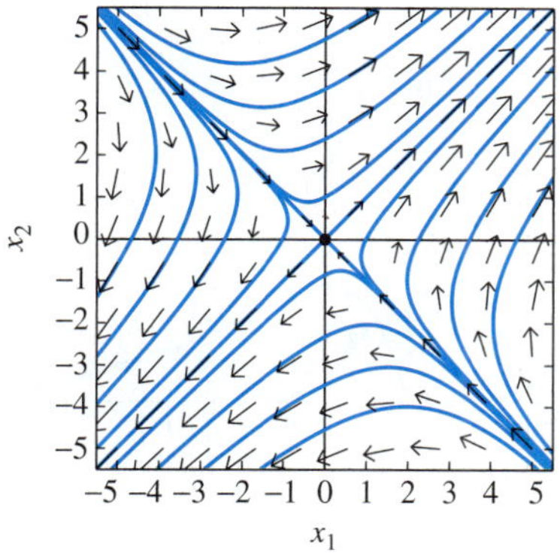

2. $x_1(t) = c_1e^{-t} + 3c_2e^{4t}$, $x_2(t) = -c_1e^{-t} + 2c_2e^{4t}$

3. General solution $x_1(t) = c_1e^{-t} + 4c_2e^{6t}$, $x_2(t) = -c_1e^{-t} + 3c_2e^{6t}$
Particular solution $x_1(t) = \frac{1}{7}(-e^{-t} + 8e^{6t})$,
$x_2(t) = \frac{1}{7}(e^{-t} + 6e^{6t})$.

4. $x_1(t) = c_1e^{-2t} + c_2e^{5t}$, $x_2(t) = -6c_1e^{-2t} + c_2e^{5t}$

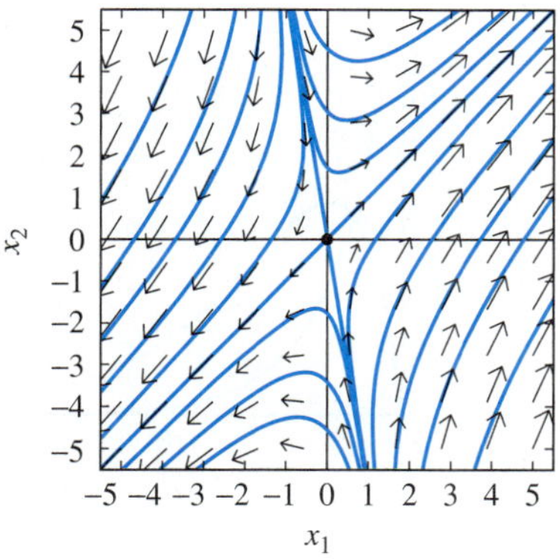

5. $x_1(t) = c_1e^{-t} + 7c_2e^{5t}$, $x_2(t) = c_1e^{-t} + c_2e^{5t}$

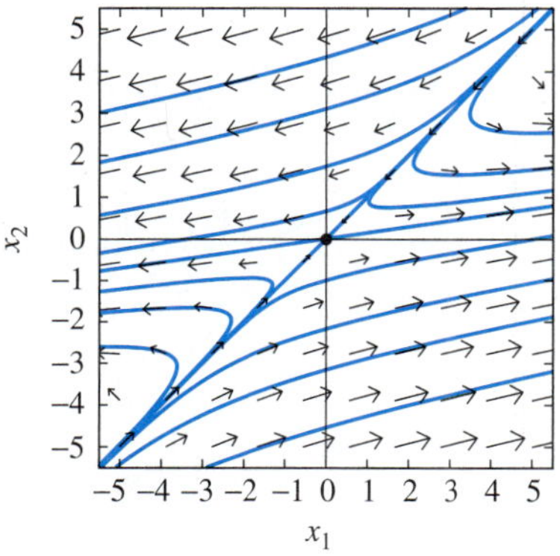

6. General solution $x_1(t) = 5c_1e^{3t} + c_2e^{4t}$, $x_2(t) = -6c_1e^{3t} - c_2e^{4t}$
Particular solution $x_1(t) = -5e^{3t} + 6e^{4t}$, $x_2(t) = 6e^{3t} - 6e^{4t}$.

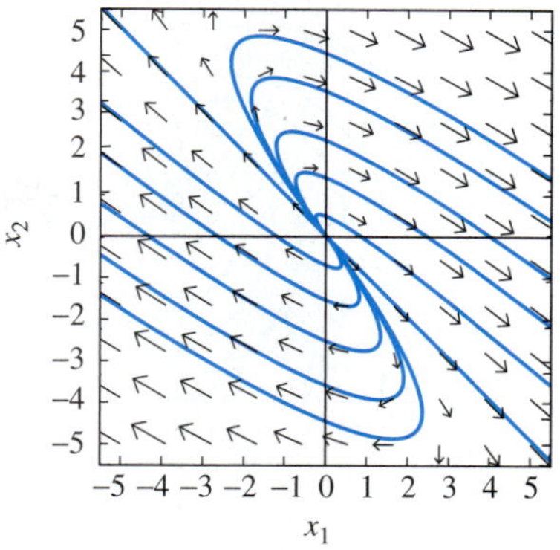

7. $x_1(t) = c_1e^{t} + 2c_2e^{-9t}$, $x_2(t) = c_1e^{t} - 3c_2e^{-9t}$

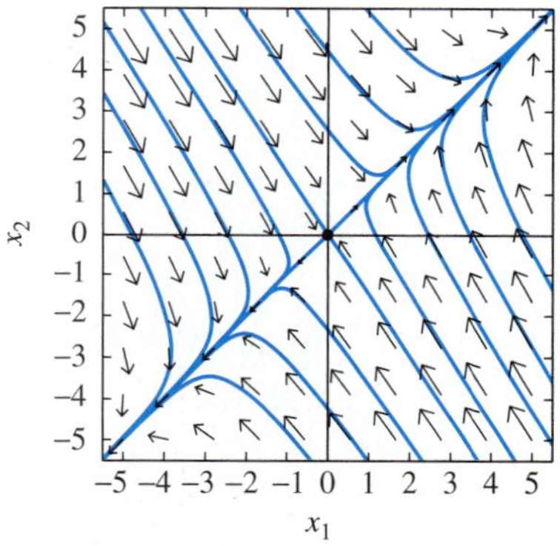

8. $x_1(t) = 5c_1 \cos 2t + 5c_2 \sin 2t$,
$x_2(t) = (c_1 - 2c_2)\cos 2t + (2c_1 + c_2)\sin 2t$

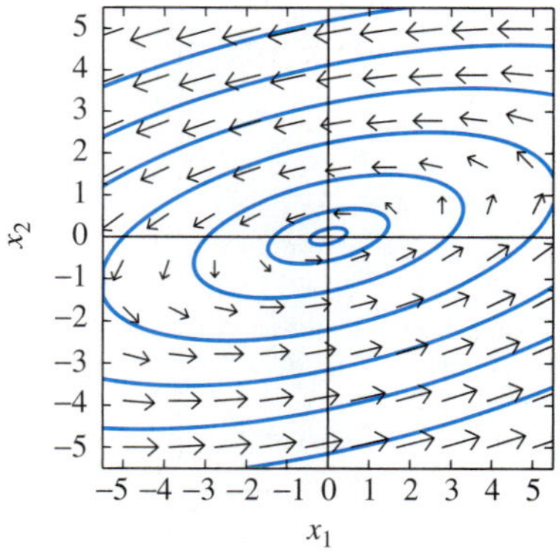

9. General solution $x_1(t) = 5c_1 \cos 4t + 5c_2 \sin 4t$,
$x_2(t) = c_1(2\cos 4t + 4\sin 4t) + c_2(2\sin 4t - 4\cos 4t)$.
Particular solution $x_1(t) = 2\cos 4t - \frac{11}{4}\sin 4t$,
$x_2(t) = 3\cos 4t + \frac{1}{2}\sin 4t$

10. $x_1(t) = -2c_1 \cos 3t - 2c_2 \sin 3t$,
$x_2(t) = (3c_1 + 3c_2)\cos 3t + (3c_2 - 3c_1)\sin 3t$

11. General solution $x_1(t) = e^{t}(c_1 \cos 2t - c_2 \sin 2t)$,
$x_2(t) = e^{t}(c_1 \sin 2t + c_2 \cos 2t)$
Particular solution $x_1(t) = -4e^{t}\sin 2t$, $x_2(t) = 4e^{t}\cos 2t$

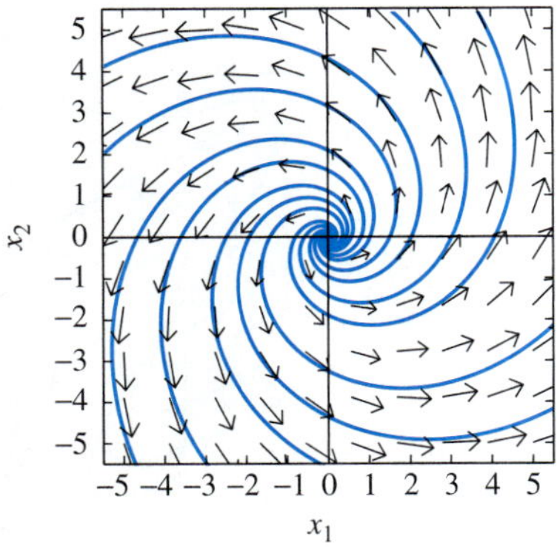

12. $x_1(t) = e^{2t}(-5c_1 \cos 2t - 5c_2 \sin 2t)$,
$x_2(t) = e^{2t}[(c_1 + 2c_2)\cos 2t + (-2c_1 + c_2)\sin 2t]$

13. $x_1(t) = 3e^{2t}(c_1 \cos 3t - c_2 \sin 3t)$,
$x_2(t) = e^{2t}[(c_1 + c_2)\cos 3t + (c_1 - c_2)\sin 3t]$

14. $x_1(t) = e^{3t}(c_1 \cos 4t + c_2 \sin 4t)$,
$x_2(t) = e^{3t}(c_1 \sin 4t - c_2 \cos 4t)$

15. $x_1(t) = 5e^{5t}(c_1 \cos 4t - c_2 \sin 4t)$,
$x_2(t) = e^{5t}[(2c_1 + 4c_2)\cos 4t + (4c_1 - 2c_2)\sin 4t]$

16. $x_1(t) = c_1 e^{-10t} + 2c_2 e^{-100t}$, $x_2(t) = 2c_1 e^{-10t} - 5c_2 e^{-100t}$

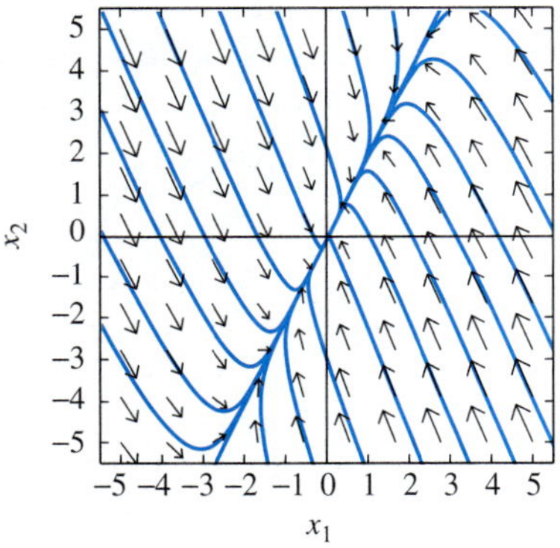

17. $x_1(t) = c_1 e^{9t} + c_2 e^{6t} + c_3$,
$x_2(t) = c_1 e^{9t} - 2c_2 e^{6t}$,
$x_3(t) = c_1 e^{9t} + c_2 e^{6t} - c_3$

18. $x_1(t) = c_1 e^{9t} + 4c_3$,
$x_2(t) = 2c_1 e^{9t} + c_2 e^{6t} - c_3$,
$x_3(t) = 2c_1 e^{9t} - c_2 e^{6t} - c_3$

19. $x_1(t) = c_1 e^{6t} + c_2 e^{3t} + c_3 e^{3t}$,
$x_2(t) = c_1 e^{6t} - 2c_2 e^{3t}$,
$x_3(t) = c_1 e^{6t} + c_2 e^{3t} - c_3 e^{3t}$

20. $x_1(t) = c_1 e^{9t} + c_2 e^{6t} + c_3 e^{2t}$,
$x_2(t) = c_1 e^{9t} - 2c_2 e^{6t}$,
$x_3(t) = c_1 e^{9t} + c_2 e^{6t} - c_3 e^{2t}$

21. $x_1(t) = 6c_1 + 3c_2 e^{t} + 2c_3 e^{-t}$,
$x_2(t) = 2c_1 + c_2 e^{t} + c_3 e^{-t}$,
$x_3(t) = 5c_1 + 2c_2 e^{t} + 2c_3 e^{-t}$

22. $x_1(t) = c_2 e^{t} + c_3 e^{3t}$,
$x_2(t) = c_1 e^{-2t} - c_2 e^{t} - c_3 e^{3t}$,
$x_3(t) = -c_1 e^{-2t} + c_3 e^{3t}$

23. $x_1(t) = c_1 e^{2t} + c_3 e^{3t}$,
$x_2(t) = -c_1 e^{2t} + c_2 e^{-2t} - c_3 e^{3t}$,
$x_3(t) = -c_2 e^{-2t} + c_3 e^{3t}$

24. $x_1(t) = c_1 e^{t} + c_2(2\cos 2t - \sin 2t) + c_3(\cos 2t + 2\sin 2t)$
$x_2(t) = -c_1 e^{t} - c_2(3\cos 2t + \sin 2t) + c_3(\cos 2t - 3\sin 2t)$
$x_3(t) = c_2(3\cos 2t + \sin 2t) + c_3(3\sin 2t - \cos 2t)$

25. $x_1(t) = c_1 + e^{2t}[(c_2 + c_3)\cos 3t + (-c_2 + c_3)\sin 3t]$,
$x_2(t) = -c_1 + 2e^{2t}(-c_2 \cos 3t - c_3 \sin 3t)$,
$x_3(t) = 2e^{2t}(c_2 \cos 3t + c_3 \sin 3t)$

26. $x_1(t) = 4e^{3t} - e^{-t}(4\cos t - \sin t)$,
$x_2(t) = 9e^{3t} - e^{-t}(9\cos t + 2\sin t)$,
$x_3(t) = 17e^{-t}\cos t$

27. $x_1(t) = 15e^{-0.2t}$, $x_2(t) = 15(e^{-0.2t} - e^{-0.4t})$.
The maximum amount ever in tank 2 is $x_2(5 \ln 2) = 3.75$ lb.

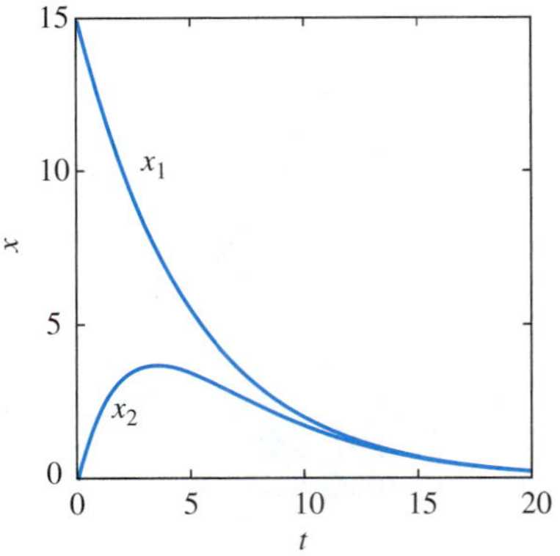

28. $x_1(t) = 15e^{-0.4t}$, $x_2(t) = 40(-e^{-0.4t} + e^{-0.25t})$.
The maximum amount ever in tank 2 is about 6.85 lb.

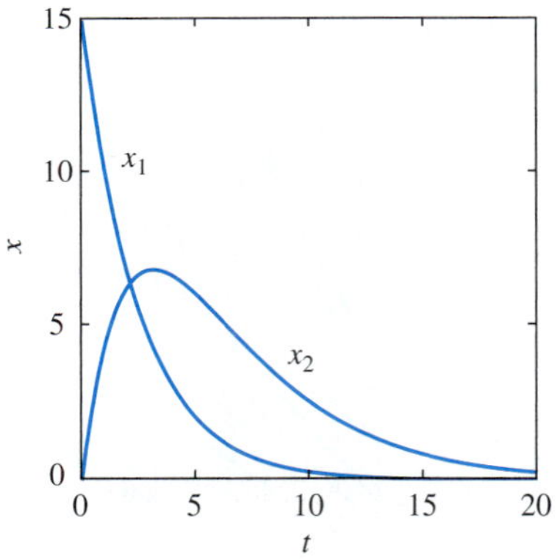

29. $x_1(t) = 10 + 5e^{-0.6t}$, $x_2(t) = 5 - 5e^{-0.6t}$

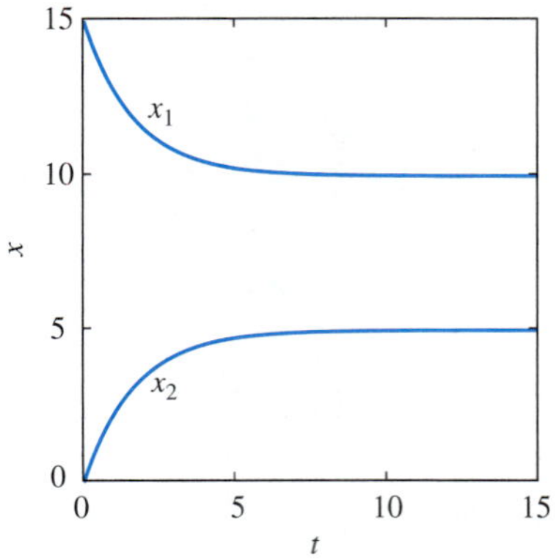

30. $x_1(t) = 5c_1 + c_2e^{-0.65t}$, $x_2(t) = 8c_1 - c_2e^{-0.65t}$.

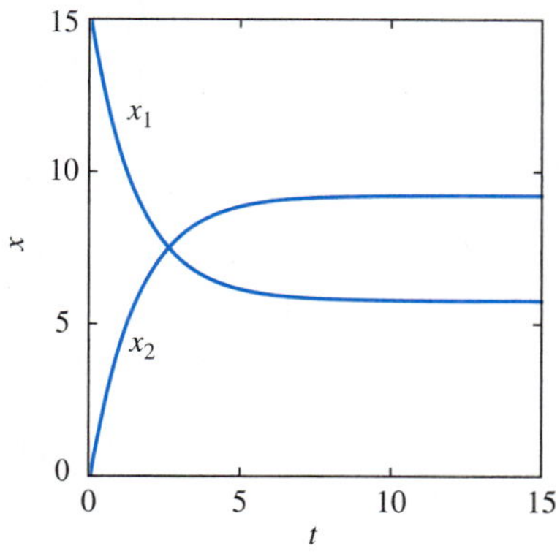

31. $x_1(t) = 27e^{-t}$,
$x_2(t) = 27e^{-t} - 27e^{-2t}$,
$x_3(t) = 27e^{-t} - 54e^{-2t} + 27e^{-3t}$.
The maximum amount of salt ever in tank 3 is $x_3(\ln 3) = 4$ pounds.

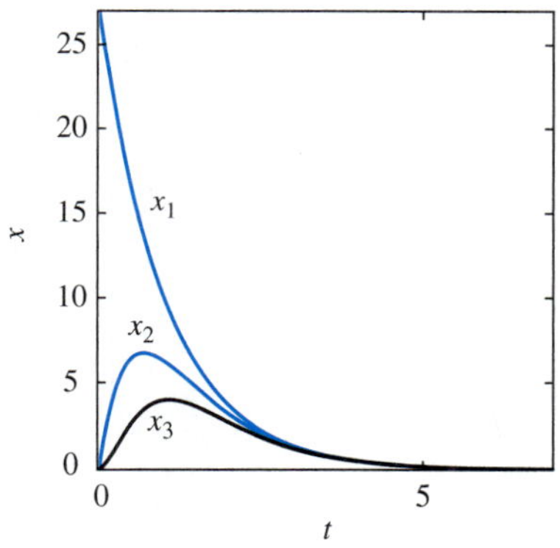

32. $x_1(t) = 45e^{-3t}$,
$x_2(t) = -135e^{-3t} + 135e^{-2t}$,
$x_3(t) = 135e^{-3t} - 270e^{-2t} + 135e^{-t}$.
The maximum amount of salt ever in tank 3 is $x_3(\ln 3) = 20$ pounds.

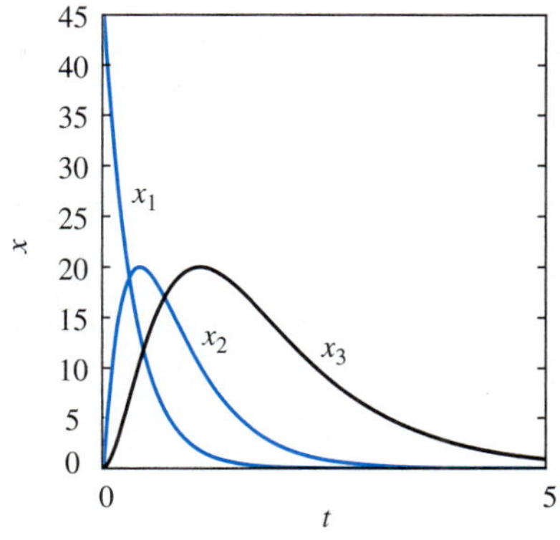

33. $x_1(t) = 45e^{-4t}$,
$x_2(t) = 90e^{-4t} - 90e^{-6t}$,
$x_3(t) = -270e^{-4t} + 135e^{-6t} + 135e^{-2t}$.
The maximum amount of salt ever in tank 3 is $x_3\left(\frac{1}{2}\ln 3\right) = 20$ pounds.

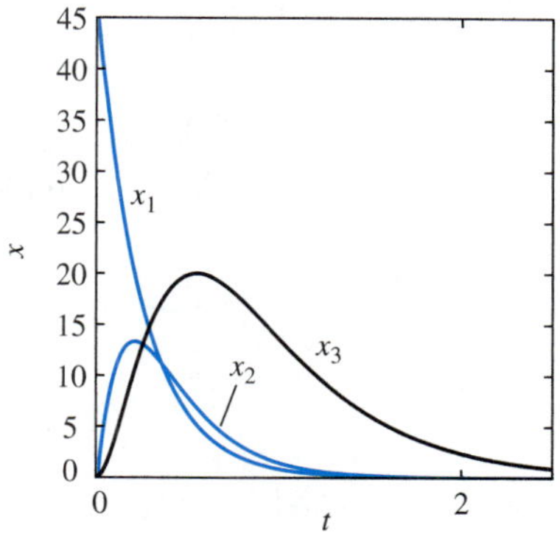

34. $x_1(t) = 40e^{-3t}$,
$x_2(t) = 60e^{-3t} - 60e^{-5t}$,
$x_3(t) = -150e^{-3t} + 75e^{-5t} + 75e^{-t}$.
The maximum amount of salt ever in tank 3 is $x_3(\frac{1}{2}\ln 5) \approx 21.4663$ pounds.

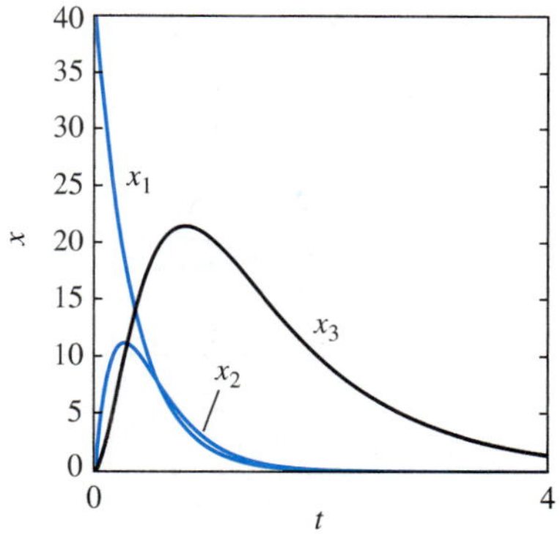

35. $x_1(t) = 10 - \frac{1}{7}\left(55e^{-18t} - 216e^{-11t}\right)$,
$x_2(t) = 3 - \frac{1}{7}\left(165e^{-18t} - 144e^{-11t}\right)$,
$x_3(t) = 20 + \frac{1}{7}\left(220e^{-18t} - 360e^{-11t}\right)$.
The limiting amounts of salt in tanks 1, 2, and 3 are 10 lb, 3 lb, and 20 lb.

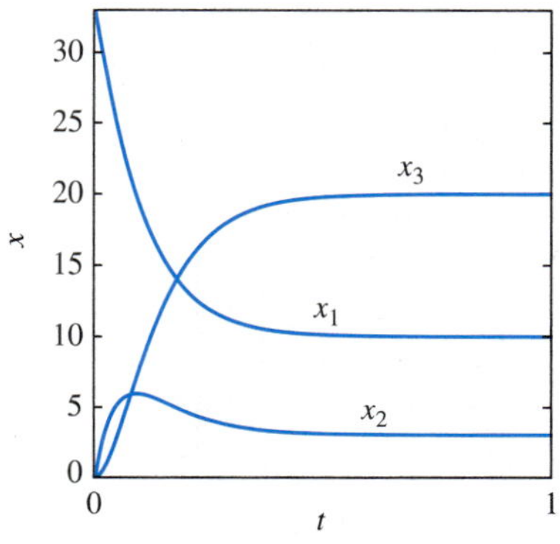

36. $x_1(t) = 4 + e^{-3t/5}[14\cos(3t/10) - 2\sin(3t/10)]$,
$x_2(t) = 10 - e^{-3t/5}[10\cos(3t/10) - 10\sin(3t/10)]$,
$x_3(t) = 4 - e^{-3t/5}[4\cos(3t/10) + 8\sin(3t/10)]$.
The limiting amounts of salt in tanks 1, 2, and 3 are 4 lb, 10 lb, and 4 lb.

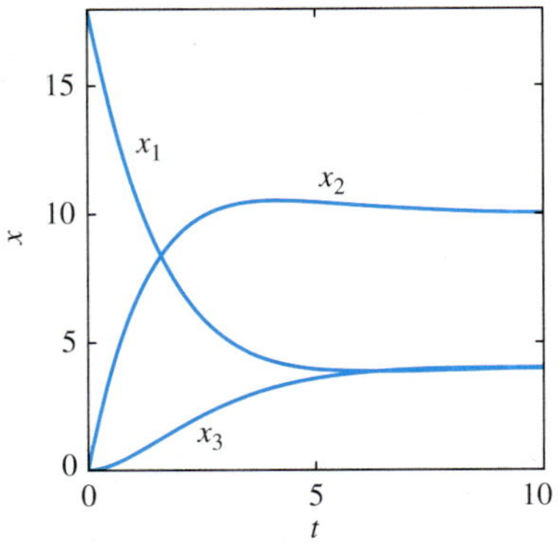

37. $x_1(t) = 30 + e^{-3t}[25\cos(t\sqrt{2}) + 10\sqrt{2}\sin(t\sqrt{2})]$,
$x_2(t) = 10 - e^{-3t}[10\cos(t\sqrt{2}) - \frac{25}{2}\sqrt{2}\sin(t\sqrt{2})]$,
$x_3(t) = 15 - e^{-3t}[15\cos(t\sqrt{2}) + \frac{45}{2}\sqrt{2}\sin(t\sqrt{2})]$.
The limiting amounts of salt in tanks 1, 2, and 3 are 30 lb, 10 lb, and 15 lb.

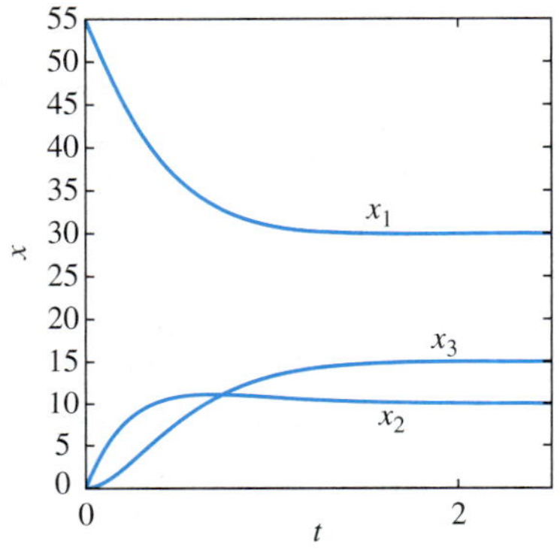

38. $x_1(t) = c_1e^t$,
$x_2(t) = -2c_1e^t + c_2e^{2t}$,
$x_3(t) = 3c_1e^t - 3c_2e^{2t} + c_3e^{3t}$,
$x_4(t) = -4c_1e^t + 6c_2e^{2t} - 4c_3e^{3t} + c_4e^{4t}$

39. $x_1(t) = 3c_1e^t + c_4e^{-2t}$,
$x_2(t) = -2c_1e^t + c_3e^{2t} - c_4e^{-2t}$,
$x_3(t) = 4c_1e^t + c_2e^{-t}$,
$x_4(t) = c_1e^t$

40. $x_1(t) = c_1e^{2t}$,
$x_2(t) = -3c_1e^{2t} + 3c_2e^{-2t} - c_4e^{-5t}$,
$x_3(t) = c_3e^{5t}$,
$x_4(t) = -c_2e^{-2t} - 3c_3e^{5t}$

41. $x_1(t) = 2e^{10t} + e^{15t} = x_4(t)$, $x_2(t) = -e^{10t} + 2e^{15t} = x_3(t)$

42. $\mathbf{x}(t) = c_1\begin{bmatrix} 3 \\ -1 \\ 2 \end{bmatrix} + c_2\begin{bmatrix} 1 \\ 1 \\ 1 \end{bmatrix}e^{2t} + c_3\begin{bmatrix} 2 \\ -3 \\ 1 \end{bmatrix}e^{5t}$

43. $\mathbf{x}(t) = c_1\begin{bmatrix} 3 \\ -1 \\ 5 \end{bmatrix}e^{-2t} + c_2\begin{bmatrix} 1 \\ 1 \\ 1 \end{bmatrix}e^{4t} + c_3\begin{bmatrix} 1 \\ -1 \\ 3 \end{bmatrix}e^{8t}$

44. $\mathbf{x}(t) = c_1\begin{bmatrix} 3 \\ -2 \\ 2 \end{bmatrix}e^{-3t} + c_2\begin{bmatrix} 7 \\ 1 \\ 5 \end{bmatrix}e^{6t} + c_3\begin{bmatrix} 5 \\ -3 \\ 3 \end{bmatrix}e^{12t}$

45. $\mathbf{x}(t) = c_1\begin{bmatrix} 1 \\ 1 \\ 1 \\ -1 \end{bmatrix}e^{-3t} + c_2\begin{bmatrix} 1 \\ 2 \\ -1 \\ 1 \end{bmatrix} + c_3\begin{bmatrix} 2 \\ 1 \\ 1 \\ 1 \end{bmatrix}e^{3t} + c_4\begin{bmatrix} 1 \\ -1 \\ 2 \\ -1 \end{bmatrix}e^{6t}$

46. $\mathbf{x}(t) =$
$c_1\begin{bmatrix} 3 \\ 2 \\ -1 \\ 1 \end{bmatrix}e^{-4t} + c_2\begin{bmatrix} 1 \\ 2 \\ 2 \\ -1 \end{bmatrix}e^{2t} + c_3\begin{bmatrix} 1 \\ 1 \\ -1 \\ 1 \end{bmatrix}e^{4t} + c_4\begin{bmatrix} 3 \\ -2 \\ 3 \\ -3 \end{bmatrix}e^{8t}$

47. $\mathbf{x}(t) =$
$c_1\begin{bmatrix} 2 \\ 2 \\ 1 \\ -1 \end{bmatrix}e^{-3t} + c_2\begin{bmatrix} 1 \\ 2 \\ -1 \\ 1 \end{bmatrix}e^{3t} + c_3\begin{bmatrix} 2 \\ 1 \\ 1 \\ 1 \end{bmatrix}e^{6t} + c_4\begin{bmatrix} 1 \\ -1 \\ 2 \\ -1 \end{bmatrix}e^{9t}$

48. $\mathbf{x}(t) =$
$c_1\begin{bmatrix} 1 \\ 2 \\ -1 \\ 2 \end{bmatrix}e^{16t} + c_2\begin{bmatrix} 2 \\ 5 \\ 1 \\ -1 \end{bmatrix}e^{32t} + c_3\begin{bmatrix} 3 \\ -1 \\ 1 \\ 2 \end{bmatrix}e^{48t} + c_4\begin{bmatrix} 1 \\ 1 \\ 2 \\ -3 \end{bmatrix}e^{64t}$

49. $\mathbf{x}(t) = c_1\begin{bmatrix} 1 \\ 0 \\ 3 \\ 1 \\ 1 \end{bmatrix}e^{-3t} + c_2\begin{bmatrix} 0 \\ 3 \\ 0 \\ -1 \\ 1 \end{bmatrix} + c_3\begin{bmatrix} 1 \\ 7 \\ 1 \\ 1 \\ 1 \end{bmatrix}e^{3t} +$
$c_4\begin{bmatrix} 0 \\ 1 \\ 0 \\ 1 \\ 1 \end{bmatrix}e^{6t} + c_5\begin{bmatrix} 2 \\ 0 \\ 5 \\ 2 \\ 1 \end{bmatrix}e^{9t}$

50. $\mathbf{x}(t) = c_1\begin{bmatrix} 0 \\ 1 \\ 1 \\ 1 \\ 0 \\ 1 \end{bmatrix}e^{-7t} + c_2\begin{bmatrix} 1 \\ 0 \\ 0 \\ 0 \\ 1 \\ 1 \end{bmatrix}e^{-4t} + c_3\begin{bmatrix} 0 \\ 1 \\ 0 \\ 1 \\ 0 \\ 1 \end{bmatrix}e^{3t} +$
$c_4\begin{bmatrix} 0 \\ 0 \\ 1 \\ 0 \\ 1 \\ 0 \end{bmatrix}e^{5t} + c_5\begin{bmatrix} 1 \\ 1 \\ 0 \\ 0 \\ 0 \\ 1 \end{bmatrix}e^{9t} + c_6\begin{bmatrix} 0 \\ 0 \\ 1 \\ -1 \\ -1 \\ 0 \end{bmatrix}e^{11t}$

Section 7.4

1. The natural frequencies are $\omega_0 = 0$ and $\omega_1 = 2$. In the degenerate natural mode with "frequency" $\omega_0 = 0$ the two masses move linearly with $x_1(t) = x_2(t) = a_0 + b_0t$. At frequency $\omega_1 = 2$ they oscillate in opposite directions with equal amplitudes.

2. The natural frequencies are $\omega_1 = 1$ and $\omega_2 = 3$. In the natural mode with frequency ω_1, the two masses m_1 and m_2 move in the same direction with equal amplitudes of oscillation. At frequency ω_2 they move in opposite directions with equal amplitudes.

3. The natural frequencies are $\omega_1 = 1$ and $\omega_2 = 2$. In the natural mode with frequency ω_1, the two masses m_1 and m_2 move in the same direction with equal amplitudes of oscillation. In the natural mode with frequency ω_2 they move in opposite directions with the amplitude of oscillation of m_1 twice that of m_2.

4. The natural frequencies are $\omega_1 = 1$ and $\omega_2 = \sqrt{5}$. In the natural mode with frequency ω_1, the two masses m_1 and m_2 move in the same direction with equal amplitudes of oscillation. At frequency ω_2 they move in opposite directions with equal amplitudes.

5. The natural frequencies are $\omega_1 = \sqrt{2}$ and $\omega_2 = 2$. In the natural mode with frequency ω_1, the two masses m_1 and m_2 move in the same direction with equal amplitudes of oscillation. At frequency ω_2 they move in opposite directions with equal amplitudes.

6. The natural frequencies are $\omega_1 = \sqrt{2}$ and $\omega_2 = \sqrt{8}$. In the natural mode with frequency ω_1, the two masses m_1 and m_2 move in the same direction with equal amplitudes of oscillation. In the natural mode with frequency ω_2 they move in opposite directions with the amplitude of oscillation of m_1 twice that of m_2.

7. The natural frequencies are $\omega_1 = 2$ and $\omega_2 = 4$. In the natural mode with frequency ω_1, the two masses m_1 and m_2 move in the same direction with equal amplitudes of oscillation. At frequency ω_2 they move in opposite directions with equal amplitudes.

8. $x_1(t) = 2\cos t + 3\cos 3t - 5\cos 5t$,
$x_2(t) = 2\cos t - 3\cos 3t + \cos 5t$.
We have a superposition of three oscillations, in which the two masses move (1) in the same direction with frequency $\omega_1 = 1$ and equal amplitudes; (2) in opposite directions with frequency $\omega_2 = 3$ and equal amplitudes; (3) in opposite directions with frequency $\omega_3 = 5$ and with the amplitude of motion of m_1 being 5 times that of m_2.

9. $x_1(t) = 5\cos t - 8\cos 2t + 3\cos 3t$,
$x_2(t) = 5\cos t + 4\cos 2t - 9\cos 3t$.
We have a superposition of three oscillations, in which the two masses move (1) in the same direction with frequency $\omega_1 = 1$ and equal amplitudes; (2) in opposite directions with frequency $\omega_2 = 2$ and with the amplitude of motion of m_1 being twice that of m_2; (3) in opposite directions with frequency $\omega_3 = 3$ and with the amplitude of motion of m_2 being 3 times that of m_1.

10. $x_1(t) = \cos 2t - 15\cos 4t + 14\cos t$,
$x_2(t) = \cos 2t + 15\cos 4t + 16\cos t$.
We have a superposition of three oscillations, in which the two masses move (1) in the same direction with frequency $\omega_1 = 1$ and with the amplitude of motion of m_2 being 8/7 times that of m_1; (2) in the same direction with frequency $\omega_2 = 2$ and equal amplitudes; (3) in opposite directions with frequency $\omega_3 = 4$ and equal amplitudes.

11. (a) The natural frequencies are $\omega_1 = 6$ and $\omega_2 = 8$. In mode 1 the two masses oscillate in the same direction with frequency $\omega_1 = 6$ and with the amplitude of motion of m_1 being twice that of m_2. In mode 2 the two masses oscillate in opposite directions with frequency $\omega_2 = 8$ and with the amplitude of motion of m_2 being 3 times that of m_1.
(b) $x(t) = 2\sin 6t + 19\cos 7t$, $y(t) = \sin 6t + 3\cos 7t$
We have a superposition of (only two) oscillations, in which the two masses move (1) in the same direction with frequency $\omega_1 = 6$ and with the amplitude of motion of m_1 being twice that of m_2; (2) in the same direction with frequency $\omega_3 = 7$ and with the amplitude of motion of m_1 being 19/3 times that of m_2.

12. The system's three natural modes of oscillation have (1) natural frequency $\omega_1 = \sqrt{2}$ with amplitude ratios $1:0:-1$; (2) natural frequency $\omega_2 = \sqrt{2+\sqrt{2}}$ with amplitude ratios $1:-\sqrt{2}:1$; (3) natural frequency $\omega_3 = \sqrt{2-\sqrt{2}}$ with amplitude ratios $1:\sqrt{2}:1$.

13. The system's three natural modes of oscillation have (1) natural frequency $\omega_1 = 2$ with amplitude ratios $1:0:-1$; (2) natural frequency $\omega_2 = \sqrt{4+2\sqrt{2}}$ with amplitude ratios $1:-\sqrt{2}:1$; (3) natural frequency $\omega_3 = \sqrt{4-2\sqrt{2}}$ with amplitude ratios $1:\sqrt{2}:1$.

15. $x_1(t) = \frac{2}{3}\cos 5t - 2\cos 5\sqrt{3}\,t + \frac{4}{3}\cos 10t$,
$x_2(t) = \frac{4}{3}\cos 5t + 4\cos 5\sqrt{3}\,t - \frac{16}{3}\cos 10t$.
We have a superposition of two oscillations with the natural frequencies $\omega_1 = 5$ and $\omega_2 = 5\sqrt{3}$ and a forced oscillation with frequency $\omega = 10$. In each of the two natural oscillations the amplitude of motion of m_2 is twice that of m_1, while in the forced oscillation the amplitude of motion of m_2 is four times that of m_1.

20. $x_1'(t) = -v_0$, $x_2'(t) = 0$, $x_3'(t) = v_0$ for $t > \pi/2$

21. $x_1'(t) = -v_0$, $x_2'(t) = 0$, $x_3'(t) = 2v_0$ for $t > \pi/2$

22. $x_1'(t) = -2v_0$, $x_2'(t) = v_0$, $x_3'(t) = v_0$ for $t > \pi/2$

23. $x_1'(t) = 2v_0$, $x_2'(t) = 2v_0$, $x_3'(t) = 3v_0$ for $t > \pi/2$

24. (a) $\omega_1 \approx 1.0293$ Hz; $\omega_2 \approx 1.7971$ Hz.
(b) $v_1 \approx 28$ mi/h; $v_2 \approx 49$ mi/h

27. $\omega_1 = 2\sqrt{10}$, $v_1 \approx 40.26$ (ft/s (about 27 mi/h),
$\omega_2 = 5\sqrt{5}$, $v_2 \approx 71.18$ ft/s (about 49 mi/h)

28. $\omega_1 \approx 6.1311$, $v_1 \approx 39.03$ ft/s (about 27 mi/h),
$\omega_2 \approx 10.3155$, $v_2 \approx 65.67$ ft/s (about 45 mi/h)

29. $\omega_1 \approx 5.0424$, $v_1 \approx 32.10$ ft/s (about 22 mi/h),
$\omega_2 \approx 9.9158$, $v_2 \approx 63.13$ ft/s (about 43 mi/h)

Section 7.5

1. Repeated eigenvalue $\lambda = -3$, eigenvector $\mathbf{v} = \begin{bmatrix} 1 & -1 \end{bmatrix}^T$;
$x_1(t) = (c_1 + c_2 + c_2 t)e^{-3t}$, $x_2(t) = (-c_1 - c_2 t)e^{-3t}$

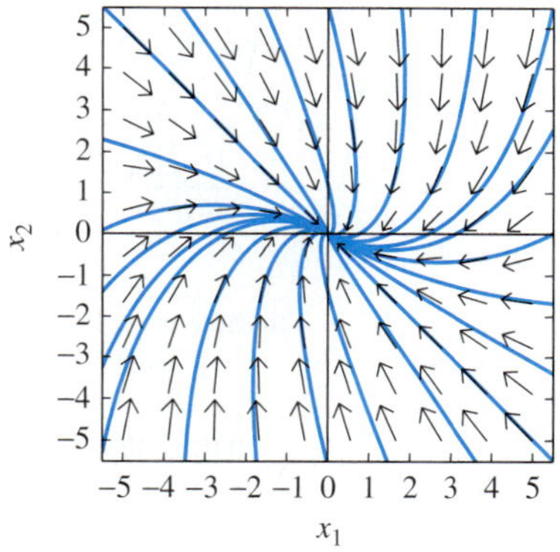

2. Repeated eigenvalue $\lambda = 2$, single eigenvector $\mathbf{v} = \begin{bmatrix} 1 & 1 \end{bmatrix}^T$;
$x_1(t) = (c_1 + c_2 + c_2 t)e^{2t}$, $x_2(t) = (c_1 + c_2 t)e^{2t}$

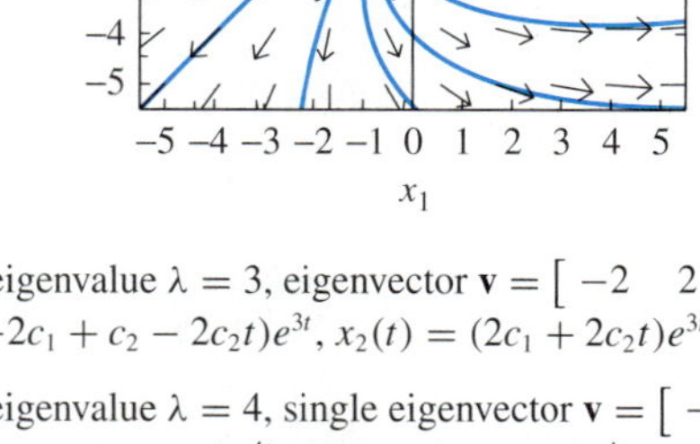

3. Repeated eigenvalue $\lambda = 3$, eigenvector $\mathbf{v} = \begin{bmatrix} -2 & 2 \end{bmatrix}^T$;
$x_1(t) = (-2c_1 + c_2 - 2c_2 t)e^{3t}$, $x_2(t) = (2c_1 + 2c_2 t)e^{3t}$

4. Repeated eigenvalue $\lambda = 4$, single eigenvector $\mathbf{v} = \begin{bmatrix} -1 & 1 \end{bmatrix}^T$;
$x_1(t) = (-c_1 + c_2 - c_2 t)e^{4t}$, $x_2(t) = (c_1 + c_2 t)e^{4t}$

5. Repeated eigenvalue $\lambda = 5$, eigenvector $\mathbf{v} = \begin{bmatrix} 2 & -4 \end{bmatrix}^T$;
$x_1(t) = (2c_1 + c_2 + 2c_2 t)e^{5t}$, $x_2(t) = (-4c_1 - 4c_2 t)e^{5t}$

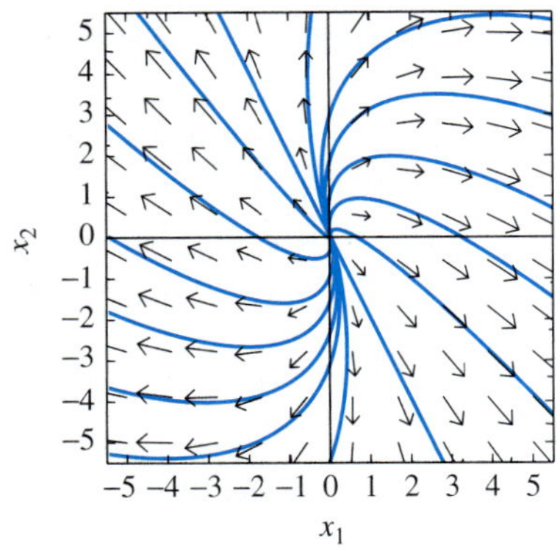

6. Repeated eigenvalue $\lambda = 5$, single eigenvector $\mathbf{v} = \begin{bmatrix} -4 & 4 \end{bmatrix}^T$; $x_1(t) = (-4c_1 + c_2 - 4c_2t)e^{5t}$, $x_2(t) = (4c_1 + 4c_2t)e^{5t}$

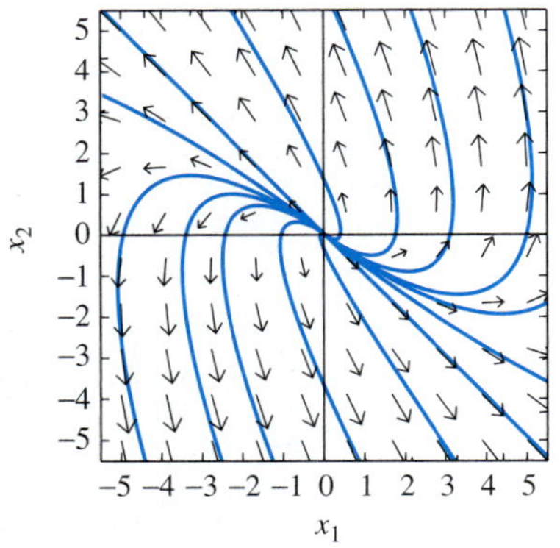

7. Eigenvalues $\lambda = 2, 2, 9$ with three linearly independent eigenvectors; $x_1(t) = c_1e^{2t} + c_2e^{2t}$, $x_2(t) = c_1e^{2t} + c_3e^{9t}$, $x_3(t) = c_2e^{2t}$

8. Eigenvalues $\lambda = 7, 13, 13$ with three linearly independent eigenvectors; $x_1(t) = 2c_1e^{7t} - c_3e^{13t}$, $x_2(t) = -3c_1e^{7t} + c_3e^{13t}$, $x_3(t) = c_1e^{7t} + c_2e^{13t}$

9. Eigenvalues $\lambda = 5, 5, 9$ with three linearly independent eigenvectors; $x_1(t) = c_1e^{5t} + 7c_2e^{5t} + 3c_3e^{9t}$, $x_2(t) = 2c_1e^{5t}$, $x_3(t) = 2c_2e^{5t} + c_3e^{9t}$

10. Eigenvalues $\lambda = 3, 3, 7$ with three linearly independent eigenvectors; $x_1(t) = 5c_1e^{3t} - 3c_2e^{3t} + 2c_3e^{7t}$, $x_2(t) = 2c_1e^{3t} + c_3e^{7t}$, $x_3(t) = c_2e^{3t}$

11. Triple eigenvalue $\lambda = -1$ of defect 2;
$x_1(t) = (-2c_2 + c_3 - 2c_3t)e^{-t}$,
$x_2(t) = (c_1 - c_2 + c_2t - c_3t + \frac{1}{2}c_3t^2)e^{-t}$,
$x_3(t) = (c_2 + c_3t)e^{-t}$

12. Triple eigenvalue $\lambda = -1$ of defect 2;
$x_1(t) = e^{-t}(c_1 + c_3 + c_2t + \frac{1}{2}c_3t^2)$
$x_2(t) = e^{-t}(c_1 + c_2t + \frac{1}{2}c_3t^2)$,
$x_3(t) = e^{-t}(c_2 + c_3t)$

13. Triple eigenvalue $\lambda = -1$ of defect 2;
$x_1(t) = (c_1 + c_2t + \frac{1}{2}c_3t^2)e^{-t}$,
$x_2(t) = (2c_2 + c_3 + 2c_3t)e^{-t}$,
$x_3(t) = (c_2 + c_3t)e^{-t}$

14. Triple eigenvalue $\lambda = -1$ of defect 2;
$x_1(t) = e^{-t}(5c_1 + c_2 + c_3 + 5c_2t + c_3t + \frac{5}{2}c_3t^2)$,
$x_2(t) = e^{-t}(-25c_1 - 5c_2 - 25c_2t - 5c_3t - \frac{25}{2}c_3t^2)$,
$x_3(t) = e^{-t}(-5c_1 + 4c_2 - 5c_2t + 4c_3t - \frac{5}{2}c_3t^2)$

15. Triple eigenvalue $\lambda = 1$ of defect 1;
$x_1(t) = (3c_1 + c_3 - 3c_3t)e^t$,
$x_2(t) = (-c_1 + c_3t)e^t$, $x_3(t) = (c_2 + c_3t)e^t$

16. Triple eigenvalue $\lambda = 1$ of defect 1;
$x_1(t) = e^t(3c_1 + 3c_2 + c_3)$
$x_2(t) = e^t(-2c_1 - 2c_3t)$,
$x_3(t) = e^t(-2c_2 + 2c_3t)$

17. Triple eigenvalue $\lambda = 1$ of defect 1;
$x_1(t) = (2c_1 + c_2)e^t$, $x_2(t) = (-3c_2 + c_3 + 6c_3t)e^t$,
$x_3(t) = -9(c_1 + c_3t)e^t$

18. Triple eigenvalue $\lambda = 1$ of defect 1;
$x_1(t) = e^t(-c_1 - 2c_2 + c_3)$,
$x_2(t) = e^t(c_2 + c_3t)$,
$x_3(t) = e^t(c_1 - 2c_3t)$

19. Double eigenvalues $\lambda = -1$ and $\lambda = 1$, with four linearly independent solutions;
$x_1(t) = c_1e^{-t} + c_4e^t$,
$x_2(t) = c_3e^t$,
$x_3(t) = c_2e^{-t} + 3c_4e^t$,
$x_4(t) = c_1e^{-t} - 2c_3e^t$

20. Eigenvalue $\lambda = 2$ with multiplicity 4 and defect 3;
$x_1(t) = (c_1 + c_3 + c_2t + c_4t + \frac{1}{2}c_3t^2 + \frac{1}{6}c_4t^3)e^{2t}$,
$x_2(t) = (c_2 + c_3t + \frac{1}{2}c_4t^2)e^{2t}$,
$x_3(t) = (c_3 + c_4t)e^{2t}$, $x_4(t) = c_4e^{2t}$

21. Eigenvalue $\lambda = 1$ with multiplicity 4 and defect 2;
$x_1(t) = (-2c_2 + c_3 - 2c_3t)e^t$, $x_2(t) = (c_2 + c_3t)e^t$,
$x_3(t) = (c_2 + c_4 + c_3t)e^t$, $x_4(t) = (c_1 + c_2t + \frac{1}{2}c_3t^2)e^t$

22. Eigenvalue $\lambda = 1$ with multiplicity 4 and defect 2;
$x_1(t) = (c_1 + 3c_2 + c_4 + c_2t + 3c_3t + \frac{1}{2}c_3t^2)e^t$,
$x_2(t) = -(2c_2 - c_3 + 2c_3t)e^t$, $x_3(t) = (c_2 + c_3t)e^t$,
$x_4(t) = -(2c_1 + 6c_2 + 2c_2t + 6c_3t + c_3t^2)e^t$

23. $\mathbf{x}(t) = c_1\mathbf{v}_1e^{-t} + (c_2\mathbf{v}_2 + c_3\mathbf{v}_3)e^{3t}$ with $\mathbf{v}_1 = \begin{bmatrix} 1 & -1 & 2 \end{bmatrix}^T$, $\mathbf{v}_2 = \begin{bmatrix} 4 & 0 & 9 \end{bmatrix}^T$, $\mathbf{v}_3 = \begin{bmatrix} 0 & 2 & 1 \end{bmatrix}^T$

24. $\mathbf{x}(t) = c_1\mathbf{v}_1e^{-t} + (c_2\mathbf{v}_2 + c_3\mathbf{v}_3)e^{3t}$ with $\mathbf{v}_1 = \begin{bmatrix} 5 & 3 & -3 \end{bmatrix}^T$, $\mathbf{v}_2 = \begin{bmatrix} 4 & 0 & -1 \end{bmatrix}^T$, $\mathbf{v}_3 = \begin{bmatrix} 2 & -1 & 0 \end{bmatrix}^T$

25. $\mathbf{x}(t) = \left[c_1\mathbf{v}_1 + c_2(\mathbf{v}_1t + \mathbf{v}_2) + c_3\left(\frac{1}{2}\mathbf{v}_1t^2 + \mathbf{v}_2t + \mathbf{v}_3\right)\right]e^{2t}$ with $\mathbf{v}_1 = \begin{bmatrix} -1 & 0 & -1 \end{bmatrix}^T$, $\mathbf{v}_2 = \begin{bmatrix} -4 & -1 & 0 \end{bmatrix}^T$, and $\mathbf{v}_3 = \begin{bmatrix} 1 & 0 & 0 \end{bmatrix}^T$

26. $\mathbf{x}(t) = \left[c_1\mathbf{v}_1 + c_2(\mathbf{v}_1t + \mathbf{v}_2) + c_3\left(\frac{1}{2}\mathbf{v}_1t^2 + \mathbf{v}_2t + \mathbf{v}_3\right)\right]e^{3t}$ with $\mathbf{v}_1 = \begin{bmatrix} 0 & 2 & 2 \end{bmatrix}^T$, $\mathbf{v}_3 = \begin{bmatrix} 2 & 1 & -3 \end{bmatrix}^T$, and $\mathbf{v}_3 = \begin{bmatrix} 1 & 0 & 0 \end{bmatrix}^T$

27. $\mathbf{x}(t) = [c_1\mathbf{v}_1 + c_2(\mathbf{v}_1t + \mathbf{v}_2) + c_3\mathbf{v}_3]e^{2t}$ with $\mathbf{v}_1 = \begin{bmatrix} -5 & 3 & 8 \end{bmatrix}^T$, $\mathbf{v}_2 = \begin{bmatrix} 1 & 0 & 0 \end{bmatrix}^T$, $\mathbf{v}_3 = \begin{bmatrix} 1 & 1 & 0 \end{bmatrix}^T$

28. $\mathbf{x}(t) = \left[c_1\mathbf{v}_1 + c_2(\mathbf{v}_1t + \mathbf{v}_2) + c_3\left(\frac{1}{2}\mathbf{v}_1t^2 + \mathbf{v}_2t + \mathbf{v}_3\right)\right]e^{2t}$ with $\mathbf{v}_1 = \begin{bmatrix} 119 & -289 & 0 \end{bmatrix}^T$, $\mathbf{v}_2 = \begin{bmatrix} -17 & 34 & 17 \end{bmatrix}^T$, and $\mathbf{v}_3 = \begin{bmatrix} 1 & 0 & 0 \end{bmatrix}^T$

29. $\mathbf{x}(t) = [c_1\mathbf{v}_1 + c_2(\mathbf{v}_1t + \mathbf{v}_2)]e^{-t} + [c_3\mathbf{v}_3 + c_4(\mathbf{v}_3t + \mathbf{v}_4)]e^{2t}$ with $\mathbf{v}_1 = \begin{bmatrix} 1 & -3 & -1 & -2 \end{bmatrix}^T$, $\mathbf{v}_2 = \begin{bmatrix} 0 & 1 & 0 & 0 \end{bmatrix}^T$, $\mathbf{v}_3 = \begin{bmatrix} 0 & -1 & 1 & 0 \end{bmatrix}^T$, $\mathbf{v}_4 = \begin{bmatrix} 0 & 0 & 2 & 1 \end{bmatrix}^T$

30. $\mathbf{x}(t) = [c_1\mathbf{v}_1 + c_2(\mathbf{v}_1t + \mathbf{v}_2)]e^{-t} + [c_3\mathbf{v}_3 + c_4(\mathbf{v}_3t + \mathbf{v}_4)]e^{2t}$, with $\mathbf{v}_1 = \begin{bmatrix} 0 & 1 & -1 & -3 \end{bmatrix}^T$, $\mathbf{v}_2 = \begin{bmatrix} 0 & 0 & 1 & 2 \end{bmatrix}^T$, $\mathbf{v}_3 = \begin{bmatrix} -1 & 0 & 0 & 0 \end{bmatrix}^T$, $\mathbf{v}_4 = \begin{bmatrix} 0 & 0 & 3 & 5 \end{bmatrix}^T$

31. $\mathbf{x}(t) = [c_1\mathbf{v}_1 + c_2(\mathbf{v}_1t + \mathbf{v}_2) + c_3\left(\frac{1}{2}\mathbf{v}_1t^2 + \mathbf{v}_2t + \mathbf{v}_3\right) + c_4\mathbf{v}_4]e^t$ with
$\mathbf{v}_1 = \begin{bmatrix} 42 & 7 & -21 & -42 \end{bmatrix}^T$,
$\mathbf{v}_2 = \begin{bmatrix} 34 & 22 & -10 & -27 \end{bmatrix}^T$, $\mathbf{v}_3 = \begin{bmatrix} 1 & 0 & 0 & 0 \end{bmatrix}^T$,
$\mathbf{v}_4 = \begin{bmatrix} 0 & 1 & 3 & 0 \end{bmatrix}^T$

32. $\mathbf{x}(t) = (c_1\mathbf{v}_1 + c_2\mathbf{v}_2)e^{2t} + (c_3\mathbf{v}_3 + c_4\mathbf{v}_4 + c_5\mathbf{v}_5)e^{3t}$ with
$\mathbf{v}_1 = \begin{bmatrix} 8 & 0 & -3 & 1 & 0 \end{bmatrix}^T$, $\mathbf{v}_2 = \begin{bmatrix} 1 & 0 & 0 & 0 & 3 \end{bmatrix}^T$,
$\mathbf{v}_3 = \begin{bmatrix} 3 & -2 & -1 & 0 & 0 \end{bmatrix}^T$,
$\mathbf{v}_4 = \begin{bmatrix} 2 & -2 & 0 & -3 & 0 \end{bmatrix}^T$, $\mathbf{v}_5 = \begin{bmatrix} 1 & -1 & 0 & 0 & 3 \end{bmatrix}^T$

33. $\mathbf{x}_1(t) = \begin{bmatrix} \cos 4t & \sin 4t & 0 & 0 \end{bmatrix}^T e^{3t}$,
$\mathbf{x}_2(t) = \begin{bmatrix} -\sin 4t & \cos 4t & 0 & 0 \end{bmatrix}^T e^{3t}$,
$\mathbf{x}_3(t) = \begin{bmatrix} t\cos 4t & t\sin 4t & \cos 4t & \sin 4t \end{bmatrix}^T e^{3t}$,
$\mathbf{x}_4(t) = \begin{bmatrix} -t\sin 4t & t\cos 4t & -\sin 4t & \cos 4t \end{bmatrix}^T e^{3t}$

34. $\mathbf{x}_1(t) = \begin{bmatrix} \sin 3t \\ 3\cos 3t - 3\sin 3t \\ 0 \\ \sin 3t \end{bmatrix} e^{2t}$,

$\mathbf{x}_2(t) = \begin{bmatrix} -\cos 3t \\ 3\sin 3t + 3\cos 3t \\ 0 \\ -\cos 3t \end{bmatrix} e^{2t}$,

$\mathbf{x}_3(t) = \begin{bmatrix} 3\cos 3t + t\sin 3t \\ (3t-10)\cos 3t - (3t+9)\sin 3t \\ \sin 3t \\ t\sin 3t \end{bmatrix} e^{2t}$,

$\mathbf{x}_4(t) = \begin{bmatrix} -t\cos 3t + 3\sin 3t \\ (3t+9)\cos 3t + (3t-10)\sin 3t \\ -\cos 3t \\ -t\cos 3t \end{bmatrix} e^{2t}$

35. $x_1(t) = x_2(t) = v_0(1 - e^{-t})$; $\lim_{t\to\infty} x_1(t) = \lim_{t\to\infty} x_2(t) = v_0$

36. $x_1(t) = v_0(2 - 2e^{-t} - te^{-t})$,
$x_2(t) = v_0(2 - 2e^{-t} - te^{-t} - \frac{1}{2}t^2e^{-t})$;
$\lim_{t\to\infty} x_1(t) = \lim_{t\to\infty} x_2(t) = 2v_0$

In Problems 37, 39, 41, 43, and 45 we give a nonsingular matrix **Q** and a Jordan-form matrix **J** such that $\mathbf{A} = \mathbf{QJQ}^{-1}$. Any scalar multiple of **Q** will do the same job.

37. $\mathbf{Q} = \begin{bmatrix} 1 & 4 & -2 \\ -1 & 0 & 9 \\ 2 & 9 & 0 \end{bmatrix}$, $\mathbf{J} = \begin{bmatrix} -1 & 0 & 0 \\ 0 & 3 & 0 \\ 0 & 0 & 3 \end{bmatrix}$

39. $\mathbf{Q} = \begin{bmatrix} 1 & 4 & -1 \\ 0 & 1 & 0 \\ 1 & 0 & 0 \end{bmatrix}$, $\mathbf{J} = \begin{bmatrix} 2 & 1 & 0 \\ 0 & 2 & 1 \\ 0 & 0 & 2 \end{bmatrix}$

41. $\mathbf{Q} = \begin{bmatrix} 0 & -5 & 0 \\ 1 & 3 & 0 \\ 1 & 8 & 1 \end{bmatrix}$, $\mathbf{J} = \begin{bmatrix} 2 & 0 & 0 \\ 0 & 2 & 1 \\ 0 & 0 & 2 \end{bmatrix}$

43. $\mathbf{Q} = \begin{bmatrix} -1 & 0 & 0 & 0 \\ 3 & -1 & -1 & 2 \\ 1 & 0 & 1 & 0 \\ 2 & 0 & 0 & 1 \end{bmatrix}$, $\mathbf{J} = \begin{bmatrix} -1 & 1 & 0 & 0 \\ 0 & -1 & 0 & 0 \\ 0 & 0 & 2 & 1 \\ 0 & 0 & 0 & 2 \end{bmatrix}$

45. $\mathbf{Q} = \begin{bmatrix} -30 & 432 & 360 & 0 \\ -7 & 72 & 228 & 0 \\ 9 & -216 & -108 & 0 \\ 30 & -432 & -288 & 12 \end{bmatrix}$, $\mathbf{J} = \begin{bmatrix} 1 & 0 & 0 & 0 \\ 0 & 1 & 1 & 0 \\ 0 & 0 & 1 & 1 \\ 0 & 0 & 0 & 1 \end{bmatrix}$

Section 7.6

The format for the first eight answers is this: $(x(t), y(t))$ at $t = 0.2$ by the Euler method, by the improved Euler method, by the Runge–Kutta method, and finally the actual values.

1. (0.8800, 2.5000), (0.9600, 2.6000), (1.0027, 2.6401), (1.0034, 2.6408)
2. (0.8100, −0.8100), (0.8200, −0.8200), (0.8187, −0.8187), (0.8187, −0.8187)
3. (2.8100, 2.3100), (3.2200, 2.6200), (3.6481, 2.9407), (3.6775, 2.9628)
4. (3.3100, −1.6200), (3.8200, −2.0400), (4.2274, −2.4060), (4.2427, −2.4205)
5. (−0.5200, 2.9200), (−0.8400, 2.4400), (−0.5712, 2.4485), (−0.5793, 2.4488)
6. (−1.7600, 4.6800), (−1.9200, 4.5600), (−1.9029, 4.4995), (−1.9025, 4.4999)
7. (3.1200, 1.6800), (3.2400, 1.7600), (3.2816, 1.7899), (3.2820, 1.7902)
8. (2.1600, −0.6300), (2.5200, −0.4600), (2.5320, −0.3867), (2.5270, −0.3889)
9. At $t = 1$ we obtain $(x, y) = (3.99261, 6.21770)$ ($h = 0.1$) and (3.99234, 6.21768) ($h = 0.05$); the actual value is (3.99232, 6.21768).
10. At $t = 1$ we obtain $(x, y) = (1.31498, 1.02537)$ ($h = 0.1$) and (1.31501, 1.02538) ($h = 0.05$); the actual value is (1.31501, 1.02538).
11. At $t = 1$ we obtain $(x, y) = (-0.05832, 0.56664)$ ($h = 0.1$) and (−0.05832, 0.56665) ($h = 0.05$); the actual value is (−0.05832, 0.56665).
12. We solved $x' = y$, $y' = -x + \sin t$, $x(0) = y(0) = 0$. With $h = 0.1$ and also with $h = 0.05$ we obtain the actual value $x(1.0) \approx 0.15058$.
13. Runge–Kutta, $h = 0.1$: about 1050 ft in about 7.7 s
14. Runge–Kutta, $h = 0.1$: about 1044 ft in about 7.8 s
15. Runge–Kutta, $h = 1.0$: about 83.83 mi in about 168 s
16. At 40°: 5.0 s, 352.9 ft; at 45°: 5.4 s, 347.2 ft; at 50°: 5.8 s, 334.2 ft (all values approximate)
17. At 39.0° the range is about 352.7 ft. At 39.5° it is 352.8; at 40°, 352.9; at 40.5°, 352.6; at 41.0°, 352.1.
18. Just under 57.5°
19. Approximately 253 ft/s
20. Maximum height: about 1005 ft, attained in about 5.6 s; range: about 1880 ft; time aloft: about 11.6 s
21. Runge–Kutta with $h = 0.1$ yields these results:
(a) 21400 ft, 46 s, 518 ft/s; (b) 8970 ft, 17.5 s; (c) 368 ft/s (at $t \approx 23$).

Chapter 8

Section 8.1

1. $\mathbf{\Phi}(t) = \begin{bmatrix} e^t & e^{3t} \\ -e^t & e^{3t} \end{bmatrix}$, $\mathbf{x}(t) = \frac{1}{2}\begin{bmatrix} 5e^t + e^{3t} \\ -5e^t + e^{3t} \end{bmatrix}$

2. $\mathbf{\Phi}(t) = \begin{bmatrix} 1 & e^{4t} \\ 2 & -2e^{4t} \end{bmatrix}$, $\mathbf{x}(t) = \frac{1}{4}\begin{bmatrix} 3 + 5e^{4t} \\ 6 - 10e^{4t} \end{bmatrix}$

3. $\mathbf{\Phi}(t) = \begin{bmatrix} 5\cos 4t & -5\sin 4t \\ 2\cos 4t + 4\sin 4t & 4\cos 4t - 2\sin 4t \end{bmatrix}$,
$\mathbf{x}(t) = \frac{1}{4}\begin{bmatrix} -5\sin 4t \\ 4\cos 4t - 2\sin 4t \end{bmatrix}$

4. $\mathbf{\Phi}(t) = e^{2t}\begin{bmatrix} 1 & 1+t \\ 1 & t \end{bmatrix}$, $\mathbf{x}(t) = e^{2t}\begin{bmatrix} 1+t \\ t \end{bmatrix}$

5. $\mathbf{\Phi}(t) = \begin{bmatrix} 2\cos 3t & -2\sin 3t \\ -3\cos 3t + 3\sin 3t & 3\cos 3t + 3\sin 3t \end{bmatrix}$,
$\mathbf{x}(t) = \frac{1}{3}\begin{bmatrix} 3\cos 3t - \sin 3t \\ -3\cos 3t + 6\sin 3t \end{bmatrix}$

6. $\mathbf{\Phi}(t) = e^{5t}\begin{bmatrix} \cos 4t - 2\sin 4t & 2\cos 4t + 2\sin 4t \\ 2\cos 4t & 2\sin 4t \end{bmatrix}$,
$\mathbf{x}(t) = 2e^{5t}\begin{bmatrix} \cos 4t + \sin 4t \\ \sin 4t \end{bmatrix}$

7. $\mathbf{\Phi}(t) = \begin{bmatrix} 6 & 3e^t & 2e^{-t} \\ 2 & e^t & e^{-t} \\ 5 & 2e^t & 2e^{-t} \end{bmatrix}$, $\mathbf{x}(t) = \begin{bmatrix} -12 + 12e^t + 2e^{-t} \\ -4 + 4e^t + e^{-t} \\ -10 + 8e^t + 2e^{-t} \end{bmatrix}$

8. $\mathbf{\Phi}(t) = \begin{bmatrix} 0 & e^t & e^{3t} \\ e^{-2t} & -e^t & -e^{3t} \\ -e^{-2t} & 0 & e^{3t} \end{bmatrix}$, $\mathbf{x}(t) = \begin{bmatrix} e^t \\ -e^t + e^{-2t} \\ -e^{-2t} \end{bmatrix}$

9. $e^{\mathbf{A}t} = \begin{bmatrix} 2e^{3t} - e^t & -2e^{3t} + 2e^t \\ e^{3t} - e^t & -e^{3t} + 2e^t \end{bmatrix}$

10. $e^{\mathbf{A}t} = \begin{bmatrix} -2+3e^{2t} & 3-3e^{2t} \\ -2+2e^{2t} & 3-2e^{2t} \end{bmatrix}$

11. $e^{\mathbf{A}t} = \begin{bmatrix} 3e^{3t}-2e^{2t} & -3e^{3t}+3e^{2t} \\ 2e^{3t}-2e^{2t} & -2e^{3t}+3e^{2t} \end{bmatrix}$

12. $e^{\mathbf{A}t} = \begin{bmatrix} -3e^{t}+4e^{2t} & 4e^{t}-4e^{2t} \\ -3e^{t}+3e^{2t} & 4e^{t}-3e^{2t} \end{bmatrix}$

13. $e^{\mathbf{A}t} = \begin{bmatrix} 4e^{3t}-3e^{t} & -4e^{3t}+4e^{t} \\ 3e^{3t}-3e^{t} & -3e^{3t}+4e^{t} \end{bmatrix}$

14. $e^{\mathbf{A}t} = \begin{bmatrix} -8e^{t}+9e^{2t} & 6e^{t}-6e^{2t} \\ -12e^{t}+12e^{2t} & 9e^{t}-8e^{2t} \end{bmatrix}$

15. $e^{\mathbf{A}t} = \begin{bmatrix} 5e^{2t}-4e^{t} & -10e^{2t}+10e^{t} \\ 2e^{2t}-2e^{t} & -4e^{2t}+5e^{t} \end{bmatrix}$

16. $e^{\mathbf{A}t} = \begin{bmatrix} -9e^{t}+10e^{2t} & 15e^{t}-15e^{2t} \\ -6e^{t}+6e^{2t} & 10e^{t}-9e^{2t} \end{bmatrix}$

17. $e^{\mathbf{A}t} = \frac{1}{2}\begin{bmatrix} e^{4t}+e^{2t} & e^{4t}-e^{2t} \\ e^{4t}-e^{2t} & e^{4t}+e^{2t} \end{bmatrix}$

18. $e^{\mathbf{A}t} = \dfrac{1}{2}\begin{bmatrix} e^{2t}+e^{6t} & -e^{2t}+e^{6t} \\ -e^{2t}+e^{6t} & e^{2t}+e^{6t} \end{bmatrix}$

19. $e^{\mathbf{A}t} = \frac{1}{5}\begin{bmatrix} 4e^{10t}+e^{5t} & 2e^{10t}-2e^{5t} \\ 2e^{10t}-2e^{5t} & e^{10t}+4e^{5t} \end{bmatrix}$

20. $e^{\mathbf{A}t} = \dfrac{1}{5}\begin{bmatrix} e^{5t}+4e^{15t} & -2e^{5t}+2e^{15t} \\ -2e^{5t}+2e^{15t} & 4e^{5t}+e^{15t} \end{bmatrix}$

21. $e^{\mathbf{A}t} = \begin{bmatrix} 1+t & -t \\ t & 1-t \end{bmatrix}$

22. $e^{\mathbf{A}t} = \begin{bmatrix} 1+6t & 4t \\ -9t & 1-6t \end{bmatrix}$

23. $e^{\mathbf{A}t} = \begin{bmatrix} 1+t & -t & -t-t^2 \\ t & 1-t & t-t^2 \\ 0 & 0 & 1 \end{bmatrix}$

24. $e^{\mathbf{A}t} = \begin{bmatrix} 1+3t & 0 & -3t \\ 5t+18t^2 & 1 & 7t-18t^2 \\ 3t & 0 & 1-3t \end{bmatrix}$

25. $e^{\mathbf{A}t} = \begin{bmatrix} e^{2t} & 5te^{2t} \\ 0 & e^{2t} \end{bmatrix}$, $\mathbf{x}(t) = e^{\mathbf{A}t}\begin{bmatrix} 4 \\ 7 \end{bmatrix}$

26. $e^{\mathbf{A}t} = \begin{bmatrix} e^{7t} & 0 \\ 11te^{7t} & e^{7t} \end{bmatrix}$, $\mathbf{x}(t) = e^{\mathbf{A}t}\begin{bmatrix} 5 \\ -10 \end{bmatrix}$

27. $e^{\mathbf{A}t} = \begin{bmatrix} e^{t} & 2te^{t} & (3t+2t^2)e^{t} \\ 0 & e^{t} & 2te^{t} \\ 0 & 0 & e^{t} \end{bmatrix}$, $\mathbf{x}(t) = e^{\mathbf{A}t}\begin{bmatrix} 4 \\ 5 \\ 6 \end{bmatrix}$

28. $e^{\mathbf{A}t} = \begin{bmatrix} e^{5t} & 0 & 0 \\ 10te^{5t} & e^{5t} & 0 \\ (20t+150t^2)e^{5t} & 30te^{5t} & e^{5t} \end{bmatrix}$, $\mathbf{x}(t) = e^{\mathbf{A}t}\begin{bmatrix} 40 \\ 50 \\ 60 \end{bmatrix}$

29. $e^{\mathbf{A}t} = \begin{bmatrix} 1 & 2t & 3t+6t^2 & 4t+6t^2+4t^3 \\ 0 & 1 & 6t & 3t+6t^2 \\ 0 & 0 & 1 & 2t \\ 0 & 0 & 0 & 1 \end{bmatrix} e^{t}$,

$\mathbf{x}(t) = e^{\mathbf{A}t}\begin{bmatrix} 1 \\ 1 \\ 1 \\ 1 \end{bmatrix}$

30. $e^{\mathbf{A}t} = e^{3t}\begin{bmatrix} 1 & 0 & 0 & 0 \\ 6t & 1 & 0 & 0 \\ 9t+18t^2 & 6t & 1 & 0 \\ 12t+54t^2+36t^3 & 9t+18t^2 & 6t & 1 \end{bmatrix}$,

$\mathbf{x}(t) = e^{\mathbf{A}t}\begin{bmatrix} 1 \\ 1 \\ 1 \\ 1 \end{bmatrix}$

33. $\mathbf{x}(t) = \begin{bmatrix} c_1\cosh t + c_2\sinh t \\ c_1\sinh t + c_2\cosh t \end{bmatrix}$

35. $e^{\mathbf{A}t} = \begin{bmatrix} e^{3t} & 4te^{3t} \\ 0 & e^{3t} \end{bmatrix}$

36. $e^{\mathbf{A}t} = e^{t}\begin{bmatrix} 1 & 2t & 3t+4t^2 \\ 0 & 1 & 4t \\ 0 & 0 & 1 \end{bmatrix}$

37. $e^{\mathbf{A}t} = \begin{bmatrix} e^{2t} & 3e^{2t}-3e^{t} & 13e^{2t}-(13+9t)e^{t} \\ 0 & e^{t} & 3te^{t} \\ 0 & 0 & e^{t} \end{bmatrix}$

38. $e^{\mathbf{A}t} = \begin{bmatrix} e^{5t} & 4e^{10t}-4e^{5t} & 16e^{10t}-(16+50t)e^{5t} \\ 0 & e^{10t} & 4e^{10t}-4e^{5t} \\ 0 & 0 & e^{5t} \end{bmatrix}$

39. $e^{\mathbf{A}t} =$
$\begin{bmatrix} e^{t} & 3te^{t} & 12e^{2t}-(12+9t)e^{t} & (51+18t)e^{t}-(51-36t)e^{2t} \\ 0 & e^{t} & 3e^{2t}-3e^{t} & 6e^{t}-(6-9t)e^{2t} \\ 0 & 0 & e^{2t} & 3te^{2t} \\ 0 & 0 & 0 & e^{2t} \end{bmatrix}$

40. $e^{\mathbf{A}t} =$
$\begin{bmatrix} e^{2t} & 4te^{2t} & (4t+8t^2)e^{2t} & 100e^{3t}-(100+96t+32t^2)e^{2t} \\ 0 & e^{2t} & 4te^{2t} & 20e^{3t}-(20+16t)e^{2t} \\ 0 & 0 & e^{2t} & 4e^{3t}-4e^{2t} \\ 0 & 0 & 0 & e^{3t} \end{bmatrix}$

Section 8.2

1. $x(t) = \frac{7}{3}$, $y(t) = -\frac{8}{3}$

2. $x(t) = \frac{1}{8}(1+12t)$, $y(t) = -\frac{1}{4}(5+4t)$

3. $x(t) = \frac{1}{756}(864e^{-t}+4e^{6t}-868+840t-504t^2)$,
$y(t) = \frac{1}{756}(-864e^{-t}+3e^{6t}+861-882t+378t^2)$

4. $x(t) = \frac{1}{84}(99e^{5t}-8e^{-2t}-7e^{t})$, $y(t) = \frac{1}{84}(99e^{5t}+48e^{-2t}-63e^{t})$

5. $x(t) = \frac{1}{3}(-12-e^{-t}-7te^{-t})$, $y(t) = \frac{1}{3}(-6-7te^{-t})$

6. $x(t) = -\frac{1}{256}(91+16t)e^{t}$, $y(t) = \frac{1}{32}(25+16t)e^{t}$

7. $x(t) = \frac{1}{410}(369e^{t}+166e^{-9t}-125\cos t-105\sin t)$,
$y(t) = \frac{1}{410}(369e^{t}-249e^{-9t}-120\cos t-150\sin t)$

8. $x(t) = \frac{1}{3}(17\cos t+2\sin t)$, $y(t) = \frac{1}{3}(3\cos t+5\sin t)$

9. $x(t) = \frac{1}{4}(\sin 2t+2t\cos 2t+t\sin 2t)$, $y(t) = \frac{1}{4}t\sin 2t$

10. $x(t) = \frac{1}{13}e^{t}(4\cos t-6\sin t)$, $y(t) = \frac{1}{13}e^{t}(3\cos t+2\sin t)$

11. $x(t) = \frac{1}{2}(1-4t+e^{4t})$, $y(t) = \frac{1}{4}(-5+4t+e^{4t})$

12. $x(t) = t^2$, $y(t) = -t^2$

13. $x(t) = \frac{1}{2}(1+5t)e^{t}$, $y(t) = -\frac{5}{2}te^{t}$

14. $x(t) = \frac{1}{8}(-2+4t-e^{4t}+2te^{4t})$, $y(t) = \frac{1}{2}t(-2+e^{4t})$

15. (a) $x_1(t) = 200(1-e^{-t/10})$, $x_2(t) = 400(1+e^{-t/10}-2e^{-t/20})$
(b) $x_1(t) \to 200$ and $x_2(t) \to 400$ as $t \to +\infty$
(c) Tank 1: about 6 min 56 s; tank 2: about 24 min 34 s

16. (a) $x_1(t) = 600(1-e^{-t/20})$, $x_2(t) = 300(1+e^{-t/10}-2e^{-t/20})$
(b) $x_1(t) \to 600$ and $x_2(t) \to 300$ as $t \to \infty$
(c) Tank 1: about 8 min 7 sec; tank 2: about 17 min 13 sec

17. $x_1(t) = 102-95e^{-t}-7e^{5t}$, $x_2(t) = 96-95e^{-t}-e^{5t}$

18. $x_1(t) = 68-110t-75e^{-t}+7e^{5t}$, $x_2(t) = 74-80t-75e^{-t}+e^{5t}$

19. $x_1(t) = -70-60t+16e^{-3t}+54e^{2t}$,
$x_2(t) = 5-60t-32e^{-3t}+27e^{2t}$

20. $x_1(t) = 3e^{2t}+60te^{2t}-3e^{-3t}$, $x_2(t) = -6e^{2t}+30te^{2t}+6e^{-3t}$

21. $x_1(t) = -e^{-t}-14e^{2t}+15e^{3t}$, $x_2(t) = -5e^{-t}-10e^{2t}+15e^{3t}$

22. $x_1(t) = -10e^{-t}-7te^{-t}+10e^{3t}-5te^{3t}$,
$x_2(t) = -15e^{-t}-35te^{-t}+15e^{3t}-5te^{3t}$

23. $x_1(t) = 3+11t+8t^2$, $x_2(t) = 5+17t+24t^2$

24. $x_1(t) = 2+t+\ln t$, $x_2(t) = 5+3t-\dfrac{1}{t}+3\ln t$

25. $x_1(t) = -1+8t+\cos t-8\sin t$,
$x_2(t) = -2+4t+2\cos t-3\sin t$

26. $x_1(t) = 3\cos t-32\sin t+17t\cos t+4t\sin t$,
$x_2(t) = 5\cos t-13\sin t+6t\cos t+5t\sin t$

27. $x_1(t) = 8t^3+6t^4$, $x_2(t) = 3t^2-2t^3+3t^4$

28. $x_1(t) = -7 + 14t - 6t^2 + 4t^2 \ln t$,
$x_2(t) = -7 + 9t - 3t^2 + \ln t - 2t \ln t + 2t^2 \ln t$

29. $x_1(t) = t \cos t - (\ln \cos t)(\sin t)$,
$x_2(t) = t \sin t + (\ln \cos t)(\cos t)$

30. $x_1(t) = \frac{1}{2}t^2 \cos 2t$, $x_2(t) = \frac{1}{2}t^2 \sin 2t$

31. $x_1(t) = (9t^2 + 4t^3)e^t$, $x_2(t) = 6t^2 e^t$, $x_3(t) = 6te^t$

32. $x_1(t) = (44 + 18t)e^t + (-44 + 26t)e^{2t}$,
$x_2(t) = 6e^t + (-6 + 6t)e^{2t}$, $x_3(t) = 2te^{2t}$

33. $x_1(t) = 15t^2 + 60t^3 + 95t^4 + 12t^5$, $x_2(t) = 15t^2 + 55t^3 + 15t^4$,
$x_3(t) = 15t^2 + 20t^3$, $x_4(t) = 15t^2$

34. $x_1(t) = 4t^3 + (4 + 16t + 8t^2)e^{2t}$,
$x_2(t) = 3t^2 + (2 + 4t)e^{2t}$,
$x_3(t) = (2 + 4t + 2t^2)e^{2t}$, $x_4(t) = (1 + t)e^{2t}$

Section 8.3

1. $\mathbf{x}(t) = \frac{1}{2}\begin{bmatrix} e^{-t} + e^{3t} & -e^{-t} + e^{3t} \\ -e^{-t} + e^{3t} & e^{-t} + e^{3t} \end{bmatrix}$

2. $\mathbf{x} = \frac{1}{5}\begin{bmatrix} 2e^{-t} + 3e^{4t} & -3e^{-t} + 3e^{4t} \\ -2e^{-t} + 2e^{4t} & 3e^{-t} + 2e^{4t} \end{bmatrix}$

3. $\mathbf{x}(t) = \frac{1}{7}\begin{bmatrix} 3e^{-t} + 4e^{6t} & -4e^{-t} + 4e^{6t} \\ -3e^{-t} + 3e^{6t} & 4e^{-t} + 3e^{6t} \end{bmatrix}$

4. $\mathbf{x}(t) = \frac{1}{7}\begin{bmatrix} e^{-2t} + 6e^{5t} & -e^{-2t} + e^{5t} \\ -6e^{-2t} + 6e^{5t} & 6e^{-2t} + e^{5t} \end{bmatrix}$

5. $\mathbf{x}(t) = \frac{1}{6}\begin{bmatrix} -e^{-t} + 7e^{5t} & 7e^{-t} - 7e^{5t} \\ -e^{-t} + e^{5t} & 7e^{-t} - e^{5t} \end{bmatrix}$

6. $\mathbf{x}(t) = \begin{bmatrix} -5e^{3t} + 6e^{4t} & -5e^{3t} + 5e^{4t} \\ 6e^{3t} - 6e^{4t} & 6e^{3t} - 5e^{4t} \end{bmatrix}$

7. $\mathbf{x}(t) = \frac{1}{5}\begin{bmatrix} 2e^{-9t} + 3e^{t} & -2e^{-9t} + 2e^{t} \\ -3e^{-9t} + 3e^{t} & 3e^{-9t} + 2e^{t} \end{bmatrix}$

8. $\mathbf{x}(t) = \frac{1}{2}\begin{bmatrix} 2\cos 2t + \sin 2t & -5\sin 2t \\ \sin 2t & 2\cos 2t - \sin 2t \end{bmatrix}$

9. $\mathbf{x}(t) = \frac{1}{4}\begin{bmatrix} 4\cos 4t + 2\sin 4t & -5\sin 4t \\ 4\sin 4t & 4\cos 4t - 2\sin 4t \end{bmatrix}$

10. $\mathbf{x}(t) = \frac{1}{3}\begin{bmatrix} 3\cos 3t - 3\sin 3t & -2\sin 3t \\ 9\sin 3t & 3\cos 3t + 3\sin 3t \end{bmatrix}$

11. $\mathbf{x}(t) = e^{t}\begin{bmatrix} \cos 2t & -\sin 2t \\ \sin 2t & \cos 2t \end{bmatrix}$

12. $\mathbf{x}(t) = \frac{e^{2t}}{2}\begin{bmatrix} 2\cos 2t - \sin 2t & -5\sin 2t \\ \sin 2t & 2\cos 2t + \sin 2t \end{bmatrix}$

13. $\mathbf{x}(t) = \frac{e^{2t}}{3}\begin{bmatrix} 3\cos 3t + 3\sin 3t & -9\sin 3t \\ 2\sin 3t & 3\cos 3t - 3\sin 3t \end{bmatrix}$

14. $\mathbf{x}(t) = e^{3t}\begin{bmatrix} \cos 4t & -\sin 4t \\ \sin 4t & \cos 4t \end{bmatrix}$

15. $\mathbf{x}(t) = \frac{e^{5t}}{4}\begin{bmatrix} 4\cos 4t + 2\sin 4t & -5\sin 4t \\ 4\sin 4t & 4\cos 4t - 2\sin 4t \end{bmatrix}$

16. $\mathbf{x}(t) = \frac{1}{9}\begin{bmatrix} 4e^{-100t} + 5e^{-10t} & -2e^{-100t} + 2e^{-10t} \\ -10e^{-100t} + 10e^{-10t} & 5e^{-100t} + 4e^{-10t} \end{bmatrix}$

17. $\mathbf{x}(t) =$
$$\frac{1}{6}\begin{bmatrix} 3 + e^{6t} + 2e^{9t} & -2e^{6t} + 2e^{9t} & -3 + e^{6t} + 2e^{9t} \\ -2e^{6t} + 2e^{9t} & 4e^{6t} + 2e^{9t} & -2e^{6t} + 2e^{9t} \\ -3 + e^{6t} + 2e^{9t} & -2e^{6t} + 2e^{9t} & 3 + e^{6t} + 2e^{9t} \end{bmatrix}$$

18. $\mathbf{x}(t) =$
$$\frac{1}{18}\begin{bmatrix} 16 + e^{6t} + 2e^{9t} & -4 + 4e^{9t} & -4 + 4e^{9t} \\ -4e^{6t} + 4e^{9t} & 1 + 9e^{6t} + 8e^{9t} & 1 - 9e^{6t} + 8e^{9t} \\ -4e^{6t} + 4e^{9t} & 1 - 9e^{6t} + 8e^{9t} & 1 + 9e^{6t} + 8e^{9t} \end{bmatrix}$$

19. $\mathbf{x}(t) = \frac{1}{3}\begin{bmatrix} 2e^{3t} + e^{6t} & -e^{3t} + e^{6t} & -e^{3t} + e^{6t} \\ -e^{3t} + e^{6t} & 2e^{3t} + e^{6t} & -e^{3t} + e^{6t} \\ -e^{3t} + e^{6t} & -e^{3t} + e^{6t} & 2e^{3t} + e^{6t} \end{bmatrix}$

20. $\mathbf{x}(t) =$
$$\frac{1}{6}\begin{bmatrix} 3e^{2t} + e^{6t} + 2e^{9t} & -2e^{6t} + 2e^{9t} & -3e^{2t} + e^{6t} + 2e^{9t} \\ -2e^{6t} + 2e^{9t} & 4e^{6t} + 2e^{9t} & -2e^{6t} + 2e^{9t} \\ -3e^{2t} + e^{6t} + 2e^{9t} & -2e^{6t} + 2e^{9t} & 3e^{2t} + e^{6t} + 2e^{9t} \end{bmatrix}$$

21. $\mathbf{x}(t) = e^{-3t}\begin{bmatrix} 1 + t & t \\ -t & 1 - t \end{bmatrix}$

22. $\mathbf{x}(t) = e^{2t}\begin{bmatrix} 1 + t & -t \\ t & 1 - t \end{bmatrix}$

23. $\mathbf{x}(t) = e^{3t}\begin{bmatrix} 1 - 2t & -2t \\ 2t & 1 + 2t \end{bmatrix}$

24. $\mathbf{x}(t) = e^{4t}\begin{bmatrix} 1 - t & -t \\ t & 1 + t \end{bmatrix}$

25. $\mathbf{x}(t) = e^{5t}\begin{bmatrix} 1 + 2t & t \\ -4t & 1 - 2t \end{bmatrix}$

26. $\mathbf{x}(t) = e^{5t}\begin{bmatrix} 1 - 4t & -4t \\ 4t & 1 + 4t \end{bmatrix}$

27. $\mathbf{x}(t) = \begin{bmatrix} e^{2t} & 0 & 0 \\ e^{2t} - e^{9t} & e^{9t} & -e^{2t} + e^{9t} \\ 0 & 0 & e^{2t} \end{bmatrix}$

28. $\mathbf{x}(t) = \begin{bmatrix} -2e^{7t} + 3e^{13t} & -2e^{7t} + 2e^{13t} & 0 \\ 3e^{7t} - 3e^{13t} & 3e^{7t} - 2e^{13t} & 0 \\ -e^{7t} + e^{13t} & -e^{7t} + e^{13t} & e^{13t} \end{bmatrix}$

29. $\mathbf{x}(t) = \begin{bmatrix} 7e^{5t} - 6e^{9t} & -3e^{5t} + 3e^{9t} & -21e^{5t} + 21e^{9t} \\ 0 & e^{5t} & 0 \\ 2e^{5t} - 2e^{9t} & -e^{5t} + e^{9t} & -6e^{5t} + 7e^{9t} \end{bmatrix}$

30. $\mathbf{x}(t) = \begin{bmatrix} 5e^{3t} - 4e^{7t} & -10e^{3t} + 10e^{7t} & 12e^{3t} - 12e^{7t} \\ 2e^{3t} - 2e^{7t} & -4e^{3t} + 5e^{7t} & 6e^{3t} - 6e^{7t} \\ 0 & 0 & e^{3t} \end{bmatrix}$

31. $e^{\mathbf{A}t} = \frac{1}{4}\begin{bmatrix} -e^{-t} + 5e^{3t} & e^{-t} - e^{3t} \\ -5e^{-t} + 5e^{3t} & 5e^{-t} - e^{3t} \end{bmatrix}$,
$\mathbf{x}(t) = \begin{bmatrix} -e^{-t} - 14e^{2t} + 15e^{3t} \\ -5e^{-t} - 10e^{2t} + 15e^{3t} \end{bmatrix}$

32. With $e^{\mathbf{A}t}$ as in Problem 31,
$\mathbf{x}(t) = \begin{bmatrix} (-10 - 7t)e^{-t} + (10 - 5t)e^{3t} \\ (-15 - 35t)e^{-t} + (15 - 5t)e^{3t} \end{bmatrix}$.

33. $e^{\mathbf{A}t} = \begin{bmatrix} 1 + 3t & -t \\ 9t & 1 - 3t \end{bmatrix}$, $\mathbf{x}(t) = \begin{bmatrix} 3 + 11t + 8t^2 \\ 5 + 17t + 24t^2 \end{bmatrix}$

34. With $e^{\mathbf{A}t}$ as in Problem 33, $\mathbf{x}(t) = \begin{bmatrix} 2 + t + \ln t \\ 5 + 3t - \frac{1}{t} + 3\ln t \end{bmatrix}$.

35. $e^{\mathbf{A}t} = \begin{bmatrix} \cos t + 2\sin t & -5\sin t \\ \sin t & \cos t - 2\sin t \end{bmatrix}$,
$\mathbf{x}(t) = \begin{bmatrix} -1 + 8t + \cos t - 8\sin t \\ -2 + 4t + 2\cos t - 3\sin t \end{bmatrix}$

36. With $e^{\mathbf{A}t}$ as in Problem 35,
$\mathbf{x}(t) = \begin{bmatrix} 3\cos t - 32\sin t + 17t\cos t + 4t\sin t \\ 5\cos t - 13\sin t + 6t\cos t + 5t\sin t \end{bmatrix}$.

37. $e^{\mathbf{A}t} = \begin{bmatrix} 1 + 2t & -4t \\ t & 1 - 2t \end{bmatrix}$, $\mathbf{x}(t) = \begin{bmatrix} 8t^3 + 6t^4 \\ 3t^2 - 2t^3 + 3t^4 \end{bmatrix}$

38. With $e^{\mathbf{A}t}$ as in Problem 37,
$\mathbf{x}(t) = \begin{bmatrix} -7 + 14t - 6t^2 + 4t^2 \ln t \\ -7 + 9t - 3t^2 + \ln t - 2t\ln t + 2t^2 \ln t \end{bmatrix}$.

39. $e^{\mathbf{A}t} = \begin{bmatrix} \cos t & -\sin t \\ \sin t & \cos t \end{bmatrix}$, $\mathbf{x}(t) = \begin{bmatrix} t\cos t - (\ln\cos t)(\sin t) \\ t\sin t + (\ln\cos t)(\cos t) \end{bmatrix}$

40. $e^{\mathbf{A}t} = \begin{bmatrix} \cos 2t & -2\sin 2t \\ \sin 2t & \cos 2t \end{bmatrix}$, $\mathbf{x}(t) = \begin{bmatrix} \frac{1}{2}t^2\cos 2t \\ \frac{1}{2}t^2 \sin 2t \end{bmatrix}$

41. $\mathbf{x}(t) = \begin{bmatrix} -9e^{-t} + 10e^{3t} & -2e^{-t} + 2e^{3t} & 4e^{-t} - 4e^{3t} \\ 9e^{-t} - 9e^{3t} & 2e^{-t} - e^{3t} & -4e^{-t} + 4e^{3t} \\ -18e^{-t} + 18e^{3t} & -4e^{-t} + 4e^{3t} & 8e^{-t} - 7e^{3t} \end{bmatrix}$

42. $\mathbf{x}(t) =$
$$\begin{bmatrix} -5e^{-2t}+6e^{3t} & -10e^{-2t}+10e^{3t} & -20e^{-2t}+20e^{3t} \\ -3e^{-2t}+3e^{3t} & -6e^{-2t}+7e^{3t} & -12e^{-2t}+12e^{3t} \\ 3e^{-2t}+-3e^{3t} & 6e^{-2t}-6e^{3t} & 12e^{-2t}-11e^{3t} \end{bmatrix}$$

43. $\mathbf{x}(t) = \frac{1}{2}e^{2t}\begin{bmatrix} -t^2-8t+2 & 4t^2+34t & t^2+8t \\ -2t & 8t+2 & 2t \\ -t^2 & 4t^2+2t & t^2+2 \end{bmatrix}$

44. $\mathbf{x}(t) = \frac{1}{2}e^{3t}\begin{bmatrix} 4t+2 & -2t & 2t \\ 2t^2+2t & -t^2+2 & t^2 \\ 2t^2-6t & -t^2+4t & t^2-4t+2 \end{bmatrix}$

45. $\mathbf{x}(t) = \begin{bmatrix} e^{-t} & te^{-t} & te^{-t} & -2te^{-t} \\ -3e^{-t}+(3-2t)e^{2t} & (1-3t)e^{-t} & (1-3t)e^{-t}-e^{2t} & 2(3t-1)e^{-t}+(2-t)e^{2t} \\ -e^{-t}+(1+2t)e^{2t} & -te^{-t} & -te^{-t}+e^{2t} & 2te^{-t}+te^{2t} \\ -2e^{-t}+2e^{2t} & -2te^{-t} & -2te^{-t} & 4te^{-t}+e^{2t} \end{bmatrix}$

46. $\mathbf{x}(t) = \frac{1}{2}e^{t}\begin{bmatrix} 48t^2+68t+2 & -18t^2-24t & 6t^2+8t & 36t^2+60t \\ 7t^2+44t & -3t^2-18t+2 & t^2+6t & 6t^2+38t \\ -21t^2-20t & 9t^2+6t & -3t^2-2t+2 & -18t^2-18t \\ -42t^2-54t & 18t^2+18t & -6t^2-6t & -36t^2-48t+2 \end{bmatrix}$

47. $\mathbf{x}(t) = \begin{bmatrix} \cos t \\ \cos t \end{bmatrix} + i\begin{bmatrix} -\sin 3t \\ \sin 3t \end{bmatrix}$

There are two natural modes—one in which the two masses move in the same direction with frequency $\omega_1 = 1$ and with equal amplitudes, and one in which they move in opposite directions with frequency $\omega_2 = 3$ and with equal amplitudes.

48. $\mathbf{x}(t) = \begin{bmatrix} \cos t \\ \cos t \end{bmatrix} + i\begin{bmatrix} -\sin\sqrt{5}\,t \\ \sin\sqrt{5}\,t \end{bmatrix}$

There are two natural modes—one in which the two masses move in the same direction with frequency $\omega_1 = 1$ and with equal amplitudes, and one in which they move in opposite directions with frequency $\omega_2 = \sqrt{5}$ and with equal amplitudes.

49. $\mathbf{x}(t) = \begin{bmatrix} \cos\sqrt{2}\,t \\ \cos\sqrt{2}\,t \end{bmatrix} + i\begin{bmatrix} -\sin 2t \\ \sin 2t \end{bmatrix}$

There are two natural modes—one in which the two masses move in the same direction with frequency $\omega_1 = \sqrt{2}$ and with equal amplitudes, and one in which they move in opposite directions with frequency $\omega_2 = 2$ and with equal amplitudes.

50. $\mathbf{x}(t) = \begin{bmatrix} \cos 2t \\ \cos 2t \end{bmatrix} + i\begin{bmatrix} -\sin 4t \\ \sin 4t \end{bmatrix}$

There are two natural modes—one in which the two masses move in the same direction with frequency $\omega_1 = 2$ and with equal amplitudes, and one in which they move in opposite directions with frequency $\omega_2 = 4$ and with equal amplitudes.

Chapter 9

Section 9.1

1. 9.1.14 **2.** 9.1.16 **3.** 9.1.19 **4.** 9.1.13

5. 9.1.12 **6.** 9.1.18 **7.** 9.1.15 **8.** 9.1.17

9. Equilibrium solutions $x(t) \equiv 0, \pm 2$. The critical point $(0, 0)$ in the phase plane looks like a center, whereas the points $(\pm 2, 0)$ look like saddle points.

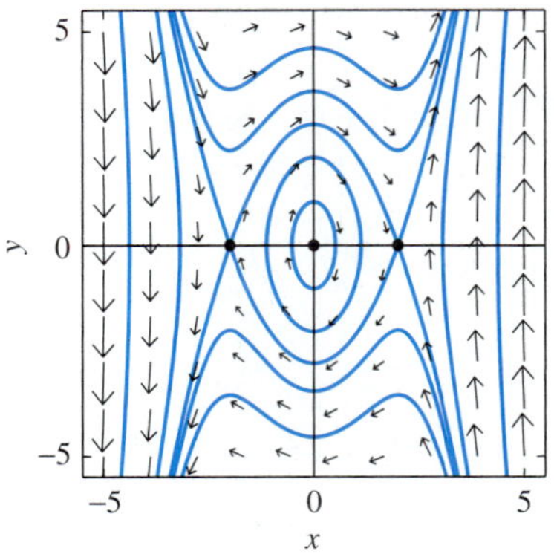

10. Equilibrium solution $x(t) \equiv 0$. The critical point $(0, 0)$ in the phase plane looks like a spiral sink.

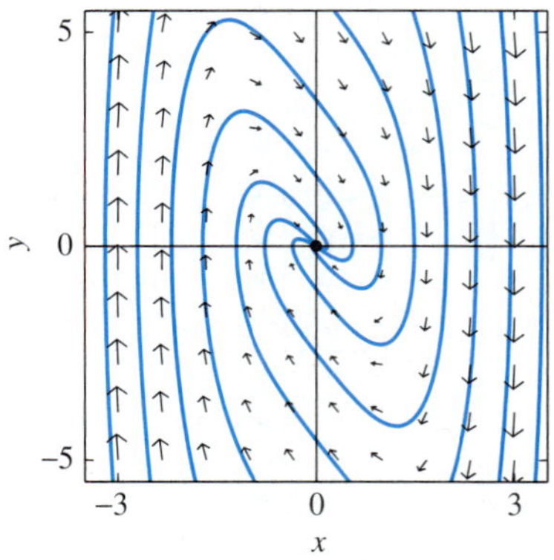

11. Equilibrium solutions $x(t) \equiv \ldots, -2\pi, -\pi, 0, \pi, 2\pi, \ldots$. The phase portrait shown in the solutions manual suggests that the critical point $(n\pi, 0)$ in the phase plane is a spiral sink if n is even, but is a saddle point if n is odd.

12. Equilibrium solution $x(t) \equiv 0$. The critical point (0, 0) in the phase plane looks like a spiral source, with the solution curves emanating from this source spiraling outward toward a closed curve trajectory.

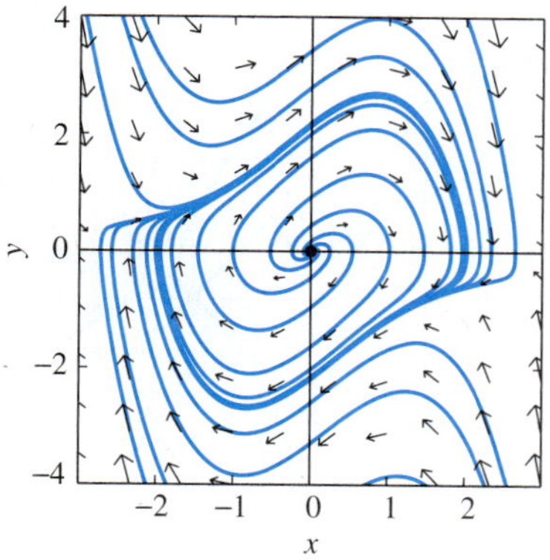

13. Solution $x(t) = x_0e^{-2t}$, $y(t) = y_0e^{-2t}$. The origin is a stable proper node similar to the one illustrated in Fig. 9.1.5.

14. Solution $x(t) = x_0e^{2t}$, $y(t) = y_0e^{-2t}$. The origin is an unstable saddle point.

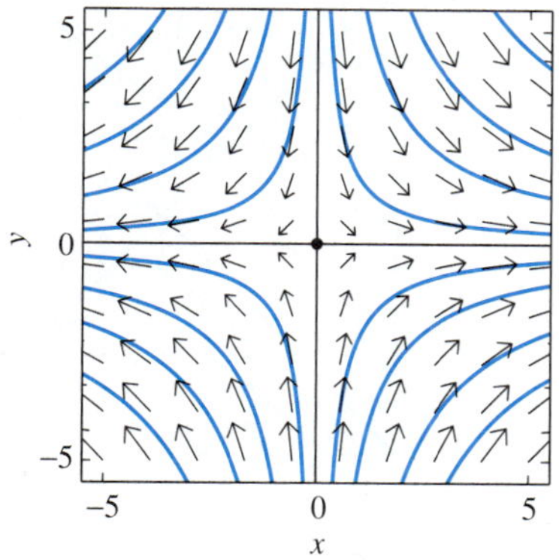

15. Solution $x(t) = x_0e^{-2t}$, $y(t) = y_0e^{-t}$. The origin is a stable node.

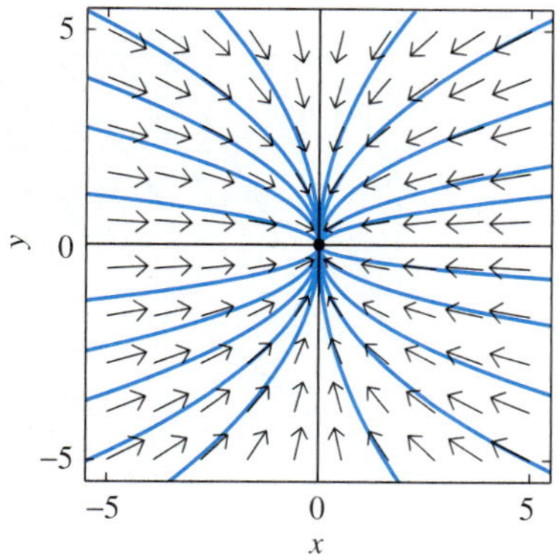

16. Solution $x(t) = x_0e^{t}$, $y(t) = y_0e^{3t}$. The origin is an unstable improper node.

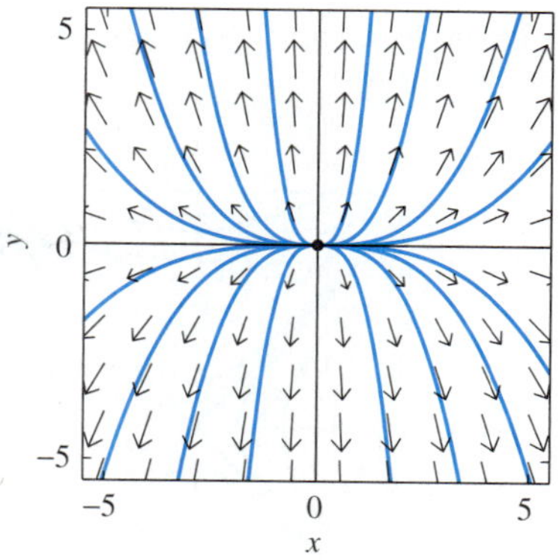

17. Solution $x(t) = A\cos t + B\sin t$, $y(t) = B\cos t - A\sin t$. The origin is a stable center.

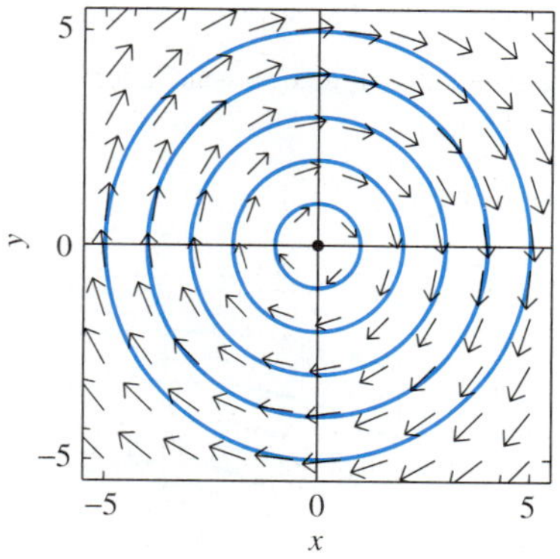

18. Solution $x(t) = A\cos 2t + B\sin 2t$, $y(t) = -2B\cos 2t + 2A\sin 2t$. The origin is a stable center.

19. Solution $x(t) = A\cos 2t + B\sin 2t$, $y(t) = B\cos 2t - A\sin 2t$. The origin is a stable center.

20. Solution $x(t) = e^{-2t}(A\cos t + B\sin t)$, $y(t) = e^{-2t}[(-2A + B)\cos t - (A + 2B)\sin t]$. The origin is a stable spiral point.

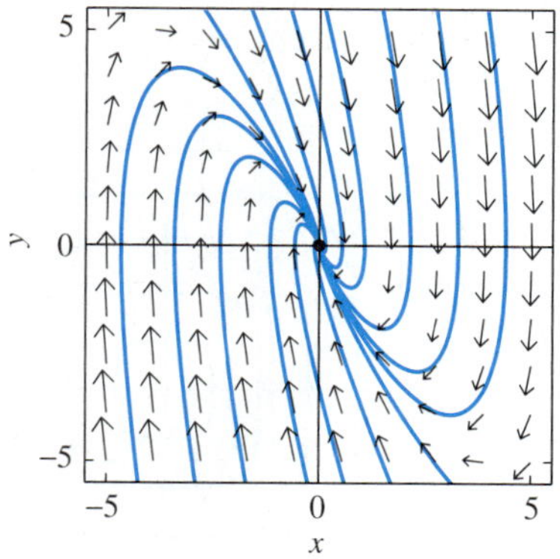

23. The origin and the circles $x^2 + y^2 = C > 0$; the origin is a stable center.

24. The origin and the hyperbolas $y^2 - x^2 = C$; the origin is an unstable saddle point.

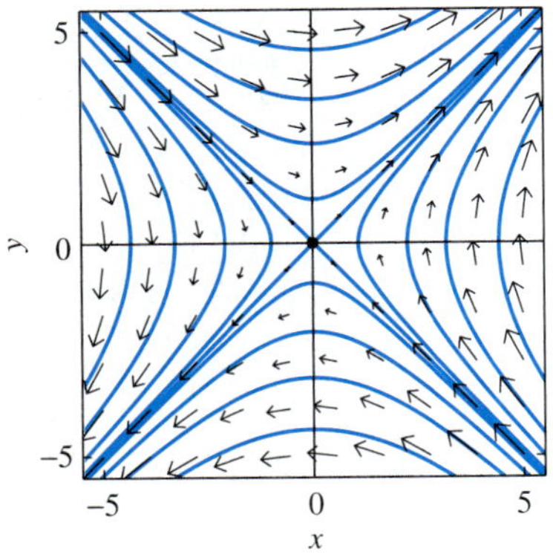

25. The origin and the ellipses $x^2 + 4y^2 = C > 0$; the origin is a stable center.

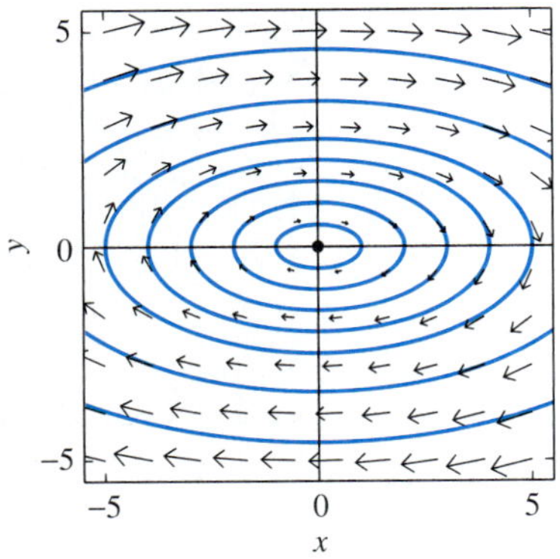

26. The origin and the ovals of the form $x^4 + y^4 = C > 0$; the origin is a stable center.

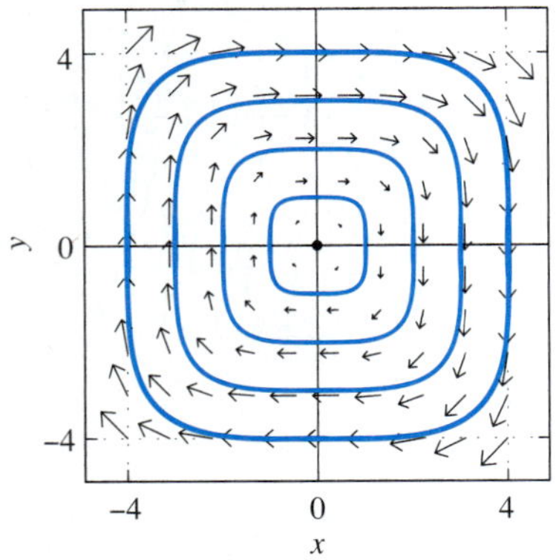

Section 9.2

1. Asymptotically stable node

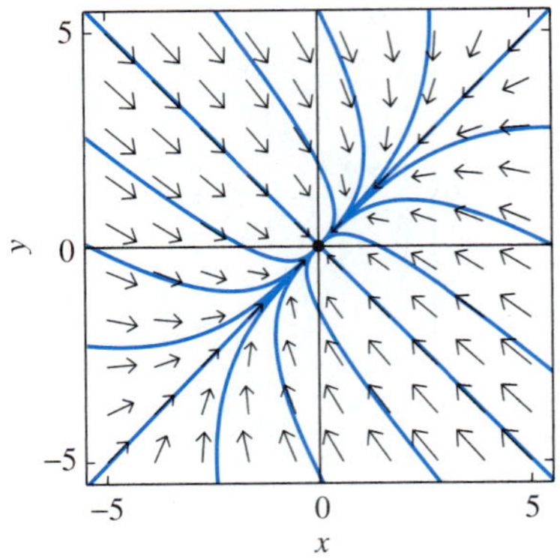

2. Unstable improper node

3. Unstable saddle point

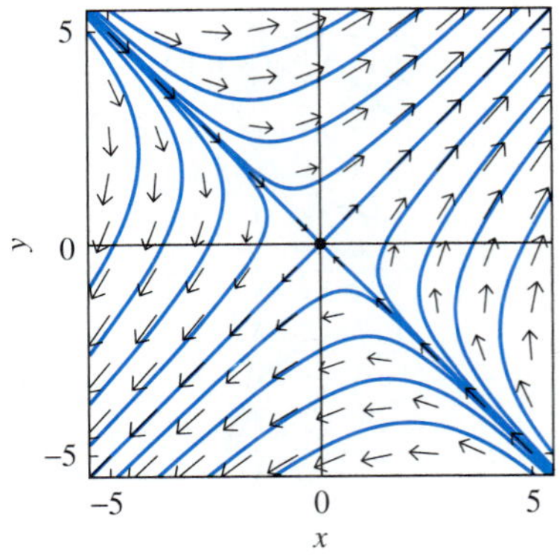

4. Unstable saddle point

5. Asymptotically stable node

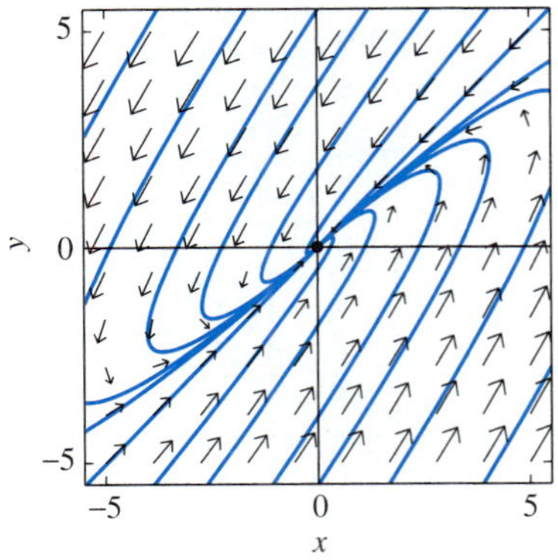

6. Unstable node

7. Unstable spiral point

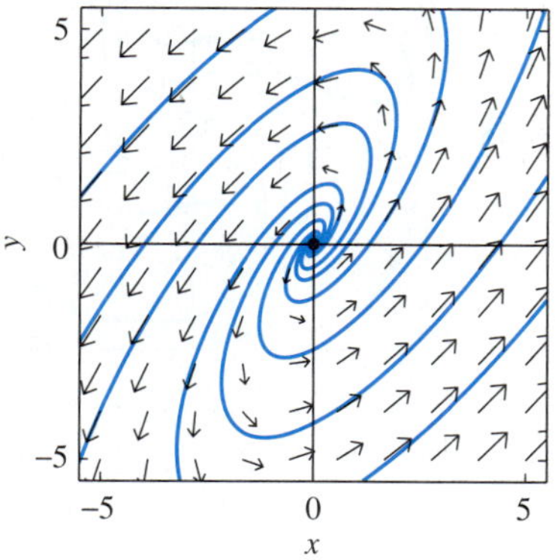

8. Asymptotically stable spiral point

9. Stable, but not asymptotically stable, center

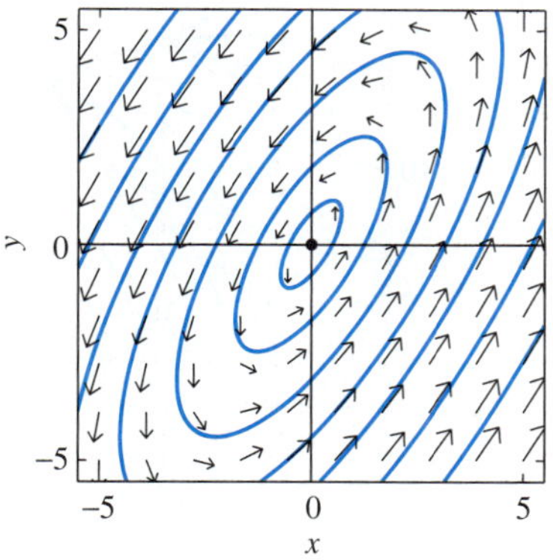

10. Stable, but not asymptotically stable, center

11. Asymptotically stable node: (2, 1)

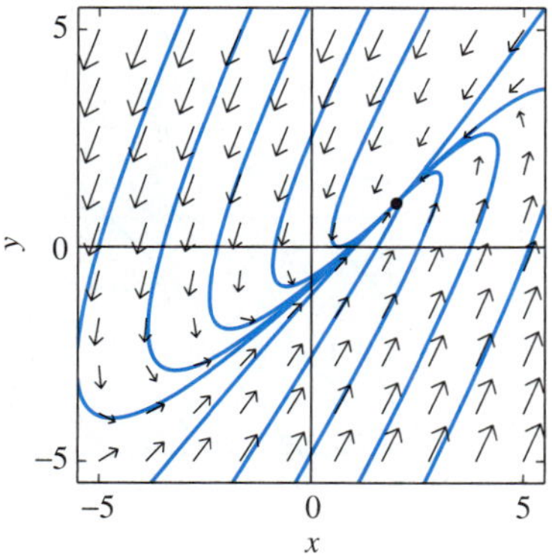

12. Unstable improper node: (2, −3)

13. Unstable saddle point: (2, 2)

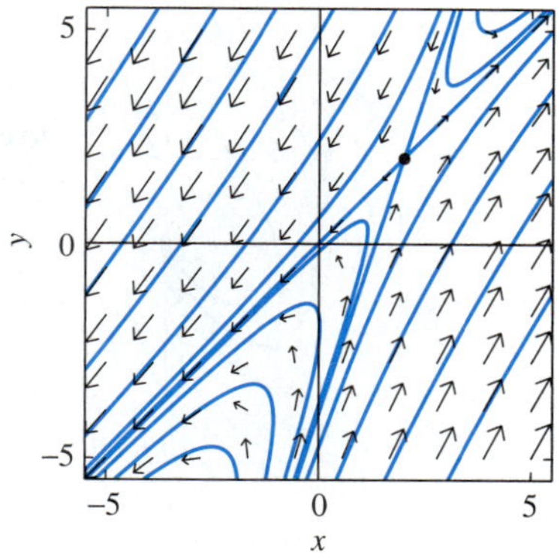

14. Unstable saddle point: (3, 4)

15. Asymptotically stable spiral point: (1, 1)

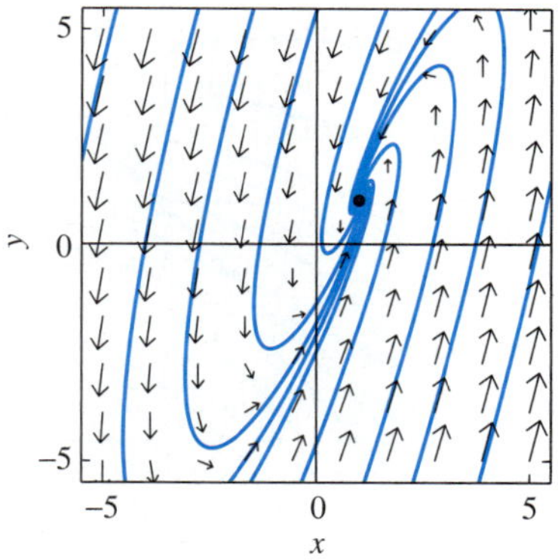

16. Unstable spiral point: (3, 2)

17. Stable center: $\left(\frac{5}{2}, -\frac{1}{2}\right)$

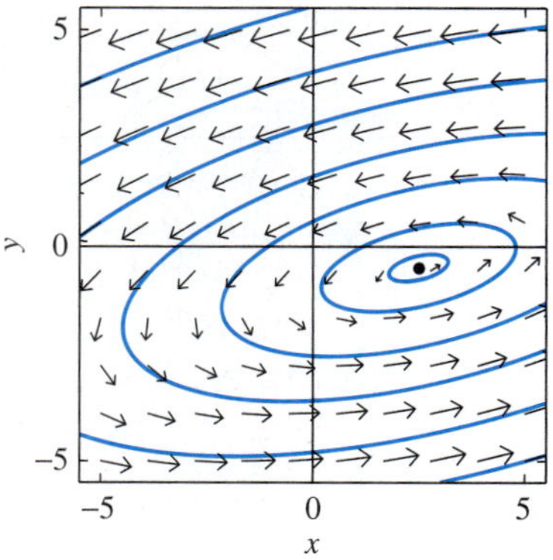

18. Stable, but not asymptotically stable, center: (−2, −1)

19. (0, 0) is a stable node. Also, there is a saddle point at (0.67, 0.40).

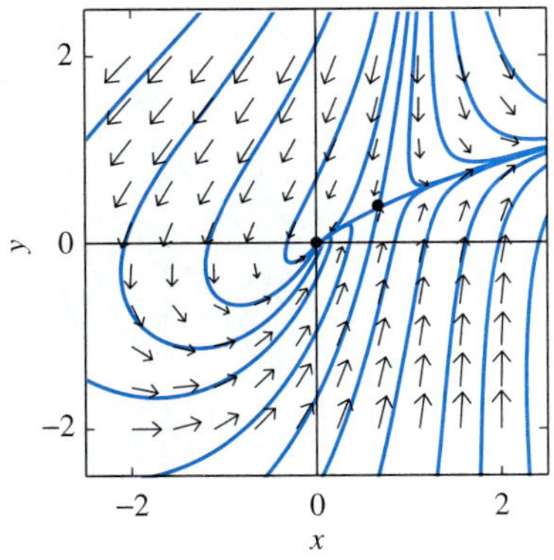

20. (0, 0) is an unstable node. Also, there is a saddle point at (−1, −1) and a spiral sink at (−2.30, −1.70).

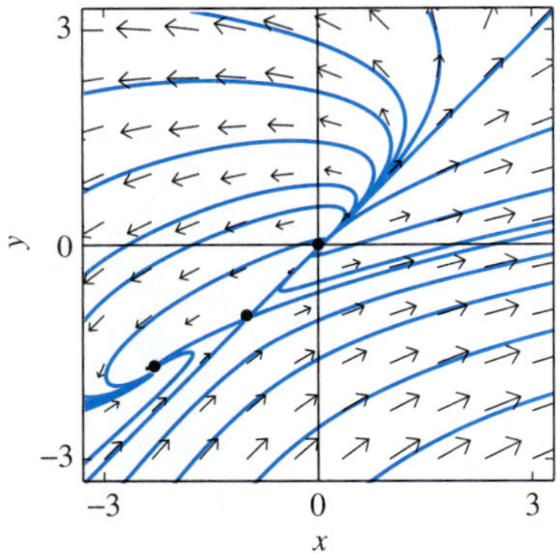

21. (0, 0) is an unstable saddle point. Also, there is a spiral sink at (−0.51, −2.12).

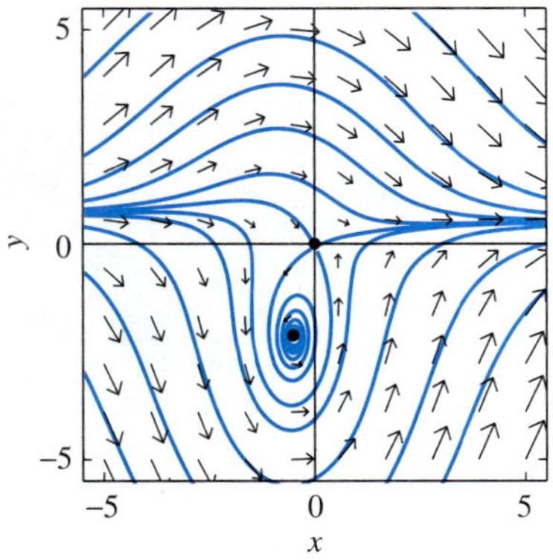

22. (0, 0) is an unstable saddle point. Also, there are nodal sinks at (±0.82, ±5.06) and nodal sources at (±3.65, ∓0.59).

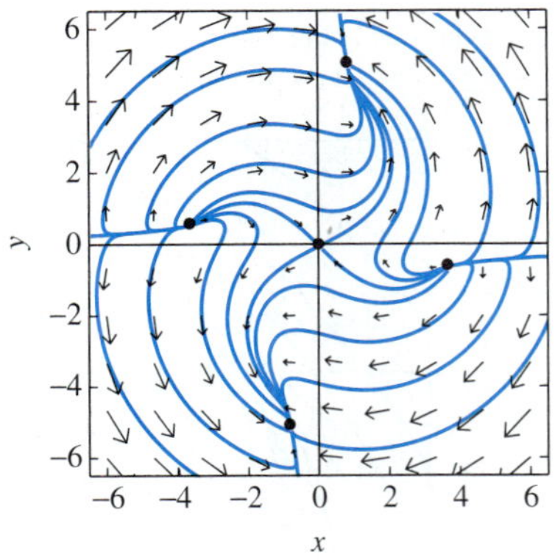

23. (0, 0) is a spiral sink. Also, there is a saddle point at (−1.08, −0.68).

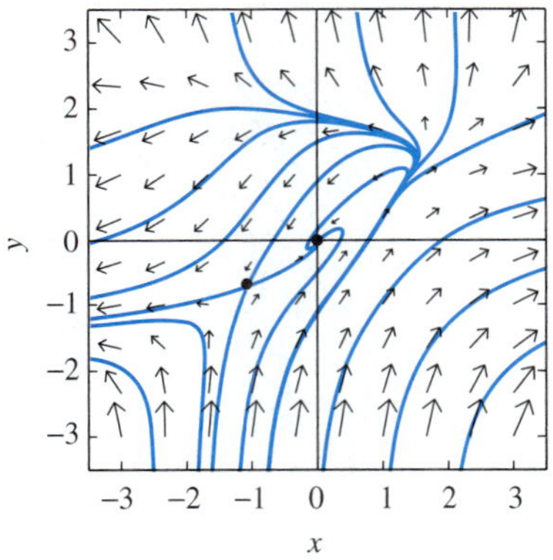

24. (0, 0) is an spiral source. No other critical points are visible.

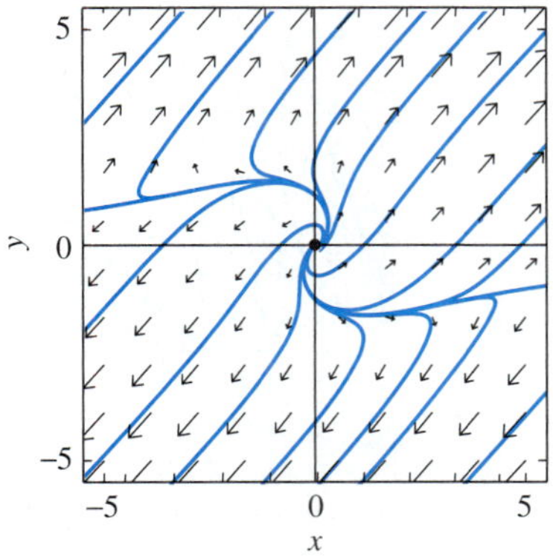

25. Theorem 2 implies only that (0, 0) is a stable sink—either a node or a spiral point. The phase portrait for $-5 \leq x, y \leq 5$ shows also a saddle point at (0.74, −3.28) and spiral sink at (2.47, −0.46). The origin looks like nodal sink in a second phase portrait for $-0.2 \leq x, y \leq 0.2$, which also reveals a second saddle point at (0.12, 0.07).

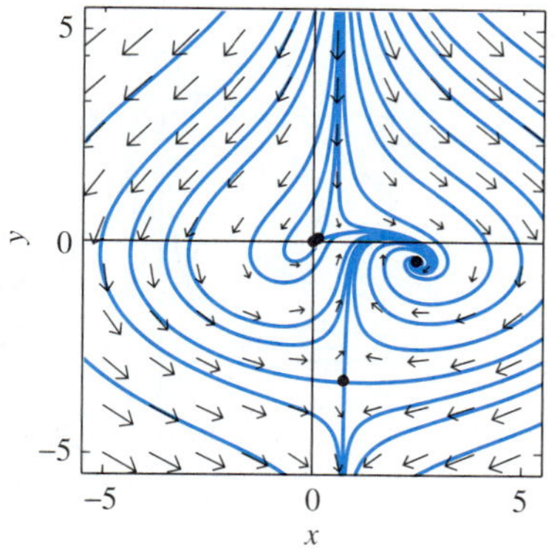

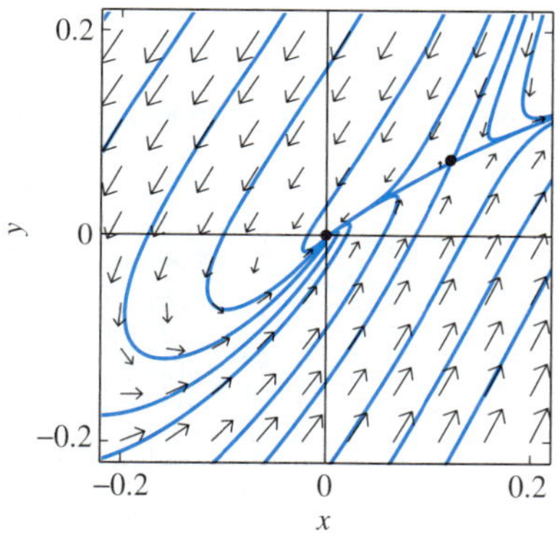

26. Theorem 2 implies only that (0, 0) is an unstable source. The phase portrait for $-3 \leq x, y \leq 3$ shows also saddle points at (0.20, 0.25) and (−0.23, −1.50), as well as a nodal sink at (2.36, 0.58).

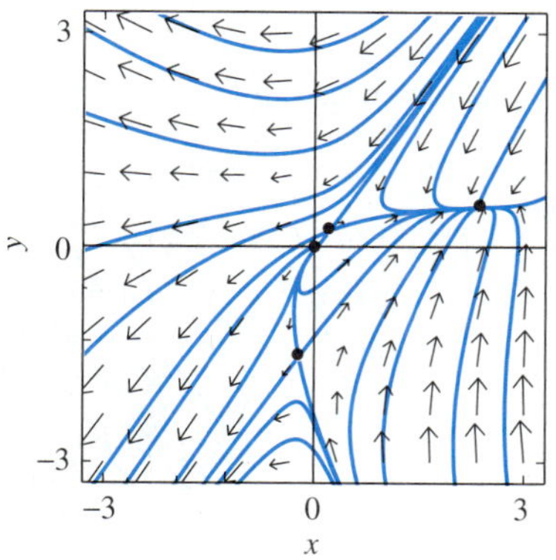

27. Theorem 2 implies only that (0, 0) is a center or a spiral point, but does not establish its stability. The phase portrait for $-2 \leq x, y \leq 2$ shows also saddle points at (−0.25, −0.51) and (−1.56, 1.64), plus a nodal sink at (−1.07, −1.20). The origin looks like a likely center in a second phase portrait for $-0.6 \leq x, y \leq 0.6$.

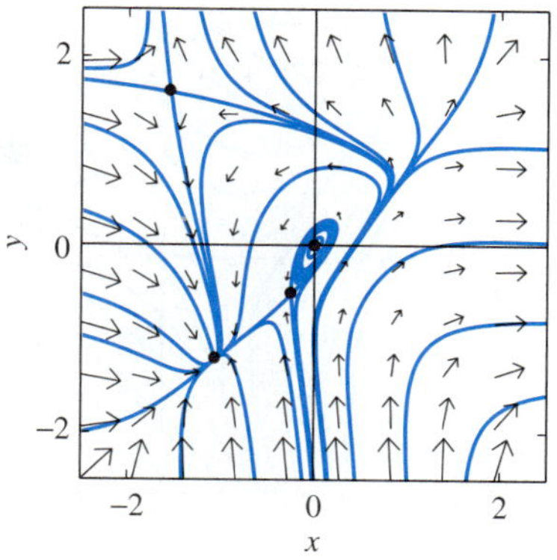

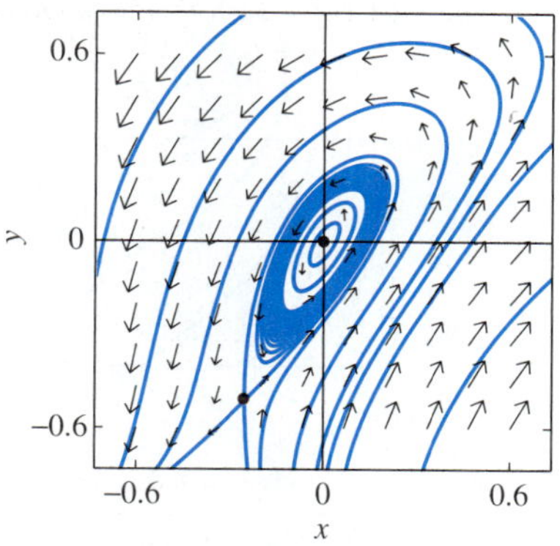

28. Theorem 2 implies only that (0, 0) is a center or a spiral point, but does not establish its stability (though in the phase portrait it looks like a likely center). The phase portrait for $-0.25 \leq x \leq 0.25$, $-1 \leq y \leq 1$ also shows saddle points at (0.13, 0.63) and (−0.12, −0.47).

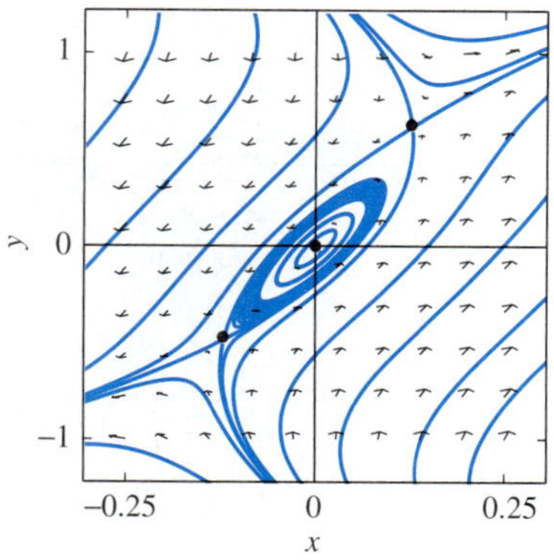

29. There is a saddle point at (0, 0). The other critical point (1, 1) is indeterminate, but looks like a center in the phase portrait.

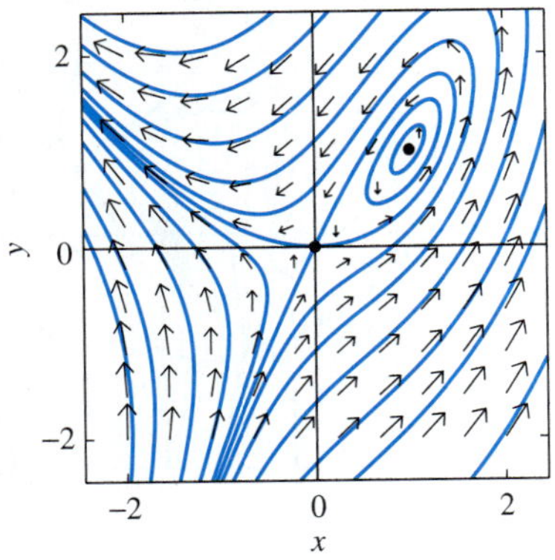

30. There is a saddle point at (1, 1) and a spiral sink at (−1, 1).

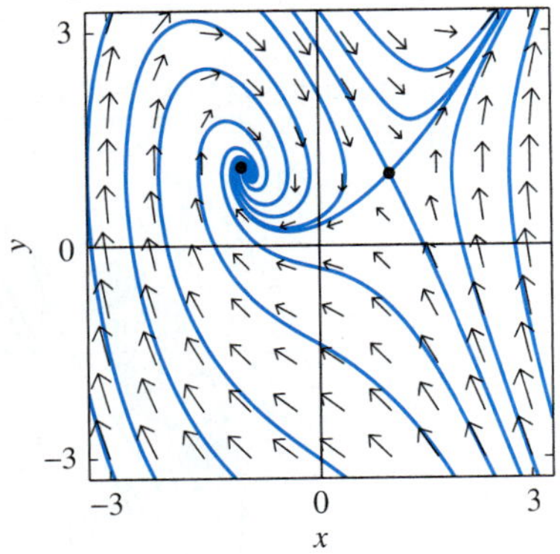

31. There is a saddle point at (1, 1) and a spiral sink at (−1, −1).

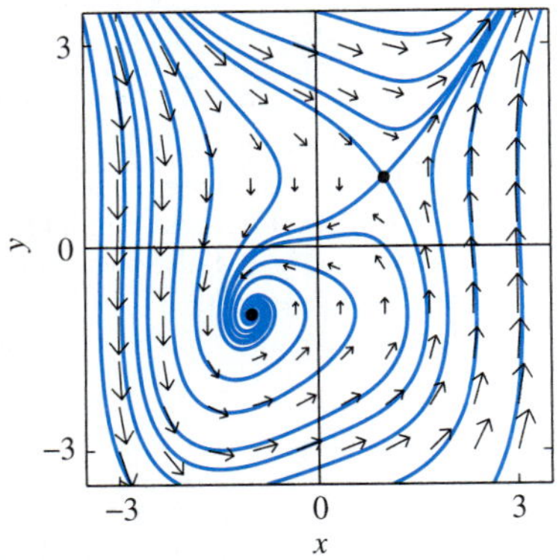

32. There is a saddle point at (2, 1) and a spiral sink at (−2, −1).

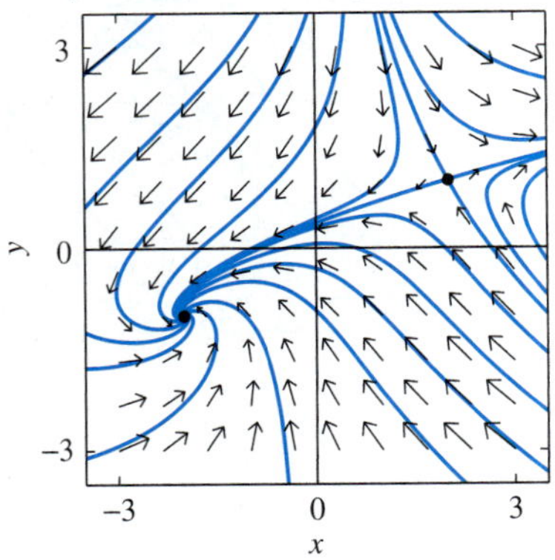

37. Note that the differential equation is homogeneous.

Section 9.3

1. Linearization at (0, 0): $x' = 200x$, $y' = -150y$; phase plane portrait:

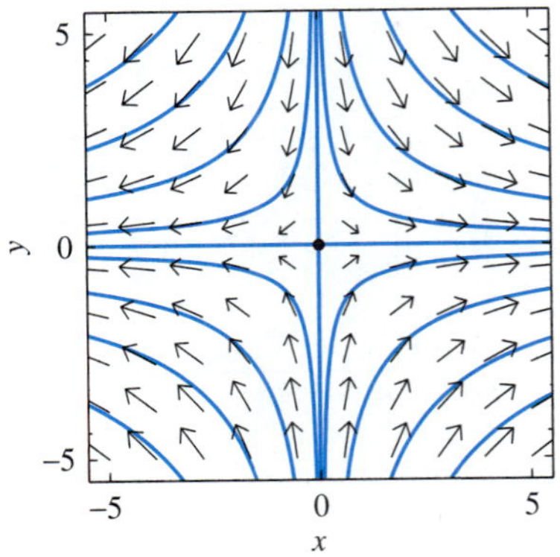

Linearization at (75, 50): $u' = -300v$, $v' = 100u$; phase plane portrait:

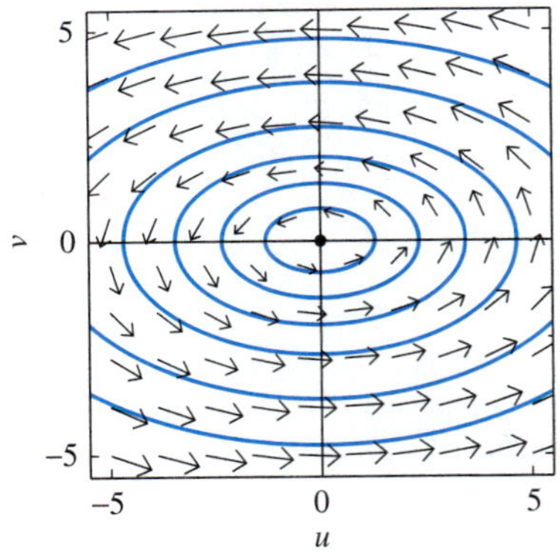

5. The characteristic equation is $\lambda^2 + 45\lambda + 126 = 0$.

7. The characteristic equation is $(-24 - \lambda)^2 - 2 \cdot (18)^2 = 0$. Phase plane portrait:

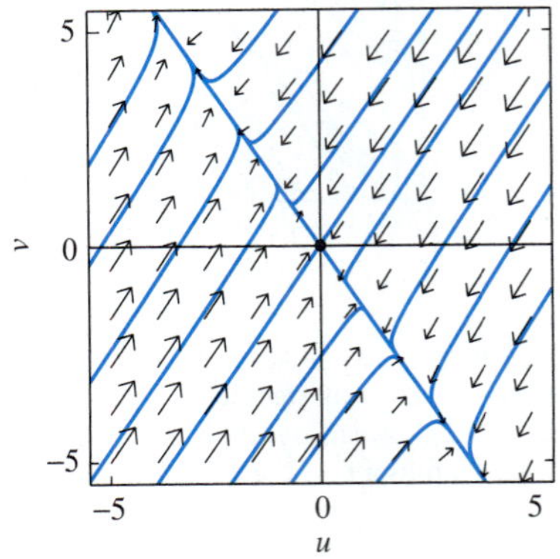

Phase plane portrait for the nonlinear system in Problems 4–7:

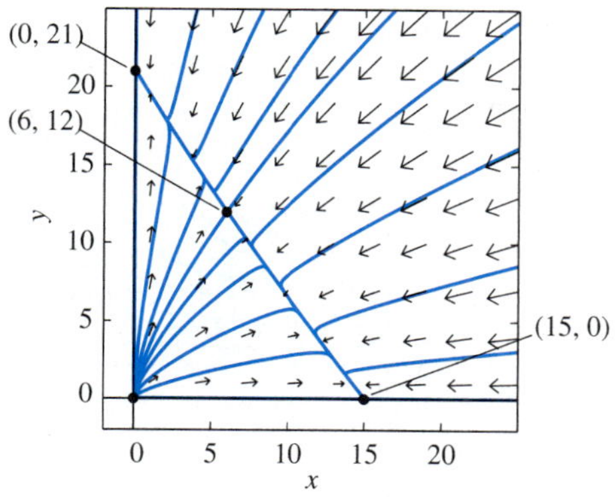

9. The characteristic equation is $\lambda^2 + 58\lambda - 120 = 0$.

10. The characteristic equation is $(\lambda + 36)(\lambda + 18) - 576 = 0$. Phase plane portrait:

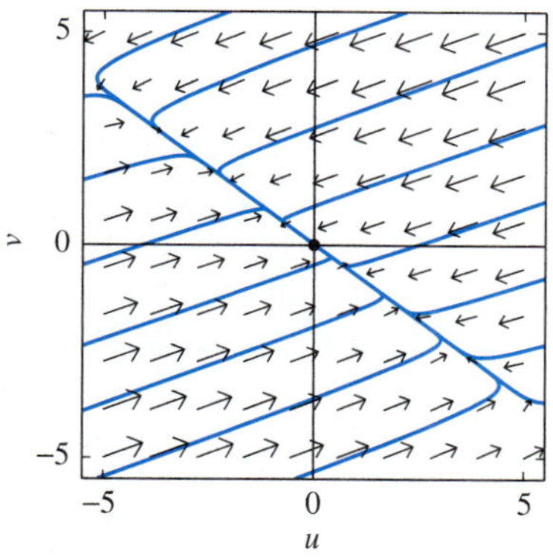

Phase plane portrait for the nonlinear system in Problems 8 through 10:

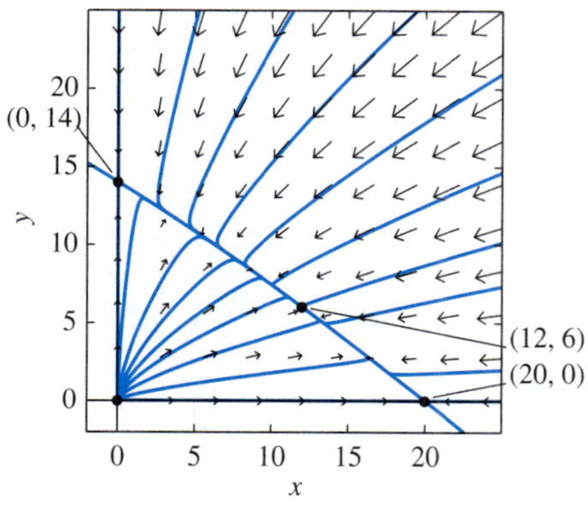

12. The characteristic equation is $\lambda^2 + 2\lambda - 15 = 0$.

13. The characteristic equation is $\lambda^2 + 2\lambda + 6 = 0$. Phase plane portrait:

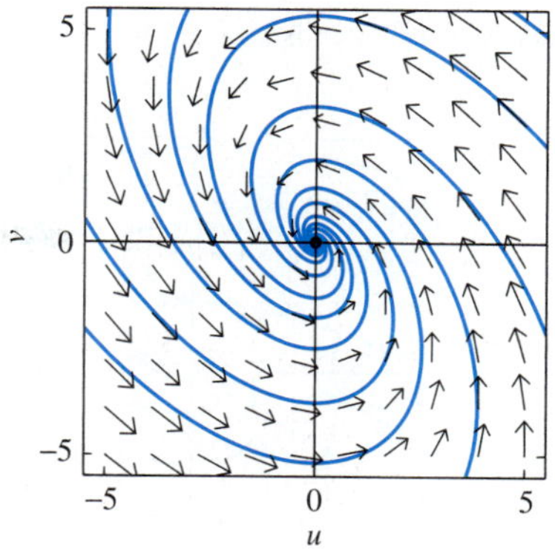

15. The characteristic equation is $\lambda^2 + 2\lambda - 24 = 0$.

17. The characteristic equation is $\lambda^2 - 4\lambda + 6 = 0$. Phase plane portrait:

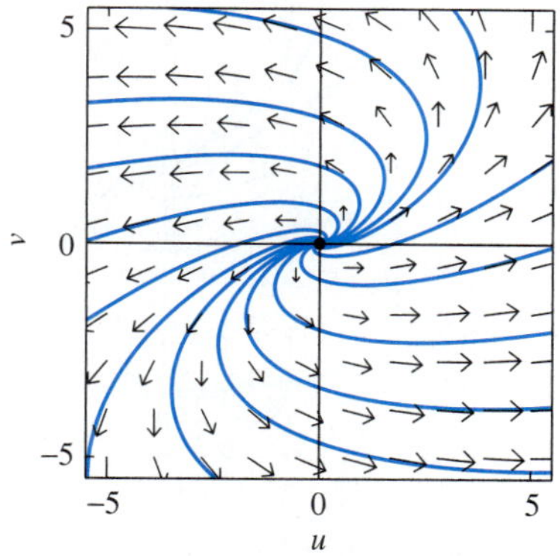

19. The characteristic equation is $\lambda^2 + 10 = 0$. Phase plane portrait:

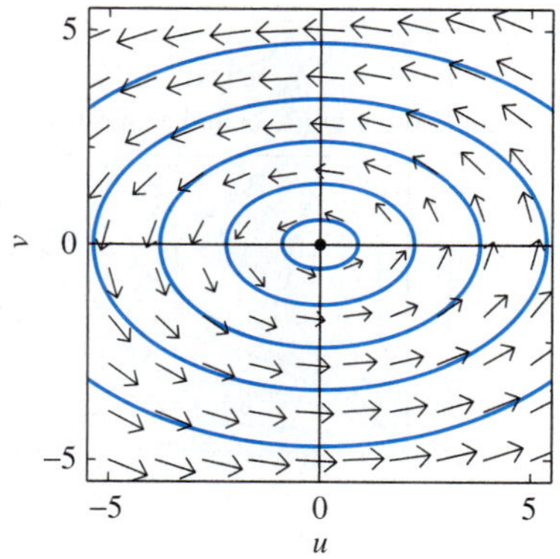

21. The characteristic equation is $\lambda^2 - \lambda - 6 = 0$.

22. The characteristic equation is $\lambda^2 - 5\lambda + 10 = 0$. Phase plane portrait:

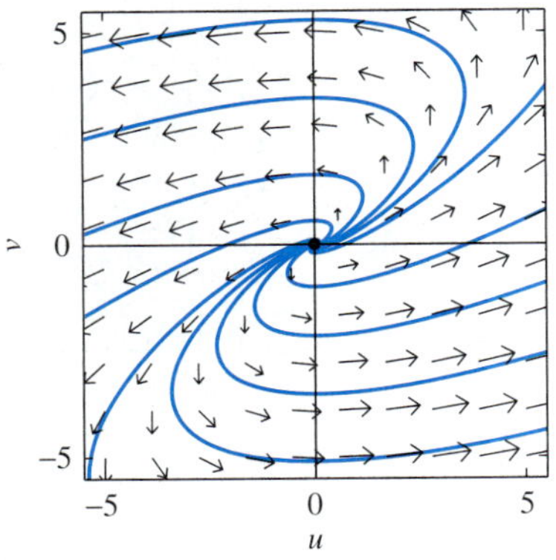

24. The characteristic equation is $\lambda^2 + 5\lambda - 14 = 0$.

25. The characteristic equation is $\lambda^2 + 5\lambda + 10 = 0$. Phase plane portrait:

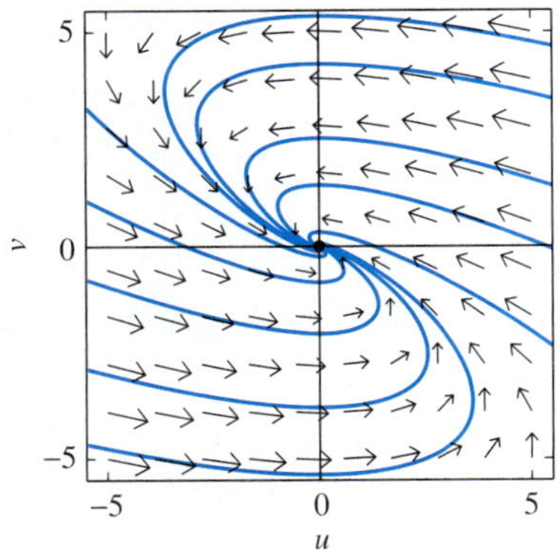

26. Naturally growing populations in competition
Critical points: nodal source (0, 0) and saddle point (3, 2)
Nonzero coexisting populations $x(t) \equiv 3$, $y(t) \equiv 2$

27. Naturally declining populations in cooperation
Critical points: nodal sink (0, 0) and saddle point (3, 2)
Nonzero coexisting populations $x(t) \equiv 3$, $y(t) \equiv 2$

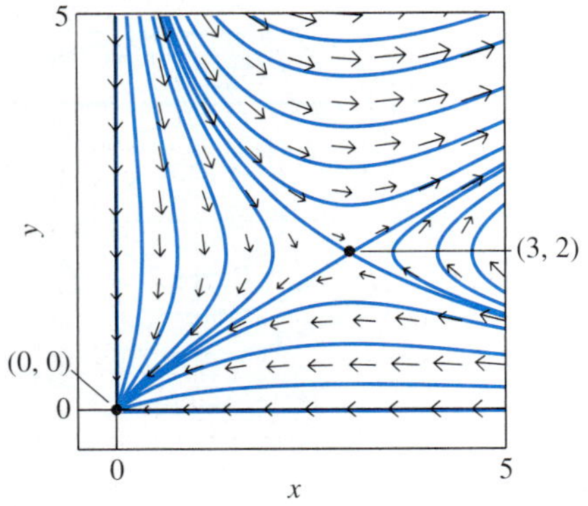

28. Naturally declining predator, naturally growing prey population
Critical points: saddle point (0, 0) and apparent stable center (4, 8)
Nonzero coexisting populations $x(t) \equiv 4$, $y(t) \equiv 8$

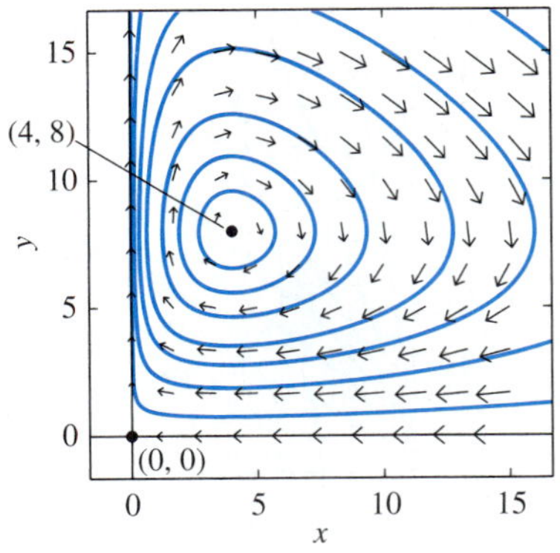

29. Logistic and naturally growing populations in competition
Critical points: nodal source (0, 0), nodal sink (3, 0), and saddle point (2, 2)
Nonzero coexisting populations $x(t) \equiv 2$, $y(t) \equiv 2$

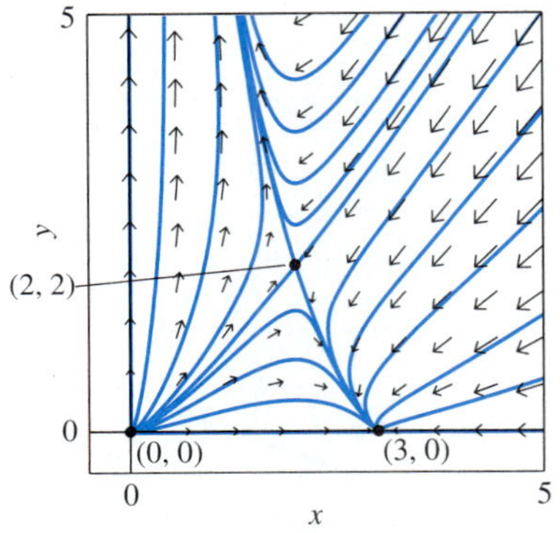

30. Logistic and naturally declining populations in cooperation
Critical points: saddle point (0, 0), nodal sink (3, 0), and saddle point (5, 4)
Nonzero coexisting populations $x(t) \equiv 5$, $y(t) \equiv 4$

31. Logistic prey, naturally declining predator population
Critical points: saddle points (0, 0) and (3, 0), spiral sink (2, 4)
Nonzero coexisting populations $x(t) \equiv 2$, $y(t) \equiv 4$

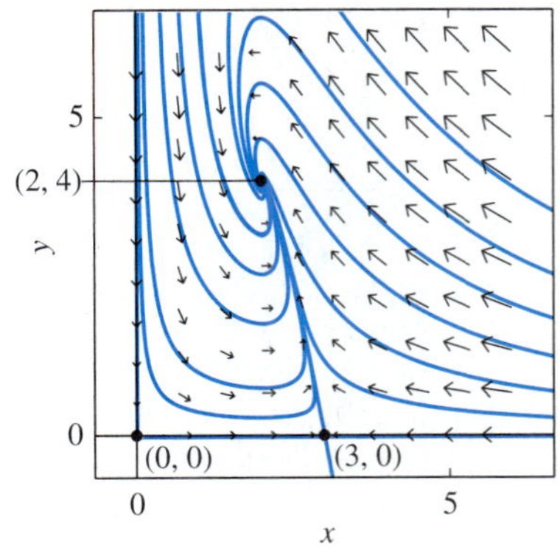

32. Logistic populations in cooperation
Critical points: nodal source (0, 0), saddle points (10, 0) and (0, 20), nodal sink (30, 60)
Nonzero coexisting populations $x(t) \equiv 30$, $y(t) \equiv 60$

33. Logistic prey and predator populations
Critical points: nodal source $(0, 0)$, saddle points $(15, 0)$ and $(0, 20)$, nodal sink $(4, 22)$
Nonzero coexisting populations $x(t) \equiv 4$, $y(t) \equiv 22$

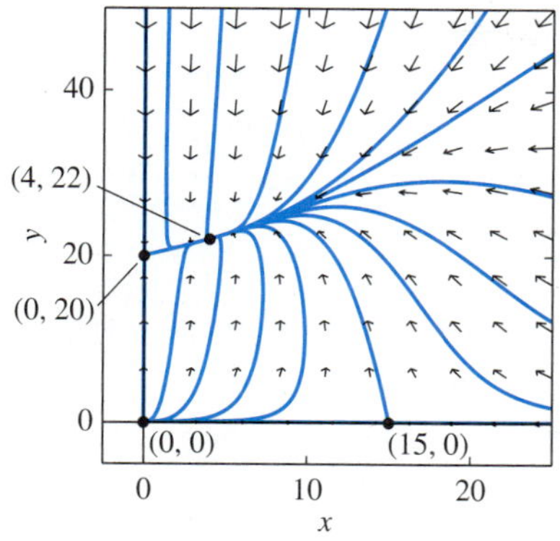

34. Logistic prey and predator populations
Critical points: nodal source $(0, 0)$, saddle points $(15, 0)$ and $(0, 5)$, spiral sink $(10, 10)$
Nonzero coexisting populations $x(t) \equiv 10$, $y(t) \equiv 10$

Section 9.4

1. Eigenvalues: $-2, -3$; stable node

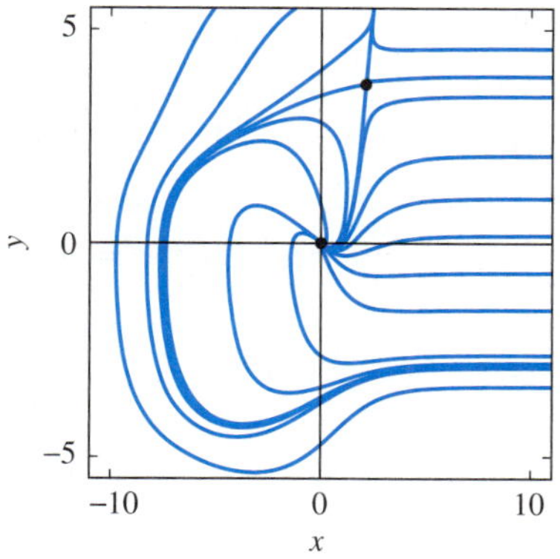

2. Eigenvalues: $1, 3$; unstable node

3. Eigenvalues: $-3, 5$; unstable saddle point

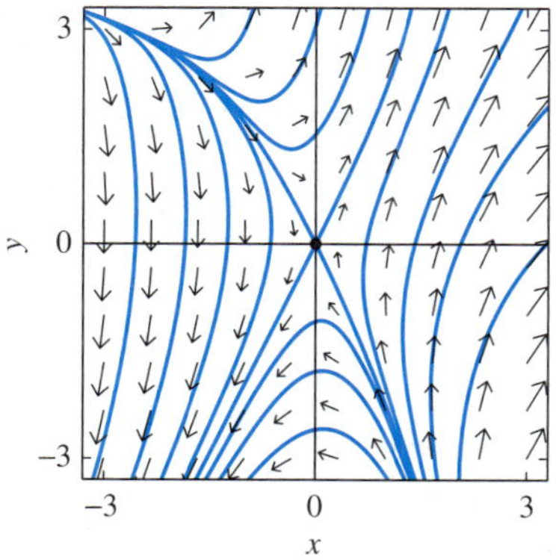

4. Eigenvalues: $-1 \pm 2i$; stable spiral point

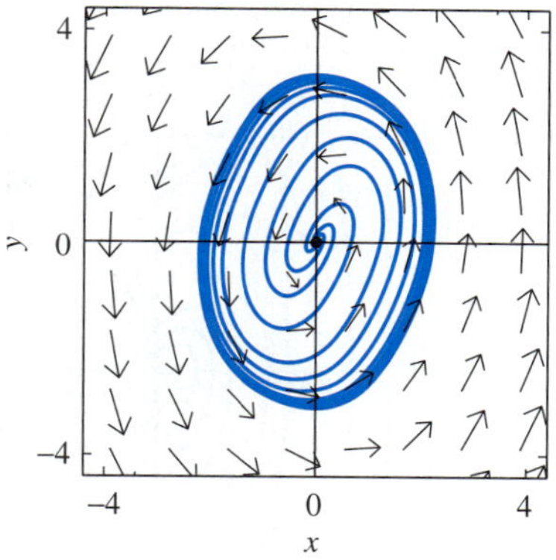

5. Critical points: $(0, n\pi)$, where n is an integer; an unstable saddle point if n is even, a stable spiral point if n is odd

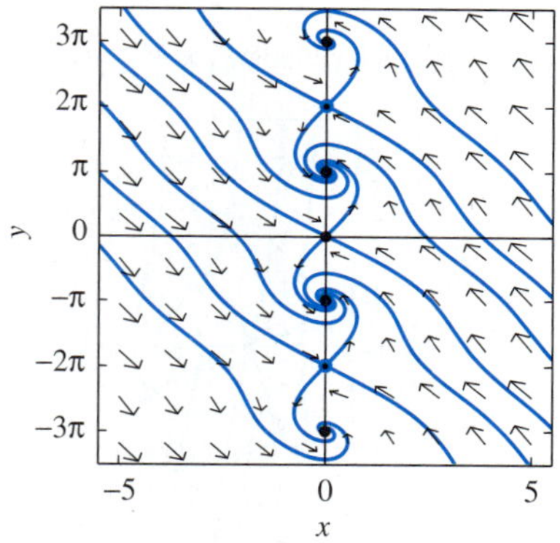

6. Critical points: $(n, 0)$, where n is an integer; an unstable saddle point if n is even, a stable spiral point if n is odd

7. Critical points: $(n\pi, n\pi)$, where n is an integer; an unstable saddle point if n is even, a stable spiral point if n is odd

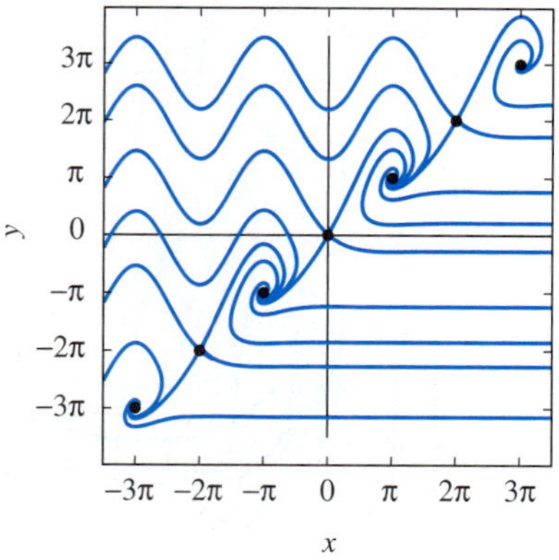

8. Critical points: $(n\pi, 0)$, where n is an integer; an unstable node if n is even, an unstable saddle point if n is odd

9. If n is odd, then $(n\pi, 0)$ is an unstable saddle point.

10. If n is odd, then $(n\pi, 0)$ is a stable node.

11. $(n\pi, 0)$ is a stable spiral point.

12. Unstable saddle points at $(2, 0)$ and $(-2, 0)$, a stable center at $(0, 0)$

13. Unstable saddle points at $(2, 0)$ and $(-2, 0)$, a stable spiral point at $(0, 0)$

14. Stable centers at $(2, 0)$ and $(-2, 0)$, an unstable saddle point at $(0, 0)$

15. A stable center at (0, 0) and an unstable saddle point at (4, 0)

16. Stable centers at (2, 0), (0, 0), and (−2, 0), unstable saddle points at (1, 0) and (−1, 0)

17. (0, 0) is a spiral sink.

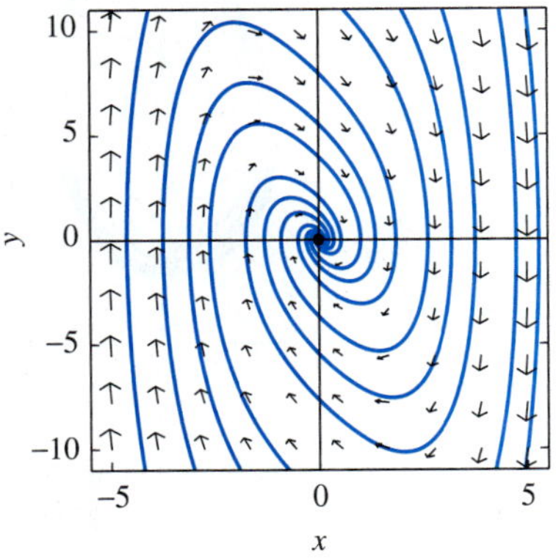

18. (0, 0) is a spiral sink; the points (±2, 0) are saddle points.

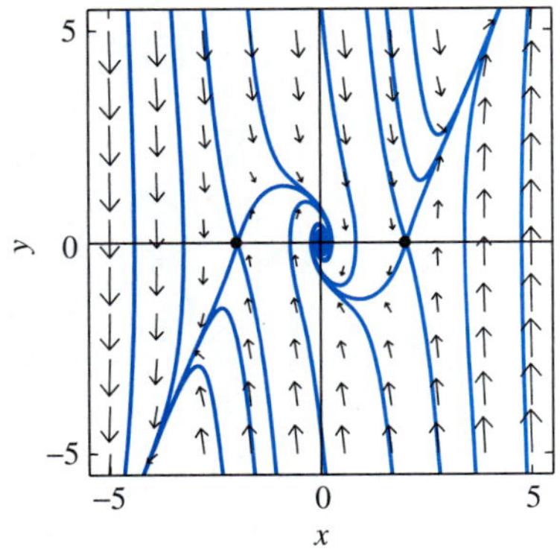

19. (0, 0) is a spiral sink.

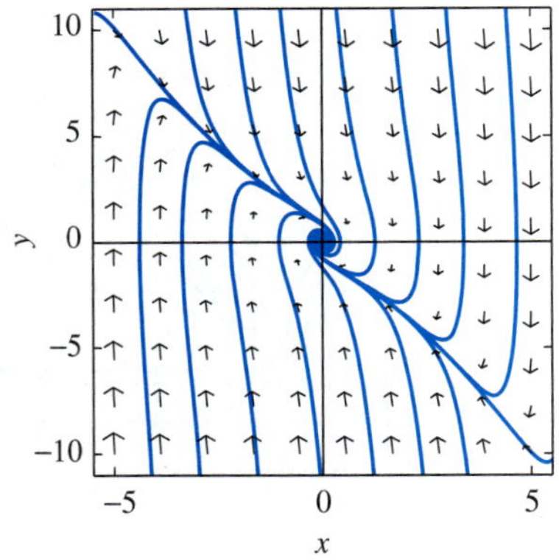

20. $(n\pi, 0)$ is a spiral sink if n is even, a saddle point if n is odd.

Chapter 10

Section 10.1

1. $1/s^2,\ s > 0$
2. $2/s^3,\ s > 0$
3. $e/(s-3),\ s > 3$
4. $s/(s^2+1),\ s > 0$
5. $1/(s^2-1),\ s > 1$
6. $\frac{1}{2}[1/s - s/(s^2+4)],\ s > 0$
7. $(1-e^{-s})/s,\ s > 0$
8. $(e^{-s} - e^{-2s})/s,\ s > 0$
9. $(1 - e^{-s} - se^{-s})/s^2,\ s > 0$
10. $(s - 1 + e^{-s})/s^2,\ s > 0$
11. $\frac{1}{2}\sqrt{\pi}s^{-3/2} + 3s^{-2},\ s > 0$
12. $(45\pi - 192s^{3/2})/(8s^{7/2}),\ s > 0$
13. $s^{-2} - 2(s-3)^{-1},\ s > 3$
14. $3\sqrt{\pi}/(4s^{5/2}) + 1/(s+10),\ s > 0$
15. $s^{-1} + s(s^2-25)^{-1},\ s > 5$
16. $(s+2)/(s^2+4),\ s > 0$
17. $\cos^2 2t = \frac{1}{2}(1+\cos 4t);\ \frac{1}{2}\left[s^{-1} + s/(s^2+16)\right],\ s > 0$
18. $3/(s^2+36),\ s > 0$
19. $s^{-1} + 3s^{-2} + 6s^{-3} + 6s^{-4},\ s > 0$
20. $1/(s-1)^2,\ s > 1$
21. $(s^2-4)/(s^2+4)^2,\ s > 0$
22. $\frac{1}{2}\left[s/(s^2-36) - s^{-1}\right]$
23. $\frac{1}{2}t^3$
24. $2\sqrt{t/\pi}$
25. $1 - \frac{8}{3}t^{3/2}\pi^{-1/2}$
26. e^{-5t}
27. $3e^{4t}$
28. $3\cos 2t + \frac{1}{2}\sin 2t$
29. $\frac{5}{3}\sin 3t - 3\cos 3t$
30. $-\cosh 2t - \frac{9}{2}\sinh 2t$
31. $\frac{3}{5}\sinh 5t - 10\cosh 5t$
32. $2u(t-3)$
37. $f(t) = 1 - u(t-a)$. Your figure should indicate that the graph of f contains the point $(a, 0)$, but not the point $(a, 1)$.
38. $f(t) = u(t-a) - u(t-b)$. Your figure should indicate that the graph of f contains the points $(a, 1)$ and $(b, 0)$, but not the points $(a, 0)$ and $(b, 1)$.
39. Figure 10.2.8 shows the graph of the unit staircase function.

Section 10.2

1. $x(t) = 5\cos 2t$
2. $x(t) = 3\cos 3t + \frac{4}{3}\sin 3t$
3. $x(t) = \frac{2}{3}(e^{2t} - e^{-t})$
4. $x(t) = \frac{1}{2}(7e^{-3t} - 3e^{-5t})$
5. $x(t) = \frac{1}{3}(2\sin t - \sin 2t)$
6. $x(t) = \frac{1}{3}(\cos t - \cos 2t)$
7. $x(t) = \frac{1}{8}(9\cos t - \cos 3t)$
8. $x(t) = \frac{1}{9}(1 - \cos 3t)$
9. $x(t) = \frac{1}{6}(2 - 3e^{-t} + e^{-3t})$
10. $x(t) = \frac{1}{4}(2t - 3 + 12e^{-t} - 9e^{-2t})$
11. $x(t) = 1,\ y(t) = -2$
12. $x(t) = \frac{2}{9}(e^{2t} - e^{-t} - 3te^{-t}),\ y(t) = \frac{1}{9}(e^{2t} - e^{-t} + 6te^{-t})$
13. $x(t) = -\left(2/\sqrt{3}\right)\sinh\left(t/\sqrt{3}\right)$,
$y(t) = \cosh\left(t/\sqrt{3}\right) + \left(1/\sqrt{3}\right)\sinh\left(t/\sqrt{3}\right)$
14. $x(t) = \frac{1}{4}(2t - 3\sin 2t),\ y(t) = -\frac{1}{8}(2t + 3\sin 2t)$
15. $x(t) = \frac{1}{3}\left(2 + e^{-3t/2}[\cos(rt/2) + r\sin(rt/2)]\right)$,
$y(t) = \frac{1}{21}\left(28 - 9e^{t} + 2e^{-3t/2}[\cos(rt/2) + 4r\sin(rt/2)]\right)$, where $r = \sqrt{3}$
16. $x(t) = \cos t + \sin t,\ y(t) = e^t - \cos t,\ z(t) = -2\sin t$
17. $f(t) = \frac{1}{3}(e^{3t} - 1)$
18. $f(t) = \frac{3}{5}(1 - e^{-5t})$
19. $f(t) = \frac{1}{4}(1 - \cos 2t) = \frac{1}{2}\sin^2 t$
20. $f(t) = \frac{1}{9}(6\sin 3t - \cos 3t + 1)$
21. $f(t) = t - \sin t$
22. $f(t) = \frac{1}{9}(-1 + \cosh 3t)$
23. $f(t) = -t + \sinh t$
24. $f(t) = \frac{1}{2}(e^{-2t} - 2e^{-t} + 1)$

Section 10.3

1. $24/(s-\pi)^5$
2. $\frac{3}{4}\sqrt{\pi}\,(s+4)^{-5/2}$
3. $3\pi/[(s+2)^2+9\pi^2]$
4. $\sqrt{2}\,(2s+5)/(4s^2+4s+17)$
5. $\frac{3}{2}e^{2t}$
6. $(t-t^2)e^{-t}$
7. te^{-2t}
8. $e^{-2t}\cos t$
9. $e^{3t}\left(3\cos 4t+\frac{7}{2}\sin 4t\right)$
10. $\frac{1}{36}e^{2t/3}\left(8\cos\frac{4}{3}t-5\sin\frac{4}{3}t\right)$
11. $\frac{1}{2}\sinh 2t$
12. $2+3e^{3t}$
13. $3e^{-2t}-5e^{-5t}$
14. $2+e^{2t}-3e^{-t}$
15. $\frac{1}{25}(e^{5t}-1-5t)$
16. $\frac{1}{125}[e^{2t}(5t-2)+e^{-3t}(5t+2)]$
17. $\frac{1}{16}(\sinh 2t-\sin 2t)$
18. $e^{4t}\left(1+12t+24t^2+\frac{32}{3}t^3\right)$
19. $\frac{1}{3}(2\cos 2t+2\sin 2t-2\cos t-\sin t)$
20. $\frac{1}{32}[e^{2t}(2t-1)+e^{-2t}(2t+1)]$
21. $\frac{1}{2}e^{-t}(5\sin t-3t\cos t-2t\sin t)$
22. $\frac{1}{64}e^{t/2}[(4t+8)\cos t+(4-3t)\sin t]$
27. $\frac{1}{4}e^{-3t}(8\cos 4t+9\sin 4t)$
28. $\frac{1}{4}(1-2e^{2t}+e^{4t})$
29. $\frac{1}{8}(-6t+3\sinh 2t)$
30. $\frac{1}{10}[2e^{-t}-e^{-2t}(2\cos 2t+\sin 2t)]$
31. $\frac{1}{15}(6e^{2t}-5-e^{-3t})$
32. $\frac{1}{2}(\cosh t+\cos t)$
33. $x(t)=r(\cosh rt\sin rt-\sinh rt\cos rt)$ where $r=1/\sqrt{2}$
34. $\frac{1}{2}\sin 2t+\frac{1}{3}\sin 3t$
35. $\frac{1}{16}(\sin 2t-2t\cos 2t)$
36. $\frac{1}{50}[2e^{2t}+(10t-2)\cos t-(5t+14)\sin t]$
37. $\frac{1}{50}[(5t-1)e^{-t}+e^{-2t}(\cos 3t+32\sin 3t)]$
38. $\frac{1}{510}e^{-3t}(489\cos 3t+307\sin 3t)+\frac{1}{170}(7\cos 2t+6\sin 2t)$
39.

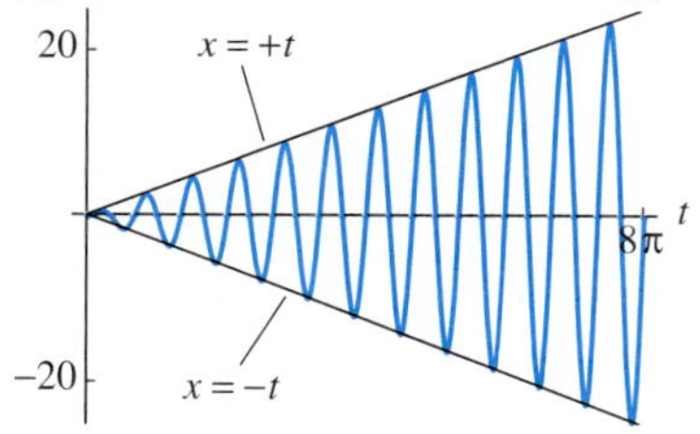

Section 10.4

1. $\frac{1}{2}t^2$
2. $(e^{at}-at-1)/a^2$
3. $\frac{1}{2}(\sin t-t\cos t)$
4. $2(t-\sin t)$
5. te^{at}
6. $(e^{at}-e^{bt})/(a-b)$
7. $\frac{1}{3}(e^{3t}-1)$
8. $\frac{1}{4}(1-\cos 2t)$
9. $\frac{1}{54}(\sin 3t-3t\cos 3t)$
10. $(kt-\sin kt)/k^3$
11. $\frac{1}{4}(\sin 2t+2t\cos 2t)$
12. $\frac{1}{5}[1-e^{-2t}(\cos t+2\sin t)]$
13. $\frac{1}{10}(3e^{3t}-3\cos t+\sin t)$
14. $\frac{1}{3}(\cos t-\cos 2t)$
15. $6s/(s^2+9)^2,\ s>0$
16. $(2s^3-24s)/(s^2+4)^3,\ s>0$
17. $(s^2-4s-5)/(s^2-4s+13)^2,\ s>0$
18. $\dfrac{2(3s^2+6s+7)}{(s+1)^2(s^2+2s+5)^2},\ s>0$
19. $\frac{1}{2}\pi-\arctan s=\arctan(1/s),\ s>0$
20. $\frac{1}{2}\ln(s^2+4)-\ln s,\ s>0$
21. $\ln s-\ln(s-3),\ s>3$
22. $\ln(s+1)-\ln(s-1),\ s>1$
23. $-(2\sinh 2t)/t$
24. $2(\cos 2t-\cos t)/t$
25. $(e^{-2t}+e^{3t}-2\cos t)/t$
26. $(e^{-2t}\sin 3t)/t$
27. $2(1-\cos t)/t$
28. $\frac{1}{8}(t\sin t-t^2\cos t)$
29. $(s+1)X'(s)+4X(s)=0;\ x(t)=Ct^3e^{-t},\ C\neq 0$
30. $X(s)=A/(s+3)^3;\ x(t)=Ct^2e^{-3t},\ C\neq 0$
31. $(s-2)X'(s)+3X(s)=0;\ x(t)=Ct^2e^{2t},\ C\neq 0$
32. $(s^2+2s)X'(s)+(4s+4)X(s)=0;$
$x(t)=C(1-t-e^{-2t}-te^{-2t}),\ C\neq 0$
33. $(s^2+1)X'(s)+4sX(s)=0;\ x(t)=C(\sin t-t\cos t),\ C\neq 0$
34. $x(t)=Ce^{-2t}(\sin 3t-3t\cos 3t),\ C\neq 0$

Section 10.5

1. $f(t)=u(t-3)\cdot(t-3)$

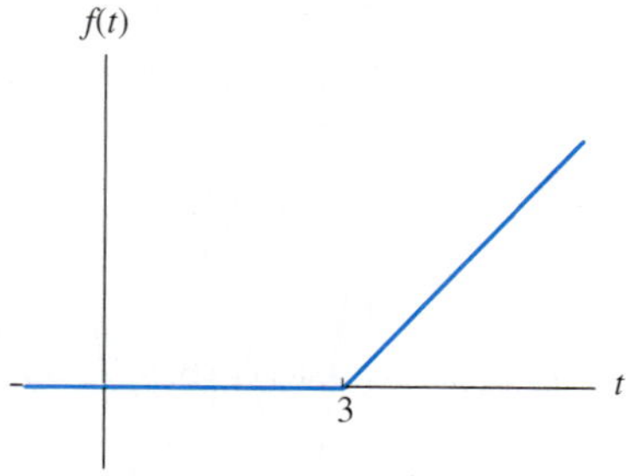

2. $f(t)=(t-1)u(t-1)-(t-3)u(t-3)$

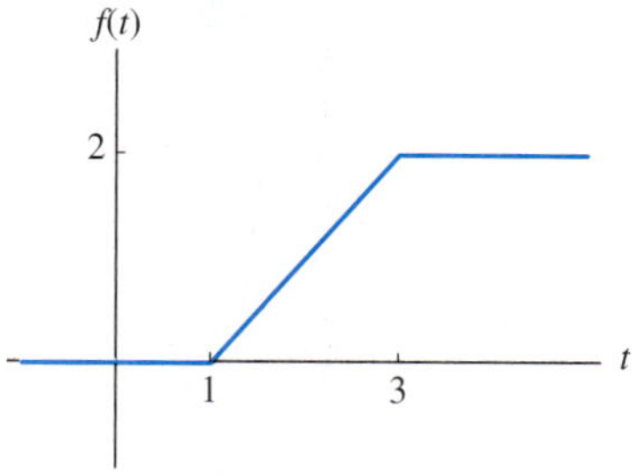

3. $f(t)=u(t-1)\cdot e^{-2(t-1)}$

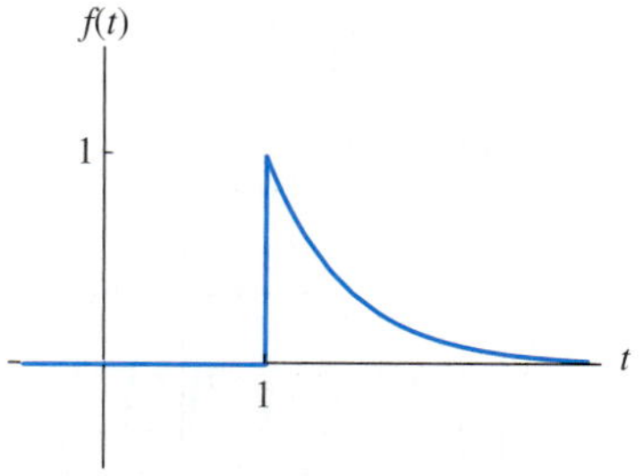

4. $f(t)=e^{t-1}u(t-1)-e^2e^{t-2}u(t-2)$

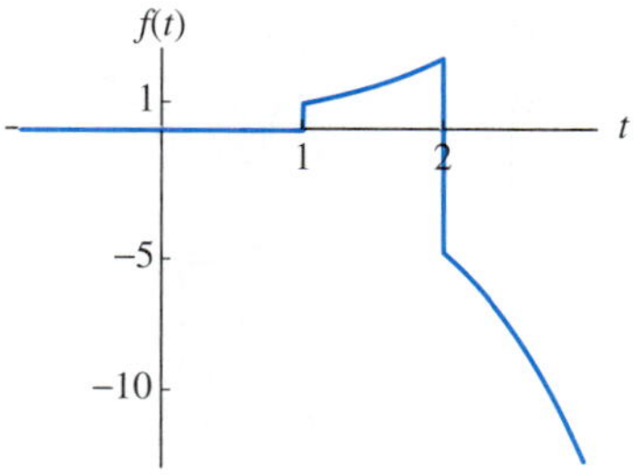

5. $f(t) = u(t-\pi)\cdot\sin(t-\pi) = -u(t-\pi)\sin t$

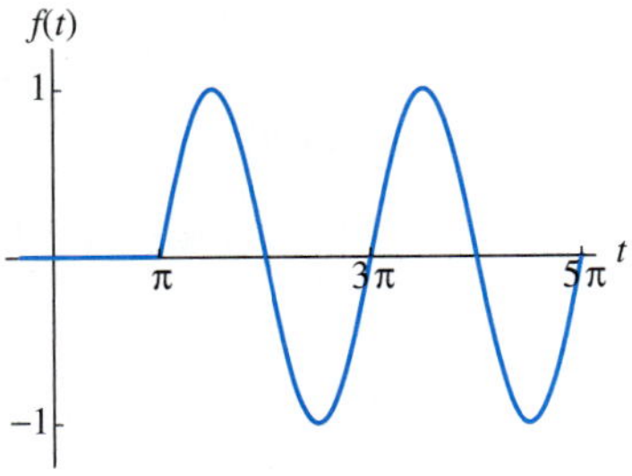

6. $f(t) = u(t-1)\cdot\cos\pi(t-1) = -u(t-1)\cos\pi t$

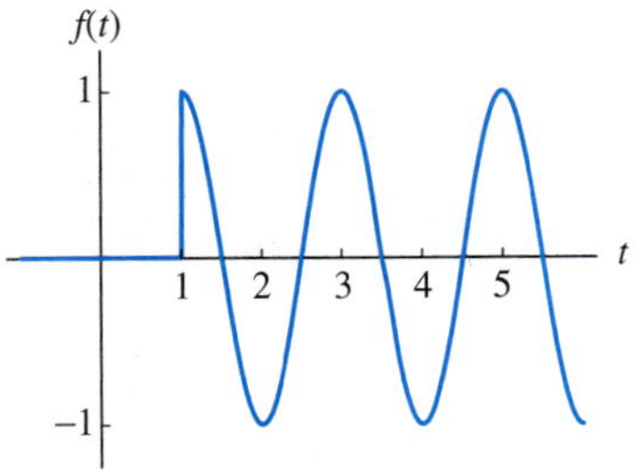

7. $f(t) = \sin t - u(t-2\pi)\sin(t-2\pi) = [1-u(t-2\pi)]\sin t$

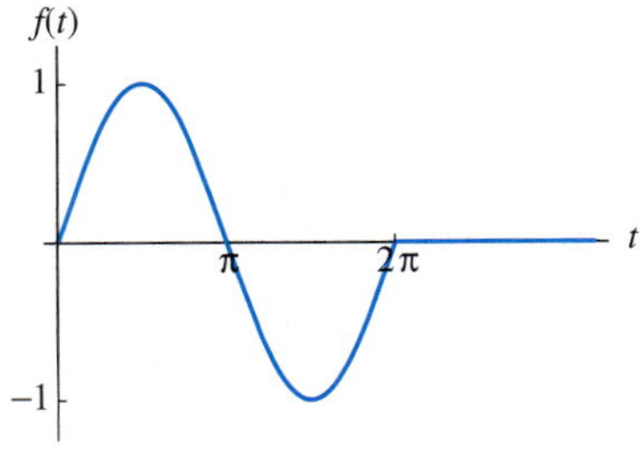

8. $f(t) = \cos\pi t - u(t-2)\cos\pi(t-2) = [1-u(t-2)]\cos\pi t$

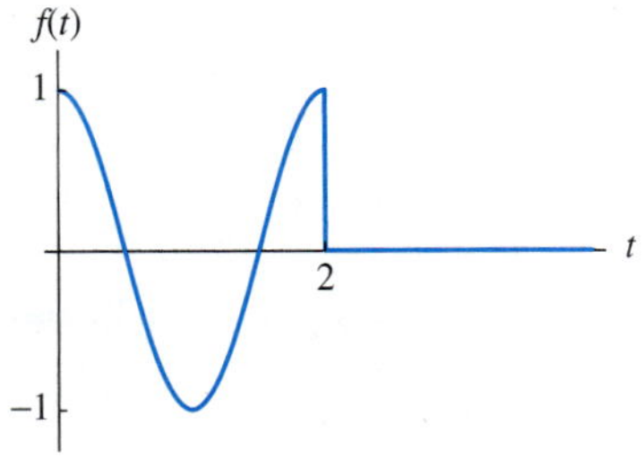

9. $f(t) = \cos\pi t + u(t-3)\cos\pi(t-3) = [1-u(t-3)]\cos\pi t$

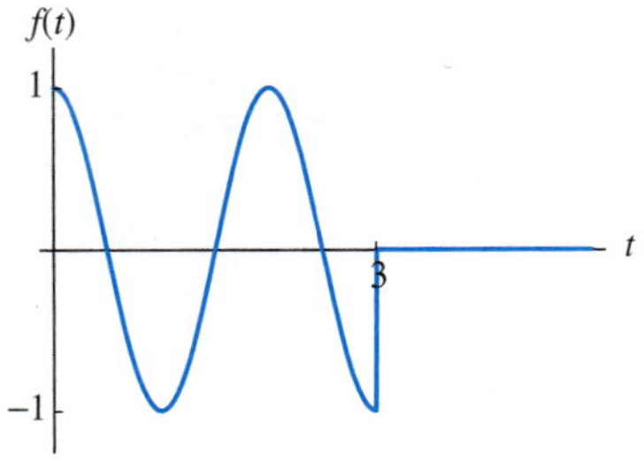

10. $f(t) = 2u(t-\pi)\cos 2(t-\pi) - 2u(t-2\pi)\cos 2(t-2\pi)$
$= 2[u(t-\pi)-u(t-2\pi)]\cos 2t$

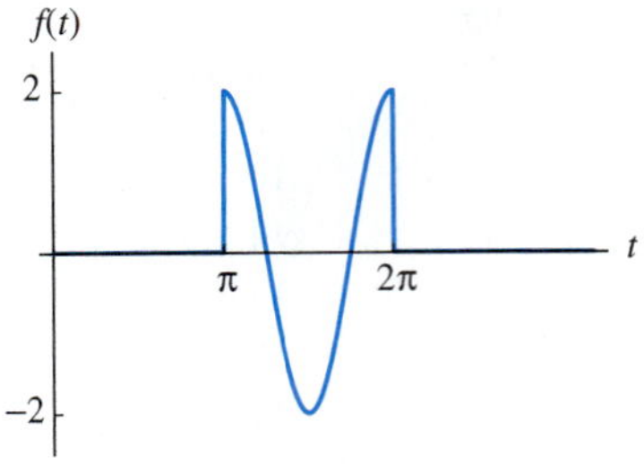

11. $f(t) = 2[1-u_3(t)]$; $F(s) = 2(1-e^{-3s})/s$

12. $F(s) = (e^{-s}-e^{-4s})/s$

13. $F(s) = (1-e^{-2\pi s})/(s^2+1)$

14. $F(s) = s(1-e^{-2s})/(s^2+\pi^2)$

15. $F(s) = (1+e^{-3\pi s})/(s^2+1)$

16. $F(s) = 2(e^{-\pi s}-e^{-2\pi s})/(s^2+4)$

17. $F(s) = \pi(e^{-2s}+e^{-3s})/(s^2+\pi^2)$

18. $F(s) = 2\pi(e^{-3s}+e^{-5s})/(4s^2+\pi^2)$

19. $F(s) = e^{-s}(s^{-1}+s^{-2})$

20. $F(s) = (1-e^{-s})/s^2$

21. $F(s) = (1-2e^{-s}+e^{-2s})/s^2$

28. $F(s) = (1-e^{-as}-ase^{-as})/[s^2(1-e^{-2as})]$

31. $x(t) = \frac{1}{2}[1-u(t-\pi)]\sin^2 t$

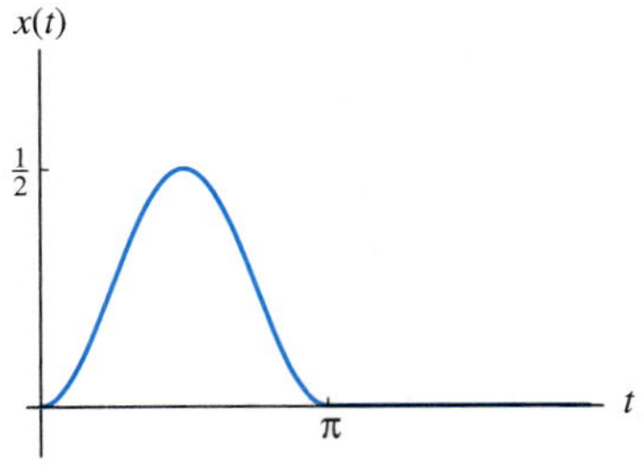

32. $x(t) = g(t) - u(t-2)g(t-2)$, where $g(t) = \frac{1}{12}(3-4e^{-t}+e^{-4t})$

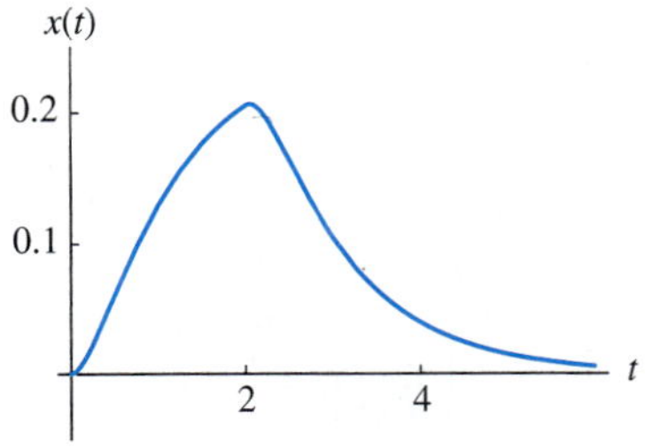

33. $x(t) = \frac{1}{8}[1-u(t-2\pi)]\left(\sin t - \frac{1}{3}\sin 3t\right)$

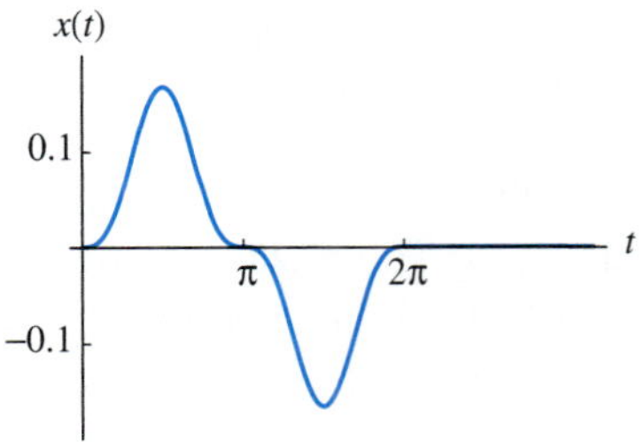

34. $x(t) = g(t) - u(t-1)[g(t-1) + h(t-1)]$, where $g(t) = t - \sin t$ and $h(t) = 1 - \cos t$

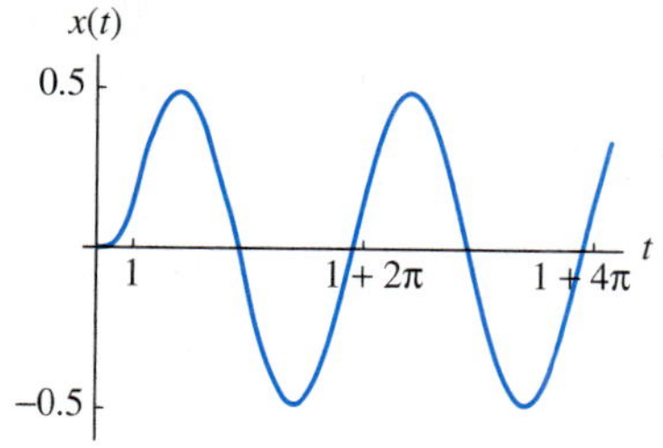

35. $x(t) = \frac{1}{4}\{-1 + t + (t+1)e^{-2t} + u(t-2)\left[1 - t + (3t-5)e^{-2(t-2)}\right]\}$

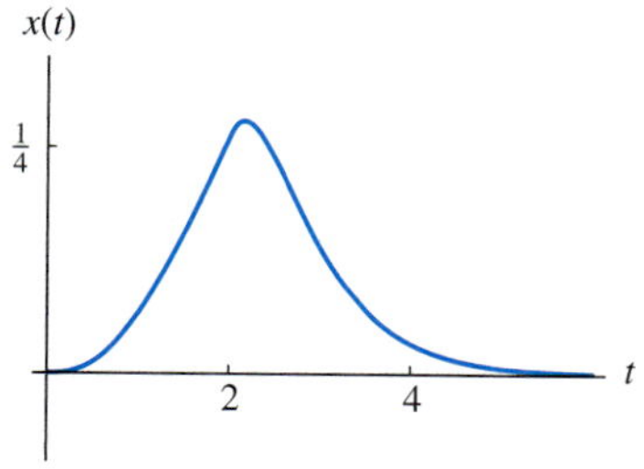

36. $x(t) = 2|\sin t|\sin t$

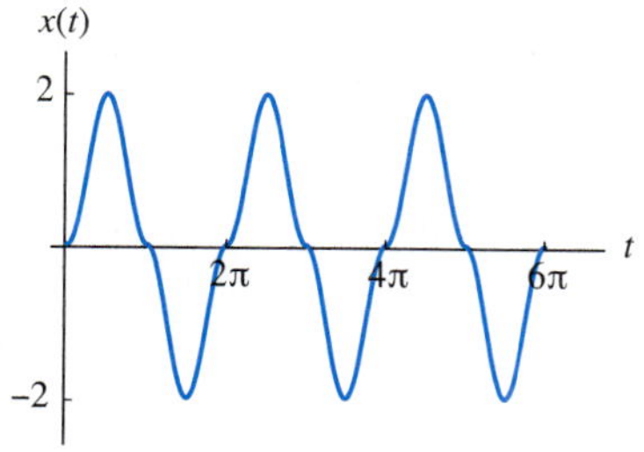

37. $x(t) = g(t) + 2\sum_{n=1}^{\infty}(-1)^n u(t - n\pi)g(t - n\pi)$, where $g(t) = 1 - \frac{1}{3}e^{-t}(3\cos 3t + \sin 3t)$

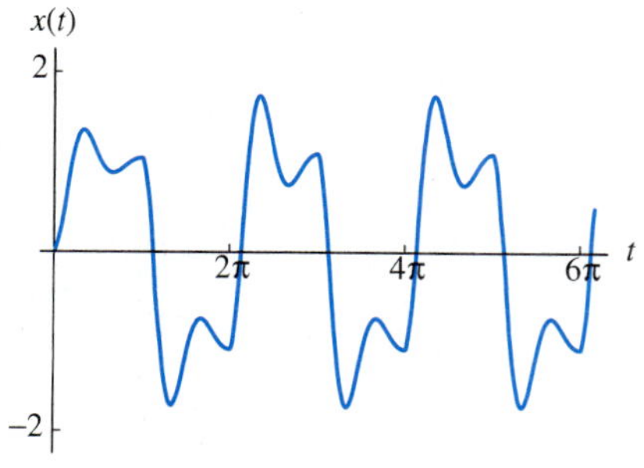

Chapter 11

Section 11.1

1. $y(x) = c_0\left(1 + x + \frac{x^2}{2} + \frac{x^3}{3!} + \cdots\right) = c_0 e^x$; $\rho = +\infty$

2. $y(x) = c_0\left(1 + \frac{4x}{1!} + \frac{4^2x^2}{2!} + \frac{4^3x^3}{3!} + \frac{4^4x^4}{4!} + \cdots\right) = c_0 e^{4x}$; $\rho = \infty$

3. $y(x) = c_0\left(1 - \frac{3x}{2} + \frac{(3x)^2}{2!2^2} - \frac{(3x)^3}{3!2^3} + \frac{(3x)^4}{4!2^4} - \cdots\right) = c_0 e^{-3x/2}$; $\rho = +\infty$

4. $y(x) = c_0\left(1 - \frac{x^2}{1!} + \frac{x^4}{2!} - \frac{x^6}{3!} + \cdots\right) = c_0 e^{-x^2}$; $\rho = \infty$

5. $y(x) = c_0\left(1 + \frac{x^3}{3} + \frac{x^6}{2!3^2} + \frac{x^9}{3!3^3} + \cdots\right) = c_0 \exp\left(\frac{1}{3}x^3\right)$; $\rho = +\infty$

6. $y(x) = c_0\left(1 + \frac{x}{2} + \frac{x^2}{4} + \frac{x^3}{8} + \frac{x^4}{16} + \cdots\right) = \frac{2c_0}{2 - x}$; $\rho = 2$

7. $y(x) = c_0(1 + 2x + 4x^2 + 8x^3 + \cdots) = \frac{c_0}{1 - 2x}$; $\rho = \frac{1}{2}$

8. $y(x) = c_0\left(1 + \frac{x}{2} - \frac{x^2}{8} + \frac{x^3}{16} - \frac{5x^4}{128} + \cdots\right) = c_0\sqrt{1 + x}$; $\rho = 1$

9. $y(x) = c_0(1 + 2x + 3x^2 + 4x^3 + \cdots) = \frac{c_0}{(1 - x)^2}$; $\rho = 1$

10. $y(x) = c_0\left(1 - \frac{3x}{2} + \frac{3x^2}{8} + \frac{x^3}{16} + \frac{3x^4}{128} + \cdots\right) = c_0(1 - x)^{3/2}$; $\rho = 1$

11. $y(x) = c_0\left(1 + \frac{x^2}{2!} + \frac{x^4}{4!} + \frac{x^6}{6!} + \cdots\right) + c_1\left(x + \frac{x^3}{3!} + \frac{x^5}{5!} + \frac{x^7}{7!} + \cdots\right) = c_0 \cosh x + c_1 \sinh x$; $\rho = +\infty$

12. $y(x) = c_0\left(1 + \frac{(2x)^2}{2!} + \frac{(2x)^4}{4!} + \frac{(2x)^6}{6!} + \cdots\right) + \frac{c_1}{2}\left((2x) + \frac{(2x)^3}{3!} + \frac{(2x)^5}{5!} + \frac{(2x)^7}{7!} + \cdots\right) = c_0 \cosh 2x + \frac{c_1}{2}\sinh 2x$; $\rho = \infty$

13. $y(x) = c_0\left(1 - \frac{(3x)^2}{2!} + \frac{(3x)^4}{4!} - \frac{(3x)^6}{6!} + \cdots\right) + \frac{c_1}{3}\left(3x - \frac{(3x)^3}{3!} + \frac{(3x)^5}{5!} - \frac{(3x)^7}{7!} + \cdots\right) = c_0 \cos 3x + \frac{1}{3}c_1 \sin 3x$; $\rho = +\infty$

14. $y(x) = x + c_0\left(1 - \frac{x^2}{2!} + \frac{x^4}{4!} - \frac{x^6}{6!} + \cdots\right) + (c_1 - 1)\left(x - \frac{x^3}{3!} + \frac{x^5}{5!} - \frac{x^7}{7!} + \cdots\right) = x + c_0 \cos x + (c_1 - 1)\sin x$; $\rho = \infty$

15. $(n+1)c_n = 0$ for all $n \geqq 0$, so $c_n = 0$ for all $n \geqq 0$.

16. $2nc_n = c_n$ for all $n \geqq 0$, so $c_n = 0$ for all $n \geqq 0$.

17. $c_0 = c_1 = 0$ and $c_{n+1} = -nc_n$ for $n \geqq 1$, thus $c_n = 0$ for all $n \geqq 0$.

18. $c_n = 0$ for all $n \geqq 0$

19. $(n+1)(n+2)c_{n+2} = -4c_n$; $y(x) = \frac{3}{2}\left[(2x) - \frac{(2x)^3}{3!} + \frac{(2x)^5}{5!} - \frac{(2x)^7}{7!} + \cdots\right] = \frac{3}{2}\sin 2x$

20. $(n+1)(n+2)c_{n+2} = 4c_n$; $y(x) = 2\left[1 + \frac{(2x)^2}{2!} + \frac{(2x)^4}{4!} + \frac{(2x)^6}{6!} + \cdots\right] = 2\cosh 2x$

21. $n(n+1)c_{n+1} = 2nc_n - c_{n-1}$; $y(x) = x + x^2 + \frac{x^3}{2!} + \frac{x^4}{3!} + \frac{x^5}{4!} + \cdots = xe^x$

22. $n(n+1)c_{n+1} = -nc_n + 2c_{n-1}$; $y = e^{-2x}$

23. As $c_0 = c_1 = 0$ and $(n^2 - n + 1)c_n + (n-1)c_{n-1} = 0$ for $n \geqq 2$, $c_n = 0$ for all $n \geqq 0$

Section 11.2

1. $c_{n+2} = c_n$; $y(x) = c_0 \sum_{n=0}^{\infty} x^{2n} + c_1 \sum_{n=0}^{\infty} x^{2n+1} = \dfrac{c_0 + c_1 x}{1 - x^2}$; $\rho = 1$

2. $c_{n+2} = -\frac{1}{2}c_n$; $\rho = 2$;
$y(x) = c_0 \sum_{n=0}^{\infty} \dfrac{(-1)^n x^{2n}}{2^n} + c_1 \sum_{n=0}^{\infty} \dfrac{(-1)^n x^{2n+1}}{2^n}$

3. $(n+2)c_{n+2} = -c_n$;
$y(x) = c_0 \sum_{n=0}^{\infty} \dfrac{(-1)^n x^{2n}}{n!2^n} + c_1 \sum_{n=0}^{\infty} \dfrac{(-1)^n x^{2n+1}}{(2n+1)!!}$; $\rho = +\infty$

4. $(n+2)c_{n+2} = -(n+4)c_n$; $\rho = 1$;
$y(x) = c_0 \sum_{n=0}^{\infty} (-1)^n (n+1) x^{2n} + \frac{1}{3} c_1 \sum_{n=0}^{\infty} (-1)^n (2n+3) x^{2n+1}$

5. $3(n+2)c_{n+2} = nc_n$; $\rho = \sqrt{3}$; $y(x) = c_0 + c_1 \sum_{n=0}^{\infty} \dfrac{x^{2n+1}}{(2n+1)3^n}$

6. $(n+1)(n+2)c_{n+2} = (n-3)(n-4)c_n$; $\rho = \infty$;
$y(x) = c_0(1 + 6x^2 + x^4) + c_1(x + x^3)$

7. $3(n+1)(n+2)c_{n+2} = -(n-4)^2 c_n$;
$y(x) = c_0\left(1 - \dfrac{8x^2}{3} + \dfrac{8x^4}{27}\right) + c_1\left(x - \dfrac{x^3}{2} + \dfrac{x^5}{120} + 9\sum_{n=3}^{\infty} \dfrac{(-1)^n[(2n-5)!!]^2 x^{2n+1}}{(2n+1)!\,3^n}\right)$

8. $2(n+1)(n+2)c_{n+2} = (n-4)(n+4)c_n$; $y(x) = c_0(1 - 4x^2 + 2x^4) + c_1\left(x - \dfrac{5x^3}{4} + \dfrac{7x^5}{32} + \sum_{n=3}^{\infty} \dfrac{(2n-5)!!(2n+3)!!x^{2n+1}}{(2n+1)!2^n}\right)$

9. $(n+1)(n+2)c_{n+2} = (n+3)(n+4)c_n$; $\rho = 1$;
$y(x) = c_0 \sum_{n=0}^{\infty} (n+1)(2n+1)x^{2n} + \dfrac{c_1}{3} \sum_{n=0}^{\infty} (n+1)(2n+3)x^{2n+1}$

10. $3(n+1)(n+2)c_{n+2} = -(n-4)c_n$;
$y(x) = c_0\left(1 + \dfrac{2x^2}{3} + \dfrac{x^4}{27}\right) + c_1\left(x + \dfrac{x^3}{6} + \dfrac{x^5}{360} + 3\sum_{n=3}^{\infty} \dfrac{(-1)^n(2n-5)!!x^{2n+1}}{(2n+1)!3^n}\right)$

11. $5(n+1)(n+2)c_{n+2} = 2(n-5)c_n$;
$y(x) = c_1\left(x - \dfrac{4x^3}{15} + \dfrac{4x^5}{375}\right) + c_0\left(1 - x^2 + \dfrac{x^4}{10} - \dfrac{x^6}{750} - 15\sum_{n=4}^{\infty} \dfrac{(2n-7)!!\,2^n x^{2n}}{(2n)!5^n}\right)$

12. $c_2 = 0$; $(n+2)c_{n+3} = c_n$;
$y(x) = c_0\left(1 + \sum_{n=1}^{\infty} \dfrac{x^{3n}}{2 \cdot 5 \cdots (3n-1)}\right) + c_1 \sum_{n=0}^{\infty} \dfrac{x^{3n+1}}{n!3^n}$

13. $c_2 = 0$; $(n+3)c_{n+3} = -c_n$;
$y(x) = c_0 \sum_{n=0}^{\infty} \dfrac{(-1)^n x^{3n}}{n!3^n} + c_1 \sum_{n=0}^{\infty} \dfrac{(-1)^n x^{3n+1}}{1 \cdot 4 \cdots (3n+1)}$

14. $c_2 = 0$; $(n+2)(n+3)c_{n+3} = -c_n$;
$y(x) = c_0\left(1 + \sum_{n=1}^{\infty} \dfrac{(-1)^n x^{3n}}{3^n \cdot n! \cdot 2 \cdot 5 \cdots (3n-1)}\right) + c_1 \sum_{n=0}^{\infty} \dfrac{(-1)^n x^{3n+1}}{3^n \cdot n! \cdot 1 \cdot 4 \cdots (3n+1)}$

15. $c_2 = c_3 = 0$; $(n+3)(n+4)c_{n+4} = -c_n$;
$y(x) = c_0\left(1 + \sum_{n=1}^{\infty} \dfrac{(-1)^n x^{4n}}{4^n \cdot n! \cdot 3 \cdot 7 \cdots (4n-1)}\right) + c_1 \sum_{n=0}^{\infty} \dfrac{(-1)^n x^{4n+1}}{4^n \cdot n! \cdot 5 \cdot 9 \cdots (4n+1)}$

16. $y(x) = x$

17. $y(x) = 1 + x^2$

18. $y(x) = 2\sum_{n=0}^{\infty} \dfrac{(-1)^n (x-1)^{2n}}{n!\,2^n}$; converges for all x

19. $y(x) = \frac{1}{3}\sum_{n=0}^{\infty} (2n+3)(x-1)^{2n+1}$; converges if $0 < x < 2$

20. $y(x) = 2 - 6(x-3)^2$; converges for all x

21. $y(x) = 1 + 4(x+2)^2$; converges for all x

22. $y(x) = 2x + 6$

23. $2c_2 + c_0 = 0$; $(n+1)(n+2)c_{n+2} + c_n + c_{n-1} = 0$ for $n \geqq 1$;
$y_1(x) = 1 - \dfrac{x^2}{2} - \dfrac{x^3}{6} + \cdots$; $y_2(x) = x - \dfrac{x^3}{6} - \dfrac{x^4}{12} + \cdots$

24. $y_1(x) = 1 + \dfrac{x^3}{3} + \dfrac{x^5}{5} + \dfrac{x^6}{45} + \cdots$;
$y_2(x) = x + \dfrac{x^3}{3} + \dfrac{x^4}{6} + \dfrac{x^5}{5} + \cdots$

25. $c_2 = c_3 = 0$, $(n+3)(n+4)c_{n+4} + (n+1)c_{n+1} + c_n = 0$ for $n \geqq 0$; $y_1(x) = 1 - \dfrac{x^4}{12} + \dfrac{x^7}{126} + \cdots$;
$y_2(x) = x - \dfrac{x^4}{12} - \dfrac{x^5}{20} + \cdots$

26. $y(x) = c_0\left(1 - \dfrac{x^6}{30} + \dfrac{x^9}{72} + \cdots\right) + c_1\left(x - \dfrac{x^7}{42} + \dfrac{x^{10}}{90} + \cdots\right)$

27. $y(x) = 1 - x - \dfrac{x^2}{2} + \dfrac{x^3}{3} - \dfrac{x^4}{24} + \dfrac{x^5}{30} + \dfrac{29x^6}{720} - \dfrac{13x^7}{630} - \dfrac{143x^8}{40320} + \cdots$;
$y(0.5) \approx 0.4156$

28. $y(x) = c_0\left(1 - \dfrac{x^2}{2} + \dfrac{x^3}{6} + \cdots\right) + c_1\left(x - \dfrac{x^3}{6} + \dfrac{x^4}{12} + \cdots\right)$

29. $y_1(x) = 1 - \dfrac{1}{2}x^2 + \dfrac{1}{720}x^6 + \cdots$;
$y_2(x) = x - \dfrac{1}{6}x^3 - \dfrac{1}{60}x^5 + \cdots$

30. $y(x) = c_0\left(1 - \dfrac{x^2}{2} + \dfrac{x^3}{6} + \cdots\right) + c_1\left(x - \dfrac{x^2}{2} + \dfrac{x^4}{18} + \cdots\right)$

33. The following figure shows the interlaced zeros of the 4th and 5th Hermite polynomials.

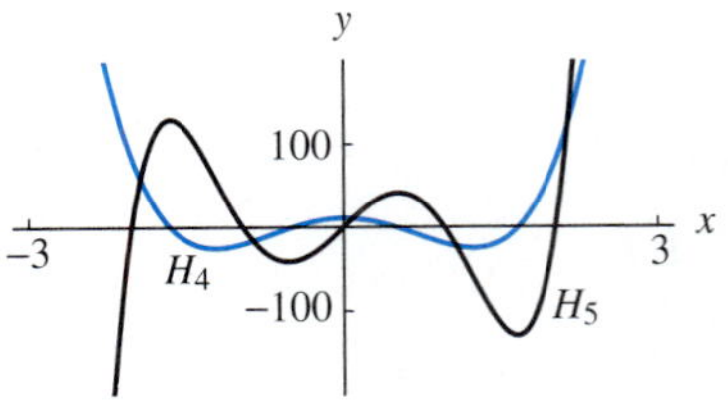

34. The figure below results when we use $n = 40$ terms in each summation. But with $n = 50$ we get the same picture as Fig. 11.2.3 in the text.

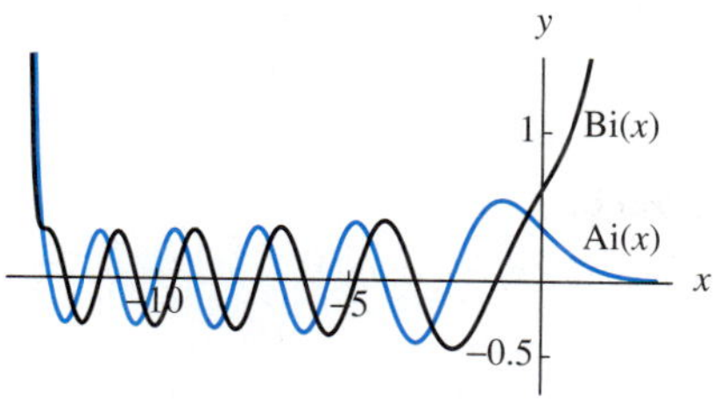

Section 11.3

1. Ordinary point
2. Ordinary point
3. Irregular singular point
4. Irregular singular point
5. Regular singular point; $r_1 = 0$, $r_2 = -1$
6. Regular singular point; $r_1 = 1$, $r_2 = -2$
7. Regular singular point; $r = -3, -2$
8. Regular singular point; $r = \frac{1}{2}, -3$
9. Regular singular point $x = 1$
10. Regular singular point $x = 1$
11. Regular singular points $x = 1, -1$
12. Irregular singular point $x = 2$
13. Regular singular points $x = 2, -2$
14. Irregular singular points $x = 3, -3$
15. Regular singular point $x = 2$
16. Irregular singular point $x = 0$, regular singular point $x = 1$
17. $y_1(x) = \cos\sqrt{x}$, $y_2(x) = \sin\sqrt{x}$
18. $y_1(x) = \sum_{n=0}^{\infty} \dfrac{x^n}{n!(2n+1)!!}$, $y_2(x) = x^{-1/2} \sum_{n=0}^{\infty} \dfrac{x^n}{n!\,(2n-1)!!}$
19. $y_1(x) = x^{3/2}\left(1 + 3\sum_{n=1}^{\infty} \dfrac{x^n}{n!\,(2n+3)!!}\right)$,
$y_2(x) = 1 - x - \sum_{n=2}^{\infty} \dfrac{x^n}{n!\,(2n-3)!!}$
20. $y_1(x) = x^{1/3} \sum_{n=0}^{\infty} \dfrac{(-1)^n 2^n x^n}{n! \cdot 4 \cdot 7 \cdots (3n+1)}$,
$y_2(x) = \sum_{n=0}^{\infty} \dfrac{(-1)^n 2^n x^n}{n! \cdot 2 \cdot 5 \cdots (3n-1)}$
21. $y_1(x) = x\left(1 + \sum_{n=1}^{\infty} \dfrac{x^{2n}}{n! \cdot 7 \cdot 11 \cdots (4n+3)}\right)$,
$y_2(x) = x^{-1/2}\left(1 + \sum_{n=1}^{\infty} \dfrac{x^{2n}}{n! \cdot 1 \cdot 5 \cdots (4n-3)}\right)$
22. $y_1(x) = x^{3/2}\left(1 + \sum_{n=1}^{\infty} \dfrac{(-1)^n x^{2n}}{n! \cdot 9 \cdot 13 \cdots (4n+5)}\right)$,
$y_2(x) = x^{-1}\left(1 + \sum_{n=1}^{\infty} \dfrac{(-1)^{n-1} x^{2n}}{n! \cdot 3 \cdot 7 \cdots (4n-1)}\right)$
23. $y_1(x) = x^{1/2}\left(1 + \sum_{n=1}^{\infty} \dfrac{x^{2n}}{2^n \cdot n! \cdot 19 \cdot 31 \cdots (12n+7)}\right)$,
$y_2(x) = x^{-2/3}\left(1 + \sum_{n=1}^{\infty} \dfrac{x^{2n}}{2^n \cdot n! \cdot 5 \cdot 17 \cdots (12n-7)}\right)$
24. $y_1(x) = x^{1/3}\left(1 + \sum_{n=1}^{\infty} \dfrac{(-1)^n x^{2n}}{2^n \cdot n! \cdot 7 \cdot 13 \cdots (6n+1)}\right)$,
$y_2(x) = 1 + \sum_{n=1}^{\infty} \dfrac{(-1)^n x^{2n}}{2^n \cdot n! \cdot 5 \cdot 11 \cdots (6n-1)}$
25. $y_1(x) = x^{1/2} \sum_{n=0}^{\infty} \dfrac{(-1)^n x^n}{n! \cdot 2^n} = x^{1/2} e^{-x/2}$,
$y_2(x) = 1 + \sum_{n=1}^{\infty} \dfrac{(-1)^n x^n}{(2n-1)!!}$
26. $y_1(x) = x^{1/2} \sum_{n=0}^{\infty} \dfrac{x^{2n}}{n! \cdot 2^n} = x^{1/2} \exp\left(\frac{1}{2}x^2\right)$,
$y_2(x) = 1 + \sum_{n=1}^{\infty} \dfrac{2^n x^{2n}}{3 \cdot 7 \cdots (4n-1)}$
27. $y_1(x) = \dfrac{1}{x}\cos 3x$, $y_2(x) = \dfrac{1}{x}\sin 3x$

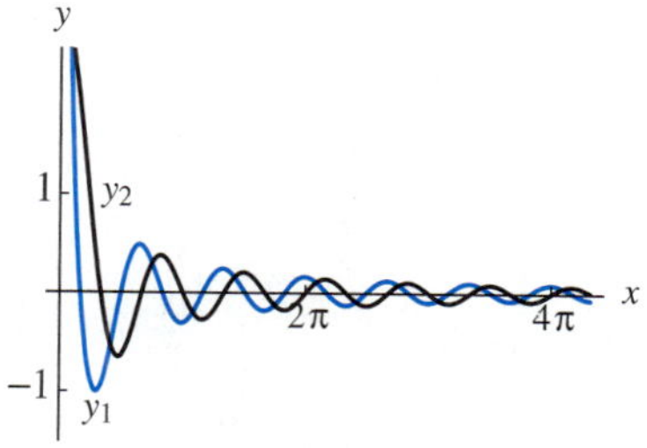

28. $y_1(x) = \dfrac{1}{x}\cosh 2x$, $y_2(x) = \dfrac{1}{x}\sinh 2x$

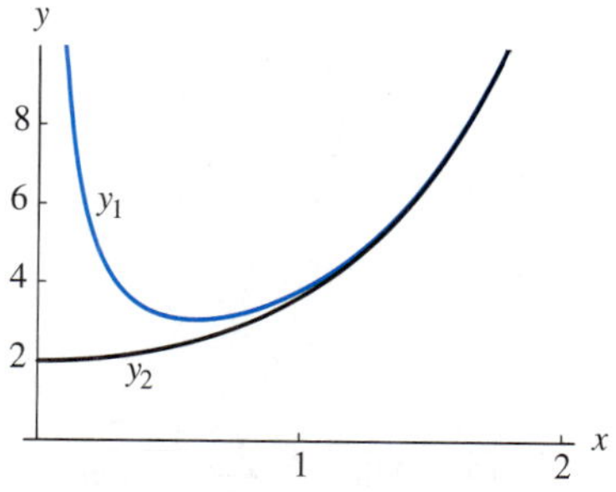

29. $y_1(x) = \dfrac{1}{x}\cos\dfrac{x}{2}$, $y_2(x) = \dfrac{1}{x}\sin\dfrac{x}{2}$

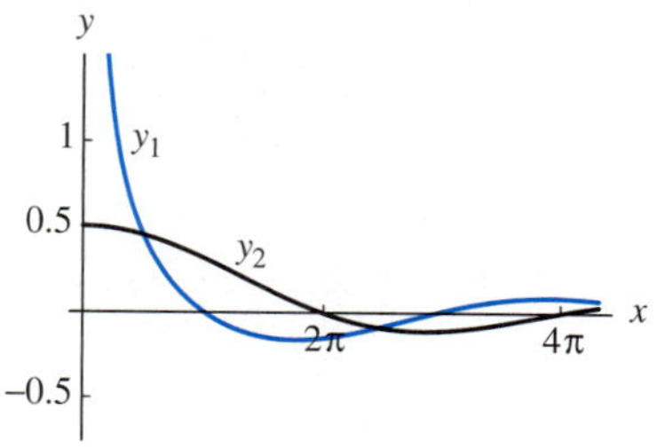

30. $y_1(x) = \cos x^2$, $y_2(x) = \sin x^2$

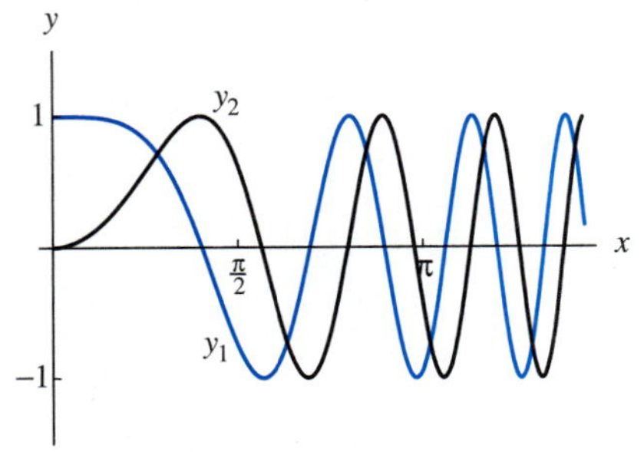

31. $y_1(x) = x^{1/2}\cosh x$, $y_2(x) = x^{1/2}\sinh x$

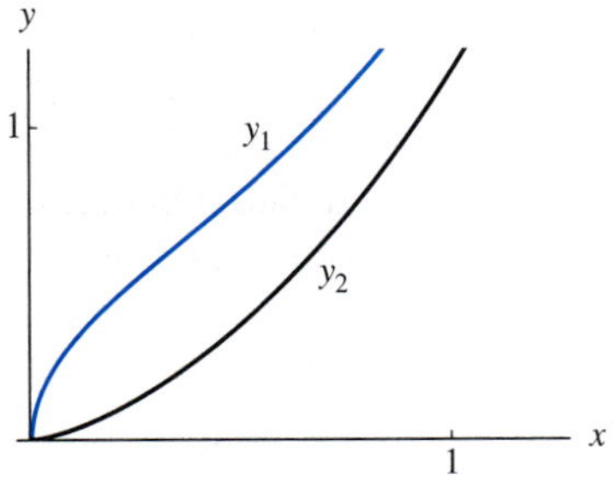

32. $y_1(x) = x + \dfrac{x^2}{5}$,

$$y_2(x) = x^{-1/2}\left(1 - \frac{5x}{2} - \frac{15x^2}{8} - \frac{5x^3}{48} + \cdots\right)$$

33. $y_1(x) = x^{-1}\left(1 + 10x + 5x^2 + \dfrac{10x^3}{9} + \cdots\right)$,

$$y_2(x) = x^{1/2}\left(1 + \frac{11x}{20} - \frac{11x^2}{224} + \frac{671x^3}{24192} + \cdots\right)$$

34. $y_1(x) = x\left(1 - \dfrac{x^2}{42} + \dfrac{x^4}{1320} + \cdots\right)$,

$$y_2(x) = x^{-1/2}\left(1 - \frac{7x^2}{24} + \frac{19x^4}{3200} + \cdots\right)$$

Section 11.4

5. $J_4(x) = \dfrac{1}{x^2}(x^2 - 24)J_0(x) + \dfrac{8}{x^3}(6 - x^2)J_1(x)$

10. 3

11. $x^2 J_1(x) + xJ_0(x) - \int J_0(x)\,dx + C$

12. $(x^3 - 4x)J_1(x) + 2x^2 J_0(x) + C$

13. $(x^4 - 9x^2)J_1(x) + (3x^3 - 9x)J_0(x) + 9\int J_0(x)\,dx + C$

14. $-xJ_1(x) + \int J_0(x)\,dx + C$

15. $2xJ_1(x) - x^2 J_0(x) + C$

16. $3x^2 J_1(x) + (3x - x^3)J_0(x) - 3\int J_0(x)\,dx + C$

17. $(4x^3 - 16x)J_1(x) + (8x^2 - x^4)J_0(x) + C$

18. $-2J_1(x) + \int J_0(x)\,dx + C$

19. $y(x) = x\,[c_1 J_0(x) + C_2 Y_0(x)]$

20. $y(x) = \dfrac{1}{x}\,[c_1 J_1(x) + c_2 Y_1(x)]$

21. $y(x) = x\left[c_1 J_{1/2}(3x^2) + c_2 J_{-1/2}(3x^2)\right]$

22. $y(x) = x^3\left[c_1 J_2(2x^{1/2}) + c_2 Y_2(2x^{1/2})\right]$

23. $y(x) = x^{-1/3}\left[c_1 J_{1/3}\left(\frac{1}{3}x^{3/2}\right) + c_2 J_{-1/3}\left(\frac{1}{3}x^{3/2}\right)\right]$

24. $y(x) = x^{-1/4}\left[c_1 J_0(2x^{3/2}) + c_2 Y_0(2x^{3/2})\right]$

25. $y(x) = x^{-1}\,[c_1 J_0(x) + c_2 Y_0(x)]$

26. $y(x) = x^2\left[c_1 J_1(4x^{1/2}) + c_2 Y_1(4x^{1/2})\right]$

27. $y(x) = x^{1/2}\left[c_1 J_{1/2}(2x^{3/2}) + c_2 J_{-1/2}(2x^{3/2})\right]$

28. $y(x) = x^{-1/4}\left[c_1 J_{3/2}\left(\frac{2}{5}x^{5/2}\right) + c_2 J_{-3/2}\left(\frac{2}{5}x^{5/2}\right)\right]$

29. $y(x) = x^{1/2}\left[c_1 J_{1/6}\left(\frac{1}{3}x^3\right) + c_2 J_{-1/6}\left(\frac{1}{3}x^3\right)\right]$

30. $y(x) = x^{1/2}\left[c_1 J_{1/5}\left(\frac{4}{5}x^{5/2}\right) + c_2 J_{-1/5}\left(\frac{4}{5}x^{5/2}\right)\right]$

Appendix A

1. $y_0 = 3$, $y_1 = 3 + 3x$, $y_2 = 3 + 3x + \frac{3}{2}x^2$,
$y_3 = 3 + 3x + \frac{3}{2}x^2 + \frac{1}{2}x^3$,
$y_4 = 3 + 3x + \frac{3}{2}x^2 + \frac{1}{2}x^3 + \frac{1}{8}x^4$; $y(x) = 3e^x$

3. $y_0 = 1$, $y_1 = 1 - x^2$, $y_2 = 1 - x^2 + \frac{1}{2}x^4$,
$y_3 = 1 - x^2 + \frac{1}{2}x^4 - \frac{1}{6}x^6$,
$y_4 = 1 - x^2 + \frac{1}{2}x^4 - \frac{1}{6}x^6 + \frac{1}{24}x^8$; $y(x) = \exp(-x^2)$

5. $y_0 = 0$, $y_1 = 2x$, $y_2 = 2x + 2x^2$,
$y_3 = 2x + 2x^2 + \frac{4}{3}x^3$,
$y_4 = 2x + 2x^2 + \frac{4}{3}x^3 + \frac{2}{3}x^4$; $y(x) = e^{2x} - 1$

7. $y_0 = 0$, $y_1 = x^2$, $y_2 = x^2 + \frac{1}{2}x^4$, $y_3 = x^2 + \frac{1}{2}x^4 + \frac{1}{6}x^6$,
$y_4 = x^2 + \frac{1}{2}x^4 + \frac{1}{6}x^6 + \frac{1}{24}x^8$; $y(x) = \exp(x^2) - 1$

9. $y_0 = 1$, $y_1 = (1 + x) + \frac{1}{2}x^2$, $y_2 = (1 + x + x^2) + \frac{1}{6}x^3$,
$y_3 = \left(1 + x + x^2 + \frac{1}{3}x^3\right) + \frac{1}{24}x^4$;
$y(x) = 2e^x - 1 - x = 1 + x + x^2 + \frac{1}{3}x^3 + \cdots$

11. $y_0 = 1$, $y_1 = 1 + x$, $y_2 = (1 + x + x^2) + \frac{1}{3}x^3$,
$y_3 = (1 + x + x^2 + x^3) + \frac{2}{3}x^4 + \frac{1}{3}x^5 + \frac{1}{9}x^6 + \frac{1}{63}x^7$;
$y(x) = \dfrac{1}{1 - x} = 1 + x + x^2 + x^3 + x^4 + x^5 + \cdots$

12. $y_0 = 1$, $y_1 = 1 + \frac{1}{2}x$, $y_2 = 1 + \frac{1}{2}x + \frac{3}{8}x^3 + \frac{1}{8}x^3 + \frac{1}{64}x^4$,
$y_3 = 1 + \frac{1}{2}x + \frac{3}{8}x^2 + \frac{5}{16}x^3 + \frac{13}{64}x^4 + \cdots$; $y(x) = (1 - x)^{-1/2}$

13. $\begin{bmatrix} x_0 \\ y_0 \end{bmatrix} = \begin{bmatrix} 1 \\ -1 \end{bmatrix}$, $\begin{bmatrix} x_1 \\ y_1 \end{bmatrix} = \begin{bmatrix} 1 + 3t \\ -1 + 3t \end{bmatrix}$,

$\begin{bmatrix} x_2 \\ y_2 \end{bmatrix} = \begin{bmatrix} 1 + 3t + \frac{1}{2}t^2 \\ -1 + 5t - \frac{1}{2}t^2 \end{bmatrix}$,

$\begin{bmatrix} x_3 \\ y_3 \end{bmatrix} = \begin{bmatrix} 1 + 3t + \frac{1}{2}t^2 + \frac{1}{3}t^3 \\ -1 + 5t - \frac{1}{2}t^2 + \frac{5}{6}t^3 \end{bmatrix}$

14. $\mathbf{x}(t) = \begin{bmatrix} e^t + te^t \\ e^t \end{bmatrix}$

16. $y_3(1) \approx 0.350185$

INDEX

Boldface page numbers indicate where terms are defined.

30. $$\int \cos au \cos bu \, du = \frac{\sin(a-b)u}{2(a-b)} + \frac{\sin(a+b)u}{2(a+b)} + C \quad \text{if } a^2 \neq b^2$$

31. $$\int \sin au \cos bu \, du = -\frac{\cos(a-b)u}{2(a-b)} - \frac{\cos(a+b)u}{2(a+b)} + C \quad \text{if } a^2 \neq b^2$$

32. $$\int \sin^n u \, du = -\frac{1}{n} \sin^{n-1} u \cos u + \frac{n-1}{n} \int \sin^{n-2} u \, du$$

33. $$\int \cos^n u \, du = \frac{1}{n} \cos^{n-1} u \sin u + \frac{n-1}{n} \int \cos^{n-2} u \, du$$

34. $$\int \tan^n u \, du = \frac{1}{n-1} \tan^{n-1} u - \int \tan^{n-2} u \, du \quad \text{if } n \neq 1$$

35. $$\int \cot^n u \, du = -\frac{1}{n-1} \cot^{n-1} u - \int \cot^{n-2} u \, du \quad \text{if } n \neq 1$$

36. $$\int \sec^n u \, du = \frac{1}{n-1} \sec^{n-2} u \tan u + \frac{n-2}{n-1} \int \sec^{n-2} u \, du \quad \text{if } n \neq 1$$

37. $$\int \csc^n u \, du = -\frac{1}{n-1} \csc^{n-2} u \cot u + \frac{n-2}{n-1} \int \csc^{n-2} u \, du \quad \text{if } n \neq 1$$

38. $$\int u \sin u \, du = \sin u - u \cos u + C$$

39. $$\int u \cos u \, du = \cos u + u \sin u + C$$

40. $$\int u^n \sin u \, du = -u^n \cos u + n \int u^{n-1} \cos u \, du$$

41. $$\int u^n \cos u \, du = u^n \sin u - n \int u^{n-1} \sin u \, du$$

FORMS INVOLVING $\sqrt{u^2 \pm a^2}$

42. $$\int \sqrt{u^2 \pm a^2} \, du = \frac{u}{2} \sqrt{u^2 \pm a^2} \pm \frac{a^2}{2} \ln \left| u + \sqrt{u^2 \pm a^2} \right| + C$$

43. $$\int \frac{du}{\sqrt{u^2 \pm a^2}} = \ln \left| u + \sqrt{u^2 \pm a^2} \right| + C$$

FORMS INVOLVING $\sqrt{a^2 - u^2}$

44. $$\int \sqrt{a^2 - u^2} \, du = \frac{u}{2} \sqrt{a^2 - u^2} + \frac{a^2}{2} \sin^{-1} \frac{u}{a} + C$$

45. $$\int \frac{\sqrt{a^2 - u^2}}{u} \, du = \sqrt{a^2 - u^2} - a \ln \left| \frac{a + \sqrt{a^2 - u^2}}{u} \right| + C$$

Table of Integrals

ELEMENTARY FORMS

1. $\displaystyle\int u\,dv = uv - \int v\,du$

2. $\displaystyle\int u^n\,du = \frac{1}{n+1}u^{n+1} + C \quad \text{if } n \neq -1$

3. $\displaystyle\int \frac{du}{u} = \ln|u| + C$

4. $\displaystyle\int e^u\,du = e^u + C$

5. $\displaystyle\int a^u\,du = \frac{a^u}{\ln a} + C$

6. $\displaystyle\int \sin u\,du = -\cos u + C$

7. $\displaystyle\int \cos u\,du = \sin u + C$

8. $\displaystyle\int \sec^2 u\,du = \tan u + C$

9. $\displaystyle\int \csc^2 u\,du = -\cot u + C$

10. $\displaystyle\int \sec u\,\tan u\,du = \sec u + C$

11. $\displaystyle\int \csc u\,\cot u\,du = -\csc u + C$

12. $\displaystyle\int \tan u\,du = \ln|\sec u| + C$

13. $\displaystyle\int \cot u\,du = \ln|\sin u| + C$

14. $\displaystyle\int \sec u\,du = \ln|\sec u + \tan u| + C$

15. $\displaystyle\int \csc u\,du = \ln|\csc u - \cot u| + C$

16. $\displaystyle\int \frac{du}{\sqrt{a^2 - u^2}} = \sin^{-1}\frac{u}{a} + C$

17. $\displaystyle\int \frac{du}{a^2 + u^2} = \frac{1}{a}\tan^{-1}\frac{u}{a} + C$

18. $\displaystyle\int \frac{du}{a^2 - u^2} = \frac{1}{2a}\ln\left|\frac{u+a}{u-a}\right| + C$

TRIGONOMETRIC FORMS

19. $\displaystyle\int \sin^2 u\,du = \frac{1}{2}u - \frac{1}{4}\sin 2u + C$

20. $\displaystyle\int \cos^2 u\,du = \frac{1}{2}u + \frac{1}{4}\sin 2u + C$

21. $\displaystyle\int \tan^2 u\,du = \tan u - u + C$

22. $\displaystyle\int \cot^2 u\,du = -\cot u - u + C$

23. $\displaystyle\int \sin^3 u\,du = -\frac{1}{3}(2 + \sin^2 u)\cos u + C$

24. $\displaystyle\int \cos^3 u\,du = \frac{1}{3}(2 + \cos^2 u)\sin u + C$

25. $\displaystyle\int \tan^3 u\,du = \frac{1}{2}\tan^2 u + \ln|\cos u| + C$

26. $\displaystyle\int \cot^3 u\,du = -\frac{1}{2}\cot^2 u - \ln|\sin u| + C$

27. $\displaystyle\int \sec^3 u\,du = \frac{1}{2}\sec u\,\tan u + \frac{1}{2}\ln|\sec u + \tan u| + C$

28. $\displaystyle\int \csc^3 u\,du = -\frac{1}{2}\csc u\,\cot u + \frac{1}{2}\ln|\csc u - \cot u| + C$

29. $\displaystyle\int \sin au\,\sin bu\,du = \frac{\sin(a-b)u}{2(a-b)} - \frac{\sin(a+b)u}{2(a+b)} + C \quad \text{if } a^2 \neq b^2$

(Continued on Rear Endpaper)